						VIII B (18)
						2 He 4.00260 Helium

			III B (13)	IV B (14)	V B (15)	VI B (16)	VII B (17)	
			5 B 10.811^b Boron	6 C 12.01115 Carbon	7 N 14.0067 Nitrogen	8 O 15.9994^{b,c} Oxygen	9 F 18.9984 Fluorine	10 Ne 20.179 Neon
VIII A (10)	I B (11)	II B (12)	13 Al 26.98154^b Aluminum	14 Si 28.086 Silicon	15 P 30.97376 Phosphorus	16 S 32.064^b Sulfur	17 Cl 35.453 Chlorine	18 Ar 39.948 Argon
28 Ni 58.69 Nickel	29 Cu 63.546^b Copper	30 Zn 65.377 Zinc	31 Ga 69.72 Gallium	32 Ge 72.59 Germanium	33 As 74.9216 Arsenic	34 Se 78.96 Selenium	35 Br 79.904 Bromine	36 Kr 83.80 Krypton
46 Pd 106.4 Palladium	47 Ag 107.868 Silver	48 Cd 112.40 Cadmium	49 In 114.82 Indium	50 Sn 118.69 Tin	51 Sb 121.75 Antimony	52 Te 127.60 Tellurium	53 I 126.9045 Iodine	54 Xe 131.30 Xenon
78 Pt 195.09 Platinum	79 Au 196.9665 Gold	80 Hg 200.59 Mercury	81 Tl 204.38 Thallium	82 Pb 207.19^b Lead	83 Bi 208.9804 Bismuth	84 Po Pollonium	85 At Astatine	86 Rn Radon

63 Eu 151.96 Europium	64 Gd 157.25 Gadolinium	65 Tb 158.9254 Terbium	66 Dy 162.50 Dysprosium	67 Ho 164.9304 Holmium	68 Er 167.26 Erbium	69 Tm 168.9342 Thulium	70 Yb 173.04 Ytterbium	71 Lu 174.97 Lutetium
95 Am Americium	96 Cm Curium	97 Bk Berkelium	98 Cf Californium	99 Es Einsteinium	100 Fm Fermium	101 Md Mendelevium	102 No Nobelium	103 Lw Lawrencium

BASIC INORGANIC CHEMISTRY

BASIC INORGANIC CHEMISTRY / THIRD EDITION

F. ALBERT COTTON

W. T. Doherty-Welch Foundation
Distinguished Professor of Chemistry
Texas A and M University
College Station, Texas, USA

GEOFFREY WILKINSON

Emeritus Professor of Inorganic Chemistry
Imperial College of Science, Technology, and Medicine
London SW7 2AY
England

PAUL L. GAUS

Professor of Chemistry
The College of Wooster
Wooster, Ohio, USA

JOHN WILEY & SONS, INC.
NEW YORK · CHICHESTER · BRISBANE · TORONTO · SINGAPORE

ACQUISITIONS EDITOR	Nedah Rose
MARKETING MANAGER	Catherine Faduska
PRODUCTION EDITOR	Deborah Herbert
TEXT DESIGNER	Karin Kincheloe
MANUFACTURING MANAGER	Susan Stetzer
COVER ILLUSTRATION	Roy Wiemann
ILLUSTRATION	Rosa Bryant

This book was set in 10×12 New Baskerville by General Graphic Services and printed and bound by Hamilton Printing. The cover was printed by Phoenix Color Corp.

Recognizing the importance of preserving what has been written, it is a policy of John Wiley & Sons, Inc. to have books of enduring value published in the United States printed on acid-free paper, and we exert our best efforts to that end.

The paper on this book was manufactured by a mill whose forest management programs include sustained yield harvesting of its timberlands. Sustained yield harvesting principles ensure that the number of trees cut each year does not exceed the amount of new growth.

Library of Congress Cataloging in Publication Data:
Cotton, F. Albert (Frank Albert), 1930–
  Basic inorganic chemistry / F. Albert Cotton, Geoffrey Wilkinson, Paul L. Gaus.—3rd ed.
      p.   cm.
  Includes index.
  ISBN 0-471-50532-3
  1. Chemistry, Inorganic.   I. Wilkinson, Geoffrey, Sir, 1921–
II. Gaus, Paul L.   III. Title.
QD141.2.C69   1995
546—dc20   94-20754
                                                              CIP

Preface

The goals for *Basic Inorganic Chemistry* remain essentially unchanged in the third edition: to teach the basics of inorganic chemistry with a primary emphasis on facts, and then to use the student's growing factual knowledge as a basis for discussing the important principles of periodicity in structure, bonding, and reactivity. Too often, we believe, have students been taught the overarching principles first, while facts have been given only secondary or sporadic emphasis. Two simple examples serve to illustrate this complaint. Although students are made to absorb elaborate theories for trends in the boiling points among various liquids, too many students do not know the boiling point of a single substance (other than water) to within $+/- 2$ °C. As a more sophisticated example consider the number of our students who can write a paragraph on the participation of *d* orbitals in the chemistry of silicon but who cannot write equations for the hydrolysis of the halides of silicon, germanium, tin, and lead, much less cite (let alone explain) the periodic trends that are found among these reactions. This book is meant for teachers who wish to avoid such errors in emphasis.

As in the second edition, we have emphasized the primary facts of inorganic chemistry, and we have organized the facts of chemical structure and reactivity (while presenting the pertinent theories) in a way that emphasizes the descriptive approach to the subject. The chemistry of the elements and their compounds is organized by classes of substances and types of reactions. Periodicity in structure and reactivity is emphasized.

This text can be used in a one-semester course that does not require physical chemistry (as taught traditionally in the United States) as a prerequisite. The principles generally encountered in the first year of college are reviewed in Chapter 1, and the book could be used in any inorganic course for which at least concurrent enrollment in sophomore organic chemistry was anticipated.

Important new material has been added to the text. This material includes a better introduction to inorganic chemistry, improved treatment of atomic orbitals and properties (such as electronegativity), new approaches to the depiction of ionic structures, nomenclature for transition metal compounds, quantitative approaches to acid–base chemistry, expanded and unified treatment of the periodicity in structure and reactivity among the main group elements, Wade's rules for boranes and carboranes, the chemistry of important new classes of substances (such as fullerenes and silenes), and a new chapter on the inorganic solid state. Material on symmetry elements, operations, and point groups has been put into an appendix. The glossary of terms has been updated.

Strategic additions or modifications have been made to most of the chapters, largely incorporating recent discoveries or additional examples that highlight periodicity in structure and reactivity. New Study Questions have been added throughout, and the Supplementary Readings lists have been brought up to date. A Solutions Manual will be available.

A number of important appendices have been added. These include symmetry operations and point groups, the full form of the hydrogen-like atomic orbital wave functions, and values for the various atomic properties, including ionization enthalpies, ionic radii, electron attachment enthalpies, and electronegativities.

The authors are grateful for the number and quality of suggestions made by teachers who have used the previous editions and by those who reviewed the manuscript for the third edition:

Donald Gaines, University of Wisconsin-Madison; Lawrence Kool, Boston College; Derek Davenport and Richard Walton, Purdue University; William Myers, University of Richmond; K. J. Balkus, University of Texas-Dallas; David C. Finster, Wittenberg University; Brice Bosnich, University of Chicago; J. H. Espenson, Iowa State University of Science and Technology; D. T. Haworth, Marquette University; John Nelson, University of Nevada-Reno; Phillip Davis, University of Tennessee-Martin

P.L.G. wishes to dedicate the Third Edition to his parents, Robert L. and Ollie M. Gaus, and to thank his wife Madonna and his daughters Laura and Amy for their prayers and support.

October 1994

F. Albert Cotton
College Station, Texas

Geoffrey Wilkinson
London, England

Paul L. Gaus
Wooster, Ohio

Preface

to the Second Edition

The principal goals in *Basic Inorganic Chemistry, Second Edition* are to set down the primary facts of inorganic chemistry in a clear and accurate manner, and to organize the facts of chemical structure and reactivity (while presenting the pertinent theories) in a way that emphasizes the descriptive approach to the subject. The chemistry of the elements and their compounds is organized by classes of substances and types of reactions, and periodicity in structure and reactivity is emphasized.

This text can be used in a one-semester course that does not require physical chemistry (as taught traditionally in the United States) as a prerequisite. The principles generally encountered in the first year of college are reviewed in Chapter 1, and the book could be used in any inorganic course for which at least concurrent enrollment in sophomore organic chemistry was anticipated. A glossary has been added to help make this second edition more useful in interdisciplinary settings.

Although the organization of the second edition is essentially unchanged from the first edition, some chapters have been revised considerably, and others have been rewritten entirely. There are, for instance, new sections on geometry and bonding in molecules and complex ions, boron chemistry, mechanisms of reactions of coordination compounds, electronic spectroscopy, and catalysis. The chapter on bioinorganic chemistry has been thoroughly revised and updated. The topics of structure, reactivity, and periodicity have been uniformly emphasized throughout the descriptive chapters. Bonding theories are developed in Chapter 3 (including an intuitive treatment of delocalized molecular orbital approaches), and these are applied in subsequent chapters wherever useful, and especially in the end-of-chapter exercises.

The end-of-chapter exercises have been revised and organized into three groups. *Review* questions are straightforward, and require only that the student recall the material in the chapter. *Additional Exercises* generally require application of important principles or additional thought by the student. *Questions from the Literature of Inorganic Chemistry* refer the student to specific journal articles that are germane to the topic at hand. Thus the study guides, supplementary readings, and study questions range in scope from a straightforward review of the chapter to the sort of professional literature on which the science is based. A separate solutions manual, containing detailed answers for each of the study questions, is also available.

The study guides at the end of certain chapters give some idea, to the student and the instructor, of the goals of, organization in, and prerequisites for a given chapter. Chapter 1 constitutes a review of the principles that are normally encountered in the first college year, and that are of use in the present text. Chapters 2 through 8 contain much of what is essential for complete comprehension of later chapters. Chapters 9 through 22 may be covered selectively, at the instructor's discretion, depending on the constraints of time. Chapter 23 is an important prerequisite for the material in Chapters 24 through 27, which are optional. Chapters 28 and 29 will be helpful to the discussion of the material in Chapter 30.

We are grateful for the efforts of those who reviewed the first edition, prior to its revision: Dr. Robert Parry, University of Utah; Dr. Richard Treptow, Chicago State University; and Prof. Glen Rodgers, Allegheny College.

We also gratefully acknowledge the very fine efforts of those who critiqued the revised edition: David Goodgame, Margaret Goodgame, Richard Treptow, Glen Rodgers, and Robert Parry. These reviewers made useful and substantial comments on the text, and have contributed significantly to its accuracy and clarity. Jeannette Stiefel was very helpful in editing the manuscript.

We would be pleased to correspond with teachers and to receive comments regarding the text. Suggestions for new journal articles to be used in *Questions from the Literature of Inorganic Chemistry* would be welcomed. Please address correspondence to P. L. Gaus.

Finally, P.L.G. wishes especially to acknowledge the help, encouragement, and support of his family: Madonna, Laura, and Amy, and to dedicate the revised edition to his parents.

October, 1986

F. Albert Cotton
College Station, Texas

Geoffrey Wilkinson
London, England

Paul L. Gaus
Wooster, Ohio

Preface

to the First Edition

Those who aspire not to guess and divine, but to discover and know, who propose not to devise mimic and fabulous worlds of their own, but to examine and dissect the nature of this very world itself, must go to facts themselves for everything.

F. Bacon, 1620

There are already several textbooks of inorganic chemistry that treat the subject in considerably less space than our comprehensive text, *Advanced Inorganic Chemistry*. Moreover, most of them include a great deal of introductory theory, which we omitted from our larger book because of space considerations. The net result is that these books contain very little of the real content of inorganic chemistry—namely, the actual facts about the properties and behavior of inorganic compounds.

Our purpose in *Basic Inorganic Chemistry*, is to meet the needs of teachers who present this subject to students who do not have the time or perhaps the inclination to pursue it in depth, but who may also require explicit coverage of basic topics such as the electronic structure of atoms and elementary valence theory. We therefore introduce material of this type, in an elementary fashion, and present only the main facts.

The point, however, is that this book does present the facts, in a systematic way. We have a decidedly Baconian philosophy about all chemistry, but particularly inorganic chemistry. We are convinced that inorganic chemistry *sans* facts (or nearly so), as presented in other books, is like a page of music with no instrument to play it on. One can appreciate the sound of music without knowing anything of musical theory, although of course one's appreciation is enhanced by knowing some theory. However, a book of musical theory, even if it is illustrated by audible snatches of themes and a few chord progressions, is quite unlike the hearing of a real composition in its entirety.

We believe that a student who has read a book on "inorganic chemistry" that consists almost entirely of theory and so-called principles, with but sporadic mention of the hard facts (only when they "nicely" illustrate the "principles") has not, in actual fact, had a course in inorganic chemistry. We deplore the current trend toward this way of teaching students who are not expected to specialize in the subject, and believe that even the nonspecialist ought to get a straight dose of the subject as it really is—"warts and all." This book was written to encourage the teaching of inorganic chemistry in a Baconian manner.

At the end of each chapter, there is a study guide. Occasionally this includes a few remarks on the scope and purpose of the chapter to help the student place it in the context of the entire book. A supplementary reading list is included in all chapters. This consists of relatively recent articles in the secondary (monograph and review) literature, which will be of interest to those who wish to pursue the subject matter in more detail. In some instances there is little literature of this kind available. However, the student—and the instructor—will find more detailed treatments of all the elements and classes of compounds, as well as further references, in our *Advanced Inorganic Chemistry,* fourth edition, Wiley, 1984, and in *Comprehensive Inorganic Chemistry,* J. C. Bailar, Jr., H. J. Emeléus, R. S. Nyholm, and A. F. Trotman-Dickinson, Eds., Pergamon, 1973.

F. Albert Cotton
Geoffrey Wilkinson

Contents

BASIC INORGANIC CHEMISTRY

Part 1

FIRST
PRINCIPLES

Chapter 1

SOME PRELIMINARIES

1-1   A Description of Inorganic Chemistry

Inorganic chemistry embraces all of the elements. Consequently, it ranges from the border of traditional organic chemistry (primarily the chemistry of carbon, specifically when bound to hydrogen, nitrogen, oxygen, sulfur, the halogens and a few other elements such as selenium and arsenic) to the borders of physical chemistry, which is the study of the physical properties and quantitative behavior of matter. Inorganic chemistry is not only concerned with molecular substances similar to those encountered in organic chemistry but is also concerned with the wider varieties of substances that include atomic gases, solids that are nonmolecular extended arrays, air sensitive (and hydrolytically sensitive) compounds, and compounds that are soluble in water and other polar solvents, as well as those that are soluble in nonpolar solvents. In short, inorganic chemistry encompasses a greater variety of substances than does organic chemistry.

A further difference between organic and inorganic chemistry is that whereas the atoms in organic substances principally have a maximum coordination number of 4 (e.g., CH_4 and NR_4^+), those in inorganic substances have coordination numbers frequently exceeding four (indeed, as high as 14), and exhibit a variety of oxidation states. Some simple examples include PF_5, which has the shape of a trigonal bipyramid, $W(CH_3)_6$, an organometallic compound that has six tungsten-to-carbon bonds, and $[Nd(H_2O)_9]^{3+}$, in which neodymium achieves a coordination number of 9.

The inorganic chemist thus faces the problem of ascertaining the structures, properties, and reactivities of an extraordinary range of materials, with widely differing properties and with exceedingly complicated patterns of structure and reactivity. We must hence be concerned with a great many methods of synthesis, manipulation, and characterization of inorganic compounds.

In accounting for the existence and in describing the behavior of inorganic materials, we shall need to use certain aspects of physical chemistry, notably thermodynamics, electronic structures of atoms, molecular bonding theories, and reaction kinetics. Some of these essential aspects of physical chemistry are reviewed later in this chapter. The rest of Part I of the text deals with atomic and molecular structure, chemical bonding, and other principles necessary for an understanding of the structure and properties of inorganic substances of all classes.

This book emphasizes the three most important aspects of inorganic chemistry: the structures, properties, and reactivities of the various inorganic substances. In doing so, one of the central themes to be found throughout the book is the periodic relationships that exist among the types of substances, their structures, and their reactivities.

Classes of Inorganic Substances

In the broadest sense, the materials that we shall discuss can be grouped into four classifications: elements, ionic compounds, molecular compounds, and polymers or network solids. The following brief list is presented to show the complicated variety of substances that are encountered in a discussion of inorganic chemistry. Greater detail is presented in the appropriate chapters to follow.

1. *The elements.* The elements have an impressive variety of structures and properties. Thus they can be
    (a) Either atomic (Ar, Kr) or molecular (H_2, O_2) gases.
    (b) Molecular solids (P_4, S_8, C_{60}).
    (c) Extended molecules or network solids (diamond, graphite).
    (d) Solid (W, Co) or liquid (Hg, Ga) metals.
2. *Ionic compounds.* These compounds are always solids at standard temperature and pressure. They include
    (a) Simple ionic compounds, such as NaCl, which are soluble in water or other polar solvents.
    (b) Ionic oxides that are insoluble in water (e.g., ZrO_2) and mixed oxides such as spinel ($MgAl_2O_4$), the various silicates [e.g., $CaMg(SiO_3)_2$], and so on.
    (c) Other binary halides, carbides, sulfides, and similar materials. A few examples are AgCl, SiC, GaAs, and BN, some of which should be better considered to be network solids.
    (d) Compounds containing polyatomic (so-called complex) ions, such as $[SiF_6]^{2-}$, $[Co(NH_3)_6]^{3+}$, $[Fe(CN)_6]^{3-}$, $[Fe(CN)_6]^{4-}$, and $[Ni(H_2O)_6]^{2+}$.
3. *Molecular compounds.* These compounds may be solids, liquids, or gases, and include, for example,
    (a) Simple, binary compounds, such as PF_3, SO_2, OsO_4, and UF_6.
    (b) Complex metal-containing compounds, such as $PtCl_2(PMe_3)_2$ and $RuH(CO_2Me)(PPh_3)_3$.
    (c) Organometallic compounds that characteristically have metal-to-carbon bonds. Some examples are $Ni(CO)_4$, $Zr(CH_2C_6H_5)_4$, and $U(C_8H_8)_2$.
4. *Network solids, or polymers.* Examples of these substances (discussed in Chapter 32), include the numerous and varied inorganic polymers and superconductors. One example of the latter has the formula $YBa_2Cu_3O_7$.

Classes of Inorganic Structures

The structures of the majority of organic substances are derived from the tetrahedron. Their predominance occurs because the maximum valence for carbon, as well as for most of the other elements (with the obvious exception of hydrogen) that are commonly bound to carbon in simple organic substances, is four. A much more complicated structural situation arises for inorganic substances since, as we have already mentioned, atoms may form many more than four bonds. It is therefore commonplace to find atoms in inorganic substances forming five, six, seven, and more bonds. The geometries of inorganic substances are, therefore, very much more elaborate and diverse than those of organic substances.

It is particularly fascinating to note that the tetrahedron, on which the geometry of organic compounds is based, is the simplest of the five regular polyhedra, otherwise known as the Platonic solids, which are shown below.

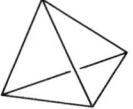
Tetrahedron
    Faces: 4 equilateral triangles
  Vertices: 4
    Edges: 6

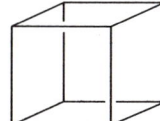
Cube
    Faces: 6 squares
  Vertices: 8
    Edges: 12

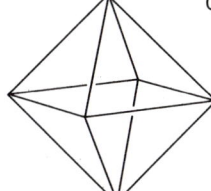
Octahedron
    Faces: 8 equilateral triangles
  Vertices: 6
    Edges: 12

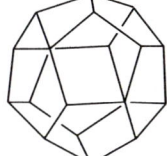
Dodecahedron
    Faces: 12 regular pentagons
  Vertices: 20
    Edges: 30

Icosahedron
    Faces: 20 equilateral triangles
  Vertices: 12
    Edges: 30

Since the days of Plato, it has been recognized that these five polyhedra constitute the complete set of regular polyhedra, which satisfy the following criteria.

1. The faces are all some regular polygon (equilateral triangle, square, or regular pentagon).
2. The vertices are all equivalent.
3. The edges are all equivalent.

Each of Plato's regular polyhedra is now known to form the basis for the structures of important classes of inorganic substances.

The structures of inorganic substances are often also based on many less regular polyhedra, such as the trigonal bipyramid, the trigonal prism, and so on, as well as on opened versions of regular and irregular polyhedra, in which one or more vertices are missing.

Clearly, structural inorganic chemistry presents a diverse array of possibilities. The student is encouraged to explore the remaining pages of the text for examples.

Classes of Inorganic Reactions

For the preponderance of organic reactions, it is appropriate to ascertain and discuss the mechanism by which the reaction proceeds. For many inorganic reactions, however, an understanding of the precise mechanism is either unnecessary or impossible. This happens for two principal reasons. First, unlike the situation for most organic substances, the bonds in inorganic compounds are often labile. Consequently, a variety of bond-making and bond-breaking events is likely during the course of an inorganic reaction. Under such circumstances, a reaction becomes capable of giving numerous products. Moreover, inorganic reactions often are conducted under circumstances, for example, vigorous stirring of a heterogeneous mixture at high temperature and pressure, that make elucidation of mechanism impossible or, at least, impractical.

For these two reasons, inorganic reactions are often best described only in terms of the overall outcome of the reaction. This approach is known as "descriptive inorganic chemistry." It should thus be readily appreciated that, although every reaction can be described in terms of the nature and identity of the products in relation to those of the reactants, not every reaction can be assigned a mechanism.

For purposes of descriptive inorganic chemistry, most reactions can be assigned to one or more of the following classes, which will be defined more thoroughly at the appropriate points in the text discussion:

1. Acid–base (neutralization).
2. Addition.
3. Elimination.
4. Oxidation–reduction (redox).
5. Insertion.
6. Substitution (displacement).
7. Rearrangement (isomerization).
8. Metathesis (exchange).
9. Solvolysis.
10. Chelation.
11. Cyclization and condensation.
12. Nuclear reactions.

At the most detailed level in our understanding of an inorganic reaction, we seek to prepare a complete reaction profile, from reactants, through any intermediates or transition states, to products. This requires intimate knowledge of the kinetics and/or thermodynamics of a reaction, as well as an appreciation of the influence of structure and bonding on reactivity. In the chapters that follow, we present this type of detail, and organize the facts so as to illustrate the periodic manner in which the structures, properties, and reactivities of inorganic substances vary.

But, first, in the rest of Chapter 1, we present a review of fundamental concepts of physical chemistry.

1-2 Thermochemistry

Standard States

To have universally recognized and understood values for energy changes in chemical processes, it is first necessary to define standard states for all substances.

The standard state for any substance is that phase in which it exists at 25 °C (298.15 K) and 1-atm (101.325 N m^{-2}) pressure. Substances in solution are at unit concentration.

Heat Content or Enthalpy

Virtually all physical and chemical changes either produce or consume energy. Generally, this energy takes the form of heat. The gain or loss of heat may be attributed to a change in the "heat content" of the substances taking part in the process. *Heat content* is called **enthalpy**, symbolized H. The change in heat content is called the **enthalpy change** ΔH, which is defined in Eq. 1-2.1.

$$\Delta H = (H \text{ of products}) - (H \text{ of reactants}) \tag{1-2.1}$$

For the case in which all products and reactants are in their standard states, the enthalpy change is designated $\Delta H°$, the **standard enthalpy change** of the process. For example, although the formation of water from H_2 and O_2 cannot actually be carried out at an appreciable rate at standard conditions, it is nevertheless useful to know, through indirect means, that the standard enthalpy change for Reaction 1-2.2 is highly negative.

$$H_2(g, 1 \text{ atm}, 25 °C) + \tfrac{1}{2} O_2(g, 1 \text{ atm}, 25 °C) = H_2O(\ell, 1 \text{ atm}, 25 °C)$$

$$\Delta H° = -285.7 \text{ kJ mol}^{-1} \quad (1\text{-}2.2)$$

The heat contents of all elements in their standard states are arbitrarily defined to be zero for thermochemical purposes.

The Signs of ΔH Values

In Eq. 1-2.2, $\Delta H°$ has a negative value. The heat content of the products has a lower value than that of the reactants, and heat is released to its surroundings by the process. This constitutes an **exothermic** process ($\Delta H < 0$). When heat is absorbed from the surroundings by the process ($\Delta H > 0$), it is called **endothermic.** The same convention will apply to changes in free energy ΔG, which will be discussed shortly.

Standard Heats (Enthalpies) of Formation

The standard enthalpy change for any reaction can be calculated if the **standard heat of formation** ΔH_f° of each reactant and product is known. It is therefore useful to have tables of ΔH_f° values, in units of kilojoules per mole (kJ mol^{-1}). The ΔH_f° value for a substance is the ΔH° value for the process in which 1 mol of that substance is produced in its standard state from elements, each in its standard state. Equation 1-2.2 describes such a process, and the ΔH° given for that reaction is actually the standard enthalpy of formation of liquid water, $\Delta H_f^\circ[H_2O(\ell)]$.

Take, as an example, the reaction shown in Eq. 1-2.3.

$$LiAlH_4(s) + 4\ H_2O(\ell) = LiOH(s) + Al(OH)_3(s) + 4\ H_2(g)$$
$$\Delta H^\circ = -599.6 \text{ kJ} \quad (1\text{-}2.3)$$

The standard enthalpy change for Reaction 1-2.3 may be calculated from Eq. 1-2.4.

$$\Delta H^\circ = \Delta H_f^\circ[LiOH(s)] + \Delta H_f^\circ[Al(OH)_3(s)]$$
$$- 4\ \Delta H_f^\circ[H_2O(\ell)] - \Delta H_f^\circ[LiAlH_4(s)] \quad (1\text{-}2.4)$$

Other Special Enthalpy Changes

Aside from formation of compounds from the elements, there are several other physical and chemical processes of special importance for which the ΔH or ΔH° values are frequently required. Among these are the process of melting (for which we specify the enthalpy of fusion ΔH_{fus}°), the process of vaporization (for which we specify the enthalpy of vaporization ΔH_{vap}°), and the process of sublimation (for which we specify the enthalpy of sublimation ΔH_{sub}°).

We also specially designate the enthalpy changes for ionization processes that produce cations or anions by loss or gain of electrons, respectively.

Ionization Enthalpies

The process of ionization by loss of electron(s), as in Reaction 1-2.5, is of particular interest.

$$Na(g) = Na^+(g) + e^-(g) \qquad \Delta H_{ion}^\circ = 502 \text{ kJ mol}^{-1} \qquad (1\text{-}2.5)$$

For many atoms, the enthalpies of removal of a second, third, and so on, electron are also of chemical interest. These enthalpies are known for most elements. For example, the first three ionization enthalpies of aluminum are given in Reactions 1-2.6–1-2.8.

$$Al(g) = Al^+(g) + e^- \qquad \Delta H^\circ = 577.5 \text{ kJ mol}^{-1} \qquad (1\text{-}2.6)$$

$$Al^+(g) = Al^{2+}(g) + e^- \qquad \Delta H^\circ = 1817 \text{ kJ mol}^{-1} \qquad (1\text{-}2.7)$$

$$Al^{2+}(g) = Al^{3+}(g) + e^- \qquad \Delta H^\circ = 2745 \text{ kJ mol}^{-1} \qquad (1\text{-}2.8)$$

The overall ionization enthalpy for formation of the $Al^{3+}(g)$ ion is then the sum of the single ionization enthalpies, as shown in Reaction 1-2.9.

$$Al(g) = Al^{3+}(g) + 3\ e^- \qquad \Delta H° = 5140\ kJ\ mol^{-1} \qquad (1\text{-}2.9)$$

Ionization enthalpies may also be defined for molecules, as in Eq. 1-2.10.

$$NO(g) = NO^+(g) + e^- \qquad \Delta H° = 890.7\ kJ\ mol^{-1} \qquad (1\text{-}2.10)$$

Note that for molecules and atoms the ionization enthalpies are always positive; energy must be expended to detach electrons. Also, the increasing magnitudes of successive ionization steps, as shown previously for aluminum, are completely general; the more positive the system becomes, the more difficult it is to ionize it further.

Electron Attachment Enthalpies

Consider Reactions 1-2.11 to 1-2.13.

$$Cl(g) + e^- = Cl^-(g) \qquad \Delta H° = -349\ kJ\ mol^{-1} \qquad (1\text{-}2.11)$$

$$O(g) + e^- = O^-(g) \qquad \Delta H° = -142\ kJ\ mol^{-1} \qquad (1\text{-}2.12)$$

$$O^-(g) + e^- = O^{2-}(g) \qquad \Delta H° = 844\ kJ\ mol^{-1} \qquad (1\text{-}2.13)$$

The $Cl^-(g)$ ion forms exothermically. The same is true of the other halide ions. Observe that the formation of the oxide ion $O^{2-}(g)$ requires first an exothermic and then an endothermic step. This is understandable because the O^- ion, which is already negative, will tend to resist the addition of another electron.

In most of the chemical literature, the negative of the enthalpy change for processes such as Eqs. 1-2.11 to 1-2.13 is called the electron affinity (A) for the atom. In this book, however, we shall use only the systematic notation illustrated previously: we shall speak of the enthalpy changes (ΔH_{EA}) that accompany the attachment of electrons to form specific ions.

Direct measurement of ΔH_{EA} values is difficult, and indirect methods tend to be inaccurate. To give an idea of their magnitudes, some of the known values (with those that are estimates in parentheses) are listed in kilojoules per mole:

H	−73											
Li	−58	Be	(+60)	B	(−30)	C	−120	N	(+10)	O	−142	F	−328
Na	(−50)					Si	(−135)	P	(−75)	S	−200	Cl	−349
										Se	(−160)	Br	−324
												I	−295

Bond Energies

Consider homolytic cleavage of the HF molecule as in Reaction 1-2.14.

$$HF(g) = H(g) + F(g) \qquad \Delta H_{298} = 566\ kJ\ mol^{-1} \qquad (1\text{-}2.14)$$

The enthalpy requirement of this process has a simple, unambiguous significance. It is the energy required to break the H—F bond. It can be called the "H—F bond energy," and we can, if we prefer, think of 566 kJ mol^{-1} as the energy released when the H—F bond is formed: a perfectly equivalent and equally unambiguous statement.

Consider, however, the stepwise cleavage of the two O—H bonds in water, as in Eqs. 1-2.15 and 1-2.16.

$$H_2O(g) = H(g) + OH(g) \qquad \Delta H_{298} = 497 \text{ kJ mol}^{-1} \qquad (1\text{-}2.15)$$

$$OH(g) = H(g) + O(g) \qquad \Delta H_{298} = 421 \text{ kJ mol}^{-1} \qquad (1\text{-}2.16)$$

These two processes of breaking the O—H bonds one after the other have different energies. Furthermore, the overall homolytic cleavage of the two O—H bonds, as in Eq. 1-2.17,

$$H_2O(g) = 2\,H(g) + O(g) \qquad \Delta H_{298} = 918 \text{ kJ mol}^{-1} \qquad (1\text{-}2.17)$$

has an associated enthalpy change that is the sum of those for the individual steps (Eq. 1-2.15 + Eq. 1-2.16). How then can we define the O—H bond energy? It is customary to take the mean of the two values for Reactions 1-2.15 and 1-2.16, which is one half of their sum: $918/2 = 459$ kJ mol^{-1}. We then speak of a mean O—H bond energy, a quantity that we must remember is somewhat artificial; we cannot know the actual enthalpies of either step if we know only their mean.

When we consider molecules containing more than one kind of bond, the problem of defining bond energies becomes even more subtly troublesome. For example, we can consider that the total enthalpy change for Reaction 1-2.18

$$H_2N\text{—}NH_2(g) = 2\,N(g) + 4\,H(g) \qquad \Delta H_{298} = 1724 \text{ kJ mol}^{-1} \qquad (1\text{-}2.18)$$

consists of the sum of the N—N bond energy E_{N-N}, and four times the mean N—H bond energy E_{N-H}. But is there any unique and rigorous way to divide the total enthalpy needed for Reaction 1-2.18 (1724 kJ mol^{-1}) into these component parts? The answer is no. Instead we take the following practical approach.

We know, from experiment, the enthalpy change for Reaction 1-2.19.

$$NH_3(g) = N(g) + 3\,H(g) \qquad \Delta H_{298} = 1172 \text{ kJ mol}^{-1} \qquad (1\text{-}2.19)$$

Thus we can determine that the mean N—H bond energy (E_{N-H}) is

$$E_{N-H} = \frac{1172}{3} = 391 \text{ kJ mol}^{-1} \qquad (1\text{-}2.20)$$

If we make the *assumption* that this value can be transferred to H$_2$NNH$_2$, then we can evaluate the N—N bond energy according to Eq. 1-2.21.

$$E_{N-N} + 4\,E_{N-H} = 1724 \text{ kJ mol}^{-1}$$

$$E_{N-N} = 1724 - 4\,E_{N-H}$$

$$= 1724 - 4(391)$$

$$= 160 \text{ kJ mol}^{-1} \qquad (1\text{-}2.21)$$

By proceeding in this way it is possible to build up a table of bond energies. These values can then be used to calculate the enthalpies of *forming* molecules from their constituent gaseous atoms. The success of this approach indicates that the energy of the bond between a given pair of atoms is somewhat independent of the molecular environment in which that bond occurs. This assumption is only approximately true, but true enough that the approach can be used in understanding and interpreting many chemical processes.

Thus far only single bonds have been considered. What about double and triple bonds? The bond energy increases as the bond order increases, in all cases. The increase is not linear, however, as shown in Fig. 1-1. A list of some generally useful bond energies is given in Table 1-1.

1-3   **Free Energy and Entropy**

The direction in which a chemical reaction will go and the point at which equilibrium will be reached depend on two factors: (1) The tendency to give off energy; exothermic processes are favored. (2) The tendency to attain a state that is statistically more probable, crudely describable as a "more disordered" state.

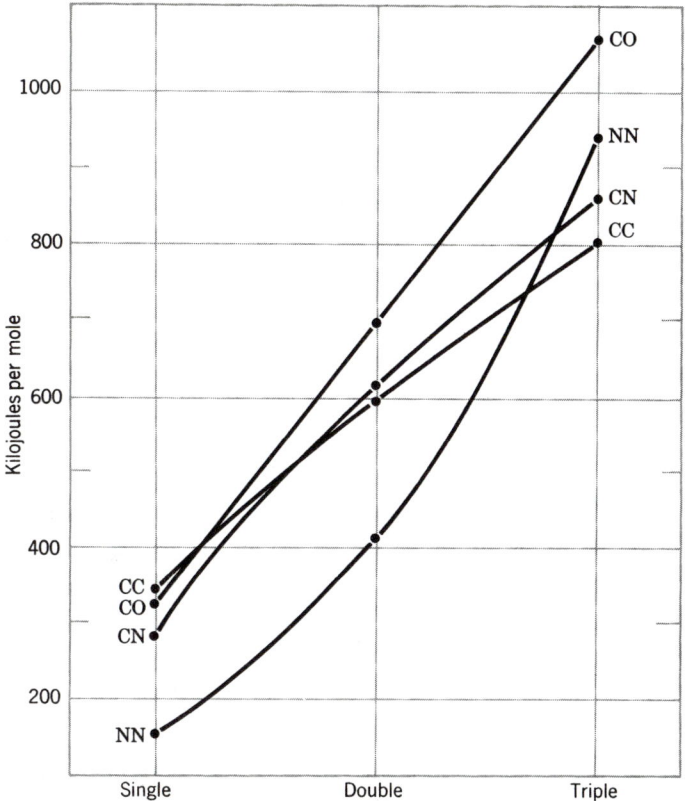

Figure 1-1   The variation of the bond energy with bond order for CC, NN, CN, and CO bonds.

Table 1-1   Some Average Thermochemical Bond Energies at 25 °C (in kJ mol^{-1})

	H	C	Si	Ge	N	P	As	O	S	Se	F	Cl	Br	I
						A. Single bond energies								
H	436	416	323	289	391	322	247	467	347	276	566	431	366	299
C		356	301	255	285	264	201	336	272	243	485	327	285	213
Si			226	—	335	—	—	368	226	—	582	391	310	234
Ge				188	256	—	—	—	—	—	—	342	276	213
N					160	~200	—	201	—	—	272	193	—	—
P						209	—	~340	—	—	490	319	264	184
As							180	331	—	—	464	317	243	180
O								146	—	—	190	205	—	201
S									226	—	326	255	213	—
Se										172	285	243	—	—
F											158	255	238	—
Cl												242	217	209
Br													193	180
I														151

B. Multiple bond energies

C=C 598	C=N 616	C=O 695	N=N 418
C≡C 813	C≡N 866	C≡O 1073	N≡N 946

We already have a measure of the energy change of a system: the magnitude and sign of ΔH.

The statistical probability of a given state of a system is measured by its *entropy*, denoted S. The greater the value of S, the more probable (and, generally, more disordered) is the state. Thus we can rephrase the two statements made in the first paragraph as follows: The likelihood of a process occurring increases as (1) ΔH becomes more negative, or (2) ΔS becomes more positive.

Only in rare cases (an example being racemization)

$$2\ d\text{-}[\text{Co(en)}_3]^{3+} = d\text{-}[\text{Co(en)}_3]^{3+} + \ell\text{-}[\text{Co(en)}_3]^{3+} \tag{1-3.1}$$

$$(\text{en} = \text{ethylenediamine})$$

does a reaction have $\Delta H = 0$. In such a case, the direction and extent of reaction depend solely on ΔS. In the case where $\Delta S = 0$, ΔH would alone determine the extent and direction of reaction. However, both cases are exceptional and it is, therefore, necessary to know how these two quantities combine to influence the direction and extent of a reaction. Thermodynamics provides the necessary relationship, which is

$$\Delta G = \Delta H - T\,\Delta S \tag{1-3.2}$$

in which T represents the absolute temperature in kelvins (K).

The letter G stands for the *free energy*, which is measured in kilojoules per mole (kJ mol^{-1}). The units of entropy are joules per kelvin per mol (J K^{-1} mol^{-1}), but for use with ΔG and ΔH in kilojoules per mole (kJ mol^{-1}), ΔS must expressed as kilojoules per kelvin per mol (kJ K^{-1} mol^{-1}).

1-4   Chemical Equilibrium

For any chemical reaction,

$$aA + bB + cC + \cdots = kK + lL + mM + \cdots \tag{1-4.1}$$

the position of equilibrium, for given temperature and pressure, is expressed by the equilibrium constant K. This is defined as follows:

$$K = \frac{[K]^k[L]^l[M]^m \cdots}{[A]^a[B]^b[C]^c \cdots} \tag{1-4.2}$$

where [A], [B], and so on, represent the thermodynamic *activities* of A, B, and so on. For reactants in solution, the activities are approximated by the concentrations in moles per liter so long as the solutions are not too concentrated. For gases, the activities are approximated by the pressures. For a pure liquid or solid phase X, the activity is defined as unity. Therefore, $[X]^x$ can be omitted from the expression for the equilibrium constant.

1-5 $\Delta G°$ As a Predictive Tool

For any reaction, the position of the equilibrium at 25 °C is determined by the value of $\Delta G°$. The parameter $\Delta G°$ is defined in a manner similar to that for $\Delta H°$, namely, Eq. 1-5.1,

$$\Delta G° = \sum \Delta G_f°(\text{products}) - \sum \Delta G_f°(\text{reactants}) \tag{1-5.1}$$

which similarly applies only at 25 °C (298.15 K). In terms of enthalpy and entropy we also have Eq. 1-5.2, at 25 °C:

$$\Delta G° = \Delta H° - 298.15 \, \Delta S° \tag{1-5.2}$$

where $\Delta S°$, the standard entropy change for the reaction, is defined as the difference between the sum of the absolute entropies of the products and the sum of the absolute entropies of the reactants.

$$\Delta S° = \sum S°(\text{products}) - \sum S°(\text{reactants}) \tag{1-5.3}$$

The standard against which we tabulate entropy for any substance is the perfect crystalline solid at 0 K, for which the absolute entropy is taken to be zero.

The following relationship exists between ΔG and the equilibrium constant, K:

$$\Delta G = -RT \ln K \tag{1-5.4}$$

where R is the gas constant and has the value

$$R = 8.314 \, \text{J K}^{-1} \, \text{mol}^{-1} \tag{1-5.5}$$

in units appropriate to this equation. At 25 °C we have

$$\Delta G° = -5.69 \log K_{298.15} \tag{1-5.6}$$

For a reaction with $\Delta G° = 0$, the equilibrium constant is unity. The more negative the value of $\Delta G°$ the more the reaction proceeds in the direction written, that is, to produce the substances on the right and consume those on the left.

When $\Delta G°$ is considered as the net result of enthalpy ($\Delta H°$) and entropy ($\Delta S°$) contributions, a number of possibilities must be considered. Reactions that proceed as written, that is, from left to right, have $\Delta G° < 0$. There are three main ways this can happen.

1. Both $\Delta H°$ and $\Delta S°$ favor the reaction. That is, $\Delta H° < 0$ and $\Delta S° > 0$.
2. The parameter $\Delta H°$ favors the reaction while $\Delta S°$ does not, but $\Delta H°$ (<0) has a greater absolute value than $T\Delta S°$, thus giving a net negative $\Delta G°$.
3. The parameter $\Delta H°$ (>0) disfavors the reaction, but $\Delta S°$ is positive and sufficiently large so that $T\Delta S°$ has a larger absolute magnitude than $\Delta H°$.

There are actual chemical reactions that belong to each of these categories.

Case 1 is fairly common. The formation of carbon monoxide (CO) from the elements is an example:

$$\tfrac{1}{2} O_2(g) + C(s) = CO(g)$$
$$\Delta G° = -137.2 \text{ kJ mol}^{-1}$$
$$\Delta H° = -110.5 \text{ kJ mol}^{-1}$$
$$T\Delta S° = 26.7 \text{ kJ mol}^{-1} \tag{1-5.7}$$

as are a host of combustion reactions, for example,

$$S(s) + O_2(g) = SO_2(g)$$
$$\Delta G° = -300.4 \text{ kJ mol}^{-1}$$
$$\Delta H° = -292.9 \text{ kJ mol}^{-1}$$
$$T\Delta S° = 7.5 \text{ kJ mol}^{-1} \tag{1.5.8}$$

$$C_4H_{10}(g) + \tfrac{13}{2} O_2(g) = 4\, CO_2(g) + 5\, H_2O(g)$$
$$\Delta G° = -2705 \text{ kJ mol}^{-1}$$
$$\Delta H° = -2659 \text{ kJ mol}^{-1}$$
$$T\Delta S° = 46 \text{ kJ mol}^{-1} \tag{1-5.9}$$

The reaction used in industrial synthesis of ammonia is an example of case 2.

$$N_2(g) + 3\, H_2(g) = 2\, NH_3(g)$$
$$\Delta G° = -16.7 \text{ kJ mol}^{-1}$$
$$\Delta H° = -46.2 \text{ kJ mol}^{-1}$$
$$T\Delta S° = -29.5 \text{ kJ mol}^{-1} \tag{1-5.10}$$

The negative entropy term can be attributed to the greater "orderliness" of a product system that contains only 2 mol of independent particles compared with the reactant system in which there are 4 mol of independent molecules.

Case 3 is the rarest. Examples are provided by substances that dissolve endothermically to give a saturated solution greater than 1 M in concentration. This happens with sodium chloride (NaCl).

$$NaCl(s) = Na^+(aq) + Cl^-(aq)$$

$$\Delta G^\circ = -2.7$$

$$\Delta H^\circ = +1.9$$

$$T\,\Delta S^\circ = +4.6 \tag{1-5.11}$$

Note that the ΔG° value does not *necessarily* predict the *actual result* of a reaction, but only the result that corresponds to the attainment of equilibrium at 25 °C. This value tells what is *possible,* but not what will actually *occur.* Thus, none of the first four reactions cited, which all have $\Delta G^\circ < 0$, actually occurs to a detectable extent at 25 °C simply on mixing the reactants. Activation energy and/or a catalyst (see page 23) must be supplied. Moreover, there are many compounds that are perfectly stable in a practical sense with positive values of ΔG°_f. These compounds do not spontaneously decompose into the elements, although that would be the equilibrium situation. Common examples are benzene, CS_2, and hydrazine (H_2NNH_2).

The actual occurrence of a reaction requires not only that ΔG° be negative but that the *rate* of the reaction be appreciable.

1-6   Temperature Dependence of the Equilibrium Constant

The equilibrium constant for a reaction depends on temperature. That dependence is determined by ΔH°, and the dependence can be used to determine ΔH° in the following way. If the value of the equilibrium constant is known to be K_1 at T_1 and K_2 at T_2, then we have Eqs. 1-6.1 and 1-6.2.

$$\ln K_1 = -\frac{\Delta H^\circ}{RT_1} + \frac{\Delta S^\circ}{R} \tag{1-6.1}$$

$$\ln K_2 = -\frac{\Delta H^\circ}{RT_2} + \frac{\Delta S^\circ}{R} \tag{1-6.2}$$

By subtracting Eqs. 1-6.1 and 1-6.2 we have Eq. 1-6.3:

$$\ln K_1 - \ln K_2 = \ln\frac{K_1}{K_2} = -\frac{\Delta H^\circ}{R}\left(\frac{1}{T_1} - \frac{1}{T_2}\right) \tag{1-6.3}$$

which allows us to calculate ΔH° if we can measure the equilibrium constant at two different temperatures. In practice, one secures greater accuracy by measuring the equilibrium constant at several different temperatures and plotting ln

K versus $1/T$. Such a plot should be a straight line with a slope of $-(\Delta H^\circ/R)$, assuming that ΔH° is constant over the temperature range employed.

1-7   Electrochemical Cell Potentials

Although it is true that the direction and extent of a reaction are indicated by the sign and magnitude of ΔG°, this is not generally an easy quantity to measure. There is one class of reactions, redox reactions in solution, that frequently allows straightforward measurement of ΔG°. The quantity actually measured is the potential difference ΔE (in volts, V), between two electrodes. Under the proper conditions, this can be related to ΔG° beginning with the following equation:

$$\Delta E = \Delta E^\circ - \frac{RT}{n\mathscr{F}} \ln Q \qquad (1\text{-}7.1)$$

The parameter ΔE° is the so-called standard potential, which will be discussed more fully. The number of electrons in the redox reaction as written is n, and $\mathscr{F}$ is the faraday, 96,486.7 C mol^{-1}.

The expression Q has the same algebraic form as the equilibrium constant for the reaction, into which the actual activities that exist when ΔE is measured are inserted. Clearly, when each concentration equals unity, the $\log Q = \log 1 = 0$ and the measured ΔE equals ΔE°, which is the standard potential for the cell.

To illustrate, the reaction between zinc and hydrogen ions may be used.

$$\text{Zn}(s) + 2\,\text{H}^+(aq) = \text{Zn}^{2+}(aq) + \text{H}_2(g) \qquad (1\text{-}7.2)$$

For this, $n = 2$ and Q has the form

$$\frac{A_{\text{H}_2} \cdot A_{\text{Zn}^{2+}}}{A_{\text{H}^+}^2} \qquad (A_{\text{Zn}} = 1) \qquad (1\text{-}7.3)$$

The symbol A_X represents the thermodynamic activity of X. For dilute gases, the activity is equal to the pressure, and for dilute solutions, the activity is equal to the concentration. At higher pressures or concentrations, correction factors (called activity coefficients) are necessary. In these cases the activity is not equal to pressure or concentration. We shall assume here that the activity coefficients can be ignored, so that the actual pressures and concentrations may be used.

Now, suppose the reaction of interest is allowed to run until equilibrium is reached. The numerical value of Q is then equal to the equilibrium constant, K. Moreover, at equilibrium there is no longer any tendency for electrons to flow from one electrode to the other: $\Delta E = 0$. Thus, we have

$$0 = \Delta E^\circ - \frac{RT}{n\mathscr{F}} \ln K \qquad (1\text{-}7.4)$$

or

$$\Delta E^\circ = \frac{RT}{n\mathscr{F}} \ln K \qquad (1\text{-}7.5)$$

However, we already know that

$$\Delta G^\circ = -RT \ln K \qquad (1\text{-}7.6)$$

Therefore, we have a way of relating cell potentials to ΔG° values, that is,

$$\frac{n\mathscr{F}}{RT}\Delta E^\circ = -\frac{1}{RT}\Delta G^\circ \qquad (1\text{-}7.7)$$

or

$$\Delta G^\circ = -n\mathscr{F}\,\Delta E^\circ \qquad (1\text{-}7.8)$$

Just as ΔG° values for a series of reactions may be added algebraically to give ΔG° for a reaction that is the sum of those added so, too, may ΔE° values be combined. But, remember that it is $n\,\Delta E^\circ$, not simply ΔE°, which must be used for each reaction. The factor $\mathscr{F}$ will, of course, cancel out in such a computation. For example, take the sum of Eqs. 1-7.9 and 1-7.10:

$$(n = 2)$$
$$\mathrm{Zn(s)} + 2\,\mathrm{H^+(aq)} = \mathrm{Zn^{2+}(aq)} + \mathrm{H_2(g)} \qquad \Delta E_1^\circ = +0.763 \qquad (1\text{-}7.9)$$

$$(n = 2)$$
$$2\,\mathrm{Cr(aq)^{3+}} + \mathrm{H_2(g)} = 2\,\mathrm{Cr^{2+}(aq)} + 2\,\mathrm{H^+(aq)} \qquad \Delta E_2^\circ = -0.408 \qquad (1\text{-}7.10)$$

$$(n = 2)$$
$$\mathrm{Zn(s)} + 2\,\mathrm{Cr^{3+}(aq)} = \mathrm{Zn^{2+}(aq)} + 2\,\mathrm{Cr^{2+}(aq)} \qquad \Delta E_3^\circ = +0.355 \qquad (1\text{-}7.11)$$

The correct relationship for the potential of the net reaction 1-7.11 is

$$2\,\Delta E_3^\circ = 2\,\Delta E_1^\circ + 2\,\Delta E_2^\circ \qquad (1\text{-}7.12)$$

In this example, we have added balanced equations to give a balanced equation. This automatically ensures that the coefficient n is the same for each ΔE° value. However, in dealing with electrode potentials (see the next section) instead of potentials of balanced reactions the cancellation is not automatic, as we shall learn presently.

Signs of ΔE° Values

Physically, there is no absolute way to associate algebraic signs with measured ΔE° values. Yet, a convention must be adhered to since, as illustrated previously, the signs of some are opposite to those of others.

Negative values of ΔG° correspond to reactions for which the equilibrium state favors products, that is, reactions that proceed in the direction written. Therefore, reactions that "go" also have positive ΔE° values. The reduction of Cr^{3+} by elemental zinc ($E^\circ = +0.355$ V) therefore goes as written in the previous example.

Half-Cells and Half-Cell (or Electrode) Potentials

Any complete, balanced chemical reaction can be artificially separated into two "half-reactions." Correspondingly, any complete electrochemical cell can be separated into two hypothetical half-cells. The potential of the actual cell, ΔE°, can then be regarded as the algebraic sum of the two half-cell potentials. In the three previously cited reactions, there are a total of three distinct half-cells. Let us consider first the reaction of zinc and $H^+(aq)$.

$$Zn(s) = Zn^{2+}(aq) + 2\ e^- \qquad E_1^\circ = +0.763\ V$$

$$\underline{2\ H^+(aq) + 2\ e^- = H_2(g) \qquad E_2^\circ = 0.000\ V}$$

$$Zn(s) + 2\ H^+(aq) = Zn^{2+}(aq) + H_2(g) \qquad E^\circ = +0.763\ V \qquad (1\text{-}7.13)$$

The half-cells E_1° and E_2° must be chosen to give the sum +0.763 V. The only solution to this or any similar problem is to assign an arbitrary *conventional* value to one such half-cell potential. All others will then be determined. The conventional choice is to assign the hydrogen half-cell a standard potential of zero. The zinc half-cell reaction as written will then have $E^\circ = +0.763$ V. In an exactly analogous way we get

$$Cr^{3+}(aq) + e^- = Cr^{2+}(aq) \qquad E^\circ = -0.408\ V \qquad (1\text{-}7.14)$$

These two half-cell potentials may then be used directly to calculate the standard potential for reduction of Cr^{3+} by $Zn(s)$.

$$Zn(s) = Zn^{2+}(aq) + 2\ e^- \qquad E^\circ = +0.763\ V$$

$$\underline{2\ e^- + 2\ Cr^{3+}(aq) = 2\ Cr^{2+}(aq) \qquad E^\circ = -0.408\ V}$$

$$Zn(s) + 2\ Cr^{3+}(aq) = Zn^{2+}(aq) + 2\ Cr^{2+}(aq) \qquad E^\circ = +0.355\ V \qquad (1\text{-}7.15)$$

Since each reaction involves the same number of electrons, the factor n in the expression $\Delta G^\circ = -n\mathscr{F}E^\circ$ is the same in this case and will cancel out.

When two half-cell reactions are added to give a third half-cell reaction, the n values will not be able to cancel out and must be explicitly employed in the computation. For example,

$$Cl^- + 3\ H_2O = ClO_3^- + 6\ H^+ + 6\ e^- \qquad E^\circ = -1.45 \qquad 6\ E_1^\circ = -8.70\ V$$

$$\underline{e^- + \tfrac{1}{2}\ Cl_2 = Cl^- \qquad\qquad\qquad\qquad E^\circ = +1.36 \qquad 1\ E_2^\circ = +1.36\ V}$$

$$\tfrac{1}{2}\ Cl_2 + 3\ H_2O = ClO_3^- + 6\ H^+ + 5\ e^- \qquad E^\circ = -1.47 \qquad 5\ E_3^\circ = -7.34\ V$$

where it should be emphasized that the correct relationship between the half-cell potentials is given in Eq. 1-7.16:

$$5\ E_3^\circ = 6\ E_1^\circ + 1\ E_2^\circ$$

$$E_3^\circ \neq E_1^\circ + E_2^\circ \qquad (1\text{-}7.16)$$

Thus, the correct value of E_3° (−1.47 V) is nowhere near the simple sum of $E_1^\circ + E_2^\circ$ (−0.09 V).

Tables of Half-Cell or Electrode Potentials

The International Union of Pure and Applied Chemistry has agreed that half-cell and electrode potentials shall be written as reductions and the terms "half-cell potential" or "electrode potential" shall mean values carrying the sign appropriate to the reduction reaction. For example, the zinc electrode reaction is tabulated as

$$Zn^{2+}(aq) + 2\ e^- = Zn(s) \qquad E° = -0.763\ V \qquad (1\text{-}7.17)$$

Zinc is said to have an electrode potential of *minus* 0.763 V.

This convention is most easily remembered by noting that a half-cell reaction with a *negative* potential is *electron* rich. When two half-cells are combined to produce a complete electrolytic cell, the electrode having the more negative standard half-cell potential will be, physically, the negative electrode (electron source) if the cell is to be operated as a battery.

A list of some important standard half-cell or electrode potentials is given in Table 1-2.

1-8   Kinetics

It is primarily through the study of the kinetics of a reaction that one gains insight into the mechanism of the reaction. In kinetics experiments, the rate of a reaction is studied as a function of the concentrations of each of the reactants and products. Activities or pressures may be employed in place of concentration. The rate of a reaction is also studied as a function of reaction conditions: temperature, solvent polarity, catalysis, and the like. A kinetic study begins with the determination of the rate law for the reaction. It is assumed that the correct stoichiometry has already been determined.

The Rate Law

This is an algebraic equation, determined experimentally for each reaction, which tells how the rate of reaction (units = concentration $\times$ time^{-1}) depends on the concentrations of reactants and products, other things, such as temperature, being fixed. For example, it has been shown that Reaction 1-8.1:

$$4\ HBr(g) + O_2(g) = 2\ H_2O(g) + 2\ Br_2(g) \qquad (1\text{-}8.1)$$

has the rate law Eq. 1-8.2:

$$\frac{d[O_2]}{dt} = -k[HBr][O_2] \qquad (1\text{-}8.2)$$

The rate of Reaction 1-8.1 (expressed as the decrease in the concentration of O_2 as a function of time) is proportional to the first power of the HBr concentration and to the first power of the oxygen concentration. Note that the rate law is not derived from the stoichiometry of the reaction; four equivalents of HBr are consumed in the stoichiometric equation, but the HBr concentration is only featured to the first power in the rate law. Although a total of five molecules must react to complete the process of Reaction 1-8.1, the rate law implies that the slowest or rate-determining step in the process is one that engages only one O_2 molecule and one HBr molecule.

Table 1-2   Some Half-Cell Reduction Potentials

Reaction Equation			$E°$ (V)
$Li^+ + e^-$	$=$	Li	-3.04
$Cs^+ + e^-$	$=$	Cs	-3.02
$Rb^+ + e^-$	$=$	Rb	-2.99
$K^+ + e^-$	$=$	K	-2.92
$Ba^{2+} + 2\,e^-$	$=$	Ba	-2.90
$Sr^{2+} + 2\,e^-$	$=$	Sr	-2.89
$Ca^{2+} + 2\,e^-$	$=$	Ca	-2.87
$Na^+ + e^-$	$=$	Na	-2.71
$Mg^{2+} + 2\,e^-$	$=$	Mg	-2.34
$\frac{1}{2}H_2 + e^-$	$=$	H^-	-2.23
$Al^{3+} + 3\,e^-$	$=$	Al	-1.67
$Zn^{2+} + 2\,e^-$	$=$	Zn	-0.76
$Fe^{2+} + 2\,e^-$	$=$	Fe	-0.44
$Cr^{3+} + e^-$	$=$	Cr^{2+}	-0.41
$H_3PO_4 + 2\,H^+ + 2\,e^-$	$=$	$H_3PO_3 + H_2O$	-0.20
$Sn^{2+} + 2\,e^-$	$=$	Sn	-0.14
$H^+ + e^-$	$=$	$\frac{1}{2}H_2$	0.00
$Sn^{4+} + 2\,e^-$	$=$	Sn^{2+}	0.15
$Cu^{2+} + e^-$	$=$	Cu^+	0.15
$S_4O_6^{2-} + 2\,e^-$	$=$	$2\,S_2O_3^{2-}$	0.17
$Cu^{2+} + 2\,e^-$	$=$	Cu	0.34
$Cu^+ + e^-$	$=$	Cu	0.52
$\frac{1}{2}I_2 + e^-$	$=$	I^-	0.53
$H_3AsO_4 + 2\,H^+ + 2\,e^-$	$=$	$H_3AsO_3 + H_2O$	0.56
$O_2 + 2\,H^+ + 2\,e^-$	$=$	H_2O_2	0.68
$Fe^{3+} + e^-$	$=$	Fe^{2+}	0.76
$\frac{1}{2}Br_2 + e^-$	$=$	Br^-	1.09
$IO_3^- + 6\,H^+ + 6\,e^-$	$=$	$I^- + 3\,H_2O$	1.09
$IO_3^- + 6\,H^+ + 5\,e^-$	$=$	$\frac{1}{2}I_2 + 3\,H_2O$	1.20
$\frac{1}{2}Cl_2 + e^-$	$=$	Cl^-	1.36
$\frac{1}{2}Cr_2O_7^{2-} + 7\,H^+ + 3\,e^-$	$=$	$Cr^{3+} + \frac{7}{2}H_2O$	1.36
$MnO_4^- + 8\,H^+ + 5\,e^-$	$=$	$Mn^{2+} + 4\,H_2O$	1.52
$Ce^{4+} + e^-$	$=$	Ce^{3+}	1.61
$H_2O_2 + 2\,H^+ + 2\,e^-$	$=$	$2\,H_2O$	1.77
$\frac{1}{2}S_2O_8^{2-} + e^-$	$=$	SO_4^{2-}	2.05
$O_3 + 2\,H^+ + 2\,e^-$	$=$	$O_2 + H_2O$	2.07
$\frac{1}{2}F_2 + e^-$	$=$	F^-	2.85
$\frac{1}{2}F_2 + H^+ + e^-$	$=$	HF	3.03

This reaction is called a second-order reaction because the sum of the exponents on the concentration terms of the rate law is two. The reaction is further said to be first order in each reactant.

The other common type of reaction, kinetically speaking, is the first-order reaction. The decomposition of N_2O_5 according to Eq. 1-8.3 is an example:

$$2\,N_2O_5(g) = 4\,NO_2(g) + O_2(g) \tag{1-8.3}$$

$$\frac{d[N_2O_5]}{dt} = -k[N_2O_5] \tag{1-8.4}$$

The first-order rate law implies certain useful regularities. Equation 1-8.4 can be rearranged and integrated as follows:

$$\frac{d[N_2O_5]}{[N_2O_5]} = -kdt$$

$$d\{\ln[N_2O_5]\} = -kdt \tag{1-8.5}$$

$$\ln\frac{[N_2O_5]_t}{[N_2O_5]_0} = -kt$$

where $[N_2O_5]_0$ denotes the initial reactant concentration that is employed at the start of a kinetics experiment, and $[N_2O_5]_t$ denotes the concentration that is found after some time t.

An equivalent expression can be given for any substance that disappears in first-order fashion, namely, Eq. 1-8.6.

$$\frac{[X]_t}{[X]_0} = e^{-kt} \tag{1-8.6}$$

For the particular case where one-half of the original quantity of reactant has disappeared, we have

$$[X]_t = \tfrac{1}{2}[X]_0 \tag{1-8.7}$$

so that Eq. 1-8.5 becomes

$$\ln \tfrac{1}{2} = -kt_{1/2} \tag{1-8.8}$$

or

$$t_{1\,2} = \frac{1}{k}\ln 2 = \frac{0.693}{k} \tag{1-8.9}$$

Thus the half-life $t_{1/2}$ of a first-order process is inversely proportional to the rate constant k. The higher the rate constant, the faster is the reaction, and the shorter is the half-life.

The Effect of Temperature on Reaction Rates

The rates of chemical reactions increase with increasing temperature. Generally, the dependence of the rate constant k on temperature T (in kelvins, K) follows the Arrhenius equation, at least over moderate temperature ranges ($\sim$ 100 K).

$$k = Ae^{-E_a/RT} \tag{1-8.10}$$

The coefficient A is called the frequency factor and E_a is called the activation energy. The higher the activation energy the slower the reaction at any given temperature. By plotting $\log k$ against T the value of E_a (as well as A) can be determined. These E_a values are often useful in interpreting the reaction mechanism.

An alternative approach to interpreting the temperature dependence of re-action rates, especially for reactions in solution, is based on the so-called absolute reaction rate theory. In essence, this theory postulates that in the rate-determining step, the reacting species A and B combine reversibly to form an "activated complex" $AB^{\ddagger}$, which can then decompose into products. Thus the following pseudoequilibrium constant is written

$$K^{\ddagger} = \frac{[AB^{\ddagger}]}{[A][B]} \tag{1-8.11}$$

The activated complex $AB^{\ddagger}$ is treated as a normal molecule except that one of its vibrations is considered to have little or no restoring force, which allows dissoci-ation into products. The frequency ν with which dissociation to products takes place is assumed to be given by equating the "vibrational" energy $h\nu$ to thermal energy $\mathbf{k}T$. Thus we write

$$-\frac{d[A]}{dt} = \nu[AB^{\ddagger}] = \left(\frac{\mathbf{k}T}{h}\right)[AB^{\ddagger}] \tag{1-8.12}$$

The measurable rate constant is defined by

$$-\frac{d[A]}{dt} = k[A][B] \tag{1-8.13}$$

so that we have

$$k = \left(\frac{\mathbf{k}T}{h}\right)\frac{[AB^{\ddagger}]}{[A][B]} = \frac{\mathbf{k}T}{h}K^{\ddagger} \tag{1-8.14}$$

The formation of this activated complex is governed by thermodynamic con-siderations similar to those of ordinary chemical equilibria. Thus we have

$$\Delta G^{\ddagger} = -RT \ln K^{\ddagger} \tag{1-8.15}$$

and, therefore,

$$k = \left(\frac{\mathbf{k}T}{h}\right)e^{-\Delta G^{\ddagger}/RT} \tag{1-8.16}$$

Furthermore, since

$$\Delta G^{\ddagger} = \Delta H^{\ddagger} - T\Delta S^{\ddagger} \tag{1-8.17}$$

we obtain

$$k = \left(\frac{\mathbf{k}T}{h}\right)e^{\Delta S^{\ddagger}/R}\, e^{-\Delta H^{\ddagger}/RT} \tag{1-8.18}$$

By taking the logarithm of both sides of Eq. 1-8.18, we obtain Eq. 1-8.19.

$$\ln k = \text{constant} + \Delta S^{\ddagger}/R - \Delta H^{\ddagger}/RT \qquad (1\text{-}8.19)$$

A graph of $\ln k$ versus $1/T$ should be a straight line with a slope related to $\Delta H^{\ddagger}$ and an intercept related to $\Delta S^{\ddagger}$. Thus the activation enthalpies and entropies can be determined from a study of the dependence of the rate constant on temperature.

This absolute rate theory approach is entirely consistent with the Arrhenius approach. From standard classical thermodynamics, we have Eq. 1-8.20.

$$E = \Delta H + RT \qquad (1\text{-}8.20)$$

Making the appropriate substitution into Eq. 1-8.18, we get Eq. 1-8.21.

$$k = (\mathbf{k}T/h)e^{\Delta S^{\ddagger}/R}e^{-(E-RT)/RT}$$

$$= \left(\frac{e\mathbf{k}T}{h}\right)e^{\Delta S^{\ddagger}/R}e^{-E_a/RT} \qquad (1\text{-}8.21)$$

Thus we see that the Arrhenius factor is a function of the entropy of activation.

Reaction Profiles

The course of a chemical reaction, as described in the absolute reaction rate theory, can be conveniently depicted in a graph of free energy versus the *reaction coordinate*. The latter is simply the pathway along which the changes in various interatomic distances progress as the system passes from reactants to activated complex to products. A representative graph is shown in Fig. 1-2 for the unimolecular decomposition of formic acid.

The Effect of Catalysts

A catalyst is a substance that causes a reaction to proceed more rapidly to equilibrium. It does not change the value of the equilibrium constant, and it does not itself undergo any net change. In terms of the absolute reaction rate theory, the role of a catalyst is to lower the free energy of activation $\Delta G^{\ddagger}$. Some catalysts do this by simply assisting the reactants to attain basically the same activated complex as they do in the absence of a catalyst. However, most catalysts appear to provide a different sort of pathway, in which they are temporarily bound, and which has a lower free energy.

An example of acid catalysis, in which protonated intermediates play a role, is provided by the catalytic effect of protonic acids on the decomposition of formic acid. Figure 1-3, when compared with Fig. 1-2 (the uncatalyzed reaction pathway), shows how the catalyst modifies the reaction pathway so that the highest value of the free energy that must be reached is diminished.

Catalysis may be either homogeneous or heterogeneous. In the previous example it is homogeneous. The strong acid is added to the solution of formic acid and the whole process proceeds in the one liquid phase. On the other hand, es-

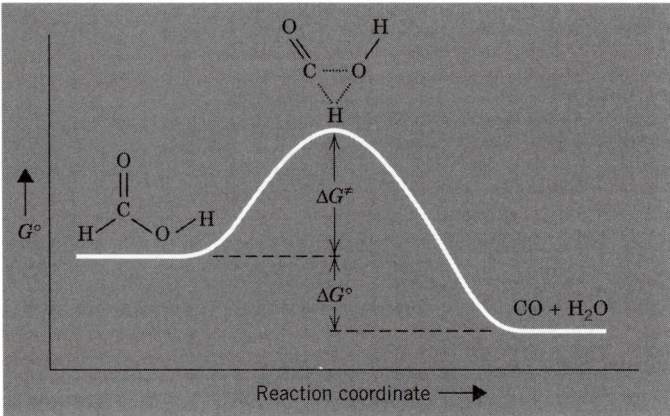

Figure 1-2   The free energy profile for the decomposition of formic acid. The free energy of activation is $\Delta G^{\ddagger}$. The standard free energy change for the overall reaction is ΔG°.

pecially in the majority of industrially important reactions, the catalyst is a solid surface and the reactants, either as gases or in solution, flow over the surface. Many reactions can be catalyzed in more than one way, and in some cases both homogeneously and heterogeneously.

The hydrogenation of alkenes affords an example where both heterogeneous and homogeneous catalyses are effective. The simple, uncatalyzed reaction shown in Reaction 1-8.22

$$RCH{=}CH_2 + H_2 \longrightarrow RCH_2CH_3 \qquad (1\text{-}8.22)$$

is impractically slow unless very high temperatures are used, which give rise to other difficulties, such as the expense and difficulty of maintaining the temper-

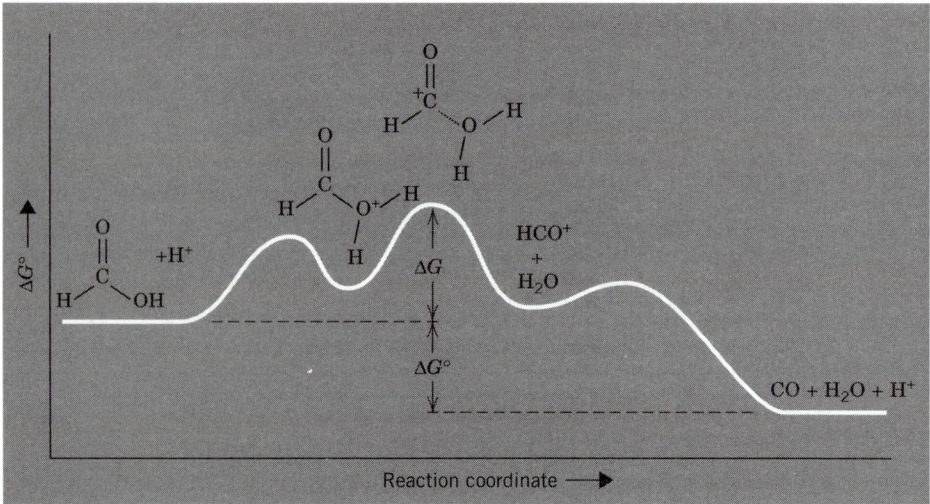

Figure 1-3   The free energy profile for the acid catalyzed decomposition of formic acid. The parameter ΔG° is the same as in Fig. 1-2, but $\Delta G^{\ddagger}$ is now smaller.

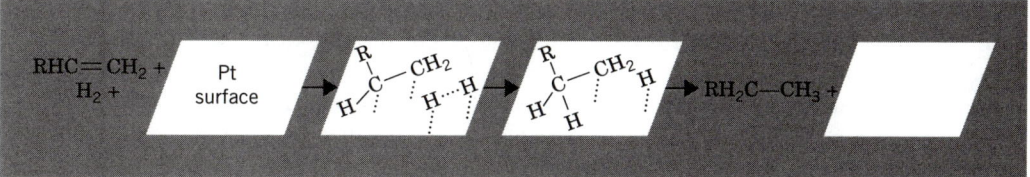

Figure 1-4   A sketch of how a suitable platinum surface can catalyze alkene hydrogenation by binding and bringing together the reactants.

ature and the occurrence of other, undesired reactions. If the gases are allowed to come in contact with certain forms of noble metals (e.g., platinum) supported on high surface area materials (e.g., silica or alumina) catalysis occurs. It is believed that both reactants are absorbed by the metal surface, possibly with dissociation of the hydrogen, as indicated in Fig. 1-4. Homogeneous catalysis (one of many examples to be discussed in detail in Chapter 30) proceeds somewhat similarly but entirely on one metal ion that is present in solution as a complex.

1-9   Nuclear Reactions

Although chemical processes essentially depend on how the electrons in atoms and molecules interact with each other, both the internal nature of nuclei and changes in nuclear composition (nuclear reactions) play an important role in the study and understanding of chemical processes. Conversely, the study of nuclear processes constitutes an important area of applied chemistry, particularly inorganic chemistry.

Atomic nuclei consist of a certain number (N), of protons (p) called the *atomic number*, and a certain number of neutrons (n). The masses of these particles are each approximately equal to one mass unit, and the total number of nucleons (protons and neutrons) is called the *mass number A*. The two numbers N and A completely designate a given nuclear species (neglecting the excited states of nuclei). It is the number of protons, that is, the atomic number, which identifies the *element*. For a given N, the different values of A, resulting from different numbers of neutrons, are responsible for the existence of different *isotopes* of that element.

When it is necessary to specify a particular isotope of an element, the mass number is placed as a left superscript. Thus the isotopes of hydrogen are ^1H, ^2H, and ^3H. In this one case, separate symbols and names are generally used for the less common isotopes $^2H = D$ (deuterium) and $^3H = T$ (tritium).

All isotopes of an element have the same chemical properties except where the mass differences alter the exact magnitudes of reaction rates and thermodynamic properties. These mass effects are virtually insignificant for elements other than hydrogen where the percentage variation in the masses of the isotopes is uniquely large.

Most elements are found in nature as a mixture of two or more isotopes. Tin occurs as a mixture of nine isotopes from ^{112}Sn (0.96%) through the most abundant isotopes ^{118}Sn (24.03%) to ^{124}Sn (5.94%). A few common elements that are terrestrially monoisotopic are ^{27}Al, ^{31}P, and ^{55}Mn. Because the exact masses of protons and neutrons differ, and neither is precisely equal to 1 atomic mass unit (amu), and for other reasons to be mentioned later, the masses of nuclei are not equal to their mass numbers. The actual atomic mass of ^{55}Mn, for example, is 54.9381 amu.

Usually, the isotopic composition of an element is constant all over the earth and thus its practical atomic weight, as found in the usual tables, is invariant. In a few instances, lead being most conspicuous, isotopic composition varies from place to place because of the different parentage of the element in radioactive species of higher atomic number. Also, for elements that do not occur in nature, the atomic weight depends on which isotope or isotopes are made in nuclear reactions. In tables, it is customary to give these elements the mass number of the longest lived isotope known.

Spontaneous Decay of Nuclei

Only certain nuclear compositions are stable indefinitely. All others spontaneously decompose by emitting α particles (2p2n) or β particles (positive or negative electrons) or by capture of a 1s electron. Emission of high energy photons (γ rays) generally accompanies nuclear decay. Alpha emission reduces the atomic number by two and the mass number by four. An example is

$$^{238}\text{U} \longrightarrow\ ^{234}\text{Th} + \alpha \qquad (1\text{-}9.1)$$

Beta decay advances the atomic number by one unit without changing the mass number. In effect, a neutron becomes a proton. An example is

$$^{60}\text{Co} \longrightarrow\ ^{60}\text{Ni} + \beta^- \qquad (1\text{-}9.2)$$

These decay processes follow first-order kinetics (page 21) and are insensitive to the physical or chemical conditions surrounding the atom. The half-lives are unaffected by temperature, which is an important distinction from first-order chemical reactions. In short, the half-life of an unstable isotope is one of its fixed, characteristic properties.

All elements have some unstable (i.e., radioactive) isotopes. Of particular importance is the fact that some elements have no stable isotopes. No element with atomic number 84 (polonium) or higher has *any* stable isotope. Some, for instance, U and Th, are found in substantial quantities in nature because they have at least one very long-lived isotope. Others, for instance, Ra and Rn, are found only in small quantities in a steady state as intermediates in radioactive decay chains. Others, for instance, At and Fr, have no single isotope stable enough to be present in macroscopic quantities. There are also two other elements, Tc and Pm, which do not have a stable isotope or one sufficiently long lived to have any detectable quantities of these elements occur in nature. Both are recovered from fission products.

Nuclear Fission

Many of the heaviest nuclei can be induced to break up into two fragments of intermediate size. This process is called nuclear fission. The stimulus for this is the capture of a neutron by the heavy nucleus. This capture creates an excited state that splits. In the process, several neutrons and a great deal of energy are re-

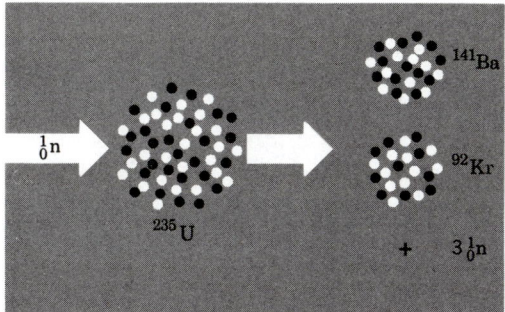

Figure 1-5   A schematic equation for a typical nuclear fission process.

leased. Because the process generates more neutrons than are required to stimulate it, a chain reaction is possible. Each individual fission can lead to an average of more than one subsequent fission. Thus, the process can become self-sustaining (nuclear reactor) or even explosive (atomic bomb). A representative example of a nuclear fission process (shown schematically in Fig. 1-5) is the following:

$$^{235}\text{U} + \text{n} \rightarrow {}^{141}\text{Ba} + {}^{92}\text{Kr} + 3\text{n}$$

Mass number	235	1	141	92	3	
Atomic number	92	0	56	36	0	
Neutrons	143	1	85	56	3	(1-9.3)

Nuclear Fusion

In principle, very light nuclei can combine to form heavier ones and release energy as they do so. Such processes are the main source of the energy generated in the sun and other stars. These processes also form the basis of the hydrogen bomb. At present, engineering research is underway to adapt nuclear fusion processes to the controlled, sustained generation of energy, but practical results cannot be expected in the near future.

Nuclear Binding Energies

The reason that fission and fusion processes are sources of nuclear energy can be understood by referring to a plot of the binding energy per nucleon as a function of mass number (Fig. 1-6). Binding energy is figured by subtracting the actual nuclear mass from the sum of the individual masses of the constituent neutrons and protons and converting that mass difference into energy using Einstein's equation, $E = mc^2$. The usual unit for nuclear energies is 1 million electron we have: volts (MeV), which is equal to 96.5×10^6 kJ mol^{-1}.

For example, for ^{12}C we have:

1.	Actual mass	12.000000 amu
2.	6 × proton mass	6.043662 amu
3.	6 × neutron mass	6.051990 amu
	(2) + (3) − (1) =	0.095652 amu

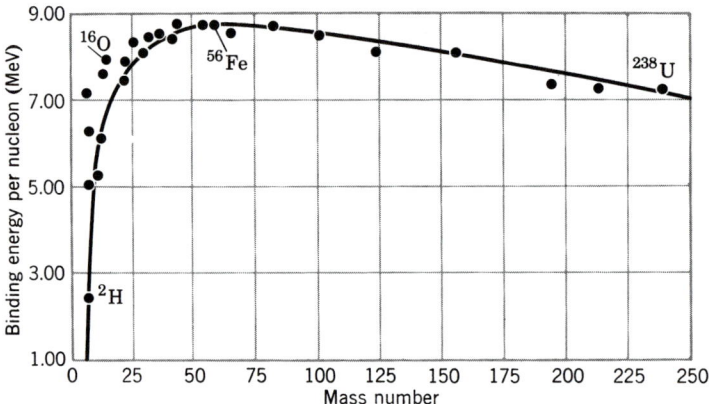

Figure 1-6   The binding energy of nucleons as a function of mass number.

One amu = 931.4 MeV. Hence,

$$\text{Total binding energy} = (931.4)(0.095652) = 89.09 \text{ MeV}$$

$$\text{Binding energy per nucleon} = (89.09)/12 = 7.42 \text{ MeV}$$

Since the formation of nuclei of intermediate masses releases more energy per nucleon than the formation of very light or very heavy ones, energy will be released when very heavy nuclei split (fission) or when very light ones coalesce (fusion).

Nuclear Reactions

The chemist, for a variety of purposes, will often require a particular isotope that is not available in nature, or even an element not found in nature. These isotopes or elements can be made in nuclear reactors. In general, they are formed when the nucleus of a particular isotope of one element captures one or more particles (α-particles or neutrons) to form an unstable intermediate. This intermediate decays, ejecting one or more particles, to give the product. The more common changes are indicated in Fig. 1-7.

A convenient shorthand for writing nuclear reactions is illustrated below for the process used to prepare an isotope of astatine.

$$^{209}\text{Bi} \, (\alpha, 2n)^{211}\text{At}$$

This equation says that ^{209}Bi captures an α-particle, and the resulting nuclear species, which is not isolable, promptly emits two neutrons to give the astatine isotope of mass number 211. The mass number increases by 4 (for α) minus 2 (for 2n) = 2 units and the atomic number increases by 2 units due to the two protons in the α-particle. Another representative nuclear reaction is

$$^{209}\text{Bi} \, (n, \gamma)^{210}\text{Bi} \longrightarrow {}^{210}\text{Po} + \beta \qquad (1\text{-}9.4)$$

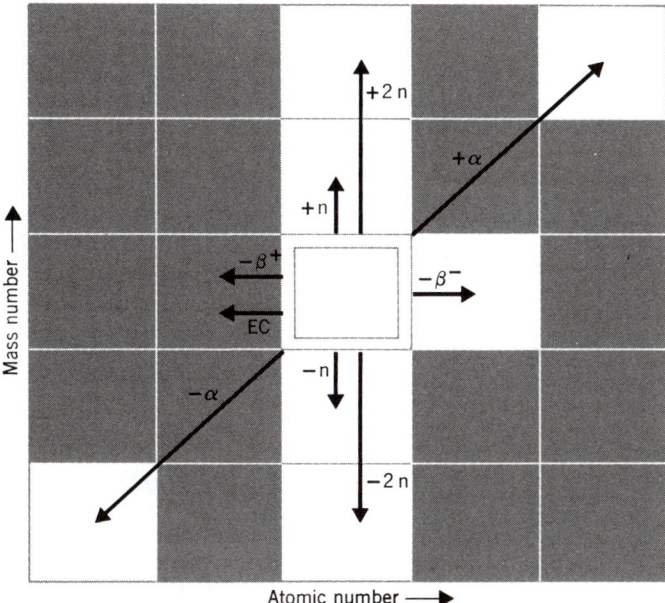

Figure 1-7   A chart showing how the more important processes of capture and ejection of particles change the nuclei (EC = electron capture).

1-10   Units

There is now an internationally accepted set of units for the physical sciences. It is called the SI (for *Système International*) units. Based on the metric system, it is designed to achieve maximum internal consistency. However, since it requires the abandonment of many familiar units and numerical constants in favor of new ones, its adoption in practice will take time. In this book, we shall take a middle course, adopting some SI units (e.g., joules for calories) but retaining some non-SI units (e.g., angstroms, Å).

The SI Units

The SI system is based on the following set of defined units:

Physical Quantity	Name of Unit	Symbol for Unit
Length	meter	m
Mass	kilogram	kg
Time	second	s
Electric current	ampere	A
Temperature	kelvin	K
Luminous intensity	candela	cd
Amount of substance	mole	mol

Multiples and fractions of these are specified using the following prefixes:

Multiplier	Prefix	Symbol
10^{-1}	deci	d
10^{-2}	centi	c
10^{-3}	milli	m
10^{-6}	micro	μ
10^{-9}	nano	n
10^{-12}	pico	p
10^{-15}	femto	f
10	deka	da
10^2	hecto	h
10^3	kilo	k
10^6	mega	M
10^9	giga	G
10^{12}	tera	T

In addition to the defined units, the system includes a number of derived units. The following table lists the main units.

Physical Quantity	Name of Unit	Symbol	Basic Units
Force	newton	N	$= \text{kg m s}^{-2}$
Work, energy, quantity of heat	joule	J	$= \text{N m or kg m}^2 \text{ s}^{-2}$
Power	watt	W	$= \text{J s}^{-1}$
Electric charge	coulomb	C	$= \text{A s}$
Electric potential	volt	V	$= \text{WA}^{-1}, \text{ kg m}^2 \text{ s}^{-3} \text{ A}^{-1}, \text{ or J/C}$
Electric capacitance	farad	F	$= \text{A s V}^{-1}$
Electric resistance	ohm	Ω	$= \text{V A}^{-1}$
Frequency	hertz	Hz	$= \text{s}^{-1}$
Magnetic flux	weber	Wb	$= \text{V s}$
Magnetic flux density	tesla	T	$= \text{Wb m}^{-2}$
Inductance	henry	H	$= \text{V s A}^{-1}$

Units to Be Used in This Book

Energy

Joules and kilojoules will be used. Much of the chemical literature to date employs calories, kilocalories, electron volts and, to a limited extent, wavenumbers (cm^{-1}). Conversion factors are given below.

Bond Lengths

The angstrom (Å) will be used. This is defined as 10^{-8} cm. The nanometer (10 Å) and picometer (10^{-2} Å) will also be used. The C—C bond length in diamond has the value:

$$
\begin{array}{rl}
1.54 & \text{angstroms} \\
0.154 & \text{nanometers} \\
154 & \text{picometers}
\end{array}
$$

Pressure

Atmospheres (atm) and Torr (1/760 atm) will be used.

Some Useful Conversion Factors and Numerical Constants

Conversion Factors

1 calorie (cal)	$= 4.184$ joules (J)
1 electron volt per molecule	$= 96.485$ kilojoules per mole (kJ mol^{-1})
(eV/molecule)	$= 23.06$ kilocalories per mole (kcal mol^{-1})
1 kilojoule per mole	$= 83.54$ wavenumbers (cm^{-1})
(kJ/mol^{-1})	
1 atomic mass unit (amu)	$= 1.6605655 \times 10^{-27}$ kilogram (kg)
	$= 931.5016$ mega electron volt (MeV)

Important Constants

Avogadro's number	N_A	$= 6.022045 \times 10^{23}$ mol^{-1}
Electron charge	e^-	$= 4.8030 \times 10^{-10}$ abs esu
		$= 1.6021892 \times 10^{-19}$ C
Electron mass	m_e	$= 9.1091 \times 10^{-31}$ kg
		$= 0.5110$ MeV
Proton mass	m_p	$= 1.6726485 \times 10^{-27}$ kg
		$= 1.007276470$ amu
Gas constant	R	$= 8.31441$ J mol^{-1} K^{-1}
		$= 1.9872$ cal mol^{-1} K^{-1}
		$= 0.08206$ L atm mol^{-1} K^{-1}
Ice point		$= 273.15$ K
Molar volume		$= 22.414 \times 10^3$ cm^3 mol^{-1}
		$= 2.2414 \times 10^{-2}$ m^3 mol^{-1}
Planck's constant	h	$= 6.626176 \times 10^{-34}$ J s
		$= 6.626176 \times 10^{-27}$ erg s
Boltzmann's constant	k	$= 1.380662 \times 10^{-23}$ J K^{-1}
Rydberg constant	$\mathcal{R}$	$= 1.097373177 \times 10^7$ m^{-1}
Speed of light	c	$= 2.99792458 \times 10^8$ m s^{-1}
Bohr radius	a_0	$= 0.52917706 \times 10^{-10}$ m
Other numbers	π	$= 3.14159$
	e	$= 2.7183$
	ln 10	$= 2.3026$

Coulombic Force and Energy Calculations in SI Units

Although SI units do, for the most part, lead to simplification, one computation that is important to inorganic chemistry becomes slightly more complex. We explain that point in detail here. It traces back to the concept of the dielectric constant ϵ, which relates the intensity of an electric field induced within a substance D to the intensity of the field applied E by the equation

$$D = \epsilon E \qquad (1\text{-}10.1)$$

The same parameter appears in the Coulomb equation for the force F between two charges q_1 and q_2, which are separated by a distance d and immersed in a medium with a dielectric constant ϵ.

$$F = \frac{q_1 \times q_2}{\epsilon d^2} \qquad (1\text{-}10.2)$$

In the old centimeter-gram-second (cgs) system of units, which the SI system replaces, units and magnitudes were defined so that ϵ was a dimensionless quantity and for a vacuum $\epsilon_0 = 1$.

For reasons that we shall not pursue here, Coulomb's law of electrostatic force, in SI units, must be written

$$F = \frac{q_1 \times q_2}{4\pi\epsilon d^2} \tag{1-10.3}$$

The charges are expressed in coulombs (C), the distance in meters (m), and the force is obtained in newtons (N). Therefore, ϵ is no longer dimensionless and has units $C^2 \, m^{-1} \, J^{-1}$. Moreover, the dielectric constant of a vacuum (the permittivity, as it should formally be called) is no longer unity. It is, instead,

$$\epsilon_0 = 8.854 \times 10^{-12} \, C^2 \, m^{-1} \, J^{-1} \tag{1-10.4}$$

Thus, to calculate a coulomb energy E in joules (J) we must employ the expression

$$E = \frac{q_1 \times q_2}{4\pi\epsilon d} \tag{1-10.5}$$

with all quantities being as defined for the Coulombic force.

STUDY GUIDE

Study Questions

1. Define the terms exothermic and endothermic. What are the signs of ΔH for each type of process?

2. How is the standard enthalpy of formation of a substance defined? Write the balanced chemical equation that applies to $\Delta H_f^\circ[CF_3SO_3H]$.

3. Write balanced chemical equations that apply to each of the following enthalpy changes:

   (a) $\Delta H_{sub}^\circ[H_2O]$ (b) $\Delta H_{vap}^\circ[C_6H_6]$

   (c) $\Delta H_{EA}[Cl(g)]$ (d) $\Delta H_{ion}[Na(g)]$

4. Write an equation that can be used to define the mean S—F bond energy in SF_6. How is this value likely to be related in magnitude to the energy of the process $SF_6(g) \rightarrow SF_5(g) + F(g)$?

5. Prepare graphical representations of the relationships between

   (a) ΔG and T; four separate possibilities depending on the signs of ΔH and ΔS.

   (b) k and T, using $\Delta H^\ddagger$ and $\Delta S^\ddagger$.

6. Give a qualitative definition of entropy.

7. Against what standard are the absolute entropies of substances tabulated?

8. Prepare graphs of the concentration of reactant A as a function of time if A disappears in first-order fashion. How should the data for such a first-order reaction be plotted in order to obtain a straight line relationship?

9. Answer as in Question 8, but for a second-order disappearance of reactant A.

10. What elements might have negative electron attachment enthalpies? What is the meaning of a negative sign for the electron attachment enthalpy?

11. The N—N bond energy in F_2NNF_2 is only about 80 kJ mol^{-1} compared to 160 kJ mol^{-1} in H_2NNH_2. Suggest a reason.

12. Predict the signs of the entropy changes for the following processes:

    (a) $H_2O(\ell) \rightarrow H_2O(g)$

    (b) $P_4(g) + 10\ F_2(g) \rightarrow 4\ PF_5(g)$

    (c) $I_2(s) + Cl_2(g) \rightarrow 2\ ICl(g)$

    (d) $BF_3(g) + NH_3(g) \rightarrow H_3NBF_3(g)$

    (e) $CO_2(g) \rightarrow CO_2(s)$

13. Use the data of Table 1-1 to estimate ΔH_f° values for the following molecules:

    (a) $HNCl_2$ (b) CF_3SF_3 (c) Cl_2NNH_2

14. What do you suppose is the main thermodynamic reason why the following reaction has an equilibrium constant > 1?

$$BCl_3(g) + BBr_3(g) \longrightarrow BCl_2Br(g) + BClBr_2(g)$$

15. Use the data of Table 1-1 to predict the enthalpy change for the reaction $CO + H_2O \rightarrow CO_2 + H_2$.

16. What is the value of the equilibrium constant for a reaction that has $\Delta G^\circ = 0$? Draw the reaction profile for such a system.

17. The following data are available for the forward direction of an equilibrium system: $\Delta G^\circ = -50$ kJ mol^{-1} and $\Delta G^\ddagger = 20$ kJ mol^{-1}. What is the activation free energy for the reverse direction of the equilibrium? Prepare a reaction profile, showing the relative magnitude of each of these three quantities.

18. The conversion of diamond into graphite is a thermodynamically favorable (spontaneous) process, and yet one does not expect a diamond to change into graphite. Why?

19. Determine the standard cell potentials for the following redox reactions:

    (a) The oxidation of lithium by chlorine

    (b) The reduction of Ce^{4+} by iodide.

20. Use Eq. 1-6.3 to ascertain the relative values of K_1 and K_2 for an exothermic reaction, assuming that $T_1 > T_2$. Answer also for an endothermic reaction. Explain the consequences in terms of the Principle of Le Châtelier.

Chapter 2

THE ELECTRONIC STRUCTURE OF ATOMS

2-1 Introduction

The term **electronic structure,** when used with respect to an atom, refers to the number and the distribution of electrons about a central nucleus. The nucleus can be considered to consist of the proper number of protons and neutrons, depending on the mass number and atomic number of the isotope in question. It is reasonable, for our purposes, to take this simplistic view of the nucleus. Apart from electrostatic repulsions between nuclei, all of the major interactions between atoms in normal chemical reactions (or in the structures of elemental and compound substances) involve the electrons. Ultimately, we would like to be able to use our understanding of the electronic structures of atoms to describe the structures and reactivities of molecules and ions.

A complete description of the electronic structure of an atom would include more than just the number and the spatial distribution of electrons within the atom. Nevertheless, most of what we would like to know about electronic structure is dictated by these two properties. Once the spatial distribution of the electrons is known, other important properties follow. For instance, the energies, ionization enthalpies, sizes, and magnetic properties of atoms all depend on the number and arrangement of the electrons within the atom.

Much of the experimental work on the electronic structures of atoms done prior to 1913 involved measuring those frequencies of electromagnetic radiation that are absorbed or emitted by atoms. It was found to be characteristic of atoms that they absorb or emit only certain sharply defined frequencies of electromagnetic radiation. The exact pattern of emission frequencies was found to be characteristic of each particular element, with the emission or absorption patterns being more complex for the heavier elements. Although the emission and absorption spectra for most of the elements were known before the turn of the century, a suitable theory was not then available for even the simplest case: the hydrogen atom.

The atomic emission spectrum for atomic hydrogen (Fig. 2-1) was found to consist of several series of **lines,** or spectroscopic emissions. Within each series, the lines become increasingly closely spaced, until they converge at a limiting value. It was Rydberg who recognized that these emission lines for the hydrogen atom had wavenumbers $\bar{v}$ (equal to v/c, where v is the frequency of the emission line and c is the speed of light) that conformed to the relationship shown in Eq. 2-1.1.

$$\bar{v} = \mathscr{R}\left(\frac{1}{m^2} - \frac{1}{n^2}\right) \quad \begin{array}{l} m = 1,\ 2,\ 3,\ 4,\ \ldots \\ n = (m+1),\ (m+2),\ (m+3),\ \ldots \end{array} \quad (2\text{-}1.1)$$

Thus each emission of light from the hydrogen atom occurs at a precise value of $\bar{v}$, in units of reciprocal centimeters (cm^{-1}). The various lines are then each found at specific locations in the spectrum (i.e., at specific values of $\bar{v}$) depending on the values of the integers m and n.

The integer $m = 1$ gives rise to the Lyman series (Fig. 2-1) for which the convergence limit is 109,678 cm^{-1}. When $m = 2$, the Balmer series arises, and so on. Two more well-defined series of lines appear at lower energies (i.e., in the IR portion of the spectrum), but are not shown in Fig. 2-1. In addition to deducing how the integers m and n could be used in Eq. 2-1.1 to generate the spectroscopic emission pattern for atomic hydrogen, Rydberg also empirically determined that the constant $\mathscr{R}$ = 109,678 cm^{-1}. Although it seems straightforward now, the accomplishment of Rydberg was remarkable. The existence of a quantitative description, Eq. 2-1.1, of the spectroscopic lines for atomic hydrogen made it quite clear that the pattern of lines was significant. The pattern was clear, but the meaning was not.

The meaning was clarified in 1913 by the Danish physicist Niels Bohr, who realized that the Rydberg equation could not be explained in terms of the strictly classical theories then in use. Bohr reasoned that if only discrete frequencies could be emitted or absorbed by an atom, then only discrete energies were possible for the electrons in that atom. Bohr broke with the tradition of classical physics and proposed that the electron could revolve indefinitely about the proton in orbits of fixed radii. According to classical physics, this should be impossible; the electron should spiral inward towards the nucleus, emitting a continuum of frequencies before crashing into the nucleus.

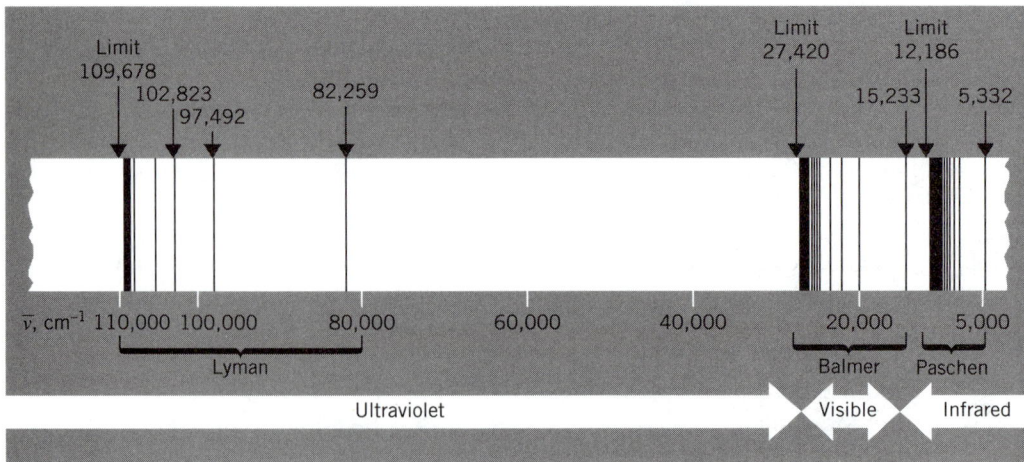

Figure 2-1 The emission spectrum of atomic hydrogen as recorded on a strip of film. Each line represents an emission frequency. Three series of lines are shown. Within each series, the lines converge to a limit. Two more well-defined series occur for atomic hydrogen in the infrared (IR) region, but they are not shown here.

Bohr's theory for the electronic structure of the hydrogen atom was founded on the assumption that for each discrete orbit, the angular momentum of the electron must be quantized according to Eq. 2-1.2

$$mvr = nh/2\pi \qquad (2\text{-}1.2)$$

where n is an integer, m and v are the mass and velocity of the electron, respectively, r is the radius of the orbit, and h is Planck's constant. According to the Bohr theory, the electron traveling in an orbit with radius r would possess an angular momentum mvr, which depended on the quantum number n. These two proposals—stable orbits and the quantization of angular momentum (and hence of the radius and energy of the electron)—were in conflict with and utterly outside of the accepted physical theory of the time. However, by using these assumptions and by treating the rest of the problem in a perfectly traditional way, Bohr was able to show that *allowed* orbits were those with radii given by Eq. 2-1.3.

$$r = \frac{n^2 h^2 \epsilon_0}{\pi m Z e^2} \qquad (2\text{-}1.3)$$

The requirement that mvr can take only those values that are multiples of $h/2\pi$ means that only certain values of r (those given by Eq. 2-1.3) are allowed. Electrons within orbits with discrete radii then must have energies that are quantized according to Eq. 2-1.4.

$$E = -\frac{mZ^2 e^4}{8n^2 h^2 \epsilon_0^2} \qquad (2\text{-}1.4)$$

The letter Z is the nuclear charge and is equal to 1 for the hydrogen atom.

The most exciting support for Bohr's theory was that the collection of constants other than the quantum number n in Eq. 2-1.4 is numerically equal to the value for $\mathcal{R}$, which Rydberg had determined empirically. In short, Bohr had obtained Eq. 2-1.5.

$$E = -\frac{\mathcal{R}}{n^2} \qquad (2\text{-}1.5)$$

The explanation for each series of spectroscopic lines in the spectrum for atomic hydrogen was now at hand (see Fig. 2-2). An electron would have lowest (most negative) energy when in the orbit for which $n = 1$. The radius of this orbit ($a_0 = 0.529$ Å) can be calculated using Eq. 2-1.3. Each higher value of the quantum number yields a correspondingly larger and less stable orbit, that is, one with a less negative energy. If an electron is excited to an orbit with higher energy ($n \geq 2$) and returns to the ground state ($n = 1$), discrete energies equal to $\mathcal{R}[(1/1^2) - (1/n^2)]$ are emitted. In this case, the Lyman series of spectroscopic lines (Fig. 2-1) is observed. The other series arise when the electron drops from upper levels to those with $n = 2$ (Balmer series) and $n = 3$ (Paschen series), as shown in Fig. 2-2.

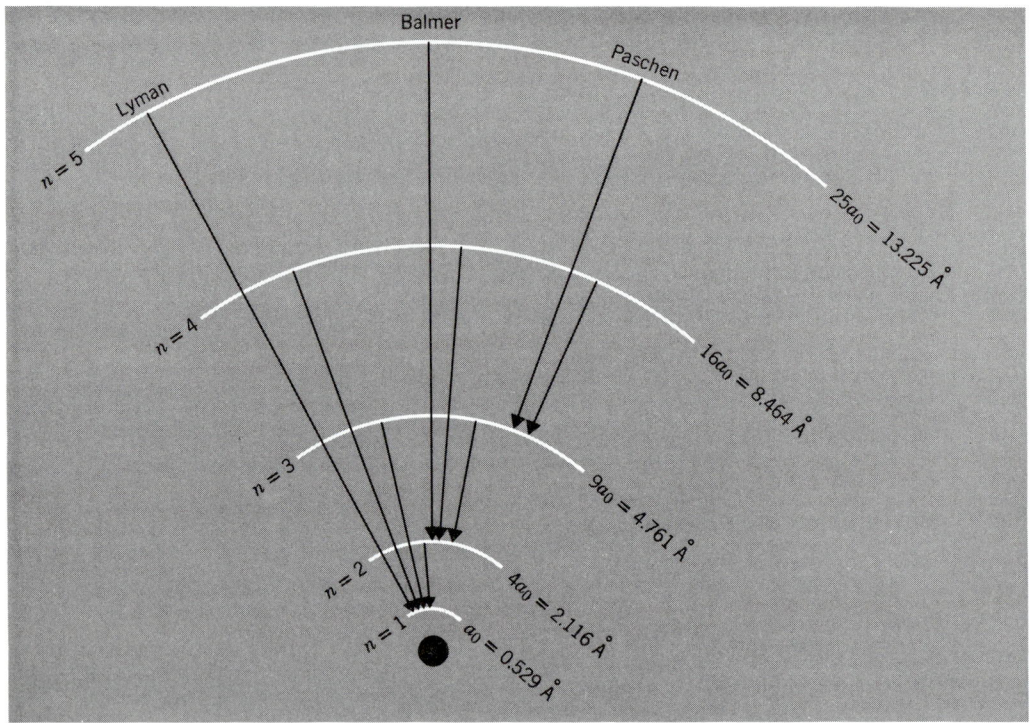

Figure 2-2   A diagram of the Bohr orbits and the corresponding energies for an electron in the hydrogen atom. Each arc represents a portion of an orbit. The transitions that give rise to the three series of spectroscopic lines of Fig. 2-1 are indicated.

In developing his theory, Bohr made use of Planck's earlier postulate (also in conflict with classical physics), which stated that electromagnetic radiation is itself quantized. According to Planck, each quantum of electromagnetic radiation of frequency ν has energy given by Eq. 2-1.6.

$$E = h\nu \tag{2-1.6}$$

The Bohr model was refined by Arnold Sommerfeld, who showed that finer features of the hydrogen spectrum, which were observed on application of a magnetic field, could be accounted for if elliptical, as well as circular orbits were used. This gave another quantum number that dictated the ellipticity of the orbits.

In spite of the success of the Bohr–Sommerfeld quantum theories for the hydrogen atom, the theories had to be abandoned for a number of reasons. First, the approach could not be applied successfully to the interpretation of spectra for atoms more complex than hydrogen. Perhaps more important, later work showed that electrons cannot be regarded as discrete particles with both precisely defined positions and velocities. It is not that the quantum approach was wrong, but that the electrons cannot be described adequately by the simplistic notion that they are only particles. In fact, it became evident that electrons also possess the same wavelike properties that Planck had already ascribed (Eq. 2-1.6) to photons. We now know that this wave–particle duality (both wave and particle characteristics are necessary for a full description) is typical of all matter, not just photons and electrons.

2-2 Wave Mechanics

In 1924, the French physicist Louis Victor de Broglie suggested that all matter could exhibit wavelike properties. For particles, such as electrons or nucleons, that travel with velocity v, de Broglie proposed the important matter–wave relationship shown in Eq. 2-2.1:

$$\lambda = h/mv \qquad (2\text{-}2.1)$$

Matter with mass m and velocity v (properties associated with particles) has a wavelength λ (a property associated with waves). While all matter in motion would then have an associated waveform, the wavelength is meaningful to spectroscopists only when m is small. De Broglie's proposal was substantiated a few years later when the two Americans Clinton J. Davisson and L. H. Germer found experimental evidence that electrons do behave in a wavelike manner. They demonstrated that a beam of electrons is diffracted by a crystal in much the same way as a beam of X-rays. The wavelength that Davisson and Germer determined for the electrons was just that predicted by Eq. 2-2.1.

Concurrent with these developments was the proposal by the Viennese physicist Erwin Schrödinger stating that the electron should be described in a way that would emphasize its wave nature. The Schrödinger **wave equation,** shown in its most general form in Eq. 2-2.2

$$\mathcal{H}\Psi = E\Psi \qquad (2\text{-}2.2)$$

represented a new method—**wave mechanics**—for describing the behavior of subatomic particles. Wave mechanics leads to the same energy levels of the electron in the hydrogen atom that Bohr obtained. In addition, it gives a better account of other properties of atomic hydrogen, but most important, it can give a correct account of more complex atoms as well.

The method of wave mechanics, as expressed in Eq. 2-2.2, is the method of operator algebra. The operator $\mathcal{H}$, called the Hamiltonian operator, prescribes a series of mathematical operations that are to be performed on the wave function, Ψ. The wave function, Ψ, is a mathematical expression that describes or defines the electron in terms of its wave properties. If the electron is accurately described by the wave function, Ψ, then Ψ is said to be a proper wave function (an eigenfunction) for the Hamiltonian operator. According to the dictates of operator algebra, this will happen only when the mathematical manipulations prescribed by the Hamiltonian operator give the wave function back, unchanged, save for multiplication by the constant E. The constant, E, is the energy that the electron would have if it were to be described as, or behave according to, the wave function, Ψ. Although there is only one set of wave functions that can exactly satisfy the Schrödinger equation, and thus correspond exactly to those energies E actually possessed by the electron in the various states of the atom, the wave functions are not easy to determine. Instead, it is necessary to devise various trial wave functions and test them. A comparison of the energies observed

Table 2-1   Some Hydrogen-like Wave Functions $\Psi = R(r)\Theta(\theta)\Phi(\phi)$ Factored Into Radial $[R(r)]$ and Angular $[\Theta(\theta)\Phi(\phi)]$ Components[a]

Orbital	$R(r)$	$\Theta(\theta)\Phi(\phi)$
$1s$	$N_c \left(\dfrac{Z}{a_0} \right)^{3/2} e^{-br}$	$\left(\dfrac{1}{4\pi} \right)^{1/2}$
$2p_y$	$N_c \left(\dfrac{Z}{a_0} \right)^{3/2} (ar)e^{-br}$	$\left(\dfrac{3}{4\pi} \right)^{1/2} \sin\theta \sin\phi$
$3d_{xy}$	$N_c \left(\dfrac{Z}{a_0} \right)^{3/2} (ar)^2 e^{-br}$	$\left(\dfrac{15}{4\pi} \right)^{1/2} \sin\theta \cos\theta \sin\phi$

[a]The factors a, b, and N_c depend variously and in part on one or more of the quantum numbers n, ℓ, and m_ℓ. The value a_0 is the Bohr radius. The correspondence between spherical polar coordinates and the more familiar Cartesian coordinates (x, y, and z) is discussed in the text.

spectroscopically (Fig. 2-1) with those calculated from a trial set of wave functions (each corresponding to an energy level E) gives an indication of how closely the trial wave functions match the true wave properties of the electron.

For a system as simple as the hydrogen atom, the wave functions can be determined precisely. Some of these functions are given in abbreviated form in Table 2-1, where they are factored into their various components in the polar coordinates r, θ, and ϕ. The polar coordinates r, θ, and ϕ correspond to the Cartesian coordinates in the following ways: $x = r \sin\theta \cos\phi$, $y = r \sin\theta \sin\phi$, and $z = r \cos\theta$. The numbers a, b, and N_c take different values depending on the quantum numbers at hand.

A more complete listing of the wave functions for the hydrogen atom is given in Appendix IIA. The wave functions are three dimensional, and contain the three quantum numbers n, ℓ, and m_ℓ as integers. In addition, each electron is characterized by the spin quantum number m_s, equal to the quantity either plus or minus one half. One electron is made distinct from another because of a difference in the value of at least one of these four quantum numbers. We shall have more to say about quantum numbers in Section 2-3. First, it is important to understand the meaning of the wave functions themselves.

While it is difficult to assign a physical meaning to the wave function itself, its square gives us a measure of the electron density in the various regions about the nucleus. A three-dimensional plot of the values of Ψ^2 centered on the nucleus gives us an indication of those regions about the nucleus where the electron, if it were behaving as the waveform, Ψ, would be most densely distributed. According to this interpretation of Ψ^2, the electron is regarded as a smeared out distribution of negative charge whose density varies from place to place according to the magnitude of Ψ^2. We have a situation in which the electron is smeared about the nucleus in a way that varies with the distance, as governed by the radial portion of the wave function, $R(r)$, and in different angular patterns, as governed by the angular portion of the wave function, $\Theta(\theta)\Phi(\phi)$. If the electron is thought of as "being" the wave function, then the electron can be said to be distributed into an **orbital** (a term borrowed from the Bohr concept of orbits), which has a size dictated by the function $R(r)$, a shape dictated by the function $\Theta(\theta)\Phi(\phi)$, and an energy, E, which can be calculated by using Eq. 2-2.2.

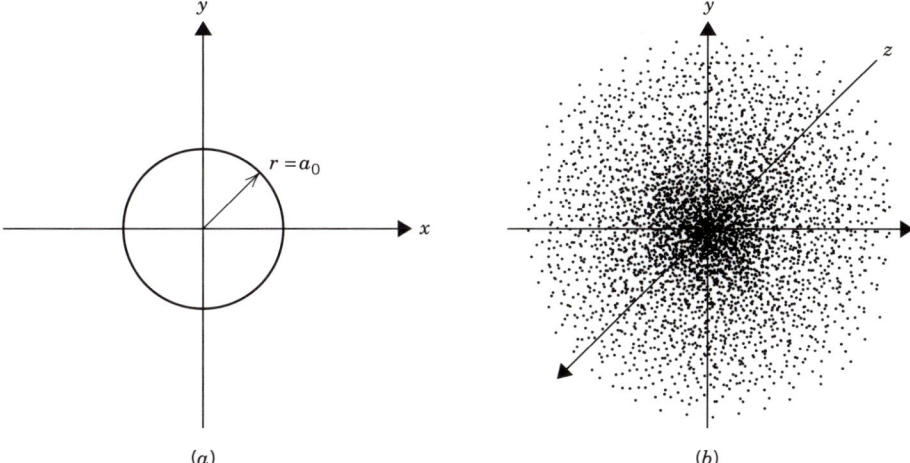

(a)   (b)

Figure 2-3 (*a*) The precise circular path of an electron in the first Bohr orbit, for which the radius is $a_0 = 0.529$ Å. (*b*) The electron density pattern for the comparable atomic orbital, drawn so that the stippling intensity corresponds to the value of the function $[R(r)]^2$ for the $1s$ atomic orbital.

It is instructive to compare the Bohr result with that of wave mechanics. The exact (precisely defined) radius for the electron in the first Bohr orbit is $a_0 = 0.529$ Å. This is shown in Fig. 2-3(*a*) as a circle having a radius equal to a_0. The wave mechanical result for the same orbital is shown in Fig. 2-3(*b*) as an electron density (or stippling) pattern, where the electron density (as indicated by the intensity of the stippling) is dependent on the value of the function $[R(r)]^2$.

The electron density in the same orbital is shown in a different way in Fig. 2-4. Here we have plotted the value of the function $4\pi r^2[R(r)]^2$ as a function of r, the distance from the nucleus. This function represents the electron density that is encountered within each spherical shell of thickness dr, as the distance r from the nucleus increases incrementally from the value r to the value $r + dr$. The function $4\pi r^2[R(r)]^2$ has reached its maximum at precisely the value of Bohr's first orbit: $r = 0.529$ A. The correspondence between the two theories in this respect is reassuring.

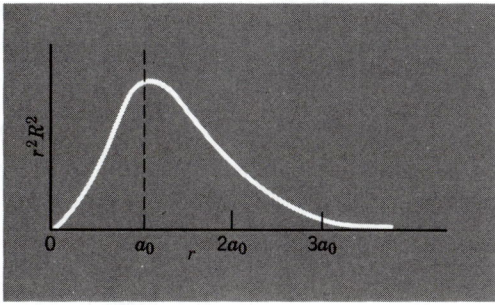

Figure 2-4 A comparison of the radial density distribution function $r^2R(r)^2$, which has maximum value at $r = a_0$, and the Bohr radius, where $r = a_0$, exclusively.

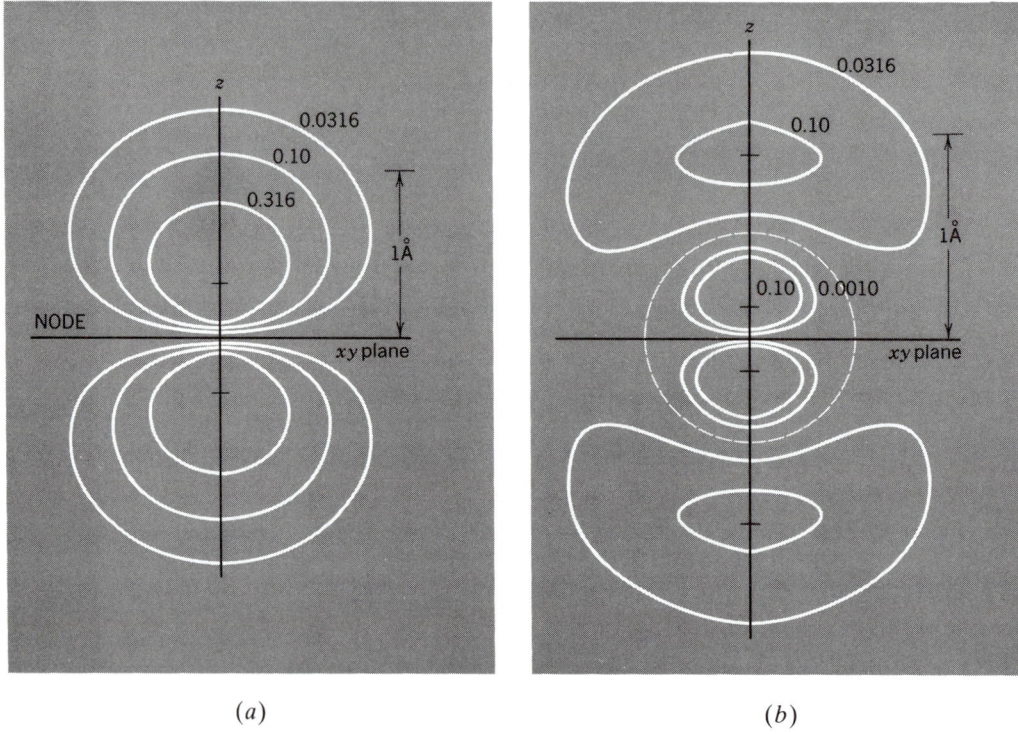

(a)   (b)

Figure 2-5   Contour maps of some orbitals, showing both shape and "internal" gradients in electron density. (a) The $2p$ orbital. (b) The $3p$ orbital. The contour lines are drawn at the points where some fraction (arbitrarily 0.0316, 0.10, and 0.316) of the maximum electron density has been reached. The maxima are indicated by the bars on the axes. [Reprinted with permission from E. A. Ogryzlo and G. B. Porter, *J. Chem. Educ.*, *40*, 256–261 (1963).]

Clearly, it is difficult to represent both the shape and size of an orbital on one graph. Where a cross section of an orbital is shown, one sees radial changes in electron density quite readily, but one loses a sense of the three-dimensional "roundness" that orbitals have. Where shape is shown (look ahead to Fig. 2-6), one loses the ability to show attenuations in electron density as a function of $R(r)$. The best solution to this dilemma is shown in Fig. 2-5, where contour and shape are shown simultaneously for two orbitals that we shall discuss shortly.

In some cases it is not efficient to show the full contour diagram for an orbital. The orbital is simply drawn as an **enclosure surface,** inside of which a majority (arbitrarily, >95%) of the electron density is known to reside. Thus the shape of each orbital may be drawn as in Fig. 2-6. Although these shapes are constructed from Ψ^2 (which must be everywhere positive), each lobe of the orbital is given the sign of the original wave function Ψ. What is not shown in the enclosure surfaces of Fig. 2-6 is the gradation in electron density that is contained in the function $R(r)$.

2-3   Atomic Orbitals in Wave Mechanics

We now consider the entire set of orbitals for the electron in the hydrogen atom. The orbital designation and the unique set of the quantum numbers n, ℓ, and m_ℓ that gives rise to each one are listed in Table 2-2. The shape of each type of

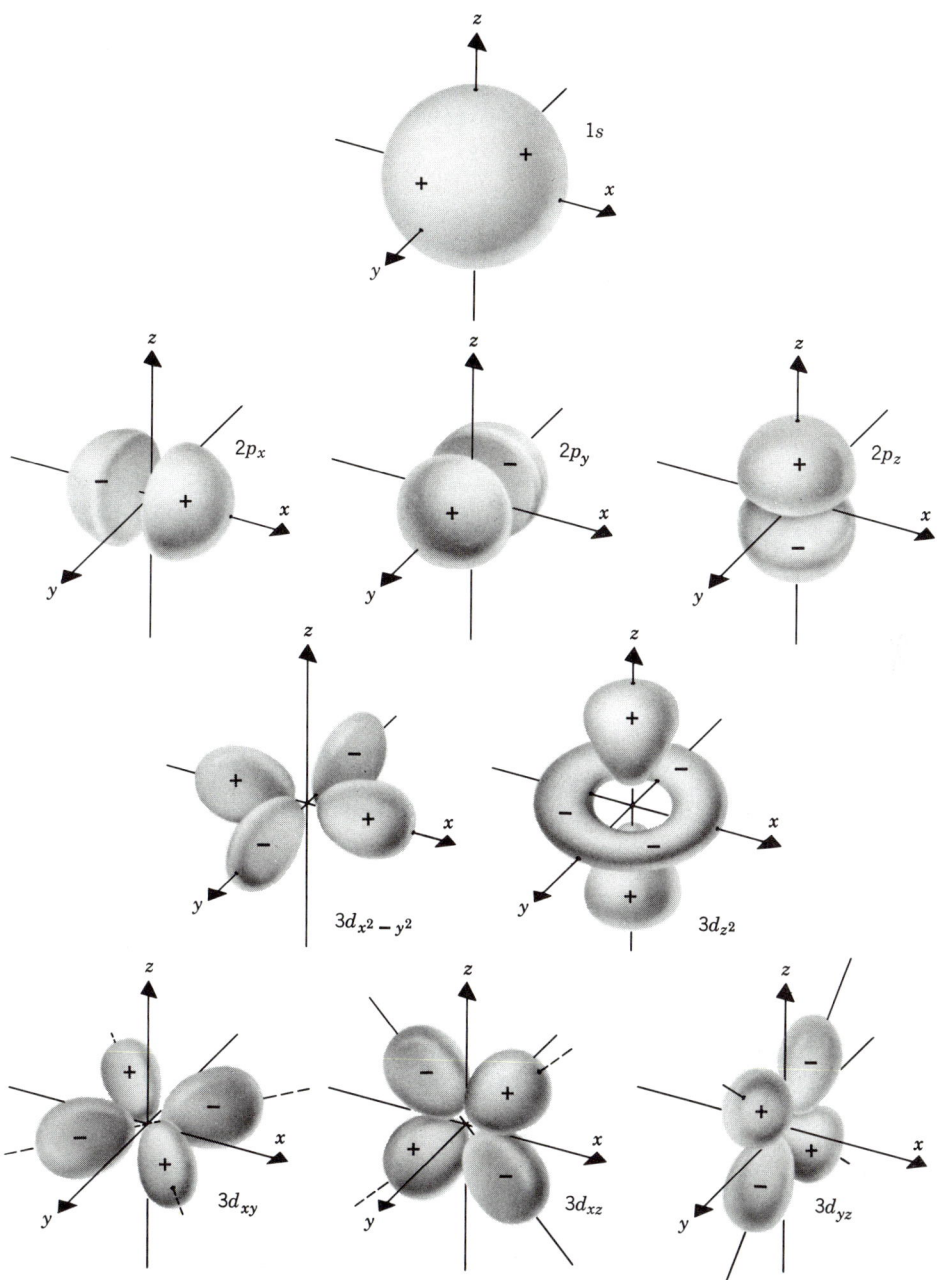

Figure 2-6   Atomic orbitals for the hydrogen atom, drawn as enclosure surfaces as described in the text. The lobes are derived from Ψ^2 and must be everywhere positive. The lobes have been given the signs, however, of the original wave functions Ψ, as this is information that becomes important when considering bonding via overlap of such orbitals.

orbital is shown in Fig. 2-6, and the radial dependences, $R(r)$ and $r^2 R(r)^2$, are graphed in Fig. 2-7. The principal quantum number n may take integral values from 1 to ∞, although values larger than 7 are spectroscopically and chemically unimportant. It is the value of this quantum number n that determines the size and energy of the orbital. For a given value of n, the quantum number ℓ may

Table 2-2   Quantum Numbers and Atomic Orbital Designations

Shell	n	ℓ	m_ℓ	Orbital
K	1	0	0	$1s$
L	2	0	0	$2s$
	2	1	$-1, 0, +1$	$2p$
M	3	0	0	$3s$
	3	1	$-1, 0, +1$	$3p$
	3	2	$-2, -1, 0, +1, +2$	$3d$
N	4	0	0	$4s$
	4	1	$-1, 0, +1$	$4p$
	4	2	$-2, -1, 0, +1, +2$	$4d$
	4	3	$-3, -2, -1, 0, +1, +2, +3$	$4f$
O	5	0	0	$5s$
—	—	—	—	—
—	—	—	—	—

take values $0, 1, 2, 3, \ldots, (n-1)$. It is this quantum number that determines the shape of the orbital. A letter designation is used for each orbital shape: s, when $\ell = 0$; p, when $\ell = 1$; d, when $\ell = 2$; f, when $\ell = 3$; followed alphabetically by the letter designations g, h, and so on. Finally, for any one orbital shape, the quantum number m_ℓ may take integral values from $-\ell$ to $+\ell$. This latter quantum number governs the orientation of the orbital. Once the electron for the hydrogen atom is placed into one specific orbital, the values of the three quantum numbers n, ℓ, and m_ℓ are known. In addition, the electron may have a value for the spin quantum number (m_s) of $+\frac{1}{2}$ or $-\frac{1}{2}$.

s Orbitals

Every s orbital has quantum number $\ell = 0$ and is spherically symmetrical. The smallest such orbital, the $1s$ orbital, has its maximum electron density closest to the nucleus, as in Fig. 2-7(b). Hence, this is the most stable orbital for the electron of the hydrogen atom. The sign of the $1s$ wave function is everywhere positive, as shown in Fig. 2-7(a). Beginning with the $2s$ orbital, there are positive and negative values for a wave function, a change occurring each time that the function $R(r)$ crosses the abscissa [Fig. 2-7(a)]. These changes in sign for the function $R(r)$ correspond to **nodes** in the functions $r^2 R(r)^2$—values of r where the electron density becomes zero. Notice from Fig. 2-7(b) that as the value of n increases, the maximum in the radial electron density shifts farther from the nucleus. Thus an orbital gets larger as the principal quantum number n increases. Correspondingly, the energy of the electron in such an orbital becomes less negative, meaning that the electron is less strongly bound.

p Orbitals

For each p orbital, the quantum number ℓ equals 1, and the shape is that shown in Fig. 2-6. Three values of the quantum number m_ℓ are possible $(-1, 0, +1)$, representing each of the three possible orientations in space. There is a node at the nucleus for each p orbital because the p-type wave function [Fig. 2-7(a)] has the value zero at the nucleus. The sign of the p orbital therefore changes at the nu-

A   B

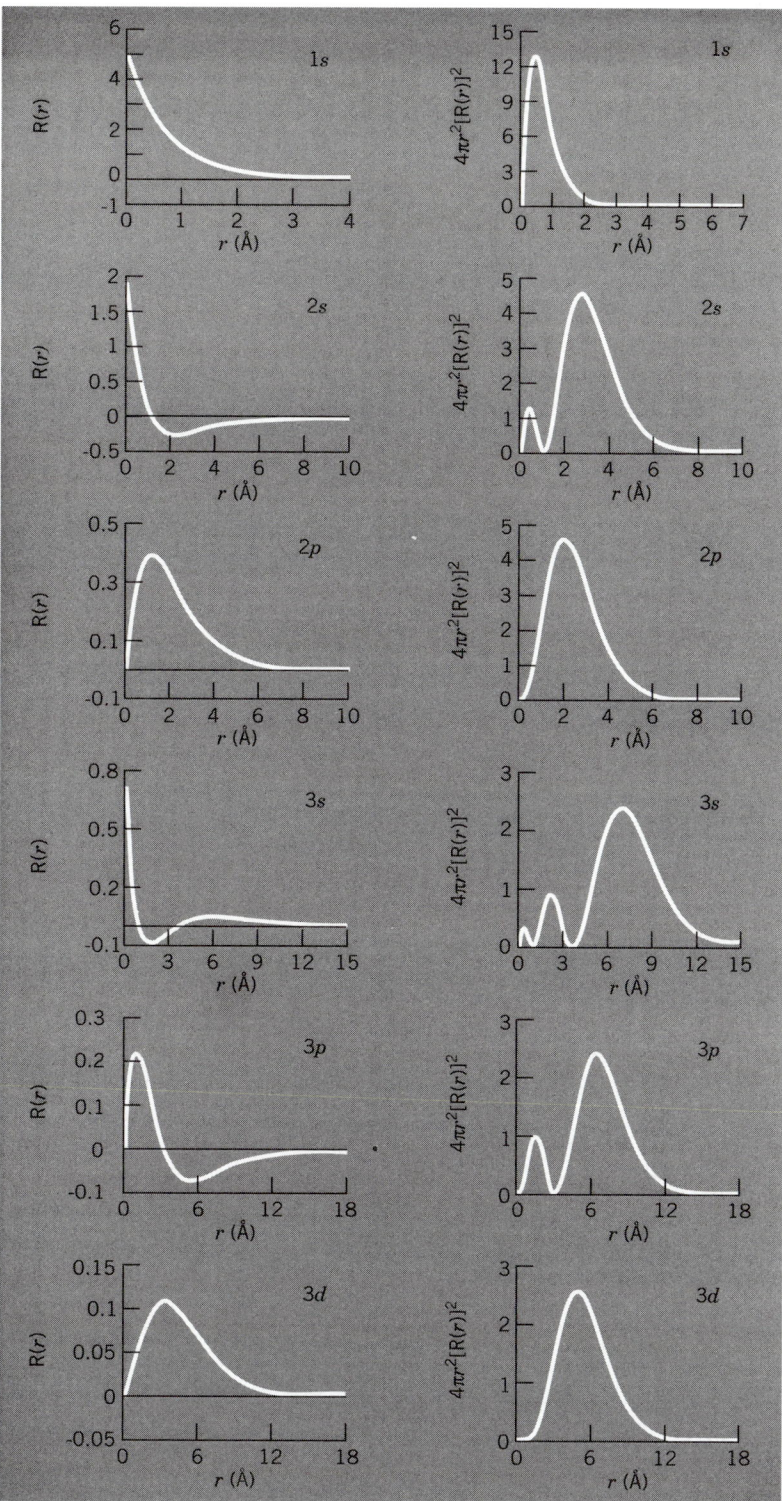

Figure 2-7 Plots for the hydrogen-like wave functions of (*a*) the radial function $R(r)$ versus r, the distance from the nucleus; and (*b*) the probability distribution function $4\pi r^2[R(r)]^2$ versus r, the distance from the nucleus. Note that the range of values for the two axes have been changed from one graph to another.

cleus. The $2p$ orbitals have no other nodes, but beginning with $3p$ there are additional radial nodes, as shown in Fig. 2-7. As was true for the s orbitals, the size of a p orbital depends on the principal quantum number in the order $2p$, $3p$, $4p$, and so on. This can be seen by comparing the positions of the largest maxima in the graphs of $r^2R(r)^2$ in Fig. 2-7(b).

d Orbitals

Each set of d orbitals consists of five members whose shapes are shown in Fig. 2-6. The five members arise because there are five possible values for the quantum number m_ℓ (-2, -1, 0, $+1$, $+2$). Within each lobe of the d orbitals, the radial electron density changes as shown in Fig. 2-7. The following features are important. The d_{z^2} orbital is symmetrical about the z axis. The d_{xz}, d_{yz}, and d_{xy} orbitals are alike, except that they have their lobes in the xz, yz, and xy planes, respectively. The $d_{x^2-y^2}$ orbital has the same shape as the d_{xy}, but the former orbital is rotated by 45° about the z axis so that its lobes lie on the x and y axes instead of between the x and y axes. The d orbitals appear only when the principal quantum number n has risen to the value of three or greater.

f Orbitals

For each value of the principal quantum number $n \geq 4$, there is a set of seven f orbitals for which $\ell = 3$. Within this set of seven orbitals, the quantum number m_ℓ takes the seven values -3, -2, -1, 0, $+1$, $+2$, $+3$. The f orbitals play an important role in chemical behavior only for compounds of the lanthanides and actinides. The typical shapes of these orbitals are given in Fig. 2-8.

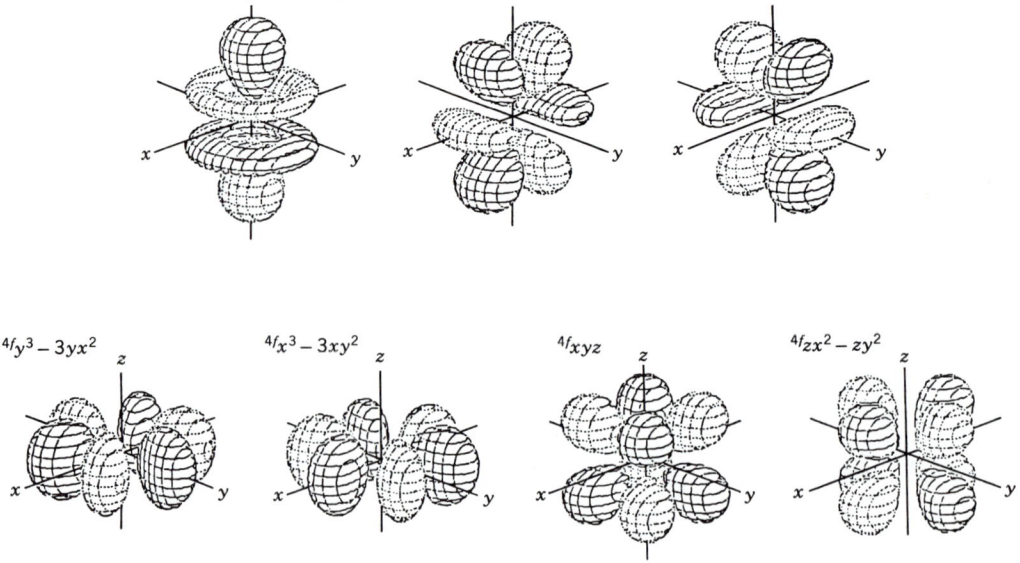

Figure 2-8   Shapes of the seven f orbitals. Solid lines represent positive amplitude and dotted lines negative amplitude of the wave function. [Taken from Q. Kikuchi and K. Suzuki, *J. Chem. Educ.*, **1985**, *62*, 206–209 and used with permission.]

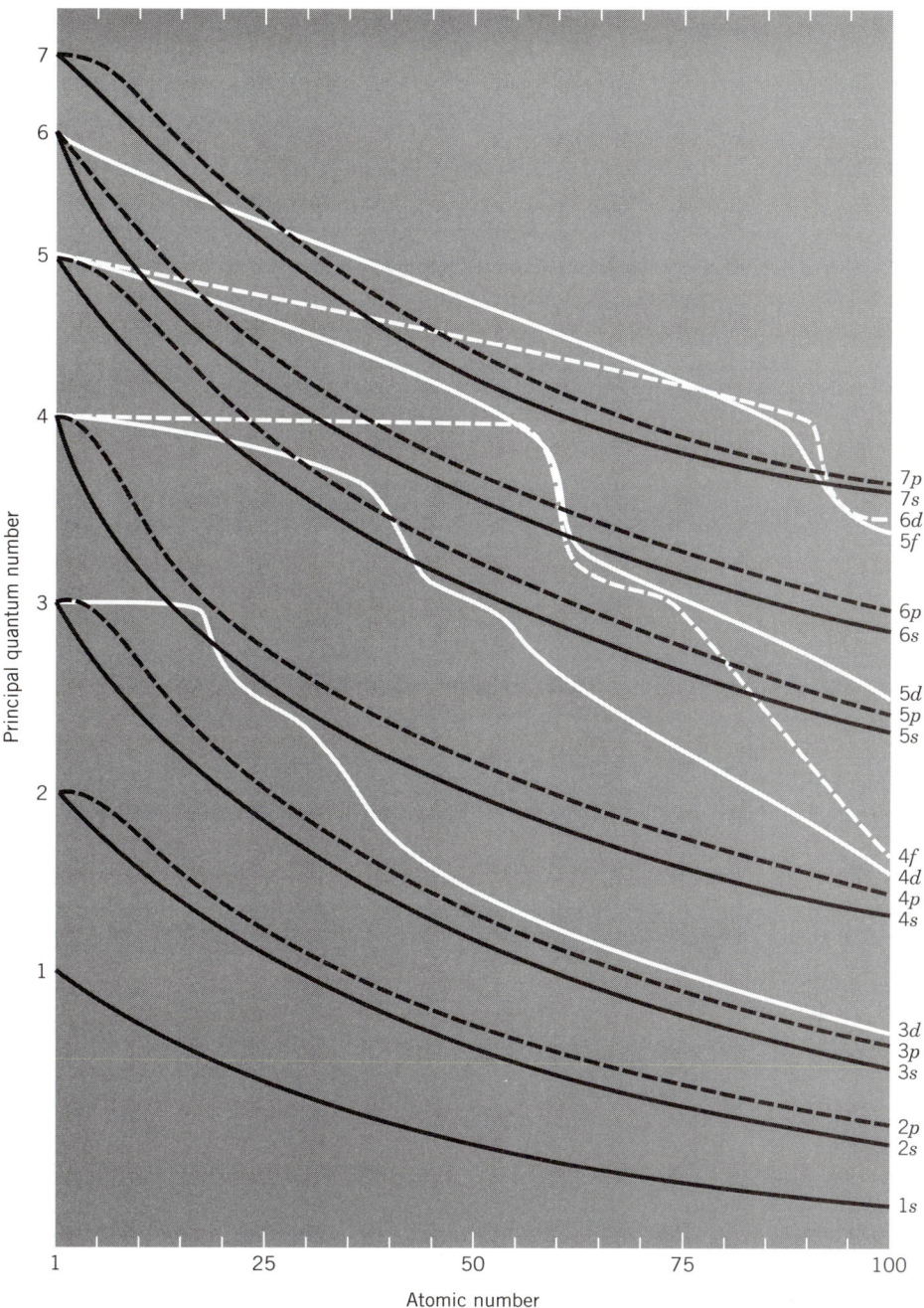

Figure 2-9 The relative energies of the atomic orbitals as a function of atomic number.

Energy Levels in the Hydrogen Atom

For the hydrogen atom, the order of increasing energy for the atomic orbitals is determined only by the principal quantum number n. The energy of the electron is the same regardless of whether it is in an s, p, d, or f orbital, as long as the principal quantum number n is the same. This is shown in Fig. 2-9, where for the hydrogen atom (atomic number 1), the energies of the s, p, d, and f orbitals converge at a value that depends only on n. This is obviously not true for atoms with more than one electron. For these more complex atoms, the hydrogen-like atomic orbitals must be modified to reflect the pattern of energies shown in Fig. 2-9.

2-4   Structures of Atoms with Many Electrons

By using the experimental data of Fig. 2-9, it is possible to arrange the atomic orbitals in multielectron atoms in the correct energy order for building up the electronic structures for atoms having an atomic number greater than 1. This ordering of atomic orbitals for multielectron atoms is shown in Fig. 2-10. Note that the p orbitals retain their threefold degeneracy, the d orbitals their fivefold degeneracy, and the f orbitals their sevenfold degeneracy. Before writing the electron configurations for multielectron atoms, it is useful to understand the reasons for the energy ordering given in Fig. 2-10. Why, for a given value of the principal quantum number n, are the atomic orbitals used in the order s, p, d, f, and so on? The answer to this question can be found in part by examining the repulsions among the electrons of a multielectron atom and the relative amplitudes of the radial portions of the atomic orbital wave functions. Our goal in examining the latter is to compare the extent to which the radial portions of the various wave functions "penetrate to the nucleus," which is the topic of the next section.

Radial Penetration of the Wave Functions

An electron in an atomic orbital that has appreciable electron density close to the nucleus is stabilized by close interaction with the positive charge of the nucleus, in the same way that the most stable Bohr orbit is the one with the smallest radius. One can evaluate the stability of an electron in various orbitals by comparing the radial electron density functions $4\pi r^2 [R(r)]^2$, as is done in Fig. 2-11. For any given value of the principal quantum number n, it is the s-type orbital that most has appreciable electron density close to the nucleus, followed in order by p-, d-, and f-type orbitals. The relative order of stability for the various orbital types is, then, $s > p > d > f$, and so on, for any one value of n. This is the ordering the orbitals are given in Fig. 2-10. It should be noted, though, from Figs. 2-9 and 2-10 that at crucial points there is a crossing of the ns and $(n-1)d$ energy levels. This will become important in writing the electron configurations of the elements.

Electron Configurations

The electrons are assigned to an atom by placing them into the various atomic orbitals according to three rules:

1. *The **aufbau principle**.* The electron configurations are built up from the bottom, using the lowest energy orbitals first.
2. *Hund's rule.* Where orbitals are available in degenerate sets, maximum spin multiplicity is preserved; that is, electrons are not paired until each orbital in a degenerate set has been half-filled.

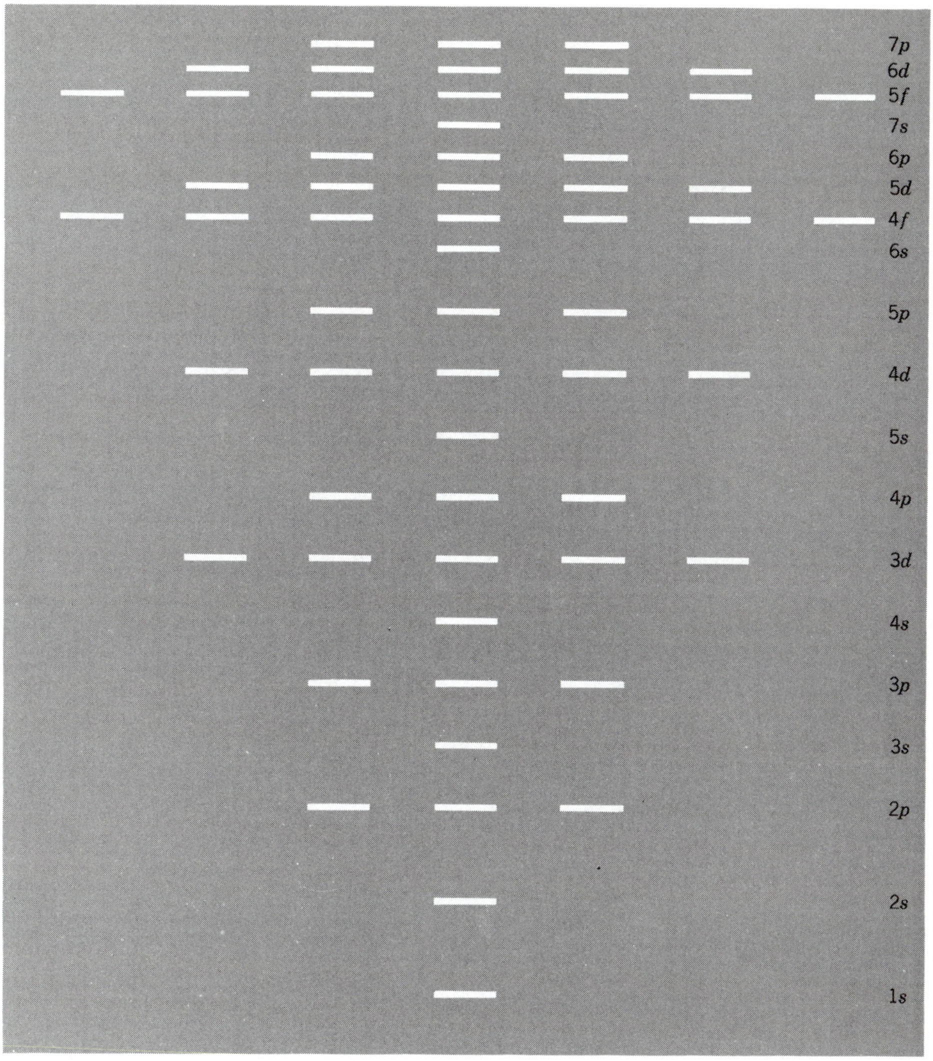

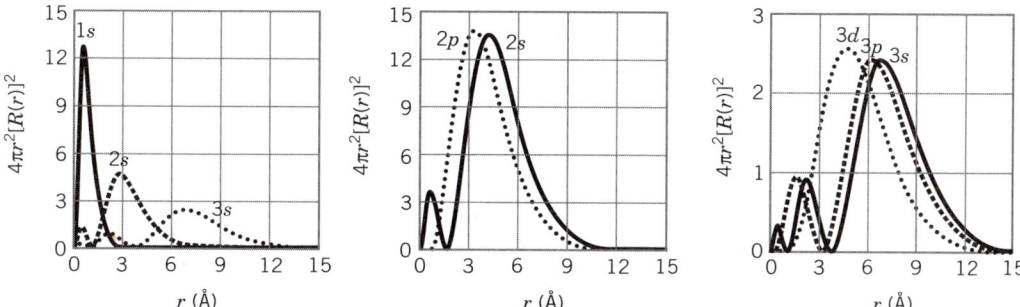

Figure 2-10 The order in which the atomic orbitals are used in building up the electron configurations of many-electron atoms. The orbitals are used in sequence, from the bottom, in accordance with the aufbau principle, Hund's rule, and the Pauli exclusion principle.

Figure 2-11 A comparison of the radial density distribution functions, $4\pi r^2 [R(r)]^2$, showing the relative penetration of various orbitals.

Periodic trends in electron configuration

Group =	IA (1)	IIA (2)	IIIA (3)	IVA (4)	VA (5)	VIA (6)	VIIA (7)	VIII (8)	VIII (9)	VIII (10)	IB (11)	IIB (12)	IIIB (13)	IVB (14)	VB (15)	VIB (16)	VIIB (17)	VIIIB (18)	Closed Shell
Period = 1	1 H $1s^1$																	2 He $1s^2$	K
2	3 Li $2s^1$	4 Be $2s^2$											5 B $2p^1$	6 C $2p^2$	7 N $2p^3$	8 O $2p^4$	9 F $2p^5$	10 Ne $2p^6$	L
3	11 Na $3s^1$	12 Mg $3s^2$											13 Al $3p^1$	14 Si $3p^2$	15 P $3p^3$	16 S $3p^4$	17 Cl $3p^5$	18 Ar $3p^6$	M
4	19 K $4s^1$	20 Ca $4s^2$	21 Sc $4s^2 3d^1$	22 Ti $4s^2 3d^2$	23 V $4s^2 3d^3$	24 Cr $4s^1 3d^5$	25 Mn $4s^2 3d^5$	26 Fe $4s^2 3d^6$	27 Co $4s^2 3d^7$	28 Ni $4s^2 3d^8$	29 Cu $4s^1 3d^{10}$	30 Zn $4s^2 3d^{10}$	31 Ga $4p^1$	32 Ge $4p^2$	33 As $4p^3$	34 Se $4p^4$	35 Br $4p^5$	36 Kr $4p^6$	N
5	37 Rb $5s^1$	38 Sr $5s^2$	39 Y $5s^2 4d^1$	40 Zr $5s^2 4d^2$	41 Nb $5s^1 4d^4$	42 Mo $5s^1 4d^5$	43 Tc $5s^2 4d^5$	44 Ru $5s^1 4d^7$	45 Rh $5s^1 4d^8$	46 Pd $5s^0 4d^{10}$	47 Ag $5s^1 4d^{10}$	48 Cd $5s^2 4d^{10}$	49 In $5p^1$	50 Sn $5p^2$	51 Sb $5p^3$	52 Te $5p^4$	53 I $5p^5$	54 Xe $5p^6$	O
6	55 Cs $6s^1$	56 Ba $6s^2$	57 La $6s^2 5d^1$	72 Hf $6s^2 5d^2$	73 Ta $6s^2 5d^3$	74 W $6s^2 5d^4$	75 Re $6s^2 5d^5$	76 Os $6s^2 5d^6$	77 Ir $6s^2 5d^7$	78 Pt $6s^1 5d^9$	79 Au $6s^1 5d^{10}$	80 Hg $6s^2 5d^{10}$	81 Tl $6p^1$	82 Pb $6p^2$	83 Bi $6p^3$	84 Po $6p^4$	85 At $6p^5$	86 Rn $6p^6$	P
7	87 Fr $7s^1$	88 Ra $7s^2$	89 Ac $7s^2 6d^1$	104	105	106	107	108	109										

s-Block Elements (Groups IA–IIA)
d-Block Elements (Groups IIIA–IIB)
p-Block Elements (Groups IIIB–VIIIB)

f-Block Elements

58 Ce $6s^2$ $6d^0 4f^2$	59 Pr $6s^2$ $5d^0 4f^3$	60 Nd $6s^2$ $5d^0 4f^4$	61 Pm $6s^2$ $5d^0 4f^5$	62 Sm $6s^2$ $5d^0 4f^6$	63 Eu $6s^2$ $5d^0 4f^7$	64 Gd $6s^2$ $5d^1 4f^7$	65 Tb $6s^2$ $5d^0 4f^9$	66 Dy $6s^2$ $5d^0 4f^{10}$	67 Ho $6s^2$ $5d^0 4f^{11}$	68 Er $6s^2$ $5d^0 4f^{12}$	69 Tm $6s^2$ $5d^0 4f^{13}$	70 Yb $6s^2$ $5d^0 4f^{14}$	71 Lu $6s^2$ $5d^1 4f^{14}$
90 Th $7s^2$ $6d^2 5f^0$	91 Pa $7s^2$ $6d^1 5f^2$	92 U $7s^2$ $6d^1 5f^3$	93 Np $7s^2$ $6d^1 5f^4$	94 Pu $7s^2$ $6d^0 5f^6$	95 Am $7s^2$ $6d^0 5f^7$	96 Cm $7s^2$ $6d^1 5f^7$	97 Bk $7s^2$ $6d^0 5f^9$	98 Cf $7s^2$ $6d^0 5f^{10}$	99 Es $7s^2$ $6d^0 5f^{11}$	100 Fm $7s^2$ $6d^0 5f^{12}$	101 Md $7s^2$ $6d^0 5f^{13}$	102 No $7s^2$ $6d^0 5f^{14}$	103 Lw $7s^2$ $6d^1 5f^{14}$

Figure 2-12 Periodic trends in electron configuration, showing the outermost or differentiating electrons for each element.

3. *The **Pauli exclusion principle**.* No two electrons may have the same set of four quantum numbers. Where two electrons occupy the same orbital, they must have opposite spins: $m_s = +\frac{1}{2}$ for one electron and $m_s = -\frac{1}{2}$ for the second electron. Because the spin quantum number m_s can take only one of two values, an orbital can house at most two electrons.

If these rules, which we shall examine in more detail in Section 2-6, are followed, the electron configuration that is specified is the ground-state configuration. Other electron configurations are possible, but they represent excited state configurations. A discussion of the ground-state electron configurations of the elements follows. The reader should refer to Fig. 2-12.

Elements of Period One
The following are electron configurations for the two elements of row one:

H $1s^1$
He $1s^2$

For both atoms, the principal quantum number, n, equals one. Row one of the periodic table is completed with the element He, because the only orbital ($1s$) in the first or **K** shell (where $n = 1$) becomes filled with two electrons. The electrons of subsequent atoms must begin using orbitals of the next shell, where $n = 2$.

Elements of Period Two
The eight elements of this row make use of the four orbitals with principal quantum number $n = 2$. The following are the ground-state electron configurations for the elements:

Li $1s^2 2s^1$
Be $1s^2 2s^2$
B $1s^2 2s^2 2p^1$
C $1s^2 2s^2 2p^2$
N $1s^2 2s^2 2p^3$
O $1s^2 2s^2 2p^4$
F $1s^2 2s^2 2p^5$
Ne $1s^2 2s^2 2p^6$

Notice that with the element boron, the $2p$ orbitals begin to be used, eventually holding six electrons with the completion of the row at Ne. It is the **L** shell that becomes filled at Ne. It is upon this neon core [Ne] that the electron configurations for the elements of row three are built.

Elements of Period Three
Beginning with sodium, orbitals with principal quantum number $n = 3$ are used.

Na $[Ne]3s^1$
Mg $[Ne]3s^2$
Al $[Ne]3s^2 3p^1$
Si $[Ne]3s^2 3p^2$
P $[Ne]3s^2 3p^3$

S $[Ne]3s^23p^4$

Cl $[Ne]3s^23p^5$

Ar $[Ne]3s^23p^6$

The row is completed with argon, where the three $3p$ orbitals are filled with two electrons each. In fact, it is characteristic of all elements in Group VIIIB(18) that they complete a row of the periodic table and have the filled outermost electron configuration np^6. Notice also that in row two it was the element boron where the appropriate np orbitals first were used. Here, in row three, this is true of the element aluminum. Both elements are in Group IIIB(13) and have the outermost electronic configuration that is characteristic of all elements of Group IIIB(13): np^1. The $3d$ atomic orbitals were not used for the electron configurations of the elements of row three. Notice also from Fig. 2-10 that the $3d$ orbitals are not yet next in line to be used. The next orbital that is available is the $4s$ orbital, and it is the first to be used after the argon core [Ar] in writing the electron configurations for elements of row four. The shell that is completed with the third-row element argon is the **M shell.**

Elements of Period Four

The fourth row of the periodic table begins with potassium, which has the characteristic outermost electron configuration of all elements in Group IA(1): ns^1.

K	$[Ar]4s^1$		Sc	$[Ar]4s^23d^1$
Ca	$[Ar]4s^2$		Ti	$[Ar]4s^23d^2$
			V	$[Ar]4s^23d^3$
$\leftarrow$			Cr	$[Ar]4s^13d^5$
			Mn	$[Ar]4s^23d^5$
Ga	$[Ar]4s^23d^{10}4p^1$		Fe	$[Ar]4s^23d^6$
Ge	$[Ar]4s^23d^{10}4p^2$		Co	$[Ar]4s^23d^7$
As	$[Ar]4s^23d^{10}4p^3$		Ni	$[Ar]4s^23d^8$
Se	$[Ar]4s^23d^{10}4p^4$		Cu	$[Ar]4s^13d^{10}$
Br	$[Ar]4s^23d^{10}4p^5$		Zn	$[Ar]4s^23d^{10}$
Kr	$[Ar]4s^23d^{10}4p^6$			

The $4s$ orbital becomes filled at the element calcium, which has the outermost configuration typical of all elements in Group IIA(2): ns^2. The two portions of the main group elements are interrupted with the 10 elements scandium through zinc, where the previously unused $3d$ orbitals become available. The series of elements from scandium to zinc is 10 elements in length because the five d orbitals, holding 2 electrons each, require 10 electrons to be filled. After zinc, the row is completed with 6 elements having outermost electron configurations featuring successive use of the three $4p$ orbitals.

The orderly pattern of filling of the d orbitals seems to be interrupted at the elements chromium and copper. In these cases a $4s$ electron is "borrowed" in order to obtain either a half-filled d orbital set (Cr) or a completely filled d orbital set (Cu). In each case, this leads to a greater stability because of the half-filled or filled d orbital set. The same anomaly takes place for Mo [also of Group VIA(6)] and for the other elements of Group IB(11), Ag and Au.

Elements of Period Five

The elements of period five, beginning with rubidium and ending with xenon, follow the same pattern of electron configurations as that for the preceding period four. The valence orbitals in question are now, in order of use, the $5s$, $4d$, and $5p$ orbitals. The $5d$ and the $4f$ orbital sets are not used at this time. As was true for chromium and copper in the first transition series, anomalies occur in the regular filling of the d orbitals at the elements molybdenum and silver.

Elements of Period Six

Period six of the periodic table is composed of 32 elements from cesium (55) to radon (86). The $6s$ orbital is filled at barium. The $5d$ orbital set begins to be used with lanthanum, but the series is immediately interrupted by 14 elements. In this series of 14 elements, as well as in those immediately below them, the sevenfold degenerate f orbitals are used, two electrons eventually being distributed into each orbit. Only then is the use of the d orbitals resumed at hafnium. The row is ended with the usual p-block elements, in this case thallium through radon.

There is an important reason why the ns orbital for any row n is used before the $(n-1)d$ or the $(n-2)f$ orbitals. The radial portion of the wave function for an s orbital is characteristically closer to the nucleus than d and f orbitals. Hence, the $(n-1)d$ orbital is higher in energy than the ns orbital for certain elements (see Fig. 2-9). Consequently, the $3d$ orbitals are not used in row three, but in row four of the periodic table. Similarly, it is not until row six that the $4f$ and the $5d$ orbitals are used.

Elements of Period Seven

The elements of this period complete the periodic table. The short-lived elements 104–109 have now been detected. The $7s$-block elements francium and radium are followed by the second series of f-block elements, for which the $5f$ and $6d$ orbital energies are similar. It is not necessary to be concerned with the exact arrangement of electrons in these f and d orbitals because two or more different configurations differ so little in energy that the exact configuration in the ground state of the free atom has little to do with the chemical properties of the element in its compounds.

2-5   The Periodic Table

More than a century ago chemists began to search for a tabular arrangement of the elements that would group together those with similar chemical properties and also arrange them in some logical sequence. The sequence was generally the order of increasing atomic weights. As is well known, these efforts culminated in the type of periodic table devised by Mendeleev, in which the elements were arranged in horizontal rows with row lengths chosen so that like elements would form vertical columns.

It was Moseley who showed that the proper sequence criterion was not atomic weight but atomic number (although the two are only rarely out of register). It then followed that vertical columns contained chemically similar elements, as well as electronically similar atoms. All of Chapter 8 is devoted to a discussion of the practical chemical aspects of the periodic table. Since we have just studied how the electron configurations of atoms are built up, it is now appropriate to point out that these configurations lead logically to the same periodic arrangement that Mendeleev deduced from strictly chemical observations.

The vertical columns of the periodic table on the inside of the front cover and elsewhere in this text are labeled in two fashions. First, we give a traditional column (or group) designation using Roman numerals I–VIII, with letters A or B. Second, and parenthetically, we give the newest group designations adopted by the International Union of Pure and Applied Chemistry: Arabic numerals, 1–18.

To build up a periodic table based on similarities in electron configuration, a convenient point of departure is to require all atoms with outer ns^2np^6 configurations to fall in a column. It is convenient to place this column at the extreme right, and to include also He ($1s^2$). This column thus contains those elements called the **noble gases:** He, Ne, Ar, Kr, Xe, Rn.

If the elements that have a single electron in the ns orbitals are placed in the Group IA(1) column at the extreme left of the table, the remaining pattern of the table is established. The elements of Group IA(1) are called the **alkali metals.** The ionization enthalpies of the single s electrons in the valence shell of these elements is low, and the +1 cations of these elements are readily formed. The chemistry of these elements is mostly that of these +1 cations. Each of them is followed by one of the elements of Group IIA(2), which have the characteristic ns^2 configuration. These elements (Be, Mg, Ca, Sr, Ba, and Ra) are called the **alkaline earth metals,** and characteristically form +2 cations.

Now, if we return to the noble gas column and begin to work back from right to left, it is clear that we shall get columns of elements with outer electron configurations ns^2np^5, ns^2np^4, . . . , ns^2np^1. The ns^2np^5 elements F, Cl, Br, I, and At are called the **halogens** (meaning salt formers). Those with the ns^2np^4 configurations are O, S, Se, Te, and Po; they are given the family name **chalcogens.** The other three columns, that is, ns^2np^3 (N, P, As, Sb, and Bi), ns^2np^2 (C, Si, Ge, Sn, and Pb), and ns^2np^1 (B, Al, Ga, In, and Tl), have no commonly accepted trivial group names, although use of the term **pnictogens** for elements of Group VB(15) is on the increase.

Thus far we have developed a rational arrangement for nearly one half of the elements. These elements, which involve outer shells consisting solely of s and p electrons, are called the **main group elements.** In particular, they are either the **s-block elements** or the **p-block elements.** Most of the remaining elements are called **d-block elements.** They occupy the central region of the periodic table, between the two main blocks. Their occurrence at this position is due to the filling of the $(n-1)d$ orbitals. Their electron configurations have already been discussed. With the exception of Group IIB(12) (Zn, Cd, and Hg), these elements are also called the **transition elements.** Their common characteristic is that either the neutral atom, or some important ion it forms, has an incomplete set of d electrons. The set of elements with incomplete $3d$ shells is called the **first transition series,** and those with partial $4d$ and $5d$ shells are called the **second** and **third transition series,** respectively.

The elements Zn, Cd, and Hg have unique properties. While they resemble the alkaline earths in giving no oxidation state higher than +2, they differ because the configuration immediately underlying their valence orbitals is a rather polarizable nd^{10} shell instead of a more tightly bound noble gas shell. Their chemistry will be taken up in Chapter 22.

Finally, the 14 elements between La and Hf, in which the $4f$ orbitals are being filled, are placed at the bottom of the table, to avoid making it excessively wide. These elements are called the **lanthanides** because of their chemical resemblance to lanthanum. A somewhat similar set of elements, called the **actinides,** have partially filled $5f$ orbitals. These elements are placed beneath the corresponding lanthanides. These two series are collectively called the *f*-**block elements.**

2-6 Hund's Rule, Electron Configurations, and Effective Nuclear Charge

Thus far, for atoms in which there are partly filled p or d shells, the electron configurations have simply been written p^n or d^n. However, it is possible, and important, to specify them in greater detail. For instance, for the configuration p^2, there are 15 distinct ways of assigning quantum numbers to the two electrons in the three degenerate orbitals. All of the corresponding orientations of the two electrons are available to the atom, but only one assignment is most stable. The ground states for the p^n configurations are illustrated in Fig. 2-13.

There are two important features of the pattern shown there:

1. Within a set of degenerate orbitals (in this case the p_x, p_y, and p_z), the electrons make use of different orbitals so long as it is possible.

2. Parallel spins (same value of m_s) are used until the Pauli exclusion principle requires pairing of spins.

The first of these features is partly a consequence of the charge of the electrons. The electrons can minimize the repulsive forces among themselves by occupying different p orbitals. This is true because the p orbitals occupy regions of space along different axes. Repulsive forces among the p electrons are thus minimized when the electrons are distributed as far from one another as possible. The second feature arises because pairing of spins before it is required by the Pauli exclusion principle leads to a less stable arrangement. Consequently, we have

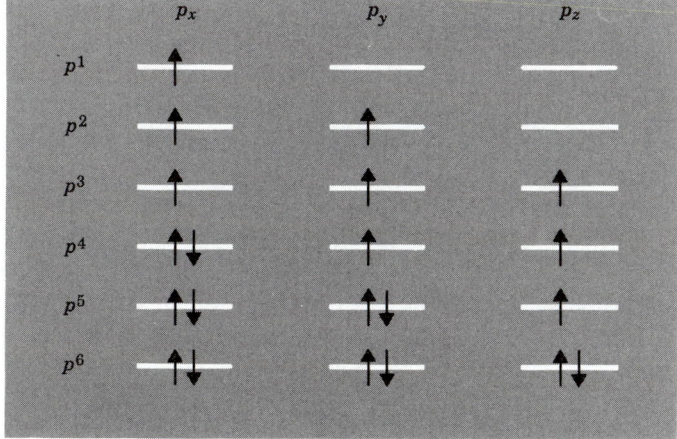

Figure 2-13 Application of the principle of maximum spin multiplicity in the filling of the degenerate set of $2p$ orbitals for the configurations $2p^1$–$2p^6$.

Hund's rule: The most stable electronic state (among several that are possible within a degenerate set of orbitals) is that state with **maximum spin multiplicity:** the one with the largest number of unpaired electron spins. This rule immediately implies the spreading out of electrons into as much of the space surrounding the nucleus as is possible. Hund's rule also implies that this spreading out of electron density leads to extra stability not only for the individual electron at hand, but also for the ensemble of electrons that make up a multielectron atom. It is important to understand how this stability arises. Part of the explanation involves the concept of **effective nuclear charge.**

An electron will occupy that orbital, of all those yet unoccupied on an atom, where the nucleus is most effective at offering positive charge to stabilize the electron. It was J. C. Slater who realized that the effective nuclear charge, Z^*, which is felt by an electron, is not the actual charge Z of the nucleus of the atom. Rather, the amount of nuclear charge actually felt by an electron depends on the type of orbital in which the electron is housed, and on the ability of other electrons in more penetrating orbitals to screen (or shield) the electron in question from the nucleus.

We have already pointed out that, among the orbitals having the same value of the principal quantum number, the s orbital is the most stable. Furthermore, any atomic orbital that places appreciable electron density between the nucleus and a second orbital is said to penetrate the region of space occupied by the second orbital. To the extent that this happens, the electron is more stable in the more penetrating orbital. The relative extent to which the various orbitals penetrate the electron clouds of other orbitals is $s > p > d > f >$ and so on. Thus, for any given principal quantum number n, an electron will experience the greatest effective nuclear charge when housed in an s orbital, then a p orbital, and so on. This finding was already cited as the reason for the order of orbital filling among the elements. But what of the atoms of the p block of the periodic table, where the last electron is placed in every case into an orbital of the same type? Consider the elements of row two, beginning with B and proceeding to Ne. Here, with each successive proton that is added in making the next element, there is added an electron into the $2p$ orbitals. Each new electron is added in accordance with Hund's rule. Also, each new electron experiences a new and different effective nuclear charge. At B the new electron is added into one of the p orbitals, say the p_x orbital. The new electron that is added for C must now go into another of the $2p$ orbitals, say the p_y. But the p_y orbital is perpendicular to the p_x orbital, and the p_y orbital is poorly screened from the nuclear charge by the p_x orbital. Consequently, the effective nuclear charge for the last electron in C is higher than that for B. It is the geometry and the orientation of the p orbitals that makes them poor at shielding one another from the nucleus. Consider the next element, N. The third p electron that is added to make this element is poorly screened from the growing nuclear charge because the other two p electrons that are already there lie at 90° to this last one. Thus the effective nuclear charge for the differentiating electron of nitrogen is even higher. Where screening of an electron is poor, the effective nuclear charge is correspondingly high. Thus Hund's rule: Electrons spread out into a degenerate set of orbitals in order to experience this maximum effective nuclear charge.

This view is admittedly simplistic. There are other factors (such as the quantum mechanical exchange energy associated with a set of electrons with parallel spins) that influence the energies of the various electron configurations. These other issues need not detain us in the following limited discussion of shielding and effective nuclear charge.

J. C. Slater proposed an empirical constant that represents the cumulative extent to which the other electrons of an atom shield (or screen) any particular electron from the nuclear charge. Thus Slater's screening constant σ is used in Eq. 2-6.1.

$$Z^* = Z - \sigma \qquad (2\text{-}6.1)$$

Here, Z is the atomic number of the atom, and hence is equal to the actual number of protons in the atom. The parameter Z^* is the **effective nuclear charge,** which according to Eq. 2-6.1 is smaller than Z, since the electron in question is screened (shielded) from Z by an amount σ. We found that in cases for which screening is small, the effective nuclear charge Z^* is large. Conversely, an electron that is well shielded (large value for σ in Eq. 2-6.1) from the nuclear charge Z experiences a small effective nuclear charge Z^*.

The value of σ for any one electron in a given electron configuration (i.e., in the presence of the other electrons of the atom in question) is calculated using a set of empirical rules developed by Slater. According to these rules, the value of σ for the electron in question is the cumulative total provided by the various other electrons of the atom. The other electrons of the atom each add an intrinsically different contribution to the value of σ as follows:

If the electron in question resides in an s or p orbital,

1. All electrons in principal shells higher than the electron in question contribute zero to σ.
2. Each electron in the same principal shell contributes 0.35 to σ.
3. Electrons in the $(n-1)$ shell each contribute 0.85 to σ.
4. Electrons in deeper shells each contribute 1.00 to σ.

If the electron in question resides in a d or f orbital,

1. All electrons in principal shells higher than the electron in question contribute zero to σ.
2. Each electron in the same principal shell contributes 0.35 to σ.
3. All inner-shell electrons [i.e., $(n-1)$ and lower] uniformly contribute 1.00 to σ.

To illustrate the application of these empirical rules, let us estimate the effective nuclear charge for one of the outer electrons (a $2p$ electron) of the fluorine atom, which has a configuration of $1s^2 2s^2 2p^5$. The inner shell $1s^2$ contributes $2 \times 0.85 = 1.70$ to σ. Each of the electrons with $n = 2$ (other than the one under consideration) contributes 0.35. Therefore, we have

$$\sigma = 2 \times 0.85 + 6 \times 0.35$$
$$= 1.70 + 2.10 = 3.80$$

and

$$Z^* = 9 - 3.80 = 5.20$$

Values of Z^* for some other elements are listed in Table 2-3.

Table 2-3 Calculation of Effective Nuclear Charge Z^*, According to Slater's Eq. 2-6.1.

n	Z	σ	Z^*
1 (H)	1	0	1.00
(He)	2	0.35	1.65
2 (Li)	3	1.70	1.30
(Be)	4	2.05	1.95
(B)	5	2.40	2.60
(C)	6	2.75	3.25
(N)	7	3.10	3.90
(O)	8	3.45	4.55
(F)	9	3.80	5.20
(Ne)	10	4.15	5.85

There are trends among the elements for the effective nuclear charge experienced by the last or **differentiating electron** of an element. In the next section we will illustrate this more completely by comparing other physical properties of the elements. Still, it is useful to pause long enough to compare the values of Z^* across a row. As is shown in Table 2-3, there is a steady increase (by 0.65 units) in Slater's Z^* across each row of the periodic table.

Other trends in Z^* are less meaningful. Also, values of Slater's Z^* become increasingly less reliable for the heavier elements, or in comparisons down a group of the periodic table. Modifications have been made, and more accurate effective nuclear charges have been estimated for all of the elements. Regardless of the particular set of values for Z^* that one adopts, the important conclusion for our purposes is that the effective nuclear charge increases continually from left to right across the rows of the periodic table, because of imperfect shielding.

2-7   Periodic Trends in the Properties of the Elements

There is an overall harmony among the properties of the elements and their electronic structures. In fact, the periodic trends that we shall discuss can be traced in part to differences in the orbitals in which the electrons are housed. The concepts that are to be used in establishing this harmony obey the following hierarchy:

1. The different interpenetrations of the atomic orbitals can be judged from an evaluation of the size of the orbitals, $R(r)^2$, and the orientations of the orbitals, $\Theta(\theta)^2\Phi(\phi)^2$.

2. Because of these different penetrations and orientations of the atomic orbitals, the orbitals are used in the sequence shown in Fig. 2-10.

3. Because of these different penetrations and orientations of the atomic orbitals, the valence electrons of the atoms experience different effective nuclear charges, as illustrated in Table 2-3.

4. Properties such as first ionization enthalpy also follow trends that reflect the different electron configurations in any period or group of the periodic table.

Ionization Enthalpy

The periodic trends in **first ionization enthalpies** (ΔH_{ion}) for the elements H to Rn are shown in Fig. 2-14. There are three major trends that merit comment. First, the maxima occur at the noble gases and the minima occur at the alkali metals. This finding is easily understandable, since the closed-shell configurations of the noble gases are very stable and resist disruption, either to form chemical bonds or to become ionized. In the alkali metal atoms, there is an electron outside the preceding noble gas configuration. This electron is well shielded from the attraction of the nucleus; it is therefore relatively easy to remove. In terms of effective nuclear charge, this comparison proceeds as follows. The outermost ($2p$) electron of a neon atom experiences an effective nuclear charge of $Z^* = 5.85$. On the other hand, the outermost ($3s$) electron of a sodium atom experiences an effective nuclear charge of only $11 - (8 \times 0.85 + 2 \times 1) = 2.20$. Thus the relative values of Z^* for Ne and Na correctly correspond to the relative values of their ionization enthalpies, 2080 and 496 kJ mol^{-1}, respectively.

Second, from each alkali metal (ns^1 configuration—minima in Fig. 2-14) across a row of the periodic table to the next noble gas (closed-shell electron configuration—maxima in Fig. 2-14) there is an overall increase in ionization enthalpies. The increase is not perfectly regular, as will be discussed shortly. Nevertheless, the trend is obvious; across any row of the periodic table there occurs an increase, generally, in the first ionization enthalpies. This trend conforms to a similar pattern noted previously for effective nuclear charge. The effective nuclear charge grows across a row because of the cumulative effects of imperfect shielding by orbitals of the same principal quantum number. As the effective nuclear charge increases, so does the energy necessary to ionize the atom.

Third, the increase just discussed is not smooth. Instead, there are two well-defined jogs that occur at corresponding positions in each series, that is, from Li

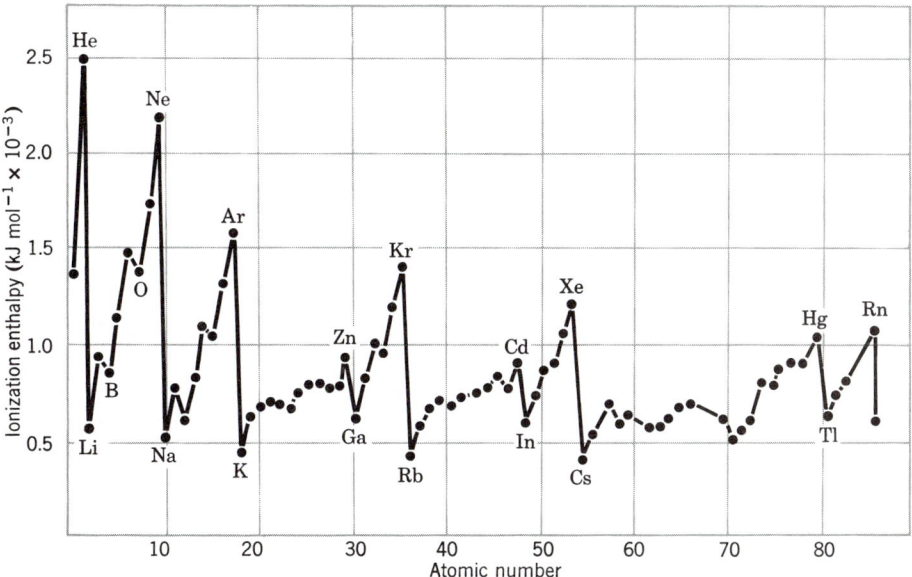

Figure 2-14 Periodic trends in the first ionization enthalpies, ΔH_{ion}. Values for the first ionization enthalpies of the elements are also listed in Appendix IIB.

to Ne, from Na to Ar and, with some differences due to intervention of the transition elements, in subsequent periods of the table. In each case, the ionization enthalpy drops from the s^2 to the s^2p configuration and again from the s^2p^3 to the s^2p^4 configuration. The explanation becomes apparent if the facts are stated in a slightly different way, where the elements of the Li to Ne period are used as an example. The ionization enthalpies of B, C, and N increase regularly but they are all lower than values that would be extrapolated from Li and Be. This occurs because p electrons are less penetrating than s electrons. These electrons are, therefore, more shielded and more easily removed than extrapolation from the behavior of s electrons would predict. Again, the ionization enthalpies of O, F, and Ne increase regularly, but all are lower than would be expected by extrapolation from B, C, and N. This occurs because the $2p$ shell is half-full at N, and each of the additional $2p$ electrons enters an orbital already singly occupied. These electrons are partly repelled by the electron already present in the same orbital, and are thus less tightly bound.

Atomic Radii

It is necessary to distinguish among at least three different types of radii that might be listed for the elements. The **single-bond covalent radius** r_{cov} of an element represents the typical contribution by that element to the length of a predominantly covalent bond. Values for r_{cov} are estimated from the known lengths of covalent bonds involving any particular element, in the absence of multiple bonds. For instance, the covalent radius for fluorine is taken to be one half the internuclear distance in the homonuclear diatomic F_2. Typical values for r_{cov} are listed in Fig. 2-15, along with two other important types of radii, r_{ion} and r_{vdw}. The values of **van der Waals radii** (r_{vdw}) are obtained from the nonbonded distance of closest approach between atoms that are in contact with, but not bonded to, one another. The sum of the van der Waals radii of two atoms is thus the shortest distance we expect to find, in the structure of a solid compound, between two immediately adjacent atoms (either in the same or a different molecule or ion) that are not bonded to one another, nor otherwise constrained to be close.

The values in Fig. 2-15 for the **ionic radii** r_{ion} represent radii that are assigned to the various ions of the elements as they are found in predominantly ionic compounds. In particular, the values in Fig. 2-15 are those compiled by Shannon and Prewitt. These are the most widely accepted values currently available. The values of Shannon and Prewitt for r_{ion} are additive and internally consistent. Nevertheless, numerous assumptions were required to derive r_{ion} values from interatomic data gathered on structures of ionic solids. First, it is obvious from the nature of the atomic orbital wave functions that no ion or atom has a single, precisely defined radius. The only way that radii can be assigned to individual ions is to determine how closely two ions actually approach one another in solid compounds, and then to assume that such a distance is equal to the sum of the radii of the two ions. It is then additionally necessary to decide what portions of the interionic distance one should assign to the cation and the anion. This presents a dilemma, since the ionic radius of an element can be expected to depend on numerous other factors such as the oxidation state of the element, the number of nearest neighbor atoms in the structure of the ionic compound, the arrangement of the ions in the solid, the identity of the other ions in the substance, and the degree of covalency in the substance. The particular values of r_{ion} that are

Radii (in pm), listed in the order: r_{cov} / r_{vdw} / r_{ion}

(100 pm = 1Å)

IA (1)	IIA (2)	IIIA (3)	IVA (4)	VA (5)	VIA (6)	VIIA (7)	VIII (8)	VIII (9)	VIII (10)	IB (11)	IIB (12)	IIIB (13)	IVB (14)	VB (15)	VIB (16)	VIIB (17)	VIIIB (18)
1 H / 37 / 130 / 208(1−)																	2 He / 54 / 140 / ⋯
3 Li / 140 / 180 / 90(1+)	4 Be / 120 / ⋯ / 59(2+)											5 B / 83 / ⋯ / ⋯	6 C / 77 / 170 / ⋯	7 N / 73 / 155 / ⋯	8 O / 70 / 140 / 126(2−)	9 F / 54 / 135 / 119(1−)	10 Ne / ⋯ / 154 / ⋯
11 Na / 154 / 230 / 116(1+)	12 Mg / 148 / 170 / 86(2+)											13 Al / 130 / 190 / 68(3+)	14 Si / 117 / 210 / 54(4+)	15 P / 110 / 185 / 212(3−)	16 S / 103 / 185 / 170(2−)	17 Cl / 97 / 180 / 167(1−)	18 Ar / ⋯ / 192 / ⋯
19 K / 198 / 280 / 152(1+)	20 Ca / 178 / ⋯ / 114(2+)	21 Sc / 160 / ⋯ / 89(3+)	22 Ti / ⋯ / ⋯ / 100(2+)	23 V / 131 / ⋯ / ⋯	24 Cr	25 Mn	26 Fe	27 Co / 125 / ⋯ / 83(2+)	28 Ni / 125 / ⋯ / 83(2+)	29 Cu / 128 / 140 / 91(1+)	30 Zn / 120 / 140 / 88(2+)	31 Ga / 130 / 190 / 76(3+)	32 Ge / 122 / ⋯ / 67(4+)	33 As / 121 / ⋯ / ⋯	34 Se / 117 / 200 / 184(2−)	35 Br / 114 / 190 / 187(1−)	36 Kr / 110 / 200 / ⋯
37 Rb / 220 / ⋯ / 166(1+)	38 Sr / 192 / ⋯ / 132(2+)	39 Y / 178 / ⋯ / 104(3+)	40 Zr / 134 / ⋯ / 80(4+)	41 Nb	42 Mo	43 Tc	44 Ru	45 Rh	46 Pd / ⋯ / 160 / 100(2+)	47 Ag / 144 / 170 / 129(1+)	48 Cd / 148 / 160 / 99(2+)	49 In / 148 / 190 / 94(3+)	50 Sn / 140 / 220 / 83(4+)	51 Sb / 141 / ⋯ / 90(3+)	52 Te / 137 / 220 / 207(2−)	53 I / 133 / 200 / 206(1−)	54 Xe / 130 / 220 / ⋯
55 Cs / 265 / ⋯ / 181(1+)	56 Ba / 218 / ⋯ / 149(2+)	57 La / ⋯ / ⋯ / 115(3+)	72 Hf	73 Ta	74 W	75 Re	76 Os	77 Ir	78 Pt / ⋯ / 175 / 94(2+)	79 Au / 144 / 170 / 151(1+)	80 Hg / 150 / 150 / 116(2+)	81 Tl / 144 / 200 / 103(3+)	82 Pb / 144 / ⋯ / 92(4+)	83 Bi / 152 / ⋯ / 117(3+)	84 Po	85 At	86 Rn
87 Fr	88 Ra	89 Ac / 187 / ⋯ / 111(3+)	104	105	106	107	108	109									

Figure 2-15 Periodic trends in atomic and ionic radii. For each element, the top value is the average single-bond covalent radius r_{cov}; the middle value is the average van der Waals radius r_{vdw}; the bottom value is the "Shannon and Prewitt" ionic radius r_{ion} for the oxidation state that is specified in parentheses, as described in the text. Each radius is given in picometers (pm), one angstrom (Å) being equal to 100 pm.

tabulated in Fig. 2-15 are those for a common oxidation state of the element (designated in parentheses) and were determined by using compounds for which the number of nearest neighbor ions (coordination number) of the particular element is six. More complete lists of r_{ion} values are presented in Appendix IIC and in tables as needed throughout the remainder of the text. We will have more to say on the subject in Section 4-5.

Since van der Waals forces (or intermolecular forces) are generally weaker than intramolecular bonds, the value of r_{vdw} for an atom is always larger than r_{cov}. Thus the simple contact (nonbonded) distance between molecules in liquid or solid molecular substances is greater than the distance between atoms covalently bonded to one another. On the other hand, values of r_{cov} are greater than r_{ion} for cations because cations are formed by removal of electron(s) from the atom. Conversely, anions are larger than their parent atoms, since they are formed from the latter by addition of electrons. Before saying more about these radii, their sources, and their uses, let us consider the periodic trends in the various radii.

Two trends in the values tabulated in Fig. 2-15 need to be mentioned. First, down any particular group of the periodic table, the radii of the elements increase by large amounts due to the successive use (with each new row) of orbitals having principal quantum number n one higher than the last. For any group of the periodic table, the size of the atoms increases as the quantum number n increases, or as one descends the group. The size of the atoms increases in spite of increasing effective nuclear charge because of the greater importance of placing electrons into higher level shells.

Second, across a row of the table, there is a progressive decrease in the size of the atoms within molecules (r_{cov}), as well as a decrease in the volume requirement of atoms between molecules (r_{vdw}). This decrease in size takes place in spite of the obvious fact that additional electrons become added with each new element! This demonstrates the importance of the imperfect shielding among the orbitals. As the effective nuclear charge grows across a row, the sizes of the atoms decrease.

Electron Attachment Enthalpies

The enthalpy change ΔH_{EA} that accompanies addition of an electron(s) provides a measure of the willingness of an atom to form anions. Where these enthalpy changes are negative, formation of the anion is favorable (exothermic). For example, the electron configuration of the halogens allows addition of an electron to form the uninegative ions.

$$X(g) + e^- \longrightarrow X^-(g) \qquad (2\text{-}7.1)$$

Values for electron attachment enthalpies of the elements are listed in Appendix IID.

Where positive values of ΔH_{EA} arise, an atom resists formation of the anion. In fact, for many elements, electron attachment enthalpies must be estimated because the normal chemistry of an element might entail formation of cations rather than anions. The alkaline earth elements, for instance, have positive electron attachment enthalpies, reflecting the tendency of these elements to form +2 cations rather than anions. The electron attachment enthalpies of the noble gases are similarly positive, reflecting the stability of the closed-shell configurations of these elements.

Where addition of a second electron is known to be common (i.e., the chalcogens, which form dinegative anions, such as O^{2-}), the addition of the first electron is typically favorable. The addition of the second electron involves increasing electronic repulsions, making the overall process unfavorable from the standpoint of ΔH_{EA}. Still, there is a rich chemistry of the stable oxides and sulfides, and so on, and more must be considered in assessing the stability of a particular anion. These complexities, coupled with the difficulties in measuring electron attachment enthalpies, make a discussion of periodic trends in electron attachment enthalpies less straightforward. Our interest in them is in their contribution to he next topic: electronegativities.

Electronegativities

Electronegativity (χ) is an empirical measure of the tendency of an atom in a molecule to attract electrons. (Chi, χ, is conventionally used for electronegativity, as well as for magnetic susceptibility.) It will, naturally, vary with the oxidation state of the atom, and for a number of reasons the numerical values that have been assigned should not be taken too literally. It is useful only as a semiquantitative notion.

It should be stressed that electronegativity is not the same as the enthalpy of electron attachment ΔH_{EA}, although the two are related. R. S. Mulliken has shown that reasonable values of χ can be calculated from the average of the negative of the electron attachment enthalpy $(-\Delta H_{EA})$ and the ionization enthalpy (ΔH_{ion}). That is, electronegativities are determined in part by the tendency of an atom to gain additional electron density and by its tendency to retain the electron density it already has. A complete electronegativity scale cannot be established using this approach, however, because electron attachment enthalpies are not available for all of the elements.

Alternative ways of computing electronegativities have been suggested. The first general method was proposed by Pauling. He suggested that if two atoms A and B had the same electronegativity, the strength of the A—B bond would be equal to the geometric mean of the A—A and B—B bond energies, since the electrons in the bond would be equally shared in purely covalent bonds in all three cases. He observed, however, that for the majority of A—B bonds the energy exceeds that geometric average because, in general, two different atoms have different electronegativities, and there is an ionic contribution to the bond in addition to the covalent one. He proposed that the "excess" A—B bond energies could be used as an empirical basis to determine electronegativity differences. For instance, the H—F bond energy is 566 kJ mol^{-1}, whereas the H—H and F—F bond energies are 436 and 158 kJ mol^{-1}, respectively. Their geometric mean is $(158 \times 436)^{1/2} = 262$ kJ mol^{-1}. The difference Δ is 304 kJ mol^{-1}. He then found that to get a consistent set of electronegativities, so that $\chi_A - \chi_B = (\chi_C - \chi_B) - (\chi_C - \chi_A)$, and so on, the electronegativity differences would have to obey the equation

$$\chi_A - \chi_B = 0.102 \, \Delta^{1/2} \qquad (2\text{-}7.2)$$

Pauling originally assigned the most electronegative of the elements, fluorine, $\chi = 4.00$. From these data, we could calculate

$$\chi_H = 4.00 - 0.102(304)^{1/2} = 2.22 \tag{2-7.3}$$

Another method of calculating electronegativities is that of Allred and Rochow. It has the advantage of being more easily applied to a larger number of the elements. The rationale is that an atom will attract electron density in a chemical bond according to Coulomb's law (Chapter 1), as shown in Eq. 2-7.4

$$\text{Force} = \frac{(Z^* e)(e)}{4\pi r^2 \epsilon_0} \tag{2-7.4}$$

where Z^* is the effective nuclear charge, e is the charge of the electron, and r is the mean radius of the electron, essentially r_{cov}. Equation 2-7.4 is the basis for the empirically adjusted electronegativities, which are given by Eq. 2-7.5.

$$\chi = 0.359 \frac{Z^*}{r^2} + 0.744 \tag{2-7.5}$$

The numerical constants were chosen to bring the range of values for electronegativity into accord with those of Pauling.

Values for the three different electronegativities are listed in Fig. 2-16. The variation of these values with position in the periodic table is reasonable. The atoms with the highest electronegativities are those with the smallest radii and the highest effective nuclear charges (e.g., F). The larger radii correspond to the lower electronegativities (e.g., Cs).

A more recent scale of electronegativities has been developed by L. C. Allen for the representative (i.e., nontransitional) elements. Accordingly, the "spectroscopic electronegativity" χ_{spec} is calculated as in Eq. 2-7.6

$$\chi_{spec} = \frac{m\epsilon_p + n\epsilon_s}{m + n} \tag{2-7.6}$$

where m and n are the number of p and s electrons, respectively, and ϵ_p and ϵ_s are the corresponding average one-electron ionization enthalpies (averaged over all multiplicities) of an atom. Precise values of ϵ_p and ϵ_s can be determined using high-resolution spectroscopic data for each element. Thus, electronegativity is the average one-electron ionization enthalpy of all s and p electrons in the valence shell of an atom. The resulting values of χ_{spec} are given in Table 2-4. Comparison with the electronegativity values presented in Fig. 2-16 shows that Allen's values are not substantially different from those of the others. The Allen method of calculating electronegativities is intuitively satisfying since the "tendency of an atom to attract electrons to itself in a molecule" ought to be related to the average one-electron valence shell ionization enthalpy of an atom.

Allen also suggested that χ_{spec}, as calculated in Eq. 2-7.6, constitutes the so-called "third dimension" of the periodic table. This finding is depicted in Fig. 2-

Figure 2-16 Electronegativities of the elements.

I	II	III	IV	II	II	II	II	II	II	II	II	III	IV	III	II	I	
H 2.20																	He
Li 0.97 / 0.98	Be 1.47 / 1.57 / 1.46											B 2.01 / 2.04 / 2.01	C 2.50 / 2.55 / 2.63	N 3.07 / 3.04 / 2.33	O 3.50 / 3.44 / 3.17	F 4.10 / 3.98 / 3.91	Ne
Na 1.01 / 0.93 / 0.93	Mg 1.23 / 1.31 / 1.32											Al 1.47 / 1.61 / 1.81	Si 1.74 / 1.90 / 2.44	P 2.06 / 2.19 / 1.81	S 2.44 / 2.58 / 2.41	Cl 2.83 / 3.16 / 3.00	Ar
K 0.91 / 0.82 / 0.80	Ca 1.04 / 1.00	Sc 1.20 / 1.36	Ti 1.32 / 1.54	V 1.45 / 1.63	Cr 1.56 / 1.66	Mn 1.60 / 1.55	Fe 1.64 / 1.83	Co 1.70 / 1.88	Ni 1.75 / 1.91	Cu 1.75 / 1.90 / 1.36	Zn 1.66 / 1.65 / 1.49	Ga 1.82 / 1.81 / 1.95	Ge 2.02 / 2.01	As 2.20 / 2.18 / 1.75	Se 2.48 / 2.55 / 2.23	Br 2.74 / 2.96 / 2.76	Kr
Rb 0.89 / 0.82	Sr 0.99 / 0.95	Y 1.11 / 1.22	Zr 1.22 / 1.33	Nb 1.23	Mo 1.30 / 2.16	Tc 1.36	Ru 1.42	Rh 1.45 / 2.28	Pd 1.35 / 2.20	Ag 1.42 / 1.93 / 1.36	Cd 1.46 / 1.69 / 1.4	In 1.49 / 1.78 / 1.80	Sn 1.72 / 1.96	Sb 1.82 / 2.05 / 1.65	Te 2.01	I 2.21 / 2.66 / 2.56	Xe
Cs 0.86 / 0.79	Ba 0.97 / 0.89	*La 1.03 / 1.10	Hf 1.23	Ta 1.33	W 1.40 / 2.36	Re 1.46	Os 1.52	Ir 1.55 / 2.20	Pt 1.44 / 2.28	Au 1.42 / 2.54	Hg 1.44 / 2.00	Tl 1.44 / 2.04	Pb 1.55 / 2.33	Bi 1.67 / 2.02	Po 1.76	At 1.96	Rn
Fr 0.86	Ra 0.97	**Ac 1.00															

III	IV	II	II	II	II	II	II	II	II	III	IV	III	II	I
*La 1.03 / 1.10	Ce 1.06 / 1.12	Pr 1.07 / 1.13	Nd 1.07 / 1.14	Pm 1.07	Sm 1.07 / 1.17	Eu 1.01	Gd 1.11 / 1.20	Tb 1.10	Dy 1.10 / 1.22	Ho 1.10 / 1.23	Er 1.11 / 1.24	Tm 1.11 / 1.25	Yb 1.06	Lu 1.14 / 1.27
**Ac 1.00	Th 1.11	Pa 1.14	U 1.22 / 1.38	Np 1.22 / 1.36	Pu 1.22 / 1.28	Am	Cm	Bk	Cf	Es	Fm	Md		

~1.2 (estimated)

Values in bold type are calculated using the approach of A. L. Allred and E. G. Rochow. *J. Inorg. Nucl. Chem.*, **1958**, *5*, 264. Values in italics are estimated by Pauling's method [A. L. Allred, *J. Inorg Nucl. Chem.*, **1961**, *17*, 215]. Values in Roman type are obtained using Mulliken's method [H. O. Pritchard and H. A. Skinner, *Chem. Rev.*, **1955**, *55*, 715]. Roman numerals at the top give the oxidation states used for the Pauling-type values.

Table 2-4   The Allen Electronegativities χ_{spec}, Determined by Using Eq. 2-7.6. The Values of Pauling χ_P and of Allred and Rochow $\chi_{A\,\&\,R}$ are Listed for Comparison

Atom	χ_{spec}	χ_P	$\chi_{A\,\&\,R}$
H	2.300	2.20	2.20
Li	0.912	0.98	0.97
Be	1.576	1.57	1.47
B	2.051	2.04	2.01
C	2.544	2.55	2.50
N	3.066	3.04	3.07
O	3.610	3.44	3.50
F	4.193	3.98	4.10
Ne	4.787		
Na	0.869	0.93	1.01
Mg	1.293	1.31	1.23
Al	1.613	1.61	1.47
Si	1.916	1.90	1.74
P	2.253	2.19	2.06
S	2.589	2.58	2.44
Cl	2.869	3.16	2.83
Ar	3.242		
K	0.734	0.82	0.91
Ca	1.034	1.00	1.04
Ga	1.756	1.81	1.82
Ge	1.994	2.01	2.02
As	2.211	2.18	2.20
Se	2.424	2.55	2.48
Br	2.685	2.96	2.74
Kr	2.966		
Rb	0.706	0.82	0.89
Sr	0.963	0.95	0.99
In	1.656	1.78	1.49
Sn	1.824	1.96	1.72
Sb	1.984	2.05	1.82
Te	2.158	2.10	2.01
I	2.359	2.66	2.21
Xe	2.582		

17, where the value of χ_{spec} is plotted in the vertical dimension of the otherwise traditional periodic table for the s- and p-block elements. This very useful result gives us an elegant and new visual perspective of an important atomic property. For instance, Fig. 2-17 shows that metalloids fall between elements having low values of χ_{spec} (metals) and those having high values of χ_{spec} (nonmetals). The periodic trends in electronegativity along a horizontal row and down a particular group of the periodic table are also evident in Fig. 2-17. The substantial resistance to ionization of the noble gas elements is well explained by their relatively high χ_{spec} values (see Fig. 2-17). One drawback to the Allen electronegativities is that Eq. 2-7.6 cannot readily be applied to the calculation of the electronegativity of a transition element.

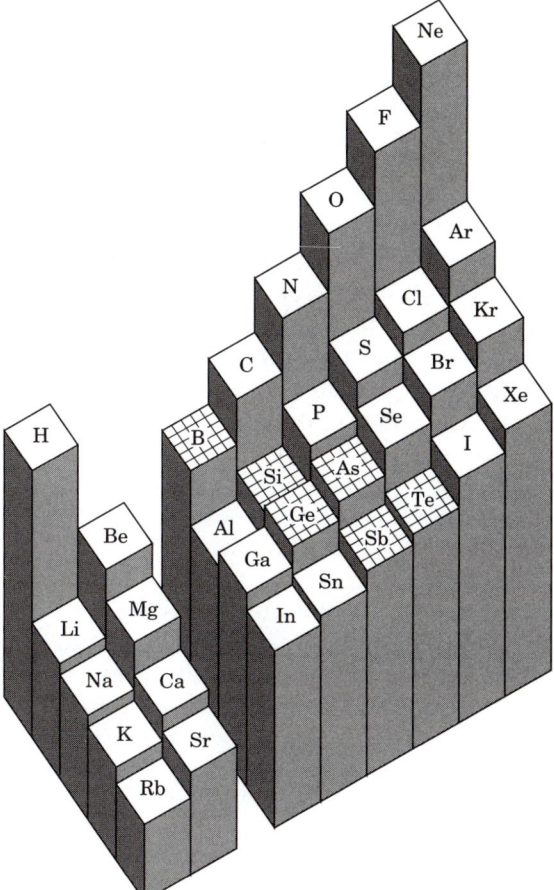

Figure 2-17   Values of the spectroscopic electronegativity (χ_{spec}, as determined using Eq. 2-7.6) as a function of the position of an element in the periodic table. The elements of the metalloid band are designated with cross hatching. [Reprinted with permission from L. C. Allen, *J. Am. Chem. Soc.*, *111*, 9003–9014 (1989). Copyright © (1989) American Chemical Society.]

2-8   Magnetic Properties of Atoms and Ions

Any atom, ion, or molecule that has one or more unpaired electrons is *paramagnetic*. This means that it, or any material in which it is found, will be attracted into a magnetic field. In cases where paramagnetic atoms or ions are very close together they interact cooperatively, and other, more or less intense and more complicated forms of magnetism, ferromagnetism, and antiferromagnetism (in particular) are observed. These forms are not discussed here, but see Chapter 32. Substances that do not contain unpaired electrons (with certain exceptions that need not concern us here) are *diamagnetic*. This means they are weakly repelled by a magnetic field. Thus, the measurement of paramagnetism affords a powerful tool for detecting the presence and number of unpaired electrons in chemical elements and compounds.

The full power of magnetic measurements comes from the fact that the magnitude of the *magnetic susceptibility*, which is a measure of the force exerted by the magnetic field on a unit mass of the specimen, is related to the number of unpaired electrons present per unit weight—and hence, per mole.

Actually, the paramagnetism of a substance containing unpaired electrons receives a contribution from the orbital motion of the unpaired electrons as well as from their spins. However, there are important cases where the spin contribution is so predominant that measured susceptibility values can be interpreted in terms of how many unpaired electrons are present. This correlation is best expressed by using a quantity called the *magnetic moment,* μ, which may be calculated from the measured susceptibility per mole χ_M. It is best to use χ_M^{corr}, where a correction has been applied to the measured χ_M to allow for the diamagnetic effect, which is always present, and which may be estimated from measurements on similar substances that lack the atom or ion that has the unpaired electrons.

Curie's Law

It was shown by Pierre Curie that for most paramagnetic substances, the magnetic susceptibility varies inversely with absolute temperature. In other words, the product $\chi_M^{corr} \times T$ is a constant, called the Curie constant for the substance. From the theory of electric and magnetic polarization it can be shown that, if the paramagnetic susceptibility is due to the presence of individual, independent paramagnetic atoms or ions within the substance, each with a magnetic dipole moment, μ, the following equation holds true:

$$\mu = 2.84\sqrt{\chi_M^{corr}T} \tag{2-8.1}$$

Clearly, this expression incorporates Curie's law.

Now, from the quantum theory for atoms (and ions) it can also be shown that the magnetic moment due entirely to the spins of n unpaired electrons on the atom or ion is given by

$$\mu = 2\sqrt{S(S+1)} \tag{2-8.2}$$

where S equals the sum of the spins of all the unpaired electrons (i.e., $n \times \frac{1}{2}$). From Eq. 2-8.2, it can easily be calculated that for one to five unpaired electrons the magnetic moments should be those shown in Table 2-5. The unit for atomic magnetic moments is the Bohr magneton (BM).

Table 2-5   Spin-Only Magnetic Moments μ (in BM)

Number of Unpaired Electrons (n)	S^a	μ(BM)
1	$\frac{1}{2}$	1.73
2	1	2.83
3	$\frac{3}{2}$	3.87
4	2	4.90
5	$\frac{5}{2}$	5.92

[a]The total spin quantum number $S = n \times \frac{1}{2}$.

To illustrate the application of these ideas, consider copper(II) sulfate, $CuSO_4 \cdot 5\,H_2O$. From the magnetic susceptibility the magnetic moment is found to be 1.95 BM. This value is only a little higher than the calculated value for one unpaired electron, and the discrepancy can be attributed to the contribution made by orbital motion of the electron. Thus the magnetic properties of $CuSO_4 \cdot 5\,H_2O$ are in accord with the presence of a Cu^{2+} ion that should have a $[Ar]3d^9$ configuration with one unpaired electron. For comparison, $MnSO_4 \cdot 4\,H_2O$ has a magnetic moment of 5.86 BM, which is approximately the number expected for a Mn^{2+} ion with the electron configuration $[Ar]3d^5$.

STUDY GUIDE

Scope and Purpose

This chapter covers fundamental principles of atomic structure, wave mechanics for atoms, and the periodic table. These topics are important in subsequent discussions of structure, bonding, and reactivity. Additional help with these important topics is available in the works listed under **Supplementary Reading.** The student should master the material sufficiently to be able to give ready answers to the **Study Questions** listed under "A. Review." More demanding exercises are listed under "B. Additional Exercises."

Study Questions

A. Review

1. The emission lines of the hydrogen atom come in related sets. What is the form of the equations for these sets? An equation of this type is named for whom?

2. What were the two bold postulates made by Bohr that allowed him to derive an equation for the energies of the electron in a hydrogen atom?

3. Write and explain the meaning of the equation relating the energy and frequency of radiation. What is the constant in it called?

4. What does the term Bohr radius mean?

5. What is de Broglie's equation for the wavelength associated with a moving particle of mass m and velocity v? What physical effect first showed directly that the wave character of the electron really exists?

6. State the relationship between the Bohr orbit with $n = 1$ and the wave mechanical orbital with $n = 1$ for the hydrogen atom.

7. Specify the set of quantum numbers used to describe an orbital and state what values of each are possible.

8. State the quantum numbers for each of the following orbitals: $1s$, $2s$, $2p$, $4d$, $4f$.

9. Draw diagrams of each of the following orbitals: $1s$, $2p_x$, $2p_y$, $2p_z$, $3d_{z^2}$, $3d_{xy}$, $3d_{yz}$, $3d_{zx}$, $3d_{x^2-y^2}$.

10. State the Pauli exclusion principle in the form relevant to atomic structure. Show how it leads to the conclusion that in a given principal shell there can be only two s, six p, ten d, and fourteen f electrons.

11. What does the term penetration mean, and why is it important in understanding the relative energies of the s, p, d, and f electrons with the same principal quantum number?

12. Define each of the following: alkali metals; alkaline earth metals; halogens; noble

gases; main group elements; d-block elements; f-block elements; lanthanides; transition elements.

13. What is Hund's first rule? Show how it is used to specify in detail the electron configurations of the elements from Li to Ne.

14. Why is the first ionization enthalpy of the oxygen atom lower than that of the nitrogen atom?

15. How is the magnetic moment of a substance containing an ion with unpaired electrons (e.g., $CuSO_4 \cdot 5\,H_2O$) related to its magnetic susceptibility at various temperatures if the substance follows Curie's law?

16. How is the magnetic moment μ related to the number of unpaired electrons if the magnetism is due solely to the electron spins? Calculate μ for an ion with three unpaired electrons.

17. R. S. Mulliken showed that electronegativity is related to both ΔH_{EA} and ΔH_{ion}. What is the relationship he gave?

18. What are the particular physical properties on which each of the following electronegativity scales is based?

    (a) Pauling's χ_P

    (b) Allred and Rochow's $\chi_{A \& R}$

    (c) Allen's χ_{spec}

19. Make a list of the factors that can influence the ionic radius, r_{ion}, of an element.

B. Additional Exercises

1. The He^+ ion is a one-electron system similar to hydrogen, except that $Z = 2$. Calculate the wavenumbers (in cm^{-1}) for the first and last lines in each of the three spectroscopic series corresponding to those discussed for the hydrogen atom.

2. The first ionization enthalpy for Li is 520 kJ mol^{-1}. This value corresponds to complete removal of the electron from the nucleus, and is achieved when $n = \infty$. From this value, calculate the effective charge felt by the $2s$ electron of Li. Why is this less than the actual charge of +3?

3. A consistent set of units that may be used in de Broglie's Eq. 2-2.1 is: λ in cm, mass in g, velocity in cm s^{-1}, and h (Planck's constant) in g cm^2 s^{-1} (or erg s). What is the wavelength in cm and in Å of (a) an electron traveling at 10^6 cm s^{-1}, a velocity typical in the electron microscope? (b) a baseball or cricket ball thrown at 10^3 cm s^{-1}? Assume that mass equals 2.00×10^2 g.

4. Consider the ground-state electron configurations of the atoms with the following atomic numbers: 7, 20, 26, 32, 37, 41, 85, 96. Calculate the total spin quantum number S for each, as well as its magnetic moment in Bohr magnetons.

5. As noted for Fig. 2-1, there are three series of lines in the emission spectrum for hydrogen. Calculate the position of the series limit for each.

6. Explain the trend in the ionization enthalpies illustrated in Fig. 2-14 for the noble gases.

7. If the wavelength of an electron is 6.0 Å, what is its velocity ($m = 9.1 \times 10^{-28}$ g)?

8. Prepare a graph of the effective nuclear charge Z^* versus Z using the data of Table 2-3. Explain any trends.

9. Use Eq. 2-7.5 and values for χ and r_{cov} found elsewhere in the chapter to estimate Z^* for elements 19, 20, 31, and 32. Explain the trends.

10. Explain the differences between the functions $R(r)$, $R(r)^2$, and $r^2[R(r)]^2$, using, for example, the $1s$ orbital for hydrogen.

11. Prepare dot density patterns similar to that of Fig. 2-3 for the following orbitals: $2p_x$, $3s$, and $3d_{xy}$. Both shape and the function $r^2[R(r)]^2$ [Fig. 2-7(b)] must be considered.

12. Use the Bohr theory to calculate the following values for the one-electron helium ion (He^+) for which $Z = 2$.

    (a) The first, second, and third orbit radii.

    (b) The second ionization enthalpy of helium (i.e., ΔH_{ion} for He^+).

    (c) The energy of the electron in the first, second, and third orbits.

13. Based on experimentally determined magnetic susceptibilities at 20 °C, the magnetic moments for the following substances have been calculated.

$MnSO_4 \cdot 4\,H_2O$	$\mu_{eff} = 5.85$ BM
$CuSO_4 \cdot 5\,H_2O$	$\mu_{eff} = 1.94$ BM
$(NH_4)_2Fe(SO_4)_2 \cdot 6\,H_2O$	$\mu_{eff} = 5.50$ BM
$[Cr(NH_3)_6](NO_3)_3$	$\mu_{eff} = 3.69$ BM
$[Cu(NH_3)_4]SO_4 \cdot 3\,H_2O$	$\mu_{eff} = 1.71$ BM
$[Co(NH_3)_6]Cl_3$	$\mu_{eff} = -0.01$ BM

Use these data to deduce the number of unpaired electrons on the transition metal ions in these substances.

SUPPLEMENTARY READING

Adamson, A. W., "Domain Representations of Orbitals," *J. Chem. Educ.,* **1965,** *42,* 141.

Atkins, P. W., *Molecular Quantum Mechanics,* Oxford University Press, New York, 1983.

Barrow, G. M., *Physical Chemistry,* 5th ed., McGraw-Hill, New York, 1988.

Berry, R. S., "Atomic Orbitals," *J. Chem. Educ.,* **1966,** *43,* 283.

Cohen, I. and Bustard, T., "Atomic Orbitals: Limitations and Variations," *J. Chem. Educ.,* **1966,** *43,* 187.

Gerloch, M., *Orbitals, Terms and States,* Wiley, New York, 1986.

Goodisman, J., *Contemporary Quantum Chemistry,* Plenum, New York, 1977.

Guillemin, V., *The Story of Quantum Mechanics,* Scribner, New York, 1968.

Herzberg, G., *Atomic Spectra and Atomic Structure,* Dover Publications, 1944.

Johnson, R. C. and Rettew, R. R., "Shapes of Atoms," *J. Chem. Educ.,* **1965,** *42,* 145.

Karplus, M. and Porter, R. N., *Atoms and Molecules: An Introduction for Students of Physical Chemistry,* Benjamin, Menlo Park, CA, 1970.

Kikuchi, Q. and Suzuki, K., "Orbital Shape Representations," *J. Chem. Educ.,* **1985,** *62,* 206.

Ogryzlo, E. A. and Porter, G. B., "Contour Surfaces for Atomic and Molecular Orbitals," *J. Chem. Educ.,* **1963,** *40,* 256.

Perlmutter-Hayman, B., "The Graphical Representation of Hydrogen-Like Functions," *J. Chem. Educ.,* **1969,** *46,* 428.

Powell, R. E., "The Five Equivalent *d* Orbitals," *J. Chem. Educ.,* **1968,** *45,* 1.

Price, W. C., Chissick, S. S., and Ravensdale, T., Eds., *Wave Mechanics, The First 50 Years,* Butterworths, London, 1973.

Pritchard, H. O. and Skinner, H. A., "Electronegativity Scales," *Chem. Rev.,* **1955,** *55,* 745.

Verkade, John G., *A Pictorial Approach to Molecular Bonding,* Springer-Verlag, New York, 1986.

Chapter 3

STRUCTURE AND BONDING IN MOLECULES

3-1   Introduction

Modern techniques, such as X-ray crystallography and spectroscopy, have made it possible for us to determine the structures of molecules and complex ions with great accuracy. As information about structure has increased, so has our understanding of bonding. The more powerful bonding theories have allowed us to make detailed predictions and comparisons regarding not only structure, but also spectroscopy, reactivity, and so on. The simpler bonding theories, although known to be incomplete and only partially accurate, have still been useful because of the lessons that they have provided about electronic structures in molecules. This is especially true of the localized bonding theories, which will be discussed shortly. Later in this chapter, we shall develop more sophisticated, delocalized bonding theories.

The material of the previous chapter (Chapter 2) is important here, because the electrons (and the orbitals in which they are housed) are the focus of any discussion of bonding. We shall show how orbitals interact to provide new locations for the electrons within molecules, and we shall be concerned with how this leads to the bonding of atoms in molecules and complex ions. Once we have established the types of orbital interactions that generally take place within molecules and complex ions, we shall have also gained insight into matters of structure, spectroscopy, and reactivity. To organize the subject, three main types of bonding are considered:

1. Covalent bonding between atom pairs (two-center bonds).
2. Delocalized (multicenter) covalent bonding.
3. Ionic bonding.

The first two types of bonding are discussed in this chapter, while ionic bonding and related topics are considered in Chapter 4. In addition, a few special forms of bonding are discussed elsewhere, such as metallic bonding (Section 8-6), the hydrogen bond (Section 9-3), and ligand field theory (Chapter 23).

There is surely no bonding that is literally and completely *ionic* but, for practical purposes, a great many compounds can be treated to a reasonable approximation as if the attractive forces were just the electrostatic attractions between ions of opposite charge. The treatment of these substances, for example, NaCl, MgO, $NiBr_2$, and the like, takes a different form from that used for covalent bonding, where electron sharing between atoms is considered the dominant factor. Therefore, it is appropriate to discuss covalent bonding (this chapter) separately from ionic bonding (Chapter 4).

The student, while studying the material in Section 3-2, should keep in mind that three separate theories are presented somewhat simultaneously. These theories are the Lewis electron-pair bond theory, the hybridization theory, and the valence shell electron-pair repulsion (VSEPR) theory. Although there is a satisfying correspondence among the three theories, each constitutes a separate approach and addresses a distinctly different aspect of the localized bond problem. Since each of these theories has advantages, as well as limitations, we must learn to move readily from one theory to the next, depending on which bonding or structural features we seek to explain.

3-2   The Localized Bond Approach

The simplest view of bonding in any molecule or complex ion is (1) the electrons that are involved in bonding remain localized between pairs of atoms, and (2) the bonding in the whole structure is the sum of the individual bonds between pairs of atoms. The approach is useful because of its simplicity and because it is easy to represent in molecular diagrams. In fact, the Lewis concept of a localized electron-pair bond is so much a part of the modern vernacular that it would be difficult to imagine working without it. As we shall see, however, the idea that electrons always remain localized between atom pairs has important limitations.

Lewis Concepts

It was the American chemist, G. N. Lewis, who first recognized that bonding between atoms involves the sharing of electrons. According to the Lewis definition, one covalent bond between two atoms results from the sharing of a pair of electrons between the atoms. Such a pair of **bonding electrons** is considered to be localized or fixed between the two atoms, and the bond is represented by a line connecting the atoms. Electrons that are not shared between atoms are localized as **lone-pair** electrons on one or another atom(s) within a molecule. The electronic structure of the entire molecule is represented by the sum of all of the bonding pairs and the lone pairs of electrons. Based on these concepts, it is possible to represent the electronic structure of a molecule in diagramatic form. Such representations are called Lewis diagrams. Some chemical intuition is needed in drawing the **Lewis diagram** for a molecule or ion. The Lewis diagram for a molecule or ion represents an approximate arrangement of atoms and the location of all valence electrons within the structure. The familiar result can, with experience, be quickly written down for any of a number of classes of substances. The utility of this approach is obvious.

Once the Lewis diagram has been correctly written for a substance, the Lewis approach can be extended with the use of hybridization theory, and with the VSEPR theory, to account for subtle aspects of geometry. These three concepts (i.e., the Lewis diagram, hybridization, and VSEPR theory) in unison become extraordinarily powerful as an approach to structure and bonding. Eventually, however, the concepts fail because of the limitations of viewing the electrons in a strictly localized way. Resonance can be added to the paradigm, but this represents only a temporary (although historically important) "fix." This localized approach to bonding is useful because of its simplicity.

Lewis Diagrams

When drawing a Lewis diagram for a molecule or complex ion, only the valence electrons of the atoms are used. The Lewis diagram is complete when the atoms have been connected properly and the valence electrons have been distributed within the structure as either bonding or lone electron pairs. It may be necessary to look up the actual structure or to make an educated guess about the placement of atoms within the molecule or complex ion. Some chemical intuition goes a long way here, and a little experience is required. Under most circumstances, a simple and symmetrical geometry is correct. Atoms that are present only once within a substance tend to reside at the center of the structure. Metals tend to be central atoms. Oxygen is commonly, and hydrogen is nearly always, peripheral. Once the positions of the atoms have been set down, the distribution of electrons into the diagram is considered. In simple cases, the valence electrons are arranged so as to give an octet of electrons to each nonhydrogen atom, although exceptions are common, especially for atoms from rows three and below of the periodic table. Some examples follow.

For many substances the number of valence electrons is just sufficient to provide an octet for each nonhydrogen atom. These are **saturated** systems, and the Lewis diagrams can be written using single bonds exclusively. Examples are CH_4, NH_3, H_2O, and HF.

Unsaturated substances are those where the number of valence electrons that are available within a molecule or complex ion is not sufficient to allow the Lewis diagram to be written using single bonds only. Then, the use of **multiple bonds** between selected atoms is required to complete the octet for each atom in the structure. As examples containing a double bond, consider NO_3^-, acetone, or SO_2.

A triple bond (or two double bonds) is necessary when there is extensive unsaturation, as in CO_2 or thiocyanate ion.

In each of these last examples, unsaturation requires the use of multiple bonds in order to maintain an octet of electrons for each atom, without using more than the number of valence electrons that are actually available.

For some **electron deficient** molecules, all of the available valence electrons are used before an **octet** is achieved for each nonhydrogen atom. The Lewis diagrams are written so as to reflect this electron deficiency, although more complete molecular orbital approaches give a better description of electron deficient molecules. The molecules that feature this electron deficiency usually involve the elements boron, beryllium, or sometimes aluminum. An example is BeH_2.

$$H—Be—H$$

Unsaturated systems are different from electron deficient ones. In the former, an octet is achieved through multiple bonding. In the latter case, the Lewis diagram is properly written with less than an octet of electrons for certain atoms.

For molecules or ions involving atoms beyond row two of the periodic table, the octet rule does not necessarily apply. These larger atoms may acquire more than an octet of electrons. This is called **valence shell expansion,** and it is made possible by the availability of valence level d orbitals on these atoms. Examples include PCl_5, BrF_3, and XeF_2.

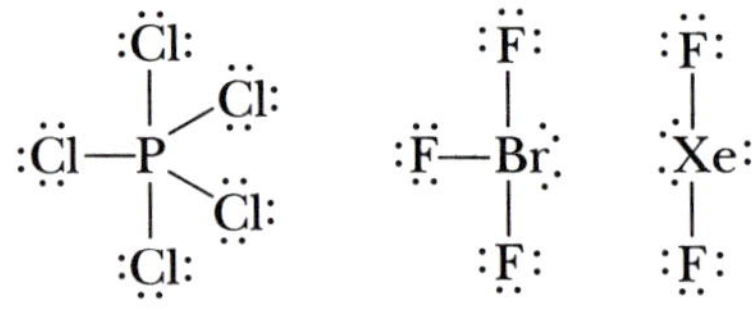

In a preliminary fashion, we might also mention the coordination compounds, which feature a central metal atom bonded to other groups. The groups that are bonded to the central metal atom are called **ligands.** Examples of coordination compounds are $Ni(CO)_4$, $[Co(NH_3)_6]^{3+}$, and $Pt(NH_3)_2Cl_2$. The Lewis diagrams for simple coordination compounds of the transition metals may be written without taking into consideration the presence of the $(n-1)$ d electrons of the metal. The bonds are considered to be **coordinate covalent bonds** in which both electrons of the metal–ligand bond are supplied by the ligand. The ligands are considered to be simply **Lewis bases** (electron-pair donors), and the metal centers are considered to be **Lewis acids** (electron-pair acceptors). The octet rule does not apply. Instead, the ligands add enough electrons to those of the metal to bring the total for the metal to that of the next noble gas: 18 valence electrons in all. Hence, the octet rule is replaced by the **18-electron rule** because of the additional 10 electrons of the d orbitals in any transition series. Some examples of coordination compounds that obey the 18-electron rule are $Ni(CO)_4$ and $[Co(NH_3)_6]^{3+}$.

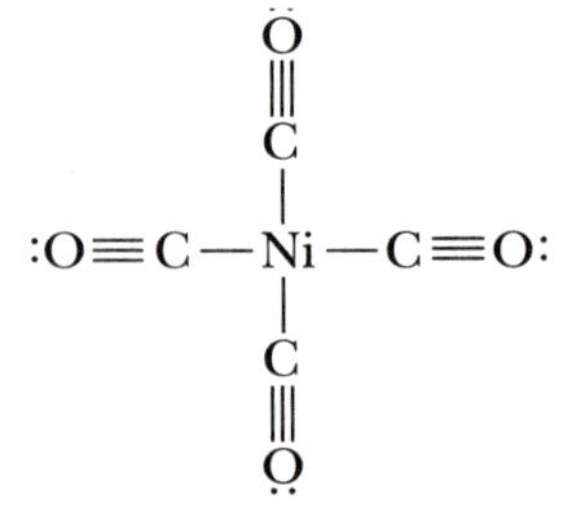

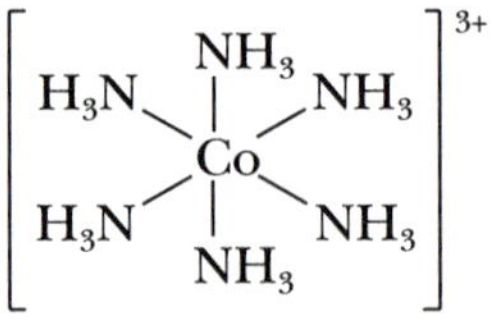

In each case the metal electrons are not listed in the Lewis diagram, but they are counted towards the 18-electron total. Note also that many transition metal compounds have other than the closed-shell, 18-electron total, and they are still perfectly stable. We shall have more to say about this in later chapters. For now it is

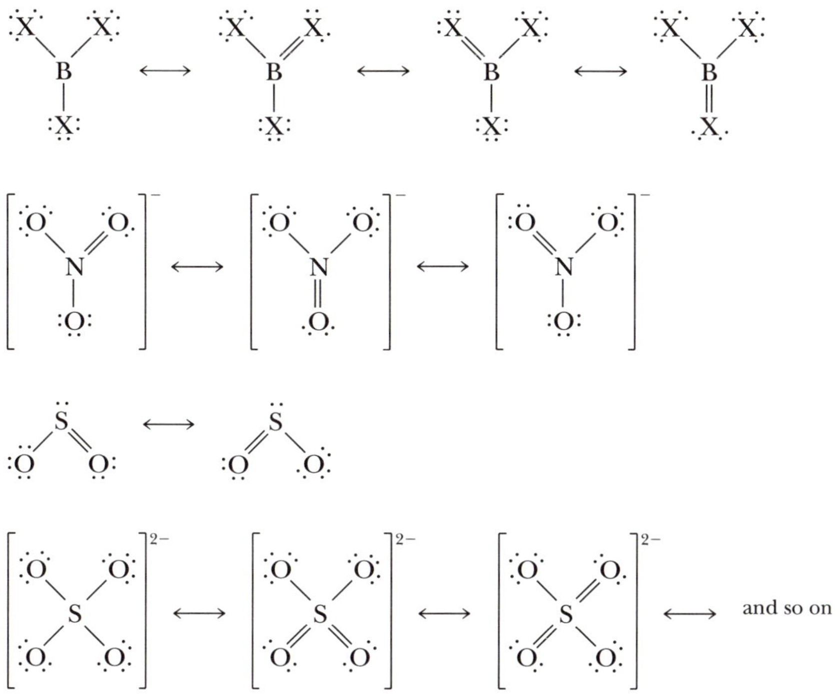

Figure 3-1 Resonance forms.

interesting to note that the octet rule finds only limited application, being replaced by the 18-electron rule when considering the coordination compounds formed by metals.

Resonance

In many of the previously written Lewis diagrams, it would have been possible to have arranged the electrons about the fixed nuclei in different (but each perfectly proper) ways. In fact, the Lewis description of the bonding in a molecule is not complete until all contributing possibilities have been written down. The overall result is delocalization of electrons within the structure through the recognition that other Lewis diagrams may be equally valid. This is **resonance,** and it is equivalent to the molecular orbital concept of delocalization. Figure 3-1 shows contributing resonance forms for the molecules and ions that were discussed previously.

Let us consider in detail the planar AB_3 systems such as BF_3, NO_3^-, and CO_3^{2-}. If we try to write a diagram for such a molecule or ion, in which each atom acquires an octet of electrons, we obtain Structure 3-I.

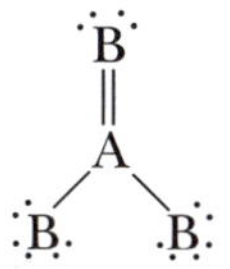

3-I

This representation implies that there are two A—B single bonds and one A=B double bond, whereas experimental data show conclusively that all A—B bonds and all B—A—B angles are equal. To bring theory and experiment into accord, the former is modified by the postulate that Structure 3-I alone does not describe the actual molecule but is only one of three equivalent, hypothetical structures, 3-I to 3-III. The real molecule has an electron distribution corresponding to the average of these three contributing structures, and is said to be a *resonance hybrid* of them. The double-headed arrow is used to indicate that the structures are mixing to give a resonance hybrid.

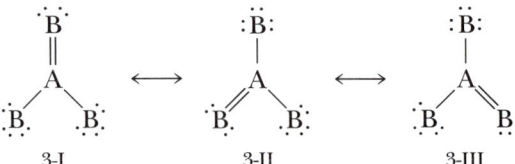

Care is required to avoid misinterpretation of the resonance concept. At no instant does the molecule actually have any one of the canonical structures. Each of these implies that one bond is stronger, and thus presumably is shorter than the other two, whereas all three bonds are always entirely equivalent. The canonical structures have no real existence, in any way or sense, but their average corresponds to the actual structure.

The concept of resonance can be justified from an energy point of view. It can be shown that a resonance hybrid must have a lower energy, that is, be more stable, than any single contributing structure. This concept explains why the molecule exists in the hybrid structure rather than any one of the contributing structures.

One particular type of resonance requires special mention, namely, *ionic–covalent resonance*. We pointed out in Section 2-7 that a bond between unlike atoms (A—B) is always more or less stronger than the average of the A—A and B—B bond strengths. This was used for calculating electronegativity differences, on the basis that an ionic or polar contribution to the bond made it stronger than the purely covalent bond alone. Actually, the situation is a little different, because it is resonance rather than simple additivity that Pauling invoked to account for the extra bond energy.

If A is more electronegative than B, the A—B bond can be represented by a resonance hybrid of Structures 3-IVa and 3-IVb.

$$A—B \longleftrightarrow \; :\!\ddot{A}\!:^- \; B^+$$
$$\text{3-IVa} \qquad \text{3-IVb}$$

As we explained, the actual A—B bond will then (1) combine the properties of both contributing structures, and (2) be more stable than either one alone. Thus, the actual A—B bond will be polar to an extent depending on how much Structure 3-IVb contributes to the average structure. The increased strength of the bond, when compared with the strength expected for a purely covalent bond, will be proportional to the square of the electronegativity difference, since that difference determines the importance of Structure 3-IVb compared with Structure 3-IVa.

When the Lewis diagram for a molecule or complex ion has been written correctly, one has accounted for all of the atoms and valence electrons of the structure. However, more needs to be known about the bonding in these structures. If an electron-pair bond involves sharing of electrons between atoms, then how is this sharing accomplished? Which orbitals are involved on the two atoms, and which orbitals in the molecule? Why does the sharing of the electrons in a bond lead to stability? What geometry should one assign to the molecule overall, and what particular bond angles and lengths result? Obviously, a wide variety of bond angles occur, not just the 90° angle at which the atomic p orbitals are disposed on any one atom. It quickly becomes apparent that atomic orbitals must be modified in such a way as to allow for the correct angles in molecules. Just as was true for atoms, orbitals must be provided for each electron, whether it is a member of a lone pair or a bonding pair. These orbitals (and the lone and bonding pairs that they house) must be arranged about each atom in the correct orientations, namely, those that are in agreement with geometry. An approach powerful enough to allow geometry to be predicted is what we seek.

In the localized bond approach, the answers to the questions just posed are obtained by employing either the **hybridization** or the **VSEPR** theory. Hybridization allows for the "construction" of new orbitals on atoms, so that the bonding in a molecule is made to be consistent with its known geometry. On the other hand, without any consideration of the orbitals involved in bonding, the VSEPR theory allows the best geometry for a molecule or polyatomic ion to be predicted. We begin with hybridization in which it is assumed that bonding arises because of the overlap of orbitals (a concept that will be reinforced with molecular orbital theory) and that the proper set of orbitals for any atom in the structure can be deduced by knowing the number of groups (atoms plus lone pairs) which occupy the space around that particular atom. Let us begin the discussion with the simplest case: linear BeH_2.

Hybridization

In BeH_2, for example, it is perfectly satisfactory, for nearly any purpose, to consider that there is one electron pair localized between each adjacent pair of atoms. Thus, we have the simple, familiar representation, H:Be:H. An electron-pair bond of the type indicated can be thought of as arising from the overlap of two orbitals, one from each of the atoms bonded, with the electrons concentrated in the region of overlap between the atoms. In the case of BeH_2, which is linear, this raises the question of how to account for the linearity. In answering that question, two new concepts, the valence state and hybridization, are introduced.

The beryllium atom has the electron configuration $1s^2 2s^2$. Thus its valence shell has only one occupied orbital and the electrons are paired. On the other hand, if it is to form two bonds by sharing one electron with each of two other atoms, it must first be put into a state where each electron is in a different orbital, and each spin is uncoupled from the other and, thus, is ready to be paired with the spin of an electron on the atom to which the bond is to be formed. When the atom is in this condition, it is said to be in a valence state.

For the particular case of BeH_2, the valence state of lowest energy is obtained by promoting one of the electrons from the $2s$ orbital to one of the $2p$ orbitals, and decoupling their spins. This process requires the expenditure of about 323 kJ mol^{-1}.

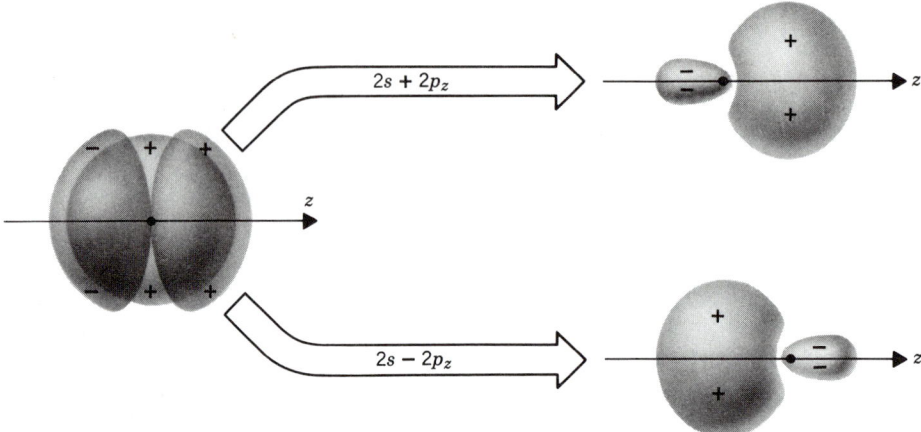

Figure 3-2   The formation of two equivalent *sp* hybrid orbitals from linear combinations of a $2s$ and a $2p_z$ atomic orbital. The dots on the z axis represent the position of the atom on which the hybrid is constructed. The two orientations of the hybrids (180° with respect to one another) result from the different sign used in the two linear combinations.

Although the promotion of the Be atom to the valence state prepares it to form two bonds to the H atoms, it does not provide an explanation or a reason why the molecule should be linear, rather than bent. The $2s$ orbital of Be has the same amplitude in all directions. Therefore, whichever of the $2p$ orbitals is used to form one Be—H bond, the other bond in which the $2s$ orbital is used could make any angle with it, insofar as overlap of the H $1s$ and Be $2s$ orbitals is concerned. However, the preference for a linear structure can be attributed to the fact that if a $2s$ and $2p$ orbital are mixed so as to form two *hybrid* (i.e., mixed) orbitals, better total overlap with the H $1s$ orbitals can be obtained. The results of mixing the $2s$ and $2p_z$ orbitals are shown in Fig. 3-2.

Each of the hybrid orbitals has a large positive lobe concentrated in a particular direction and is, therefore, able to overlap very strongly with an orbital on another atom located at an appropriate distance in that direction. Actual calculations show that the extent of overlap thus obtained is greater than that obtain-

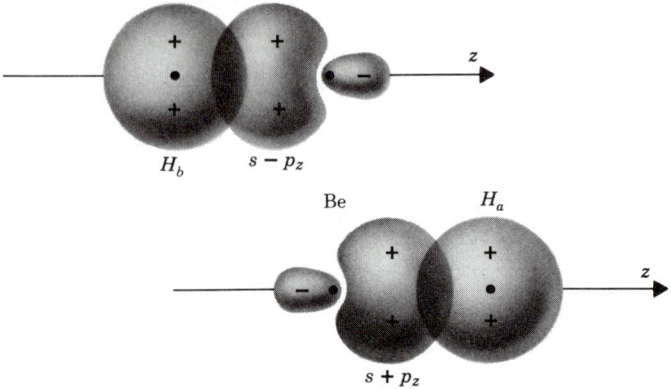

Figure 3-3   The formation of covalent bonds in BeH_2 by overlap of *sp* hybrid orbitals on Be with $1s$ atomic orbitals on H.

able by using either a pure $2s$ or pure $2p_z$ orbital. This overlap is not difficult to see without calculation, if we note that one half of the p_z orbital is found in the $+z$ direction and one half is in the $-z$ direction. The $2s$ orbital is uniformly distributed in all directions. The hybrid orbitals, however, are each strongly concentrated in just one direction.

The linearity of the BeH_2 molecule suggests the use of the hybrid orbitals. Figure 3-2 shows that the sp hybrids are oriented in the $+z$ and $-z$ directions because of the spatial properties of the s and p orbitals themselves. The best Be to H overlaps are then obtained by placing the H atoms along the $+z$ and $-z$ axes (Fig. 3-3). The correctness of sp hybridization for Be is affirmed by the known linearity of the molecule; the best geometry is the one that disperses the two bonding pairs as far from one another as possible: $180°$.

The hybrid orbitals just described are called sp hybrids, to indicate that they are formed from one s orbital and from one p orbital. There are also other ways of mixing s and p orbitals to obtain hybrid orbitals. The element boron forms many compounds, among which are the simple BX_3 substances that aptly illustrate the next important case of hybridization.

The boron atom has a ground-state electron configuration $1s^2 2s^2 2p$. To form three bonds it must first be promoted to a valence state based on a configuration $2s 2p_x 2p_y$, in which the three valence electrons have decoupled their spins. The

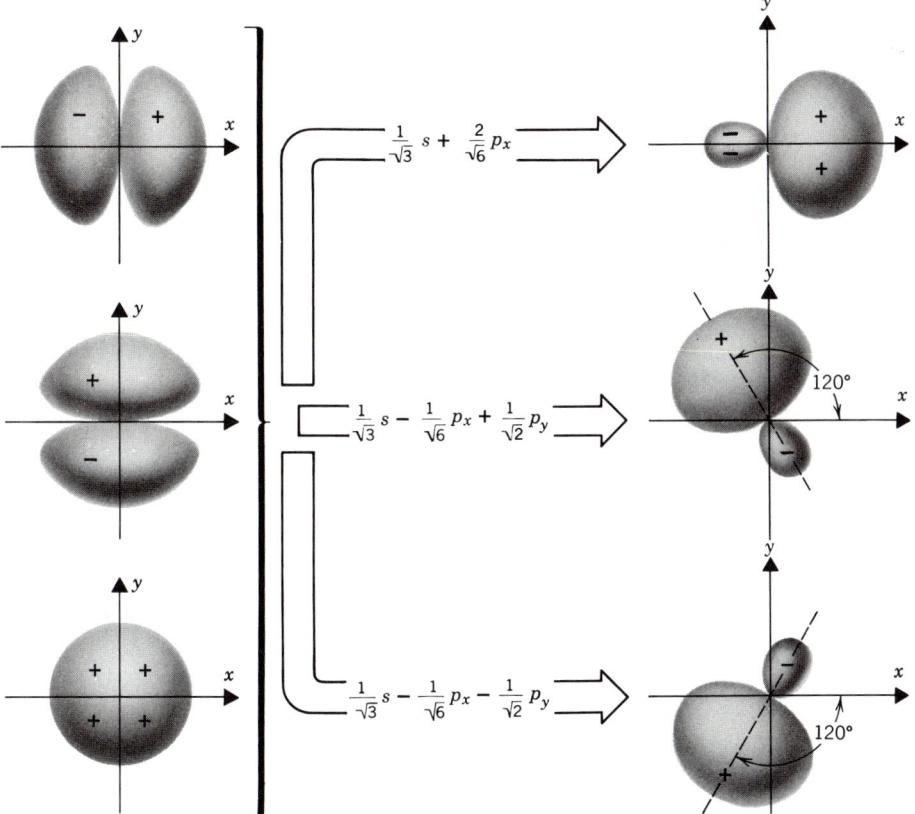

Figure 3-4 The formation of three equivalent sp^2 hybrid orbitals.

choice of $2p_x$ and $2p_y$ is arbitrary; any two $2p$ orbitals would be satisfactory. The ability of the central atom to form three bonds is now taken care of, but the question of securing maximum overlap must be dealt with. Again, it develops straightforwardly that by mixing the s and the two p orbitals equally, hybrid orbitals, called sp^2 hybrids, can be formed. These hybrids give superior overlap in certain definite directions, as is shown in Fig. 3-4. The three hybrid orbitals lie in the xy plane, and their maxima lie along the lines that are 120° apart. Thus, the BX_3 molecules have a planar, triangular structure.

The next type of hybridization that we shall discuss is the last one in which only s and p orbitals are involved. Let us consider how the carbon atom combines with four hydrogen atoms to form methane. Again, promotion from a ground state ($1s^2 2s^2 2p^2$), which does not have a sufficient number of unpaired electrons, to the valence state ($2s2p_x2p_y2p_z$) is required first. Then, the four orbitals of the valence state are mixed to give a set of four equivalent orbitals, each of which is called an sp^3 hybrid, as shown in Fig. 3-5. The hybrid orbitals of the sp^3 set are directed towards the vertices of a tetrahedron. Note that this geometry arises exclusively and directly from the algebra of hybridization. The geometry also happens to be that which most disperses the four C—H bonding pairs of electrons as far from one another as is possible.

In summary, an atom that has only s and p orbitals in its valence shell can form three types of hybrid orbitals, depending on the number of electrons available to form bonds:

sp hybrids give a linear molecule

sp^2 hybrids give a plane triangular molecule

sp^3 hybrids give a tetrahedral molecule

When d orbitals as well as s and p orbitals are available, the following important sets of hybrids, each illustrated in Fig. 3-6, can arise.

1. d^2sp^3, *Octahedral hybridization.* When the $d_{x^2-y^2}$ and d_{z^2} orbitals are combined with an s orbital and a set of p_x, p_y, and p_z orbitals, a set of equivalent orbitals with lobes directed to the vertices of an octahedron can be formed.

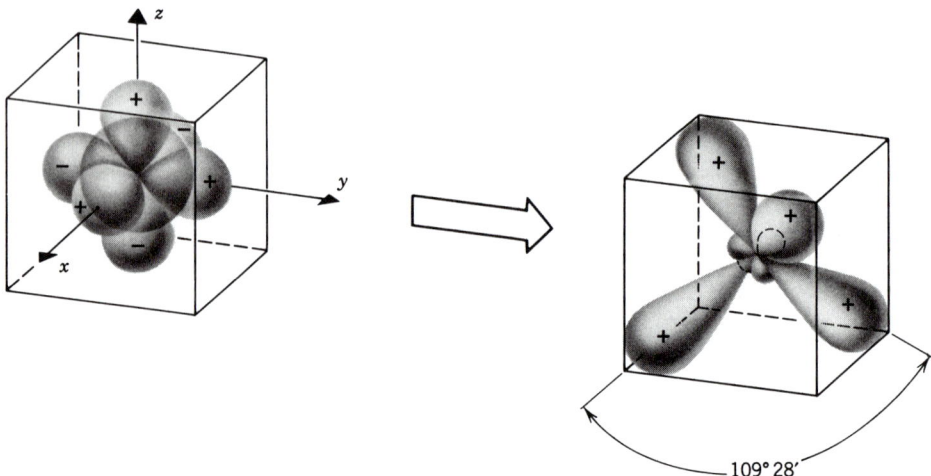

Figure 3-5 The formation of four equivalent sp^3 hybrid orbitals. A tetrahedron is defined by the four alternate corners of a cube to which the four hybrid orbitals are directed.

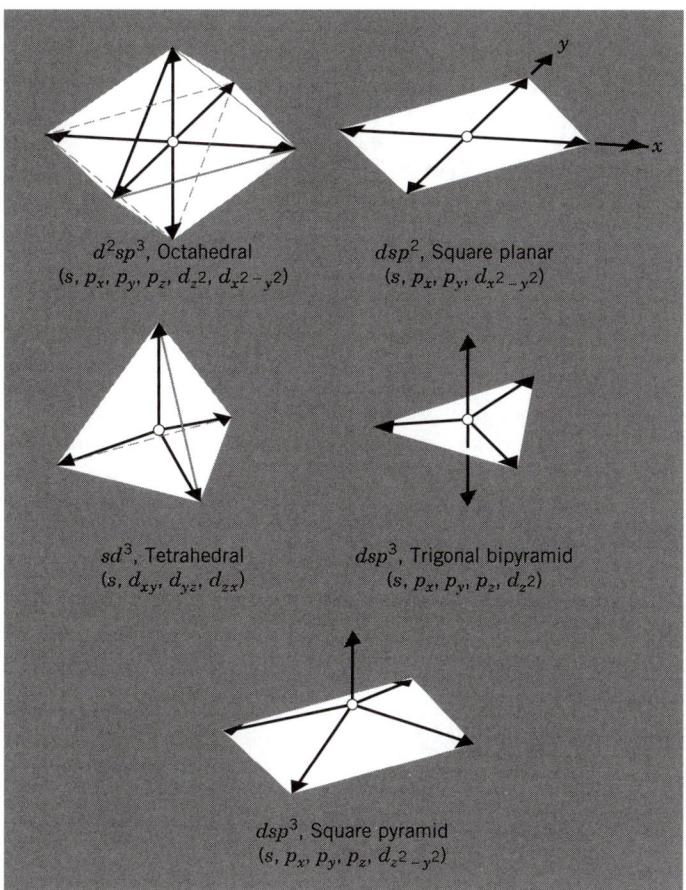

d^2sp^3, Octahedral
(s, p_x, p_y, p_z, d_{z^2}, $d_{x^2-y^2}$)

dsp^2, Square planar
(s, p_x, p_y, $d_{x^2-y^2}$)

sd^3, Tetrahedral
(s, d_{xy}, d_{yz}, d_{zx})

dsp^3, Trigonal bipyramid
(s, p_x, p_y, p_z, d_{z^2})

dsp^3, Square pyramid
(s, p_x, p_y, p_z, $d_{z^2-y^2}$)

Figure 3-6 Five important hybridization schemes involving d orbitals. Arrows show the direction in which the hybrid orbitals point within each different set.

2. dsp^2, *Square planar hybridization.* A $d_{x^2-y^2}$ orbital, an s orbital, and p_x and p_y orbitals can be combined to give a set of equivalent hybrid orbitals with lobes directed to the corners of a square in the xy plane.

3. sd^3, *Tetrahedral hybridization.* An s orbital and the set d_{xy}, d_{yz}, d_{zx} may be combined to give a tetrahedrally directed set of orbitals.

4. dsp^3, *Trigonal bipyramidal hybridization.* The orbitals s, p_x, p_y, p_z, and d_{z^2} may be combined to give a nonequivalent set of five hybrid orbitals directed to the vertices of a trigonal bipyramid.

5. dsp^3, *Square pyramidal hybridization.* The orbitals s, p_x, p_y, p_z, and $d_{x^2-y^2}$ may be combined to give a nonequivalent set of five hybrid orbitals directed to the vertices of a square pyramid.

The use of hybridized orbitals to explain bonding and correlate structures has become less common in recent years, giving way to the more general use of molecular orbital (MO) theory. The main reasons for this are that the MO approach lends itself more readily to quantitative calculations employing digital computers and because, with such calculations, it is possible to account for molecular spectra more easily. Nevertheless, the concept of hybrid orbitals retains certain advantages of simplicity and, in many instances, affords a very easy way to correlate and "explain" molecular structures.

Valence Shell Electron-Pair Repulsion (VSEPR) Theory

There is a very natural correlation between the orientation of the bonds (valences) to an atom and the spatial requirements of the bonding and nonbonding (lone-pair) electrons that reside at, and hence occupy the space surrounding, that atom. Electron pairs, whether in bonds to other atoms or in lone-pair orbitals on the atom in question, will tend to stay as far apart from one another as possible, to minimize repulsions among the various pairs. Thus the geometry at any atom in a molecule or polyatomic ion is dictated by the need of each electron pair to have as great a distance as possible separating it from other electron pairs residing on that atom. The electron pairs residing on an atom thus repel each other. Furthermore, it is assumed that lone-pair to lone-pair repulsions are most severe, followed by lone-pair to bonding-pair repulsions, and that bonding-pair to bonding-pair repulsions are the least significant of the three. This is sensible, since bonding pairs of electrons are confined to the relatively smaller space between nuclei, where they are constrained by interaction with two nuclei, whereas lone pairs of electrons fall under the attractive influence of only a single nucleus. Lone pairs are thus considered to require more room in the space immediately surrounding an atom than bonding pairs, and they are more repulsive towards other electron pairs residing on an atom than are bonding pairs. The angles between the various valences at an atom are then said to become adjusted so as to minimize the repulsions among the valence shell electron pairs. Additionally, it is found that the repulsive influence on adjacent electron pairs from electrons in a multiple bond is larger than that from the electron pair of a single bond. This difference obviously arises from the greater electron density that resides along the bond axis when multiple bonding is present. Finally, in geometries for which there is a difference between axial and equatorial positions on a polyhedron (namely, the trigonal bipyramid to be discussed shortly), the equatorial positions are favored by lone electron pairs over the axial positions. This broadly constitutes the approach of VSEPRs, as developed principally by R. J. Gillespie.

First, before applying VSEPR arguments to explain the geometries of molecules and polyatomic ions, it will be convenient to define a quantity known as the "occupancy." Second, once we know how many groups (whether atoms or lone pairs) are needed to "occupy" the space around an atom, then we can deduce the best prototype geometrical arrangement of those groups. Finally, starting with the prototype geometry, we can analyze electron-pair repulsions to explain small deviations in angles from those of the prototype.

For accounting purposes, it is convenient to define a quantity known as the occupancy for an atom. For structures AB_xE_y (where A is the central atom), x is the number of other atoms B bound to A, and y is the number of lone electron pairs E residing on atom A. The sum $(x + y)$ is what we shall call, for want of a better word, the **occupancy** of atom A. The space surrounding atom A is said to be *occupied* by $(x + y)$ other atoms or lone pairs. The occupancy of N in $:NH_3$ is four, for example. The occupancy for an atom is defined so that it is independent of the presence of multiple bonds; whether atoms B are singly or multiply

Table 3-1   The Separate Correspondence between Occupancy $(x + y)$ and Either Prototype Geometry or Hybridization of the Central Atom (A) in the Structures $AB_x E_y{}^a$

Occupancy $(x + y)$	Prototype Geometry	Hybridization
Two	Linear	sp
Three	Triangular	sp^2
Four	Tetrahedral	sp^3, sd^3
	Square (planar)	dsp^2
Five	Square pyramidal	dsp^3
	Trigonal bipyramidal	dsp^3
Six	Octahedral	d^2sp^3

^{a}The "central" atom (i.e., the one for which geometry is being considered) is designated atom A. Other atoms bonded to A are designated "B," whereas lone pairs of electrons on atom A are designated "E."

bonded to atom A, each B still occupies only one position in the space surrounding atom A.

Table 3-1 lists the occupancies (atoms plus lone pairs) and the corresponding geometries that best minimize electron-pair repulsions for each situation. Those hybridizations of the central atom that separately give a particular geometry are also listed in Table 3-1, although it should be remembered that VSEPR theory should be applied without reference to hybridization. (It is only convenient to list the two together in Table 3-1 because of the close correspondence in result that is often seen when applying the two theories.) Figure 3-7 shows the prototype shapes for the various molecules $AB_x E_y$. For the formulas AB_2, AB_3, AB_4, AB_5, and AB_6, in which there are no lone pairs, the molecular shapes are regular polyhedra: linear (AB_2), trigonal planar (AB_3), tetrahedral (AB_4), trigonal bipyramidal (AB_5), and octahedral (AB_6). Subgroups of these regular geometries are obtained for formulas with lone pairs E at the central atom. Although the positions of the lone pairs are specified, the geometry of the molecules is defined by the positions of the atoms A and B only. Thus, for the formula AB_3E, the four sp^3 hybrid orbitals of atom A are arranged roughly in the shape of a tetrahedron, but the molecule is said to be pyramidal. The following are specific examples of each structural type. The student should refer to Figs. 3-8 to 3-12.

Examples

In the following examples, the three localized bonding theories just mentioned have been applied in roughly the following fashion. First determine the proper Lewis diagram for the molecule or polyatomic ion. These diagrams are given in Figs. 3-8 to 3-12. Next, having determined the occupancy value for the atom of interest, deduce the atom's hybridization. Multiple bonds are then invoked in electronically unsaturated systems, using unhybridized p or d orbitals. Also, once occupancy has been determined, a prototype geometry can be chosen, and VSEPR theory can be used to explain deviations from the prototype. Although only the salient steps in this type of analysis are given in the examples that follow, the student is encouraged to work out all of the details for each example, starting with the Lewis diagram, and arriving eventually at a hybridization, a de-

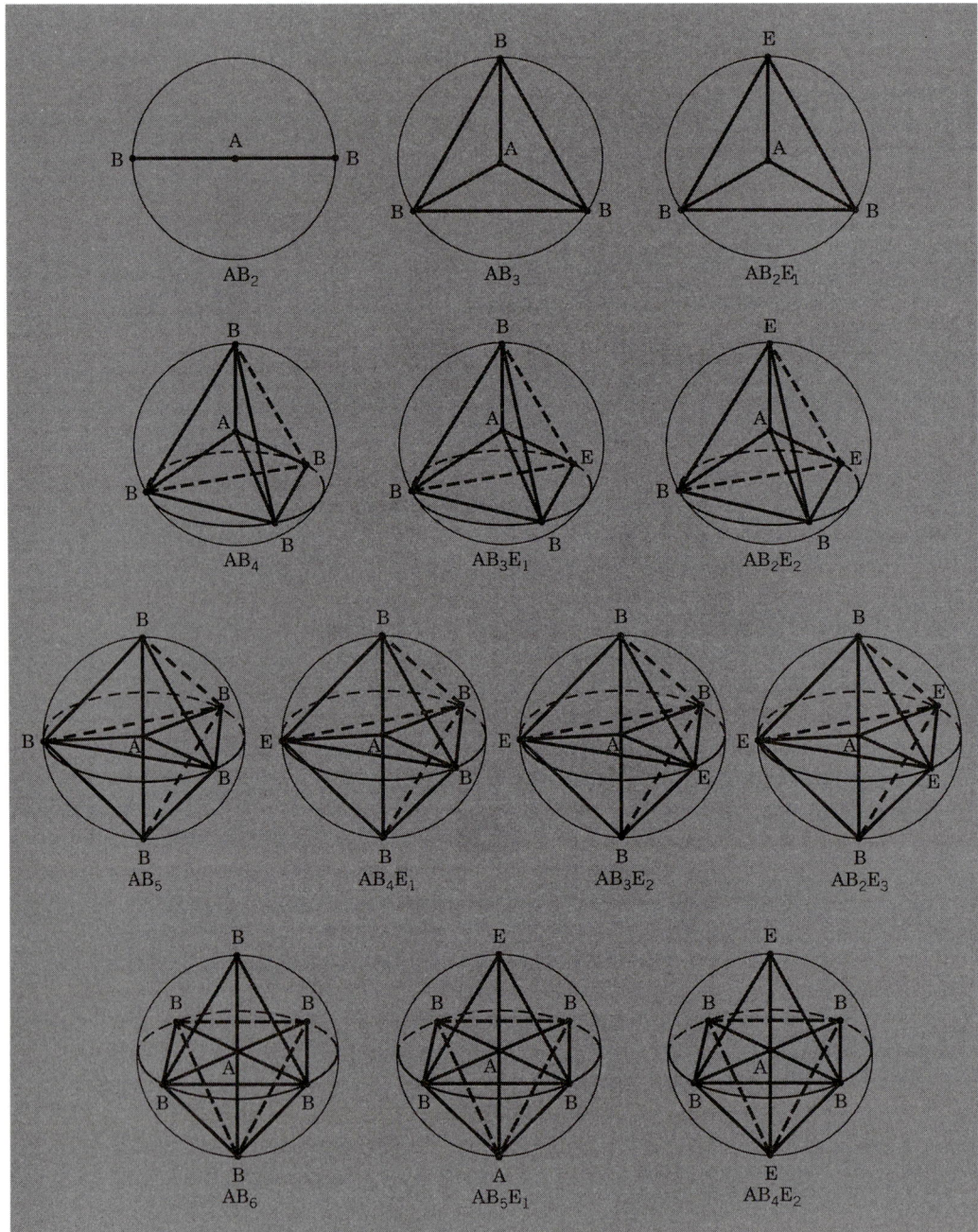

Figure 3-7   Idealized geometries for structures having the formulas AB_xE_y, where A is a central atom, B are peripheral atoms, and E are lone pairs residing on A.

scription of the multiple bonds, and VSEPR adjustments of the prototype geometries.

AB_2

BeH_2. This molecule has been discussed previously. It only remains to point out that the unhybridized p orbitals on Be are perpendicular to the molecular axis as well as to one another.

AB_2

1. H—Be—H $\angle$ HBeH = 180°

2. Ö=C=Ö sp hybridized C
                       $\angle$ OCO = 180°

3. [N̈=N=N̈]⁻ sp hybridized central N
                           $\angle$ NNN = 180°

ABE

1. :C≡O: sp hybridized C and O

2. [:C≡N̈]⁻

Figure 3-8 Examples of sp hybridization in structures AB_xE_y, where occupancy $(x + y) = 2$.

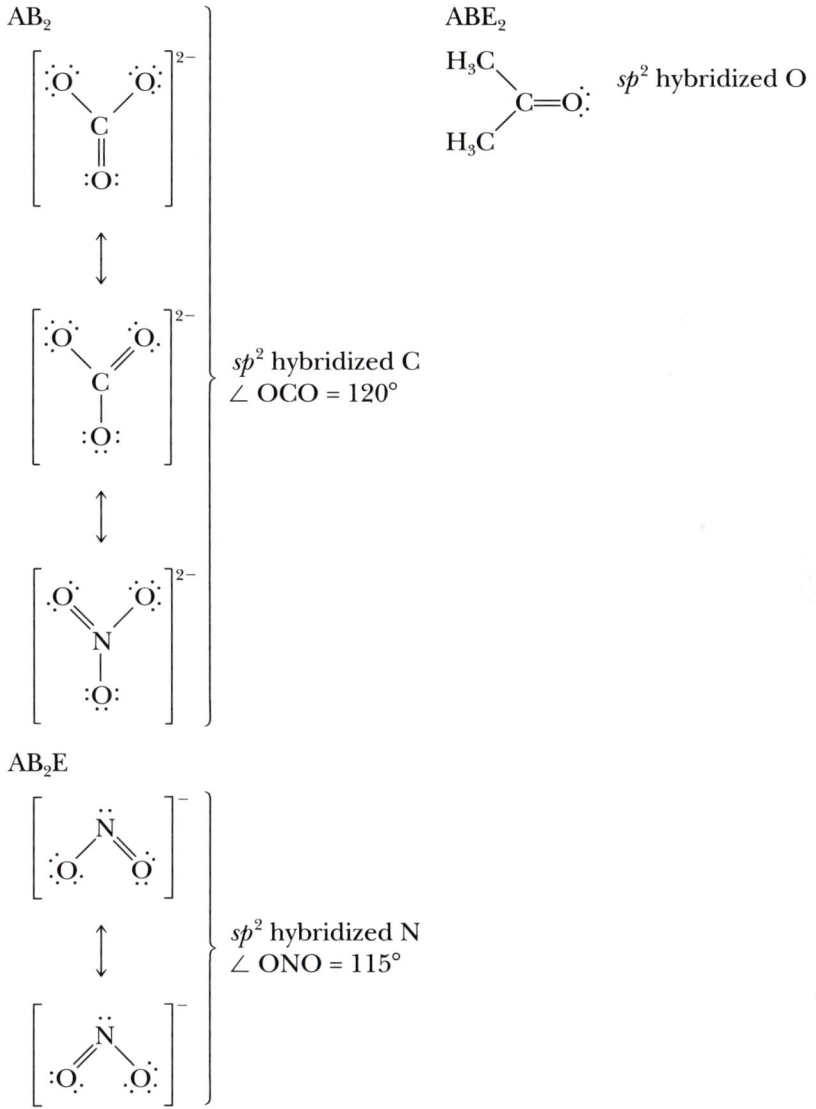

AB_2

sp^2 hybridized C
$\angle$ OCO = 120°

ABE_2

sp^2 hybridized O

AB_2E

sp^2 hybridized N
$\angle$ ONO = 115°

Figure 3-9 Examples of sp^2 hybridization in structures AB_xE_y, where occupancy $(x + y) = 3$.

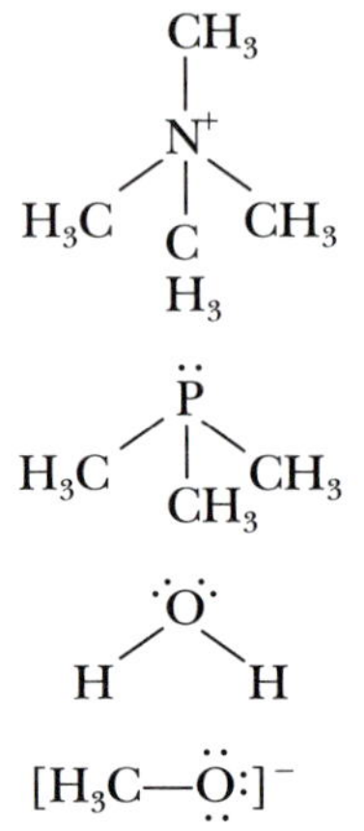

AB$_4$

$$CH_3$$
$$|$$
$$N^+$$
$$H_3C \diagup \; | \; \diagdown CH_3$$
$$C$$
$$H_3$$

sp^3 hybridized N
$\angle$ CNC = 109°

AB$_3$E

$$\ddot{P}$$
$$H_3C \diagup | \diagdown CH_3$$
$$CH_3$$

sp^3 hybridized P
$\angle$ CPC = 99°

AB$_2$E$_2$

$$\ddot{\overset{..}{O}}$$
$$H \diagup \diagdown H$$

sp^3 hybridized O
$\angle$ HOH = 104°

ABE$_3$

$[H_3C—\overset{..}{\underset{..}{O}}:]^-$

sp^3 hybridized O
(although see text)

Figure 3-10 Examples of sp^3 hybridization in structures AB$_x$E$_y$, where occupancy $(x + y) = 4$.

CO$_2$. The central carbon is sp hybridized, as shown in Fig. 3-8, and the molecule is linear. The unhybridized p orbitals of carbon are involved in π bonding with the atomic p orbitals of oxygen, as shown in Fig. 3-13. The two π bond systems are perpendicular to one another because the unhybridized atomic p orbitals of carbon are oriented 90° to one another. The π bonds each involve two regions of overlap (above and below the O—C—O bond axis). The σ bond system involves overlap of sp hybrids on carbon with sp^2 hybrids on oxygen. The σ bond system lies along the internuclear axis of the molecule, while the π bond system has a node along the internuclear axis. The azide anion N$_3^-$ is completely analogous to CO$_2$.

ABE
CO. Carbon monoxide contains a triple bond: one σ and two mutually perpendicular π bonds. There is a lone pair of electrons on each atom, housed in an sp hybrid. The π-bond system is illustrated in Fig. 3-13. It is sp hybridization that leaves two π atomic orbitals available on both C and O for the formation of these π bonds. Other examples that are **isostructural** (have the same structures) and **isoelectronic** (have the same electron configurations) are the ions CN$^-$ and NO$^+$.

For both systems described (namely, cases AB$_2$ and ABE) the atoms or lone pairs that occupy the space about an atom are disposed 180° from one another. This occurs because only two groups must be accommodated at the atom in question, that is, occupancy [the quantity $(x + y)$ in the cases AB$_x$E$_y$] is two. Linear geometry and sp hybridization always result under these circumstances. Other examples include alkynes (—C≡C—), nitriles (R—C≡N:), isonitriles (R—N≡C:), metal carbonyls (M—C≡O:), and cyanate (:N≡C—O$^-$).

AB$_3$
CO$_3^{2-}$. The sp^2 hybridization of carbon in the carbonate ion allows for the use of one unhybridized p atomic orbital on carbon in the formation of one π bond.

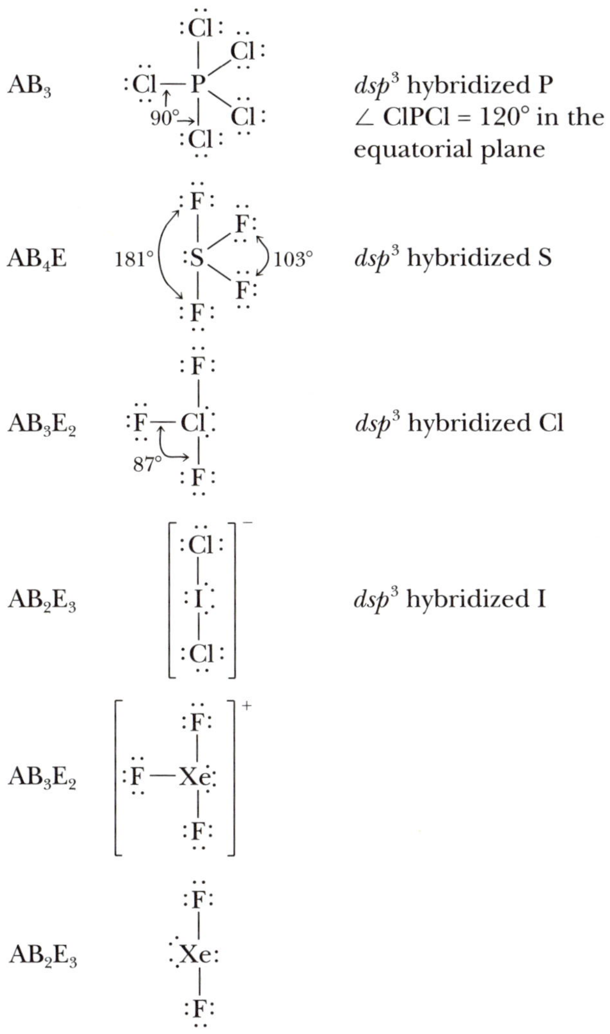

AB$_3$	$:\overset{\cdot\cdot}{Cl}:$ ... $\overset{\cdot\cdot}{Cl}:$ $:\overset{\cdot\cdot}{Cl}-P$ $\overset{90°}{\to}$ $:\overset{\cdot\cdot}{Cl}:$ $:\overset{\cdot\cdot}{Cl}:$	dsp^3 hybridized P $\angle$ ClPCl = 120° in the equatorial plane
AB$_4$E	181° $\Big($ $:\overset{\cdot\cdot}{S}$ $\Big)$ 103° :F: :F: :F: :F:	dsp^3 hybridized S
AB$_3$E$_2$	:F: $:\overset{\cdot\cdot}{F}-\overset{\cdot\cdot}{Cl}:$ 87° :F:	dsp^3 hybridized Cl
AB$_2$E$_3$	$\left[\begin{array}{c}:\overset{\cdot\cdot}{Cl}: \\ :\overset{\cdot}{\underset{\cdot}{I}}: \\ :\overset{\cdot\cdot}{Cl}:\end{array}\right]^{-}$	dsp^3 hybridized I
AB$_3$E$_2$	$\left[\begin{array}{c}:\overset{\cdot\cdot}{F}: \\ :\overset{\cdot\cdot}{F}-Xe: \\ :\overset{\cdot\cdot}{F}:\end{array}\right]^{+}$	
AB$_2$E$_3$	:F: $:\overset{\cdot\cdot}{Xe}:$:F:	

Figure 3-11 Examples of dsp^3 hybridization in structures AB$_x$E$_y$, where occupancy $(x + y) = 5$.

As shown for each resonance form in Fig. 3-9, π bonding can take place between the central carbon and any one of the three equivalent oxygen atoms. Three resonance forms are required to show this delocalization of the one π bond. For any one resonance form, the π bond is illustrated in Fig. 3-13. The ion has trigonal planar geometry; the oxygen atoms are dispersed 120° with respect to one another in order to minimize repulsions among the electrons of the C—O bonds. This geometry is typical of other structures (e.g., BF$_3$ and SO$_3$) having occupancies of three for the central atoms. In the examples that follow, however, because of the presence of lone pairs E, or nonequivalent substituents B, the perfect 120° angles are not observed.

Carbonyls. For organic carbonyls (R$_2$C=O), or for the acyls [R(X)C=O] and the formyls [H$_2$C=O, X$_2$C=O, and M—C(H)=O (a metal formyl)], the

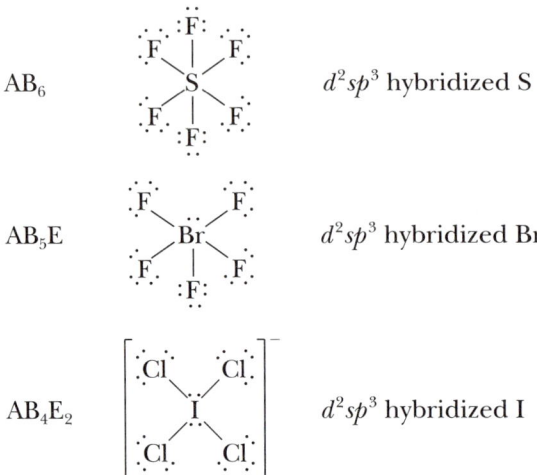

AB$_6$ d^2sp^3 hybridized S

AB$_5$E d^2sp^3 hybridized Br

AB$_4$E$_2$ d^2sp^3 hybridized I

Figure 3-12 Examples of d^2sp^3 hybridization in structures AB$_x$E$_y$, where occupancy $(x + y) = 6$.

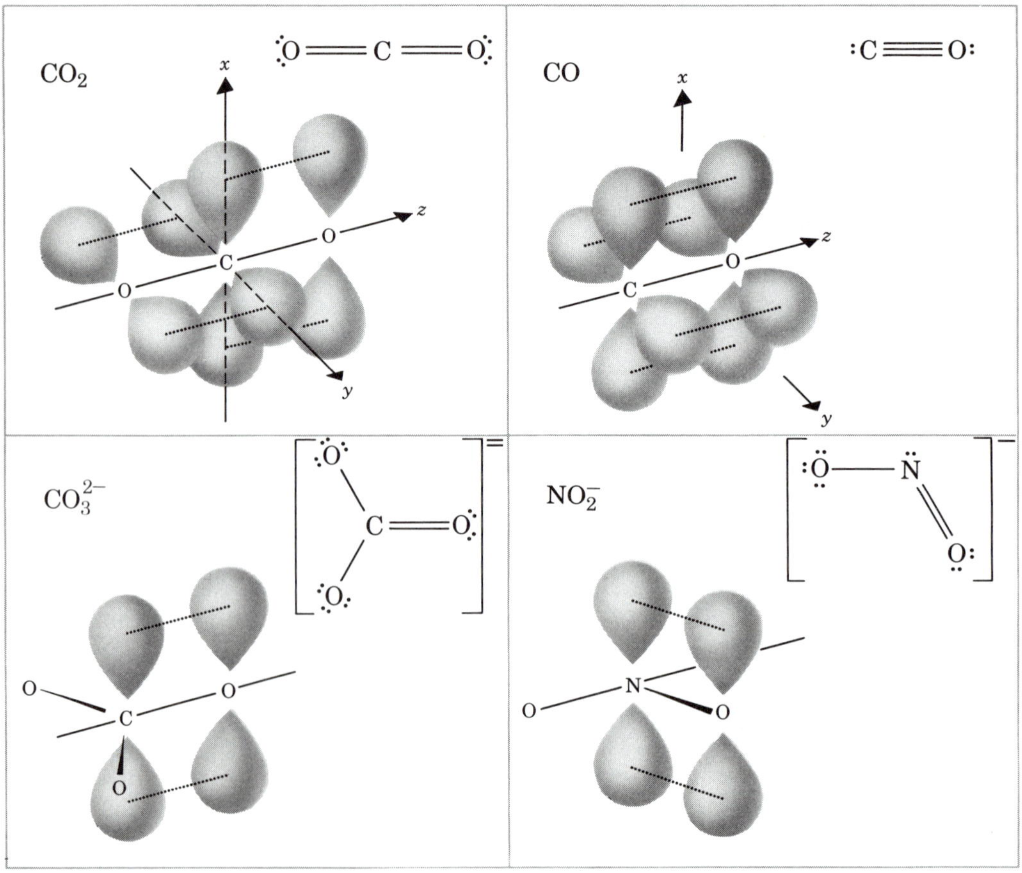

CO$_2$

CO

CO$_3^{2-}$

NO$_2^-$

Figure 3-13 Examples of the formation of π bonds via overlap of unhybridized (atomic) p orbitals.

carbon atoms can be taken to be sp^2 hybridized, with occupancy equal to three. The groups that are bound to the central carbon are nonequivalent, and the idealized geometry of 120° is altered due to repulsions among the bonding electrons at carbon. Consider the formyls $H_2C\!=\!O$ and $X_2C\!=\!O$ shown in Structures 3-V.

116°	111°	108°
3-Va	3-Vb	3-Vc

The electrons of the C=O double bond require the greatest room in these structures. Consequently, the HCH and XCX bond angles collapse from the normal 120° for sp^2 hybridization to those values listed in Structures 3-V. Repulsions from the C=O double-bond system become balanced by repulsions between the electrons of the two C—H or C—X bonds, and the resulting angles reflect the willingness of the electrons of the C—H or the C—X groups to approach one another in either $H_2C\!=\!O$ or $X_2C\!=\!O$. Obviously, the more electronegative groups X allow for a closer approach to one another by the C—X bonds. This occurs because the electron density in the C—X bonds is farther out towards the X extremities of the C—X bond (and collapse of the XCX bond angle is less troublesome) for atoms X with the higher electronegativities.

AB_2E

NO_2^-. The nitrite anion is planar, and sp^2 hybridization is consistent with the occupancy of three for the central nitrogen atom. One π bond is present in each of the contributing resonance forms. As shown in Fig. 3-13, this π-bond system lies above and below the plane of the ion, and is perpendicular to it. Although the central nitrogen is sp^2 hybridized, the ONO angle is not a perfect 120° because of the larger volume requirement of the lone-pair electrons. The bonding electron pairs move closer to one another in response to repulsion from the lone pair of nitrogen. The larger lone-pair–bonding-pair repulsion is balanced by the less intense bonding-pair–bonding-pair interaction once the ONO angle has collapsed from the idealized 120° to the actual 115° found in the ion. This result is shown in Structure 3-VIa. Removal of one electron from the nitrite anion gives the neutral radical NO_2, shown in Structure 3-VIb. Here the ONO angle opens to the value 132° because only a lone electron, not a pair, is housed on the nitrogen. Now the most severe repulsion is between the electrons of the NO bonds, and the ONO angle can become larger without encountering restrictions from a full lone pair of electrons on nitrogen. For the cation NO_2^+ the central nitrogen has occupancy equal to two and, as shown in Structure 3-VIc, the geometry is linear.

115°	132°	
3-VIa	3-VIb	3-VIc

ABE_2

Simple examples in this category include O_2 and NO^-. Otherwise, we must look to terminal atoms for more examples.

It is not necessary to specify a hybridization for terminal atoms. Whatever the bonding scheme, the geometry is linear by definition. The presence of lone pairs can be inferred from the Lewis diagrams, and the presence of single or multiple bonds can be inferred from the length of the bond. There are terminal atoms, though, where it is instructive to examine the hybridization. Such a case is the terminal oxygen of a carbonyl group in aldehydes or ketones. The occupancy formula for such an oxygen atom is ABE_2, and a double bond to C is typical. The hybridization of such an oxygen is said to be sp^2, and two of these hybrid orbitals are used to house the two lone pairs on oxygen. The other sp^2 hybrid forms a σ bond to carbon by overlap with an sp^2 hybrid from carbon. The σ bond to carbon and the two lone pairs of oxygen lie in a plane. The π bond is perpendicular to this plane, above and below it. It is not necessary or proper to speak of the geometry at such a terminal atom, because it lies on the periphery of the molecule. It is helpful, though, to realize that the method of determining occupancy gives a hybridization that is consistent with the number of π bonds to the atom.

AB_4

In addition to the many organic compounds having sp^3 hybridized carbon, there are important AB_4 examples among inorganic systems where the occupancy is also four. The best examples are the tetraoxides of the main group elements (general formula AO_4^{n-}) and the tetracoordinated compounds of the transition metals that contain a central metal and four ligands [e.g., $Ni(CO)_4$]. The geometry for main group atoms A is always tetrahedral. When A is a transition metal, the ligands can be arranged either in tetrahedral fashion (sp^3 hybridization) or in square planar fashion (dsp^2 hybridization), as shown in Fig. 3-7 and in Table 3-1. It is the number of d electrons that determines which of these two geometries is preferred, although the d electrons are not considered in writing the Lewis diagram. More will be said about this in subsequent chapters. For now, we shall restrict our attention to the oxy anions AO_4^{n-} and the transition metal systems ML_4^{n+}.

AO_4^{n-}. The familiar ions phosphate (PO_4^{3-}), sulfate (SO_4^{2-}), and perchlorate (ClO_4^-) are isostructural and isoelectronic. The central atom is tetrahedrally surrounded by four oxygen atoms and an octet is achieved for all atoms in the ions when single bonds are used exclusively. As already shown for SO_4^{2-} in Fig. 3-1, there can be additional π bonding that increases the electron density at the central atom. This π bonding involves the use of empty d orbitals on the central atom, as shown in Fig. 3-14. Former lone-pair electrons of oxygen are shared with the central atom through $d\pi$–$p\pi$ overlap. This requires a rehybridization of the terminal oxygen atoms from sp^3 (A—O groups) to sp^2 (A=O groups). The terminal oxygen atoms are said to be π donors and the central atom A is said to be a π acceptor. The double-bond system is most evident when the central atom has the highest electronegativity (e.g., ClO_4^-). For elements of rows three and below of the periodic table $d\pi$–$p\pi$ bonding is prominent because of the presence of valence-level d orbitals on these atoms. For rows one and two of the periodic table, the d orbitals are not found in the valence levels. These orbitals are consequently too high in energy to be of use in bonding.

The availability of empty d orbitals also plays a role in the chemistries of third-row compounds, as illustrated by two examples with this same structure,

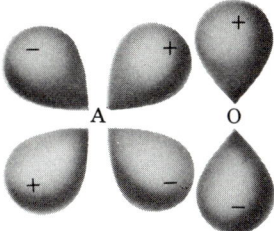

Figure 3-14 An example of $d\pi–p\pi$ bonding. An empty d orbital on a central atom accepts electron density from a filled p orbital of another atom.

AB_4. Both CCl_4 and the corresponding third-row compound of silicon ($SiCl_4$) are tetrahedral, with sp^3 hybridized central atoms. The carbon analog is stable towards attack by simple nucleophiles, such as water, while the Si compound is not. The empty d orbitals in the valence shell of Si provide the needed site for nucleophilic attack, and the larger size of the central Si atom facilitates the hydrolysis shown in Reaction 3-2.1.

$$SiCl_4 + 4\ ROH \longrightarrow Si(OR)_4 + 4\ HCl \qquad (3\text{-}2.1)$$

One could add the silicates SiO_4^{4-} and the tetrahedral XeO_4, which complete the isoelectronic series of tetraoxides of row three, to the discussion of structures AB_4. It is more common, though, for silicate structures to occur in polymeric form (as discussed in Chapter 15) rather than as discrete anions. Xenon tetraoxide is an explosively unstable gas.

ML_4^{n+}. Transition metal compounds that are four coordinate may adopt either tetrahedral or square planar geometry, depending on the number of d electrons that reside at the metal. Square planar geometry is common for d^8 systems, such as in Structures 3-VIIa and 3-VIIb.

3-VIIa

3-VIIb

Tetrahedral geometry is found for d^{10} systems as in Structures 3-VIIIa and 3-VIIIb.

3-VIIIa

3-VIIIb

AB₃E

Of the structures with occupancy four and one lone pair, the most familiar are the amines (:NR₃), the phosphines (:PR₃), the arsines (:AsR₃), and the stibines (:SbR₃). All are pyramidal, and can serve as Lewis bases by reason of the one lone pair on the central atom. In fact, these compounds serve as useful ligands for coordination to metals. The halides (e.g., :NX₃) should also be considered here.

As a class, the molecules may be taken to have sp^3 hybridized central atoms and roughly pyramidal molecular geometries. The lone pair of electrons causes deviations from the ideal 109.5° angles expected for perfect sp^3 hybrid sets. The HAH angle is smallest in the molecules :AH₃ where the central atom A is the largest.

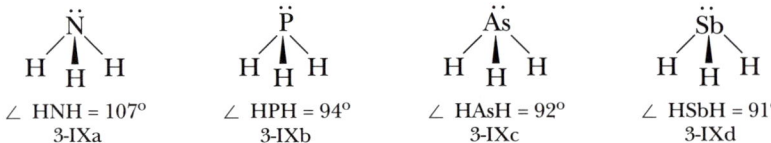

∠ HNH = 107°   ∠ HPH = 94°   ∠ HAsH = 92°   ∠ HSbH = 91°
3-IXa   3-IXb   3-IXc   3-IXd

The XAX angle is smallest in the molecules :AX₃ where the atoms X are most electronegative.

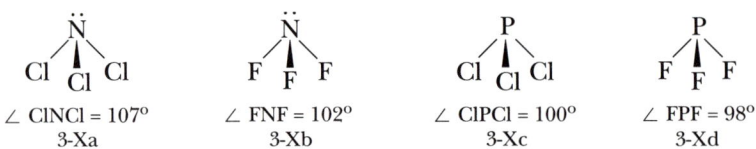

∠ ClNCl = 107°   ∠ FNF = 102°   ∠ ClPCl = 100°   ∠ FPF = 98°
3-Xa   3-Xb   3-Xc   3-Xd

Presumably, the A—X bonding electrons are polarized towards the electronegative atoms X, so that a decrease of the XAX angle is less troublesome where X is more electronegative.

The angles noted in Structures 3-Xa to 3-Xd may indicate that the choice of sp^3 hybridization is inappropriate for some of the examples given. After all, angles close to 90° may indicate, if anything, a lack of hybridization for the central atoms Sb and As. The fully delocalized MO treatment, which is presented later in this chapter, offers a more satisfactory explanation of the bonding in such systems.

AB₂E₂

This familiar case includes the dihydrides of Group VIB(16): H₂O, H₂S, H₂Se, and H₂Te, as shown in Structures 3-XIa to 3-XId.

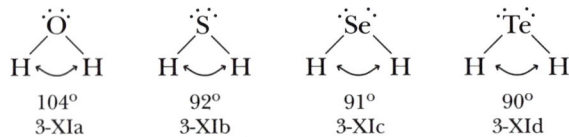

104°   92°   91°   90°
3-XIa   3-XIb   3-XIc   3-XId

The two lone pairs provide the most severe repulsions, and the space that they require in order to minimize this repulsion is achieved by a decrease in the HAH angle. This result is most readily accomplished for the larger central atoms Se and Te. In the latter cases it is inappropriate to consider that the central atom is simply sp^3 hybridized. Other cases where this AB₂E₂ structure arises include the alcohols (ROH) and ethers (ROR).

ABE₃

It is unnecessary and inappropriate to assign a hybridization for a "central" atom, such as A, in the case ABE$_3$. It is neither possible nor necessary to know the positions of the lone electron pairs. It is certain that the electron pairs are as far from one another as is possible, but it is a matter of theory, not fact, to speculate about the orbital arrangement for those electrons. (In contrast, it is possible to speak with certainty about the positions of atoms.) Nevertheless, we have grown accustomed to speaking of the oxygen of alkoxides (RO$^-$) as being sp^3 hybridized, for instance, because this does provide maximum room for each of the three lone pairs of oxygen. It also correctly accounts for the single remaining bond to carbon in the octet of oxygen. One must examine the energy of the entire ion (including three electron pairs somewhat localized on the oxygen and a bonding pair somewhat localized between the carbon and the oxygen) before deciding if the best bond between oxygen and carbon is provided by overlap of two sp^3 hybrid orbitals. In short, a more delocalized bonding theory may prove better.

AB₅

This case begins the series in Fig. 3-11 in which the occupancy at a central atom is five. Where no lone-pair electrons reside at the central atom [PCl$_5$, Fe(CO)$_5$, or CuCl$_5^{3-}$], the geometry is a perfect trigonal bipyramid. As illustrated in Structure 3-XII:

3-XII

the two axial positions in this polyhedron are not equivalent to the three equatorial positions. The **axial** groups are positioned above and below the triangular plane, while the three **equatorial** groups are positioned in the triangular plane. As shown in Figs. 3-6 and 3-7, there is one form of dsp^3 hybridization that gives this orbital arrangement.

An occupancy of five is also accommodated by square pyramidal geometry and the other type of dsp^3 hybridization. In this case, however, it is the $d_{x^2-y^2}$ orbital that is required, as shown in Fig. 3-6. An interesting example of this geometrical difference is given by the compounds studied by R. R. Holmes, and shown in Structures 3-XIIIa and 3-XIIIb.

3-XIIIa

3-XIIIb

The Si atom of Structure 3-XIIIa is at the center of a square pyramid, while that of Structure 3-XIIIb is trigonal bipyramidal.

AB_4E

SF_4 has the structure given in Fig. 3-11. This structure is derived from that of the trigonal bipyramid, with the lone electron pair of S occupying an equatorial position. This structure is preferred because there is a close (~90°) interaction between this lone pair and only two axial bonding pairs. The other bonding pairs are at a relatively distant 128°. Had the lone electron pair of S been put into an axial position, there would have been three close 90° interactions with bonding pairs in the equatorial positions. This structure would clearly be less stable.

AB_3E_2

ClF_3 has the distorted planar T shape shown in Fig. 3-11. The axial FClF angle is not 180° because the two equatorial lone pairs push the two axial fluorines back from their formal positions. The ion $[XeF_3]^+$ is T shaped.

AB_2E_3

The ICl_2^- ion is linear. Axial placement of the two chlorines allows the three lone pairs of I to be accommodated in the relatively "roomy" equatorial plane. Similarly, xenon difluoride is a linear molecule with three equatorial lone pairs at the central xenon atom. The only other possible geometry for such a system with an occupancy of five would be to place the two fluorine atoms adjacent to one another, giving a bent geometry. This latter case is less favored because it would result in one lone pair having two 90° interactions with other lone pairs. The existing structure is one in which each lone pair suffers only two close interactions (120°) with other equatorial lone pairs.

AB_6

Finally, we consider the cases with occupancies of six and d^2sp^3 hybridization. The AB_6 system is represented by a host of transition metal compounds with octahedral or pseudooctahedral geometries. Further examples will be given in the following chapters. Silicon hexafluoride (SF_6) is a good example of a main group nonmetal compound with octahedral geometry.

AB_5E

The lone pair on Br in the BrF_5 molecule gives a square pyramidal geometry, although the orbital arrangement is still roughly that of an octahedron. Unlike the trigonal bipyramid, all positions on the octahedron are equivalent, and placement of the lone pair is not an issue. This is not true of the next example, however.

AB_4E_2

The ion ICl_4^- is planar because two lone pairs on the central iodine are placed opposite (180°) one another. The other possibility is less stable because it would involve placement of lone pairs at 90° to one another.

Table 3-2 Single-Bond Covalent Radii (in Å)

H	0.28	C	0.77	N	0.70	O	0.66	F	0.64
		Si	1.17	P	1.10	S	1.04	Cl	0.99
		Ge	1.22	As	1.21	Se	1.17	Br	1.14
		Sn	1.40	Sb	1.41	Te	1.37	I	1.33

3-3 Bond Lengths and Covalent Radii

If we consider a single bond between like atoms, say Cl—Cl, we can define the single-bond covalent radius of the atom as one half of the bond length. Thus the Cl—Cl distance (1.988 Å) yields a covalent radius of 0.99 Å for the chlorine atom. In a similar way, radii for other atoms (e.g., 0.77 Å for carbon by taking one half the C—C bond length in diamond) are obtained. It is then gratifying to find that the lengths of heteronuclear bonds can often be predicted with useful accuracy. For example, from Table 3-2 we can predict the following bond lengths, in angstroms, which agree pretty well with the measured values given in parentheses:

C—Si	1.94 (1.87)	P—Cl	2.09 (2.04)
C—Cl	1.76 (1.77)	Cl—Br	2.13 (2.14)

The agreement cannot be expected to be perfect, since bond properties (including length) vary somewhat with the environment.

Multiple bonds are always shorter than corresponding single bonds. This is illustrated by bonds between nitrogen atoms:

$$N\equiv N \ (1.10 \text{ Å}) \qquad N=N \ (1.25 \text{ Å}) \qquad N—N \ (1.45 \text{ Å})$$

Consequently, double- and triple-bond radii can also be defined. For the elements C, N, and O, which form most of the multiple bonds, the double- and triple-bond radii are approximately 0.87 and 0.78 times the single-bond radii, respectively.

The hybridization of an atom affects its covalent radius; since s orbitals are more contracted than p orbitals, the radius decreases with increasing s character. Carbon has the following single-bond radii:

$$C(sp^3), 0.77 \text{ Å} \qquad C(sp^2), 0.73 \text{ Å} \qquad C(sp), 0.70 \text{ Å}$$

When there is a great difference in the electronegativities (Section 2-7) of two atoms, the bond length is usually less than the sum of the covalent radii, sometimes by a considerable amount. Thus, from Table 3-2, the C—F and Si—F distances are calculated to be 1.44 and 1.81 Å, whereas the actual distances in CF_4 and SiF_4 are 1.32 and 1.54 Å. In the case of the C—F bond it is believed that the shortening can be attributed to ionic–covalent resonance, which strengthens and, hence, shortens (by 0.12 Å) the bond. For SiF_4 only part of the very pronounced shortening can be thus explained. Much of it is thought to be due to π bonding using filled fluorine $p\pi$ and empty silicon $d\pi$ orbitals.

3-4 Molecular Packing: van der Waals Radii

When molecules pack together in the liquid and solid states, their approach to one another is limited by short-range repulsive forces, which result from overlapping of the diffuse outer regions of the electron clouds around the atoms.

Table 3-3   van der Waals Radii of Nonmetallic Atoms (in Å)

H	1.1–1.3					He	1.40
N	1.5	O	1.40	F	1.35	Ne	1.54
P	1.9	S	1.85	Cl	1.80	Ar	1.92
As	2.0	Se	2.00	Br	1.95	Kr	1.98
Sb	2.2	Te	2.20	I	2.15	Xe	2.18

Radius of a methyl group, 2.0 Å
Half-thickness of an aromatic ring, 1.85 Å

The actual distance apart at which any two molecules would come to rest is determined by the equalization of attractive and repulsive forces. There are also weak, short-range attractive forces between molecules that result from permanent dipoles, dipole–induced dipole, and so-called London forces. The latter arise from interaction between fluctuating dipoles whose time-average value in any one molecule is zero.

Collectively, all these attractive and repulsive forces that are neither ionic nor covalent are called *van der Waals forces*.

For the vast majority of molecules we find that both the attractive and repulsive forces are of roughly constant magnitude. Thus the distances between molecules in condensed phases do not vary a great deal. Consequently, it is possible to compile a list of van der Waals radii, which give the typical internuclear distances between nearest neighbor atoms in different molecules in condensed phases. The van der Waals radii for some common atoms are listed in Table 3-3.

van der Waals radii are much larger than covalent radii and are roughly constant for isoelectronic species. Thus, in crystalline Br_2, the covalent radius of Br is 1.15 Å, whereas the van der Waals radius (one-half of the shortest intermolecular Br⋯Br distance) is 1.95 Å. The latter differs little from the Kr⋯Kr packing distance of 1.98 Å in solid Kr, since Br when bonded to another atom is isoelectronic with the Kr atom.

3-5   The Delocalized Approach to Bonding: Molecular Orbital Theory

The MO theory description of the chemical bond involves the simple and broadly applicable idea that a chemical bond can exist when outer orbitals on different atoms overlap so as to concentrate electron density between the atomic cores. The criterion of net positive overlap of atomic orbitals is of unparalleled usefulness as a qualitative guide, and indicates whether bonding will actually occur. Consequently, the examination of these overlaps will be our first consideration.

Overlap of Orbitals

If two atoms approach each other closely enough for one orbital on each atom to have appreciable amplitude in a region of space common to both of them, the orbitals are said to overlap. The magnitude of the overlap may be positive, negative, or zero, according to the properties of the orbitals concerned. Examples of these three cases are illustrated in Fig. 3-15.

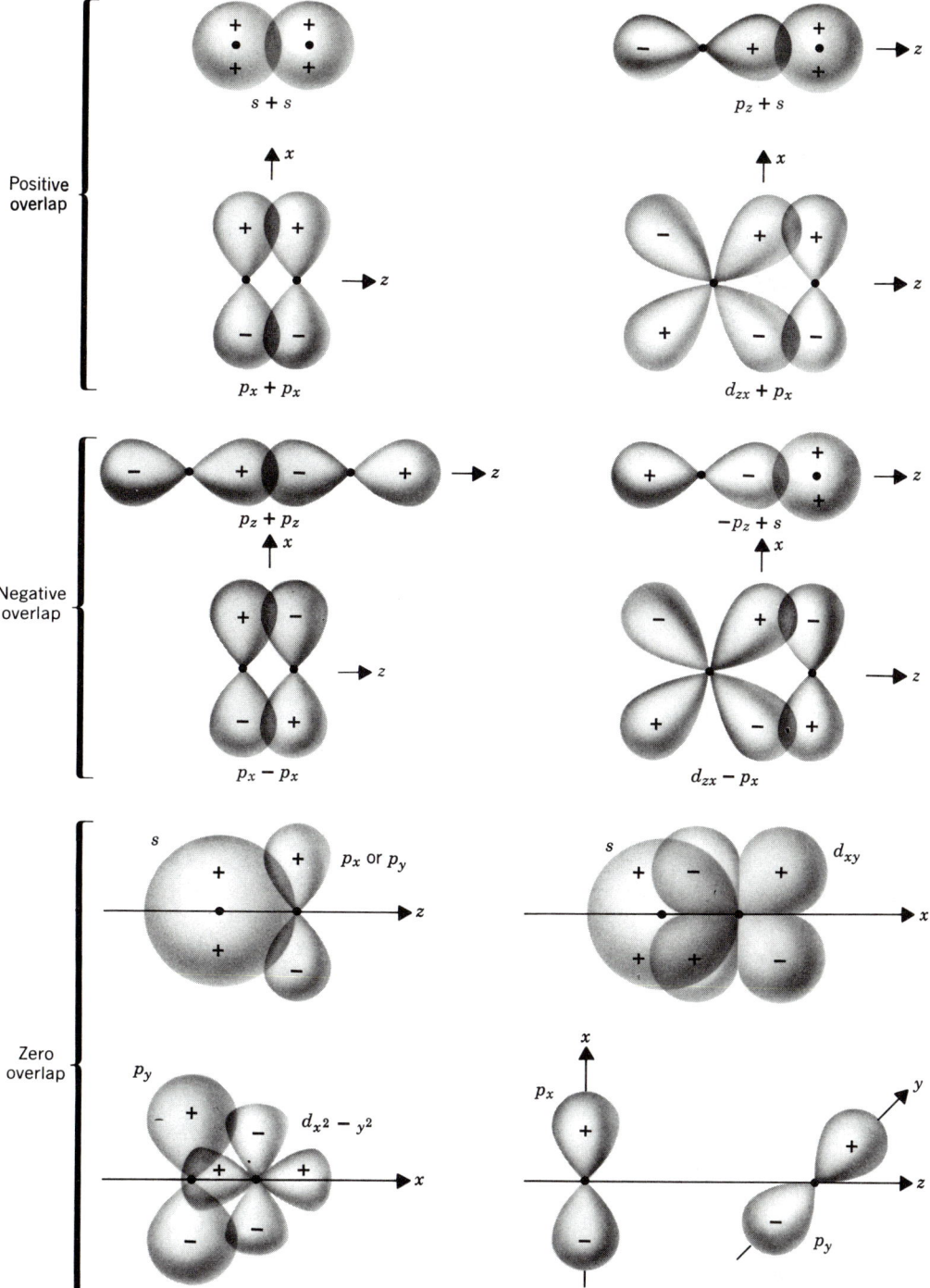

Figure 3-15 Some common types of orbital interaction leading to positive, negative, and zero overlap.

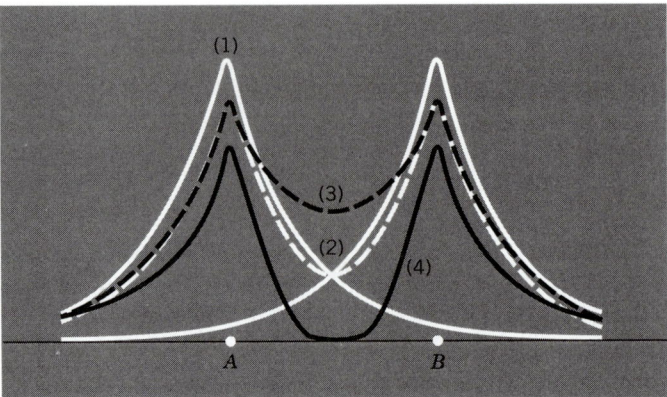

Figure 3-16   Electron density distributions for the one-electron H_2^+ ion, with H_A at point A and H_B at point B. (1) For each atom, taken separately, the solid white curve represents either ϕ_A^2 or ϕ_B^2. (2) The broken white curve represents the simple sum $(\phi_A^2 + \phi_B^2)/2$. (3) The broken black curve represents the bonding function $(\phi_A + \phi_B)^2/2$. (4) The solid black curve represents the antibonding function $(\phi_A - \phi_B)^2/2$.

Overlap has a positive sign when the superimposed regions of the two orbitals have the same sign: both + or both −. Overlap has a negative sign when the superimposed regions of the two orbitals have opposite signs. Precisely zero overlap results when there are precisely equal regions of overlap with opposite signs.

The physical reason for the validity of the overlap criterion is straightforward. In a region where two orbitals ϕ_1 and ϕ_2 have positive overlap, the electron density is higher than the mere sum of the electron densities of the two separate orbitals. That is, $(\phi_1 + \phi_2)^2$ is greater than $\phi_1^2 + \phi_2^2$, by $2\phi_1\phi_2$. More electron density is shared between the two atoms. The attraction of both nuclei for these electrons is greater than the mutual repulsion of the nuclei. A net attractive force or bonding interaction therefore results.

This interaction is shown in Fig. 3-16 for the H_2^+ ion. The full light lines (1) show the electron distributions in the $1s$ orbitals for each atom, ϕ_A^2 and ϕ_B^2. The light dash line (2) shows the simple average of these, $(\phi_A^2 + \phi_B^2)/2$. If these two orbitals are brought together with the same sign, they give a positive overlap and the electron density will be given by $(\phi_A + \phi_B)^2/2$. This is shown as line (3) which lies above line (2) throughout the region between the nuclei. In other words, the electron becomes concentrated between the nuclei where it is simultaneously attracted to both of them and the H_2^+ ion is more stable than $H^+ + H$ or $H + H^+$.

Clearly, in the case of negative overlap, shared electron density is reduced and internuclear repulsion increases. This causes a net repulsive or *antibonding* interaction between the atoms. This is also illustrated for H_2^+ in Fig. 3-16. The electron density distribution given by $(\phi_A - \phi_B)^2/2$ is shown by the solid curve (4). The electron density is now much lower everywhere between the nuclei, actually reaching zero at the midpoint.

When the net overlap is zero there is neither an increase nor a decrease in shared electron density and, therefore, neither a repulsive nor an attractive interaction. This situation is described as a *nonbonding* interaction.

Diatomic Molecules: H_2 and He_2

Once the sign and magnitude of the overlap between a particular pair of orbitals are known, the result, in terms of the energy of interaction, may be expressed in an energy-level diagram. This is best explained by using an example: the hydrogen molecule (H_2). Each atom has only one orbital, namely, its $1s$ orbital, which is stable enough to be used in bonding. Thus we examine the possible ways in which the two $1s$ orbitals ϕ_1 and ϕ_2 may overlap as two H atoms approach each other.

There are only two possibilities, as illustrated in Fig. 3-17. If the two $1s$ orbitals are combined with positive overlap, a bonding interaction results. The positively overlapping combination $\phi_1 + \phi_2$ can be regarded as an orbital in itself, which is called a *molecular orbital* (MO), and is denoted Ψ_b. The subscript b stands for *bonding*. Similarly, the negatively overlapping combination $\phi_1 - \phi_2$ also constitutes a molecular orbital Ψ_a, where the subscript a stands for antibonding.

Now imagine that two hydrogen atoms approach each other so that the molecular orbital Ψ_b is formed. An MO, like an atomic orbital, is subject to the exclusion principle, which means that it may be occupied by no more than two electrons, and then only if these two electrons have opposite spins. A bond will be formed if we assume that the two electrons present, one from each H atom, pair their spins, and occupy Ψ_b. The energy of the system will decrease as r, the internuclear distance, decreases following the curve labeled b in Fig. 3-18. At a certain value of the internuclear distance, r_e, the energy will reach a minimum and then begin to rise again, very steeply. At the minimum the attractive force due to the sharing of the electrons just balances the forces due to repulsions between particles of like charge. At shorter distances the repulsive forces increase very rapidly. It is this rapid increase in repulsive forces at short distances that causes the H_2 molecule (and all other molecules) to have a minimum energy at a particular internuclear distance and prevents the atoms from coalescing. This minimum energy, relative to the energy of the completely separated ($r = \infty$) atoms is called the bond energy and is denoted E_b in Fig. 3-18.

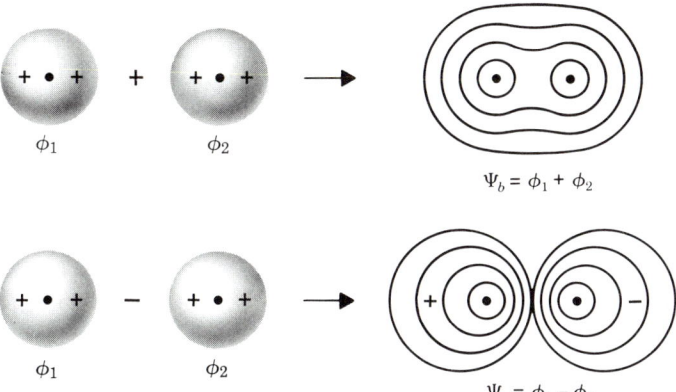

$\Psi_b = \phi_1 + \phi_2$

$\Psi_a = \phi_1 - \phi_2$

Figure 3-17   The $1s$ orbitals (ϕ_1 and ϕ_2) on two hydrogen or helium atoms may combine to form either a bonding MO, Ψ_b, or an antibonding MO, Ψ_a. The sign of Ψ_b is everywhere positive. The sign of Ψ_a changes between the nuclei; a nodal plane exists here because the value of Ψ_a is zero at the midpoint between the atoms.

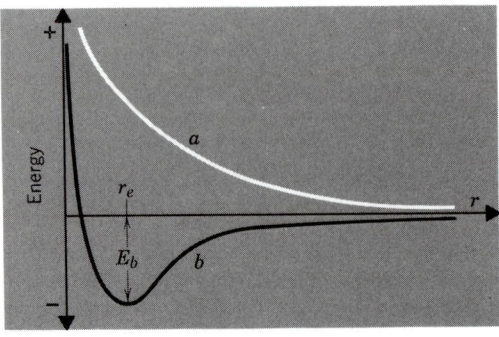

Figure 3-18   The variation of the energy associated with the antibonding orbital Ψ_a (a–the solid white curve) and the bonding orbital Ψ_b (b–the solid black curve) as a function of the distance, r, between the two atoms. The equilibrium internuclear distance, r, corresponds to the minimum in curve b. Here the stability associated with the bond is maxi-

Now, if the two H atoms approach each other so as to form the antibonding orbital Ψ_a with both electrons occupying that orbital, the energy of the system would vary, as shown in curve a. The energy would continuously increase, because at all values of r the interaction is repulsive.

We may now consider the possible formation of an He_2 molecule by using the same basic considerations, represented in Figs. 3-17 and 3-18, as for the H_2 molecule. Again, only the $1s$ orbitals are stable enough to be potentially useful in bonding. The He atom differs from the H atom in having two electrons, and this is crucial because in the He_2 molecule there are then four electrons. This means that Ψ_b and Ψ_a must each be occupied by an electron pair. Therefore, whatever stabilization results from the occupation of Ψ_b, it is offset (actually outweighed) by the antibonding effect of the electrons in Ψ_a. The result is that no net, appreciable bonding occurs and the He atoms are more stable apart than together.

Homonuclear Diatomics in General

The foregoing explanation of why H_2 is a stable molecule and He_2 is not, when coupled with the previous results concerning orbital overlaps, provides all the essential features needed to discuss the bonding in all homonuclear diatomic molecules. We shall explicitly consider the molecules that might be formed by the elements of the first short period, that is, Li_2, Be_2, . . . , F_2, Ne_2.

Before we do so, however, we introduce a different type of energy-level diagram from that in Fig. 3-18—one more suitable to molecules with many MO's. Instead of trying to represent the energy as a function of internuclear distance, we select one particular distance, r_e (or the estimated value thereof). The energies of the MO's at that distance are then shown in the center of the diagram. The energies of the atomic orbitals are shown for the separate atoms on each side of the diagram. The presence of electrons in the orbitals can then be represented by dots (or sometimes arrows). For H_2 and He_2 the appropriate diagrams are shown in Fig. 3-19.

Similar diagrams can be used when the two atomic orbitals are not of identical energy, in which case the appearance will be as shown in Fig. 3-20. Two important features must be emphasized for this case. (1) The more the two atomic orbitals differ in energy to begin with, the less they interact and the smaller are the potential bonding energies. (2) While the MO's Ψ_a and Ψ_b in Fig. 3-19 contain equal contributions from ϕ_1 and ϕ_2, this is not true when ϕ_1 and ϕ_2 differ in energy. In that case, Ψ_b has more ϕ_2 than ϕ_1 character while, conversely, Ψ_a has a preponderance of ϕ_1 character. When ϕ_1 and ϕ_2 differ very greatly in en-

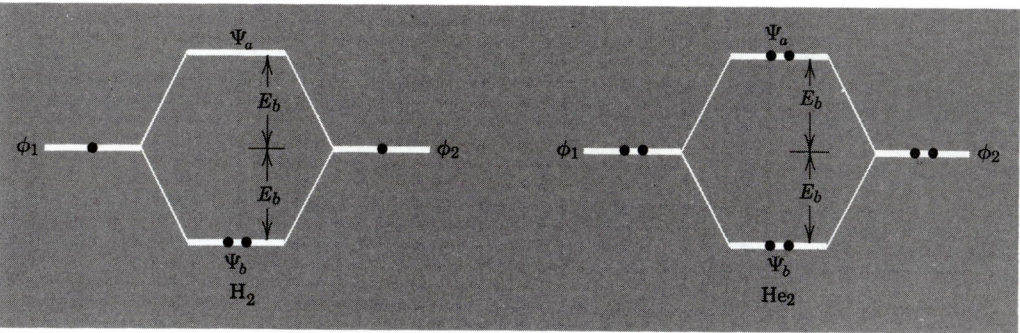

Figure 3-19 The MO energy-level diagrams for the H_2 and He_2 molecules. The orbitals marked ϕ_1 and ϕ_2 are the contributing $1s$ atomic orbitals on either two H or two He atoms. The MO's marked Ψ_a and Ψ_b correspond to those diagramed in Fig. 3-17.

ergy, the interaction becomes so small that Ψ_a is virtually identical in form and energy with ϕ_1 and Ψ_b with ϕ_2, as is shown in Fig. 3-20(b).

Diagrams of this type can be used to show the formation of bonding and antibonding MO's from any two atomic orbitals, or from two entire sets of atomic orbitals. We are interested here in the interactions of the entire set of $2s2p_x2p_y2p_z$ orbitals on one atom with the equivalent set on another.

If we define the internuclear axis as the z axis, we first note that only certain overlaps can be nonzero:

$2s$	with	$2s'$
$2s$	with	$2p'_z$
$2p_z$	with	$2s'$
$2p_z$	with	$2p'_z$
$2p_x$	with	$2p'_x$
$2p_y$	with	$2p'_y$

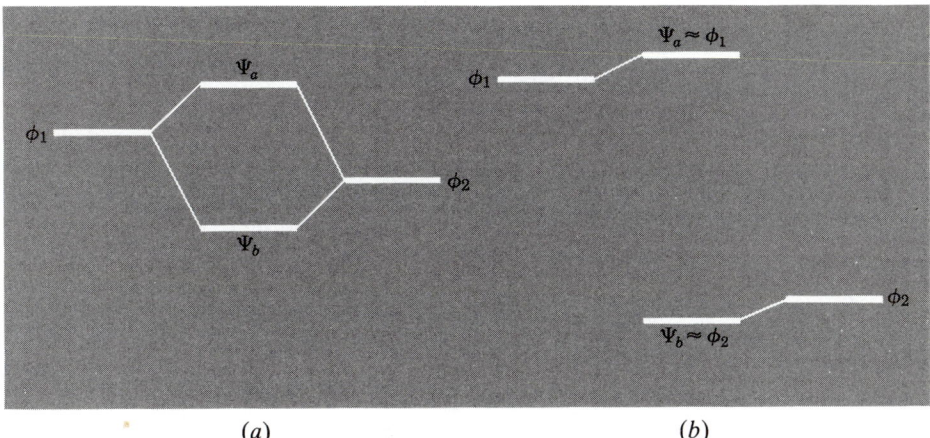

(a)   (b)

Figure 3-20 The MO energy-level diagrams for cases where the interacting atomic orbitals ϕ_1 and ϕ_2 initially differ in energy. In (b) the energy difference between ϕ_1 and ϕ_2 is so great that, even were the symmetry correct, little overlap is possible. As a result, the MO's are only slightly different in either energy or shape from the initial atomic orbitals.

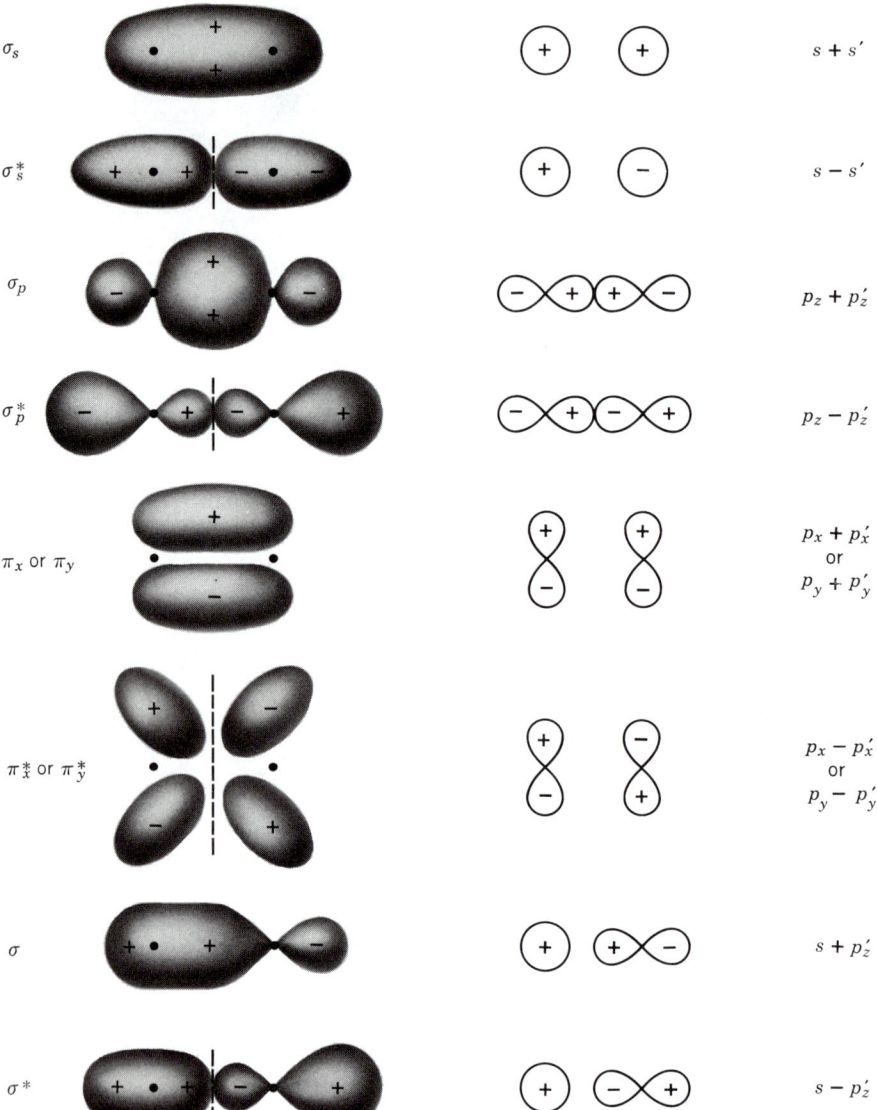

Figure 3-21   Diagrams showing each of the types of overlap that are important in a diatomic molecule. In the right-most column are given those algebraic combinations of orbitals on two adjacent atoms that lead to either bonding or antibonding MO's. Diagrams of these orbital combinations are given in the adjacent column. In each case, it is the z axis that is taken to be the internuclear axis and, by convention, the positive z direction for each atom is that which points towards the other atom. The sign for each lobe of an orbital is the sign of the original wave function, although the orbital is drawn from the square of the wave function. The algebraic sign for each combination in the right-hand column is chosen to give either a bonding or antibonding interaction. The resulting MO's are given the designations listed in the left-most column, where the σ and π notation conforms to that explained in the text, and * indicates an antibonding MO. The approximate shapes of the MO's are given by the shaded figures. Each antibonding MO is characterized by a nodal plane perpendicular to the internuclear axis, as indicated by the dashed lines.

All the remaining 10 (e.g., $2s$ with $2p'_x$, $2p_x$ with $2p'_y$, etc.) are rigorously zero and need not be further considered.

Figure 3-21 shows the overlaps just mentioned in more detail, and indicates how the resulting MO's are symbolized. The first four types of overlap, whether positive (to give a bonding MO) or negative (to give an antibonding MO) give rise to MO's that are designated σ. The $p_x \pm p_x$ and $p_y \pm p_y$ overlaps give rise to orbitals designated π. The last two, $s \pm p'_z$, also give σ molecular orbitals. The basis for this notation will now be explained.

σ, π, and δ Notation

If we view a MO between two atoms along the direction of the bond, that is, we look at it end-on, the following possibilities must be considered, as shown in Fig. 3-22.

(a) We shall see a wave function that has the same sign, either + or −, all the way around. In other words, as we trace a circle about the bond axis, no change in sign occurs throughout the entire circle. An MO of this kind is called a σ (sigma) MO. Such an MO can only be formed by overlap

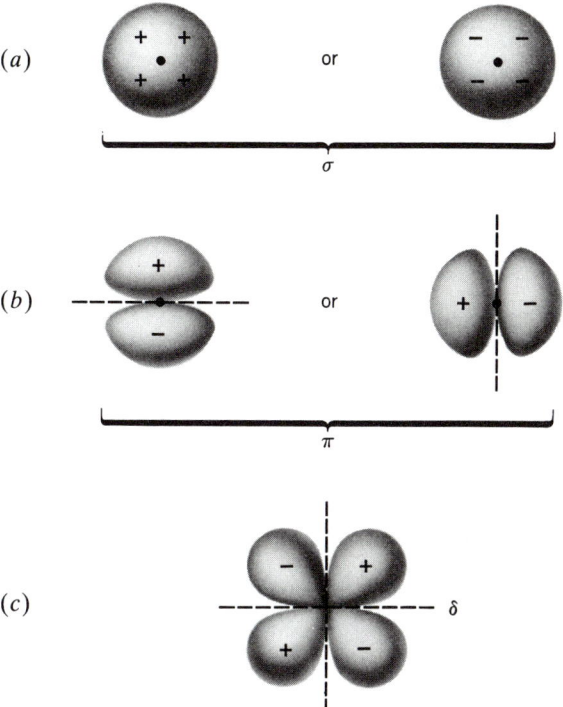

Figure 3-22 Characteristics of σ, π, and δ molecular orbitals as seen along (down) the internuclear axis, such that the first atom eclipses the second. As shown in the two examples of (a), σ molecular orbitals are not broken by any nodal planes that include the internuclear axis. Two examples of π molecular orbitals are shown in (b). These orbitals possess one plane that includes the internuclear axis. The δ-type MO of part (c) is formed by two d-type atomic orbitals placed face to face. These δ molecular orbitals possess two nodal planes that include the internuclear axis.

(either + or −) of two atomic orbitals that have the same property with respect to the axis in question. Thus these atomic orbitals can also be designated σ. Only the s and p_z orbitals in the sets we are using have this property. The symbol σ is used because σ is the letter s in the Greek alphabet, and a σ MO is analogous to an atomic s orbital, although it need not be formed from atomic s orbitals.

(b) We may see a wave function that is separated into two regions of opposite sign. With respect to the entire MO, there is a *nodal plane*. Precisely in this plane the wave function has an amplitude of zero, over the entire length of the bond. The symbol π, the Greek letter p, is used because this type of MO is analogous to an atomic p orbital. As shown in Fig. 3-21, it can be formed by overlap of two suitably oriented p orbitals. In the simple case of a diatomic molecule, or any other linear molecule, π orbitals always come in pairs because there are always two similar p orbitals, p_x and p_y, on each atom. The orbitals are equivalent to each other and thus two equivalent π bonding MO's and two equivalent π antibonding MO's are formed.

(c) Although we shall not encounter this possibility until much later, when we discuss certain transition metal compounds, there are MO's that have two nodal planes. These are called δ, the Greek letter d, orbitals. The δ molecular orbitals cannot be formed with s and p atomic orbitals, but the overlap of suitable atomic d orbitals (e.g., two d_{xy} or two $d_{x^2-y^2}$ orbitals) will form a δ molecular orbital.

Antibonding orbitals shall be designated with an asterisk: σ*, π*, and δ*.

The F_2 Molecule

We now consider energy-level diagrams for specific homonuclear diatomic molecules formed from the elements of row two in the periodic table. First we consider F_2. Each fluorine atom has the electron configuration $1s^2 2s^2 2p^5$. The $1s$ electrons are so close to the nucleus and so much lower in energy than the valence shell that they play no significant role in bonding; this is almost always true of so-called inner-shell electrons. Thus only the $2s$ and $2p$ orbitals and their electrons need be considered. (Recall, as well, that only valence electrons are considered in drawing a Lewis diagram.)

For a fluorine atom, the effective nuclear charge is high, and the energy difference between the $2s$ and $2p$ atomic orbitals is great. For this reason, in the F_2 molecule, the $2s$ orbital of one fluorine atom interacts only slightly with the $2p_z$ orbital on the other fluorine atom. The symmetry is proper for overlap, as shown in Fig. 3-21, but the energy difference between the two orbitals is so great that overlap is not effective. This is illustrated in Fig. 3-20(b). As a result, there is no contribution to bonding from interaction of these two orbitals. Thus only $2s$–$2s$, $2p_x$–$2p_x$, $2p_y$–$2p_y$, and $2p_z$–$2p_z$ interactions need be considered, and the diagram of Fig. 3-23 is obtained. The internuclear axis is the z axis.

In Fig. 3-23 the π and π* molecular orbitals are each doubly degenerate. These orbitals are formed by p_x to p_x and p_y to p_y overlap, so that the π molecular orbitals differ only in their orientation around the internuclear (z) axis. The overlap is positive for π_x $(2p_x + 2p_x)$ and negative for π_x^* $(2p_x - 2p_x)$, and similarly for π_y and π_y^*. The orbitals in Fig. 3-23 have the shapes designated in Fig. 3-21.

For F_2 there is a total of $(7 + 7) = 14$ valence electrons that must occupy these MO's in keeping with the *aufbau* principle, Hund's rule, and the Pauli exclusion

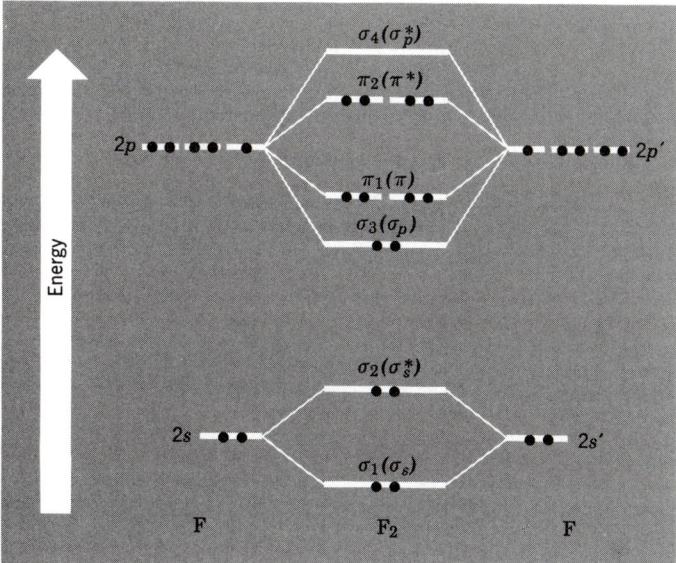

Figure 3-23   An MO diagram for the fluorine molecule (F_2). Atomic orbitals of each fluorine atom are listed on the left and the right. The MO's that result are listed in the center. The σ molecular orbitals are each singly degenerate, and are given the arbitrary designations $\sigma_1 \cdots \sigma_4$. The π_1 and π_2 levels are each doubly degenerate. The parenthetical MO designations correspond to those given in Fig. 3-21.

principle. By adding electrons in this fashion, we get the occupation shown in Fig. 3-23. For all pairs of electrons except those of σ_p, the stability gained for a bonding pair is offset by an antibonding pair of electrons. Hence, only the electron pair in σ_p gives a net bonding effect, and we conclude that the F_2 molecule has a single bond, in agreement with the Lewis diagram.

In general, **bond order** is defined in MO theory in just this way. If we take the number of electron pairs in bonding molecular orbitals (n_b) and subtract the number of pairs in antibonding molecular orbitals (n_a), we have the bond order, namely, $n_b - n_a$.

The Li_2 Molecule

The diagram is somewhat different for the Li_2 molecule because the $2s$ and $2p$ atomic orbital separation is smaller in the Li atom. Consequently, the $2s$ orbital of one atom is close in energy to the $2p$ orbital of the second atom. Their overlap cannot be ignored. The diagram that shows this is Fig. 3-24, where the internuclear axis is again taken to be the z axis. As a result of s to p'_z and p_z to s' interactions, the molecular orbitals σ_2 and σ_3 have both p_z and s character. Thus there is an upward displacement of σ_3 so that it lies above π_x and π_y. Although this has practically no importance for the stability of Li_2, it will become important as we proceed to molecules with more electrons. For Li_2 the two valence electrons occupy σ_1, and the bond order is one. It is a weak bond because the overlap of such diffuse $2s$ atomic orbitals is poor. A computer-drawn electron density map for this bonding electron pair, which is represented as $[\sigma_1]^2$, is shown in Fig. 3-25. This quantitative representation of the overlap should be compared with that of the purely schematic depictions of Figs. 3-17 and 3-21.

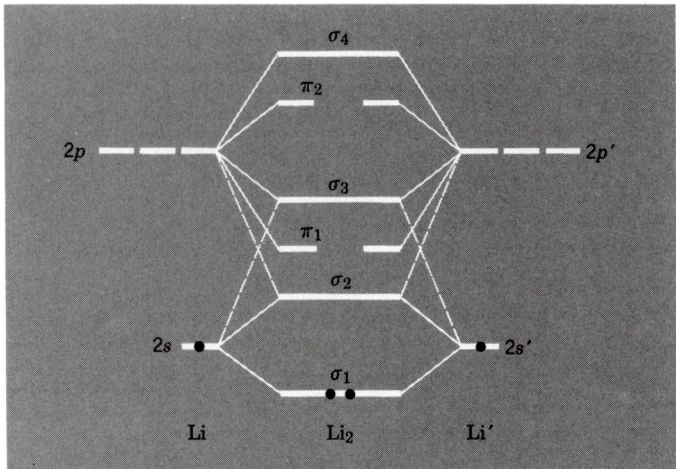

Figure 3-24   An MO diagram for the dilithium molecule
(Li_2). As in F_2, the σ molecular orbitals are singly degenerate,
and the π molecular orbitals are each doubly degenerate.
However, because of s–p mixing, as discussed in the text, the resulting MO's are not strictly those shown in Fig. 3-21.

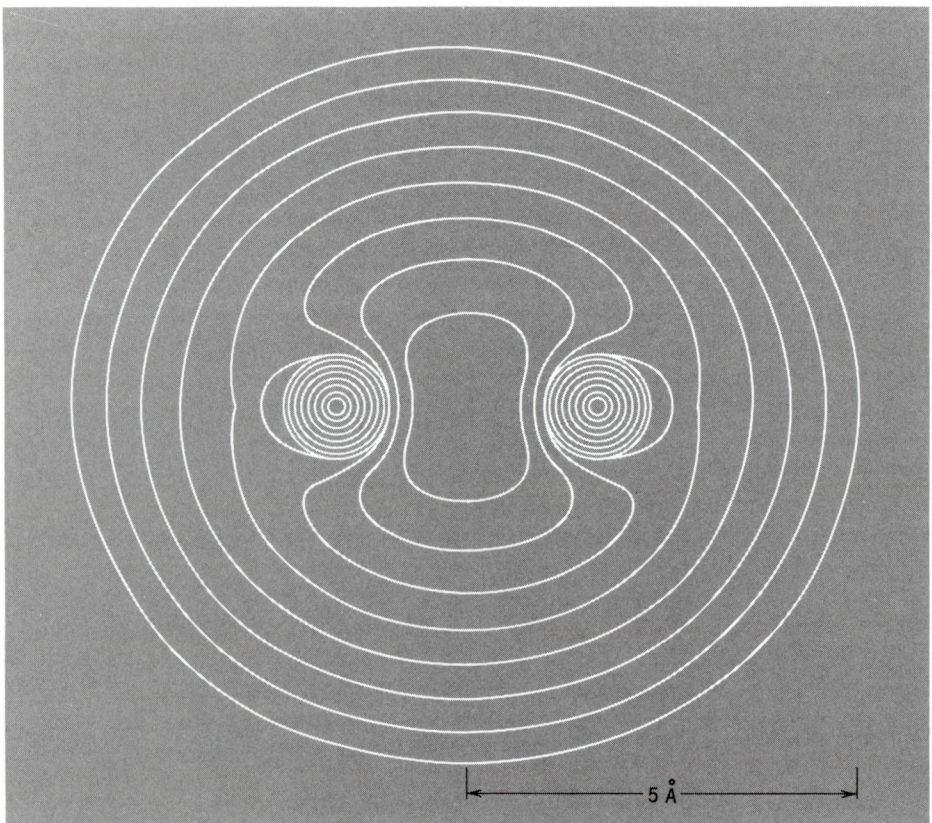

Figure 3-25   Electron density contours for the filled bonding molecular orbital, σ_1, in
Li_2. Each new contour line from the outside in represents an increase in electron density by a factor of two. The atoms are located at the positions of highest electron density.

The Complete Series

We can now consider the entire series of molecules from Li_2 to F_2. The progressive changes in orbital energies and electron populations from one extreme to the other are shown in Fig. 3-26, along with bond distances and energies. The dilithium molecule has the longest and weakest bond of all because it is only a single bond formed by overlap of two fairly diffuse $2s$ atomic orbitals. The Li atoms are large, and the effective nuclear charge is low.

The beryllium atom has the ground-state electron configuration $1s^2 2s^2$. Four valence electrons are to be considered for the Be_2 molecule, and these electrons are assigned as shown in Fig. 3-26. Because the bond order is zero, there is no stable Be_2 molecule.

For B_2 there are six electrons to occupy the MO's. The last two enter the doubly degenerate π_1 level according to Hund's rule. The B_2 molecule is, therefore, paramagnetic with two unpaired spins. The bond order is one because the σ_1 and σ_2 pairs cancel one another, leaving one net bond due to $[\pi_1]^2$. The bond distance is shorter and the bond energy is higher than in Li_2 because of the smaller size of the B atoms.

For C_2 the π_1 orbitals are only slightly lower in energy than σ_3, but they are low enough to give the ground-state electron configuration $[\sigma_1]^2[\sigma_2]^2[\pi_1]^4$, as shown in Fig. 3-26. The C_2 molecule is diamagnetic and has a considerably shorter and stronger bond than does B_2. The excited state $[\sigma_1]^2[\sigma_2]^2[\pi_1]^3[\sigma_3]^1$ for C_2 lies only about 10 kJ mol^{-1} above the ground state.

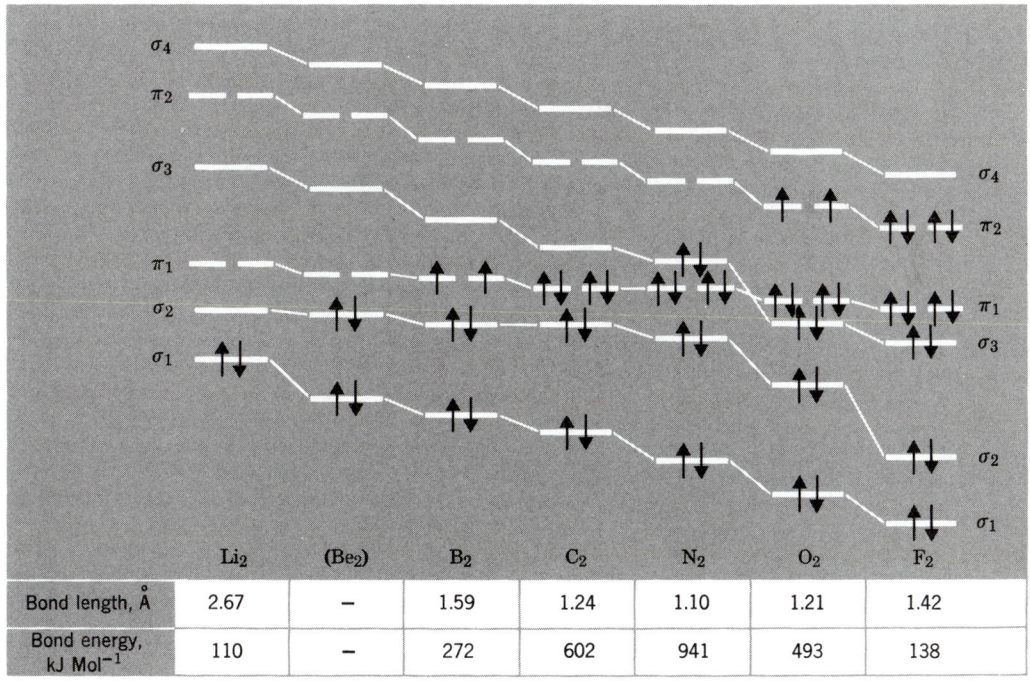

	Li_2	(Be_2)	B_2	C_2	N_2	O_2	F_2
Bond length, Å	2.67	–	1.59	1.24	1.10	1.21	1.42
Bond energy, kJ Mol^{-1}	110	–	272	602	941	493	138

Figure 3-26   The MO energy-level diagrams for the diatomic molecules from Li_2 to F_2 showing the changes in MO energies, electron configurations, bond lengths, and bond energies. For molecules with high effective nuclear charge (e.g., F_2, O_2, and perhaps N_2), the MO's are essentially those of Fig. 3-21, as designated in Fig. 3-23. For molecules with low effective nuclear charge, $s-p$ mixing is extensive, as discussed in the text.

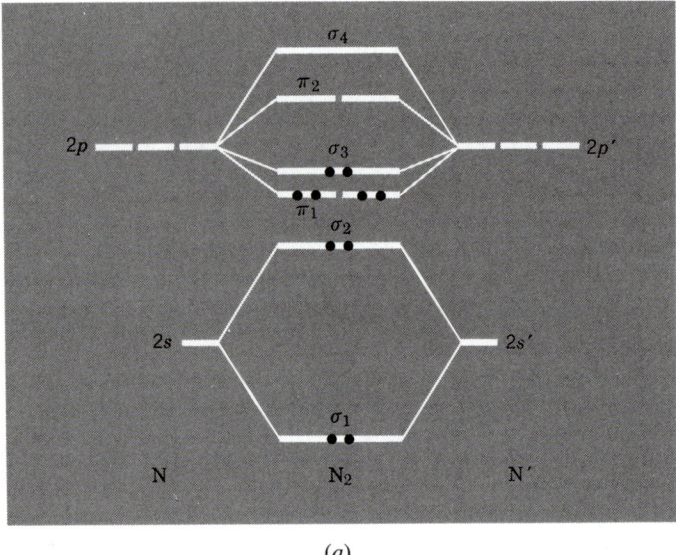

(a)

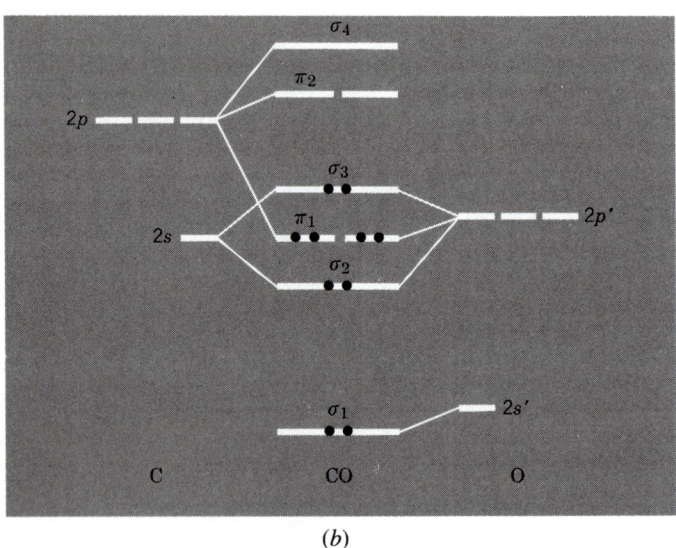

(b)

Figure 3-27   Molecular orbital diagrams for (*a*) the dinitrogen molecule and (*b*) the carbon monoxide molecule. Both have the same number of electrons, but the MO diagrams are different because of the different starting energies for the atomic orbitals of C and O.

The N_2 molecule has the highest bond order (three), the shortest bond, and the strongest bond of any molecule in the series. The bond order of three is in agreement with the Lewis diagram that has a triple bond, :N≡N:.

With the O_2 molecule, bond order and bond strength begin to decrease since, following N_2, only antibonding MO's remain to be occupied. For O_2, the two additional electrons enter the doubly degenerate π_2 level, which is antibonding. The bond order is two. The electrons in the π_2 level are unpaired, and this accords with the fact that the substance is paramagnetic with two unpaired electrons. The correct prediction of this by simple MO theory is in contrast with the difficulty of explaining it in Lewis terms. The Lewis approach correctly requires a double bond, but not the presence of two unpaired electrons.

The Ne_2 molecule is not stable, and the reason for this is clear. The bond order would be zero, because all MO's through σ_4 in Fig. 3-26 would be filled by the 16 valence electrons of two Ne atoms.

Heteronuclear Diatomic Molecules

The extension of the MO method for homonuclear diatomic molecules to include heteronuclear diatomic molecules, such as CO and NO, is not difficult. It depends on making allowance for the different effective nuclear charges of the two atoms. This is shown in Fig. 3-27 where the isoelectronic molecules N_2 and CO are contrasted. There are two important features to be noted in this comparison. First, all orbitals of the oxygen atom lie at lower energies than the corresponding ones of the carbon atom, because oxygen has the higher effective nuclear charge. This finding is in keeping with Fig. 2-14, which indicates that the first ionization enthalpy of O is several hundred kilojoules per mole greater than that of C. Second, the $2s$–$2p$ energy separation is greater for O than for C. The resulting MO diagram for CO [Fig. 3-27(b)] emphasizes the overlap of the carbon $2s$ atomic orbital with that atomic orbital of O closest to it in energy, the oxygen $2p$ atomic orbital. This s–p mixing is not prominent in the MO diagram for N_2, Fig. 3-27(a).

For these reasons, the MO's for CO are significantly different from those for N_2. The highest filled MO for N_2 is σ_3 of Fig. 3-27(a). This orbital is essentially σ_p of Fig. 3-21. Because it is an orbital of high bonding character, loss of an electron (to form N_2^+) weakens the N—N bond. In CO, however, the highest filled MO [σ_3 of Fig. 3-27(b)] is slightly antibonding in character. Hence, the CO$^+$ ion has a slightly stronger bond than does CO.

Another important heteronuclear diatomic molecule is nitric oxide (NO). Since N and O differ by only one atomic number, the energy-level diagram for NO is rather similar to that of N_2. The additional electron of NO must occupy the antibonding π_2 orbital of Fig. 3-27(a). Because π_2 is antibonding, the last electron of NO is easily removed to form NO$^+$, which then has a stronger bond than the neutral NO. The electronic structure of NO might equally well have been derived qualitatively by removing one electron from the configuration of the O_2 molecule.

3-6 Molecular Orbital Theory for Polyatomic Molecules

Linear Triatomics: BeH$_2$

The MO method can be generalized to larger molecules. To illustrate, let us consider the simplest linear triatomic molecule BeH$_2$. Let us choose the z axis as the molecular axis. We first note that only σ molecular orbitals can be formed because the hydrogen atoms have only their $1s$ orbitals to use in bonding. These orbitals are themselves of σ character with respect to any axis that passes through the nucleus, and therefore they can contribute only to σ molecular orbitals. Then, on the Be atom, only the $2s$ and $2p_z$ orbitals can participate in bonding. The p_x and p_y orbitals, which have π character and zero overlap with any σ orbital, will not play a role in bonding in the BeH_2 molecule.

The $2s$ orbital of beryllium can combine with the two $1s$ orbitals of the hydrogen atoms to form bonding and antibonding MO's, as is shown in Fig. 3-28.

The $2p_z$ orbital of beryllium also combines with the hydrogen $1s$ orbitals, as is shown in Fig. 3-28, to form bonding and antibonding σ molecular orbitals. In these, the $1s$ orbitals are *out of phase* with each other.

The important points to remember about these four σ molecular orbitals are the following:

1. In each bonding MO, electron density is large and continuous between adjacent atoms, while in the antibonding MO's there is a node between each adjacent pair of nuclei.
2. In each bonding MO, the wave function indicates that an electron pair occupying it is *spread out* over the whole molecule, and is shared by all of the atoms, not just a particular adjacent pair. In other words, in MO's electrons are *delocalized* over the whole extent of the MO.

The MO treatment of the bonding in BeH_2 can be expressed in terms of an energy-level diagram, as shown in Fig. 3-29. The main features here are that the

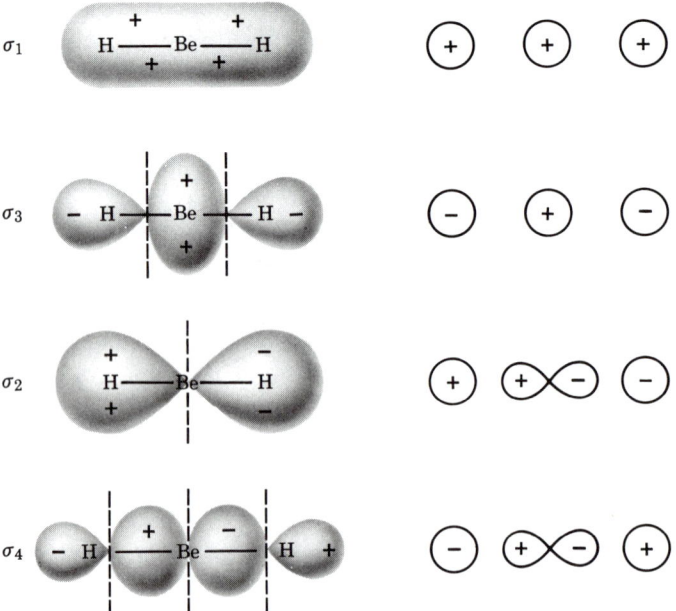

Figure 3-28 The four σ molecular orbitals for the linear BeH_2. The dashed vertical lines are nodal planes perpendicular to the molecular axis. The notation for orbitals $\sigma_1 \cdots \sigma_4$ conforms to that used in Fig. 3-29. Those p atomic orbitals of Be that are perpendicular to the molecular axis have nonbonding interactions with the hydrogen $1s$ orbitals.

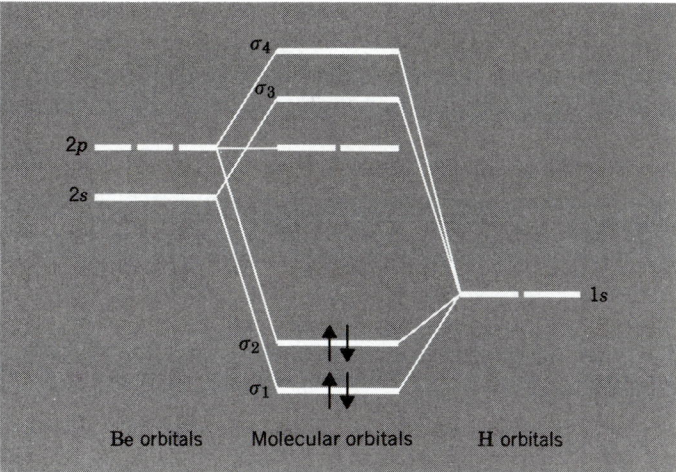

Figure 3-29 An MO energy-level diagram for BeH_2. The MO designations σ_1–σ_4 correspond to those of Fig. 3-28. The two atomic p orbitals of Be that are unchanged in energy lie at right angles to the internuclear axis and have nonbonding interactions with the $1s$ atomic orbitals of the hydrogen atoms.

hydrogen $1s$ orbitals lie at much lower energy (~400 kJ mol^{-1}) than the beryllium $2s$ orbital and that the p_x and p_y orbitals of Be carry over completely unchanged into the center column, because they do not overlap with any other orbitals. The four valence electrons, $2s^2$ from Be and $1s$ from each H, occupy σ_1 and σ_2. The total bond order of the Be—H bonds is two. Since each Be—H pair participates equally in the molecule, this is equivalent to saying that there are two equivalent B—H single bonds.

Trigonal Planar Molecules: AB$_3$

A particularly important and more general application of MO theory in polyatomic molecules deals with π bonding in planar systems. One important group is the symmetrical compounds of the general formula AB_3. Examples include BF_3, CO_3^{2-}, and NO_3^-. If these trigonal planar systems are oriented so that the central atom is at the origin of the coordinate system and the molecular or ionic plane coincides with the xy plane, then the π-bond system will be formed entirely by the p_z atomic orbitals of the four atoms. The π-bond system must then have a node in the xy plane. It must also be equally dispersed over the three A—B bonds, in agreement with the resonance result discussed previously. Consistent with this, the MO approach involves overlap that encompasses the whole structure.

　　There are three linear combinations of p_z atomic orbitals from the separate peripheral B atoms of AB_3 which are important to the discussion. We shall not discuss the methods that are used to deduce these particular linear combinations, except to point out that these three particular linear combinations have been chosen to provide the best total, positive overlap with the atomic p_z orbital of the central atom A. These three linear combinations are termed **group or-**

bitals, as illustrated in Fig. 3-30. The linear combinations [group orbitals (GO)] represented in Fig. 3-30 arise from three distinct arrangements of the p_z atomic orbitals at the vertices of the AB_3 triangle. Only one group orbital of Fig. 3-30 has non-zero overlap with the p_z atomic orbital of the central atom A as shown in Fig. 3-31. The other two interactions are nonbonding (e.g., π_{2a} and π_{2b}) and the resulting MO's appear in the energy-level diagram of Fig. 3-32 with energies that are unchanged. One of the interactions shown in Fig. 3-31 is bonding, and this leads to the π_1 molecular orbital shown in Fig. 3-32. Its antibonding counterpart is listed in Fig. 3-32 as π_1^*.

For each of the species BF_3, CO_3^{2-}, or NO_3^-, there are only 6 electrons to occupy the π molecular orbitals of Fig. 3-32. (The other 18 valence electrons occupy the various σ orbitals of the xy plane.) The six π electrons are distributed as is shown in Fig. 3-32. The 4 electrons in the two degenerate orbitals (π_{2a} and π_{2b}) neither contribute to nor detract from the stability of the π-bond system because they are nonbonding. Thus, the π-bonding stability is provided entirely by the

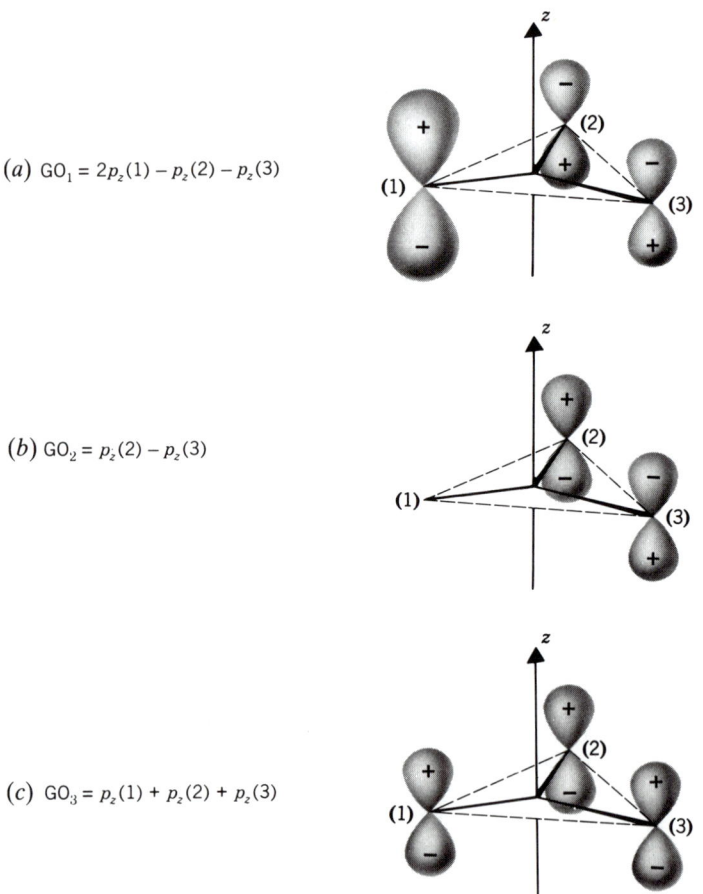

(a) $GO_1 = 2p_z(1) - p_z(2) - p_z(3)$

(b) $GO_2 = p_z(2) - p_z(3)$

(c) $GO_3 = p_z(1) + p_z(2) + p_z(3)$

Figure 3-30   Three combinations of p_z orbitals from the three outer B atoms of a planar AB_3 molecule. Each combination (called a group orbital, GO) is multicentered, and each is constructed to be used as a group in overlapping with the p_z atomic orbital of the central atom A, as shown in Fig. 3-31.

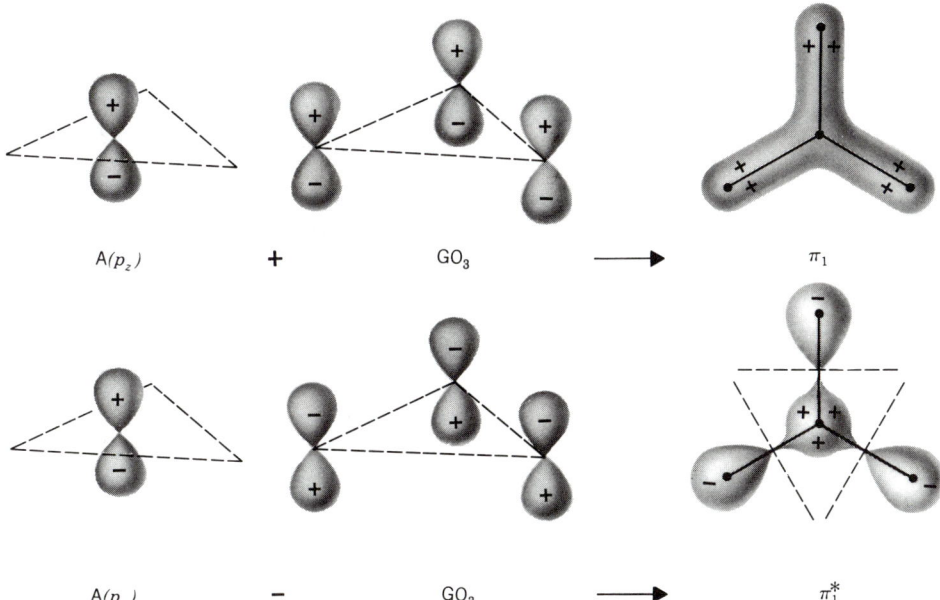

Figure 3-31 Diagrams showing how bonding (π_1) and antibonding (π_1^*) molecular orbitals are formed in an AB_3 molecule using the p_z orbital of the central atom A and a group orbital (GO_3 of Fig. 3-30) from the outer atoms B. The MO's themselves, at the right, are viewed from above. The MO's change sign in the molecular plane, as do the p_z atomic orbitals from which they are formed. In addition, π_1^* has three nodal planes perpendicular to the molecular plane.

one electron pair in the π_1 molecular orbital. The total π bond order of one is equally distributed over the three equivalent AB regions so that the net π bond per AB group is one-third. The conclusion here is that one π bond is delocalized over three AB atom pairs. The same conclusion was reached previously through the concept of resonance.

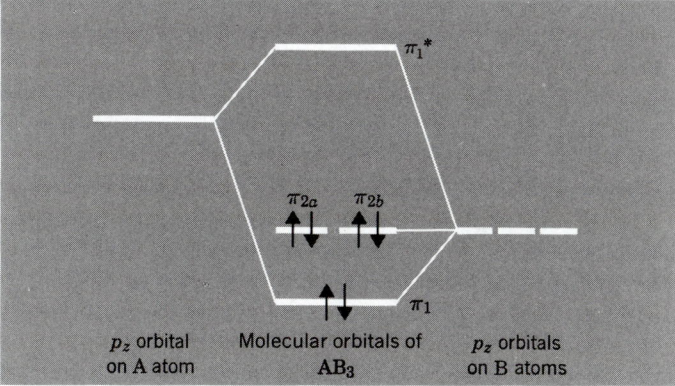

Figure 3-32 The MO energy-level diagram for the π-bond system in a planar, symmetrical AB_3 molecule. The symbols π_1 and π_1^* correspond to those used in Fig. 3-31. The nonbonding orbitals π_{2a} and π_{2b} are essentially GO_1 and GO_2 of Fig. 3-30.

3-7   Multicenter Bonding in Electron Deficient Molecules

In some molecules, there are not enough electrons to allow at least one electron-pair bond between each adjacent pair of atoms. Examples of molecules displaying this type of electron deficiency are shown in Structures 3-XIV and 3-XV.

3-XIV   3-XV

In Structure 3-XIV and in the Al_2C_6 skeleton of Structure 3-XV there are eight adjacent pairs of atoms, but there are only six pairs of electrons available for bonding. Eight bonds are required for the normal distribution of two-center, two-electron bonds. Clearly, this is not possible for Structures 3-XIV and 3-XV. [Note that Structures 3-XIV and 3-XV are the actual structures for molecules with empirical formulas BH_3 and $Al(CH_3)_3$.]

Both Structures 3-XIV and 3-XV present the same problem for a bonding description. We shall concentrate on Structure 3-XIV, since it is less cumbersome. We could try to account for Structure 3-XIV by invoking a resonance description, namely, the canonical forms 3-XIVa and 3-XIVb.

3-XIVa   3-XIVb

This would imply that, in each B- - - -H- - - -B bridge, one electron pair is shared between or distributed over two B- - - -H bonds, giving each bridging BH group a bond order of one half. The electron deficiency is obvious, but the lack of formal bonds in each resonance form seems somewhat artificial. The remaining B—H bonds of the terminal BH groups are adequately described as normal two-centered, two-electron bonds. An analogous description could be used for the central Al—C bonds in Structure 3-XV.

Another way to describe the bonding in Structure 3-XIV is to use an MO treatment that encompasses only the bridging system. The terminal BH groups are handled separately as localized electron-pair bonds, so that within each starting BH_2 unit there are ordinary B—H bonds formed using sp^3 hybrid orbitals on the B atoms. If these two BH_2 units are brought together, as shown in Fig. 3-33(a), so as to make the $H_2B \cdots BH_2$ sets of atoms coplanar, the remaining two sp^3 hybrid orbitals on each B atom point toward each other. Now, if the remaining two hydrogen atoms are placed in their proper bridging positions, as shown in Fig. 3-33(b), each of the $1s$ atomic orbitals of these hydrogen atoms overlaps with two of the sp^3 orbitals from the B atoms. In this way an orbital is formed that extends over each B- - - -H- - - -B unit. There are two such orbitals. Each is three centered, has no nodes, and is, therefore, capable of bonding all three atoms together. Since each boron atom and each bridging hydrogen atom supplies one electron, there are four electrons to be distributed into the two

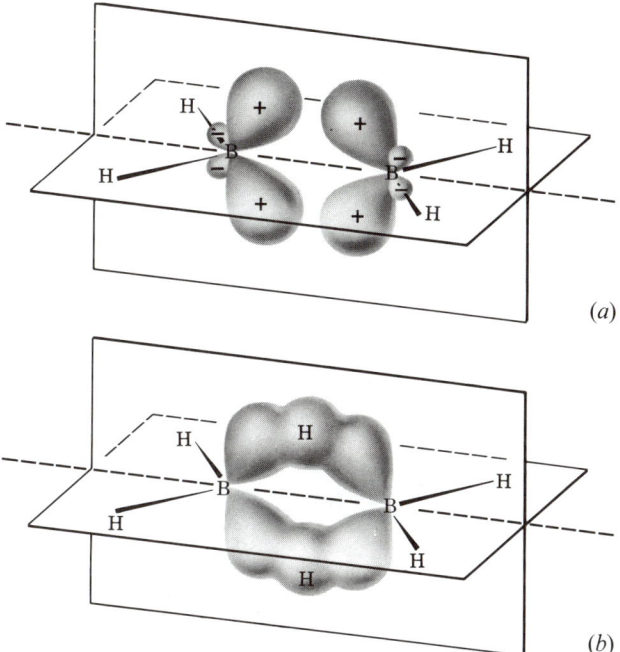

Figure 3-33   The formation of 3c–2e bonds in B_2H_6. The orientation of two coplanar BH_2 groups, with sp^3 hybrids on B atoms, is shown in (*a*). When the bridging H atoms are placed as in (*b*), continuous overlap within each B—H—B arch results in two separate 3c–2e bonds.

three-centered orbitals. Thus one electron pair can be used for each three-centered orbital. In this way we establish a type of bond called a three-center, two-electron bond, abbreviated 3c–2e. Since one electron pair is shared between three atoms instead of two, 3c–2e bonds have about one-half the strength of the normal two-center, two-electron (2c–2e) bond. This is equivalent to the bond order of one-half obtained in the resonance treatment.

To appreciate and utilize more fully the concept of 3c–2e bonding, it is necessary to examine it in more detail. Suppose we consider only the sp^3 hybrid orbital on each B atom and the $1s$ orbital of the bridging H atom. These three atomic orbitals can be combined into three MO's, as shown in Fig. 3-34. One of these (Ψ_b) is a bonding orbital; it is the same one already discussed. There is also an antibonding orbital Ψ_a, which has a node between each adjacent pair of atoms. The third orbital Ψ_n has the signs of the two sp^3 orbitals out of phase and cannot have any net overlap with the hydrogen $1s$ orbital. It is a *nonbonding* orbital.

We can now draw an energy-level diagram that expresses these results, as shown in Fig. 3-35. By placing an electron pair in Ψ_b, the bonding MO, we have a complete picture of the 3c–2e bonding situation.

In the case of $Al_2(CH_3)_6$, Structure 3-XV, the 3c–2e bridge bonding can be described in a very similar fashion. Each Al atom provides sp^3 hybrid orbitals, as do the boron atoms in B_2H_6. Instead of the $1s$ orbital of the H atom, we now have the large positive lobe of a carbon sp^3 orbital at the center.

$\Psi_a = sp^3(B_1) - 1s(H) + sp^3(B_2)$

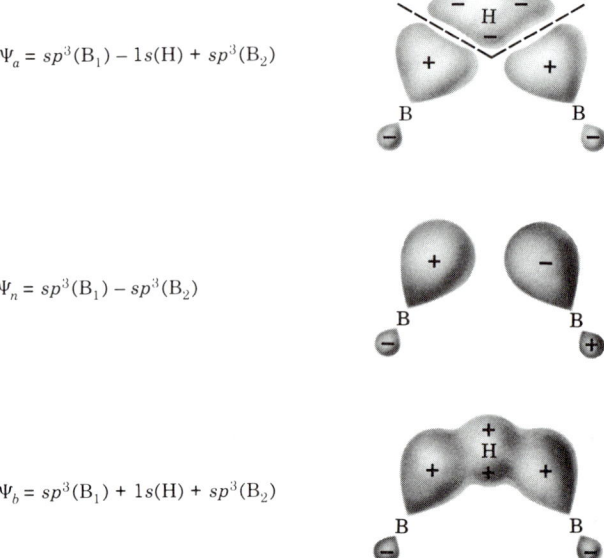

$\Psi_n = sp^3(B_1) - sp^3(B_2)$

$\Psi_b = sp^3(B_1) + 1s(H) + sp^3(B_2)$

Figure 3-34   The formation of three distinct three-center MO's in a B—H—B bridge system.

The energy-level diagram in Fig. 3-35 can also be applied to the interesting case of the three-center, four-electron (3c–4e) bond. In the FHF⁻ ion, which is symmetrical (although most hydrogen bonds are weaker and unsymmetrical), each F atom supplies a σ orbital and an electron pair. Thus a set of orbitals essentially similar to that in the BHB system is used, and an energy-level diagram, which is essentially like that in Fig. 3-35, is applicable. However, there are now two electron pairs. One pair occupies Ψ_b and the other Ψ_n. The pair in Ψ_n has no significant effect on the bonding because Ψ_n is a nonbonding orbital. The net result is that here, too, the bond orders are one half.

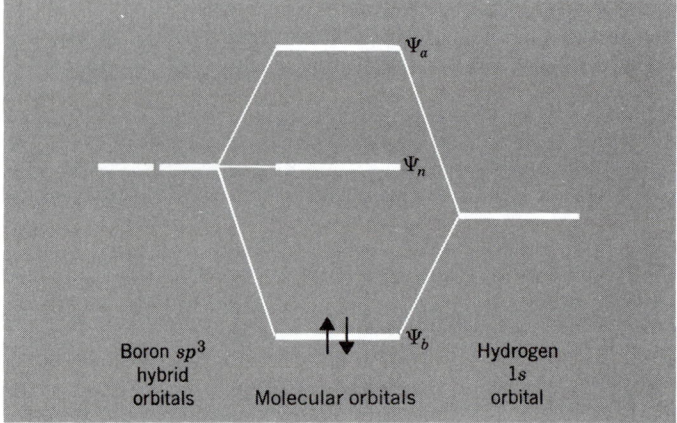

Figure 3-35   An energy-level diagram for the three MO's of Fig. 3-34 that are formed in a three-center B—H—B bridge bond of B_2H_6.

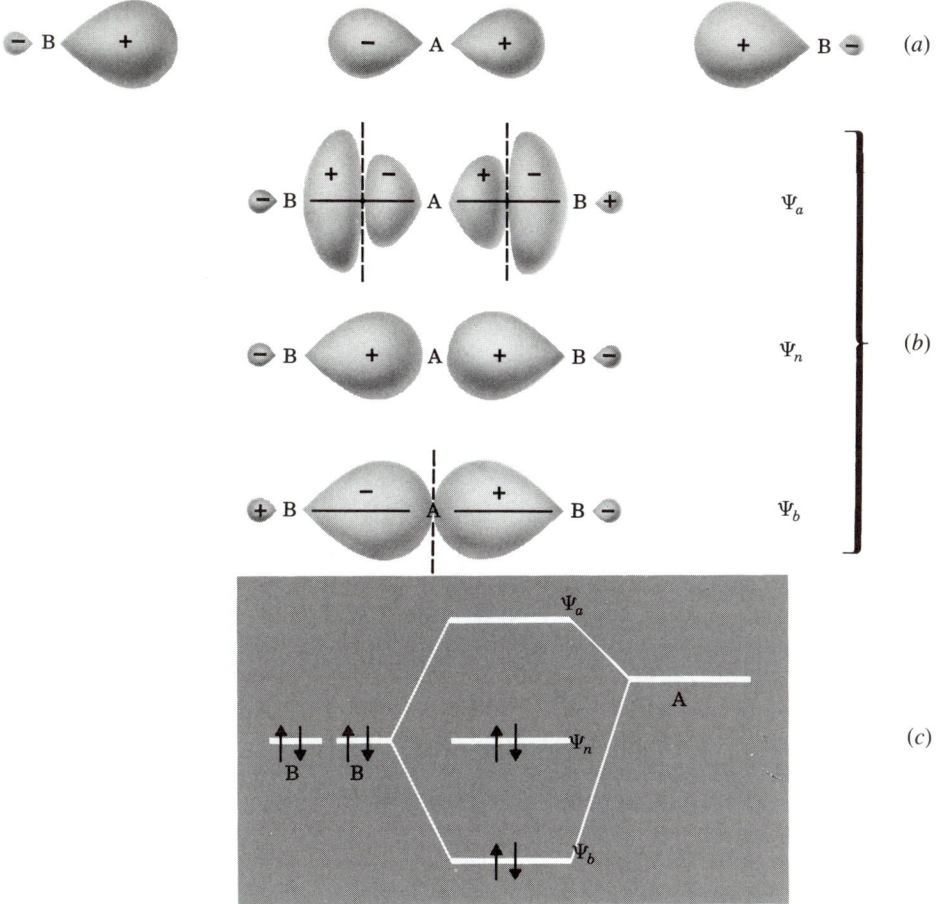

Figure 3-36   The formation of three-center orbitals in a B—A—B system, where the central atom A uses a p orbital. The orbitals that are used are shown in (a). The shapes of the MO's formed are shown in (b). An energy-level diagram showing the occupation of the orbitals for a 3c–4e bond is shown in (c).

One other type of (3c–4e) bonding must also be discussed since it is essential to the discussion of molecular shapes. Suppose we have a set of three atoms, B—A—B, most probably linear but possibly bent to some extent, such that the central atom uses a p orbital rather than an s orbital. The situation is shown in Fig. 3-36(a). Again, it is possible to form three multicenter orbitals, as shown in Fig. 3-36(b). The result turns out to be very similar to that already seen where the central atom uses an s orbital, in that bonding Ψ_b, nonbonding Ψ_n, and antibonding Ψ_a orbitals are formed and the energy-level diagram is analogous, as is shown in Fig. 3-36(c).

The interesting result, as seen in either Fig. 3-35 or 3-36, is that even if two electron pairs are available, the A—B bonds will have orders of only one-half, because one electron pair occupies the nonbonding orbital Ψ_n. Here we are dealing with an orbitally deficient system rather than an electron deficient one. If the central atom in either case had an additional σ-type atomic orbital the system would be equivalent to that in BeH_2 and two bonds, each of order one, could be formed.

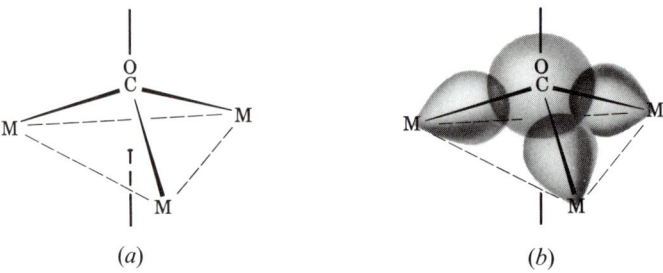

Figure 3-37   (*a*) The orientation of a CO group over three metal atoms as found in some metal carbonyl compounds. (*b*) The overlap of three metal atom orbitals with themselves and with the σ_3 orbital [see Fig. 3-27(*b*)] of CO to form a four-center orbital. It is this orbital (and the electron pair that occupies it) that is responsible for binding of the CO group in this "capping" position over the three metal atoms. The electron pair originally filled σ_3 of the CO molecule.

Multicenter bonding can occur in larger groups of atoms. There are, for example, compounds in which a single CO ligand lies perpendicularly over the center of a triangular set of metal atoms, as shown in Fig. 3-37(*a*). In such cases, the best and simplest way to describe the bonding is in terms of a four-center, two-electron (4c–2e) bond. If one orbital on each metal atom is directed toward the carbon lone-pair orbital of CO [σ_3 in Fig. 3-27(*b*)], there will be mutual overlap of all four orbitals, as shown in Fig. 3-37(*b*), and the resulting four-center orbital will be occupied by the electron pair initially in σ_3 of the CO molecule.

STUDY GUIDE

Scope and Purpose

A brief and qualitative introduction to the twin subjects of structure and bonding has been given. These will be important topics in subsequent discussions of the reactions and properties of compounds. Bonding theory and molecular structure should not be taken as ends in themselves, but only as important tools in understanding the actual properties and reactivities of chemical compounds. Of the Study Questions, those marked "A. Review" should be used by the student as a self-study guide to mastery of the material in the text. More challenging exercises are listed under "B. Additional Exercises." Questions marked "C. Questions from the Literature of Inorganic Chemistry," require the use of specific journal articles.

Study Questions

A. Review

1. Why are the sign and magnitude of overlap between orbitals on adjacent atoms good indications of whether and how strongly the atoms are bonded?

2. Show with drawings how an *s* orbital, each of the three *p* orbitals, and each of the five *d* orbitals on one atom would overlap with the *s* orbital, one of the *p* orbitals, and any

two of the d orbitals on another atom close to it. Characterize each overlap as positive, negative, or exactly zero.

3. Draw an energy-level diagram for the interaction of two atoms each with an s orbital. Show how the MO's would be occupied if the two atoms in question were H atoms and if they were He atoms. What conclusions are to be drawn about the formation of a bond in each case?

4. When a bond is formed between two atoms, they are drawn together. What limits their internuclear distance so that they do not coalesce?

5. State the defining characteristics of σ, π, and δ molecular orbitals.

6. What is meant by a node? A nodal plane?

7. How is bond order defined for a diatomic molecule in MO theory?

8. Show with an energy-level diagram why the C_2 molecule has a bond order of 2 and no unpaired electrons, but has a low-lying excited state in which there are two unpaired electrons.

9. Show how the electronic structure of the NO molecule can be inferred from that of O_2. Explain why NO^+ has a stronger bond than NO itself.

10. True or False: The set of valence shell orbitals ($2s$, $2p$) for N are of higher energy than those for C. Explain the reason for your answer.

11. Write the electron configurations for the ground states and the valence states of Be, B, C, and N atoms, so that each one can form the maximum number of 2c–2e (ordinary electron-pair) bonds.

12. What are the three important types of hybrid orbitals that can be formed by an atom with only s and p orbitals in its valence shell? Describe the molecular geometry that each of these produces.

13. State the geometric arrangement of bonds produced by each of the following sets of hybrid orbitals: dsp^2, d^2sp^3, dsp^3. For each one state explicitly which d and p orbitals are required for each geometric arrangement.

14. Explain in detail, using both the MO approach and the resonance theory, why the NO bonds in NO_3^- have a bond order of $1\frac{1}{3}$.

15. Why is the use of hybrid orbitals preferable to the use of single atomic orbitals in forming bonds? Illustrate.

16. What does the term "electron deficient molecule" mean?

17. Why does B_2H_6 not have the same kind of structure as C_2H_6? Draw the structure that B_2H_6 does have and describe the nature of the two sorts of BH bonds therein.

18. Using the VSEPR model, predict the structures of the following ions and molecules: BeF_2, CH_2, OF_2, PCl_4^+, SO_2, ClF_2^+, BrF_3, BrF_5, SbF_5, ICl_4^-.

19. Why are the Kr $\cdots$ Kr and intermolecular Br $\cdots$ Br distances in the solid forms of the two elements practically identical? How would you expect the Br $\cdots$ Br distances in solid CBr_4 to be related to the above distances?

20. Although substances such as $OPCl_3$, SO_3^{2-}, and ClO_4^- are electronically saturated, the X—O bond lengths found in these compounds are shorter than would be expected for purely single bonds. Explain how P—O, S—O, or Cl—O double-bond character can arise in these three examples.

21. Explain why there is no $2s$–$2p$ mixing in the MO energy-level diagram for F_2, whereas $2s$–$2p$ mixing is evident in the MO energy-level diagram for Li_2.

B. Additional Exercises

1. For the series of diatomics O_2^+, O_2, O_2^-, and O_2^{2-}, determine from an MO energy-level diagram how the bond lengths will vary and how many unpaired electrons each should have.

2. The ionization enthalpies for H and F are given in Fig. 2-14. Draw an MO energy-level diagram for the HF molecule. How is the polarity of the molecule indicated in this diagram?

3. Draw Lewis diagrams and predict the structures of $(CH_3)_2S$ (dimethyl sulfide) and $(CH_3)_2SO$ (dimethyl sulfoxide). How will the CSC bond angles differ?

4. Draw Lewis diagrams for each of the series OCO, NNO, ONO⁻, NCO⁻, NNN⁻, and NCN²⁻. What is the same in all of these systems?

5. Describe the bonding in Al_2Br_6.

6. Determine the hybridization for the central atom in $OPCl_3$, OSF_4, and OIF_5. Predict the fine points of geometry using VSEPR theory.

7. Predict the geometry of gaseous GeF_2, and explain your reasoning.

8. Choose a reasonable geometry for seven-coordinate iodine in IF_7.

9. Draw Lewis diagrams for O_3 and SO_2. For ozone, $\angle OOO = 117°$. For sulfur dioxide, $\angle OSO = 120°$. Explain.

10. The molecules CO_2, $HgCl_2$, and $(CN)_2$ are linear. Draw Lewis diagrams and assign hybridizations for each atom.

11. Consider the series CO_3^{2-}, NO_3^-, and SO_3. What geometry do you predict for each?

12. Construct an MO energy-level diagram for NO, NO^+, and NO^-. Determine the bond order and the number of unpaired electrons in each.

13. Describe the geometry and the hybridization in $[PtCl_6]^{2-}$.

14. Draw a qualitatively correct energy-level diagram for the CO_2 molecule. Show that it accounts correctly for the presence of double bonds.

15. Sketch the π bonds found in NO_3^-, SO_3, and NO^+.

16. From among the following molecules and ions:

Al_2Cl_6	$SnCl_2$	SO_3^{2-}	BF_3
B_2H_6	SO_2	$OPCl_3$	PF_6^-

list the ones that (a) are coordinatively saturated, (b) are electronically unsaturated, (c) contain bridging atoms, and (d) are electron deficient.

17. In the following series of electronically saturated and isoelectronic ions, the observed X—O bond distances have been determined:

Ion	X—O Bond Length (Å)
SiO_4^{4-}	1.63
PO_4^{3-}	1.54
SO_4^{2-}	1.49
ClO_4^-	1.46

Compare these data to the proper sum of radii from Fig. 2-15 to determine which of the above ions has significant $p\pi–d\pi$ bond character.

18. Use VSEPR theory to predict the bond angles in each of the following:

(a) OF_2
(b) SO_2
(c) ClF_2^+
(d) BrF_3
(e) BrF_5
(f) SbF_5
(k) XeF_2
(l) BF_3
(m) NO_3^-
(n) NO_2^-
(o) NO_2^+
(p) PCl_4^+

(g) ICl_4^- (q) PF_6^-

(h) $OPCl_3$ (r) PCl_3

(i) OSF_4 (s) PCl_5

(j) OIF_5

19. Identify the molecules and ions in the preceding question that are

(a) Electronically saturated.

(b) Coordinatively saturated.

C. Questions from the Literature of Inorganic Chemistry

1. The structure of the pentafluorotellurate monoanion has been determined by X-ray techniques. (See S. H. Mastin, R. R. Ryan, and L. B. Asprey, *Inorg. Chem.*, **1970,** *9,* 2100–2103.) What is the oxidation state of Te in this anion? Draw the Lewis diagram for the anion. Determine the occupancy and the formula AB_xE_y for Te, and explain any deviations from ideal geometry using VSEPR theory.

2. Consider the compound $(CH_3)_3SnCl \cdot 2,6\text{-}(CH_3)_2C_5H_3NO$, whose structure has been determined by X-ray techniques. (See A. L. Rheingold, S. W. Ng, and J. J. Zuckerman, *Organometallics*, **1984,** *3,* 233–237.) Determine a hybridization for each atom in the structure. This compound can be considered to be an adduct of which Lewis acid and which Lewis base?

3. Sulfate becomes bound to four $Al(CH_3)_3$ fragments in the dianion $[Al_4(CH_3)_{12}SO_4]^{2-}$. (See R. D. Rogers and J. L. Atwood, *Organometallics*, **1984,** *3,* 271–274.) Using the bond angles and lengths as a guide, decide if the $Al(CH_3)_3$ groups alter the SO_4^{2-} group significantly upon formation of the $[Al_4(CH_3)_{12}SO_4]^{2-}$ product. Explain.

4. The compound $SnCl_2$ can serve either as a Lewis acid (electron-pair acceptor) or as a Lewis base (electron-pair donor). In fact, it can do both, simultaneously. (See C. C. Hsu and R. A. Geanangel, *Inorg. Chem.*, **1980,** *19,* 110–119.) Draw a Lewis diagram for $SnCl_2$. Give an example where the Sn atom serves as (a) an electron-pair donor, (b) an electron-pair acceptor, and (c) both a donor and an acceptor. In each of these cases, list the occupancy and the hybridization at the Sn atom. Is there a change in either hybridization or geometry when $SnCl_2$ serves as a Lewis base or a Lewis acid?

SUPPLEMENTARY READING

Atkins, P. W., *Molecular Quantum Mechanics,* Oxford University Press, Oxford, 1983.

Ballhausen, C. J. and Gray, H. B., *Molecular Orbital Theory,* Benjamin, Menlo Park, CA, 1964.

Ballhausen, C. J. and Gray, H. B., *Molecular Electronic Structures,* Benjamin–Cummings, Menlo Park, CA, 1980.

Burdett, J. K., *Molecular Shapes, Theoretical Models of Inorganic Stereochemistry,* Wiley-Interscience, New York, 1980.

Cartmell, E. and Fowles, G. W. A., *Valency and Molecular Structure,* Butterworths, London, 1966.

Champion, A., *Chemical Bonding,* McGraw-Hill, New York, 1964.

Coulson, C. A., *Valence,* Oxford University Press, New York, 1964.

Coulson, C. A., *The Shape and Structure of Molecules,* 2nd ed., revised by R. McWeeny, Clarendon, Oxford, 1982.

DeKock, R. L. and Gray, H. B., *Chemical Structure and Bonding*, Benjamin–Cummings, Menlo Park, CA, 1980.

Ebsworth, E. A. V., Rankin, D. W. H., and Cradock, S., *Structural Methods in Inorganic Chemistry*, Blackwell Scientific Publications, Oxford, 1987.

Ferguson, J. E., *Stereochemistry and Bonding in Inorganic Chemistry*, Prentice-Hall, Englewood Cliffs, NJ, 1974.

Gillespie, R. J., *Molecular Geometry*, Van Nostrand-Reinhold, London, 1972.

Gimarc, B. M. *Molecular Structure and Bonding*, Academic, NY, 1979.

Goodisman, J., *Contemporary Quantum Chemistry*, Plenum, New York, 1977.

Karplus, M. and Porter, R. N., *Atoms and Molecules: An Introduction for Students of Physical Chemistry*, Benjamin, Menlo Park, CA, 1970.

McWeeny, R., *Coulson's Valence*, Oxford University Press, Oxford, 1979.

Müller, U., *Structural Inorganic Chemistry*, Wiley, NY, 1993.

Murrell, J. N., Kettle, S. F., Tedder, J. M., *The Chemical Bond*, 2nd ed., Wiley, NY, 1985.

Verkade, J. G., *A Pictorial Approach to Molecular Bonding*, Springer-Verlag, New York, 1986.

Wade, K., *Electron Deficient Compounds*, Nelson, London, 1971.

Wahl, A. C., "Electron Density Maps," *Science*, **1966**, *151*, 961.

Chapter 4

IONIC SOLIDS

4-1   Introduction

A great many inorganic solids, and even a few organic ones, can usefully be thought of as consisting of a three-dimensional array of ions. This ionic model can be developed in further detail in two main ways.

First, it is assumed that the energy of this array of ions can be treated as the sum of the following contributions:

1. Coulombic (electrostatic) attractive and repulsive energies.
2. Additional repulsive energy that results from repulsion between the overlapping outer electron density of adjacent ions.
3. A variety of minor energy terms, mainly van der Waals and zero point vibrational energy.

The important point here is that no explicit account is taken of covalent bonding. This is doubtless an oversimplification in *every* case, but evidently in many substances the pure ionic description leads to fairly accurate estimates of the enthalpies of formation of the compounds. There probably is a certain approximate compensation so that covalent bond energy, which may actually be present, arises at the expense of a nearly equal amount of coulomb energy. Thus, so long as the covalence is small, the error involved in assuming that one form of energy exactly offsets the other is an acceptable approximation.

Second, the main features of the structures of ionic solids can be understood by treating these substances as efficiently packed arrays of ions. To be efficient, the packing of ions in the structure of an ionic compound must maximize the number of contacts between oppositely charged ions, while simultaneously keeping ions of the same sign as far apart as possible. A set of radii (r_{ion}, as defined in Chapter 2) for the different ions, together with a geometrical and electrostatic analysis, can enable us to understand why, for example, NaCl, CsCl, and CuCl all have different structures.

4-2   The Lattice Energy of Sodium Chloride

We begin by considering how to calculate the enthalpy of forming a solid ionic compound from a dilute gaseous collection of the constituent ions. For definiteness, we shall first consider a specific example, NaCl. X-ray study shows that the atoms are arranged as in Fig. 4-1. If we assume that the atoms are in fact the ions Na^+ and Cl^-, the energy of the array can be calculated in the following way.

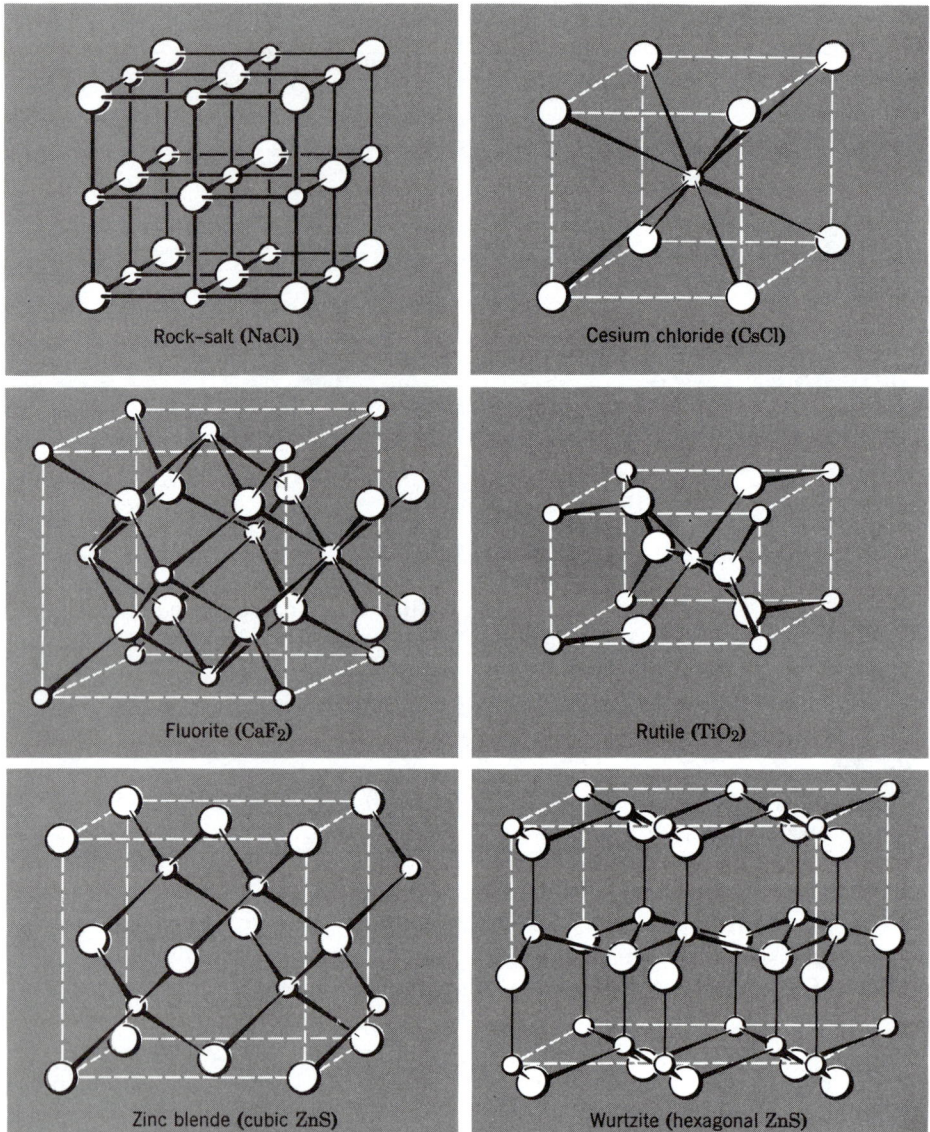

Rock-salt (NaCl)

Cesium chloride (CsCl)

Fluorite (CaF$_2$)

Rutile (TiO$_2$)

Zinc blende (cubic ZnS)

Wurtzite (hexagonal ZnS)

Figure 4-1 Six important ionic structures. Small circles denote metal cations and large circles denote anions.

The shortest Na$^+$—Cl$^-$ distance is called r_0. The electrostatic energy between two neighboring ions is given by Eq. 4-2.1.

$$E \,(\text{joules}) = \frac{e^2}{4\pi\epsilon_0 r_0} \qquad (\epsilon_0 = 8.854 \times 10^{-12}\ \text{C}^2\ \text{m}^{-1}\ \text{J}^{-1}) \qquad (4\text{-}2.1)$$

where e is the electron charge in coulombs, and ϵ_0 is the dielectric constant of a vacuum, as defined in Chapter 1.

Each Na$^+$ ion is surrounded by six Cl$^-$ ions at the distance r_0 (in meters) giving an energy term $6e^2/4\pi\epsilon_0 r_0$. The next closest neighbors to a given Na$^+$ ion are

12 Na^+ ions which, by simple trigonometry, lie $\sqrt{2}r_0$ away. Thus, another energy term, with a minus sign because it is repulsive, is $-12e^2/\sqrt{2}r_0 4\pi\epsilon_0$. By repeating this sort of procedure, successive terms are found, which lead to the expression:

$$E = \frac{1}{4\pi\epsilon_0}\left(\frac{6e^2}{r_0} - \frac{12e^2}{\sqrt{2}r_0} + \frac{8e^2}{\sqrt{3}r_0} - \frac{6e^2}{2r_0} + \cdots\right)$$

$$= \frac{e^2}{4\pi\epsilon_0 r_0}\left(6 - \frac{12}{\sqrt{2}} + \frac{8}{\sqrt{3}} - \frac{6}{2} + \frac{24}{\sqrt{5}} - \cdots\right) \qquad (4\text{-}2.2)$$

The parenthetical expression in Eq. 4-2.2 is an infinite series. It eventually converges to a single value because the electrostatic interactions at great distances become unimportant.

It is possible to derive a general formula for the infinite series and to find the numerical value to which it converges. That value is characteristic of the structure and independent of the particular ions present. It is called the *Madelung constant* (M_{NaCl}) for the NaCl structure. This constant is actually an irrational number, whose value can be given to as high a degree of accuracy as needed, for example, 1.747 . . . or 1.747558 . . . , or better. Madelung constants for many common ionic structures have been evaluated, and a few are given in Table 4-1 for illustrative purposes. The structures themselves (see Fig. 4-1) will be discussed presently.

A unique Madelung constant is defined only for those structures in which all ratios of interatomic vectors are fixed by symmetry. For the rutile structure there are two crystal dimensions that can vary independently. There is a different Madelung constant for each ratio of the two independent dimensions.

When a mole (N ions of each kind, where N is Avogadro's number) of sodium chloride is formed from the gaseous ions, the total electrostatic energy released is given by

$$E_e = NM_{NaCl}\left(\frac{e^2}{4\pi\epsilon_0 r_0}\right) \qquad (4\text{-}2.3)$$

This is true because the expression for the electrostatic energy of one Cl^- ion would be the same as that for an Na^+ ion. If we were to add the electrostatic energies for the two kinds of ions, the result would be twice the true electrostatic energy because each pairwise interaction would have been counted twice.

The electrostatic energy given by Eq. 4-2.3 is not the actual energy released in the process

$$Na^+(g) + Cl^-(g) = NaCl(s) \qquad (4\text{-}2.4)$$

Table 4-1   Madelung Constants for Several Structures

Structure Type	M
NaCl	1.74756
CsCl	1.76267
CaF_2	5.03878
Zinc blende	1.63805
Wurtzite	1.64132

Real ions are not rigid spheres. The equilibrium separation of Na^+ and Cl^- in NaCl is fixed when the attractive forces are exactly balanced by repulsive forces. The attractive forces are Coulombic and strictly follow a $1/r^2$ law. The repulsive forces are more subtle and follow an inverse r^n law, where n is >2 and varies with the nature of the particular ions. We can write, in a general way, that the total repulsive energy per mole at any value of r is

$$E_{rep} = \frac{NB}{r^n} \tag{4-2.5}$$

where B is a constant.

At the equilibrium distance, the net energy U for Reaction 4-2.4 is determined by contributions from both repulsive (Eq. 4-2.5) and attractive (Eq. 4-2.3) forces. This is given by Eq. 4-2.6.

$$U = -NM_{NaCl}\left(\frac{e^2}{4\pi\epsilon_0 r}\right) + \frac{NB}{r^n} \tag{4-2.6}$$

where the algebraic signs are chosen in accord with the convention that the attractive forces produce an exothermic term, and the repulsive forces produce an endothermic term.

The constant B can now be eliminated if we recognize that at equilibrium (where $r = r_0$) the energy U is, by definition, at a minimum. Hence, the derivative of U with respect to r, evaluated at $r = r_0$, must equal zero. Differentiating Eq. 4-2.6 we get Eq. 4-2.7:

$$\left(\frac{dU}{dr}\right)_{r=r_0} = \frac{NM_{NaCl}e^2}{4\pi\epsilon_0 r_0^2} - \frac{nNB}{r_0^{n+1}} = 0 \tag{4-2.7}$$

which can be rearranged and solved for B

$$B = \frac{e^2 M_{NaCl}}{4\pi\epsilon_0 n} r_0^{n-1} \tag{4-2.8}$$

When the result of Eq. 4-2.8 is substituted into Eq. 4-2.6, we obtain Eq. 4-2.9.

$$U = \frac{NM_{NaCl}e^2}{4\pi\epsilon_0 r_0}\left(1 - \frac{1}{n}\right) \tag{4-2.9}$$

The value of n can be estimated to be 9.1 from the measured compressibility of NaCl.

In a form suitable for calculating numerical results (in kJ mol^{-1}) by using r_0 in Angstroms, Eq. 4-2.9 becomes

$$U = -1389 \frac{M_{NaCl}}{r_0}\left(1 - \frac{1}{n}\right) \tag{4-2.10}$$

and inserting appropriate values of parameters we obtain Eq. 4-2.11.

$$U = -1389\frac{1.747}{2.82}\left(1-\frac{1}{9.1}\right)$$

$$U = -860 + 95 = -765 \text{ kJ mol}^{-1} \tag{4-2.11}$$

Notice that the repulsive energy (95 kJ mol^{-1}) equals only about 11% of the attractive (Coulombic) energy (860 kJ mol^{-1}). The total is, therefore, not very sensitive to the value of n. A small error in the estimation of n from compressibility data is not highly significant. For instance, if a value of $n = 10$ had been used, an error of only 9 kJ mol^{-1} (or 1.2%) would have been made.

4-3   Generalization of the Lattice Energy Calculation

As mentioned in Section 4-2, the Madelung constant is determined solely by the geometry of the structure. For an ionic structure that is the same as that of NaCl, but where each ion has a charge of ±2 (as in MgO), the Madelung constant for NaCl can still be used. It is only necessary to modify Eq. 4-2.9 to replace the quantity $-e^2$ with the appropriate charges. For MgO, this would be $(2e)(-2e)$. In general, Eq. 4-2.9 becomes Eq. 4-3.1.

$$U = \frac{NM_{\text{NaCl}}Z^2e^2}{4\pi\epsilon_0 r_0}\left(1-\frac{1}{n}\right) \tag{4-3.1}$$

Equation 4-3.1 may be used for any structure whose Madelung constant is M_{NaCl}, and where the ions have the charges Z^+ and Z^-.

The value of n can be estimated for alkali halides by using the average of the following numbers:

He   5   Kr   10
Ne   7   Xe   12
Ar   9

where the noble gas symbol denotes the noble gas-like electron configuration of the ion. Thus, for LiF, an average of the He and Ne values $(5 + 7)/2 = 6$ would be used.

4-4   The Born–Haber Cycle: Experimental Approaches to Lattice Energies

One test of whether an ionic model is a useful description of a substance is the ability of the model to produce an accurate value for the enthalpy of formation of the substance. It is not possible, though, to measure the enthalpy of Reaction 4-2.4 or its reverse. It is not experimentally feasible to do so because NaCl does not vaporize to give Na$^+$ and Cl$^-$. Rather, it vaporizes to give NaCl(g) and, depending on temperature, a number of aggregates, $(\text{NaCl})_x(g)$, which dissociate at very high temperatures into atoms.

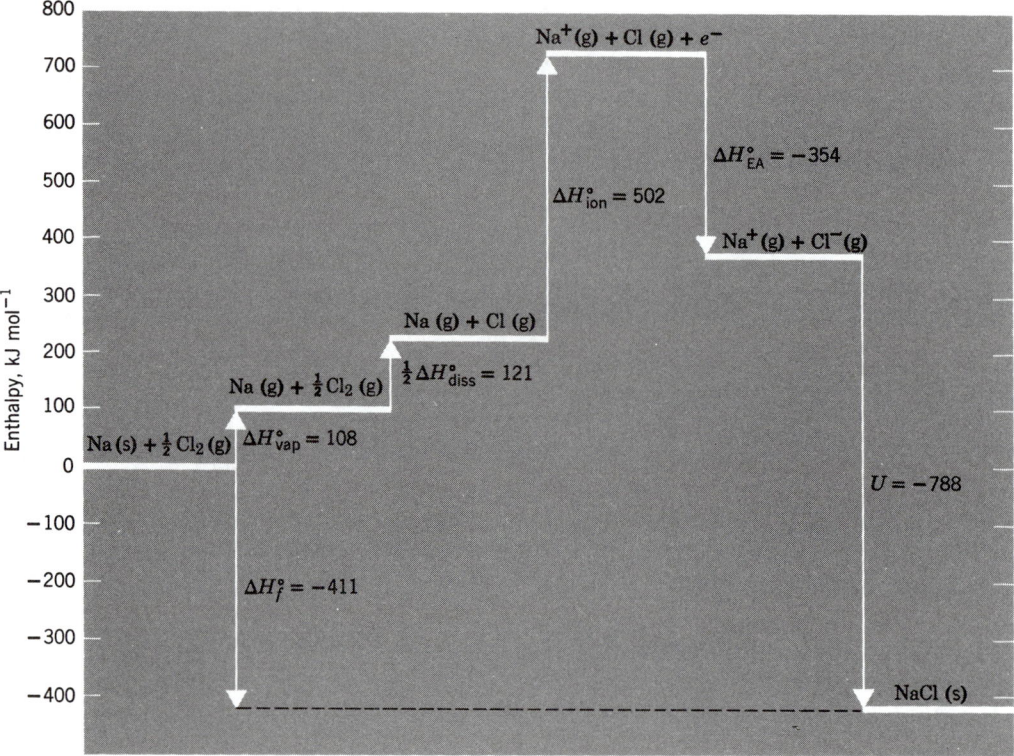

Figure 4-2 The Born–Haber cycle for NaCl.

To circumvent this problem, the Born–Haber thermodynamic cycle is used. This is illustrated in Fig. 4-2. The cycle is useful because the formation of NaCl(s) from the elements according to Reaction 4-4.1:

$$Na(s) + \tfrac{1}{2} Cl_2(g) \longrightarrow NaCl(s) \tag{4-4.1}$$

can be broken down into a series of steps. If the enthalpies of these steps are added algebraically, the result must equal the enthalpy for Reaction 4-4.1, which is the enthalpy of formation (ΔH_f°) for NaCl(s).

$$\Delta H_f^\circ = \Delta H_{vap}^\circ + \tfrac{1}{2} \Delta H_{diss}^\circ + \Delta H_{EA}^\circ + \Delta H_{ion}^\circ + U \tag{4-4.2}$$

Each term in Eq. 4-4.2 corresponds to a step in the cycle shown in Fig. 4-2. The enthalpy terms correspond to the vaporization of sodium (ΔH_{vap}°), the dissociation of Cl_2 into gaseous atoms (ΔH_{diss}°), electron attachment to Cl(g) to give $Cl(g)^-$ (ΔH_{EA}°), the first ionization enthalpy of a gaseous sodium atom (ΔH_{ion}°), and the formation of NaCl(s) from gaseous ions (U).

Any one of the enthalpies in Eq. 4-4.2 can be calculated if the others are known. For NaCl all the enthalpies except U have been measured independently. The following summation can thus be made:

$$\Delta H_f^\circ \quad -411$$

$$-\Delta H_{vap}^\circ \quad -108$$

$$-\tfrac{1}{2}\Delta H_{diss}^\circ = -121$$

$$-\Delta H_{EA}^\circ \quad 349$$

$$-\Delta H_{ion}^\circ \quad -502$$

$$U \quad = -784 \text{ kJ mol}^{-1} \tag{4-4.3}$$

The result is within 1% of the value of U obtained in very precise calculations. This good agreement supports (but does not prove) the idea that the ionic model for NaCl is a useful one.

4-5   Ionic Radii

In a manner similar in principle to that in which covalent radii were estimated, it is possible to assign radii to ions. The internuclear distance d between two ions in an ionic structure is assumed to be equal to the sum of the radii of the ions:

$$d = r^+ + r^- \tag{4-5.1}$$

By comparing distances in different compounds with an ion in common, it can first be shown that the radii of ions are substantially constant. For example, the difference in the radii of K^+ and Na^+ can be evaluated in four different halides.

$$
\begin{aligned}
r_{K^+} - r_{Na^+} &= d_{KF} - d_{NaF} &= 0.35 \text{ Å} \\
&= d_{KCl} - d_{NaCl} &= 0.33 \text{ Å} \\
&= d_{KBr} - d_{NaBr} &= 0.32 \text{ Å} \\
&= d_{KI} - d_{NaI} &= 0.30 \text{ Å}
\end{aligned}
$$

Actually, the apparent trend as the halide ion size increases is a real effect that can be understood in terms of packing considerations, but we shall not discuss that topic further. Suffice it to say that if $(r_{K^+} - r_{Na^+})$ is substantially constant, it is reasonable to assume that r_{K^+} and r_{Na^+} are themselves substantially constant.

It is easy to work out extensive sets of sums and differences of ionic radii. Then, provided that the actual radius of any one ion can be evaluated, the radii of all of the ions will be determined. Although this problem has no rigorous solution, Pauling proposed a practical one, namely, that for two ions with the same noble gas configuration, say Na^+ and F^-, the ratio of the radii should be inversely proportional to the ratio of the nuclear charges felt by the outer electrons.

The nuclear charges that are felt by the outer electrons are the effective nuclear charges defined in Section 2-6. For Na^+ and F^-, the effective nuclear charge is given by $Z^* = Z - \sigma$. In each case, the value of the screening constant σ is 4.15. This value is the same for the isoelectronic Ne. The calculations develop as follows: $Z^*(Na^+) = 11.00 - 4.15 = 6.85$, while $Z^*(F^-) = 9.00 - 4.15 = 4.85$. Hence, according to Pauling's proposal, the ratio of the ionic radii should be given by Eq. 4-5.2.

$$\frac{r_{Na^+}}{r_{F^-}} = \frac{4.85}{6.85} = 0.71 \qquad (4\text{-}5.2)$$

Since the internuclear distance in NaF is 2.31 Å, we have Eq. 4-5.3.

$$r_{Na^+} + r_{F^-} = 2.31 \text{ Å} \qquad (4\text{-}5.3)$$

Treating the ratio (Eq. 4-5.2) and the sum (Eq. 4-5.3) as a pair of simultaneous equations in two unknowns, we obtain the individual radii:

$$r_{F^-} = 1.35 \text{ Å}$$

$$r_{Na^+} = 0.96 \text{ Å}$$

We have outlined Pauling's method of determining ionic radii because it straightforwardly shows the two principal steps in any procedure for estimating such radii: (1) making radii additive and (2) finding a way to divide up the sums of cation and anion radii into separate, individual radii. However, since the first efforts by Pauling and others in the 1920s to determine useful sets of radii, a great deal of sophisticated work has gone into this activity, and many tabulations have appeared. Today, there is a widely used, extensive set of radii, where the coordination number is taken into account. These are the Shannon and Prewitt radii mentioned previously in Section 2-7. These radii are listed in Appendix IIC.

4-6   Geometries of Crystal Lattices

Figure 4-1 shows six of the most important structures formed by essentially ionic substances. All of these structures have a common qualitative feature: The ions are packed to maximize the contacts between those of opposite charge and to minimize repulsions between those of the same charge. In a three-dimensional sense, ions of opposite charge alternate. The nearest neighbors of one ion are ions of opposite charge. However, this qualitative idea alone does not account for all of the features that can be seen in Fig. 4-1. For AB-type compounds we see four structure types. Consider first those of NaCl and CsCl. The difference is that in the NaCl structure each cation has six nearest neighbor anions, whereas in the CsCl structure each cation has eight such neighbors. We say that the *coordination numbers* of the cations are six and eight, respectively. In both the zinc blende and wurzite structures the cation has a coordination number of only four. Again, for AB_2-type compounds there is a fluorite structure where the cation coordination number is eight and a rutile structure where it is six. Why does a particular AB or AB_2 compound adopt one and not another of these structures?

The answer lies partly in a consideration of the relative sizes of the ions. Anions are almost always larger than cations, since the net excess of nuclear

Table 4-2 Values of the Pauling Ionic Radii (in Å)[a]

Main Group Elements

Li^+	0.60	Be^{2+}	0.31	Al^{3+}	0.50					H^-	2.08
Na^+	0.96	Mg^{2+}	0.65	Ga^{3+}	0.62	Sn^{4+}	0.71			O^{2-}	1.40
K^+	1.33	Ca^{2+}	0.99	In^{3+}	0.81	Pb^{4+}	0.84			S^{2-}	1.84
Rb^+	1.48	Sr^{2+}	1.13	Tl^{3+}	0.95			Pb^{2+}	1.21	Se^{2-}	1.98
Cs^+	1.69	Ba^{2+}	1.35							Te^{2-}	2.21

F^-	1.35
Cl^-	1.81
Br^-	1.95
I^-	2.16

Transition Metal Ions

Ti^{4+}	0.68	Fe^{3+}	0.53	Mn^{2+}	0.80		
Zr^{4+}	0.80	Cr^{3+}	0.55	Fe^{2+}	0.75		
Ce^{4+}	1.01			Co^{2+}	0.72		
				Ni^{2+}	0.69		

Others

Zn^{2+}	0.74
Cd^{2+}	0.97
Hg^{2+}	1.10

[a]A more satisfactory list of values for ionic radii has since been prepared by Shannon and Prewitt, as explained in Section 2-7, and given in Appendix IIC.

133

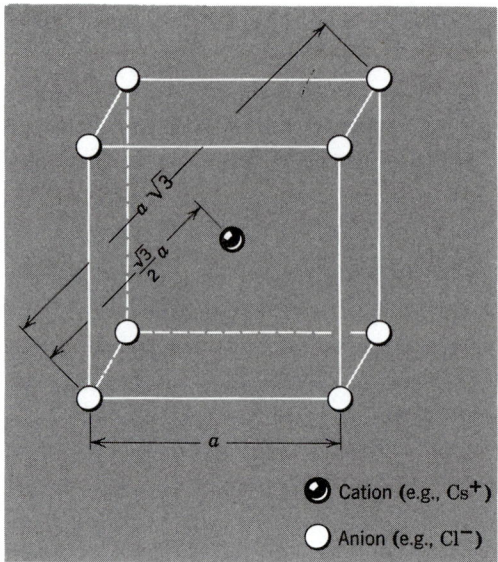

Figure 4-3   The geometry of the crystal lattice for CsCl.

charge on cations draws their electron clouds in, while the excess of negative charge on anions causes the electron clouds to expand. The optimum arrangement should allow the maximum number of oppositely charged ions to be neighbors without unduly squeezing together ions of the same charge. Thus the greater the ratio of cation to anion size, the higher the coordination number of the cation can—and should—be. That is why the relatively large Cs^+ ion surrounds itself with eight Cl^- ions, but for the smaller Na^+ ion there are only six.

It is possible to treat this idea in a semiquantitative way, by finding for each structure that ratio (r^-/r^+) for which the anions just touch one another while making contact with the cation. We shall call this situation perfect packing. For the CsCl structure, the relevant geometric relations (Fig. 4-3) are as follows.

First, the anions just touch one another along the edge a of the cube. The radii of the two anions, therefore, combine to give the length of the cube edge, as in Eq. 4-6.1.

$$2\,r^- = a \tag{4-6.1}$$

Second, the cation touches each anion along the body diagonal of the cube, which has length $a\sqrt{3}$. The cation–anion distance is therefore one-half this distance, as in Eq. 4-6.2.

$$r^+ + r^- = \frac{\sqrt{3}}{2}a \tag{4-6.2}$$

Equations 4-6.1 and 4-6.2 define the geometric requirements for perfect packing of ions in the CsCl-type crystal lattice. Both equations are satisfied for values of r^+ and r^- such that $r^-/r^+ = 1.37$. Similar considerations suggest that perfect pack-

Table 4-3   Radius Ratios r^-/r^+ for Several Crystal Structures, and the Resulting Coordination Number of the Cation

Structure Type	Ideal Values[a]	Cation Coordination Number
CsCl	1.37	8
NaCl	2.44	6
ZnS	4.44	4

[a]These values correspond to perfect packing, that is, they give a perfect match between the size of the anion and that of the cation.

ing for octahedral coordination number six is achieved when $r^-/r^+ = 2.44$. Also, the tetrahedral coordination number four is preferred when $r^-/r^+ = 4.44$. This is summarized in Table 4-3.

It should be stressed, however, that the foregoing analysis, which is based on ion sizes, is only a part of the picture. It works best for compounds that are most truly ionic (namely, alkali and alkaline earth halides, oxides, and sulfides) but even some of these compounds do not obey predictions based solely on the radius ratio. Coordination numbers often are lower than expected for compounds in which the ions are highly polarizable [e.g., copper(I) and zinc compounds].

In a case where the cation is very small relative to the anion ($r^-/r^+ \gg 4.44$), it will be impossible to achieve good cation–anion contact, even when anion–anion contacts are very close. Thus, ionic salts of this type are relatively unstable. Salts of the small cations Li^+, Be^{2+}, Al^{3+}, and Mg^{2+} with large polyatomic anions (e.g., ClO_4^-, CO_3^{2-}, NO_3^-, O_2^-) or even monatomic anions, such as Cl^-, Br^-, and I^-, are cases in point. The consequences of this are threefold.

1. In some cases, the anhydrous compounds are unstable relative to hydrates in which the cations surround themselves with water molecules. Thus $Mg(ClO_4)_2$ is a powerful absorbant for water, and lithium perchlorate forms a stable hydrate ($LiClO_4 \cdot 3\ H_2O$), whereas the other alkali metal perchlorates do not.

2. In other cases, the result of the bad packing is thermal instability. Thus, the large polyatomic anion decomposes to leave behind a smaller one that can pack better with the small cation. Examples are

$$Li_2CO_3 \longrightarrow Li_2O + CO_2 \tag{4-6.3}$$

$$2\ NaO_2 \longrightarrow Na_2O + \tfrac{3}{2}\ O_2 \tag{4-6.4}$$

$$4\ Be(NO_3)_2 \longrightarrow Be_4O(NO_3)_6 + N_2O_4 + \tfrac{1}{2}\ O_2 \tag{4-6.5}$$

3. The solubility relations are related to point (1). Thus $LiClO_4$ is about 10 times as soluble as $NaClO_4$ which, in turn, is about 10^3 times as soluble as $KClO_4$, $RbClO_4$, and $CsClO_4$. This trend is due partly to the solvation enthalpies of the cations decreasing as they increase in size, but is enhanced by the fact that poor packing of the small Li^+ and Na^+ cations with the large ClO_4^- ions decreases the intrinsic stability of the crystals.

4-7   Structures of Ionic Substances Based on Close Packing of Anions

Close Packing of Spheres

The structures of many inorganic substances can usefully be regarded as essentially infinite arrays of spheres that are packed efficiently into three-dimensional space, that is, to occupy the least possible volume. This statement is true both of the metals to be discussed in Section 8-6, and of those ionic substances for which the anions are considerably larger than the cations. In the latter case, large spherical anions are arranged in space in a so-called close-packed pattern to be described shortly. We will show that the close packing of spherical anions in three dimensions results in the creation of specific types, numbers, and arrangements of interstices (or holes) between the anions, into which the relatively smaller cations fit. Thus, instead of defining the unit cell, an alternate description of the structure of an ionic substance can be given by stating the shapes and numbers of holes (between anions) in which the cations are found to reside. To use this approach, it is necessary to understand the close-packed structures that arise from the stacking of layers of spheres on top of one another.

Figure 4-4 shows the close packing of spheres in a single layer. The pattern that is produced is an array of contiguous equilateral triangles. A second layer of spheres can now be laid down over the first. There is only one way that this can be accomplished so as to use space most efficiently: Atoms in layer B are placed so that they nestle into the depressions between spheres in layer A. This stacking is called the close packing of spheres in two layers, and it is depicted in Fig. 4-5(a). It is also shown in Structure 4-I, in which the pattern is represented by only the centers of the spheres, plus connecting lines between the spheres in each separate layer. Four things should be noted in Fig. 4-5(a). First, two types of interstices (or holes) are created between layers A and B. Each hole has the shape of either an octahedron or a tetrahedron. Second, as seen in Structure 4-I, only one half of the depressions between spheres in layer A become covered by spheres in layer B. (This finding will shortly become important, when we place a third layer on top of the first two.) Third, twice as many tetrahedral holes exist between layers A and B as octahedral holes. Fourth, as shown in Fig. 4-5(a), the octahedral and tetrahedral holes between layers A and B are arranged in a regular pattern, as follows: All tetrahedral holes share edges (but not faces) with adjacent tetrahedral holes. Thus part of Fig. 4-5(a) can be described by saying that we have drawn two tetrahedra sharing an edge. Also, all octahedral holes share edges with adjoining octahedral holes, but they share faces only with adja-

Figure 4-4   Close packing of spheres in a single plane.

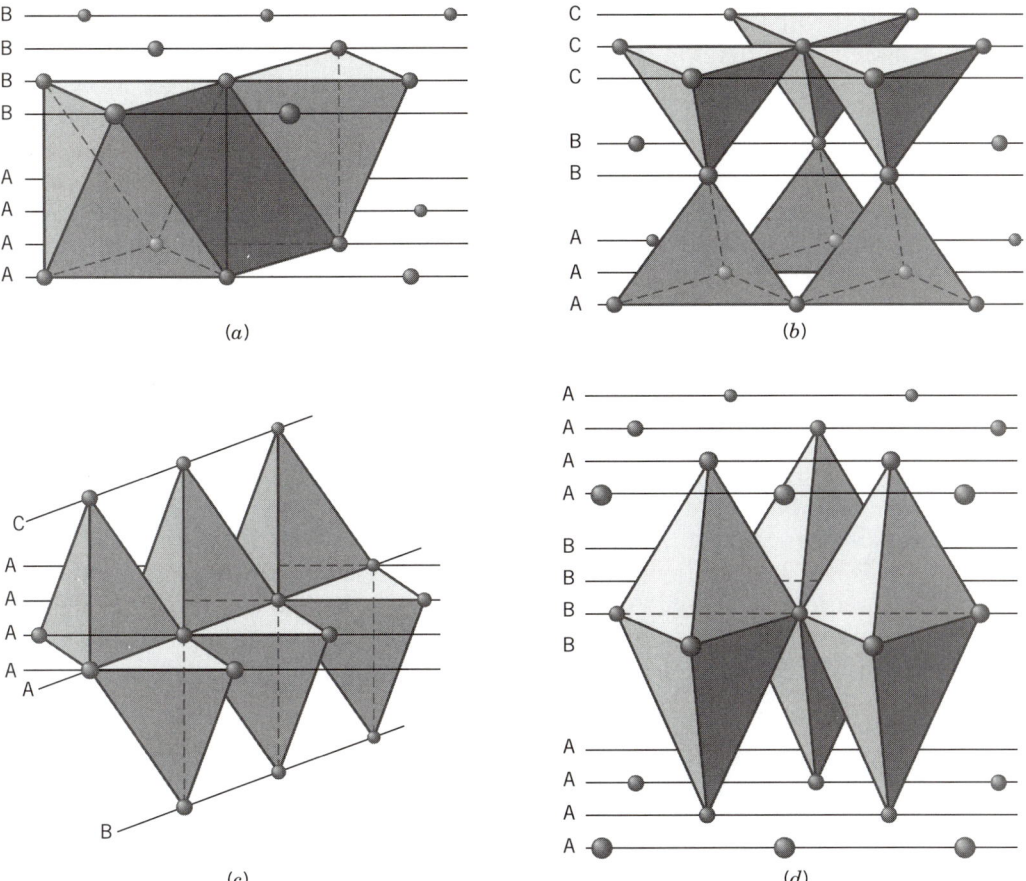

Figure 4-5   Diagrams showing the shapes and arrangements of interstices or holes formed between spheres in close-packed arrays. (*a*) One of the octahedral and two of the tetrahedral holes formed between two parallel layers of close-packed spheres. (*b*) One view of the tetrahedral holes formed by three parallel layers of spheres in the cubic close-packed structure. Atoms in layers C and B form one set of tetrahedra, whereas those in layers A and B form another set. (*c*) An alternate view of the cubic close-packed array of spheres, emphasizing the way tetrahedral holes formed by spheres in layers C and A share edges with tetrahedral holes formed by spheres from layers A and B. (*d*) Three layers of spheres from the hexagonal close-packed structure, showing how tetrahedral holes formed by spheres from layer B and the top A layer share faces with those formed from spheres in layer B and the bottom A layer.

cent tetrahedral holes. That is, octahedral holes are never found to share a face with nearby octahedral holes in the pattern created by close packing of spheres in two layers. Therefore, Fig. 4-5(*a*) shows an octahedral hole sharing a face with one of the two tetrahedral holes, which in turn shares only an edge with the adjoining tetrahedron. It is interesting to note that, in order to define the close packing of spheres in two layers, we need only define the arrangement of the octahedral and tetrahedral holes. The positions of the spheres are taken to coincide with the vertices of the holes.

To continue the build up of a close-packed structure in three dimensions, only a third layer of close-packed spheres needs to be added to the two layers of Fig. 4-5(*a*). We will see that there are two distinct ways to place a third layer on top of the first two. In each case, the spheres of layer C nestle into depressions between the spheres in layer B. But, as shown in Structure 4-I, these depressions are of two types: those residing over octahedral holes found between layers A and B, and those residing over tetrahedral holes found between layers A and B. The arrangement of the spheres (and hence, the shapes and arrangements of the holes created between the spheres) in these first three layers can then continue to create the entire structure. The first possibility gives rise to **cubic close packing,** whereas the second gives us **hexagonal close packing.** Let us now examine each of these in detail.

Cubic Close Packing of Spheres

When the third layer of spheres is placed on top of layers A and B so that the spheres in layer C lie over the octahedral holes created by layers A and B, then a **cubic close packed (ccp)** array of spheres is formed. This pattern, when viewed from the top, as in Structure 4-II, does not superimpose spheres of layer C with spheres of layer A. Therefore, the three layers are geometrically distinct from one another, and the pattern is described as **ABC.** If a fourth layer is added so as to coincide with the first layer, we have the pattern **ABCABCABC** . . . , and so on. The traditional view of a ccp array, emphasizing the cubic symmetry of the unit cell, is given in Structure 4-III.

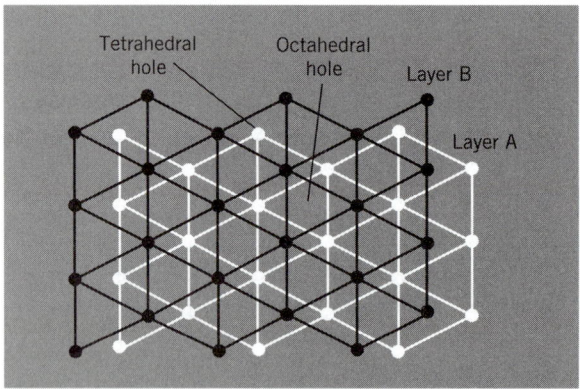

4-I

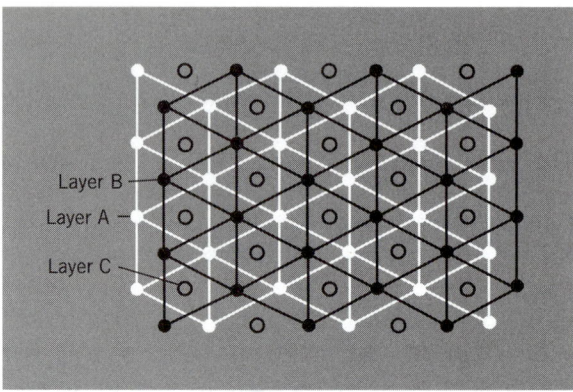

4-II

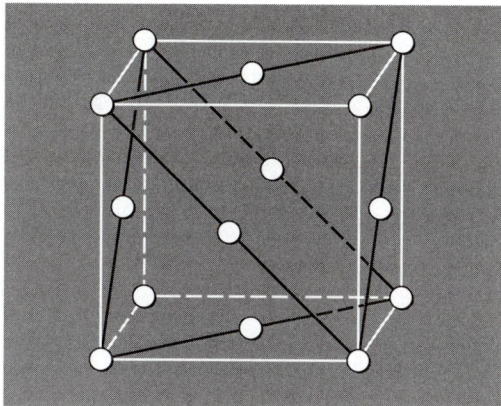

4-III

As an alternative to specifying the positions of the spheres, the cubic close-packed structure can be described in terms of the arrangement of octahedral and tetrahedral holes. For instance, in the ccp structure, tetrahedral holes in adjoining layers are oriented between parallel layers A, B, and C, as shown in Fig. 4-5(b). Here, the tetrahedral holes created by spheres in layers B and C are staggered with respect to tetrahedral holes created between layers B and A. Furthermore, these two layers of tetrahedral holes share spheres in layer B as common vertices.

A slightly different view of the ccp array is given in Fig. 4-5(c). Here, other groups of tetrahedral holes between adjoining layers share edges. Thus adjacent tetrahedra in the ccp structure variously share either vertices [Fig. 4-5(b)] or edges [Fig. 4-5(c)]. There is, however, no sharing of faces between tetrahedra in the ccp structure. All tetrahedral holes share faces only with adjacent octahedral holes in the ccp structure. This is one of the differences between the ccp structure and the other close-packing possibility: hexagonal close packing.

Hexagonal Close Packing of Spheres
The **hexagonal close-packed (hcp)** array is formed when the third layer of spheres is placed on top of layers A and B so that spheres in layer C lie over the tetrahedral holes created by layers A and B. When viewed from the top, this pattern superimposes spheres in layer C over those in layer A. Since this arrangement makes the first and third layers equivalent, the stacking pattern may be simply depicted **AB**. When repeated indefinitely, we have **ABABAB** . . . , and so on. A portion of an hcp array is shown in Fig. 4-5(d). Here we see that, unlike ccp, tetrahedral holes formed by the first and second layers of spheres do share faces with tetrahedral holes formed by the second and third layers of spheres.

Another useful distinction between the ccp and hcp structures can be seen by examining the different ways in which edges and faces of adjoining octahedra are shared. A different view of the ccp structure is given in Fig. 4-6(a). This view places the ABC planes at a 45° angle to the horizontal plane. As shown in Fig. 4-6(a), in the ccp structure, adjoining octahedra share, at most, an edge. Furthermore, adjoining octahedra in the ccp structure never share faces. In contrast, as shown in Fig. 4-6(b), in the hcp structure, octahedra share faces as well as edges.

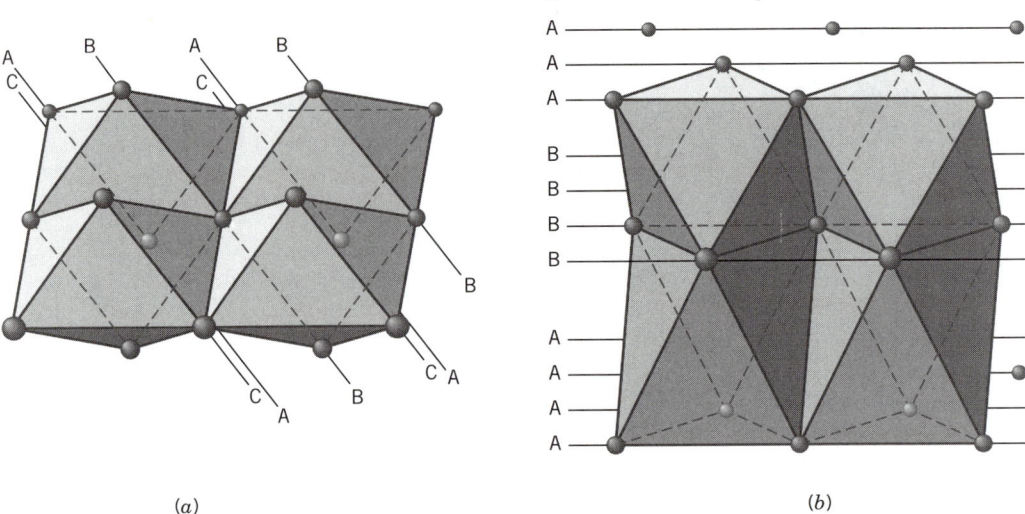

(a)   (b)

Figure 4-6   Diagrams showing the arrangements of the octahedral holes in the two close-packed structures. (a) Adjacent octahedral holes in the cubic close-packed structure, emphasizing the fact that octahedral holes share edges but not faces with one another in this structure. The parallel planes of spheres (A, B, and C) are oriented at a 45° angle to the horizontal plane. (b) Four octahedral holes in the hexagonal close-packed structure, emphasizing the fact that octahedral holes in this structure share both edges and faces with one another.

Ionic Substances

The structures of many ionic substances can be elegantly described using the close-packed structures defined previously. Since the anions are typically larger than the cations, we generally find (1) that the anions adopt either the ccp or the hcp structure, and (2) that the cations occupy particular octahedral or tetrahedral interstices. Consider NaCl, whose ionic lattice was illustrated in Fig. 4-1. An equally correct but alternative description of the structure of NaCl is to say that the chloride anions adopt a cubic close-packed array, with a sodium cation residing in each of the octahedral holes. Table 4-4 lists similar descriptions of the structures of ionic substances, chiefly halides and oxides, using the close-packed

Table 4-4   A Description of the Structures of Ionic Substances Using the Concept of Close Packing of Anions

Formula	Structure of the Anions	Location of the Cations
CdI_2	hcp	Cations occupy octahedral holes in every other layer
$CdCl_2$	ccp	Cations occupy octahedral holes in every other layer
NaCl	ccp	Cations occupy all octahedral holes in every layer
BI_3	ccp	Cations occupy two-thirds of the octahedral holes in every other layer
Al_2O_3	hcp	Cations occupy two-thirds of the octahedral holes in every layer
$CaTiO_3$	hcp	Fe^{2+} and Ti^{4+} ions jointly occupy two-thirds of the octahedral holes
Na_2S	ccp	Cations occupy all of the tetrahedral holes[a]

[a]This is the so-called antifluorite structure. The structure of CaF_2 (fluorite) can be regarded as being formed from a ccp array of *cations*, with anions occupying all of the tetrahedral holes.

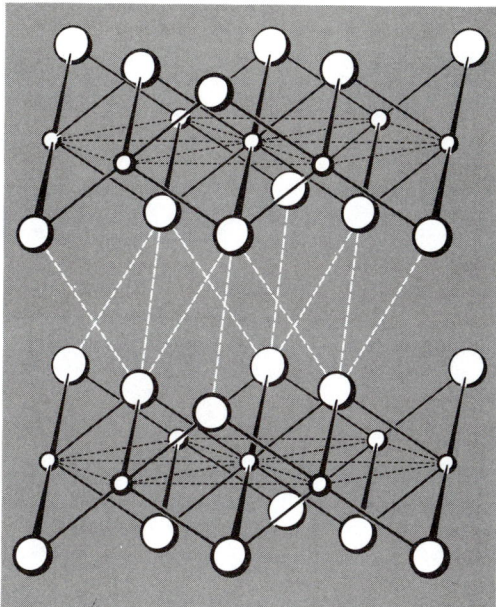

Figure 4-7   A portion of the CdI_2 structure.
Small spheres represent metal cations.

approach. As a further example, consider CdI_2, whose structure is given in Fig. 4-7. This structure is adopted by a number of MX_2 compounds. The anions are hcp and the metal ions occupy octahedral holes, but only in every other layer. The compound $CdCl_2$ has a ccp array of anions, again with every other layer of octahedral holes fully occupied by cations. Substances having this type of a layered structure, with every other layer of octahedral holes unoccupied by cations, are often flakey crystalline solids, making cleavage along the vacant planes easy. In the BI_3 structure, which is adopted by many MX_3 compounds, every other layer of octahedral holes in a ccp array of anions is partially occupied by cations.

Corundum, the α form of Al_2O_3, has an hcp array of oxide ions with two-thirds of the octahedral holes occupied by cations, but not in a layered fashion. This important structure is adopted by many other M_2O_3 compounds. Some examples are Fe_2O_3, V_2O_3, and Rh_2O_3.

4-8   Mixed-Metal Oxides

There are a large number of metal oxides, of great scientific and technical importance, which are essentially ionic substances. Many contain two or more different kinds of metal ions. These oxides tend to adopt one of a few basic, general structures, the names of which are derived from the first compound (or an important one) found to have that structure.

The Spinel Structure

Spinel is a mineral ($MgAl_2O_4$). The structure is based on a ccp array of oxide ions, with Mg^{2+} ions in a set of tetrahedral holes and Al^{3+} ions in a set of octahe-

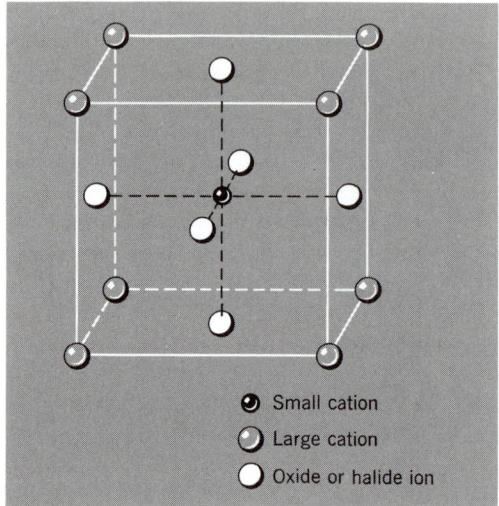

● Small cation
◐ Large cation
○ Oxide or halide ion

Figure 4-8   The perovskite structure.

dral holes. Many substances of the types $M^{2+}(M^{3+})_2O_4$, $M^{4+}(M^{2+})_2O_4$, and $M^{6+}(M^{+})_2O_4$ have this structure. More highly charged cations tend to prefer the octahedral holes so that in $M^{4+}(M^{2+})_2O_4$ compounds the octahedral holes are occupied by all the M^{4+} ions and one-half of the M^{2+} ions.

The Ilmenite Structure

Ilmenite is the mineral $FeTiO_3$. Its structure is closely related to the corundum structure except that the cations are of two kinds. In ilmenite the cations are Fe^{2+} and Ti^{4+}, but many substances with the ilmenite structure have cations with charges of (+1, +5) or (+3, +3).

The Perovskite Structure

Perovskite is the mineral $CaTiO_3$. Its structure, shown in Fig. 4-8, is based on a ccp array of oxide ions together with large cations, similar in size to the oxide ion. The smaller cations lie in octahedral holes formed entirely by oxide ions. Again, the individual cation charges are not important so long as their sum is +6. The structure is adopted by many fluorides with cations of disparate sizes, such as $KZnF_3$.

STUDY GUIDE

Scope and Purpose

The fundamental aspects of the bonding within ionic compounds have been discussed. In developing the model of ionic bonding, a complete lack of covalency has been assumed. In addition, we have treated the simplest cases, those involving spherical ions. Where compounds are not perfectly ionic, or where complex, nonspherical ions are to be packed into a crystal lattice, the description of the bonding is not quite so straightforward. Still the principle of most efficient packing requires the same sorts of electrostatic and geometric considerations as have been presented here. The study questions marked "A. Review" require a straightforward understanding of the material presented in the chapter. Those study questions under "B. Additional Exercises" require application.

Study Questions

A. Review

1. What are the two main contributions to the cohesive energy of an ionic solid?

2. What is a Madelung constant? Why can the same Madelung constant be used for seemingly different substances?

3. What is n in the Born expression for the non-coulomb repulsive energy? What are typical values for n?

4. Use Fig. 4-2 as a guide and write out balanced chemical equations for each step in the Born–Haber cycle for CrN, KF, and MgO.

5. What proposal did Pauling make to estimate the ratio of the radii of certain anion–cation pairs?

6. Define the coordination number of a cation in a crystal lattice. Why are we more concerned with the coordination number of the cation than the anion?

7. Describe a close-packed layer of spheres.

8. Show, with drawings, the two different ways to stack three close-packed layers of spheres.

9. Explain the difference between cubic and hexagonal close packing.

10. How are the ilmenite and corundum structures related?

11. What is the name of the mineral whose formula is $MgAl_2O_4$? What other cation charges can exist in mixed oxides of this structure?

12. For the perovskite structure to occur, what must be true of the sizes of the cations?

13. How are the fluorite and antifluorite structures related to one another?

14. As a means of becoming acquainted with the hole arrangements in the ccp and hcp arrays, prepare simple drawings of the following structures:
    (a) Two tetrahedra sharing an edge.
    (b) Two tetrahedra sharing a face.
    (c) Two octahedra sharing an edge.
    (d) Two octahedra sharing a face.

15. How are the corundum, ilmenite, and perovskite structures related?

B. Additional Exercises

1. Consider a line of alternating cations and anions. Evaluate the Madelung constant to within 1%.

2. Design a cycle of the Born–Haber type to evaluate the enthalpy of the reaction $NH_3(g) + H^+(g) \rightarrow NH_4^+(g)$.

3. For MgO, which has the NaCl structure, the unit cell edge is 4.21 Å. Use the approach of Pauling to evaluate the radii of Mg^{2+} and O^{2-}. Compare your results with those of Table 4-2.

4. What is the coordination number of each atom in the hexagonal close-packed structure?

5. Use a Born–Haber cycle to calculate the energy of electron attachment to $O(g)$ to form $O^{2-}(g)$. You will need the information in Problem 4, part A, as well as the following: ΔH_f° of $MgO(s) = -602$ kJ mol^{-1}; ΔH_{vap} of $Mg = 150.2$ kJ mol^{-1}; ΔH_{diss} of $O_2 = 497.4$ kJ mol^{-1}; $\Delta H_{ion}(1) + \Delta H_{ion}(2)$ for $Mg = 2188$ kJ mol^{-1}.

6. Table 4-3 lists the ratios r^-/r^+ for *perfect* packing in structures having coordination numbers of four, six, and eight. In practice, a range of values for these ratios is observed within a series of compounds having the same structure. The typical ranges for the three coordination numbers are $(4.44 - 2.44)$ for coordination number four; $(2.44 - 1.37)$ for coordination number six; 1.37 and below for coordination number eight. With this in mind, predict coordination numbers of four, six, or eight for the compounds NaF, KBr, and LiCl.

7. Why do you think the value of n in the Born repulsion expression can be estimated from compressibility data, namely, the change in volume suffered by a substance for each unit change in pressure?

8. Both NaH and LiH adopt a structure with H^- anions forming a ccp array and cations occupying every octahedral hole. The length of the unit cell edge of NaH is 4.88 Å, whereas that of LiH is 4.08 Å. Use the approach of Pauling to estimate the ionic radii of Na^+, Li^+, and H^-, and compare your results to those in Table 4-2.

9. Consider the formulas and the structures of Na_2S, $CdCl_2$, and NaCl, as listed in Table 4-4. Deduce the number of octahedral and tetrahedral holes per anion, in the ccp structure.

10. How can the structure of $CdCl_2$ be used to construct the lattice of NaCl?

11. Draw the unit cell for perovskite, emphasizing the octahedral hole into which a small cation fits.

12. Redraw the unit cell of perovskite (Fig. 4-8) choosing the center of the cube to be the large cation, rather than the small cation. This produces a completely valid, alternate unit cell. What is the shape of the hole in which the large cation now resides?

13. Lithium fluoride (LiF) adopts the NaCl structure, whereas CsI adopts the CsCl structure. Use the radius ratio approach to explain this.

14. Use Pauling's approach to estimate the ionic radii of Li^+ and F^-, given that LiF adopts the NaCl structure, with a unit cell edge of 4.02 Å.

15. Compare the unit cell of zinc blende (Fig. 4-1) with that of diamond [Fig. 8-2(a)]. What similarities are there between these two structures?

16. Use Pauling's method to estimate the ionic radii of Cs^+ and I^-, given that CsI adopts the CsCl structure, with a unit cell edge of 4.56 Å.

17. The distance between a Cs^+ and a Br^- ion in CsBr is 3.72 Å. Knowing that this substance adopts the CsCl structure, determine the unit cell edge of CsBr.

18. Based on the data of Table 4-3, determine the radius ratios (r^-/r^+, using Pauling's values from Table 4-2) for LiH, LiF, CsI, and CsBr, and assign expected coordination numbers for the cations in these compounds.

19. Study the unit cell drawings found in the chapter for NaCl, zinc blende, perovskite, rutile, CsCl, and fluorite. Next, go through the exercise of assigning each atom to one of the following locations of the unit cell: corner, edge, face, or internal to the unit cell. Now deduce the empirical formula of each substance based on the structure of the unit cell, and the number and locations (corner, edge, face, or internal) of the various atoms. *Hint:* Realize that, in an ionic lattice, each corner atom is shared by eight contiguous unit cells. Each corner atom therefore contributes only one-eighth to a given unit cell. Likewise each edge atom is shared by four unit cells, and therefore contributes only one-fourth to any one unit cell. Every face atom similarly contributes only one-half to each unit cell, being shared in ionic substances by two adjoining unit cells. Atoms that reside within the unit cell (internal atoms) are assigned completely to that unit cell, since no other unit cell shares its contribution.

SUPPLEMENTARY READING

Adams, D. M., *Inorganic Solids,* Wiley, New York, 1974.

Dasent, W. E., *Inorganic Energetics,* 2nd ed., Cambridge University Press, London, 1982.

Galasso, F. S., *Structure and Properties of Inorganic Solids,* Pergamon Press, Elmsford, NY, 1970.

Greenwood, N. N., *Ionic Crystals, Lattice Defects and Non-Stoichiometry,* Butterworths, London, 1968.

Hannay, N. B., *Solid-State Chemistry,* Prentice-Hall, Englewood Cliffs, NJ, 1967.

Johnson, D. A., *Some Thermodynamic Aspects of Inorganic Chemistry,* 2nd ed., Cambridge University Press, London, 1982.

Krebs, H., *Fundamentals of Inorganic Crystal Chemistry,* McGraw-Hill, New York, 1968.

Müller, U., *Structural Inorganic Chemistry,* Wiley, NY, 1993.

O'Keeffe, M. and Navrotsky, A., Eds., *Structure and Bonding in Crystals,* Vols. I and II, Academic, New York, 1981.

Wells, A. F., *Structural Inorganic Chemistry,* 5th ed., Oxford University Press, London, 1984.

Chapter 5

THE CHEMISTRY OF SELECTED ANIONS

5-1   Introduction

We have thus far discussed covalent bonding and some of the characteristics of simple ionic compounds, that is, compounds consisting mainly of monatomic cations (e.g., Na^+ or Ca^{2+}) and monatomic anions (e.g., F^- or O^{2-}). However, much of inorganic chemistry deals with ionic compounds of more elaborate types. In these types, either the cation, or the anion, or both of them are polyatomic species, within which there are bonds and stereochemical relationships analogous to those within the uncharged polyatomic species that we call molecules.

The next two chapters consider the properties of anions and cations in more detail, with particular, though not exclusive, reference to the more complex polyatomic members of each group. The chemistry of cations is generally called *coordination chemistry* and is discussed in Chapter 6. Here, the general properties of anions, as well as the specific chemistry of some of the more important ones, are outlined.

One term that must be defined here, in a preliminary way, is *ligands* (although the subject will be covered in detail in Chapter 6). When an anion (or other group) is bonded to a metal ion, it is called a ligand.

We may classify anions as follows:

1. Simple anions, such as O^{2-}, F^-, or CN^-.
2. Discrete oxo anions, such as NO_3^- or SO_4^{2-}.
3. Polymeric oxo anions, such as silicates, borates, or condensed phosphates.
4. Complex halide anions (e.g., TaF_6^-) and anionic complexes containing multibasic anions (e.g., oxalate). An example of an oxalate is $[Co(C_2O_4)_3]^{3-}$.

Some of these, such as the oxide ion O^{2-}, or most silicate anions, can exist only in the solid state. Others, such as chloride ion (Cl^-), can also exist in aqueous solution. Furthermore, some elements that form anions (notably the halogens, O, and S) may be bound to other elements by covalent bonds as in PCl_3, CS_2, or NO_2.

More complex anions, such as dithiocarbamate ($R_2NCS_2^-$) or acetylacetonate ($CH_3COCHCOCH_3^-$), which occur mainly in coordination compounds, are discussed in Chapter 6. The compounds involving carbanions such as CH_3^-, $C_6H_5^-$, or $C_5H_5^-$ (Chapter 29) are described separately, since they constitute a very different class of compounds. Hydride (H^-) and complex hydrido ions (BH_4^- and AlH_4^-) are also more conveniently treated separately (Chapters 9, 12, and 13). The most extensive, important, and varied classes of anions are those containing oxygen, and we discuss them first.

5-2   The Oxide, Hydroxide, and Alkoxide Ions

Oxides

The nature of several important oxide lattices has been discussed in Chapter 4. Discrete O^{2-} ions exist in many oxides but the ion cannot exist in aqueous solutions owing to the hydrolytic reaction

$$O^{2-}(s) + H_2O = 2\ OH^-(aq) \qquad K > 10^{22} \qquad (5\text{-}2.1)$$

As an example, consider Eq. 5-2.2.

$$CaO(s) + H_2O \longrightarrow Ca^{2+}(aq) + 2\ OH^- \qquad (5\text{-}2.2)$$

Thus only those ionic oxides that are insoluble in water are inert to it. When insoluble in water, they usually dissolve in dilute acids, as in Reaction 5-2.3:

$$MgO(s) + 2\ H^+(aq) \longrightarrow Mg^{2+}(aq) + H_2O \qquad (5\text{-}2.3)$$

Ionic oxides function as **basic anhydrides;** they react with water to produce aqueous metal hydroxides (Reaction 5-2.2) or with acids to produce water (Reaction 5-2.3).

In contrast, the covalent oxides of the nonmetals are usually acidic in water.

$$N_2O_5 + H_2O \longrightarrow 2\ H^+(aq) + 2\ NO_3^-(aq) \qquad (5\text{-}2.4)$$

When insoluble in water, as for some of the oxides of less electropositive metals, these **acidic anhydrides** still generally dissolve in base.

$$Sb_2O_5(s) + 2\ OH^- + 5\ H_2O \longrightarrow 2\ Sb(OH)_6^- \qquad (5\text{-}2.5)$$

Basic and acidic oxides will often combine directly, as in Reaction 5-2.6.

$$\underset{\text{Base}}{Na_2O} + \underset{\text{Acid}}{Si_2O} \xrightarrow{\text{fusion}} Na_2SiO_3 \qquad (5\text{-}2.6)$$

Amphoteric oxides behave as bases towards strong acids and as acids toward strong bases. An example is ZnO, as in Reactions 5-2.7 and 5-2.8.

$$ZnO + 2\ H^+(aq) \longrightarrow Zn^{2+} + H_2O \qquad (5\text{-}2.7)$$

$$ZnO + 2\ OH^- + H_2O \longrightarrow Zn(OH)_4^{2-} \qquad (5\text{-}2.8)$$

Some relatively inert oxides dissolve neither in acid nor in base. Examples are N_2O, CO, and MnO_2. When MnO_2 does react with concentrated hydrochloric acid, it is a redox reaction, not an acid–base reaction, which takes place because the Mn^{4+} ion is unstable and reacts with Cl^-, as in Reaction 5-2.9.

$$4\, H^+ + 2\, Cl^- + MnO_2 \longrightarrow Mn^{2+} + 2\, H_2O + Cl_2 \qquad (5\text{-}2.9)$$

Some elements form several oxides. For chromium, the most stable oxide is chromium(III) oxide (Cr_2O_3), which is formed when the metal or other oxides are heated in air. It is amphoteric, as described in the section below on hydrous oxides. The oxide with chromium in the highest oxidation state is chromium(VI) oxide (CrO_3), which is an acidic anhydride.

$$CrO_3 + H_2O \longrightarrow \underset{\text{Chromic acid}}{H_2CrO_4} \qquad (5\text{-}2.10)$$

In contrast, chromium(II) oxide (CrO) is a basic anhydride.

$$CrO + H_2O \longrightarrow \underset{\text{(unstable)}}{Cr(OH)_2} \qquad (5\text{-}2.11)$$

It is typical of all elements capable of forming several oxides that the oxide with the element in the highest formal oxidation state is most acidic, while that with the element in the lowest formal oxidation state is most basic.

Hydroxides

Discrete OH^- ions exist only in the hydroxides of the more electropositive elements such as Na or Ba. For such ionic materials, dissolution in water results in the formation of aquated metal ions and hydroxide ions, as in Reaction 5-2.12

$$M^+OH^-(s) + n\, H_2O \longrightarrow M^+(aq) + OH^-(aq) \qquad (5\text{-}2.12)$$

and the substance is a strong base.

In the limit of an extremely covalent M—O bond, dissociation will occur to varying degrees according to Reaction 5-2.13

$$MOH + n\, H_2O \rightleftharpoons MO^-(aq) + H_3O^+(aq) \qquad (5\text{-}2.13)$$

and the substance must be considered an acid.

Amphoteric hydroxides are those in which there is the possibility of either kind of dissociation, the one of Reaction 5-2.14 being favored by a strong acid

$$M—O—H + H^+ = M^+ + H_2O \qquad (5\text{-}2.14)$$

whereas dissociation according to Reaction 5-2.15

$$M—O—H + OH^- = MO^- + H_2O \qquad (5\text{-}2.15)$$

is favored by a strong base, because the formation of water

$$H^+ + OH^- = H_2O \qquad K_{25\,°C} = 10^{14} \tag{5-2.16}$$

is so highly favored. Similarly, the hydrolytic reactions of many metal ions, which are often written as in Reaction 5-2.17

$$M^{n+} + H_2O = (MOH)^{(n-1)+} + H^+ \tag{5-2.17}$$

can be more realistically written as acid dissociations of the aquo ions, as in Reaction 5-2.18

$$M(H_2O)_x^{n+} = [M(H_2O)_{x-1}(OH)]^{(n-1)+} + H^+ \tag{5-2.18}$$

The higher the positive charge on the metal, the more acidic are the hydrogen atoms of the coordinated water molecules.

The OH$^-$ ion has the ability to form bridges between metal ions. Thus, there are various compounds of the transition metals containing hydroxo bridges between pairs of metal atoms, as in Structure 5-I. Although bridges of the type 5-I are most common, there are also triply bridging hydroxo groups as in Structure 5-II.

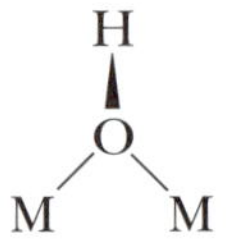

5-I    5-II

Hydrous Oxides

Many so-called metal hydroxides do not have discrete hydroxide ions in the lattice of the crystalline compound. This is because the compounds are actually hydrous metal oxides, or oxides with varying degrees of hydration. Hydroxo bridges are involved in the early stages of the precipitation of hydrous metal oxides. In the case of Fe^{3+}, precipitation of Fe$_2$O$_3 \cdot n$ H$_2$O—commonly, but incorrectly, written Fe(OH)$_3$—proceeds through the following stages on adding OH$^-$

$$[Fe(H_2O)_6]^{3+} \longrightarrow [Fe(H_2O)_5OH]^{2+} \tag{5-2.19}$$
$$\text{pH} < 0 \qquad\qquad 0 < \text{pH} < 2$$

$$\longrightarrow [(H_2O)_4Fe(OH)_2Fe(H_2O)_4]^{4+} \tag{5-2.20}$$
$$\sim 2 < \text{pH} < \sim 3$$

$$\longrightarrow \text{colloidal Fe}_2\text{O}_3 \cdot x\,\text{H}_2\text{O} \tag{5-2.21}$$
$$\sim 3 < \text{pH} < \sim 5$$

$$\longrightarrow \text{Fe}_2\text{O}_3 \cdot n\,\text{H}_2\text{O ppt} \tag{5-2.22}$$
$$\text{pH} \sim 5$$

Similar behavior is exhibited by chromium. The hydrous oxide (Cr$_2$O$_3 \cdot n$ H$_2$O) is precipitated from chromium(III) solution by aqueous ammonia. The hydrous

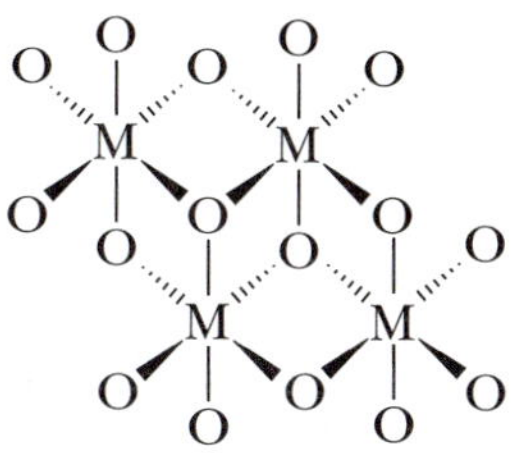

Figure 5-1 An important type of tetrameric structure for $M(OR)_4$ alkoxides. The circles represent entire alkoxide groups.

oxide is amphoteric, reacting not only with acid, as in Reaction 5-2.23,

$$Cr_2O_3 \cdot n\,H_2O + acid \longrightarrow [Cr(H_2O)_6]^{3+} \qquad (5\text{-}2.23)$$

but also with bases to form polymeric chromite ions, $[CrO_2]_x^{n-} \cdot y\,H_2O$.

Alkoxides

The alkoxide ions (RO^-) are analogous to the hydroxide ion. These ions are stronger bases than OH^-, and are therefore hydrolyzed immediately, as in Reaction 5-2.24.

$$RO^- + H_2O = OH^- + ROH \qquad (5\text{-}2.24)$$

Many alkoxides formally analogous to the hydroxides are known [e.g., $Ti(OH)_4$ and $Ti(OR)_4$]. The alkoxides are often polymeric owing to the occurrence of bridging RO^- groups similar to Structures 5-I and 5-II. For example, the structure shown in Fig. 5-1 is a common one for $M(OR)_4$ compounds, where the metal prefers a coordination number of six and the R group is not too large. Note that the structure of Fig. 5-1 contains all three types of RO^- groups: non-bridging (or terminal), doubly bridging, and triply bridging.

Very bulky alkyl or aryl oxides can give complexes with unusually low coordination numbers, for example, square $Cr(py)_2(OAr)_2$, where $Ar = 2,4,6\text{-}t\text{-}BuC_6H_2$. Mixed alkoxides with two or more metals have been much studied since they give mixed oxides on thermal decomposition.

5-3 Oxo Anions

Oxo Anions of Carbon

Both carbonate (CO_3^{2-}) and bicarbonate (HCO_3^-) ions exist in crystalline ionic solids and in neutral or alkaline solutions. There are many naturally occurring carbonates, some of which are very important, such as limestone $(CaCO_3)$.

The ions (Structures 5-III and 5-IV) are planar. In carbonate, because of delocalized π bonding (Section 3-6), the bond lengths are equal, and the bond angles are 120°. The carbonate ion constitutes an AB_3 system.

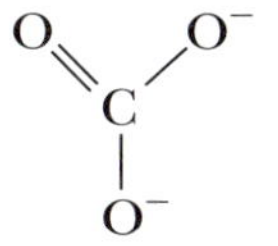

5-III 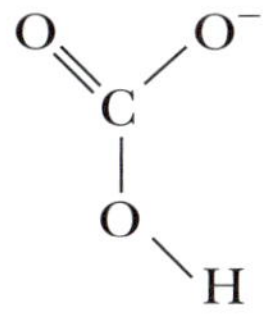 5-IV 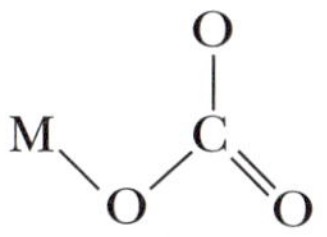 5-V 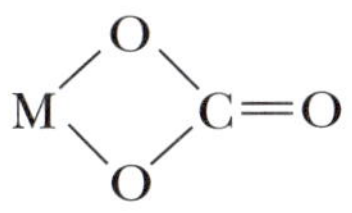 5-VI

The soluble carbonates, such as those of the alkali metals, form solutions that are basic due to the hydrolysis shown in Reaction 5-3.1.

$$CO_3^{2-} + H_2O = HCO_3^- + OH^- \qquad (5\text{-}3.1)$$

The majority of the carbonates are insoluble in water, the principal exceptions being salts of the alkali metals, or of Tl^+ or NH_4^+. When insoluble carbonates are precipitated from aqueous solution, the precipitates are frequently and variously contaminated with hydroxide. This contamination is especially true for the transition metal ions, which have a great affinity for hydroxide.

Like the other oxo anions discussed here, carbonate can act as a ligand, for example, in $[Co(NH_3)_5CO_3]^+$, forming one bond to the metal, as in Structure 5-V. It can also form two bonds to a metal (Structure 5-VI), as in $[Co(NH_3)_4CO_3]^+$.

Oxalate ($C_2O_4^{2-}$) gives insoluble salts with +2 ions such as Cu^{2+}. It is frequently found as a ligand, usually forming two bonds to the same cation, as in $[Cr(C_2O_4)_3]^{3-}$, but it can also act as a bridge.

The *carboxylate* anions have several ways in which they can behave as ligands, as distinct from ionic behavior, in, say, sodium acetate. The main possibilities are Structures 5-VII to 5-IX. The type of structure shown in Structure 5-VIII is quite common and occurs in $Na[UO_2(RCO_2)_3]$. Symmetrical bridging (Structure 5-IX) occurs in the binuclear carboxylates $M_2(CO_2R)_4$ of Cu^{II}, Cr^{II}, Mo^{II}, and Rh^{II}, where four carboxylato bridges are formed.

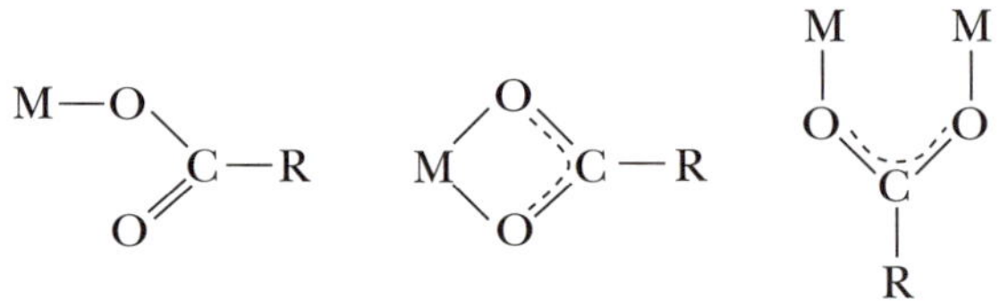

5-VII   5-VIII   5-IX

Oxo Anions of Nitrogen

Nitrite (NO_2^-) occurs normally as an anion only in $NaNO_2$ or KNO_2. It can act as a ligand in several ways (Structures 5-X, 5-XI, and 5-XII):

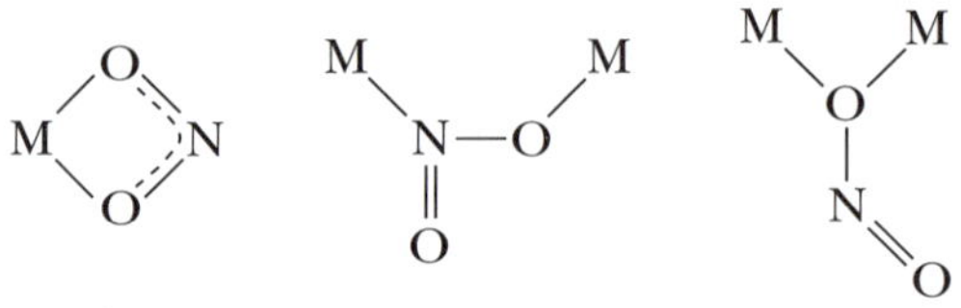

5-X   5-XI   5-XII

The occurrence of a particular form can often be deduced from infrared (IR) spectra. Finally, there are two tautomers: *nitrito* (M—ONO) and *nitro* (M—NO_2). Such tautomers occur for organic compounds. The first inorganic example was discovered by S. M. Jørgensen in 1894 when he isolated the tautomers $[Co(NH_3)_5ONO]Cl_2$ and $[Co(NH_3)_5NO_2]Cl_2$. The nitro isomer is always the more stable one.

Nitrates are made by dissolving the metals, oxides, or hydroxides in HNO_3. The crystalline salts are frequently hydrated and soluble in water. Alkali metal nitrates give nitrites on strong heating; others decompose to the metal oxides, water, and nitrogen oxides.

Like nitrite, nitrate may bond in several ways in complexes (see Structures 5-XIII to 5-XVI). The symmetrical Structure 5-XVI is quite common. Nitrate ion is a relatively weak ligand in aqueous solutions but cations of charge +3 or more are often complexed in solution as MNO_3^{2+}.

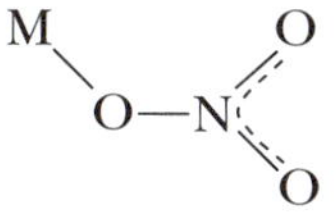

5-XIII 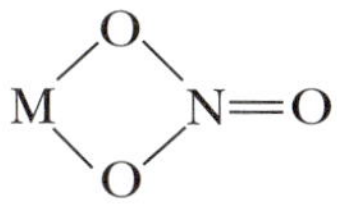   5-XIV 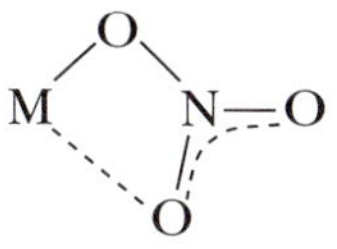   5-XV 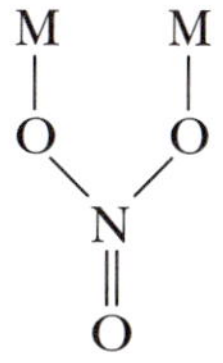   5-XVI

Oxo Anions of Phosphorus

The most important oxo anions of phosphorus are those of P^V. These anions are derived from orthophosphoric acid (H_3PO_4), which is properly written $O{=}P(OH)_3$. Such orthophosphates have tetrahedral PO_4 groups, and are known in one form or another (i.e., PO_4^{3-}, HPO_4^{2-}, or $H_2PO_4^-$) for most metal ions. Some are of practical importance, for example, ammonium phosphate fertilizers, alkali metal phosphate buffers in analysis, and the like. Natural phosphorus minerals are all orthophosphates and a major one is *fluoroapatite*, $Ca_9(PO_4)_6 \cdot CaF_2$. Hydroxy apatites, partly carbonated, make up the mineral part of teeth. The precipitation of insoluble phosphates from 3–6 M HNO_3 is a characteristic of the +4 ions of Ce, Th, Zr, and U. Phosphates also form complexes in aqueous solution with many of the metal ions.

Arsenates generally resemble phosphates and the salts are often isomorphous. However, *antimony* differs in giving crystalline antimonates of the type $KSb(OH)_6$.

Oxo Anions of Sulfur

The common oxo anions of sulfur are sulfite, SO_3^{2-} (pyramidal, see Structure 5-XVII); bisulfite, HSO_3^- (also pyramidal, see Structure 5-XVIII); sulfate, SO_4^{2-} (tetrahedral, see Structure 5-XIX); and bisulfate, HSO_4^- (tetrahedral, see Structure 5-XX).

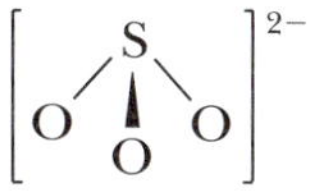

5-XVII 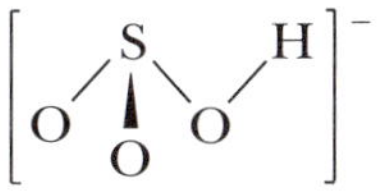   5-XVIII 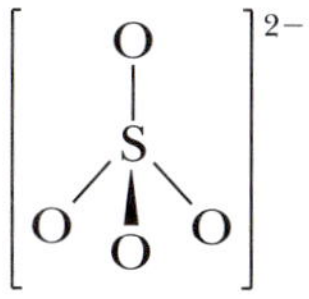   5-XIX 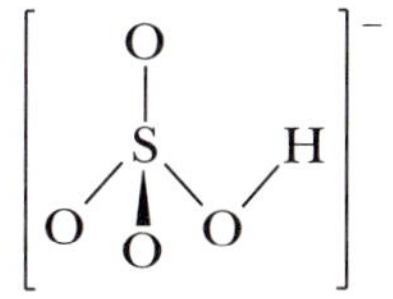   5-XX

The sulfate ion forms many complexes in which it may coordinate to the metal ion through one oxygen atom (Structure 5-XXI), through two oxygen atoms (Structure 5-XXII), or it may serve as a bridge between two metal atoms (Structure 5-XXIII).

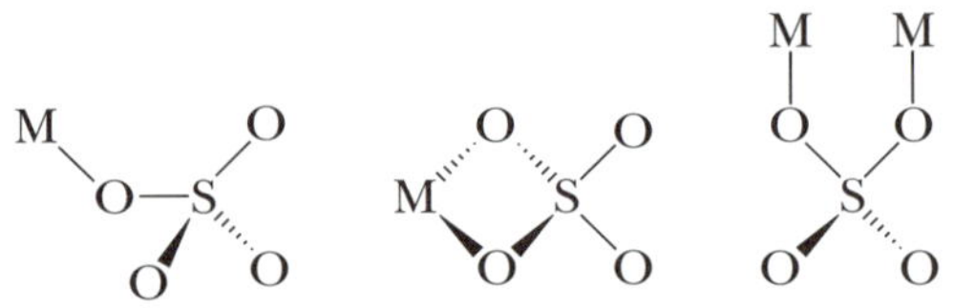

5-XXI   5-XXII   5-XXIII

Selenates are generally similar to the salts of SO_4^{2-} or HSO_4^- and are often isomorphous with them. *Tellurates* are invariably octahedral as in Hg_3TeO_6 or $K[TeO(OH)_5]\cdot H_2O$, and the parent acid is best regarded as $Te(OH)_6$.

Oxo Anions of the Halogens

Chlorates, bromates, and iodates are pyramidal ions (XO_3^-), known almost exclusively in alkali metal salts.

Iodates of +4 ions, Ce, Zr, Hf, Th, and so on, can be precipitated from 6 M HNO_3 and provide a useful separation of these elements.

The most important *perhalate ion* (XO_4^-) is the perchlorate ion (ClO_4^-). It forms soluble salts with virtually all metal ions except the larger alkali ions, K^+, Rb^+, and Cs^+. It is often used to precipitate salts of other large +1 cations, for example, $[Cr(en)_2Cl_2]^+$, where en is ethylenediamine. This is highly inadvisable for organometallic ions such as (η^5-C_5H_5)$_2Fe^+$, as these compounds are often treacherously explosive. It is safer to employ $CF_3SO_3^-$, BF_4^-, or PF_6^- ions. The perchlorate ion has only a small tendency to serve as a ligand and is often used to minimize complex formation. It does, however, have some ability to coordinate, and a few perchlorate complexes are known.

Perbromate ion is a laboratory curiosity. Periodates are of two types: tetrahedral IO_4^- ion and the octahedral ions $IO_2(OH)_4^-$ and $IO_3(OH)_3^{2-}$. Perbromates and periodates are chiefly important as oxidants.

Oxo Ions of the Transition Metals

Tetrahedral oxo anions (MO_4^{n-}) are formed by V^V, Cr^{VI}, Mo^{VI}, W^{VI}, Mn^{VI}, Mn^{VII}, Tc^{VII}, Re^{VII}, Fe^{VI}, Ru^{VII}, and Os^{VII} and can exist in solutions and in crystalline salts. They are not of general utility as anionic ligands. The best known are the permanganate (MnO_4^-) and chromate (CrO_4^{2-}) ions that are widely used as oxidants, but not as anions. We consider their chemistry elsewhere under the appropriate elements.

5-4   Polynuclear Oxo Anions

The oxo anions just discussed have two, three, or four oxygen atoms attached to a central atom to give a discrete anion. However, it is possible for one or more of these oxygen atoms to be shared between two atoms to give an ion with a bridge oxygen. One example of the simplest type is dichromate (Structure 5-XXIV), which is formed from CrO_4^- on acidification. It is essentially two tetrahedra sharing one oxygen atom.

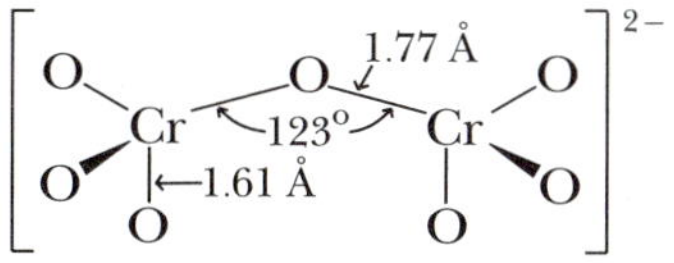

$$
\left[
\begin{array}{c}
\text{structure}
\end{array}
\right]^{2-}
$$

5-XXIV

Silicates and Borates

Silicates are built up on the basis of sharing oxygen atoms of tetrahedral SiO_4 units. *Borates,* which are rather similar, are built up from planar BO_3, or less commonly from tetrahedral BO_4 units. Linking of such units can produce small groups, such as $O_3SiOSiO_3^{6-}$ or $O_2BOBO_2^{4-}$. However, cyclic (Structure 5-XXV), infinite chain (Structure 5-XXVI), and sheet structures can be formed by appropriate oxygen sharing, and are of preeminent importance for silicates. The charges on the anions can be ascertained by regarding nonbridging oxygen atoms as being derived from an —OH group by loss of H^+.

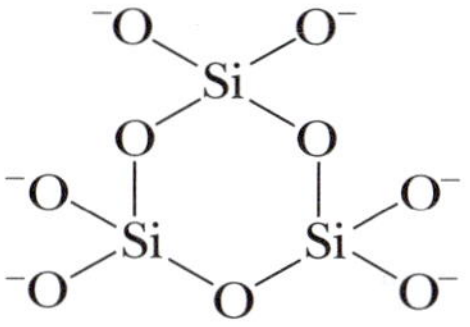

Ring anion
$Si_3O_9^{6-}$
5-XXV

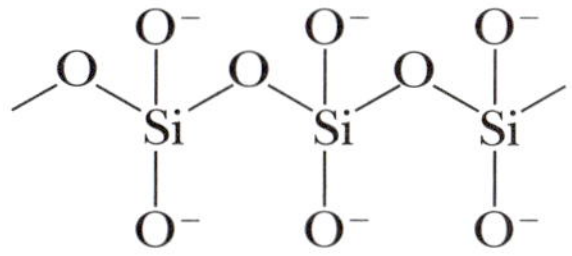

Infinite chain anion (pyroxene)
$(SiO_3^{2-})_n$
5-XXVI

Figure 5-2 shows an infinite sheet of SiO_4 units tetrahedrally linked in a two-dimensional network. The stoichiometry is $(Si_2O_5^{2-})_n$.

In silicate or borate structures, the specific nature of the cations or even their charges are relatively unimportant, so long as the total positive charge is equivalent to the total negative charge. Thus, for the pyroxene structure, which occurs in many minerals, we can have $MgSiO_3$, $CaMg(SiO_3)_2$, $LiAl(SiO_3)_2$, and so on. The cations lie between the chains so that their specific identity is of minor importance in the structure, so long as the required positive charge is supplied. Similarly, for sheet anions, the cations lie *between* sheets. Such substances could be expected to cleave readily. This is found to be so in *micas*, which are sheet silicates.

The final extension to complete sharing of oxygen atoms of each tetrahedron leads, of course, to the structure of SiO_2, *silica*. However, if some of the formally Si^{4+} "ions" are replaced by Al^{3+}, then the framework must have a negative charge—and positive counterions must be distributed through it. Such *framework minerals* are called *aluminosilicates*. They are among the most diverse, widespread, and useful natural silicate minerals. Many synthetic aluminosilicates can be made, and several are manufactured industrially for use as ion exchangers (when wet) and "molecular sieves" (when dry).

Among the most important framework aluminosilicates are the *zeolites*. Their chief characteristic is the openness of the $[(Al, Si)O_2]_n$ framework (Figs. 5-3 and

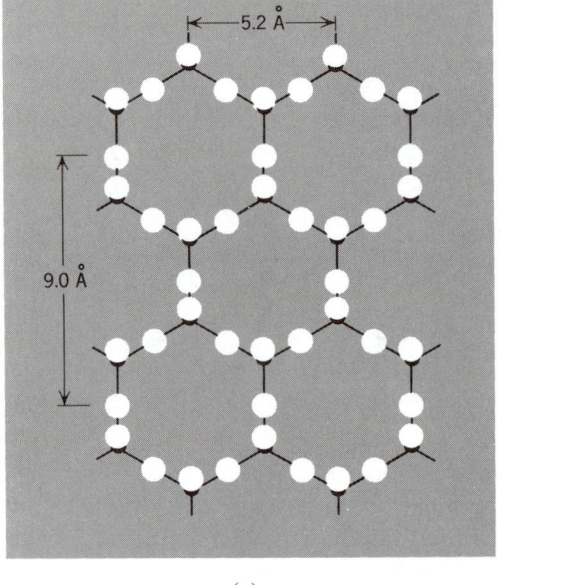

(a)   (b)

Figure 5-2 (a) The hexagonal arrangement of linked SiO_4 tetrahedra giving an infinite sheet of composition ($Si_2O_5^{2-}$), where ● = Si and ○ = O. The Si atoms are coplanar, and each is substantially eclipsed by a terminal (nonlinking) oxygen. (b) The tetrahedral arrangement for each Si atom in (a). The sheet is characterized by three planes: one containing the capping (terminal) O atoms that eclipse each Si in (a), a second plane containing each Si atom, and a third plane formed by the network of bridging O atoms.

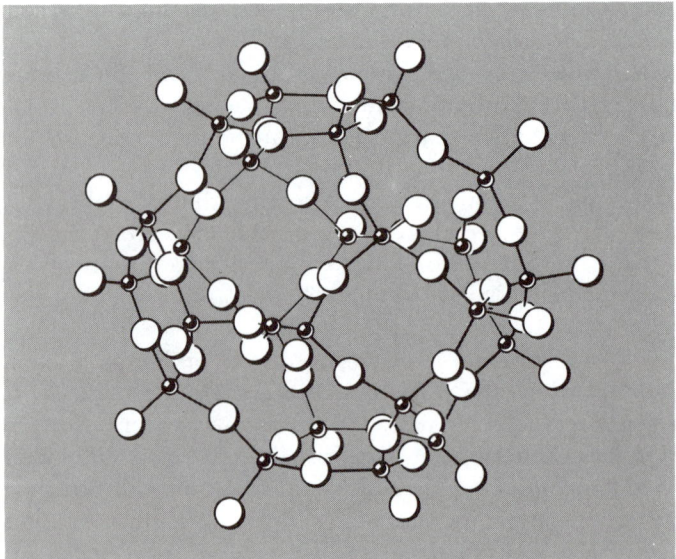

Figure 5-3 The arrangement of AlO_4 and SiO_4 tetrahedra that gives the cubooctahedral cavity in some zeolites and felspathoids. The ● represents Si or Al.

Figure 5-4 Model of a zeolite (edingtonite) showing the channels in the structure. The spheres represent oxygen atoms. The Si and Al atoms lie at the centers of O_4 tetrahedra and cannot be seen. Such a tetrahedron is most easily recognized at the lower right-hand corner of the model.

5-4). The composition is always of the type $M_{x/n}[(AlO_2)_x(SiO_2)_y]\cdot z\,H_2O$ where n is the charge of the metal cation M^{n+}, which is usually Na^+, K^+, or Ca^{2+}, and z is the number of moles of water of hydration, which is highly variable. The openness of these structures results in the formation of channels and cavities of different sizes ranging from 2 to 11 Å in diameter. Molecules of appropriate sizes may thus be trapped in the holes, and it is this property that makes possible their use as selective absorbents. Such zeolites are called "molecular sieves." Zeolites are also used as supports for metals or metal complexes used in heterogeneous catalytic reactions. The zeolites used are mainly synthetic. For example, slow crystallization under precisely controlled conditions of a sodium aluminosilicate gel of proper composition gives the crystalline compound $Na_{12}[(AlO_2)_{12}\text{-}(SiO_2)_{12}]\cdot 27\,H_2O$. This hydrated form can be used as a cation exchanger in basic solution.

In the hydrate all the cavities contain water molecules. In the anhydrous state, which is obtained by heating in vacuum to about 350 °C, the same cavities may be occupied by other molecules brought into contact with the zeolite, providing such molecules are able to squeeze through the aperatures connecting cavities. Molecules within the cavities then tend to be held there by attractive forces of electrostatic and van der Waals types. Thus the zeolite will be able to absorb and strongly retain molecules just small enough to enter the cavities. Those too large to enter will not be absorbed at all, and it will weakly absorb very small molecules or atoms that can enter but also leave easily. For example, straight-chain hydrocarbons but not branched-chain or aromatic ones may be absorbed.

Some *germanates* corresponding to silicates are known, but Ge, Sn, and Pb usually form octahedral anions, $[M(OH)_6]^{2-}$. *Borates* do not form frameworks and are ring or chain polymeric anions. The most common boron mineral, *borax* ($Na_2B_4O_7 \cdot 10\ H_2O$), contains an anion with the Structure 5-XXVII.

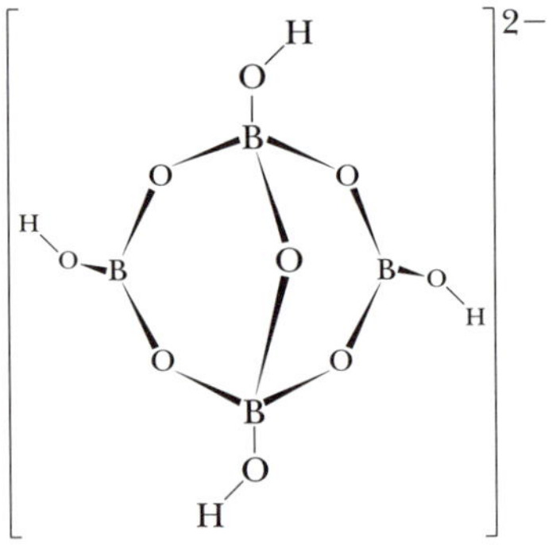

5-XXVII

Polymeric or Condensed Phosphates

Orthophosphate anions can also be linked by oxygen bridges. Three types of building blocks occur (Structures 5-XXVIII to 5-XXX). The resulting polymeric anions are called metaphosphates if they are cyclic (Structure 5-XXXI) or polyphosphates if they are linear (Structure 5-XXXII). Sodium salts of condensed phosphates are widely used as water softeners, since they form soluble

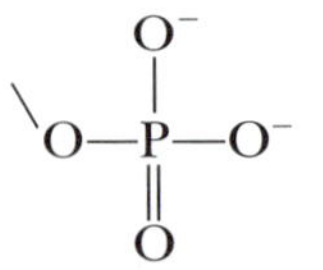

$PO_{3.5}^{2-}$	PO_3^-	$PO_{2.5}$
End Unit	Middle unit	Branching unit
5-XXVIII	5-XXIX	5-XXX

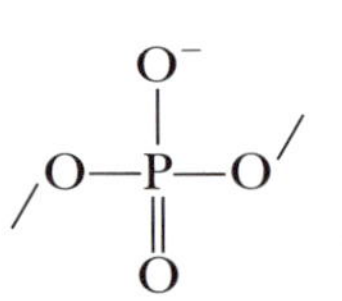

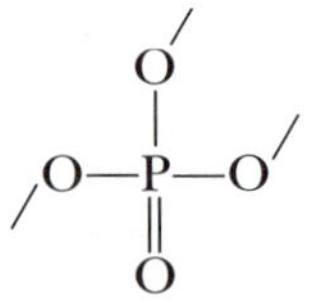

complexes with calcium and other metals. The use of phosphates has led to some ecological problems, since they also act as fertilizers and in lakes can lead to abnormally high growths of algae.

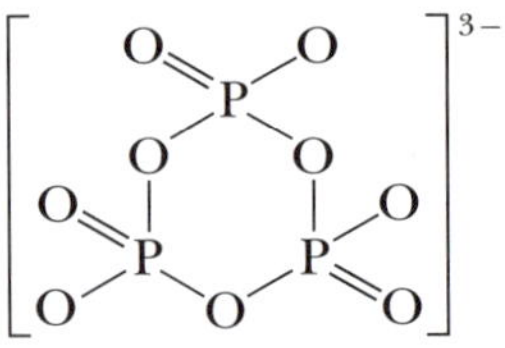

$P_3O_9^{3-}$ $P_3O_{10}^{5-}$
5-XXXI   5-XXXII

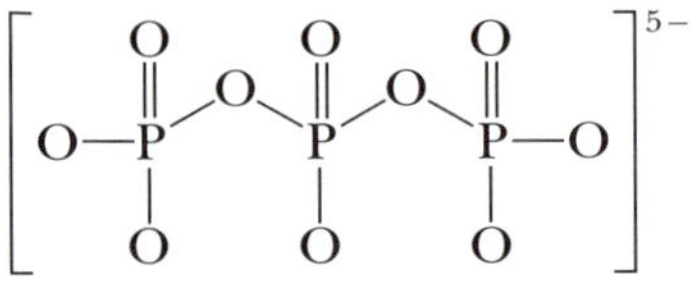

Condensed phosphates are usually prepared by dehydration of orthophosphates under various conditions of temperature (300–1200 °C) and also by appropriate dehydration of hydrated species as, for example,

$$(n-2)\mathrm{NaH_2PO_4} + 2\,\mathrm{Na_2HPO_4} \xrightarrow{\text{heat}} \underset{\text{Polyphosphate}}{\mathrm{Na}_{n+2}\mathrm{P}_n\mathrm{O}_{3n+1}} + (n-1)\mathrm{H_2O} \quad (5\text{-}4.1)$$

$$n\,\mathrm{NaH_2PO_4} \xrightarrow{\text{heat}} \underset{\text{Metaphosphate}}{(\mathrm{NaPO_3})_n} + n\,\mathrm{H_2O} \quad (5\text{-}4.2)$$

They can also be prepared by controlled addition of water to P_4O_{10}. The resulting complex mixtures of anions can be separated by ion exchange or chromatography.

The most important *cyclic* phosphate is *tetrametaphosphate*, which can be prepared by heating copper nitrate with slightly more than an equimolar amount of H_3PO_4 (75%) slowly to 400 °C. The sodium salt can be obtained by treating a solution of the copper salt with Na_2S. Slow addition of P_4O_{10} to ice water gives about 75% of the P as tetrametaphosphate. Condensed arsenates exist only in the solid state, and are rapidly hydrolyzed by water.

Polyanions of the Transition Metals

Next we look at the *transition metal polyanions*. Although we cannot discuss them in detail, the oxo anions of V^V, Nb^V, Ta^V, Mo^{VI}, and W^{VI} form extensive series of what are called *isopoly* and *heteropoly* anions. Both are built up by sharing oxygen atoms in MO_6 octahedra, where corners and edges, but not faces, may be shared. An example is shown in Fig. 5-5.

Isopoly anions, which contain only the element and oxygen, have stoichiometries such as $Nb_6O_{19}^{8-}$ and $Mo_7O_{24}^{6-}$. In heteropoly anions an additional

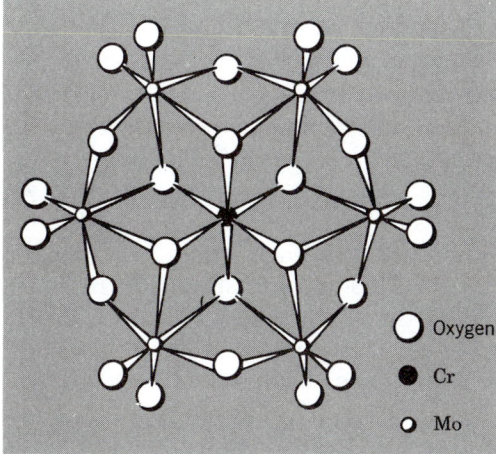

Figure 5-5   The structure of $[\mathrm{CrMo_6O_{24}H_6}]^{3-}$. The hydrogen atoms are probably bound to oxygen atoms of the central octahedron.

metal or nonmetal atom is present. One example is $[Co_2^{II}W_{12}O_{42}]^{8-}$. The heteropoly salt ammonium phosphomolybdate, $(NH_4)_3[P^VMo_{12}O_{40}]$, is used in the determination of phosphorus while the large silicotungstate anion is sometimes used for precipitation of large +1 cations.

5-5   Halogen-Containing Anions

Ionic Halides

Most halides of metals in +1, +2, and +3 oxidation states are predominantly ionic in character. Of course, there is a uniform gradation from halides that are for all practical purposes purely ionic, through those of intermediate character, to those that are essentially covalent. Covalent halides and the preparation of halides are discussed in Chapter 20.

Many metals show their highest oxidation state in the fluorides. For very high oxidation states, which are formed notably with transition metals (e.g., WF_6 or OsF_6), the compounds are generally gases, volatile liquids, or solids closely resembling the covalent fluorides of the nonmetals. The question as to whether a metal fluoride will be ionic or molecular cannot be reliably predicted, and the distinction between the types is not always sharp.

Fluorides in high oxidation states are often hydrolyzed by water, for example,

$$4\ RuF_5 + 10\ H_2O \longrightarrow 3\ RuO_2 + RuO_4 + 20\ HF \qquad (5\text{-}5.1)$$

The driving force for such reactions results from the high stability of the oxides and the low dissociation of HF in aqueous solution.

The halides of the alkali and alkaline earth elements (with the exception of Be) most of the lanthanides, and a few halides of the d-group metals and actinides can be considered as mainly ionic materials. As the charge/radius ratio of the metal ions increases, however, covalence increases. Consider, for instance, the sequence KCl, $CaCl_2$, $ScCl_3$, $TiCl_4$. Potassium chloride is completely ionic, but $TiCl_4$ is an essentially covalent molecular compound. Similarly, for a metal with variable oxidation state, the lower halides will tend to be ionic, whereas the higher ones will tend to be covalent. As examples we can cite $PbCl_2$ and $PbCl_4$, and UF_4, which is an ionic solid, while UF_6 is a gas.

Most ionic halides dissolve in water to give hydrated metal ions and halide ions. However, the lanthanide and actinide elements in the +3 and +4 oxidation states form fluorides insoluble in water. Fluorides of Li, Ca, Sr, and Ba are also sparingly soluble. Lead gives a sparingly soluble salt PbClF, which can be used for gravimetric determination of F^-. The chlorides, bromides, and iodides of Ag^I, Cu^I, Hg^I, and Pb^{II} are also insoluble. The solubility through a series of mainly ionic halides of a given element, $MF_n \rightarrow MI_n$, may vary in either order. In cases where all four halides are essentially ionic, the solubility order will be iodide > bromide > chloride > fluoride, since the governing factor will be the lattice energies, which increase as the ionic radii decrease. This order is found among the alkali, alkaline earth, and lanthanide halides. On the other hand, if covalence is fairly important, it can invert the trend, making the fluoride most and the iodide least soluble, as in the cases of Ag^+ and Hg_2^{2+} halides.

Halide Complex Anions

Complex halogeno anions, especially of fluoride and chloride, are of considerable importance. Halogeno anions may be formed by interaction of a metallic or nonmetallic halide acting as a Lewis acid toward the halide acting as a base:

$$AlCl_3 \ + \ Cl^- \ = \ AlCl_4^- \tag{5-5.2}$$

$$FeCl_3 \ + \ Cl^- \ = \ FeCl_4^- \tag{5-5.3}$$

$$BF_3 \ \ + \ F^- \ = \ BF_4^- \tag{5-5.4}$$

$$PF_5 \ \ + \ F^- \ = \ PF_6^- \tag{5-5.5}$$

Many such halogeno anions can be formed in aqueous solution. The relative affinities of F^-, Cl^-, Br^-, and I^- for a given metal ion are not fully understood. For crystalline materials, lattice energies are important. For BF_4^-, BCl_4^-, and BBr_4^-, the last two of which are known only in crystalline salts of large cations, lattice energies are governing. In considering the stability of the complex ions *in solution,* it is important to recognize that (a) the stability of the complex involves not only the bond strength of the M—X bond, but also its stability relative to the stability of ion–solvent bonds, and (b) in general an entire series of complexes will exist, $M^{n+}(aq)$, $MX^{(n-1)+}(aq)$, $MX_2^{(n-2)+}(aq)$, ..., $MX_x^{(n-x)+}(aq)$, where x is the maximum coordination number of the metal ion. These two points, of course, apply to all types of complexes in solution.

   Generally, the stability decreases in the series F > Cl > Br > I, but with some metal ions the order is the opposite: F < Cl < Br < I. This problem is one of several involving acid–base interactions (see Chapter 7 for a discussion). It is to be emphasized that all complex fluoro "acids," such as HBF_4 and H_2SiF_6, are *necessarily strong*, since the proton can be bound *only* to a solvent molecule.

   Halogeno anions are important in several ways. These anions are involved in many important reactions in which Lewis acids, particularly $AlCl_3$ and BF_3, take part; one example is the Friedel–Crafts reaction. For several elements, they are among the most accessible source materials; platinum as chloroplatinic acid, $(H_3O^+)_2PtCl_6$ and potassium chloroplatinite, K_2PtCl_4, are good examples. Large or undeformable anions like BF_4^- or PF_6^- can be used to obtain sparingly soluble salts of appropriate cations. Finally, halide complex formation can be used for separations with anion-exchange resins. To take an extreme example, Co^{2+} and Ni^{2+} can be separated by passing a strong HCl solution through an anion-exchange column. The Co^{2+} ion readily forms $CoCl_3^-$ and $CoCl_4^{2-}$, whereas nickel does not give chloro complexes in aqueous solutions. Effective separation usually depends on properly exploiting the *difference* in complex formation between two cations *both* of which have some tendency to form anionic halide complexes.

Pseudohalides

Pseudohalides are substances containing two or more atoms that have halogen-like properties. Thus cyanogen (NC—CN) gives the *cyanide* ion (CN^-) and shows halogen-like behavior. Compare

$$Cu^{2+} + 2 \ CN^- = CuCN + \tfrac{1}{2}(CN)_2 \tag{5-5.6}$$

$$Cu^{2+} + 2 \ I^- = CuI + \tfrac{1}{2} I_2 \tag{5-5.7}$$

Other pseudohalide ions are *cyanate* (OCN⁻) and *thiocyanate* (SCN⁻). These ions are formed, respectively, from CN⁻ by oxidation, for example by PbO, and by fusing KCN with S_8. Their Ag^+ salts, like those of the halides, are insoluble in water.

The pseudohalide ions are very good ligands. For cyanate and thiocyanate there are two binding possibilities—through N or through O or S. For OCN, most nonmetals seem to be *N*-bonded in covalent compounds, such as $P(NCO)_3$, while the corresponding thiocyanates are *S*-bonded.

Cyanate and the more numerous thiocyanate complexes usually have stoichiometries similar to the analogous halide complexes.

Cyanide is somewhat different in that the formation of cyanide *complexes* is restricted to transition metal *d*-block elements and Zn, Cd, and Hg. This suggests that π acceptor bonding is important in the binding of CN⁻ to the metal, which is almost invariably through carbon. The π acceptor character of CN⁻ is not nearly so high as for CO, RNC, or similar ligands (Chapter 28). This is clearly reasonable in view of its negative charge. Indeed, CN⁻ is a strong nucleophile, so that back-bonding need not be invoked to explain the stability of its complexes with metals in +2 and +3 oxidation states. However, CN⁻ does have the ability to stabilize metal ions in low oxidation states, for example, $[Ni(CN)_4]^{4-}$. Here, some acceptance of electron density into π* orbitals of CN⁻ is likely.

The majority of cyanide complexes are anionic, typical examples being $[Fe^{II}(CN)_6]^{4-}$, $[Ni(CN)_4]^{2-}$, and $[Mo(CN)_8]^{3-}$. In contrast to similar halide complexes, the free acids of many cyano anions are known, for example, $H_4[Fe(CN)_6]$ and $H_3[Rh(CN)_6]$. The reason is that the proton can be stabilized in hydrogen bonds between the cyano anions, that is, $M-CN \cdots H \cdots NC-M$.

5-6   The Sulfide and Hydrosulfide Anions

Only the alkalis and alkaline earths form sulfides that contain the S^{2-} ion. Only these sulfides dissolve in water. Although S^{2-} is not as extensively hydrolyzed as O^{2-}, nevertheless essentially only SH⁻ ions are present in aqueous solutions owing to the low second dissociation constant of H_2S. The S^{2-} ion is present in strongly alkaline solution, but it cannot be detected in solution less alkaline than 8 *M* NaOH owing to the reaction

$$S^{2-} + H_2O = SH^- + OH^- \qquad K \sim 1 \qquad (5\text{-}6.1)$$

Polysulfide ions S_n^{2-} are formed when solutions of alkali sulfides are boiled with sulfur. Salts can be crystallized. The ions contain kinked chains of sulfur atoms as illustrated by the S_4^{2-}, Structure 5-XXXIII.

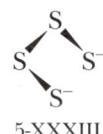

5-XXXIII

Polysulfide ions (and also ions such as thiomolybdate, MoS_4^{2-}) are important ligands for transition metals, forming, for example $[Pt(S_5)_3]^{2-}$, which has *d* and *ℓ* isomers, as shown in Fig. 19-1.

STUDY GUIDE

Scope and Purpose

The structures and chemistries of a number of important classes of anions have been presented. These anions are of particular interest to the discussion of co-ordination chemistry in Chapter 6, because the anions are important both as ligands and as counterions. Further details for each of the systems discussed in this chapter are available in later sections of this book. As usual, the study questions under "A. Review" are intended as a guide to the student.

Study Questions

A. Review

1. Why does the ion O^{2-} exist only in ionic lattices?
2. List the ways in which OH^- can act as a ligand.
3. List the elements that form oxoanions.
4. Many oxoanions can act as ligands in more than one way. Give the ways for
   (a) CO_3^{2-}, (b) SO_4^{2-}, (c) NO_3^-, (d) $CH_3CO_2^-$, (e) NO_2^-.
5. Draw the structures of $Cr_2O_7^{2-}$, $Si_2O_7^{6-}$, and $B_2O_5^{4-}$.
6. How are two-dimensional silicate networks built up?
7. What is the composition of zeolites? What are molecular sieves?
8. How do the oxoanions of Ge, Sn, and Pb differ from silicates?
9. Draw structures for cyclic and linear condensed phosphates.
10. What is meant by the terms iso- and heteropoly anions?

B. Additional Exercises

1. Compare the properties of the oxides of Mg, B, Si, and Sb^V. What are their formulas, and which are acidic and/or basic?
2. Why is the oxide of an element most acidic in the highest oxidation state?
3. Titanium ethoxide is a tetramer, $[Ti(OC_2H_5)_4]_4$. Write a plausible structure for this molecule. Write a balanced equation for its reaction with water.
4. Compare the Lewis diagrams for the simpler oxoanions of S, Se, and Te.
5. What are the structures of the anions in $K_3B_3O_6$, CaB_2O_4, and $KB_5O_8 \cdot 4H_2O$?
6. Draw Lewis diagrams and discuss the nature of the multiple bonding in SO_4^{2-}, NO_3^-, and ClO_4^-. What orbitals are involved in the overlap that leads to π bonding in each case?
7. Draw an MO energy-level diagram for CN^-. What is the highest occupied MO? What is the lowest unoccupied MO? Draw the lowest unoccupied MO and show how it is involved in π bonding with a metal d-type orbital for metal cyanides.
8. Besides cyanide, what other pseudohalides might enter into π bonding with metals? Let the Lewis diagrams for these pseudohalides guide your thinking. Remember to consider d orbitals on atoms other than metals.
9. Draw Lewis diagrams for typical halate and perhalate anions XO_3^- and XO_4^-, respectively, where X = halogen.
10. Predict the products of the reaction of the complexes $[Co(NH_3)_4CO_3]^+$ and $[Co(NH_3)_5CO_3]^+$ with acid.
11. Predict the product upon treating aqueous chromium(III) ion with ammonia.

12. The mineral chromite ($FeCr_2O_4$) can be formed by fusing which two simple, anhydrous oxides?

C. Questions from the Literature of Inorganic Chemistry

1. The following questions should be answered by consulting the paper by A. F. Reid and M. J. Sienko, *Inorg. Chem.*, **1967**, *6*, 531–524.

   (a) Write balanced chemical equations for the solid state reactions used to synthesize $ScTiO_3$ and $ScVO_3$ (two methods).

   (b) What is the oxidation state of Sc in Sc_2O_3 and in the mixed metal oxides $ScVO_3$ and $ScTiO_3$? Based on magnetic susceptibility data available in this article, what oxidation state should be assigned to V in $ScVO_3$ and to Ti in $ScTiO_3$?

   (c) What is the electron configuration (d^n) for the V and Ti ions in $ScVO_3$ and $ScTiO_3$, respectively?

   (d) What is the likely crystal structure for $ScVO_3$ and $ScTiO_3$? How have the authors reached this conclusion?

SUPPLEMENTARY READING

Further details concerning individual anions and classes of anions can be found later in this book and in the following useful references.

Cotton, F. A. and Wilkinson, G., *Advanced Inorganic Chemistry*, 5th ed., Wiley-Interscience, New York, 1988.

Latimer, W. M. and Hildebrand, J. H., *Reference Book of Inorganic Chemistry*, 3rd ed., Macmillan, New York, 1951.

Purcell, K. F. and Kotz, J. C., *Inorganic Chemistry*, Saunders, Philadelphia, PA, 1977.

Wells, A. F., *Structural Inorganic Chemistry*, 5th ed., Clarendon, Oxford, 1984.

Chapter 6

COORDINATION CHEMISTRY

6-1   Introduction

In coordination compounds, metals are surrounded by groups that are called ligands. The types of groups that may surround a metal atom or ion are greatly varied, but they may be broadly considered to be of two types: ligands that bond to metal atoms or ions through carbon atoms, and ligands that do not. The former are involved in organometallic compounds, and we postpone discussion of them until Chapters 28–30. The branch of inorganic chemistry concerned with the remaining combined behavior of cations and their ligands is called coordination chemistry. There is, of course, no sharp dividing line between coordination chemistry and the chemistry of covalent molecules, including organometallic compounds. Nor, in the other extreme, is there a clear distinction between the chemistry of coordination compounds and that of ionic solids. It is, however, traditional and convenient, in discussions of coordination compounds, to view the central metal as a cation, and to view the ligands as Lewis bases.

A few examples will help to illustrate this classification. We traditionally consider that CH_4 and SF_6 are covalent substances, while treating BH_4^- and AlF_6^{3-} as if they were coordination compounds, formally derived from B^{3+} + $4 H^-$ and $Al^{3+} + 6 F^-$, respectively. In terms of fundamental electronic properties, these distinctions would not be easy to defend. Similarly, metal–ligand bonding in $Na_3[AlF_6]$ and $AlF_3(s)$ cannot be qualitatively very different, even though we traditionally call the former a coordination compound (and AlF_6^{3-} a complex ion), and the latter an ionic compound.

The main justification for classifying many substances as coordination compounds is that their chemistry can conveniently be described in terms of a central cation M^{n+}, about which a great variety of ligands L, L′, L″, and so on, may be placed in an essentially unlimited number of combinations. The overall charge on the resulting complex $[ML_xL'_yL''_z\ldots]$ is determined by the charge on M, and the sum of the charges on the ligands. For example, the Pt^{2+} ion forms a great many complexes, studies of which have provided much of our basic knowledge of coordination chemistry. Examples of its complexes, all of which can be interconverted by varying the concentrations of the different ligands are

$$[Pt(NH_3)_4]^{2+} \quad [Pt(NH_3)_3Cl]^+ \quad [Pt(NH_3)_2Cl_2] \quad [Pt(NH_3)Cl_3]^- \quad [PtCl_4]^{2-}$$

For complexes of Pt^{2+} the four ligands lie at the vertices of a square with the Pt^{2+} ion at the center. Thus, structurally, four of the five complexes in this series are unambiguously:

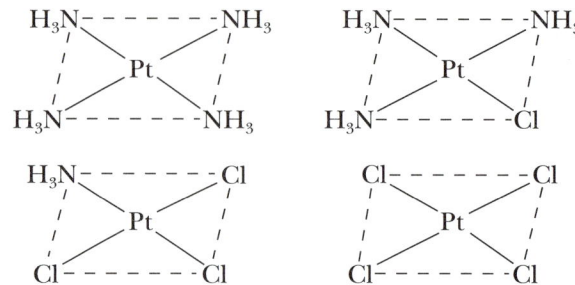

Notice that the structure of the middle member of the series $[Pt(NH_3)_2Cl_2]$ is ambiguous from the formula. Two isomers (cis and trans) are possible and both are well known.

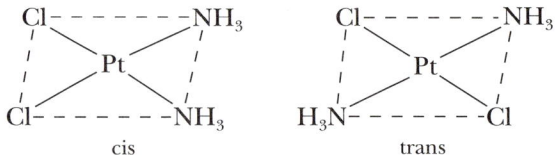

This is one of the simplest examples of the occurrence of isomers among coordination compounds. A number of other important cases will be discussed in Section 6-4.

The fundamental and classical investigations in coordination chemistry were carried out between about 1875 and 1915 by the Danish chemist S. M. Jørgensen (1837–1914) and the Swiss chemist Alfred Werner (1866–1919). When they began their studies the nature of coordination compounds was a huge puzzle, which the contemporary ideas of valence and structure could not accommodate. How, for example, could a stable metal salt (e.g., MCl_n) combine with a group of stable, independently existing molecules (e.g., x NH_3) to form a compound $M(NH_3)_xCl_n$ with wholly new properties? How were bonds formed? What was the structure? Jørgensen and Werner prepared thousands of new compounds, seeking to find regularities and relationships that would suggest answers to these questions. Finally, Werner developed the concept of ligands surrounding a central metal ion—the concept of a coordination complex—and deduced the geometrical structures of many of them. His structure deductions were based on the study of isomers such as those just discussed. In this very instance he reasoned that the arrangement had to be planar to give the two isomers; a tetrahedral structure could not account for their existence. Werner received the Nobel prize in Chemistry for his work in 1913.

6-2   Structures of Coordination Compounds

Coordination Numbers and Coordination Geometries

The term coordination number has already been introduced (Chapter 4) in discussing the packing of ions in crystal lattices. The term is also widely applied to the coordination compounds that are formed between a central metal (a cation or a zero-valent metal) and its ligands. Thus, whether one discusses an array of ions in a crystal lattice or a discrete complex ion (coordination compound), the **coordination number** is the number of groups that immediately surround the metal. In addition to the number of ligands surrounding a metal, it is important to know the arrangement of the ligands: the **coordination geometry.** There is a definite correspondence between coordination geometry and coordination number. The relationship is more complicated than that previously discussed (Chapter 3) between geometry and occupancy in compounds AB_xE_y, because for coordination compounds the number of d electrons can significantly influence geometry. We now discuss the most common coordination numbers, and under each, the most common coordination geometries.

Coordination Number Two

This coordination number is relatively rare, occurring mainly with the $+1$ cations of Cu, Ag and Au, and with Hg^{2+}. The coordination geometry is linear. Examples include the ions $[H_3N—Ag—NH_3]^+$, $[NC—Ag—CN]^-$, and $[Cl—Au—Cl]^-$. Such complexes are typically unstable towards the addition of further ligands as in Reaction 6-2.1.

$$[Cu(CN)_2]^- + 2\ CN^- \longrightarrow [Cu(CN)_4]^{3-} \qquad (6\text{-}2.1)$$

Coordination number two can also be stabilized for other metals by use of bulky ligands such as the bis(triphenylsilylamido) anion, for instance in $Fe[N(SiPh_3)_2]_2$, whose coordination geometry is linear.

Coordination Number Three

The most important geometries for complexes with coordination number three are the trigonal plane and the trigonal pyramid. Examples are the planar HgI_3^- and $[Cu(CN)_3]^{2-}$ and the pyramidal $SnCl_3^-$. The latter can be considered to be derived from the Lewis acid $SnCl_2$ and the Lewis base (ligand) Cl^-, as in Reaction 6-2.2.

$$SnCl_2 + Cl^- \longrightarrow Cl \overset{\overset{\displaystyle \cdot\cdot}{Sn}}{\underset{Cl}{\diagup \ \big\downarrow \ \diagdown}} Cl^- \qquad (6\text{-}2.2)$$

In some cases where the empirical formula might suggest three coordination (e.g., $AlCl_3$, $FeCl_3$, and $PtCl_2PR_3$), there exist, instead, dinuclear structures in which two ligands are shared so as to give each metal center an effective coordination number of four. Two such examples are shown in Structures 6-I and 6-II:

6-I 6-II

Coordination Number Four

This coordination number is very important, since it gives either tetrahedral or square planar coordination geometries. Tetrahedral complexes predominate, being formed almost exclusively by nontransition metals and by transition metals other than those near the right of the d block. The variety of compounds that adopt the tetrahedral geometry is striking. Examples include $Li(H_2O)_4^+$, BeF_4^{2-}, BH_4^-, $AlCl_4^-$, $CoBr_4^{2-}$, ReO_4^-, and $Ni(CO)_4$. Tetrahedral geometry is preferred for valence electron configurations d^0 or d^{10}, as well as for d^n configurations where square planar geometry (or coordination number expansion to an octahedron) is not favored by the number of d electrons. It is the d^8 electron configuration that characteristically leads to square planar geometry. Thus, it is common for complexes of the ions Ni^{2+}, Pd^{2+}, Pt^{2+}, Rh^+, Ir^+, and Au^{3+}. This geometry is also common for complexes of the d^9 ion, Cu^{2+}. The special preference of the d^8 metal ions for the square planar geometry occurs because this requires only one d orbital to be used in forming the four metal–ligand σ bonds (namely, the $d_{x^2-y^2}$ orbital), which has lobes pointing towards the ligands. It is then possible for the four electron pairs of the metal ion to occupy the remaining four d orbitals without being repelled by the electron pairs that form the metal–ligand bonds. For the d^9 case, only one electron has to be placed in the high energy $d_{x^2-y^2}$ orbital.

Coordination Number Five

This coordination number is less common than four or six, but is still very important. The two most symmetrical coordination geometries are the trigonal bipyramid (Structure 6-III) and the square pyramid (Structure 6-IV).

6-III   6-IV

These two geometries (previously discussed in Chapter 3) usually differ little in energy, and one may become converted into the other by small changes in bond angles. Consequently, many five-coordinate complexes do not have either structure precisely, but a structure that is intermediate between the two. Moreover, even those that do have one or the other structure in the crystal may become stereochemically nonrigid in solution, with the ligands interchanging positions rapidly, as explained in Section 6-6. Another interesting illustration of the similar stabilities of the two geometries for coordination number five is afforded by the $[Ni(CN)_5]^{3-}$ ion. This ion forms one crystalline salt in which both geometries are found.

Coordination Number Six

This coordination number is enormously important, since nearly all cations form six-coordinate complexes. Practically all of these have one geometrical form, the octahedron (Structure 6-V). It is essential to recognize that the octahedron is an extremely symmetrical figure, even though some of the stylized ways of drawing it might not show this clearly. All six ligands, and all six M—L bonds, are equivalent in a regular octahedral ML_6 complex.

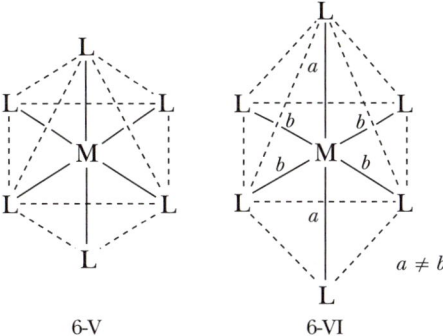

6-V   6-VI

As with other prototype geometries, we continue to describe complexes as "octahedral" even when different kinds of ligands are present and, hence, the full symmetry of the true octahedron cannot be retained. Even in cases where all ligands are chemically the same, octahedra are often distorted, either by electronic effects inherent in the metal ion or by forces in the surroundings. A compression or elongation of one L—M—L axis relative to the other two is called a *tetragonal* distortion (Structure 6-VI), whereas a complete breakdown of the equality of the axes gives a *rhombic* distortion (Structure 6-VII). If the octahedron is compressed or elongated on an axis connecting the centers of two opposite triangular faces, the distortion is called *trigonal* (Structure 6-VIII).

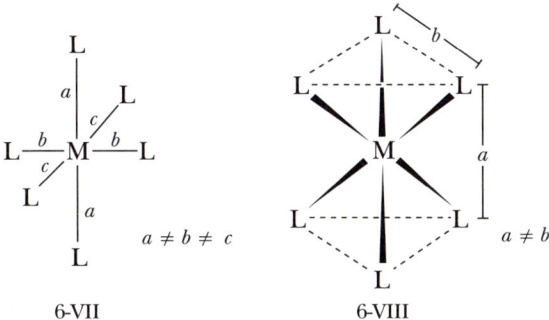

6-VII   6-VIII

There are a few cases in which six ligands lie at the vertices of a trigonal prism (Structure 6-IX). The prism is related to the octahedron in a simple way: If one triangular face of an octahedron is rotated 60° relative to the one opposite to it, a prism is formed. The superior stability of the octahedron compared with the prism has at least two causes. The most evident is steric: On the average, the octahedron allows the ligands to stay further away from each other than the prism does for any given M—L distance. It is also likely that in most cases the metal ion can form stronger bonds to an octahedral set of ligands. The cases where a trigonal prism is found mostly involve either a set of six sulfur atoms, which may interact directly with each other to stabilize the prism, or some sort of rigid cage ligand, which forces the prismatic arrangement.

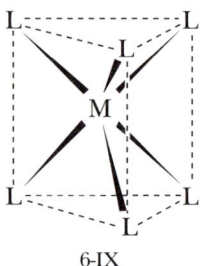

6-IX

Higher Coordination Numbers

Coordination numbers of seven, eight, and nine are not infrequently found for some of the larger cations. In each of these cases there are several geometries that generally do not differ much in stability. Thus complexes with high coordination numbers are characteristically stereochemically nonrigid (Section 6-6).

For seven coordination there are three fairly regular geometries: (1) the pentagonal bipyramid (Structure 6-X), (2) an arrangement derived from the octahedron by spreading one face to make room for the seventh ligand (Structure 6-XIa), and (3) an arrangement similarly derived from a trigonal prism (Structure 6-XIb).

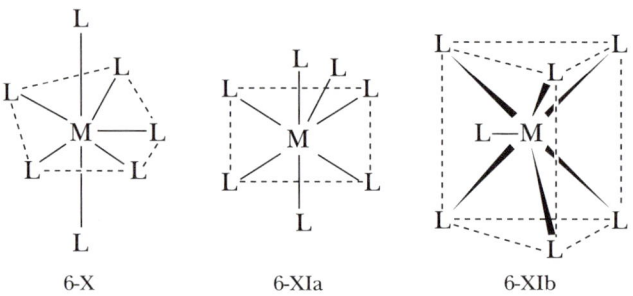

6-X   6-XIa   6-XIb

Coordination number eight also has three important geometries, all of which are shown in Fig. 6-1. The cube itself is rare, since by distorting to either the antiprism or the triangular dodecahedron, interligand repulsions can be diminished while still maintaining close M—L contacts.

For nine coordination the only symmetrical arrangement is that shown in Fig. 6-2. This is observed in many lanthanide compounds in the solid state.

Types of Ligands

The majority of ligands are anions or neutral molecules that can be thought of as electron-pair donors. Common ligands are F^-, Cl^-, Br^-, CN^-, NH_3, H_2O, CH_3OH, and OH^-. When ligands such as these donate one electron pair to one metal atom they are called *monodentate* (literally, one-toothed) ligands. The five complexes of Pt^{2+} mentioned in the Introduction contain only monodentate ligands, Cl^- and NH_3.

Ligands that contain two or more atoms, each of which can simultaneously form a two-electron donor bond to the same metal ion, are called *bidentate* li-

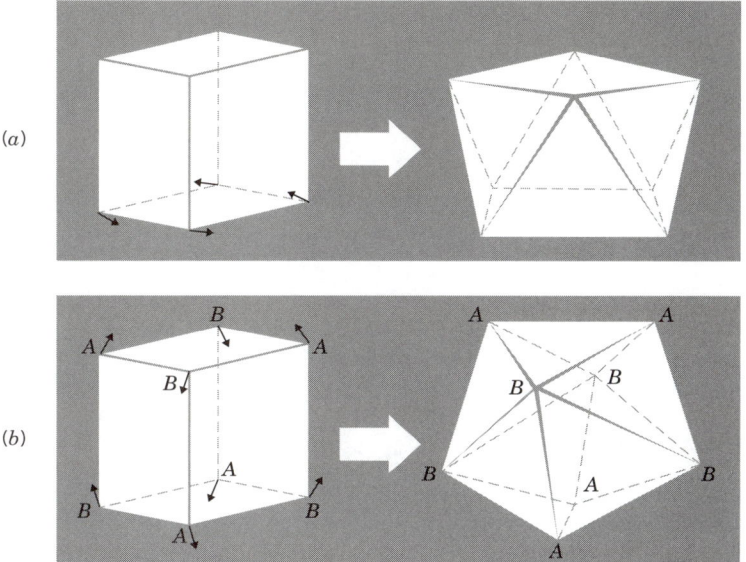

Figure 6-1 Important geometries for eight-coordinate complexes: The cube and its two principal distortions (*a*) to produce a square antiprism, and (*b*) to produce a dodecahedron.

gands. These ligands are also called *chelate* (from the Greek for claw) ligands since they appear to grasp the cation between the two or more donor atoms.

Bidentate Ligands

The most common of the polydentate ligands are bidentate, that is, having two possible points of attachment to a metal ion. Neutral bidentate ligands include the following: diamines, diphosphines, and diethers, all of which form five-membered rings with a metal atom.

$$H_2N \diagdown \qquad \diagup NH_2$$
$$CH_2CH_2$$

Ethylenediamine (en)

$$(C_6H_5)_2P \diagdown \qquad \diagup P(C_6H_5)_2$$
$$CH_2CH_2$$

Bis(diphenylphosphino)ethane (diphos or dppe)

$$(CH_3)_2P \diagdown \qquad \diagup P(CH_3)_2$$
$$CH_2CH_2$$

Bis(dimethylphosphino)ethane (dmpe)

$$\ddot{O} \qquad \ddot{O}$$
$$H_3C \diagup \quad CH_2CH_2 \diagup \quad CH_3$$

Glyme

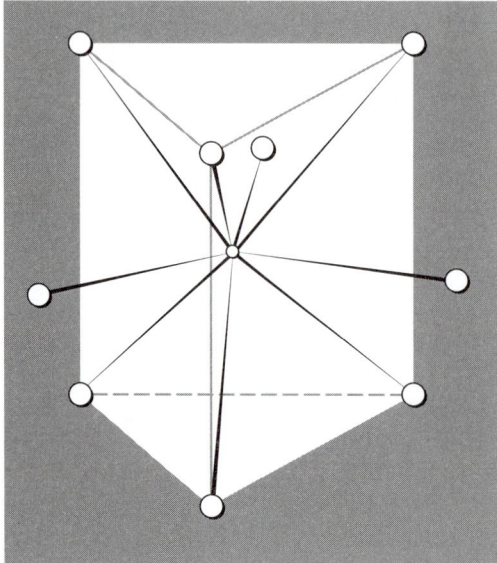

Figure 6-2   The structure of many nine-coordinate complexes. Six ligands (three each, top and bottom) define the trigonal prism that is capped above each rectangular face by one of three "equatorial" ligands.

Two important aromatic amines form five-membered rings with the metal.

2,2′-Bipyridine (bpy)

1,10-Phenanthroline (phen)

The anion of acetylacetone, acetylacetonate (acac), forms a six-membered ring when coordinated to a metal,

Acetylacetonate (acac)

whereas a number of other common anions may form four-membered rings with a metal (although these are often also monodentate).

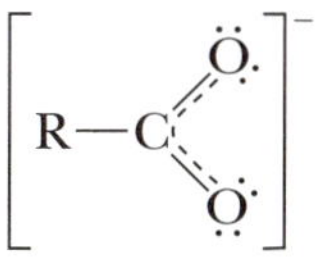

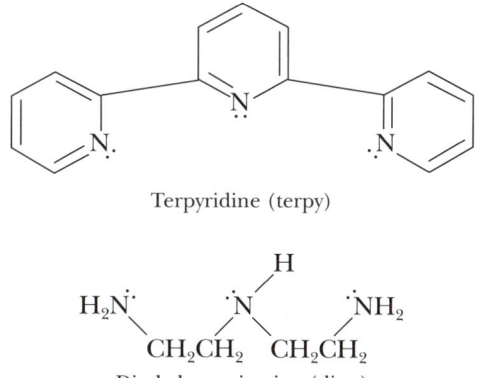

Carboxylates 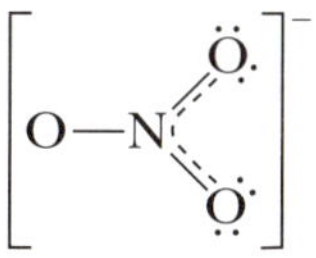   Nitrate

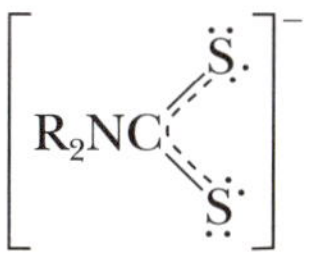

Dithiocarbamates 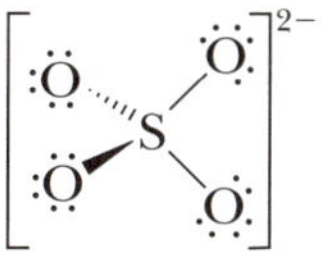   Sulfate

Tridentate Ligands

Two of the most important tridentate ligands are triamines.

Terpyridine (terpy)

Diethylene triamine (dien)

Tetradentate Ligands

There are many important tetradentate ligands. First, we have the bis(dimethylglyoximato) system. It consists of two closely coupled bidentate units that form a planar chelate, locked into planarity by two strong hydrogen bonds.

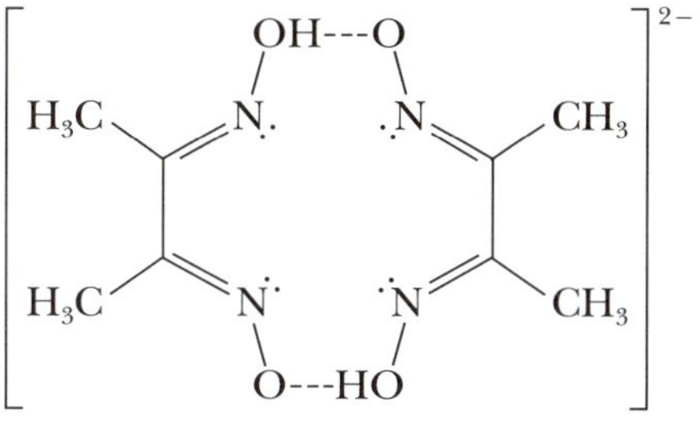

Bis(dimethylglyoximato) (dmgH)

An important open chain tetramine is triethylenetetramine:

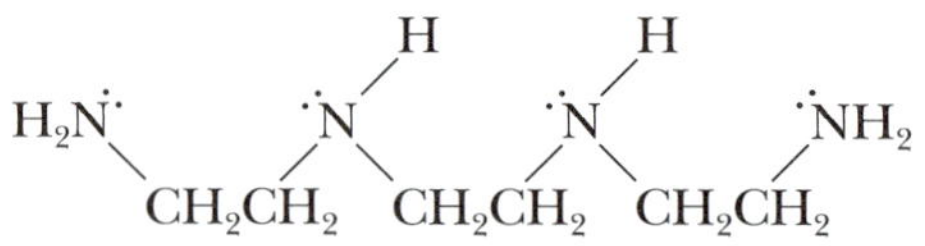

Triethylenetetramine (trien)

In addition, there are open chain, anionic tetradentate ligands. The following Schiff base, which is derived from acetylacetone and ethylenediamine (otherwise known as acacen) is an important example. Perhaps more important are the many "biological" macrocyclic ligands, such as porphyrin (Structure 6-XII) and its derivatives, phthalocyanine (Structure 6-XIII), and a host of similar molecules

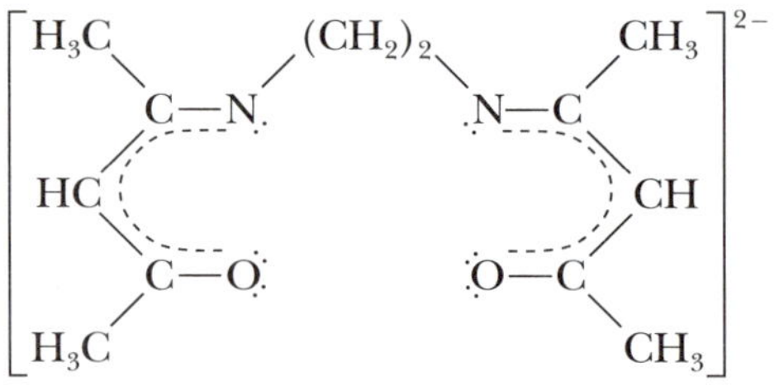

Acacen

that can be readily synthesized (e.g., Structure 6-XIV).

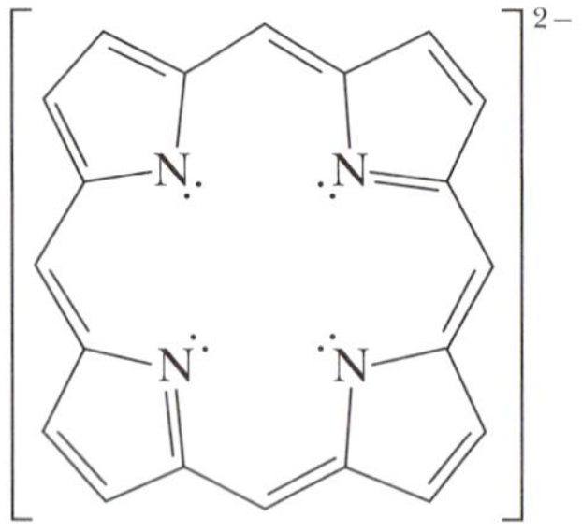

6-XII 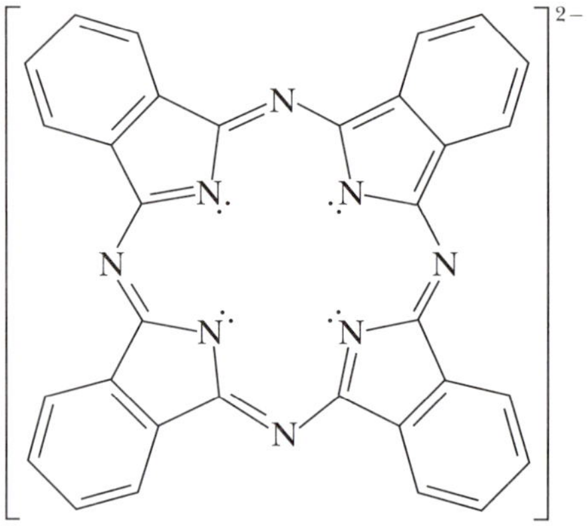   6-XIII 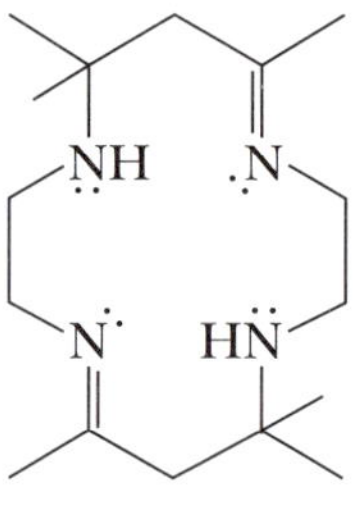   6-XIV

There are also the tripod ligands that favor the formation of trigonal bipyramidal complexes, as shown in Structure 6-XVa. An example is the molecule $N[CH_2CH_2P(C_6H_5)_2]_3$, which coordinates as seen in Structure 6-XVb.

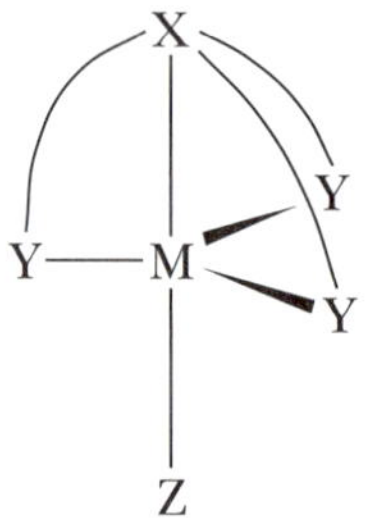

6-XVa 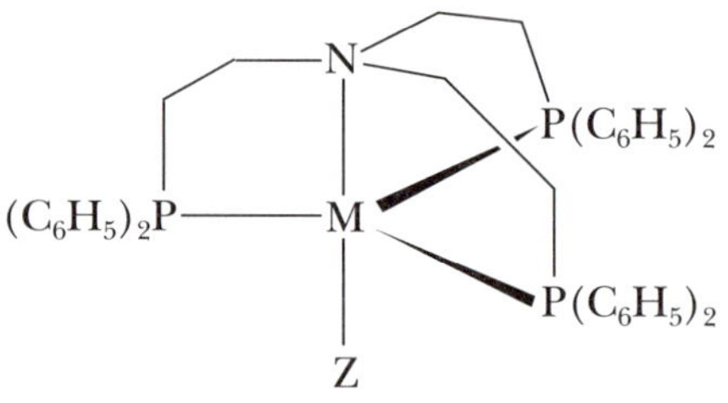   6-XVb

Isomerism in Coordination Compounds

Geometrical Isomerism

One reason coordination chemistry can become quite complicated is that there are many ways in which isomers can arise. We have already observed that square complexes of the type ML_2X_2 can exist as cis and trans isomers. Other important forms of geometrical isomerism are illustrated in Structures 6-XVI to 6-XIX. Isomers of octahedral complexes that are of particular importance are the trans (Structure 6-XVI) and cis (Structure 6-XVII) isomers of the ML_4X_2 species and the facial (Structure 6-XVIII) and meridional (Structure 6-XIX) isomers of ML_3X_3 species.

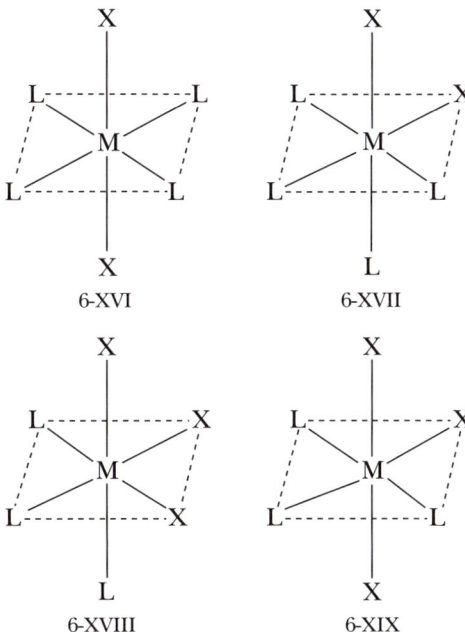

Optical Isomerism

Optical isomers are molecules that are mirror images of each other that cannot be superimposed. Since they cannot be superimposed, they are not identical, even though all their internal distances and angles are identical. These isomers also react identically unless the reactant is also one of a pair of optical isomers. Their most characteristic difference, which gives rise to the term *optical*, is that each one causes the plane of polarization of plane-polarized light to be rotated, but in opposite directions.

Two molecules that are optical isomers in this sense are called enantiomorphs. Their existence was first recognized among organic compounds when a tetrahedral carbon atom was bonded to four different groups, as in lactic acid.

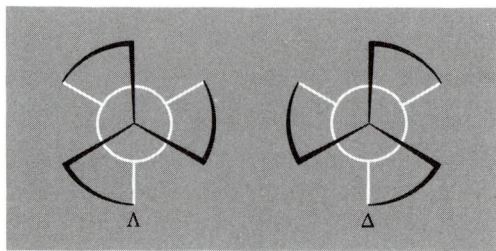

Figure 6-3   Diagrams of tris-chelate complexes showing how the absolute configurations are defined according to twist. The two optical isomers are oriented to show their mirror image relationship. The view for each is along the axis of threefold symmetry.

One of Werner's accomplishments was to recognize that enantiomorphs exist for certain types of octahedral complexes. He prepared and resolved these compounds and used this result to support his hypothesis that the coordination geometry was indeed octahedral. Among the most important enantiomorphous octahedral complexes are those that contain two or three bidentate ligands. The enantiomorphs of a $M(L-L)_2X_2$ complex are shown as Structures 6-XX and 6-XXI. Those of the $M(L-L)_3$ type are Structures 6-XXII and 6-XXIII.

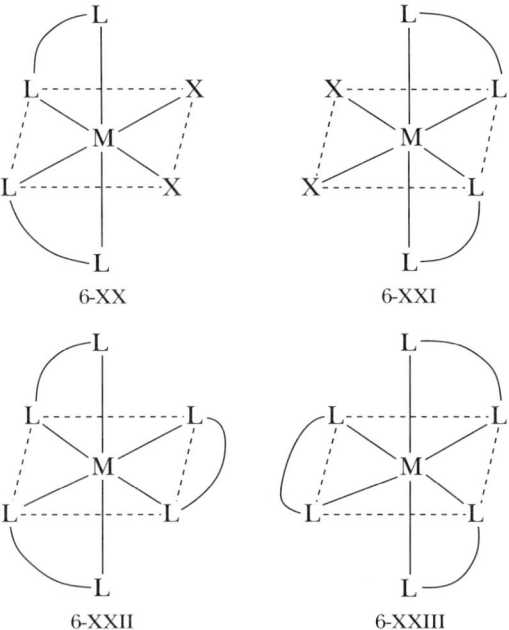

For the latter, which are called tris-chelate complexes, another useful way to regard them is shown in Fig. 6-3, where the view is perpendicular to one pair of opposite triangular faces of the octahedron. Viewed in this way, the molecules have the appearance of helices, like a ship's propellor, with the twist of the helix being opposite in the two cases. Figure 6-3 also defines a notation for the absolute configurations: Λ (Greek capital lambda) for laevo or left; Δ (Greek capital delta) for dextro or right.

Ionization Isomerism
Compounds that have the same empirical formula may still differ with respect to which anions are coordinated to the metal and which are present as

counterions within the crystal lattice. Such isomers yield different ions when dissolved, as illustrated by Reactions 6-2.3 and 6-2.4.

$$[Co(NH_3)_4Cl_2]NO_2 \longrightarrow [Co(NH_3)_4Cl_2]^+ + NO_2^- \qquad (6\text{-}2.3)$$

$$[Co(NH_3)_4Cl(NO_2)]Cl \longrightarrow [Co(NH_3)_4Cl(NO_2)]^+ + Cl^- \qquad (6\text{-}2.4)$$

The two reactants $[Co(NH_3)_4Cl_2]NO_2$ and $[Co(NH_3)_4Cl(NO_2)]Cl$ are ionization isomers. Consider also the three ionization isomers shown in Reactions 6-2.5 to 6-2.7.

$$[Co(en)_2(NO_2)Cl]SCN \longrightarrow [Co(en)_2(NO_2)Cl]^+ + SCN^- \qquad (6\text{-}2.5)$$

$$[Co(en)_2(NO_2)SCN]Cl \longrightarrow [Co(en)_2(NO_2)SCN]^+ + Cl^- \qquad (6\text{-}2.6)$$

$$[Co(en)_2(SCN)Cl]NO_2 \longrightarrow [Co(en)_2(SCN)Cl]^+ + NO_2^- \qquad (6\text{-}2.7)$$

In these illustrations the square brackets are used to enclose the metal atom and all the ligands that are directly bound to it, namely, those groups that reside in the first coordination shell. This use of square brackets is a way of making this distinction in formulas when necessary and will be found in the research literature. These brackets can, however, be omitted when no confusion would arise, as in $Co(NH_3)_3Cl_3$.

The concept of ionization isomerism provides the key to understanding many simple but otherwise puzzling observations. For example, there are three different substances of the composition $CrCl_3 \cdot 6\,H_2O$. One is violet and is $[Cr(H_2O)_6]Cl_3$; it does not lose water over H_2SO_4 and all Cl^- is immediately precipitated by Ag^+ from a fresh solution. The complex $[Cr(H_2O)_5Cl]Cl_2 \cdot H_2O$ is green; it loses one H_2O over H_2SO_4 and only two thirds of its chlorine content is precipitated promptly. The complex $[Cr(H_2O)_4Cl_2]Cl \cdot 2\,H_2O$, which is also green, loses two H_2O over H_2SO_4 and only one third of its chlorine content is promptly precipitated.

Linkage Isomerism

Some ligands can bind in more than one way, and often isomeric complexes with different modes of binding can be isolated. The oldest example is the isomeric pair

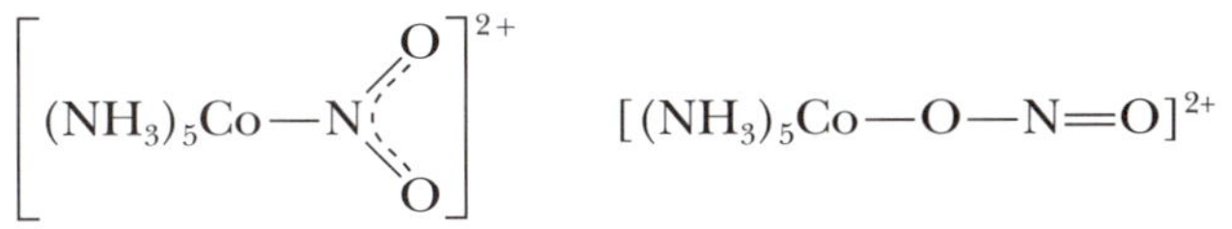

Other ligands prone to give linkage isomers, or at least to bind in different ways in different compounds, are thiocyanate, SCN^- (which may use either S or N as the donor atom) and the sulfoxides, $R_2S{=}O$ (which may use either S or O as the donor). A ligand that can bind in two ways is called an ambidentate ligand.

Coordination Isomerism

In compounds where both the cation and anion are complex, the distribution of ligands can vary, giving rise to isomers. The following are examples:

$$[Co(NH_3)_6][Cr(CN)_6] \quad \text{and} \quad [Cr(NH_3)_6][Co(CN)_6]$$
$$[Cr(NH_3)_6][Cr(SCN)_6] \quad \text{and} \quad [Cr(NH_3)_4(SCN)_2][Cr(NH_3)_2(SCN)_4]$$
$$[Pt^{II}(NH_3)_4][Pt^{IV}Cl_6] \quad \text{and} \quad [Pt^{IV}(NH_3)_4Cl_2][Pt^{II}Cl_4]$$

6-3   Nomenclature for Coordination Compounds

The names of coordination compounds are written by employing rules established by the International Union of Pure and Applied Chemistry (IUPAC). Our discussion of the various rules for nomenclature will be aided by the following four examples, which will be discussed in the context of the pertinent rules.

Example A. $Na[PtCl_3(NH_3)]$
Sodium trichloroammineplatinate(II)

Example B. $K_2[CuBr_4]$
Potassium tetrabromocuprate(II)

Example C.   *trans*-$[Co(en)_2(I)(H_2O)](NO_3)_2$
trans-Iodoaquabis(ethylenediamine)cobalt(III) nitrate

Example D.   *mer*-$Ru(PPh_3)_3Cl_3$
mer-Trichlorotris(triphenylphosphine)ruthenium(III)

RULE 1   The names of neutral coordination compounds are given without spaces. For coordination compounds that are ionic (i.e., the coordination sphere serves as either the cation or the anion of an ionic substance), the cation is named first and separated by a space from the anion, as is customary for all ionic substances. No spaces are used within the name of the coordination ion.

Thus in Examples A and B, the cations sodium and potassium are named first and separated by a space from the names of the anions. In Example C, the portion with the coordination sphere is the cation, and is therefore named first, followed by the name of the counter anion, nitrate. Moreover, there are no spaces within the names of the coordination anions in Examples A and B, or in the name of the coordination cation in Example C. In Example D, since the coordination sphere is a neutral compound, the name is given entirely without spaces.

RULE 2   The name of the coordination compound (whether neutral, cationic, or anionic) begins with the names of the ligands. The metal is listed next, followed in parentheses by the oxidation state of the metal.

In all four examples, regardless of the charge (or lack of charge) on the coordination sphere, the ligands are first named as a set, followed by the metal, and last, the oxidation state of the metal. The latter is always enclosed in parentheses, and can be deduced from the overall charges on the ligands and the net charge on the coordination sphere.

RULE 3   When more than one of a given ligand is bound to the same metal atom or ion, the number of such ligands is designated by the following prefixes:

2 di	6 hexa	10 deca
3 tri	7 hepta	11 undeca
4 tetra	8 octa	12 dodeca
5 penta	9 nona	

However, when the name of the ligand in question already contains one of these prefixes, then a prefix from the following list is used instead:

2 bis	6 hexakis
3 tris	7 heptakis
4 tetrakis	8 octakis
5 pentakis	9 ennea

In Example A, since the name of the Cl⁻ ligand (chloro) does not itself contain a prefix, we are free to use the prefix tri to designate three such ligands. However, Examples C and D are good illustrations of the use of bis and tris to designate two and three ligands, respectively, each of which already contains the prefix di or tri. That is to say, it is the occurrence of the prefix di in ethylene*di*-amine that requires the use of the prefix bis to designate two ethylenediamine ligands in Example C. Moreover, it is the occurrence of the prefix tri in *tri*phenylphosphine that requires the use of the prefix tris, designating three such phosphine ligands in Example D.

RULE 4   With the exceptions to be noted shortly, neutral ligands are given the same name as the uncoordinated molecule, but with spaces omitted. Specific examples are

$(CH_3)_2SO$	dimethylsulfoxide (DMSO)
$(NH_2)_2CO$	urea
C_5H_5N	pyridine
terpy	terpyridine
bpy	2,2′-bipyridine
SO_2	sulfurdioxide
N_2	dinitrogen
O_2	dioxygen
PCl_3	trichlorophosphine
PPh_3	triphenylphosphine
$P(OCH_3)_3$	trimethylphosphite
$OP(CH_3)_3$	trimethylphosphineoxide

There are, however, some neutral molecules which, when serving as ligands, are given special names. These are

NH_3	ammine
H_2O	aqua
NO	nitrosyl
CO	carbonyl
CS	thiocarbonyl

In Examples A–D, the names of the neutral ligands are ammine, ethylene-diamine, aqua, and triphenylphosphine. These neutral ligands are distinguished from anionic ones by the fact that the latter are given names that end in "o," according to Rule 5.

RULE 5   Anionic ligands are given names that end in the letter "o." When the name of the free, uncoordinated anion ends in "ate," the ligand name is changed to end in "ato." Some examples are

$CH_3CO_2^-$ (acetate)	acetato
SO_4^{2-} (sulfate)	sulfato
CO_3^{2-} (carbonate)	carbonato
acac	acetylacetonato

When the name of the free, uncoordinated anion ends in "ide," the ligand name is changed to end in "ido." Some examples are

N^{3-} (nitride)	nitrido
N_3^- (azide)	azido
NH_2^- (amide)	amido
$(CH_3)_2N^-$ (dimethylamide)	dimethylamido

When the name of the free, uncoordinated anion ends in "ite," the ligand name is changed to end in "ito." Some examples are

NO_2^- (nitrite)	nitrito
SO_3^{2-} (sulfite)	sulfito
ClO_3^- (chlorite)	chlorito

Certain anionic ligands are given special names, all ending in "o":

CN^-	cyano
F^-	fluoro
Cl^-	chloro
Br^-	bromo
I^-	iodo
O^{2-}	oxo
O_2^{2-}	peroxo
O_2^-	superoxo
OH^-	hydroxo
H^-	hydrido
CH_3O^-	methoxo

Organic groups, although implicitly considered to be anions, are given their regular names, without an "o" ending. Some examples are

CH_3 (Me)	methyl
C_2H_5 (Et)	ethyl
C_3H_7 (Pr)	propyl
C_6H_5 (Ph)	phenyl

In Examples A–D, the only anionic ligands are chloro and iodo. The nitrate anion in Example C is not named as a ligand (i.e., it does not end in "o") because it is not coordinated to the metal ion, serving only as the counterion.

RULE 6 The ligands are named in groups, according to charge. All anionic ligands are named first. Neutral ligands are named in the second group, and in rare cases where they occur, cationic ligands are named last. Within each charge group, the ligands are named in alphabetical order, ignoring the prefixes that are used to designate the number of each ligand.

In Example A, the chloro ligands, being anions, are named before the ammine ligand, which is neutral. In Example C, the anionic iodo ligand is named first. Then the neutral ligands are named, aqua coming before ethylenediamine. In Example D, the anionic chloro ligands are named before the neutral triphenylphosphine ligands. (Notice that the prefixes bis and tris were ignored for purposes of assessing alphabetical order.)

RULE 7 When the coordination entity is either neutral or cationic (as in Examples D and C, respectively), the usual name of the metal is used, followed in parentheses by the oxidation state of the metal. However, when the coordination entity is an anion, the name of the metal is altered to end in "ate." This is done for some metals by simply changing the ending "ium" to "ate":

scandium	scandate
titanium	titanate
chromium	chromate
zirconium	zirconate
niobium	niobate
ruthenium	ruthenate
rhodium	rhodate
palladium	palladate
rhenium	rhenate

For other metals, the name is given the ending "ate":

manganese	manganate
cobalt	cobaltate
nickel	nickelate
molybdenum	molybdate
tantalum	tantalate
tungsten	tungstate
platinum	platinate

Finally, the names of some metals are based on the Latin name of the element:

iron	ferrate
copper	cuprate
silver	argentate
gold	aurate

RULE 8 Optical isomers are designated by the symbols Δ or Λ. Geometrical isomers are designated by *cis* or *trans* and *mer* or *fac,* the latter two standing for meridional or facial, respectively.

RULE 9 Bridging ligands are designated with the prefix μ-. When there are two bridging ligands of the same kind, the prefix di-μ- is used. Bridging ligands are listed in order with the other ligands, according to Rule 6, and set off between hyphens. An important exception arises when the molecule is symmetrical, and a more compact name can be given by listing the bridging ligand first. The following examples illustrate Rule 9:

$$[(NH_3)_5Co—NH_2—Co(NH_3)_4(H_2O)]Cl_5$$

Pentaamminecobalt(III)-μ-amidotetraammineaquacobalt(III) chloride

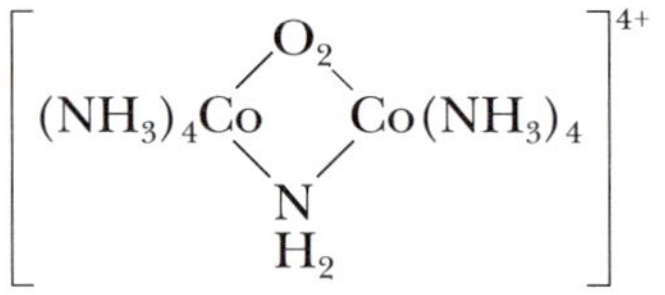

Tetraamminecobalt(III)-μ-amido-μ-superoxotetraamminecobalt(III)

The bridging —O_2— group in the above example is named for the superoxide anion O_2^-, because physical data suggest the −1 charge.

$$[(NH_3)_5Cr—OH—Cr(NH_3)_5]Br_5$$

μ-Hydroxobis[pentaamminechromium(III)] bromide

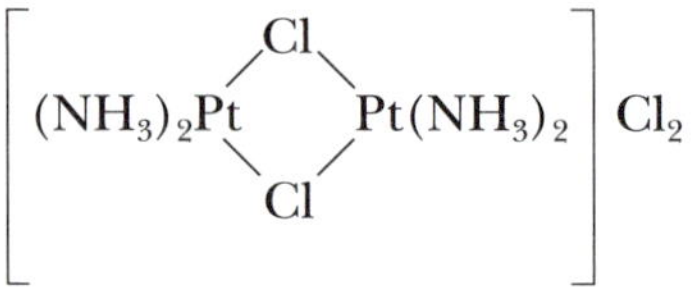

Di-μ-chlorobis[diammineplatinum(II)] chloride

RULE 10   Ligands that are capable of linkage isomerism are given specific names for each mode of attachment. Common examples are

—SCN⁻	thiocyanato
—NCS⁻	isothiocyanato
—SeCN⁻	selenocyanato
—NCSe⁻	isoselenocyanato
—NO₂⁻	nitro
—ONO⁻	nitrito

RULE 11   Compounds that are hydrated (contain a fixed and crystallographically distinct number of water molecules of hydration in the crystalline solid) are so designated as the last step in constructing the name of a coordination compound:

· H_2O	monohydrate
· 1.5 H_2O	sesquihydrate
· 2 H_2O	dihydrate
· 3 H_2O	trihydrate, and so on

The designation of hydration is customarily set off from the name of the compound by a space.

Some further examples are now presented. Note that in the chemical formula of a substance, the metal is listed first (not last as in the name), and that it is a common (though not universal) practice to enclose the coordination sphere in square brackets.

1. $$[Co(NH_3)_5CO_3]Cl$$
   Carbonatopentaamminecobalt(III) chloride

2. $$[Cr(H_2O)_4Cl_2]Cl$$
   Dichlorotetraaquachromium(III) chloride

3. $$K_2[OsCl_5N]$$
   Potassium pentachloronitridoosmate(VI)

4. $$[Ph_4As][PtCl_2(H)(CH_3)]$$
   Tetraphenylarsonium dichlorohydridomethylplatinate(II)

5. $$Mo(Ph_2PCH_2CH_2PPh_2)_2(N_2)_2$$
   Bis(1,2-diphenylphosphinoethane)bis(dinitrogen)molybdenum(0)

6. $$K_3[Fe(CN)_5NO]\cdot2\,H_2O$$
   Potassium pentacyanonitrosylferrate(II) dihydrate

6-4   The Stability of Coordination Compounds

Equilibrium Constants for the Formation of Complexes in Solution

The formation of complexes in aqueous solution is a matter of great importance not only in inorganic chemistry but also in biochemistry, analytical chemistry, and in a variety of applications. The extent to which a cation combines with ligands to form complex ions is a thermodynamic problem and can be treated in terms of appropriate expressions for equilibrium constants.

Suppose we put a metal ion M and some monodentate ligand L together in solution. If we assume that no insoluble products or any species containing more than one metal ion are formed, then equilibrium expressions of the following sort will describe the system:

$$M + L = ML \qquad K_1 = \frac{[ML]}{[M][L]}$$

$$ML + L = ML_2 \qquad K_2 = \frac{[ML_2]}{[ML][L]}$$

$$ML_2 + L = ML_3 \qquad K_3 = \frac{[ML_3]}{[ML_2][L]}$$

$$\vdots \qquad \vdots \qquad \vdots \qquad \vdots$$

$$ML_{N-1} + L = ML_N \qquad K_N = \frac{[ML_N]}{[ML_{N-1}][L]} \tag{6-4.1}$$

There will be N such equilibria, where N represents the maximum coordination number of the metal ion M for the ligand L. The parameter N may vary from one ligand to another. For instance, Al^{3+} forms $AlCl_4^-$ and AlF_6^{3-}, and Co^{2+} forms $CoCl_4^{2-}$ and $Co(NH_3)_6^{2+}$, as the highest complexes with the ligands indicated.

Another way of expressing the equilibrium relations is the following:

$$M + L = ML \qquad \beta_1 = \frac{[ML]}{[M][L]}$$

$$M + 2L = ML_2 \qquad \beta_1 = \frac{[ML]}{[M][L]}$$

$$M + 3L = ML_3 \qquad \beta_3 = \frac{[ML_3]}{[M][L]^3}$$

$$\vdots \qquad\qquad \vdots \quad \vdots \qquad\qquad \vdots$$

$$M + NL = ML_N \qquad \beta_N = \frac{[ML_N]}{[M][L]^N} \qquad (6\text{-}4.2)$$

Since there can be only N independent equilibria in such a system, it is clear that the K_i's and the β_i's must be related. The relationship is indeed rather obvious. Consider, for example, the expression for β_3. Let us multiply both numerator and denominator by $[ML][ML_2]$, and then rearrange slightly:

$$\beta_3 = \frac{[ML_3]}{[M][L]^3} \cdot \frac{[ML][ML_2]}{[ML][ML_2]}$$

$$= \frac{[ML]}{[M][L]} \cdot \frac{[ML_2]}{[ML][L]} \cdot \frac{[ML_3]}{[ML_2][L]}$$

$$= K_1 K_2 K_3 \qquad (6\text{-}4.3)$$

It is not difficult to see that this kind of relationship is perfectly general, namely:

$$\beta_k = K_1 K_2 K_3 \ldots K_k = \prod_{i=1}^{i=k} K_i \qquad (6\text{-}4.4)$$

The K_i's are called the *stepwise formation constants* (or stepwise stability constants), and the β_i's are called the *overall formation constants* (or overall stability constants); each type has its special convenience.

The set of stepwise formation constants (K_i's) provide particular insight into the species present as a function of concentrations. With only a few exceptions, there is generally a slowly descending progression in the values of the K_i's in any particular system. This is illustrated by the data for the Cd^{2+}—NH_3 system, where the ligands are uncharged, and by the Cd^{2+}—CN^- system where the ligands are charged.

$$Cd^{2+} + NH_3 = [Cd(NH_3)]^{2+} \qquad K = 10^{2.65}$$

$$[Cd(NH_3)]^{2+} + NH_3 = [Cd(NH_3)_2]^{2+} \qquad K = 10^{2.10}$$

$$[Cd(NH_3)_2]^{2+} + NH_3 = [Cd(NH_3)_3]^{2+} \qquad K = 10^{1.44}$$

$$[Cd(NH_3)_3]^{2+} + NH_3 = [Cd(NH_3)_4]^{2+} \qquad K = 10^{0.93} \;\; (\beta_4 = 10^{7.12}) \quad (6\text{-}4.5)$$

$$Cd^{2+} + CN^- = [Cd(CN)]^+ \qquad K = 10^{5.48}$$

$$[Cd(CN)]^+ + CN^- = [Cd(CN)_2] \qquad K = 10^{5.12}$$

$$[Cd(CN)_2] + CN^- = [Cd(CN)_3]^- \qquad K = 10^{4.63}$$

$$[Cd(CN)_3]^- + CN^- = [Cd(CN)_4]^{2-} \qquad K = 10^{3.55} \ (\beta_4 = 10^{18.8}) \quad (6\text{-}4.6)$$

Thus, as ligand is added to the solution of metal ion, ML first forms more predominantly than any other complex in the series. As addition of ligands is continued, the ML_2 concentration rises rapidly, while the ML concentration drops. Then ML_3 becomes dominant, ML and ML_2 become unimportant, and so on, until the highest complex ML_N is formed to the nearly complete exclusion of all others at very high ligand concentrations.

A steady decrease in K_i as i increases is almost always observed, although occasional exceptions occur because of unusual steric or electronic effects. The principal reason for the decrease is statistical. At any given step, say from ML_n to ML_{n+1}, there is a certain probability for the complexes ML_n to gain another ligand, and a different probability for ML_{n+1} to lose a ligand. As n increases, there are more ligands to be lost and fewer places $(N - n)$ in the coordination shell to accept additional ligands. For a series of steps ML to $ML_2, \ldots , ML_5$ to ML_6, the magnitude of log K_i tends to decrease by about 0.5 at each step for statistical reasons alone.

Many methods of chemical analysis and separation are based on the formation of complexes in solution, and accurate values for formation constants are helpful. For example, different transition metal ions can be selectively determined by complexation with the hexadentate chelate $EDTA^{4-}$, shown in Fig. 6-4. By adjusting the concentration of $EDTA^{4-}$ and the pH, one ion can be complexed while another ion (which is simultaneously in solution) is not complexed. This is the basis for the determination of Th^{4+} in the presence of divalent cations. The analysis is made possible by the large difference in formation constants for the $EDTA^{4-}$ complexes of the 4+ and 2+ cations. The $EDTA^{4-}$ ligand is less selective among ions of like charge, but the addition of CN^- allows the determination of the alkaline earth cations in the presence of the cations of Zn,

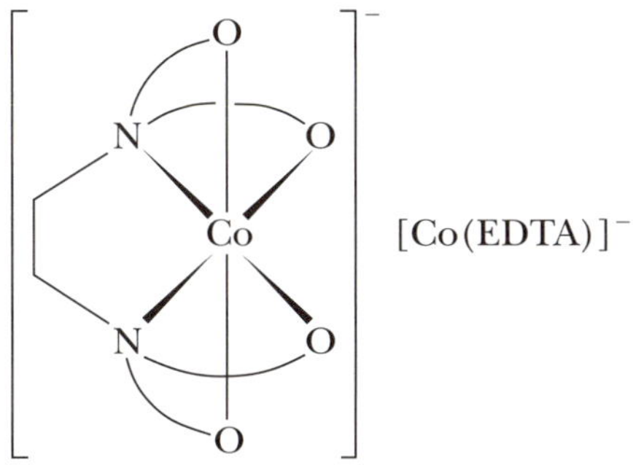

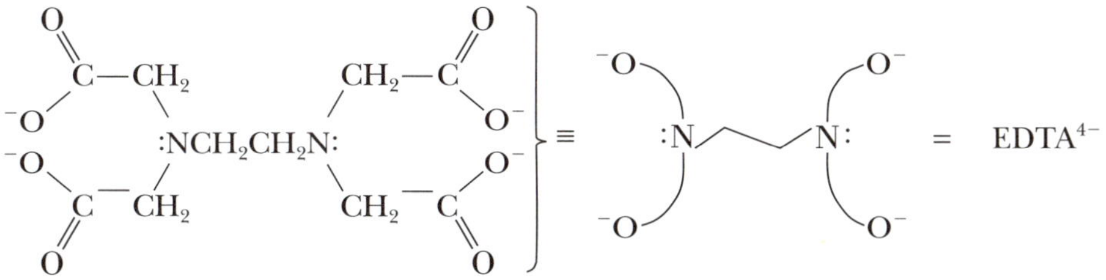

Figure 6-4   The chelation of a metal by the hexadentate ligand $EDTA^{4-}$.

Cd, Cu, Co, and Ni, because the latter form more stable complexes with CN^- than with $EDTA^{4-}$.

The Chelate Effect

As a general rule, a complex containing one (or more) five- or six-membered chelate rings is more stable (has a higher formation constant) than a complex that is as similar as possible but lacks some or all of the chelate rings. A typical illustration is

$$Ni^{2+}(aq) + 6\ NH_3(aq) =$$

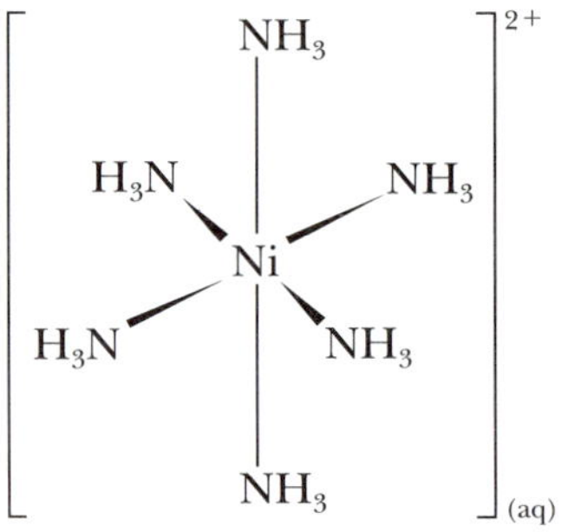

$$\beta_6 = 10^{8.6}$$

$$Ni^{2+}(aq) + 3\ H_2NCH_2CH_2NH_2(aq) =$$

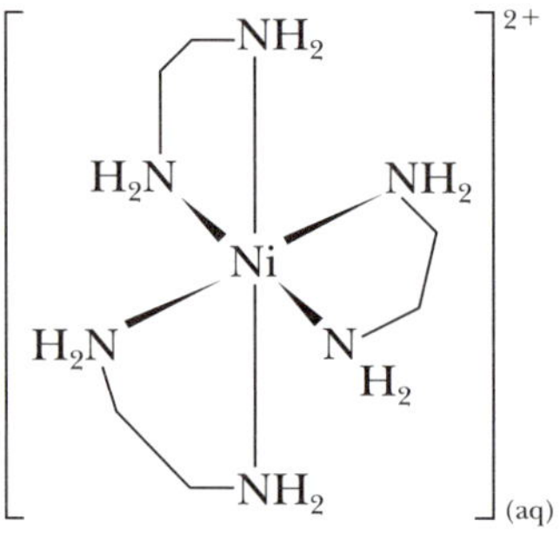

$$\beta_3 = 10^{18.3}$$

The complex with three chelate rings is about 10^{10} times more stable. Why should this be true? As with all questions concerning thermodynamic stability, we are dealing with free energy changes ($\Delta G°$) and we first look at the contributions of enthalpy and entropy, to see if one or the other is the main cause of the difference.

We can more directly compare these two reactions by combining them in the equation

$$Ni(NH_3)_6^{2+}(aq) + 3(en)(aq) = Ni(en)_3^{2+}(aq) + 6\ NH_3(aq) \qquad (6\text{-}4.7)$$
$$(en = \text{ethylenediamine})$$

for which

$$K = 10^{9.7}$$

$$\Delta G° = -RT \ln K = -67\ \text{kJ mol}^{-1} = \Delta H° - T\ \Delta S°$$

$$\Delta H° = -12\ \text{kJ mol}^{-1}$$

$$-T\ \Delta S° = -55\ \text{kJ mol}^{-1}$$

It is evident that both enthalpy and entropy favor the chelate complex, but the entropy contribution is far more important. Data for a large number of these reactions, with many different metal ions and ligands, show that enthalpy contributions to the chelate effect are sometimes favorable, sometimes unfavorable, but always relatively small. The general conclusion is that *the chelate effect is essentially an entropy effect.* The reason for this is as follows.

The nickel ion is coordinated by six H_2O molecules. In each of the first two reactions, these six H_2O molecules are liberated when the nitrogen ligands become coordinated. On that score, the two processes are equivalent. However, in one case *six* NH_3 molecules *lose* their freedom at the same time, and there is no net change in the number of particles. In the other case, only *three* en molecules lose their freedom, and thus there is a net increase of 3 mol of individual molecules. The reaction with three en causes a much greater increase in disorder than does that with six NH_3 molecules and, therefore, $\Delta S°$ is more positive (more favorable) in the former case than in the latter. It is easy to see that this reasoning is general for all such comparisons of a chelate with a nonchelate process.

Another way to state the matter is to visualize a chelate ligand with one donor atom attached to a metal ion. The other donor atom cannot then get very far away, and the probability of it, too, becoming attached is greater than if it were in an entirely independent molecule, with access to the entire volume of solution. Thus the chelate effect weakens as ring size increases. The effect is greatest for five- and six-membered rings, becomes marginal for seven-membered rings, and is unimportant thereafter. When the ring to be formed is large, the probability of the second donor atom attaching itself promptly to the same metal atom is no longer large as compared with its encountering a different metal atom, or as compared with the dissociation of the first donor atom before the second one makes contact.

6-5   Reactivity of Coordination Compounds

Virtually all of transition metal chemistry and a great deal of the rest of inorganic chemistry could be included under this title, taken in its broadest sense. Only three aspects will be covered in this and the following sections: substitution, electron-transfer, and isomerization reactions. Additional aspects of reactivity will be discussed in Chapters 28–30, under organometallic compounds. A detailed correlation of structure, bonding, and reactivity will have to be postponed until Chapter 23 and later, when a discussion of bonding in coordination compounds can be developed.

Substitution Reactions of Octahedral Complexes

The ability of a complex to engage in reactions that result in replacing one or more ligands in its coordination sphere (by other ligands in solution, for instance) is called its lability. Those complexes for which such substitution reactions are rapid are called **labile,** whereas those for which such substitution reactions proceed slowly (or not at all) are called **inert.** We note that these terms are kinetic terms, because they reflect rates of reaction. These terms should not be confused with the thermodynamic terms **stable** and **unstable,** which refer to the tendency of species to exist (as governed by the equilibrium constants K or β)

under equilibrium conditions. A simple example of this distinction is provided by the $[Co(NH_3)_6]^{3+}$ ion, which will persist for months in an acid medium because of its kinetic inertness (slow reactivity) despite the fact that it is thermodynamically unstable, as shown by the large equilibrium constant ($K \sim 10^{25}$) for Reaction 6-5.1.

$$[Co(NH_3)_6]^{3+} + 6\,H_3O^+ \longrightarrow [Co(H_2O)_6]^{3+} + 6\,NH_4^+ \qquad (6\text{-}5.1)$$

In contrast, the overall formation constant ($\beta_4 = 10^{22}$) for Reaction 6-5.2 indicates that the thermodynamic stability of $[Ni(CN)_4]^{2-}$ is high.

$$Ni^{2+} + 4\,CN^- \longrightarrow [Ni(CN)_4]^{2-} \qquad (6\text{-}5.2)$$

Nevertheless, the rate of exchange of CN^- ligands with excess CN^- in solution is immeasurably fast by ordinary techniques. The complex $[Ni(CN)_4]^{2-}$ is both thermodynamically stable and kinetically labile; the terms are not contradictory. In other words, it is not required that there be any relationship between thermodynamic stability and kinetic lability. Of course this lack of any necessary relation between thermodynamics and kinetics is generally found in chemistry, but its appreciation here is especially important.

A practical definition of the terms labile and inert can be given. Inert complexes are those whose substitution reactions have half-lives longer than a minute. Such reactions are slow enough to be studied by classical techniques where the reagents are mixed and changes in absorbance, pH, gas evolution, and so on, are followed directly by the observer. Data can be taken conveniently for such reactions. Labile complexes are those that have half-lives for a reaction under a minute. Special techniques are required for collecting data during such reactions, as they may appear to be finished within the time of mixing.

In the first transition series, virtually all octahedral complexes save those of Cr^{III} and Co^{III}, and sometimes Fe^{II}, are normally labile; that is, ordinary complexes come to equilibrium with additional ligands (including water) so rapidly that the reactions appear instantaneous by ordinary techniques of kinetic measurement. Complexes of Co^{III} and Cr^{III} ordinarily undergo substitution reactions with half-lives of hours, days, or even weeks at 25 °C.

Two extreme mechanistic possibilities may be considered for any ligand substitution process or for any single step in a series of substitution reactions. First, there is the dissociative (D) mechanism in which the ligand to be replaced dissociates from the metal center and the vacancy in the coordination sphere is taken by the new ligand. This mechanism is shown in Reaction 6-5.3:

$$[L_5MX] \xrightarrow{\text{slow}} X + \underset{\substack{\text{Five-coordinate} \\ \text{intermediate}}}{[L_5M]} \xrightarrow[\text{fast}]{+Y} [L_5MY] \qquad (6\text{-}5.3)$$

where L represents a nonlabile ligand, X is the leaving ligand, and Y is the entering ligand. The important feature of such a mechanism is that the first step (dissociation of the leaving group) is rate determining. Once formed by cleavage of the bond to the leaving group, X, the five-coordinate intermediate will react with the new ligand, Y, almost immediately. This mechanism for ligand substitution is comparable to the S_N1 mechanism in organic systems, because the formation of the intermediate with reduced coordination number is unimolecular, as well as rate determining.

The other extreme possibility for ligand substitution is the addition–elimination mechanism, or the associative (A) mechanism. In this case the new ligand, Y, directly attacks the original complex to form a seven-coordinate intermediate in the rate-determining step, as shown in Reaction 6-5.4.

$$[L_5MX] + Y \xrightarrow{\text{slow}} \left[L_5M \diagdown{\overset{X}{\diagup}}_{Y} \right] \xrightarrow{\text{fast}} [L_5MY] + X \qquad (6\text{-}5.4)$$

After rate-determining association between the entering ligand Y and the metal complex, the leaving group X is lost in a fast step. The rate-determining step is bimolecular for the mechanism shown by Reaction 6-5.4.

Unfortunately, these two extreme mechanisms are just that—extremes—and observed mechanisms are seldom so simple. Instead of a five- or seven-coordinate intermediate, a transition state may be reached in which some degree of bond breaking accompanies a given degree of bond making. The interchange of the ligands X and Y could be accomplished mostly by breaking the bond to the leaving group (interchange–dissociative, I_d) or by making the bond to the entering group (interchange–associative, I_a), but in each case both ligands are bound to the metal to one extent or another. Figure 6-5 presents reaction profiles for each of the four mechanistic cases just mentioned.

To complicate matters further, the rate law that is determined for a reaction from kinetic data cannot be used to identify the mechanism for that reaction. This is so because additional steps in the overall substitution may take place, obscuring the simple first- and second-order rate laws that are expected for unimolecular and bimolecular processes, respectively. The three most important cases that illustrate this sort of complication are (1) solvent intervention, (2) ion-pair formation, and (3) conjugate-base formation.

1. Solvent intervention.   Many reactions of complexes have been studied in solvents that are themselves ligands. Water, for instance, is a respectable ligand, and is present in aqueous solution in high and effectively constant concentration ($\sim 55.5\ M$). The substitution of X by Y might take place by the sequence of Reactions 6-5.5 and 6-5.6.

$$[L_5MX] + H_2O \longrightarrow [L_5MH_2O] + X \qquad \text{(Slow)} \qquad (6\text{-}5.5)$$

$$[L_5MH_2O] + Y \longrightarrow [L_5MY] + H_2O \qquad \text{(Fast)} \qquad (6\text{-}5.6)$$

A simple first-order rate law would be observed, and yet either Reaction 6-5.5 or 6-5.6 could proceed by an A (or I_a) or a D (or I_d) mechanism.

Intervention of the solvent in Reaction 6-5.5 obscures the molecularity of the rate-determining step; the reaction will necessarily be observed to be first order because of the high and constant concentration of the entering ligand, H_2O.

2. Ion-pair formation.   When the reacting complex and the entering ligand are both ions, especially when both have high charges, ion pairs (or outer-sphere complexes, as they are sometimes called) will form, as in Reaction 6-5.7.

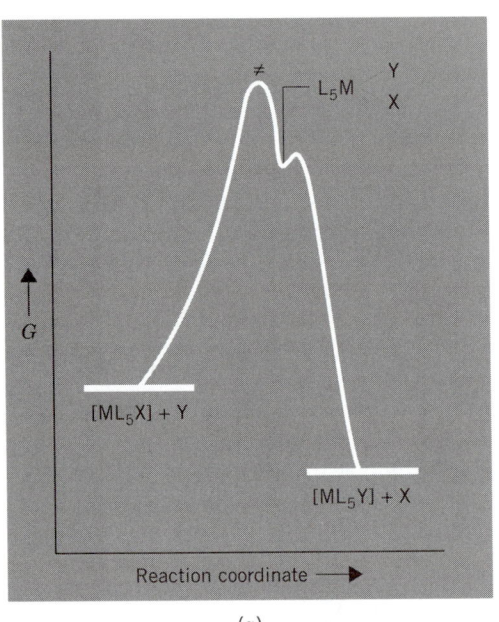

(a)

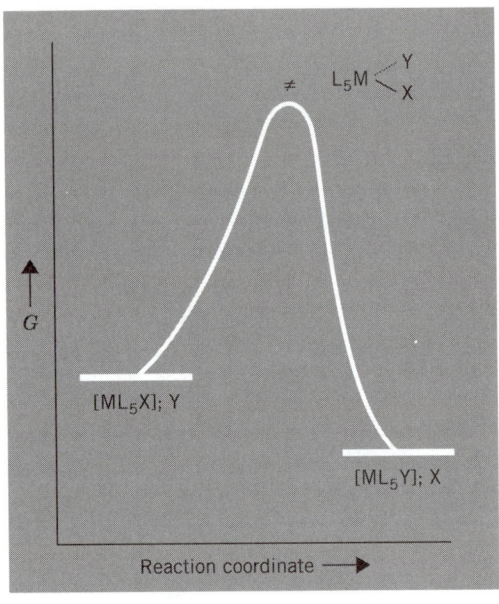

(b)

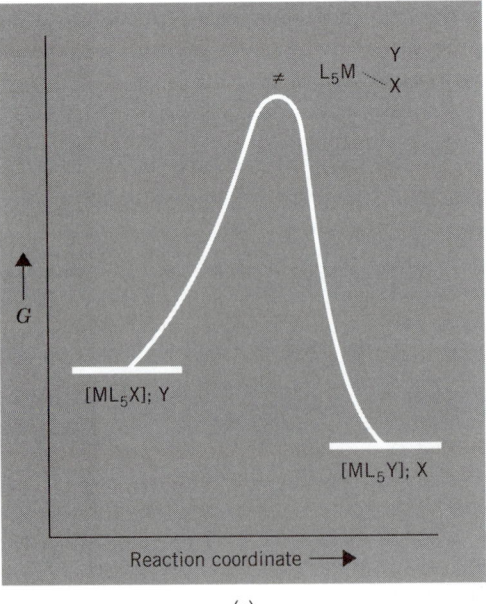

(c)

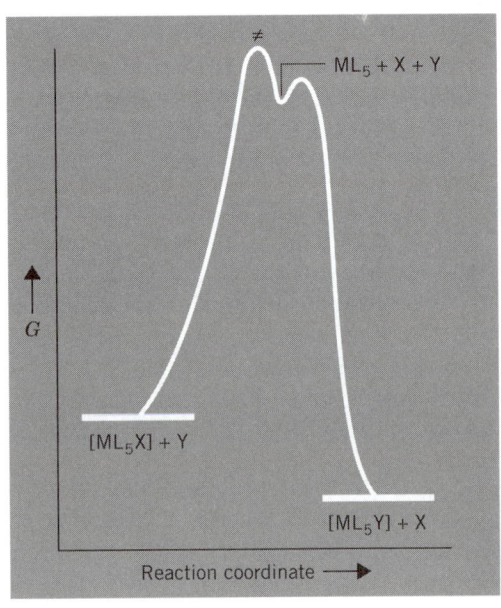

(d)

$$[L_5MX]^{n+} + Y^{m-} = \{[L_5MX]Y\}^{n-m} \qquad (6\text{-}5.7)$$

In the product of Reaction 6-5.7, the entering ligand Y has been stabilized at the outer edge of the coordination sphere of the complex $[L_5MX]^{n+}$ primarily by electrostatics. In cases where charges on ions are not involved, an entering group Y may be bound at the periphery of the metal complex through, for instance, hydrogen bonding. Outer-sphere or ion-pair equilibrium constants K_{os} are generally in the range 0.05–40, depending on the charges on the ions and on their effective radii. Where ion pairs (or neutral outer-sphere complexes) are featured as intermediates in the reaction path that leads to ligand substitution, then observed rate laws will be second order, whether or not the mechanism at the rate-determining step involves associative or dissociative activation.

3. Conjugate-base formation. When experimental rate laws contain [OH$^-$], there is the question whether OH$^-$ actually attacks the metal in a true associative fashion, or whether it appears in the rate law through operation of the mechanism shown in Eqs. 6-5.8 and 6-5.9.

$$[Co(NH_3)_5Cl]^{2+} + OH^- = [Co(NH_3)_4(NH_2)Cl]^+ + H_2O \quad \text{(Fast)} \quad (6\text{-}5.8)$$

$$[Co(NH_3)_4(NH_2)Cl]^+ \xrightarrow[\text{then } H^+]{+Y^-} [Co(NH_3)_5Y]^{2+} + Cl^- \quad \text{(Slow)} \quad (6\text{-}5.9)$$

In this conjugate-base (CB) mechanism, the hydroxide first deprotonates a ligand (usually NH$_3$) forming the conjugate base, here leading to the NH$_2^-$ ligand. It is then the conjugate base of the original metal complex that reacts with the incoming ligand, as in Eq. 6-5.9.

Water Exchange in Aqua Ions

Since many reactions in which complexes are formed occur in aqueous solution, one of the most fundamental reactions to be studied and understood is that in which the water ligands in the aqua ion $[M(H_2O)_n]^{m+}$ are displaced from the first coordination shell by other ligands. Included here is the simple case in which the new ligand is another water molecule, the water-exchange reaction.

A partial survey of results is given in Fig. 6-6. Not shown here are systems where the water exchange is characteristically slow: Cr^{3+}, Co^{3+}, Rh^{3+}, Ir^{3+}, and Pt^{2+}. These five typically inert aqua ions have exchange rate constants in the range 10^{-3}–10^{-6} s^{-1}. Those ions included in Fig. 6-6 are broadly considered to be

---

Figure 6-5 The four general mechanisms for ligand substitution in the complexes [ML$_5$X], where L are nonlabile ligands, X is the leaving ligand, and Y is the entering ligand. (*a*) The associative (A) mechanism in which an intermediate of expanded coordination number is formed first through rate-determining entry of the ligand Y. (*b*) The interchange-associative mechanism (I$_a$), in which the transition state is reached mostly through formation of the bond (M$\cdots$Y) to the entering ligand. The notations [ML$_5$X];Y and [ML$_5$Y];X for the reactants and products, respectively, represent outer-sphere complexes (or ion pairs) as formed in Reaction 6-5.7. (*c*) The interchange–dissociative mechanism (I$_d$), in which the transition state is reached mostly through breaking the bond (M$\cdots$X) to the leaving ligand. Again, the reactants and products are outer-sphere complexes (or ion pairs) as featured in Reaction 6-5.7. (*d*) The dissociative or D mechanism in which an intermediate of reduced coordination number is formed first through rate-determining cleavage of the bond to the leaving group.

labile, but a range of 10^{10} in lability is covered. It is convenient to divide the ions into four classes, depending on these rate constants for water exchange:

Class I. Rate constants for water exchange exceed 10^8 s^{-1} for ions that fall into this class. The exchange process is as fast here as is allowed by diffusion within the solvent, that is, these are diffusion-controlled reactions. Ions that fall into this class include those of Group IA(1), Group IIA(2) (except Be and Mg), Group IIB(12) (except Zn^{2+}), and Cr^{2+} and Cu^{2+} from the first transition series.

Class II. Ions that fall into this class have water-exchange rate constants in the range 10^4–10^8 s^{-1}. These include many of the 2+ ions of the first transition series (excepting V^{2+}, which is slower and Cr^{2+} and Cu^{2+}, which are in Class I), and the 3+ ions of the lanthanides.

Class III. Water exchange rate constants cover the range 1–10^4 s^{-1} for ions in this class: Be^{2+}, Al^{3+}, Ga^{3+}, V^{2+}, and some others.

Class IV. These are the ions mentioned previously that are inert, having rate constants for water exchange in the range 10^{-3}–10^{-6} s^{-1}.

There are a number of important trends that should be noted in the data of Fig. 6-6. First consider either of the series of ions in Groups IA(1), IIA(2), IIB(12), or IIIB(13), where partially filled d orbitals are not featured. In each of these series, the exchange rate constant decreases as the size of the ion decreases, that is, exchange rates are lower for the smaller ions. We expect that the leaving ligands will be more tightly bound by ions of smaller size because the smaller ions (of those with a given charge) are the ones with the higher charge densities. The data of Fig. 6-6 indicate, then, that a dissociative process (D or I$_d$) operates in water exchange; dissociation of the leaving group is slower (smaller rate constants) where the leaving group is bound more tightly (to a smaller ion).

Such simple correlations of rate and size do not work for ions of the transition series, where the number of d electrons can influence reactivity. Compare, for instance, Cr^{2+}, Ni^{2+}, and Cu^{2+}, which have similar radii, but different reactivities. Also, the inertness of Co^{3+} is completely out of line with ionic size. More will be said later about these ions. For now, it is useful to note that transition metal ions that are typically inert include those with d^6 electron configurations (Co^{3+}, Rh^{3+}, and Ir^{3+}) and those with d^3 electron configurations (Cr^{3+}). The characteristically labile ions include the d^4 (Cr^{2+}) and d^9 (Cu^{2+}) systems.

Anation Reactions

An important reaction of the aqua ions is the addition of an anion, as in Reaction 6-5.10:

$$[M(H_2O)_6]^{n+} + X^- \longrightarrow [M(H_2O)_5X]^{(n-1)+} + H_2O \qquad (6\text{-}5.10)$$

Such reactions are especially germane to the synthesis of new complexes starting with the simple aqua ions. Anation reactions may also be considered to include reactions in which coordinated water in the substituted complexes [ML$_5$H$_2$O]$^{n+}$ is replaced by an incoming anion. In either case, two remarkably general observations have been made concerning the rates at which water ligands are replaced by anions:

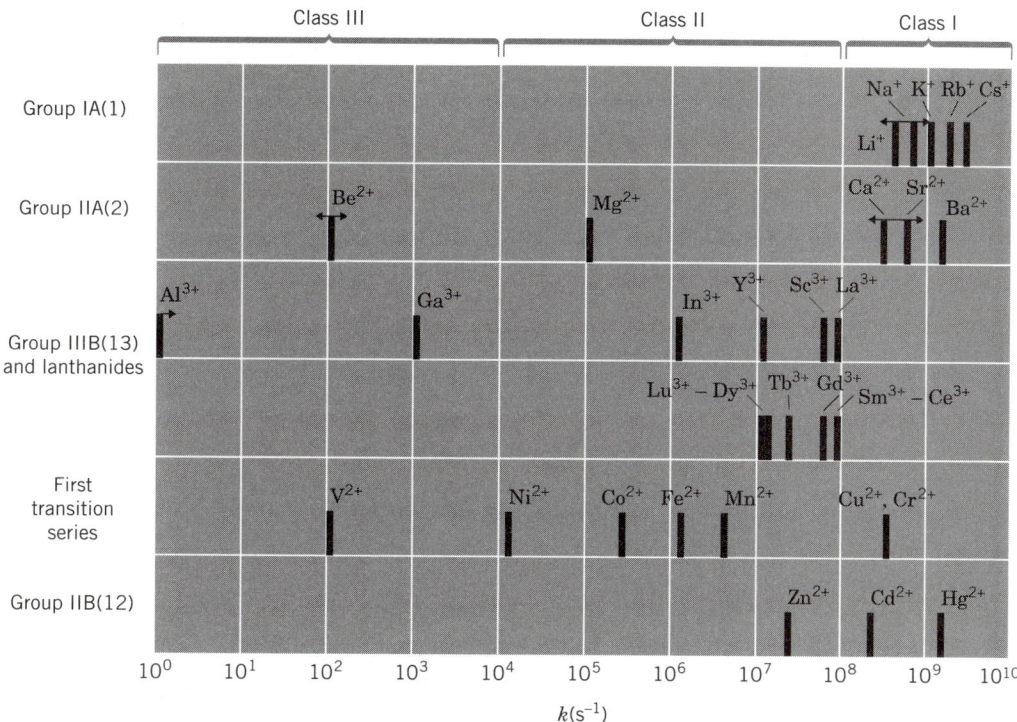

Figure 6-6   Characteristic rate constants (s^{-1}) for substitution of inner-sphere water ligands in various aqua ions. [Adapted from M. Eigen, *Pure Appl. Chem.*, **1963**, *6*, 105, with revised data kindly provided by M. Eigen. See also H. P. Bennetto and E. F. Caldin, *J. Chem. Soc. A*, **1971**, 2198.]

1. For a given aqua ion and a series of entering monoanions X^- (or a separate series of dianions), the rate constants for anation show little or no dependence ($<$ a factor of 10) on the identity of the entering ligand.

2. Rate constants for anation of a given aqua ion are practically the same (perhaps ~10 times slower) as the rate constant for water exchange for that aqua ion.

The most reasonable explanation for these observations is that the overall process involves the following three steps:

$$[M(H_2O)_6]^{n+} + X^- \;\overset{K_{os}}{=}\; \{[M(H_2O)_6]X\}^{(n-1)+} \tag{6-5.11}$$

$$\{[M(H_2O)_6]X\}^{(n-1)+} \;\overset{k_0}{\longrightarrow}\; \{[M(H_2O)_5]X\}^{(n-1)+} + H_2O \tag{6-5.12}$$

$$\{[M(H_2O)_5]X\}^{(n-1)+} \;\overset{\text{very}}{\underset{\text{fast}}{\longrightarrow}}\; [M(H_2O)_5X]^{(n-1)+} \tag{6-5.13}$$

In the first step an outer-sphere complex (here an ion pair) is formed with an equilibrium constant K_{os} (Reaction 6-5.11). A coordinated water molecule is then lost (Reaction 6-5.12) with rate constant k_0, a rate constant that should be close to that for water exchange in the parent aqua ion. In the third step, which is very fast, and may not be distinct from the second step, the entering ligand X^-

slips into the coordination spot vacated by the water ligand. The most appropriate rate law for the overall sequence of reactions 6-5.11 to 6-5.13 is given by Eq. 6-5.14:

$$\text{rate} = k_{\text{obs}}[\text{M}(\text{H}_2\text{O})_6^{n+}][\text{X}^-] \qquad (6\text{-}5.14)$$

Experimentally, one expects to observe second-order kinetics where such a mechanism operates, and the observed second-order rate constant k_{obs} should be equal to the product $K_{\text{os}}k_0$. Values for K_{os} can be estimated and factored out of the experimentally determined k_{obs}, yielding k_0. When this is done, for any of a number of anation reactions, the values for k_0 closely resemble those for the simple water exchange in $[\text{M}(\text{H}_2\text{O})_6]^{n+}$. This is taken to be evidence that the mechanism for anation also involves dissociative activation. When coupled with a lack of dependence on the identity of the entering ligand (as long as ions of like charge are compared), this argument is convincing.

Aquation Reactions

Complexes that are present in aqueous solution are susceptible to aquation or hydrolysis reactions in which a ligand is replaced by water. Even where other entering ligands Y are part of an overall reaction, it appears that there are few reactions in which the leaving ligand X is not first replaced by water. Thus solvent intervention is a key feature in substitutions of X by Y, and aquation of the ligand X in $[\text{ML}_5\text{X}]$ is a reaction of fundamental importance.

Our discussion will emphasize aquation of the ligand X in amine complexes of Co^{III}, as seen in Reaction 6-5.15, where A represents an amine-type ligand such as NH_3.

$$[\text{CoA}_5\text{X}]^{n+} + \text{H}_2\text{O} \longrightarrow [\text{CoA}_5\text{OH}_2]^{(n+1)+} + \text{X}^- \qquad (6\text{-}5.15)$$

The rate law observed for such aquation reactions is a two-term rate law, shown in Eq. 6-5.16.

$$\text{rate} = k_a[\text{CoA}_5\text{X}^{n+}] + k_b[\text{CoA}_5\text{X}^{n+}][\text{OH}^-] \qquad (6\text{-}5.16)$$

The first term, involving the acid hydrolysis rate constant k_a, predominates at low pH, where $[\text{OH}^-]$ is low. The second term, involving the base hydrolysis rate constant k_b, predominates at high pH. The two-term rate law is an indication that two paths for aquation are possible, an acid hydrolysis and a base hydrolysis reaction path. At intermediate values of pH, both paths will be available. In general, k_b is approximately 10^4 times k_a, and it is often true that complexes that are inert under acidic conditions become labile in the presence of bases. The amines of Co^{III}, for instance, are so labile towards substitution in aqueous base that they generally decompose in that medium through rapid, successive substitutions leading to hydroxides and hydrous metal oxides.

Acid Hydrolysis. The general equation for acid hydrolysis is Reaction 6-5.15. The ligand undergoing substitution is replaced in the first coordination sphere by the entering ligand, water. Since the entering ligand is present in high and effectively constant concentration, the rate law does not contain $[\text{H}_2\text{O}]$, and tells us nothing about the order of the reaction with respect to water. The rate law is, in fact, simply a first-order rate law, as shown in Eq. 6-5.17:

$$\text{rate} = k_a[\text{CoA}_5\text{X}^{n+}] \qquad (6\text{-}5.17)$$

and the observed rate constant is always a simple, first-order rate constant, k_a. For these reasons, the rate law itself does not provide the means for deciding whether the reactions proceed by D or A mechanisms. The means for determining mechanism must be sought elsewhere. Hundreds of specific reactions have been studied, and although numerous exceptions exist, most acid hydrolysis reactions of octahedral complexes appear to proceed through dissociative processes (D or I_d). Some of the evidence that supports this conclusion comes from the study of (1) leaving group effects, (2) steric effects, and (3) charge effects.

The effect of the leaving ligand on the acid hydrolysis rate constant k_a can be seen in the data of Table 6-1. C. H. Langford (and later A. Haim) has pointed out that the dependence is linear. This is shown in Fig. 6-7, and is called a linear free energy relationship. The rate constant k_a for Reaction 6-5.15 is seen to be linearly dependent on the equilibrium constant K_a for Reaction 6-5.18.

$$[\text{CoA}_5\text{OH}_2]^{3+} + \text{X}^- = \text{CoA}_5\text{X}^{2+} + \text{H}_2\text{O} \qquad K_a \qquad (6\text{-}5.18)$$

Where the equilibrium constant K_a is largest, the anion X^- (here NCS^- or F^-) is most tightly bound to cobalt in the complex $\text{CoA}_5\text{X}^{2+}$. As can be seen in the data of Table 6-1, these are also the systems that are least labile (have the smallest k_a). The most labile complexes (large k_a) have the least tightly bound anions X^- (small K_a). One concludes that the strength of the bond to the leaving group is important in controlling the rate of the reaction. Furthermore, that the slope in Fig. 6-7 is exactly 1.0, as pointed out by Langford, indicates that the nature of X^- in the transition state is the same as its nature in the products of the reaction: a solvated anion. In other words, one must completely cleave the bond to the leaving group in order to reach the transition state for the reaction. A dissociative mechanism is suggested, and I_d is the most reasonable proposal since no five-coordinate intermediate has been detected.

Linear free energy relationships have been observed for other reactions. A plot similar to that of Fig. 6-7 can be constructed from the data for Reaction 6-5.19.

$$[\text{IrA}_5\text{X}]^{2+} + \text{H}_2\text{O} = [\text{IrA}_5\text{OH}_2]^{3+} + \text{X}^- \qquad (6\text{-}5.19)$$

The slope for such a plot is 0.9, and the rate constants k_a follow the trend $\text{NO}_3^- > \text{I}^- > \text{Br}^- > \text{Cl}^-$. The smaller slope indicates a less complete requirement for breaking the bond to the leaving group before the transition state is reached.

The dissociative nature of the acid hydrolysis mechanisms for octahedral complexes is also indicated by studies of steric effects. The data obtained by R. G. Pearson for Reaction 6-5.20 are typical.

$$[\text{Co}(\text{A}{-}\text{A})_2\text{Cl}_2]^+ + \text{H}_2\text{O} \longrightarrow [\text{Co}(\text{A}{-}\text{A})_2\text{Cl}(\text{OH}_2)]^{2+} + \text{Cl}^- \quad (6\text{-}5.20)$$

Table 6-1   Rate Constants k_a for the Reactions

$$[CoA_5X]^{2+} + H_2O \longrightarrow [CoA_5OH_2]^{3+} + X^-$$

and Equilibrium Constants K_a for the Reactions

$$[CoA_5OH_2]^{3+} + X^- = [CoA_5X]^{2+} + H_2O$$

X^-	k_a (s^{-1})	K_a (M^{-1})
NCS$^-$	5.0×10^{-10}	470
F$^-$	8.6×10^{-8}	20
H$_2$PO$_4^-$	2.6×10^{-7}	7.4
Cl$^-$	1.7×10^{-6}	1.25
Br$^-$	6.3×10^{-6}	0.37
I$^-$	8.3×10^{-6}	0.16
NO$_3^-$	2.7×10^{-5}	0.077

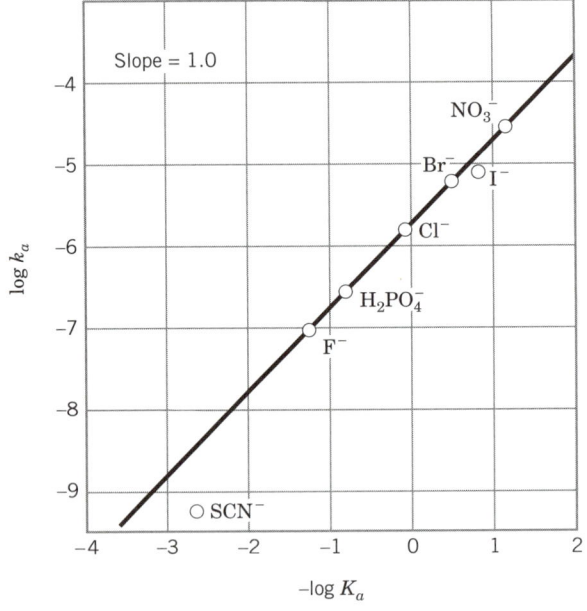

$$CoA_5X^{2+} + H_2O \xrightarrow{k_a} CoA_5H_2O^{3+} + X^-$$

$$CoA_5OH_2{}^{3+} + X^- \underset{\rightleftharpoons}{\overset{K_a}{}} CoA_5X^{2+} + H_2O$$

Figure 6-7   A plot of log k_a (the acid hydrolysis rate constant for Reaction 6-5.15) versus $-$log K_a (the equilibrium constant for Reaction 6-5.18). The slope of the plot is 1.0, indicating that factors controlling the strength of the Co—X bond (as measured by K_a) also influence the lability of the Co—X bond (as measured by k_a).

The ligands A—A in Reaction 6-5.20 are bidentate diamines that have been substituted in the carbon chains to provide increased crowding in the coordination sphere of the cobalt reactant. The data in Table 6-2 are typical of those for reactions that exhibit steric acceleration; the complexes having the larger ligands (A—A) react more quickly. Dissociative activation is indicated. No five-coordinate intermediate has been detected, so an I_d mechanism is assigned.

Charge effects also indicate dissociative activation for substitution reactions of octahedral complexes of cobalt. Compare, for instance,

$$[Co(NH_3)_5Cl]^{2+} \qquad (k_a = 6.7 \times 10^{-6}\ s^{-1})$$

with

$$trans\text{-}[Co(NH_3)_4Cl_2]^+ \qquad (k_a = 1.8 \times 10^{-3}\ s^{-1})$$

Table 6-2   Acid Hydrolysis Rate Constants for Aquation of the First Chloride Ligand in the Complexes $trans\text{-}[Co(A\text{—}A)_2Cl_2]^+$. A—A Represents a Bidentate Diamine Ligand with Increasing Substitution in the Carbon Chain[a]

A—A	k_a (s^{-1})
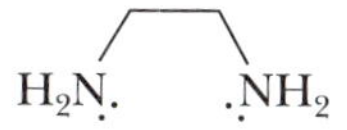 H$_2$N. .NH$_2$	3.2×10^{-5}
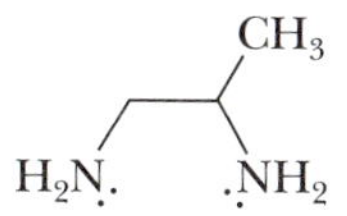 CH$_3$ / H$_2$N. .NH$_2$	6.2×10^{-5}
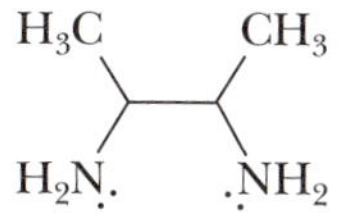 H$_3$C CH$_3$ / H$_2$N. .NH$_2$	4.2×10^{-3}
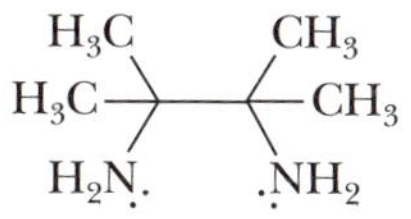 H$_3$C CH$_3$ / H$_3$C— —CH$_3$ / H$_2$N. .NH$_2$	3.3×10^{-2}

[a]From the work of R. G. Pearson, C. R. Boston, and F. Basolo, *J. Am. Chem. Soc.*, **1953**, *75*, 3089.

Where the charge on the cobalt reactant is higher, the rate of separation of the anion Cl$^-$ is slower.

There are exceptions, but the majority of octahedral complexes appear to undergo substitution through mechanisms that involve dissociation of the leaving group as a predominant step. However, the extreme D mechanism should be assigned only to those rare systems where a five-coordinate intermediate can be detected.

Base Hydrolysis. Aquation reactions of octahedral complexes of Co^{III} that take place in basic solution display the rate law shown in Eq. 6-5.21.

$$\text{rate} = k_b[\text{CoA}_5\text{X}^{n+}][\text{OH}^-] \qquad (6\text{-}5.21)$$

This is simply the second-order term in the general rate law, Eq. 6-5.16. The second-order rate term in Eq. 6-5.16 predominates in basic solution, so that one observes simple second-order kinetics (Eq. 6-5.21).

The interpretation of a term of the type $k_b[CoA_5X^{n+}][OH^-]$ in a rate law for base hydrolysis has long been disputed. It could, of course, be interpreted as representing a genuine associative (A) process: OH^- being a nucleophile. However, the possibility of a CB mechanism (Reactions 6-5.8 and 6-5.9) must be considered. There are arguments on both sides, and it is possible that the mechanism for base hydrolysis may vary for different complexes. Studies of base hydrolysis of Co^{III} complexes suggest that, for these complexes, the CB mechanism is the reasonable one.

As already mentioned, base hydrolysis of Co^{III} complexes is generally much faster than acid hydrolysis because $k_a < k_b$ in Eq. 6-5.16. This, in itself, provides evidence against a simple A mechanism. Therefore, this reaction favors the CB mechanism, because there is no reason to expect OH^- to be uniquely capable of attack on the metal. In the reactions of square complexes, OH^- turns out to be a distinctly inferior nucleophile toward Pt^{II}.

The CB mechanism, of course, requires that the reacting complex have at least one protonic hydrogen atom on a nonleaving ligand, and that the rate of reaction of this hydrogen be fast compared with the rate of ligand displacement. It has been found that the rates of proton exchange in many complexes subject to rapid base hydrolysis are, in fact, some 10^5 times faster than the hydrolysis itself {e.g., in $[Co(NH_3)_5Cl]^{2+}$ and $[Co(en)_2NH_3Cl]^{2+}$}. Such observations are in keeping with the CB mechanism but afford no positive proof of it.

If the CB mechanism is correct, there is the question of why the conjugate base so readily dissociates to release the ligand X. In view of the very low acidity of coordinated amines, the concentration of the conjugate base is a very small fraction of the total concentration of the complex. Thus, its reactivity is enormously greater, by a factor far in excess of the mere ratio of k_b/k_a. It can be estimated that the ratio of the rates of aquation of $[Co(NH_3)_4NH_2Cl]^+$ and $[Co(NH_3)_5Cl]^{2+}$ must be greater than 10^6. Two features of the conjugate base have been considered in efforts to account for this reactivity. First, there is the obvious charge effect. The conjugate base has a charge that is one unit less positive than the complex from which it is derived. Although it is difficult to construct a rigorous argument, it seems entirely unlikely that the charge effect, in itself, can account for the enormous rate difference involved. It has been proposed that the amide ligand could labilize the leaving group X by a combination of electron repulsion in the ground state and a π-bonding contribution to the stability of the five-coordinate intermediate, as is suggested in Fig. 6-8.

Attack on Ligands. There are some reactions where ligand exchange does not involve the breaking of metal–ligand bonds; instead, bonds within the ligands themselves are broken and reformed. One well-known case is the aquation of a carbonato complex according to Reaction 6-5.22.

$$[Co(NH_3)_5OCO_2]^+ + 2\,H_3{}^*O^+ \longrightarrow$$
$$[Co(NH_3)_5(H_2O)]^{3+} + 2\,H_2{}^*O + CO_2 \quad (6\text{-}5.22)$$

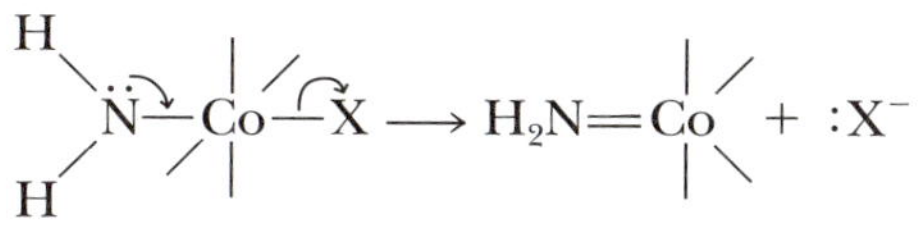

(a)   (b)

Figure 6-8 A diagram showing how an amide group can promote the dissociation of the trans ligand X through (a) electronic repulsion in the ground state and (b) stabilization of the five-coordinate intermediate via π bonding.

When isotopically labeled water ($H_2{}^*O$) is used, it is found that no $^*O$ gets into the coordination sphere of the cobalt during aquation.

The most likely path for this aquation involves proton attack on the oxygen atom bonded to cobalt (Structure 6-XXIV). This attack is followed by elimination of CO_2 and protonation of the hydroxo complex, as in Reaction 6-5.23.

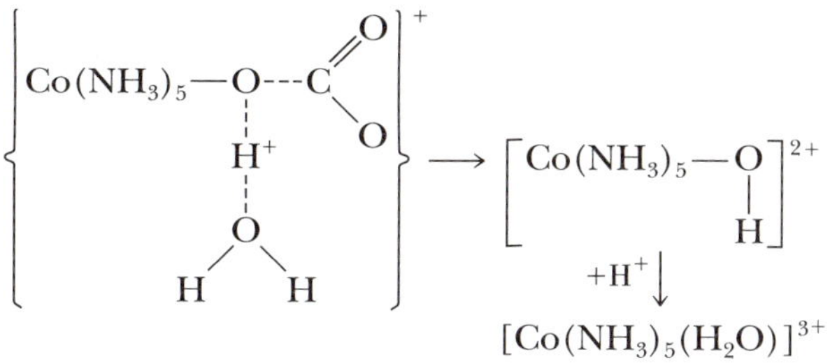

Transition state
6-XXIV

As another example, consider the reaction of nitrite with the pentaammineaquacobalt(III) ion, as in Reaction 6-5.24.

$$[CoA_5(^*OH_2)]^{3+} + NO_2^- \longrightarrow [CoA_5(N^*OO)]^{2+} + H_2O \qquad (6\text{-}5.24)$$

Isotopic labeling studies show that the oxygen of the aqua ligand is one of the oxygen atoms that is found in the nitro ligand. This remarkable result can be explained by the sequence of Reactions 6-5.25 to 6-5.27:

$$2\,NO_2^- + 2\,H^+ = N_2O_3 + H_2O \qquad (6\text{-}5.25)$$

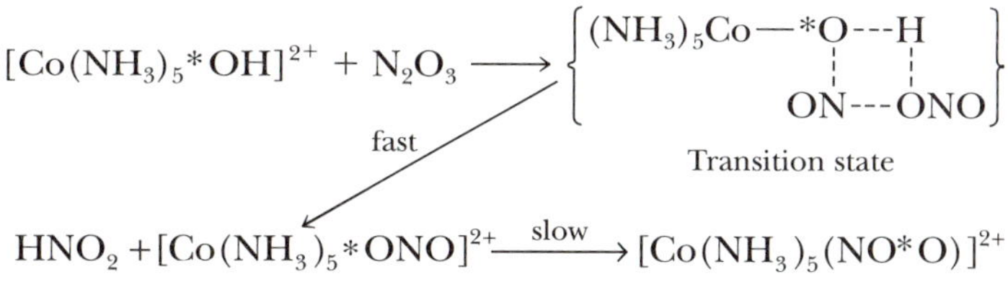

$$[Co(NH_3)_5{}^*OH]^{2+} + N_2O_3 \longrightarrow \left\{ \begin{array}{c} (NH_3)_5Co\!-\!^*O\text{---}H \\ \\ ON\text{---}ONO \end{array} \right\} \qquad (6\text{-}5.26)$$

fast

Transition state

$$HNO_2 + [Co(NH_3)_5{}^*ONO]^{2+} \xrightarrow{\text{slow}} [Co(NH_3)_5(NO^*O)]^{2+} \qquad (6\text{-}5.27)$$

In the transition state, it is an O—H bond that is broken, not a Co—O bond. The oxygen of the aqua ligand is, therefore, retained in the nitro ligand.

Substitution Reactions in Square Complexes

For square complexes, the mechanistic problem is more straightforward and better understood. One might expect that four-coordinate complexes would be more likely than octahedral complexes to react by associative pathways because the coordination sphere is less congested to start with. Extensive studies of Pt^{II} complexes have shown that this is true.

For reactions in aqueous solution, of the type shown in Reaction 6-5.28,

$$PtL_3X + Y \longrightarrow PtL_3Y + X \tag{6-5.28}$$

L represents nonlabile ligands, X is the leaving ligand, and Y is the entering ligand. Reactions of the type shown in Reaction 6-5.28 are found to have the rate law shown in Eq. 6-5.29.

$$\text{rate} = k_1[PtL_3X] + k_2[PtL_3X][Y] \tag{6-5.29}$$

Such a two-term rate law indicates that two paths are available for reaction: a first-order path characterized by the rate constant k_1 and a second-order path characterized by the rate constant k_2. It is believed that the second path (k_2) proceeds through a genuine associative (A) mechanism in which Y is added to the Pt center to form a five-coordinate intermediate. The first path (k_1) represents a two-step process in which X is first replaced by solvent (water) in the rate-determining step. (This k_1 path involves, then, solvent intervention, and must, for the same reasons previously discussed for aquation reactions of octahedral complexes, obey first-order kinetics.) The k_1 path is completed when the intervening water ligand is, in turn, replaced by Y. The two paths indicated by the rate law (Eq. 6-5.29) are shown in Fig. 6-9. Both paths appear to involve associative activation, and A or I_a mechanisms for each step of either the k_1 or the k_2 paths are assigned. Some of the evidence for this follows.

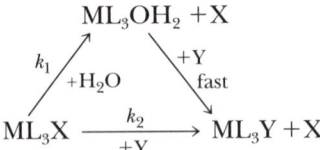

Figure 6-9 The two reaction paths for ligand substitution in square complexes, as indicated by the two-term rate law, Eq. 6-5.29. The k_1 path involves rate-determining formation of the aquated intermediate $[PtL_3(OH_2)]^{n+}$ and subsequent, rapid substitution of the aqua ligand by Y. Both of these steps in the k_1 path appear to involve associative activation. The k_2 path involves direct replacement of X by Y, also via associative activation.

Charge Effects

Consider the series of Pt^{II} complexes with charges varying from +1 to −2:

$$[Pt(NH_3)_3Cl]^+ \qquad [Pt(NH_3)_2Cl_2] \qquad [Pt(NH_3)Cl_3]^- \qquad [PtCl_4]^{2-}$$

The observed rate constants k_1 (for aquation in water solvent) vary only by a factor of two. This is a remarkably small variation, given the large differences in charge among the complexes. The Pt—Cl bond breaking should be more difficult in the complexes with the higher positive charge. Also, complexes with the higher positive charge should favor approach of the nucleophile. Since neither of these trends is observed, an associative process is indicated in which both Pt—Cl bond breaking and Pt—OH_2 bond making are of comparable importance.

Steric Effects

Steric acceleration is observed for substitution reactions of octahedral complexes, and this was taken to be evidence for a dissociative nature in such reactions. For square complexes, substitution reactions are retarded by steric crowding at the metal center. This is taken to be evidence that the entering ligand Y must approach the metal center in order to reach the transition state. This result is consistent with either an associative (A) or an interchange (I_a) mechanism.

Entering Ligand Effects

The second-order rate constant k_2 in Eq. 6-5.29 is strongly dependent on the nature of the entering ligand. A reactivity series can be established in which the entering ligands Y are placed in order depending on the value of k_2:

$$F^- \sim H_2O \sim OH^- < Cl^- < Br^- \sim NH_3 \sim \text{alkenes} < C_6H_5NH_2 < C_5H_5N$$
$$< NO_2^- < N_3^- < I^- \sim SCN^- \sim R_3P$$

This is essentially the order of nucleophilicity towards Pt^{II} that is expected for these ligands, and an associative mechanism is indicated.

Stereochemistry

A general representation of the stereochemical course of substitution reactions of square complexes is given in Fig. 6-10. Carefully note that this process is entirely stereospecific: cis and trans starting materials lead, respectively, to cis and trans products. Whether any of the three intermediate configurations possess enough stability to be regarded as actual intermediates rather than merely phases of the activated complex remains uncertain.

Nonlabile Ligands: The trans Effect

A particular feature of substitution at square complexes is the important role played by nonlabile ligands that are trans to the leaving ligand. Consider Reaction 6-5.30.

$$[PtLX_3] + Y \longrightarrow [PtLX_2Y] + X \qquad (6\text{-}5.30)$$

Any one of three labile ligands X can be replaced by the entering ligand Y. Furthermore, the ligand X that is replaced can be either cis or trans to L, lead-

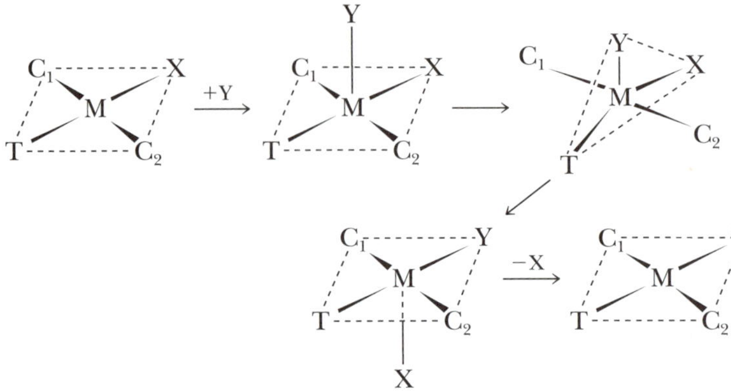

Figure 6-10 The steric course of ligand substitution in square complexes, and the structure (trigonal bipyramidal) of the five-coordinate intermediate. The ligands C_1 and C_2 that occupy axial positions in the intermediate trigonal bipyramid are the ligands that are cis to the leaving group X in the reactant. The ligand T in the reactant is the strongest trans director, as it lies trans to the leaving ligand X. The entering ligand Y, the leaving ligand X, and the trans ligand T, share the equatorial positions of the trigonal bipyramidal intermediate. The new ligand Y in the product occupies the coordination position that was vacated by the leaving ligand X.

ing to cis or trans orientation of Y with respect to L in the product. It has been found that the relative proportions of cis and trans products varies appreciably with the nature of the ligand L. Ligands L that strongly favor substitution to give trans products in reactions such as Eq. 6-5.30 are said to be strong trans directors. A fairly extensive series of ligands L may be arranged in order with respect to their tendency to be strong trans directors:

$$H_2O, OH^-, NH_3, py\ (NC_5H_5) < Cl^-, Br^- < SCN^-, I^-, NO_2^-,$$
$$C_6H_5^- < SC(NH_2)_2, CH_3^- < H^-, PR_3 < C_2H_4, CN^-, CO$$

This is also known as the *trans effect* series. It is to be emphasized that the trans effect is here defined solely as a kinetic phenomenon. It is the effect of the ligand L on the rate of substitution in the position trans to itself. A strong trans director (a ligand high in the trans effect series) promotes more rapid substitution of the ligand trans to itself than it does of the ligand cis to itself.

The trans effect has proved very useful in rationalizing known synthetic procedures and in devising new ones. As an example, we consider the synthesis of the cis and trans isomers of $[Pt(NH_3)_2Cl_2]$. The synthesis of the cis isomer is accomplished by treatment of the $[PtCl_4]^{2-}$ ion with ammonia, as in Reaction 6-5.31.

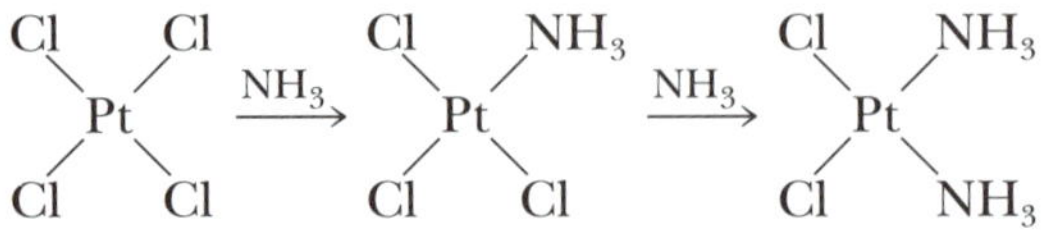

$$(6\text{-}5.31)$$

Since Cl^- has a greater *trans* directing influence than NH_3, substitution of NH_3 into $[Pt(NH_3)Cl_3]^-$ is least likely to occur in the position *trans* to the NH_3 already present. Thus, the *cis* isomer is favored. The trans isomer is synthesized by treating $[Pt(NH_3)_4]^{2+}$ with Cl^-, as in Reaction 6-5.32.

$$
\underset{H_3N}{\overset{H_3N}{\diagdown}}\!\!Pt\!\!\underset{NH_3}{\overset{NH_3}{\diagup}} \xrightarrow{Cl^-} \underset{H_3N}{\overset{H_3N}{\diagdown}}\!\!Pt\!\!\underset{Cl}{\overset{NH_3}{\diagup}} \xrightarrow{Cl^-} \underset{H_3N}{\overset{Cl}{\diagdown}}\!\!Pt\!\!\underset{Cl}{\overset{NH_3}{\diagup}}
\tag{6-5.32}
$$

In this case the intermediate is disposed to give the trans isomer because of the greater trans effect of Cl^-. The first Cl^- directs the second Cl^- to the trans position.

All theorizing about the trans effect must recognize the fact that since it is a kinetic phenomenon, depending on activation energies, the stabilities of *both* the ground state and the activated complex are relevant. The activation energy can be affected by changes in one or the other of these energies or by changes in both.

The earliest attempt to explain the trans effect was the so-called polarization theory of Grinberg, which is primarily concerned with effects in the ground state. This theory deals with a postulated charge distribution, as shown in Fig. 6-11. The primary charge on the metal ion induces a dipole in the ligand (L), which in turn induces a dipole in the metal. The orientation of this dipole on the metal is such as to repel negative charge in the trans ligand X. Hence, X is less attracted by the metal atom because of the presence of L. This theory would lead to the expectation that the magnitude of the trans effect of L and its polarizability should be monotonically related, and for some ligands in the trans effect series (e.g., H^-, $I^- > Cl^-$), such a correlation is observed. In effect, this theory says that the trans effect is attributable to a ground-state weakening of the bond to the ligand that is to be displaced.

An alternative theory of the trans effect was developed with special reference to the activity of ligands such as phosphines, CO, and alkenes, which are known to be strong π acids (see Chapter 28 for further details). This model attributes their effectiveness primarily to their ability to stabilize a five-coordinate transition state or intermediate. This model is, of course, only relevant if the reactions are bimolecular; there is good evidence that this is true in the vast majority of, if not all, cases. Figure 6-12 shows how the ability of a ligand to withdraw metal $d\pi$ electron density into its own empty π or π^* orbitals could enhance the stability

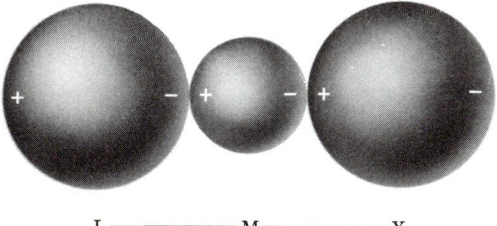

Figure 6-11   The arrangement of dipoles along the trans L—M—X axis according to the polarization theory of the *trans* effect.

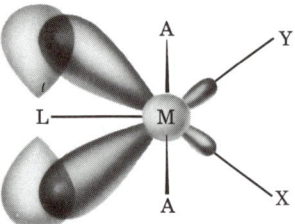

Figure 6-12 The postulated activated complex (a five-coordinate trigonal bipyramid) for reaction of Y with *trans*-MA$_2$LX to displace X.

of a species in which both the incoming ligand Y and the outgoing ligand X are simultaneously bound to the metal atom.

Recently, evidence has been presented to show that even in cases where stabilization of a five-coordinate activated complex may be important, there is still a ground-state effect: a weakening and polarization of the trans bond. In the anion $[C_2H_4PtCl_3]^-$ the Pt—Cl bond trans to ethylene is slightly longer than the cis bonds, the Pt—*trans*-Cl stretching frequency is lower than the average of the two Pt—*cis*-Cl frequencies, and there is evidence that the *trans*-Cl atom is more ionically bonded.

The consensus among workers in the field, in each case, for the entire series of ligands whose trans effect has been studied, is that both the ground-state bond weakening and the activated-state stabilizing roles may be involved to some extent. For a hydride ion or a methyl group it is probable that we have the extreme of pure, ground-state bond weakening. With the alkenes the ground-state effect may play a secondary role compared with activated-state stabilization, although the relative importance of the two effects in such instances remains a subject for speculation, and further studies are needed.

Electron-Transfer Reactions

These are oxidation–reduction (redox) reactions in which an electron passes from one complex to another. Electron-transfer reactions may involve substitution of one or more ligands in the first- or inner-coordination spheres of either reactants or products, but this is not necessary. An example of an electron-transfer reaction is given by Eq. 6-5.33

$$Fe^{2+}(aq) + Ce^{4+}(aq) \longrightarrow Fe^{3+}(aq) + Ce^{3+}(aq) \tag{6-5.33}$$

in which the aqua ion of Ce^{IV} is reduced by the aqua ion of Fe^{II}.

An electron-transfer reaction may take place so that there is actually no net chemical change, as in Reaction 6-5.34.

$$[*Fe(CN)_6]^{4-} + [Fe(CN)_6]^{3-} = [*Fe(CN)_6]^{3-} + [Fe(CN)_6]^{4-} \tag{6-5.34}$$

Reactions such as Reaction 6-5.34 are called **self-exchange** reactions. Self-exchange reactions can only be followed by using isotopic tracers or certain mag-

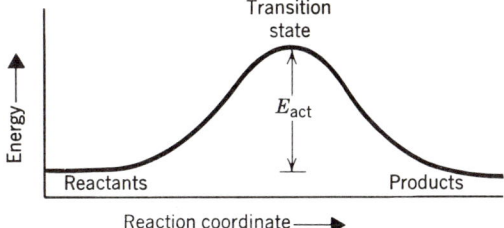

Figure 6-13 Free energy versus reaction co-ordinate for a self-exchange reaction. The pro-file is symmetrical because the reactants and products are identical. For other electron-transfer reactions known as cross reactions, the products are at a lower energy than reac-tants, in proportion to the electrochemical po-tential ($\Delta G = -n\mathscr{F}E$) for the reaction.

netic resonance techniques. These reactions are of interest because there is no change in free energy as a consequence of reaction, and the free energy profile (Fig. 6-13) is symmetrical.

There are two well-established general mechanisms for electron-transfer re-actions. In the first, called an **outer-sphere** electron-transfer mechanism, only the outer, or solvent, coordination spheres of the two metal complexes are displaced during the reaction. No substitution of the ligands in the inner-coordination spheres of either reactant is needed in order for electron transfer to take place. (There are required changes in metal–ligand bond lengths, however.) In the sec-ond mechanism, called **inner-sphere** electron transfer, the inner-coordination sphere of one reactant must first undergo substitution to accept a new ligand. The new ligand must serve, once substitution has taken place, to bridge the two metal centers together. This bridging ligand is bound to the inner-coordination spheres of both metal centers.

The Outer-Sphere Mechanism

This mechanism is certain to be correct when both complexes participating in the reaction undergo ligand substitution reactions more slowly than they par-ticipate in electron-transfer reactions. An example is the reaction shown in Reaction 6-5.35.

$$[Fe^{II}(CN)_6]^{4-} + [Ir^{IV}Cl_6]^{2-} \rightleftharpoons [Fe^{III}(CN)_6]^{3-} + [Ir^{III}Cl_6]^{3-} \quad (6\text{-}5.35)$$

where both reactants are "inert" towards substitution ($t_{1/2} > 1$ ms), but the redox reaction is fast ($k = 10^5$ L mol^{-1} s^{-1}). Clearly, the electron-transfer process is not constrained to wait for substitution to take place or it would be itself as slow as substitution. The outer-sphere mechanism is also correct when no ligand is ca-pable of serving as a bridging ligand.

The two steps of a general outer-sphere mechanism can be illustrated using Reaction 6-5.35. There is a preequilibrium, characterized by the constant K_{os}, in which an outer-sphere complex (or ion pair) is formed, as in Reaction 6-5.36.

$$[Fe(CN)_6]^{4-} + [IrCl_6]^{2-} = [Fe(CN)_6]^{4-}/[IrCl_6]^{2-} \qquad K_{os} \qquad (6\text{-}5.36)$$

$$[Fe(CN)_6]^{4-}/[IrCl_6]^{2-} \longrightarrow [Fe(CN)_6]^{3-} + [IrCl_6]^{3-} \qquad k_{et} \qquad (6\text{-}5.37)$$

This encounter (outer-sphere complex or ion pair) between the reactants brings them to within the internuclear separation required for electron transfer. The electron-transfer step (Reaction 6-5.37) takes place within this outer-sphere complex, only after metal–ligand bond lengths have been altered enough to allow the electron transfer to take place adiabatically (without further change in energy). R. Marcus recognized that the electron transfer should be adiabatic, because electron motion should be faster than nuclear motion. In other words, the electron transfer takes place quickly, once internuclear distances have become appropriately adjusted. For the complex that is being oxidized, metal–ligand distances in the activated complex must generally become shorter, because of the higher oxidation state that is to exist on the metal upon oxidation. The complex being reduced must achieve longer metal–ligand bond distances in the activated complex, in anticipation of the lower oxidation state that develops at the metal upon reduction.

Self-Exchange. Some self-exchange reactions that are believed to proceed by outer-sphere mechanisms are listed in Table 6-3. (The second-order rate laws that one usually observes for such reactions do not in themselves indicate an

Table 6-3   Rate Constants for Some Self-Exchange Reactions that Proceed via Outer-Sphere Mechanisms

Reactants	Rate Constants ($L\ mol^{-1}\ s^{-1}$)
$[Fe(bpy)_3]^{2+}$, $[Fe(bpy)_3]^{3+}$ $[Mn(CN)_6]^{3-}$, $[Mn(CN)_6]^{4-}$ $[Mo(CN)_8]^{3-}$, $[Mo(CN)_8]^{4-}$ $[W(CN)_8]^{3-}$, $[W(CN)_8]^{4-}$ $[IrCl_6]^{2-}$, $[IrCl_6]^{3-}$ $[Os(bpy)_3]^{2+}$, $[Os(bpy)_3]^{3+}$	10^4–10^6
$[Fe(CN)_6]^{3-}$, $[Fe(CN)_6]^{4-}$ $[MnO_4]^-$, $[MnO_4]^{2-}$	7.4×10^2 3×10^3
$[Co(en)_3]^{2+}$, $[Co(en)_3]^{3+}$ $[Co(NH_3)_6]^{2+}$, $[Co(NH_3)_6]^{3+}$ $[Co(C_2O_4)_3]^{3-}$, $[Co(C_2O_4)_3]^{4-}$	$\sim 10^{-4}$

outer-sphere mechanism; one also observes second-order kinetics for most inner-sphere electron-transfer processes.)

The range covered by these rate constants is very large, extending from 10^{-4} up to, perhaps, the very high rate constants typical of processes that are slowed only by the ability of the reactants to diffuse through the solvent ($\sim 10^9$). It is possible to account qualitatively for the observed variation in rate constants in terms of the different amounts of energy required to change the metal–ligand bond distances from their initial values to those needed in the transition state. For the case of self-exchange reactions, the transition state must be symmetrical; the two halves of the activated complex must be identical. The lengthening of metal–ligand bonds that is required of the complex undergoing reduction is equal to the shortening of the metal–ligand bonds that is required of the complex undergoing oxidation. After all, self-exchange simply transforms one reactant into the other, with no net chemical change (Fig. 6-13). Furthermore, it can be shown that an unsymmetrical transition state would correspond to a higher activation energy and, therefore, would not lie along the preferred reaction path.

In the seven fastest reactions of Table 6-3 there is very little difference in the metal–ligand bond lengths in the two reacting complexes. Thus, very little energy of bond stretching and bond compressing is needed to achieve the symmetrical transition state. For the MnO_4^-/MnO_4^{2-} pair the bond length difference is somewhat greater, and for the last three reactions there is a considerable difference between the two reactants in metal–ligand bond distance.

Cross Reactions. Electron-transfer reactions between dissimilar complexes (e.g., Reactions 6-5.33 and 6-5.35) are called **cross reactions.** For cross reactions there is a net decrease in free energy, and the free energy profile is not symmetrical. A linear free energy relationship exists for such reactions, and the faster reactions tend to be those for which the free energy change is most favorable. Marcus and Hush derived the relationship shown in Eq. 6-5.38.

$$k_{12} = [k_{11}k_{22}K_{12}f]^{1/2} \qquad (6\text{-}5.38)$$

This equation allows the calculation of the rate constant for a cross reaction (k_{12}) from the two appropriate self-exchange rate constants (k_{11} and k_{22}) and the equilibrium constant for the overall cross reaction (K_{12}). The constant f in Eq. 6-5.38 is a statistical and steric factor that is usually about 1. The linear free energy relationship arises because the rate of reaction (as measured by k_{12}) depends on the net free energy change of the reaction (as measured by K_{12}). In fact, it is a general result that the faster cross reactions are those with the larger equilibrium constants. Thus rate constants for cross reactions are generally higher than those for the comparable self-exchanges.

As a specific example, consider the cross reaction shown in Eq. 6-5.39.

$$[Fe(CN)_6]^{4-} + [Mo(CN)_8]^{3-} \longrightarrow [Fe(CN)_6]^{3-} + [Mo(CN)_8]^{4-} \quad (6\text{-}5.39)$$

for which k_{12} is sought. The equilibrium constant K_{12} for Reaction 6-5.39 is 1.0×10^2, and the electrochemical potential is $E = 0.12$ V. The self-exchange reactions that apply are given in Eqs. 6-5.40 and 6-5.41.

$$[Fe(CN)_6]^{4-} + [Fe(CN)_6]^{3-} \longrightarrow [Fe(CN)_6]^{3-} + [Fe(CN)_6]^{4-} \qquad k_{11} \quad (6\text{-}5.40)$$

$$[Mo(CN)_8]^{3-} + [Mo(CN)_8]^{4-} \longrightarrow$$
$$[Mo(CN)_8]^{4-} + [Mo(CN)_8]^{3-} \qquad k_{22} \quad (6\text{-}5.41)$$

Values for the self-exchange rate constants are $k_{11} = 7.4 \times 10^2$ L mol^{-1} s^{-1} and $k_{22} = 3.0 \times 10^4$ L mol^{-1} s^{-1}. Substitution of these values into Eq. 6-5.38, and using a value for f of 0.85 yields the prediction that k_{12} should be about 4×10^4 L mol^{-1} s^{-1}. The value that is obtained experimentally is 3×10^4 L mol^{-1} s^{-1}.

The Inner-Sphere (or Ligand-Bridged) Mechanism

Ligand-bridged transition states have been shown to occur in a number of reactions, mainly through the elegant experiments devised by H. Taube and his students. He has demonstrated that the following general reaction occurs:

$$[Co(NH_3)_5X]^{2+} + Cr^{2+}(aq) + 5\ H^+$$
$$= [Cr(H_2O)_5X]^{2+} + Co^{2+}(aq) + 5\ NH_4^+ \quad (6\text{-}5.42)$$

$$(X = F^-, Cl^-, Br^-, I^-, SO_4^{2-}, NCS^-, N_3^-, PO_4^{3-}, P_2O_7^{4-},$$
$$CH_3CO_2^-, C_3H_7CO_2^-, \text{crotonate, succinate, oxalate, maleate})$$

The significance and success of these experiments rest on the following facts. The Co^{III} complex is not labile, while the Cr^{II} aqua ion is. In the products, the $[Cr(H_2O)_5X]^{2+}$ ion is not labile, whereas the Co^{II} aqua ion is. It is found that the transfer of X from $[Co(NH_3)_5X]^{2+}$ to $[Cr(H_2O)_5X]^{2+}$ is quantitative. The most reasonable explanation for these facts is a mechanism such as that illustrated in Reaction 6-5.43.

$$Cr^{II}(H_2O)_6^{2+} + Co^{III}(NH_3)_5Cl^{2+} \longrightarrow [(H_2O)_5Cr^{II}ClCo^{III}(NH_3)_5]^{4+}$$

$$\updownarrow \text{electron transfer}$$

$$Cr(H_2O)_5Cl^{2+} + Co(NH_3)_5(H_2O)^{2+} \longleftarrow [(H_2O)_5Cr^{III}ClCo^{II}(NH_3)_5]^{4+}$$

$$\downarrow H^+, H_2O$$

$$Co(H_2O)_6^{2+} + 5\ NH_4^+$$

Since all Cr^{III} species, including $[Cr(H_2O)_6]^{3+}$ and $Cr(H_2O)_5Cl^{2+}$, are substitution inert, the quantitative production of $Cr(H_2O)_5Cl^{2+}$ must imply that electron transfer ($Cr^{II} \rightarrow Co^{III}$) and Cl^- transfer from Co to Cr are mutually interdependent acts, neither of which is possible without the other. Postulation of the binuclear, chloro-bridged intermediate appears to be the only chemically credible way to explain this phenomenon. As implied by Reaction 6-5.42, many ligands can serve as ligand bridges in inner-sphere reactions.

In reactions between Cr^{2+} and CrX^{2+} and between Cr^{2+} and $Co(NH_3)_5X^{2+}$, which are inner sphere, the rates decrease as X is varied in the order $I^- > Br^- > Cl^- > F^-$. This seems reasonable if ability to "conduct" the transferred electron is associated with polarizability of the bridging group, and it appeared that this order might even be considered diagnostic of the mechanism. However, the opposite order is found for the $Fe^{2+}/Co(NH_3)_5X^{2+}$ and for the $Eu^{2+}/Co(NH_3)_5X^{2+}$ reactions. Moreover, the $Eu^{2+}/Cr(H_2O)_5X^{2+}$ reactions give the order first mentioned, thus showing that the order is not simply a function of the reducing ion used. The order must, of course, be determined by the relative stabilities of transition states with different X, and the variation in reactivity order has been rationalized on this basis.

There are now a number of cases (e.g., those of $Co(NH_3)_5X^{2+}$ with $[Co(CN)_5]^{3-}$, where $X = F^-$, CN^-, NO_3^-, and NO_2^-, and that of Cr^{2+} with $[IrCl_6]^{2-}$) in which the electron transfer is known to take place by both inner- and outer-sphere pathways.

6-6   **Stereochemical Nonrigidity**

No molecule is strictly rigid in the sense that all the interatomic distances and bond angles are fixed at one precise set of values. On the contrary, all molecules, even at absolute zero, constantly execute a set of vibrations, such that all of the atoms oscillate with amplitudes of a few tenths of an angstrom, about their average positions. In this sense, no molecule is rigid, but there are many molecules that undergo rapid deformational rearrangements of a much greater amplitude, in which atoms actually change places with each other. Such rearrangements are found among an enormous variety of compounds, including inorganic molecules, such as PF_5, metal carbonyls, organometallic compounds, and organic molecules. Molecules that behave in this way are said to be stereochemically nonrigid. The recognition of stereochemical nonrigidity and its study is only possible by nuclear magnetic resonance (NMR) spectroscopy. Let us consider one of the earliest inorganic examples, PF_5.

Five-Coordinate Complexes: PF_5

This molecule is known to have a trigonal bipyramidal structure. It would be expected that the fluorine (^{19}F) NMR spectrum would show a complex multiplet of relative intensity two for the axial fluorine atoms and another of intensity three for the equatorial ones. The multiplets would result from coupling of each type of fluorine to those of the other type, and from coupling of both types to the phosphorus atom that has a spin of one half. In fact, only a sharp doublet is seen, indicating that, as far as NMR can tell, all five fluorine atoms are equivalent; the doublet structure results from their coupling to the phosphorus atom.

   This result is due to the axial and equatorial fluorine atoms changing places with one another so rapidly (>10,000 times per second) that the NMR spectrometer cannot sense the two different environments and records all five of them at a single frequency, which is the weighted average of those frequencies for each environment. However, the splitting of the fluorine resonance into a doublet by the phosphorus atom is maintained which indicates that the exchange of places occurs without breaking the P—F bonds.

   The generally accepted explanation for the rapid exchange of axial and equatorial fluorine atoms in PF_5 was suggested by R. S. Berry and is shown in Fig. 6-14. This rearrangement pathway has two main stages. First, there is a concerted motion of the two axial F atoms and two of the equatorial ones, so that these four atoms come into the same plane and define a square. All four atoms are now equivalent to each other, and the entire set of five atoms defines a square pyramid. Second, a trigonal bipyramidal arrangement is now recovered. There are two equally probable ways for this to happen. In one, the same F atoms that were initially axial can return to axial positions. This would do nothing to cause exchange. However, if the other diagonally opposite pair of F atoms, which were initially equatorial, move to axial positions (while the other two, which were initially axial necessarily become equatorial), an exchange of positions involving all but one of the F atoms is accomplished. The same process can now be repeated so that the equatorial F atom that did not exchange the first time becomes exchanged. If this process is repeated indefinitely, all F atoms will constantly pass back and forth between axial and equatorial positions.

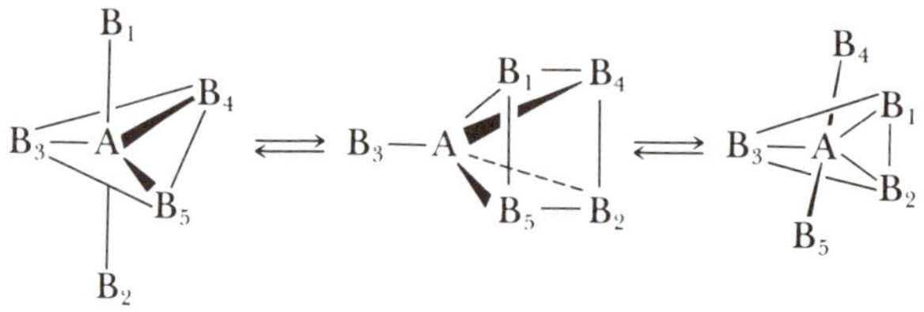

Figure 6-14   A simple mechanism that interchanges axial and equatorial ligands of a tbp by passage through an sp intermediate.

Note that the molecules that exist immediately before and after the rearrangement steps (or after any number of steps) are chemically identical. These molecules differ only in the interchange of indistinguishable nuclei; the process causes no net chemical change and has $\Delta H° = \Delta S° = \Delta G° = 0$. Molecules of this type are by far the most common and important stereochemically nonrigid molecules and are called *fluxional molecules.*

An important fact about the process occurring in PF_5 is that it consists of a rearrangement of one of the more symmetrical forms of five coordination [the trigonal bipyramid (tbp)] into the other [the square pyramid (sp)], and then back to an equivalent version of the first in which some ligands have changed places. This type of process has been called a polytopal rearrangement, because the two different arrangements of the ligand set are polytopes.

For coordination number five, the tbp and sp arrangements seldom differ greatly in energy, so that whichever one is the preferred arrangement in a given substance, the other one can provide a low-energy pathway for averaging the ligand environments. As a general rule, five-coordinate species are fluxional, even at very low temperatures.

Polytopal rearrangements are generally facile for complexes with coordination numbers higher than six as well. This occurs because while one symmetrical structure may be somewhat more stable than any other, the other arrangements are only a few kilojoules less stable, and with ordinary thermal energies available, they provide accessible intermediates for rearrangement. For example, consider an eight-coordinate complex with dodecahedral structure. The eight ligands are not all equivalent but fall into two sets, the A's and the B's, as shown in Fig. 6-1. It is easy to see how the dodecahedron could be converted by relatively slight changes in interatomic distances into either a cubic or a square antiprismatic intermediate from which a new dodecahedron with the A's and B's interchanged would be recovered.

Six-Coordinate Complexes: Racemization of tris-Chelate Complexes

Octahedral complexes are generally not fluxional. That is, even when cis and trans isomers of MX_4Y_2 complexes interconvert, they do so by ligand dissociation and recombination rather than by any intramolecular rearrangement. However, in a few cases it has been shown that intramolecular rearrangement by way of a twist does occur. These are mostly tris-chelate species, where the process studied is racemization.

As previously stated, these exist in enantiomeric configurations, Λ and Δ (Fig. 6-3). At various rates, depending on the metal ion involved and the experimental conditions, these can interconvert. A sample consisting entirely of one enantiomer will eventually racemize, that is, become a mixture of both in equal quantities. Possible pathways for racemization fall into two broad classes: (1) those without breaking of metal–ligand bonds, and (2) those with bond rupture.

Two possible pathways without bond rupture are the trigonal (or Bailar) twist and the rhombic (Ray–Dutt) twist, shown as (a) and (b) in Fig. 6-15. Many dissociative (bond-rupture) type pathways may be imagined; one is shown as (c) in Fig. 6-15. It appears that racemization most often occurs via some pathway with bond rupture, although in a few cases there is evidence for the trigonal twist.

Notice in Fig. 6-3 that if the top part of the Λ isomer is twisted relative to the bottom one half by 60°, the molecule will reach a trigonal prismatic intermediate structure, and can then become the Δ isomer. This sort of process, shown in Fig. 6-15(a), is in general not facile and is rapid only in cases where the chelate ligands have a relatively short distance between their donor atoms (a small "bite"). Since the distance to be spanned is shorter in the eclipsed trigonal prismatic intermediate than in the octahedral structure, such ligands cause the two structures to be closer in stability, so the prism becomes a thermally accessible intermediate or transition state.

Fluxional behavior will be mentioned again later in discussing metal carbonyls (Chapter 28) and organometallic compounds (Chapter 29).

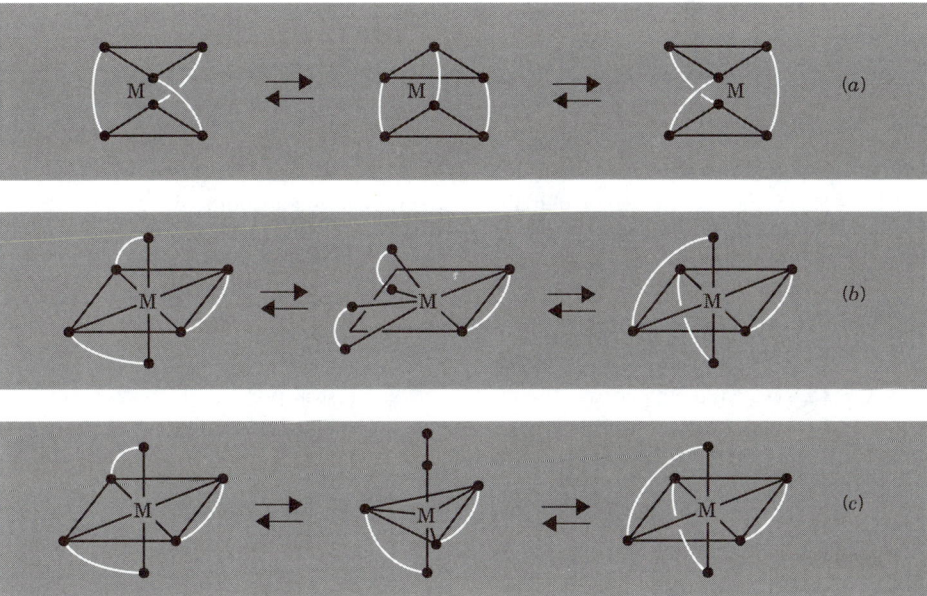

Figure 6-15   Three possible paths for racemization of a tris-chelate complex. (a) The trigonal twist. (b) The rhombic twist. (c) One of many paths involving metal–ligand bond rupture.

STUDY GUIDE

Scope and Purpose

An overview has been presented of the structures, nomenclature, reactivities, stabilities, and so on, of coordination compounds. We shall return to these topics in later chapters where theories of bonding will be added to the discussion. For now, we have confined the discussion to compounds that are traditionally considered to be coordination compounds; organometallic compounds will be covered in Chapters 28–30. The extensive material on mechanisms of reactions may be considered by the instructor to be optional, without much loss in continuity when moving on into later chapters. The general aspects of substitution reactions may, however, be useful, and we encourage some treatment of the differences between associative and dissociative processes.

Study Questions

A. Review

1. For each coordination number from two to nine, mention the principal geometrical arrangement (or arrangements).

2. What does each of the following abbreviations stand for: tbp, sp, *fac*, *mer*?

3. What is meant by tetragonal, rhombic, and trigonal distortion of an octahedron?

4. What do the terms *chelate* and *polydentate* mean?

5. What are the structures of the following ligands: acetylacetonate, ethylenediamine, diethylenetriamine, $EDTA^{4-}$?

6. Show with drawings the enantiomorphs of $M(L—L)_2X_2$ and $M(L—L)_3$ type complexes.

7. Give one example of each of the following types of isomers: ionization isomers, linkage isomers, coordination isomers.

8. Write the names of each of the following: $[Co(NH_3)_4(en)]Cl_3$, $[Cr(en)Cl_4]^-$, $[Pt(acac)NH_3Cl]$, $[Ru(NH_3)_5N_2](NO_3)_2$, $KFeCl_4$.

9. What are the two principal sets of equilibrium constants (K_i's and β_i's) for expressing the formation of a series of complexes, ML, ML_2, ML_3, and so on? How are they related?

10. Except in rare cases, how do the magnitudes of the constants K_i vary with increasing i? What is the underlying reason for this, regardless of the charges?

11. What is meant by the *chelate effect*? Give an example.

12. For what ring sizes is the chelate effect most important? How do you explain it?

13. Explain the difference between kinetic inertness (or lability) and thermodynamic stability (or instability).

14. What are the two limiting mechanisms for ligand exchange?

15. Explain how solvent intervention, ion-pair formation, and conjugate-base formation can affect the observed rate law.

16. Why does the rate law tell us nothing as to the true order of an aquation (acid hydrolysis) reaction carried out in aqueous solution?

17. True or false: the high rate of basic hydrolysis of $[Co(NH_3)_5Cl]^{2+}$ is attributable to the exceptional ability of OH^- to attack the cobalt ion nucleophilically. If false, give an alternative explanation of the high rate.

18. Why do many square complexes have two-term rate laws for ligand replacement reactions?

19. What is meant by the term *trans effect*?

20. Discuss the two general mechanisms for electron-transfer reactions.

21. Describe the type of reaction and the reasoning used by Taube to prove that certain electron-transfer reactions must occur by way of a bridged intermediate.

22. What is meant by a fluxional molecule? What is the experimental evidence that proves PF_5 to be one?

B. Additional Exercises

1. Show with drawings how axial–equatorial exchange in a square pyramidal complex AB_5 could occur via a tbp intermediate.

2. Draw all the isomers of an octahedral complex having four different monodentate ligands. Indicate optical isomers.

3. Show how the experimental determination of the number of isomers of $[Co(NH_3)_4Cl_2]^+$ would enable you to show that the coordination geometry is octahedral, not trigonal prismatic.

4. Why do you think most species, such as $AlCl_3$, $[CuCl_3]^-$, $Pt(NH_3)_2Cl^+$, are not actually such three-coordinate monomers but, instead, dimerize?

5. Suppose you prepared $[Co(en)_2Cl_2]^+$. Ignoring possible ring conformation effects, how many isomers, geometric and optical, could be formed?

6. Write the proper names for the complexes
   (a) $K_3[Co(C_2O_4)_3]$ (b) $Fe(CO)_5$
   (c) $[Co(NH_3)_5Cl]Cl_2$ (d) $[Co(en)_3]_2(SO_4)_3$
   (e) $Na_2[PtCl_4]$ (f) $[Ru(NH_3)_5(N_2)]Cl_2$
   (g) $Na[Fe(CO)_4H]$ (h) $K_2[Fe(CN)_5NO] \cdot 2 H_2O$
   (i) $K[(NH_3)_5Co{-}NC{-}Co(CN)_5]$

7. Draw all the possible isomers of the dinuclear complex $L_2X_2M(\mu\text{-}X)_2ML_2X_2$, where L is a ligand that cannot be a bridge.

8. Assign an inner-sphere or an outer-sphere mechanism for the following reactions, and draw out the details of the reaction sequence:

$$(a) \quad \left[Co(NH_3)_5N \bigcirc - \bigcirc N \right]^{3+} + [Fe(CN)_5OH_2]^{3-}$$

The Co^{III} reactant is substitution inert, while the Fe^{II} reactant is substitution labile. The products are $Co^{2+}(aq)$, which is substitution labile, and

$$\left[Fe(CN)_5N \bigcirc - \bigcirc N \right]^{2-}$$

which is substitution inert.

(b) $[Co(NH_3)_6]^{3+} + [Fe(CN)_5OH_2]^{3-}$

The hexaammine of cobalt is substitution inert and the Fe^{II} reactant is substitution labile. The products are $Co^{2+}(aq)$ and $[Fe(CN)_5OH_2]^{2-}$.

9. For $[PtX_4]^{2-}$ complexes both ligand exchange rate and thermodynamic stability increase in the order X = Cl < Br < I < CN. Explain why these observations are not inconsistent.

10. Using the trans effect sequence given in the text, devise rational procedures for selectively synthesizing each of the three isomers of $[Pt(py)NH_3NO_2Cl]$.

11. If application of the Marcus equation were to be made in order to predict the electron-transfer rate constants k_{12} for the following cross reactions, list the self-exchange reactions for which self-exchange rate constants k_{11} and k_{22} would be needed.
    (a) $[Fe(CN)_6]^{4-} + [Co(en)_3]^{3+}$
    (b) $Fe^{3+}(aq) + [Cr(phen)_3]^{2+}$
    (c) $[Rh(phen)_3]^{3+} + [Ru(phen)_3]^{2+}$

12. Predict the value for the second-order rate constants k_{12} for the following cross reactions, assuming in each case that f in the Marcus equation equals 0.8.
    (a) $[Fe(CN)_6]^{4-} + [MnO_4]^-$ where $K_{12} = 2.5 \times 10^3$.
    (b) $[Mo(CN)_8]^{4-} + [IrCl_6]^{2-}$ where $K_{12} = 1.5 \times 10^2$,
        $k_{11} = 3.0 \times 10^4 \text{ L mol}^{-1} \text{ s}^{-1}$ and $k_{22} = 2.3 \times 10^5 \text{ L mol}^{-1} \text{ s}^{-1}$.

13. Give the proper name for each of the following compounds:
(a) $Pt(NH_3)_2Cl_2$	(m)	$[Cr(H_2O)_6]Cl_2$
(b) $[Rh(NH_3)_5Cl]Cl_2$	(n)	$[Co(en)_3]_2(SO_4)_3$
(c) $[Co(NH_3)_6](NO_3)_3$	(o)	$Na[HB(OCH_3)_3]$
(d) $[Co(H_2O)_4]SO_4$	(p)	$[Pt(py)_4][PtCl_4]$
(e) $[Co(NH_3)_4(OH_2)_2](BF_4)_3$	(q)	$Na_2[PdCl_6]$
(f) $[Fe(H_2O)_6]Br_2$	(r)	$(NEt_4)_3[Cr(CN)_6]$
(g) $Na_3[Fe(CN)_6]\cdot 2\,H_2O$	(s)	$[Ni(phen)_3](ClO_4)_2$
(h) $Na_4[Fe(CN)_6]$	(t)	$[Co(NH_3)_5NO_2]SO_4$
(i) $Ni(CO)_4$	(u)	$[Co(en)_2(Cl)(NO_2)]SCN$
(j) $[Cu(NH_3)_4]SO_4$	(v)	$[N(CH_3)_4][W(CO)_5Cl]$
(k) $[Pt(en)_2](ClO_4)_2$	(w)	$[Cr(H_2O)_6]Cl_3$
(l) $Co(NH_3)_2(Cl)(Br)(CH_3CO_2)$	(x)	$Pt(acac)(NH_3)(Cl)$

14. Go back through the compounds of Questions 6 and 13. Identify each compound that can display (a) geometrical isomerism, (b) linkage isomerism, and (c) optical isomerism.

15. Draw the structure of each of the following coordination compounds or ions. Then draw the structure for each geometrical, linkage, or optical isomer that is possible:
    (a) *cis*-Dichlorotetraaquachromium(III) chloride
    (b) Potassium pentachloronitroosmate(IV)
    (c) *mer*-Trihydridotris(triphenylphosphine)ruthenium(III)
    (d) Potassium trioxalatocobaltate(III)
    (e) Chloropentaamminecobalt(III) nitrate
    (f) Tris(ethylenediamine)cobalt(III) nitrate
    (g) Sodium tetrabromoplatinate(II)
    (h) Pentaamminedinitrogenruthenium(II) chloride
    (i) Sodium pentacyanonitrosylferrate(II) dihydrate
    (j) Tetraammineaquacobalt(III)-μ-cyanobromotetraamminecobaltate(III)

16. The pentacyanocobaltate(II) ion is a catalyst for the conversion of $[Co(CN)_5\text{—}NCS]^{3-}$ to $[Co(CN)_5\text{—}SCN]^{3-}$, by an inner-sphere electron-transfer mechanism. Show all of the necessary steps for this reaction.

17. The alkali metal cations fall into Class 1, undergoing water exchange exceedingly

rapidly. Discuss the evidence that suggests an I_d mechanism. (*Note:* It will be instructive at this point to preview the material presented in Table 10-1 and Section 10-7, which clearly demonstrates that the water ligands in the aqua ions become less tightly bound in the series $Li^+ > Na^+ > K^+ > Rb^+ > Cs^+$.)

18. List and define all of the rate and equilibrium constants, plus the cross reaction and each self-exchange reaction, that must be known in order to use the Marcus equation to predict the rate constant of the following outer-sphere electron-transfer reaction.

$$[Fe(CN)_6]^{4-} + [Co(en)_3]^{3+} \longrightarrow [Fe(CN)_6]^{3-} + [Co(en)_3]^{2+}$$

19. Sketch the structure of the following:
    (a) *mer*-Bromochloroisothiocyanatotris(triphenylphosphine)rhodium(III)
    (b) All possible linkage, geometrical, and optical isomers of $[Co(en)_2(NO_2)Cl]^+$.

20. Write the chemical equations for
    (a) Water exchange in the hexaaquanickel(II) ion.
    (b) Acid hydrolysis of $[Co(NH_3)_5Cl]^{2+}$.
    (c) Base hydrolysis of $[Co(NH_3)_5Cl]^{2+}$.
    (d) Self-exchange of $[IrCl_6]^{2-,3-}$.
    (e) Acid hydrolysis of $[Co(NH_3)_5CO_3]^+$.
    (f) $[Co(NH_3)_5OH_2]^{3+} + NO_2^-$
    (g) *cis*-$Pt(PEt_3)_2(CN)(Cl) + H_2O$
    (h) $[Co(NH_3)_5I]^{2+} + [Cr(H_2O)_6]^{2+} + 5\ H^+$

21. Show the mechanisms that explain why the following reactions occur far more rapidly than would be true for simple substitution or ligand replacement:
    (a) $[Co(NH_3)_5H_2O]^{3+} + NO_2^-$
    (b) $[Co(NH_3)_5CO_3]^+ + H_3O^+$
    (c) $[Cr(NH_3)_5NCS]^{2+} + Hg^{2+}$

22. Briefly explain how each of the following classic "obstacles" in kinetics serves to obscure the molecularity of a reaction:
    (a) Solvent intervention, as in the aquation (acid hydrolysis) of $[Rh(NH_3)_5Cl]^{2+}$.
    (b) Ion-pair formation, as in anation of $[Co(NH_3)_5OH_2]^{3+}$.
    (c) Conjugate base formation, as in base hydrolysis of $[Co(NH_3)_5Br]^{2+}$.

23. Contrast the general trends for substitution reactions of (1) octahedral and (2) square complexes regarding:
    (a) Leaving group effects.
    (b) Charge effects.
    (c) Steric effects.

24. Explain the mechanistic significance of the observed two-term rate laws for
    (a) Substitution reactions for square complexes, that is, $k_{obs} = k_1 + k_2[Y]$.
    (b) Aquation reactions of octahedral complexes, that is, $k_{obs} = k_a + k_b[OH^-]$.

25. Show the steps of the electron-transfer mechanisms that account for the following reactions. Note that in each of these reactions, one of the reactant metal ions is labeled, making its identification among the products possible.
    (a) $[*Co(NH_3)_5—NCS]^{2+} + 5\ CN^- + Co^{2+}(aq) \longrightarrow [Co(CN)_5—SCN]^{2+} + 5\ NH_3 + *Co^{2+}(aq)$
    (b) $[*Cr(H_2O)_5—SCN]^{2+} + Cr^{2+}(aq) \longrightarrow [Cr(H_2O)_5—NCS]^{2+} + *Cr^{2+}(aq)$

26. Explain why assignment of an inner-sphere electron-transfer mechanism for each reaction in Question 25 requires (a) an ambidentate ligand, (b) an inert transition metal reactant, (c) a labile transition metal reactant, and (d) an inert transition metal product.

27. Summarize the types of data that indicate a predominantly dissociative mechanism for substitution reactions of octahedral compounds.

28. Summarize the types of data that indicate a predominantly associative mechanism for substitution reactions of square compounds.

29. What mechanistic interpretation can we give to the following data for the reaction

$$Pt(dien)Br^+ + Y^- \longrightarrow Pt(dien)Y^+ + Br^-$$

Y	$k_2(M^{-1}s^{-1}) \times 10^4$
OH^-	1
Cl^-	8.8
py	33
I^-	2300

30. What mechanistic interpretation should we give to the following data for the reaction

$$trans\text{-}Pt(py)_2Cl_2 + {}^{36}Cl^- \longrightarrow trans\text{-}Pt(py)_2\,{}^{36}ClCl + Cl^-$$

Solvent	$k_1(s^{-1}) \times 10^5$
$(CH_3)_2SO$	38
H_2O	3.5
C_2H_5OH	1.4

31. Why is the first reaction below 100 times faster than the second?

$$Co(en)_2Cl_2^+ + H_2O \longrightarrow Co(en)_2(Cl)OH_2^{2+} + Cl^-$$

$$Co(en)_2(Cl)OH_2^{2+} + H_2O \longrightarrow Co(en)_2(OH_2)_2^{3+} + Cl^-$$

32. Direct aquation of $trans$-Co(trien)Cl_2^+ produces cis-Co(trien)(Cl)(OH$_2$)$^{2+}$. In the presence of Hg^{2+}, we get $HgCl_2$, plus $trans$-Co(trien)(Cl)(OH$_2$)$^{2+}$, which subsequently undergoes isomerization to the cis product. Explain these observations with a detailed mechanistic proposal.

33. A ligand-bridged intermediate has been observed in the following reaction. Write out a likely mechanism for the process.

$$(H_2O)_5Cr\!-\!NCS^{2+} + Hg^{2+} \longrightarrow Cr(H_2O)_6^{3+} + Hg\!-\!SCN^+$$

34. The five-coordinate compound $PF_4(NEt_2)$ is fluxional at room temperature, ^{19}F NMR spectroscopy indicating that all fluorine atoms in the molecule are made equivalent on the NMR time scale by a rapid polytopal rearrangement. At -85 °C, however, NMR indicates the presence of two distinct types of fluorine atoms, in a ratio of 1:1. Explain.

35. At -22 °C, PCl_2F_3 is fluxional, all fluorine atoms being indistinguishable by NMR spectroscopy. However, at -143 °C, NMR spectroscopy clearly indicates the presence of two distinct types of fluorine atoms, in an F_{axial} to $F_{equatorial}$ ratio of 2:1. Explain.

36. At -74 °C, NMR spectroscopy indicates that the four fluorine atoms in $PF_4[NMe_2]$ are equivalent to one another. Is the molecule fluxional or not at this temperature? Explain.

37. Interpret the following data by showing the fundamental mechanistic steps that are indicated, and by labeling the slopes and intercepts properly in terms of specific rate constants for each step of the mechanism, for the reaction:

$$Pt(dien)Br^+ + Y^- \longrightarrow Pt(dien)Y^+ + Br^-$$

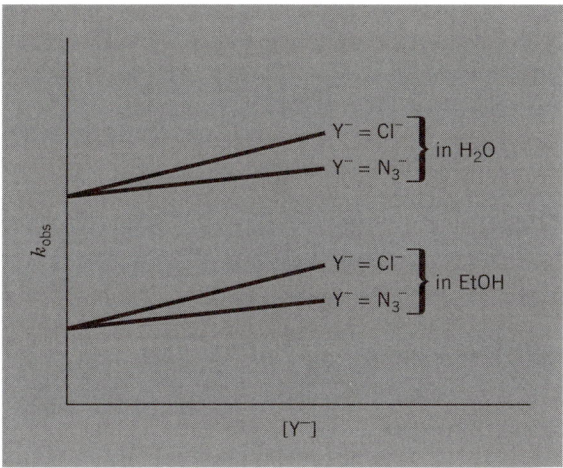

C. Questions from the Literature of Inorganic Chemistry

1. Consider the paper by L. R. Carey, W. E. Jones, and T. W. Swaddle, *Inorg. Chem.*, **1971**, *10*, 1566–1570, dealing with the mechanisms of aquation reactions of $[Cr(H_2O)_5X]^{2+}$ complexes.

    (a) What evidence do the authors cite for discounting the D mechanism?

    (b) For which reactions is an I_a mechanism suspected? An I_d mechanism?

    (c) List the principal evidence that the authors cite for these mechanistic assignments.

    (d) How do these mechanistic results differ from those for the cobalt(III) ammines?

2. Consider the papers by C. Shea and A. Haim, *J. Am. Chem. Soc.*, **1971**, *93*, 3055–3056, and *Inorg. Chem.*, **1973**, *12*, 3013–3015.

    (a) Draw the structures (including oxidation states of the metals and overall charge on the complexes) of the reactants, intermediates, and products of the redox reactions reported in these papers. Include the intermediates for both adjacent and remote attack. Realize that the compounds that are abbreviated $CrNCS^{2+}$, $CrSCN^{2+}$, and Cr^{2+} are probably six-coordinate complexes.

    (b) How were the yields of $CrSCN^{2+}$ and $CrNCS^{2+}$ determined? Is there likely to be any uncatalyzed isomerization of these forms of linkage of the SCN^- ligands in these products?

    (c) What is the reaction that the authors call the "Cr^{2+}-catalyzed isomerization of $CrSCN^{2+}$ to $CrNCS^{2+}$"? How does this isomerization take place, and what must be true of the labilities of the Cr^{2+} and Cr^{3+} complexes?

    (d) How do the reductants Cr^{2+} and $[Co(CN)_5]^{3-}$ differ in reactions with $[(NH_3)_5Co—NCS]^{2+}$ and in reactions with $[(NH_3)_5Co—SCN]^{2+}$?

    (e) Why is the isomer $[(NH_3)_5Co—NCS]^{2+}$ only susceptible to remote attack by the reductant?

3. Consider the paper: A. J. Miralles, A. P. Szecsy, and A. Haim, *Inorg. Chem.*, **1982**, *21*, 697–699.

    (a) To what fundamental steps in the reaction sequence do the authors ascribe each of the spectroscopic changes that take place upon mixing reactants?

    (b) What two consecutive redox reactions take place in these systems? What are their mechanisms? What are their relative rates? Why was an ion pair observable only for the first?

(c) How have the authors decided upon the orientation of the reactants within the ion pairs that precede the first outer-sphere electron-transfer process?

4. Consider the paper by J. L. Burmeister, *Inorg. Chem.*, **1964,** *3*, 919–920. Propose a mechanism for the synthesis [reaction (1)] of $[Co(CN)_5—SCN]^{3-}$.

SUPPLEMENTARY READING

Atwood, J. D., *Inorganic and Organometallic Reaction Mechanisms,* Brooks/Cole, Monterey, CA, 1985.

Basolo, F. and Johnson, R. C., *Coordination Chemistry,* Benjamin, Menlo Park, CA, 1964.

Basolo, F. and Pearson, R. G., *Mechanisms of Inorganic Reactions,* 2nd ed., Wiley, New York, 1967.

Benson, D., *Mechanisms of Inorganic Reactions in Solution,* McGraw-Hill, New York, 1968.

Cannon, R. D., *Electron Transfer Reactions,* Butterworths, London, 1980.

Edwards, J. O., "Inorganic Reaction Mechanisms," Parts I and II, Vols. 13 and 17, *Progress in Inorganic Chemistry,* Wiley-Interscience, New York, 1970 and 1972.

Edwards, J. O., *Inorganic Reaction Mechanisms,* Benjamin, Menlo Park, CA, 1964.

Langford, C. H. and Gray, H. B., *Ligand Substitution Processes,* Benjamin, Menlo Park, CA, 1984.

Lippard, S. J., "An Appreciation of Henry Taube," Vol. 30, *Progress in Inorganic Chemistry. An Appreciation of Henry Taube,* Vol. 30, Wiley-Interscience, New York, 1983.

Martell, A. E., Ed., *Coordination Chemistry,* Vols. 1 and 2, Van Nostrand–Reinhold, New York, 1971 and 1978.

Martell, A. E. and Motekaitis, R. J., *Determination and Use of Stability Constants,* VCH Publishers, Weinheim, 1989.

Sykes, A. G., *Kinetics of Inorganic Reactions,* Pergamon Press, Elmsford, NY, 1966.

Taube, H., *Electron Transfer Reactions of Complex Ions in Solution,* Academic, New York, 1970.

Twigg, M. V., Ed., *Mechanisms of Inorganic and Organometallic Reactions,* Vols. 1 and 2, Plenum, New York, 1982 and 1984.

Wilkins, R. G., *The Study of Kinetics and Mechanism of Reactions of Transition Metal Complexes,* Allyn and Bacon, Boston, 1974.

Chapter 7

SOLVENTS, SOLUTIONS, ACIDS, AND BASES

The majority of chemical reactions and many measurements of properties are carried out in a solvent. The properties of the solvent are crucial to the success or failure of the study. For the inorganic chemist, water has been the most important solvent, and it will continue to be, but many other solvents have been tried and found useful. A few of them, and the concepts that influence the choice of a solvent, are discussed here. Closely connected with the properties of solvents is the behavior of acids and bases. In this chapter some fundamental concepts concerning acids and bases are also presented.

7-1   Solvent Properties

Properties that chiefly determine the utility of a solvent are

1. The temperature range over which it is a liquid.
2. Its dielectric constant.
3. Its donor and acceptor (Lewis acid–base) properties.
4. Its protonic acidity or basicity.
5. The nature and extent of autodissociation.

The first two are of rather obvious import and need not detain us long. The others will merit discussion in subsequent sections.

Liquid Range

Solvents that are liquid at room temperature and 1-atm pressure are most useful because they are easily handled, but it is also desirable that measurements or reactions be feasible at temperatures well above and below room temperature. As Table 7-1 shows, N,N-dimethylformamide (DMF), propane-1,2-diol carbonate, and acetonitrile are especially good in this respect.

Dielectric Constant

The ability of a liquid to dissolve ionic solids depends strongly, although not exclusively, on its *dielectric constant*, ϵ. The force (F) of attraction between cations

Table 7-1   Properties of Some Useful Solvents[a]

Name	Abbreviation	Formula	Liquid Range (°C)	ϵ/ϵ_0
Water		H_2O	0 to 100	82
Acetonitrile		CH_3CN	−45 to 82	38
N,N-Dimethylformamide	DMF	$HC(O)N(CH_3)_2$	−61 to 153	38
Dimethyl sulfoxide	DMSO	$(CH_3)_2SO$	18 to 189	47
Nitromethane		CH_3NO_2	−29 to 101	36
Sulfolane		 (ring) SO_2	28 to 285	44
Propane-1,2-diol carbonate		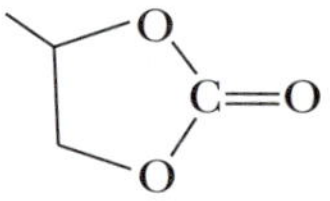 (ring) $C{=}O$	−49 to 242	64
Hexamethylphosphoramide	HMP	$OP[N(CH_3)_2]_3$	8 to 230	30
Glycol dimethyl ether	Glyme	$CH_3OCH_2CH_2OCH_3$	−58 to 83	3.5
Tetrahydrofuran	THF	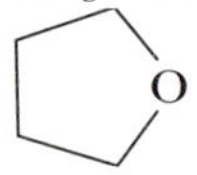 (ring) O	−65 to 66	7.6
Dichloromethane		CH_2Cl_2	−97 to 40	9
Ammonia		NH_3	−78 to −33	23 (−50 °C)
Sulfuric acid		H_2SO_4	10 to 338	100
Hydrogen fluoride		HF	−83 to 20	84 (0 °C)
Hydrogen cyanide		HCN	−14 to 26	107

[a] In this table, instead of the absolute value of ϵ, we give the ratio of ϵ to ϵ_0, with the latter being the value for a vacuum. In subsequent sections the term "dielectric constant" refers to this ratio.

and anions immersed in a medium of dielectric constant ϵ is inversely proportional to ϵ, as in Eq. 7-1.1.

$$F = \frac{q^+ q^-}{4\,\pi\epsilon r^2} \tag{7-3.8}$$

Thus, water ($\epsilon = 82\epsilon_0$ at 25 °C, where ϵ_0 is for a vacuum) reduces the attractive force nearly to 1% of its value in the absence of a solvent. Solvents with lower dielectric constants are less able to reduce the attractive forces (F in Eq. 7-1.1) between dissolved cations and anions. Such solvents are, therefore, less able to dissolve ionic substances.

7-2   Donor and Acceptor Properties: Solvent Polarity

The ability of a solvent to keep a given solute in solution depends considerably on its ability to solvate the dissolved particles, that is, to interact with them in a quasichemical way. For ionic solutes, there are both cations and anions to be solvated. Commonly, the cations are smaller [e.g., $Ca(NO_3)_2$, $FeCl_3$] and the solvation of the cations is of prime importance. The solvation of simple cations is essentially the process of forming complexes in which the ligands are solvent molecules. The order of coordinating ability toward typical cations for some common solvents is

$$DMSO > DMF \approx H_2O > \text{acetone} \approx (CH_3CHCH_2)O_2CO \approx CH_3CN$$
$$> (CH_2)_4SO_2 > CH_3NO_2 > C_6H_5NO_2 \gg CH_2Cl_2$$

Acceptor properties are usually manifested less specifically. The positive ends of the solvent molecule dipoles will orient themselves toward the anions.

Note that in general the dielectric constant and the ability to solvate ions are related properties, which tend to increase simultaneously, but there is no quantitative correlation. The more polar the molecules of a solvent the higher its dielectric constant tends to be (although the extent of hydrogen bonding also plays a very important role); at the same time, the more polar a molecule the better able it is to use its negative and positive regions to solvate cations and anions, respectively.

7-3 Protic Solvents

These solvents contain ionizable protons and are more or less acidic. Examples are H_2O, HCl, HF, H_2SO_4, and HCN. Even ammonia, which is usually considered a base, is a protic solvent and can supply H^+ to stronger bases. Protic solvents characteristically undergo autodissociation.

Autodissociation of Protic Solvents

For some of the examples just mentioned, the autodissociation reactions can be written in the simplest way as follows:

$$2\,H_2O = H_3O^+ + OH^- \tag{7-3.1}$$

$$2\,HCl = H_2Cl^+ + Cl^- \tag{7-3.2}$$

$$2\,HF = H_2F^+ + F^- \tag{7-3.3}$$

$$2\,H_2SO_4 = H_3SO_4^+ + HSO_4^- \tag{7-3.4}$$

$$2\,NH_3 = NH_4^+ + NH_2^- \tag{7-3.5}$$

In each of Reactions 7-3.1 to 7-3.5, autodissociation involves proton transfer between two solvent molecules to give the protonated solvent cation and the deprotonated solvent anion.

The significance of autodissociation is that solutes encounter not only the molecules of the solvent but the cations and anions that form in the autodissociation process. The autodissociations of several of the acid solvents are discussed in detail in Section 7-11. Here, we give a closer examination of the processes in water and ammonia. These simple equations do not consider the further solvation of the primary products of autodissociation in detail, and this is important.

Water

A more general equation for the autodissociation of water is

$$(n + m + 1)H_2O = [H(H_2O)_n]^+ + [HO(H_2O)_m]^- \tag{7-3.6}$$

For the hydrogen ion, $[H(H_2O)_n]^+$, there is strong association of H^+ with one water molecule to give H_3O^+, a pyramidal ion (Structure 7-I) isoelectronic with NH_3. This ion is observed in a number of crystalline compounds. In water it is further solvated. Another species actually observed in crystals is the $H_5O_2^+$ ion (Structure 7-II). Probably the $H_9O_4^+$ ion (Structure 7-III) is the largest well-

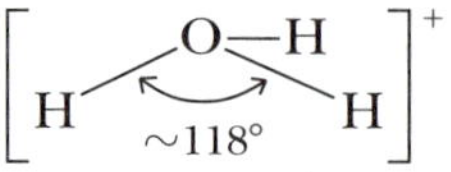

$$\left[\begin{array}{c} H \\[0.5em] \overset{O-H}{\underset{\sim 118°}{\diagup \diagdown}} \\ H \end{array} H \right]^+ \qquad \left[\begin{array}{c} H \quad H \\ \diagdown \; | \\ O\text{---}H\text{---}O-H \\ \diagdown \\ H \end{array} \right]^+$$

7-I 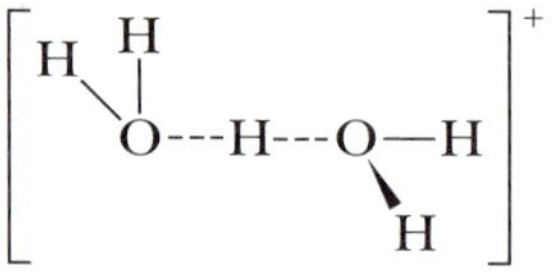   7-II

$$\left[\begin{array}{c} \qquad\qquad OH_2 \\ H_2O \diagdown \quad O-H \diagup \\ \qquad H \diagdown \quad \diagup \\ \qquad\quad H \\ \qquad\qquad OH_2 \end{array} \right]^|$$

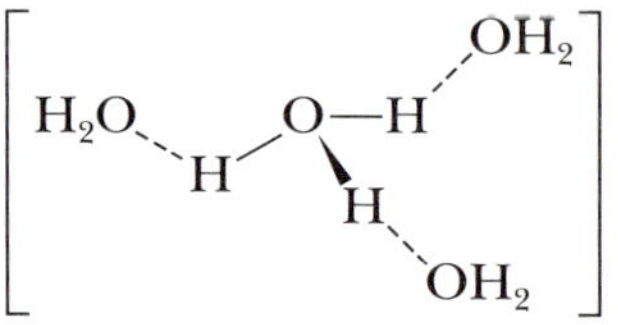

7-III

defined species. The extent of autodissociation (Eq. 7-3.6) is slight, as shown by the small value for the equilibrium constant, Eq. 7-3.7.

$$K'_{25\,°C} = \frac{[H^+][OH^-]}{[H_2O]} = (1.0 \times 10^{-14})/55.56 \tag{7-3.7}$$

In practice, the essentially constant 55.56 M concentration of H_2O molecules is omitted (because it *is* constant), and the constant $K_{25\,°C} = [H^+][OH^-] = 1.0 \times 10^{-14}$ is used.

Liquid Ammonia

Liquid ammonia is a colorless liquid that is useful as a solvent over the temperature range −78 to −33 °C. Its autodissociation (Eq. 7-3.5) is less than that of water:

$$K_{-50\,°C} = [NH_4^+][NH_2^-] = 10^{-30} \tag{7-3.8}$$

Here too, autodissociation involves proton transfer between two solvent molecules to form the conjugate base of ammonia (NH_2^-) and the conjugate acid of ammonia (NH_4^+).

7-4   Aprotic Solvents

There are three broad classes of aprotic solvents:

   1. Nonpolar, or very weakly polar, nondissociated liquids, which do not solvate strongly.   Examples are carbon tetrachloride (CCl_4) and hydrocarbons. Because of low polarity, low dielectric constants, and poor donor power, these are not powerful solvents except for other nonpolar substances. Their main value, when they can be used, is that they play a minimal role in the chemistry of reactions carried out therein.
   2. Nonionized but strongly solvating (generally polar) solvents.   Examples of this type are acetonitrile (CH_3CN), *N,N*-dimethylformamide (DMF), di-

methyl sulfoxide (DMSO), tetrahydrofuran (THF), and sulfur dioxide (SO_2). These substances have in common the facts that they are aprotic, that no autodissociation equilibria are known to occur, and that they strongly solvate ions. In other respects they differ. Some are high boiling (DMSO), others are low boiling (SO_2); some have high dielectric constants (DMSO, 45) while others are of low polarity (THF, 7.6). For the most part, they solvate cations best by using negatively charged oxygen atoms, but SO_2 has pronounced acceptor ability, and solvates anions and other Lewis bases effectively. For example, the molecular adduct $(CH_3)_3N \rightarrow SO_2$ can be isolated.

   3. Highly polar, autoionizing solvents.   Some of these solvents are interhalogen compounds, such as BrF_3 and IF_5, whose structures were discussed in Chapter 3. Examples of their autoionizations are given in Eqs. 7-4.1 and 7-4.2.

$$2\ BrF_3 = BrF_2^+ + BrF_4^- \tag{7-4.1}$$

$$2\ IF_5 = IF_4^+ + IF_6^- \tag{7-4.2}$$

Another example is trichlorophosphine oxide:

$$2\ Cl_3PO = Cl_2PO^+ + Cl_4PO^- \tag{7-4.3}$$

which undergoes autoionization through Cl^- transfer.

7-5   Molten Salts

These salts represent a kind of extreme of aprotic, autoionizing solvents. In them ions predominate over neutral molecules which, in some cases, are of negligible concentration. The alkali metal halides and nitrates are among the "totally" ionic molten salts, whereas others (e.g., molten halides of zinc, tin, and mercury) contain many molecules, as well as ions.

   Low melting points are often achieved with either mixtures or by using halides of alkylammonium ions. Thus an appropriate mixture of $LiNO_3$, $NaNO_3$, and KNO_3 has a melting point as low as 160 °C and $(C_2H_5)_2H_2NCl$ has a melting point of 215 °C. There are even molten salts that are liquids at and below room temperature. These are formed by mixing $AlCl_3$ with compounds such as that shown below.

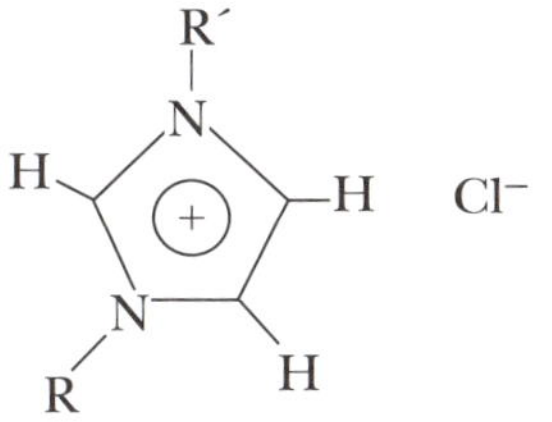

   Examples of important reactions carried out in molten salts are the following preparations of low-valent metal salts.

$$CdCl_2 + Cd \xrightarrow{\text{liquid } AlCl_3} Cd_2[AlCl_4]_2 \tag{7-5.1}$$

$$Re_3Cl_9 \xrightarrow{\text{liquid } (C_2H_5)_2H_2NCl} [(C_2H_5)_2H_2N]_2[Re_2Cl_8] \tag{7-5.2}$$

The industrial production of aluminum is carried out by electrolysis of a solution of Al_2O_3 in molten Na_3AlF_6.

7-6   Solvents for Electrochemical Reactions

A good solvent for electrochemical reactions must meet several criteria. First, electrochemical reactions involve ionic substances, so that a dielectric constant of 10 or better is desirable. Second, the solvent must have a wide range of voltage over which it is not oxidized or reduced, so its own electrode reactions will not take precedence over those of interest.

Water is a widely useful solvent for electrochemistry. Because of its high dielectric constant and solvating ability, it dissolves many electrolytes. Its intrinsic conductance is suitably low. Its range of redox stability is fairly wide, as shown by the following potentials, although its reduction is often a limitation.

$$O_2 + 4\,H^+(10^{-7}\,M) + 4\,e^- = 2\,H_2O \qquad E° = +0.82\,V \qquad (7\text{-}6.1)$$

$$H^+(10^{-7}\,M) + \ e^- = \tfrac{1}{2}\,H_2 \qquad E° = -0.41\,V \qquad (7\text{-}6.2)$$

Acetonitrile, CH_3CN, is widely used for solutes such as organometallic compounds or salts containing large alkylammonium ions, which are insufficiently soluble in water. It is stable over a wide range of voltages.

N,N-Dimethylformamide, $HC(O)N(CH_3)_2$, is similar to CH_3CN but is easier to reduce. Dichloromethane and nitromethane are sometimes used for organic solutes. Molten salts are also useful.

7-7   Purity of Solvents

Although it is obvious that a solvent should be pure if reproducible and interpretable results are to be obtained, it is not always obvious what subtle forms of contamination can occur. Of particular importance are water and oxygen. Oxygen is slightly soluble in virtually all solvents, and saturated solutions are formed on brief exposure to air, for example, when pouring. Oxygen can be partially removed by bubbling nitrogen through the liquid, but only repeated freezing and pumping on a vacuum line can completely remove it. Certain organic solvents, especially ethers, react with oxygen on long exposure to air, forming peroxides. The solvents can best be purified of peroxides by distillation from reductants (e.g., hydrides) or by passage through "molecular sieves" (Section 5-4).

Water also dissolves readily in solvents exposed to the air or to glass vessels that have not been baked dry. It is important to recognize that even small quantities of H_2O on a weight percentage basis can be important. For example, acetonitrile, which contains only 0.1% by weight of water, is about 0.04 M in H_2O, so that the properties of 0.1 M solutions can be *seriously* influenced by the *"trace"* of water.

7-8   Definitions of Acids and Bases

The concepts of acidity and basicity are so pervasive in chemistry that acids and bases have been defined many times and in various ways. One definition, probably the oldest, is so narrow as to pertain only to water as solvent. According to this definition, acids and bases are sources of H^+ and OH^-, respectively. A somewhat broader, but closely allied definition, which is applicable to all protonic solvents, is that of Brønsted and Lowry.

Brønsted–Lowry Definition

An acid is a substance that supplies protons and a base is a proton acceptor. Thus, in water, any substance that increases the concentration of hydrated protons (H_3O^+) above that due to the autodissociation of the water is an acid, and any substance that lowers it is a base. Any solute that supplies hydroxide ions (OH^-) is a base, since these combine with protons to reduce the H_3O^+ concentration. However, other substances, such as sulfides, oxides, or anions of weak acids (F^- or CN^-), are also bases.

Solvent System Definition

This definition can be applied in all cases where the solvent has a significant autoionization reaction, whether protons are involved or not. Some examples are

$$2\,H_2O = H_3O^+ + OH^- \tag{7-8.1}$$

$$2\,NH_3 = NH_4^+ + NH_2^- \tag{7-8.2}$$

$$2\,H_2SO_4 = H_3SO_4^+ + HSO_4^- \tag{7-8.3}$$

$$2\,OPCl_3 = OPCl_2^+ + OPCl_4^- \tag{7-8.4}$$

$$2\,BrF_3 = BrF_2^+ + BrF_4^- \tag{7-8.5}$$

A solute that increases the cationic species natural to the solvent is an acid; one that increases the anionic species is a base. Thus, for the BrF_3 solvent, a compound such as BrF_2AsF_6, which dissolves to give BrF_2^+ and AsF_6^- ions, is an acid, while $KBrF_4$ is a base. If solutions of acid and base are mixed, a neutralization reaction, producing a salt and solvent molecules, takes place.

$$\underbrace{BrF_2^+ + AsF_6^-}_{\text{Acid}} + \underbrace{K^+ + BrF_4^-}_{\text{Base}} = \underbrace{K^+ + AsF_6^-}_{\text{Salt}} + \underbrace{2\,BrF_3}_{\text{Solvent}} \tag{7-8.6}$$

Even for protonic solvents this is a broader and more useful definition, because it explains why acid or base character is not an absolute property of the solute. Rather, the acid or base character of a substance can only be specified in relation to the solvent used. For example, in water, CH_3CO_2H (acetic acid) is an acid.

$$CH_3CO_2H + H_2O = H_3O^+ + CH_3CO_2^- \tag{7-8.7}$$

In the sulfuric acid solvent system, CH_3CO_2H is a base.

$$H_2SO_4 + CH_3CO_2H = CH_3CO_2H_2^+ + HSO_4^- \tag{7-8.8}$$

As another example, urea, $H_2NC(O)NH_2$, which is essentially neutral in water, is an acid in liquid ammonia:

$$NH_3 + H_2NC(O)NH_2 = NH_4^+ + H_2NC(O)NH^- \tag{7-8.9}$$

The Lux–Flood Definition

Consider Reaction 7-8.10:

$$CaO + H_2O \longrightarrow Ca(OH)_2 \tag{7-8.10}$$

in which CaO serves as a basic anhydride. Then consider Reaction 7-8.11:

$$CO_2 + H_2O \longrightarrow H_2CO_3 \tag{7-8.11}$$

in which CO_2 serves as an acidic anhydride. In these two cases, CaO and CO_2 are first allowed to react with water, and the hydration products are readily recognized as a base, $Ca(OH)_2$, and an acid, H_2CO_3. Furthermore, Reaction 7-8.12

$$Ca(OH)_2 + H_2CO_3 \longrightarrow CaCO_3 + H_2O \tag{7-8.12}$$

is readily recognized as a neutralization reaction in which a salt (plus solvent) is formed. The salt ($CaCO_3$) may be prepared directly, without intervention of solvent, as in Reaction 7-8.13.

$$CaO + CO_2 \longrightarrow CaCO_3 \tag{7-8.13}$$

It is natural to continue to regard Reaction 7-8.13 as an acid–base reaction. Some other examples of direct reactions between acidic and basic oxides are given in Reactions 7-8.14 and 7-8.15.

$$CaO + SiO_2 \longrightarrow CaSiO_3 \tag{7-8.14}$$

$$3\, Na_2O + P_2O_5 \longrightarrow 2\, Na_3PO_4 \tag{7-8.15}$$

The general principle involved in such processes was recognized by Lux and Flood, who proposed that an acid be defined as an oxide ion acceptor and a base as an oxide ion donor. Thus, in Reactions 7-8.13 to 7-8.15, the bases CaO and Na_2O donate oxide ions to the acids CO_2, SiO_2, and P_2O_5, to form the ions CO_3^{2-}, SiO_3^{2-}, and PO_4^{3-}.

The Lux–Flood concept of acids and bases is very useful in dealing with high temperature, anhydrous systems, such as those encountered in the oxide chemistries of ceramics and metallurgy. Furthermore, the Lux–Flood definition has a direct relation to the aqueous chemistry of acids and bases because the bases are oxides (basic anhydrides) that react with water as in Reaction 7-8.16

$$Na_2O + H_2O \longrightarrow 2\, Na^+ + 2\, OH^- \tag{7-8.16}$$

and the acids are oxides (acidic anhydrides) that react with water as in Reaction 7-8.17.

$$P_2O_5 + 3\, H_2O \longrightarrow 2\, H_3PO_4 \tag{7-8.17}$$

The Lewis Definition

One of the most general (and useful) of all definitions was proposed by G. N. Lewis. He defined an acid as an electron-pair acceptor and a base as an electron-pair donor. This definition includes the Brønsted–Lowry definition as a special case, since the proton can be regarded as an electron-pair acceptor and the base, be it OH^-, NH_2^-, HSO_4^-, and so on, as an electron-pair donor. Consider, for example, Reaction 7-8.18.

$$H^+ + :OH^- = H:OH \tag{7-8.18}$$

The Lewis definition covers a great many systems where protons are not involved at all, however. The reaction between NH_3 and BF_3 is an acid–base reaction.

$$\underset{\substack{\text{Lewis} \\ \text{base}}}{H_3N:} \quad + \quad \underset{\substack{\text{Lewis} \\ \text{acid}}}{BF_3} \quad \longrightarrow \quad H_3N:BF_3 \tag{7-8.19}$$

In the Lewis sense, all of the usual ligands can be regarded as bases and all metal ions can be regarded as acids. The degree of affinity of a metal ion for ligands can be termed its Lewis acidity, and the tendency of a ligand to become bound to a metal ion can be regarded as a measure of its Lewis basicity.

Base and acid strengths in the Lewis sense are not fixed, inherent properties of the species concerned, but vary somewhat with the nature of the partner. That is, the order of base strength of a series of Lewis bases may change when the type of acid with which they are allowed to combine changes. We discuss this in Section 7-9.

Observe that, for a given donor or acceptor atom, basicity or acidity can be influenced greatly by the nature of the substituents. Substituent influence can be either electronic or steric in origin.

Electronic Effects

The electronegativity of substituents exercises an obvious effect. Thus base strength and acid strength are affected oppositely, as the following examples show.

$$\text{Base strength} \quad (CH_3)_3N > H_3N > F_3N$$
$$\text{Acid strength} \quad (CH_3)_3B < H_3B < F_3B$$

The more electron withdrawing (electronegative) the substituent the more it enhances Lewis acidity and diminishes Lewis basicity.

However, more subtle electronic effects can also be important. On simple electronegativity grounds the following order of acid strengths would be predicted: $BF_3 > BCl_3 > BBr_3$. Experimentally, just the opposite is found. This finding can be understood when the existence of π interactions in the planar molecules is taken into account, and when it is noted that, after the Lewis acid has combined with a base, the BX_3 group becomes pyramidal and the boron atom no longer interacts with the π electrons of the X atoms. Simple calculations indicate that the B—X π interactions will decrease in strength in the order $F \gg Cl > Br$. Therefore, BF_3 is a weaker Lewis acid than BCl_3 because the planar BF_3 molecule is stabilized to a greater extent than BCl_3 by B—X π bonding. Borate esters, $B(OR)_3$, are also surprisingly weak Lewis acids for the same reason.

Steric Effects

There may be several kinds of steric effects. For the following three bases (Structures 7-IV to 7-VI) base strength toward the proton increases slightly from

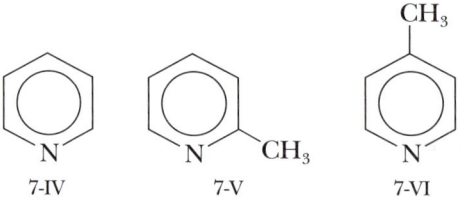

| 7-IV | 7-V | 7-VI |

Structure 7-IV to 7-V and is virtually the same for Structures 7-V and 7-VI, as is expected from the ordinary inductive effect of a methyl group. However, with respect to $B(CH_3)_3$, the order of basicity is

$$7\text{-IV} \approx 7\text{-VI} \gg 7\text{-V}$$

This results from the steric hindrance between the ortho methyl group of the base and the methyl groups of $B(CH_3)_3$. For the same reason quinuclidine, (Structure 7-VII) is a far stronger base toward $B(CH_3)_3$ than is triethylamine, (Structure 7-VIII):

| 7-VII | 7-VIII |

A different sort of steric effect results as the bulk on the boron atom in a BR_3 acid is increased. Since, as we stated previously, the BR_3 molecule goes from planar to pyramidal when it interacts with the base, the R groups must be squeezed into considerably less space. As the R groups increase in size, this effect strongly opposes the formation of the $A:BR_3$ compound, thus effectively decreasing the acidity of BR_3.

7-9   "Hard" and "Soft" Acid Base (HSAB) Concepts

It has been known for a long time that metal ions can be sorted into two groups according to their preference for various ligands. Let us consider the ligands formed by the elements of Groups VB(15), VIB(16), and VIIB(17). For Group VB(15) we might take a homologous series, such as R_3N, R_3P, R_3As, R_3Sb, and for Group VIIB(17) we take the anions themselves, F^-, Cl^-, Br^-, and I^-. For type (a) metals, complexes are more stable with the lightest ligands and less stable as each group is descended. For the type (b) elements the trend is just the opposite. This is summarized as follows:

Complexes of type (a) metal	Ligands			Complexes of type (b) metal
Strongest	R_3N	R_2O	F^-	Weakest
↑	R_3P	R_2S	Cl^-	
	R_3As	R_2Se	Br^-	↓
Weakest	R_3Sb	R_2Te	I^-	Strongest

Type (a) metal ions include principally:

1. Alkali metal ions.
2. Alkaline earth ions.
3. Lighter and more highly charged ions, for example,

$$Ti^{4+}, Fe^{3+}, Co^{3+}, Al^{3+}$$

Type (b) metal ions include principally:

1. Heavier transition metal ions, such as

$$Hg_2^{2+}, Hg^{2+}, Pt^{2+}, Pt^{4+}, Ag^+, Cu^+$$

2. Low-valent metal ions, such as the formally zero-valent metals in metal carbonyls.

This empirical ordering proved very useful in classifying, and to some extent predicting, the relative stabilities of complexes. Later, Pearson observed that it might be possible to generalize the correlation to include a broader range of acid–base interactions. He noted that the type (a) metal ions (acids) were small, compact, and not very polarizable, and that they preferred ligands (bases) that were also small and less polarizable. He called these acids and bases "hard." Conversely, the type (b) metal ions, and the ligands they prefer, tend to be larger and more polarizable; he described these acids and bases as "soft." The empirical relationship could then be expressed, qualitatively, by the statement that *hard acids prefer hard bases and soft acids prefer soft bases.* Although the point of departure for the "hard and soft" terminology was the concept of polarizability, other factors undoubtedly enter into the problem. There is no unanimity among chemists as to the detailed nature of "hardness" and "softness," but clearly Coulombic attraction will be of importance for hard–hard interactions while covalence will be quite significant for soft–soft interactions. The participation of both electrostatic and covalent forces in acid–base interactions will be considered in Section 7-10.

7-10   The Drago-Wayland Equation for Quantitatively Estimating the Strength of Lewis Acid-Base Interactions

In an attempt to account quantitatively for the enthalpy of formation of a Lewis acid–base adduct, Drago and his students proposed Eq. 7-10.1.

$$-\Delta H_{AB} = E_A \times E_B + C_A \times C_B + W \qquad (7\text{-}10.1)$$

The parameter ΔH_{AB} is the (normally exothermic) enthalpy of combining a Lewis acid A with a Lewis base B to give the adduct B $\rightarrow$ A. The form of Eq. 7-10.1 is based on the notion that for each acid–base interaction there will be both electrostatic and covalent components to the dative bond. Drago further *postulated* that the tendency of an individual acid or base to contribute either to electrostatic or covalent interaction with *any* partner is a fixed characteristic that can be defined quantitatively. Thus, each acid or base is said to have a characteristic value E_A or E_B, respectively, which is indicative of the normal contribution of that acid or base to the **electrostatic** component of any dative bond that it forms. Similarly, the contribution of an acid or a base to the **covalent** component of any dative bond is measured by the parameter C_A or C_B, respectively. Thus the electrostatic contribution in Eq. 7-10.1 to the total enthalpy of adduct formation is the quantity $E_A \times E_B$, and the covalent contribution to the enthalpy of adduct formation is the quantity $C_A \times C_B$. The factor W in Eq. 7-10.1 is usually zero; it is used only when there is suspected to be a constant contribution to the enthalpies of reaction for a particular acid (or base), such contribution being independent of the base (or acid) reacting.

The Drago–Wayland equation states that the enthalpy of adduct formation will be a large negative number (and the dative bond of the acid–base adduct will be strong) in cases where either the electrostatic term $E_A \times E_B$ or the covalent term $C_A \times C_B$ is large. This is tantamount to requiring that the acid A and the base B be properly matched in bonding characteristics. Thus, A and B are most likely to form a strong bond when both contribute to electrostatic interaction (i.e., have large E values) or both contribute to covalent bonding (i.e., have large C values). A mismatch, where one partner prefers electrostatic and the other covalent interactions, is disfavored since both $E_A \times E_B$ and $C_A \times C_B$ will be small.

Drago's values for the E and C parameters for a variety of acids and bases are given in Table 7-2. In establishing these parameters, it was necessary to assign arbitrary values for the four parameters identified in Table 7-2. After that, data-fitting procedures were used to arrive at the other values.

Table 7-2   Drago's Parameters for Estimating the Strength of Acid–Base Interactions[a]

Acids	E_A	C_A
$(C_2H_5)_3Al$	12.5	2.04
$(CH_3)_3B$	5.79	1.57
$BF_3(g)$	12.19	0.81
$(CH_3)_3Al$	17.32	0.94
I_2	1.00[b]	1.00[b]
$(CH_3)_3Ga$	13.83	0.40
SO_2	1.11	0.74
$(CH_3)_3In$	13.19	0.37

Bases	E_B	C_B
$(CH_3)_3N$	1.19	11.20
$(C_2H_5)_3N$	1.29	10.83
$NH(CH_3)_2$	1.33	8.47
$(CH_3)_2S$	0.57	6.49
$(C_2H_5)_2S$	0.55	7.40[b]
$(CH_3)_3P$	1.11	6.51
NC_5H_5 (py)	1.30	6.69
NH_2CH_3	1.50	5.63
NH_3	1.48	3.32
$(CH_3)_2SO$	1.36	2.78
$(CH_3)_2NOCH_3$	1.32[b]	2.48
CH_3CN	0.90	1.34
$(C_2H_5)_2O$	1.08	3.08

[a]When used in Eq. 7-10.1, these parameters provide an estimate of the enthalpy of adduct formation (in kcal mol^{-1}) for the Lewis acid–base pair, B → A. The data were taken from R. S. Drago, N. Wong, C. Bilgrien, and G. C. Vogel, *Inorg. Chem.*, **1987**, *26*, 9–14.

[b]One of four parameters whose numerical values are assigned arbitrarily. Iterative data-fitting procedures are then used to determine a consistent set of values for other substances.

As an example of the use of the Drago–Wayland equation, consider the two adducts formed between trimethylaluminum and either trimethylamine or trimethylphosphine, as in Reactions 7-10.2 and 7-10.3.

$$(CH_3)_3Al + :N(CH_3)_3 \longrightarrow (CH_3)_3Al:N(CH_3)_3 \qquad (7\text{-}10.2)$$

$$(CH_3)_3Al + :P(CH_3)_3 \longrightarrow (CH_3)_3Al:P(CH_3)_3 \qquad (7\text{-}10.3)$$

For the trimethylamine adduct, the Drago–Wayland equation gives us the following prediction:

$$-\Delta H = 17.32 \times 1.19 + 0.94 \times 11.20 = 31.14 \text{ kcal mol}^{-1}$$

Thus the enthalpy of Reaction 7-10.2 is -31.14 kcal mol^{-1}. Correspondingly, for the trimethylphosphine adduct we get:

$$-\Delta H = 17.32 \times 1.11 + 0.94 \times 6.51 = 25.35 \text{ kcal mol}^{-1}$$

and the enthalpy of Reaction 7-10.3 is -25.35 kcal mol^{-1}. Hence, trimethylaluminum is found to form a more stable adduct with trimethylamine than with trimethylphosphine.

The Drago–Wayland equation has some advantages over the simple HSAB approach, because it uses more parameters in order to arrive at a more quantitative understanding of the acid–base interaction. In cases where a detailed comparison and understanding of relative acid–base strengths is required, Drago's approach should be used. Also, the Drago–Wayland equation clearly provides a quantitative assessment of the relative importance of electrostatic versus covalent bonding in acid–base adducts.

7-11   Some Common Protic Acids

Sulfuric Acid (H$_2$SO$_4$)

This acid is of enormous industrial importance and is manufactured in larger quantities than any other. The preparation first requires the burning of sulfur to SO_2. Oxidation of SO_2 to SO_3 must then be catalyzed either homogeneously by oxides of nitrogen (lead chamber process) or heterogeneously by platinum (contact process). Sulfuric acid is ordinarily sold as a 98% mixture with water (18 M). The pure substance is obtained as a colorless liquid by addition of sufficient SO_3 to react with the remaining H_2O. The solid and liquid are built of SO_4 tetrahedra linked by hydrogen bonds.

Addition of further SO_3 to 100% H_2SO_4 gives *fuming sulfuric acid* or *oleum,* which contains polysulfuric acids, such as pyrosulfuric acid ($H_2S_2O_7$), and, with more SO_3, $H_2S_3O_{10}$, and $H_2S_4O_{13}$.

Sulfuric acid is not a very strong oxidizing agent, but it is a powerful dehydrating agent for carbohydrates and other organic substances, often degrading the former to elemental carbon.

$$C_nH_{2n}O_n \xrightarrow{\;H_2SO_4\;} n\,C + H_2SO_4 \cdot n\,H_2O \qquad (7\text{-}11.1)$$

The equilibria in pure H_2SO_4 are complex. Besides self-ionization

$$2\ H_2SO_4 = H_3SO_4^+ + HSO_4^- \qquad K_{10\ ^\circ C} = 1.7 \times 10^{-4}\ mol^2\ kg^2 \quad (7\text{-}11.2)$$

there are hydration–dehydration equilibria, such as

$$2\ H_2SO_4 = H_3O^+ + HS_2O_7^- \qquad\qquad\qquad\qquad (7\text{-}11.3)$$

$$2\ H_2SO_4 = H_2O + H_2S_2O_7 \qquad\qquad\qquad\qquad (7\text{-}11.4)$$

$$H_2SO_4 + H_2S_2O_7 = H_3SO_4^+ + HS_2O_7^- \qquad \text{and so on} \qquad (7\text{-}11.5)$$

Nitric Acid (HNO_3)

The normally available, concentrated acid is about 70% by weight HNO_3 in water. It is colorless when pure but is often yellow as a result of photochemical decomposition, which gives NO_2.

$$2\ HNO_3 \xrightarrow{\ hv\ } 2\ NO_2 + H_2O + \tfrac{1}{2}O_2 \qquad\qquad (7\text{-}11.6)$$

Red, "fuming" nitric acid is essentially 100% HNO_3, which contains additional NO_2.

The pure acid is a colorless liquid or solid that must be stored below 0 °C to avoid thermal decomposition according to the same reaction as Eq. 7-11.6 for photochemical decomposition. In the pure liquid the following equilibria occur:

$$2\ HNO_3 = H_2NO_3^+ + NO_3^- \qquad\qquad\qquad (7\text{-}11.7)$$

$$H_2NO_3^+ = NO_2^+ + H_2O \qquad\qquad\qquad (7\text{-}11.8)$$

While aqueous HNO_3 below 2 M concentration is not strongly oxidizing, the concentrated acid is a very powerful oxidizing agent. It will attack nearly all metals except for Au, Pt, Rh, and Ir and a few others that quickly become passivated (covered with a resistant oxide film), such as Al, Fe, and Cu.

Aqua Regia

A mixture of about three volumes of HCl to one volume of HNO_3, prepared from the concentrated aqueous acids, is known as aqua regia. It contains free Cl_2 and ClNO, and is, therefore, a powerful oxidizing agent. It readily dissolves even Au and Pt, owing to the ability of Cl^- to stabilize the Au^{3+} and Pt^{4+} cations by forming the complexes $AuCl_4^-$ and $PtCl_6^{2-}$. Aqua regia may also be used to dissolve certain difficultly soluble salts because of its combined oxidizing and coordinating abilities. For example, HgS dissolves in aqua regia both because the sulfide is oxidized to sulfur, and because the mercury(II) ion is stabilized by formation of the complex ion, $HgCl_4^{2-}$.

Perchloric Acid ($HClO_4$)

This acid is normally available in concentrations 70–72% by weight. The pure substance, which can be obtained by vacuum distillation in the presence of the dehydrating agent $Mg(ClO_4)_2$, is stable at 25 °C for only a few days, decomposing to give off Cl_2O_7. Both the pure and the concentrated aqueous acid react explosively with organic matter. The ClO_4^- ion is a very weak ligand, and perchloric acid and alkali perchlorates are, therefore, of use in preparing solutions in which complexing of cations is to be minimized.

Solid perchlorate complexes of transition metals can be dangerously explosive. For synthetic purposes, ClO_4^- is best replaced by $CF_3SO_3^-$ (trifluoromethane sulfonate), which is commonly known as triflate. The acid (bp 162 °C) is very hygroscopic and is a superacid (Section 7-13), with H_0 −13.8.

The Hydrohalic Acids (HCl, HBr, and HI)

These three acids are similar but differ markedly from hydrofluoric acid, which we describe later in this section. The pure compounds are pungent gases at 25 °C but are highly soluble in water to give strongly acidic solutions. One molar solutions are virtually 100% dissociated. For aqueous HBr, and especially HI, their reactivity as simple acids is complicated by the reducing character of the Br^- and I^- ions.

Only HCl (bp −85 °C) has been much studied as a pure liquid. Its autoionization according to Eq. 7-11.9 is small,

$$3 \text{ HCl} = H_2Cl^+ + HCl_2^- \tag{7-11.9}$$

but many organic and some inorganic compounds dissolve to give conducting solutions. A number of compounds containing the $[Cl{-}H{-}Cl]^-$ and $[Br{-}H{-}Br]^-$ ions have been isolated.

Hydrofluoric Acid (HF)

In dilute aqueous solution HF is a weak acid,

$$HF + H_2O = H_3O^+ + F^- \qquad K_{25 \, °C} = 7.2 \times 10^{-5} \tag{7-11.10}$$

This is mainly due to the formation of strong hydrogen-bonded ion pairs, such as $F^-{-}{-}{-}^+[H{-}OH_2]$. The aqueous acid readily attacks glass and silica because the stable SiF_6^{2-} ion can be formed:

$$6 \text{ HF(aq)} + SiO_2 = 2 \, H_3O^+ + SiF_6^{2-} \tag{7-11.11}$$

and it is used commercially to etch glass.

In contrast to the aqueous solution, liquid HF (bp 19.5 °C) is one of the strongest acids known. The principal self-ionization equilibria are

$$2 \text{ HF} = H_2F^+ + F^- \tag{7-11.12}$$

$$F^- + n \text{ HF} = HF_2^- + H_2F_3^- + H_3F_4^- \qquad \text{and so on} \tag{7-11.13}$$

There are a few substances that act as solvent–system acids towards liquid HF, namely, as fluoride ion acceptors. Through F^- transfer, they serve to increase the concentration of the solvent cation H_2F^+. An example is SbF_5, which operates as in Eq. 7-11.14.

$$2\,HF + SbF_5 = H_2F^+ + SbF_6^- \tag{7-11.14}$$

Liquid HF has a dielectric constant (84 at 0 °C) comparable to that of water, and it is an excellent solvent for a wide range of inorganic and organic compounds.

7-12 Some Rules Concerning the Strengths of Oxy Acids

Acids consisting of a central atom surrounded by O atoms and OH groups, $XO_n(OH)_m$ are very common, including H_2SO_4, H_3PO_4, and HNO_3. For these acids there are two important generalizations:

1. The ratio of successive dissociation constants (K_n/K_{n-1}) is 10^{-4}–10^{-5}, (which is equivalent to $pK_{n-1} - pK_n = 4.5 \pm 0.5$, where $pK = -\log K$).
2. The magnitude of K_1 depends on n, the number of additional oxygen atoms besides those in OH groups. The more of these, the greater the acid strength, according to:

n	K_1	Acid Strength
3	Very, very large	Very strong
2	~10^2	Strong
1	10^{-2}–10^{-3}	Medium
0	$10^{-7.5}$–$10^{-9.5}$	Weak

The basis for these rules, and their general validity, lies in the delocalization of the charge of the anions. For a given initial dissociation,

$$XO_n(OH)_m = [XO_{n+1}(OH)_{m-1}]^- + H^+ \tag{7-12.1}$$

the greater the number $(n+1)$ of oxygen atoms in the conjugate base $[XO_{n+1}(OH)_{m-1}]^-$, the more the negative charge of the anion can be spread out, and thus the more stable is the anion. For instance, the negative charge in nitrate is dispersed uniformly, as in Structure 7-IX, in the same way that the π bond in an AB_3 system is delocalized via resonance. For cases where there are many

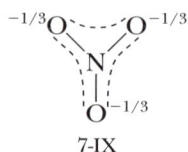

7-IX

oxygen atoms and only a single proton (e.g., $HClO_4$) delocalization of the negative charge in the conjugate base is very effective, and the dissociation of a proton from the parent acid is very favorable. In contrast, when $n = 0$, there is practically no delocalization of negative charge in the anion, and dissociation of a proton from the parent "acid" is not favorable. An example is $Te(OH)_6$, which is not appreciably acidic, because there is little charge delocalization in $Te(OH)_5O^-$.

The steady decrease in the values of K_1, K_2, K_3, and so on, occurs because after each dissociation, there is an increased negative charge that lessens the tendency of the next proton to depart.

Apparent exceptions to rule (2) turn out not to have simple $XO_n(OH)_m$ type structures. For example, phosphorous acid (H_3PO_3) would have $K_1 \approx 10^{-8}$ if it were $P(OH)_3$. In fact, the value of K_1 is about 10^{-2}, which should mean that it has $n = 1$. It actually does belong in that group since its structure is $HPO(OH)_2$, with one hydrogen atom directly attached to P. Similarly, hypophosphorous acid (H_3PO_2) has $K_1 \approx 10^{-2}$ and its actual structure is $H_2PO(OH)$.

Carbonic acid also deviates from expectation, but for a different reason. For $CO(OH)_2$ we expect $K_1 \approx 10^{-2}$, whereas the measured value is approximately 10^{-6}. This occurs because much of the solute in a solution of *carbonic acid* is present as loosely hydrated CO_2 and not as $CO(OH)_2$. When a correction is made for this, the true dissociation constant of $CO(OH)_2$ is found to be about 2×10^{-4}, which is close to the expected range.

7-13   Superacids

There are a number of liquids that are considerably more acidic, by as much as 10^6–10^{10} times, than concentrated aqueous solutions of so-called very strong acids, such as nitric and sulfuric acids. These are called superacids, and in recent years a great deal of new chemistry has been found to occur in these media. Superacid systems are necessarily nonaqueous, since the acidity of any aqueous system is limited by the fact that the strongest acid that can exist in the presence of water is H_3O^+. Any stronger acid simply transfers its protons to H_2O to form H_3O^+.

To measure superacidity it is necessary to define a scale that goes beyond the normal pH scale and is defined in terms of an experimental measurement. The usual one is the Hammett acidity function H_0, defined as follows:

$$H_0 = pK_{BH^+} - \log \frac{[BH^+]}{[B]}$$

where B is an indicator base, BH^+ is its protonated form, and pK_{BH^+}, is $-\log K$ for dissociation of BH^+. The ratio $[BH^+]/[B]$ can be measured spectrophotometrically. By employing bases with very low basicities (very negative pK values), the H_0 scale may be extended to the very negative values appropriate to the superacids. The H_0 scale becomes identical to the pH scale in dilute aqueous solution. Crudely, H_0 values can be thought of as pH values extending below $pH = 0$.

The first superacid systems to be studied quantitatively were very concentrated solutions of H_2SO_4. Pure H_2SO_4 has $H_0 = -12$; it is thus about 10^{12} times more acidic that 1 M aqueous H_2SO_4. When SO_3 is added, to produce oleum, H_0 can reach about -15.

Hydrofluoric acid has H_0 of about -11, and the acidity is increased to about -12 on the addition of fluoride ion acceptors such as SbF_5.

Superacid media that have found wide application are obtained on addition of AsF_5 or SbF_5 to fluorosulfonic acid (HSO_3F). Pure fluorosulfonic acid has $H_0 = -15$ and is useful because of its wide liquid range, from -89 to $+164\ °C$, its ease of purification, and the fact that it does not attack glass, provided it is free of HF. The self-ionization of HSO_3F is

$$2 \, HSO_3F = H_2SO_3F^+ + SO_3F^- \tag{7-13.2}$$

and any additive that increases the concentration of $H_2SO_3F^+$ increases the acidity. The addition of about 10 mol % of SbF_5 to HSO_3F increases $-H_0$ to about 19. A 1:1 molar mixture of HSO_3F and SbF_5 is colloquially known as "magic acid," although the additional SbF_5 beyond about 10% does little to increase the acidity.

The ability of SbF_5 to increase the acidity of HSO_3F is due mainly to the equilibrium

$$2 \, HSO_3F + SbF_5 = H_2SO_3F^+ + SbF_5(SO_3F)^- \tag{7-13.3}$$

through which the concentration of the solvent cation $H_2SO_3F^+$ is increased.

Superacid media have been used in many ways. The most obvious is to protonate molecules not normally thought of as bases, for instance, aromatic hydrocarbons. Thus, fluorobenzene in HF/SbF_5 or HSO_3F/SbF_5 produces the ion (Structure 7-X).

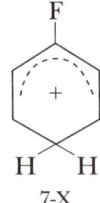

7-X

Many other cationic species that would be immediately destroyed by even the weakest of bases can be prepared in and isolated from superacid media. These include carbonium ions as in Reaction 7-13.4:

$$(CH_3)_3COH \xrightarrow{\text{superacid}} (CH_3)_3C^+ + H_3O^+ \tag{7-13.4}$$

and halogen cations as in Reaction 7-13.5:

$$I_2 \xrightarrow{\text{superacid}} I_2^+ \text{ and or } I_3^+ \tag{7-13.5}$$

It is also possible to prepare some remarkable polynuclear cations of sulfur, selenium, and tellurium, such as S_4^{2+}, S_8^{2+}, Se_4^{2+}, and Te_4^{2+}.

STUDY GUIDE

Scope and Purpose

We have presented an overview of the properties of solvents and of acids and bases that are important to the material in subsequent chapters. It is also intended that the various definitions of acids and bases that are set down in this chapter will find appropriate use in describing reactions. The principal goals of the student are to learn to recognize the various types and descriptions of acids and bases, and to develop a ready appreciation for the definition or description that most suits a particular reaction. The relative strengths of acids and bases and the relative extents of autoionization, and so on, are also important concepts and trends that have been discussed.

Study Questions

A. Review

1. Name some properties that determine the utility of a solvent.
2. What is the principal effect of the dielectric constant?
3. What is the relationship between donor and/or acceptor ability of a solvent and its ability to function as a solvent?
4. Name four protic solvents besides water.
5. Discuss the autodissociation of water and the forms of the hydrated proton.
6. In liquid NH_3 what are the species characteristic of acids? And bases?
7. Describe the three classes of aprotic solvents, mentioning examples of each.
8. Name an important industrial process that employs a molten salt as a solvent.
9. What two properties are generally important in a solvent for electrochemical reactions?
10. Name two common impurities in solvents and indicate how they can be removed.
11. State the Brønsted–Lowry definition of acids and bases.
12. Discuss the solvent system definition and show how it includes the Brønsted–Lowry definition as a special case.
13. Why is acetic acid not an acid in H_2SO_4?
14. To what sort of systems does the Lux–Flood concept apply? Give a representative equation.
15. State the Lewis definition of acids and bases and write three equations that illustrate it, including one that involves a protonic acid.
16. Why is F_3N a much weaker base than H_3N?
17. Why is BBr_3 a stronger acid than BF_3?
18. Describe the origin of the concept of hard and soft acids and bases.
19. Write the type of equation used to account for the combined effect of both electrostatic and covalent forces in acid–base interactions.
20. What are the main properties of each of the following common acids? H_2SO_4, HNO_3, $HClO_4$, HF.
21. Rank the following acids in order of their strengths: $HClO_2$, $HClO_3$, $HClO_4$, H_2SeO_3, H_3AsO_4, $HMnO_4$, H_2SeO_4. Explain your reasoning.
22. What is the definition of the Hammett acidity function (H_0)?
23. Why does the addition of SbF_5 to HSO_3F cause H_0 to become more negative?
24. What are the four parameters that are used in the Drago–Wayland equation to estimate the enthalpy of adduct formation, and what does each of the four quantities represent?

B. Additional Exercises

1. Consider acetic acid as a solvent. Its dielectric constant is about 10. What is its mode of self-ionization likely to be? Name some substances that will be acids and some that will be bases in acetic acid. Will it be a better or poorer solvent than H_2O for ionic compounds?

2. State whether each of the following would act as an acid or a base in liquid HF.

$$BF_3, SbF_5, H_2O, CH_3CO_2H, C_6H_6$$

In each case write an equation, or equations, to show the basis for your answer.

3. Dimethyl sulfoxide is a very good solvent for polar and ionic materials. Why?

4. Why are only superacids good solvents for species such as I_2^+, Se_4^{2+}, S_8^{2+}, and so on? How would they react with less acidic solvents, such as H_2O or HNO_3?

5. Why do you think phosphines (R_3P) and phosphine oxides (R_3PO) differ considerably in their base properties?

6. Which member of each pair would you expect to be the more stable? (1) $PtCl_4^{2-}$ or PtF_4^{2-}. (2) $Fe(H_2O)_6^{3+}$ or $Fe(PH_3)_6^{3+}$. (3) $F_3B:thf$ or $Cl_3B:thf$. (4) $(CH_3)_3B:PCl_3$ or $(CH_3)_3B:P(CH_3)_3$. (5) $(CH_3)_3Al:pyridine$ or $(CH_3)_3Ga:pyridine$. (6) $Cl_3B:NCCH_3$ or $(CH_3)_3B:NCCH_3$.

7. In terms of the HSAB concept, which end of the SCN^- ion would you expect to coordinate to Cr^{3+}? To Pt^{2+}?

8. Estimate pK_1 values for H_2CrO_4, $HBrO_4$, $HClO$, H_5IO_6, and HSO_3F.

9. Write equations for the probable main self-ionization equilibria in liquid HCN.

10. Aluminum trifluoride (AlF_3) is insoluble in HF, but dissolves when NaF is present. When BF_3 is passed into the solution, AlF_3 is precipitated. Account for these observations using equations.

11. What change in hybridization is necessary when the following serve as Lewis acids: BF_3, $AlCl_3$, and $SnCl_2$?

12. Balance the equation for the oxidation of Au by aqua regia.

13. Draw the Lewis diagram and predict the structure for SiF_6^{2-}.

14. Write equations representing the autoionization of the following solvents, and classify the process as hydrogen ion or halide ion transfer: (a) HCl, (b) HNO_3, (c) $OPCl_3$, (d) HF. Identify which species in these systems are the solvent's conjugate acid and the solvent's conjugate base.

15. Draw the Lewis diagrams for all species involved in Reaction 7-13.2. Discuss this equilibrium (a) in terms of the solvent system definition of acids and bases and (b) in terms of the Lewis definition of acids and bases.

16. Boric acid, $B(OH)_3$, acts as an acid in water, but does not do so via ionization of a proton. Rather, it serves as a Lewis acid towards OH^-. Explain with the use of a balanced equation.

17. The parameter K_1 is about 10^{-2} for the three acids H_3PO_4 (orthophosphoric acid), H_3PO_3 (phosphorous acid), and H_3PO_2 (hypophosphorous acid). Use this information to draw the Lewis diagram for each.

18. Use the HSAB theory to predict which of these two adducts should be the more stable adduct, and then explain your choice, both in terms of σ- and π-bonding effects.

$$H_3N:BBr_3 \quad or \quad F_3N:BF_3$$

19. Give a good definition for each of the following three terms, and illustrate each with an example. (a) solvent–system base, (b) autoionization, and (c) amphoterism.

20. Discuss the following two acid–base reactions from the solvent–system point of view.

$$2\,PCl_5 + 2\,TiCl_4 = 2\,PCl_4^+ + Ti_2Cl_{10}^{2-}$$
$$PCl_5 + NbCl_5 = NbCl_6^- + PCl_4^+$$

21. Using the most appropriate acid–base theory, identify the acids and bases in the following reactions.

(a) $SiO_2 + Na_2O = Na_2SiO_3$

(b) $B(OR)_3 + NaH = Na[HB(OR)_3]$

(c) $N_2O_5 + H_2O = 2\ HNO_3$

(d) $Cl_3PO + Cl^- = Cl_4PO^-$

(e) $Li_3N + 2\ NH_3 = 3\ Li^+ + 3\ NH_2^-$

(f) $2\ HF + PF_5 = H_2F^+ + PF_6^-$

(g) $\frac{1}{2} Al_2Cl_6 + PF_3 = Cl_3Al{:}PF_3$

(h) $BF_3 + 2\ ClF = Cl_2F^+ + BF_4^-$

(i) $NOF + ClF_3 = NO + ClF_4^-$

(j) $XeOF_4 + XeO_3 = 2\ XeO_2F_2$

(k) $XeO_3 + OH^- = HXeO_4^-$

(l) $SiO_2 + 2\ XeF_6 = 2\ XeOF_4 + SiF_4$

(m) $PCl_5 + ICl = PCl_4^+ + ICl_2^+$

(n) $10\ S + 4\ NH_3 = S_6^{2-} + S_4N^- + 3\ NH_4^+$

22. Use the Drago–Wayland equation to compare the enthalpy of adduct formation for (a) I_2 plus $(C_2H_5)_2O$, with (b) I_2 plus $(C_2H_5)_2S$.

23. Use the Drago–Wayland equation to rank, according to decreasing stability, the various adducts that can be formed from among the following acids and bases.

Acids	$B(CH_3)_3$, $Al(CH_3)_3$, and $Ga(CH_3)_3$
Bases	$(CH_3)_3N$, $(C_2H_5)_3N$, $(CH_3)_2S$, $(CH_3)_3P$, and $(C_2H_5)_2O$

24. Use the values of pK_1 provided below to deduce the structures of the following oxo acids.

Acid	pK_1
H_3PO_4	2
HNO_3	−3
$HClO_4$	−8
H_5IO_6	2
H_3PO_3	2
H_3PO_2	2

SUPPLEMENTARY READING

Barton, A. F. M., *Handbook of Solubility Parameters and Other Cohesive Parameters*, CRC Press, Cleveland, OH, 1983.

Bell, R. P., *The Proton in Chemistry*, 2nd ed., Chapman & Hall, London, 1973.

Burger, K., *Ionic Solvation and Complex Formation Reactions in Non-Aqueous Solvents*, Elsevier, New York, 1983.

Drago, R. S., "A Modern Approach to Acid–Base Chemistry," *J. Chem. Educ.*, **1974**, *51*, 300.

Gillespie, R. J., "Fluorosulfonic Acid and Related Superacid Media," *Acc. Chem. Res.*, **1968**, *1*, 202–209.

Gillespie, R. J., "The Chemistry of Superacid Systems," *Endeavour*, **1973**, *32*, 541.

Gutmann, V., *The Donor–Acceptor Approach to Molecular Interactions*, Plenum, New York, 1978.

Ho, Tse-Lok, *Hard and Soft Acid and Base Principles in Organic Chemistry*, Academic, New York, 1977.

Jensen, W. B., *The Lewis Acid–Base Concepts. An Overview,* Wiley, New York, 1980.

Lagowski, J. J., Ed., *The Chemistry of Non-Aqueous Solvents,* Vols. 1 and 2, Academic, New York, 1966 and 1967.

Luder, W. F. and Zuffanti, S., *The Electronic Theory of Acids and Bases,* 2nd ed., Dover, New York, 1961.

Olah, G. A., Surya Prakash, G. K., and Sommer, J., *Superacids,* Wiley-Interscience, New York, 1985.

Pearson, R. G., Ed., *Hard and Soft Acids and Bases,* Dowden, Hutchinson, and Ross, Stroudsburg, PA, 1973.

Seaborg, G. T., "The Research Style of G. N. Lewis," *J. Chem. Educ.,* **1984,** *61,* 93.

Chapter 8

THE PERIODIC TABLE AND THE CHEMISTRY OF THE ELEMENTS

8-1 Introduction

Inorganic chemistry has often been said to comprise a vast collection of unrelatable facts in contrast to organic chemistry, where there appears to be a much greater measure of systematization and order. This statement is in part true, since the subject matter of inorganic chemistry is far more diverse and complicated and the rules for chemical behavior are often less well established. The subject matter is complicated because even among elements of similar electronic structure, such as Li, Na, K, Rb, and Cs, Group IA(1), differences arise because of differences in the size of atoms, ionization potentials, hydration, solvation energies, or the like. Some of these differences may be quite subtle (e.g., those that enable the human cell and other living systems to discriminate among Li, Na, and K). In short, every element behaves in a different way.

Organic chemistry deals with many compounds that are formed by a *few* elements, namely, carbon in sp, sp^2, or sp^3 hybridization states, along with H, O, N, S and the halogens, and less commonly B, Si, Se, P, Hg, and so on. The chemistry is mainly one of molecular compounds that are liquids or solids commonly soluble in nonpolar solvents, distillable, or crystallizable and normally stable to, though combustible in, air or oxygen.

Inorganic chemistry, by contrast, deals with many compounds formed by *many* elements. It involves the study of the chemistry of more than 100 elements that can form compounds as gases, liquids, or solids, whose reactions may be (or may have to be) studied at very low or very high temperatures. The compounds may form ionic, extended-covalent, or molecular crystals and their solubility may range from essentially zero in all solvents to high solubility in alkanes; they may react spontaneously and vigorously with water or air. Furthermore, while organic compounds almost invariably follow the octet rule with a maximum coordination number and a maximum valence of 4 for all elements, inorganic compounds may have coordination numbers up to 14 with those of 4, 5, 6, and 8 being especially common, and valence numbers from −2 to +8. Finally, there are types of bonding in inorganic compounds that have no parallel in organic chemistry, where σ and $p\pi$–$p\pi$ multiple bonding normally prevail.

Although various concepts help to bring order and systematics into inorganic chemistry, the oldest and still the most meaningful concept of order is the periodic table of the elements. As we pointed out in Chapter 2, the order in the periodic table depends on the electronic structures of the gaseous atoms. By successively adding electrons to the available energy levels, we can build up the pattern of the electronic structures of the elements from the lightest to the heaviest one currently known, element 109. Moreover, on the basis of the electron configurations, the elements can be arranged in the conventional long form of the periodic table.

However, the periodic table can also be constructed solely on the basis of the chemical properties of the elements, and one of its chief uses is to provide a compact mnemonic device for correlating chemical facts. In this chapter, the periodic table is discussed from the chemical, instead of the theoretical, aspect. In effect, the kinds of chemical observations that originally stimulated chemists such as Mendeleev to devise the periodic table are examined here. Now, in addition, we are able to correlate such facts, and to interpret them, in terms of the electronic structures of the atoms.

Heavier Elements

Elements with atomic numbers 104–109 have now been discovered, and those through 106, namely Dubnium (Db, 104), Joliotium (Jl, 105) and Rutherfordium (Rf, 106) have been independently confirmed. In general, the elements beyond number 100, Fermium, have been made only a few atoms at a time. The detection of element 109 (mass number 266) is based on the observation of three decay events (or atoms) after 10 days of bombarding a bismuth target ($^{209}Bi_{83}$) with $^{58}Fe_{26}$. The half-life of isotope-266 of element 109 is about 3.4 ms. Two isotopes of element 108 are claimed on the basis of three decay events from long-term bombardment of a lead ($^{208}Pb_{82}$ and $^{207}Pb_{82}$) target with $^{58}Fe_{26}$. The half-life of isotope-265 of element 108 is about 1.8 ms, whereas the half-life of isotope-264 of element 108 is only 76 μs. Only three atoms of isotope-265 of element 108 have been synthesized, and only one atom of isotope-264 of element 108 has been detected. Element-107 was synthesized by fusion of bismuth-209 and chromium-54, giving in one trial 14, and in another trial 15, atoms of isotope-262, and 9 atoms of isotope-261. Attempts to synthesize element 110 using either lead-208 and nickel-64 or bismuth-209 and cobalt-59 have been unsuccessful.

PART A

THE NATURE AND TYPES OF THE ELEMENTS

8-2   Monatomic Elements: He, Ne, Ar, Kr, Xe, and Rn

The noble gases, with their closed-shell electronic structures, are necessarily monatomic. In the vapor state, mercury ($5d^{10}6s^2$) is also monatomic. However, liquid mercury, despite its relatively high vapor pressure and solubility in water and other solvents, has appreciable electrical conductivity and is bright and metallic in appearance. This occurs because the $6p$ orbitals are available to participate in metallic bonding.

8-3 Diatomic Molecules: H_2, N_2, O_2, F_2, Cl_2, Br_2, and I_2

For the halogens, the formation of a single electron-pair bond in a diatomic molecule completes the octet. For nitrogen and oxygen, multiple bonding allows for a simple diatomic molecule. The diatomic molecules P_2 and S_2 are stable at elevated temperatures, but not at 25 °C. In dihydrogen, formation of a single bond completes the $1s$ shell for each atom.

8-4 Discrete Polyatomic Molecules: P_4, S_n, Se_8, and Fullerenes (C_n)

For the second-row and heavier elements, $p\pi$–$p\pi$ bonding of the type found in N_2 and O_2 is less effective. The formation by phosphorus and sulfur of the normal number of single electron-pair bonds as expected from their electronic structures (namely, three and two, respectively) leads either to discrete molecules or to chain structures, which are more stable than the diatomics.

White phosphorus has tetrahedral P_4 molecules (Structure 8-I) with the P—P distance 2.21 Å, and the P—P—P angles are, of course, 60°. The small angle implies that the molecule might be strained. Strain in this sense would mean that the total energy of the six P—P bonds in the P_4 molecule is less than the total energy of six P—P bonds formed by P atoms having normal bond angles (90°–100°). Current theoretical work indicates that such strain energy is not very large, although it is sufficient to make white phosphorus less stable than black phosphorus, where all P—P—P bond angles are normal. White phosphorus is also much more reactive than the black allotrope. The molecules As_4 and Sb_4 are also formed on condensation from vapor but for them the tetrahedral structure is still less stable, readily transforming to the black phosphorus type of structure.

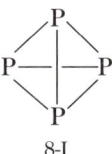

8-I

Sulfur has a profusion of allotropes; these contain multiatom sulfur rings. The largest ring thus far known is S_{20}. The allotropes are referred to as cyclohexasulfur, cyclooctasulfur, and the like. Chains occur in *catenasulfur* (S). The thermodynamically most stable form is orthorhombic sulfur (Fig. 8-1), which contains S_8 rings.

8-5 Elements with Extended Structures

In some elements, atoms form 2, 3, or 4 single covalent bonds to each other to give chains, planes, or three-dimensional networks (extended structures). The most important elements that do this are

C^a	P^a	S^a
Si	As	Se^a
Ge	Sb	Te
Sn^b	Bi	

(*a*) Also molecular   (*b*) Also metallic

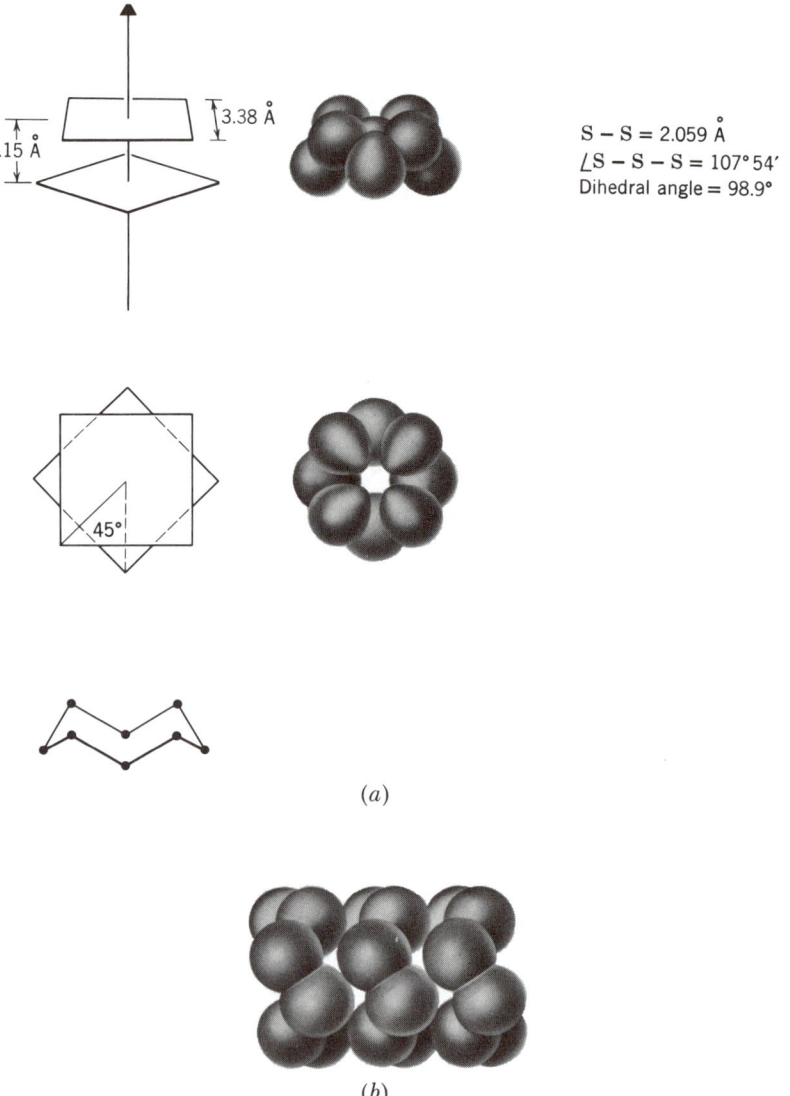

$S - S = 2.059 \text{ Å}$
$\angle S - S - S = 107°54'$
Dihedral angle $= 98.9°$

(a)

(b)

Figure 8-1   The structure of orthorhombic sulfur. (*a*) The cyclic S_8 molecule. (*b*) Stacking of S_8 molecules in the solid.

Some of these have allotropes of either molecular or metallic types. Those with metallic allotropes are discussed here, as are the ones that form extended, three-dimensional, covalent networks. First, we discuss boron, which forms limited networks based on variously linked 12-atom units.

Elemental boron has several allotropes, all based on B_{12} icosahedra (Structure 8-II). In the α-rhombohedral allotrope, the B_{12} units are packed as "spheres" in roughly cubic closest packing. The icosahedral units are linked weakly together. The β-rhombohedral form of boron also has icosahedral units linked in a complicated way. A tetragonal form of the element has B_{12} units arranged in layers that are linked through B—B bonds. The latter, obtained by crystallization of liquid boron, is the thermodynamically stable form of the element. Its structure accounts for the high melting point (2250 ± 50 °C) and for

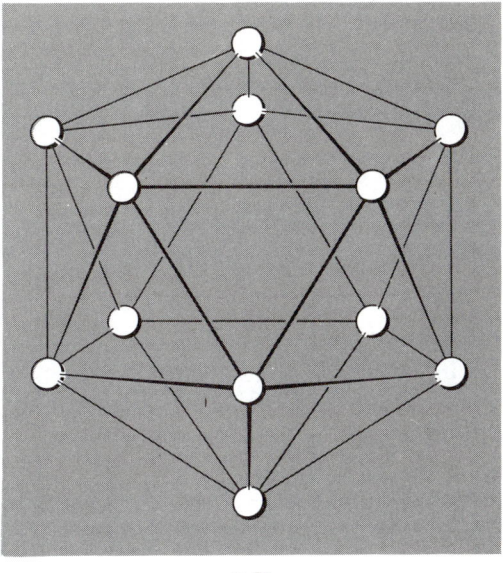

8-II

the chemical inertness of boron. The element is properly considered to be a met-alloid, and its weakly linked B_{12} structure gives it properties intermediate be-tween those of the molecular and the metallic substances.

The Group IVB(14) elements all have the diamond structure shown in Fig. 8-2. This structure has a cubic unit cell, but it can, for some purposes, be viewed as a stacking of puckered, infinite layers. All atoms in the diamond struc-ture are equivalent, each being surrounded by a perfect tetrahedron of four other atoms. Each atom forms a localized two-electron bond to each of its neigh-

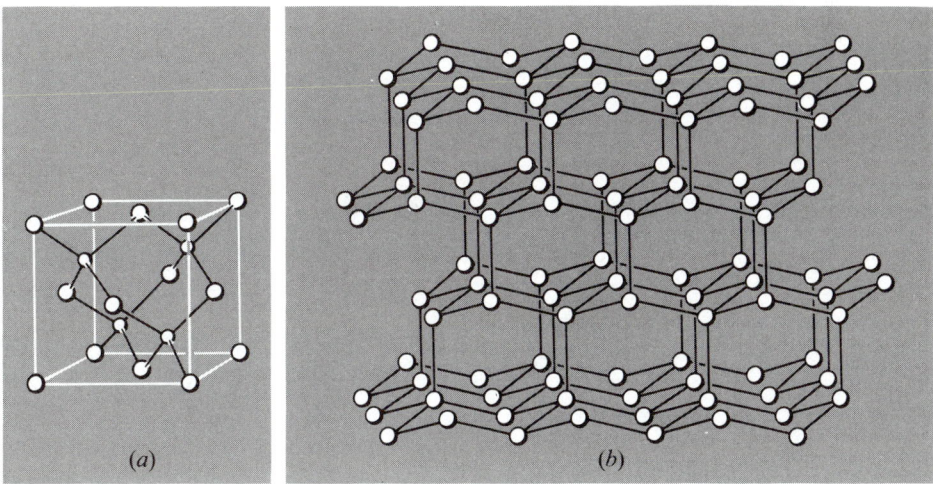

Figure 8-2   The diamond structure seen from two points of view. (*a*) The conven-tional unit cell. (*b*) A view showing how layers are stacked; these layers run perpendicu-lar to the body diagonals of the cube. Remember, however, that this is not a layer struc-ture; its properties are the same in all directions.

bors. The extended, three-dimensional, covalent network structure clearly accounts for the extreme hardness of diamond.

Silicon and *germanium* normally have the diamond structure. *Tin* has the diamond structure, but it also displays the equilibrium shown in Reaction 8-5.1.

$$\alpha - \text{Sn} \underset{}{\overset{18\ ^\circ\text{C}}{\rightleftharpoons}} \beta - \text{Sn} \tag{8-5.1}$$

"Gray"	"White"
diamond	distorted cp
$d^{20} = 5.75$	$d^{20} = 7.31$

The white allotrope has a more efficient, near-ideal, closest-packing (cp) structure, and this accounts for the higher density of the white (β-Sn) allotrope, compared to the α-Sn allotrope, which has the diamond structure (d^{20} in Reaction 8-5.1 represents density, in grams per cubic centimeter, at 20 °C).

Carbon also exists as *graphite,* which has the layer structure shown in Fig. 8-3. The separation of the layers (3.35 Å) is approximately the sum of the van der Waals radii for C and indicates that the forces between the layers should be weak. This accounts for the softness and lubricity of graphite, since the layers can easily slip over one another. Each C atom is surrounded by only three neighbors; after forming one σ bond with each neighbor, each C atom still has one electron and these electrons are paired up into a system of π bonds, as shown in Structure 8-III. Resonance makes all bonds equivalent so that the C—C bond distances are all 1.415 Å.

8-III

This is a little longer than the C—C distance in benzene where the bond order is 1.5 and corresponds to a C—C bond order in graphite of about 1.33. Since $p\pi$–$p\pi$ multiple bonding is clearly involved, the other Group IVB(14) elements cannot form this type of structure. The continuous π system in each layer makes possible good electrical conductance, especially in directions parallel to the layers. The conductance in these directions is 10^4 times greater than it is in the direction perpendicular to the layers. In diamond the conductance is only 10^{-18} of that in the graphite layers. For the elements Si, Ge, and Sn in their diamond-type structures the conductances steadily increase until at tin it is comparable to that within the graphite layers. This is an excellent illustration of increasing metallic character as a group is descended.

Recently, it has been found that carbon exists in a third allotropic form in which there are large, spheroidal C_n molecules, with $n = 60$ or more. Those with $n = 60, 70, 76,$ and 78 have structures that are more or less definitely established,

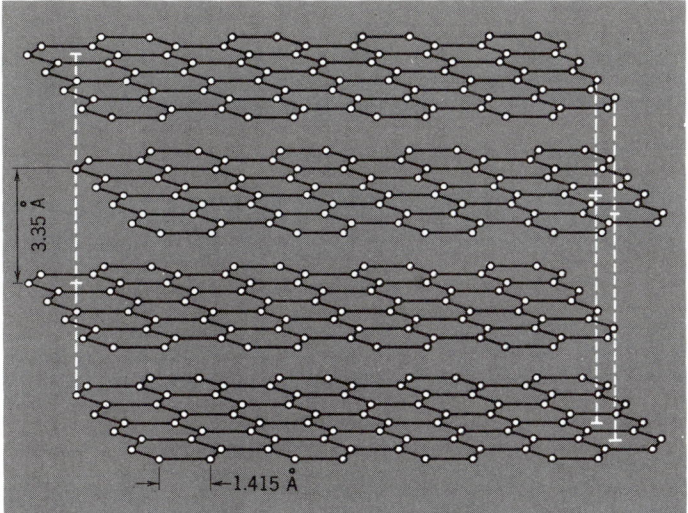

Figure 8-3 The normal structure of graphite.

as shown in Fig. 8-4. The C_{60} molecules were first recognized and christened "buckminsterfullerene" after the American architect–engineer F. Buckminster (Bucky) Fuller, who was best known for designing hemispherical geodesic domes, consisting of pentagonal and hexagonal faces. More commonly, these molecules are now called fullerenes and most informally, buckyballs. The C_{60} buckyball has the same form as a soccer ball.

Large scale preparation of fullerenes is achieved by vaporization of graphite in an electric arc or by a plasma discharge. However, these methods produce mixtures that are not easy to separate. The lower ones, especially C_{60} and C_{70}, can be separated from the many larger ones by extracting them into hexane, benzene, or toluene, when they form magenta solutions. The C_{100} to C_{250} fullerenes dissolve in high-boiling solvents, such as 1,2,4-$C_6H_3Cl_3$ (bp 214 °C), while the completely insoluble residues are thought to contain fullerenes as large as C_{400}. Separation of C_{60} and C_{70} from each other is very difficult on more than a small (~30 mg) scale.

Because of their spheroidal shape, the fullerenes C_{60} and C_{70} form very disordered crystals and the determination of their structures was not easily accomplished. However, by use of derivatives and with the help of NMR and other spectroscopic data, the structures of the smaller fullerenes are well established. In all of them, each carbon atom has three neighboring carbon atoms and forms, formally, two single and one double bond. C_{60} has 32 faces, 20 of which are hexagons and 12 of which are pentagons. C_{60} also seems to be the most stable of the fullerenes. This is, in part, due to the fact that there is considerable delocalization of the electrons in the formal double bonds, and in fact the two kinds of C—C distances are very close, with values of about 1.40 and 1.50 Å.

The fullerenes are the least stable of the carbon allotropes. Graphite, being the most stable, is assigned a standard heat of formation (ΔH_f°) of precisely zero, while diamond has $\Delta H_f^\circ = 2.9$ kJ mol^{-1}. C_{60} has $\Delta H_f^\circ = 38.1$ kJ mol^{-1}.

In Group VB(15), *phosphorus* has numerous polymorphs. The common red form, which may be a mixture of forms, has not been structurally characterized. Black phosphorus, obtained by heating white phosphorus under pressure, has

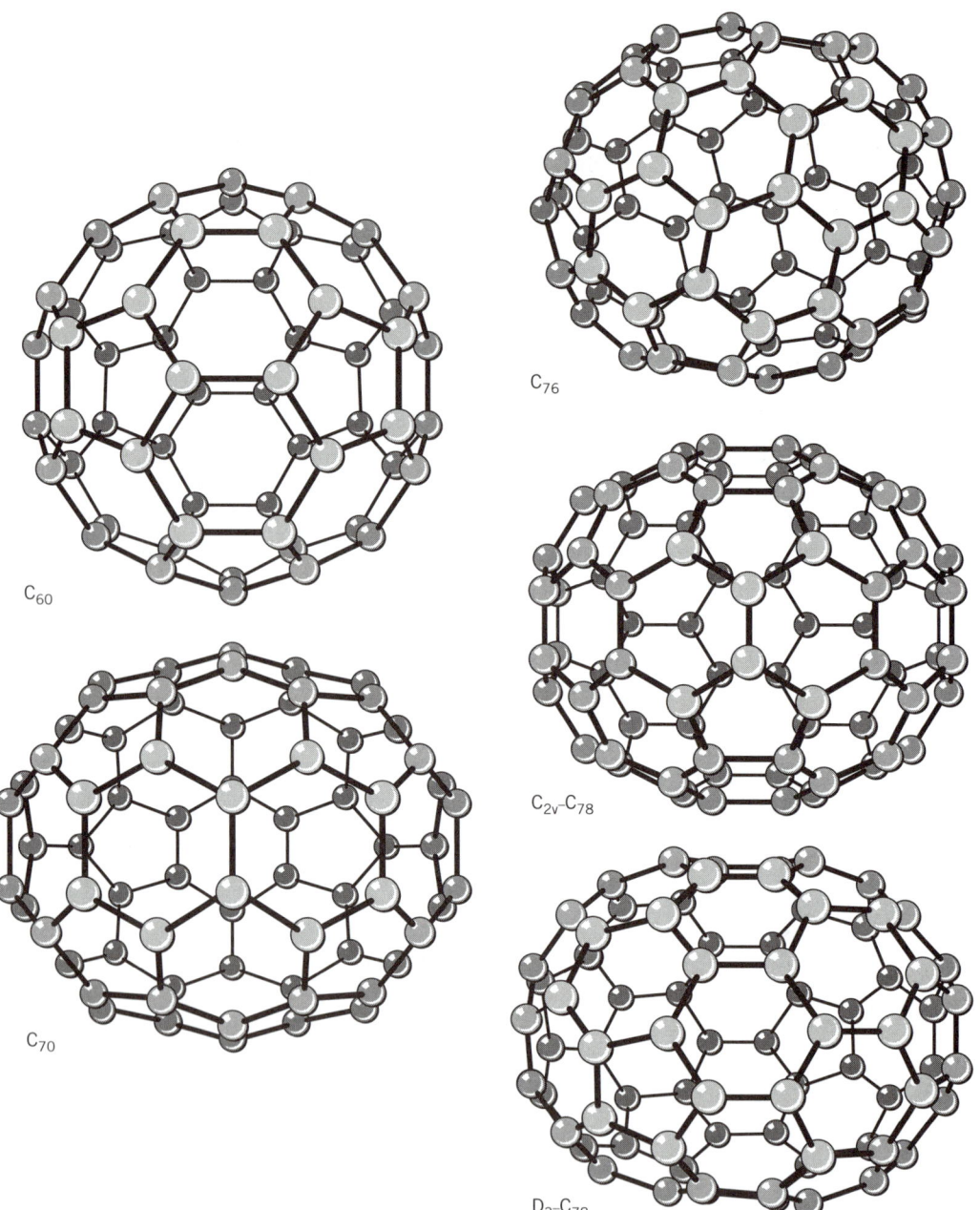

C_{60}

C_{70}

C_{76}

C_{2v}–C_{78}

D_3–C_{78}

Figure 8-4   The structures of some of the smaller fullerenes, C_{60}, C_{70}, C_{76}, and two isomers of C_{78}. [Reprinted with permission from F. Diederich and R. L. Whetten, "Beyond C_{60}: The Higher Fullerenes," *Acc. Chem. Res., 25*, 119–126 (1992). Copyright © (1992) American Chemical Society.]

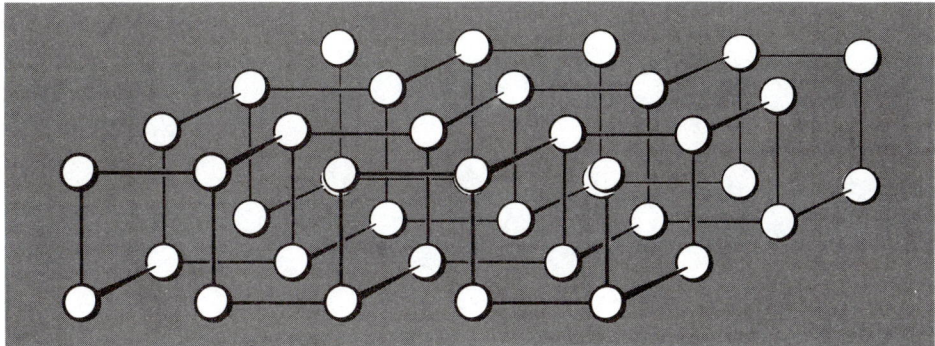

Figure 8-5   The arrangement of atoms in the double layers found in crystalline black phosphorus.

the structure shown in Fig. 8-5. Each phosphorus atom is bound to three neighbors by single bonds, 2.17–2.20 Å long. The double layers thus formed are stacked with an interlayer distance of 3.87 Å. As is true for graphite, the layer structure of black P leads to flakiness of the crystals. It also accounts for the lack of reactivity, for example, to air, compared to P_4.

Arsenic, antimony, and *bismuth* all form crystals whose structures are similar to that of black phosphorus. However, they are bright and metallic in appearance and have resistivities that are comparable to those of metals such as Ti or Mn. Clearly, structure alone does not fix the properties of a substance. In As, Sb, and Bi, the larger atomic orbitals lead to the formation of energy bands rather than purely localized bonds. This gives rise to increasing metallic character.

The chain form of *sulfur* (*catenasulfur*) is the main component of the so-called plastic sulfur obtained when molten sulfur is poured into water. It can be drawn into long fibers that contain helical chains of sulfur atoms. It slowly transforms to orthorhombic S_8.

The stable form of *selenium,* gray, metal-like crystals obtained from melts, contains infinite spiral chains. There is evidently weak interaction of a metallic nature between neighboring atoms of different chains, but in the dark the electrical conductivity of selenium is not comparable to that of true metals (resistivity 2×10^{11} μΩ cm). However, it is notably photoconductive, and is hence used in photoelectric devices. Selenium is also essential to the process of xerography.

Tellurium is isomorphous with gray Se, although it is silvery white and semimetallic (resistivity 2×10^5 μΩ cm). The resistivity of S, Se, and Te has a negative temperature coefficient, usually considered a characteristic of nonmetals.

8-6   Metals

The majority of the elements are metals. These elements have many physical properties different from those of other solids, notably: (1) high reflectivity; (2) high electrical conductance, decreasing with increasing temperature; (3) high thermal conductance; and (4) mechanical properties such as strength and ductility. There are three basic metal structures: *cubic* and *hexagonal close packed* (illustrated in Section 4-7) and *body-centered cubic, bcc* (Fig. 8-6). In *bcc* packing each atom has only 8 instead of 12 nearest neighbors, although there are 6 next nearest neighbors that are only about 15% further away. It is only 92% as dense an

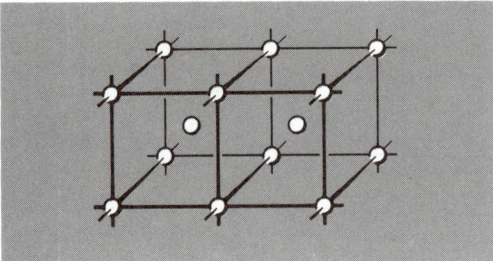

Figure 8-6   A body-centered cubic (bcc) structure.

arrangement as the *hcp* and *ccp* structures. The distribution of these three structure types, *hcp*, *ccp*, and *bcc*, in the periodic table is shown in Fig. 8-7. The majority of the metals deviate slightly from the ideal structures, especially those with *hcp* structures. For the *hcp* structure the ideal value of c/a, where c and a are the hexagonal unit-cell edges, is 1.633. All metals having this structure have a smaller c/a ratio (usually 1.57 – 1.62) except zinc and cadmium.

The characteristic physical properties of metals as well as the high coordination numbers (either 12 nearest neighbors, or 8 plus 6 more that are not too remote) suggest that the *bonding in metals* is different from that in other substances. There is no ionic contribution, and it is also impossible to have two-electron covalent bonds between all adjacent pairs of atoms, since there are neither

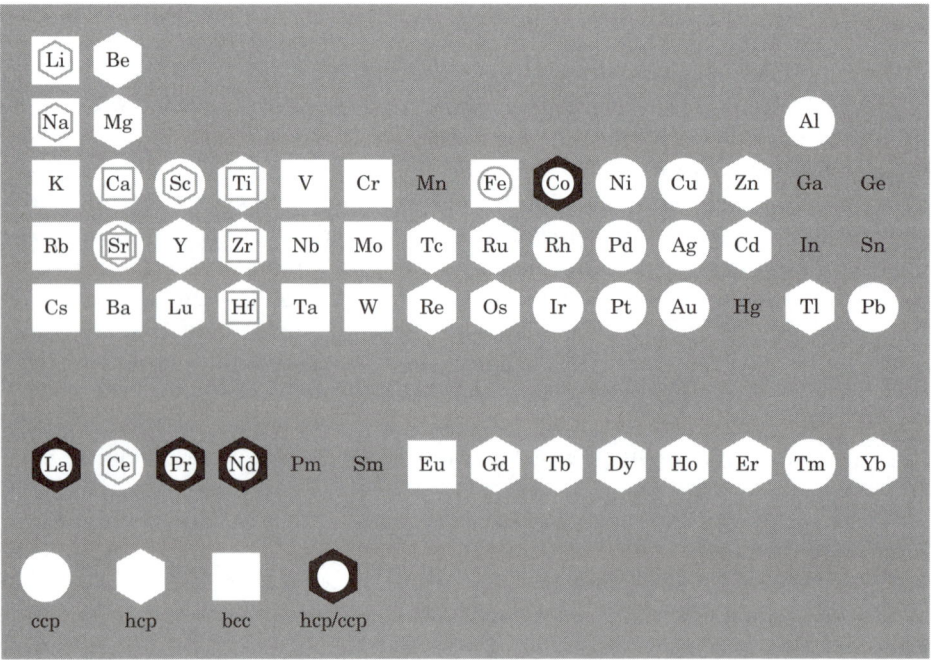

Figure 8-7   The occurrence of hexagonal close-packed (hcp), cubic close-packed (ccp), and body-centered cubic (bcc) structures among the elements. Where two or more symbols are used, the largest represents the stable form at 25 °C. The symbol labeled hcp/ccp signifies a mixed . . . *ABCABABCAB* . . . type of close packing, with overall hexagonal symmetry. [Adapted with permission from H. Krebs, *Grundzuge der Anorganishen Kristallchemie,* F. Enke Verlag, *1968.*]

sufficient electrons nor sufficient orbitals. An explanation of the characteristic properties of metals is given by the so-called band theory. This is very mathematical but the principle can be illustrated.

Imagine an array of atoms so far apart that their atomic orbitals do not interact. Now suppose this array contracts. The orbitals of neighboring atoms begin to overlap and interact with each other. So many atoms are involved that at the actual distances in metals, the interaction forms essentially continuous energy bands that spread through the metal (Fig. 8-8). The electrons in these bands are *completely delocalized*. Observe also that some bands may overlap. In Fig. 8-8, where Na is used as an illustration, the 3s and 3p bands overlap.

The energy bands can also be depicted as in Fig. 8-9. Here energy is plotted horizontally, and the envelope indicates on the vertical the number of electrons that can be accommodated at each value of the energy. Shading is used to indicate filling of the bands.

Completely filled or completely empty bands, as shown in Fig. 8-9(*a*), do not permit net electron flow and the substance is an *insulator*. Covalent solids can be discussed from this point of view (though it is unnecessary to do so) by saying that all electrons occupy low-lying bands (equivalent to the bonding orbitals) while the high-lying bands (equivalent to antibonding orbitals) are entirely empty. Metallic conductance occurs when there is a partially filled band, as in Fig. 8-9(*b*); the transition metals, with their incomplete sets of *d* electrons, have partially filled *d* bands and this accounts for their high conductances. Overlapping bands, as in Na, are illustrated in Fig. 8-9(*c*).

Cohesive Energies of Metals

The strength of binding among the atoms in metals can be measured by the *enthalpies of atomization* (Fig. 8-10). Cohesive energy maximizes with elements having partially filled *d* shells, that is, with the transition metals. However, it is particularly with the elements near the middle of the second and third transition series, especially Nb—Ru and Hf—Ir, that the energies are largest, reaching 837 kJ mol^{-1} for tungsten. It is noteworthy that these large cohesive energies are principally due to the structure of the metals where high coordination numbers are involved. For an *hcp* or *ccp* structure, there are 6 bonds per metal atom (since each of the 12 nearest neighbors has a half-share in each of the 12 bonds). Each bond, even when cohesive energy is 800 kJ mol^{-1}, has an energy of only 133 kJ mol^{-1}, roughly one half of the C—C bond energy in diamond where each carbon atom has only four neighbors, but there are three times as many of them.

PART B

THE CHEMISTRY OF THE ELEMENTS IN RELATION TO THEIR POSITION IN THE PERIODIC TABLE

We can now proceed to a more detailed commentary on the chemical reactivity and types of compounds formed by the elements. The periodic table forms the basis for the discussion, starting with the simplest chemistry, namely, that of hydrogen, and proceeding to the heaviest elements.

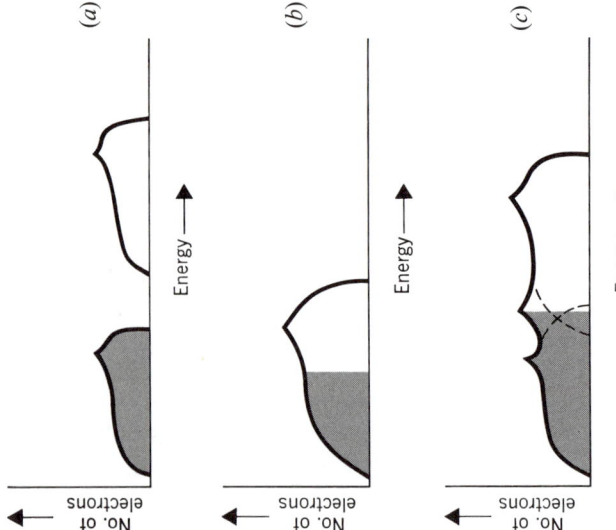

Figure 8-9 Envelopes of energy bands, with shading to indicate filling. (*a*) an insulator, (*b*) a metallic conductor, (*c*) overlapping conduction bands as in Na.

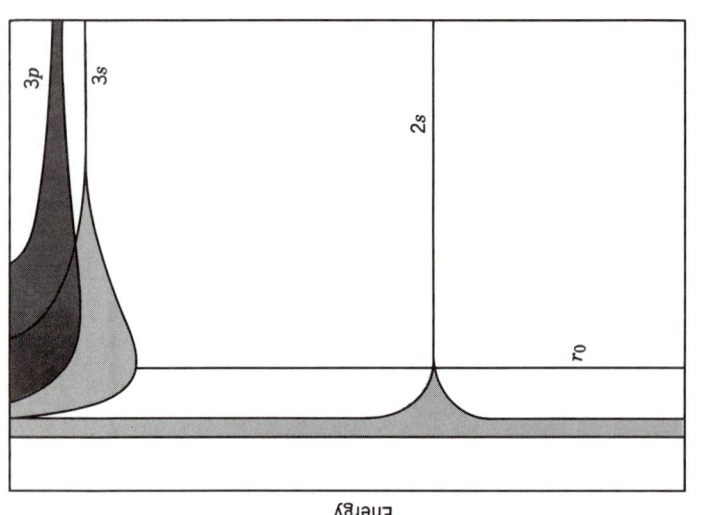

Figure 8-8 Energy bands of sodium as a function of internuclear distance. The actual equilibrium distance is represented by r_0. [Reproduced by permission from J. C. Slater, *Introduction to Chemical Physics*, McGraw-Hill, New York, 1939.]

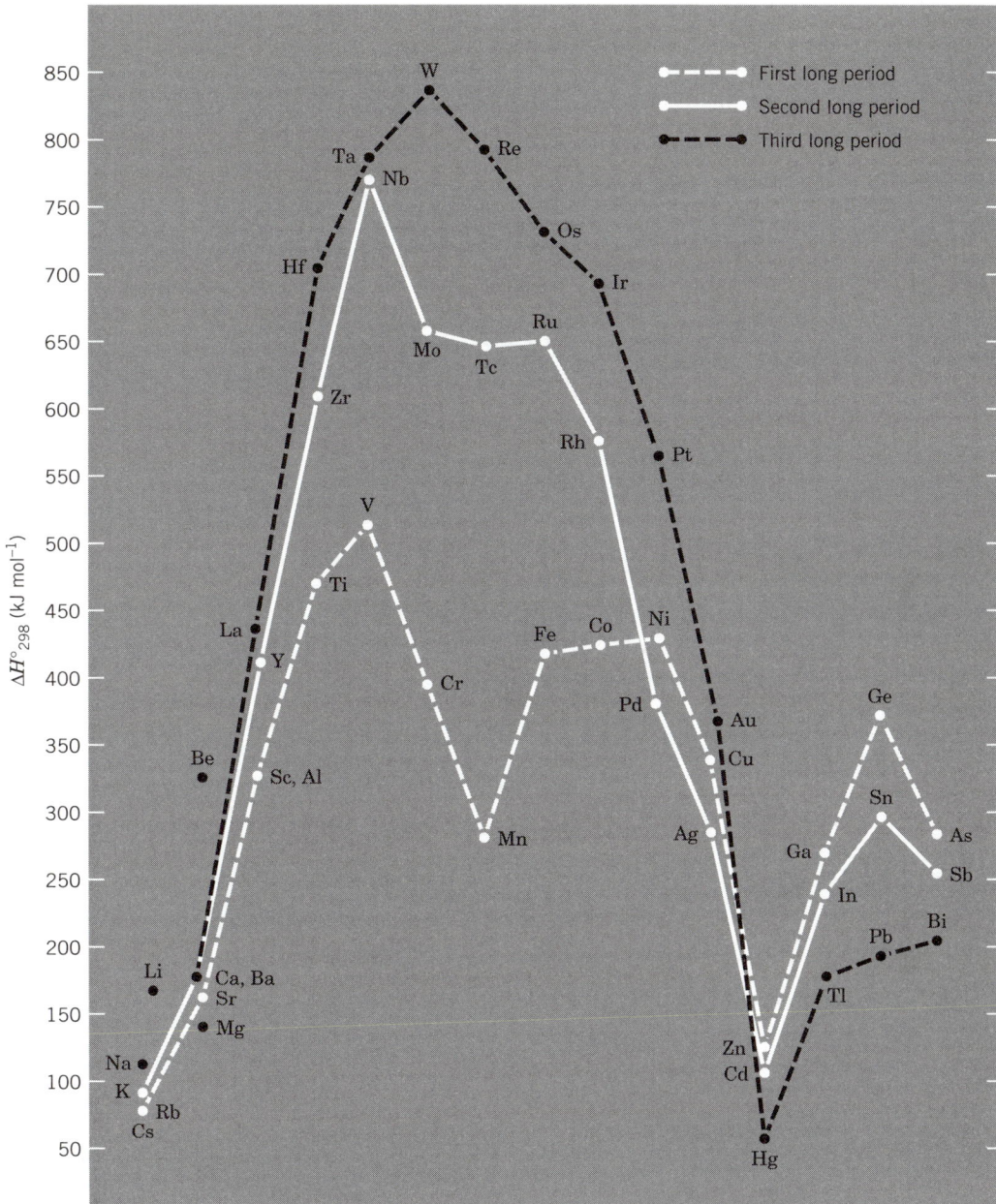

Figure 8-10   Heats of atomization of metals, ΔH°_{298} for $M(s) \rightarrow M(g)$. [Reproduced by permission from W. E. Dasent, *Inorganic Energetics,* Second Edition, Cambridge University Press, New York, 1982.]

8-7   Hydrogen: 1s¹

The chemistry of hydrogen depends on three electronic processes:

1. *Loss of the 1s valence electron.* This forms merely the proton, H⁺. Its small size, $r \sim 1.5 \times 10^{-13}$ cm, relative to atomic sizes $r \sim 10^{-8}$ cm, and its charge result in a unique ability to distort the electron clouds surrounding other

atoms. The proton *never* exists as such except in gaseous ion beams. It is invariably associated with other atoms or molecules. Although the hydrogen ion in water is commonly written as H^+, it is actually H_3O^+ or $H(H_2O)_n^+$.

2. *Acquisition of an electron.* The H atom can acquire an electron forming the *hydride ion* H^- with the He $1s^2$ structure. This ion exists only in crystalline hydrides of the most electropositive metals (e.g., NaH or CaH_2).

3. *Formation of an electron pair bond.* Nonmetals and even many metals can form covalent bonds to hydrogen.

The chemistry of hydrogen-containing substances depends greatly on the nature of the other elements and groups in the compound. The extent to which the compounds dissociate in polar solvents, acting as acids in the general way shown in Reaction 8-7.1:

$$HX \rightleftarrows H^+ + X^- \tag{8-7.1}$$

is particularly dependent on the nature of X.

The electronic structure and coordination number of the whole molecule are also important. Consider BH_3, CH_4, NH_3, OH_2, and FH. The first acts as a Lewis acid, and dimerizes instantly to B_2H_6; CH_4 is unreactive and neutral; NH_3 has a lone pair and is a base; H_2O with two lone pairs can act as a base or as a very weak acid; HF, a gas, is a much stronger though still weak acid in aqueous solution.

All H—X bonds necessarily have some polar character with the dipole oriented $\overset{\delta+}{H}$—$\overset{\delta-}{X}$ or $\overset{\delta-}{H}$—$\overset{\delta+}{X}$. The term "hydride" is usually given to those compounds in which the negative end of the dipole is on hydrogen (e.g., in SiH_4, $\overset{\delta+}{Si}$—$\overset{\delta-}{H}$). However, although HCl as $\overset{\delta+}{H}$—$\overset{\delta-}{Cl}$ is a strong acid in aqueous solution, nevertheless, it is a gas and is properly termed a covalent hydride.

8-8   Helium ($1s^2$) and the Noble Gases (ns^2np^6)

The second element *helium* (He), with $Z = 2$, has the closed $1s$ shell; its very small size leads to some physical properties that are unique to liquid helium. The physical properties of the other noble gases vary systematically with size. Although the first ionization energies are high, which is consistent with their chemical inertness, the values fall steadily as the size of the atom increases. The ability to enter into chemical combination with other atoms should increase with decreasing ionization potential and decreasing energy of promotion to states with unpaired electrons; that is, $ns^2np^6 \rightarrow ns^2np^5(n+1)s$. The threshold of chemical activity is reached at Kr, but few compounds have been isolated. The reactivity of Xe is much greater, and many compounds of Xe with O and F are known (Chapter 21). The reactivity of Rn is presumably still greater than that of the other noble gases, but since the longest-lived isotope (^{222}Rn) has a half-life of only 3.825 days, only limited tracer studies can be made.

8-9   Elements of the First Short Period

The third element *lithium* (Li), with $Z = 3$, has the structure $1s^22s$. With increasing Z, electrons enter the $2s$ and $2p$ levels until the closed-shell configuration

$1s^2 2s^2 2p^6$ is reached at neon. The seven elements Li to F constitute the first members of the *groups* of elements.

Although these elements have many properties in common with the heavier elements of their respective groups, which is to be expected in view of the similarity in the outer-electronic structures of the *gaseous* atoms, they nevertheless show highly individual behavior in many important respects. We have already seen that O_2 and N_2 form diatomic molecules, whereas their congeners, S and P, form polyatomic molecules or chains. Indeed, the differences between the chemistry of B, C, N and O, and Al, Si, P and S, and the heavier members of these groups is sufficiently striking that in many ways it is not useful to regard the elements of the first short period as prototypes for their congeners. The closest analogies between the elements of the first short period and the heavier elements of particular groups occur for Li and F, followed by Be.

The increase in nuclear charge and consequent changes in the extranuclear structure result in extremes of physical and chemical properties. Figure 8-11 gives the first ionization enthalpies. The low ionization enthalpy for *lithium* is in accord with facile loss of an electron to form the Li^+ ion, which occurs both in solids and in solution. It accords with the high reactivity of lithium with oxygen, nitrogen, water, and many other elements.

For *beryllium* (Be), the first (899 kJ mol^{-1}) and especially the second (1757 kJ mol^{-1}) ionization enthalpies are sufficiently high that total loss of both electrons to give Be^{2+} does not occur even with the most electronegative elements. Thus in BeF_2, the Be—F bonds have appreciable covalent character. The ion in aqueous solution, $[Be(H_2O)_4]^{2+}$, is very strongly aquated and undergoes hydrolysis quite readily to give species with Be(OH) bonds.

For the succeeding elements, the absence of any simple cations under any conditions is to be expected from the high ionization enthalpies. Note that the values (Fig. 8-11) for B, C, and N increase regularly but that they are lower than the values that would be predicted by extrapolation from Li and Be. This arises because p electrons are less penetrating than s electrons; they are, therefore, shielded by the s electrons and are removed more easily. Another discontinuity occurs between N and O. This occurs because the $2p$ shell is half full, that is,

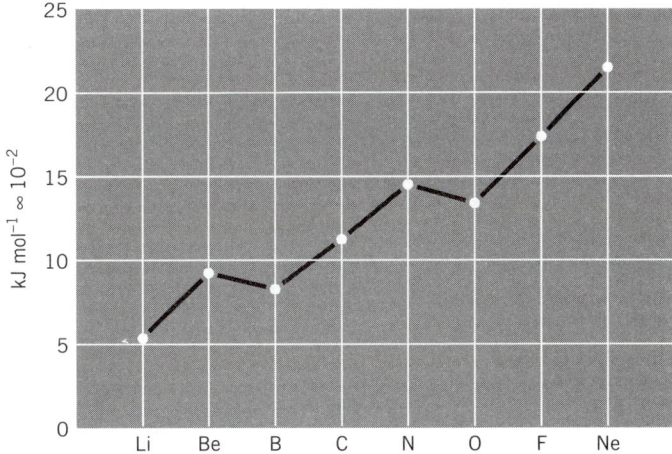

Figure 8-11   First ionization enthalpies of the elements Li to Ne. See also Fig. 2-14.

$p_x p_y p_z$ at N. The p electrons added in O, F, and Ne thus enter p orbitals that are already singly occupied. Hence, they are partly repelled by the p electron already present in the same orbital and are less tightly bound.

The electron attachment enthalpies (Section 1-2) become increasingly more negative from Li to F, and the electronegativities (Sections 1-2 and 2-7) of the elements increase from Li to F.

Boron $(2s^2 2p^1)$ has no simple ion chemistry associated with cations of the type B^{n+}. Rather, it is bound covalently in all of its compounds, as in oxoanions, organoboron compounds, or hydrides.

Anion formation first appears for carbon, which forms C_2^{2-} and some other polyatomic ions, although the existence of C^{4-} is uncertain. N^{3-} ions are stable in nitrides of highly electropositive elements. Oxide (O^{2-}) and fluoride (F^-) are common in solids, but observe that O^{2-} ions cannot exist in aqueous solutions. Compare

$$O^{2-} + H_2O = 2\,OH^- \qquad K > 10^{22} \qquad\qquad (8\text{-}9.1)$$

$$F^- + H_2O = HF + OH^- \qquad K = 10^{-7} \qquad\qquad (8\text{-}9.2)$$

Carbon is a true nonmetal and its chemistry is dominated by single, double, and triple bonds to itself or to nitrogen, oxygen, and a few other elements. What distinguishes carbon from other elements is its unique ability to form chains of carbon–carbon bonds (called catenation) in compounds—as distinct from the element itself.

Nitrogen as nitrogen gas (N_2) is relatively unreactive because of the great strength of the $N{\equiv}N$ bond and its electronic structure. Nitrogen compounds are covalent, usually involving three single bonds, although multiple bonds, such as $C{\equiv}N$ or $Os{\equiv}N$, can exist. With electropositive elements, ionic nitrides containing N^{3-} may be formed.

The diatomic molecule *oxygen* has two unpaired electrons and consequently is very reactive. There is an extensive chemistry with covalent bonds as in $(CH_3)_2C{=}O$, $(C_2H_5)_2O$, CO, SO_3, and the like. However, well-defined oxide ions O^{2-}, O_2^-, and O_2^{2-} exist in crystalline solids. Hydroxide ions (OH^-) exist both in solids and in solutions, although in hydroxylic solvents the OH^- ion is doubtlessly hydrated via hydrogen bonds.

Fluorine is extremely reactive due largely to the low bond energy in F_2. This is a result, in part, of repulsions between nonbonding electrons. Ionic compounds containing F^- ions and covalent compounds containing X—F bonds are well established. Owing to the high electronegativity of fluorine, such covalent bonds are generally quite polar in the sense X^+—F^-.

Covalent Bonds

A few points may be mentioned here.

1. Note that Be, B, and C have fewer unpaired electrons in their ground states than the number of electron-pair bonds they normally form. This has been explained previously in terms of promotion to valence states.

2. The elements of the first short period obey the octet rule. Since they have only four orbitals ($2s$, $2p_x$, $2p_y$, $2p_z$) in their valence shells, there are never more than eight electrons in their valence shells. This means that the maximum number of electron-pair bonds is four. The octet rule breaks down for elements in the second short period.

For example, phosphorus $(3s^2 3p^3 3d^0)$ can be excited to a valence state $3s^1 3p^3 3d^1$ with an expenditure of energy so modest that the heat of formation of the two additional bonds will more than compensate for it. On the other hand, promotion of N $(2s^2 2p^3)$ to any state with five unpaired electrons, such as $2s^1 2p^3 3d^1$, would require more promotional energy than could be recovered by the extra bond formation energy.

For C, promotion from $2s^2 2p^2$ to $2s 2p^3$ gives the valence of four. For N $(2s^2 2p^3)$ only three of the five electrons can possibly be unpaired, in O only two, and in F only one. Hence, these elements are limited to valences of three, two, and one. On the other side of C, that is, in Li, Be, and B, the valences are less than four because of lack of electrons to occupy the orbitals, so that by electron sharing alone these can show valences of only one, two, and three, respectively.

3. Where there are fewer electrons than are required to fill the energetically useful orbitals, as in trivalent boron compounds [e.g., BCl_3, BF_3, and $B(CH_3)_3$], there is a strong tendency to utilize these orbitals by combining with compounds that have an excess of electrons. Examples of these compounds are those of trivalent nitrogen [e.g., NH_3, $N(CH_3)_3$, etc.] or oxygen [e.g., H_2O, $(C_2H_5)_2O$, etc.] that have unshared electron pairs. The former are thus acceptors of electrons (Lewis acids) and the latter are donors of electrons (Lewis bases). The formation of a dative bond is shown in Fig. 8-12.

Notice that while nitrogen compounds have only one unshared pair ($:NR_3$), oxygen compounds have two ($:\ddot{O}R_2$); normally only one of the electron pairs is used and only in a very few cases does oxygen form four bonds. Beryllium compounds with two empty orbitals usually fill these by forming compounds with two donor molecules (BeX_2L_2).

Note that such donor–acceptor behavior is not confined to elements of the first short period, but is quite general. Adducts may be formed between compounds whenever one compound has empty orbitals and the other has unshared electron pairs.

Compounds of many elements may act as acceptors, but donors are commonly compounds of trivalent N, P, and As and compounds of divalent O and S. However, a very important class of donors are the halide and pseudohalide ions, and ions such as hydride (H^-) and carbanions (e.g., CH_3^- or $C_6H_5^-$). Some representative examples are

$$BF_3 + F^- = BF_4^- \tag{8-9.3}$$

$$PF_5 + F^- = PF_6^- \tag{8-9.4}$$

$$AlCl_3 + Cl^- = AlCl_4^- \tag{8-9.5}$$

$$PtCl_4 + 2\ Cl^- = PtCl_6^{2-} \tag{8-9.6}$$

$$Ni(CN)_2 + 2\ CN^- = Ni(CN)_4^{2-} \tag{8-9.7}$$

$$Co(NCS)_2 + 2\ SCN^- = Co(NCS)_4^{2-} \tag{8-9.8}$$

$$BH_3 + H^- = BH_4^- \tag{8-9.9}$$

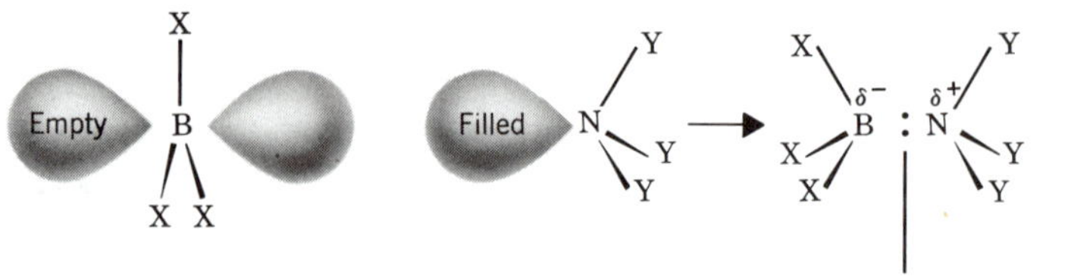

Figure 8-12   The formation of a dative bond between boron in a BX_3 acceptor and nitrogen in an NY_3 donor.

$$Al(CH_3)_3 + CH_3^- = Al(CH_3)_4^- \qquad (8\text{-}9.10)$$

Lewis-base behavior is also shown by some transition metal compounds, as we discuss later. One example is the compound $(\eta^5\text{-}C_5H_5)_2ReH$, which is as strong a base to protons as NH_3. The reason why some atoms succeed in increasing their coordination numbers from three to four but seldom from two to four can be understood if we consider the polar nature of the dative bond. The donor and acceptor molecules are both electrically neutral. When the bond is formed, the donor atom has, in effect, lost negative charge rendering it positive. It has only half-ownership of an electron pair that formerly belonged to it entirely. Conversely, the acceptor atom now has extra negative charge. This would be true for complete sharing of the electron pair (Structure 8-IV). Lesser polarity is introduced if the electron pair remains more the property of the donor atom than the acceptor (Structure 8-V), in which case we indicate only charges $\delta+$ and $\delta-$ on the atoms.

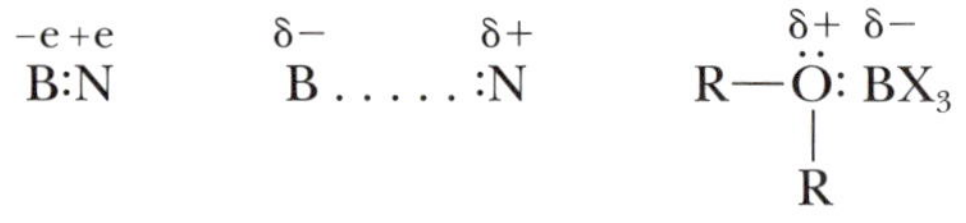

$$
\begin{array}{ccc}
\overset{-e\ +e}{B\!:\!N} & \overset{\delta-\qquad\quad \delta+}{B\ldots\ldots:\!N} & \overset{\delta+\ \ \delta-}{R\!-\!\overset{\cdot\cdot}{O}\!:\,BX_3} \\
& & | \\
& & R \\
\text{8-IV} & \text{8-V} & \text{8-VI}
\end{array}
$$

This charge separation can be achieved only by doing work against Coulomb forces, which we must assume is more than compensated by the bond energy when a stable system results. However, if we take a case where one donor bond has been formed (Structure 8-VI), then the second unshared pair on oxygen is further restrained by the positive charge that arises on O from the dative bond already formed. There is thus much more Coulombic work to be done in forming a second dative bond—enough apparently to make this process energetically unfavorable. Steric hindrance between the first acceptor and a second would also militate against addition of a second acceptor. Note that this electrostatic argument is basically the same as that used to explain relative dissociation constants in polyfunctional acids.

8-10   The Elements of the Second Short Period

The elements of the second short period are Na, Mg, Al, Si, P, S, Cl, and Ar. Although their outer-electronic structures are similar to those of the corresponding elements in the first short period, their chemistry differs considerably. In particular, the chemistry of Si, P, S, and Cl is largely different from their corresponding partners in the first short period. The elements of the second short period, however, give a better guide to the chemistry of the heavier elements in their respective groups than the elements of the first short period that start the groups. This is especially the case for the nonmetallic elements for the following reasons:

1. It is not generally favorable to form $p\pi$–$p\pi$ multiple bonds such as $Si{=}Si$, $Si{=}O$, or $P{=}P$. Most likely, this occurs because, in order to approach close enough to get good overlap of $p\pi$ atomic orbitals, the heavier atoms would encounter large repulsive forces due to overlapping of their filled inner shells. The small, compact inner shell of the elements from the second row (i.e., just $1s^2$) does not produce this repulsion.

The result is that, as we have seen, the nature of the elements of the second short period is strikingly different from that of the elements of the first short period. As a striking example, consider the vast chemistry of carbon associated with multiple bonds such as $C{=}C$, $C{\equiv}C$, $C{=}O$, and $C{=}N$. In contrast, silicon displays less tendency to form multiple bonds of this type, and although many compounds containing double bonds (e.g., $Si{=}Si$, $Si{=}P$, $Ge{=}Ge$, $Ge{=}C$, and $P{=}P$) are known, the substances are stable only when sterically encumbered by very bulky substituents so that the double bonds are *kinetically* stable. An example is $[(CH_3)_3Si]_3CP{=}PC[Si(CH_3)_3]_3$. Here, we also note that whereas CO_2 is a gas, SiO_2 (quartz) is an infinite polymer or network substance.

2. Although in certain types of compounds of P, S, and Cl, such as Cl_3PO, Cl_2SO, SO_2, ClO_4^-, and ClO_2, there is some multiple bonding, this bonding occurs by an entirely different mechanism involving d orbitals. The low-lying $3d$ orbitals can be utilized not only for $p\pi$–$d\pi$ multiple bonding, but also for additional bond formation. The octet rule now no longer holds rigorously and is indeed commonly violated.

3. The possibility of using the $3d$ orbitals then allows promotion to valence states leading to formation of five or six bonds. Hence, there are compounds such as PCl_5, SF_6, and the ion SiF_6^{2-}, in which we have five- and six-coordination. For silicon, even where there is some analogy with carbon chemistry, as in compounds with single bonds, the reactions and mechanisms operating in silicon chemistry may be vastly different. A simple example is the unreactivity of CCl_4 toward H_2O, whereas $SiCl_4$ is instantly hydrolyzed.

4. The shapes of molecules and the nature of the bonds also differ. Recall the discussion from Chapter 3 concerning VSEPR geometries.

5. Even the cation- and anion-forming elements differ. Thus while beryllium forms only $[Be(H_2O)_4]^{2+}$, the magnesium ion is $[Mg(H_2O)_6]^{2+}$, and there are substantial differences between the chemistry of Li and Na. Aluminum is an electropositive metal totally different from boron, although in certain covalent compounds there are some similarities. For Group VIIB(17), the Cl—Cl bond strength is actually higher than that of F_2, and Cl_2 is much less reactive. In addition, solid chlorides commonly have structures that are quite different from those of the corresponding fluorides. The structures of ionic chlorides are much closer to those given by sulfides.

8-11   The Remainder of the Nontransition Elements

The remainder of the nontransition elements have many important periodic trends in the physical and chemical properties and in the structures of the elements. As already pointed out, a thorough discussion of chemical periodicity is complicated by the fact that the elements in the first short period (row two of the periodic table) do not uniformly serve as reliable guides to the behavior of the other elements of their respective periodic table groups. This is in spite of the fact that the elements of the first short period each reside at the top of their group. Even Li and F, which do serve as reasonable guides to the chemistry of the other elements of their respective groups, show significant differences. Much better guides to the chemistry of the elements of the periodic table groups are the elements Na through Ar, which constitute the second short period, or row three of the periodic table. Thus there are major disparities between the chemistry of nitrogen and phosphorus in Group VB(15), followed by more regular trends in chemistry on descending the rest of Group VB(15). Likewise, carbon is not very similar to silicon, whereas the remaining elements of Group IVB(14) display a uniform trend towards increasing metallic character on descending the group from Ge (a metalloid) to Sn and Pb, both of which are metals.

The best way to outline the numerous periodic aspects of the chemistry of the nontransition elements [Groups IA(1), IIA(2), and Groups IIIB(13) through VIIIB(18)] is to state carefully:

First, the differences between each element in row two and the remaining elements in the same group.

Second, the regular variations thereafter, upon descending the group.

Among the regular variations to be discussed here and at greater length later are

1. Metallic character for the elements.
2. Properties of the oxides of the elements, including
   (a) Ionic versus covalent character.
   (b) Acidic versus basic character.
3. Properties of the halides of the elements, including
   (a) Molecular versus ionic character.
   (b) Ease of hydrolysis of the halide derivative.
4. Trends in electrovalence and covalence among the elements.
5. Trends in structure, especially coordination numbers among unassociated molecules and complex ions, as well as the tendency to form aggregates in the solid state so as to increase the effective coordination number of an element.
6. Properties of the hydride derivatives of the elements, especially volatility and reactivity.
7. Tendency for catenation among the compounds of the elements.
8. The relative importance of $p\pi–p\pi$ versus $d\pi–p\pi$ (or even $d\pi–d\pi$) bonding in compounds of the elements.
9. The general strength of covalent bonds to particular elements.
10. The relative importance of low-valent versus high-valent oxidation states of the elements.

The main features of periodicity are outlined briefly in this chapter. Greater discussion of these points is undertaken in the chapters of Part 2 of the book (Chapters 9–22, The Main Group Elements).

Group IA(1)

All the elements of Group IA(1) (Table 8-1) are highly electropositive giving +1 ions. Of all the groups in the periodic table, these metals most clearly show the effect of increasing size and mass on chemical properties. Thus, as examples, the following properties *decrease* from Li to Cs: (a) melting points and heat of sublimation of the metals; (b) lattice energies of salts except those with very small anions (because of irregular radius ratio effects); (c) effective hydrated radii and hydration energies; and (d) strength of covalent bonds in M_2 molecules.

Groups IIA(2) and IIB(12)

Some properties of the elements in Groups IIA(2) and IIB(12) are given in Table 8-2. The elements Ca, Sr, Ba, and Ra are also highly electropositive forming +2 ions. Systematic group trends are again shown, for example, by increasing insolubilities of sulfates, increasing thermal stabilities of carbonates or nitrates, and decreasing hydration energies of the ions in solution.

The elements Zn, Cd, and Hg are in Group IIB(12) and have two s electrons outside filled d shells, since they follow Cu, Ag, and Au, respectively, after the first, second, and third transition series elements. The chemistries of Zn and Cd are quite similar, but the polarizing power of the M^{2+} ions is larger than would be predicted by comparing the radii with those of the Mg to Ra group. This can be associated with the greater ease of distortion of the filled d shell compared with the noble gas shell of the Mg to Ra ions. Both Zn and Cd are quite electropositive, resembling Mg in their chemistry, although there is a greater tendency to form complexes with NH_3, halide ions, and CN^-.

Mercury is unique. It has a high *positive* potential, and the Hg^{2+} ion does *not* resemble Zn^{2+} or Cd^{2+}. For example, the formation constants for, say, halide ions are orders of magnitude greater than for Cd^{2+}. Mercury also readily forms the dimercury ion, which has a metal–metal bond ($^+Hg-Hg^+$).

Groups IIIA(3) and IIIB(13)

Some properties of the elements in Groups IIIA(3) and IIIB(13) are given in Table 8-3. This group is quite large, since it contains the Group IIIA(3) elements,

Table 8-1   Some Properties of Group IA(1) Elements

Element	Electron Configuration	mp (°C)	Ionic Radius (Å)	$E°$ (V)[a]	Ionization Enthalpy (kJ mol^{-1})
Li	$[He]2s^1$	180	0.90	−3.0	520
Na	$[Ne]3s^1$	98	1.16	−2.7	496
K	$[Ar]4s^1$	64	1.52	−2.9	419
Rb	$[Kr]5s^1$	39	1.66	−3.0	403
Cs	$[Xe]6s^1$	29	1.81	−3.0	376
Fr[b]	$[Rn]7s^1$	—	—	—	—

[a]For $M^+(aq) + e^- = M(s)$.

[b]All isotopes are radioactive with short half-lives.

Table 8-2 Some Properties of Group IIA(2) and IIB(12) Elements

Element	Electron Configuration	mp (°C)	Ionic Radius M^{2+} (Å)	$E°$ (V)^a	ΔH(kJ mol^{-1}) for $M(g) \rightarrow M^{2+}(g) + 2\,e^-$
Be	[He]$2s^2$	1280	0.59	−1.85	2657
Mg	[Ne]$3s^2$	650	0.86	−2.37	2188
Ca	[Ar]$4s^2$	840	1.14	−2.87	1735
Sr	[Kr]$5s^2$	770	1.32	−2.89	1609
Ba	[Xe]$6s^2$	725	1.49	−2.90	1463
Ra	[Rn]$7s^2$	700	1.62	−2.92	1484
Zn	[Ar]$3d^{10}4s^2$	420	0.88	−0.76	2632
Cd	[Kr]$4d^{10}5s^2$	320	0.92	−0.40	2492
Hg	[Xe]$4f^{14}5d^{10}6s^2$	−39	1.16	+0.85	2805

^{a}For $M^{2+}(aq) + 2\,e^- = M(s)$.

Sc, Y, La, and Ac, and the Group IIIB(13) elements, Al, Ga, In, and Tl. In addition, all of the lanthanide elements could be included, since their chemistry is similar to that of the Group IIIA(3) elements.

However, we consider the lanthanides separately because of their special position in the periodic table. Notice that in the Sc to Ac group the three valence electrons are d^1s^2 compared with s^2p^1 for the Al to Tl group. Despite this occupancy of the d levels, the elements show no transition metal-like chemistry. These elements are highly electropositive metals, and their chemistry is primarily one of the +3 ions that have the noble gas configuration.

Scandium with the smallest ionic radius has chemical behavior intermediate between that of Al, which has a considerable tendency to covalent bond formation, and the mainly ionic natures of the heavier elements.

The elements Ga, In, and Tl, like Al are borderline between ionic and covalent in compounds, even though the metals are quite electropositive and they form M^{3+} ions.

Table 8-3 Some Properties of the Group IIIA(3) and IIIB(13) Elements

Element	Electron Configuration	mp (°C)	Ionic Radius (Å)^a	$E°$ (V)^b
Sc	[Ar]$3d^14s^2$	1540	0.89	−1.88
Y	[Kr]$4d^15s^2$	1500	1.04	−2.37
La	[Xe]$5d^16s^2$	920	1.15	−2.52
Ac^c	[Rn]$6d^17s^2$	1050	1.11	−2.6
Al	[Ne]$3s^23p^1$	660	0.68	−1.66
Ga	[Ar]$3d^{10}4s^24p^1$	30	0.76	−0.53
In	[Kr]$4d^{10}5s^25p^1$	160	0.94	−0.34
Tl	[Xe]$4f^{14}5d^{10}6s^26p^1$	300	1.03	+0.72

^{a}For M^{3+}.

^{b}For $M^{3+}(aq) + 3\,e^- = M(s)$.

^{c}Isotopes are all radioactive.

The +1 state becomes progressively more stable as the group is descended, and for Tl the Tl^I–Tl^{III} relationship is a dominant factor of the chemistry. The occurrence of an oxidation state two units below the group valence is sometimes attributed to the *inert pair effect,* which first makes itself evident here. It could be considered to apply in the low reactivity of Hg, but it is more pronounced still in Groups IVB(14) and VB(15). The term refers to the resistance of a pair of *s* electrons to be lost or to participate in covalent bond formation. Thus Hg is difficult to oxidize, allegedly because it contains only an inert pair $(6s^2)$, Tl forms Tl^I rather than Tl^{III} because of the inert pair in the valence shell $(6s^26p)$, and so on. The concept of the inert pair tells us little, if anything, about the ultimate reasons for the stability of lower oxidation states. It is a useful label.

Group IVB(14)

Some properties of the Group IVB(14) elements are given in Table 8-4. Note that we restrict our attention to Group IVB(14), since Group IVA(4) comprises the transition metals Ti, Zr, and Hf, whose chemistry we shall consider separately. This pattern holds true for the remaining Groups VB(15)–VIIB(17).

There is no more striking an example of the enormous discontinuity in properties between the elements of the first and second short periods (followed by a relatively smooth change toward metallic character for the remaining members of the group) than that provided by Group IVB(14). Carbon is nonmetallic, as is silicon, but little of the chemistry of silicon can be inferred from that of carbon. Germanium is much like silicon, although it shows much more metallic behavior in its chemistry. Tin and lead are metals, and both have some metal-like chemistry, especially in the divalent state.

The main chemistry in the IV oxidation state for all the elements is essentially one that involves covalent bonds and molecular compounds. Typical examples are $GeCl_4$ and $PbEt_4$. There is a decrease in the tendency to *catenation,* which is a feature of carbon chemistry, in the order C ≫ Si > Ge ≃ Sn ≃ Pb. This is partly due to the diminishing strength of the C—C, Si—Si, and the like, bonds (Table 8-4). Generally, the strengths of covalent bonds to other atoms decrease in going from C to Pb.

Now let us look at the *divalent state.* Although in CO the *oxidation* state of C is *formally* taken to be two, this *is* only a formalism and carbon uses more than two valence electrons in bonding. True divalence is found only in *carbenes* (e.g., $:CF_2$), and these species are very reactive due to the accessibility of the sp^2 hybridized lone pair. The divalent compounds of the other Group IVB(14) ele-

Table 8-4   Some Properties of Group IVB(14) Elements

Element	Electron Configuration	mp (°C)	Covalent Radius (Å)	Self-Bond Energy (kJ mol⁻¹)
C	$[He]2s^22p^2$	>3550	0.77	356
Si	$[Ne]3s^23p^2$	1410	1.17	210–250
Ge	$[Ar]3d^{10}4s^24p^2$	940	1.22	190–210
Sn	$[Kr]4d^{10}5s^25p^2$	232	1.40	105–145
Pb	$[Xe]4f^{14}5d^{10}6s^26p^2$	327	1.44	—

ments can be regarded as carbene-like in the sense that they are angular with a lone pair and can readily undergo an oxidative addition reaction (see also Chapter 30) to give two new bonds to the element, for example,

$$
\begin{array}{c}
R \\
\diagdown \\
 C^{II}: + X - Y = \\
\diagup \\
R
\end{array}
\qquad
\begin{array}{c}
R \quad\; X \\
\diagdown\!\!\diagup \\
\overset{IV}{C} \\
\diagup\!\!\diagdown \\
R \quad\; Y
\end{array}
\tag{8-11.1}
$$

The increase in stability of the divalent state cannot be attributed to ionization energies as they are very similar in all cases. Factors that doubtless govern the relative stabilities are (a) promotion energies, (b) bond strengths in covalent compounds, and (c) lattice energies in ionic compounds.

For CH_4, the factor that stabilizes CH_4 relative to $CH_2 + H_2$ despite the much higher promotional energy required in forming CH_4 is the great strength of C—H bonds. If we now have a series of reactions

$$
MX_2 + X_2 = MX_4 \tag{8-11.2}
$$

in which the M—X bond energies are decreasing, as they do from Si $\rightarrow$ Pb, then it is possible that bond energy may become too small to compensate for the M^{II}—M^{IV} promotion energy, and MX_2 becomes the more stable compound.

The change in this group is shown by the reactions:

$$
GeCl_2 + Cl_2 = GeCl_4 \qquad \text{(Very rapid at 25 °C)} \tag{8-11.3}
$$

$$
SnCl_2 + Cl_2 = SnCl_4 \qquad \text{(Slow at 25 °C)} \tag{8-11.4}
$$

$$
PbCl_2 + Cl_2 = PbCl_4 \qquad \text{(Only under forcing conditions)} \tag{8-11.5}
$$

In addition, $PbCl_4$ decomposes readily, while $PbBr_4$ and PbI_4 do not exist, probably because of the reducing power of Br^- and I^-.

It is difficult to give any rigorous argument on lattice energy effects, since there is no evidence for the existence of M^{4+} ions and Pb^{2+} ions are found in only a few compounds.

Group VB(15)

Some properties of the Group VB(15) elements are given in Table 8-5. Like nitrogen, phosphorus is essentially covalent in all its chemistry, but arsenic, antimony, and bismuth show increasing tendencies to cationic behavior. Although electron gain to achieve the electronic structure of the next noble gas is con-

Table 8-5   Some Properties of Group VB(15) Elements

Element	Electron Configuration	mp (°C)	Covalent Radius (Å)	Ionic Radius (Å)
P	$[Ne]3s^23p^3$	44	1.10	2.12 (P^{3-})
As	$[Ar]3d^{10}4s^24p^3$	814 (36 atm)	1.21	—
Sb	$[Kr]4d^{10}5s^25p^3$	603	1.41	0.90 (Sb^{3+})
Bi	$[Xe]4f^{14}5d^{10}6s^26p^3$	271	1.52	1.17 (Bi^{3+})

ceivable (as in N^{3-}), considerable energies are involved so that anionic compounds are rare. Similarly, loss of valence electrons is difficult because of high ionization energies. There are no +5 ions and even the +3 ions are not simple, being SbO^+ and BiO^+. Bismuth trifluoride (BiF_3) seems predominantly ionic.

The increasing metallic character is shown by the oxides that change from acidic for phosphorus to basic for bismuth, and by halides that have increasing ionic character.

Group VIB(16)

Table 8-6 gives some properties of the Group VIB(16) elements. The atoms of this group form compounds that feature:

1. The chalcogenide ions (e.g., S^{2-} or Se^{2-}) in salts of highly electropositive elements.
2. Two electron-pair bonds, as in H_2S or $SeCl_2$.
3. Anions containing one bond, as in HS^-.
4. Monocations containing three covalent bonds, as in sulfonium cations (R_3S^+).
5. Compounds in which the Group VIB(16) element has the IV or VI oxidation state, with four, five, or six covalent bonds, as for $SeCl_4$, SeF_5^-, and TeF_6.

We have already pointed out that, from top to bottom in the group, atomic size increases and electronegativity decreases. Also, the general trend down the group is for

1. Decreasing stability of the hydrides (H_2E).
2. Increasing metallic character of the elements themselves.
3. Increasing tendency to form anionic complexes such as $SeBr_6^{2-}$, $TeBr_6^{2-}$, and PoI_6^{2-}.

Group VIIB(17)

Some properties of the Group VIIB(17) elements are given in Table 8-7. The halogen atoms are only one electron short of the noble gas configuration, and the elements form the anion X^- or a single covalent bond. Their chemistry is completely nonmetallic. The changes in behavior with increasing size are progressive and, with the exception of the Li–Cs group, there are closer similarities within this group than in any other in the periodic table.

Table 8-6 Some Properties of Group VIB(16) Elements

Element	Electron Configuration	mp (°C)	Covalent Radius (Å)	Ionic (X^{2-}) Radius (Å)
S	$[Ne]3s^23p^4$	119	1.03	1.70
Se	$[Ar]3d^{10}4s^24p^4$	217	1.17	1.84
Te	$[Kr]4d^{10}5s^25p^4$	450	1.37	2.07
Po	$[Xe]4f^{14}5d^{10}6s^26p^4$	254	—	2.30

Table 8-7 Some Properties of Group VIIB(17) Elements

Element	Electron Configuration	mp (°C)	bp (°C)	Radius X^- (Å)	Covalent Radius (Å)
F	$[He]2s^2 2p^5$	−233	−118	1.19	0.54
Cl	$[Ne]3s^2 3p^5$	−103	−34.6	1.67	0.97
Br	$[Ar]3d^{10}4s^2 4p^5$	−7.2	58.8	1.82	1.14
I	$[Kr]4d^{10}5s^2 5p^5$	113.5	184.3	2.06	1.33
At^a	$[Xe]4f^{14}5d^{10}6s^2 6p^5$	—	—	—	—

^{a}All isotopes are radioactive with short half-lives.

The halogens can form compounds in higher formal oxidation states, mainly in halogen fluorides, such as ClF_3, ClF_5, BrF_5, and IF_7 and oxo compounds.

No evidence exists for cationic behavior with ions of the type X^+. However, Br_2^+, I_2^+, Cl_3^+, and Br_3^+ and several iodine cations are known. When a halogen forms a bond to another atom more electronegative than itself (e.g., ICl) the bond will be polar with a positive charge on the heavier halogen.

8-12 The Transition Elements of the *d* and *f* Blocks

The transition elements may be strictly defined as those that, *as elements*, have partly filled *d* or *f* shells. We adopt a broader definition and also include elements that have partly filled *d* or *f* shells *in compounds*. This means that we treat the *coinage metals*, Cu, Ag, and Au, as transition metals, since Cu^{II} has a $3d^9$ configuration, Ag^{II} has a $4d^9$ configuration, and Au^{III} has a $5d^8$ configuration. Appropriately, we also consider these elements as transition elements because their chemical behavior is quite similar to that of other transition elements.

There are thus 61 transition elements, counting the heaviest ones through atomic number 109. Those through number 104 have certain common properties:

1. They are all metals.
2. They are practically all hard, strong, high-melting, high-boiling metals that conduct heat and electricity well.
3. They form alloys with one another and with other metallic elements.
4. Many of them are sufficiently electropositive to dissolve in mineral acids, although a few are "noble"; that is, they have such low electrode potentials that they are unaffected by simple acids.
5. With very few exceptions, they exhibit variable valence, and their ions and compounds are colored in one if not all oxidation states.
6. Because of partially filled shells they form at least some paramagnetic compounds.

This large number of transition elements is subdivided into three main groups: (1) the main transition elements or *d*-block elements, (2) the lanthanide elements, and (3) the actinide elements.

The main transition group or *d* block includes those elements that have partially filled *d* shells only. Thus, the element Sc, with the outer-electron configuration $4s^2 3d$, is the lightest member. The eight succeeding elements (the *first transition series*) Ti, V, Cr, Mn, Fe, Co, Ni, and Cu, all have partly filled $3d$ shells either in the ground state of the free atom (all except Cu) or in one or more of their chemically important ions (all except Sc). At Zn the configuration is $3d^{10}4s^2$, and this element forms no compound in which the $3d$ shell is ionized, nor does this ionization occur in any of the next nine elements. It is not until we come to yttrium, with a ground-state outer-electron configuration $5s^2 4d$, that we meet the next transition element. The following eight elements, Zr, Nb, Mo, Tc, Ru, Rh, Pd, and Ag, all have partially filled $4d$ shells, whether in the free element (all but Ag) or in one or more of the chemically important ions (all but Y). This group of nine elements constitutes the *second transition series.*

Again, a sequence of elements follows in which there are never d-shell vacancies under chemically significant conditions until we reach the element La, with an outer-electron configuration in the ground state of $6s^2 5d$. Now, if the pattern we have observed twice before were to be repeated, there would follow 8 elements with enlarged, but not complete, sets of $5d$ electrons. This does not happen, however. The $4f$ shell now becomes slightly more stable than the $5d$ shell and, through the next 14 elements, electrons enter the $4f$ shell until, at Lu, it becomes filled. Lutetium thus has the outer-electron configuration $4f^{14}5d6s^2$. Since both La and Lu have partially filled d shells and no other partially filled shells, it might be argued that both of them should be considered as d-block elements. For chemical reasons, however, it would be unwise to classify them in this way, since all of the 15 elements La ($Z = 57$) through Lu ($Z = 71$) have very similar chemical and physical properties, those of La being in a sense prototypal; hence, these elements are called the *lanthanides.*

The shielding of one f electron by another from the effects of the nuclear charge is quite weak because of the shapes of the f orbitals. Hence, with increasing atomic number and nuclear charge, the effective nuclear charge experienced by each $4f$ electron increases. This increase causes a shrinkage in the radii of the atoms or ions as one proceeds from La to Lu (see Table 26-1). This accumulation of successive shrinkages is called the *lanthanide contraction.* It has a profound effect on the radii of subsequent elements, which are smaller than might have been anticipated from the increased mass. Thus Zr^{4+} and Hf^{4+} have almost identical radii despite the atomic numbers of 40 and 72, respectively.

For practical purposes, the *third transition series* begins with Hf, having the ground-state outer-electron configuration $6s^2 5d^2$, and embraces the elements Ta, W, Re, Os, Ir, Pt, and Au, all of which have partially filled $5d$ shells in one or more chemically important oxidation states as well as (excepting Au) in the neutral atom.

Continuing on from Hg, which follows Au, we come via the noble gas Rn and the radioelements Fr and Ra, to Ac, with the outer-electron configuration $7s^2 6d$. Here, by analogy to what happened at La, we might expect that in the following elements electrons would enter the $5f$ orbitals, producing a lanthanide-like series of 15 elements. What actually occurs is not as simple. Although, immediately following La, the $4f$ orbitals become decisively more favorable than the $5d$ orbitals for the electrons entering in the succeeding elements, there is apparently not so great a difference between the $5f$ and $6d$ orbitals until later. Thus, for the elements immediately following Ac, and their ions, there may be electrons in the $5f$ or $6d$ orbitals, or both. Since it appears that later on, after 4 or 5 more electrons have been added to the Ac configuration, the $5f$ orbitals do become the more stable, and since the elements from about Am on show moderately homologous chemical behavior, it has become accepted practice to call the 15 elements beginning with Ac the *actinide elements*.

There is an important distinction, based on electronic structures, between the three classes of transition elements. For the *d*-block elements the partially filled shells are *d* shells: 3*d*, 4*d*, or 5*d*. These *d* orbitals project well out to the periphery of the atoms and ions so that the electrons occupying them are strongly influenced by the surroundings of the ion and, in turn, are able to influence the environments very significantly. Thus, many of the properties of an ion with a partly filled *d* shell are quite sensitive to the number and arrangement of the *d* electrons present. In marked contrast to this, the 4*f* orbitals in the lanthanide elements are rather deeply buried in the atoms and ions. The electrons that occupy them are largely screened from the surroundings by the overlying shells (6*s*, 5*p*) of electrons, and therefore reciprocal interactions of the 4*f* electrons and the surroundings of the atom or the ion are of relatively little chemical significance. This explains why the chemistry of all the lanthanides is so homologous, whereas there are seemingly erratic and irregular variations in chemical properties as one passes through a series of *d*-block elements. The behavior of the actinide elements lies between those of the two types described previously, because the 5*f* orbitals are not as well shielded as are the 4*f* orbitals, although they are not as exposed as are the *d* orbitals in the *d*-block elements.

STUDY GUIDE

Scope and Purpose

We have examined the periodic table and the positions of the elements in it, taking the opportunity to compare and contrast the properties of the elements in their uncombined states, as well as the various tendencies of the elements to form particular types of compounds. The student should especially note the highly useful and systematic manner in which the electron configurations of the elements correlate with the positions of the elements in the periodic table, and with the properties of the elements and their typical compounds.

Study Questions

A. Review

1. Which elements are (at 25 °C and 1 atm pressure)
   (a) gases   (b) liquids   (c) solids melting below 100°C?
2. Why is white phosphorus much more chemically reactive than black phosphorus?
3. Draw the structure of the most stable form of sulfur.
4. Draw the structure for carbon in (a) diamond and (b) graphite. What is the nature of C—C bonding in the two allotropes?
5. Write down the electronic structures of the elements of the first short period, then answer the following questions.
   (a) What is the first ionization energy of Li (approximately)?
   (b) Why does Be not form a 2+ ion in solids?
   (c) Why is there a discontinuity between the ionization energy of N and O?

(d) How do the electron attachment energies vary from Li to F?

(e) Which of the elements can form anions?

6. Why is dinitrogen normally unreactive?

7. What is the octet rule? Why does it apply strictly only to elements of the first short period?

8. What are Lewis acids and Lewis bases? Give two examples of each.

9. Why is there no silicon analog of graphite?

10. What are the main trends in properties of the alkali metals?

11. List the elements of Groups IIA(2) and IIB(12). Compare their main chemical features.

12. Give the electronic structures of

$$Sc \quad and \quad Ti$$
$$Y \quad and \quad Zr$$
$$La \quad and \quad Hf$$

Why are there 14 other elements between La and Hf?

13. How do the following elements attain the noble gas configuration?

(a) N   (b) S

14. Why are Cu, Ag, and Au considered as transition metals?

15. List the common features of transition metals.

16. What are the main groups of transition metals? Write out their names and give the electronic structures of the first, the middle, and the last.

17. What is an icosahedron? For which element is it the most characteristic structural feature?

18. What are the principal properties and structural types of the metals?

19. On what electronic processes does the chemistry of hydrogen depend? Explain.

20. Why is carbon unique in forming chains of single bonds in compounds?

21. What is the lanthanide contraction and what is its main effect?

22. What are the actinide elements and what relation do they bear to the lanthanide elements?

B. Additional Exercises

1. Use MO theory to explain the bonding in N_2, O_2, and F_2. Why is oxygen paramagnetic?

2. Why is the bond energy of F_2 much less than that of Cl_2?

3. Correlate the Lewis diagrams in the compounds "BH_3," CH_4, NH_3, OH_2, and HF with their chemistries.

4. Predict the products of the following acid–base reactions:

(a) $BF_3 + F^-$ (e) $Na_2O + H_2O$

(b) $BF_3 + N(CH_3)_3$ (f) $SO_2 + H_2O$

(c) $Ni(CN)_2 + CN^-$ (g) $SO_3 + H_2O$

(d) $AlCl_3 + Cl^-$

5. Make diagrams of the $d\pi$–$p\pi$ bonds in Cl_3PO, Cl_2SO, SO_2, SO_4^{2-}, ClO_2, ClO_4^-, and PO_4^{3-}. Start by drawing the Lewis diagram for each and then designate the $p\pi$-donor atoms (and their donor orbitals) and the $d\pi$-acceptor atoms (and their acceptor orbitals).

6. Why is CH_2 unstable while $PbCl_2$ is stable? Compare also the stabilities of $GeCl_2$ and $SnCl_2$.

7. Why are the chemical consequences of partially filled d orbitals so much more pronounced for the d-block elements than the consequences of partially filled f orbitals

for the *f*-block elements?

8. Preview the material of Sections 12-3 and 13-3, and use this information together with the material of Chapter 8 to make a list of periodic trends in chemical properties among the **oxides** of the elements of Group IIIB(13).

9. Repeat Question 8 in part B for the **halides** of the elements of Group IIIB(13), using Chapter 8 and the material of Sections 12-4 and 13-4.

10. Repeat Question 8 in part B for the **oxides** of the elements of Group IVB(14), using the material of Chapter 8 and Sections 14-3, 14-4, and 15-5.

11. Repeat Question 8 in part B for the **hydrides** of the elements of Group VB(15), using the material of Chapter 8 plus that of Sections 16-4 and 17-3.

12. Repeat Question 8 in part B for the **halides** of the elements of Group VB(15), using information from Chapter 8 plus Sections 16-8 and 17-4.

SUPPLEMENTARY READING

Baum, R. M., "Flood of Fullerene Discoveries Continues," *Chem. Eng. News,* **June 1, 1992,** 25–33.

Cotton, S. A. and Hart, F. A., *The Heavy Transition Elements,* Wiley, New York, 1975.

Cowley, A. H., "Stable Compounds with Double Bonding Between the Heavier Main Group Elements," *Acc. Chem. Res.,* **1984,** *17,* 386.

Cox, P. A., *The Elements. Their Origin, Abundance and Distribution,* Oxford University Press, New York, 1989.

Donohue, J., *The Structures of the Elements,* Wiley, New York, 1974.

Emsley, J., *The Elements,* 2nd ed., Clarendon Press, Oxford, 1989.

Hammond, G. S. and Kuck, V. J., Eds., "Fullerenes. Synthesis, Properties and Chemistry of Large Carbon Clusters," ACS Symposium Series, American Chemical Society, Washington DC, 1992.

Hermann, G., "Synthesis of the Heaviest Chemical Elements—Results and Perspectives," *Ang. Chem. Int. Ed. Eng.,* **1988,** *27,* 1417–1592.

Hoffman, D. C., "The Heaviest Elements," *Chem. and Eng. News,* May 2, **1994,** p.24.

Kroto, H. W., "C$_{60}$: Buckminsterfullerene, The Celestial Sphere That Fell to Earth," *Angew. Chem. Int. Ed. Eng.,* **1992,** *31,* 111–246.

Kroto, H. W., Allaf, A. W., and Balm, S. P., "C$_{60}$: Buckminsterfullerene," *Chem. Rev.,* **1991,** *91,* 1213–1235.

McLafferty, F. W., Ed., "Special Issue on Buckminsterfullerenes," *Acc. Chem. Res.,* **1992,** *26(3),* 98–175.

Parish, R. V., *The Metallic Elements,* Longman, New York, 1977.

Powell, P. and Timms, P., *The Chemistry of the Non-Metals,* Chapman & Hall, London, 1974.

Raabe, G. and Michl, J., "Multiple Bonds to Silicon," *Chem. Rev.,* **1985,** *85,* 419.

Seaborg, G. T. and Loveland, W. D., *The Elements Beyond Uranium,* Wiley-Interscience, New York, 1990.

Smalley, R. E., in "Atomic and Molecular Chemistry," E. R. Bernstein, Ed., Elsevier, Amsterdam, 1990.

Steudel, R., *Chemistry of the Non-Metals,* Walter de Gruyter, Berlin, 1977.

Troyer, R., "The Third Form of Carbon; A New Era in Chemistry," *Interdiscip. Sci. Rev.,* **1992,** *17,* 161–170.

Part 2

THE MAIN GROUP ELEMENTS

Chapter 9

HYDROGEN

9-1 Introduction

Hydrogen (not carbon) forms more compounds than any other element. For this and other reasons, many aspects of hydrogen chemistry are treated elsewhere in this book. Protonic acids and the aqueous hydrogen ion have already been discussed in Chapter 7. This chapter examines certain topics that most logically should be considered at this point.

Three isotopes of hydrogen are known: ^1H, ^2H (deuterium or D), and ^3H (tritium or T). Although isotope effects are greatest for hydrogen, justifying the use of distinctive names for the two heavier isotopes, the chemical properties of H, D, and T are essentially identical, except in matters such as rates and equilibrium constants of reactions. The normal form of the element is the diatomic molecule; the various possibilities are H_2, D_2, T_2, HD, HT, DT.

Naturally occurring hydrogen contains 0.0156% deuterium, while tritium (formed continuously in the upper atmosphere in nuclear reactions induced by cosmic rays) occurs naturally in only minute amounts that are believed to be of the order of 1 in 10^{17} and is radioactive (β^-, 12.4 years).

Deuterium, as D_2O, is separated from water by fractional distillation or electrolysis and is available in ton quantities for use as a moderator in nuclear reactors. Deuterium oxide is also useful as a source of deuterium in deuterium-labeled compounds.

Molecular hydrogen is a colorless, odorless gas (fp 20.28 K) virtually insoluble in water. It is most easily prepared by the action of dilute acids on metals such as Zn or Fe, and by electrolysis of water.

Industrially, hydrogen is obtained by the so-called steam re-forming of methane or light petroleum over a promoted nickel catalyst at about 750 °C. The process is complicated, but the main reaction, illustrated with methane, is given in Reaction 9-1.1.

$$CH_4 + H_2O = CO + 3\ H_2 \tag{9-1.1}$$

The mixtures of CO and H_2 that are produced in Reaction 9-1.1 are called synthesis gas, or "syngas." Synthesis gas can now also be produced from trash, sewage, sawdust, scrap wood, newspapers, and so on. The production of syngas from coal is termed "coal gasification."

When desired, the proportion of hydrogen in synthesis gas mixtures can be increased by use of Reaction 9-1.2.

$$CO + H_2O = CO_2 + H_2 \qquad \Delta H = -42\ kJ\ mol^{-1} \tag{9-1.2}$$

This is the water–gas shift reaction, which proceeds either at relatively high temperatures (280–350 °C) using an iron–chromate type catalyst, or at lower temperatures using copper-containing catalysts. The carbon dioxide side product is removed by scrubbing with arsenite solution or ethanolamine, from which it is recovered for other uses, such as the manufacturing of dry ice. The remaining small amounts of CO and CO_2 impurities (which may act as unwanted poisons in subsequent chemical uses of the hydrogen) are catalytically converted to methane (which is usually innocuous) according to Reactions 9-1.3 and 9-1.4.

$$CO + 3\ H_2 = CH_4 + H_2O \tag{9-1.3}$$

$$CO_2 + 4\ H_2 = CH_4 + 2\ H_2O \tag{9-1.4}$$

In addition to its use for hydrogen and carbon monoxide production, synthesis gas is used directly in large-scale catalyzed syntheses of methanol (Chapter 30), and of higher alcohols (e.g., ethanol), as shown in Reaction 9-1.5:

$$2\ CO + 4\ H_2 \longrightarrow CH_3CH_2OH + H_2O \tag{9-1.5}$$

or 2-ethylhexanol.

Hydrogen is not an exceptionally reactive element at low temperatures, because the bond dissociation energy of the molecule is considerably endothermic.

$$H_2 = 2\ H \qquad \Delta H = 434.1\ kJ\ mol^{-1} \tag{9-1.6}$$

Hydrogen burns in air to form water, and it will react explosively with oxygen and with halogens under certain conditions. At high temperatures, hydrogen gas will reduce many oxides to lower oxides, as in Reaction 9-1.7.

$$2\ MO_2 + H_2 \longrightarrow M_2O_3 + H_2O \tag{9-1.7}$$

It is also useful for the complete reduction of many metal oxides to the metals, as shown in Reactions 9-1.8 through 9-1.10.

$$MO_2 + 2\ H_2 \longrightarrow M + 2\ H_2O \tag{9-1.8}$$

$$M_2O_3 + 3\ H_2 \longrightarrow 2\ M + 3\ H_2O \tag{9-1.9}$$

$$MO + H_2 \longrightarrow M + H_2O \tag{9-1.10}$$

In the presence of iron or ruthenium catalysts at high temperature and pressure, H_2 will react with N_2 to produce NH_3. With electropositive metals and most nonmetals, hydrogen forms hydrides, as we shall discuss in Section 9-6. Hydrogen serves as a reducing or hydrogen-transfer agent for a variety of organic and inorganic substances, but a catalyst is required in most cases. The reduction of alkenes to alkanes by hydrogen over Pt or Ni is a typical example.

9-2   The Bonding of Hydrogen

The chemistry of hydrogen depends mainly on the three electronic processes discussed in Chapter 8: (1) loss of a valence electron to give H^+, (2) acquisition of an electron to give H^-, and (3) formation of a single covalent bond, as in CH_4.

However, hydrogen has additional unique bonding features. The nature of the proton and the complete absence of any shielding of the nuclear charge by electron shells allow other forms of chemical activity that are either unique to hydrogen or particularly characteristic of it. Some of these are the following, which we shall discuss in some detail subsequently.

1. The formation of numerous compounds, often nonstoichiometric, with metallic elements. These compounds are generally called hydrides but cannot be regarded as simple saline hydrides (Section 9-6).

2. The formation of hydrogen bridge bonds in electron-deficient compounds (e.g., Structure 9-I) and transition metal complexes (e.g., Structure 9-II).

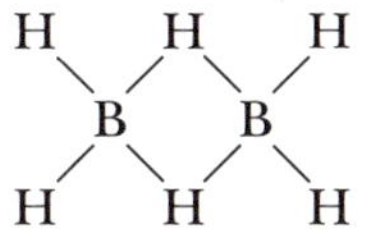

9-I 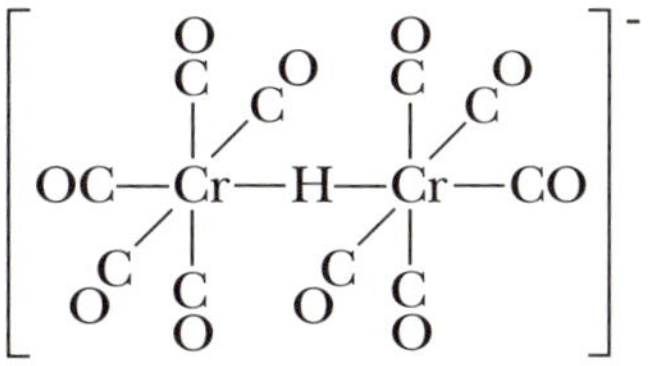   9-II

A classic example of bridge bonds is provided by diborane (Structure 9-I) and related compounds (Chapter 12). The electronic nature of such bridge bonds was discussed in Chapter 3.

3. The hydrogen bond is important not only because it is essential to an understanding of much other hydrogen chemistry, but also because it is one of the most intensively studied examples of intermolecular attraction. Hydrogen bonds generally dominate the chemistry of water, aqueous solutions, hydroxylic solvents, and OH-containing species and are of crucial importance in biological systems, since they are responsible for the linking of polypeptide chains in proteins and the base pairs of nucleic acids.

9-3 The Hydrogen Bond

When hydrogen is bonded to another atom X, mainly F, O, N, or Cl, such that the X—H bond is quite polar, with H bearing a partial positive charge, it can interact with another negative or electron-rich atom Y, to form what is called a hydrogen bond (H bond), written as

$$X—H\text{-}\text{-}\text{-}Y$$

Although the details are subject to variation, and controversy, it is generally considered that typical hydrogen bonds are due largely to electrostatic attraction of H and Y. The X—H distance becomes slightly longer, but this bond remains essentially a normal two-electron bond. The H---Y distance is generally much longer than that of a normal covalent H—Y bond.

In the case of the very strongest hydrogen bonds, the $X \cdots Y$ distance becomes quite short and the X—H and Y---H distances come close to being equal. In these cases there are presumably covalent and electrostatic components in both the X—H and Y—H bonds.

Experimental evidence for hydrogen bonding came first from comparisons of the physical properties of hydrogen compounds. The apparently abnormally high boiling points of NH_3, H_2O, and HF (Fig. 9-1) are classic examples which imply association of these molecules in the liquid phase. Other properties such as heats of vaporization provided further evidence for association. Although physical properties reflecting association are still a useful tool in detecting hydrogen bonding, the deeper understanding of H bonds and the determination of their parameters comes from X-ray or neutron diffraction of solids, and from other techniques, notably ion cyclotron resonance, NMR, IR, and Raman spectroscopies, and calorimetry.

Structural evidence for hydrogen bonds is provided by the X···Y distances, which are shorter than the expected van der Waals contact when a hydrogen bond exists. For instance, in crystalline $NaHCO_3$ there are four kinds of O···O distances between HCO_3^- ions with values of 3.12, 3.15, 3.19, and 2.55 Å. The first three are about equal to twice the van der Waals radius of oxygen, but the last one indicates a hydrogen bond, O—H---O. When an X—H group enters into hydrogen bonding, the X—H stretching band in the IR spectrum is lowered in frequency, broadened, and increased in integrated intensity.

The enthalpies of hydrogen bonds are relatively small in most instances: 20–30 kJ mol^{-1}, as compared with covalent bond enthalpies of 200 kJ mol^{-1}, and up. Nevertheless, these weak bonds can have a profound effect on the properties and chemical reactivity of substances in which they occur. This effect is clearly seen from Fig. 9-1, where water would boil at about −100 °C instead of +100 °C if hydrogen bonds did not play their role. Obviously, life itself (as we know it) depends on the existence of such weak hydrogen bonds.

However, there are also strong and very strong hydrogen bonds mainly involving O and F atoms in cations and anions. The enthalpies are in the ranges 50–100 and greater than 100 kJ mol^{-1}, respectively. The best example of a very strong, short bond is that in the FHF$^-$ anion, where the proton is centered between the F atoms that are only 2.26-Å apart. Similar very short bonds are found in [HOHOH]$^-$, HCl_2^-, $H_5O_2^+$, $(CH_3OH)_2H^+$, and $H_3F_2^-$. An unusual example involving nitrogen is the anion $[(OC)_5CrCN—H—NCCr(CO)_5]^-$.

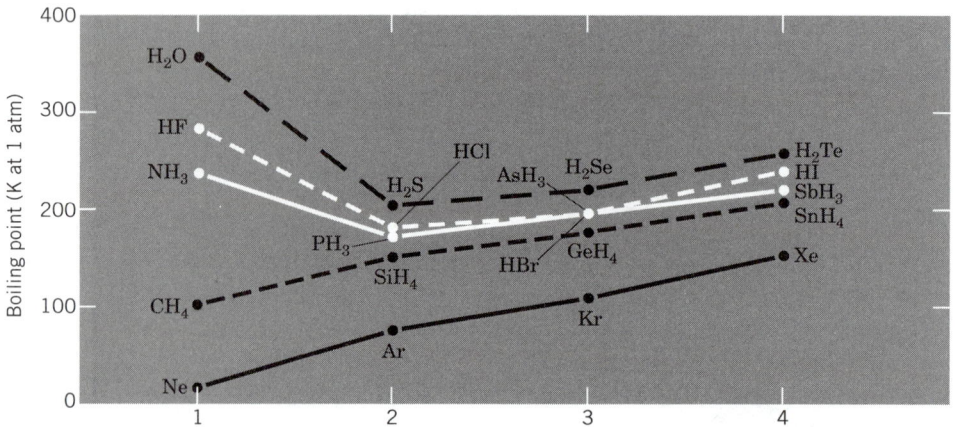

Figure 9-1   Periodic trends in the boiling points of some molecular hydrides, with a comparison to the noble gases.

Finally, we can note a rather similar, though different, type of interaction between hydrogen atoms bound to carbon atoms of ligands (e.g., CH_3 and other alkyl type ligands) and the transition metals to which these ligands are complexed, namely, C—H$\cdots$M. Such bonds are called agostic, and are identified by a sometimes significant lengthening of the C—H bond. A few additional examples of normal hydrogen bonds involving carbon (C—H$\cdots$C, C—H$\cdots$M, and C—H$\cdots$Cl) and N—H$\cdots$M have also been characterized.

9-4   Ice and Water

The structure of water is very important since it is the medium in which so much chemistry, including the chemistry of life, takes place. The structure of ice is of interest for clues about the structure of water. There are nine known modifications of ice, the stability of each depending on temperature and pressure. The ice formed in equilibrium with water at 0 °C and 1 atm is called ice I and has the structure shown in Fig. 9-2. There is an infinite array of oxygen atoms, each tetrahedrally surrounded by four others with hydrogen bonds linking each pair.

The structural nature of liquid water is still controversial. The structure is not random, as found in liquids consisting of more-or-less spherical nonpolar molecules; instead, it is highly structured owing to the persistence of hydrogen bonds. Even at 90 °C only a few percent of the water molecules appear not to be hydrogen bonded. Still, there is considerable disorder, or randomness, as befits a liquid.

In an attractive, though not universally accepted, model of liquid water the liquid consists at any instant of an imperfect network, very similar to the network of ice I, but differing in that (a) some interstices contain water molecules that do not belong to the network but, instead, disturb it; (b) the network is patchy and does not extend over long distances without breaks; (c) the short-range ordered regions are constantly disintegrating and re-forming (they are "flickering clus-

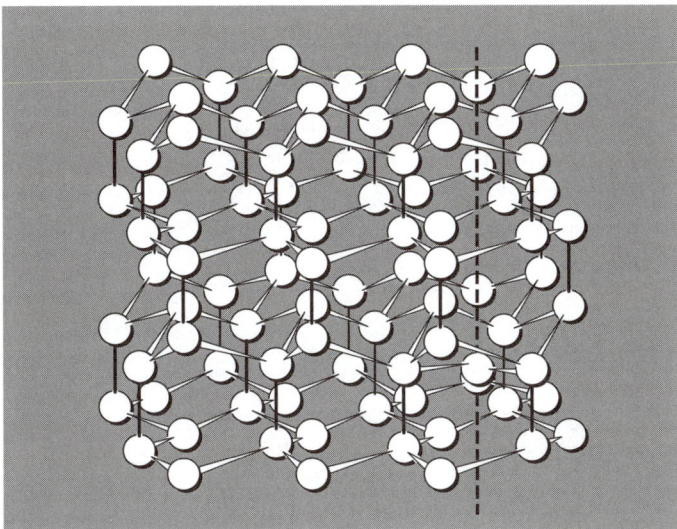

Figure 9-2   The structure of ice I. Only the oxygen atoms are shown. The O$\cdots$O distances are 2.75 Å.

ters"); and (d) the network is slightly expanded compared with ice I. The fact that water has a slightly higher density than ice I may be attributed to the presence of enough interstitial water molecules to offset the expansion and disordering of the ice I network. This model of water receives support from X-ray scattering studies.

9-5   Hydrates and Water Clathrates

Solids that consist of molecules of a compound together with water molecules are called *hydrates*. The majority contain discrete water molecules either bound to cations through the oxygen atom or bound to anions or other electron-rich atoms through hydrogen bonds, or both, as is shown in Fig. 9-3. In many cases when the hydrate is heated above 100 °C, the water can be driven off leaving the *anhydrous* compound. However, there are many cases where something other than, or in addition to, water is driven off. For example, many hydrated chlorides give off HCl and a basic or oxo chloride is left.

$$ScCl_3 \cdot 6H_2O \xrightarrow{\text{Heat}} ScOCl + 2\,HCl(g) + 5\,H_2O(g) \quad (9\text{-}5.1)$$

Water also forms materials called *gas hydrates*, which are actually a type of inclusion or clathrate compound. A clathrate (from the Latin clathratus, meaning "enclosed or protected by crossbars or gratings") is a substance in which one component, the host molecule, crystallizes with an open structure that contains holes or channels in which atoms or small molecules of the second component, the guest molecule, can be trapped. Many substances other than water, for example *p*-quinol, $C_6H_4(OH)_2$, urea, and $Fe(acac)_3$, can form inclusion compounds, and a great variety of small molecules can be trapped.

There are two common gas hydrate structures and both are cubic. In one, the unit cell contains 46 molecules of H_2O connected to form six medium-size and two small cages. This structure is adopted when atoms (Ar, Kr, or Xe) or relatively small molecules (e.g., Cl_2, SO_2, or CH_3Cl) are used, generally at pressures greater than 1 atm for the gases. Complete filling of only the medium cages by atoms or molecules (X) would give a composition $X \cdot 7.67\ H_2O$, while complete filling of all eight cages would lead to $X \cdot 5.76\ H_2O$. In practice, complete filling of all cages of one or both types is seldom attained. These formulas, therefore, represent limiting rather than observed compositions; for instance, the usual formula for chlorine hydrate is $Cl_2 \cdot 7.30\ H_2O$. The second structure, often

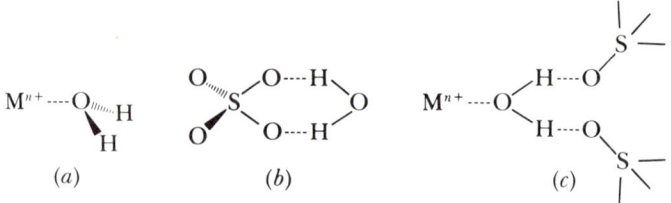

Figure 9-3   Three principal ways in which water molecules are bound in hydrates: (*a*) through oxygen to cations, (*b*) through hydrogen to anions, and (*c*) a combination of the preceding two.

formed in the presence of larger molecules of liquid substances (and thus sometimes called the *liquid hydrate* structure) such as chloroform and ethyl chloride, has a unit cell containing 136 water molecules with 8 large cages and 16 smaller ones. The anesthetic effect of substances such as chloroform may be due to the formation of liquid hydrate crystals in brain tissue. The methane clathrate occurs in vast quantities under arctic permafrost.

A third notable class of clathrate compounds, *salt hydrates,* is formed when tetraalkylammonium or sulfonium salts crystallize from aqueous solution with high water content, for example, $[(C_4H_9)_4N]C_6H_5CO_2 \cdot 39.5H_2O$ or $[(C_4H_9)_3S]F \cdot 20 H_2O$. The structures of these substances are very similar to the gas and liquid hydrate structures in a general way, although they are different in detail. These structures consist of frameworks constructed mainly of hydrogen-bonded water molecules, but apparently also include the anions (e.g., F^-) or parts of the anions (e.g., the O atoms of the benzoate ion). The cations and parts of the anions (e.g., the C_6H_5C part of the benzoate ion) occupy cavities in an incomplete and random way.

9-6 Hydrides

Although all compounds of hydrogen could be termed hydrides, not all hydrogen-containing compounds display "hydridic" character. In general, **hydridic substances** are those that either react as hydride ion (H^-) donors or clearly contain anionic hydrogen. Thus it is necessary to distinguish hydridic substances (e.g., NaH) from those that are either neutral (e.g., CH_4) or acidic (e.g., HCl). This distinction between hydrogen-containing substances that are hydridic, neutral, or acidic runs roughly parallel to the bonding considerations mentioned in Section 9-2; that is, hydrogen may be bound in its compounds essentially as (or serve, on reaction, as a source of) H^-, $H\cdot$, or H^+, respectively. It is also, at times, convenient to classify the compounds of hydrogen as being

1. Either ionic or covalent.
2. Either stoichiometric or nonstoichiometric.
3. Either binary or complex.

Among the strictly binary hydrides, Figure 9-4 gives a general idea of the types of compounds formed by hydrogen.

H																	He
Li	Be											B	C	N	O	F	Ne
Na	Mg											Al	Si	P	S	Cl	Ar
K	Ca	Sc	Ti	V	Cr	Mn	Fe	Co	Ni	Cu	Zn	Ga	Ge	As	Se	Br	Kr
Rb	Sr	Y	Zr	Nb	Mo	Tc	Ru	Rh	Pd	Ag	Cd	In	Sn	Sb	Te	I	Xe
Cs	Ba	La–Lu	Hf	Ta	W	Re	Os	Ir	Pt	Au	Hg	Tl	Pb	Bi	Po	At	Rn
Fr	Ra	Ac		U,Pu													
Saline hydrides		Transition metal hydrides								Borderline hydrides		Covalent hydrides					

Figure 9-4 A classification of the binary hydrides. For the transition elements, in addition to the simple binary hydrides, complex molecules or ions containing M—H bonds are also known.

Covalent Hydrides

The principal covalent hydrides of the nontransition elements will be discussed more completely in the appropriate chapters that remain. Briefly, the covalent hydrides include

1. Neutral, binary XH_4 compounds of Group IVB(14), for example, CH_4.
2. Somewhat basic, binary XH_3 compounds of Group VB(15), for example, NH_3 and PH_3.
3. Weakly acidic or amphoteric, binary XH_2 compounds of Group VIB(16), for example, H_2S and H_2O.
4. Strongly acidic, binary HX compounds of Group VIIB(17), for example, HCl and HI.
5. Numerous covalent hydrides of boron, to be discussed in Chapter 12.
6. Hydridic, complex compounds of hydrogen, two examples of which are $LiAlH_4$ and $NaBH_4$, which serve as powerful reducing agents despite the fact that the Al—H and B—H bonds in these substances are essentially covalent in nature.

The latter two compounds provide an interesting illustration of covalent hydrides that are hydridic. First of all, although the two compounds are ionic (being Li^+ and Na^+ salts), the tetrahedral anions in these salts contain essentially covalent bonds to hydrogen. Furthermore, the tetrahydroaluminate and tetrahydroborate anions are each hydridic, being formed by the action of LiH on Al_2Cl_6 in ether, as in Reaction 9-6.1:

$$8 \, LiH + Al_2Cl_6 \longrightarrow 2 \, LiAlH_4 + 6 \, LiCl \qquad (9\text{-}6.1)$$

and by the action of NaH on diborane, as in Reaction 9-6.2.

$$2 \, NaH + B_2H_6 \longrightarrow 2 \, NaBH_4 \qquad (9\text{-}6.2)$$

Also, each of the above tetrahydro anions is a powerful hydrogen-transfer agent, as shown in Reactions 9-6.3 and 9-6.4.

$$2 \, LiAlH_4 + 2 \, SiCl_4 \longrightarrow 2 \, SiH_4 + 2 \, LiCl + Al_2Cl_6 \qquad (9\text{-}6.3)$$

$$I_2 + 2 \, NaBH_4 \longrightarrow B_2H_6 + 2 \, NaI + H_2 \qquad (9\text{-}6.4)$$

In Reaction 9-6.3, we have the reduction of $SiCl_4$ by $LiAlH_4$ to give silane (SiH_4), whereas in Reaction 9-6.4 we have the classic synthesis of diborane (B_2H_6) by reduction of I_2 using sodium borohydride.

Saline Hydrides

The most electropositive elements, the alkali metals and the larger of the alkaline earth metals, react directly with dihydrogen to form stoichiometric hydrides having considerable ionic character. These compounds are called the saline (saltlike) hydrides. Those of the heavier metals are truly hydridic substances, since they are properly considered to contain metal cations and H^- ions. However, due to the small size and high charge density of the ions of the smaller metals [Be and Mg in Group IIA(2) and Li in Group IA(1)], their hydrides have more covalent character, and BeH_2 is best described as a covalent polymer having Be—H—Be bridges.

The saline hydrides are ionic substances, as shown by the facts that (a) they conduct electricity when molten, and (b) when dissolved and electrolyzed in molten halides, the saline hydrides evolve dihydrogen at the positive electrode (anode), where oxidation of H^- takes place. The ionic nature of the saline hydrides is further indicated by their structures. The ionic radius of H^- lies between that of F^- and Cl^-, and the alkali metal hydrides, LiH to CsH, all adopt the NaCl structure. The structure of MgH_2 is the same as that of rutile (Chapter 4), whereas CaH_2, SrH_2, and BaH_2 adopt a type of $PbCl_2$ structure having a slightly distorted hcp array.

The saline hydrides are all prepared by direct interaction of the metals with elemental hydrogen at 300–700 °C, as shown in Reactions 9-6.5 and 9-6.6.

$$2\,M(\ell) + H_2(g) \longrightarrow 2\,MH(s) \tag{9-6.5}$$

$$M(\ell) + H_2(g) \longrightarrow MH_2(s) \tag{9-6.6}$$

The rates for Reaction 9-6.5 are in the order Li > Cs > K > Na. The products of Reactions 9-6.5 and 9-6.6 are white crystalline solids when they are pure, but are usually gray owing to traces of the metals from which they were made.

All of the saline hydrides decompose thermally to give the metal and hydrogen, although lithium hydride alone is stable to its melting point (688 °C). Also, only LiH is unreactive at moderate temperatures towards oxygen or chlorine. Because of its relative unreactivity, LiH finds practical use only in the synthesis of $LiAlH_4$, as in Reaction 9-6.1.

Since they are hydridic, the saline hydrides (except LiH) are quite reactive with water and air, as shown in Reactions 9-6.7 and 9-6.8.

$$MH(s) + H_2O \longrightarrow H_2(g) + MOH(aq) \tag{9-6.7}$$

$$MH_2(s) + H_2O \longrightarrow H_2(g) + M(OH)_2(aq) \tag{9-6.8}$$

The saline hydrides are powerful reducing or hydrogen-transfer agents, as shown in Reactions 9-6.1 and 9-6.2, as well as by Reactions 9-6.9 through 9-6.11.

$$NaH + B(OCH_3)_3 \longrightarrow Na[HB(OCH_3)_3] \tag{9-6.9}$$

$$4\,NaH + TiCl_4 \longrightarrow Ti^0 + 4\,NaCl + 2\,H_2 \tag{9-6.10}$$

$$NaH + ROH \longrightarrow NaOR + H_2 \tag{9-6.11}$$

Transition Metal Hydrides

The transition metal hydrides are extremely diverse in their structures and properties. The wide variety of transition metal compounds that contain M—H bonds includes stoichiometric binary anions such as $[ReH_9]^{2-}$ and $[FeH_6]^{4-}$; complex stoichiometric substances with essentially covalent bonds to hydrogen, such as $HMn(CO)_5$ (to be discussed in Chapter 28) and $Re_2H_8(PR_3)_4$; as well as non-stoichiometric compounds formed by the direct reaction of hydrogen with various transition metals, as described below.

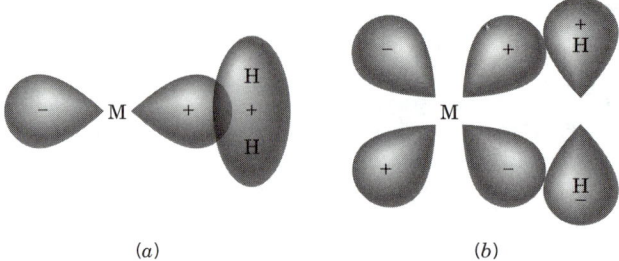

(a)   *(b)*

Figure 9-5   The two components of the metal–dihydrogen bond in so-called "side-on" dihydrogen complexes. (*a*) Donation of electron density from the σ-bonding molecular orbital of dihydrogen into an empty σ orbital of the metal, and (*b*) "back-donation" of π-electron density from a filled *d* orbital of the metal into the σ* antibonding orbital of the H—H linkage. Both components of the M—H$_2$ bond weaken the H—H linkage.

Hydrogen reacts with many transition metals or their alloys on heating to give exceedingly complicated substances. They are black or grayish-black, non-stoichoimetric solids, typical formulas being LaH$_{2.87}$, YbH$_{2.55}$, TiH$_{1.7}$, and ZrH$_{1.9}$. Under conditions of excess hydrogen, limiting compositions may be achieved, but in any given preparation, numerous structural phases may be present, each with its own stoichiometric composition.

A satisfactory theoretical understanding of these substances has not yet been developed. Whether the hydrogen is bound in the metal lattice in its hydridic, protonic, or molecular form is not known. The most straightforward view of the compounds is that hydrogen atoms are located in regular interstices between the metal atoms, and these substances are therefore sometimes termed the *interstitial hydrides.* In this fashion, the element palladium (and to a lesser extent Pt) can absorb very large volumes of hydrogen, and thus can be used to purify hydrogen.

Uniquely, uranium forms a well-defined, stoichiometric hydride by the rapid and exothermic reaction of the metal with hydrogen at 250–300 °C to yield a pyrophoric black powder, as in Reaction 9-6.12.

$$U + \tfrac{3}{2}\,H_2 \longrightarrow UH_3 \tag{9-6.12}$$

Uranium hydride (UH$_3$) is of importance chemically because it is often more suitable for the synthesis of uranium compounds than the metal. For example, with water it yields UO$_2$, whereas with Cl$_2$ and H$_2$S it yields UCl$_4$ and US$_2$, respectively.

Dihydrogen (H$_2$) as a Ligand

Within the past decade, it has been shown that the H$_2$ molecule can behave as a ligand and occupy a place in the coordination sphere around a metal atom. This happens only under special circumstances, with metals in low oxidation states. The H$_2$ molecule takes a "side-on" orientation with respect to the metal, and the bonding is accomplished by a combination of (a) weak donation of the bonding electrons of the H$_2$ molecule to an empty σ orbital of the metal atom and (b) ac-

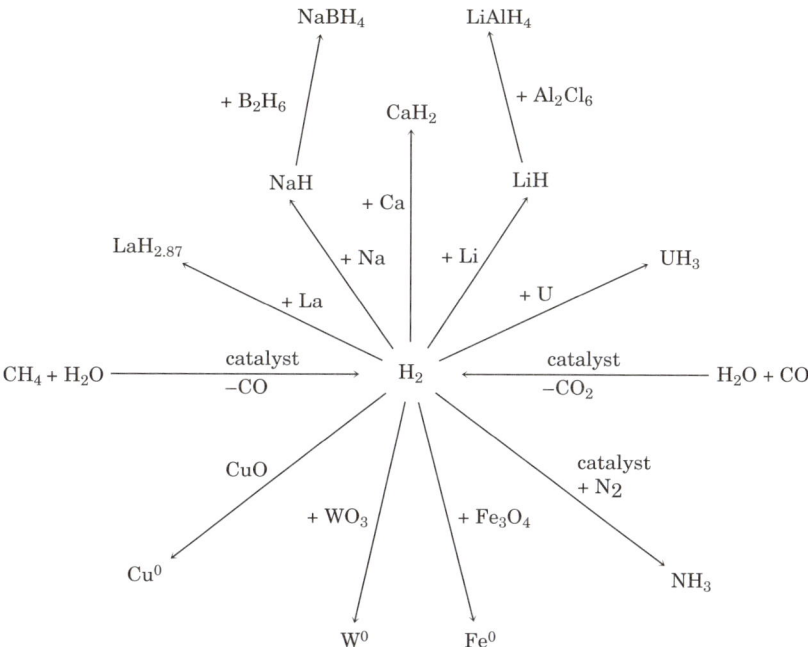

Figure 9-6   Some reactions of hydrogen.

ceptance of electrons from a filled orbital of the metal atom into the σ^* anti-bonding orbital of H_2, as shown in Fig. 9-5. Clearly, each of these parts of the bonding mechanism weakens and lengthens the H—H bond. Thus, unless conditions are very delicately balanced, the system tends toward a conventional dihydride, as represented by Structure 9-IV.

9-7   Reaction Summary

As a study aid, the various reactions of hydrogen are illustrated in Fig. 9-6 Rather than being a comprehensive list of reactions of hydrogen, Fig. 9-6 is intended only to be an overview of the important types of reactions that hydrogen is capable of undergoing.

STUDY GUIDE

Study Questions

A. Review

1. What are the three isotopes of hydrogen called? What are their approximate natural abundances? Which one is radioactive?

2. What is the chief large-scale use for D_2O?

3. What is one thing that helps to explain the relatively low reactivity of elemental hydrogen?

4. What are the three principal electronic processes that lead to formation of compounds by the hydrogen atom?

5. When a hydrogen bond is symbolized by X—H---Y, what do the solid and dashed lines represent? Which distance is shorter?

6. How does hydrogen-bond formation affect the properties of HF, H_2O, and NH_3? Compared with what?

7. What is the usual range of enthalpies of a hydrogen bond?

8. Describe the main features of the structure of ice I. How is the structure of water believed to differ from that?

9. In what two principal ways is water bound in salt hydrates?

10. Can it safely be assumed that whenever a salt hydrate is heated at 100–120 °C the corresponding anhydrous salt will remain?

11. What is the true nature of so-called chlorine hydrate ($Cl_2 \cdot 7.3\ H_2O$)?

12. What is a saline hydride? What elements form them? Why are they believed to contain cations and H^- ions?

13. Define and cite examples of the different types of hydrogen-containing compounds that are discussed in this chapter, listing the distinguishing electronic, structural, and reactive characteristics of each class.

14. Which are the types of metals that react directly with hydrogen to form (a) ionic and (b) interstitial hydrides?

15. Give an explanation of the structural role of water in each of the following types of compounds, together with an example of a specific chemical substance for each type.
    (a) A hydrated compound.
    (b) A hydrous compound.
    (c) A gas hydrate.
    (d) A liquid hydrate.
    (e) A salt hydrate.

16. How could a nonstoichiometric hydride be made? What metal might one use? How could the hydridic character of the product be demonstrated?

B. Additional Exercises

1. Suggest a means of preparing pure HD.

2. It is believed that the shortest H bonds become symmetrical. How must the conventional description (X—H---Y) be modified to cover this situation?

3. Which H bond would you expect to be stronger, and why?
   S—H---O   or   O—H---S

4. Prepare a qualitative Born–Haber cycle to explain why only the most electropositive elements form saline hydrides.

5. Complete and balance the following reactions featuring hydrides:

    (a) $CaH_2 + H_2O \rightarrow$

    (b) $B_2H_6 + NaH \rightarrow$

    (c) $SiCl_4 + LiAlH_4$ to give silane, SiH_4

    (d) $Al_2Cl_6 + LiH$ to give $LiAlH_4$

6. The boiling points of the hydrogen halides follow the trend $HF(20 °C) > HCl(-85 °C) < HBr(-67 °C) < HI(-36 °C)$. Explain.

7. The three different aspects of the chemistry of hydrogen can be illustrated by the reactivity of water with NaH, CH_4, and HCl. Explain.

8. Compare the bonding in "BH_3" and BCl_3. Why is BCl_3 monomeric and "BH_3" dimeric?

9. Suggest a synthesis of H_2Se and H_2S; of $NaBH_4$ and $LiAlH_4$; of HCl and HI; of NaH and CaH_2.

10. Prepare an MO description of the linear and symmetrical hydrogen bond in $[F—H—F]^-$ using the $1s$ atomic orbital on the central hydrogen atom and ligand group orbitals (formed from appropriately oriented $2p$ atomic orbitals) on the two fluorine atoms. Prepare the MO energy-level diagram that accompanies these three MO's and add the proper number of electrons to it. What is the bond order in each F—H half?

11. Finish and balance the following equations:

    (a) $CaH_2 + H_2O$

    (b) $K^0 + C_2H_5OH$

    (c) $KH + C_2H_5OH$

    (d) $UH_3 + H_2O$

    (e) $UH_3 + H_2S$

    (f) $UH_3 + HCl$

    (*Hint:* Dihydrogen is a product of all of these reactions.)

12. Suggest a two-step synthesis of lithium aluminum hydride ($LiAlH_4$), using only elements and Al_2Cl_6. Repeat this for $NaBH_4$, using B_2H_6.

13. Write balanced equations for the reaction of dihydrogen with sodium, B_2H_6, calcium, lithium, nitrogen, oxygen, and uranium.

14. Write balanced equations representing the steam re-forming of ethane, reduction of Fe_2O_3 by hydrogen, reaction of CaH_2 with water, and the water–gas shift reaction.

15. Draw the unit cell for NaH. What is the coordination number of Na^+ in this structure?

16. Review the material of Chapter 4 and draw out a Born–Haber cycle for NaH. After considering each step of the cycle, explain what two steps in the cycle give sodium (and the other alkali metal hydrides) an advantage over other metals in the formation of an ionic hydride as opposed to a covalent hydride.

17. The gallium analog of $LiAlH_4$, namely, $LiGaH_4$, is thermally unstable, decomposing to LiH and elements. Write a balanced equation to represent this. Why do you expect that the same reaction for $LiAlH_4$ is not observed?

SUPPLEMENTARY READING

Alefield, G. and Volkl, J., Eds., *Hydrogen in Metals,* Springer, New York, 1978.

Attwood, J. L., Davies, J. E. D., and McNichol, D. D., Eds., *Inclusion Compounds,* Academic, London, 1991.

Bau, R., Ed., *Transition Metal Hydrides*, ACS Advances in Chemistry Series No. 167, American Chemical Society, Washington, DC, 1978.

Berecz, E. and Balla-Achs, M., *Gas Hydrates*, Elsevier, Amsterdam, 1983.

Cotton, F. A. and Wilkinson, G., *Advanced Inorganic Chemistry, Fifth Edition*, Wiley-Interscience, New York, 1988, Chapter 3.

Evans, E. A., *Tritium and Its Compounds, Second Edition*, Halstead-Wiley, New York, 1974.

Franks, F., Ed., *Water, A Comprehensive Treatise*, Vol. 1, Plenum, New York, 1972.

Hibbert, F. and Emsley, J., "Hydrogen Bonding and Chemical Reactivity," *Adv. Phys. Org. Chem.*, **1990**, *26*, 255.

Libowitz, G. C., *The Solid State Chemistry of Binary Hydrides*, Benjamin, Menlo Park, CA, 1965.

Lutz, H. D., "Bonding and Structure in Solid Hydrates," *Structure and Bonding*, Vol. 69, Springer-Verlag, New York, 1988.

Muetterties, E. L., Ed., *Transition Metal Hydrides*, Dekker, New York, 1971.

Moore, R. A., Ed., *Water and Aqueous Solutions: Structures, Thermodynamics, and Transport Processes*, Wiley, New York, 1972.

Pimentel, G. C. and McClellan, A. L., *The Hydrogen Bond*, Freeman, San Francisco, 1960.

Schuster, P., Ed., *Hydrogen Bonds*, Springer, New York, 1983.

Shaw, B. L., *Inorganic Hydrides*, Pergamon Press, Elmsford, NY, 1967.

Snoeyink, V. L. and Jenkins, D., *Water Chemistry*, Wiley, New York, 1980.

Vinogradov, S. N., *Hydrogen Bonding*, Van Nostrand-Reinhold, New York, 1971.

Wiberg, E. and Amberger, E., *Hydrides*, Elsevier, New York, 1971.

Chapter 10

THE GROUP IA(1) ELEMENTS: LITHIUM, SODIUM, POTASSIUM, RUBIDIUM, AND CESIUM

10-1 Introduction

Sodium and potassium are abundant (2.6 and 2.4%, respectively) in the lithosphere. There are vast deposits of rock salt (NaCl) and $KCl \cdot MgCl_2 \cdot 6\,H_2O$ (carnallite) resulting from evaporation of lagoons over geologic time. The Great Salt Lake of Utah and the Dead Sea in Israel are examples of evaporative processes at work today. The elements Li, Rb, and Cs have much lower abundances and occur in a few silicate minerals.

The element Fr has only very short-lived isotopes that are formed in natural radioactive decay series or in nuclear reactors. Tracer studies show that the ion behaves as expected from the position it holds in Group IA(1).

Sodium and its compounds are of great importance. The metal, as Na—Pb alloy, is used to make tetraalkylleads (Section 29-9), and there are other industrial uses. The hydroxide, carbonate, sulfate, tripolyphosphate, and silicate are among the top 50 industrial chemicals.

Potassium salts, usually sulfate, are used in fertilizers. The main use for Li is as a metal in the synthesis of lithium alkyls (Section 29-3).

Both Na^+ and K^+ are of physiological importance in animals and plants; cells probably differentiate between Na^+ and K^+ by some type of complexing mechanism. Lithium salts are used in the treatment of certain mental disorders.

Some properties of the elements were given in Table 8-1. The low ionization enthalpies and the fact that the resulting M^+ ions are spherical and of low polarizability leads to a chemistry of +1 ions. The high second ionization enthalpies preclude the formation of +2 ions. Despite the essentially ionic nature of Group IA(1) compounds, some degree of covalent bonding can occur. The diatomic molecules of the elements (e.g., Na_2) are covalent. In some chelate and organometallic compounds, the M—O, M—N, and M—C bonds have a slight covalent nature. The tendency to covalency is greatest for the ion with the greatest polarizing power, that is, Li^+. The charge/radius ratio for Li^+, which is similar to that for Mg^{2+}, accounts for the similarities in their chemistry, where Li^+ differs from the other members.

Some other ions that have +1 charge and radii similar to those of the alkalis may have similar chemistry. The most important are

1. Ammonium and substituted ammonium ions. The solubilities and crystal structures of salts of NH_4^+ resemble those of K^+.
2. The Tl^+ ion can resemble either Rb^+ or Ag^+; its ionic radius is similar to that of Rb^+, but it is more polarizable.
3. Spherical +1 complex ions, such as $(\eta^5\text{-}C_5H_5)_2Co^+$ (Chapter 29).

10-2 Preparation and Properties of the Elements

Both Li and Na are obtained by electrolysis of fused salts or of low-melting eutectics such as $CaCl_2$ + NaCl. Because of their low melting points and ready vaporization K, Rb, and Cs cannot readily be made by electrolysis, but are obtained by treating molten chlorides with Na vapor. The metals are purified by distillation. The elements Li, Na, K, and Rb are silvery but Cs has a golden-yellow cast. Because there is only one valence electron per metal atom, the binding energies in the close-packed metal lattices are relatively weak. Hence, the metals are very soft with low melting points. The Na—K alloy, with 77.2% K, has a melting point of −12.3 °C.

The elements Li, Na, or K may be dispersed on various solid supports, such as Na_2CO_3 or kieselguhr, by melting. They are used as catalysts for various reactions of alkenes, notably the dimerization of propene to 4-methyl-1-pentene. Dispersions in hydrocarbons result from high-speed stirring of a suspension of the melted metal. These dispersions may be poured in air, and they react with water with effervescence. They may be used where sodium shot or lumps would react too slowly.

The metals are highly electropositive (Table 8-1) and react directly with most other elements and many compounds on heating. Lithium is usually the least, and Cs the most, reactive.

Lithium is only slowly attacked by water at 25 °C and will not replace the weakly acidic hydrogen in $C_6H_5C{\equiv}CH$, whereas the other alkali metals will do so. However, Li is uniquely reactive with N_2 (slowly at 25 °C, but rapidly at 400 °C) forming a ruby-red crystalline nitride (Li_3N). Like Mg, which gives Mg_3N_2, lithium can be used to absorb N_2.

With water, Na reacts vigorously, K inflames, and Rb and Cs react explosively; large lumps of Na may also react explosively. The elements Li, Na, and K can be handled in air although they tarnish rapidly. The others must be handled under Ar.

A fundamental difference, which is attributable to cation size, is shown by the reaction with O_2. In air (or O_2) at 1 atm the metals burn. Lithium gives only Li_2O with a trace of Li_2O_2. Sodium normally gives the peroxide, Na_2O_2, but it will take up further O_2 under pressure and heat to give the superoxide, NaO_2. The elements K, Rb, and Cs form the superoxides MO_2. The increasing stability of the per- and superoxides as the size of the alkali ions increases is a typical example of the stabilization of larger anions by larger cations through lattice-energy effects, as is explained in Section 4-6.

The metals react with alcohols to give the alkoxides, and Na or K in ethanol or *tert*-butanol is commonly used in organic chemistry as a reducing agent and a source of the nucleophilic RO^- ions.

Sodium and the other metals dissolve with much vigor in mercury. Sodium amalgam (Na/Hg) is a liquid when low in sodium, but is solid when rich in sodium. It is a useful reducing agent and can be used for aqueous solutions.

10-3 Solutions of the Metals in Liquid Ammonia and Other Solvents

The Group IA(1) metals, and to a lesser extent Ca, Sr, Ba, Eu, and Yb, are soluble in ammonia giving solutions that are blue when dilute. These solutions conduct electricity and the main current carrier is the solvated electron. While the lifetime of the solvated electron in water is very short, in very pure liquid ammonia it may be quite long (<1% decomposition per day).

In *dilute solutions* the main species are metal ions (M^+) and electrons, which are both solvated. The broad absorption around 15,000 Å, which accounts for the common blue color, is due to the solvated electrons. Magnetic and electron spin resonance studies show the presence of individual electrons, but the decrease in paramagnetism with increasing concentration suggests that the electrons can associate to form diamagnetic electron pairs. Although there may be other equilibria, the data can be accommodated by the equilibria

$$\text{Na(s) (dispersed)} \rightleftharpoons \text{Na (in solution)} \rightleftharpoons \text{Na}^+ + \text{e}^- \quad (10\text{-}3.1)$$

$$2\,\text{e}^- \rightleftharpoons \text{e}_2^{2-} \quad (10\text{-}3.2)$$

The most satisfactory models of the solvated electron assume that the electron is not localized but is "smeared out" over a large volume so that the surrounding molecules experience electronic and orientational polarization. The electron is trapped in the resultant polarization field, and repulsion between the electron and the electrons of the solvent molecules leads to the formation of a cavity within which the electron has the highest probability of being found. In ammonia this is estimated to be approximately 3.0–3.4 Å in diameter; this cavity concept is based on the fact that solutions are of much lower density than the pure solvent, that is, they occupy far greater volume than that expected from the sum of the volumes of metal and solvent.

As the concentration of metal increases, metal ion clusters are formed. Above 3 M concentration, the solutions are copper colored with a metallic luster. Physical properties, such as their exceedingly high electrical conductivities, resemble those of liquid metals. More is said about this in Section 10-7.

The metals are also soluble to varying degrees in other amines, hexamethylphosphoramide, $OP(NMe_2)_3$, and in ethers such as THF or diglyme, giving blue solutions.

The ammonia and amine solutions are widely used in organic and inorganic synthesis. (Lithium in methylamine or ethylenediamine can reduce aromatic rings to cyclic monoalkenes.) Sodium in liquid ammonia is the most widely used of such reducing agents. The blue solution is moderately stable at temperatures where ammonia is still a liquid, but reaction to give the amide (Reaction 10-3.3),

$$\text{Na} + \text{NH}_3(\ell) = \text{NaNH}_2 + \tfrac{1}{2}\,\text{H}_2 \quad (10\text{-}3.3)$$

can occur photochemically and is catalyzed by transition metal salts. Thus sodium amide is prepared by treatment of Na with ammonia in the presence of a trace of iron(III) chloride. Primary and secondary amines react similarly, giving alkylamides (Reaction 10-3.4),

$$Li(s) + CH_3NH_2(\ell) \longrightarrow LiNHCH_3(s) + \tfrac{1}{2} H_2 \qquad (10\text{-}3.4)$$

and dialkylamides (Reaction 10-3.5), respectively.

$$Li(s) + (C_2H_5)_2NH(\ell) \longrightarrow LiN(C_2H_5)_2(s) + \tfrac{1}{2} H_2 \qquad (10\text{-}3.5)$$

The lithium dialkylamides are used to make compounds with $M-NR_2$ bonds (Section 24-7).

The formation of the amides of K, Rb, and Cs is reversible owing to the favorable potential for half-reaction 10-3.6.

$$e^- + NH_3 = NH_2^- + \tfrac{1}{2} H_2 \qquad K = 5 \times 10^4 \qquad (10\text{-}3.6)$$

The similar reactions for Li and Na are irreversible, owing to the insolubility of the latter amides in ammonia:

$$Na^+(am) + e^-(am) + NH_3(\ell) = NaNH_2(s) + \tfrac{1}{2} H_2 \qquad K = 3 \times 10^9 \quad (10\text{-}3.7)$$

where am denotes a solution in ammonia.

COMPOUNDS OF THE GROUP IA(1) ELEMENTS

10-4 Binary Compounds

The metals of Group IA(1) react directly with most other elements to give binary compounds or alloys. Many of these compounds are described under the appropriate element. The most important are the oxides (M_2O), peroxides (M_2O_2), and superoxides (MO_2). Although all three types can be obtained for each alkali metal, indirect methods are often required. The direct reactions of the metals with an excess of O_2 give different products, depending on the metal: lithium predominantly forms the oxide, along with traces of the peroxide; sodium preferentially forms the peroxide, with traces of the oxide; potassium, rubidium, and cesium form superoxides.

All three types of compounds between oxygen and an alkali metal are readily hydrolyzed:

$$\text{Oxides} \qquad M_2O + H_2O \quad = 2\,M^+ + 2\,OH^- \qquad\qquad (10\text{-}4.1)$$

$$\text{Peroxides} \qquad M_2O_2 + 2\,H_2O = 2\,M^+ + 2\,OH^- + H_2O_2 \qquad (10\text{-}4.2)$$

$$\text{Superoxides} \qquad 2\,MO_2 + 2\,H_2O = O_2 + 2\,M^+ + 2\,OH^- + H_2O_2 \quad (10\text{-}4.3)$$

10-5 Hydroxides

These are white, very deliquescent crystalline solids: NaOH (mp 318 °C) and KOH (mp 360 °C). The solids and their aqueous solutions absorb CO_2 from the atmosphere. Hydroxides are freely soluble exothermically in water and in alcohols and are used whenever strong alkali bases are required.

10-6 Ionic Salts

Salts of virtually all acids are known; they are usually colorless, crystalline, ionic solids. Color arises from colored anions, except where defects induced in the lattice (e.g., by radiation) may cause *color centers,* through electrons being trapped in holes (cf. ammonia solutions cited previously).

The properties of a number of *lithium compounds* differ from those of the other Group IA(1) elements but resemble those of Mg^{2+} compounds. Many of these anomalous properties arise from the very small size of Li^+ and its effect on lattice energies, as explained in Section 4-6. In addition to examples cited there, we note that LiH is stable to approximately 900 °C, while NaH decomposes at 350 °C. The compound Li_3N is stable, whereas Na_3N does not exist at 25 °C. Lithium hydroxide decomposes at red heat to Li_2O, whereas the other hydroxides MOH sublime unchanged; LiOH is also considerably less soluble than the other hydroxides. The carbonate (Li_2CO_3) is thermally less stable relative to Li_2O and CO_2 than are other alkali metal carbonates. The solubilities of Li^+ salts resemble those of Mg^{2+}. Thus LiF is sparingly soluble (0.27 g/100 g H_2O at 18 °C) and is precipitated from ammoniacal NH_4F solutions; LiCl, LiBr, LiI and, especially $LiClO_4$ are soluble in ethanol, acetone, and ethyl acetate; LiCl is soluble in pyridine.

The alkali metal salts are generally characterized by high melting points, by electrical conductivity of the melts, and by ready solubility in water. These salts are seldom hydrated when the anions are small, as in the halides, because the hydration energies of the ions are insufficient to compensate for the energy required to expand the lattice. The Li^+ ion has a large hydration energy, and it is often hydrated in its solid salts when the same salts of other alkalis are unhydrated, e.g., $LiClO_4 \cdot 3 H_2O$. For salts of *strong* acids, the Li salt is usually the *most* soluble in water of the alkali metal salts, whereas for *weak* acids the Li salts are usually *less* soluble than those of the other elements.

There are few important *precipitation reactions* of the ions. One example is the precipitation by methanolic solutions of 4,4′-diaminodiphenylmethane (L) of Li and Na salts (e.g., NaL_3Cl). Generally, the larger the M^+ ion the more numerous are its insoluble salts. Thus Na has few insoluble salts; the mixed Na—Zn and Na—Mg uranyl acetates [e.g., $NaZn(UO_2)_3(CH_3CO_2)_9 \cdot 6 H_2O$], which may be precipitated almost quantitatively from dilute acetic acid solutions, are useful for analysis. Salts of the heavier ions, K^+, Rb^+, and Cs^+, with large anions such as ClO_4^-, $[PtCl_6]^{2-}$, $[Co(NO_2)_6]^{3-}$, and $B(C_6H_5)_4^-$, are relatively insoluble and form the basis for gravimetric analysis.

10-7 Solvation and Complexation of Alkali Metal Ions

Hydration of Alkali Metal Cations

For alkali metal cations, as well as for others, solvation must be considered from two points of view. First, each ion in solution possesses a primary solvation shell (termed hydration shell when the solvent is water), which is the number of solvent (water) molecules directly coordinated to the metal ion, as described for ligands in Chapter 6. The discussion of water exchange rates in Section 6-5 (in particular Fig. 6-6) concerned precisely this first solvation, or coordination layer. Second, there is also the overall solvation number, which is the total number of solvent molecules on which the ion exercises a substantial restraining influence. Thus, although the first solvation shell or coordination sphere of a solvated metal ion is the most important, other layers of solvent molecules are organized and influenced by the cation. As an example, consider the aqueous lithium cation which, as shown in Table 10-1, has a hydration number of about 25. This means that a total of 25 water molecules operate in aqueous solution under the restraining influence of the cation's positive charge to such an extent as to be considered bound to the cation.

In the case of Li^+, a primary coordination number of four tetrahedrally arranged water molecules is observed in numerous crystalline salts, and a similar arrangement of four water molecules probably exists in aqueous solutions. The ions Na^+ and K^+ may also have fourfold primary hydration in aqueous solutions. The primary hydration numbers of Rb^+ and Cs^+ are probably equal to six. However, as mentioned earlier, electrostatic forces operate beyond the primary hydration sphere of an ion, and additional layers of water molecules are bound to metal ions in water solution. These successive layers of bound solvent molecules are collectively termed the secondary solvation (hydration) layers. The extent of the secondary solvation layers appears to vary *inversely* with the size of the bare ion, that is, inversely with the size of the crystal radii of the ions. Thus as the crystal radii increase, the total hydration numbers, the hydrated radii, and the hydration energies all decrease. Apparently, the greater charge density of the smaller cation (i.e., Li^+) produces a greater organizing influence on secondary hydration layers in Li^+(aq) than is the case for the successively larger monocations of the Group IA(1) metals. As a result, the aqueous lithium cation is effectively larger than that of sodium, and so on, as shown in Table 10-1. Correspondingly, as hydrated radii decrease, the ionic mobilities of the aqueous alkali metal ions are found to increase, as shown in Table 10-1.

These trends play a role in the behavior of the alkali metal cations in ion exchange materials and in their passage through cell walls and other biological membranes, although doubtless other factors than size and hydration numbers are also important. In a cation exchange resin, two cations compete for attachment at anionic sites on the resin, as shown by equilibrium 10-7.1:

Table 10-1   Data on the Hydration of Aqueous Group IA(I) Ions

	Li^+	Na^+	K^+	Rb^+	Cs^+
Ionic radii[a] (Å)	0.90	1.16	1.52	1.66	1.81
Approximate hydrated radii (Å)	3.40	2.76	2.32	2.28	2.28
Approximate hydration numbers[b]	25.3	16.6	10.5	10.0	9.9
Hydration enthalpies (kJ mol^{-1})	519	406	322	293	264
Ionic mobilities[c]	33.5	43.5	64.6	67.5	68

[a]Values by Shannon and Prewitt as listed in Appendix IIC, for coordination number 6.

[b]From transference data.

[c]At 18 °C and infinite dilution.

$$A^+(aq) + [B^+R^-](s) = B^+(aq) + [A^+R^-](s) \qquad (10\text{-}7.1)$$

where R represents the solid resin and A^+ and B^+ are the cations. The value of the equilibrium constants for such equilibria can be measured quite accurately, and the order of preference of the alkali cations is usually $Li^+ < Na^+ < K^+ < Rb^+ < Cs^+$, although irregular behavior does occur in some cases. The usual order may be explained if we assume that the bonding force that holds the cation to the anionic site on the resin is essentially electrostatic, and that under ordinary conditions, the ions within the waterlogged resin are hydrated approximately to the same extent as they are in dilute aqueous solution. Then the ion with the smallest hydrated radius (which is the one with the largest "naked" radius) will be able to approach most closely to the negative site of attachment on the resin. Hence, according to Coulomb's law, this ion will be held most strongly.

Complexation of Cations by Crowns and Cryptates

Alkali metal cations may be brought into solution in solvents other than water by use of two types of special complexation ligands: *crown ethers* and *cryptates*. Ethers, polyethers, and especially cyclic polyethers are particularly suited to solvate Na^+ and other alkali metal cations. Examples are tetrahydrofuran (THF), the "glyme" solvents [which are linear polyethers such as $CH_3O(CH_2CH_2O)_nCH_3$], and the macrocyclic *crown ethers*. Five of the more common crown ethers are shown in Structures 10-I through 10-V, along with their customary names. In such crown ethers, the number of oxygen atoms and the total number of atoms

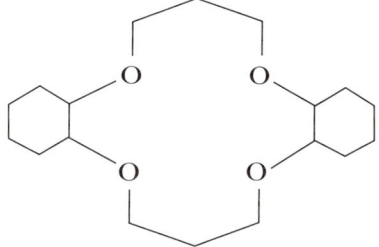

dicyclohexyl-14-crown-4

10-I

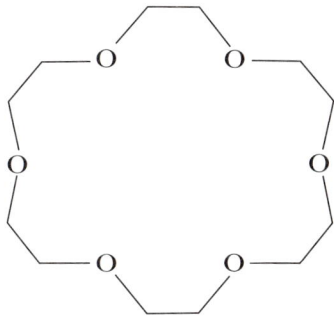

18-crown-6

10-II

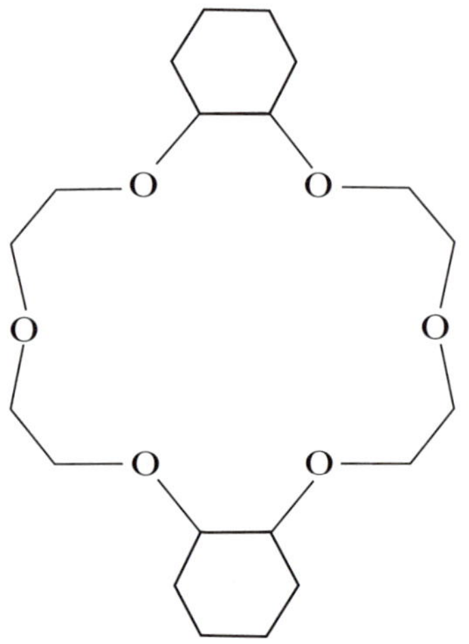

dicyclohexyl 18-crown-6

10-III

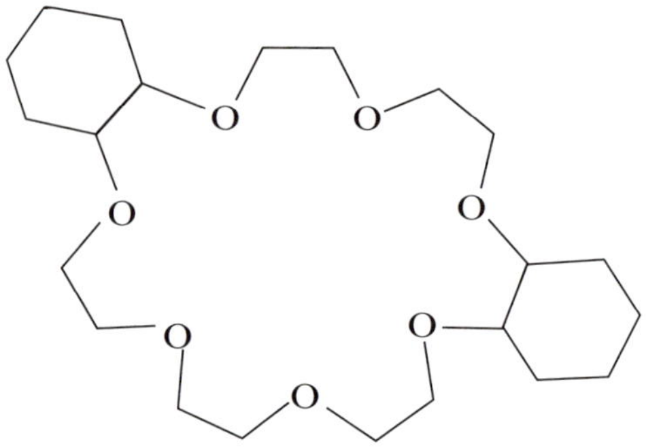

dicyclohexyl 21-crown-7

10-IV

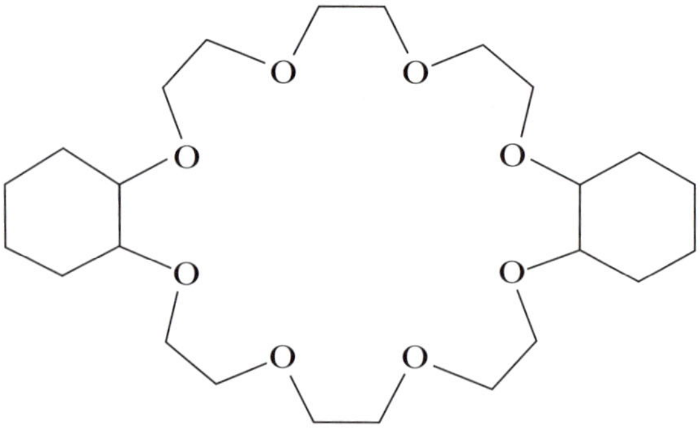

dicyclohexyl 24-crown-8

10-V

in the ring are both specified within the name of the polyether. As an example, 18-crown-6 is a symmetrical cyclic polyether containing 6 oxygen atoms and a total of 18 ring atoms, as shown in Structure 10-II. The name of crown ether 10-III is dicyclohexyl-18-crown-6.

   The bonding of an alkali metal cation within the cavity of a cyclic polyether is largely electrostatic, and a close match between the size of the cation and the

10-VI
2,2,2-crypt

size of the crown is important if the cation is to be tightly bound in the cavity created by the oxygen donor atoms. For 18-crown-6, the binding constants increase in the order $Li^+ < Na^+$, $Cs^+ < Rb^+ < K^+$. In other words, the strongest binding is achieved by K, principally because this ion possesses the best match in size to the cavity of 18-crown-6. In comparison, Rb^+ is preferentially bound by the larger dicyclohexyl-21-crown-7, and Cs^+ by dicyclohexyl-24-crown-8. In contrast, the small Li^+ ion finds its greatest binding with small crown ethers such as dicyclohexyl-14-crown-4. In each of these cases, the size ratio of the cation to that of the crown's cavity is in the optimum range of about 0.80–0.97. Obviously, a cation radius/crown cavity size ratio greater than 1 would be undesirable, since the crown ring would then be too small to surround the cation effectively.

Other factors have been found to influence the stability of a crown ether complex with an alkali metal cation. First of all, the greater the number of oxygen atoms in the crown ring, the greater the magnitude of the ion–dipole interaction. Binding is enhanced in cases where the crown donor oxygen atoms are coplanar. Also, for greatest affinity, the crown ether should not be sterically hindered, and the oxygen atoms should be symmetrically placed around the ring. Finally, for maximum binding to a given alkali metal cation, the crown ring should not contain electron-withdrawing substituents, which would decrease the basicity of the oxygen atoms.

The *cryptates* are even more potent and selective agents for binding alkali metal ions (and others). However, they differ from the crown ethers in two ways. First, they incorporate nitrogen as well as oxygen donor atoms, as shown in Structure 10-VI. Second, the cryptates are polycyclic, and hence are able more fully to surround a metal cation, thereby taking greater advantage of the chelate effect mentioned in Section 6-4. The cryptate shown in Structure 10-VI is called 2,2,2-crypt (often abbreviated C_{222}), and the structure of a representative complex is shown in Fig. 10-1.

Alkali Metal Anions

When a solution of sodium in ethylamine is cooled in the presence of 2,2,2-crypt, the compound shown in Fig. 10-2, $[Na(2,2,2\text{-crypt})]^+Na^-$, which is stable only below -10 °C, crystallizes. This fascinating compound is one of a number of known *sodide* (i.e., Na^- containing) compounds. Although Reaction 10-7.2

$$2\, Na(g) = Na^+(g) + Na^-(g) \qquad (10\text{-}7.2)$$

is endothermic by 438 kJ mol^{-1}, the lattice energy for the formation of the crys-

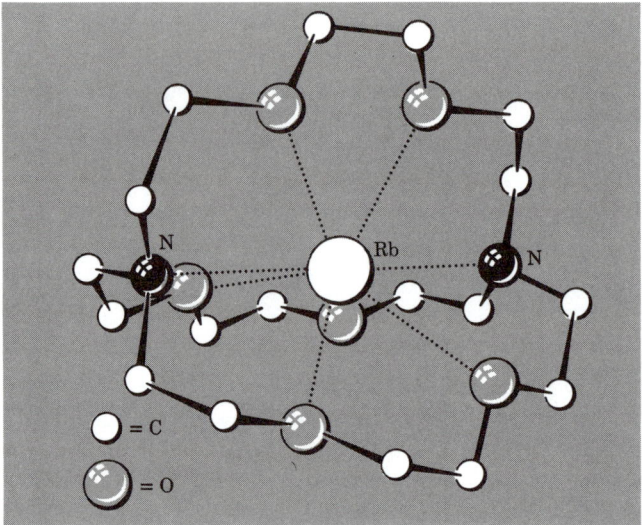

Figure 10-1 The structure of the cation in the thio-cyanate salt [Rb(2,2,2-crypt)]SCN·H₂O. [Reproduced by permission from M. R. Truter, *Chem. Br.* **1971**, 203.]

talline sodide compound and the complexation of the sodium cation by the cryptate overcome this endothermicity, thereby stabilizing the sodide (Na⁻) ion. Other less stable *alkalides* have been prepared by J. L. Dye and co-workers, for example, the potasside [K(2,2,2-crypt)]⁺K⁻, and similar cesides. The alkalides are brown or gold-brown solids that are extremely air and water sensitive, thermally unstable, diamagnetic solids.

The structure of the sodide shown in Fig. 10-2 warrants comment. The crystal structure is best described as alternating layers of [Na(2,2,2-crypt)]⁺ and Na⁻ ions in what is essentially a hcp array, as described in Chapter 4. The unusually large cryptated cations form a hcp array in which the octahedral sites are occupied by sodide ions. Furthermore, this structure is nearly identical to that of the simple cryptated salt [Na(2,2,2-crypt)]⁺I⁻, which contains the common iodide anion. The sodide anion in [Na(2,2,2-crypt)]⁺Na⁻ is located as far as possible from the negative oxygen and nitrogen atoms of the cryptate, and the shortest distance between sodide ions in the same layer is 8.83 Å. The separation between adjacent layers of sodide ions is 11.0 Å, and the distance between the Na⁻ and the Na⁺ ions is 7.06 Å.

Interestingly, a similar series of *electrides* is known. These are black, paramagnetic solids that have the general formula [M(crypt)]⁺e⁻, and which adopt structures similar to those of the alkalides. In electrides, it is the electrons rather than the alkali metal anions that are held in the cavities formed by the cryptated metal cations. For instance, in the case of the electride [Cs(crypt)]⁺e⁻, the electrons are located in cavities of diameter 2.4 Å between the cryptated cations.

Encapsulated Metal Ions in Biology

Naturally occurring small cyclic polypeptides can also act to encapsulate metal ions. These cyclic polypeptides play a role in transporting alkali and alkaline

10-VII

earth ions across membranes in living systems. More is presented on this topic in Chapter 31. Perhaps the best known examples of such cyclic polypeptides are valinomycin (Structure 10-VII) and nonactin (shown in Fig. 10-3 as the potassium complex).

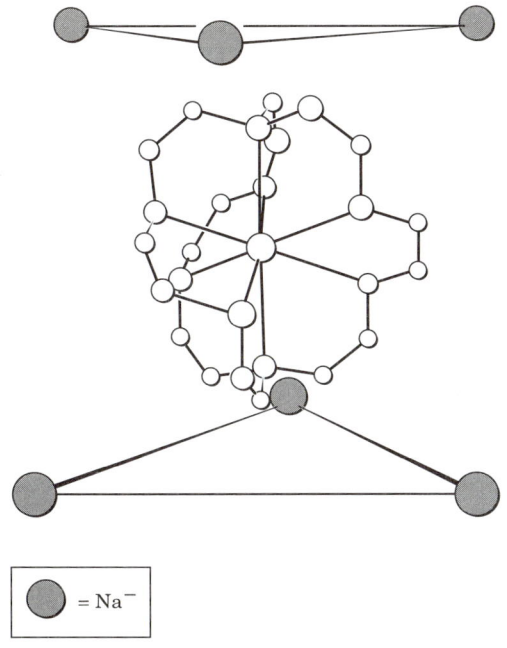

$= Na^-$

Figure 10-2   Part of the unit cell of the crystalline sodide $[Na(2,2,2\text{-crypt})]^+Na^-$ showing a single sodium cation at the center of the 2,2,2-crypt ligand and the six nearest neighbor Na^- (sodide) anions. [Reprinted in part with permission from F. J. Tehan, B. L. Barnett, and J. L. Dye, *J. Am. Chem. Soc., 96*, 7203–7208 (1974). Copyright © (1974) American Chemical Society.]

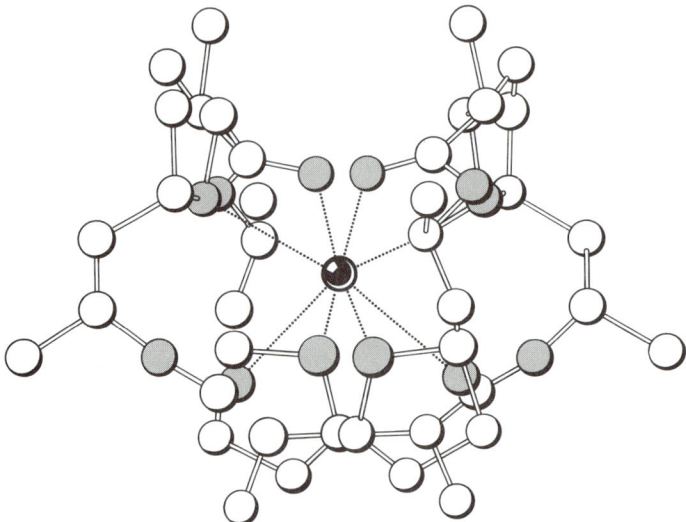

Figure 10-3   The structure of the nonactin complex of K$^+$. [Reproduced by permission from D. A. Fenton, *Chem. Soc. Rev.*, **1977**, *6*, 325–343.]

10-8   Organometallic Compounds

Lithium Alkyls and Aryls

One of the most important areas of chemistry for the Group IA(1) elements is that of their organic compounds. This is especially true of Li, whose alkyls and aryls find extensive use as alkylating and arylating agents. Organolithium compounds resemble Grignard reagents in their reactions, although the lithium reagents are generally more reactive.

Lithium alkyls and aryls are best prepared as in Reaction 10-8.1

$$C_2H_5Cl + 2\ Li \longrightarrow C_2H_5Li + LiCl \tag{10-8.1}$$

using alkyl or aryl chlorides in benzene or petroleum solvents. Methyllithium may also be prepared at low temperatures in hexane as insoluble white crystals from the exchange between butyllithium and methyl iodide.

$$C_4H_9Li + CH_3I \longrightarrow CH_3Li(s) + C_4H_9I \tag{10-8.2}$$

Organolithium compounds all react rapidly with oxygen and water, and are usually spontaneously flammable in air.

Organolithium compounds are among the few alkali metal compounds that have properties—solubility in hydrocarbons and other nonpolar liquids and high volatility—typical of covalent substances. These compounds are generally liquids or low melting solids, and molecular association is an important structural feature. For example, in the crystals of methyllithium (Fig. 10-4), the lithium atoms are associated in a tetrahedral unit with methyl groups symmetrically capping each triangular face of the Li$_4$ tetrahedron. A similar aggregation occurs for lithium alkoxides (LiOR) and dialkylamides (LiNR$_2$).

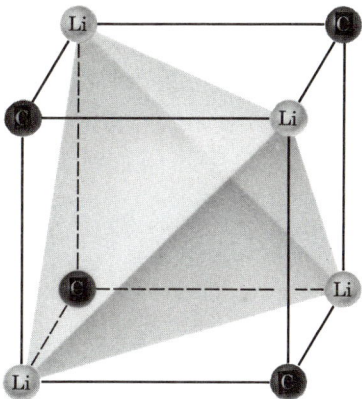

Figure 10-4   The structure of solid $(CH_3Li)_4$, showing the tetrahedral arrangement of Li atoms and the
face-capping positions of the methyl groups. The
structure may be regarded to be roughly that of a
cube.

In solution, the lithium alkyls are also aggregated, but the extent of aggregation depends on the solvent and the steric nature of the organic group. It is
not surprising, then, that the wide variations in reactivities of Li alkyls depend on
these differences in aggregation and other ion pairing interactions. An example
is benzyllithium, which is monomeric in THF and reacts as a benzylating agent
some 10^4 times as fast as methylation by the tetrameric methyllithium.

Organosodium and Organopotassium Compounds

These compounds are all appreciably ionic and are not soluble to any extent in
hydrocarbons. They are exceedingly reactive, being sensitive to air and water.
Although alkyl and aryl derivatives can be prepared *in situ* for use as reactive intermediates, they are seldom isolated.

Some of the most important compounds are those formed by the more
acidic hydrocarbons such as cyclopentadiene (Reaction 10-8.3),

$$3\ C_5H_6 + 2\ Na \longrightarrow 2\ C_5H_5^-\ Na^+ + C_5H_8 \qquad (10\text{-}8.3)$$

and acetylenes (Reaction 10-8.4).

$$RC\equiv CH + Na \longrightarrow RC\equiv C^-Na^+ + \tfrac{1}{2}\,H_2 \qquad (10\text{-}8.4)$$

Reactions 10-8.3 and 10-8.4 are best performed using sodium dispersed in THF,
glyme, or DMF. The ionic products of Reactions 10-8.3 and 10-8.4 are useful as
reagents for the synthesis of transition metal organometallic derivatives.

10-9   Other Alkali Metal Compounds

A large number of alkali metal compounds that are commonly volatile and soluble in hydrocarbon or ether solvents is known. The most important of these are

the alkyls and aryls that were discussed in Section 10-8. These compounds have much in common, however, especially the tendency to aggregate into dimers, tetramers, hexamers, and so on, with the following classes of compounds, where R = alkyl or aryl:

Alkoxides	MOR
Amides	MNHR, MNR$_2$
Phosphides	MPHR, MPR$_2$
Thiolates	MSR

Such compounds have been extensively studied recently because, if the R group is very bulky, the alkali metal compound can be used to make transition metal complexes with very low coordination numbers.

Some typical syntheses are given in Reactions 10-9.1 and 10-9.2, which should be compared with the syntheses of the alkyls (Section 10-8).

$$i\text{-}Pr_2NH + n\text{-}BuLi \longrightarrow i\text{-}Pr_2NLi + C_4H_{10} \qquad (10\text{-}9.1)$$

$$2\,t\text{-}BuOH + 2\,Na \longrightarrow 2\,t\text{-}BuONa + H_2 \qquad (10\text{-}9.2)$$

Reaction 10-9.1 illustrates the utility of alkyl lithiums as deprotonating agents; the resulting dialkyl amides can similarly act as strong bases.

A characteristic feature, especially of lithium compounds [although not restricted to them, since $(NaO\text{-}t\text{-}Bu)_6$ is a hexamer both in the solid and in benzene], is aggregation, as discussed for the alkyls in Section 10-8. The extent of aggregation typically depends on the compound, the nature of the attached groups, and on the solvent.

Other important compounds of the alkali metals include those with the transition metal carbonylates (Chapter 28), which are made in THF solvent by reactions such as 10-9.3 through 10-9.5.

$$Mn_2(CO)_{10} + 2\,Na/Hg \longrightarrow 2\,NaMn(CO)_5 \qquad (10\text{-}9.3)$$

$$Co_2(CO)_8 + 2\,Na/Hg \longrightarrow 2\,NaCo(CO)_4 \qquad (10\text{-}9.4)$$

$$Cr(CO)_6 + 2\,Na \longrightarrow Na_2Cr(CO)_5 + CO \qquad (10\text{-}9.5)$$

10-10 Reaction Summary

As a study aid, and in order to compare the chemistry of lithium with that of the other members of the group, the reactions of the Group IA(1) metals are listed in Fig. 10-5(a–c). Rather than being a comprehensive list of reactions, Fig. 10-5 is meant to be only an overview of the important types of reactions that the alkali metals typically undergo. The student should note the metal ion precipitation reactions, as well as the differences between lithium (and to some extent sodium) and the other members of the group.

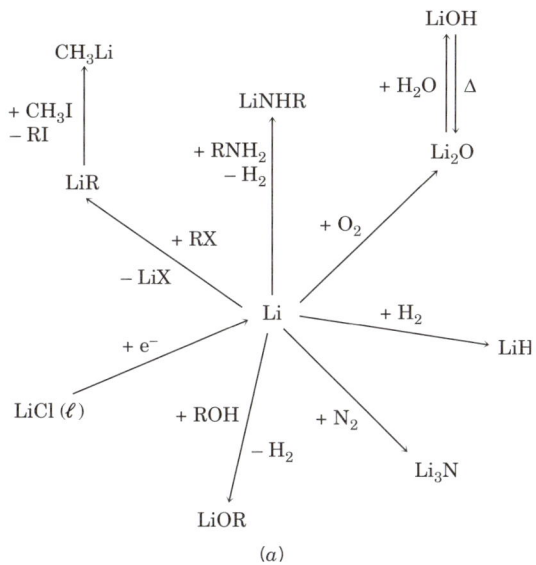

(a)

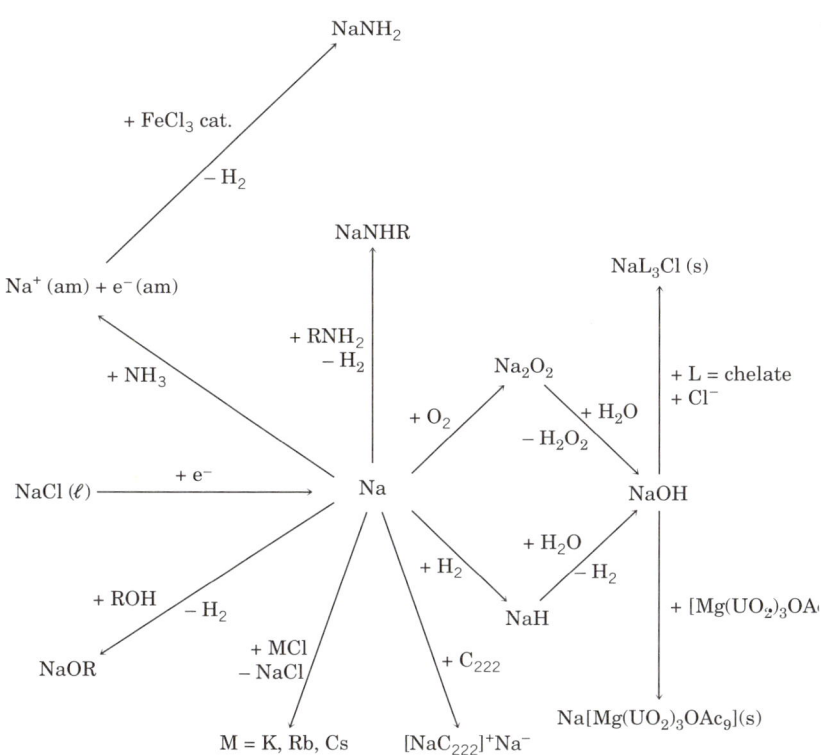

(b)

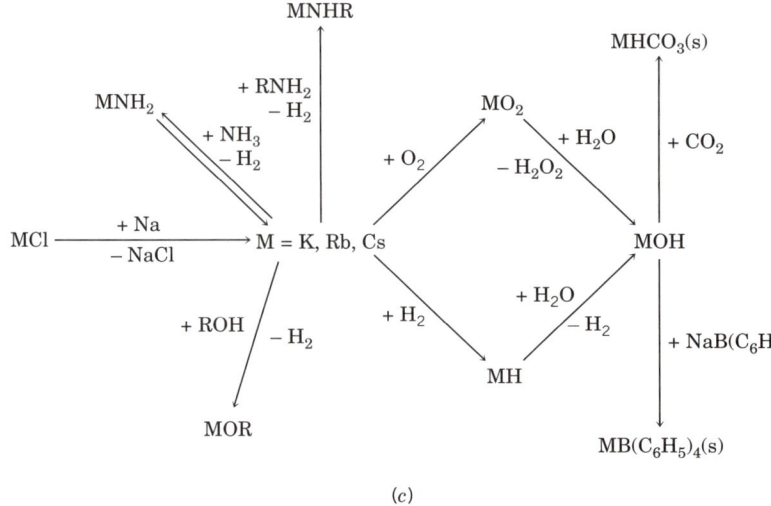

(c)

Figure 10-5 Reactions of the alkali metals.

STUDY GUIDE

Study Questions

A. Review

1. Why are the alkali metals soft and volatile?
2. Why are they highly electropositive?
3. Write down the electronic structure of francium.
4. Why are the first ionization energies of the Group IA(1) atoms low?
5. Why does the chemical reactivity of the metals increase from Li to Cs?
6. What other ions have properties similar to the alkali metal ions?
7. How does the charge-radius ratio of Li^+ differ from those of the other Group IA(1) ions? List some consequences of this difference.
8. How do the reactivity and the nature of the products vary from Li to Cs when the alkali metals react with oxygen?
9. What is the nature of the solutions of alkali metals in liquid ammonia? What is the chief reaction by which they decompose?
10. How would you make lithium hydride? Why is it more stable than NaH?
11. Draw the crystal structures of NaCl and CsCl. Why do they differ?
12. Why is sodium peroxide a useful oxidizing agent in aqueous solution?
13. In what order are the M^+ ions eluted from a cation exchange resin column?
14. Why is LiF almost insoluble in water, whereas LiCl is soluble, not only in water, but in acetone?
15. What is (a) a crown ether, (b) a cryptate?
16. Why are lithium salts commonly hydrated and those of the other alkali ions usually anhydrous?
17. How would you extinguish a sodium fire in the laboratory?

B. Additional Exercises

1. Vapors of the alkali metals contain about 1% diatomic molecules. Discuss the bonding in such molecules using the MO approach. Why do the dissociation energies of the diatomic alkali metal molecules decrease with increasing Z?

2. Anhydrous KOH in THF is one of the strongest known bases and will deprotonate exceedingly weak acids. Why? (Compare the solvation of KOH by water and by THF.)

3. The formation constant for the 1:1 complex between K^+ and cyclohexyl-18-crown-6 is much larger than the values for the other alkali metal cations. Estimate from this the size of the "hole" available for the cations in this ligand.

4. Why is there so little variation in the standard potentials for reduction of the Group IA(1) cations?

5. Which ligand would you expect more favorably to complex with K^+, cyclohexyl-18-crown-6 or 2,2,2-crypt? Why?

6. Write balanced chemical equations for the electrolysis of (a) NaCl in water, (b) molten NaCl, (c) tetraethylammonium chloride in water, (d) molten tetraethylammonium chloride.

7. Complete and balance equations for the following reactions involving the metals and the ions of Group IA(1).

   (a) $KCl + Na$ (h) $RbO_2 + H_2O$

   (b) $Li + N_2$ (i) $Li_2O + H_2O$

   (c) $Na + O_2$ (j) $KOH + CO_2$

   (d) $Cs + O_2$ (k) $K^+ + B(C_6H_5)_4^-$

   (e) $K + C_2H_5OH$ (l) $Li + ClC_6H_5$

   (f) $Li + HN(C_2H_5)_2$ (m) $C_4H_9Li + CH_3I$

   (g) $Li + HN(SiMe_3)_2$ (n) $CH_3Li + [W(CO)_5Cl]^-$

8. If a crown ether were to be modified by replacing some or all oxygen atoms with sulfur, would such a complexing agent favor K^+ or Ag^+? Explain.

9. Make a thorough list of all of the ways in which the structure and reactivity of lithium and its compounds differ from those of the other alkali metals.

10. Why do alkoxides, amides, and alkyls of lithium [as opposed to other metals of Group IA(1)] have largely covalent rather than ionic nature?

11. Suggest a reason why 14-crown-4 is able to catalyze reactions of $LiCH_3$ in organic solvents.

12. Why is butyllithium more reactive in hexane as an R^- donor (nucleophile) than CH_3Li?

13. Make a careful drawing of each of the following:

    (a) $Li^+(aq)$

    (b) $[Na(2,2,2\text{-crypt})]^+$

    (c) 24-crown-8

    (d) Methyllithium (solid state)

14. Give balanced equations for the reaction of sodium with diethylamine, hydrogen, ethanol, water, and oxygen. Repeat for lithium and for potassium.

15. Lithium hydride adopts the NaCl-type structure, having a unit cell edge of 5.36 Å. Use this information and the effective nuclear charges for each ion to determine the Pauling radius of Li^+ and H^-, as described in Chapter 4.

16. Write out the Born–Haber cycle for the formation of KH.

17. Write out the Born–Haber cycle for the formation of Na_2O_2.

18. Suggest the product on reaction of BuLi and $HN(SiMe_3)_2$.

C. Questions from the Literature of Inorganic Chemistry

1. Consider the paper by H. K. Frendsdorf, *J. Am. Chem. Soc.,* **1971,** *93,* 600–606, and references cited therein, regarding the stability constants of cyclic polyether complexes with alkali cations.

    (a) Draw the structures of the crown ethers in Table II of this paper.

    (b) What relationship exists between stability constants for the complexes in methanol, cation radius, and ring size of the various crown ethers?

    (c) How do the potassium complexes of nonactin and valinomycin compare with the potassium complexes of 24-crown-8 and 30-crown-10, as inferred with stability constants?

    (d) Why are the stability constants for crown ether–alkali metal complexes in water lower than stability constants in methanol?

2. Consider the paper by B. Van Eck, Dinh Le Long, D. Issa, and J. L. Dye, *Inorg. Chem.,* **1982,** *21,* 1966–1970.

    (a) The analysis of the crystalline alkalides that are featured in this work was performed by reacting the samples with water. Write a balanced chemical equation for the reaction that takes place.

    (b) The H_2 evolved during analysis was compared with the total titratable base that was present after reaction with water. Why? For K^+ crypt-2,2,2·Na^-, how many equivalents of titratable base are released per equivalent of hydrogen upon reaction with water?

    (c) Why are the sodides the easiest crystals to prepare and the most stable of the alkalide compounds?

3. Consider the work: E. C. Alyea, D. C. Bradley, and R. G. Copperthwaite, *J. Chem. Soc., Dalton Trans.,* **1972,** *1580–1584.*

    (a) Draw Lewis diagrams for the lithium derivatives of $[N(SiMe_3)_2]^-$, which are used as reagents in this paper.

    (b) Suggest a synthesis of the lithium bis(trimethylsilyl)amido reagents, $Li[N(SiMe_3)_2]$.

    (c) What is the likely coordination geometry of the metal complexes of Table 2?

    (d) What reactions were used to synthesize the complexes of Table 2? Write balanced chemical equations.

    (e) What π delocalizations do the authors mention involving the N and Si atoms of the silylamide ligands? Show with orbital diagrams how π overlap within the $(Si)_2N$—M framework may take place. To what extent is Sc^{3+} believed to be involved in such π bonding? Why?

    (f) Show, with orbital diagrams, both the M π donation and π acceptance that the authors discuss. For which metals is each form of π bonding apparent?

    (g) What would be the likely reaction of such ML_3 complexes with water?

SUPPLEMENTARY READING

Addison, C. C., *The Chemistry of the Liquid Alkali Metals,* Wiley, New York, 1984.

Bach, R. O., *Lithium. Current Applications in Science, Medicine, and Technology,* Wiley, New York, 1985.

Borgstedt, H. V. and Matthews, C. K., *Applied Chemistry of the Alkali Metals,* Plenum, New York, 1987.

Christensen, J. J., Eatough, D. J., and Izatt, R. M., "The Synthesis and Ion Binding of

Synthetic Multidentate Macrocyclic Compounds," *Chem. Rev.,* **1974,** *74,* 351–384.

Dietrich, B., "Coordination-Chemistry of Alkali and Alkaline-Earth Cations with Macrocyclic Ligands," *J. Chem. Educ.,* **1985,** *63,* 954.

Dunitz, J. D. et al., Eds., *Structure and Bonding,* Vol. 16, Springer-Verlag, Berlin, 1973.

Dye, J. L., "Electrides, Negatively Charged Metal Ions, and Related Phenomena," in *Progress in Inorganic Chemistry,* Vol. 32, S. Lippard, Ed., Wiley-Interscience, New York, 1984.

Fenton, D. E., "Across the Living Barrier," *Chem. Soc. Rev.,* **1977,** *16,* 325–343.

Kapoor, P. N. and Mehrotra, R. C., "Coordination Compounds of the Alkali and Alkaline Earth Elements with Covalent Characteristics," *Coord. Chem. Rev.,* **1974,** *14,* 1.

Langer, A. W., Ed., "Polyamine-Chelated Alkali Metal Compounds," in *Advances in Chemistry Series,* No. 130; American Chemical Society, Washington, DC, 1974.

Oliver, J. P., "Organoalkalimetal Compounds," in *International Review of Science, Inorganic Chemistry Series 2,* Vol. 4, B. J. Aylett, Ed., Butterworths, London, 1979, pp. 1–40.

Schade, C. and Schleyer, P. von R., "Structures of Organo Alkali Metal Compounds," *Adv. Organomet. Chem.,* **1987,** *27,* 169.

Schrauzer, G. N. and Klippel, K. F., Eds., *Lithium in Biology and Medicine,* VCH Publishers, Weinheim, 1991.

The Alkali Metals, Spec. Publ. No. 22, The Chemical Society, London, 1967.

Wakefield, B. J., *The Chemistry of Organolithium Compounds,* Pergamon Press, New York, 1974.

Wakefield, B. J., *Organolithium Methods,* Academic, London, 1988.

Wardell, J. L., *Comprehensive Organometallic Chemistry,* Vol. 1, Chapter 2, Pergamon Press, Oxford, 1981.

Chapter 11

THE GROUP IIA(2) ELEMENTS: BERYLLIUM, MAGNESIUM, CALCIUM, STRONTIUM, AND BARIUM

11-1 Introduction

Beryllium occurs in the mineral *beryl*, $Be_3Al_2(SiO_3)_6$. Compounds of Be are exceedingly toxic, especially if inhaled, whereby they cause degeneration of lung tissue similar to miners' silicosis; they must be handled with great care. This element has only minor technical importance.

The elements Mg, Ca, Sr, and Ba are widely distributed in minerals and in the sea. There are substantial deposits of limestone ($CaCO_3$), dolomite ($CaCO_3 \cdot MgCO_3$), and carnallite ($KCl \cdot MgCl_2 \cdot 6\,H_2O$). Less abundant are strontianite ($SrCO_3$) and barytes ($BaSO_4$). All isotopes of radium are radioactive. The isotope ^{226}Ra, α, 1600 years, which occurs in the ^{238}U decay series, was first isolated by Pierre and Marie Curie from the uranium ore, pitchblende. It was collected from solutions by coprecipitation with $BaSO_4$ and the nitrates subsequently fractionally crystallized. Its use in cancer therapy has been supplanted by other forms of radiation.

The positions of the Group IIA(2) elements and of the related Group IIB(12) (Zn, Cd, and Hg) elements in the periodic table and some of their properties have been given in Chapter 8.

The atomic radii are smaller than those of the Li to Cs group as a result of the increased nuclear charge (cf. Table 4-2). The number of bonding electrons in the metals is now two, so that these have higher melting and boiling points and densities. The ionization enthalpies are higher than those of Group IA(1) atoms and their enthalpies of vaporization are higher. Nevertheless, the high lattice energies and high hydration energies of M^{2+} ions compensate for these increases. The metals are hence electropositive with high chemical reactivities and standard electrode potentials. Born–Haber cycle calculations show that MX compounds would be unstable in the sense that the following reactions should have very large negative enthalpies:

$$2\,MX = M + MX_2 \qquad (11\text{-}1.1)$$

Covalency and Stereochemistry for Beryllium

In the case of beryllium, because of its exceptionally small atomic radius and high enthalpies of ionization and sublimation, the lattice or hydration energies are insufficient to provide for complete charge separation to give a simple Be^{2+} cation in beryllium-containing compounds. (Recall the material in Chapter 4 on Born–Haber cycles for ionic compounds.) Consequently, although the oxides and fluorides of the other elements of Group IIA(2) (except perhaps Mg) are ionic, BeF_2 and BeO show evidence of covalent character. Also, covalent compounds with bonds from Be to C are quite stable. In these respects, Be resembles Zn. Note that to form two covalent bonds, promotion of Be from the $2s^2$ to the $2s^1 2p^1$ electron configuration is required. Thus BeX_2 molecules should be linear. Since such molecules are coordinatively unsaturated, they exist only in the gas phase. In condensed phases, at least threefold, and more commonly fourfold (maximum) coordination is achieved in the following ways.

1. Polymerization may occur through bridging groups, such as H, F, Cl, or CH_3, giving chain polymers of the type $[BeF_2]_n$, $[BeCl_2]_n$, and $[Be(CH_3)_2]_n$, as shown in Fig. 11-1. The coordination of Be in these chains is not exactly tetrahedral. For instance, the internal Cl—Be—Cl angles in $[BeCl_2]_n$ are 98.2°, which means the $Be(\mu_2\text{-}Cl)_2Be$ units are somewhat elongated in the direction of the chain axis. In contrast, the C—Be—C angles in $[Be(CH_3)_2]_n$ are 114°. These distortions from the ideal tetrahedral angle for a four-coordinate Be atom are dependent on the nature of the bridging group, and are related to the presence or absence of lone pairs on the bridging atoms.

Other important examples of bridging to Be atoms include the following. As already noted, in the gas phase at high temperature, the halides are linear molecules, X—Be—X. At low temperatures, however, the chloride exists in appreciable amounts (~20% at 560 °C) as a dimer, $[BeCl_2]_2$, in which Be is presumably three coordinate. Interestingly, in compounds of the type $(M^I)_2(Be_4Cl_{10})$ (M = K, Rb, Tl, NO, or NH_4), the anion (Structure 11-I) resembles a portion of the $[BeCl_2]_n$ chain.

11–I

11–II

2. Alkoxides, $[Be(OR)_2]_n$, usually have associated structures with both μ_2-bridging and terminal OR groups. For example, $[Be(OCH_3)_2]_n$ is a high polymer, insoluble in hydrocarbon solvents. On the other hand, the *tert*-butoxy derivative is less condensed, being only a trimer $[Be(O\text{-}t\text{-}Bu)_2]_3$ (Structure 11-II), which is soluble in hydrocarbon solvents. Only when the alkoxide is bulky are monomers obtained with two-coordinate Be, as in Structure 11-III.

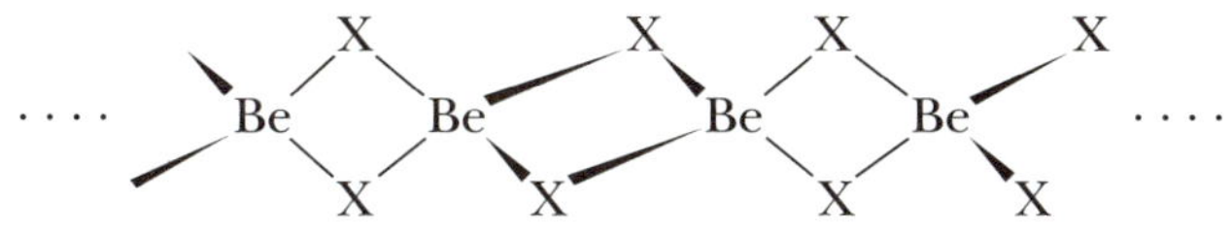

Figure 11-1 The infinite chain structure of BeX_2 compounds ($X = F$, Cl, or CH_3), whereby each Be atom achieves a coordination number of four.

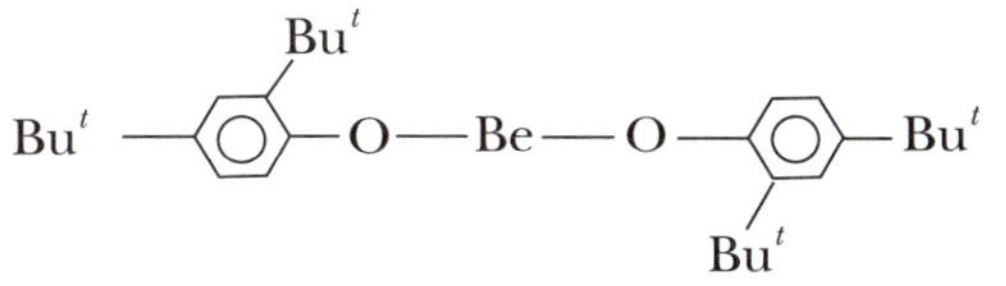

11–III

Another coordinatively unsaturated Be compound containing bulky organic groups is the two-coordinate beryllium alkyl $Be(t\text{-}Bu)_2$, which reacts with *tert*-butyllithium in dry pentane at room temperature, as in Reaction 11-1.2:

$$Li\text{-}t\text{-}Bu + Be(t\text{-}Bu)_2 \longrightarrow Li[Be(t\text{-}Bu)_3] \tag{11-1.2}$$

to give the three-coordinate lithium tri-*tert*-butylberyllate, $Li[Be(t\text{-}Bu)_3]$, in which the donor carbon atoms of the three *tert*-butyl groups are arranged in a trigonal planar coordination geometry around the Be atom. Further aspects of the organochemistry of the Group IIA(2) elements are given in Chapter 29.

3. By functioning as Lewis acids, many Be compounds obtain maximum coordination of the metal atom. The chloride $BeCl_2$ reacts with donor solvents to form four-coordinate etherates such as $BeCl_2(OEt_2)_2$. Interaction with anions gives complex ions such as BeF_4^{2-}. The aqua ion is four-coordinate, $[Be(H_2O)_4]^{2+}$. In chelate compounds, such as the acetylacetonate, $Be(acac)_2$, four approximately tetrahedral bonds are formed, with four equal C–O bonds and four equal Be–O bonds.

4. Beryllium also achieves tetrahedral four coordination in compounds such as BeO and BeS, the structures of which are often those of the corresponding Zn derivatives. Thus low-temperature BeO has the ZnO–wurtzite structure, whereas BeS adopts the ZnS–zinc blende structure (Fig. 4-1). The most stable $Be(OH)_2$ polymorph has the $Zn(OH)_2$ structure. It also may be noted that Be with F gives compounds that are often isomorphous with oxygen compounds of silicon. An example is $NaBeF_3$, which is isomorphous with $CaSiO_3$. In addition, there are five different corresponding forms of Na_2BeF_4 and Ca_2SiO_4.

Magnesium

The second member of Group IIA(2) (Mg) is intermediate in behavior between Be and the remainder of the group whose chemistry is entirely ionic in nature. The Mg^{2+} ion has a high polarizing ability, and there is, consequently, a decided tendency for its compounds to have nonionic behavior, although not as much as for Be. Magnesium, therefore, readily forms bonds to carbon, as discussed in Chapter 29. Like $Be(OH)_2$, $Mg(OH)_2$ is only sparingly soluble in H_2O, whereas the hydroxides of the other members of Group IIA(2) are water soluble and highly basic.

Calcium, Strontium, Barium, and Radium

The elements Ca, Sr, Ba, and Ra form a closely related group in which the chemical and physical properties change systematically with increasing size. Examples are increases from Ca to Ra in (a) the electropositive nature of the element (cf. $E°$ values, Table 8-2); (b) hydration energies of salts; (c) insolubility of most salts, notably sulfates; and (d) thermal stabilities of carbonates and nitrates. As in Group IA(1), and as explained in Section 4-6, the larger Group IIA(2) cations can stabilize large anions such as O_2^{2-}, O_2^-, and I_3^-.

Because of similarity in charge and radius, the 2+ ions of the lanthanides (Section 26-5) resemble the Sr to Ra ions. Thus Eu, which forms an insoluble sulfate ($EuSO_4$), sometimes occurs in Group IIA(2) minerals.

11-2   Beryllium and Its Compounds

The metal, obtained by Ca or Mg reduction of $BeCl_2$, or by Mg reduction of BeF_2, is very light and has been used for windows in X-ray apparatus. The absorption of electromagnetic radiation in general depends on the electron density in matter, and Be has the lowest stopping power per unit of mass thickness of all constructional materials, hence its utility as a nonabsorbing surface, or window.

Beryllium metal is relatively unreactive compared to other members of its group, especially in its massive state, where it does not react with water at red heat, and it does not react with air below 600 °C. It can be ignited in air only when finely powdered, to give BeO and Be_3N_2. Beryllium does not react directly with hydrogen; consequently BeH_2 must be prepared by less direct methods, such as reduction of $BeCl_2$ in ether by LiH, or pyrolysis of Be(t-Bu)$_2$. Of the Group IIA(2) elements, only Be reacts with aqueous bases (NaOH or KOH) to liberate hydrogen and form the beryllate ion, $[Be(OH)_4]^{2-}$. The latter is also formed when beryllium hydroxide is dissolved is aqueous alkali. Thus beryllium metal and the hydroxide are chemically similar to aluminum and Al(OH)$_3$. The hydroxide, Be(OH)$_2$, has several polymorphs, the most stable of which is crystallized when boiled solutions of $BeCl_2$ and OH$^-$ are cooled.

Beryllium metal is unreactive towards cold, concentrated HNO$_3$, due to passivation. However, Be does react with concentrated solutions of noncomplexing acids (Chapter 7) to form the tetraaqua ion, $[Be(H_2O)_4]^{2+}$, crystalline salts of which may be readily obtained. The water ligands in such salts are more strongly bound than is typical of other divalent cations. For instance, $[Be(H_2O)_4]Cl_2$ does not lose H_2O over strong dessicants such as P_2O_5. The stability of Be complexes with ligands containing nitrogen or other donors is lower than that of complexes possessing oxygen donor ligands. Thus $[Be(NH_3)_4]Cl_2$ is thermally stable, but rapidly hydrolyzed to the tetraaqua ion.

The firm attachment of the H_2O molecules in $[Be(OH_2)_4]^{2+}$ causes a weakening of the O—H bonds. This means that the aqua ion is acidic, as shown in Reaction 11-2.1.

$$[Be(H_2O)_4]^{2+} = [Be(H_2O)_3OH]^+ + H^+ \qquad (11\text{-}2.1)$$

Thus aqueous solutions of beryllium salts are extensively hydrolyzed and are acidic. In fact, the $[Be(OH_2)_3(OH)]^+$ ion is itself unstable, and quickly trimerizes to give the $[Be_3(OH)_3(H_2O)_x]^{3+}$ ion.

The tetrafluoroberyllate ion, $[BeF_4]^{2-}$, is formed in fluoride-containing solutions. It is also obtained by dissolving BeO or $Be(OH)_2$ in concentrated aqueous fluoride solutions, or in nonaqueous melts of acid fluorides such as NH_4HF_2. The tetrafluoroberyllate anion behaves in crystals much like SO_4^{2-}; thus $PbBeF_4$ and $PbSO_4$ have similar structures and solubilities.

The white crystalline oxide BeO is obtained on ignition of Be or its compounds in air. It resembles Al_2O_3 in being highly refractory (mp 2570 °C). The high-temperature form (>800 °C) is exceedingly inert and dissolves readily only in a hot syrup of concentrated H_2SO_4 and $(NH_4)_2SO_4$. More reactive forms of BeO dissolve in hot aqueous alkali or fused $KHSO_4$.

Beryllium fluoride (BeF_2) is obtained as a glassy, hygroscopic mass by thermal decomposition of $(NH_4)_2BeF_4$. On a small scale, the chloride and the bromide are best obtained by direct interaction of the elements in a hot tube. Otherwise, $BeCl_2$ may be prepared by passing CCl_4 over BeO at 800 °C, or at 600–800 °C as in Reaction 11-2.2.

$$BeO + C + Cl_2 \longrightarrow BeCl_2 + CO \qquad (11\text{-}2.2)$$

As noted earlier, $BeCl_2$ forms long chains in the crystal, and this compound and the similar methyl derivative, $[Be(CH_3)_2]_n$, are cleaved by donor molecules or ions to give, for example in ethers, adducts of the type $BeCl_2(OR_2)_2$. (Such Lewis acid behavior is also typical of Al, Mg, and Zn halides and alkyls.) Beryllium chloride also dissolves exothermically in H_2O, and the salt $[Be(H_2O)_4]Cl_2$ can be obtained from aqueous hydrochloric acid solutions. In melts with alkali halides, chloroberyllate ions, $[BeCl_4]^{2-}$, are formed, although this ion, unlike the tetrafluoro ion, does not exist in aqueous solution.

The most unusual oxygen-containing complexes of Be have the formula $Be_4O(O_2CR)_6$ and are formed by refluxing $Be(OH)_2$ with carboxylic acids. These white crystalline compounds are soluble in nonpolar organic solvents, such as alkanes, but are insoluble in water and lower alcohols. In solution, the compounds are un-ionized and monomeric. They have the structure illustrated in Fig. 11-2. The central oxygen atom is tetrahedrally surrounded by the four Be atoms (this

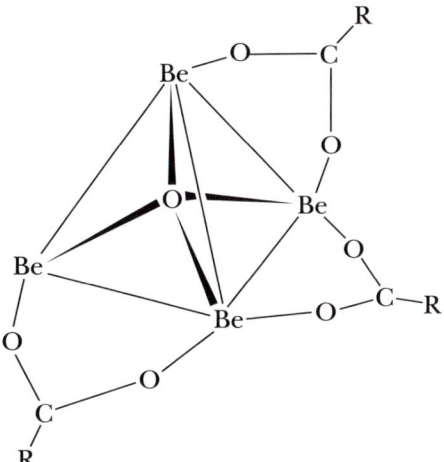

Figure 11-2 The structure of the "basic carboxylate" complexes $Be_4O(O_2CR)_6$. Only three of the six carboxylate groups are shown.

being one of the few cases, excepting solid oxides, in which oxygen is four coordinated), and each Be atom is tetrahedrally surrounded by four oxygen atoms.

Note that Be and its compounds are exceedingly poisonous. Inhalation of Be or Be compounds can cause serious respiratory disease, and soluble compounds may produce dermatitis on contact with the skin. Great precautions should be taken in handling either elemental Be or its compounds.

11-3   The Remaining Elements of Group IIA(2) and Their Properties

Magnesium

Magnesium is produced in several ways. Two important sources are dolomite rock and seawater, which contains 0.13% Mg. Dolomite is first calcined to give a CaO/MgO mixture from which the calcium can be removed by ion exchange using seawater. The equilibrium is favorable because the solubility of $Mg(OH)_2$ is lower than that of $Ca(OH)_2$

$$Ca(OH)_2 \cdot Mg(OH)_2 + Mg^{2+} \longrightarrow 2\,Mg(OH)_2 + Ca^{2+} \qquad (11\text{-}3.1)$$

The most important processes for obtaining the metal are (a) the electrolysis of fused halide mixtures (e.g., $MgCl_2 + CaCl_2 + NaCl$) from which the least electropositive metal (Mg) is deposited, and (b) the reduction of MgO or of calcined dolomite ($MgO \cdot CaO$). The latter is heated with ferrosilicon:

$$CaO \cdot MgO + FeSi = Mg + \text{silicates of Ca and Fe} \qquad (11\text{-}3.2)$$

and the Mg is distilled out. Magnesium oxide can be heated with coke at 2000 °C and the metal deposited by rapid quenching of the high-temperature equilibrium that lies well to the right.

$$MgO + C \rightleftharpoons Mg + CO \qquad (11\text{-}3.3)$$

Magnesium is grayish white and has a protective surface oxide film. Thus despite the favorable potential, it is not attacked by water unless it is amalgamated. Magnesium, however, is readily soluble in dilute acids. It is used in light constructional alloys and for the preparation of Grignard reagents (Section 29-5) by interaction with alkyl or aryl halides in ether solution. It is essential to life because it occurs in chlorophyll (cf. Section 31-4).

Calcium

Calcium, strontium, and barium are made only on a relatively small scale by reduction of the halides with Na. These elements are soft and silvery, resembling Na in their reactivities, although they are somewhat less reactive. Calcium is used for the reduction to the metal of actinide and lanthanide halides and for the preparation of CaH_2, which is a useful reducing agent.

11-4   Binary Compounds

Oxides

The oxides (MO) are white, high-melting crystalline solids, with NaCl-type lattices. They are obtained by calcining the carbonates. Calcium oxide, for instance, is made on a vast scale for the cement industry, as in Reaction 11-4.1.

$$CaCO_3 \longrightarrow CaO + CO_2(g) \qquad \Delta H° = 178.1 \text{ kJ mol}^{-1} \qquad (11\text{-}4.1)$$

Magnesium oxide is relatively inert, especially after ignition at high temperatures, but the other oxides react with H_2O, evolving heat, to form the hydroxides. They absorb CO_2 from the air. Magnesium hydroxide is insoluble in water ($\sim 1 \times 10^{-4}$ g L^{-1} at 20 °C) and can be precipitated from Mg^{2+} solutions; it is a much weaker base than the Ca to Ra hydroxides, although it has no acidic properties and unlike $Be(OH)_2$ is insoluble in an excess of hydroxide. The Ca to Ra hydroxides are all soluble in water, increasingly so with increasing atomic number [$Ca(OH)_2$, ~ 2 g L^{-1}; $Ba(OH)_2$, ~ 60 g L^{-1} at ~ 20 °C], and all are strong bases.

Halides

The anhydrous halides can be made by dehydration (Section 20-3) of the hydrated salts. Both Mg and Ca halides readily absorb water. The ability to form hydrates, as well as the solubilities in water, decrease with increasing size, and Sr, Ba, and Ra halides are normally anhydrous. This is because the hydration energies decrease more rapidly than the lattice energies with increasing size of M^{2+}. All the halides appear to be essentially ionic.

The fluorides vary in solubility in the reverse order, that is, Mg < Ca < Sr < Ba, because of the small size of the F^- relative to the M^{2+} ion. The lattice energies decrease unusually rapidly because the large cations make contact with one another without at the same time making contact with the F^- ions.

Other Compounds

The metals, like the alkalis, react with many other elements. Compounds such as phosphides, silicides, or sulfides are mostly ionic and are hydrolyzed by water.

Calcium *carbide,* obtained by reduction of the oxide with carbon in an electric furnace, is an acetylide $Ca^{2+}C_2^{2-}$. It can be employed as a source of acetylene:

$$Ca^{2+}C_2^{2-} + 2\,H_2O \longrightarrow Ca(OH)_2 + HC{\equiv}CH \qquad (11\text{-}4.2)$$

The binary hydrides MH_2 are ionic, apart from MgH_2, which is more covalent in nature. The compound CaH_2 reacts smoothly with water and is used as a drying agent for organic solvents and gases.

11-5   Oxo Salts, Ions, and Complexes

All the elements form *oxo salts;* those of Mg and Ca are often hydrated. The carbonates are all rather insoluble in water and the solubility products decrease with increasing size of M^{2+}; $MgCO_3$ is used in stomach powders to absorb acid. The same solubility order applies to the *sulfates;* magnesium sulfate which, as Epsom salt ($MgSO_4 \cdot 7\,H_2O$), is used as a mild laxative in "health" salts, is readily soluble in water. Calcium sulfate has a hemihydrate $2\,CaSO_4 \cdot H_2O$ (plaster of Paris) which readily absorbs more water to form the very sparingly soluble $CaSO_4 \cdot 2\,H_2O$ (gypsum), while Sr, Ba, and Ra sulfates are insoluble and anhydrous. Barium sulfate is accordingly used for "barium meals" as it is opaque to X-rays and provides a suitable shadow in the stomach. The *nitrates* of Sr, Ba, and Ra are also anhydrous and the last two can be precipitated from cold aqueous solution by the addition of fuming nitric acid. *Magnesium perchlorate* is used as a drying agent, but contact with organic materials must be avoided because of the hazard of explosions.

For water, acetone, and methanol solutions, NMR studies have shown that the coordination number of Mg^{2+} is six, although in ammonia it appears to be five. The $[Mg(H_2O)_6]^{2+}$ ion is not acidic and in contrast to $[Be(H_2O)_4]^{2+}$ can be dehydrated fairly readily. It occurs in a number of crystalline salts.

Only Mg^{2+} and Ca^{2+} show any appreciable tendency to form *complexes* and in solution, with a few exceptions, these are of oxygen ligands. The compounds $MgBr_2$, MgI_2, and $CaCl_2$ are soluble in alcohols and polar organic solvents. Adducts such as $MgBr_2(OEt_2)_2$ and $MgBr_2(THF)_4$ can be obtained.

Oxygen chelate complexes, among the most important being those with ethylenediaminetetraacetate (EDTA) type ligands, readily form in alkaline aqueous solution. For example,

$$Ca^{2+} + EDTA^{4-} = [Ca(EDTA)]^{2-} \qquad (11\text{-}5.1)$$

The cyclic polyethers and related nitrogen compounds form strong complexes whose salts can be isolated. The complexing of calcium by $EDTA^{4-}$ and by polyphosphates is of importance not only for removal of Ca^{2+} from water, but also for the volumetric estimation of Ca^{2+}.

Both Mg^{2+} and Ca^{2+} have important biological roles (Chapter 31). The tetrapyrrole systems in chlorophyll form an exception to the rule that complexes of Mg (and the other elements) with nitrogen ligands are weak.

11-6 Summary of Group Trends for the Elements of Group IIA(2)

By using the list of periodic chemical properties developed in Section 8-11, together with the information given in this chapter, we can summarize the periodic trends in the chemical properties of the elements of Group IIA(2).

1. Beryllium
   (a) Forms covalent compounds almost exclusively, even with the most electronegative elements.
   (b) Does not form ionic compounds containing simple Be^{2+} ions, but does readily achieve a maximum coordination number of four, through formation of complex ions such as BeF_4^{2-} and $Be(H_2O)_4^{2+}$, in which the Be-to-ligand bonds possess considerable covalent character.
   (c) Forms a series of organo derivatives, BeR_2 and $[BeR_3]^-$, which contain covalent Be—C bonds.
   (d) The oxide and especially the hydroxide are amphoteric, reacting either with acids or aqueous OH^-.
   (e) The halides are covalent polymers that are readily hydrolyzed or cleaved by donors.
   (f) The hydride is a covalent polymer.

2. Magnesium
   (a) Forms ionic substances that have partial covalent character.
   (b) Forms many ionic substances containing the uncomplexed Mg^{2+} ion, and forms numerous coordination compounds having a maximum coordination number of six.
   (c) Forms an important series of organo derivatives, namely, the Grignard reagents $RMgX$ and the dialkyls MgR_2, both of which are discussed in Chapter 29.
   (d) The oxide is basic, and the hydroxide is only weakly basic compared to the lower members of the group. Also, the hydroxide, unlike $Be(OH)_2$, does not dissolve in aqueous hydroxide.
   (e) The halides are essentially ionic.
   (f) The hydride is only partially covalent.

3. Calcium, Strontium, and Barium

   (a) Form only ionic substances.
   (b) Do not form covalent bonds as in the alkyls of magnesium.
   (c) The oxides are basic, and the hydroxides are strong bases, the solubility increasing with atomic number.
   (d) The halides are crystalline ionic substances that are readily hydrated.
   (e) The hydrides are ionic and powerfully hydridic, as discussed in Chapter 9, and illustrated in Fig. 9-4.

STUDY GUIDE

Study Questions

A. Review

1. Name the important minerals of the Group IIA(2) elements.
2. Why do these metals have higher melting points than the alkali metals?
3. Why does beryllium tend to form covalent compounds?
4. Why do linear molecules X—Be—X exist only in the gas phase?
5. Which compound, when dissolved in water, would give the most acid solution, $BeCl_2$ or $CaCl_2$?
6. Draw the structures of $BeCl_2$ and $CaCl_2$ in the solid state.
7. How is magnesium made?
8. What are the properties of the hydroxides, $M(OH)_2$?
9. How do the solubilities of (a) hydroxides, (b) chlorides, and (c) sulfates vary in Group IIA(2)?
10. What and where are the Dolomites from which $MgCO_3 \cdot CaCO_3$ gets its name?
11. What is an important fact about beryllium compounds from a safety point of view?
12. Compare the physical properties of Be, Mg, Ca, and Sr.
13. Do the alkaline earth cations form many complexes? Which cations tend most to do so and what are the best complexing agents?
14. What are the main types of compounds formed by the alkaline earth elements? Are they generally soluble in water?

B. Additional Exercises

1. Beryllium readily forms a compound of stoichiometry $Be_4O(CO_2CH_3)_6$. Write a likely structure for this compound.

2. Write a balanced chemical equation for the synthesis of hydrogen peroxide using barium oxide.

3. Why do you think that the usual coordination numbers for Be^{2+} and Mg^{2+} are four and six, respectively?

4. The hydroxide of beryllium (actually a hydrous metal oxide) is a white, gelatinous substance that is amphoteric. Write balanced chemical equations for its reaction with H^+ and with OH^-.

5. Why does the increase in the number of valence electrons for the alkaline earth metals over that for the alkali metals give the alkaline earths higher melting points, higher boiling points, and higher densities?

6. Write balanced equations for one method of preparation of each of the metals of this group.

7. Describe the bonding in the chainlike $[Be(CH_3)_2]_n$.

8. Describe the bonding in $BeCl_2(g)$ and $[BeCl_2]_n(s)$.

9. Sketch a likely structure for $(BeCl_2)_2$, based on information provided in this chapter.

10. What type of compound does one expect on dissolution of $BeCl_2$ in donor solvents? Give two examples, with equations.

11. Write equations for each of the following reactions:
    (a) Reduction of $BeCl_2$ with magnesium.
    (b) Ignition of finely powered Be in air.
    (c) Dissolution of Be in aqueous KOH.
    (d) Dissolution of $Be(OH)_2$ in aqueous KOH.
    (e) Hydrolysis of $Be(NH_3)_4Cl_2$.
    (f) Ligand substitution in the tetraaquaberyllate ion by excess aqueous fluoride ion.
    (g) A nonaqueous synthesis of the tetrafluoroberyllate ion.
    (h) Thermal decomposition of $(NH_4)_2BeF_4$.
    (i) Hydrolysis of $BeCl_2$.
    (j) Reaction of beryllium hydroxide in refluxing acetic acid.

12. Unlike the aqua ion of beryllium, $Mg^{2+}(aq)$ has coordination number six. Also, the aqua ion of Mg^{2+} undergoes more rapid water exchange (Chapter 6) and does not ionize a proton as shown for $[Be(H_2O)_4]^{2+}$ in Reaction 11-2.1. Explain these differences based on a comparison of the properties of Be and Mg.

C. Problems from the Literature of Inorganic Chemistry

1. Consider the paper by R. Aruga, *Inorg. Chem.*, **1980**, *19*, 2895–2896.
    (a) What are the three series or behaviors in stability constants that are listed in the introduction?
    (b) How is each series distinguished?
    (c) Into which series do the Group IIA(2) cation complexes of iminodiacetate fit? of thiosulfate? of sulfate? of malate?
    (d) For which behavior (series) is entropy an important factor in determining the stability of the complexes?

2. Answer the following questions concerning lithium tri-*tert*-butylberyllate after reading the article by J. R. Wermer, D. F. Gaines, and H. A. Harris, *Organometallics*, **1988**, *7*, 2421–2422.

(a) What are the important structural facts for the title compound in sofaras the Be atom is concerned? Concerning the Li atom?

(b) What facts about the structure and physical properties of the title compound indicate a covalent nature for the bonding of Li in this compound?

SUPPLEMENTARY READING

Bell, N. A., "Beryllium Halides and Complexes," *Adv. Inorg. Chem. Radiochem.,* **1972,** *14,* 225.

Boynton, R. S., *Chemistry and Technology of Lime and Limestone,* 2nd ed., Wiley, New York, 1980.

Dietrich, B., "Coordination-Chemistry of Alkali and Alkaline-Earth Cations with Macrocyclic Ligands," *J. Chem. Educ.,* **1985,** *63,* 954.

Everest, D. A., *The Chemistry of Beryllium,* Elsevier, Amsterdam, 1964.

Hughes, M. N. and Birch, N. J., "IA and IIA Cations in Biology," *Chem. Br.,* **1982,** 196–198.

Poonia, N. S. and Bajag, A. V., "Complexes of the Group II Elements," *Coord. Chem. Rev.,* **1988,** *87,* 55.

Skilleter, D. N., "Properties, Uses and Toxicity of Beryllium," *Chem. Br.,* **1990,** 26.

Sobota, P., "MgCl$_2$ Reactions and Complexes," *Polyhedron,* **1992,** *11,* 715.

Spiro, T. G., Ed., *Calcium in Biology,* Wiley-Interscience, New York, 1983.

Wacker, W. E. C., *Magnesium and Man,* Harvard University Press, Cambridge, MA, 1980.

Chapter 12

BORON

12-1 Introduction

The principal ores of boron are borates such as:

Ulexite $\{NaCa[B_5O_6(OH)_6] \cdot 5\,H_2O\}$

Borax $\{Na_2[B_4O_5(OH)_4] \cdot 8\,H_2O\}$

Colemanite $\{Ca_2[B_3O_4(OH)_3]_2 \cdot 2\,H_2O\}$

Kernite $\{Na_2[B_4O_5(OH)_4] \cdot 2\,H_2O\}$

The structures of borate minerals are complex and diverse, but they characteristically contain trigonal BO_3 or tetrahedral BO_4 units in large boron–oxygen anions. Some oxygen atoms in borate minerals are monoprotonated to give hydroxyl groups, while others are diprotonated to give waters of hydration. The cations in such minerals are usually alkali or alkaline earth cations. The structure of the borate anion in borax is shown in Structure 5-XXVII. Borax occurs in large deposits in the Mojave desert of California, which is the major source of boron.

No ionic compounds involving simple B^{3+} cations are formed because the ionization enthalpies for boron are so high that lattice energies or hydration enthalpies cannot offset the energy required for formation of a cation. Boron does form three covalent bonds using sp^2 hybrid orbitals in a trigonal plane. All such BX_3 compounds are coordinatively unsaturated and act as strong Lewis acids; interaction with Lewis bases (molecules or ions) gives tetrahedral adducts such as $BF_3 \cdot O(C_2H_5)_2$, BF_4^-, and $B(C_6H_5)_4^-$. The formation of such Lewis acid–base adducts requires a change to sp^3 hybridization for boron.

Another major feature of boron chemistry is the preponderance of compounds consisting of boron atoms in closed polyhedra or in open, basketlike arrangements. Often the structures are seen to be derivatives or fragments of the icosahedron. The frameworks of such molecules may include atoms other than boron (e.g., carbon) and many of those with carbon (the carboranes) form complexes with transition metals.

Among the Group IIIB(13) elements, it is the chemistry of boron that is unique. The chemistry of boron has only a few features in common with that of aluminum. The main resemblances to silicon and differences from the more metallic aluminum are as follows:

1. The oxide B_2O_3 and $B(OH)_3$ are acidic. The compound $Al(OH)_3$ is a basic hydroxide, although it shows weak amphoteric properties by dissolving in strong NaOH.

2. Borates and silicates are built on similar structural principles with sharing of oxygen atoms so that complicated chain, ring, or other structures result.

3. The halides of B and Si (except BF_3) are readily hydrolyzed. The Al halides are solids and only partly hydrolyzed by water. All act as Lewis acids.

4. The hydrides of B and Si are volatile, spontaneously flammable, and readily hydrolyzed. Aluminum hydride is a polymer, $(AlH_3)_n$.

12-2   Isolation of the Element

Boron forms a number of allotropes that are difficult to purify because of the high melting points of the solids (e.g., 2250 °C for the β-rhombohedral form) and because of the corrosive nature of the liquid. Boron is made in 95–98% purity as an amorphous powder by reduction of the oxide B_2O_3 with Mg

$$B_2O_3 + 3\,Mg \longrightarrow 2\,B + 3\,MgO \qquad (12\text{-}2.1)$$

followed by washing of the powder with NaOH, HCl, and HF. Other electropositive metals may be used in place of Mg. Purer forms of the element are available from the reductions of boron trihalides with zinc at 900 °C, as in Reaction 12-2.2

$$2\,BCl_3 + 3\,Zn \longrightarrow 3\,ZnCl_2 + 2\,B \qquad (12\text{-}2.2)$$

or from reductions with hydrogen over hot tantalum metal as a catalyst, as in Reaction 12-2.3

$$2\,BX_3 + 3\,H_2 \longrightarrow 6\,HX + 2\,B \qquad (12\text{-}2.3)$$

The several allotropes of crystalline boron all have structures built up of B_{12} icosahedra (Structure 8-II), one form differing from another by the manner in which these icosahedra are packed into the unit cell.

Crystalline boron is very inert and is attacked only by hot concentrated oxidizing agents. Amorphous boron is more reactive. With ammonia for instance, amorphous boron at white heat gives $(BN)_x$, a slippery white solid with a layer structure resembling that of graphite, but with hexagonal rings of alternating B and N atoms.

12-3   Oxygen Compounds of Boron

Almost all of the naturally occurring forms of boron are the oxygen-containing borate minerals, which are mentioned in the introduction to this chapter and in Section 5-4. In addition, there are many types of organic derivatives containing boron–oxygen bonds, the chief examples being those containing trigonal boron: the orthoborates, $B(OR)_3$; the acyl borates, $B(OCOR)_3$; the peroxo borates, $B(OOR)_3$; and the boronic acids, $RB(OH)_2$, all of which are best considered to be derivatives of boric acid. We consider first the borate-containing compounds.

Crystalline Borates

Many borates occur naturally, usually in hydrated form. Anhydrous borates may be made by fusion of boric acid and metal oxides. The hydrated borates may be precipitated from aqueous solution. The stoichiometries of borates (e.g., $KB_5O_8 \cdot 4\,H_2O$, $Na_2B_4O_7 \cdot 10\,H_2O$, CsB_2O_4, and $Mg_3B_7O_{13}Cl$) give little idea of the structure of the anions in these substances. The main structural principles of the borates are similar to those for silicates: cyclic or linear polyoxo anions, formed by the linking together of BO_3 and/or BO_4 units shared by oxygen atoms.

In crystalline borates, the most common structural units are those shown in Fig. 12-1. Notice that the skeletal boron–oxygen units may be protonated to varying degrees, and that the boron atoms are either tetrahedral or trigonal. Recall also the structure of the borate anion in borax, Structure 5-XXVII. The largest discrete borate anion known is $B_{10}O_{21}^{12-}$, which consists of two tetraborate units linked by two BO_3 triangles.

In *anhydrous borates,* the BO_3^{3-} and $B_3O_6^{3-}$ ions are common, as is the infinite chain anion $(BO_2)_n^{n-}$, which occurs in $Ca(BO_2)_2$. Planar BO_3 units are linked in three dimensions in the mineral tourmaline. Also common in minerals are networks of $B_6O_{12}^{6-}$ and $B_3O_6^{3-}$ linked by shared oxygen atoms.

Hydrated borates contain polyoxo anions in the crystal, with the following important structural features.

1. Both BO_3 and tetrahedral BO_4 groups are present, the number of BO_4 units being equal to the charge on the anion. Thus $KB_5O_8 \cdot 4\,H_2O$ has one BO_4 and four BO_3, whereas $Ca_2B_6O_{11} \cdot 7\,H_2O$ has four BO_4 and two BO_3 groups.

2. Anions that do not have BO_4 groups, such as metaborate, $B_3O_6^{3-}$, or metaboric acid, $B_3O_3(OH)_3$, hydrate rapidly and lose their original structures. Also, although certain complex borates can be crystallized from solution, this does not constitute evidence for the existence of these ions in solution, since other less complex polyoxo anions can readily combine during the crystallization process.

3. Certain discrete as well as chain-polymer borate anions can be formed by the linking of two or more rings by shared tetrahedral boron atoms.

Examples of many of these structural features are illustrated in Fig. 12-1.

Boric Acid

The acid $B(OH)_3$ can be obtained as white needles either from borates, or by hydrolysis of boron trihalides. The $B(OH)_3$ units are linked together by hydrogen bonds to form infinite layers of nearly hexagonal symmetry. The layers are 3.12 Å apart, and the crystals are readily cleaved along interlayer planes. Some reactions of boric acid are given in Fig. 12-2.

When heated, boric acid loses water stepwise to form one of three forms of metaboric acid, HBO_2. If $B(OH)_3$ is heated below 130 °C, the so-called form-III is obtained, which has a layer structure in which B_3O_3 rings are joined by hydrogen bonding. On continued heating of form-III of HBO_2, between 130 and 150 °C, HBO_2-II is formed. This compound has a more complex structure containing both BO_4 tetrahedra and B_2O_5 groups in chains linked by hydrogen bonds. Finally, on heating of form-II above 150 °C, cubic HBO_2 (form-I) is obtained, in which all boron atoms are four coordinate. Complete fusion of boric acid gives the oxide B_2O_3 as a glass. The melt readily dissolves metal oxides to give borate glasses. It also reacts with silica to give the borosilicate glass known as Pyrex.

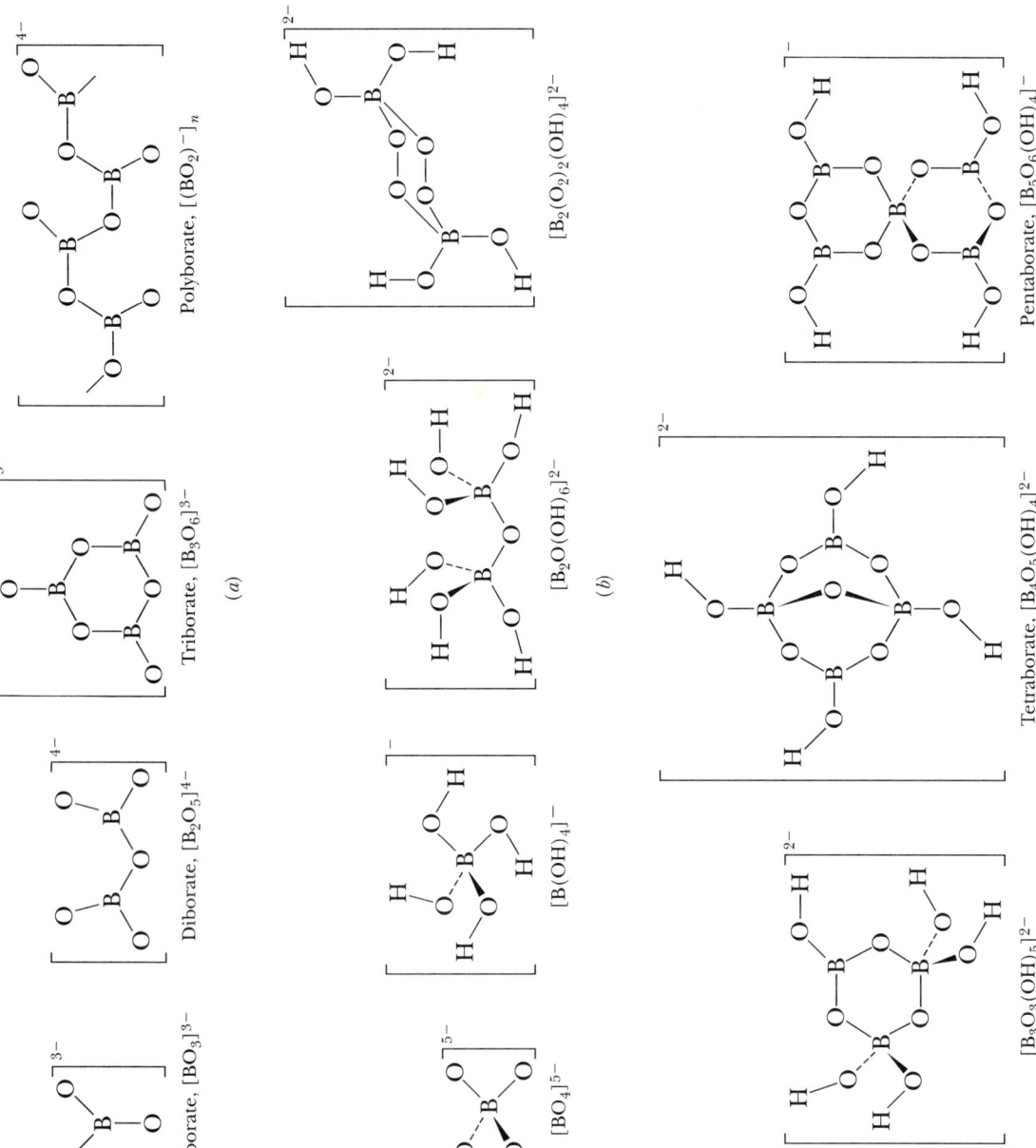

Figure 12-1   The structure of borate anions in boron-containing minerals. (*a*) Anions containing boron in planar BO₃ units. (*b*) Anions containing boron in tetrahedral BO₄ units. (*c*) Anions containing boron in both planar BO₃ and tetrahedral BO₄ units.

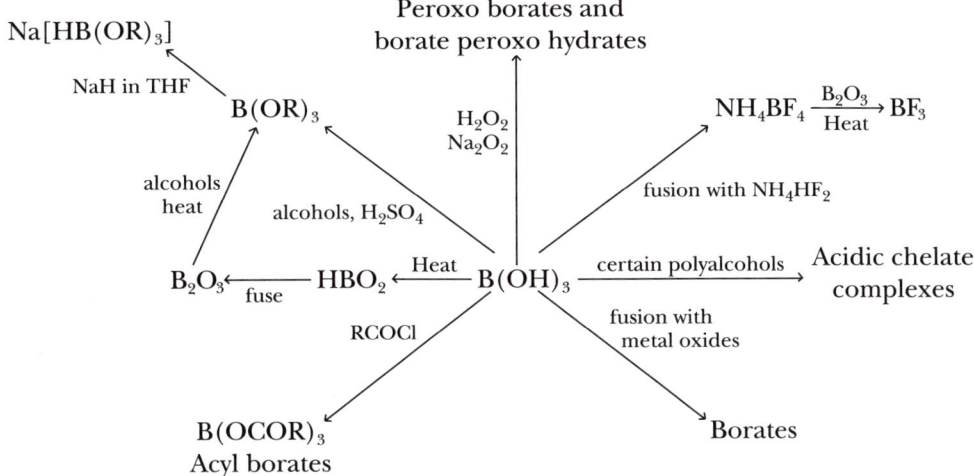

Figure 12-2   Some reactions of boric acid.

Boric acid is readily converted to alkyl or aryl orthoborates, $B(OR)_3$, by condensation with alcohols in the presence of sulfuric acid. These compounds are usually colorless liquids that are converted, in ether solvents, by alkali metal hydrides to the very useful reducing agents $[HB(OR)_3]^-$. The reactivity and selectivity of the latter as reducing agents can be controlled by changing the R groups.

Treatment of boric acid with sodium peroxide leads to peroxoborates, variously formulated as $NaBO_3 \cdot 4\,H_2O$ or $NaBO_2 \cdot H_2O_2 \cdot 3\,H_2O$, which are extensively used in washing powders because they afford H_2O_2 in solution. For example, in solution there is the equilibrium shown in Reaction 12-3.1:

$$[B(OH)_3(O_2H)]^- + H_2O = [B(OH)_4]^- + H_2O_2 \qquad (12\text{-}3.1)$$

Boric Acid and Borate Ions in Solution

Boric acid is moderately soluble in water, where it acts as a weak Lewis acid towards OH^-

$$B(OH)_3 + H_2O = B(OH)_4^- + H^+ \qquad pK = 9.0 \qquad (12\text{-}3.2)$$

The $B(OH)_4^-$ ion occurs in many borate-type minerals, but most borates (especially those formed by fusion of boric acid with metal oxides) have complex structures such as the ring anion (Structure 12-I). Boric acid and borates form very stable complexes with 1,2-diols, as in Structure 12-II.

The concentration of boric acid can be determined by complexation with a diol such as glycerol (Reaction 12-3.3)

$$B(OH)_3 + 2 \quad \underset{\wedge\ \wedge}{\overset{HO\ OH}{\underset{|}{\overset{|}{C}}-\underset{}{\overset{|}{C}}}} \longrightarrow H_3O^+ + \left[\begin{array}{c} \overset{\backslash|}{C}-O \qquad O-\overset{|/}{C} \\ | \qquad\qquad B \qquad | \\ \underset{/|}{C}-O \qquad O-\underset{|\backslash}{C} \end{array} \right]^- + 2\,H_2O \qquad (12\text{-}3.3)$$

followed by titration with NaOH.

As noted previously, in concentrated solutions of boric acid, polyoxo borate anions are also present. These are formed, for example, as in equilibrium 12-3.4.

$$2\,B(OH)_3 + B(OH)_4^- = B_3O_3(OH)_4^- + 3\,H_2O \qquad (12\text{-}3.4)$$

Equilibria between various borate anions is rapidly established in aqueous solution, as shown by rapid exchange between $B(OH)_3$ labeled with ^{18}O and borates. The larger polyoxo anions $B_5O_6(OH)_4^-$ and $B_4O_5(OH)_4^{2-}$ are formed at higher pH. In dilute solutions, however, depolymerization to the mononuclear species occurs. Thus when borax is dissolved in dilute solution, $B(OH)_4^-(aq)$ is formed.

12-4   The Halides of Boron

Trihalides

Boron trifluoride is a pungent, colorless gas (bp -101 °C) that is obtained by heating B_2O_3 with NH_4BF_4, or with CaF_2 and concentrated H_2SO_4. It is commercially available in tanks.

Boron trifluoride is one of the strongest Lewis acids known and reacts readily with most Lewis bases, such as ethers, alcohols, amines, or water to give adducts, and with F^- to give the tetrafluoroborate ion, BF_4^-. The diethyletherate, $(C_2H_5)_2OBF_3$, a viscous liquid, is available commercially. Unlike the other halides, BF_3 is only partially hydrolyzed by water:

$$4\,BF_3 + 6\,H_2O = 3\,H_3O^+ + 3\,BF_4^- + B(OH)_3 \qquad (12\text{-}4.1)$$

$$BF_4^- + H_2O = [BF_3OH]^- + HF \qquad (12\text{-}4.2)$$

Because of this, and its potency as a Lewis acid, BF_3 is widely used to promote various organic reactions. Examples are

1. Ethers or alcohols + acids → esters + H_2O or ROH.
2. Alcohols + benzene → alkylbenzenes + H_2O.
3. Polymerization of alkenes and alkene oxides such as propylene oxide.
4. Friedel–Crafts-like acylations and alkylations.

In (1) and (2) the effectiveness of BF_3 must depend on its ability to form an adduct with one or both of the reactants, thus lowering the activation energy of the rate-determining step in which H_2O or ROH is eliminated by breaking of C—O bonds. In reactions of type (4), intermediates may be characterized at low temperatures. Thus the interaction of benzene and C_2H_5F proceeds as in

Reaction 12-4.3. It is clear that BF_3 is not actually a catalyst, since it must be present in stoichiometric amount and is consumed in removing HF as HBF_4.

$$C_2H_5F + BF_3 \longrightarrow [C_2H_5^{\delta+}\text{---}F\text{--}^{\delta-}BF_3] \xrightarrow{C_6H_6} \left[\left\langle\!\!\left\langle \bigcirc \right\rangle\!\!\right\rangle \!\!\begin{array}{c} H \\ \\ C_2H_5 \end{array} \right]^+ + BF_4^-$$

$$\left\langle\!\!\left\langle \bigcirc \right\rangle\!\!\right\rangle\!\!-C_2H_5 + HBF_4 \qquad (12\text{-}4.3)$$

Fluoroboric acid solutions are formed on dissolving $B(OH)_3$ in aqueous HF

$$B(OH)_3 + 4\,HF = H_3O^+ + BF_4^- + 2\,H_2O \qquad (12\text{-}4.4)$$

The commercial solutions contain 40% acid. Fluoroboric acid is a strong acid and cannot, of course, exist as HBF_4. The ion is tetrahedral and fluoroborates resemble the corresponding perchlorates in their solubilities and crystal structures. Like ClO_4^- and PF_6^-, the anion has a low tendency to act as a ligand toward metal ions in *aqueous* solution. In nonaqueous media, there is evidence for complex formation.

Boron trichloride (bp 12 °C) and the *bromide* (bp 90 °C) are obtained by direct interaction at elevated temperatures. They fume in moist air and are violently hydrolyzed by water.

$$BCl_3 + 3\,H_2O = B(OH)_3 + 3\,HCl \qquad (12\text{-}4.5)$$

The rapid hydrolysis supports other evidence that these halides are stronger Lewis acids than BF_3.

Reactions of the Trihalides of Boron

As already mentioned, the boron trihalides are Lewis acids, and they readily react with Lewis bases to form adducts. Two other important reactions that we shall consider are halide exchange among the trihalides themselves, and the elimination of HX from adducts of the trihalides when an acidic hydrogen is available.

Formation of Adducts with Lewis Bases

Even the weakest of bases will form adducts with the trihalides of boron. Ethers, amines, phosphines, alcohols, anions, carbon monoxide, and the like all form adducts by donation of an electron pair to boron. The rehybridization of boron that accompanies adduct formation results in a loss in BX double-bond character, as shown in Fig. 12-3. When the Lewis donor is trimethylamine, the enthalpy change for adduct formation, as in Reaction 12-4.6,

$$BX_3 + :N(CH_3)_3 \longrightarrow X_3B{:}N(CH_3)_3 \qquad (12\text{-}4.6)$$

is most negative for BBr_3 and least negative for BF_3. We would expect that the higher electronegativity of fluorine should enhance the stability of the trimethyl-

sp² Boron   sp³ Boron

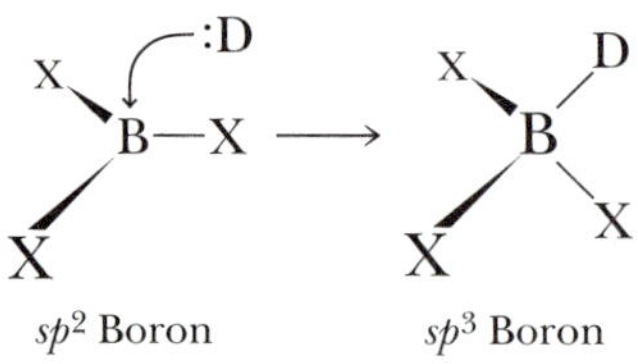

Figure 12-3 The reaction of a trigonal trihalide of boron with a Lewis base (:D) to give a tetrahedral adduct. The rehybridization of boron that is required when :D disrupts the B—X π bond in the BX$_3$ reactant.

amine adduct with BF$_3$. Since the enthalpy of adduct formation is least favorable with BF$_3$, however, it is concluded that the loss in BX double-bond character upon rehybridization to form an adduct is greater with BF$_3$ than in the other trihalides. From this we can conclude that the double-bond character in the trihalides follows the order BF$_3$ > BCl$_3$ > BBr$_3$, a trend opposite to that expected from the electronegativities of the halides. (Recall that the double bond in BX$_3$ results from donation of π-electron density from X into an empty $2p$ atomic orbital of an sp^2-hybridized boron atom. The π-bond system in these sorts of molecules was discussed in Section 3-6.) Evidently the π-bond system in BF$_3$ is especially strong because of effective overlap between the boron $2p$ and the fluorine $2p$ atomic orbitals—overlap that is effective because of the closely matched energies and sizes of the orbitals. The $3p$ and $4p$ atomic orbitals of Cl and Br have the proper symmetry for π overlap with the $2p$ atomic orbital of B in the compounds BX$_3$, but the π overlap is less effective because the energies and sizes of the π-donor orbitals ($3p$ for chlorine and $4p$ for bromine) are not well matched to those of the π-acceptor ($2p$) orbital of boron.

Halide Exchange Reactions Among the Boron Trihalides

Mixtures of two different trihalides of boron undergo exchange of halide as illustrated in Reaction 12-4.7.

$$BCl_3 + BBr_3 = BCl_2Br + BBr_2Cl \tag{12-4.7}$$

The position of equilibrium in Reaction 12-4.7 lies mostly to the left, but small amounts of the exchange products can be detected spectroscopically. The trifluoride undergoes halide exchange less readily than BBr$_3$ and BCl$_3$. No intermediates have been detected, but it is reasonable to propose that the exchange involves the type of dimeric structure shown in Structure 12-III. Such a dimer would be similar to Al$_2$Cl$_6$ (Structure 6-I).

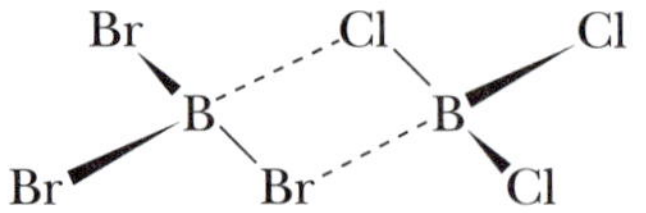

12-III

Equilibria of the type illustrated by Reaction 12-4.7 are established rapidly, and only small amounts of the exchange products can be detected. Attempts to isolate the exchange products from such systems are not successful because of the facile nature of the equilibria. Thus, no pure mixed halide of boron is known. A concerted mechanism, as illustrated in Structure 12-III, would be consistent with all of the facts as long as the new, bridging BX bonds that form the dimer are weak.

Elimination Reactions of BX_3 Adducts

When an acidic hydrogen is present in an adduct of BX_3, elimination of HX is possible, as illustrated in the following reactions. Hydrolysis of BCl_3 by alcohols involves adduct formation followed by elimination of HCl as in Reaction 12-4.8.

$$BCl_3 + C_2H_5OH \longrightarrow Cl_2B\!-\!OC_2H_5 + HCl \qquad (12\text{-}4.8)$$

Stepwise addition and elimination eventually leads to complete solvolysis as in Reaction 12-4.9:

$$BCl_3 + 3\,C_2H_5OH \longrightarrow B(OC_2H_5)_3 + 3\,HCl \qquad (12\text{-}4.9)$$

The dimethylamine adduct of BCl_3 undergoes elimination of HCl to give an aminoborane as in Reaction 12-4.10.

$$(CH_3)_2NH{:}BCl_3 \longrightarrow (CH_3)_2N\!-\!BCl_2 + HCl \qquad (12\text{-}4.10)$$

Aminoboranes, and in particular, the nature of the BN bond in aminoboranes, will be discussed in Section 12-6.

Subhalides of Boron

A number of interesting subhalides of boron, in which the proportion of halogen to boron is less than 3:1, are known. The best characterized are

1. The gaseous monohalides BF and BCl.
2. The so-called monohalides of Cl, Br, and I: B_nCl_n ($n = 8$, 9, 10, or 11); B_nBr_n ($n = 7$, 8, 9, or 10); and B_nI_n ($n = 8$ or 9).
3. The diboron tetrahalides, B_2X_4, X = F, Cl, Br, or I, although the last one has been little studied.
4. Certain other fluorides: B_3F_5, B_8F_{12}, and $B_{14}F_{18}$.

All of these require special synthetic techniques to avoid reactions with air, water, and even hydrocarbon or silicone greases. For instance, the diboron tetrahalides are pyrophoric, water sensitive, and thermally unstable either at room temperature (B_2F_4, B_2Cl_4, or B_2Br_4) or at the melting point (B_2I_4). The compounds have been characterized by mass spectrometry, and by IR and Raman spectroscopies in the solid, liquid, and gas. In some cases (see below), structures have been determined by X-ray crystallography.

Boron monochloride, BCl, is produced when B_2Cl_4 is passed rapidly through a quartz tube at 1000 °C. It is also obtained by electric discharge of B_2Cl_4 at a copper electrode, at liquid nitrogen temperature. Boron monofluoride, BF, is produced by passing BF_3 gas over boron at 1950-2000 °C.

Diboron tetrafluoride (B_2F_4) is made by fluorination of B_2Cl_4 with either SbF_3 or TiF_4. Also, condensation of BF together with BF_3 converts about 25% of the BF to B_2F_4. Triboron pentafluoride (B_3F_5) is obtained by condensing BF with B_2F_4. Disproportionation of liquid B_3F_5 at -30 °C gives B_8F_{12}, as in Reaction 12-4.11:

$$4 \, B_3F_5 \longrightarrow 2 \, B_2F_4 + B_8F_{12} \tag{12-4.11}$$

which may then be separated from B_2F_4 by fractional distillation.

Diboron tetrachloride (B_2Cl_4) is made from BCl_3 by radiofrequency discharge in the presence of mercury, as in Reaction 12-4.12.

$$2 \, BCl_3 + 2 \, Hg \longrightarrow B_2Cl_4 + Hg_2Cl_2 \tag{12-4.12}$$

This compound can also be made by condensation of gaseous B_2O_2 with BCl_3 at -196 °C, as in Reaction 12-4.13.

$$2 \, B_2O_2 + 4 \, BCl_3 \longrightarrow 2 \, B_2O_3(s) + 3 \, B_2Cl_4 \tag{12-4.13}$$

Condensation of BCl with BCl_3 at -196 °C gives B_4Cl_4. Boron monochloride is produced when B_2Cl_4 is rapidly passed through a quartz tube at 1000 °C. It is also obtained from B_2Cl_4 by electric discharge from copper electrodes at liquid nitrogen temperatures. The thermal decomposition of B_2Cl_4 at temperatures between 100 and 450 °C gives a mixture of B_nCl_n compounds in which $n = 8, 9, 10$, or 11, from which the individual compounds may be separated. The relative amounts of the various compounds produced by this method depend on the temperature used. Recently, B_9X_9 molecules (X = H, Cl, Br, or I, but not F) have been prepared by oxidation of $B_9X_9^{2-}$ ions using sulfuryl chloride in CH_2Cl_2, starting with the salts $[n\text{-}Bu_4N]_2[B_9X_9]$.

Diboron tetrabromide (B_2Br_4) is made by radiofrequency discharge of $BBr_3(g)$ in the presence of Hg, or by treating B_2Cl_4 with excess BBr_3. It decomposes readily at room temperature to produce B_7Br_7, B_9Br_9, and $B_{10}Br_{10}$. Reaction of B_8Cl_8 with Al_2Br_6 in BBr_3 solvent at 100 °C affords B_8Br_8. Through silent electric discharge, B_2Br_4 gives B_9Br_9 and BBr_3.

Diboron tetraiodide (B_2I_4) is obtained by radiofrequency discharge of BI_3. It is also produced (among other compounds) by reacting I_2 with $Zr(BH_4)_4$ (Fig. 12-7). Above its melting point (92–94 °C), B_2I_4 decomposes to give B_9I_9 and B_8I_8.

In the low-temperature crystal, both B_2F_4 and B_2Cl_4 are planar, with XBX angles close to 120°. In the case of B_2Cl_4, the planarity in the crystal is evidently due to crystal packing forces that overcome the steric considerations that would otherwise favor a staggered conformation. This can be deduced from the fact that in the liquid and the gas, the staggered conformation (Structure 12-IV) is preferred, with a barrier to rotation about the B—B bond of 1.85 kcal mol^{-1}. There has been some disagreement over the liquid and gas phase structures of B_2F_4, but recently Raman and IR analysis suggested that it is planar in the liquid and gas too, with an exceedingly small barrier (< 1.1 kcal mol^{-1}) to rotation about the B—B bond. This is in agreement with theoretical calculations.

12-IV

In both B_2Cl_4 and B_2F_4, the B—X bonds are somewhat shorter than is expected from the sum of the single-bond covalent radii. This suggests a delocalized π-bond system as in Structure 12-V.

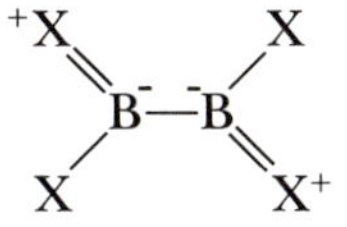

$$\begin{array}{c} {}^{+}X \diagdown \qquad X \\ B^{\cdot\cdot}\!\!-\!\!B \\ X \diagup \qquad X^{+} \end{array}$$

12-V

Such a π-bond system, when conjugated across the B—B bond, should favor planar geometry. Evidently, in the case of B_2Cl_4 (though not for B_2F_4), this is outweighed by steric considerations that should favor the staggered form.

The tetrahedral structure of B_4Cl_4 (Structure 12-VI) has been determined by X-ray crystallography and by IR and Raman spectroscopy in the gas phase. Similarly, a dodecahedron forms the basis for the structure of B_8Cl_8 (Structure 12-VII), whereas B_9Cl_9 (Structure 12-VIII) is based on a tricapped trigonal prism.

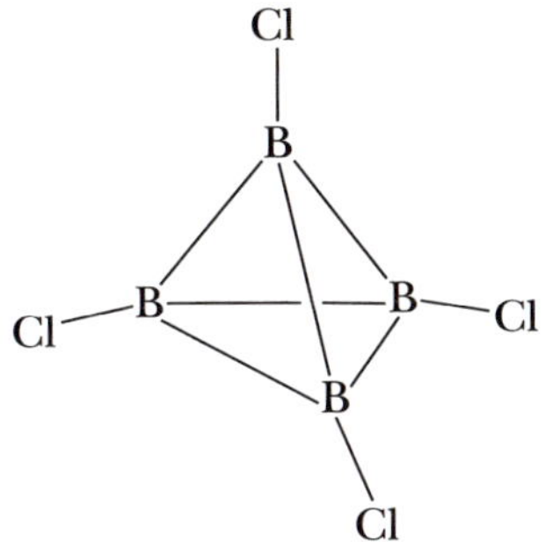

12-VI

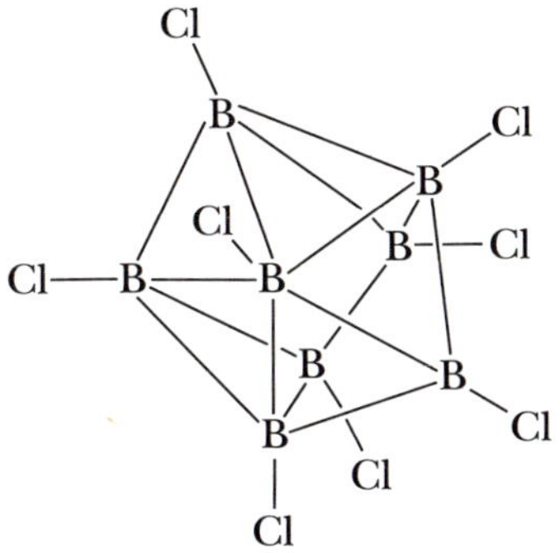

12-VII

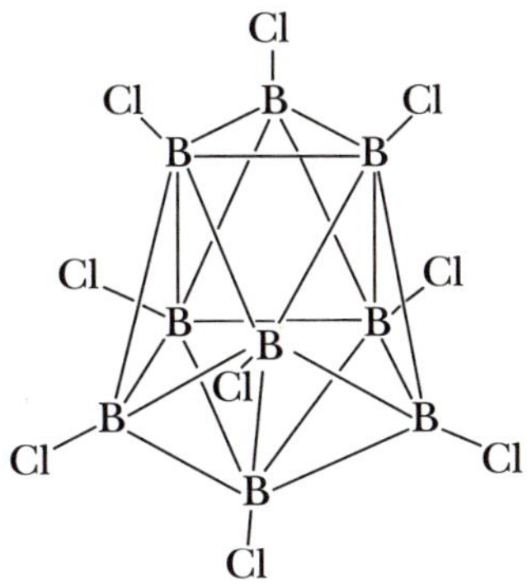

12-VIII

12-5 The Hydrides of Boron

Boranes

Boron forms an extensive series of molecular hydrides called *boranes*. Typical boranes are B_2H_6, B_4H_{10}, B_9H_{15}, $B_{10}H_{14}$, and $B_{20}H_{16}$. Boranes were first prepared between 1912 and 1936 by Alfred Stock who developed vacuum line techniques to handle these reactive materials. Stock's original synthesis (the reaction of Mg_3B_2 with acid) is now superseded for all but B_6H_{10}. Most syntheses now involve thermolysis of B_2H_6 under varied conditions, often in the presence of hydrogen.

The properties of some boranes are listed in Table 12-1. The nomenclature that is used for boranes is straightforward: the number of boron atoms is indicated by the prefix, and the number of hydrogen atoms is indicated parenthetically. For example, B_4H_{10} is *tetra*borane(*10*).

Diborane(6)

Diborane(6) (B_2H_6) is a gas (bp −92.6 °C) that is spontaneously flammable in air and instantly hydrolyzed by H_2O to H_2 and boric acid. It is obtained virtually quantitatively in ether, at room temperature, by the reaction of sodium borohydride with BF_3, as in Reaction 12-5.1.

$$3\,NaBH_4 + 4\,BF_3 \longrightarrow 2\,B_2H_6 + 3\,NaBF_4 \tag{12-5.1}$$

Laboratory quantities may be prepared by oxidation of sodium borohydride by iodine in diglyme, as in Eq. 12-5.2.

$$2\,NaBH_4 + I_2 \longrightarrow B_2H_6 + 2\,NaI + H_2 \tag{12-5.2}$$

Industrial quantities are prepared at high temperatures by reduction of BF_3 with sodium hydride.

$$2\,BF_3 + 6\,NaH \longrightarrow B_2H_6 + 6\,NaF \tag{12-5.3}$$

Borane (BH_3) has only a transient existence in the thermal decomposition of diborane.

$$2\,B_2H_6 = BH_3 + B_3H_9 \tag{12-5.4}$$

Reactions of diborane are discussed later in Section 12-5. Note that diborane is an extremely versatile reagent for the synthesis of organoboranes, which in turn are very useful intermediates in organic synthesis. Diborane is also a powerful reducing agent for some functional groups, for example, aldehydes and organic nitriles.

Higher Boranes

The heavier boranes (e.g., B_6H_{10}) are mainly liquids whose flammability in air decreases with increasing molecular weight. One of the most important is decaborane ($B_{10}H_{14}$), a solid (mp 99.7 °C) that is stable in air and only slowly hydrolyzed by water. It is obtained by heating B_2H_6 at 100 °C and is an important starting material for the synthesis of the $B_{10}H_{10}^{2-}$ anion and carboranes discussed later.

Table 12-1   Important Properties of Some Boranes

Formula	Name	Melting Point (°C)	Boiling Point (°C)	Reaction with Air (at 25°C)	Thermal Stability	Reaction with Water
B_2H_6	Diborane(6)	−164.85	−92.59	Spontaneously flammable	Fairly stable at 25 °C	Instant hydrolysis
B_4H_{10}	Tetraborane(10)	−120	18	Not spontaneously flammable if pure	Decomposes fairly rapidly at 25 °C	Hydrolysis in 24 h
B_5H_9	Pentaborane(9)	−46.6	48	Spontaneously flammable	Stable at 25 °C; slow decomposition 150°C	Hydrolyzed only on heating
B_5H_{11}	Pentaborane(11)	−123	63	Spontaneously flammable	Decomposes very rapidly at 25 °C	Rapid hydrolysis
B_6H_{10}	Hexaborane(10)	−62.3	108	Spontaneously flammable	Slow decomposition at 25 °C	Hydrolyzed only on heating
B_6H_{12}	Hexaborane(12)	−82.3	80–90	—	Liquid stable few hours at 25 °C	Quantitative, to give B_4H_{10}, $B(OH)_3$, H_2
B_8H_{12}	Octaborane(12)	−20	—	—	Decomposes above −20 °C	—
B_8H_{18}	Octaborane(18)	—	—	—	Unstable	—
B_9H_{15}	Enneaborane(15)	2.6	—	Stable	—	—
$B_{10}H_{14}$	Decaborane(14)	99.7	213 (extrap.)	Very stable	Stable at 150 °C	Slow hydrolysis

Structure and Bonding in the Boranes

The structures of the boranes are unlike those of other hydrides, such as those of carbon, and are unique. A few of them are shown in Fig. 12-4. Observe that in none are there sufficient electrons to allow the formation of conventional two-electron bonds between all adjacent pairs of atoms (2c–2e bonds). There is thus the problem of electron deficiency. It was to rationalize the structures of boranes that the earliest of the various concepts of multicenter bonding (Chapter 3) were first developed.

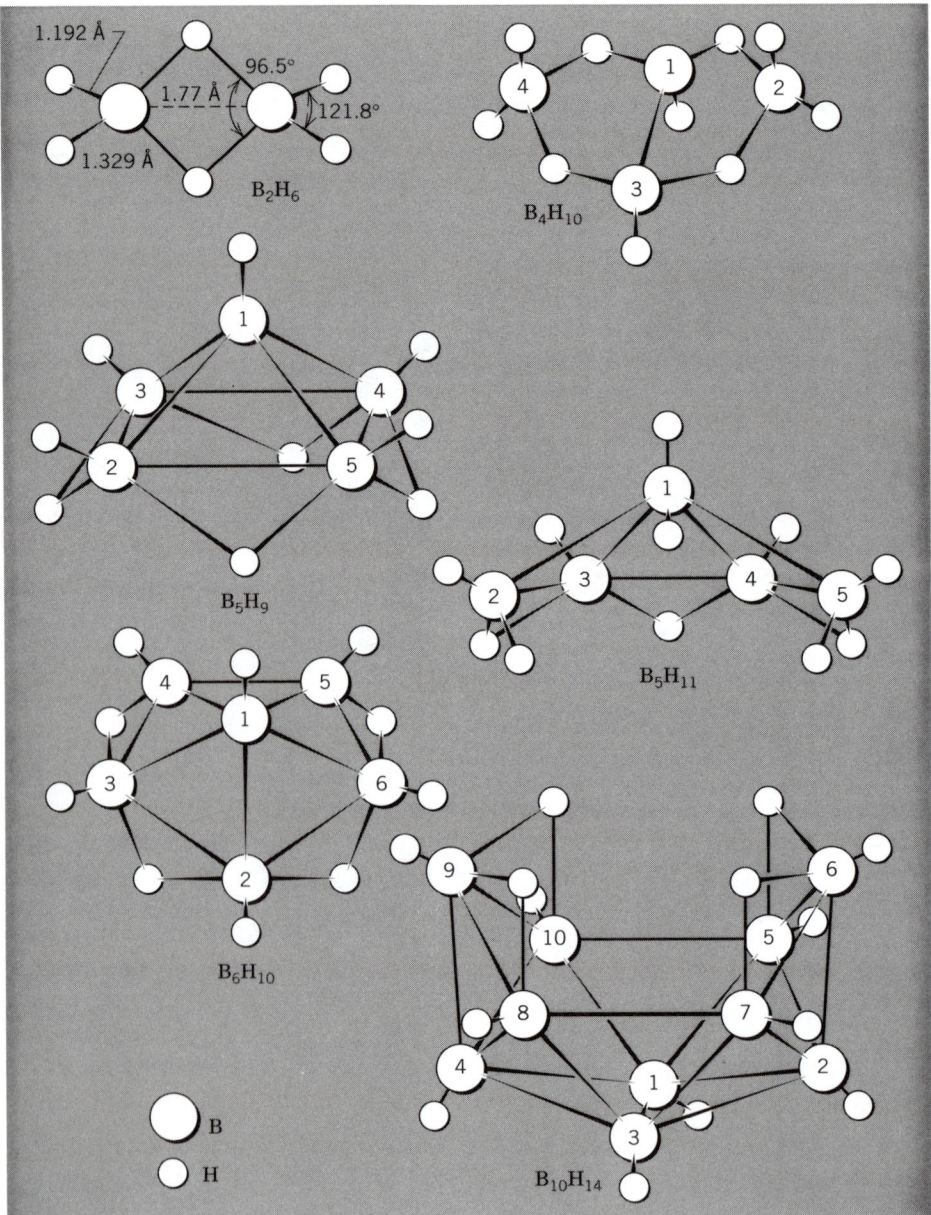

Figure 12-4   The structures of some boranes.

For diborane itself (3c–2e) bonds are required to explain the B—H—B bridges. The terminal B—H bonds may be regarded as conventional (2c–2e) bonds. Thus, each boron atom uses two electrons and two roughly sp^3 orbitals to form (2c–2e) bonds to two hydrogen atoms. The boron atom in each BH_2 group still has one electron and two hybrid orbitals for use in further bonding. The plane of the two remaining orbitals is perpendicular to the BH_2 plane. When two such BH_2 groups approach each other, as is shown in Fig. 12-5, with hydrogen atoms also lying in the plane of the four empty orbitals, two B—H—B (3c–2e) bonds are formed. The total of four electrons required for these bonds is provided by the one electron carried by each H atom and by each BH_2 group.

We have just seen that two structure–bonding elements are used in B_2H_6, namely, (2c–2e) BH groups and (3c–2e) BHB groups. To account for the structures and bonding of the higher boranes, these elements, as well as three others, are required. The three others are (2c–2e) BB groups, (3c–2e) open BBB groups, and (3c–2e) closed BBB groups. These five structure–bonding elements are conveniently represented in the following way:

Terminal (2c–2e) boron–hydrogen bond   B—H

(3c–2e) Hydrogen bridge bond

(2c–2e) Boron–boron bond   B—B

Open (3c–2e) B—B—B bond

Closed (3c–2e) boron bond

By using these five elements, W. N. Lipscomb was able to develop "semitopological" descriptions of the structures and bonding in all of the boranes. The scheme is capable of elaboration into a comprehensive, semipredictive tool for correlating all the structural data. Figure 12-6 shows a few examples of its use to depict known structures.

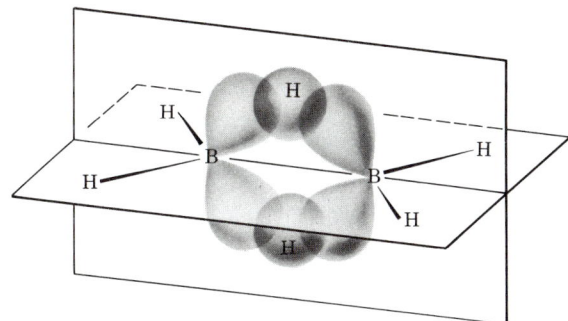

Figure 12-5   A diagram showing the formation of two bridging (3c–2e) B—H—B bonds in diborane.

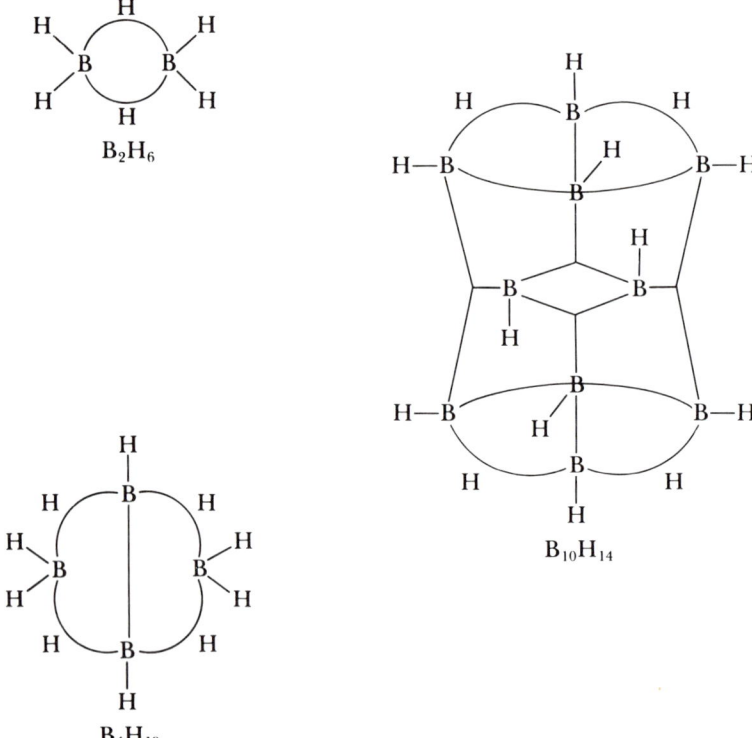

Figure 12-6 Valence descriptions of some electron deficient boranes using Lipscomb's "semitopological" scheme.

The semitopological scheme does not always provide the best description of bonding in the boranes, and related species such as the polyhedral borane anions and carboranes we shall discuss later. Where there is symmetry of a high order it is often better to think in terms of a highly delocalized MO description of the bonding. For instance, in B_5H_9 (Fig. 12-4), where the four basal boron atoms are equivalently related to the apical boron atom, it is *possible* to depict a resonance hybrid involving the localized $B \overset{B}{\frown} B$ and B—B elements, namely,

but it is neater and simpler to formulate a set of seven five-center MO's with the lowest three occupied by electron pairs. When one approaches hypersymmetrical species such as $B_{12}H_{12}^{2-}$, use of the full molecular symmetry in an MO treatment becomes the only practical course.

Reactions of the Boranes

The boranes undergo an impressive variety of reactions including oxidations to oxides, pyrolysis to higher boranes, attack by nucleophiles and electrophiles, reduction to borane anions, and reactions with bases such as OH^- and NH_3. In some cases it is useful to view at least the substitutions as being either reactions of terminal BH groups or of bridging BHB groups. We shall restrict our attention to three illustrative systems: diborane(6), B_2H_6; pentaborane(9), B_5H_9; and decaborane(14), $B_{10}H_{14}$.

Diborane(6), B_2H_6. Controlled pyrolysis of diborane leads to most of the higher boranes. Reaction with oxygen is extremely exothermic.

$$B_2H_6 + 3\,O_2 \longrightarrow B_2O_3 + 3\,H_2O \qquad \Delta H = -2160 \text{ kJ mol}^{-1} \quad (12\text{-}5.5)$$

Reaction of diborane with water is instantaneous.

$$B_2H_6 + 6\,H_2O \longrightarrow 2\,B(OH)_3 + 6\,H_2 \qquad (12\text{-}5.6)$$

Diborane is also hydrolyzed by weaker acids (e.g., alcohols), as in Reaction 12-5.7.

$$B_2H_6 + 6\,ROH \longrightarrow 2\,B(OR)_3 + 6\,H_2 \qquad (12\text{-}5.7)$$

Reaction with HCl replaces a terminal H with Cl

$$B_2H_6 + HCl \longrightarrow B_2H_5Cl + H_2 \qquad (12\text{-}5.8)$$

and reaction with chlorine gives the trichloride, as in Reaction 12-5.9.

$$B_2H_6 + 6\,Cl_2 \longrightarrow 2\,BCl_3 + 6\,HCl \qquad (12\text{-}5.9)$$

The electron deficient 3c–2e BHB bridges are sites of nucleophilic attack. Small amines such as NH_3, CH_3NH_2, and $(CH_3)_2NH$ give unsymmetrical cleavage of diborane, as in Reaction 12-5.10.

$$B_2H_6 + 2\,NH_3 \longrightarrow [H_2B(NH_3)_2]^+[BH_4]^- \qquad (12\text{-}5.10)$$

The boronium ion products, $[H_2BL_2]^+$, are tetrahedral, and can undergo substitution by other bases, as in Reaction 12-5.11.

$$[H_2B(NH_3)_2]^+ + 2\,PR_3 \longrightarrow [H_2B(PR_3)_2]^+ + 2\,NH_3 \qquad (12\text{-}5.11)$$

Large amines, such as $(CH_3)_3N$ and pyridine, give symmetrical cleavage of diborane, as in Reaction 12-5.12.

$$B_2H_6 + 2\,N(CH_3)_3 \longrightarrow 2\,H_3B{\leftarrow}N(CH_3)_3 \qquad (12\text{-}5.12)$$

The amine borane products from symmetrical cleavage of diborane are Lewis base adducts of BH_3. Amine boranes will be discussed more in Section 12-6.

Reduction of diborane can be accomplished with sodium

$$2\,B_2H_6 + 2\,Na \longrightarrow NaBH_4 + NaB_3H_8 \qquad (12\text{-}5.13)$$

or with sodium borohydride

$$B_2H_6 + NaBH_4 \longrightarrow NaB_3H_8 + H_2 \qquad (12\text{-}5.14)$$

Reductions of diborane with sodium borohydride can also lead to higher borane anions, as in Reaction 12-5.15.

$$2\,NaBH_4 + 5\,B_2H_6 \longrightarrow Na_2B_{12}H_{12} \qquad (12\text{-}5.15)$$

The polyhedral borane anion $B_{12}H_{12}^{2-}$ has icosahedral geometry. Other polyhedral borane anions and carboranes will be discussed shortly.

Pentaborane(9), B_5H_9. Pentaborane(9) has the structure shown in Fig. 12-4. The apical boron is bonded to a single terminal hydrogen atom, while each of the four basal borons is bonded to one terminal hydrogen atom and to two bridging hydrogen atoms. The relative electron deficiency in the basal plane is illustrated by reaction of B_5H_9 with base, as in Reaction 12-5.16.

$$B_5H_9 + NaH \longrightarrow Na^+B_5H_8^- + H_2 \qquad (12\text{-}5.16)$$

Higher boranes are even more acidic than B_5H_9. The anionic product of Reaction 12-5.16 (Structure 12-IX) is fluxional (Section 6-6) due to rapid cycling

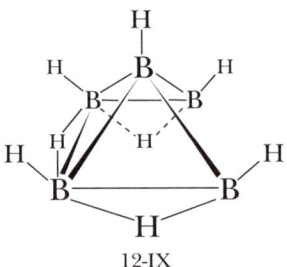

12-IX

of bridging hydrogen atoms. Thus the basal boron atoms are indistinguishable using NMR techniques, as are the bridging hydrogen atoms.

The relatively electron-rich apical BH group of pentaborane(9) is susceptible to attack by electrophiles, as in Reaction 12-5.17.

$$B_5H_9 + I_2 \longrightarrow B_5H_8I + HI \qquad (12\text{-}5.17)$$

Decaborane(14), $B_{10}H_{14}$. The structure of decaborane(14) is shown in Fig. 12-4. Four electron deficient bridging BHB groups cap the top of this icosahedral fragment, making this part of the molecule the preferred site for attack by nucleophiles. As for pentaborane(9), it is the bridging hydrogen atoms that are acidic.

$$B_{10}H_{14} + OH^- \longrightarrow B_{10}H_{13}^- + H_2O \qquad (12\text{-}5.18)$$

Reduction by sodium converts two of the bridging hydrogen atoms at the top of the molecule to terminal hydrogen atoms.

$$B_{10}H_{14} + 2\,Na \longrightarrow Na_2B_{10}H_{14} \tag{12-5.19}$$

Nucleophiles react to give 6,9-disubstituted products as in Reactions 12-5.20 to 12-5.22:

$$B_{10}H_{14} + 2\,CN^- \longrightarrow B_{10}H_{12}(CN)_2^{2-} + H_2 \tag{12-5.20}$$

$$B_{10}H_{14} + 2\,CH_3CN \longrightarrow B_{10}H_{12}(NCCH_3)_2 + H_2 \tag{12-5.21}$$

$$B_{10}H_{14} + 2\,PR_3 \longrightarrow B_{10}H_{12}(PR_3)_2 + H_2 \tag{12-5.22}$$

In contrast to reactions with nucleophiles, decaborane(14) reacts with electrophiles to give 2,4- or 1,3-disubstituted products. An example is shown in Reaction 12-5.23.

$$B_{10}H_{14} + I_2 \longrightarrow 2,4\text{-}I_2B_{10}H_{12} + H_2 \tag{12-5.23}$$

Charge distribution calculations using MO theory indicate that considerable excess negative charge should be assigned to boron atoms 1, 2, 3, and 4, with positive charge assigned to the electron deficient positions elsewhere in the molecule. It is thus gratifying that experiments show consistently that only positions 1, 2, 3, and 4 can be substituted electrophilically.

The Tetrahydroborate Ion (BH$_4^-$)

The tetrahydroborate ion (BH$_4^-$) is the simplest of a number of borohydride anions. This ion is of great importance as a reducing agent and as a source of H$^-$ ion both in inorganic and organic chemistry; derivatives such as [BH(OCH$_3$)$_3$]$^-$ and [BH$_3$CN]$^-$ are also useful, the latter because it can be used in acidic solutions.

Borohydrides of many metals have been made and some representative syntheses are:

$$4\,NaH + B(OCH_3)_3 \xrightarrow{\sim 250\,^\circ C} NaBH_4 + 3\,NaOCH_3 \tag{12-5.24}$$

$$NaH + B(OCH_3)_3 \xrightarrow{\text{THF}} NaBH(OCH_3)_3 \tag{12-5.25}$$

$$2\,LiH + B_2H_6 \xrightarrow{\text{ether}} 2\,LiBH_4 \tag{12-5.26}$$

$$AlCl_3 + 3\,NaBH_4 \xrightarrow{\text{heat}} Al(BH_4)_3 + 3\,NaCl \tag{12-5.27}$$

$$UF_4 + 2\,Al(BH_4)_3 \longrightarrow U(BH_4)_4 + 2\,AlF_2BH_4 \tag{12-5.28}$$

The most important salt is NaBH$_4$. This is a white crystalline solid, which is stable in dry air, and nonvolatile. It is insoluble in diethyl ether but dissolves in H$_2$O, THF, and ethyleneglycol ethers from which it can be crystallized.

Many borohydrides are ionic, containing the tetrahedral BH$_4^-$ ion. However, BH$_4^-$ can serve as a ligand, interacting more or less covalently with metal ions, by bridging hydrogen atoms. Thus in (Ph$_3$P)$_2$CuBH$_4$ there are two Cu—H—B bridges, whereas in Zr(BH$_4$)$_4$, each BH$_4$ forms three bridges to Zr, as shown in Fig. 12-7. These M—H—B bridges are (3c–2e) bonding systems.

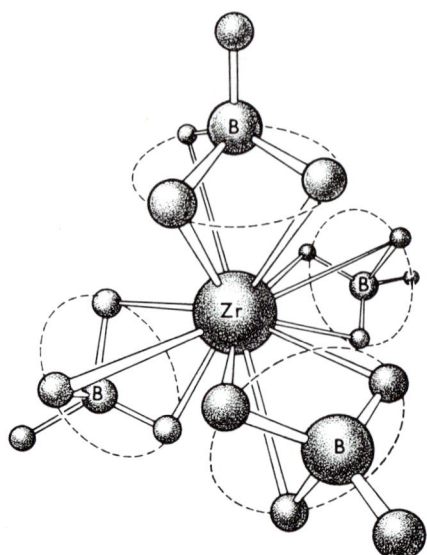

Figure 12-7   The structure of
$Zr(BH_4)_4$. [Taken from Bird, P. H. and
Churchill, M. R., *J. Chem. Soc., Chem.
Commun.*, **1967,** 403. Used with permis-
sion.]

Polyhedral Borane Anions and Carboranes

The polyhedral borane anions have the formula $B_nH_n^{2-}$. The carboranes may be
considered to be *formally* derived from $B_nH_n^{2-}$ by replacement of BH^- by the isoelectronic and isostructural CH. Thus two replacements lead to neutral molecules ($B_{n-2}C_2H_n$). Carboranes or derivatives with $n = 5$ to $n = 12$ are known, in
some of which two or more isomers may be isolated. Sulfur and phosphorus derivatives can also be obtained, with PH^+, for example, replacing CH or BH^-.

Geometrically, there are three broad classes of boranes or carboranes.

1. Those in which the boron or boron–carbon framework forms a regular
    polyhedron. These are called *closo* (Greek for cage) compounds.
2. Those in which the boron or boron–carbon framework has the structure
    of a regular polyhedron with one vertex missing. These are called *nido*
    (nest) compounds.
3. Those in which the boron or boron–carbon framework has the structure
    of a regular polyhedron with two vertices missing. These are termed
    arachno (spider web) compounds.

A systematic method for counting electrons and for organizing structures in
these and other classes of compounds will be presented in Section 12-7. For now,
examples of some important *closo* and *nido* borane anions and carboranes are
presented in Fig. 12-8. Structures of some other *nido* and *arachno* compounds
have already been presented, for example *nido*-B_5H_9 and *arachno*-B_4H_{10} in Fig.
12-4.

$\boldsymbol{B_nH_n^{2-}}$ **Ions.** The most stable and best studied ions are $B_{10}H_{10}^{2-}$ and $B_{12}H_{12}^{2-}$,
which can be synthesized by the reactions

$$B_{10}H_{14} + 2\,R_3N \xrightarrow{\ 150\,°C\ } 2\,(R_3NH)^+ + B_{10}H_{10}^{2-} + H_2 \qquad (12\text{-}5.29)$$

$$6\,B_2H_6 + 2\,R_3N \xrightarrow{\ 150\,°C\ } 2\,(R_3NH)^+ + B_{12}H_{12}^{2-} + 11\,H_2 \qquad (12\text{-}5.30)$$

The most important general reaction of the anions is attack by electrophilic reagents such as Br^+, $C_6H_5N_2^+$, and RCO^+, in strongly acid media. The $B_{10}H_{10}^{2-}$ ion is more susceptible to substitution than $B_{12}H_{12}^{2-}$.

$B_{n-2}C_2H_n$ Carboranes. The most important carboranes are 1,2- and 1,7-di-carba-*closo*-dodecaborane ($B_{10}C_2H_{12}$) and their *C*-substituted derivatives. The 1,2 isomer may be obtained by the reactions

$$B_{10}H_{14} + 2\,R_2S = B_{10}H_{12}(R_2S)_2 + H_2 \qquad (12\text{-}5.31)$$

$$B_{10}H_{12}(R_2S)_2 + RC{\equiv}CR' = 1,2\text{-}B_{10}H_{10}C_2RR' + 2\,R_2S + H_2 \quad (12\text{-}5.32)$$

On heating at 450 °C the 1,2 isomer rearranges to the 1,7 isomer.

Derivatives may be obtained from $B_{10}C_2H_{12}$ by replacement of the CH hydrogen atoms by Li. The dilithio derivatives react with many other reagents (Scheme 12-1) where a self-explanatory abbreviation is used for $B_{10}H_{10}C_2$.

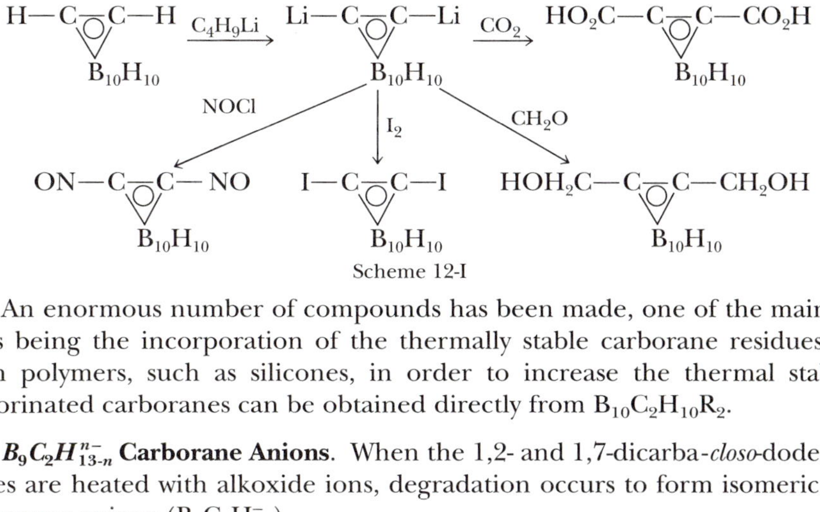

Scheme 12-I

An enormous number of compounds has been made, one of the main motives being the incorporation of the thermally stable carborane residues into high polymers, such as silicones, in order to increase the thermal stability. Chlorinated carboranes can be obtained directly from $B_{10}C_2H_{10}R_2$.

$B_9C_2H_{13-n}^{n-}$ Carborane Anions. When the 1,2- and 1,7-dicarba-*closo*-dodecaboranes are heated with alkoxide ions, degradation occurs to form isomeric *nido*-carborane anions ($B_9C_2H_{12}^-$).

$$B_{10}C_2H_{12} + C_2H_5O^- + 2\,C_2H_5OH = B_9C_2H_{12}^- + B(OC_2H_5)_3 + H_2 \qquad (12\text{-}5.33)$$

This removal of a BH^{2+} unit from $B_{10}C_2H_{12}$ may be interpreted as a nucleophilic attack at the most electron-deficient boron atoms of the carborane. Molecular orbital calculations show that the C atoms in carboranes have considerable electron-withdrawing power. The most electron-deficient B atoms are those adjacent to carbon. In 1,2-$B_{10}C_2H_{12}$ these will be in positions three and six while in 1,7-$B_{10}C_2H_{12}$ they will be at positions two and three.

While alkoxide ion attack produces only $B_9C_2H_{12}^-$, use of the very strong base NaH forms the $B_9C_2H_{11}^{2-}$ ions.

$$B_9C_2H_{12}^- + NaH = Na^+ + B_9C_2H_{11}^{2-} + H_2 \qquad (12\text{-}5.34)$$

The structures of the isomeric $B_9C_2H_{11}^{2-}$ ions are shown in Fig. 12-9. The $B_9C_2H_{11}^{2-}$

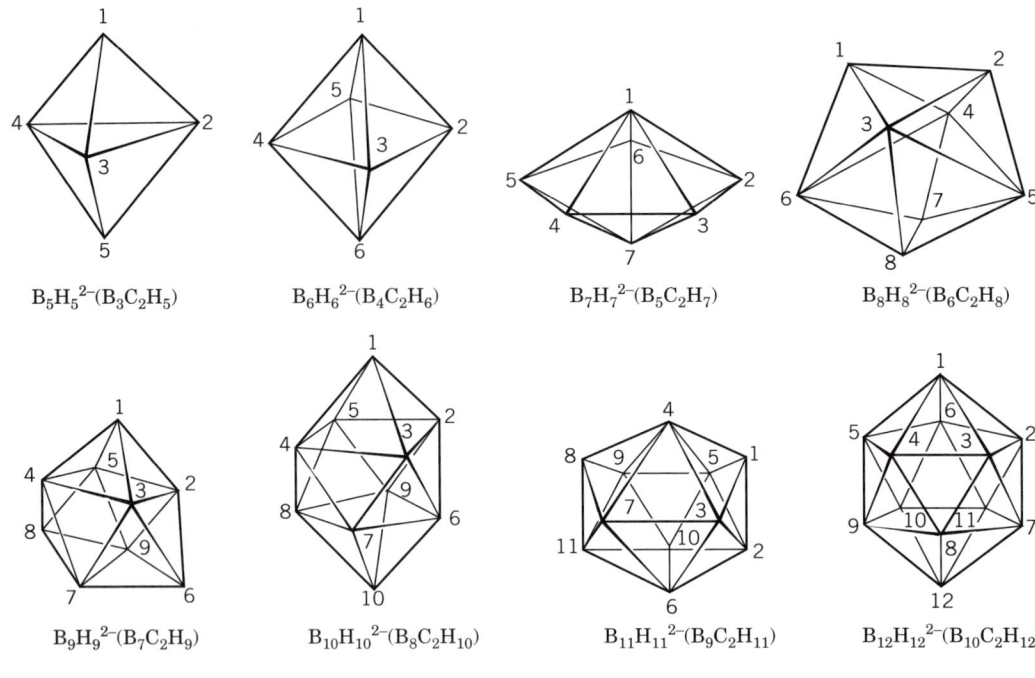

$B_5H_5^{2-}(B_3C_2H_5)$

$B_6H_6^{2-}(B_4C_2H_6)$

$B_7H_7^{2-}(B_5C_2H_7)$

$B_8H_8^{2-}(B_6C_2H_8)$

$B_9H_9^{2-}(B_7C_2H_9)$

$B_{10}H_{10}^{2-}(B_8C_2H_{10})$

$B_{11}H_{11}^{2-}(B_9C_2H_{11})$

$B_{12}H_{12}^{2-}(B_{10}C_2H_{12})$

(a)

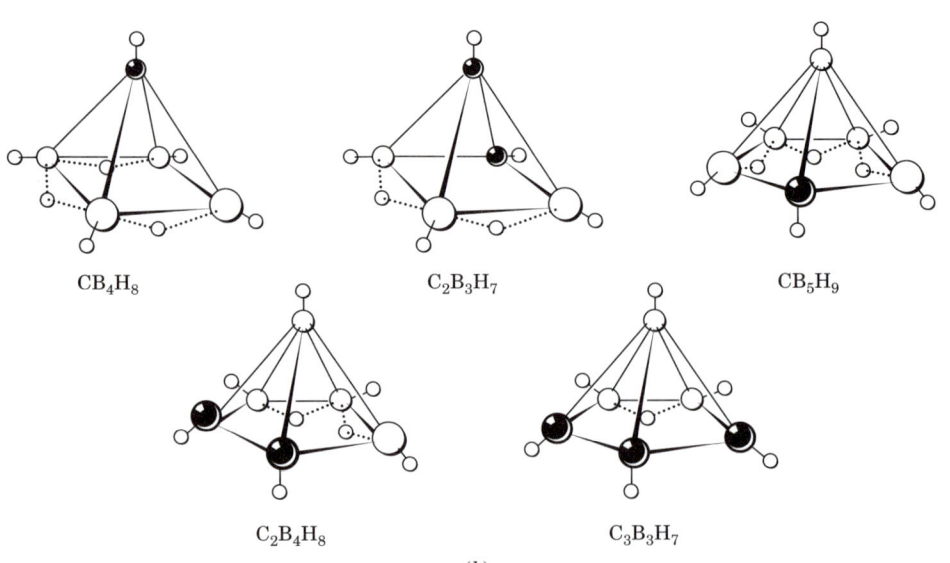

CB_4H_8

$C_2B_3H_7$

CB_5H_9

$C_2B_4H_8$

$C_3B_3H_7$

(b)

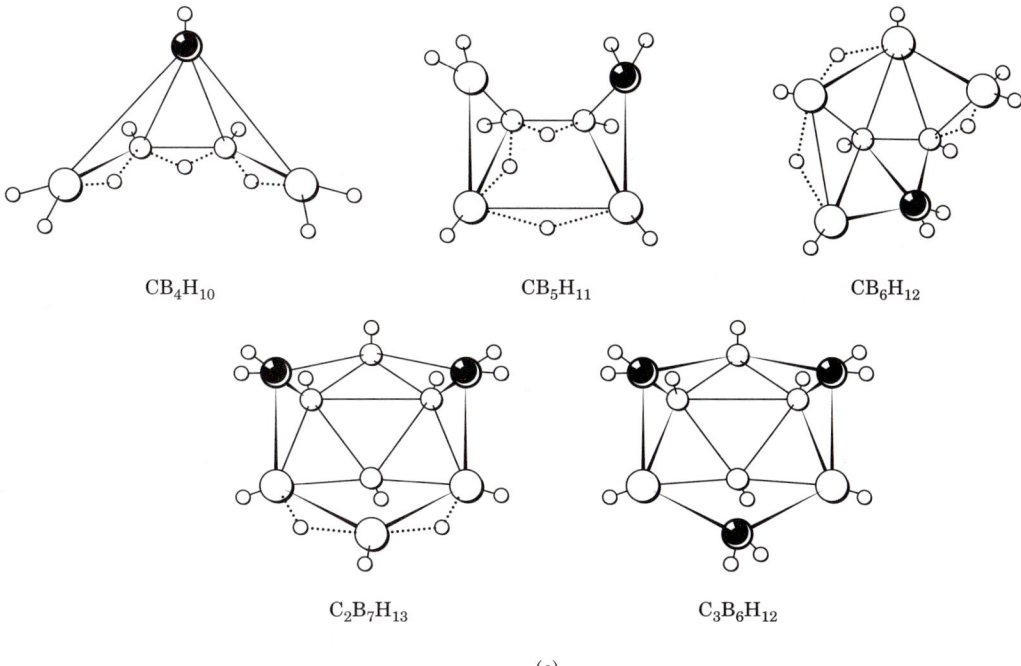

CB_4H_{10} CB_5H_{11} CB_6H_{12}

$C_2B_7H_{13}$ $C_3B_6H_{12}$

(c)

Figure 12-8   The structures of selected boranes and carboranes. (*a*) The triangulated regular polyhedra, which serve as the framework structures for the *closo* borane anions, $B_nH_n^{2-}$, and for the neutral, isoelectronic carboranes, $C_2B_{n-2}H_n$. Conventional numbering schemes are indicated. Each vertex B or C atom is bonded to a terminal H atom, which is not shown. (*b*) Selected *nido* carboranes. Note the presence of both terminal B—H (or C—H) groups and bridging B—H—B groups in these *nido* carboranes, as well as in the *nido* boranes shown in Fig. 12-4. (*c*) Selected *arachno* carboranes. Note, for the *arachno* carboranes shown here, as well as for the *arachno* boranes shown in Fig. 12-4, the additional presence of boron atoms bound to two terminal H atoms, namely, BH_2 groups.

ions are very strong bases and readily acquire H^+ to give $B_9C_2H_{12}^-$. These, in turn, can be protonated to form the neutral *nido*-carboranes $B_9C_2H_{13}$, which are strong acids.

$$B_9C_2H_{11}^{2-} \underset{}{\overset{H^+}{\rightleftharpoons}} B_9C_2H_{12}^- \underset{}{\overset{H^+}{\rightleftharpoons}} B_9C_2H_{13} \qquad (12\text{-}5.35)$$

Heating $B_9C_2H_{13}$ gives yet another *closo*-carborane ($B_9C_2H_{11}$) with loss of hydrogen.

Metal Complexes of Carborane Anions.   The open pentagonal faces of the $B_9C_2H_{11}^{2-}$ ions (Fig. 12-9) were recognized by M. F. Hawthorne in 1964 to bear a strong resemblance structurally and electronically to the cyclopentadienyl ion ($C_5H_5^-$). The latter forms strong bonds to transition metals, as we discuss in Chapter 29.

Interaction of $Na_2B_9C_2H_{11}$ with metal compounds such as those of Fe^{2+} or Co^{3+} thus leads to species isoelectronic with ferrocene, $(C_5H_5)_2Fe$, or the cobalticinium ion, $(C_5H_5)_2Co^+$, namely, $(B_9C_2H_{11})_2Fe^{2-}$ and $(B_9C_2H_{11})_2Co^-$, respectively. The iron complex undergoes reversible oxidation like ferrocene:

$$[(C_5H_5)_2Fe^{III}]^+ + e^- = [(C_5H_5)_2Fe^{II}]^0 \qquad (12\text{-}5.36)$$

$$[(B_9C_2H_{11})_2Fe]^- + e^- = [(B_9C_2H_{11})_2Fe]^{2-} \qquad (12\text{-}5.37)$$

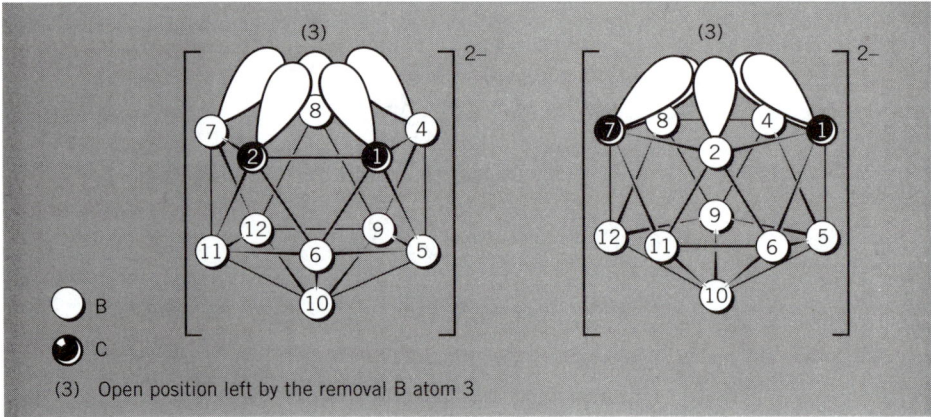

Figure 12-9 The isomeric *nido*-1,2- and *nido*-1,7-carborane anions ($B_9C_2H_{11}^{2-}$).

The formal nomenclature for the $B_9C_2H_{11}^{2-}$ ion and its complexes is unwieldy and the trivial name "*dicarbollide*" ion was proposed (from the Spanish *olla* for pot, referring to the potlike shape of the B_9C_2 cage).

The structures of two types of bis(dicarbollide) metal complexes are shown in Fig. 12-10. While some complexes have a symmetrical "sandwich" structure [Fig. 12-10(a)] others have the metal disposed unsymmetrically.

Finally, comparable with η^5-$C_5H_5Mn(CO)_3$ (Chapter 29), there are mixed complexes with only one dicarbollide unit and other ligands such as CO, $(C_6H_5)_4C_4$, and C_5H_5 [Fig. 12-10(b)].

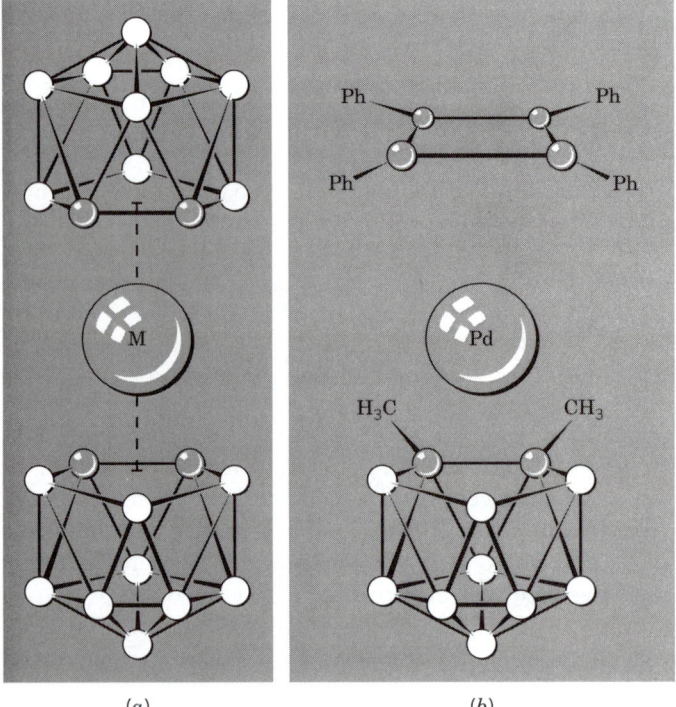

(*a*)　　　　　　　　(*b*)

Figure 12-10 (*a*) The general structure of bis(dicarbollide) metal complexes. (*b*) An example of a mono(dicarbollide) complex of palladium.

12-6 **Boron–Nitrogen Compounds**

Here we describe three types of B—N compounds, each of which is analogous to C—C compounds, but with some differences. Since the covalent radius and the electronegativity of carbon are each intermediate between those of B and N, it is to be expected that C—C compounds will be similar to, but less polar than, their isoelectronic B—N counterparts. We shall consider amine boranes (analogous to alkanes), aminoboranes (analogous to alkenes), and borazines (analogous to benzenes).

Amine Boranes

Amine boranes are Lewis acid–base adducts containing a boron–nitrogen donor bond. Both boron and nitrogen are typically tetrahedral, and the B—N bond length is comparable to the C—C bond lengths found in simple alkanes such as ethane. Amine boranes are formed by symmetrical cleavage of diborane or by reaction of ammonium salts as in Reaction 12-6.1.

$$[H_3NR]Cl + LiBH_4 \longrightarrow RH_2N{\rightarrow}BH_3 + LiCl + H_2 \qquad (12\text{-}6.1)$$

The B—N bond strength varies from one adduct to another. The weakest B—N bonds are represented as in Structure 12-X, where an arrow indicates a slight donor → acceptor interaction. More complete sharing of nitrogen electrons with boron is represented by Structure 12-XI, which is expected to be polar.

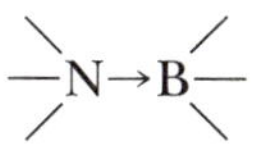

$$-\underset{\diagup}{\overset{\diagdown}{N}}{\rightarrow}\underset{\diagdown}{\overset{\diagup}{B}}- \qquad -\underset{\diagup}{\overset{\diagdown}{\overset{+}{N}}}-\underset{\diagdown}{\overset{\diagup}{\overset{-}{B}}}-$$

12-X 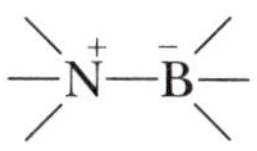 12-XI

Steric hindrance can prevent the formation of some adducts, for example, 2,6-dimethylpyridine with trimethylborane. Diadducts can be obtained, as in Reaction 12-6.2.

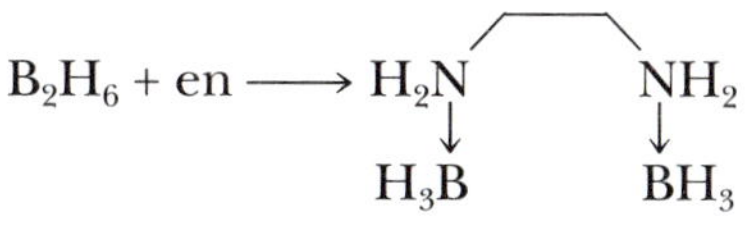

$$B_2H_6 + en \longrightarrow \underset{H_3B}{\overset{}{H_2N}} \overset{\frown}{} \underset{BH_3}{\overset{}{NH_2}} \qquad (12\text{-}6.2)$$

The chief reaction of amine boranes is elimination either of HX or of RH, to give aminoboranes.

Aminoboranes

Aminoboranes are B—N compounds that are analogous to alkenes. Boron is trigonal in aminoboranes, and the three substituents at boron are planar, or very nearly so. Two resonance forms may be written, Structures 12-XII and 12-XIII.

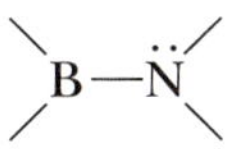

$$\underset{\diagup}{\overset{\diagdown}{B}}-\ddot{\underset{\diagdown}{\overset{\diagup}{N}}} \longleftrightarrow \overset{-}{\underset{\diagup}{\overset{\diagdown}{B}}}{=}\overset{+}{\underset{\diagdown}{\overset{\diagup}{N}}}$$

12-XII 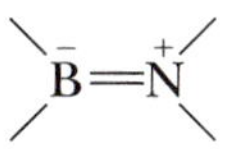 12-XIII

In most aminoboranes there is something less than a full double bond between boron and nitrogen, but in some cases high barriers to rotation about the B—N bond indicate that the B—N bond order exceeds 1.0. The multiple bond in aminoboranes is formed by overlap of atomic p orbitals, as shown in Fig. 12-11. Average rotational barriers are lower in bis(amino)boranes, indicating competition between the two nitrogen π donors for the empty p orbital of boron. Rotational barriers are lower still for tris(amino)boranes.

Aminoboranes are synthesized by reduction of ammonium salts with tetrahydroborate reagents as in Reaction 12-6.1 (followed by dehydrohalogenation of the intermediate amine borane), or by treatment of certain aminoboranes with Grignard reagents, as in Reaction 12-6.3.

$$R_2\ddot{N}{-}BCl_2 + 2\ R'MgX \longrightarrow R_2\ddot{N}{-}BR'_2 + 2\ MgXCl \qquad (12\text{-}6.3)$$

The sequence of Reactions 12-6.4 to 12-6.6 serves as a useful example.

$$(CH_3)_2HN{:} + BCl_3 \longrightarrow (CH_3)_2HN{\rightarrow}BCl_3 \qquad (12\text{-}6.4)$$

$$(CH_3)_2HN{\rightarrow}BCl_3 \longrightarrow (CH_3)_2\overset{+}{N}{=}\overset{-}{B}Cl_2 + HCl \qquad (12\text{-}6.5)$$

$$(CH_3)_2\overset{+}{N}{=}\overset{-}{B}Cl_2 + 2EtMgCl \longrightarrow (CH_3)_2\overset{+}{N}{=}\overset{-}{B}Et_2 + 2\ MgCl_2 \qquad (12\text{-}6.6)$$

The chief reaction (other than substitution) of aminoboranes is condensation to cyclic systems, as in the formation of a dimer (Structure 12-XIV).

$$
\begin{array}{ccc}
R & & Cl \\
| & & | \\
R{-}N{-} & \!\!\!B & {-}Cl \\
\downarrow & \uparrow & \\
Cl{-}B & \!\!\!N & {-}R \\
| & & | \\
Cl & & R \\
\end{array}
$$

12-XIV

Condensation of aminoboranes to cyclic trimers, when accompanied by elimination of either HX or RH, leads to the borazine derivatives.

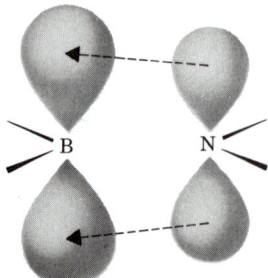

Figure 12-11 The formation of a π bond in aminoboranes as a result of donation of electrons from a filled p orbital on nitrogen to an empty p orbital on boron.

Borazines

One of the most interesting B—N compounds is *borazine*, $B_3N_3H_6$ (Structure 12-XV).

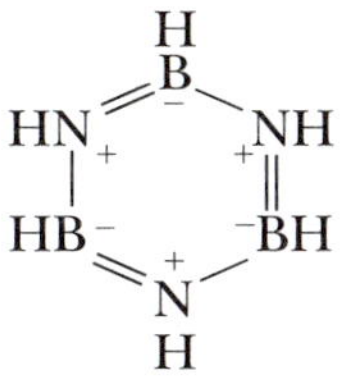

12-XV

It has an obvious formal resemblance to benzene, and the physical properties of the compounds are similar. However, borazine is much more reactive than benzene and readily undergoes addition reactions, as in Reaction 12-6.7:

$$B_3N_3H_6 + 3 HX \longrightarrow (-H_2N-BHX-)_3 \quad X = Cl, OH, OR, \text{ and so on} \quad (12\text{-}6.7)$$

which do not occur with benzene. Borazine also decomposes slowly and may be hydrolyzed to NH_3 and boric acid at elevated temperatures. As with benzene, π complexes with transition metals may be obtained (Chapter 29); thus hexamethylborazine gives compound 12-XVI:

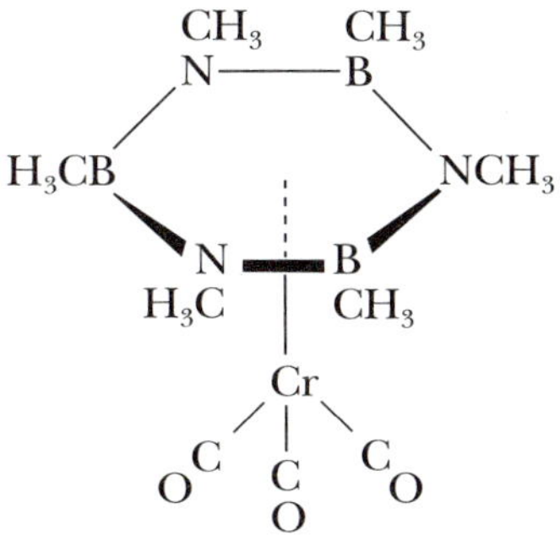

12-XVI

Borazine and substituted borazines may be synthesized by reactions such as 12-6.8 to 12-6.11.

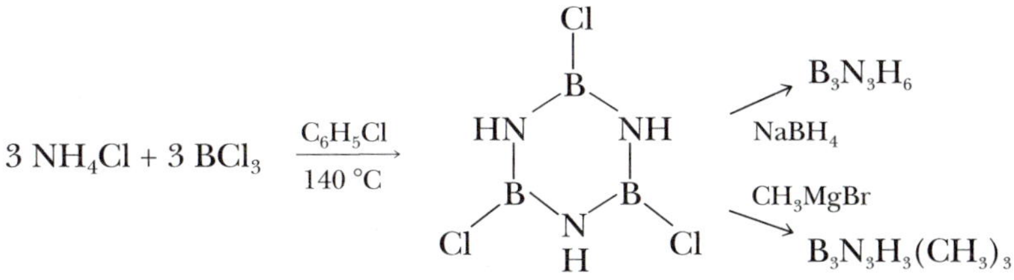

$$3 NH_4Cl + 3 BCl_3 \xrightarrow[140\ °C]{C_6H_5Cl}$$

$$\xrightarrow{NaBH_4} B_3N_3H_6 \quad (12\text{-}6.8)$$

$$\xrightarrow{CH_3MgBr} B_3N_3H_3(CH_3)_3 \quad (12\text{-}6.9)$$

$$CH_3NH + BCl_3 \xrightarrow[C_6H_5Cl]{boiling} Cl_3B \cdot NH_2CH_3 \quad (12\text{-}6.10)$$

(mp 126-128 °C)

$$3\,Cl_3B \cdot NH_2CH_3 + 6(CH_3)_3N \xrightarrow{\text{toluene}}$$

$$6(CH_3)_3NHCl +$$

(12-6.11)

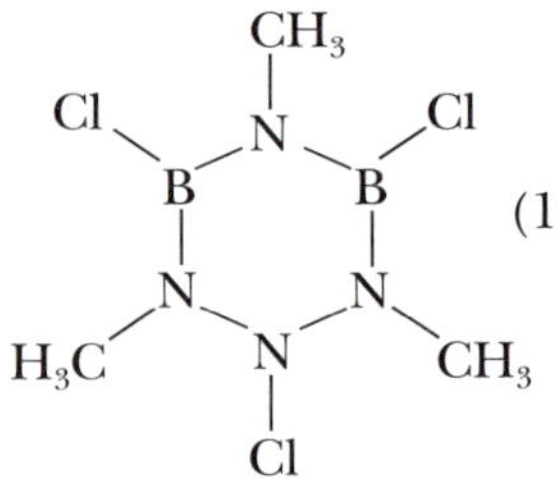

(mp 153-156 °C)

12-7 Electron Counting for Boranes and Other Framework Substances: Wade's Rules

It is now instructive to review the structures of the numerous boranes and carboranes, which, as noted earlier, fall into the *closo, nido,* and *arachno* categories. The type of structure adopted by a particular compound has been shown to be related to the number of electrons that are available in the compound for bonding within the polyhedral framework, that is, the number of "framework electrons." A way of correlating the number of framework electrons with structure was first articulated by K. Wade, hence the name "Wade's rules."

Wade's Rules as Applied to Boranes and Carboranes

We start by defining the quantity F, the number of electrons available for framework bonds, as in Eq. 12-7.1:

$$F = 3b + 4c + h + x - 2n \tag{12-7.1}$$

where

b = the number of boron atoms

c = the number of carbon atoms

h = the number of hydrogen atoms

x = the amount of **negative** charge on the ion

n = the number of vertices, that is, $b + c$

Note that x is defined so as to be a positive quantity for anions. Thus the number of valence electrons available for the framework bonds (F) is the number that remains after providing for n exo-framework (2c–2e) terminal B—H or C—H bonds.

RULE 1   When the value of F is equal to the quantity $(2n + 2)$, the substance should have a *closo* structure, that is, the framework geometry is based on an n vertex, triangulated, regular polyhedron. This result is obtained for all of the borane dianions ($B_nH_n^{2-}$), for the carborane anions ($CB_{n-1}H_n^-$), and for the neutral carboranes ($C_2B_{n-2}H_n$), since substitution of a BH^- group by the isoelectronic CH unit does not change the value of F as defined in Eq. 12-7.1.

Two examples readily illustrate this result. For $B_6H_6^{2-}$, the value of F is $3 \times 6 + 4 \times 0 + 6 + 2 - 2 \times 6 = 14$. Since the quantity $(2n + 2)$ is also equal to 14, we have identified a *closo* situation. Similarly, for the carborane $C_2B_4H_6$, we have $F = 3 \times 4 + 4 \times 2 + 6 + 0 - 2 \times 6 = 14$. The structure for both $B_6H_6^{2-}$ and $C_2B_4H_6$

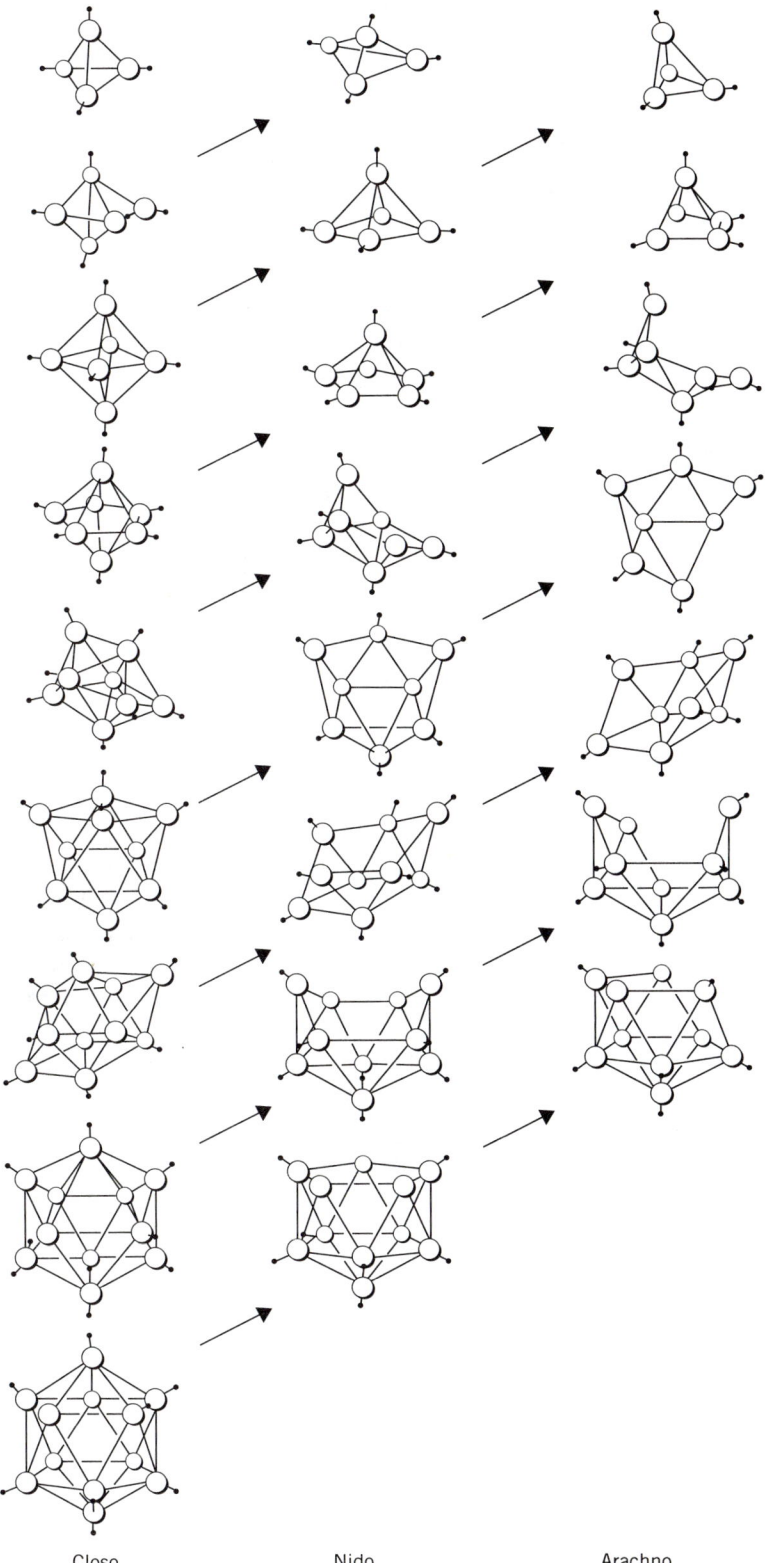

Closo	Nido	Arachno

Figure 12-12   The conversion of *closo* polyhedral borane and heterob-
orane frameworks to *nido* and *arachno* structures by removal of one or
two vertices, respectively. Bridge hydrogen atoms are not shown in the
nido and *arachno* structures, and BH$_2$ groups are not distinguished from
B—H groups in the *arachno* structures. The lines linking boron atoms
are meant merely to illustrate cluster geometry. [Reprinted with permission from R. W. Rudolph, *Acc. Chem. Res., 9,* 446 (1976). Copyright ©
(1976) American Chemical Society.]

is thus a six vertex polyhedron (namely, the octahedron), as shown in Figs. 12-8(*a*) and 12-12. In *closo* compounds, the bonds to hydrogen are only of the terminal-type B—H or C—H, 2c–2e bonds.

RULE 2   When the value of F is equal to the quantity $(2n + 4)$, the substance should have the *nido* structure, that is, an $(n + 1)$ vertex polyhedron, with one vertex missing, as illustrated in Fig. 12-12.

For example, for B_5H_9, the quantity F is equal to $3 \times 5 + 4 \times 0 + 9 + 0 - 2 \times 5 = 14$. Since this is equal to the quantity $(2n + 4)$, the structure of B_5H_9 (Fig. 12-4) is that of an $(n + 1) = 6$ vertex polyhedron, with one vertex missing. This structure is well illustrated in Fig. 12-12. A similar result is obtained for $C_2B_3H_7$. In *nido* compounds, there are B—H—B bridge bonds at those edges left open by the missing vertex atom. The other hydrogen atoms are bonded in the 2c–2e terminal fashion. It is characteristic, then, of *nido* compounds that we find two types of groups: n terminal B—H hydrogen atoms, and B—H—B bridges for the remainder.

RULE 3   When the value of F is equal to the quantity $(2n + 6)$, the compound falls into the *arachno* category, and the preferred structure is that of the $(n + 2)$ vertex polyhedron, with two vertices missing.

The compound B_4H_{10} and the ion $B_9H_{14}^-$ provide useful examples. For B_4H_{10}, the quantity F is equal to $3 \times 4 + 4 \times 0 + 10 + 0 - 2 \times 4 = 14$. This is equal to the quantity $(2n + 6)$ and, as shown in Fig. 12-4, the structure is based on an $(n + 2)$ vertex polyedron, with two adjacent vertices missing (Fig. 12-12). For $B_9H_{14}^-$, the value of F is $3 \times 9 + 4 \times 0 + 14 + 1 - 2 \times 9 = 24$, which is equal to $(2n + 6)$. The same value is obtained for CB_8H_{14}. Both have a structure (Fig. 12-12) based on a $(9 + 2) = 11$ vertex polyhedron, with two adjacent vertices missing. It is characteristic of *arachno* compounds that we find hydrogen bound in three ways: B—H or C—H terminal bonds, B—H—B bridge bonds, and BH_2 groups. The compound B_5H_{11} (Fig. 12-4) provides another example.

12-8   Descriptive Summary of Reactions

As a study aid, and as a means of summarizing the chemistry of this chapter, we present Figs. 12-13(*a–c*), which illustrate some of the key reactions for $B(OR)_3$, $B(OH)_3$, B_2H_6, BCl_3, and BF_3. The student should also note Fig. 12-1, as well as Study Question 12-3, in part B.

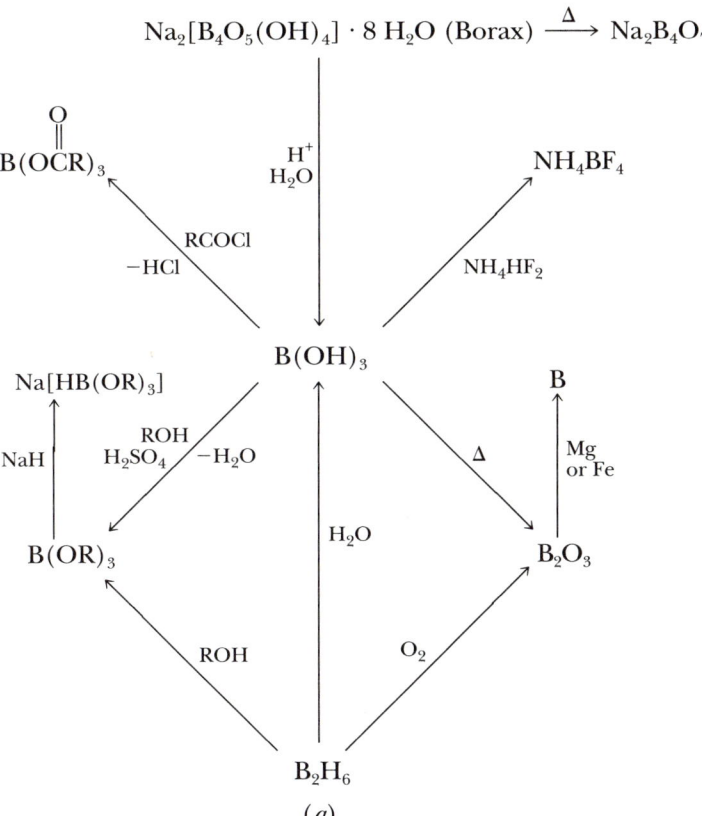

Figure 12-13a

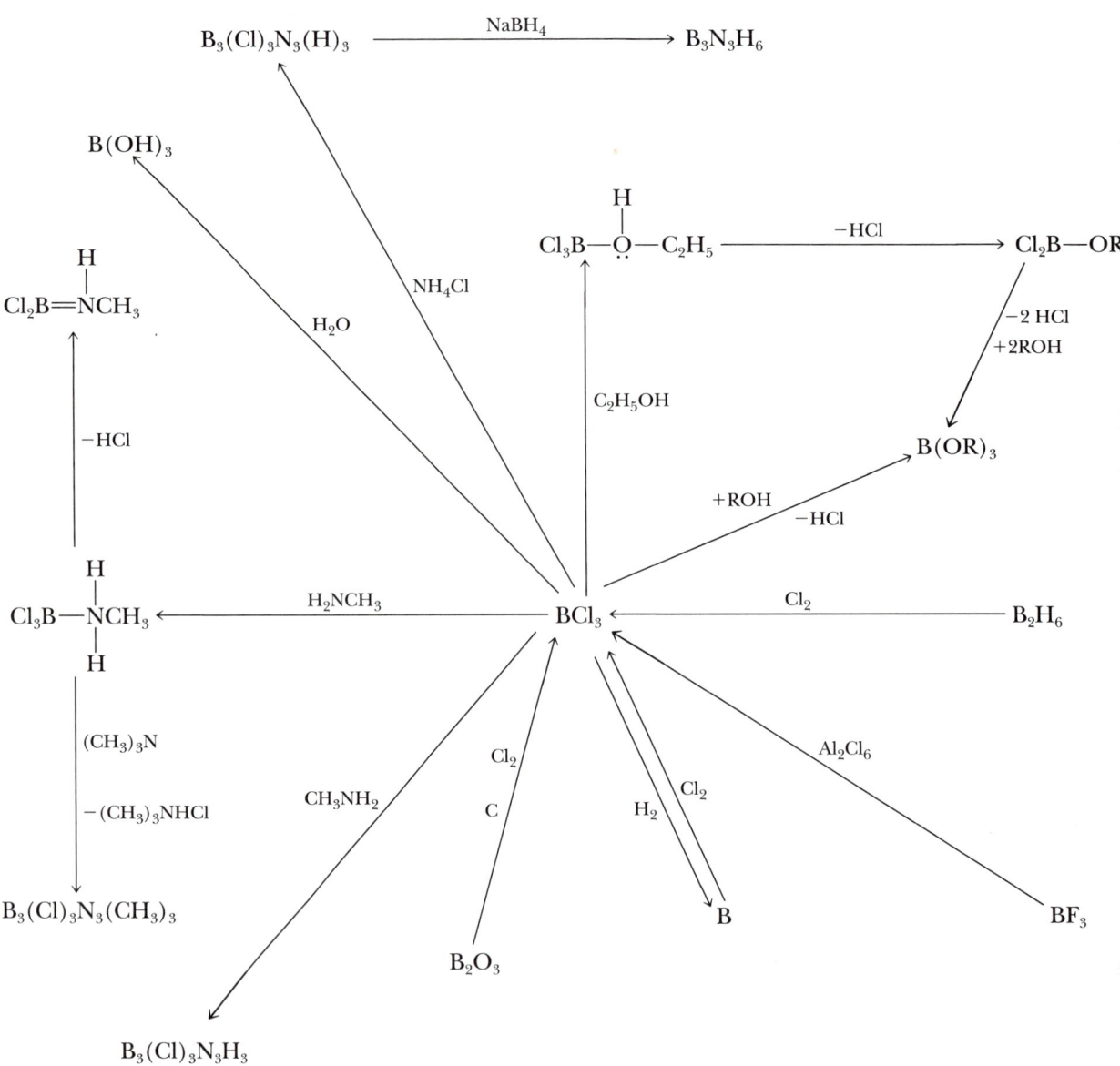

$$B_3(Cl)_3N_3(H)_3 \xrightarrow{\quad NaBH_4 \quad} B_3N_3H_6$$

(b)

Figure 12-13b

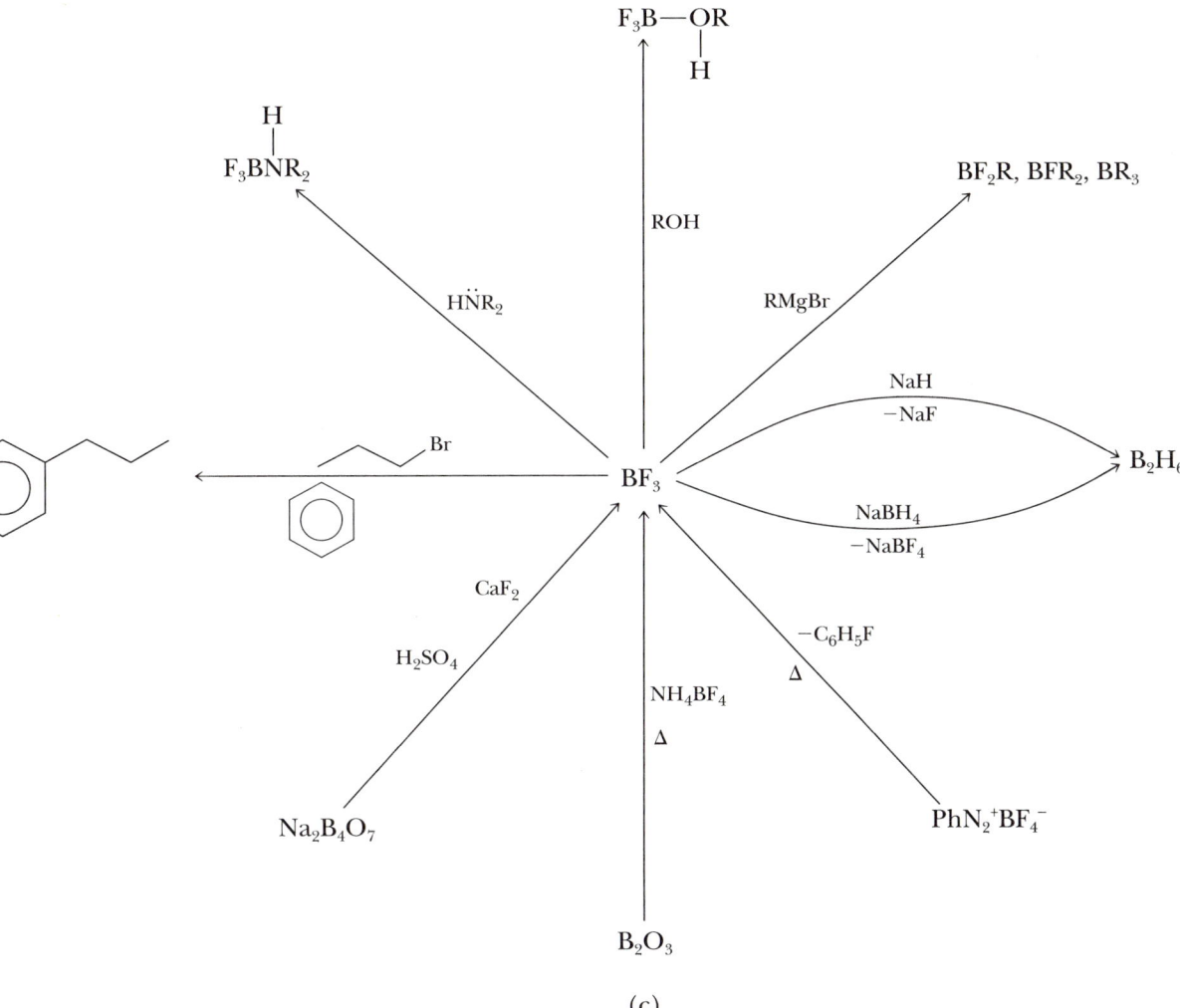

(c)

Figure 12-13c

STUDY GUIDE

Study Questions

A. Review

1. Draw the structure of the B12 unit that is found in elemental boron.
2. Draw the structures of the cyclic borate anion in $K_3B_3O_6$ and the chain borate anion in $Mg_2B_2O_5$. Indicate the hybridization at each atom in these borate anions.
3. How does boric acid ionize in water? How strong an acid is it?
4. Why is the activity of boric acid increased by the addition of glycerol?
5. How would one best prepare BF_3 in the laboratory?
6. Why is BBr_3 a better Lewis acid than BF_3?

7. Draw the structure of diborane and describe its bonding.

8. Give equations for one useful synthesis of diborane(6).

9. How is sodium borohydride (or tetrahydroborate) prepared?

B. Additional Exercises

1. Review each of the structures of the boron hydrides that have been presented in this chapter and decide which can properly be thought of as a fragment of the icosahedron.

2. The borate anion, $[B_5O_6(OH)_4]^-$, has one tetrahedral boron and four trigonal boron atoms, and has two six-membered rings. Draw the structure.

3. Review the reactions of diborane and prepare the same sort of "reaction wheel" for it that has already been prepared for boric acid (Fig. 12-2).

4. Explain why the barriers to rotation around the B—N bond in the following aminoboranes display the trend $H_2B(-NR_2) > HB(-NR_2)_2 > B(-NR_2)_3$.

5. Propose a structure for the anion $B_3H_8^-$ featured in Reactions 12-5.13 and 12-5.14.

6. Draw the structure of the disubstituted product from the reaction of decaborane(14) with CN^-.

7. What are the hybridizations and the geometries of the C and O atoms in the organic derivatives $B(OR)_3$, $B(OCOR)_3$, and $B(OOR)_3$?

8. What is the structure of the anion formed upon deprotonation of decaborane(14)? Which are the acidic hydrogens in decaborane(14) and why?

9. Predict the products of the following reactions:

   (a) $BF_3 + OEt_2$

   (b) $BF_3 + H_2O$

   (c) $BCl_3 + ROH$

   (d) $B_2H_6 + HCl$

   (e) $B_{10}H_{14} + NR_3$

   (f) $B_{10}H_{14} + I_2$

   (g) $LiH + B_2H_6$

   (h) $NH_4Cl + LiBH_4$

   (i) $(CH_3)_2N-BCl_2 + C_6H_5MgBr$

   (j) $B_3N_3H_6 + H_2O$

   (k) $B_3N_3H_6 + HBr$

   (l) $(Cl-B)_3(NH)_3 + C_2H_5MgBr$

10. Suggest a series of reactions for the synthesis of

    (a) Borazine, beginning with boron trichloride.

    (b) Decaborane(14), starting with diborane.

    (c) $[H_2B(NMe_3)_2]^+$, starting with diborane.

    (d) Diethylaminodichloroborane, starting with BCl_3.

    (e) *B*-Trichloro-*N*-trimethylborazine, starting with BCl_3.

11. Suggest a reason for the greater reactivity of borazine than benzene towards addition of HX.

12. Draw the structure of $(Ph_3P)_2CuBH_4$. Carefully show the geometry at P, Cu, and B.

13. Consider the semitopological diagram of $B_{10}H_{14}$ in Fig. 12-6. Account for all of the electrons in the molecule by listing the number of each that is involved in (a) terminal BH groups, (b) bridging BHB groups, (c) open BBB bridge groups, (d) two-electron BB bonds, and (e) closed BBB bonds.

14. Use Wade's rules to classify each of the following.

    (a) $B_6H_6^{2-}$

    (j) B_8H_{14}

    (b) $C_2B_4H_6$

    (k) $C_2B_7H_{11}$

    (c) B_5H_9

    (l) $B_{10}H_{14}^{2-}$

    (d) $C_2B_3H_7$

    (m) $B_7H_7^{2-}$

    (e) B_4H_{10}

    (n) $C_2B_8H_{10}$

    (f) $B_9H_{14}^-$ (o) $C_2B_8H_{12}$

    (g) $B_{10}H_{14}$ (p) $C_2B_9H_{11}$

    (h) B_5H_{11} (q) B_6H_{12}

    (i) B_6H_{10} (r) B_9Cl_9 and $B_9Cl_9^{2-}$

15. The spectrum of B_3F_5 suggests the presence of two types of F atoms in a ratio of 4:1 and two types of B atoms in a ratio of 2:1. Suggest a possible structure for this compound using trigonal boron atoms only.

16. The compound B_8F_{12} has four trigonal boron atoms and four tetrahedral atoms. Furthermore, there appears to be four terminal BF_2 groups and two bridging BF_2 groups. Suggest a plausible structure.

17. Write equations for each of the following reactions.

    (a) Reaction of diborane with ammonia.

    (b) Reaction of diborane with HCl.

    (c) Reduction of boron oxide by Fe.

    (d) $B(OH)_3 + CH_3COCl$

    (e) Reduction of BF_3 with NaH.

    (f) Reaction of $B_{10}H_{14}$ with I_2.

    (g) Hydrolysis of B_5H_9.

    (h) $B_2H_6 + O_2$

    (i) Hydrolysis of $BiCl_3$.

    (j) Condensation of boric acid with ethanol.

    (k) Reaction of BCl_3 with chlorine.

    (l) Reduction of diborane with sodium.

    (m) Hydrolysis of diborane.

    (n) Reaction of BCl_3 with ethanol.

    (o) Thermolysis of boric acid.

    (p) $B(OH)_3 + NH_4HF_2$

18. Write out a stepwise synthesis, starting with borax, of $Cl_2B—OC_2H_5$.

19. Show how to make the following compounds from the given starting materials.

    (a) B_2H_6 from BF_3 (g) B_9Cl_9 from B_2Cl_4

    (b) $[ClB—NH]_3$ from BCl_3 (h) $B(OC_2H_5)_3$ from B_2H_6

    (c) $B(OCH_3)_3$ from $B(OH)_3$ (i) BCl_3 from BF_3

    (d) B_2F_4 from B_2Cl_4 (j) $NaBH_4$ from B_2H_6

    (e) B_2Cl_4 from BCl_3 (k) $B_3N_3H_6$ from BCl_3

    (f) B_4Cl_4 from BCl_3

C. Problems from the Literature of Inorganic Chemistry

1. Consider the paper by R. W. Parry, R. W. Rudolph, and D. F. Shriver, *Inorg. Chem.*, **1964**, *3*, 1479–1483.

    (a) Write balanced equations for the symmetrical and unsymmetrical cleavage reactions of tetraborane(10) by a general nucleophile, L.

    (b) Write balanced equations for the symmetrical and unsymmetrical cleavage reactions of tetraborane(10) by $NaBH_4$.

    (c) Write the balanced equations for the symmetrical and unsymmetrical cleavage of tetraborane(10) by $NaBD_4$, and account for the predicted percentage of D label in the products for each case.

    (d) Why have the authors so carefully argued against "exchange" in such reactions or among the reaction products?

    (e) What are the products of the cleavage of tetraborane(10) by NH_3?

2. Consider the comparison of amine boranes and borazines made by O. T. Beachley, Jr., and B. Washburn, *Inorg. Chem.*, **1975**, *14*, 120–123.

    (a) Write balanced chemical equations to represent the reactions that were employed to synthesize

        (i) $H_2ClB \cdot N(CH_3)_3$ and $H_2BrB \cdot N(CH_3)_2H$

        (ii) $H_2CH_3B \cdot N(CH_3)_3$ and $H_2CH_3B \cdot N(CH_3)_2H$

        (iii) $H_2(CN)B \cdot N(CH_3)_3$ and $H_2(CN)B \cdot N(CH_3)_2H$

    (b) Draw the Lewis diagram for each adduct mentioned in (a).

    (c) What reaction takes place between $HgBr_2$ and (i) $H_3B \cdot N(CH_3)_3$; (ii) $H_3B_3N_3H_3$?

    (d) What reaction takes place between AgCN and (i) $H_3B_3N_3H_3$ at 0 °C (ii) $H_3B \cdot N(CH_3)_3$ at 130 °C?

    (e) What mechanistic interpretation do the authors give to the facts in (c)?

    (f) How do π- and σ-bond effects combine in the borazine ring to make the BH group sufficiently hydridic to react with $HgCl_2$?

    (g) What suggestion do the authors make to explain the facts in (d)?

3. Methylation at boron of the *closo*-carborane, $2,4$-$C_2B_5H_7$, has been studied by J. F. Ditter, E. B. Klusmann, R. E. Williams, and T. Onak, *Inorg. Chem.*, **1976**, *15*, 1063–1065.

    (a) When methylation was performed with methylchloride in the presence of an excess of $AlCl_3$, which boron atom(s) was methylated to give (i) $CH_3C_2B_5H_6$ via monomethylation (ii) $(CH_3)_2C_2B_5H_5$ via dimethylation (iii) $(CH_3)_3C_2B_5H_4$ via trimethylation?

    (b) What do the facts in (a) suggest about the relative availability of electrons (as judged by readiness to react with electrophilic reagents) at the different boron atoms in $C_2B_5H_7$?

    (c) How does its position in the cage influence the electron availability at a boron atom, according to these authors?

4. Although borazine, the inorganic analog of benzene, was known as early as 1926, a similar B—P cyclic trimer was not reported until 1987. Read the subsequent account of the compound by H. V. Rasika Dias and P. P. Power, *J. Am. Chem. Soc.*, **1989**, *111*, 144–148, and answer the following questions.

    (a) What synthetic method was used for the title compounds?

    (b) What mechanisms are proposed for formation of the compounds?

    (c) For the compound $(MesB$—$PC_6H_5)_3$, what is the significance of the planarity of the ring atoms and the six substituent carbon atoms?

    (d) What other structural data suggest a considerable amount of B—P double-bond character in the rings?

5. Read the article on synthesis of $B_5H_9^{2-}$ and B_5H_{11} by J. R. Wermer and S. G. Shore, *Inorg. Chem.*, **1987**, *26*, 1644–1645.

    (a) Write equations for the syntheses reported here of $B_5H_9^{2-}$ and B_5H_{11}, starting with B_5H_9.

    (b) Use Wade's rules to classify the structures of the above three compounds.

    (c) How do the structure and chemistry of $B_5H_9^{2-}$ compare with those of $B_5H_8^-$?

6. Read the article by T. Davan and J. A. Morrison, *Inorg. Chem.*, **1986**, *25*, 2366–2372.

    (a) What is the overall stability order found for the polyhedral boron chlorides?

(b) How does this stability order differ from that for the polyhedral borane anions, $B_nH_n^{2-}$?

(c) What difference is there between the two classes of compounds insofar as Wade's rules are concerned?

SUPPLEMENTARY READING

Adams, R. M., "Nomenclature of Inorganic Boron Compounds," *Pure Appl. Chem.,* **1972,** *30,* 683.

Brown, H. C., *Boranes in Organic Chemistry,* Cornell University Press, Ithaca, NY, 1972.

Brown, H. C., *Organic Syntheses via Boranes,* Wiley, New York, 1975.

Greenwood, N. N., *Boron,* Pergamon Press, Elmsford, NY, 1975.

Grimes, R. N., *Carboranes,* Academic, New York, 1971.

Grimes, R. N., Ed., *Metal Interactions with Boron Clusters,* Plenum, New York, 1982.

Grimes, R. N., "Carbon-Rich Carboranes and their Metal Derivatives," in *Advances in Inorganic Chemistry and Radiochemistry,* Vol. 26, H. J. Emeleus and A. G. Sharpe, Eds., Academic, pp. 55–117, 1983.

Lubman, J. F., Greenberg, A., and Williams, R. E., Eds., *Advances in Boron and the Boranes,* VCH Publishers, New York, 1988.

Massey, A. G., "The Subhalides of Boron," in *Advances in Inorganic Chemistry and Radiochemistry,* Vol. 26, H. J. Emeleus and A. G. Sharpe, Eds., Academic, New York, pp. 1–54, 1983.

Michl, J. Ed., *Chem. Rev.,* **1992,** *92(2),* 177–362. A special issue devoted to boron chemistry.

Mingos, D. M. P., "Polyhedral Skeletal Electron Pair Approach," *Acc. Chem. Res.,* **1984,** *17,* 311–319.

Morrison, J. A., "Polyhedral Boron Halides and Their Reactions," *Chem. Rev.,* **1991,** *91,* 35.

Muetterties, E. L., Ed., *The Chemistry of Boron and its Compounds,* Wiley, New York, 1967.

Muetterties, E. L., Ed., *Boron Hydride Chemistry,* Academic, New York, 1975.

Muetterties, E. L. and Knoth, W. H., *Polyhedral Boranes,* Dekker, New York, 1968.

Niedenzu, K. and Dawson, J. W., *Boron–Nitrogen Compounds,* Springer-Verlag, New York, 1965.

Wade, K., "Structure and Bonding Patterns in Cluster Chemistry," in *Advances in Inorg. Chem. and Radiochem.,* Vol. 18, H. J. Emeleus and A. G. Sharpe, Eds., Academic, New York, pp. 1–66, 1976.

Woollins, J. D., *Non-metal Rings, Cages and Clusters,* Wiley, New York, 1988.

Chapter 13

THE GROUP IIIB(13) ELEMENTS: ALUMINUM, GALLIUM, INDIUM, AND THALLIUM

13-1 Introduction

Aluminum is the commonest metallic element in the earth's crust and occurs in rocks such as felspars and micas. More accessible deposits are hydrous oxides such as bauxite ($Al_2O_3 \cdot n\,H_2O$) and cryolite (Na_3AlF_6). The elements Ga and In occur only in traces in Al and Zn ores. Thallium, also a rare element, is recovered from flue dusts from the roasting of pyrite and other sulfide ores.

Aluminum metal has many uses and some salts, such as the sulfate ($\sim 10^8$ kg/year in the USA), are made on a large scale. Gallium finds some use in solid state devices as GaAs. Thallium is used mainly as the Tl^{III} carboxylates in organic synthesis.

The position of the elements and their relation to the Sc, Y, La group is discussed in Chapter 8, where Table 8-3 gives some important properties of the elements.

The elements are more metallic than boron, and their chemistry in compounds is more ionic. Nevertheless, many of the compounds are on the borderline of ionic–covalent character. All four elements give trivalent compounds, but the univalent state becomes increasingly important for Ga, In, and Tl. For Tl the two states are about equally important and the redox system Tl^I–Tl^{III} dominates the chemistry. The Tl^+ ion is well defined in solutions.

The main reason for the existence of the univalent state is the decreasing strengths of bonds in MX_3; thus, for the chlorides, the mean bond energies are Ga(242), In(206), and Tl(153) kJ mol^{-1}. Hence, there is an increasing drive for Reaction 13-1.1 to occur.

$$MX_3 = MX + X_2 \qquad (13\text{-}1.1)$$

The compounds of MX_3 or MR_3 resemble similar BX_3 compounds in that they are Lewis acids, with strengths decreasing in the order B > Al > Ga > In ~ Tl. However, while all BX_3 compounds are planar monomers, the halides of the other elements have crystal structures in which the coordination number is increased. Coordination numbers of four occur in bridged dimers such as $Cl_2Al(\mu\text{-}Cl)_2AlCl_2$ and $(AlMe_3)_2$, whereas with bulky ligands, monomeric three-coordinate compounds may be formed, for example, $Ga(SR)_3$, where Ar =

2,4,6-t-BuC$_6$H$_2$. Adducts of the Lewis acids MX$_3$ can be five-coordinate, an example being (Me$_3$N)$_2$AlH$_3$.

Each of the elements forms an aqua ion, [M(H$_2$O)$_6$]$^{3+}$, and gives simple salts and complex compounds, where virtually all of the metals are octahedrally coordinated.

13-2   Occurrence, Isolation, and Properties of the Elements

Aluminum is prepared on a vast scale from bauxite, Al$_2$O$_3 \cdot n$ H$_2$O (n = 1–3). This is purified by dissolution in aqueous NaOH (giving Al(OH)$_4^-$), filtration to remove Fe and other insoluble hydroxides, and finally by precipitation of Al(OH)$_3 \cdot 3$ H$_2$O on cooling. The dehydrated product is dissolved in molten cryolite and the melt at 800–1000 °C is electrolyzed. Aluminum is a hard, strong, white metal. Although highly electropositive, it is nevertheless resistant to corrosion because a hard, tough film of oxide is formed on the surface. Thick oxide films are often electrolytically applied to aluminum, a process called anodizing; the fresh films can be colored by pigments. Aluminum is soluble in dilute mineral acids, but is "passivated" by concentrated HNO$_3$. If the protective effect of the oxide film is broken, for example by scratching or by amalgamation, rapid attack can occur even by water. The metal is readily attacked by hot aqueous NaOH, halogens, and various nonmetals.

The elements Ga, In, and Tl are usually obtained by electrolysis of aqueous solutions of their salts; for Ga and In this possibility arises because of large overvoltages for hydrogen evolution of these metals. These elements are soft, white, comparatively reactive metals, dissolving readily in acids. Thallium dissolves only slowly in H$_2$SO$_4$ or HCl, since the Tl^I salts formed are only sparingly soluble. Gallium, like Al, is soluble in aqueous NaOH. The elements react rapidly at room temperature (or on warming) with the halogens and with nonmetals such as sulfur.

13-3   Oxides

The only oxide of aluminum is *alumina* (Al$_2$O$_3$). However, this simplicity is compensated by the occurrence of polymorphs and hydrated materials whose nature depends on the conditions of preparation. There are two forms of anhydrous Al$_2$O$_3$: α-Al$_2$O$_3$ and γ-Al$_2$O$_3$. Other trivalent metals (e.g., Ga or Fe) form oxides that crystallize in these same two structures. Both have close-packed arrays of oxide ions but differ in the arrangement of the cations.

α-Al$_2$O$_3$ is stable at high temperatures and also indefinitely metastable at low temperatures. It occurs in nature as the mineral corundum and may be prepared by heating γ-Al$_2$O$_3$ or any hydrous oxide above 1000 °C. Gamma-Al$_2$O$_3$ is obtained by dehydration of hydrous oxides at low temperatures (~450 °C). Alpha-Al$_2$O$_3$ is hard and is resistant to hydration and to attack by acids. Gamma-Al$_2$O$_3$ readily absorbs water and dissolves in acids; the aluminas used for chromatography and conditioned to different reactivities are γ-Al$_2$O$_3$. Large quantities of α-Al$_2$O$_3$ are used in industry as a support material for heterogeneous catalysts.

There are several hydrated forms of alumina of stoichiometries from AlO·OH to Al(OH)$_3$. Addition of ammonia to a boiling solution of an aluminum salt produces a form of AlO·OH known as *boehmite*. A second form of AlO·OH occurs in nature as the mineral *diaspore*. The true *hydroxide*, $Al(OH)_3$, is obtained as a crystalline white precipitate when CO_2 is passed into alkaline "aluminate" solutions.

The oxides of Ga and In are similar, but Tl gives only brown-black Tl_2O_3, which decomposes to Tl_2O at 100 °C.

The elements form *mixed oxides* with other metals. Aluminum oxides containing only traces of other metal ions include ruby (Cr^{3+}) and blue sapphire (Fe^{2+}, Fe^{3+}, and Ti^{4+}). Synthetic ruby, blue sapphire, and white sapphire (gem-quality corundum) are manufactured in large quantities. Mixed oxides containing macroscopic proportions of other elements include the minerals *spinel* ($MgAl_2O_4$) and *crysoberyl* ($BeAl_2O_4$). The *spinal structure* (Section 4-8) is important as a prototype for many other $M^{II}M_2^{III}O_4$ compounds. Compounds such as $NaAlO_2$, which can be made by heating Al_2O_3 with sodium oxalate at 1000 °C, are also ionic mixed oxides.

13-4 Halides

All four halides of each element are known, with one exception. The compound TlI_3, obtained by adding iodine to thallium(I) iodide, is not thallium(III) iodide, but rather thallium(I) triiodide, $Tl^I(I_3)$. This situation may be compared with the nonexistence of iodides of other oxidizing cations, such as Cu^{2+} and Fe^{3+}, except that here a lower-valent compound fortuitously has the same stoichiometry as the higher-valent one. The coordination numbers of the halides are shown in Table 13-1. The fluorides of Al, Ga, and In are ionic and high melting (>950 °C), whereas the chlorides, bromides, and iodides have lower melting points. There is some correlation between melting points and coordination number, since the halides with coordination number four consist of discrete dinuclear molecules (Fig. 13-1) and the melting points are low. Thus, the three chlorides have the following melting points: $AlCl_3$, 193 °C (at 1700 mm Hg); $GaCl_3$, 78 °C; $InCl_3$, 586 °C. In the vapor, aluminum chloride is also dimeric so that there is a radical change of coordination number on vaporization. The dimer structures persist in the vapor phase at temperatures close to the boiling points but at higher temperatures dissociation occurs, giving triangular monomers analogous to the boron halides.

The covalent halides dissolve readily in nonpolar solvents such as benzene, in which they are dimeric. As Fig. 13-1 shows, the configuration of halogen atoms about each metal atom is distorted tetrahedral. The formation of such dimers is attributable to the tendency of the metal atoms to complete their octets.

Table 13-1 Coordination Numbers of Metal Atoms in Group IIIB(13) Halides

	F	Cl	Br	I
Al	6	6	4	4
Ga	6	4	4	4
In	6	6	6	4
Tl	6	6	4	

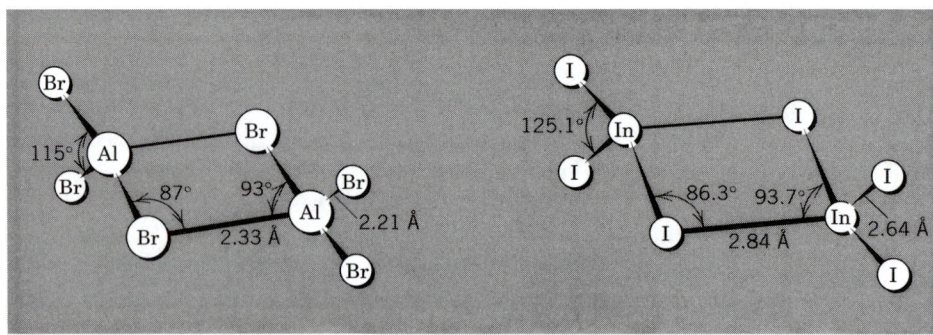

Figure 13-1   The structures of Al_2Br_6 and In_2I_6.

The thallium(III) halides vary considerably in thermal stability. Although TlF_3 is stable to 500 °C, $TlCl_3$ loses chlorine at about 40 °C forming $TlCl$, while $TlBr_3$ loses Br_2 at even lower temperatures to give first "$TlBr_2$," which is actually $Tl^I[Tl^{III}Br_4]$.

The trihalides (fluorides excepted) are strong Lewis acids, and this is one of the most important aspects of their chemistry, as well as that of other MR_3 compounds, such as the alkyls and AlH_3. Adducts are readily formed with Lewis bases (including halide ions). The dimeric halides are cleaved to give products such as $Cl_3AlN(CH_3)_3$ and $AlCl_4^-$.

Aluminum chloride and bromide especially are used as catalysts (Friedel–Crafts type) in a variety of reactions. The formation of $AlCl_4^-$ or $AlBr_4^-$ ions is essential to the catalytic action, since in this way carbonium ions are formed (Reaction 13-4.1).

$$RCOCl + AlCl_3 = RCO^+ + AlCl_4^- \quad \text{(ion pair)} \qquad (13\text{-}4.1)$$

and made available for reaction as in Reaction 13-4.2.

$$RCO^+ + C_6H_6 \longrightarrow [RCOC_6H_6]^+ \longrightarrow RCOC_6H_5 + H^+ \qquad (13\text{-}4.2)$$

13-5   The Aqua Ions, Oxo Salts, and Aqueous Chemistry

The elements Al, Ga, In, and Tl form well-defined octahedral aqua ions, $[M(H_2O)_6]^{3+}$, and many salts containing these ions are known, including hydrated halides, sulfates, nitrates, and perchlorates. Phosphates are sparingly soluble.

In aqueous solution, the octahedral ions $[M(H_2O)_6]^{3+}$ are quite acidic. For Reaction 13-5.1

$$[M(H_2O)_6]^{3+} = [M(H_2O)_5(OH)]^{2+} + H^+ \qquad (13\text{-}5.1)$$

the constants are K_a(Al), 1.12×10^{-5}; K_a(Ga), 2.5×10^{-3}; K_a(In), 2×10^{-4}; and K_a(Tl), $\sim 7 \times 10^{-2}$. Although little emphasis can be placed on the exact numbers, the orders of magnitude are important, for they show that aqueous solutions of the

M^{III} salts are subject to extensive hydrolysis. Indeed, salts of weak acids (sulfides, carbonates, cyanides, acetates, and the like) cannot exist in contact with water.

In addition to this hydrolysis reaction there is also a dimerization as in Reaction 13-5.2.

$$2\,AlOH^{2+}(aq) = [Al_2(OH)_2]^{4+}(aq) \qquad K = 600\ M^{-1}\ (30\ °C) \quad (13\text{-}5.2)$$

More complex species of the general formula $Al[Al_3(OH)_8]_m^{m+3}$ have also been postulated and some, such as $[Al_{13}O_4(OH)_{24}(H_2O)_{12}]^{7+}$ and its gallium analog, have been identified in crystalline basic salts.

An important class of aluminum salts, the *alums,* are structural prototypes and give their name to a large number of analogous salts formed by other elements. These salts have the general formula $MAl(SO_4)_2 \cdot 12\,H_2O$ in which M is practically any common univalent, monatomic cation except Li^+, which is too small to be accommodated without loss of stability of the structure. The crystals are made up of $[M(H_2O)_6]^+$, $[Al(H_2O)_6]^{3+}$, and two SO_4^{2-} ions. Salts of the same type, $M^I M^{III}(SO_4)_2 \cdot 12\,H_2O$, and having the same structures, are formed by other M^{3+} ions, including those of Ti, V, Cr, Mn, Fe, Co, Ga, In, Rh, and Ir. All such compounds are referred to as alums. The term is used so generally that those alums containing aluminum are redundantly designated aluminum alums.

Aluminum ions and complexes are environmentally important. The leaching of Al^{3+} from silicate rocks by acid rain leads to high concentrations in lakes. Such high concentrations are toxic to aquatic life. Although senile dementia (Alzheimer's disease) may have a genetic origin, a symptom is the accumulation of aluminum complexes in the brain. The Al^{3+} ion is known to bind to iron sites in human serum transferrin (Chapter 31), and citrates, which occur in blood plasma, lactates, and other complexing agents may be involved.

Thallium carboxylates, particularly the acetate and trifluoroacetate, which can be obtained by dissolution of the oxide in the acid, are extensively used in organic synthesis. The trifluoroacetate will directly thallate (cf. mercuration, Chapter 29) aromatic compounds to give aryl thallium ditrifluoroacetates [e.g., $C_6H_5Tl(OOCCF_3)_2$]. It also acts as an oxidant, for example, by converting para substituted phenols into *p*-quinones.

The hydroxides of aluminum and gallium are amphoteric:

$$Al(OH)_3(s) = Al^{3+} + 3\,OH^- \qquad K \approx 5 \times 10^{-33} \qquad (13\text{-}5.3)$$

$$Al(OH)_3(s) = AlO_2^- + H^+ + H_2O \qquad K \approx 4 \times 10^{-13} \qquad (13\text{-}5.4)$$

$$Ga(OH)_3(s) = Ga^{3+} + 3\,OH^- \qquad K \approx 5 \times 10^{-37} \qquad (13\text{-}5.5)$$

$$Ga(OH)_3(s) = GaO_2^- + H^+ + H_2O \qquad K \approx 10^{-15} \qquad (13\text{-}5.6)$$

Like the oxides, these compounds also dissolve in bases as well as in acids. By contrast, the oxides and hydroxides of In and Tl are purely basic. According to Raman spectra, the main aluminate species from pH 8 to 12 appears to be an OH bridged polymer with octahedral Al, but at pH > 13 and concentrations below 1.5 M the tetrahedral $Al(OH)_4^-$ ion is present. Above 1.5 M there is condensation to give the ion $[(HO)_3AlOAl(OH)_3]^{2-}$. This occurs in the crystalline salt $K_2[Al_2O(OH)_6]$ which has an angular Al—O—Al bridge.

13-6 Coordination Compounds

The trivalent elements form four-, five- and six-coordinate complexes, which may be cationic, like $[Al(H_2O)_6]^{3+}$ or $[Al(OSMe_2)_6]^{3+}$; neutral, for example, $AlCl_3(NMe_3)_2$; or anionic, like $[AlF_6]^{3-}$ and $[In(NCS)_6]^{3-}$.

One of the most important salts is *cryolite,* whose structure (Fig. 13-2) is adopted by many other salts that contain small cations and large octahedral anions and, with reversal of cations and anions, by many salts of the same type as $[Co(NH_3)_6]I_3$. It is closely related to the structures adopted by many compounds of the types $M_2^+[AB_6]^{2-}$ and $[XY_6]^{2+}Z_2^-$. The last two structures are essentially the fluorite (or antifluorite) structures (see Fig. 4-1), except that the anions (or cations) are octahedra whose axes are oriented parallel to the cube edges. The relationship of the two structures can be seen in Fig. 13-2, since the Na^+ ions have been indicated by both open ○ and marked ⊗ circles. If all of the marked circles (one at the center and one on each of the cube edges) in Fig. 13-2 are removed, the cryolite structure reduces to the $M_2^+[AB_6]^{2-}$ fluorite-type structure.

Many of the important octahedral complexes are those containing chelate rings. Some typical structures contain β-diketones, pyrocatechol (Structure 13-I), dicarboxylic acids (Structure 13-II), and 8-quinolinol (Structure 13-III).

13-I	13-II	13-III

The neutral complexes are soluble in organic solvents, but insoluble in water. The acetylacetonates have low melting points (<200 °C) and vaporize without de-

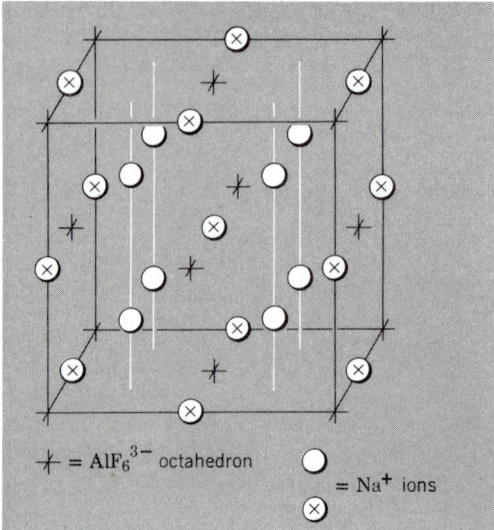

Figure 13-2 The cubic structure of cryolite (Na_3AlF_6).

composition. The anionic complexes are isolated as the salts of large univalent cations. The 8-quinolinolates are used for analytical purposes.

The four elements form *alkoxides,* but only those of aluminum and gallium are important. The isopropoxide of aluminum is widely used in organic chemistry to catalyze the reduction of aldehydes and ketones by alcohols or vice versa (Meerwein–Ponndorf–Oppenauer–Verley reactions). Alkoxides can be made by Reactions 13-6.1 and 13-6.2.

$$Al + 3\ ROH \xrightarrow[\text{catalyst, warm}]{\text{1\% HgCl}_2 \text{ as}} (RO)_3 Al + \tfrac{3}{2} H_2 \qquad (13\text{-}6.1)$$

$$AlCl_3 + 3\ RONa \xrightarrow{\text{ROH}} (RO)_3 Al + 3\ NaCl \qquad (13\text{-}6.2)$$

The *tert*-butoxide has the dimeric structure typical of $M_2(OR)_6$ compounds both in the crystalline form and in solution (Structure 13-IV). The commonly used isopropoxide has different oligomers, one of which is the tetramer shown in Structure 13-V. This compound can be regarded as an Al^{3+} ion coordinated by three $[Al(OR)_4]^-$ groups. Other alkoxides normally form dimers and trimers, but where R groups are very bulky, three-coordinate monomers can be formed.

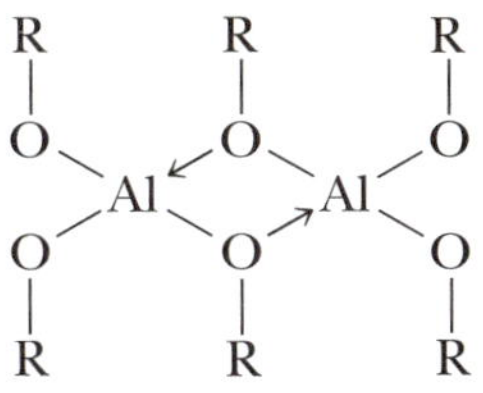

13-IV 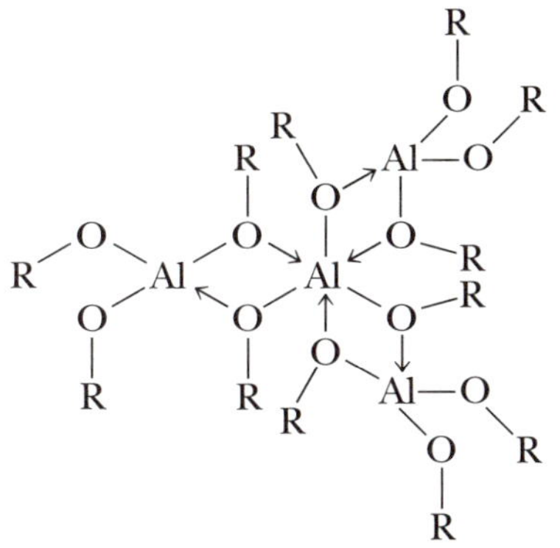   13-V

Terminal and bridging alkoxyl groups can be distinguished by nmr spectra. Other alkoxides form dimers and trimers.

13-7 Hydrides

The salts containing the tetrahedral anion AlH_4^-, which is similar in some ways to BH_4^-, are important hydrides of Al. Gallium also forms a tetrahydrido anion. The thermal and chemical stabilities of these tetrahydrido anions vary with the ability of the MH_3 groups to act as an H^- acceptor, as in Reaction 13-7.1.

$$MH_3 + H^- = MH_4^- \qquad (13\text{-}7.1)$$

The order is B > Al > Ga. Thus $LiGaH_4$ decomposes slowly even at 25 °C to LiH, Ga, and H_2 and is a milder reducing agent than $LiAlH_4$. Similarly, although BH_4^- is stable in water, the Al and Ga salts are rapidly and often explosively hydrolyzed by water.

$$MH_4^- + 4\,H_2O = 4\,H_2 + M(OH)_3 + OH^- \qquad (13\text{-}7.2)$$

The most important compound is *lithium tetrahydridoaluminate,* which is widely used in both organic and inorganic chemistry as a reducing agent. It accomplishes many otherwise tedious or difficult reductions, for example, —CO$_2$H to —CH$_2$OH. It is a nonvolatile, crystalline solid, which is white when pure but is usually gray. It is stable below 120 °C and is soluble in diethyl ether, THF, and glymes.

Both aluminum and gallium salts are made by reaction of the chloride with lithium hydride, as in Reaction 13-7.3.

$$4\,LiH + MCl_3 \xrightarrow{\;(C_2H_5)_2O\;} LiMH_4 + 3\,LiCl \qquad (13\text{-}7.3)$$

The sodium salt can be obtained by direct interactions of the elements, as in Reaction 13-7.4.

$$Na + Al + 2\,H_2 \xrightarrow[150\,°C\ 2000\ psi]{THF} NaAlH_4 \qquad (13\text{-}7.4)$$

The addition of toluene precipitates NaAlH$_4$ which can be converted to the lithium salt by recrystallization from ether in the presence of LiCl, as in Reaction 13-7.5.

$$NaAlH_4 + LiCl \xrightarrow{\;(C_2H_5)_2O\;} NaCl(s) + LiAlH_4 \qquad (13\text{-}7.5)$$

Donor Adducts of the Hydrides

There is an extensive range of complex hydrides that may be regarded as arising from the Lewis acid behavior of the MH$_3$ fragments. These adducts may be formed with donor molecules (e.g., NR$_3$ and PR$_3$) or with anions (e.g., H$^-$) as in Reaction 13-7.3 above. The various adducts are similar to the borane adducts, the stability order being B > Al > Ga. The most studied adducts are the trialkylamine alanes (alane = AlH$_3$). Trimethylamine in ether, at room temperature or below, gives both 1:1 and 1:2 adducts, as in Reactions 13-7.6 through 13-7.9.

$$Me_3N—AlCl_3 + 3\,LiH \longrightarrow Me_3N—AlH_3 + 3\,LiCl \qquad (13\text{-}7.6)$$

$$Me_3NH^+Cl^- + LiAlH_4 \longrightarrow Me_3N—AlH_3 + LiCl + H_2 \qquad (13\text{-}7.7)$$

$$3\,LiAlH_4 + AlCl_3 + 4\,NMe_3 \longrightarrow 4\,Me_3N—AlH_3 + 3\,LiCl \qquad (13\text{-}7.8)$$

$$Me_3NAlH_3 + Me_3N \longrightarrow (Me_3N)_2AlH_3 \qquad (13\text{-}7.9)$$

The monotrimethylamine alane adduct is a white, volatile, crystalline solid (mp 75 °C), that is readily hydrolyzed. It is monomeric and tetrahedral. The bis amine product of Reaction 13-7.9 is trigonal bipyramidal, with axial N atoms. Tetrahydrofuran also gives both a 1:1 and a 2:1 adduct, but ether, presumably for steric reasons, forms only a mono adduct.

Similar monoamine gallane adducts exist. These have strong Ga—H bonds, making them less sensitive to hydrolysis than are the aluminum analogs. The diamine adduct (Me$_3$N)$_2$GaH$_3$ is stable only below −60 °C.

13-8   Lower Valent Compounds

Since the outer-electron configuration is ns^2np^1, univalent compounds are, in principle, possible. Aluminum forms such species only at high temperature in the gas phase, for example,

$$AlCl_3(s) + 2\,Al(s) \rightleftharpoons 3\,AlCl(g) \qquad (13\text{-}8.1)$$

Some gallium(I) and indium(I) compounds are known. The so-called dichloride "$GaCl_2$" is actually $Ga^I[Ga^{III}Cl_4]$.

Thallium has a well-defined unipositive state. In aqueous solution it is distinctly more stable than Tl^{III}

$$Tl^{3+} + 2\,e^- = Tl^+ \qquad E° = +1.25\ V \qquad (13\text{-}8.2)$$

The Tl^+ ion is not very sensitive to pH, although the Tl^{3+} ion is extensively hydrolyzed to $TlOH^{2+}$ and the colloidal oxide, even at pH 1–2.5. The redox potential is, hence, very dependent on pH, as well as on the presence of complexing anions. For example, the presence of Cl^- stabilizes Tl^{3+} more (by formation of complexes) than Tl^+, and the potential is thereby lowered.

The colorless Tl^+ ion has a radius of 1.64 Å, comparable to those of K^+, Rb^+, and Ag^+ (1.52, 1.66, and 1.29 Å). Thus it resembles the alkali ions in some ways and the Ag^+ ion in others. It may replace K^+ in certain enzymes and has potential use as a probe for potassium. In crystalline salts, the Tl^+ ion is usually six or eight coordinate. The yellow hydroxide is unstable, giving the black oxide Tl_2O at about 100 °C. The oxide and hydroxide are soluble in water giving strongly basic solutions. These absorb CO_2 from the air, although TlOH is a weaker base than KOH. Many thallium(I) salts (e.g., Tl_2SO_4, Tl_2CO_3, or $TlCO_2CH_3$) have solubilities somewhat lower than those of the corresponding K^+ salts, but otherwise they are similar to and quite often isomorphous with them. Thallium(I) fluoride is soluble in water but the other halides are sparingly soluble. Thallium(I) chloride also resembles AgCl in being photosensitive and darkening on exposure to light, but differs in being insoluble in ammonia. All thallium compounds are exceedingly poisonous.

13-9   Summary of Periodic Trends for the Elements of Group IIIB(13)

By using the list of periodic chemical properties developed in Section 8-11, as well as properties mentioned in Chapters 12 and 13, we can now summarize the periodic trends in the properties of the elements of Group IIIB(13).

1. Boron

   (a) Forms no simple B^{3+} cation.

   (b) Forms covalent compounds almost exclusively, and all polyatomic ions have covalent bonds.

   (c) Obeys the octet rule, the maximum covalence being four.

   (d) Forms trivalent compounds that readily serve as Lewis acids.

  (e) Frequently forms polyhedral structures: boranes and borates.
  (f) Forms an oxide, B_2O_3, and a hydroxide, $B(OH)_3$, both of which are acidic.
  (g) Forms covalent halides that are readily hydrolyzed.
  (h) Forms numerous covalent hydrides, all of which are volatile, flammable, and readily hydrolyzed.
  (i) Forms a stable and important hydride anion, BH_4^-.

 2. Aluminum
  (a) Readily forms an important 3+ ion, because it is electropositive.
  (b) Is much more metallic than boron, and forms a greater number and variety of ionic substances.
  (c) Forms both molecular and ionic substances, with coordination numbers of six and higher.
  (d) Forms two oxides, only one of which is acidic.
  (e) Forms a hydroxide that is weakly amphoteric, although mostly basic.
  (f) Forms solid halides that are only partially hydrolyzable.
  (g) Forms a polymeric hydride.
  (h) Forms an anionic hydride (AlH_4^-) that is more reactive than BH_4^-.

 3. Gallium, Indium, and Thallium
  (a) Readily give the M^{3+} ion in solution, and have a rich coordination chemistry typical of metals.
  (b) Form increasingly stable lower valent compounds, especially Tl^+.
  (c) Increasingly form weaker covalent bonds on descent of the group, enhancing the formation of monovalent compounds.
  (d) Form MX_3 halides that are increasingly aggregated in the solid state (through halide ion bridges) to give coordination numbers of four, six, and higher.
  (e) Do not form important EH_4^- anions, except perhaps GaH_4^-.

STUDY GUIDE

Study Questions

A. Review

 1. What is bauxite, and how is it purified for Al production?
 2. Why is aluminum resistant to air and water, even though it is very electropositive?
 3. What are the formulas and structures of (a) corundum, (b) the mineral spinel?
 4. What is the structure of the trihalide dimers, M_2X_6? What happens to these molecules at high temperatures?
 5. What is an alum? What species are present in a crystalline alum?
 6. For cryolite, give the formula, structure, and chief industrial use.
 7. Compare the properties of B_2O_3 and Al_2O_3.
 8. How is $LiAlH_4$ prepared? Why does it explode with water, while $NaBH_4$ does not?
 9. Write equations to show that the hydroxides of Al and Ga are amphoteric.

B. Additional Exercises

1. Discuss the reasons why $Tl^{III}I_3$ is unstable relative to Tl^II_3, whereas the opposite is true for Al, Ga, and In.

2. How might one establish that the true nature of "$GaCl_2$" is actually $Ga^I[Ga^{III}Cl_4]$?

3. Interaction of Al with alcohols using $HgCl_2$ as a catalyst gives alkoxides of Al that are tetrameric in solution. Write a structure for the aluminum–isopropoxide tetramer.

4. Show, with equations, how $AlCl_3$ functions as a Friedel–Crafts catalyst.

5. Why is the Tl^+/Tl^{3+} electrochemical potential sensitive to pH and to the presence of complexing anions?

6. Explain the preference shown in Table 13-1 of six coordination for fluorides and chlorides versus four coordination for bromides and iodides.

7. Give equations for the following:

   (a) Aluminum chloride plus PF_3.

   (b) Synthesis of $LiAlH_4$ starting with elements only.

   (c) Thermal decomposition of $LiGaH_4$.

   (d) Thermal decomposition of $TlCl_3$.

   (e) Hydrolysis of Al^{3+} salts.

   (f) Hydrolysis of $GaCl_3$.

   (g) Reaction of Al with ethanol.

   (h) Thermal decomposition of Tl_2O_3.

   (i) Reaction of Al_2Cl_6 with $N(CH_3)_3$.

   (j) Amphoteric behavior by aluminum hydroxide (two equations).

   (k) Reduction of $AlCl_3$ by Al, at high temperature.

   (l) Synthesis of $(Me_3N)_2GaH_3$.

C. Questions from the Literature of Inorganic Chemistry

1. Complexes of the type $InCl_3 \cdot 3$ L and $TlX_3 \cdot 2$ L were studied by B. F. G. Johnson and R. A. Walton, *Inorg. Chem.*, **1966**, *5*, 49–53.

   (a) Write balanced equations for the reactions that were employed in the syntheses of these two types of compounds.

   (b) Suggest a structure for $TlCl_4^-$, for $TlCl_3 \cdot 2$ py, and for $InCl_3 \cdot 3$ py.

2. What evidence do the authors present for the presence of a metal–metal bond in the compound $Ga_2I_4 \cdot 2(diox)$? See J. C. Beamish, R. W. H. Small, and I. J. Worrall, *Inorg. Chem.*, **1979**, *18*, 220–223.

3. Consider the paper by E. R. Alton, R. G. Montemayer, and R. W. Parry, *Inorg. Chem.*, **1974**, *13*, 2267–2270.

   (a) Which of the Lewis bases featured in this study (:PF_3, :PCl_3, :C≡O:, or :NH_3) form complexes with the Lewis acids (i) BF_3 (ii) $AlCl_3$ (iii) $(CH_3)_3Al$?

   (b) What conclusions in reference to σ-base strength do the authors reach for PF_3 versus CO?

   (c) What is the *distortion energy* that the authors mention, and how can this concept be used to explain a higher stability for $F_3P{:}AlCl_3$ than for $F_3P{:}BF_3$?

4. Read about the synthesis of gallane and other materials in the article by A. J. Downs, M. J. Goode, and C. R. Pulham, *J. Am. Chem. Soc.*, **1989**, *111*, 1936–1937.

   (a) How was the starting material $[H_2GaCl]_2$ prepared?

   (b) How was the title compound prepared?

   (c) What reaction takes place between gallane and anhydrous HCl, and how was this used for analysis of the chemical composition of the title compound?

(d) What is the significance of the fact that the title compound reacts at low temperature with an excess of trimethylamine to give a single product, $(Me_3N)_2GaH_3$?

(e) What compound is obtained from the thermal decomposition of $(Me_3N)_2GaH_3$?

SUPPLEMENTARY READING

Carty, A. J. and Tuck, D. J., "Coordination Chemistry of Indium," *Prog. Inorg. Chem.*, **1975,** *19,* 243.

Cucinella, S., Mazzei, A., and Marconi, W., "Synthesis and Reactions of Aluminum Hydride Derivatives," *Inorg. Chim. Acta Rev.,* **1970,** *4,* 51.

Greenwood, N. N., "The Chemistry of Gallium," *Adv. Inorg. Chem. Radiochem.,* **1963,** *5,* 91.

Lee, A. G., *The Chemistry of Thallium,* Elsevier, Amsterdam, 1971.

Lee, A. G., "Coordination Chemistry of Thallium(I)," *Coord. Chem. Rev.,* **1972,** *8,* 289.

Massey, R. C. and Taylor, D., Eds., *Aluminum in Food and the Environment,* Royal Society of Chemistry, Cambridge, UK, 1989.

Olah, G. A., *Friedel–Crafts Chemistry,* Wiley, New York, 1973.

Sheka, I. A., Chans, I. S., and Mityureva, T. T., *The Chemistry of Gallium,* Elsevier, Amsterdam, 1966.

Walton, R. A., "Coordination Complexes of the Thallium(III) Halides and Their Behavior in Non-Aqueous Media," *Coord. Chem. Rev.,* **1971,** *6,* 1–25.

Chapter 14

CARBON

14-1 Introduction

There are more known compounds of carbon than of any other element except hydrogen. Most are best regarded as organic chemicals. This chapter considers certain compounds traditionally considered "inorganic." Chapter 29 discusses organometallic or, more precisely, organoelement compounds in which there are bonds to carbon such as Fe—C, P—C, Si—C, and Al—C.

The electronic structure of C in its ground state is $1s^2 2s^2 2p^2$, so that to accommodate the normal four covalence the atom must be promoted to a valence state $2s2p_x2p_y2p_z$ (see Section 3-2). The ion C^{4+} does not arise in any normal chemical process, but C^{4-} may possibly exist in some carbides of the most electropositive metals.

Some cations, anions, and radicals have been detected as transient species in organic reactions. Certain stable species of these types are known. The ions are known as *carbonium* ions [e.g., $(C_6H_5)_3C^+$] or *carbanions* [e.g., $(NC)_3C^-$]. These species can be stable only when the charge is extensively delocalized onto the attached groups.

Divalent carbon species or *carbenes* ($:CR_1R_2$) play a role in many reactions, but they are highly reactive. Carbenes can be trapped by binding to transition metals and many metal carbene compounds are known (Section 29-17).

The divalent species of some other Group IVB(14) elements, such as $:SiF_2$ or $:SnCl_2$, can be considered to have carbene-like behavior.

A unique feature of carbon is its propensity for bonding to itself in chains or rings, not only with single bonds (C—C), but also with multiple bonds (C=C or C≡C). Sulfur and silicon are the elements next most inclined to *catenation*, as this self-binding is called, but they are far inferior to carbon. The reason for the thermal stability of carbon chains is the intrinsic high strength of the C—C single bond (356 kJ mol^{-1}). The Si—Si bond (226 kJ mol^{-1}) is weaker but another important factor is that Si—O bonds (368 kJ mol^{-1}) are much stronger than C—O bonds (336 kJ mol^{-1}). Hence, given the necessary activation energy, compounds with Si—Si links are converted very exothermically into ones with Si—O bonds.

14-2 The Chemistry and Physical Properties of Diamond, Graphite, the Fullerenes, and Carbides

Diamond

Diamond differs from graphite in its physical and chemical properties because of differences in the arrangement and bonding of the atoms (Section 8-5). Diamond (3.51 g cm^{-3}) is denser than graphite (2.22 g cm^{-3}), but graphite is more stable, by 2.9 kJ mol^{-1} at 300 K and 1 atm pressure. From the densities, it follows that to transform graphite into diamond, pressure must be applied. From the thermodynamic properties of the allotropes it is estimated that they would be in equilibrium at 300 K under a pressure of about 15,000 atm. Because equilibrium is attained extremely slowly at this temperature, the diamond structure persists under ordinary conditions.

Diamonds can be produced from graphite only by the action of high pressure, and high temperatures are necessary for an appreciable rate of conversion. Naturally occurring diamonds must have been formed when those conditions were provided by geological processes.

In 1955 a successful synthesis of diamonds from graphite was reported. Although graphite can be directly converted into diamond at about 3000 K and pressures above 125 kbar, in order to obtain useful rates of conversion, a transition metal catalyst, such as Cr, Fe, or Pt, is used. It appears that a thin film of molten metal forms on the graphite, which dissolves some graphite and reprecipitates it as diamond, which is less soluble. Diamonds up to 0.1 carat (20 mg) of high industrial quality can be routinely produced at competitive prices. Some gem quality diamonds have also been made but the cost, thus far, has been prohibitive. Diamond will burn in air at 600–800 °C but its chemical reactivity is much lower than that of graphite or amorphous carbon.

Graphite

Many forms of amorphous carbon (including charcoals, certain soots, and lampblack) are all actually microcrystalline forms of graphite. The physical properties of such materials are mainly determined by the nature and extent of their surface areas. The finely divided forms, which present relatively vast surfaces with only partially saturated attractive forces, readily absorb large amounts of gases and solutes from solution. Active forms of carbon impregnated with palladium, platinum, or other metals are widely used as industrial catalysts.

An important aspect of graphite technology is the production of very strong fibers by pyrolysis, at 1500 °C or above, of oriented organic polymer fibers (e.g., those of polyacrylonitrile, polyacrylate esters, or cellulose). When incorporated into plastics the reinforced materials are light and of great strength. Other forms of graphite, such as foams, foils, or whiskers, can also be made.

The loose layered structure of graphite allows many molecules and ions to penetrate the layers to form what are called *intercalation* or *lamellar compounds*. Some of these may be formed spontaneously when the reactant and graphite are brought together. The alkali metals, halogens, and metal halides and oxides (e.g., $FeCl_3$ and MoO_3) are examples of reactants.

Fullerenes

The sootlike substances known as the fullerenes have already been introduced (Section 8-5). In the last few years there has been a remarkable explosion of papers in the chemical research literature on the fullerenes, and no doubt the topic will grow in scope as new discoveries are made. The reactions listed below represent only a portion of the emerging chemistry of the fullerenes. For this reason, the list of *Supplementary Reading* materials at the end of this chapter is more extensive than usual. The interested student is encouraged to consult not only these sources, but also the latest research and review literature, as advances in this area are expected to be unusually rapid.

The unsaturation of C_{60} is indicated by its reduction by Li in $NH_3(\ell)/$ t-BuOH (Birch reduction) to give a light cream solid composed of $C_{60}H_{36}$ and $C_{60}H_{18}$. Reaction with primary and secondary amines (e.g., n-$PrNH_2$, t-$BuNH_2$, ethylenediamine, morpholine, and n-dodecylamine) results in the multiple addition of H and NR_2 groups across the C=C double bonds to give $C_{60}H_6(NR_2)_6$. Each such addition results in the rehybridization of the carbon atoms from sp^2 to sp^3.

The first derivative structure of C_{60} was that of the remarkable osmium compound made as in Reaction 14-2.1.

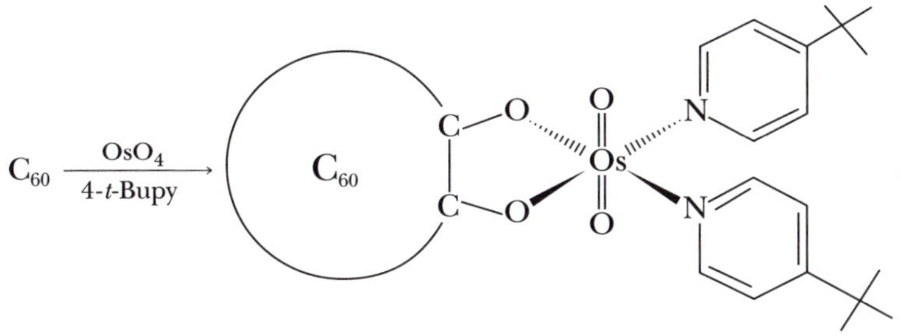

$$C_{60} \xrightarrow[\text{4-}t\text{-Bupy}]{\text{OsO}_4} \quad (14\text{-}2.1)$$

This reaction of OsO_4 is characteristic of the C=C double bonds of alkenes. Two similar pyridine (py) derivatives have been prepared by reacting either 2 equivalents of OsO_4 and 5 equivalents of py, or 1 equivalent of OsO_4 and 2.2 equivalents of py, with C_{60} in toluene, at 0 °C, giving Structures 14-Ia and 14-Ib, respectively.

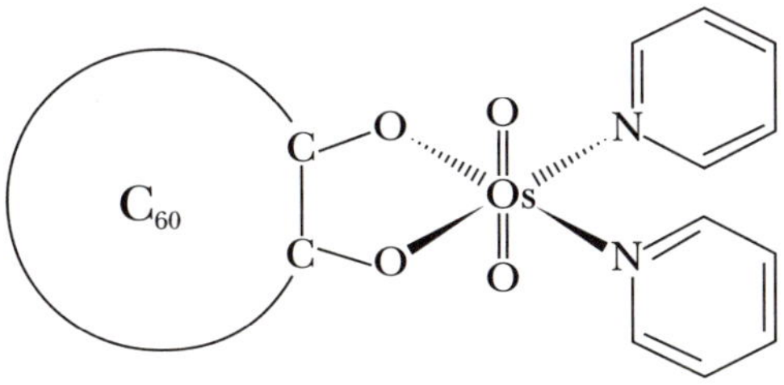

14-Ia

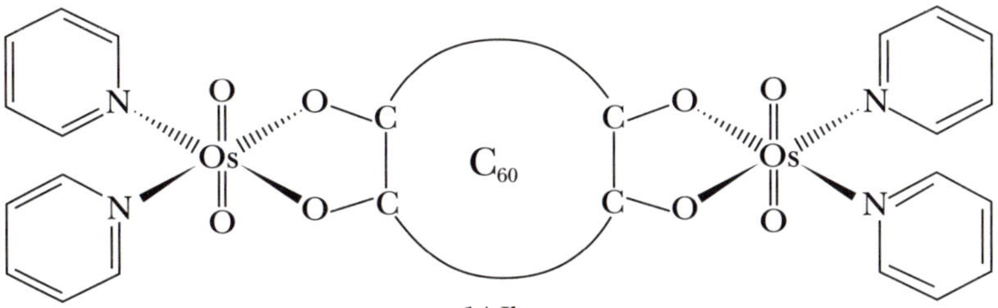

14-Ib

Such *osmylations* are typical of pyridine-activated polycyclic aromatic hydrocarbons, and underscore the "aromaticity" of C_{60}. Nevertheless, certain facts are best interpreted by regarding the C_{60} structure as a series of isolated alkenes. This is consistent with the two distinct C—C bond lengths in C_{60}, and with the fact that only small ring currents are detectable in C_{60}. The high reactivity of C_{60} is attributable to the nonplanarity of the C=C groups, which causes high-strain energy, and because each C=C double bond is attached to four electron-withdrawing groups.

That C_{60} can behave as an alkene towards transition metals is also shown by Reactions 14-2.2 and 14-2.3, in which a side-on, η^2 connection to either Pt or Ir is made by the π electrons of a C=C group, a classic bonding situation typical

$$C_{60} + (\eta^2\text{-}C_2H_4)Pt(PPh_3)_2 \longrightarrow C_2H_4 + (\eta^2\text{-}C_{60})Pt(PPh_3)_2 \quad (14\text{-}2.2)$$

$$C_{60} + Ir(CO)Cl(PPh_3)_2 \longrightarrow (\eta^2\text{-}C_{60})Ir(CO)Cl(PPh_3)_2 \quad (14\text{-}2.3)$$

of simple alkenes, as discussed further in Chapter 29. Although such compounds could be formulated as in Structure 14-IIa, analogous to the bonding of ethylene to transition metals (Section 29-12), it is probably more like Structure 14-IIb,

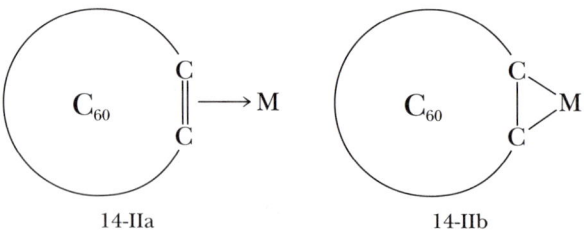

14-IIa   14-IIb

more typical of the bonding to transition metals by alkenes containing electron-withdrawing substituents, such as C_2F_4 or $C_2(CN)_4$. This formulation of the bond between C_{60} and Ir is also better for the product of Reaction 14-2.3, in that the reaction is then understood to be both an oxidation of iridium from Ir(I) to Ir(III), as well as an addition of the new C=C ligand. (See oxidative addition reactions, in Chapter 30.) The three-membered ring with single bonds (Structure 14-IIb) is likely to be the correct form of the epoxide, $C_{60}O$ (Structure 14-IIc), which is made by photochemical oxidation in benzene.

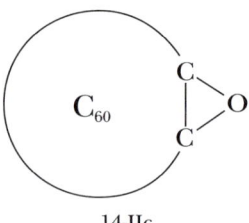

14-IIc

Reactions such as those in Equations 14-2.2 and 14-2.3 may be further understood by appreciating more about the details of the structure of C_{60}, which is composed of 20 six-membered rings interconnected with 12 five-membered rings, such that no five-membered rings share an edge with other five-membered rings. Thus we find six-membered rings fused both to other six-membered rings (6–6 fusions) and to five-membered rings (6–5 fusions), but we find five-membered rings fused only to six-membered rings. Although all carbon atoms are the same, as discussed in Chapter 8, there are two types of C—C bonds, one longer than the other by about 0.1 Å. These two types of C—C bonds appear in the sphere at regular locations, one at the 6–6 ring fusions and the other at the 6–5 ring fusions. Consequently, when C_{60} reacts by simple addition to a transition metal to form an η^2 attachment, increasing the coordination number of the metal from four to six, as in Reaction 14-2.3, the metal atom is found attached specifically to the two carbon atoms (designated C_1—C_2) of a 6–6 ring fusion. The coordinated carbon atoms C_1 and C_2 are pulled away from the C_{60} sphere, and the C_1—C_2 bond is somewhat elongated. Reaction 14-2.3 may be reversed by dissolving the product in CH_2Cl_2. Thus C_{60} behaves like tetracyanoethylene and O_2, both of which reversibly add to $IrCOCl(PPh_3)_2$, as discussed in Chapter 30. A similar reaction has been reported for C_{70}.

Partial *halogenation* of C_{60} and C_{70} may be accomplished by reaction with Cl_2 or Br_2, although the extent of halogenation is sometimes uncertain. Reaction of C_{60} with Br_2 gives $C_{60}Br_2$ and $C_{60}Br_4$. In each case, the bromination can be reversed at 150 °C, giving a quantitative recovery of bromine. Chlorination of C_{60} gives mixtures of $C_{60}Cl_n$, the average value of n being 24. The chloro derivatives are dechlorinated only at temperatures above 400 °C and are thus more stable than the bromo derivatives. The chlorine atoms of $C_{60}Cl_n$ can be replaced by OCH_3 groups, using methanolic KOH, as well as by C_6H_5 groups, in a Friedel–Crafts reaction (Section 13-4) with benzene, catalyzed by $AlCl_3$. Partially fluorinated derivatives, $C_{60}F_6$ and $C_{60}F_{42}$, have been isolated, but prolonged (12 days) interaction with F_2 gives colorless $C_{60}F_{60}$.

Anions, known as *fullerides,* are readily obtained, and these can be either diamagnetic or, like the radical $C_{60}^{\cdot-}$, paramagnetic. From bulk electrolysis, the salt $(Ph_4P)^+C_{60}^{\cdot-}(Ph_4PCl)_2$ has been obtained. The anion is also formed in THF solvent, using the tetraphenylporphyrin complex of Cr^{2+} as a reducing agent, as in Reaction 14-2.4, where TPP = tetraphenylporphyrin.

$$Cr^{II}(TPP) + C_{60} \longrightarrow [Cr^{III}(TPP)]^+C_{60}^{\cdot-} \qquad (14\text{-}2.4)$$

Reduction of C_{60}/C_{70} mixtures by Li gives red-brown solutions which, on treatment with CH_3I, gives polymethylated fullerenes with 1–24 methyl groups. Direct interaction with other alkali metals gives black materials such as $(K^+)_3C_{60}^{3-}$. Also, films of C_{60} doped with K, Rb, or Cs metal vapor can be prepared, which are superconducting and may be of value since the critical temperature for superconductivity is relatively high. For instance, T_c for Rb_nC_{60} is 30 K.

Heterofullerenes can be expected, since BN is isoelectronic with CC. As already discussed, α-BN is an analog of graphite and β-BN is an analog of diamond. Not surprisingly, then, calculations have suggested that $C_{12}B_{24}N_{24}$ should be stable. So far, laser vaporization of graphite has given $C_{58}B_2$ and $C_{59}B$.

Large metal atoms may be inserted into the center of certain fullerenes, giving compounds such as La_2C_{80} and LaC_{82}. These are obtained by the arc-vaporization of La_2O_3 and graphite, which yields solvent-extractable products. The similar LaC_{60} is not solvent extractable, but it can be sublimed. The details on such compounds are still forthcoming. Since the heavy atoms are thought to be encapsulated within the fullerene sphere, these substances have been called the "endohedral metallofullerenes." Interestingly, certain small fullerene compounds appear to be especially stable, for instance, MC_{28} (M = U, Zr, Hf, and Ti) and KC_{44}. It should be noted, however, that no one has, as yet, isolated a pure endohedral metallofullerene; the materials claimed to date have been characterized principally by mass spectrometry.

Reactions of C_{60} are listed in Fig. 14-1.

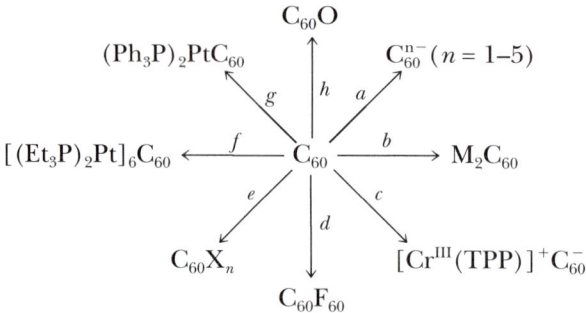

Figure 14-1   Some reactions of C_{60}: (*a*) electrochemical reduction ($E°$ depends on the solvent); (*b*) alkali metals; (*c*) $Cr^{II}(TPP)$; (*d*) F_2, 70 °C; (*e*) Cl_2, Br_2; (*f*) $Pt(Et_3P)_4$; (*g*) $Pt(C_2H_4)(PPh_3)_2$; and (*h*) C_6H_6, O_2, $h\nu$.

Carbides

Solid compounds of carbon with elements other than hydrogen are generally called carbides. However, there are quite diverse types of carbides, which may be classified as follows.

Ionic Carbides. These are formed by the most electropositive metals, such as the alkali and alkaline earth metals and aluminum. While it is a bit of an oversimplification to call them ionic, these carbides behave in many ways as though the carbon atoms were present in anionic form, for example, as C^{4-} or C_2^{2-} ions. This is particularly evident in their reactions with water, as in Reactions 14-2.5 and 14-2.6.

$$Al_4C_3 + 12\ H_2O \longrightarrow 4\ Al(OH)_3 + 3\ CH_4 \qquad (14\text{-}2.5)$$

$$CaC_2 + 2\ H_2O \longrightarrow Ca(OH)_2 + C_2H_2 \qquad (14\text{-}2.6)$$

Interstitial Carbides. The transition metals form carbides in which carbon atoms occupy tetrahedral holes in the close-packed arrays (Chapter 4) of metal atoms. Such materials are commonly very hard, electrically conducting, and have very high melting points (3000–4800 °C). Tungsten carbide (WC) is so hard that it is used to make tool bits for machining steel. The smaller metals Cr, Mn, Fe, Co, and Ni give carbides that are intermediate between typically ionic and interstitial carbides, and these are hydrolyzed by water or dilute acids.

Covalent Carbides. The metalloids, especially silicon and boron, form SiC and B_4C, which are also extremely hard, infusible, and chemically inert. Silicon carbide has a diamond-like structure (Chapters 4 and 8) in which C and Si atoms are each tetrahedrally surrounded by four of the other kind of atoms. Under the name *carborundum,* it is used in cutting tools and abrasives.

14-3   Carbon Monoxide

This colorless toxic gas (bp −190 °C) is formed when carbon is burned in a deficiency of oxygen. The following equilibrium is found at all temperatures

$$2 \, CO(g) = C(s) + CO_2(g) \qquad (14\text{-}3.1)$$

but this equilibrium is rapidly attained only at elevated temperatures. Carbon monoxide is made commercially along with hydrogen (Section 9-1) by steam reforming or partial combustion of hydrocarbons and by Reaction 14-3.2.

$$CO_2 + H_2 = CO + H_2O \qquad (14\text{-}3.2)$$

A mixture of CO and H_2 (synthesis gas) is very important commercially, being used in the hydroformylation process (Section 30-9) and for the synthesis of methanol. Carbon monoxide is also formed when carbon is used in reduction processes, for example, of phosphate rock to give phosphorus (Section 17-2) and in automobile exhausts. Carbon monoxide is also released by certain marine plants and it occurs naturally in the atmosphere.

Carbon monoxide is formally the anhydride of formic acid (HCO_2H), but this is not an important aspect of its chemistry. Although CO is an exceedingly weak base, one of its important properties is its ability to act as a ligand toward transition metals. The metal—CO bond involves a certain type of multiple bonding ($d\pi$–$p\pi$ bonding discussed in Chapter 28). The toxicity of CO arises from this ability to bind to the Fe atom in hemoglobin (Section 31-4) in the blood. Only iron and nickel react directly with CO (Chapter 28) under practical conditions.

14-4 Carbon Dioxide and Carbonic Acid

Carbon dioxide is present in the atmosphere (300 ppm), in volcanic gases, and in supersaturated solution in certain spring waters. It is released on a large scale by fermentation processes, limestone calcination, and all forms of combustion of carbon and carbon compounds. It is involved in geochemical cycles as well as in photosynthesis. In the laboratory it can be made by the action of heat or acids on carbonates. Solid CO_2 (sublimes -78.5 °C) or "dry ice" is used for refrigeration.

Carbon dioxide is the anhydride of the most important simple acid of carbon, *carbonic acid*. For many purposes, the following acid dissociation constants are given for aqueous carbonic acid:

$$\frac{[H^+][HCO_3^-]}{[H_2CO_3]} = 4.16 \times 10^{-7}$$

$$\frac{[H^+][CO_3^{2-}]}{[HCO_3^-]} = 4.84 \times 10^{-11}$$

The equilibrium quotient in the first equation is incorrect because not all the CO_2 dissolved and undissociated is present as H_2CO_3. The greater part of the dissolved CO_2 is only loosely hydrated, so that the correct first dissociation constant, using the real concentration of H_2CO_3, has the much larger value of about 2×10^{-4}, more in keeping (see Section 7-12) with the structure $(HO)_2CO$.

The rate at which CO_2 comes into equilibrium with H_2CO_3 and its dissociation products when passed into water is measurably slow. This explains why we can analytically distinguish between H_2CO_3 and the loosely hydrated $CO_2(aq)$. This slowness is of great importance in biological, analytical, and industrial chemistry.

The slow reaction can be shown by addition of a saturated aqueous solution of CO_2, on the one hand, and of dilute acetic acid, on the other, to solutions of dilute NaOH containing phenolphthalein indicator. The acetic acid neutralization is instantaneous, whereas with the CO_2 neutralization, it takes several seconds for the color to fade.

The hydration of CO_2 occurs by two paths. For pH < 8, the principal mechanism is direct hydration of CO_2 according to Eq. 14-4.1, followed by a rapid acid–base reaction to give bicarbonate:

$$CO_2 + H_2O = H_2CO_3 \qquad \text{(Slow)} \qquad (14\text{-}4.1)$$

$$H_2CO_3 + OH^- = HCO_3^- + H_2O \qquad \text{(Instantaneous)} \qquad (14\text{-}4.2)$$

The rate law for this process is first order.

$$\frac{-d[CO_2]}{dt} = k_{CO_2}[CO_2] \qquad k_{CO_2} = 0.03\ s^{-1} \qquad (14\text{-}4.3)$$

At pH > 10, the predominant reaction of CO_2 is by direct attack with OH^-, as in Reaction 14-4.4, followed by a rapid acid–base reaction to give carbonate:

$$CO_2 + OH^- = HCO_3^- \qquad \text{(Slow)} \qquad (14\text{-}4.4)$$

$$HCO_3^- + OH^- = CO_3^{2-} + H_2O \qquad \text{(Instantaneous)} \qquad (14\text{-}4.5)$$

for which the rate law is

$$\frac{-d[CO_2]}{dt} = k_{OH^-}[OH^-][CO_2] \qquad k_{OH^-} = 8500\ M^{-1}\ s^{-1} \qquad (14\text{-}4.6)$$

Because k_{OH^-} is so much larger than k_{CO_2}, it can be considered that the mechanism given by Reactions 14-4.4 and 14-4.5 represents base catalysis of the CO_2 hydrolysis mechanism given by Reactions 14-4.1 and 14-4.2. Both mechanisms operate in the pH range 8–10.

For each hydration process there is a corresponding dehydration reaction.

$$H_2CO_3 \longrightarrow H_2O + CO_2 \qquad k_{H_2CO_3} = 20\ s^{-1} \qquad (14\text{-}4.7)$$

$$HCO_3^- \longrightarrow CO_2 + OH^- \qquad k_{HCO_3^-} = 2 \times 10^{-4}\ s^{-1} \qquad (14\text{-}4.8)$$

Hence, for the overall equilibrium represented by Reaction 14-4.9

$$H_2CO_3 \rightleftharpoons CO_2 + H_2O \qquad (14\text{-}4.9)$$

the equilibrium constant can be determined to be

$$K = \frac{[CO_2]}{[H_2CO_3]} = \frac{k_{H_2CO_3}}{k_{CO_2}} = \text{about } 660 \qquad (14\text{-}4.10)$$

It follows from the large value of K in Reaction 14-4.10 that the true ionization constant (K_a) of H_2CO_3 is greater than the apparent constant, as noted previously.

14-5 Compounds with C—N Bonds; Cyanides and Related Compounds

An important area of "inorganic" carbon chemistry is that of compounds with C—N bonds. The most important species are the cyanide, cyanate, and thiocyanate ions and their derivatives.

Cyanogen, (CN)$_2$. This flammable gas (bp −21°C) is stable despite the fact that it is highly endothermic ($\Delta H_f^\circ = 297$ kJ mol^{-1}). It can be obtained by catalytic gas-phase oxidation of HCN by NO$_2$

$$2 \text{ HCN} + \text{NO}_2 \longrightarrow (\text{CN})_2 + \text{NO} + \text{H}_2\text{O} \tag{14-5.1}$$

$$\text{NO} + \tfrac{1}{2}\text{O}_2 \longrightarrow \text{NO}_2 \tag{14-5.2}$$

Cyanogen can also be obtained from CN$^-$ by aqueous oxidation using Cu^{2+} (cf. the Cu^{2+}–I$^-$ reaction):

$$\text{Cu}^{2+} + 2 \text{ CN}^- \longrightarrow \text{CuCN} + \tfrac{1}{2}(\text{CN})_2 \tag{14-5.3}$$

or acidified peroxodisulfate. Dry (CN)$_2$ is made by the reaction

$$\text{Hg(CN)}_2 + \text{HgCl}_2 \longrightarrow \text{Hg}_2\text{Cl}_2 + (\text{CN})_2 \tag{14-5.4}$$

Although pure (CN)$_2$ is stable, the impure gas may polymerize at 300–500 °C. Cyanogen dissociates into CN radicals and, like halogens, can oxidatively add to lower valent metal atoms (Chapter 30) giving dicyano complexes, for example,

$$(\text{Ph}_3\text{P})_4\text{Pd} + (\text{CN})_2 \longrightarrow (\text{Ph}_3\text{P})_2\text{Pd(CN)}_2 + 2 \text{ Ph}_3\text{P} \tag{14-5.5}$$

A further resemblance to the halogens is the disproportionation in basic solution.

$$(\text{CN})_2 + 2 \text{ OH}^- \longrightarrow \text{CN}^- + \text{OCN}^- + \text{H}_2\text{O} \tag{14-5.6}$$

Thermodynamically this reaction can occur in acid solution but is rapid only in base. A stoichiometric mixture of O$_2$ and (CN)$_2$ burns producing one of the hottest flames (~5050 K) known from a chemical reaction.

Hydrogen Cyanide. Like the hydrogen halides, HCN is a covalent, molecular substance, but is capable of dissociation in aqueous solution. It is an extremely poisonous (though less so than H$_2$S), colorless gas and is evolved when cyanides are treated with acids. Liquid HCN (bp 25.6 °C) has a very high dielectric constant (107 at 25 °C) that is due (as for H$_2$O) to association of the polar molecules by hydrogen bonding. Liquid HCN is unstable and can polymerize violently in the absence of stabilizers. In aqueous solutions polymerization is induced by ultraviolet light.

Hydrogen cyanide is thought to have been one of the small molecules in the earth's primeval atmosphere and to have been an important source or intermediate in the formation of biologically important chemicals. For example, under pressure, with traces of water and ammonia, HCN pentamerizes to adenine.

In aqueous solution, HCN is a very weak acid ($pK_{25°C} = 9.21$) and solutions of soluble cyanides are extensively hydrolyzed, but the pure liquid is a strong acid.

Hydrogen cyanide is made industrially from CH_4 and NH_3 by the reactions

$$2\,CH_4 + 3\,O_2 + 2\,NH_3 \xrightarrow[>800\,°C]{\text{catalyst}} 2\,HCN + 6\,H_2O \qquad (14\text{-}5.7)$$

$$\Delta H = -475 \text{ kJ mol}^{-1}$$

or

$$CH_4 + NH_3 \xrightarrow[Pt]{1200\,°C} HCN + 3\,H_2 \qquad \Delta H = +240 \text{ kJ mol}^{-1} \quad (14\text{-}5.8)$$

Hydrogen cyanide has many industrial uses. It may be added directly to alkenes; for example, butadiene gives adiponitrile, $NC(CH_2)_4CN$ (for nylon), in the presence of zero-valent Ni alkylphosphite catalysts that operate by oxidative–addition and transfer reactions (Chapter 30).

Cyanides. Sodium cyanide is manufactured by the fusion of calcium cyanamide with carbon and sodium carbonate.

$$CaCN_2 + C + Na_2CO_3 \longrightarrow CaCO_3 + 2\,NaCN \qquad (14\text{-}5.9)$$

The cyanide is leached with water. The $CaCN_2$ is made in an impure form contaminated with CaO, CaC_2, C, and so on, by the interaction

$$CaC_2 + N_2 \xrightarrow{\sim 1100\,°C} CaNCN + C \qquad (14\text{-}5.10)$$

The linear NCN^{2-} ion is isostructural and isoelectronic with CO_2. Cyanamide itself (H_2NCN) can be made by acidification of CaNCN. The commercial product is the dimer, $H_2NC(=NH)NHCN$, which also contains much of the tautomer containing the substituted *carbodiimide group*, $H_2N-C(=NH)-N=C=NH$. Organocarbodiimides are important synthetic reagents in organic chemistry and $CH_3N=C=NCH_3$ is stable enough to be isolated.

Sodium cyanide can also be obtained by the reaction

$$NaNH_2 + C \xrightarrow{500\text{-}600\,°C} NaCN + H_2 \qquad (14\text{-}5.11)$$

Cyanides of electropositive metals are water soluble but those of Ag^I, Hg^I, and Pb^{II} are very insoluble. The cyanide ion is of great importance as a ligand (Chapter 28), and many cyano complexes of transition metals are known (e.g., Zn, Cd, or Hg); some, like $Ag(CN)_2^-$ and $Au(CN)_2^-$, are of technical importance and others are employed analytically. The complexes sometimes resemble halogeno complexes [e.g., $Hg(CN)_4^{2-}$ and $HgCl_4^{2-}$], but other types exist. Fusion of alkali cyanides with sulfur gives the *thiocyanate* ion (SCN^-).

14-6 Compounds with C—S Bonds

Carbon disulfide (CS_2) is a very toxic liquid (bp 46 °C), usually pale yellow, and is prepared on a large scale by the interaction of methane and sulfur over silica or alumina catalysts at about 1000 °C.

$$CH_4 + 4 S = CS_2 + 2 H_2S \qquad (14\text{-}6.1)$$

In addition to its high flammability in air, CS_2 is a very reactive molecule and has an extensive chemistry, much of it organic in nature. It is used to prepare carbon tetrachloride industrially.

$$CS_2 + 3 Cl_2 \longrightarrow CCl_4 + S_2Cl_2 \qquad (14\text{-}6.2)$$

Carbon disulfide is one of the small molecules that readily undergo the "insertion reaction" (Chapter 30), where the $\overset{\displaystyle |}{\underset{\displaystyle S}{-S-C-}}$ group is inserted between

Sn—N, Co—Co, and other bonds. Thus dithiocarbamates are obtained with titanium dialkylamides.

$$Ti(NR_2)_4 + 4 CS_2 \longrightarrow Ti(S_2CNR_2)_4 \qquad (14\text{-}6.3)$$

The CS_2 molecule can also serve as a ligand, being either bound as a donor through sulfur or added oxidatively (Chapter 30) to give a three-membered ring, as in Structure 14-III.

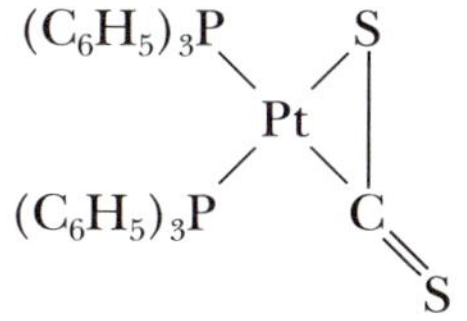

$$(C_6H_5)_3P \diagdown \underset{Pt}{\diagup} \overset{S}{\diagdown} \underset{(C_6H_5)_3P}{\diagup} \overset{\diagdown}{\underset{S}{C}}$$

14-III

Important reactions of CS_2 involve nucleophilic attacks on carbon by the ions RO^- and HS^- and by primary or secondary amines, which lead, in basic solution, respectively, to xanthates, thiocarbonates, and dithiocarbamates. For example,

$$\left. \begin{matrix} RO^- \\ HS^- \\ R_2HN \end{matrix} \right\} + \underset{S}{\overset{S}{\underset{\|}{\overset{\|}{C}}}} \longrightarrow \left\{ \begin{matrix} ROCS_2^- & \text{Xanthate} & (14\text{-}6.4) \\ CS_3^{2-} & \text{Thiocarbonate} & (14\text{-}6.5) \\ R_2NCS_2^- & \text{Dithiocarbamate} & (14\text{-}6.6) \end{matrix} \right.$$

Dithiocarbamates are normally prepared as Na salts by the action of primary or secondary amines on CS_2 in the presence of NaOH. The Zn, Mn, and Fe dithiocarbamates are used as agricultural fungicides, and Zn salts are used as accelerators in the vulcanization of rubber.

Dithiocarbamates form many complexes with metals. The CS_2^- group in dithiocarbamates, as well as in xanthates, thioxanthates, and thiocarbonates, is usually chelated (as in Structure 14-IV), but monodentate and bridging dithiocarbamates are known.

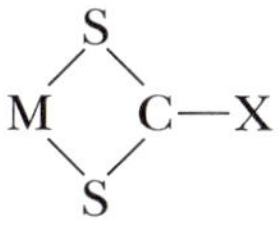

$$\underset{\text{14-IV}}{M \underset{\diagdown S \diagup}{\overset{\diagup S \diagdown}{}} C-X} \qquad \begin{matrix} X = NHR \text{ or } NR_2, \\ OR \text{ or } SR, \\ O, S, \text{ or } S-S \end{matrix}$$

On oxidation of aqueous solutions by H_2O_2, Cl_2, or $S_2O_8^{2-}$, *thiuram disulfides* are obtained, for example,

$$I_2 + 2\ (CH_3)_2NCS_2^- \longrightarrow (CH_3)_2NC\underset{\underset{S}{\|}}{-}S-S-\underset{\underset{S}{\|}}{C}N(CH_3)_2 + 2\ I^- \quad (14\text{-}6.7)$$

Thiuram disulfides, which are strong oxidants, are used as polymerization initiators (for, when heated, they give radicals) and as vulcanization accelerators. Tetraethylthiuram disulfide is "Antabuse," the agent for rendering the body allergic to ethanol.

STUDY GUIDE

Scope and Purpose

Most of the chemistry of the element carbon constitutes the field of organic chemistry. The inorganic chemist, however, is legitimately concerned with certain aspects of carbon that are very important and that have traditionally not been included in the realm of organic chemistry. These include nearly all of the chemistry of the element itself, of compounds in which carbon is combined with metals and metalloids, and much of the chemistry of the simple, binary compounds with nonmetals (oxides, cyanides, or halides). The field of organometallic chemistry, which we examine in Chapters 29 and 30, is a truly interdisciplinary one.

Study Questions

A. Review

1. The electronic structure of C in its ground state is $1s^2 2s^2 2p_x 2p_y$. Why does carbon usually form four single bonds and not two?

2. Give examples of a stable carbonium ion, a carbanion, and a free radical. What is a carbene?

3. What is meant by catenation? Why does silicon have much less tendency to catenation than carbon? Could the same be said for nitrogen?

4. Describe the synthesis and main properties of diamond.

5. What is graphite? Draw its structure and explain why its properties differ from those of diamond.

6. List ways in which CO can be made.

7. List ways in which CO_2 can be made.

8. On which side is the equilibrium in the reaction

$$CO_2(aq) + 2\ H_2O \rightleftharpoons H_3O^+ + HCO_3^-$$

9. Why does $CaCO_3$ dissolve to some extent in CO_2 saturated water? Write balanced equations for the reactions involved.

10. How could you make cyanogen in the laboratory? Write balanced equations.

11. List similarities between $(CN)_2$ and CN^- and Cl_2 and Cl^-.

12. Why are solutions of KCN in water alkaline?

13. Give the industrial synthesis and major properties of hydrogen cyanide.

14. How is CS_2 prepared? Write equations for its reaction with C_2H_5ONa in ethanol and with $(C_2H_5)_2NH$ in the presence of aqueous NaOH.

15. How would you convert $BaCO_3$ labeled with ^{13}C or ^{14}C, which is the usual source of labeled carbon compounds, to (a) $Ni(*CO)_4$, (b) $*C_2H_2$, (c) $*CH_4$, (d) $*CS_2$, and (e) $*CH_3OH$?

B. Additional Exercises

1. The C—C bond length in graphite is 1.42 Å. How does this compare with the C—C bond length in (a) diamond, (b) ethylene, and (c) benzene? What do you expect is the C—C bond order in graphite? Explain.

2. Write down the structures, the Lewis diagrams, and the MO's for the isoelectronic molecules carbon dioxide and allene. What sort of differences in chemistry do you expect?

3. Hydrogen cyanide (HCN) can give dimers, trimers, tetramers, pentamers, and polymers on polymerization. Write some plausible structures for these molecules.

4. Explain why HCN is a weak acid in aqueous solution yet as the pure liquid it is a strong acid. Recall the material of Chapter 7.

5. Zinc dithiocarbamates are dimeric. Propose a structure.

6. Draw the Lewis diagrams for each reactant and product of Reaction 14-6.3.

7. Identify the oxidizing and reducing agents in Reactions 14-5.1 and 14-5.2. Draw the Lewis diagram for each reactant and product.

C. Questions from the Literature of Inorganic Chemistry

1. Consider the paper by A. L. Balch, V. J. Catalano, and J. W. Lee, "Accumulating Evidence for the Selective Reactivity of the 6–6 Ring Fusion of C_{60}. Preparation and Structure of $(\eta^2\text{-}C_{60})Ir(CO)Cl(PPh_3)_2 \cdot 5 \, C_6H_6$," *Inorg. Chem.,* **1991,** *30,* 3980–3981.

   (a) List all of the significant structural changes to the C_{60} framework that occur upon formation of the η^2 attachment to Ir in the title compound.

   (b) How was the formation of the title compound shown to be reversible?

   (c) What conclusions do the authors reach regarding the two types of ring fusions in the C_{60} framework?

   (d) Five benzene molecules are found in the crystal. What effects do these have on the structure of the coordination compound?

2. Consider the paper by P. J. Fagan, J. C. Calabrese, and B. Malone, "A Multiply-Substituted Buckminsterfullerene (C_{60}) with Octahedral Array of Platinum Atoms," *J. Am. Chem. Soc.,* **1991,** *113,* 9408–9409.

   (a) Explain how NMR spectroscopy has been used to determine the structure of the title compound.

   (b) What structural features make this compound similar to that of Question 1C above?

SUPPLEMENTARY READING

Ansell, M. F., "Diamond Cleavage," *Chem. Ber.,* **1984,** 1017–1021.

Baum, R. M., "Flood of Fullerene Discoveries Continues," *Chem. Eng. News,* **June 1, 1992,** 25–33.

Diederich, F. and Whetten, R. L., "C_{60}: From Soot to Superconductors," *Angew. Chem. Int. Ed. Engl.,* **1991,** *30,* 678–680.

Diederich, F. and Whetten, R. L., "Beyond C_{60}: The Higher Fullerenes," *Acc. Chem. Res.,* **1992,** *25,* 119–126.

Fagan, P. J., Calabrese, J. C., and Malone, B., "Metal Complexes of Buckminsterfullerene (C_{60})," *Acc. Chem. Res.,* **1992,** *25,* 134–142.

Fischer, J. E., Heiney, P. A., and Smith, A. B., "Solid-State Chemistry of Fullerene-Based Materials," *Acc. Chem. Res.,* **1992,** *25,* 112–118.

Fleming, R. M. et al., "Preparation and Structure of the Alkali-Metal Fulleride A_4C_{60}," *Nature (London),* **1991,** *352,* 701–703.

Haddon, R. C., "Electronic Structure, Conductivity, and Superconductivity of Alkali Metal Doped C_{60}," *Acc. Chem. Res.,* **1992,** *25,* 127–133.

Hammond, G. S. and Kuck, V. J., Eds., "Fullerenes. Synthesis, Properties and Chemistry of Large Carbon Clusters," ACS Symposium Series, American Chemical Society, Washington DC, 1992.

Hare, J. P. and Kroto, H. W., "A Postbuckminsterfullerene View of Carbon in the Galaxy," *Acc. Chem. Res.,* **1992,** *25,* 106–112.

Hawkins, J. M., "Osmylation of C_{60}: Proof and Characterization of the Soccer-Ball Framework," *Acc. Chem. Res.,* **1992,** *25,* 150–156.

Johnson, R. D., Bethune, D. S., and Yannoni, C. S., "Fullerene Structure and Dynamics: A Magnetic Resonance Potpourri," *Acc. Chem. Res.,* **1992,** *25,* 169–175.

Kelty, S. P., Chen, C., and Lieber, C. M., "Superconductivity at 30 K in Cesium-Doped C_{60}," *Nature (London),* **1991,** *352,* 223–225.

Kroto, H. W., "C_{60}: Buckminsterfullerene, The Celestial Sphere That Fell to Earth," *Angew. Chem., Int. Ed. Eng.,* **1992,** *31,* 111–246.

Kroto, H. W., Allaf, A. W., and Balm, S. P., "C_{60}: Buckminsterfullerene," *Chem. Rev.,* **1991,** *91,* 1213–1235.

McElvany, S. W., Ross, M. M., and Callahan, J. H., "Characterization of Fullerenes by Mass Spectrometry," *Acc. Chem. Res.,* **1992,** *25,* 162–168.

McLafferty, F. W., Ed., "Special Issue on Buckminsterfullerenes," *Acc. Chem. Res.,* **1992,** *25(3),* 98–175.

Schwarz, H., "C_{60}-Fullerene. A Playground for Chemical Manipulations on Curved Surfaces and in Cavities," *Angew. Chem. Int. Ed. Engl.,* **1992,** *31,* 293–298.

Sleight, A. W., "Buckminsterfullerene. Sooty Superconductors," *Nature (London),* **1991,** *350,* 557–558.

Smalley, R. E., *Atomic and Molecular Crystals,* E. R. Bernstein, Ed., Elsevier, Amsterdam, 1990. A general reference for C_{60}.

Smalley, R. E., "Self-Assembly of the Fullerenes," *Acc. Chem. Res.,* **1992,** *25,* 98–105.

Troyer, R., "The Third Form of Carbon; A New Era In Chemistry," *Interdisc. Sci. Rev.,* **1992,** *17,* 161–170.

Weaver, J. H., "Fullerenes and Fullerides: Photoemission and Scanning Tunneling Microscopy Studies," *Acc. Chem. Res.,* **1992,** *25,* 143–149.

Wudl, F., "The Chemical Properties of Buckminsterfullerene (C_{60}) and the Birth and Infancy of Fulleroids," *Acc. Chem. Res.,* **1992,** *25,* 157–161.

Chapter 15

THE GROUP IVB(14) ELEMENTS: SILICON, GERMANIUM, TIN, AND LEAD

15-1 Introduction

Silicon is second only to oxygen in its natural abundance (~28% of the earth's crust) and occurs in a great variety of silicate minerals and as quartz (SiO_2).

Germanium, tin, and lead are rare elements (~10^{-3}%). Tin and lead have been known since antiquity because of the ease with which they are obtained from their ores.

Cassiterite (SnO_2) occurs mixed in granites, sands, and clays. Lead occurs mainly as *galena* (PbS).

Germanium was discovered in 1886 following the prediction of its existence by Dimitri Mendeleev. It occurs widely but in small amounts and is recovered from coal and zinc ore concentrates.

The main use of Ge, Sn, and Pb is as the metals, but alkyltin and alkyllead compounds are made on a large scale (Chapter 29).

The position of the elements in the periodic table and some general features, including the reasons for the existence of the lower II oxidation state, were discussed in Section 8-11. Some properties of the elements were given in Table 8-4.

Multiple Bonding

It was earlier thought that silicon and the remainder of the Group IVB(14) elements did not form stable $p\pi$–$p\pi$ multiple bonds, as is common for carbon. Beginning in the 1960s, however, transient intermediates with Si=C, $p\pi$–$p\pi$ bonding were discovered in thermal decomposition reactions such as 15-1.1, which takes place at 560 °C.

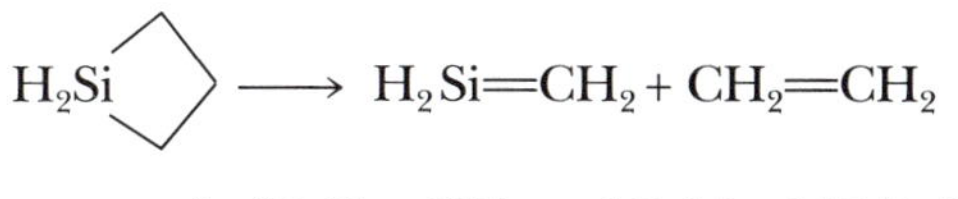

$$H_2Si \triangleright \longrightarrow H_2Si{=}CH_2 + CH_2{=}CH_2 \qquad (15\text{-}1.1)$$

Numerous such compounds ($R_2Si{=}CR_2'$ and $R_2M{=}MR_2'$) for Si, Ge, and Sn are now known to be isolable, provided that bulky groups are used, as discussed in Section 15-7.

Although stoichiometric similarities exist between the compounds of carbon and those of the remaining elements of Group IVB(14) [e.g., the pairs CO_2 and SiO_2, as well as $(CH_3)_2CO$ and $(CH_3)_2SiO$], there is no structural or chemical similarity between them. Carbon dioxide is a gas, properly written $O{=}C{=}O$, whereas SiO_2 is a giant molecule, or network substance, with each Si atom singly bonded to four adjacent oxygen atoms, giving linked SiO_4 tetrahedra, as discussed in Section 5-4. Also, reactions of the compounds of silicon and the lower elements of the group do not give products analogous to those for carbon. For example, the dehydration of alcohols gives alkenes, but the dehydration of silanols, $R_2Si(OH)_2$, is accompanied by condensation, giving $(R_2SiO)_n$ and $R_2(OH)SiOSi(OH)R_2$.

Whereas multiple bonds to carbon involve overlap of the $p\pi$–$p\pi$ variety, multiple bonding for silicon and germanium (and to a lesser extent for tin) usually arise from a $p\pi$–$d\pi$ component, especially in bonds to O and N. It is important to note that this does not usually lead to conjugation, as is so prevalent for carbon. The following structural and chemical features of silicon and germanium compounds are best explained by some degree of $p\pi$–$d\pi$ double-bond character.

1. Trisilylamine, $(H_3Si)_3N$, differs from trimethylamine, $(CH_3)_3N$, being planar rather than pyramidal, and being a very weak base. Disilylamine is also planar. These observations can be explained by supposing that nitrogen forms dative π bonds to the silicon atom, as shown in Fig. 15-1. We assume that the central nitrogen atom is sp^2 hybridized, leaving a filled $2p_z$ orbital, which overlaps appreciably with an empty silicon $3d_{xz}$ (or $3d_{yz}$) orbital. Thus a dative $p\pi \to d\pi$ bond is established, which provides additional bond strength in each Si—N linkage of the molecule. It is this additional bond strength that stabilizes the NSi_3 skeleton in a planar configuration. In contrast, for $N(CH_3)_3$, since carbon has no low-lying d orbitals, σ bonding alone determines the configuration at the AB_3E carbon atom, which is pyramidal. As an interesting comparison, consider trisilylphosphine, $(H_3Si)_3P$, which is pyramidal. Evidently phosphorus is less able than nitrogen to form a $p\pi \to d\pi$ dative bond to silicon.

2. In the vapor phase, H_3SiNCO is linear (hydrogen atoms excepted). This can be explained by the formation of a $p\pi \to d\pi$ bond between nitrogen and silicon ($H_3Si{=}N{=}C{=}O$). The corresponding carbon compound (H_3CNCO) is not linear, since carbon has no vacant, low-lying d orbitals. Interestingly, H_3GeNCO is not linear in the gas phase. Evidently, effective $p\pi$–$d\pi$ bonding occurs for Si—N, but not for Ge—N.

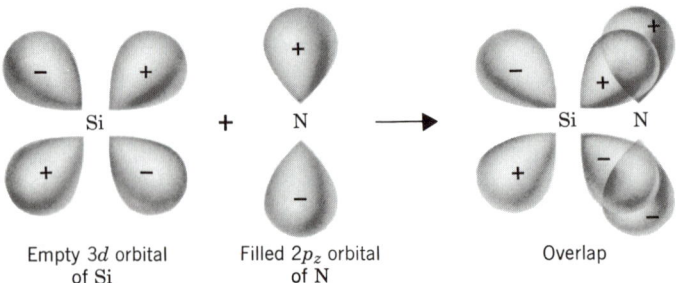

Empty $3d$ orbital of Si   Filled $2p_z$ orbital of N   Overlap

Figure 15-1   The formation of a $d\pi$–$p\pi$ bond between Si and N atoms in trisilylamine.

3. The disilyl ethers, $(R_3Si)_2O$, all have large angles at oxygen (140–180°), and both electronic and steric explanations have been suggested. Electronically, overlap between filled oxygen $p\pi$ orbitals and empty silicon $d\pi$ orbitals would improve with increasing Si—O—Si angles, and might be most effective for a linear Si—O—Si group. There may also be very strong steric factors favoring more linear structures, especially for large R groups. For instance, the angle at oxygen is 180° for $(Ph_3Si)_2O$.

4. Silanols such as $(CH_3)_3SiOH$ are stronger protonic acids than their carbon analogs, and form stronger hydrogen bonds. This is due to stabilization of the conjugate base anion by $O(p\pi) \rightarrow Si(d\pi)$ bond formation. A similar stabilization of the conjugate base anion can be invoked to explain the order of acidities (M = Si > Ge > C) in the series R_3MCO_2H.

Stereochemistry

The stereochemistry of silicon compounds and the lower members of Group IVB(14) depend on the oxidation state. Also, unlike carbon, certain compounds of these elements have five, six, seven, and eight or higher coordination.

Compounds having *oxidation state IV* are listed in Table 15-1. All of the elements form tetrahedral compounds, some of which are chiral, for example, $GeH(CH_3)C_6H_5$ (α-napthyl). Since valence shell expansion by use of outer d orbitals can occur, giving hybridizations (such as dsp^3 and d^2sp^3), five- and six-coordinate compounds are common, as shown in Table 15-1. Pentacoordination is found mainly in

(a) Anions such as MX_5^- and $MR_nX_{n-5}^-$, which are usually trigonal bipyramidal, and are stabilized in the solid state by large cations.

(b) Adducts of donor ligands with halides or substituted halides of the elements, such as $L{\rightarrow}MX_4$.

(c) For Sn, polymeric compounds R_3SnX, where X acts as a bridge in the solid state structure.

Octahedral coordination is common for all of the elements, although for ions and adducts, the preference for five or six coordination depends on delicate energy balances, and cannot be predicted.

Table 15-1   Coordination Number and Geometry of Tetravalent Compounds of the Group IVB(14) Elements

Coordination Number	Geometry	Examples
3	Trigonal (AB_3)	$(C_6H_5)_3Si^+$
4	Tetrahedral (AB_4)	SiO_2, $SiCl_4$, GeH_4, $Pb(CH_3)_4$
5	Trigonal bipyramidal (AB_5)	$(CH_3)_3SnCl(py)$, $SnCl_5^-$, SiF_5^-, $RSiF_4^-$
5	Square pyramidal (AB_5)	$[XSi(O_2C_6H_4)_2]^-$
6	Octahedral (AB_6)	SiF_6^{2-}, $[Si(acac)_3]^+$, $[Si(ox)_3]^{2-}$, GeO_2, $PbCl_6^{2-}$, *trans*-$GeCl_4(py)_2$, $Sn(S_2CNEt_2)_4$
7	Pentagonal bipyramid	$Ph_2Sn(NO_3)_2(OPPh_3)$
8	Dodecahedral	$Sn(NO_3)_4$, $Pb(O_2CCH_3)_4$

Table 15-2   Coordination Number and Geometry of Divalent Compounds of the Group IVB(14) Elements

Coordination Number	Geometry	Examples
2	Angular (AB$_2$E)	Ge(N-t-Bu$_2$)$_2$, Pb(C$_6$H$_5$)$_2$
3	Pyramidal (AB$_3$E)	SnCl$_2$·2 H$_2$O, SnCl$_3^-$, Pb(SC$_6$H$_5$)$_3^-$
4	"Seesaw" (AB$_4$E)	Pb^{II} in Pb$_3$O$_4$
		Sn(S$_2$CNR$_2$)$_2$
5	Square pyramidal (AB$_5$E)	SnO (blue-black form), PbO
6	Octahedral	PbS (NaCl type)
		GeI$_2$ (CaI$_2$ type)
7	Complex	(18-C-6)SnCl$^+$
9, 10	Complex	Pb(NO$_3$)$_2$(semicarbazone)
		Pb(O$_2$CCH$_3$)$_2$·3 H$_2$O

Compounds having *oxidation state II* are listed in Table 15-2. In many of the compounds of Sn^{II}, and to a lesser extent Ge^{II} and Pb^{II}, the lone pair of electrons on the metal atom has important structural and stereochemical consequences. First, the structures are such that the lone pairs, unlike the so-called "inert pairs," appear to occupy a bond position. Thus the SnCl$_3^-$ ion is pyramidal with a lone pair, as in NH$_3$. According to the AB$_x$E$_y$ scheme, which was discussed in Chapter 3, we would therefore consider this Sn atom to fall into the AB$_3$E classification. The lone pair not only has structural consequences, but chemical ones as well; SnCl$_3^-$ can act as a donor toward transition metals, as in the complex [Pt^{II}(SnCl$_3$)$_5$]$^{3-}$. Consider also SnCl$_2$·2H$_2$O, which contains a pyramidal SnCl$_2$OH$_2$ molecule; the second water molecule is not coordinated, and is readily lost at 80 °C. Other Sn^{II} compounds, such as SnCl$_2$ and SnS, accomplish three coordination in the solid by use of a bridging group between the metal atoms. The Sn$_2$F$_5^-$ ion consists of two SnF$_3$ groups sharing a fluorine atom.

In Ge$_5$F$_{12}$, the Ge^{II} atoms fall into the AB$_5$E classification, being square pyramidal with the lone pair occupying the sixth position. The same is true of SnO (the blue-black form) and of PbO, in which there are MO$_5$E metal atoms.

15-2   Isolation and Properties of the Elements

Silicon is obtained in the ordinary commercial form by reduction of SiO$_2$ with carbon or CaC$_2$ in an electric furnace. Similarly, Ge is prepared by reduction of GeO$_2$ with C or H$_2$. Silicon and Ge are used as semiconductors, especially in transistors. For this purpose, exceedingly high purity (<10^{-9} atom % of impurities) is essential, and special methods are required to obtain usable materials. The element is first converted to the tetrachloride, which is reduced back to the metal by hydrogen at high temperatures. After casting into rods it is *zone refined*. A rod of metal is heated near one end so that a cross-sectional wafer of molten silicon is produced. Since impurities are more soluble in the melt than they are in the solid they concentrate in the melt, and the melted zone is then caused to move slowly along the rod by moving the heat source. This carries impurities to the end. This process may be repeated. The impure end is then removed. Superpure Ge is made in a similar way.

Tin and lead are obtained by reduction of the oxide or sulfide with carbon. The metals can be dissolved in acid and deposited electrolytically to effect further purification.

Silicon is ordinarily rather unreactive. It is attacked by halogens giving tetrahalides, and by alkalis giving solutions of silicates. Silicon is not attacked by acids except hydrofluoric; presumably the stability of $[SiF_6]^{2-}$ provides the driving force here.

Germanium is somewhat more reactive than silicon and dissolves in concentrated H_2SO_4 and HNO_3. Tin and lead dissolve in several acids and are rapidly attacked by halogens. These elements are slowly attacked by cold alkali, and rapidly by hot, to form stannates and plumbites. Lead often appears to be more noble and unreactive than would be indicated by its standard potential of -0.13 V. This low reactivity can be attributed to a high overvoltage for hydrogen and also, in some instances, to insoluble surface coatings. Thus lead is not dissolved by dilute H_2SO_4 and concentrated HCl.

15-3   Hydrides: MH$_4$

These are colorless gases. Only *monosilane* (SiH$_4$) is of any importance. This spontaneously flammable gas is prepared by the action of LiAlH$_4$ on SiO$_2$ at 150–170 °C or by reduction of SiCl$_4$ with LiAlH$_4$ in an ether. Although stable to water and dilute acids, rapid base hydrolysis gives hydrated SiO$_2$ and H$_2$.

Substituted silanes with organic groups are of great importance, as are some closely related tin compounds (Chapter 29). The most important reaction of compounds with Si—H bonds, such as HSiCl$_3$ or HSi(CH$_3$)$_3$, is the Speier or hydrosilation reaction of alkenes.

$$RCH{=}CH_2 + SiHCl_3 \longrightarrow RCH_2CH_2SiCl_3 \qquad (15\text{-}3.1)$$

This reaction, which employs chloroplatinic acid as a catalyst, is commercially important for the synthesis of precursors to silicones.

15-4   Chlorides: MCl$_4$

Chlorination of the hot Group IVB(14) elements gives colorless liquids (MCl$_4$), except PbCl$_4$, which is yellow. The compound PbCl$_4$ may also be prepared by Reaction 15-4.1.

$$PbO_2 + 4\,HCl \longrightarrow PbCl_4 + 2\,H_2O \qquad (15\text{-}4.1)$$

The tetrachlorides are eventually hydrolyzed by water to hydrous oxides, but limited hydrolysis may give oxochlorides. In aqueous HCl, the tetrachlorides of Sn and Pb give chloroanions, $[MCl_6]^{2-}$.

The compound GeCl$_4$ differs from SiCl$_4$ in that the former can be distilled and separated from concentrated HCl, whereas silicon tetrachloride is immediately hydrolyzed by water.

The principal uses of SiCl$_4$ and GeCl$_4$ are in the synthesis of pure Si and Ge. Additional uses of SiCl$_4$ and SnCl$_4$ are in syntheses of organometallic compounds (Chapter 29).

15-5   Oxygen Compounds

Silica

Pure SiO_2 occurs in two forms, *quartz* and *cristobalite*. The Si is always tetrahedrally bound to four oxygen atoms but the bonds have considerable ionic character. In cristobalite the silicon atoms are placed as are the carbon atoms in diamond, with the oxygen atoms midway between each pair. In quartz, there are helices so that enantiomorphic crystals occur, and these may be easily recognized and separated mechanically.

Quartz and cristobalite can be interconverted when heated. These processes are slow because the breaking and re-forming of bonds is required and the activation energy is high. However, the rates of conversion are profoundly affected by the presence of impurities, or by the introduction of alkali metal oxides.

Slow cooling of molten SiO_2 or heating any solid form to the softening temperature gives an amorphous material that is glassy in appearance and is indeed a glass in the general sense, that is, a material with no long-range order but, instead, a disordered array of polymeric chains, sheets, or three-dimensional units.

Silica is relatively unreactive towards Cl_2, H_2, acids, and most metals at 25 °C or even at slightly elevated temperatures but is attacked by F_2, aqueous HF, alkali hydroxides, and fused carbonates.

Aqueous HF gives solutions containing fluorosilicates (e.g., $[SiF_6]^{2-}$). The *silicates* have been discussed in Section 5-4. The fusion of excess alkali carbonates with SiO_2 at about 1300 °C gives water-soluble products commercially sold as a syrupy liquid that has many uses. Aqueous sodium silicate solutions appear to contain the ion $[SiO_2(OH)_2]^{2-}$ but, depending on the pH and concentration, polymerized species are also present. In weathering of rocks and soils, "silicic acid," $Si(OH)_4$, is released in addition to $[Al(H_2O)_5(OH)]^{2+}$ and $[Al(H_2O)_4(OH)_2]^+$, and it appears that soluble silica can thereby reduce the Al levels, through formation of aluminosilicates (Chapter 5).

The basicity of the dioxides increases, with SiO_2 being purely acidic, GeO_2 less so, SnO_2 amphoteric, and PbO_2 somewhat more basic. When SnO_2 is made at high temperatures or by dissolving Sn in hot concentrated nitric acid, it is, like PbO_2, remarkably inert to attack.

Only lead forms a stable oxide containing both Pb^{II} and Pb^{IV}, namely, Pb_3O_4, which is a bright red powder known commercially as red lead. It is made by heating PbO and PbO_2 together at 250 °C. Although it behaves chemically as a mixture of PbO and PbO_2, the crystal contains $Pb^{IV}O_6$ octahedra linked in chains by sharing opposite edges. The chains are linked by Pb^{II} atoms each bound to three O atoms.

There are no true hydroxides and the products of hydrolysis of the hydrides or halides, and the like, are best regarded as hydrous oxides.

Among the most interesting and commercially valuable of silicon–oxygen compounds are the aluminosilicates, which have been mentioned earlier (Section 5-4).

15-6   Complex Compounds

Most of the complex species contain halide ions or donor ligands that are O, N, S, or P compounds.

Anionic Complexes

Silicon forms only fluoroanions, normally $[SiF_6]^{2-}$, whose high formation constant accounts for the incomplete hydrolysis of SiF_4 in water, according to Reaction 15-6.1.

$$2\,SiF_4 + 2\,H_2O = SiO_2 + [SiF_6]^{2-} + 2\,H^+ + 2\,HF \qquad (15\text{-}6.1)$$

The ion is usually made by dissolving SiO_2 in aqueous HF and is stable even in basic solution. Under selected conditions and with cations of the right size, the $[SiF_5]^-$ ion can be isolated, for example,

$$SiO_2 + HF(aq) + R_4N^+Cl^- \xrightarrow{\ CH_3OH\ } [R_4N][SiF_5] \qquad (15\text{-}6.2)$$

By contrast with $[SiF_6]^{2-}$, the $[GeF_6]^{2-}$ and $[SnF_6]^{2-}$ ions are hydrolyzed by bases; $[PbF_6]^{2-}$ ion is hydrolyzed even by water.

Although Si does not, the other elements give chloroanions, and all the elements form oxalato ions $[M(ox)_3]^{2-}$.

Cationic Complexes

The most important are those of chelating uninegative oxygen ligands, such as the acetylacetonates. An example is $[Ge(acac)_3]^+$.

The tetrahalides act as Lewis acids; $SnCl_4$ is a good Friedel–Crafts catalyst. The *adducts* are 1:1 or 1:2 but it is not always clear in the absence of X-ray evidence whether they are neutral, that is, MX_4L_2, or whether they are salts, for example, $[MX_2L_2]X_2$. Some of the best defined are the pyridine adducts, for example, *trans*-$(py)_2SiCl_4$.

Alkoxides, Carboxylates, and Oxo Salts

All four elements form alkoxides. Those of silicon [e.g., $Si(OC_2H_5)_4$] are the most important; the surface of glass or silica can also be alkoxylated. Alkoxides are normally obtained by the standard method, solvolysis of chlorides, as in Eq. 15-6.3.

$$MCl_4 + 4\,ROH + 4\,amine \longrightarrow M(OR)_4 + 4\,amine\cdot HCl \qquad (15\text{-}6.3)$$

Silicon alkoxides are hydrolyzed by water, eventually to hydrous silica. Of the carboxylates, *lead tetraacetate* is the most important, as it is used in organic chemistry as a strong but selective oxidizing agent. It is made by dissolving Pb_3O_4 in hot glacial acetic acid or by electrolytic oxidation of Pb^{II} in acetic acid. In oxidations the attacking species is probably $Pb(OOCCH_3)_3^+$, which is isoelectronic with the similar oxidant, $Tl(OOCCH_3)_3$, but this is not always so, and some oxidations are free radical in nature. The trifluoroacetate is a white solid, which will oxidize even heptane to give the $ROOCCF_3$ species, whence the alcohol ROH is obtained by hydrolysis; benzene similarly gives phenol.

Tin(IV) sulfate, $Sn(SO_4)_2\cdot 2\,H_2O$, can be crystallized from solutions obtained by oxidation of Sn^{II} sulfate; it is extensively hydrolyzed in water.

Tin(IV) nitrate is a colorless volatile solid made by interaction of N_2O_5 and $SnCl_4$; it contains bidentate NO_3^- groups giving dodecahedral coordination. The compound reacts with organic matter.

15-7   The Divalent State

Silicon

Divalent silicon species are thermodynamically unstable under normal conditions. However, several species, notably SiO and SiF_2, have been identified in high temperature reactions and trapped by chilling to liquid nitrogen temperatures. Thus at about 1100 °C and low pressures, the following reaction goes in about 99.5% yield:

$$SiF_4 + Si \rightleftharpoons 2\,SiF_2 \qquad (15\text{-}7.1)$$

Silicon difluoride (SiF_2) is stable for a few minutes at 10^{-4} cm pressure; the molecule is angular and diamagnetic. When the frozen compound warms, it gives fluorosilanes up to $Si_{16}F_{34}$.

Germanium

Germanium dihalides are stable. Germanium difluoride (GeF_2) is a white crystalline solid obtained by the action of anhydrous HF on Ge at 200 °C; it is a fluorine bridged polymer with approximately tbp coordination of Ge. Germanium dichloride ($GeCl_2$) gives salts of the $GeCl_3^-$ ion similar to those of Sn noted in the next subsection.

Tin

The most important compounds are SnF_2 and $SnCl_2$, which are obtained by heating Sn with gaseous HF or HCl. The fluoride is sparingly soluble in water and is used in fluoride-containing toothpastes. Water hydrolyzes $SnCl_2$ to a basic chloride, but from dilute acid solutions $SnCl_2 \cdot 2H_2O$ can be crystallized. Both halides dissolve in solutions containing an excess of halide ion, thus

$$SnF_2 + F^- = SnF_3^- \qquad pK \approx 1 \qquad (15\text{-}7.2)$$

$$SnCl_2 + Cl^- = SnCl_3^- \qquad pK \approx 2 \qquad (15\text{-}7.3)$$

In aqueous fluoride solutions SnF_3^- is the major species, but the ions SnF^+ and $Sn_2F_5^-$ can be detected.

The halides dissolve in donor solvents such as acetone, pyridine, or DMSO, to give pyramidal adducts, for example, $SnCl_2OC(CH_3)_2$.

The very air-sensitive tin(II) ion (Sn^{2+}) occurs in acid perchlorate solutions, which may be obtained by reduction of copper(II) perchlorate as in Reaction 15-7.4.

$$Cu(ClO_4)_2 + Sn/Hg = Cu + Sn^{2+} + 2\,ClO_4^- \qquad (15\text{-}7.4)$$

Hydrolysis gives $[Sn_3(OH)_4]^{2+}$, with $SnOH^+$ and $[Sn_2(OH)_2]^{2+}$ in minor amounts.

$$3\,Sn^{2+} + 4\,H_2O \rightleftharpoons [Sn_3(OH)_4]^{2+} + 4\,H^+ \qquad \log K = -6.77 \quad (15\text{-}7.5)$$

The trimeric, probably cyclic, ion appears to provide the nucleus of several basic tin(II) salts obtained from aqueous solutions at fairly low pH. Thus the nitrate appears to be $Sn_3(OH)_4(NO_3)_2$ and the sulfate, $Sn_3(OH)_2OSO_4$. All Sn^{II} solutions are readily oxidized by oxygen and, unless stringently protected from air, normally contain some Sn^{IV}. The chloride solutions are often used as mild reducing agents.

$$SnCl_6^{2-} + 2\,e^- = SnCl_3^- + 3\,Cl^- \qquad E° = ca.\ 0.0\ V\ (1\ M\ HCl,\ 4\ M\ Cl^-) \qquad (15\text{-}7.6)$$

Lead

Of the four elements, only lead has a well-defined low-valent cationic chemistry. The lead(II) ion (Pb^{2+}) is partially hydrolyzed in water.

$$Pb^{2+} + H_2O = PbOH^+ + H^+ \qquad \log K \approx -7.9 \qquad (15\text{-}7.7)$$

In concentrated solutions and on addition of base, polymeric ions that contain three, four, and six Pb atoms are formed. The crystalline "basic" salt

$$[Pb_6O(OH)_6]^{4+}(ClO_4^-)_4 \cdot H_2O$$

has the cluster structure in Fig. 15-2. The O atom lies at the center of the middle tetrahedron, while the OH groups lie on the faces of the outer tetrahedra.

Most lead salts are only sparingly soluble in water and some (e.g., $PbSO_4$ or $PbCrO_4$) are insoluble. The common soluble salts are $Pb(NO_3)_2$ and $Pb(CO_2CH_3)_2 \cdot 2\,H_2O$, which is incompletely ionized in water. The halides are always anhydrous and in solution they form complex species PbX^+, PbX_3^-, and so on, except for the fluoride where only PbF^+ occurs.

Silenes and Other Organo Compounds

Although for many of the elements, discussion of organo chemistry has been reserved for later chapters, it is now appropriate to mention the recent developments in divalent organo chemistry for silicon, germanium, and tin. Compounds of stoichiometry GeR_2 or SnR_2, which were known for a long time, proved to be cyclogermanes or stannanes such as $(Me_2Sn)_6$ (where $Me = CH_3$), or various other polymers with M^{II}—M^{II} bonds. Also, the silicon compounds made by the reduction of R_2SiCl_2 with Li or Na/K in THF, where R is *not* a bulky ligand, are

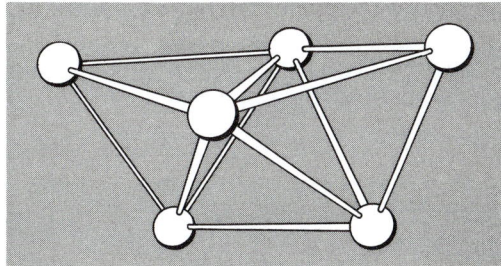

Figure 15-2 The three face-sharing tetrahedra of Pb atoms in the $Pb_6O(OH)_6^{4+}$ cluster.

cyclic polymers. These compounds are often similar to hydrocarbons, but differ in having large (~115°) Si—Si—Si angles.

However, when the alkyl or aryl R groups are very bulky, monomers (MR_2) or dimers ($R_2M\!\!=\!\!MR_2$) can be isolated. The simplest silene (Me_2Si) can be obtained only in the gas phase or in solution by thermal decomposition of the cyclic hexamer, as in Reaction 15-7.8.

$$cyclo\text{-}(Me_2Si)_6 \longrightarrow cyclo\text{-}(Me_2Si)_5 + Me_2Si \tag{15-7.8}$$

The first isolable *silene*, the yellow tetramesityl disilene, Structure 15-I

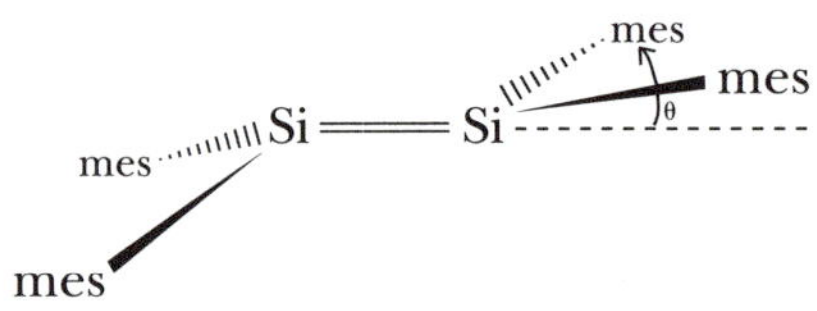

15-I

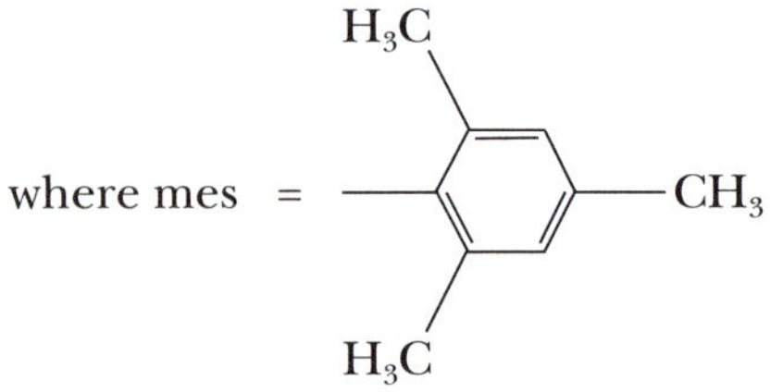

where mes =

can be made photochemically according to Reaction 15-7.9.

$$(mes)_2Si(SiMe_3)_2 \longrightarrow (mes)_2Si\!\!=\!\!Si(mes)_2 + (Me_3Si)_2 \tag{15-7.9}$$

This compound is an air sensitive, but thermally stable solid, with a slightly bent trans structure, the angle θ being 18°. In this respect, this silene differs from typical alkenes, which are planar. The $Si\!\!=\!\!Si$ bond (2.16 Å) is about 9% shorter than a Si—Si single bond. Many other disilenes can be made by the reduction of R_2SiCl_2 with Li, and some compounds have sufficiently strong double bonds to permit the existence of cis and trans isomers. There are similar $R_2Si\!\!=\!\!CR_2$ compounds containing $Si\!\!=\!\!C$ bonds. A good example is $Me_2Si\!\!=\!\!C(SiMe_3)(SiMe\text{-}t\text{-}Bu_2)$, where the $C_2Si\!\!=\!\!CSi_2$ skeleton is planar, with a $C\!\!=\!\!Si$ bond distance of 1.702 Å. Finally, compounds with $Si\!\!=\!\!N$, $Si\!\!=\!\!P$, and $Si\!\!=\!\!O$ bonds are known, for example, $t\text{-}Bu_2Si\!\!=\!\!NSi\text{-}t\text{-}Bu_3$.

The *germenes,* which can be made by the action of Grignard reagents on $GeCl_2$·dioxane in ether, and stannenes, are less stable than silenes. Distortions from planar geometry are larger than is found among the silenes. Tin and lead form highly colored monomers or dimers, but in Sn_2R_4 the Sn—Sn distance is 2.76 Å, a value closer to that of a Sn—Sn single bond.

15-8   Summary of Group Trends for the Elements of Group IVB(14)

Using the list of periodic chemical properties listed in Section 8-11, as well as properties mentioned in Chapters 14 and 15, we can now summarize the periodic trends in the properties of the elements of Group IVB(14).

1. Carbon
   (a) Is completely nonmetallic.
   (b) Has a strong tendency for catenation among its compounds.
   (c) Forms molecular (covalent) substances almost exclusively (carbides excepted).
   (d) Obeys the octet rule, the maximum covalence being four.
   (e) Forms divalent (lower valent) compounds that are unstable (as in reactive intermediates), or that exist as such only as a formality (e.g., CO).
   (f) Forms hydrides that are stable, molecular substances that are difficult to hydrolyze, but oxidize readily.
   (g) Forms stable, molecular halides that are not readily oxidized or hydrolyzed.
   (h) Forms oxides (CO and CO_2) that are acidic anhydrides.
   (i) Forms multiple bonds of the $p\pi-p\pi$ variety, which can be conjugated.

2. Silicon
   (a) Is a nonmetal.
   (b) Displays little or no tendency for catenation among its compounds.
   (c) Forms mostly covalent substances, as well as polyatomic ions and oxoanions containing covalent bonds.
   (d) Readily undergoes coordination number expansion to a maximum covalence of six, namely, SiF_5^- and SiF_6^{2-}.
   (e) Forms divalent (lower valent) compounds only rarely, an example being the unstable SiF_2.
   (f) Forms hydrides that are reactive and unstable, an example being SiH_4, which is readily hydrolyzed.
   (g) Forms molecular halides that are readily hydrolyzed.
   (h) Forms an oxide (SiO_2) that is an acidic, covalent-network substance.
   (i) Forms strong, but unconjugated multiple bonds of the $p\pi-d\pi$ variety, especially to O and N.

3. Germanium, Tin, and Lead
   (a) Are increasingly metallic on descent of the group, Ge being most like Si.
   (b) Display little catenation, since in general bond strength decreases on descending the group.
   (c) Form both covalent and ionic substances.
   (d) Form compounds with a variety of coordination numbers, six or eight being common.
   (e) Form divalent (lower valent) compounds that are increasingly stable upon descending the group.
   (f) Do not form any important covalent hydrides.
   (g) Form both high-valent (MX_4) and low-valent (MX_2) molecular halides, which are readily hydrolyzed and undergo coordination number expansion to produce, for instance, $SnCl_6^{2-}$ or $PbCl_6^{2-}$.
   (h) Display increasingly metallic character on descent of the group, as

demonstrated by the following. Whereas SiO_2 is acidic and SnO_2 is amphoteric, PbO_2 is purely a basic anhydride.

(i) Form multiple bonds of the $p\pi$–$d\pi$ variety, but less effectively on descent of the group.

STUDY GUIDE

Study Questions

A. Review

1. Why is CO_2 a gas and SiO_2 a giant molecule?
2. Explain what is meant by $d\pi$–$p\pi$ bonding.
3. Why does tin form divalent inorganic compounds more easily than silicon?
4. How is super pure Ge made from GeO_2?
5. Write balanced equations for the synthesis of SiH_4 and for its hydrolysis by aqueous KOH.
6. Why is CCl_4 unreactive to H_2O, whereas $SiCl_4$ is rapidly hydrolyzed?
7. Why is SiF_4 incompletely hydrolyzed by water?
8. Explain the nature of zeolites and of molecular sieves.
9. Why does silicon have much less tendency to form bonds to itself than does carbon?
10. How is lead tetraacetate made?
11. What is red lead?
12. What is the nature of Sn^{II} in aqueous chloride solution?

B. Additional Exercises

1. Explain why H_3SiNCS has a linear SiNCS group, whereas in H_3CNCS the CNC group is angular.
2. Why are silanols, such as $(CH_3)_3SiOH$, stronger acids than their carbon analogs?
3. List the various types of geometries among the compounds of the tetravalent Group IVB(14) elements and give examples. For each example, give the structural classification for the Group IVB(14) atom, according to the AB_xE_y scheme of Chapter 3.
4. What methods could one use to determine the nature of 1:1 and 1:2 adducts of $SnCl_4$ with neutral donors?
5. Why can Sn^{II} compounds, such as $SnCl_3^-$, act as donors (ligands) to transition metals?
6. The single-bond energies for the elements of the first and second short periods follow the trends C > Si; N < P; O < S; F < Cl. Why is the first pair in the list apparently anomalous?
7. Predict the relative π-bond strength between B and N in the two compounds bis(trimethylsilyl)aminoborane and bis(*tert*-butyl)aminoborane. Explain your answer in terms of the π orbitals that are involved.
8. Draw the π-bond system that is responsible for the planarity of trisilylamine.
9. Draw the Lewis diagrams and discuss the geometries of $SnCl_2$, $SnCl_3^-$, and $[Pt(SnCl_3)_5]^{3-}$.
10. Balance the equation for the reaction of $SiCl_4$ with $LiAlH_4$.
11. Use valence shell electron-pair repulsion (VSEPR) theory to compare the bond angles in the pyramidal ions SnF_3^-, $SnCl_3^-$, and $GeCl_3^-$.

12. Review the material of Section 8-11 plus the material of this chapter, and summarize the facts concerning the low-valent state for the elements of Group IVB(14), citing specific compounds as examples to illustrate each point.

13. Compare the reactivities of the divalent chlorides ($GeCl_2$, $SnCl_2$, and $PbCl_2$) with chlorine, and use this information to arrive at the correct order of stabilities of the divalent state for these elements.

14. Diagram the apparent extent of the π-bonding systems in H_3SiNCO and in H_3GeNCO, taking into consideration the geometries of the two.

15. Explain the planarity of disilylamine using an orbital overlap approach.

16. Offer an explanation for the relative extent of N-to-M π bonding in H_3SiNCO versus H_3GeNCO.

17. The compound $SnCl_2(C_2H_5)_2$ crystallizes as long needles, in which there are Sn—Cl—Sn bridges in one plane and ethyl groups in coordination positions perpendicular to that plane. The effective coordination number of Sn in the solid is six, but there are two Sn—Cl distances in the structure. Propose a solid state structure.

18. Choose the correct answer from among the following possibilities:
    (a) The most stable low-valent halide:
        $GeCl_2$ $SnCl_2$ $PbCl_2$
    (b) A nonexistent halide:
        $SnCl_4$ $PbCl_4$ PbI_4
    (c) A purely acidic oxide:
        PbO_2 SnO_2 SiO_2
    (d) Forms an oxoacid on treatment with HNO_3:
        P_4 Sb_4 Bi
    (e) The most stable hydride:
        NH_3 PH_3 AsH_3
    (f) The substance that is coordinatively saturated:
        CCl_4 $SiCl_4$ $PbCl_4$
    (g) The substance that is not coordinatively saturated:
        SnF_5^- CH_4 PCl_6^-

19. Explain how the following reaction demonstrates the acidity of SiO_2:

$$SiO_2 + Na_2O \longrightarrow Na_2SiO_3$$

20. List and explain three ways in which the chemistry of carbon differs from that of the other members of the group.

21. Give balanced equations for each of the following:
    (a) Production of Ge from the oxide.
    (b) Oxidation of Si by chlorine.
    (c) Dissolution of $SnCl_2$ in pyridine.
    (d) Hydrolysis of $GeCl_4$.
    (e) Hydrolysis of Sn^{4+} solutions.
    (f) Hydrolysis of SiF_4.

22. Why does the tendency towards catenation decrease on descent of Group IVB(14)? Illustrate your answer with some examples.

23. Suggest a synthesis, starting from elemental silicon and fluorine, of SiF_6^{2-}.

24. What is the main product on reaction of lead with chlorine, $PbCl_4$, $PbCl_2$, or PbOCl?

25. Suggest a synthesis of lithium bis(dimethylsilyl)amide.

26. Sketch the structures of $SnCl_2$, $SnCl_3^-$, SnF_4, and SnF_5^-.

C. Questions from the Literature of Inorganic Chemistry

1. Let the paper by R. H. Nielson and R. L. Wells, *Inorg. Chem.*, **1977**, *16*, 7–11, serve as a basis for the following questions:

    (a) What typical values for B—N rotational barriers does one expect for mono-, bis-, and tris-aminoboranes?

    (b) Why should studies of rotational barriers in these aminoboranes indicate the relative extent of π bonding between boron and an *N*-trimethylsilyl, an *N*-trimethylgermyl, and an *N*-trimethylstannyl substituent?

    (c) Both a steric and a competitive π-bonding argument can be given to explain the trends reported here. Elaborate.

2. Consider the paper by D. Kummer and T. Seshadri, *Angew. Chem. Int. Ed. Eng.*, **1975**, *14*, 699–700.

    (a) Determine the oxidation state of Si and draw the Lewis diagram for each of the Si-containing compounds mentioned in this article.

    (b) Predict the geometry for each of these compounds.

3. Compare and contrast the structure of and the bonding in two different classes of Sn^{IV} compounds as presented in

    (a) R_2SnX_2

    N. W. Alcock and J. F. Sawyer, *J. Chem. Soc., Dalton Trans.*, **1977**, 1090–1095.

    (b) $SnCl_4(PR_3)_2$

    G. G. Mather, G. M. McLaughlin, and A. Pidcock, *J. Chem. Soc., Dalton Trans.*, **1973**, 1823–1827.

4. Consider the compounds $M[CH(Me_3Si)_2]_2$, where M = Ge, Sn, or Pb, as described by J. D. Cotton, P. J. Davidson, and M. F. Lappert, *J. Chem. Soc., Dalton Trans.*, **1976**, 2275–2285.

    (a) Draw the Lewis diagram of these substances.

    (b) Explain (and give an example of) each of the four types of reactions mentioned for these substances.

5. Look up the structure of PbO (*Acta Crystallogr.*, **1961**, *14*, 1304) and describe the geometry at lead. What structural role does the "lone pair" play?

6. The structure of the $[K(18\text{-}C\text{-}6)]^+$ salt of $[(t\text{-}Bu)_3C_6H_2\text{—}SiF_4]^-$ was reported by S. E. Johnson, R. O. Day, and R. R. Holmes, *Inorg. Chem.*, **1989**, *28*, 3182. What unusual bond angles are there in this anion, and what is the apparent cause?

7. Read the first report on the structure of two stable disilenes by M. J. Fink, M. J. Michalczyk, K. J. Haller, R. West, and J. Michl, "X-ray Structure of Two Disilenes," *Organometallics*, **1984**, *3*, 793–800.

    (a) Which of these two disilenes has a structure most like an alkene?

    (b) What two principle deviations from planarity are noted for Compound *1a*?

    (c) What explanations do the authors give for the pyramidalization at Si in Compound *1a*?

    (d) Why do these distortions from planarity not occur for Compound *1b*?

SUPPLEMENTARY READING

Breck, D. W., *Molecular Sieves*, Wiley, New York, 1973.

Burger, H. and Eugen, R., "The Chemistry of Lower-Valent Silicon," *Topics in Current Chemistry*, No. 5, Springer-Verlag, Berlin, 1974.

Cowley, A. H. and Norman, N. C., "The Synthesis, Properties, and Reactivities of Stable Compounds Featuring Double Bonding Between Heavier Group 14 and 15 Elements," in *Progress in Inorganic Chemistry,* Vol. 34, Wiley-Interscience, New York, 1986.

Davidov, V. I., *Germanium,* Gordon and Breach, New York, 1966.

Donaldson, J. D., "The Chemistry of Divalent Tin," in *Progress in Inorganic Chemistry,* Vol. 8, Interscience, New York, 1967.

Drake, J. E. and Riddle, C., "Volatile Compounds of the Hydrides of Silicon and Germanium with Elements of Groups V and VI," *Q. Rev.,* **1970,** *24,* 263.

Eaborn, C., *Organosilicon Compounds,* Butterworths, London, 1960.

Ebsworth, E. A. V., *Volatile Silicon Compounds,* Pergamon Press, Elmsford, NY, 1963.

Ebsworth, E. A. V., *The Organometallic Compounds of the Group IV Elements,* G. MacDiarmid, Ed., Dekker, New York, 1968.

Glocking, F., *The Chemistry of Germanium,* Academic, New York, 1969.

Harrison, P. G. et al., Eds., *The Chemistry of Tin,* Blackie, London, and Chapman & Hall, New York, 1989.

Holmes, R. R., "The Stereochemistry of Nucleophilic Substitution at Tetracoordinated Silicon," *Chem. Rev.,* **1990,** *90,* 17–31.

Holt, M. S., Wilson, W. L., and Nelson, J. H., "The Chemistry of Transition Metal to Tin Bonds," *Chem. Rev.,* **1989,** *89,* 11.

Iler, R. K., *The Chemistry of Silicon,* Wiley, New York, 1979.

Lesbre, M., Mazerolles, P., and Satge, J., *The Organic Compounds of Germanium,* Wiley, New York, 1971.

Margrave, J. L. and Wilson, P. W., "Silicon Difluoride, Its Reactions and Properties," *Acc. Chem. Res.,* **1971,** *4,* 145.

Ng, S. W. and Zuckerman, J. J., "Where Are the Lone Pair Electrons in Subvalent Fourth-Group Compounds?" *Adv. Inorg. Chem. Radiochem.,* **1985,** *29,* 297–325.

Noll, W. et al., *Chemistry and Technology of Silicones,* Academic, New York, 1968.

Patai, S. and Rappaport, Z., *The Chemistry of Organo Silicon Compounds,* Wiley, New York, 1989.

Petz, W., "Transition Metal Complexes with Derivatives of Si^{II}, Ge^{II}, Sn^{II}, and Pb^{II}, as Ligands," *Chem. Rev.,* **1986,** *86,* 1019.

Raabe, G. and Michl, J., "Multiple Bonding to Silicon," *Chem. Rev.,* **1985,** *85,* 419–509.

Shapiro, H. and Frey, F. W., *The Organic Compounds of Lead,* Wiley, New York, 1968.

Zuckerman, J. J., Ed., "Organotin Compounds: New Chemistry and Applications," *Advances in Chemistry Series,* No. 157, American Chemical Society, Washington, DC, 1976.

Chapter 16

NITROGEN

16-1 Introduction

The nitrogen atom $(1s^2 2s^2 2p_x 2p_y 2p_z)$ can complete its valence shell in the following ways:

1. Electron gain to form the nitride ion N^{3-}; this ion is found only in saltlike nitrides of the most electropositive metals.

2. Formation of electron-pair bonds: (a) single bonds, as in NH_3, or (b) multiple bonds, as in $:N{\equiv}N:$, $-\ddot{N}{=}\ddot{N}-$, or NO_2.

3. Formation of electron-pair bonds with electron gain, as in NH_2^- or NH^{2-}.

4. Formation of electron-pair bonds with electron loss, as in the tetrahedral ammonium and substituted ammonium ions, $[NR_4]^+$.

The following structural types (recall Chapter 3) are common among those compounds of nitrogen having covalent bonds: AB_4 (as in tetrahedral NR_4^+); AB_3E (as in pyramidal NR_3); AB_2E_2 (as in bent NR_2^-); AB_3 (as in planar NO_3^-); AB_2E (as in bent $R_2C{=}N{-}OH$); and ABE (N_2). There are a few stable species in which, formally, the nitrogen valence shell is incomplete. Nitroxides, $R_2\dot{N}{=}\ddot{O}$, NO, and NO_2 are the best examples; these have unpaired electrons and are paramagnetic.

Three-Covalent Nitrogen

The molecules NR_3 are pyramidal; the bonding is best considered as involving sp^3 hybrid orbitals so that the lone pair occupies the fourth position. There are three points to note:

1. As a result of the nonbonding electron pair, all NR_3 compounds behave as Lewis bases and they give donor–acceptor complexes with Lewis acids, for example, $F_3B:N(CH_3)_3$, and they act as ligands toward transition metal ions as in, for example, $[Co(NH_3)_6]^{3+}$.

2. Pyramidal molecules ($NRR'R''$) should be chiral. Optical isomers cannot be isolated, however, because such molecules very rapidly undergo a motion known as *inversion* in which the N atom oscillates through the plane of the three R groups, much as an umbrella can turn inside out (Fig. 16-1). The energy barrier for this process is only about 24 kJ mol^{-1}.

3. There are a very few cases where three-covalent nitrogen is planar; in these cases multiple bonding is involved as we discussed for $N(SiMe_3)_3$, in

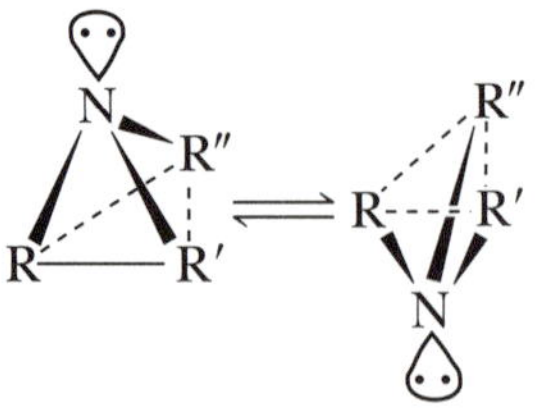

Figure 16-1 Diagram illustrating the inversion of NH_3.

Section 15-1. The N-centered triangular metal complexes such as $[NIr_3(SO_4)_6(H_2O)_3]^{4-}$ are similar.

N—N Single-Bond Energy

The N—N single bond is quite weak. If we compare the single-bond energies:

H_3C—CH_3	H_2N—NH_2	HO—OH	F—F	Units
350	160	140	150	kJ mol^{-1}

it is clear that there is a profound drop between C and N. This difference is probably attributable to the effects of repulsion between nonbonding lone pairs. The result is that, unlike carbon, nitrogen has little tendency to catenation.

Multiple Bonds

The propensity of nitrogen, like carbon, to form $p\pi$–$p\pi$ multiple bonds is a feature that distinguishes it from phosphorus and the other Group VB(15) elements. Thus nitrogen as the element is dinitrogen (N_2), with a very high bond strength and a short internuclear distance (1.094 Å), whereas phosphorus forms P_4 molecules or infinite layer structures in which there are only single bonds (Section 8-5).

Where a nitrogen atom forms one single and one double bond, nonlinear molecules result, as shown in Structures 16-I to 16-IV.

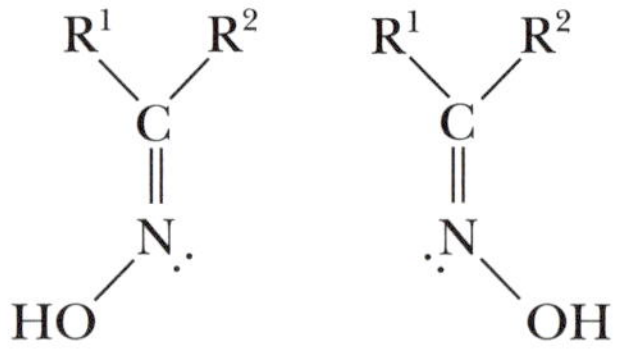

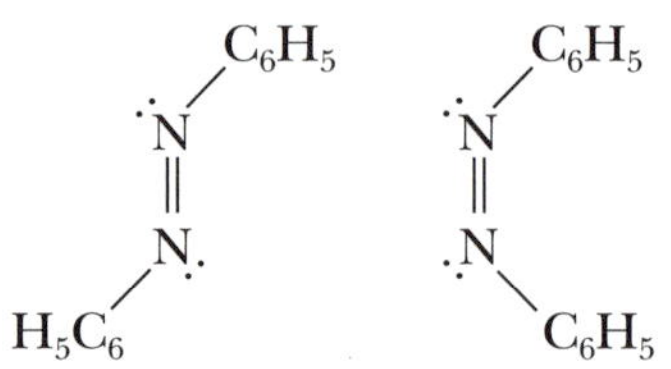

Each nitrogen atom in these structures is of the AB_2E type, uses sp^2 hybrid orbitals, and forms a π bond using the unhybridized $2p$ orbital.

In the oxo anions NO_2^- (AB_2E) and NO_3^- (AB_3), there are multiple bonds that may be formulated in either resonance or MO terms, as discussed in Chapter 3.

16-2 Occurrence and Properties of the Element

Nitrogen occurs in nature mainly as dinitrogen, N_2 (bp 77.3 K), which comprises 78% by volume of the earth's atmosphere. The isotopes ^{14}N and ^{15}N have an absolute ratio $^{14}N/^{15}N = 272.0$. Compounds enriched in ^{15}N are used in tracer studies.

The heat of dissociation of N_2 is extremely large.

$$N_2(g) = 2\,N(g) \qquad \Delta H = 944.7 \text{ kJ mol}^{-1} \qquad K_{25\,°C} = 10^{-120} \quad (16\text{-}2.1)$$

The great strength of the N≡N bond is principally responsible for the chemical inertness of N_2 and for the fact that most simple nitrogen compounds are endothermic even though they may contain strong bonds. Dinitrogen is notably unreactive in comparison with isoelectronic, triply bonded systems such as X—C≡C—X, :C≡O:, X—C≡N:, and X—N≡C:. Both —C≡C— and —C≡N groups can act as donors by using their π electrons, whereas N_2 does not. It can, however, form complexes similar to those formed by CO, although to a much more limited extent, in which there are M←N≡N: and M←C≡O: configurations (Chapter 28).

Nitrogen is obtained by liquefaction and fractionation of air. It usually contains some argon and, depending on the quality, upwards of about 30 ppm of oxygen. Spectroscopically pure N_2 is made by thermal decomposition of sodium or barium azide.

$$2\,NaN_3 \longrightarrow 2\,Na + 3\,N_2 \qquad (16\text{-}2.2)$$

The only reactions of N_2 at room temperature are with metallic Li to give Li_3N, with certain transition metal complexes, and with nitrogen fixing bacteria. These nitrogen fixing bacteria are either free living or symbiotic on the root nodules of clover, peas, beans, and the like. The mechanism by which these bacteria fix N_2 is unknown.

At elevated temperatures nitrogen becomes more reactive, especially when catalyzed. Typical reactions are

$$N_2(g) + 3\,H_2(g) = 2\,NH_3(g) \qquad K_{25\,°C} = 10^3 \text{ atm}^{-2} \qquad (16\text{-}2.3)$$

$$N_2(g) + O_2(g) = 2\,NO(g) \qquad K_{25\,°C} = 5 \times 10^{-31} \qquad (16\text{-}2.4)$$

$$N_2(g) + 3\,Mg(s) = Mg_3N_2(s) \qquad (16\text{-}2.5)$$

$$N_2(g) + CaC_2(s) = C(s) + CaNCN(s) \qquad (16\text{-}2.6)$$

16-3 Nitrides

Nitrides of electropositive metals have structures with discrete nitrogen atoms and can be regarded as ionic, for example, $(Ca^{2+})_3(N^{3-})_2$, and $(Li^+)_3N^{3-}$. Their ready hydrolysis to ammonia and the metal hydroxides is consistent with this.

Such nitrides are prepared by direct interaction or by loss of ammonia from amides on heating, for example,

$$3 \text{ Ba(NH}_2)_2 \longrightarrow \text{Ba}_3\text{N}_2 + 4 \text{ NH}_3 \qquad (16\text{-}3.1)$$

Transition metal *nitrides* are often nonstoichiometric and have nitrogen atoms in the interstices of close-packed arrays of metal atoms. Like the similar carbides or borides they are hard, chemically inert, high melting, and electrically conducting.

There are numerous covalent *nitrides* (BN, S_4N_4, P_3N_5, etc.), and their properties vary greatly depending on the element with which nitrogen is combined. These are, therefore, discussed more fully under the appropriate element.

16-4 Nitrogen Hydrides

Ammonia

Ammonia (NH_3) is formed by the action of a base on an ammonium salt.

$$\text{NH}_4\text{X} + \text{OH}^- \longrightarrow \text{NH}_3 + \text{H}_2\text{O} + \text{X}^- \qquad (16\text{-}4.1)$$

Industrially, ammonia is made by the Haber process in which the reaction

$$\text{N}_2(\text{g}) + 3 \text{ H}_2(\text{g}) = 2 \text{ NH}_3(\text{g}) \qquad \Delta H = -46 \text{ kJ mol}^{-1} \quad (16\text{-}4.2)$$
$$K_{25\,°C} = 10^3 \text{ atm}^{-2}$$

is carried out at 400–500 °C and pressures of 10^2–10^3 atm in the presence of a catalyst. Although the equilibrium is most favorable at low temperature, even with the best catalysts, elevated temperatures are required to obtain a satisfactory rate. The best catalyst is α-iron containing some oxide to widen the lattice and enlarge the active interface.

Ammonia is a colorless, pungent gas (bp −33.35 °C). The liquid has a large heat of evaporation (1.37 kJ g^{-1} at the boiling point) and can be handled in ordinary laboratory equipment. Liquid NH_3 resembles water in its physical behavior, being highly associated via strong hydrogen bonding. Its dielectric constant (~22 at −34 °C; cf. 81 for H_2O at 25 °C) is sufficiently high to make it a fair ionizing solvent. Its self-ionization has been discussed previously (Section 7-3).

Liquid NH_3 has lower reactivity than H_2O toward electropositive metals and dissolves many of them (Section 10-3).

Because $NH_3(\ell)$ has a much lower dielectric constant than water, it is a better solvent for organic compounds but generally a poorer one for ionic inorganic compounds. Exceptions occur when complexing by NH_3 is superior to that by water. Thus AgI is exceedingly insoluble in water but very soluble in NH_3. Primary solvation numbers of cations in NH_3 appear similar to those in H_2O, for example, 5.0 ± 0.2 and 6.0 ± 0.5 for Mg^{2+} and Al^{3+}, respectively.

Ammonia burns in air:

$$4 \text{ NH}_3(\text{g}) + 3 \text{ O}_2(\text{g}) = 2 \text{ N}_2(\text{g}) + 6 \text{ H}_2\text{O}(\text{g}) \qquad K_{25\,°C} = 10^{228} \quad (16\text{-}4.3)$$

Reaction 16-4.3 is thermodynamically favored under normal conditions. However, at 750–900 °C, in the presence of a platinum or a platinum-rhodium catalyst, reaction of ammonia with oxygen can be made to give NO instead of N_2, as in Eq. 16-4.4:

$$4\,NH_3 + 5\,O_2 = 4\,NO + 6\,H_2O \qquad K_{25\,°C} = 10^{168} \qquad (16\text{-}4.4)$$

thus affording a useful synthesis of NO. The latter reacts with an excess of O_2 to produce NO_2, and the mixed oxides can be absorbed in water to form nitric acid.

$$2\,NO + O_2 \longrightarrow 2\,NO_2 \qquad\qquad\qquad (16\text{-}4.5)$$

$$3\,NO_2 + H_2O \longrightarrow 2\,HNO_3 + NO \qquad \text{and so on} \qquad (16\text{-}4.6)$$

Thus the sequence in industrial utilization of atmospheric nitrogen is as follows:

$$N_2 \xrightarrow[\substack{\text{Haber}\\\text{process}}]{H_2} NH_3 \xrightarrow[\substack{\text{Ostwald}\\\text{process}}]{O_2} NO \xrightarrow{O_2+H_2O} HNO_3(aq) \qquad (16\text{-}4.7)$$

Ammonia is extremely soluble in water. Although aqueous solutions are generally referred to as solutions of the weak base NH_4OH, called *ammonium hydroxide,* undissociated NH_4OH probably does not exist. The solutions are best described as $NH_3(aq)$, with the equilibrium written as

$$NH_3(aq) + H_2O = NH_4^+ + OH^- \qquad K_{25\,°C} = \frac{[NH_4^+][OH^-]}{[NH_3]} \qquad (16\text{-}4.8)$$

$$= 1.77 \times 10^{-5} \; (p K_b = 4.75)$$

Ammonium Salts

Stable crystalline salts of the tetrahedral NH_4^+ ion are mostly water soluble. Ammonium salts generally resemble those of potassium and rubidium in solubility and structure, since the three ions are of comparable (Pauling) radii: $NH_4^+ = 1.48$ Å, $K^+ = 1.33$ Å, $Rb^+ = 1.48$ Å. Salts of strong acids are fully ionized, and the solutions are slightly acidic.

$$NH_4Cl = NH_4^+ + Cl^- \qquad K \approx \infty \qquad (16\text{-}4.9)$$

$$NH_4^+ + H_2O = NH_3 + H_3O^+ \qquad K_{25\,°C} = 5.5 \times 10^{-10} \qquad (16\text{-}4.10)$$

Thus, a 1 *M* solution will have a pH of about 4.7. The constant for the second reaction is sometimes called the hydrolysis constant; however, it may equally well be considered as the acidity constant of the cationic acid NH_4^+, and the system regarded as an acid-base system in the following sense:

$$\underset{\text{Acid}}{NH_4^+} + \underset{\text{Base}}{H_2O} = \underset{\text{Acid}}{H_3O^+} + \underset{\text{Base}}{NH_3(aq)} \qquad (16\text{-}4.11)$$

Many ammonium salts volatilize with dissociation around 300 °C, for example,

$$NH_4Cl(s) = NH_3(g) + HCl(g) \qquad \Delta H = 177 \text{ kJ mol}^{-1}$$
$$K_{25\ °C} = 10^{-16} \qquad (16\text{-}4.12)$$

$$NH_4NO_3(s) = NH_3(g) + HNO_3(g) \qquad \Delta H = 171 \text{ kJ mol}^{-1} \quad (16\text{-}4.13)$$

Salts that contain oxidizing anions may decompose when heated, with oxidation of the ammonia to N_2O or N_2, or both. For example,

$$(NH_4)_2Cr_2O_7(s) = N_2(g) + 4\,H_2O(g) + Cr_2O_3(s)$$
$$\Delta H = -315 \text{ kJ mol}^{-1} \qquad (16\text{-}4.14)$$

$$NH_4NO_3(\ell) = N_2O(g) + 2\,H_2O(g) \qquad \Delta H = -23 \text{ kJ mol}^{-1} \qquad (16\text{-}4.15)$$

Hydrazine

Hydrazine (N_2H_4) may be thought of as derived from ammonia by replacement of a hydrogen atom by the NH_2 group. It is a bifunctional base,

$$N_2H_4(aq) + H_2O = N_2H_5^+ + OH^- \qquad K_{25\ °C} = 8.5 \times 10^{-7} \quad (16\text{-}4.16)$$

$$N_2H_5^+(aq) + H_2O = N_2H_6^{2+} + OH^- \qquad K_{25\ °C} = 8.9 \times 10^{-15} \quad (16\text{-}4.17)$$

and two series of hydrazinium salts are obtainable. Those of $N_2H_5^+$ are stable in water, while those of $N_2H_6^{2+}$ are extensively hydrolyzed. Salts of $N_2H_6^{2+}$ can be obtained by crystallization from aqueous solution containing a large excess of the acid, since they are usually less soluble than the monoacid salts.

Anhydrous N_2H_4 is a fuming colorless liquid (bp 114 °C). It is surprisingly stable in view of its endothermic nature ($\Delta H_f° = 50 \text{ kJ mol}^{-1}$). It burns in air with considerable evolution of heat.

$$N_2H_4(\ell) + O_2(g) = N_2(g) + 2\,H_2O(\ell) \qquad \Delta H° = -622 \text{ kJ mol}^{-1} \quad (16\text{-}4.18)$$

Aqueous hydrazine is a powerful reducing agent in basic solution, normally being oxidized to nitrogen. Hydrazine is made by the interaction of aqueous ammonia with sodium hypochlorite.

$$NH_3 + NaOCl \longrightarrow NaOH + NH_2Cl \qquad (\text{Fast}) \quad (16\text{-}4.19)$$

$$NH_3 + NH_2Cl + NaOH \longrightarrow N_2H_4 + NaCl + H_2O \qquad (16\text{-}4.20)$$

However, there is a competing reaction that is rather fast once some hydrazine has been formed.

$$2\,NH_2Cl + N_2H_4 \longrightarrow 2\,NH_4Cl + N_2 \qquad (16\text{-}4.21)$$

To obtain appreciable yields, it is necessary to add gelatine. This sequesters heavy metal ions that catalyze the parasitic reaction; even the part per million or so of Cu^{2+} in ordinary water will almost completely prevent the formation of hydrazine if no gelatine is used. Since simple sequestering agents such as EDTA are not as beneficial as gelatine, the latter is assumed to have a catalytic effect as well.

Hydroxylamine

Hydroxylamine (NH_2OH) is a weaker base than NH_3:

$$NH_2OH(aq) + H_2O = NH_3OH^+ + OH^- \qquad K_{25\,°C} = 6.6 \times 10^{-9} \quad (16\text{-}4.22)$$

It is prepared by reduction of nitrates or nitrites either electrolytically or with SO_2, under controlled conditions. Hydroxylamine is a white unstable solid. In aqueous solution, or as its salts $[NH_3OH]Cl$ or $[NH_3OH]_2SO_4$, it is used as a reducing agent.

Azides

Sodium azide can be obtained by the reaction

$$3\,NaNH_2 + NaNO_2 \xrightarrow{175\,°C} NaN_3 + 3\,NaOH + NH_3 \qquad (16\text{-}4.23)$$

Heavy metal azides are explosive and lead or mercury azide have been used in detonation caps. The azide ion, which is linear and symmetrical, behaves rather like a halide ion and can act as a ligand in metal complexes. The pure acid (HN_3) is a dangerously explosive liquid.

16-5 Nitrogen Oxides

Dinitrogen Monoxide (Nitrous Oxide)

Nitrous oxide (N_2O) is obtained by thermal decomposition of molten ammonium nitrate.

$$NH_4NO_3 \xrightarrow{250\,°C} N_2O + 2\,H_2O \qquad (16\text{-}5.1)$$

The contaminants are NO (which can be removed by passage through ferrous sulfate solution), and 1-2% of nitrogen. Thermodynamically, nitrous oxide is unstable relative to N_2 and atomic oxygen ($\Delta G = 105$ kJ mol^{-1}), but it is kinetically stable in the absence of transition metal complexes with which it reacts by O atom transfer, giving N_2 and M=O or M—O—M bonds.

Nitrous oxide has the linear structure NNO. It is relatively unreactive, being inert to the halogens, alkali metals, and ozone at room temperature. On heating, it decomposes to N_2 and O_2. At elevated temperatures, it will react with the alkali metals and with many organic compounds. It will oxidize some low-valent transition metal complexes and itself forms the complex, $[Ru(NH_3)_5N_2O]^{2+}$. It is used as an anaesthetic.

Nitrogen Monoxide (Nitric Oxide)

Nitric oxide (NO) is formed in many reactions involving reduction of nitric acid and solutions of nitrates and nitrites. For example, with 8 M nitric acid, we have:

$$8\,HNO_3 + 3\,Cu \longrightarrow 3\,Cu(NO_3)_2 + 4\,H_2O + 2\,NO \qquad (16\text{-}5.2)$$

Reasonably pure NO is obtained by the aqueous reactions:

$$2 \text{ NaNO}_2 + 2 \text{ NaI} + 4 \text{ H}_2\text{SO}_4 \longrightarrow$$
$$\text{I}_2 + 4 \text{ NaHSO}_4 + 2 \text{ H}_2\text{O} + 2 \text{ NO} \qquad (16\text{-}5.3)$$

$$2 \text{ NaNO}_2 + 2 \text{ FeSO}_4 + 3 \text{ H}_2\text{SO}_4 \longrightarrow \text{Fe}_2(\text{SO}_4)_3 + 2 \text{ NaHSO}_4$$
$$+ 2 \text{ H}_2\text{O} + 2 \text{ NO} \qquad (16\text{-}5.4)$$

or, using molten salts,

$$3 \text{ KNO}_2(\ell) + \text{KNO}_3(\ell) + \text{Cr}_2\text{O}_3(\text{s}) \longrightarrow 2 \text{ K}_2\text{CrO}_4(\text{s}, \ell) + 4 \text{ NO} \qquad (16\text{-}5.5)$$

Nitric oxide reacts rapidly with dioxygen, as in Reaction 16-5.6:

$$2 \text{ NO} + \text{O}_2 \longrightarrow 2 \text{ NO}_2 \qquad (16\text{-}5.6)$$

but the reaction is slow under dilute conditions. Nitric oxide apparently plays a respiratory role in controlling blood pressure.

Nitric oxide is oxidized to nitric acid by strong oxidizing agents; the reaction with permanganate is quantitative and provides a method of analysis. It is reduced to N_2O by SO_2 and to NH_2OH by Cr^{2+}, in acid solution in both cases.

Nitric oxide is thermodynamically unstable and at high pressures it readily decomposes in the range 30-50 °C.

$$3 \text{ NO} \longrightarrow \text{N}_2\text{O} + \text{NO}_2 \qquad (16\text{-}5.7)$$

The NO molecule is paramagnetic with the electron configuration

$$(\sigma_1)^2(\sigma_2)^2(\sigma_3)^2(\pi)^4(\pi^*)^1$$

The electron in the π^* orbital is relatively easily lost to give the *nitrosonium ion* (NO^+), which forms many salts. Because the electron removed comes out of an antibonding orbital, the bond is stronger in NO^+ than in NO; the bond length decreases by 0.09 Å and the vibration frequency rises from 1840 cm^{-1} in NO to 2150-2400 cm^{-1} (depending on environment) in NO^+.

The ion is formed when N_2O_3 or N_2O_4 is dissolved in concentrated sulfuric acid.

$$\text{N}_2\text{O}_3 + 3 \text{ H}_2\text{SO}_4 = 2 \text{ NO}^+ + 3 \text{ HSO}_4^- + \text{H}_3\text{O}^+ \qquad (16\text{-}5.8)$$

$$\text{N}_2\text{O}_4 + 3 \text{ H}_2\text{SO}_4 = \text{NO}^+ + \text{NO}_2^+ + 3 \text{ HSO}_4^- + \text{H}_3\text{O}^+ \qquad (16\text{-}5.9)$$

The compound $\text{NO}^+\text{HSO}_4^-$, nitrosonium hydrogen sulfate, is an important intermediate in the lead-chamber process for manufacture of sulfuric acid.

Not only does the NO^+ ion react with many reducing agents, but it may be part of a reversible electrode reaction in nonaqueous solvents (e.g., CH_3CN), as in Reaction 16-5.10.

$$\text{NO}^+ + \text{e}^- = \text{NO} \qquad (16\text{-}5.10)$$

Nitric oxide forms many complexes with transition metals (Chapter 28) some of which can be considered to arise from NO^+.

Nitrogen Dioxide (NO$_2$) and Dinitrogen Tetroxide (N$_2$O$_4$)

The two oxides, NO$_2$ and N$_2$O$_4$, exist in a strongly temperature-dependent equilibrium

$$2\,NO_2 \;\rightleftharpoons\; N_2O_4 \qquad (16\text{-}5.11)$$

Brown   Colorless
paramagnetic diamagnetic

both in solution and in the gas phase. In the solid state, the oxide is wholly N$_2$O$_4$. In the liquid phase, partial dissociation occurs; it is pale yellow at the freezing point (-11.2 °C) and contains 0.01% of NO$_2$, which increases to 0.1% in the deep red-brown liquid at the boiling point, 21.15 °C. Dissociation is complete in the vapor above 140 °C. Nitrogen dioxide has an unpaired electron. The other "free radical" molecules, NO and ClO$_2$ (Section 20-4), have little tendency to dimerize, and the difference may be that in NO$_2$ the electron is localized mainly on the N atom. The dimer has three isomeric forms of which the most stable and normal form has the planar structure O$_2$N—NO$_2$. The N—N bond is rather long (1.75 Å), as would be expected from its weakness. The dissociation energy of N$_2$O$_4$ is only 57 kJ mol^{-1}.

Mixtures of the two oxides are obtained by heating metal nitrates, by oxidation of NO, and by reduction of nitric acid and nitrates by metals and other reducing agents. The gases are highly toxic and attack metals rapidly. They react with water as in Reaction 16-5.12.

$$2\,NO_2 + H_2O = HNO_3 + HNO_2 \qquad (16\text{-}5.12)$$

The nitrous acid decomposes, particularly when warmed:

$$3\,HNO_2 = HNO_3 + 2\,NO + H_2O \qquad (16\text{-}5.13)$$

Thermal decomposition of NO$_2$ takes place above 150°C according to Reaction 16-5.14:

$$2\,NO_2 \rightleftharpoons 2\,NO + O_2 \qquad (16\text{-}5.14)$$

The oxides are fairly strong oxidizing agents in aqueous solution, comparable in strength to bromine.

$$N_2O_4(g) + 2\,H^+(aq) + 2\,e^- = 2\,HNO_2(aq) \qquad E° = +1.07\ V \ (16\text{-}5.15)$$

An equilibrium mixture of the oxides, *nitrous fumes*, is used in organic chemistry as a selective oxidizing agent, the oxidation proceeding through an initial hydrogen abstraction to give HONO according to Reaction 16-5.16.

$$RH + NO_2 = R\cdot + HONO \qquad (16\text{-}5.16)$$

Liquid N$_2$O$_4$ can be used as a solvent and has been utilized to make anhydrous nitrates and nitrate complexes. Thus Cu dissolves in N$_2$O$_4$ to give Cu(NO$_3$)$_2$·N$_2$O$_4$, which loses N$_2$O$_4$ on heating to give Cu(NO$_3$)$_2$.

In solvents such as anhydrous HNO_3, N_2O_4 dissociates ionically as in Reaction 16-5.17.

$$N_2O_4 = NO^+ + NO_3^- \tag{16-5.17}$$

Dinitrogen Trioxide

Dinitrogen trioxide, N_2O_3, formally the anhydride of nitrous acid, is obtained by interaction of stoichiometric amounts of NO and O_2, or of NO and N_2O_4. It is an intensely blue liquid or a pale blue solid. The stable form has a weak N—N bond. It exists only at low temperature, and readily dissociates to give NO and NO_2 as in Reaction 16-5.18.

$$N_2O_3 = NO + NO_2 \tag{16-5.18}$$

The N_2O_3 molecule has an O_2N—NO structure in the gas phase and at low temperature, with an extremely long (1.89 Å) N—N bond (Structure 16-V) consistent with its easy dissociation.

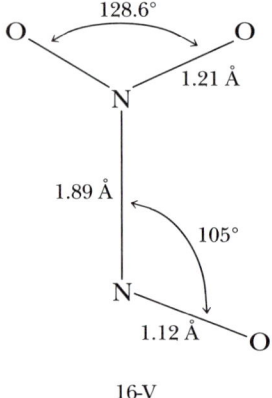

16-V

Dinitrogen Pentoxide

This oxide (N_2O_5) forms unstable colorless crystals. It is made by Reaction 16-5.19.

$$2\ HNO_3 + P_2O_5 = 2\ HPO_3 + N_2O_5 \tag{16-5.19}$$

Dinitrogen pentoxide is the anhydride of nitric acid. In the solid state it exists as the nitronium nitrate, $NO_2^+NO_3^-$.

16-6 The Nitronium Ion

Just as NO readily loses its odd electron, so does NO_2. The *nitronium* ion (NO_2^+) is involved in the dissociation of HNO_3, in solutions of nitrogen oxides in acids, and in nitration reactions of aromatic compounds. Indeed, it was studies on nitration reactions that lead to recognition of the importance of NO_2^+ as the attacking species.

The nitronium ion is formed in ionizing solvents such as H_2SO_4, CH_3NO_2, or CH_3CO_2H, by ionizations such as

$$2\ HNO_3 = NO_2^+ + NO_3^- + H_2O \qquad (16\text{-}6.1)$$

$$HNO_3 + H_2SO_4 = NO_2^+ + HSO_4^- + H_2O \qquad (16\text{-}6.2)$$

The actual nitration process can then be formulated

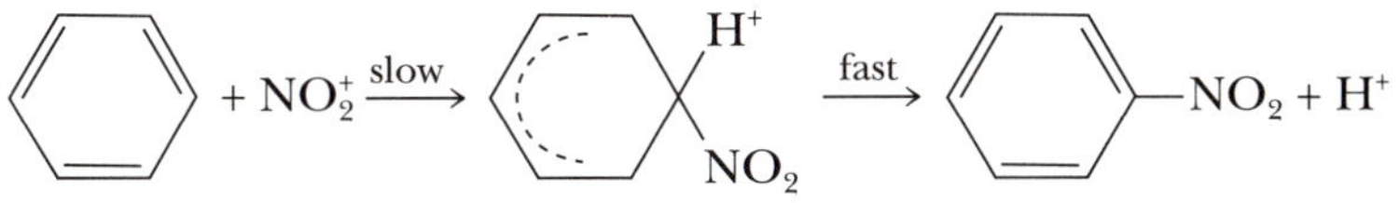

$$(16\text{-}6.3)$$

Nitronium salts can be readily isolated. These salts are thermally stable but rapidly hydrolyzed. Typical preparations are

$$N_2O_5 + HClO_4 = NO_2^+ClO_4^- + HNO_3 \qquad (16\text{-}6.4)$$

$$HNO_3 + 2\ SO_3 = NO_2^+HS_2O_7^- \qquad (16\text{-}6.5)$$

16-7   Nitrous Acid

Solutions of the weak acid HONO ($pK_a = 3.3$) are made by acidifying cold solutions of nitrites. The aqueous solution can be obtained free of salts by the reaction

$$Ba(NO_2)_2 + H_2SO_4 \longrightarrow 2\ HNO_2 + BaSO_4(s) \qquad (16\text{-}7.1)$$

The pure liquid acid is unknown, but it can be obtained in the vapor phase. Even aqueous solutions of nitrous acid are unstable and decompose rapidly when heated.

$$3\ HNO_2 \rightleftharpoons H_3O^+ + NO_3^- + 2\ NO \qquad (16\text{-}7.2)$$

Nitrites of the alkali metals are prepared by heating the nitrates with a reducing agent, such as carbon, lead, or iron. They are very soluble in water. Nitrites are very toxic but have been used for preservation of ham and other meat products; there is evidence that they can react with proteins to give carcinogenic nitrosamines.

The main use of nitrites is to generate nitrous acid for the synthesis of organic diazonium compounds from primary aromatic amines. Organic derivatives of the NO_2 group are of two types: nitrites (R—ONO) and nitro compounds (R—NO_2). Similar isomerism occurs in metal complexes where the NO_2^- ligand can be coordinated to a metal either through the nitrogen atom (i.e., the nitro ligand) or through the oxygen atom (i.e., the nitrito ligand), as has already been discussed in Chapter 6.

16-8   Nitrogen Halides

Of the binary halides we have NF_3, NF_2Cl, $NFCl_2$, and NCl_3. There are also N_2F_2, N_2F_4, and the halogen azides XN_3 (X = F, Cl, Br, I). With the exception of NF_3, the halides are reactive and some of them are explosive, for example, $NFCl_2$. Only the fluorides are important.

Nitrogen trifluoride is made by the electrolysis of NH_4F in anhydrous HF solvent, a procedure that also gives small amounts of N_2F_2. Electrolysis of molten NH_4F is the preferred method for synthesis of N_2F_2. Reaction 16-8.1,

$$NH_3 + F_2 \text{ (diluted by } N_2) \longrightarrow NF_3, N_2F_4, N_2F_2, NHF_2 \qquad (16\text{-}8.1)$$

conducted in a Cu-packed reactor, gives mixtures of fluorides. The predominant product depends on conditions, especially the F_2/NH_3 ratio.

Nitrogen trifluoride (bp -129 °C) is a very stable gas that normally is reactive only at 250–300 °C, although it reacts readily with $AlCl_3$ at 70 °C, as in Reaction 16-8.2.

$$2\, NF_3 + 2\, AlCl_3 \longrightarrow N_2 + 3\, Cl_2 + 2\, AlF_3 \qquad (16\text{-}8.2)$$

It is unreactive towards water and most other reagents at room temperature, and it is thermally stable in the absence of reducing metals. The NF_3 molecule is pyramidal, but unlike ammonia, has a very low dipole moment. Evidently, it is an extremely poor donor molecule, and does not form complexes.

Interaction of NF_3, F_2, and a strong Lewis acid, such as BF_3, AsF_5, or SbF_5, gives salts of the ion NF_4^+. Such reactions are performed at low temperature, under high pressures, with UV light, as in Reaction 16-8.3:

$$NF_3 + F_2 + BF_3 \longrightarrow NF_4^+BF_4^- \qquad (16\text{-}8.3)$$

Compounds of NF_4^+ are ionic, and other salts may be prepared similarly, namely, those of AsF_6^- and SnF_6^{2-}. The perchlorate may be prepared by low-temperature (-78 °C) metathesis in liquid HF, as in Reaction 16-8.4:

$$NF_4^+SbF_6^- + CsClO_4 \longrightarrow CsSbF_6(s) + NF_4^+ClO_4^- \qquad (16\text{-}8.4)$$

NF_4^+ is one of the strongest oxidizers known.

The oxohalides (or the nitrosyl halides), XNO, where X = F, Cl, or Br, are obtained by reaction of the halogens with NO as in Reaction 16-8.5.

$$2\, NO + X_2 \longrightarrow 2\, XNO \qquad (16\text{-}8.5)$$

All three of the nitrosyl halides are powerful oxidants, able to attack many metals. All decompose on treatment with water producing HNO_3, HNO_2, NO, and HX.

16-9   Descriptive Summary of Reactions

The chemistry of nitrogen is well organized by noting the oxidation state of nitrogen among reactants and products. As a partial summary, this is illustrated for the oxides of nitrogen in Figs. 16-2 and 16-3. The corresponding balanced chemical equations are given in Tables 16-1 and 16-2. The student is encouraged to prepare similar diagrams for the hydrides and fluorides.

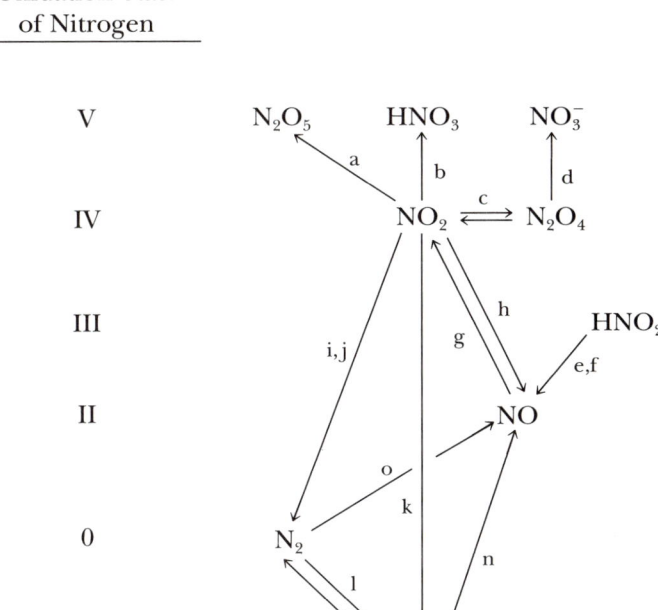

Figure 16-2 Reactions that do not involve disproportionation of the oxides of nitrogen. The oxidation state of nitrogen is indicated on the scale at left.

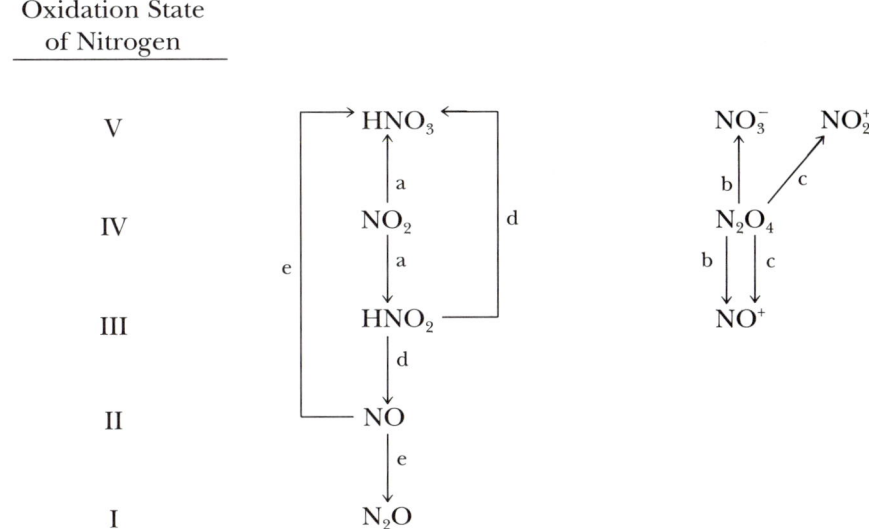

Figure 16-3 Reactions that do involve disproportionation of the oxides of nitrogen. The oxidation state of nitrogen is indicated on the scale at left.

Table 16-1   Chemical Equations for the Reactions of Fig. 16-2

(a)	$2 NO_2 + O_3 \longrightarrow N_2O_5 + O_2$
(b)	$2 NO_2 + H_2O_2 \longrightarrow 2 HNO_3$
(c)	$2 NO_2 = N_2O_4$
(d)	$N_2O_4 + xsCu \longrightarrow Cu(NO_3)_2(s)$
(e)	$2 HNO_2 + 2 HI \longrightarrow I_2 + 2 NO + 2 H_2O$
(f)	$Fe^{2+} + HNO_2 + H^+ \longrightarrow Fe^{3+} + NO + H_2O$
(g)	$2 NO + O_2 \longrightarrow 2 NO_2$
(h)	$2 Cu + NO_2 \longrightarrow Cu_2O + NO$
(i)	$C + NO_2 \longrightarrow CO_2 + \frac{1}{2}N_2$
(j)	$NO_2 + 2 H_2 \longrightarrow \frac{1}{2}N_2 + 2 H_2O$
(k)	$2 NO_2 + 7 H_2 \longrightarrow 2 NH_3 + 4 H_2O$
(l)	$N_2 + 3 H_2 \longrightarrow 2 NH_3$
(m)	$4 NH_3 + 3 O_2 \longrightarrow 2 N_2 + 6 H_2O$
(n)	$4 NH_3 + 5 O_2 \longrightarrow 4 NO + 6 H_2O$
(o)	$N_2 + O_2 \longrightarrow 2 NO$

Table 16-2   Chemical Equations for the Reactions of Fig. 16-3

(a)	$2 NO_2 + H_2O \longrightarrow HNO_3 + HNO_2$
(b)	$N_2O_4 \longrightarrow NO^+ + NO_3^-$
(c)	$N_2O_4 + 3 H_2SO_4 \longrightarrow NO^+ + NO_2^+ + 3 HSO_4^- + H_3O^+$
(d)	$3 HNO_2 \longrightarrow HNO_3 + 2 NO + H_2O$
(e)	$3 NO \longrightarrow N_2O + NO_2$

STUDY GUIDE

Study Questions

A. Review

1. Give the electronic structure of the nitrogen atom and list the ways by which the octet can be completed in forming compounds of nitrogen. Give examples.

2. Draw the Lewis diagrams and explain the geometry and hybridization at each atom in NO_2^-, NO_3^-, NO_2^+, NO^+, NO, N_2, N_3^-, FNO, and N_2O.

3. Write balanced equations for the synthesis of nitric acid from NH_3 and O_2.

4. Write equations for the action of heat on (a) $NaNO_3$, (b) NH_4NO_3, (c) $Cu(NO_3)_2 \cdot nH_2O$, (d) N_2O, and (e) N_2O_3.

5. How is hydrazine prepared?

6. Write balanced equations for three different preparations of nitric oxide.

7. How is the nitronium ion prepared? Explain its significance in the nitration of aromatic hydrocarbons.

8. In acid solution we have

$$HNO_2 + H^+ + e^- \longrightarrow NO + H_2O \qquad E^\circ = 1.0 \text{ V}$$

Write balanced equations for the reactions of nitrous acid with (a) I^-, (b) Fe^{2+}, (c) $C_2O_4^{2-}$.

9. How can NO_2^- and NO_3^- be bonded to transition metal complexes?

10. Write balanced equations for the hydrolysis of (a) calcium nitride, (b) lithium nitride, (c) dinitrogen pentoxide, and (d) dinitrogen trioxide.

11. Draw Lewis diagrams for the radicals NO and NO_2. Explain the formation of the N—N bond in N_2O_3 and in N_2O_4.

B. Additional Exercises

1. Use MO theory to compare the electronic structures of CO, N_2, CN^-, and NO^+. Why does nitrogen form complexes with metals less readily than CO?

2. Why does nitrogen form only a diatomic molecule unlike phosphorus and other elements of Group VB(15)?

3. Nitrogen trichloride is an extremely dangerous explosive oil, but NF_3 is a stable gas that reacts only above 250 °C. Explain this difference.

4. Three isomers of N_2O_4 are known. Draw likely structures for them.

5. Determine the oxidation numbers of the atoms in the molecules and ions found in the following: Reactions 16-6.1, 16-5.17, 16-5.14, 16-5.7, 16-5.1, 16-4.13, 16-4.4, and 16-2.4. Which of these are redox reactions?

6. With drawings, show how hybrid orbitals overlap in the formation of the σ-bond framework in each of the following molecules and ions: (a) N_2, (b) N_3^-, (c) NO_2^-, and (d) ClNO.

7. With drawings, show the hybrid orbitals that house lone pairs of electrons in the molecules and ions of Problem 6, in Part B. Be careful to show the geometry correctly, including the likely position of the lone electrons in the molecules and ions, and specify the type of hybrid that is used in each case.

8. With drawings, show the formation of the π-bond system in the molecules and ions of Problem 6, in part B.

9. Complete and balance the following equations:

(a) $Li + N_2$ (b) $Cu + NO_2$

(c) $C + NO_2$ (d) $H_2O_2 + NO_2$

(e) $O_3 + NO_2$ (f) $H_2 + NO_2$

(g) $HI + HNO_2$

10. Draw the Lewis diagrams for N_2F_2 and N_2F_4, each of which has a nitrogen-nitrogen linkage. Classify each nitrogen atom according to the AB_xE_y scheme of Chapter 3 and give the hybridization for each nitrogen.

11. Give the AB_xE_y classification (Chapter 3) for each oxide mentioned in Section 16-5.

12. Use the style of Figs. 16-2 and 16-3 to diagram the conversion of NO_2 into

(a) HNO_3

(b) N_2O_4

(c) N_2

(d) NH_3

13. Give diagrams for the stepwise conversion of NO_2 into

(a) HNO_2 and NO

(b) N_2 and NH_3

14. Give the principal products on reaction of each of the following:

(a) $NH_3 + O_2$ (uncatalyzed)

(b) Disproportionation of NO.

(c) Oxidation of copper by NO.

(d) Oxidation of NO_2 by ozone.

(e) Reduction of NO_2 by excess hydrogen.

(f) Disproportionation of HNO_2.

(g) The Haber process for ammonia.

(h) Hydrolysis of N_2O_3.

(i) Hydrolysis of N_2O_5.

(j) Dissolution of N_2O_4 in anhydrous HNO_3.

(k) Dimerization of NO_2.

(l) Oxidation of ammonia by air over a Pt catalyst at 750 °C.

15. Outline the synthesis of HNO_3, starting from the elements.

16. Which oxide is the anhydride of HNO_3? of HNO_2? Explain by using equations.

C. Questions from the Literature of Inorganic Chemistry

1. Hydrolysis of dinitrogen trioxide is described in the paper by G. Y. Markovits, S. E. Schwartz, and L. Newman, *Inorg. Chem.,* **1981,** *20,* 445–450.

(a) Draw the Lewis diagrams, discuss the geometry, and assign an oxidation number to each atom in the substances found in Reactions (1), (2), and (3).

(b) What evidence do the authors cite for an equilibrium in which N_2O_3 is formed from nitrous acid in acidic medium?

(c) How is Eq. (15) obtained?

(d) The authors report a value for $\Delta G_f^\circ[N_2O_3(aq)]$. How was this number calculated?

(e) Of Reactions (1), (2), (3), (9), and (13), which represent disproportionation, hydrolysis, and/or acid-base type reactions?

2. Consider the paper by K. O. Christe, C. J. Schack, and R. D. Wilson in *Inorg. Chem.,* **1977,** *16,* 849-854.

(a) What is the nature of solid SnF_4, and why is it not a good Lewis acid?

(b) What reaction takes place in liquid HF solvent between KF and SnF_4?

(c) Draw the Lewis diagrams and predict the geometries of NF_4^+, BF_4^-, $(SnF_4)_x$, SnF_5^-, and $[SnF_6]^{2-}$.

(d) What reaction takes place in liquid HF solvent between NF_4BF_4 and SnF_4?

(e) Why does NF_4SnF_5 not react with a second equivalent of NF_4BF_4 to form $(NF_4)_2SnF_6$?

3. Some reaction chemistry of NF_4^+ is reported by K. O. Christe, W. W. Wilson, and R. D. Wilson in *Inorg. Chem.,* **1980,** *19,* 1494-1498.

(a) Write balanced equations for the reactions in anhydrous HF solvent between NF_4^+ and (i) ClO_4^-, (ii) BrO_4^-, and (iii) HF_2^-.

(b) Write balanced equations for the reactions in BrF_5 solvent between NF_4^+ and (i) BrF_4^- and (ii) BrF_4O^-.

SUPPLEMENTARY READING

Bottomley, F. "Reactions of Nitrosyls," in *Reactions of Coordinated Ligands,* P. S. Braterman, Ed., Plenum, New York, 1989.

Bottomley, F. and Burns, R. C., *Treatise on Dinitrogen Fixation,* Wiley, New York, 1979.

Chatt, J. C., da C. Pina, L. M., and Richards, R. L., *New Trends in Nitrogen Fixation,* Academic, New York, 1980.

Colburn, C. B., Ed., *Developments in Inorganic Nitrogen Chemistry,* Vols. 1 and 2, Elsevier, Amsterdam, 1966 and 1973.

Dehnicke, K. and Strahl, J., "Nitrido Complexes of the Transition Metals," *Angew. Chem. Int. Ed. Eng.,* **1992,** *31,* 955–978.

Emeleus, H. J., Shreeve, J. M., and Verma, R. D., "The Nitrogen Fluorides and Some Related Compounds," *Adv. Inorg. Chem.,* **1989,** *33,* 139–196.

Griffith, W. P., "Transition-Metal Nitrido Complexes," *Coord. Chem. Rev.,* **1972,** *8,* 369–396.

Jolly, W. L., *The Inorganic Chemistry of Nitrogen,* Benjamin, New York, 1964.

Smith, P. A. S., *The Open-Chain Chemistry of Organic Nitrogen Compounds,* Vols. 1 and 2, Benjamin, New York, 1966.

Wright, A. N. and Winkler, C. A., *Active Nitrogen,* Academic, New York, 1968.

Chapter 17

THE GROUP VB(15) ELEMENTS: PHOSPHORUS, ARSENIC, ANTIMONY, AND BISMUTH

17-1 Introduction

Phosphorus occurs mainly in minerals of the *apatite* family, $Ca_9(PO_4)_6 \cdot CaX_2$; X = F, Cl, or OH, which are the main components of amorphous phosphate rock, millions of tons of which are processed annually. The elements As, Sb, and Bi occur mainly as sulfide minerals, such as *mispickel* (FeAsS) or *stibnite* (Sb_2S_3).

Some properties of the elements are given in Table 8-5, and some general features and trends are noted in Chapter 8.

The valence shells of the atoms (ns^2np^3) are similar to the electron configuration of N, but beyond the similarity in stoichiometries of compounds such as NH_3 and PH_3, there is little resemblance in the chemistry between even P and N. Phosphorus is a true nonmetal in its chemistry but As, Sb, and Bi show an increasing trend to metallic character and cationic behavior.

The principal factors responsible for the differences between nitrogen and phosphorus group chemistry are those responsible for the C to Si differences, namely, (a) the diminished ability of the second-row element to form $p\pi$-$p\pi$ multiple bonds, and (b) the possibility of utilizing the lower lying $3d$ orbitals.

The first explains features such as the fact that nitrogen forms esters O=NOR, whereas phosphorus gives $P(OR)_3$. Nitrogen oxides and oxoacids all involve multiple bonds (Section 16-1), whereas the phosphorus oxides have single P—O bonds, as in P_4O_6, and phosphoric acid is $PO(OH)_3$ in contrast to $NO_2(OH)$.

The utilization of d orbitals has three effects. First, it allows some $p\pi$-$d\pi$ bonding as in $R_3P{=}O$ or $R_3P{=}CH_2$. Thus amine oxides, R_3NO, have only a single canonical structure ($R_3N^+{-}O^-$) and are chemically reactive, while P—O bonds are shorter than expected for the sum of single–bond radii, indicating multiple bonding, and are very strong, about 500 kJ mol^{-1}. Second, there is the possibility of expansion of the valence shell, whereas nitrogen has a covalency maximum of four. Thus we have compounds such as PF_5, $P(C_6H_5)_5$, $P(OCH_3)_6^-$, and PF_6^-.

Notice that for many of the five-coordinate species, especially of phosphorus, the energy difference between the trigonal bipyramidal and square pyramidal configurations is small, and such species are usually stereochemically nonrigid (Section 6-6).

When higher coordination numbers occur for the elements in the III oxidation state, as in $[SbF_5]^{2-}$, the structures take the form of a square pyramid. As discussed in Chapter 3, AB_5E systems such as these accommodate one lone pair (E), in addition to the five peripheral atoms (B), at the central atom (A).

Finally, while trivalent nitrogen and the other elements in compounds such as $N(C_2H_5)_3$, $P(C_2H_5)_3$, and $As(C_6H_5)_3$ have lone pairs and act as donors, there is a profound difference in their donor ability toward transition metals. This follows from the fact that although NR_3 has no low-lying acceptor orbitals, the others do have such orbitals, namely, the empty d orbitals. These can accept electron density from filled metal d orbitals to form $d\pi$-$d\pi$ bonds, as we shall discuss in detail later (Section 28-15).

17-2   The Elements

Phosphorus is obtained by reduction of phosphate rock with coke and sand in an electric furnace. Phosphorus distills and is condensed under water as P_4. Phosphorus allotropes have been discussed (Section 8-4).

$$2\ Ca_3(PO_4)_2 + 6\ SiO_2 + 10\ C = P_4 + 6\ CaSiO_3 + 10\ CO \qquad (17\text{-}2.1)$$

P_4 is stored under water to protect it from air in which it will inflame. Red and black P are stable in air but will burn on heating. P_4 is soluble in CS_2, benzene, and similar organic solvents; it is very poisonous.

The elements *As, Sb,* and *Bi* are obtained as metals (Section 8-5) by reduction of their oxides with carbon or hydrogen. The metals burn on heating in oxygen to give the oxides.

All the elements react readily with halogens but are unaffected by nonoxidizing acids. Nitric acid gives, respectively, phosphoric acid, arsenic acid, antimony trioxide, and bismuth nitrate, which nicely illustrates the increasing metallic character as the group is descended.

Interaction with various metals and nonmetals gives phosphides, arsenides, and the like, which may be ionic, covalent polymers or metal-like solids. Gallium arsenide (GaAs)—one of the so-called III–V compounds of a Group IIIB(13) and a Group VB(15) element—has semiconductor properties similar to those of Si and Ge.

There are a number of ligands that consist exclusively of Group VB(15) atoms. The P_3 ring forms an η^3 attachment to metals that are also stabilized by tripod ligands (Chapter 6), as in $LCoP_3$, where L = a tripod ligand. The P_4 molecule can serve as an η^1 or an η^2 ligand, for example, in $LNi(\eta^1\text{-}P_4)$ and *trans*-$[RhCl(PPh_3)_2(\eta^2\text{-}P_4)]$. The P_2 and As_2 molecules can bind to metals in a variety of side-on and bridging attachments that resemble those of acetylene (Chapter 29).

17-3   Hydrides (EH₃)

The stability of these EH_3 gases decreases in the series NH_3, PH_3, AsH_3, SbH_3, and BiH_3. The last two in the series are very unstable thermally. The average bond energies are N—H, 391; P—H, 322; As—H, 247; and Sb—H, 255 kJ mol^{-1}.

Phosphine (PH_3) is made by the action of acids on zinc phosphide. Pure PH_3 is not spontaneously flammable, but it often inflames owing to traces of P_2H_4 or P_4 vapor. It is exceedingly poisonous. Because of its poor ability to enter into hydrogen bonding, it is not associated in the liquid state, in contrast to the behavior of ammonia. Phosphine is sparingly soluble in water, and it is a very weak base. The proton affinities of PH_3 and NH_3 differ considerably, as indicated by the relative values of $\Delta H°$ for Reactions 17-3.1 and 17-3.2.

$$PH_3(g) + H^+(g) = PH_4^+(g) \quad \Delta H° = -770 \text{ kJ mol}^{-1} \quad (17\text{-}3.1)$$

$$NH_3(g) + H^+(g) = NH_4^+(g) \quad \Delta H° = -866 \text{ kJ mol}^{-1} \quad (17\text{-}3.2)$$

Although PH_3 is the weaker base, it does react with gaseous HI to give PH_4I as unstable colorless crystals. Phosphonium iodide (PH_4I) is completely hydrolyzed by water, as in Reaction 17-3.3.

$$PH_4I(s) + H_2O = H_3O^+ + I^- + PH_3(g) \quad (17\text{-}3.3)$$

It is the low basicity of PH_3 that forces the equilibrium in Reaction 17-3.3 to lie far to the right. Phosphine is used industrially to make organophosphorus compounds (Chapter 29).

17-4   Halides (EX₃, EX₅) and Oxohalides

The trihalides, except PF_3, are obtained by direct halogenation, keeping the element in excess. An excess of the halogen gives EX_5. The trihalides are rapidly hydrolyzed by water and are rather volatile; the gaseous molecules have pyramidal structures. The chlorides and bromides, as well as PF_3 and PI_3, have molecular lattices. The compounds AsI_3, SbI_3, and BiI_3 have layer structures based on hexagonal close packing of iodine atoms with the Group VB(15) atoms in octahedral holes. Bismuth trifluoride (BiF_3) is known in two forms, in both of which Bi has the coordination number eight, while SbF_3 has an intermediate structure in which SbF_3 molecules are linked through F bridges to give each Sb^{III} a very distorted octahedral environment.

Phosphorus trifluoride is a colorless, toxic gas, made by fluorination of PCl_3. It forms complexes with transition metals similar to those formed by CO (Section 28-15). Unlike the other trihalides, PF_3 is hydrolyzed only slowly by H_2O, but it is attacked rapidly by alkalis. It has no Lewis acid properties.

Phosphorus trichloride is a low-boiling liquid that is hydrolyzed by water to give phosphorous acid. It reacts with oxygen to give $OPCl_3$. Figure 17-1 illustrates some of the important reactions of PCl_3. Many of these reactions are typical of other EX_3 compounds and also, with obvious changes in formulas, of $OPCl_3$ and other oxo halides.

Arsenic trihalides are similar to those of phosphorus. Antimony trichloride ($SbCl_3$) differs in that it dissolves in a limited amount of water to give a clear solution that, on dilution, gives insoluble oxo chlorides such as SbOCl and $Sb_4O_5Cl_2$. No simple Sb^{3+} ions exist in the solutions. Bismuth trichloride ($BiCl_3$), a white, crystalline solid, is hydrolyzed by H_2O to BiOCl, but this reaction is reversible.

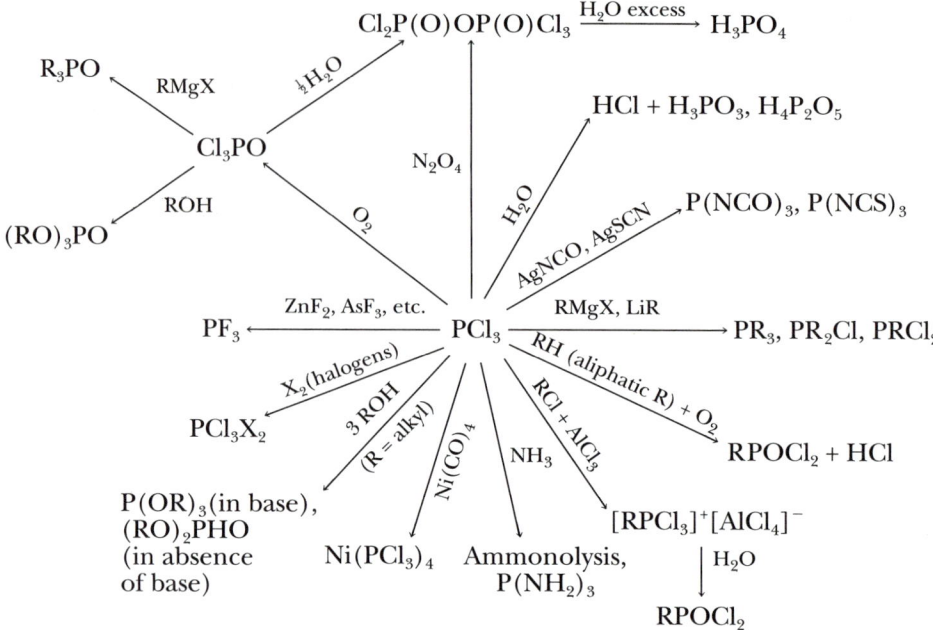

Figure 17-1   Some important reactions of PCl_3. Many of these are typical of other EX_3 and OEX_3 compounds.

$$BiCl_3 + H_2O \rightleftharpoons BiOCl + 2\ HCl \qquad (17\text{-}4.1)$$

Phosphorus pentafluoride (PF_5) is prepared by the interaction of PCl_5 with CaF_2 at 300–400 °C. It is a very strong Lewis acid and forms complexes with amines, ethers, and other bases, as well as with F^-, in which phosphorus becomes six co-ordinate. However, these organic complexes are less stable than those of BF_3 and are rapidly decomposed by water and alcohols. Like BF_3, PF_5 is a good catalyst, especially for ionic polymerization. Arsenic pentafluoride (AsF_5) is similar.

Antimony pentafluoride (SbF_5) is a viscous liquid (bp 150 °C). Its association is due to polymerization through fluorine bridging. The crystal has cyclic tetramers. Its main use is in "superacids" (Section 7-13).

The compounds AsF_5, SbF_5, and PF_5 are potent fluoride ion acceptors, forming MF_6^- ions. The PF_6^- ion is a common and convenient *noncomplexing* anion.

Phosphorus(V) chloride has a trigonal bipyramidal structure in the gas, melt, and solution in nonpolar solvents, but the solid is $[PCl_4]^+[PCl_6]^-$, and it is ionized in polar solvents like CH_3NO_2. The tetrahedral PCl_4^+ ion can be considered to arise here by transfer of Cl^- to the Cl^- acceptor, PCl_5. Therefore, it is not surprising that many salts of the PCl_4^+ ion are obtained when PCl_5 reacts with other Cl^- acceptors, namely,

$$PCl_5 + TiCl_4 \longrightarrow [PCl_4^+]_2[Ti_2Cl_{10}]^{2-} \quad \text{and} \quad [PCl_4]^+[Ti_2Cl_9]^- \quad (17\text{-}4.2)$$

$$PCl_5 + NbCl_5 \longrightarrow [PCl_4]^+[NbCl_6]^- \qquad (17\text{-}4.3)$$

Solid *phosphorus pentabromide* is also ionic, but differs, being $PBr_4^+Br^-$. Antimony forms *antimony pentachloride,* a fuming liquid which is colorless when pure, but usually yellow. While it is a powerful chlorinating agent, it is also useful for removing chloride, as in Reaction 17-4.4.

$$CuCl_2 + 2\ PhCN + 2\ SbCl_5 \longrightarrow Cu(NCPh)_2^+ + 2\ SbCl_6^- \qquad (17\text{-}4.4)$$

Arsenic does not form a pentabromide, and the pentachloride decomposes above −50 °C. The cations AsX_4^+ (X = F, Cl, Br, and I) are all known.

Phosphoryl halides are X_3PO, in which X may be F, Cl, or Br. The most important one is Cl_3PO, which is obtainable by the reactions

$$2\ PCl_3 + O_2 \longrightarrow 2\ Cl_3PO \qquad (17\text{-}4.5)$$

$$P_4O_{10} + 6\ PCl_5 \longrightarrow 10\ Cl_3PO \qquad (17\text{-}4.6)$$

The reactions of Cl_3PO are much like those of PCl_3 (Fig. 17-1). Hydrolysis by water yields phosphoric acid. Cl_3PO also has donor properties and many complexes are known, in which oxygen is the ligating atom.

The oxohalides SbOCl and BiOCl are precipitated when solutions of Sb^{III} and Bi^{III} in concentrated HCl are diluted.

17-5 Oxides

The oxides of the Group VB(15) elements clearly exemplify two important trends that are manifest to some extent in all groups of the periodic table: (1) the stability of the higher oxidation state decreases with increasing atomic number, and (2) in a given oxidation state the metallic character of the elements, and, therefore, the basicity of the oxides, increase with increasing atomic number. Thus, P^{III} and As^{III} oxides are acidic, Sb^{III} oxide is amphoteric, and Bi^{III} oxide is strictly basic.

Phosphorus pentoxide is so termed for historical reasons but its correct molecular formula is P_4O_{10} [Fig. 17-2(a)]. It is made by burning phosphorus in excess oxygen. It has at least three solid forms. Two are polymeric but one is a white, crystalline material that sublimes at 360 °C and 1 atm. Sublimation is an excellent method of purification, since the products of incipient hydrolysis, which are the commonest impurities, are comparatively nonvolatile. This form and the vapor consist of molecules in which the P atoms are at the corners of a tetrahedron with six oxygen atoms along the edges. The remaining four O atoms lie along extended threefold axes of the tetrahedron. The P—O—P bonds are single but the length of the four apical P—O bonds indicates $p\pi\text{-}d\pi$ bonding, that is, P=O.

The compound P_4O_{10} is one of the most effective drying agents known at temperatures below 100 °C. It reacts with water to form a mixture of phosphoric acids whose composition depends on the quantity of water and other conditions. It will even extract the elements of water from many other substances which are themselves considered to be good dehydrating agents; for example, it converts pure HNO_3 into N_2O_5 and H_2SO_4 into SO_3. It also dehydrates many organic compounds, for example, converting amides into nitriles.

The *trioxide* is also polymorphous: one form contains discrete molecules (P_4O_6). The structure [Fig. 17-2(b)] is similar to that of P_4O_{10} except that the four nonbridging apical oxygen atoms in the latter are missing. P_4O_6 is a colorless, volatile compound that is formed in about 50% yield when P_4 is burned in

Figure 17-2   The structure of (a) P_4O_{10} and (b) P_4O_6.

a deficit of oxygen. The compounds As_4O_6 and Sb_4O_6 are similar to P_4O_6 both structurally and in their acidic nature. The compound Bi_2O_3 and the hydroxide, $Bi(OH)_3$, precipitated from bismuth (III) solution have no acidic properties.

17-6   Sulfides

Phosphorus and sulfur combine directly above 100 °C to give several sulfides, the most important being P_4S_3, P_4S_5, P_4S_7, and P_4S_{10}. Each compound is obtained by heating stoichiometric quantities of red P and sulfur. The compound P_4S_3 is used in matches. It is soluble in organic solvents such as carbon disulfide and benzene. The compound P_4S_{10} has the same structure as P_4O_{10}. The others also have structures based on a tetrahedral group of phosphorus atoms with P—S—P bridges or apical P=S groups. P_4S_{10} reacts with alcohols:

$$P_4S_{10} + 8\ ROH \longrightarrow 4(RO)_2P(S)SH + 2\ H_2S \qquad (17\text{-}6.1)$$

to give dialkyl and diaryl dithiophosphates that form the basis of many extreme-pressure lubricants, of oil additives, and of flotation agents.

Arsenic forms As_4S_3, As_4S_4, As_2S_3, and As_2S_5 by direct interaction. The last two can also be precipitated from hydrochloric acid solutions of As^{III} and As^V by hydrogen sulfide. As_2S_3 is insoluble in water and acids but is acidic, dissolving in alkali sulfide solutions to give thio anions. As_2S_5 behaves similarly. As_4S_4, which occurs as the mineral *realgar,* has a structure with an As_4 tetrahedron.

Antimony forms Sb_2S_3 either by direct interaction or by precipitation with H_2S from Sb^{III} solutions; it dissolves in an excess of sulfide to give anionic thio complexes, probably mainly SbS_3^{3-}. Antimony trisulfide (Sb_2S_3), as well as Bi_2S_3, possess a ribbonlike polymeric structure in which each Sb atom and each S atom is bound to three atoms of the opposite kind, forming interlocking SbS_3 and SSb_3 pyramids.

Bismuth gives dark brown Bi_2S_3 on treatment of Bi^{III} solutions with H_2S; it is not acidic.

Some of the corresponding selenides and tellurides of As, Sb, and Bi have been studied intensively as semiconductors. (See Section 32-3.)

17-7 The Oxo Acids

The nature and properties of the oxoanions of the Group VB(15) elements have been discussed in Chapter 5. Here we discuss only the important acids and some of their derivatives.

Phosphorous acid is obtained when PCl_3 or P_4O_6 are hydrolyzed by water. It is a deliquescent colorless solid (mp 70 °C, $pK = 1.26$). The acid and its mono- and diesters differ from PCl_3 in that there are *four* bonds to P, one being P—H. The presence of hydrogen bound to P can be demonstrated by NMR or other spectroscopic techniques. Phosphorous acid is, hence, best written $HP(O)(OH)_2$ as in Structure 17-I. Hypophosphorous acid, H_3PO_2, has two P—H bonds (Structure 17-II). By contrast the triesters have only three bonds to phosphorus, thus being analogous to PCl_3. The trialkyl and aryl phosphites, $P(OR)_3$, have excellent donor properties toward transition metals and many complexes are known.

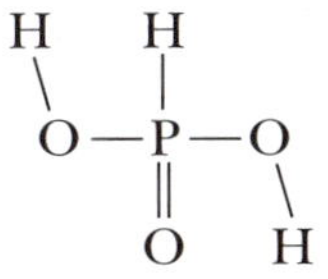

17-I 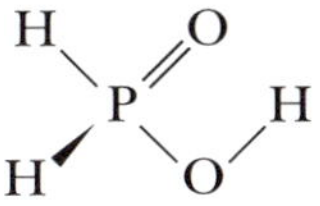   17-II

Phosphorous acid may be oxidized by chlorine or other agents to phosphoric acid, but the reactions are slow and complex. However, the triesters are quite readily oxidized and must be protected from air.

$$2(RO)_3P + O_2 = 2(RO)_3PO \qquad (17\text{-}7.1)$$

These compounds also undergo the Michaelis-Arbusov reaction with alkyl halides, forming dialkyl phosphonates:

$$P(OR)_3 + R'X \longrightarrow [(RO)_3PR']X \longrightarrow RO\underset{\underset{OR}{|}}{\overset{\overset{O}{\|}}{P}}R' + RX \qquad (17\text{-}7.2)$$

Phosphonium intermediate

Trimethylphosphite easily undergoes spontaneous isomerization to the dimethyl ester of methylphosphonic acid.

$$P(OCH_3)_3 \longrightarrow CH_3PO(OCH_3)_2 \qquad (17\text{-}7.3)$$

Orthophosphoric acid, H_3PO_4, commonly called phosphoric acid, is one of the oldest known and most important phosphorus compounds. It is made in vast quantities, usually as 85% syrupy acid, by the direct reaction of ground phosphate rock with sulfuric acid and also by the direct burning of phosphorus and subsequent hydration of P_4O_{10}. The pure acid is a colorless crystalline solid (mp 42.35 °C). It is very stable and has essentially no oxidizing properties below 350–400 °C. At elevated temperatures it is fairly reactive toward metals, which reduce it, and it will attack quartz. *Pyrophosphoric acid* is also produced:

$$2\,H_3PO_4 \longrightarrow H_2O + H_4P_2O_7 \qquad (17\text{-}7.4)$$

but this conversion is slow at room temperature.

The acid is tribasic: at 25 °C, $pK_1 = 2.15$, $pK_2 = 7.1$, $pK_3 \approx 12.4$. The pure acid and its crystalline hydrates have tetrahedral PO_4 groups connected by hydrogen bonds. Hydrogen bonding persists in the concentrated solutions and is responsible for the syrupy nature. For solutions of concentration less than about 50%, the phosphate anions are hydrogen bonded to the liquid water rather than to other phosphate anions.

Phosphates and the polymerized phosphate anions (for which the free acids are unknown) are discussed in Section 5-4. Large numbers of *phosphate esters* can be made by the reaction

$$OPCl_3 + 3\ ROH = OP(OR)_3 + 3\ HCl \qquad (17\text{-}7.5)$$

or by oxidation of trialkylphosphites. Phosphate esters, such as tributylphosphate, are used in the extraction of certain +4 metal ions (see Section 26-2) from aqueous solutions.

Phosphate esters are also of fundamental importance in living systems. It is because of this that their hydrolysis has been studied. Triesters are attacked by OH^- at P and by H_2O at C, depending on pH.

$$OP(OR)_3 \begin{cases} \xrightarrow{^{18}OH^-} OP(OR)_2(^{18}OH) + RO^- & (17\text{-}7.6) \\ \xrightarrow{H_2^{18}O} OP(OR)_2(OH) + R^{18}OH & (17\text{-}7.7) \end{cases}$$

Diesters, which are strongly acidic, are completely in the anionic form at normal (and physiological) pH values.

$$RO\overset{\displaystyle O}{\underset{\displaystyle OH}{\overset{\|}{\underset{|}{-P-}}}}OR' \rightleftharpoons R'OPO_2OR^- + H^+ \qquad K \approx 10^{-1.5} \qquad (17\text{-}7.8)$$

These diesters are thus relatively resistant to nucleophilic attack by either OH^- or H_2O, which is the reason why enzymic catalysis is indispensible if we wish to achieve useful rates of reaction.

Much remains to be learned concerning the mechanisms of most phosphate ester hydrolyses, especially the many enzymic ones. Two important possibilities are the following:

1. One-step nucleophilic displacement (S_N2) with inversion.

$$H_2O\,(\text{or } OH^-) + \begin{array}{c} O \\ \diagdown \\ \text{P}-OR \\ \diagup \diagdown \\ ^-O OR' \end{array} \longrightarrow \begin{array}{c} O \\ \diagdown \\ HO-\text{P} \\ \diagup \diagdown \\ ^-O OR' \end{array} + HOR \qquad (17\text{-}7.9)$$

2. Release of a short-lived *metaphosphate* group (PO_3^-) which rapidly recovers the four-connected orthophosphate structure.

$$-O\overset{\displaystyle O}{\underset{\displaystyle O}{-P-}}O\overset{\displaystyle O}{\underset{\displaystyle O}{-P-}}O\overset{\displaystyle O}{\overset{\|}{-P-}}OH \longrightarrow$$
$$H^+ O^- H^+$$

$$-O\overset{\displaystyle O\ O}{\underset{\displaystyle O}{POPOH}} + PO_3^- \xrightarrow{H_2O} H_2PO_4^- \qquad (17\text{-}7.10)$$

17-8 Complexes of the Group VB(15) Elements

The main aqueous chemistry of Sb^{III} is in oxalato, tartrato, and similar hydroxy acid complexes.

The $[Sb(C_2O_4)_3]^{3-}$ ion forms isolable salts and has been shown to have the incomplete pentagonal bipyramid structure (Fig. 17-3) with a lone pair at one axial position. The tartrate complexes of antimony(III) have been greatly studied, and have been used medicinally as "tartar emetic" for more than 300 years. The structure of the anion in this salt, $K_2[Sb_2(d\text{-}C_4H_2O_6)_2]\cdot3H_2O$, is shown in Fig. 17-4.

Only for bismuth is there a true cationic chemistry. Aqueous solutions contain well-defined hydrated cations, but there is no evidence for a simple aqua ion $[Bi(H_2O)_n]^{3+}$. In neutral perchlorate solutions the main species is $[Bi_6O_6]^{6+}$ or its hydrated form, $[Bi_6(OH)_{12}]^{6+}$, while $[Bi_6O_6(OH)_3]^{3+}$ is formed at a higher pH. The $[Bi_6(OH)_{12}]^{6+}$ species contains an octahedron of Bi^{3+} ions with an OH^- bridging each edge.

17-9 Phosphorus–Nitrogen Compounds

Many compounds are known with P—N and P≡N bonds. The R_2N—P bonds are particularly stable and occur widely in combination with bonds to other univalent groups, such as P—R, P—Ar, and P—halogen.

Phosphazenes are cyclic or chain compounds that contain alternating phosphorus and nitrogen atoms with two substituents on each phosphorus atom. The three main structural types are the cyclic trimer (Structure 17-III), cyclic tetramer (Structure 17-IV), and the oligomer or high polymer (Structure 17-V). The alternating sets of single and double bonds in Structures 17-III to 17-V are written for convenience but, in general, all P—N distances are found to be equal. It appears that they are of the order of about 1.5, since their lengths (1.56–1.61 Å) are appreciably shorter than expected ($\approx$1.80 Å) for P—N single bonds. Hexachlorocyclotriphosphazene, $(NPCl_2)_3$, is a key intermediate in the

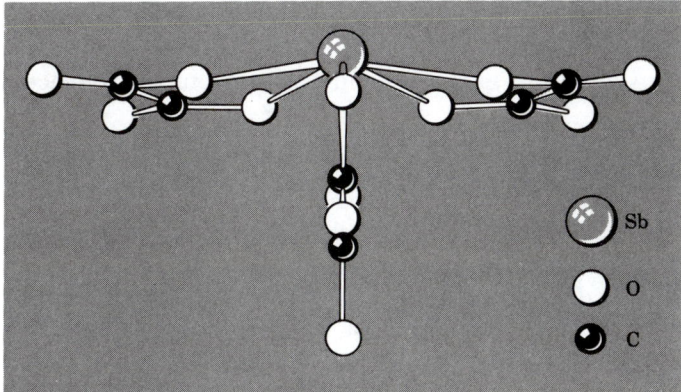

Figure 17-3 The $[Sb(C_2O_4)_3]^{3-}$ ion. Two oxalato, $C_2O_4^{2-}$, ligands are bidentate and one is monodentate. The oxygen donor atoms form a pentagonal base to the pyramid that is capped by Sb^{3+}.

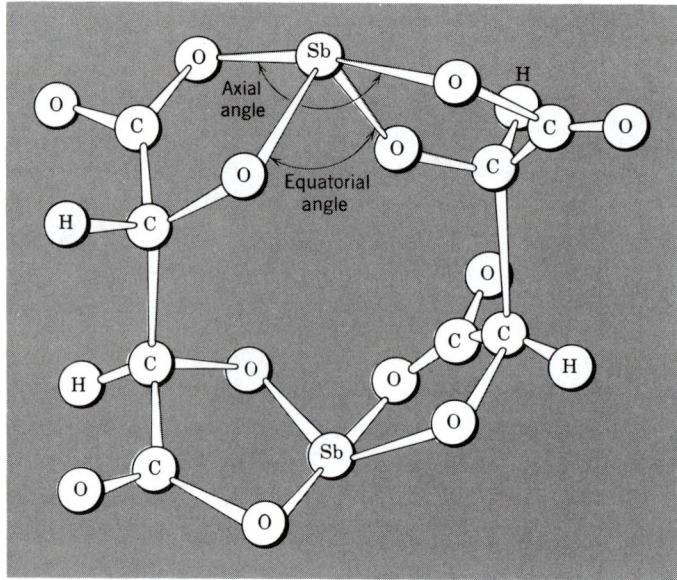

Figure 17-4   Geometry of the anion $[Sb_2(C_4H_2O_6)_2]^{2-}$. Water molecules link the anions into sheets by hydrogen bonding to carboxylate carbon atoms. [Reproduced by permission from Tapscott, R. E., Belford, R. L., and Paul, I. C., *Coord. Chem. Rev.*, **1969**, *4*, 323.]

| 17-III | 17-IV | 17-V |

synthesis of many other phosphazenes and is manufactured by Reaction 17-9.1:

$$n\,PCl_5 + n\,NH_4Cl \xrightarrow{\text{in } C_2H_2Cl_4 \text{ or } C_6H_5Cl} (NPCl_2)_n + 4n\,HCl \quad (17\text{-}9.1)$$

Reaction 17-9.1 produces a mixture of cyclic $(NPCl_2)_n$ compounds with $n = 3, 4, 5, \ldots$, as well as some low-molecular weight linear polymers. Control of the reaction conditions can give 90% yields of either the compound with $n = 3$ or 4, which can be purified by extraction, recrystallization, or sublimation.

Structures are given in Fig. 17-5 of the cyclic trimer $[NPCl_2]_3$ and the tetramer $[NPClPh]_4$. Most six-membered rings such as $[NPX_2]_3$ are planar, while the larger rings are nonplanar. The fluoroderivatives, $[NPF_2]_n$ are planar, or nearly so, when $n = 3-6$.

The majority of the reactions of phosphazenes involve replacement of the substituents at phosphorus by nucleophiles (e.g., OH, OR, NR_2, or R) to give substituted derivatives, as in Reactions 17-9.2 to 17-9.4.

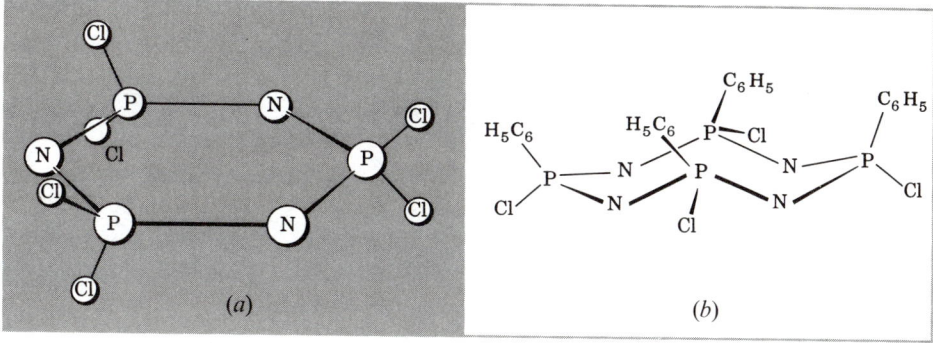

Figure 17-5   The structures of two representative cyclic phosphazenes (*a*) [NPCl$_2$]$_3$ and (*b*) all-*cis*-[NPClPh]$_4$.

$$[NPCl_2]_3 + 6\ NaOR \longrightarrow [NP(OR)_2]_3 + 6\ NaCl \qquad (17\text{-}9.2)$$

$$[NPCl_2]_3 + 6\ NaSCN \longrightarrow [NP(SCN)_2]_3 + 6\ NaCl \qquad (17\text{-}9.3)$$

$$[NPF_2]_3 + 6\ PhLi \longrightarrow [NPPh_2]_3 + 6\ LiF \qquad (17\text{-}9.4)$$

Hexachlorotriphosphazene, [NPCl$_2$]$_3$, is especially susceptible to hydrolysis as in Reaction 17-9.5.

$$[NPCl_2]_3 + 6\ H_2O \longrightarrow [NP(OH)_2]_3 + 6\ HCl \qquad (17\text{-}9.5)$$

Hexachlorotriphosphazene undergoes a ring-opening polymerization above 250 °C to give the linear polydichlorophosphazene represented in Structure 17-VI. Although the dichloro polymer is hydrolytically unstable, it is readily converted, by reactions analogous to those of the cyclic trimer, to derivatives such as Structures 17-VII and 17-VIII. The properties of such polymers depend largely on the nature of the groups attached to phosphorus. Especially stable fibers and useful elastomers are obtained when the substituents are the perfluoroalkoxy groups, such as $CF_3(CF_2)_nCH_2O$, or the amides such as $-NHCH_3$.

$$\left[\begin{array}{c} Cl \\ | \\ -N=P- \\ | \\ Cl \end{array}\right]_n \qquad \left[\begin{array}{c} OR \\ | \\ -N=P- \\ | \\ OR \end{array}\right]_n \qquad \left[\begin{array}{c} NR_2 \\ | \\ -N=P- \\ | \\ NR_2 \end{array}\right]_n$$

$$\text{17-VI} \qquad\qquad \text{17-VII} \qquad\qquad \text{17-VIII}$$

17-10   Compounds with Element–Element Double Bonds

Although N≡N double bonds abound, other Group VB(15) E═E bonds were unknown until only recently. Now we have stable compounds that contain P═P, P═As, and As═As bonds. Similar E═E or E═E' bonds involving antimony or bismuth are still unknown. The best calculations show that the HN═NH and HP═PH π-bond strengths are 256 and 150 kJ mol^{-1}, respectively. Thus the P═P π bond has considerable strength, but is weaker than the N═N π bond.

It is thermodynamics that makes obtaining compounds with E═E bonds difficult. Compounds with such bonds are unstable relative to cyclic oligomers of the type (RP)$_n$ or (RAs)$_n$. It has been found that cyclization can be thwarted by employing large R groups, partly because they diminish the rate of oligomerization, and partly because they reduce the stability of certain cyclic products.

Some of the E=E bond distances of RE=E′R′ molecules (Structure 17-IX)

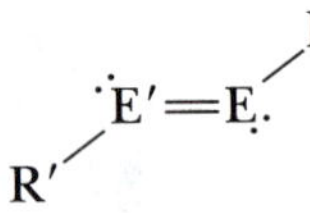

$$\overset{\displaystyle R'}{\underset{\displaystyle R'}{}}\ddot{E'}=\ddot{E}\overset{\displaystyle R}{}$$

17-IX

are listed in Table 17-1. The molecules are all planar in their X—E=E′=X portions, and the E=E′ distances are approximately 0.20 Å shorter than the corresponding E—E′ single-bond lengths.

Table 17-1   Bond Distances in Some
RE=ER Compounds (Structure 17-IX)

E	E′	R^a	R′	Distance E=E′ (Å)
P	P	Ar*	Ar*	2.034
P	P	$(Me_3Si)_3C$	$(Me_3Si)_3C$	2.014
P	As	Ar*	$(Me_3Si)_2CH$	2.124
As	As	Ar*	$(Me_3Si)_2CH$	2.224

^{a} Ar* = 2,4,6-$(Me_3C)_3C_6H_2$.

Two of the principal methods of preparation are shown in Reactions 17-10.1 and 17-10.2.

$$2\ RPCl_2 + 2\ Mg \longrightarrow RP{=}PR + 2\ MgCl_2 \qquad (17\text{-}10.1)$$

$$RECl_2 + H_2E'R' \xrightarrow{\text{base}} RE = E'R' \qquad (17\text{-}10.2)$$

17-11 Summary of Group Trends for the Elements of Group VB(15)

The list of periodic chemical properties from Section 8-11 can be used now, together with properties mentioned in Chapters 16 and 17, to summarize the periodic trends in the properties and reactivites of the elements of Group VB(15). Among these trends one finds increasing metallic character on descent of the group.

1. Nitrogen
   (a) Forms covalent compounds almost exclusively, the only important exceptions being simple nitrides, such as Li_3N.
   (b) Forms oxides that are covalent and serve as acid anhydrides.
   (c) Forms halides (fluorides predominantly) that are covalent (e.g., NF_3 and NF_4^+).
   (d) Forms hydrides that are covalent and nonhydridic.
   (e) Forms esters of the type

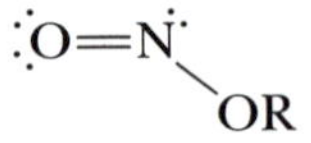

$$\ddot{O}{=}\ddot{N}\diagdown_{OR}$$

   (f) Frequently forms compounds that are electronically unsaturated, in which the unsaturation is exclusively of the $p\pi$–$p\pi$ type.
2. Phosphorus
   (a) Forms covalent substances almost exclusively, most of which are electronically saturated.

(b) Forms electronically saturated covalent oxides that serve as acidic anhydrides.

(c) Forms low-valent (PX_3) and high-valent (PX_5) molecular halides that are readily hydrolyzed.

(d) Forms a gaseous hydride, PH_3.

(e) Forms electronically saturated esters of the type $P(OR)_3$.

(f) Forms compounds that are electronically saturated, but which contain $p\pi–d\pi$ (rather than $p\pi–p\pi$) double bonding.

(g) Compounds with P=P and P=As double bonds are becoming increasingly known.

3. Arsenic, Antimony, and Bismuth

(a) Increasingly form ionic compounds rather than covalent ones on descent of the group.

(b) Rather than simple ions such as M^{3+} or M^{5+}, form oxo ions such as SbO^+ and BiO^+.

(c) Form oxides that are, on descent of the group, increasingly basic, as seen by the following trend: P and As (acidic oxides), Sb (amphoteric oxide), and Bi (basic oxide).

(d) Form halides that are ionic and increasingly aggregated in the solid state through halide bridges, giving expanded coordination numbers at the metal ion.

(e) Form increasingly weaker bonds to hydrogen.

(f) Increasingly form more stable low-valent compounds than is typical of phosphorus, for example, the oxochloride of bismuth, BiOCl.

(g) Compounds containing As=As and As=P double bonds are known, but the antimony and bismuth analogs are not.

17-12 Descriptive Summary of Reactions

Some of the important reactions of PCl_3 were given in Fig. 17-1. As a study aid, other reactions of phosphorus and its compounds are diagrammed in Figs. 17-6 and 17-7.

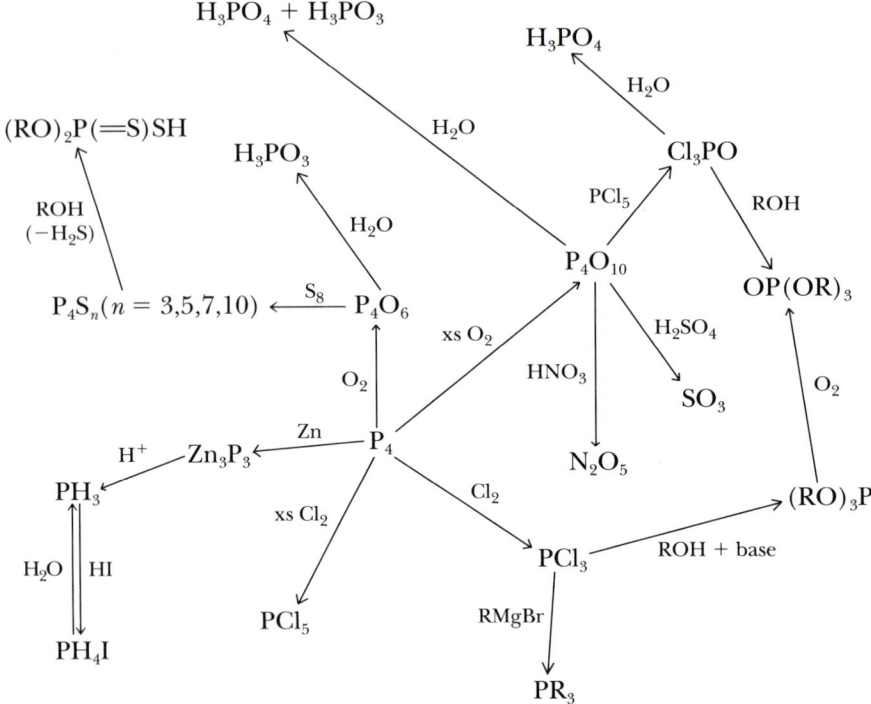

Figure 17-6 Some reactions of P_4 and its derivatives.

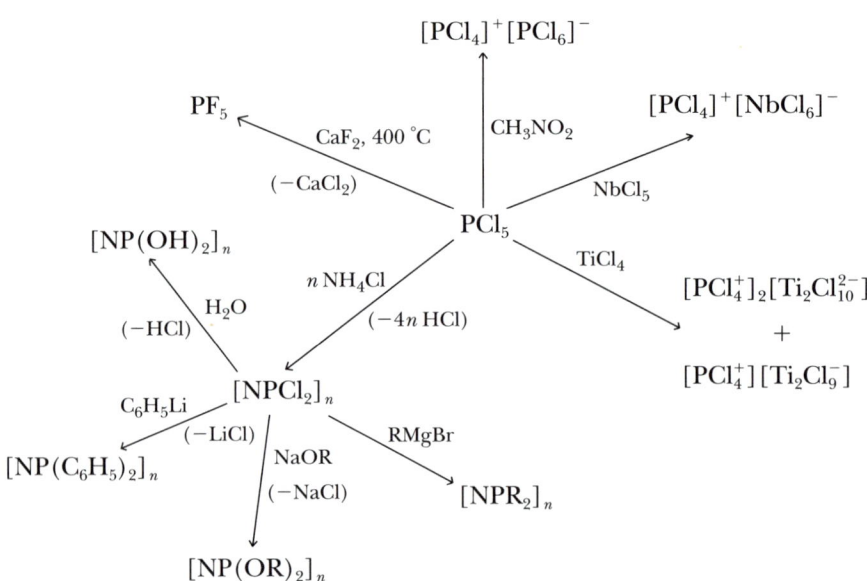

Figure 17-7 Some reactions of PCl_5.

STUDY GUIDE

Study Questions

A. Review

1. Why does phosphorus form P_4 molecules while nitrogen is N_2?
2. How are white and red phosphorus obtained from phosphate rock?
3. What are the principal factors responsible for the differences between the chemistry of nitrogen and the chemistry of phosphorus?
4. Explain the differences in (a) basicity and (b) donor ability toward transition metals of $N(CH_3)_3$ and $P(CH_3)_3$.
5. Write balanced equations for the reactions:
   (a) $P_4 + HNO_3$ (b) $AsCl_3 + H_2O$ (c) $POCl_3 + H_2O$
   (d) $P_4O_{10} + HNO_3$ (e) $P_4O_6 + H_2O$ (f) $Zn_3P + $ dilute HCl
6. How is PCl_5 made? What is its structure in solutions and in the solid state?
7. Draw the structures of P_4O_{10} and As_4O_6.
8. What happens when H_2S is passed into acidic (HCl) solution of trivalent P, As, Sb, and Bi?
9. What are the structures of (a) phosphorous acid and (b) triethylphosphite?
10. What is the Michaelis–Arbusov reaction?
11. Why is pure phosphoric acid syrupy?
12. What is the structure of "tartar emetic"?
13. What are phosphazenes and how are they made?
14. Describe the interaction of water with $SbCl_3$ and $BiCl_3$.
15. How is PF_5 prepared? Give its main chemical properties.
16. Compare the structure and properties of nitric and phosphoric acids.

B. Additional Exercises

1. Discuss the importance of $d\pi$–$p\pi$ bonding for phosphorus. Give examples, with explanations for differences between the chemistries of N and P.
2. The compound NF_3 had no donor properties at all, but PF_3 forms numerous complexes with metals, for example, $Ni(PF_3)_4$. Explain.
3. Both P and Sb form stable pentachlorides but As does not. Why?
4. Compare the oxides of N with those of P.
5. Show with drawings the formation of the π bonds in R_3PO and $R_3P{=}CH_2$. What is the geometry at P in each case?
6. Draw the Lewis diagrams and discuss the geometries in PF_3, PF_5, and PF_6^-.
7. Write balanced equations for the following reactions.
   (a) The hydrolysis of PCl_3.
   (b) Air oxidation of PCl_3.
   (c) The hydrolysis of $BiCl_3$.
   (d) A synthesis of triethylphosphine.
   (e) Oxidation of PCl_3 by F_2.
   (f) Methanolysis of trichlorophosphine oxide.
   (g) Dissolution of PCl_5 in polar solvents.
   (h) Ammonolysis of PCl_3.
   (i) The synthesis of hexachlorotriphosphazene.

8. Suggest a synthesis of $[NP(CH_3)_2]_3$ starting with PCl_5, NH_4Cl, and a Grignard reagent.

9. How many isomers are possible for the partially substituted cyclic trimer $N_3P_3F_2Cl_4$?

10. Discuss the changes in hybridization, oxidation state, and geometry (use the AB_xE_y classification scheme of Chapter 3 and VSEPR theory) that take place on forming

    (a) SbF_6^- from SbF_5 (b) PCl_4^+ from PCl_5

    (c) PCl_6^- from PCl_5 (d) $[SbF_5]^{2-}$ from SbF_5

11. Use the Lewis theory of acids and bases to discuss the reactions that are found in Problem 10, part B.

12. Beginning with PCl_5, and using two steps or fewer, list as many derivatives as can be made using the reactions of this chapter.

13. Give the chemical equation that represents each of the following reactions.

    (a) Reduction of phosphate rock by carbon and sand.

    (b) Hydrolysis of $OPCl_3$, using an excess of water.

    (c) Reaction (condensation) of $OPCl_3$ with phenol.

    (d) Oxidation of phosphorus with an excess of oxygen.

    (e) Air oxidation of $P(OC_6H_5)_3$.

    (f) Reaction of PCl_3 with C_2H_5MgBr.

    (g) Reaction of PCl_3 with CH_3OH.

    (h) $PCl_3 + AsF_3$

    (i) $PCl_5 + H_2$

14. Of P_4, Sb_4 and Bi, which is the only element that forms an oxoacid on treatment with HNO_3? Explain.

15. Although compounds such as $OPCl_3$ are properly said to be electronically saturated, the OP linkage possesses considerable double-bond character. Explain.

16. Which elements of Group VB(15) form hydrolyzable halides of both the low- and high-valent variety?

17. Which elements of Group VB(15) form an amphoteric oxide?

18. Give the products to be expected on reaction of P_4 with

    (a) A deficiency of oxygen.

    (b) An excess of oxygen.

    (c) A deficiency of Cl_2.

    (d) An excess of Cl_2.

    (e) S_8.

19. Give the principal P-containing product for each of the following:

    (a) $PCl_5 + NbCl_5$

    (b) PCl_5 dissolved in $CH_3NO_2(\ell)$

    (c) Metathesis of PCl_5 and CaF_2 at 400 °C

    (d) Thermal reaction of PCl_5 and NH_4Cl

    (e) $[NPCl_2]_3 + NaOC_2H_5$

    (f) $[NPCl_2]_3 + C_6H_5Li$

    (g) $[NPCl_2]_3 + C_6H_5MgBr$

    (h) $PCl_3 + C_6H_5MgBr$

    (i) $PCl_5 + TiCl_4$

20. Explain how the differing reactions of the M_4 elements of Group VB(15) with nitric acid are consistent with increasing metallic behavior on descent of the group.

21. Compare the oxides of phosphorus with those of nitrogen and bismuth.

22. The compound P_4S_{10} is isostructural with P_4O_{10}. It also undergoes the following alcoholysis reaction:

$$P_4S_{10} + 8\ ROH \longrightarrow 4(RO)_2P(S)SH + 2\ H_2S$$

Draw the Lewis diagram of each reactant and product, and give the occupancy notation $(AB_xE_y$, as in Chapter 3) for each distinct P, O, and S atom.

C. Questions from the Literature of Inorganic Chemistry

1. Consider the paper by B. H. Christian, R. J. Gillespie, and J. F. Sawyer, *Inorg. Chem.*, **1981**, *20*, 3410-3420.

(a) Salts of the cations $As_3S_4^+$ and $As_3Se_4^+$ have been prepared starting with As_4S_4 or As-Se alloys and using (as oxidants) the Lewis acids AsF_5 or SbF_5. Draw Lewis diagrams for the cations and anions that are formed in these reactions.

(b) What (different) products were obtained upon oxidation of As_4F_4 by $SbCl_5$, Cl_2, or Br_2? Why?

(c) How does the structure of the starting material As_4S_4 differ from its oxidized product, $As_3S_4^+$?

2. The dianion $[Sb_2OCl_6]^{2-}$ is described in a paper by M. Hall and D.B. Sowerby, *J. Chem. Soc., Chem. Commun.*, **1979**, 1134–1135.

(a) How is this dianion uniquely different from other antimony chlorides or antimony oxide chlorides?

(b) Show with drawings how each Sb^{III} center can be viewed as an AB_5E system (according to the classification of Chapter 3) in which the "sixth position" of a pseudooctahedron is occupied by a lone electron pair.

(c) Is there evidence among the structural data (either in terms of bond angles or bond lengths) for the presence of a lone pair of electrons on each Sb^{III} center? Answer in terms of VSEPR theory (Chapter 3).

3. The structure of the ion $[SbCl_5]^{2-}$ was reported by R. K. Wismer and R. A. Jacobson, *Inorg. Chem.*, **1974**, *13*, 1678-1680.

(a) Use VSEPR theory and the AB_xE_y classification that was presented in Chapter 3 to discuss the hybridizations and geometries around antimony in the compounds $SbCl_3$, $(NH_4)_2SbCl_5$, $(pyH)SbCl_4$, and $[Co(NH_3)_6][SbCl_6]$.

(b) In the crystals of K_2SbCl_5, the square-pyramidal $[SbCl_5]^{2-}$ units were found to be packed base to base. The short interion Sb-Sb distance indicates little *stereochemical effect* from a localized lone pair of electrons on Sb. Elaborate and explain.

4. Consider the work by P. Wisian-Neilson and R. H. Neilson, *J. Am. Chem. Soc.*, **1980**, *102*, 2848–2849.

(a) What problems normally arise in the syntheses of *fully alkylated* polymeric dialkylphosphazenes, $[NPR_2]_n$, starting with $[NPCl_2]_n$ polymers and using Grignard reagents?

(b) Compound 2 as reported in this work leads to fully alkylated polymers, $[NPR_2]_n$, without the problems mentioned in (a). Why? Show the elimination that must take place upon polymerization.

(c) Draw the Lewis diagrams and discuss the hybridizations and geometries around all atoms in Compounds 1 and 2 of this paper.

(d) Show at each Si, N, and P atom how a *p* or *d* orbital may become involved in a π-bond system in each Molecule 1 and 2.

(e) Elimination reactions of Compound 1 gave a cyclic tetramer, $[NP(CH_3)_2]_4$. Show the necessary elimination reactions and draw the likely structure of the cyclized product.

SUPPLEMENTARY READING

Allcock, H. R., *Phosphorus-Nitrogen Compounds,* Academic, New York, 1972.

Allcock, H. R., "Inorganic Macromolecules," *Chem. Eng. News,* **1985,** March 18, 22–36.

Corbridge, D. E. C., *The Structural Chemistry of Phosphorus,* Elsevier, Amsterdam, 1974.

Corbridge, D. E. C., *Phosphorus: An Outline of its Chemistry, Biochemistry and Technology,* 4th ed., Elsevier, Amsterdam, 1990.

Cowley, A. H. and Norman, N. C., "The Synthesis, Properties and Reactivities of Stable Compounds Featuring Double Bonding Between Heavier Group 14 and 15 Elements," in *Progress in Inorganic Chemistry,* Vol. 34, Wiley-Interscience, New York, 1986.

Doak, G. O. and Freedman, L. D., *Organometallic Compounds of Arsenic, Antimony, and Bismuth,* Wiley, New York, 1970.

Emsley, J. and Hall, D., *The Chemistry of Phosphorus,* Harper & Row, New York, 1976.

Fluck, E., "The Chemistry of Phosphine," in *Topics in Current Chemistry,* Springer-Verlag, Berlin, 1973.

Goldwhite, H., *Introduction to Phosphorus Chemistry,* Cambridge University Press, Cambridge, UK, 1981.

Mann, F. G. *Heterocyclic Derivatives of P, As, Sb, and Bi,* Wiley, New York, 1970.

McAuliffe, C. A. and Levason, W., *Phosphine, Arsine, and Stibine Complexes of the Transition Elements,* Elsevier, Amsterdam, 1979.

Regitz, M., Ed. *Multiple Bonds and Low Coordination in Phosphorus Chemistry,* G. Thieme, Stuttgart, 1990.

Toy, A. D. F., *The Chemistry of Phosphorus,* Pergamon Press, New York, 1975.

Toy, A. D. and Walsh, E. N., *Phosphorus Chemistry in Everyday Living,* 2nd ed., American Chemical Society, Washington, DC, 1987.

Walsh, E. N. et al., Eds., "Phosphorus Chemistry: Developments in American Science," ACS Symposium Series, No. 486, American Chemical Society, Washington, DC, 1992.

Woolins, J. D., *Nonmetal Rings, Cages, and Clusters,* Wiley, New York, 1988.

Chapter 18

OXYGEN

18-1 Introduction

Oxygen compounds of all the elements except He, Ne, and possibly Ar are known. Molecular oxygen (dioxygen, O_2) reacts (at room temperature or on heating) with all other elements except the halogens, a few noble metals, and the noble gases.

The chemistry of oxygen involves the completion of the octet (neon configuration) by one of the following means:

1. Electron gain to form the oxide O^{2-}.
2. Formation of two single covalent bonds, usually in bent AB_2E_2 systems, such as water and ethers.
3. Formation of a double bond, as in ABE_2 systems, such as ketones or $Cl_4Re{=}O$.
4. Formation of a single bond, as well as electron gain, as in ABE_3 systems, such as OH^- and RO^-.
5. Formation of three covalent bonds, usually in pyramidal AB_3E systems, such as H_3O^+ and R_3O^+.
6. Formation in rare cases of four covalent bonds, as, for example, in $Be_4O(CH_3CO_2)_6$.

The wide range of physical properties shown by the binary oxides of the elements is due to the broad range of bond types from essentially ionic systems to essentially covalent ones. Thus we distinguish the highly ionic oxides (such as those of the alkali and alkaline earth metals) from the completely covalent, molecular oxides, such as CO_2. There are, however, intermediate cases such as the oxides of boron, aluminum, or silicon.

Ionic Oxides

The formation of the oxide ion from molecular oxygen requires about 1000 kJ mol^{-1}:

$$\tfrac{1}{2} O_2(g) = O(g) \qquad \Delta H = 248 \text{ kJ mol}^{-1} \qquad (18\text{-}1.1)$$

$$O(g) + 2e^- = O^{2-} \qquad \Delta H = 752 \text{ kJ mol}^{-1} \qquad (18\text{-}1.2)$$

In forming an ionic metal oxide, energy must also be expended to vaporize and to ionize the metal. Thus the stability of ionic metal oxides is a consequence only of the high lattice energies that are obtained with the small and highly charged oxide ion.

Where the lattice energy is not sufficient to offset the energies for ionization, and so on, oxides with substantial covalent character are formed. Examples of oxides with some covalent character are BeO, SiO_2, and oxides of boron, such as B_2O_3.

Covalent or Molecular Oxides

Covalent or molecular oxides are compounds, such as CO_2, SO_2, SO_3, and NO_2, in which covalent bonding is dominant. Such compounds are well described by the AB_xE_y classification, as presented in Chapter 3, with some exceptions, as noted in the following subsection. Use of the p orbitals in π bonding with other atoms is an important aspect in the bonding of molecular oxides. This may be $p\pi-p\pi$ bonding as in the ketones ($R_2C{=}O$), or $p\pi-d\pi$ bonding as in phosphine oxides ($R_3P{=}O$) or linear $M{=}O{=}M$ systems.

ABE_3 Systems

Terminal oxygen atoms that bear three lone pairs of electrons are found in alkoxides (RO^-), and hydroxide (OH^-). Such oxygen atoms may be considered to be sp^3 hybridized.

AB_2E_2 Systems

The compounds that fit into this class are usually angular due to the volume requirements of two lone pairs of electrons. Examples include water, alcohols, and ethers. The oxygen atoms are considered to be sp^3 hybridized, but there are wide variations from the tetrahedral bond angles due to electronic repulsions between the two lone pairs of electrons: H_2O (104.5°) and $(CH_3)_2O$ (111°). Where the atoms bound to oxygen have d orbitals available, some $p\pi-d\pi$ character is often present in the bond to oxygen, and the B—A—B angles may be even larger, for example, the angle Si—O—Si in quartz is 142° and in H_3Si—O—SiH_3 it is greater than 150°.

A linear B—A—B situation at oxygen occurs in some AB_2E_2 systems containing transition metals (e.g., $[Cl_5Ru{-}O{-}RuCl_5]^{4-}$). The σ bonds to Ru are formed by sp hybrids on oxygen, thus leaving two pairs of π electrons on oxygen in p orbitals that are oriented perpendicular to the Ru—O—Ru axis. These filled p orbitals on oxygen interact with empty d orbitals on the Ru atoms, forming a π-bond system.

AB_3E Systems

The third example containing sp^3 hybridized oxygen atoms is that of the oxonium ions $:OH_3^+$ and $:OR_3^+$. The formation of oxonium ions is analogous to formation of ammonium ions (NH_4^+). Oxygen is less basic than nitrogen, and the oxonium ions are therefore less stable. Notice that ions of the type OH_4^{2+} are unlikely (even though $:OH_3^+$ still has a lone electron pair), because of electrostatic repulsion of the $:OH_3^+$ ion towards another proton. As for $:NR_3$, the pyramidal $:OR_3^+$ ions undergo rapid inversion.

ABE_2 Systems

Oxygen atoms of this type include those of ketones, aldehydes, and other organic carbonyls. The oxygen atoms are sp^2 hybridized and have a roughly trigonal arrangement around the oxygen of the lone pairs E and the carbonyl carbon.

The sp^2 hybridization of the carbon atom leaves one p orbital available for formation of a π bond perpendicular to the trigonal plane.

Acid–Base Properties of Oxides

Generally, the oxides of the metals are basic, whereas those of the nonmetals are acidic. There are also a number of important amphoteric oxides.

Basic Oxides

Although X-ray studies show the existence of discrete oxide ions (O^{2-}) [as well as peroxide (O_2^{2-}) and superoxide (O_2^-) to be discussed later], these ions cannot exist in aqueous solution owing to the hydrolysis reactions shown in Reactions 18-1.3 through 18-1.5.

$$O^{2-} + H_2O \longrightarrow 2\ OH^- \tag{18-1.3}$$

$$O_2^{2-} + H_2O \longrightarrow HO_2^- + OH^- \tag{18-1.4}$$

$$2\ O_2^- + H_2O \longrightarrow O_2 + HO_2^- + OH^- \tag{18-1.5}$$

Consequently, only those ionic oxides that are insoluble in water are inert to it. Ionic oxides function as *basic anhydrides*. When insoluble in water, they usually dissolve in dilute acids, as in Reaction 18-1.6.

$$MgO(s) + 2\ H^+(aq) \longrightarrow Mg^{2+} + H_2O \tag{18-1.6}$$

However, some ionic oxides (e.g., MgO) become very slow to dissolve in acids after high-temperature ignition.

Acidic Oxides

The covalent oxides of the nonmetals are usually acidic, dissolving in water to produce solutions of acids. They are termed *acid anhydrides*. An example is given in Reaction 18-1.7, in which N_2O_5 is seen to be the acid anhydride of nitric acid.

$$N_2O_5 + H_2O \longrightarrow 2\ H^+ + 2\ NO_3^- \tag{18-1.7}$$

Even when these oxides are insoluble in water (e.g., as in the case of Sb_2O_5), they will generally dissolve in bases (as in Reaction 18-1.8).

$$Sb_2O_5 + 2\ OH^- + 5\ H_2O \longrightarrow 2\ Sb(OH)_6^- \tag{18-1.8}$$

Acidic oxides will often combine directly, by fusion, with basic oxides to form salts, as in Reaction 18-1.9.

$$Na_2O + SiO_2 \xrightarrow{\text{fusion}} Na_2SiO_3 \tag{18-1.9}$$

Amphoteric Oxides

These oxides behave acidicly towards strong bases and as bases towards strong acids. The example of ZnO is illustrated in Reactions 18-1.10 and 18-1.11.

$$ZnO(s) + 2\,H^+(aq) \longrightarrow Zn^{2+}(aq) + H_2O \qquad (18\text{-}1.10)$$

$$ZnO + 2\,OH^- + H_2O \longrightarrow Zn(OH)_4^{2-} \qquad (18\text{-}1.11)$$

Other Oxides

There are other oxides, some of which are relatively inert, which dissolve in neither acids nor bases (e.g., N_2O, CO, PbO_2, and MnO_2). When MnO_2 and PbO_2 do react with acids (e.g., conc HCl) they do so by a redox rather than an acid–base reaction, as in Reaction 18-1.12.

$$MnO_2(s) + 4\,HCl \longrightarrow Mn^{2+} + 2\,Cl^- + Cl_2 + 2\,H_2O \qquad (18\text{-}1.12)$$

18-2   Occurrence, Properties, and Allotropy

Oxygen has three isotopes, ^{16}O (99.759%), ^{17}O (0.0374%), and ^{18}O (0.2039%). Fractional distillation of water allows concentrates containing up to 97 atom % ^{18}O or up to 4 atom % ^{17}O to be prepared. Oxygen-18 is used as a tracer in studying reaction mechanisms of oxygen compounds. Although ^{17}O has a nuclear spin $(\tfrac{5}{2})$, its low abundance means that even when enriched samples are used spectrum accumulation and/or the Fourier transform method are required. An example of ^{17}O resonance studies is the distinction between H_2O in a complex, for example, $[Co(NH_3)_5H_2O]^{3+}$, and solvent water.

Oxygen has two allotropes; dioxygen (O_2) and trioxygen or ozone (O_3). Dioxygen is paramagnetic in all phases and has the rather high dissociation energy of 496 kJ mol^{-1}. Simple valence bond theory predicts the electronic structure :Ö=Ö: which, though accounting for the strong bond, fails to account for the paramagnetism. However, simple MO theory (Section 3-5) readily accounts for the triplet ground state having a double bond. There are several low-lying singlet states that are important in photochemical oxidations. Like NO, which has one unpaired electron in an antibonding (π^*) MO, oxygen molecules associate only weakly, and true electron pairing to form a symmetrical O_4 species does not occur even in the solid. Both liquid and solid O_2 are pale blue.

Ozone

The action of a silent electric discharge on O_2 produces O_3 in concentrations up to 10%. Ozone gas is perceptibly blue and is diamagnetic. Pure ozone obtained by fractional liquefaction of O_2—O_3 mixtures gives a deep blue, explosive liquid. The action of UV light on O_2 produces traces of O_3 in the upper atmosphere. The maximum concentration is at an altitude of about 25 km. It is of vital importance in protecting the earth's surface from excessive exposure to UV light. Ozone decomposes exothermically, as in Reaction 18-2.1:

$$O_3 = \tfrac{3}{2}O_2 \qquad \Delta H = -142 \text{ kJ mol}^{-1} \qquad (18\text{-}2.1)$$

but it decomposes only slowly at 250°C in the absence of catalysts and UV light.

The O_3 molecule is symmetrical and bent; $\angle$O—O—O, 117°; O—O, 1.28 Å. Since the O—O bond distances are 1.49 Å in HOOH (single bond) and 1.21 Å in O_2 (~ double bond), it is apparent that the O—O bonds in O_3 must have considerable double-bond character. In terms of a resonance description, this can be accounted for as in the resonance forms of Structures 18-I and 18-II.

$$\cdot \ddot{O} \cdot \;\; \cdot \ddot{O} \cdot \longleftrightarrow \cdot \ddot{O} \cdot \;\; \cdot \ddot{O} \cdot$$

18-I   18-II

Chemical Properties of O$_2$ and O$_3$

Ozone is a much more powerful oxidizing agent than O_2 and reacts with many substances under conditions where O_2 will not. The reaction

$$O_3 + 2\,KI + H_2O \longrightarrow I_2 + 2\,KOH + O_2 \tag{18-2.2}$$

is quantitative and can be used for analysis. Ozone is used for oxidations of organic compounds and in water purification. Oxidation mechanisms probably involve free radical chain processes as well as intermediates with $-$OOH groups. In acid solution, O_3 is exceeded in oxidizing power only by F_2, the perxenate ion $[H_2XeO_6]^{2-}$, atomic oxygen, OH radicals, and a few other such species.

The following potentials indicate the oxidizing strengths of O_2 and O_3 in ordinary aqueous solution.

$$O_2 + 4\,H^+(10^{-7}\,M) + 4\,e^- = 2\,H_2O \qquad E^\circ = +0.815\ \text{V} \tag{18-2.3}$$

$$O_3 + 2\,H^+(10^{-7}\,M) + 2e^- = O_2 + H_2O \quad E^\circ = +1.65\ \text{V} \tag{18-2.4}$$

The first step in the reduction of O_2 in aprotic solvents such as DMSO and pyridine appears to be a one-electron step to give the superoxide anion:

$$O_2 + e^- = O_2^- \tag{18-2.5}$$

whereas in aqueous solution a two-electron step occurs to give HO_2^-

$$O_2 + 2\ e^- + H_2O = HO_2^- + OH^- \tag{18-2.6}$$

It can also be seen from the potential given for Reaction 18-2.3 that neutral water saturated with O_2 is a fairly good oxidizing agent. For example, although Cr^{2+} is just stable toward oxidation by pure water, in oxygen-saturated water it is rapidly oxidized. Ferrous ion (Fe^{2+}) is oxidized (slowly in acid, but more rapidly in base) to Fe^{3+} in the presence of air, although in oxygen-free water it is quite stable, as shown by the potential for Reaction 18-2.7.

$$Fe^{3+} + e^- = Fe^{2+} \quad E^\circ = +0.77\ \text{V} \tag{18-2.7}$$

Many oxidations by oxygen in acid solution are slow, but the rates of oxidation may be vastly increased by catalytic amounts of transition metal ions, especially Cu^{2+}, where a Cu^I–Cu^{II} redox cycle is involved.

The dioxygen molecule is readily soluble in organic solvents, and merely pouring these liquids in air serves to saturate them with O_2. This fact should be kept in mind when determining the reactivity of air-sensitive materials in solution in organic solvents.

Measurements of electronic spectra of alcohols, ethers, benzene, and even saturated hydrocarbons show that there is reaction of the charge-transfer type with the oxygen molecule. However, there is no true complex formation, since the heats of formation are negligible and the spectral changes are due to contact between the molecules at van der Waals distances. The classic example is that of N,N-dimethylaniline, which becomes yellow in air or oxygen but colorless again when the oxygen is removed. Such weak charge-transfer complexes make certain electronic transitions in molecules more intense; they are also a plausible first stage in photooxidations.

With certain transition metal complexes, O_2 adducts may be formed, sometimes reversibly (Section 18-7). Although the O_2 entity remains intact, the complexes may be described as having coordinated O_2^- or O_2^{2-} ions, bound to the metal in a three-membered ring or as a bridging group. Coordinated O_2 is more reactive than free O_2, and substances not directly oxidized under mild conditions can be attacked in the presence of metal complexes.

The Excited State Chemistry of Oxygen

As discussed in Chapter 3, the oxygen molecule contains two unpaired electrons in π^* molecular orbitals. This electron configuration gives rise to three electronic states, as shown in Table 18-1. The triplet state ($^3\Sigma_g^+$) is the ground state, but two excited states are also available at higher energies. These excited singlet states (especially $^1\Delta_g$) have sufficiently long lifetimes to allow them to be useful for reactions with a variety of substrates, where they cause specific oxidations, a very typical example being 1,4 addition to a 1,3-diene, as in Reaction 18-2.8.

$$\text{(diene)} + O_2(\text{Singlet}) \longrightarrow \text{(cyclic peroxide)} \qquad (18\text{-}2.8)$$

Singlet oxygen molecules may be generated either by photochemical or chemical means. The photochemical route typically employs a sensitizer, which first absorbs energy from the light source and then transfers an appropriate amount of that energy to triplet oxygen to give an oxygen molecule in an excited (singlet) state. The sensitizer molecule or ion must be in an excited triplet state for this energy transfer to be spin allowed.

The chemical generation of singlet oxygen may be accomplished as in Reactions 18-2.9 and 18-2.10:

Table 18-1 The Three Electronic States Arising from the $(\pi^*)^2$ Electron Configuration of Molecular Oxygen

State	π_a^*	π_b^*	Energy
$^1\Sigma_g^+$	↑	↓	155 kJ (~13,000 cm^{-1})
$^1\Delta_g$	↑↓		92 kJ (~8,000 cm^{-1})
$^3\Sigma_g^+$	↑	↑	0 (Ground state)

$$H_2O_2 + Cl_2 \longrightarrow 2\ Cl^- + 2\ H^+ + O_2 \tag{18-2.9}$$

$$H_2O_2 + ClO^- \longrightarrow Cl^- + H_2O + O_2 \tag{18-2.10}$$

which are accompanied by a red chemiluminescent glow.

18-3 Hydrogen Peroxide

Pure hydrogen peroxide (H_2O_2) is a colorless liquid (bp 152.1 °C, fp − 0.41 °C). It resembles water in many of its physical properties and is even more highly associated via hydrogen bonding and 40% denser than is H_2O. It has a high dielectric constant, but its utility as an ionizing solvent is limited by its strong oxidizing nature and its ready decomposition in the presence of even traces of many heavy-metal ions according to the reaction:

$$2\ H_2O_2 = 2\ H_2O + O_2 \qquad \Delta H = -99\ kJ\ mol^{-1} \tag{18-3.1}$$

In dilute aqueous solution it is more acidic than water.

$$H_2O_2 = H^+ + HO_2^- \qquad K_{20°C} = 1.5 \times 10^{-12} \tag{18-3.2}$$

The molecule H_2O_2 has a skew, chain structure (Fig. 18-1).

There are two methods for large-scale production of H_2O_2. One is by autoxidation of an anthraquinol, such as 2-ethylanthraquinol.

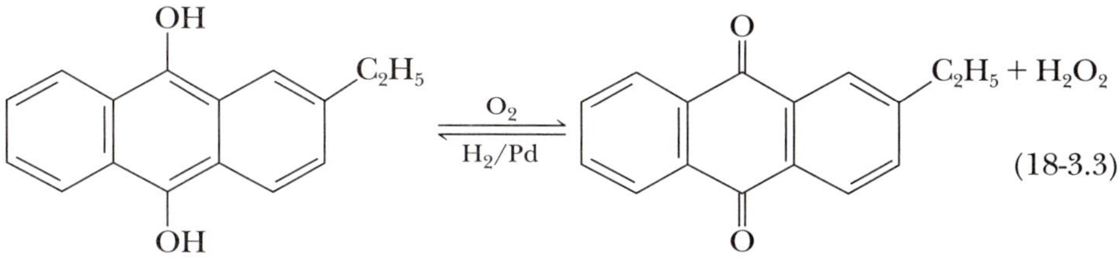

$$\tag{18-3.3}$$

The resulting quinone is reduced with H_2 gas. The H_2O_2 is obtained as a 20% aqueous solution. Only O_2, H_2 and H_2O are required as raw materials.

An older and more expensive method is electrolytic oxidation of sulfuric acid or ammonium sulfate–sulfuric acid solutions to give peroxodisulfuric acid, which is then hydrolyzed to yield H_2O_2:

$$2\ HSO_4^- \longrightarrow HO_3S{-}O{-}O{-}SO_3H + 2\ e^- \tag{18-3.4}$$

$$H_2S_2O_8 + H_2O \longrightarrow H_2SO_5 + H_2SO_4 \qquad (\text{Rapid}) \tag{18-3.5}$$

$$H_2SO_5 + H_2O \longrightarrow H_2O_2 + H_2SO_4 \qquad (\text{Slow}) \tag{18-3.6}$$

Fractional distillation can then give 90–98% H_2O_2.

The redox chemistry of H_2O_2 in aqueous solution is summarized by the potentials.

$$H_2O_2 + 2\ H^+ + 2\ e^- = 2\ H_2O \quad E° = 1.77\ V \tag{18-3.7}$$

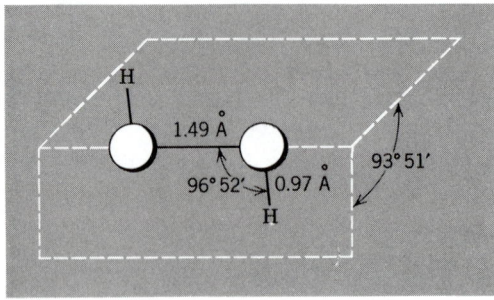

Figure 18-1 The structure of hydrogen peroxide.

$$O_2 + 2 H^+ + 2 e^- = H_2O_2 \quad E° = 0.68 \text{ V} \tag{18-3.8}$$

$$HO_2^- + H_2O + 2 e^- = 3 OH^- \quad E° = 0.87 \text{ V} \tag{18-3.9}$$

These show that H_2O_2 is a strong oxidizing agent in either acid or basic solution. It behaves as a reducing agent only toward very strong oxidizing agents such as MnO_4^-.

Dilute or 30% H_2O_2 solutions are widely used as oxidants. In acid solution, oxidations with H_2O_2 are slow, whereas in basic solution, they are usually fast. Decomposition to H_2O and O_2, which may be considered a self-oxidation, or disproportionation, occurs most rapidly in basic solution; hence an excess of H_2O_2 may best be destroyed by heating in basic solution.

Many reactions involving H_2O_2 (and also O_2) in solutions involve free radicals. Metal-ion catalyzed decomposition of H_2O_2 and other reactions form radicals of which HO_2 and OH are most important. The hydroperoxo radical (HO_2) has been detected in aqueous solutions where H_2O_2 interacts with Ti^{3+}, Fe^{2+}, or Ce^{4+} ions.

18-4   Peroxides and Superoxides

These substances are derived formally from O_2^{2-} (peroxides) and O_2^- (superoxides).

Ionic Peroxides

Ionic peroxides are formed by alkali metals, Ca, Sr, and Ba. Sodium peroxide is made commercially by air oxidation of sodium. Sodium peroxide is a yellow powder that is very hygroscopic, though thermally stable to 500 °C. It contains, according to electron spin resonance (ESR) studies, about 10% of the superoxide.

The ionic peroxides give H_2O_2 on reaction with H_2O or dilute acids. All of the ionic peroxides are powerful oxidizing agents, converting organic materials to carbonate even at moderate temperatures. Sodium peroxide will also vigorously oxidize some metals (e.g., Fe, which violently gives FeO_4^{2-}). The peroxides of the alkali metals also react with CO_2 according to Reaction 18-4.1 to give carbonates:

$$2 CO_2(g) + 2 M_2O_2 \longrightarrow 2 M_2CO_3 + O_2 \tag{18-4.1}$$

Other electropositive metals such as Mg and the lanthanides also yield peroxides; these are intermediate in character between the ionic ones and the essentially covalent peroxides of metals such as Zn, Cd, and Hg.

Many ionic peroxides form well-crystallized hydrates such as $Na_2O_2 \cdot 8H_2O$ and $M^{II}O_2 \cdot 8H_2O$. These contain discrete O_2^{2-} ions to which water molecules are hydrogen bonded, giving chains of the type shown in Structure 18-III.

$$----O_2^{2-}---(H_2O)_8----O_2^{2-}---(H_2O)_8----$$

18-III

The formation of such stable hydrates accounts for the extreme hygroscopic nature of the crystalline peroxides.

Ionic Superoxides

Ionic superoxides, MO_2, are formed by the interaction of O_2 with K, Rb, or Cs as yellow-to-orange crystalline solids. NaO_2 can be obtained by reaction of Na_2O_2 with O_2 at 300 atm and 500 °C. LiO_2 cannot be isolated. Alkaline earth, Zn, and Cd superoxides occur only in small concentrations as solid solutions in the peroxides. The O_2^- ion has one unpaired electron. Superoxides are very powerful oxidizing agents. They react vigorously with water.

$$2\ O_2^- + H_2O = O_2 + HO_2^- + OH^- \tag{18-4.2}$$

$$2\ HO_2^- = 2\ OH^- + O_2 \qquad \text{(Slow)} \tag{18-4.3}$$

The reaction with CO_2, which involves peroxocarbonate intermediates, is used for removal of CO_2 and regeneration of O_2 in closed systems (e.g., submarines). The overall reaction is

$$4\ MO_2(s) + 2\ CO_2(g) = 2\ M_2CO_3(s) + 3\ O_2(g) \tag{18-4.4}$$

18-5   Other Peroxo Compounds

There are many *organic peroxides* and *hydroperoxides*. Peroxo carboxylic acids, for example peracetic acid, $CH_3C(O)OOH$, can be obtained by the action of H_2O_2 on acid anhydrides. The peroxo acids are useful oxidants and sources of free radicals, for example by treatment with $Fe^{2+}(aq)$. Benzoyl peroxide and cumyl hydroperoxide are moderately stable and widely used where free radical initiation is required, as in polymerization reactions.

Organic peroxo compounds are also obtained by *autoxidation* of ethers, alkenes, and the like, on exposure to air. Autoxidation is a free radical chain reaction initiated by radicals generated by interaction of oxygen and traces of metals such as Cu, Co, or Fe. The attack on specific reactive C—H bonds by a radical ($X^\bullet$), first gives $R^\bullet$ and then hydroperoxides that can react further.

$$RH + X^\bullet \longrightarrow R^\bullet + HX \tag{18-5.1}$$

$$R^\bullet + O_2 \longrightarrow RO_2^\bullet \tag{18-5.2}$$

$$RO_2^\bullet + RH \longrightarrow ROOH + R^\bullet \tag{18-5.3}$$

Explosions can occur on distillation of oxidized solvents. These solvents should be washed with acidified $FeSO_4$ solution or, for ethers and hydrocarbons, passed through a column of activated alumina. Peroxides are absent when the Fe^{2+} + SCN^- reagent does not give a red color indicative of the $Fe(SCN)^{2+}$ ion.

There are also many inorganic peroxo compounds where —O— is replaced by —O—O— groups, such as peroxodisulfuric acid, $(HO)_2S(O)OOS(O)(OH)_2$, mentioned previously. Potassium and ammonium peroxodisulfates (Section 19-5) are commonly used as strong oxidizing agents in acid solution, for example to convert C into CO_2, Mn^{2+} into MnO_4^-, or Ce^{3+} into Ce^{4+}. The last two reactions are slow and normally incomplete in the absence of silver ion as a catalyst.

It is important to make the distinction between true peroxo compounds, which contain —O—O— groups, and compounds that contain H_2O_2 of crystallization, such as $2Na_2CO_3 \cdot 3H_2O_2$ or $Na_4P_2O_7 \cdot nH_2O_2$.

18-6 The Dioxygenyl Cation

The interaction of PtF_6 with O_2 gives an orange solid (O_2PtF_6) isomorphous with $KPtF_6$, which contains the paramagnetic O_2^+ ion. This reaction was of importance in that it lead N. Bartlett to treat PtF_6 with xenon (Section 21-2). A number of other salts of the O_2^+ ion are known.

It is instructive to compare the various $O_2^{n\pm}$ species, since they provide an interesting illustration of the effect of varying the number of antibonding electrons on the length and stretching frequency of a bond, as shown by the data in Table 18-2.

18-7 Dioxygen as a Ligand

Although the most common mode of reaction of molecular oxygen with transition metal complexes is oxidation (i.e., extraction of electrons from the metal or from its ligands), under appropriate circumstances the dioxygen molecule may, instead, become a ligand. Such reactions are termed **oxygenations,** because the dioxygen ligand retains its identity, whereas oxidation reactions are those in which the O_2 molecule loses its identity through reduction.

Oxygenation reactions are often reversible. That is, upon increasing temperature and/or reducing the partial pressure of O_2, the dioxygen ligand is lost by dissociation or by transfer to another acceptor (which may become oxidized). The process of reversible oxygenation plays an essential role in life processes. In humans or other higher animals, oxygen molecules are "carried" from the lungs

Table 18-2 Bond Values for Oxygen Species

Species	O—O distance Å	Number of π* Electrons	$\nu_{O—O}(cm^{-1})$
O_2^+	1.12	1	1860
O_2	1.21	2	1556
O_2^-	1.33	3	1145
O_2^{2-}	1.49	4	~770

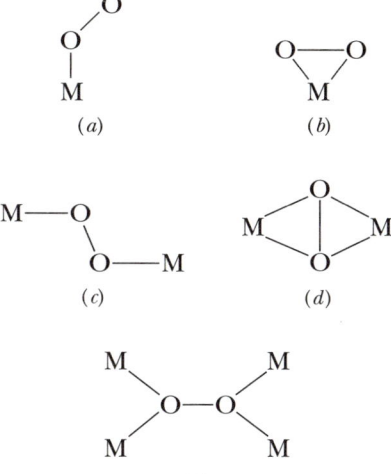

Figure 18-2   The five structural types of dioxygen ligands.

to the various tissues by hemoglobin and myoglobin molecules, in which 1:1 O_2–Fe complexes are formed. In lower animals, there are molecules such as hemerythrins and hemocyanins, that serve similar functions. More detail concerning these biological complexes will be given in Chapter 31.

Broadly speaking, there are two types of 1:1 O_2–M complexes, the "end-on" and the "slide-on" types, as shown in Fig. 18-2, types **(a)** and **(b)**. In addition, there are many 1:2 O_2—M complexes, as shown in Fig. 18-2, types **(c)** and **(d)**. The hemoglobin and myoglobin complexes are of type **(a)**, and there are a number of synthetic examples in which O_2 fills one position in an octahedral complex. Most of these can be considered to contain a coordinated superoxide ion O_2^-, and thus have an unpaired electron formally present on the coordinated dioxygen unit. Many of these complexes form reversibly.

The "side-on" complexes, type **(b)** in Fig. 18-2, are also numerous. Many are formed reversibly, as with Vaska's compound in Reaction 18-7.1.

$$(C_6H_5)_3P \quad Cl \qquad \qquad (C_6H_5)_3P \quad \diagdown O$$
$$\qquad \diagdown Ir \diagup \qquad + O_2 \; \rightleftharpoons \; \qquad Ir \qquad \qquad (18\text{-}7.1)$$
$$O^C \quad P(C_6H_5)_3 \qquad \qquad O^C \quad P(C_6H_5)_3$$
$$\qquad \qquad \qquad \qquad \qquad \qquad Cl$$

These compounds are generally best regarded as peroxide complexes, that is, compounds containing the O_2^{2-} ligand. The complexes in Fig. 18-2, types **(c)** and **(d),** are also best regarded as peroxide complexes.

18-8   Oxygen Compounds as Ligands

Water: Aqua Ligands

Hydration of transition metals has already been discussed in Chapter 6, as have the rates and thermodynamics of water ligand exchange in solution. In some cases, such as the alkali metal cations, the water ligands are weakly bound (and

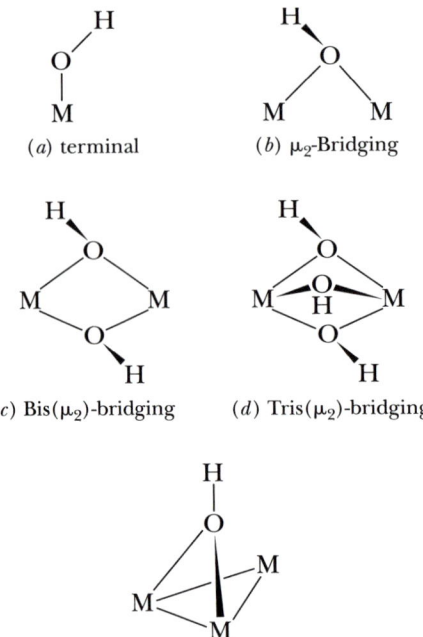

(a) terminal   (b) μ_2-Bridging

(c) Bis(μ_2)-bridging   (d) Tris(μ_2)-bridging

(e) (μ_3)-Bridging

rapidly substituted), whereas in cases such as $[Cr(H_2O)_6]^{3+}$ and $[Rh(H_2O)_6]^{3+}$, they are firmly bound and exchange with solvent water molecules only slowly (Chapter 6).

Ligand water molecules can be acidic, especially when bound to cations of high charge, giving hydroxo complexes, as in Reactions 18-8.1 and 18-8.2.

$$[Pt(NH_3)_4(H_2O)_2]^{4+} \longrightarrow [Pt(NH_3)_4(H_2O)(OH)]^{3+} + H^+ \qquad (18\text{-}8.1)$$

$$[Co(NH_3)_5(H_2O)]^{3+} \longrightarrow [Co(NH_3)_5(OH)]^{2+} + H^+ \qquad (18\text{-}8.2)$$

Hydroxide: Hydroxo Ligands

Many important hydroxo complexes are known, the hydroxo ligand serving in some cases as a simple terminal ligand, and in other cases as a bridging ligand, examples of which are shown in Fig. 18-3. Double (μ_2) bridges are most common. For complexes containing only terminal hydroxo ligands, there has been particular interest in the structural changes that are apparent when comparing the octahedral aqua ions (e.g., $[M(H_2O)_6]^{3+}$, where M = Co^{III} or Al^{III}) with the corresponding hydroxo complexes $[M(OH)_4]^-$, which are tetrahedral.

Oxide: Oxo Ligands

Oxo compounds can be of several structural types, as shown in Fig. 18-4. The multiply bonded oxo group (M=O) is found not only in oxo compounds and oxo anions of the nontransition elements (e.g., SO_4^{2-}, Chapter 5, and $Cl_3P=O$, Chapter 17), but also in transition metal compounds, such as vanadyl (V=O), uranyl (U=O), permanganate (MnO_4^-), and osmium tetroxide (OsO_4). In cases involving metals, the bond distance to oxygen (1.59–1.66 Å) corresponds to a double bond, which is best formulated as arising from $Op\pi \rightarrow Md\pi$ donation.

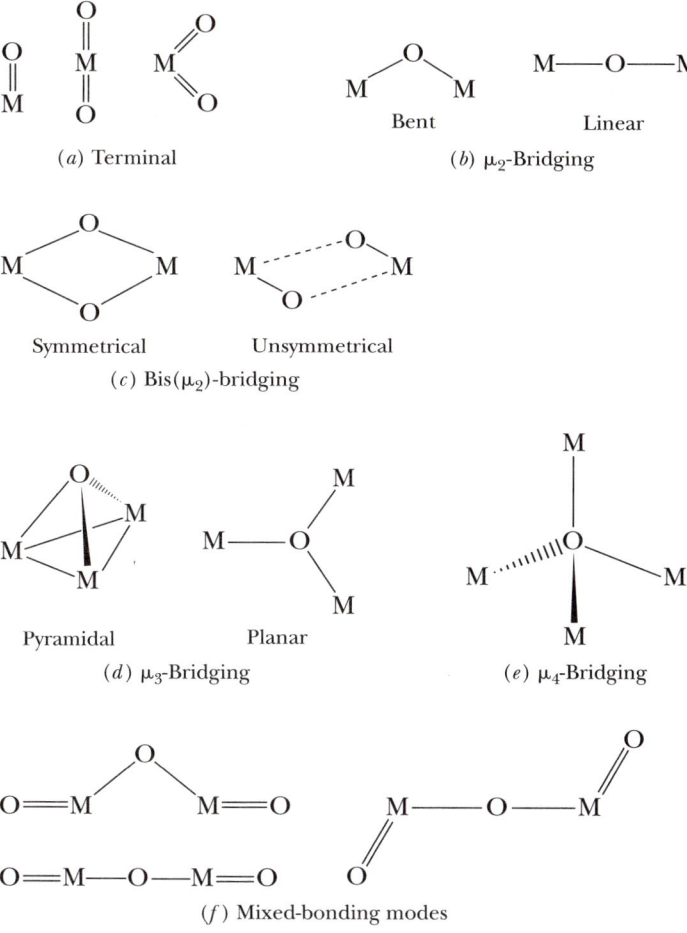

Figure 18-4   The common structural types of oxo ligands.

Thus the metal oxo complexes are most stable when the metal is in a high oxidation state. In contrast, for the oxides of the nonmetals (e.g., CO and SO_2), low oxidation states of the nonmetal are preferred.

The M=O bond is commonly affected by the nature of the group *trans* to oxygen. Donors that increase electron density on the metal tend to reduce the metal's acceptor ability, thus lowering the M=O π-bond character. Consequently, the MO stretching frequency in such complexes is found to be lower than when the oxo ligand is trans to a weak donor ligand.

8-9   Oxygen Fluorides

Most oxygen compounds are properly called oxides and, therefore, are discussed under the chemistry of the other elements. However, since fluorine is more electronegative than oxygen, it is logical to treat oxygen fluorides in this chapter. While these compounds are sometimes called fluorine oxides, it is best to call them oxygen fluorides. These compounds have been intensively studied as rocket fuel oxidizers.

Oxygen Difluoride (OF₂)

This compound can be prepared by passing fluorine rapidly through a 2% NaOH solution, by electrolysis of aqueous HF—KF solutions, or by reaction of fluorine with moist KF. It is a pale yellow, poisonous gas (bp 145 °C), which is relatively unreactive as far as this class of compounds is concerned. It can be mixed without reaction with H_2, CH_4, or CO, although an electrical spark in such mixtures will cause a violent explosion. When mixed with Cl_2, Br_2, or I_2, OF_2 will explode at room temperature. It reacts only slowly with water, as in Reaction 18-9.1, but explodes with steam. Oxygen difluoride will liberate other halogens from their acids or salts, as in Reaction 18-9.2.

$$OF_2 + H_2O \longrightarrow O_2 + 2\,HF \tag{18-9.1}$$

$$OF_2 + 4\,HX(aq) \longrightarrow 2\,X_2 + 2\,HF + H_2O \tag{18-9.2}$$

Oxygen difluoride will oxidize most metals and nonmetals, and even reacts with Xe in an electric discharge to give xenon fluorides and xenon oxide fluoride (Chapter 21).

Dioxygen Difluoride: (O₂F₂)

This compound is a yellow-orange solid (mp 109.7 K) that is made by high-voltage electric discharge on mixtures of O_2 and F_2 at low temperature and pressure. It decomposes into the elements in the gas at −50 °C, and is a potent fluorinating and oxidizing agent. Many substances explode on exposure to O_2F_2, even at low pressures.

The structure of O_2F_2 is bent, one fluorine atom being about 87° out of the plane of the other three atoms (Structure 18-IV). The O—O bond is quite short (1.217 Å) compared to the value for H_2O_2 (1.48 Å).

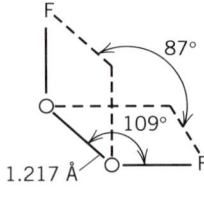

18-IV

STUDY GUIDE

Study Questions

A. Review

1. Give the electron configuration of the oxygen atom.
2. Give two examples of oxonium ions. What is their structure?
3. Describe the carbon–oxygen bond in acetone.

4. Describe the interaction with water of acidic, basic, and neutral oxides. Give two examples of each case.

5. Explain why the oxygen molecule is paramagnetic.

6. Write out the electron configurations of the two excited state singlets found in Table 18-1.

7. Describe the preparation in the laboratory of ozone.

8. How is H_2O_2 made?

9. Write balanced equations for the following reactions: (a) H_2O_2 and $KMnO_4$ in acidic solution; (b) $Fe(OH)_2$ and O_2 in basic solution; (c) sodium peroxide and CO_2; and (d) potassium superoxide and water.

10. What is the difference between oxygenation and oxidation?

B. Additional Exercises

1. Prepare MO energy-level diagrams for all of the ions O_2^{n+} that are chemically important, and determine the bond order and the expected magnetic moment (μ_{eff} in Bohr magnetons, as discussed in Chapter 2).

2. Classify the oxygen atoms in the following systems according to the AB_xE_y scheme of Chapter 3, and, where appropriate, discuss the geometry about oxygen in terms of the VSEPR theory:

   (a) O_2 and O_3 (b) O_2^- and O_2^{2-}

   (c) CH_3OH and H_2O (d) CO_2 and SO_3

   (e) H_2O_2 and OH^- (f) $(CH_3)_2O$ and CH_3CO_2H

   (g) $CH_3C(O)OOH$ (h) peroxodisulfuric acid

3. Draw the orbitals as they interact to form the π-bond systems in

   (a) ketones   (b) carbonate ion

   (c) $[Cl_5Ru-O-RuCl_5]^{4-}$ (d) ozone

   (e) triphenylphosphine oxide   (f) $H_3Si-O-SiH_3$

   (g) $OSCl_2$

4. Calculate the standard redox potential for the air oxidation of Fe^{2+} in aqueous solution.

C. Questions from the Literature of Inorganic Chemistry

1. Compare the structures and properties of two very different "reversible oxygen complexes" as reported by S. J. La Placa and J. A. Ibers, *J. Am. Chem. Soc.*, **1965**, *87*, 2581–2586, and as reported by A. L. Crumbliss and F. Basolo, *J. Am. Chem. Soc.*, **1970**, *92*, 55–60. See also L. Vaska, *Science*, **1963**, *140*, 809.

   (a) Should the oxygen ligands in these complexes be considered to be O_2, O_2^-, or O_2^{2-} ligands?

   (b) Explain how magnetic data support or conflict with your answer to (a).

   (c) What should be the approximate O—O distances in the cobalt–O_2 compounds of Crumbliss?

2. Consider the work by M. M. Morrison, J. L. Roberts, Jr., and D. T. Sawyer, *Inorg. Chem.*, **1979**, *18*, 1971–1973.

   (a) What reaction takes place between OH^- and H_2O_2 in pyridine solution?

   (b) What is formed upon electrochemical reduction of H_2O_2 in pyridine solution?

   (c) After electrochemical reduction of H_2O_2 in pyridine solution, what reaction takes place between HO_2^- and H_2O_2?

   (d) How are the reactions for (c) and (a) related?

(e) What role does solvent play in these reactions? What is different about these redox reactions in water and in pyridine?

SUPPLEMENTARY READING

Bailey, P. S., *Ozonation in Organic Chemistry*, Academic, New York, Vol. 1, 1978, Vol. 2, 1982.

Dotto, L. and Schiff, H., *The Ozone War*, Doubleday, New York, 1978.

Golodets, G. I., *Heterogeneous Catalytic Reactions Involving Oxygen*, Elsevier, Amsterdam, 1983.

Greenwood, G. and Hill, H. O. A., "Oxygen and Life," *Chem. Br.*, **1982,** 194.

Hayaishi, O., *Molecular Oxygen in Biology*, North-Holland, Amsterdam, 1974.

Hoare, P. J., *The Electrochemistry of Oxygen*, Wiley, New York, 1968.

Horvath, M., Bilitzky, L., and Huttner, J., *Ozone*, Elsevier, Amsterdam, 1985.

Martell, A. E. and Sawyer, D. T., Eds., *Oxygen Complexes and Oxygen Activation*, Plenum, New York, 1988.

Murphy, J. S. and Orr, J. R., *Ozone Chemistry and Technology*, Franklin Institute Press, Philadelphia, 1975.

Oberley, L. W., Ed., *Superoxide Dismutase*, Vol. 3. CRC Press, Boca Raton, FL, 1985.

Patai, S., Ed., *The Chemistry of the Hydroxyl Group*, Wiley-Interscience, New York, 1971.

Schaap, A. P., Ed., *Singlet Molecular Oxygen*, Wiley, New York, 1976.

Severn, D., *Organic Peroxides*, Vols. I–III, Wiley-Interscience, New York, 1972.

Spiro, T. G., *Metal Ion Activation of Oxygen*, Wiley, New York, 1983.

Toft-Sorensen, O., Ed., *Nonstoichiometric Oxides*, Academic, New York, 1981.

Valentine, J. S., "The Dioxygen Ligand in Mononuclear Group VIII Transition Metal Complexes," *Chem. Rev.*, **1973,** *73,* 235.

Vaska, L., "Dioxygen Metal Complexes," *Acc. Chem. Res.,* **1976,** *9,* 175.

Chapter 19

THE GROUP VIB(16) ELEMENTS: SULFUR, SELENIUM, TELLURIUM, AND POLONIUM

19-1 Introduction

The position of these elements in the periodic table has been discussed in Chapter 8, and some properties are listed in Table 8-6. The elements of Group VIB(16) bear little resemblance to oxygen for the following reasons:

1. Sulfur, selenium, tellurium, and polonium have lower electronegativities than oxygen; consequently, their compounds have less ionic character. The relative stabilities of their bonds to other elements are also different. In particular, the importance of hydrogen bonding is drastically lowered. Only very weak S---H—S bonds exist, and H_2S is totally different from H_2O (Chapter 7).

2. For sulfur particularly, as in other third-row elements, there is multiple $d\pi-p\pi$ bonding, but little if any $p\pi-p\pi$ bonding. The short S—O distances in SO_4^{2-} (where s and p orbitals are used in σ bonding) is a result of multiple $d\pi-p\pi$ bond character. The latter arises from the flow of electrons from filled $p\pi$ orbitals on O atoms to empty $d\pi$ orbitals on S atoms.

3. The valence for S, Se, Te, and Po atoms is not confined to 2, and d orbitals can be utilized to form more than four bonds to other elements. Examples are SF_6 and $Te(OH)_6$.

4. Sulfur has a strong tendency to catenation, equaled or exceeded only by carbon. Sulfur forms compounds for which there are no known O, Se, or Te analogs. Examples are polysulfide ions, S_n^{2-}, polythionate ions, $[O_3S—S_n—SO_3]^{2-}$, and compounds of the type XS_nX, where X = H, Cl, CN, or NR_2.

The changes in the properties of compounds on going from S to Po can be associated with the increasing size of the atoms and with the decreasing electronegativity, from top to bottom in the group. Some examples of trends in properties of compounds that arise for these reasons are

1. The decreasing stability of the hydrides H_2E.
2. The increasing tendency to form complex ions such as $SeBr_6^{2-}$.
3. The appearance of metallic properties for Te and Po atoms. Thus the oxides MO_2 are ionic and basic, reacting with HCl to give the chlorides.

19-2   Occurrence and Reactions of the Elements

Sulfur occurs widely in nature as the element, as H_2S and SO_2, in metal sulfide ores, and as sulfates [e.g., gypsum and anhydrite ($CaSO_4$), magnesium sulfate, and so on]. Sulfur is obtained on a vast scale from natural hydrocarbon gases such as those in Alberta, Canada, which contain up to 30% H_2S; this is removed by interaction with SO_2, which is obtained from burning sulfur in air.

$$S + O_2 = SO_2 \tag{19-2.1}$$

$$2\,H_2S + SO_2 = 3\,S + 2\,H_2O \tag{19-2.2}$$

Selenium and tellurium are less abundant but frequently occur as selenide and telluride minerals in sulfide ores, particularly those of Ag and Au. They are recovered from flue dusts from combustion chambers for sulfide ores.

Polonium occurs in U and Th minerals as a product of radioactive decay series. The most accessible isotope, ^{210}Po (α, 138.4 days), can be made in gram quantities by irradiation of Bi in nuclear reactors.

$$^{209}Bi(n,\,\gamma)^{210}Bi \xrightarrow{\beta^-} {}^{210}Po \tag{19-2.3}$$

The Po can be separated by sublimation on heating. It is intensely radioactive and special handling techniques are required. The chemistry resembles that of Te but is somewhat more "metallic."

The physical properties and structures of the elements have been described (Chapter 8). On melting, S_8 first gives a yellow, transparent, mobile liquid that becomes dark and increasingly viscous above about 160 °C. The maximum viscosity occurs about 200 °C, but on further heating the mobility increases until the boiling point (444.6 °C), where the liquid is dark red. The "melting point" of S_8 is actually a decomposition point. Just after melting, rings with an average of 13.8 sulfur atoms are formed and at higher temperature, still larger rings form. Then in the high viscosity region there are giant macromolecules that are probably chains with radical ends. At higher temperatures, highly colored S_3 and S_4 molecules are present to the extent of 1–3% at the boiling point. The nature of the physical changes and of the species involved are by no means fully understood.

Sulfur vapor contains S_8 and at higher temperatures S_2 molecules. The latter, like O_2, are paramagnetic with two unpaired electrons, and account for the blue color of the hot vapor.

Cyclosulfurs other than S_8, with ring sizes from S_6 to S_{20}, can be prepared by specific synthetic routes. These compounds are all unstable in solution relative to S_8, but solutions of S_8 do contain, at equilibrium, about 0.3% S_6 and 0.8% S_7, both of which are much more reactive than S_8.

The elements S, Se, and Te burn in air on heating to form the dioxides; they also react on heating with halogens, most metals, and nonmetals. They are attacked by hot oxidizing acids like H_2SO_4 or HNO_3.

In oleums (Section 7-11), S, Se, and Te dissolve to give highly colored solutions that contain cations in which the element is in a fractional oxidation state. Salts of the cations M_4^{2+}, M_8^{2+}, and M_{16}^{2+} have been obtained by selective oxidation of the elements with SbF_5 or AsF_5 in liquid HF. For example,

$$S_8 + 3\,SbF_5 = S_8^{2+} + 2\,SbF_6^- + SbF_3 \tag{19-2.4}$$

or by reactions in molten $AlCl_3$, for example,

$$7 \, Te + TeCl_4 + 4 \, AlCl_3 = 2 \, Te_4^{2+} + 4 \, AlCl_4^- \qquad (19\text{-}2.5)$$

The S_4^{2+}, Se_4^{2+}, and Te_4^{2+} ions are square (Structure 19-I) and there is probably a six π-electron quasiaromatic system. The green Se_8^{2+} ion has a ring structure (Structure 19-II). The S_{16}^{2+} and Se_{16}^{2+} ions have two M_8 rings joined together.

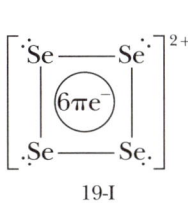

19-I

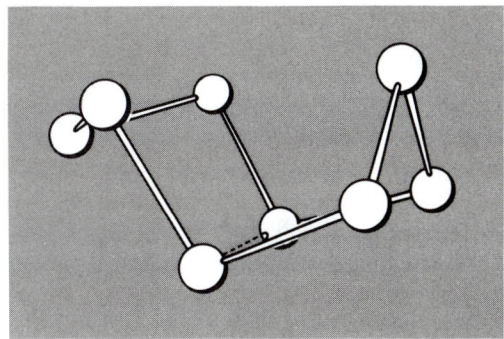

19-II

The reaction of sulfur with the double bonds of natural and synthetic rubbers (a process called vulcanization) is of great technical importance. It leads to formation of S bridges between carbon chains and, hence, to strengthening of rubber.

All reactions of S_8 must involve initial ring opening to give sulfur chains or chain compounds. Many involve nucleophilic reactants, for example,

$$S_8 + 8 \, CN^- \longrightarrow 8 \, SCN^- \qquad (19\text{-}2.6)$$

$$S_8 + 8 \, Na_2SO_3 \longrightarrow 8 \, Na_2S_2O_3 \qquad (19\text{-}2.7)$$

$$S_8 + 8(C_6H_5)_3P \longrightarrow 8(C_6H_5)_3PS \qquad (19\text{-}2.8)$$

Such reactions proceed by a series of steps such as

$$S_8 + CN^- \longrightarrow SSSSSSSSCN^- \qquad (19\text{-}2.9)$$

$$S_6—S—SCN^- + CN^- \longrightarrow S_6SCN^- + SCN^- \quad \text{and so on} \qquad (19\text{-}2.10)$$

Sulfur–sulfur bonds occur in a variety of compounds, and —S—S— bridges are especially important in certain enzymes and other proteins.

19-3   Hydrides: EH₂

These compounds are obtained by the action of acids on metal sulfides, selenides, or tellurides. The hydrides are extremely poisonous gases with revolting odors. The toxicity of H_2S far exceeds that of HCN. The thermal stability and bond strengths decrease down the series, whereas the acidity in water increases.

Hydrogen sulfide dissolves in water to give a solution about 0.1 M at 1 atm. Its dissociation constants are

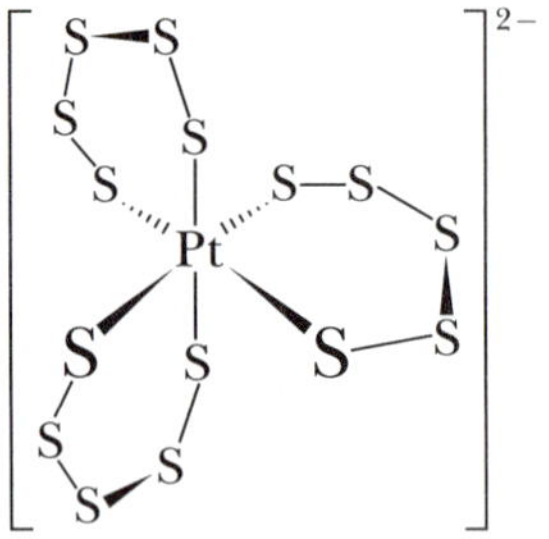

Figure 19-1   A complex ion of Pt^{IV} in which the three S_5^{2-} ligands are each bidentate.

$$H_2S + H_2O = H_3O^+ + SH^- \qquad K = 1 \times 10^{-7} \qquad (19\text{-}3.1)$$

$$SH^- + H_2O = H_3O^+ + S^{2-} \qquad K = {\sim}10^{-14} \qquad (19\text{-}3.2)$$

Owing to this small second dissociation constant, essentially only SH^- ions are present in solutions of ionic sulfides, and S^{2-} occurs only in very alkaline solutions (>8 M NaOH) as shown in Equation 19-3.3.

$$S^{2-} + H_2O = SH^- + OH^- \qquad K = {\sim}1 \qquad (19\text{-}3.3)$$

The compounds H_2S_2 to H_2S_6 are generally known as sulfanes; they contain —SS— to —SSSSS— chains. These compounds can be obtained by reactions such as

$$2\,H_2S(\ell) + S_nCl_2 = 2\,HCl(g) + H_2S_{n+2}(\ell) \qquad (19\text{-}3.4)$$

The anions of the sulfanes (polysulfides, S_n^{2-}) are also easily obtained as salts. Examples are Na_2S_5, K_2S_6, and BaS_4. In addition, the S_4^{2-} and S_5^{2-} ions can serve as chelating ligands, in complexes such as $[Pt(S_5)_3]^{2-}$, whose structure is shown in Fig. 19-1. The latter is chiral and may be resolved into enantiomorphs.

19-4   Halides and Oxohalides of Sulfur

Sulfur Fluorides

Direct fluorination of S_8 yields mainly SF_6 and traces of S_2F_{10} and SF_4. The *tetrafluoride*, SF_4 (bp $-30\ °C$), is evolved as a gas when SCl_2 is refluxed with NaF in acetonitrile at 78–80 $°C$.

$$3\,SCl_2 + 4\,NaF = SF_4 + S_2Cl_2 + 4\,NaCl \qquad (19\text{-}4.1)$$

SF_4 is extremely reactive, and instantly hydrolyzed by water to SO_2 and HF. It is a very selective fluorinating agent converting C=O and P=O groups smoothly into CF_2 and PF_2, and CO_2H and $P(O)OH$ groups into CF_3 and PF_3 groups.

Sulfur hexafluoride is very resistant to chemical attack. Because of its inertness, high dielectric strength, and molecular weight, it is used as a gaseous insulator in high-voltage generators and other electrical equipment. The low reactivity is presumably due to a combination of factors including high S—F bond strength, coordinative saturation, and steric hindrance at sulfur. The inertness of SF_6 is due to kinetic factors and not to thermodynamic stability, since its reaction with H_2O to give SO_3 and HF would be decidedly favorable ($\Delta G = -460$ kJ mol^{-1}).

Sulfur Chlorides

The chlorination of molten sulfur gives S_2Cl_2, which is an orange liquid of revolting smell. By using an excess of Cl_2, with traces of $FeCl_3$ or I_2 as catalyst, at room temperature, an equilibrium mixture of SCl_2 (~ 85%) and S_2Cl_2 is obtained. The dichloride (SCl_2) readily loses chlorine within a few hours, as in the equilibrium of Reaction 19-4.2,

$$2\ SCl_2 = S_2Cl_2 + Cl_2 \qquad (19\text{-}4.2)$$

but it can be obtained pure as a dark red liquid by fractional distillation in the presence of PCl_5, which stabilizes SCl_2.

Sulfur chlorides are solvents for sulfur, giving dichlorosulfanes up to about $S_{100}Cl_2$. These compounds are used in the vulcanization of rubber and are also useful as mild chlorinating agents.

Thionyl chloride ($SOCl_2$) is obtained by Reaction 19-4.3.

$$SO_2 + PCl_5 \longrightarrow SOCl_2 + POCl_3 \qquad (19\text{-}4.3)$$

It is a colorless fuming liquid (bp 80 °C) that is readily hydrolyzed as in Reaction 19-4.4.

$$SOCl_2 + H_2O \longrightarrow SO_2 + 2\ HCl \qquad (19\text{-}4.4)$$

Because the products of reactions such as 19-4.4 are volatile (and therefore easily removed), thionyl chloride is often used to prepare anhydrous chlorides, such as iron(III) chloride, as in Reactions 19-4.5 and 19-4.6.

$$Fe(OH)_3 + 3\ SOCl_2 \longrightarrow 3\ SO_2 + 3\ HCl + FeCl_3 \qquad (19\text{-}4.5)$$

$$FeCl_3{\cdot}6\ H_2O + 6\ SOCl_2 \longrightarrow 6\ SO_2 + 12\ HCl + FeCl_3 \qquad (19\text{-}4.6)$$

Thionyl chloride has a pyramidal structure with sulfur at the apex. Sulfur can be considered to be sp^3 hybridized, and it should be classified as an AB_3E system. The presence of one lone pair on sulfur allows thionyl chloride to act as a weak Lewis base. Some $d\pi$–$p\pi$ bonding between S and O is present.

Sulfuryl chloride (SO_2Cl_2) is obtained by Reaction 19-4.7.

$$SO_2 + Cl_2 \longrightarrow SO_2Cl_2 \qquad (19\text{-}4.7)$$

The reaction requires a catalyst such as $FeCl_3$. Sulfuryl chloride is a colorless liquid that fumes in moist air, due to hydrolysis. It finds use as a chlorinating agent for organic compounds. The structure of sulfuryl chloride may be considered to be derived from a tetrahedron.

19-5 Oxides and Oxo Acids

Table 19-1 lists the formulas and structures of the principal oxo acids of sulfur. In each case the sulfur may be considered to be roughly sp^3 hybridized and falls

Table 19-1 The Principal Oxo Acids of Sulfur

Name	Formula	Structure[a]
Acids Containing One Sulfur Atom		
Sulfurous[b]	H_2SO_3	SO_3^{2-} (in sulfites)
Sulfuric	H_2SO_4	$O{-}\overset{\displaystyle O}{\underset{\displaystyle OH}{S}}{-}OH$
Acids Containing Two Sulfur Atoms		
Thiosulfuric	$H_2S_2O_3$	$O{-}\overset{\displaystyle O}{\underset{\displaystyle OH}{S}}{-}SH$
Dithionous[c]	$H_2S_2O_4$	$HO{-}\overset{\displaystyle O}{S}{-}\overset{\displaystyle O}{S}{-}OH$
Disulfurous[c]	$H_2S_2O_5$	$HO{-}\overset{\displaystyle O}{\underset{\displaystyle O}{S}}{-}\overset{\displaystyle O}{S}{-}OH$
Dithionic	$H_2S_2O_6$	$HO{-}\overset{\displaystyle O}{\underset{\displaystyle O}{S}}{-}\overset{\displaystyle O}{\underset{\displaystyle O}{S}}{-}OH$
Disulfuric	$H_2S_2O_7$	$HO{-}\overset{\displaystyle O}{\underset{\displaystyle O}{S}}{-}O{-}\overset{\displaystyle O}{\underset{\displaystyle O}{S}}{-}OH$
Acids Containing Three or More Sulfur Atoms		
Polythionic	$H_2S_{n+2}O_6$	$HO{-}\overset{\displaystyle O}{\underset{\displaystyle O}{S}}{-}S_n{-}\overset{\displaystyle O}{\underset{\displaystyle O}{S}}{-}OH$
Peroxo Acids		
Peroxomonosulfuric	H_2SO_5	$HOO{-}\overset{\displaystyle O}{\underset{\displaystyle O}{S}}{-}OH$
Peroxodisulfuric	$H_2S_2O_8$	$HO{-}\overset{\displaystyle O}{\underset{\displaystyle O}{S}}{-}O{-}O{-}\overset{\displaystyle O}{\underset{\displaystyle O}{S}}{-}OH$

[a]In most cases the structure given is inferred from the structure of anions in salts of the acid.

[b]The acid is stable in the gas phase as $(HO)_2S{=}O$.

[c]The free acid is unknown.

into either classification AB$_3$E or AB$_4$. Extensive *dπ–pπ* bonding between oxygen and sulfur is to be expected. We approach the chemistry of the acids by considering that they are derived from hydration of the acidic anhydrides SO$_2$ or SO$_3$, or by protonation of the corresponding anions (e.g., sulfates or sulfites).

The Dioxides

The *dioxides* are obtained by burning the elements in air. Sulfur dioxide is produced when many sulfides are heated in air. Selenium and tellurium dioxides are also obtained by treating the elements with hot nitric acid to form H_2SeO_3 and 2 $TeO_2 \cdot HNO_3$, respectively, and then heating these to drive off water or nitric acid.

Sulfur dioxide is a gas with a pungent smell. The molecule is angular. Liquid SO_2 (bp -10 °C) dissolves many organic and inorganic substances and is used as a solvent for NMR studies, as well as in preparative reactions. The liquid does not undergo self-ionization, and any conductivity it may display is due to impurities.

Sulfur dioxide has lone pairs and can act as a Lewis base. However, it also acts as a Lewis acid giving complexes, for example, with amines, as in $(CH_3)_3NSO_2$ and with electron-rich transition metal complexes. In the crystalline compound $SbF_5 \cdot SO_2$, which is of interest because of the use of SO_2 as a solvent for super-acid systems (Section 7-13), the SO_2 is bound as in Structure 19-III. The bonding in Structure 19-IV differs in that the S atom is bound to the metal.

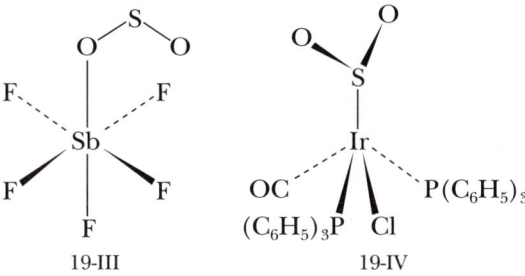

19-III   19-IV

Metal–sulfur bonding appears to be general in transition metal species. Sulfur dioxide also undergoes "insertion" reactions (Chapter 30) with metal–carbon bonds, for example,

$$RCH_2HgOAc + SO_2 \longrightarrow RCH_2SO_2HgOAc \qquad (19\text{-}5.1)$$

$$(CH_3)_4Sn + SO_2 \longrightarrow (CH_3)_3SnSO_2CH_3 \qquad (19\text{-}5.2)$$

Sulfur dioxide is quite soluble in water; such solutions, which possess acidic properties, have long been referred to as solutions of *sulfurous acid*, H_2SO_3. However, H_2SO_3 is either not present or present only in infinitesimal quantities in such solutions. The so-called hydrate, $H_2SO_3 \cdot {\sim}6H_2O$, is the gas hydrate (Section 9-5), $SO_2 \cdot 7H_2O$. The equilibria in aqueous solutions of SO_2 are best represented as

$$SO_2 + x\,H_2O = SO_2 \cdot x\,H_2O \quad \text{(hydrated } SO_2) \qquad (19\text{-}5.3)$$

$$[SO_2 \cdot x\,H_2O = H_2SO_3 \quad K \ll 1] \qquad (19\text{-}5.4)$$

$$SO_2 \cdot x\,H_2O = HSO_3^-(aq) + H_3O^+ + (x-2)H_2O \qquad (19\text{-}5.5)$$

and the first acid dissociation constant for "sulfurous acid" is properly defined as follows:

$$K_1 = \frac{[HSO_3^-][H^+]}{[\text{Total dissolved } SO_2] - [HSO_3^-] - [SO_3^{2-}]} = 1.3 \times 10^{-2}$$

Although sulfurous acid does not exist, two series of salts, the *bisulfites* (containing HSO_3^-) and the *sulfites* (containing SO_3^{2-}) are well known. The SO_3^{2-} ion in crystals is pyramidal. Only the water-soluble alkali sulfites and bisulfites are commonly encountered.

Heating solid bisulfites or passing SO_2 into their aqueous solutions affords *pyrosulfites*.

$$2\ MHSO_3 \underset{}{\overset{heat}{\rightleftharpoons}} M_2S_2O_5 + H_2O$$

$$HSO_3^-(aq) + SO_2 = HS_2O_5^-(aq)$$

Whereas pyro acids, e.g., pyrosulfuric, $H_2S_2O_7$, (Section 7-11) usually have oxygen bridges, the pyrosulfite ion has an unsymmetrical structure, O_2S-SO_3. Some important reactions of sulfites are shown in Fig. 19-2.

Solutions of SO_2 and of sulfites possess reducing properties and are often used as reducing agents.

$$SO_4^{2-} + 4\ H^+ + (x-2)H_2O + 2\ e^- = SO_2 \cdot x\ H_2O \quad E° = 0.17\ V \quad (19\text{-}5.9)$$

$$SO_4^{2-} + H_2O + 2\ e^- = SO_3^{2-} + 2\ OH^- \quad E° = -0.93\ V \quad (19\text{-}5.10)$$

Selenium dioxide is a white volatile solid; the gas consists of discrete and symmetrically bent molecules very similar to those of SO_2. In the solid state, the molecules of SeO_2 associate through $O \rightarrow Se$ donor bonds. For TeO_2, this type of association through adduct formation is so strong that the compound is not volatile.

The Trioxides

Sulfur trioxide is obtained by reaction of SO_2 with O_2, a reaction that is thermodynamically very favorable but extremely slow in the absence of a catalyst such as platinum sponge, V_2O_5, or NO. Sulfur trioxide reacts vigorously with water to form sulfuric acid. Industrially, SO_3 is absorbed in concentrated H_2SO_4 to give oleum (Section 7-11), which is then diluted. Sulfur trioxide is used as such for

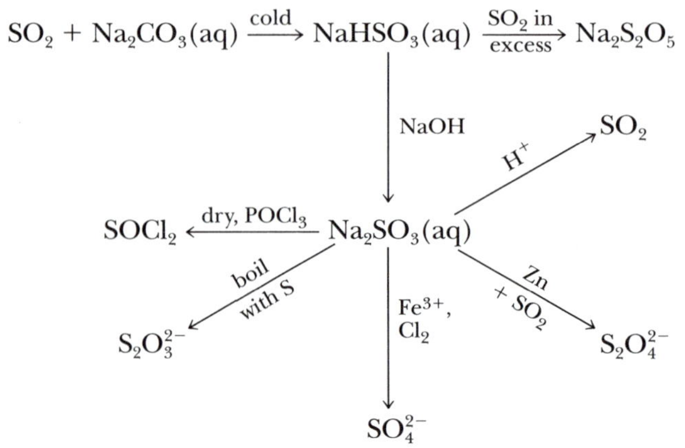

Figure 19-2   Some reactions of sulfites.

preparing sulfonated oils and alkyl arenesulfonate detergents. It is also a powerful, but generally indiscriminate, oxidizing agent.

The SO_3 molecule, in the gas phase, has a planar, triangular structure involving both $p\pi-p\pi$ and $p\pi-d\pi$ S—O bonding and forms polymers in the solid state.

Sulfuric, Selenic, and Telluric Acids

Sulfuric acid has already been discussed in Chapter 7. *Selenic acid* is similar to H_2SO_4, including the isomorphism of the hydrates and salts. It differs in being less stable, evolving oxygen above 200 °C, and being a strong but usually not kinetically fast oxidizing agent.

$$SeO_4^{2-} + 4\,H^+ + 2\,e^- = H_2SeO_3 + H_2O \quad E° = 1.15\,\text{V} \qquad (19\text{-}5.11)$$

Telluric acid, which is obtained by oxidation of Te or TeO_2 with H_2O_2 or other powerful oxidants, is very different in structure, being $Te(OH)_6$ in the crystalline form. It is a very weak dibasic acid ($K_1 \approx 10^{-7}$) and is also an oxidant. Most tellurates contain TeO_6 octahedra as in $K[TeO(OH)_5]$ or Hg_3TeO_6.

Thiosulfates

Thiosulfates are readily obtained by boiling solutions of sulfites with sulfur. The acid is unstable in aqueous solution. The alkali thiosulfates are manufactured for use in photography, where they are used to dissolve unreacted silver bromide from emulsions, by formation of the complexes $[AgS_2O_3]^-$ and $[Ag(S_2O_3)_2]^{3-}$; the thiosulfate ion also forms complexes with other metal ions.

The thiosulfate ion has the structure S—SO_3^{2-}, Structure 19-V:

$$\begin{bmatrix} \text{S} \\ | \\ \text{S} \\ \text{O}^{\diagup} \diagdown_{\text{O}}^{\text{O}} \end{bmatrix}^{2-} \qquad \begin{aligned} \text{S—S} &= 2.01\,\text{Å} \\ \text{S—O} &= 1.47\,\text{Å} \end{aligned}$$

19-V

and may be considered to be derived from sulfate by replacement of an O atom by a S atom.

Dithionites

The reduction of sulfites in aqueous solutions containing an excess of SO_2, by zinc dust, gives ZnS_2O_4. The Zn^{2+} and Na^+ salts are commonly used as powerful and rapid reducing agents in alkaline solution.

$$2\,SO_3^{2-} + 2\,H_2O + 2\,e^- = 4\,HO^- + S_2O_4^{2-} \quad E° = -1.12\,\text{V} \qquad (19\text{-}5.12)$$

In the presence of 2-anthraquinonesulfonate as a catalyst, aqueous $Na_2S_2O_4$ efficiently removes oxygen from inert gases.

The dithionite ion has the structure O_2S—SO_2^{2-}, shown in Structure 19-VI:

$$\begin{bmatrix} \text{O}^{\diagdown}\text{S—S}^{\diagup}\text{O} \\ \text{O} \qquad \text{O} \end{bmatrix}^{2-} \qquad \begin{aligned} \text{S—S} &= 2.39\,\text{Å} \\ \text{S—O} &= 1.51\,\text{Å} \end{aligned}$$

19-VI

The eclipsed conformation, shown in Structure 19-VI, is found only in salts with small cations, whereas in solution it has the staggered conformation. The S—S bond is long and weak.

Dithionates

The dithionate ion has the staggered structure O_3S—SO_3^{2-}. The ion is usually obtained by oxidation of sulfite or SO_2 solutions with manganese(IV) oxide as in Reaction 19-5.13.

$$MnO_2 + 2\ SO_3^{2-} + 4\ H^+ \longrightarrow Mn^{2+} + S_2O_6^{2-} + 2\ H_2O \qquad (19\text{-}5.13)$$

The ion itself is stable and solutions of its salts may be boiled without decomposition. It resists reaction with most oxidizing and reducing agents and is therefore a useful counterion for precipitating complex cations. The free acid may be obtained by treatment of the anion in aqueous solution with sulfuric acid. Dithionic acid is a moderately strong acid that decomposes slowly in concentrated solution or when warmed. Other salts of dithionate (e.g., BaS_2O_6) may be obtained by titration of an aqueous solution of the acid with the appropriate base [e.g., $Ba(OH)_2$]. Such salts are frequently hydrated, barium dithionate being obtained as the dihydrate, $BaS_2O_6 \cdot 2H_2O$.

Polythionates

Polythionate anions have the general formula $[O_3SS_nSO_3]^{2-}$. The corresponding acids are not stable, decomposing rapidly into S, SO_2, and sometimes SO_4^{2-}. The well-established polythionate anions are those with $n = 1$–4. They are named according to the total number of sulfur atoms and are thus called: trithionate $(S_3O_6^{2-})$, tetrathionate $(S_4O_6^{2-})$, and so on. There is evidence for anions having chains with up to 20 sulfur atoms.

Tetrathionates are obtained by treatment of thiosulfates with iodine in the reaction used in the volumetric determination of iodine.

$$2\ S_2O_3^{2-} + I_2 \longrightarrow 2\ I^- + S_4O_6^{2-} \qquad (19\text{-}5.14)$$

The structures of trithionate and tetrathionate are shown in Structures 19-VII and 19-VIII, respectively.

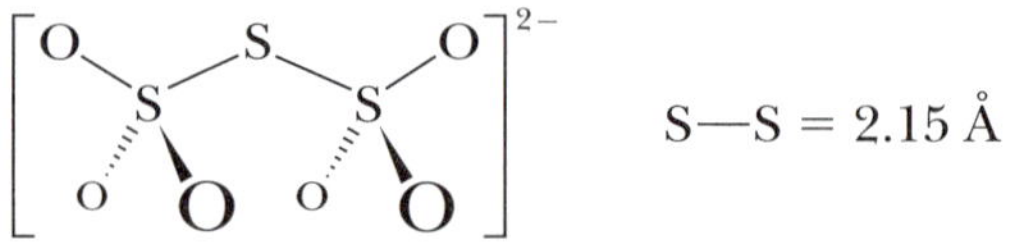

19-VII

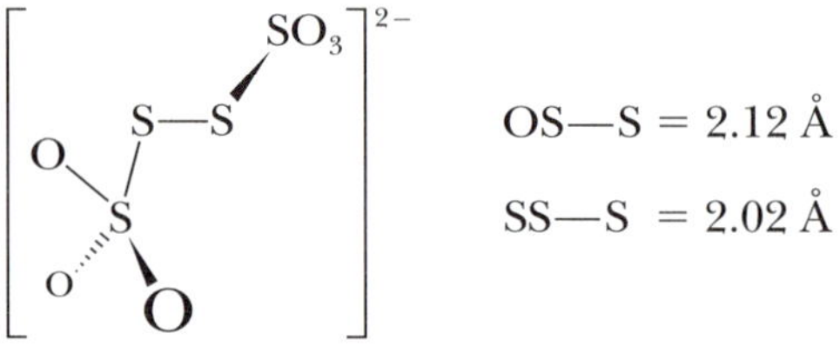

S—S = 2.15 Å

19-VIII

OS—S = 2.12 Å

SS—S = 2.02 Å

Peroxodisulfates

The NH_4^+ or Na^+ salts are obtained by electrolysis of the corresponding sulfates at low temperatures and high current densities. The $S_2O_8^{2-}$ ion has the structure $O_3S—O—O—SO_3$, with approximately tetrahedral angles about each S atom. The ion is one of the most powerful and useful of oxidizing agents.

$$S_2O_8^{2-} + 2\ e^- = 2\ SO_4^{2-} \qquad E° = 2.01\ V \qquad (19\text{-}5.15)$$

However, the reactions are complicated mechanistically. Oxidations by $S_2O_8^{2-}$ are slow and are usually catalyzed by addition of Ag^+, which is converted to Ag^{2+}, the actual oxidant.

STUDY GUIDE

Study Questions

A. Review

1. What are the principal forms in which sulfur occurs in nature?
2. Ordinary solid sulfur consists of what species? Summarize briefly what is observed when sulfur is heated from below its melting point to above its boiling point and explain the reasons for these changes.
3. What types of species are formed on dissolving S, Se, and Te in oleums or other superacids?
4. Discuss the aqueous chemistry of H_2S, SH^-, and S^{2-}.
5. What are the principal fluorides of sulfur?
6. Write equations for the preparations and for the reactions with water of thionyl chloride and sulfuryl chloride.
7. Write equations for the two most important reactions, or types of reaction, of SO_3.
8. Of what use(s) is SO_2?
9. Mention the chief similarities and differences among sulfuric, selenic, and telluric acids.
10. Give general formulas for three series of compounds that contain chains of more than two S atoms.

B. Additional Exercises

1. Compare the boiling points and the acid strengths in the series H_2X, where X = O through Te. Explain the trends.
2. Although SF_6 is unreactive, TeF_6 is hydrolyzed by water. Explain.
3. Describe the preparation and uses of SF_4 and SF_6.
4. Why is it that $SOCl_2$ can act both as a Lewis acid and as a Lewis base?
5. Predict the structure and describe the bonding in $SeOCl_2(py)_2$.
6. Unlike SO_2, SeO_2 is a solid with a chain structure. Draw a reasonable Lewis diagram for such a structure.
7. Draw Lewis diagrams for the following molecules and ions, giving the AB_xE_y classification, the hybridization, and the geometry at each sulfur atom:
    (a) $S_2O_3^{2-}$ (b) $S_2O_4^{2-}$

(c) $S_2O_6^{2-}$ (d) SO_2

(e) SO_3 (f) $SOCl_2$

(g) SO_2Cl_2 (h) SCl_2

(i) S_2Cl_2

8. The bond order of the S—O bond decreases in the series $OSF_2 > OSCl_2 > OSBr_2$. Explain.

9. Predict the structure of the adduct between $SbCl_5$ and $OPCl_3$.

10. Write a balanced equation for the dehydration of selenous acid.

11. Draw pictures representing the orbitals as they overlap in forming $p\pi$–$d\pi$ bonds in SO_3.

12. Prepare an MO energy-level diagram for the π-bond system in SO_3. Use the group orbital approach as described in Chapter 3, and construct π-molecular orbitals centered on the d atomic orbitals of S.

13. The S—O bond length in SO_4^{2-} is 1.44 Å and the S—O bond length in SO_3 is 1.42 Å. Compare these with the bond lengths given in the chapter for $S_2O_3^{2-}$, $S_2O_4^{2-}$, $S_3O_6^{2-}$, and $S_4O_6^{2-}$, and with the sum (S + S and S + O) of the S—S and the S—O covalent single-bond radii. Discuss the relative strengths of S—O and S—S bonds in these systems.

C. Questions from the Literature of Inorganic Chemistry

1. Consider the oxofluorides of Se and Te as reported by H. Oberhammer and K. Seppelt, *Inorg. Chem.*, **1979**, *18*, 2226–2229.

   (a) Draw the Lewis diagrams and discuss the hybridization and geometry (using the AB_xE_y classification and the VSEPR approach) of the following oxofluorides mentioned in this paper: SeO_2F_2, $SeOF_4$, $Se_2O_2F_8$, $Te_2O_2F_8$, $I_2O_4F_6$, F_5SOSF_5, and $F_5SeOSeF_5$.

   (b) In which compounds in (a) is $d\pi$–$p\pi$ bonding between O and Se (or Te) important? Explain.

   (c) What reason(s) do the authors give for the tendency of $SeOF_4$ to dimerize giving $Se_2O_2F_8$? Explain.

   (d) Do you suppose $TeOF_4$ is stable? Explain.

2. Consider the adducts of SO_2 described by P. G. Eller and G. J. Kubas, *Inorg. Chem.*, **1978**, *17*, 894–897.

   (a) Does SO_2 serve as a Lewis acid or as a Lewis base in forming ISO_2^-?

   (b) Other adducts XSO_2^- were not isolated, but stabilities of the adducts were studied. How?

   (c) Draw the Lewis diagram for ISO_2^- and discuss the structural data presented in the article. Classify the S atom according to the AB_xE_y system.

   (d) What is the significance of the I—S distance found in this study? Compare this I—S distance with the sum $r_{cov}(I + S)$ and the sum $r_{vdw}(I + S)$. Is this a fully covalent I—S bond?

3. Consider the paper by C. J. Schack, R. D. Wilson, and J. F. Hon, *Inorg. Chem.*, **1972**, *11*, 208-209.

   (a) Write balanced equations for every step in each synthesis of SeF_5Cl as reported in this paper.

   (b) What is the equation representing the hydrolysis in aqueous hydroxide solution of SeF_5Cl? How was this hydrolysis used to analyze the compound?

   (c) With what other compounds of S and Se do the authors suggest a similarity? On what basis are these comparisons made?

SUPPLEMENTARY READING

Ausari, M. A. and Ibers, J. A., "Soluble Metal Sulfides and Selenides," *Coord. Chem. Rev.,* **1990,** *100,* 223.

Bagnall, K. W., *The Chemistry of Se, Te, and Po,* Elsevier, Amsterdam, 1966.

Clive, D. L. J., *Modern Organo-Selenium Chemistry,* Pergamon Press, New York, 1978.

Cooper, W. C., *Tellurium,* Van Nostrand-Reinhold, New York, 1972.

Engelbrecht, E. and Sladky, F., "Selenium and Tellurium Fluorides," *Adv. Inorg. Chem. Radiochem.,* **1981,** *24,* 189.

Hargittai, I., *The Structure of Volatile Sulfur Compounds,* Kluwer, The Hague, 1985.

Heal, H. G., "Sulfur–Nitrogen Compounds," *Adv. Inorg. Chem. Radiochem.,* **1972,** *15,* 375.

Heal, H. G., *The Inorganic Heterocyclic Chemistry of Sulfur, Nitrogen, and Phosphorus,* Academic, London, 1981.

Janickis, J., "Polythionates and Selenopolythionates," *Acc. Chem. Res.,* **1969,** *2,* 316.

Meyer, B., "Elemental Sulfur," *Chem. Rev.,* **1976,** *76,* 367.

Nickless, G., Ed., *Inorganic Sulfur Chemistry,* Elsevier, Amsterdam, 1968.

Oae, S., *The Organic Chemistry of Sulfur,* Plenum, New York, 1977.

Paulmier, C. *Selenium Reagents in Organic Chemistry,* Pergamon Press, New York, 1986.

Roy, A. B. and Trudinger, P. A., *The Biochemistry of Inorganic Compounds of Sulfur,* Cambridge University Press, Cambridge, UK, 1970.

Chapter 20

THE HALOGENS: FLUORINE, CHLORINE, BROMINE, IODINE, AND ASTATINE

20-1   Introduction

With the exception of He, Ne, and Ar, all of the elements in the periodic table form halides. Ionic or covalent halides are among the most important and common compounds. They are often the easiest to prepare and are widely used as source materials for the synthesis of other compounds. Where an element has more than one valence, the halides are often the best known and most accessible compounds in all of the oxidation states. There is also an extensive and varied chemistry of organic halogen compounds; the fluorine compounds, especially where F completely replaces H, have unique properties.

The position of the elements in the periodic table is outlined in Section 2-5, and some properties are listed in Table 8-7. The element *astatine,* named for the Greek for "unstable," has no stable isotope. As far as can be ascertained by tracer studies, At behaves like I, but is perhaps somewhat less electronegative.

20-2   Occurrence, Isolation, and Properties of the Elements

Fluorine occurs widely, for example as CaF_2 (*fluorspar*), Na_3AlF_6 (*cryolite*), and $3Ca_3(PO_4)_2Ca(F,Cl)_2$ (*fluorapatite*). It is more abundant than chlorine. Fluorine was first isolated in 1886 by H. Moissan. The element is obtained by electrolysis of molten fluorides. The most commonly used electrolyte is $KF \cdot 3HF$ (mp 70–100 °C). As the electrolysis proceeds the melting point increases, but the electrolyte is readily regenerated by resaturation with HF from a storage tank. Fluorine cells are constructed of steel, Cu, or Ni–Cu alloy, which become coated with an unreactive layer of fluoride. The cathodes are steel or Cu, the anodes ungraphitized carbon. Although F_2 is often handled in metal apparatus, it can be handled in glass provided traces of HF, which attacks glass rapidly, are removed by passing the gas through anhydrous NaF or KF with which HF forms the bifluorides (MHF_2).

Fluorine is the most chemically reactive of all the elements and combines directly (often with extreme vigor), at ordinary or elevated temperatures, with all the elements other than O_2, He, Ne, and Kr. It also attacks many other compounds, breaking them down to fluorides; organic materials often inflame and burn in F_2.

The great reactivity of F_2 is in part attributable to the low dissociation energy (Table 1-1) of the F—F bond, and because reactions of atomic fluorine are strongly exothermic. The low F—F bond energy is probably due to repulsion between nonbonding electrons. A similar effect may account for the low bond energies in H_2O_2 and N_2H_4.

Chlorine occurs as NaCl, KCl, $MgCl_2$, and so on, in seawater, salt lakes, and as deposits originating from the prehistoric evaporation of salt lakes. Chlorine is obtained by electrolysis of brine. Older technology employed a mercury cathode in which the sodium dissolved.

$$Na^+ + e^- = Na \qquad (20\text{-}2.1)$$

$$Cl^- = \tfrac{1}{2}Cl_2 + e^- \qquad (20\text{-}2.2)$$

However, this process entailed a hazard because of the loss of mercury to the environment, and a newer process employing membrane cells and not requiring mercury is now common.

Chlorine is a greenish gas. It is moderately soluble in water, with which it reacts as in Reaction 20-2.3.

$$Cl_2 + H_2O = HCl + HOCl \qquad (20\text{-}2.3)$$

Bromine occurs in much smaller amounts, as bromides, along with chlorides. Bromine is obtained from brines by the reaction:

$$2\ Br^- + Cl_2 \xrightarrow{\ \text{pH} \sim 3.5\ } 2\ Cl^- + Br_2 \qquad (20\text{-}2.4)$$

It is swept out in a current of air. Bromine is a dense, mobile, dark red liquid at room temperature. It is moderately soluble in water and miscible with nonpolar solvents such as CS_2 and CCl_4.

Iodine occurs as iodide in brines and as iodate in Chile saltpeter (guano, $NaNO_3$). Various forms of marine life concentrate iodine. Production of I_2 involves either oxidizing I^- or reducing iodates to I^- followed by oxidation. An acid solution of MnO_2 is commonly used as the oxidant.

Iodine is a black solid with a slight metallic luster. At atmospheric pressure it sublimes without melting. It is readily soluble in nonpolar solvents such as CS_2 and CCl_4. Such solutions are purple, as is the vapor. In polar solvents, unsaturated hydrocarbons, and liquid SO_2, brown or pinkish-brown solutions are formed. These colors indicate the formation of weak complexes $I_2 \cdots S$ known as *charge-transfer complexes*. The bonding energy results from partial transfer of charge in the sense $I_2^- S^+$. The complexes of I_2 and also of Br_2, Cl_2, and ICl can sometimes be isolated as crystalline solids at low temperatures.

Iodine forms a blue complex with starch, in which the iodine forms linear I_5^- ions in channels in the polysaccharide amylose.

Astatine has been identified as a short-lived product in the natural radioactive decay series of uranium and thorium. The element was first obtained in quantities sufficient for study by the $^{209}Bi(\alpha,2n)^{211}At$ reaction (Chapter 1). About 20 isotopes of astatine are known, but the longest lived has a half-life of only 8.3 h. As a result, macroscopic quantities cannot normally be isolated for synthetic purposes, although a few inorganic compounds (HAt, CH_3At, AtI, AtBr, and AtCl) have been detected by mass spectrometry. Astatine appears to behave chemically about as would be expected on extrapolation of the properties of the other halogens. It is rather volatile and somewhat soluble in water. A few organic compounds, such as C_6H_5At, $C_6H_5AtCl_2$, and $C_6H_5AtO_2$ are known.

20-3 Halides

There are almost as many ways of classifying halides as there are types of halides. Binary halides may form simple molecules, or complex, infinite arrays. For ionic compounds some common types of lattices are given in Chapter 4 and some general points on halides are discussed in Section 5-5. Other types of halide compounds include oxide halides such as $VOCl_3$, hydroxy halides, and organohalides. The covalent and ionic radii are given in Table 8-7.

Preparation of Anhydrous Halides

1. Direct interaction with the elements. The halogens themselves are normally used for most elements. The compounds HF, HCl, and HBr may also be used for metals.

Direct fluorination normally gives fluorides in the higher oxidation states. Most metals and nonmetals react very vigorously with F_2; with nonmetals such as P_4, the reaction may be explosive. For rapid formation in dry reactions of *chlorides, bromides,* and *iodides* elevated temperatures are usually necessary. For metals, the reaction with Cl_2 and Br_2 may be more rapid when THF or some other ether is used as a reaction medium; the halide is then obtained as a solvate.

2. Dehydration of hydrated halides. The dissolution of metals, oxides, or carbonates in aqueous halogen acids followed by evaporation or crystallization gives hydrated halides. These can sometimes be dehydrated by heating in vacuum, but this often leads to impure products or oxohalides. Dehydration of chlorides can be effected by thionyl chloride, and halides in general can be treated with 2,2-dimethoxypropane.

$$CrCl_3 \cdot 6\,H_2O + 6\,SOCl_2 \xrightarrow{\text{reflux}} CrCl_3 + 12\,HCl + 6\,SO_2 \quad (20\text{-}3.1)$$

$$MX_n \cdot m\,H_2O \text{ in } CH_3C(OCH_3)_2CH_3 \longrightarrow$$
$$MX_n + m\,(CH_3)_2CO + 2m\,CH_3OH \quad (20\text{-}3.2)$$

The acetone and methanol products of Reaction 20-3.2 may solvate the halide products, but the solvents can easily be removed by gentle heating or at reduced pressures.

3. Treatment of oxides with other halogen compounds. Oxides may often be treated with halogen-containing compounds to replace oxygen with halogen, as in the following reactions:

$$NiO + ClF_3 \longrightarrow NiF_2 + \cdots \quad (20\text{-}3.3)$$

$$Pr_2O_3 + 6\,NH_4Cl \xrightarrow{300^\circ\,C} 3\,PrCl_3 + 3\,H_2O + 6\,NH_3 \quad (20\text{-}3.4)$$

$$Sc_2O_3 + CCl_4 \xrightarrow{600^\circ\,C} ScCl_3 + \cdots \quad (20\text{-}3.5)$$

4. Halogen exchange. Many halides react to exchange halogen with (a) elemental halogens, (b) acid halides, (c) halide salts, or (d) an excess of another halogen-containing substance. Chlorides can often be converted to either bromides (by KBr) or especially to iodides (by KI), using acetone, in which KCl is less soluble. Halogen exchange is especially important for the synthesis of fluorides from chlorides, using various metal fluorides such as CoF_3 or AsF_5. This type of replacement is used extensively in the synthesis of organic fluorine compounds, as discussed in Section 20-7.

Another fluorinating agent that has special advantages is SbF_3, which is used along with $SbCl_5$ as a catalyst in Reaction 20-3.6.

$$C_6H_5CCl_3 + SbF_3 \longrightarrow C_6H_5CF_3 + SbCl_3 \qquad (20\text{-}3.6)$$

Molecular Halides

Most of the electronegative elements, and the metals in high oxidation states, form molecular halides. These halides are gases, liquids, or volatile solids with molecules held together only by van der Waals forces. There is probably a rough correlation between increasing metal-to-halogen covalence and increasing tendency to the formation of molecular compounds. Thus the molecular halides are sometimes also called the covalent halides. The designation molecular is preferable, since it states a fact.

The formation of halide bridges between two or, less often, three other atoms is an important structural feature. Between two metal atoms, the most common situation involves two halogen atoms, but examples with one and three bridge atoms are known. Such bridges used to be depicted as involving a covalent bond to one metal atom and donation of an electron pair to the other as in Structure 20-I, but structural data show that the two bonds to each bridging halogen atom are usually equivalent as in Structure 20-II. Molecular orbital theory provides a simple, flexible formulation in which the M—X—M group is treated as a three-center, four-electron (3c–4e) group (cf. Section 3-7).

With Cl^- and Br^-, bridges are characteristically bent, whereas fluoride bridges may be either bent or linear. Thus, in BeF_2 there are infinite chains, - - - BeF_2BeF_2- - -, with bent bridges, similar to the situation in $BeCl_2$. On the other hand, transition metal pentahalides afford a notable contrast. While the pentachlorides dimerize with bent M—Cl—M bridges (Structure 20-II), the pentafluorides form cyclic tetramers with linear M—F—M bridges (Structure 20-III). The fluorides probably adopt the tetrameric structures with linear bridges, in part because the smaller size of F than of Cl would introduce excessive metal–metal repulsion in a bent bridge.

20-I   20-II   20-III

Molecular fluorides of both metals and nonmetals are usually gases or volatile liquids. Their volatility is due to the absence of intermolecular bonding other than van der Waals forces, since the polarizability of fluorine is very low and no suitable outer orbitals exist for other types of attraction. Where the central atom has suitable vacant orbitals available, and especially if the polarity of the single bonds M—F would be such as to leave a considerable charge on M, as in, say, SF_6, multiple bonding can occur using filled p orbitals of fluorine for overlap with vacant orbitals of the central atom. This multiple bonding is a major factor in the shortness and high strength of many bonds to fluorine. Because of the high electronegativity of fluorine the bonds in these compounds tend to be very polar. Because of the low dissociation energy of F_2 and the relatively high energy of many bonds to F (e.g., C—F, 486; N—F, 272; P—F, 490 kJ mol^{-1}), molecular fluorides are often formed very exothermically.

The high electronegativity of fluorine often has a profound effect on the properties of molecules in which several F atoms occur. Representative examples include (a) CF_3CO_2H is a strong acid; (b) $(CF_3)_3N$ and NF_3 have no basicity; and (c) CF_3 derivatives in general are attacked much less readily by electrophilic reagents in anionic substitutions than are CH_3 compounds. The CF_3 group may be considered as a kind of large pseudohalogen with an electronegativity about comparable to that of Cl.

A fairly general property of molecular halides is their easy hydrolysis, for example,

$$BCl_3 + 3\ H_2O \longrightarrow B(OH)_3 + 3\ H^+ + 3\ Cl^- \qquad (20\text{-}3.7)$$

$$PBr_3 + 3\ H_2O \longrightarrow HPO(OH)_2 + 3\ H^+ + 3\ Br^- \qquad (20\text{-}3.8)$$

$$SiCl_4 + 4\ H_2O \longrightarrow Si(OH)_4 + 4\ H^+ + 4\ Cl^- \qquad (20\text{-}3.9)$$

Where the maximum covalency is attained, as in CCl_4 or SF_6, the halides may be quite inert towards water. However, this is a result of kinetic and not thermodynamic factors. For instance, for CF_4, the equilibrium for hydrolysis, as in Reaction 20-3.10

$$CF_4(g) + 2\ H_2O(\ell) = CO_2(g) + 4\ HF(g) \qquad (20\text{-}3.10)$$

is thermodynamically favorable ($K_{eq} = 10^{23}$), but the rate of hydrolysis is negligible because there is no site for attack by water at carbon. The necessity for means of attack is also illustrated by the fact that SF_6 is not hydrolyzed, whereas SeF_6 and TeF_6 are hydrolyzed at 25 °C. Attack by a nucleophile (and expansion of the coordination sphere) is possible only for Se and Te, not S.

20-4 Halogen Oxides

Oxygen fluorides have been studied as potential rocket fuel oxidizers. Oxygen difluoride (OF_2) is obtained as a pale yellow gas on passing F_2 gas rapidly through a 2% NaOH solution. Dioxygen difluoride (O_2F_2) is an unstable orange-yellow solid made by the action of electric discharges on F_2–O_2 mixtures; O_2F_2 is an extremely potent oxidizing and fluorinating agent.

Chlorine oxides are reactive, and tend to explode. None can be obtained by direct reaction of Cl_2 and O_2. The *dioxide* (ClO_2) is a powerful oxidant and is used diluted with air commercially, for example, for bleaching wood pulp. It is always made "on site" by Reaction 20-4.1

$$2\ NaClO_3 + SO_2 + H_2SO_4 = 2\ ClO_2 + 2\ NaHSO_4 \qquad (20\text{-}4.1)$$

or by reduction of $KClO_3$ with moist oxalic acid at 90 °C, carbon dioxide being an additional product. Chlorine dioxide is monomeric (Structure 20-IV), but in the crystal the molecules associate loosely, pairwise as in Structure 20-V, and the solid becomes diamagnetic at low temperatures.

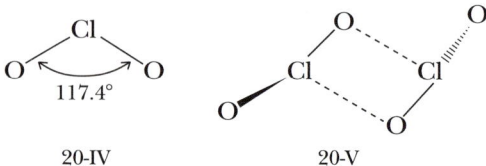

20-IV   20-V

The compound Cl_2O is used as a selective and powerful chlorinating agent for organic compounds. It can also be passed into lime water to make $Ca(OCl)_2$, a safe and useful bleaching agent.

Iodine pentoxide is the anhydride of iodic acid, and can be prepared by dehydration of iodic acid at elevated temperatures, as in Reaction 20-4.2.

$$2\ HIO_3 \underset{H_2O}{\overset{240\ °C}{\rightleftharpoons}} I_2O_5 + H_2O \qquad (20\text{-}4.2)$$
$$\text{fast}$$

Iodine pentoxide is an oxidizing agent that can be used in the determination of CO, as in Reaction 20-4.3

$$5\ CO + I_2O_5 = I_2 + 5\ CO_2 \qquad (20\text{-}4.3)$$

where the liberated iodine is determined by iodometry. Iodine pentoxide has a three-dimensional network structure with $O_2I—O—IO_2$ units linked by strong intermolecular $I \cdots O$ interactions.

20-5 The Oxo Acids

The chemistry of the halogen oxo acids is complicated. Solutions of the acids and several of the anions may be obtained by interaction of the free halogens with water or aqueous bases. In this section the term halogen refers to Cl, Br, and I only; fluorine forms only FOH as discussed in the following subsection.

Reaction of Halogens with H_2O and OH^-

The potentials and equilibrium constants necessary to understand these systems can be derived from data given in Table 20-1.

The halogens are all soluble in water to some extent. However, in such solutions there are species other than solvated halogen molecules, since a *disproportionation* reaction occurs *rapidly*.

$$X_2(g,\ell,s) = X_2(aq) \qquad\qquad K_1 \qquad (20\text{-}5.1)$$

$$X_2(aq) = H^+ + X^- + HOX \qquad K_2 \qquad (20\text{-}5.2)$$

Table 20-1   Standard Potentials (in V) for Reactions of Halogen Compounds

Reaction	Cl	Br	I
(1) $H^+ + HOX + e^- = \frac{1}{2} X_2(g,\ell,s) + H_2O$	1.63	1.59	1.45
(2) $3 H^+ + HXO_2 + 3 e^- = \frac{1}{2} X_2 (g,\ell,s) + 2 H_2O$	1.64		
(3) $6 H^+ + XO_3^- + 5 e^- = \frac{1}{2} X_2(g,\ell,s) + 3 H_2O$	1.47	1.52	1.20
(4) $8 H^+ + XO_4^- + 7 e^- = \frac{1}{2} X_2(g,\ell,s) + 4 H_2O$	1.42	1.59	1.34
(5) $\frac{1}{2} X_2(g,\ell,s) + e^- = X^-$	1.36	1.07	0.54^a
(6) $XO^- + H_2O + 2 e^- = X^- + 2 OH^-$	0.89	0.76	0.49
(7) $XO_2^- + 2 H_2O + 4 e^- = X^- + 4 OH^-$	0.78		
(8) $XO_3^- + 3 H_2O + 6 e^- = X^- + 6 OH^-$	0.63	0.61	0.26
(9) $XO_4^- + 4 H_2O + 8 e^- = X^- + 8 OH^-$	0.56	0.69	0.39

[a]Indicates that I^- can be oxidized by oxygen in aqueous solution.

The values of K_1 are Cl_2, 0.062; Br_2, 0.21; I_2, 0.0013. The values of K_2 computed from the potentials in Table 20-1 are 4.2×10^{-4} for Cl_2, 7.2×10^{-9} for Br_2, and 2.0×10^{-13} for I_2. We can also estimate from

$$\frac{1}{2} X_2 + e^- = X^- \tag{20-5.3}$$

and

$$O_2 + 4 H^+ + 4 e^- = 2 H_2O \qquad E° = 1.23 \text{ V} \tag{20-5.4}$$

that the potentials for the reactions

$$2 H^+ + 2 X^- + \frac{1}{2} O_2 = X_2 + H_2O \tag{20-5.5}$$

are -1.62 V for fluorine, -0.13 V for chlorine, 0.16 V for bromine, and 0.69 V for iodine.

Thus for saturated solutions of the halogens in water at 25 °C we have the results shown in Table 20-2. There is an appreciable concentration of HOCl in a saturated aqueous solution of Cl_2, a smaller concentration of HOBr in a saturated solution of Br_2, but only a negligible concentration of HOI in a saturated solution of I_2.

Hypohalous Acids

The colorless, very unstable gas FOH is made by passing F_2 over ice and collecting the gas in a trap. It reacts rapidly with water. The other XOH compounds are

Table 20-2   Equilibrium Concentrations in Aqueous
Solutions of the Halogens (25 °C, mol L^{-1})

	Cl_2	Br_2	I_2
Total solubility	0.091	0.21	0.0013
Concentration X_2(aq), (mol L^{-1})	0.061	0.21	0.0013
$[H^+] = [X^-] = [HOX]$	0.030	1.15×10^{-3}	6.4×10^{-6}

also unstable. They are known only in solution from the interaction of the halogen and Hg^{II} oxide.

$$2 X_2 + 2 HgO + H_2O \longrightarrow HgO \cdot HgX_2 + 2 HOX \qquad (20\text{-}5.6)$$

The hypohalous acids are very weak acids but good oxidizing agents, especially in acid solution (see Table 20-1).

The *hypohalite ions* can be produced in principle by dissolving the halogens in base according to the general reaction

$$X_2 + 2 OH^- \longrightarrow XO^- + X^- + H_2O \qquad (20\text{-}5.7)$$

and for these *rapid* reactions the equilibrium constants are all favorable: 7.5×10^{15} for Cl_2, 2×10^8 for Br_2, and 30 for I_2.

However, the hypohalite ions tend to disproportionate in basic solution to produce the *halate ions*.

$$3 XO^- = 2 X^- + XO_3^- \qquad (20\text{-}5.8)$$

For these reactions, the equilibrium constants are very favorable: 10^{27} for ClO^-, 10^{15} for BrO^-, and 10^{20} for IO^-. Thus the *actual* products obtained on dissolving the halogens in base depend on the rates at which the hypohalite ions that were initially produced undergo disproportionation. These rates vary with temperature.

The disproportionation of ClO^- is slow at and below room temperature. Thus, when Cl_2 reacts with base "in the cold," reasonably pure solutions of Cl^- and ClO^- are obtained. In hot solutions ($\sim$ 75 °C) the rate of disproportionation is fairly rapid and good yields of ClO_3^- can be secured.

The disproportionation of BrO^- is moderately fast even at room temperature. Solutions of BrO^- can only be made and/or kept at around 0 °C. At temperatures of 50–80 °C quantitative yields of BrO_3^- are obtained.

$$3 Br_2 + 6 OH^- \longrightarrow 5 Br^- + BrO_3^- + 3 H_2O \qquad (20\text{-}5.9)$$

The rate of disproportionation of IO^- is so fast that it is unknown in solution. Hence, reaction of I_2 with base gives IO_3^- quantitatively according to an equation analogous to that for Br_2.

Halous Acids

The only certain acid is *chlorous acid* ($HClO_2$). This acid is obtained in aqueous solution by treating a suspension of barium chlorite with H_2SO_4, and filtering off the $BaSO_4$. It is a relatively weak acid ($K_a \approx 10^{-2}$) and cannot be isolated. *Chlorites* ($MClO_2$) are obtained by reaction of ClO_2 with solutions of bases.

$$2 ClO_2 + 2 OH^- \longrightarrow ClO_2^- + ClO_3^- + H_2O \qquad (20\text{-}5.10)$$

Chlorites are used as bleaching agents. In alkaline solution ClO_2^- is quite stable even on boiling. In acid solutions, the decomposition is rapid and is catalyzed by Cl^-.

$$5\ HClO_2 \longrightarrow 4\ ClO_2 + Cl^- + H^+ + 2\ H_2O \qquad (20\text{-}5.11)$$

Halic Acids

Iodic acid, HIO_3, is a stable white solid obtained by oxidizing I_2 with concentrated HNO_3, H_2O_2, O_3, and so on. *Chloric* and *bromic acids* are obtained in solution by treating the barium halates with H_2SO_4.

The halic acids are strong acids and are powerful oxidizing agents. The ions XO_3^- are pyramidal, as is to be expected from the presence of an octet, with one unshared pair in the halogen valence shell, that is, an AB_3E species.

Iodates of the +4 ions of Ce, Zr, Hf, and Th can be precipitated from 6 M nitric acid to provide a useful means of separation.

Halates and Perhalates

Although disproportionation of ClO_3^- is thermodynamically very favorable,

$$4\ ClO_3^- = Cl^- + 3\ ClO_4^- \qquad K \sim 10^{29} \qquad (20\text{-}5.12)$$

the reaction occurs very slowly in solution and is not a useful preparative procedure. *Perchlorates* are prepared by electrolytic oxidation of chlorates. The properties of *perchloric acid* are discussed in Section 7-11 and perchlorates are discussed in Section 5-3.

The disproportionation of BrO_3^- to BrO_4^- and Br^- is extremely unfavorable ($K \sim 10^{-33}$). *Perbromates* can be obtained only by oxidation of BrO_3^-, preferably by F_2, in basic solution.

$$BrO_3^- + F_2 + 2\ OH^- = BrO_4^- + 2\ F^- + H_2O \qquad (20\text{-}5.13)$$

The perbromates are exceedingly powerful oxidants.

$$BrO_4^- + 2\ H^+ + 2\ e^- = BrO_3^- + H_2O \qquad E^\circ = +1.76\ V \qquad (20\text{-}5.14)$$

Solutions of $HBrO_4$ up to 6 M are stable, but decompose when stronger.

Periodates resemble tellurates in their stoichiometries. The main equilibria in acid solutions are

$$H_5IO_6 = H^+ + H_4IO_6^- \qquad K = 1 \times 10^{-3} \qquad (20\text{-}5.15)$$

$$H_4IO_6^- = IO_4^- + 2\ H_2O \qquad K = 29 \qquad (20\text{-}5.16)$$

$$H_4IO_6^- = H^+ + H_3IO_6^{2-} \qquad K = 2 \times 10^{-7} \qquad (20\text{-}5.17)$$

In aqueous solutions at 25 °C the main ion is IO_4^-. The pH-dependent equilibria are established rapidly. Kinetic studies of the hydration of IO_4^- suggest either one-step or two-step paths (Fig. 20-1), the latter being more likely. Periodic acid and its salts are used in organic chemistry as oxidants that usually react smoothly and rapidly. They are useful analytical oxidants; for example, they oxidize Mn^{2+} to MnO_4^-.

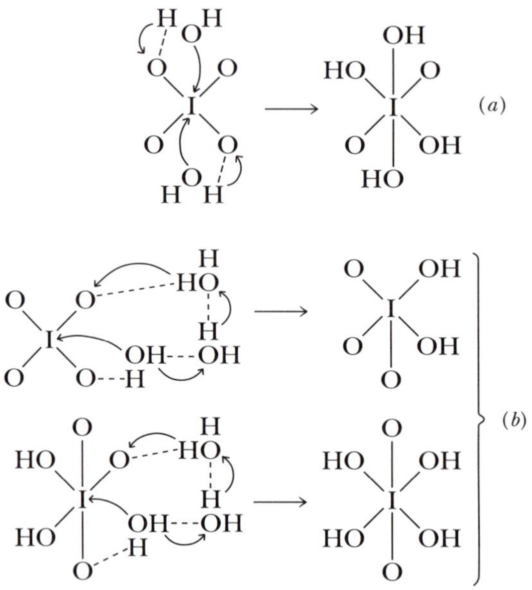

Figure 20-1 Schematic representation of (*a*) the one-step and (*b*) the two-step mechanism for the aquation of IO_4^- to $IO_2(OH)_4^-$. Dotted lines represent hydrogen bonds.

20-6 Interhalogen Compounds

The halogens form many compounds among themselves in binary combinations that may be neutral or ionic (e.g., BrCl, IF_5, Br_3^+, I_3^-). *Ternary* combinations occur only in polyhalide ions (e.g., $IBrCl^-$).

Neutral interhalogen compounds are of the type XX'_n, where *n* is an *odd* number, and X′ is always the lighter halogen when $n > 1$. Because *n* is odd, the compounds are diamagnetic; their valence electrons are present either as bonding pairs or as unshared pairs. The principles involved in the bonding are similar to those in xenon fluorides and have been discussed in Chapter 3.

Chlorine trifluoride is a liquid (bp 11.8 °C) that is commercially available in tanks. It is made by direct combination at 200–300 °C. Reaction of ClF_3 with excess Cl_2 gives *chlorine monofluoride*, which is a gas (bp −100 °C).

Bromine trifluoride, a red liquid (bp 126 °C), is also made by direct interaction.

These three substances, which are typical of all halogen fluorides, are very reactive. They react explosively with H_2O and organic substances. They are powerful fluorinating agents for inorganic compounds, and when diluted with N_2, they fluorinate organic compounds.

Interhalogen ions can be either cations or anions. Halogen fluorides react with fluoride ion acceptors, for example,

$$2\ ClF + AsF_5 = FCl_2^+AsF_6^- \tag{20-6.1}$$

or with fluoride ion donors,

$$IF_5 + CsF = Cs^+IF_6^- \tag{20-6.2}$$

It is not always clear that such products contain *discrete* ions. For instance, in $ClF_2^+SbF_6^-$ each Cl atom has two close and two distant (belonging to SbF_6^-) fluorine neighbors in a much distorted square.

The pale yellow *triiodide* ion is formed on dissolving I_2 in aqueous KI. There are numerous salts of I_3^-. Other ions are not usually stable in aqueous solution although they can be obtained in CH_3OH or CH_3CN and as crystalline salts of large cations such as Cs^+ or R_4N^+. For chlorine, the ion is formed only in concentrated solution.

$$Cl^-(aq) + Cl_2 \rightleftharpoons Cl_3^-(aq) \qquad K \approx 0.2 \qquad (20\text{-}6.3)$$

The electrical conductance of molten I_2 is ascribed to self-ionization

$$3\,I_2 \rightleftharpoons I_3^+ + I_3^- \qquad (20\text{-}6.4)$$

20-7 Organic Compounds of Fluorine

Although the halogens form innumerable organic compounds, the methods of making organic fluorine compounds and some of their unusual properties are of interest in inorganic chemistry. Fluorination of other halogen compounds by treatment with metal fluorides has been discussed in Section 20-3. These methods are expensive so that alternative cheaper methods suitable for industrial procedures have been developed.

1. Replacement of chlorine using hydrogen fluoride. Anhydrous HF is cheap and can be used to replace Cl in chloro compounds. Catalysts such as $SbCl_5$ or CrF_4, and moderate temperature and pressure are required. Examples are

$$2\,CCl_4 + 3\,HF \longrightarrow CCl_2F_2 + CCl_3F + 3\,HCl \qquad (20\text{-}7.1)$$

$$CCl_3COCCl_3 \xrightarrow{\text{HF}} CF_3COCF_3 \qquad (20\text{-}7.2)$$

2. Electrolytic replacement of hydrogen by fluorine. One of the most important laboratory and industrial methods is the electrolysis of organic compounds in liquid HF at voltages ($\sim$4.5–6) below that required for the liberation of F_2. Steel cells with Ni anodes and steel cathodes are used. Fluorination occurs at the anode. Although many organic compounds give conducting solutions in liquid HF, a conductivity additive may be required. Examples of such fluorinations are

$$(C_2H_5)_2O \longrightarrow (C_2F_5)_2O \qquad (20\text{-}7.3)$$

$$C_8H_{18} \longrightarrow C_8F_{18} \qquad (20\text{-}7.4)$$

$$(CH_3)_2S \longrightarrow CF_3SF_5 + (CF_3)_2SF_4 \qquad (20\text{-}7.5)$$

$$(C_4H_9)_3N \longrightarrow (C_4F_9)_3N \qquad (20\text{-}7.6)$$

$$CH_3CO_2H \longrightarrow CF_3CO_2F \xrightarrow{H_2O} CF_3CO_2H \qquad (20\text{-}7.7)$$

3. Direct replacement of hydrogen by fluorine. Although most organic compounds normally inflame or explode with fluorine, direct fluorination of many compounds is possible as follows.

(a) Catalytic fluorination where the reacting compound and F_2 diluted with N_2 are mixed *in the presence* of copper gauze or a cesium fluoride catalyst. An example is shown in Reaction 20-7.8.

$$C_6H_6 + 9\,F_2 \xrightarrow{\text{Cu, 265 °C}} C_6F_{12} + 6\,HF \qquad (20\text{-}7.8)$$

(b) The reaction of the substrate in the solid state, over a long period of time with F_2 (diluted with He), at low temperature. It is important to allow heat, generated in the exothermic reaction (overall for replacement of H by F, ~ 420 kJ mol^{-1}), which could lead to C—C bond breaking, to be efficiently dissipated. The replacement reaction proceeds by several steps, each less exothermic than the C—C average bond strength, so that, provided the reaction time allows separate completion of individual steps, fluorination without degradation is possible. Examples of materials that can be fluorinated in this way are polystyrene, anthracene, phthalocyanine, carboranes, and so on.

(c) Inorganic fluorides, such as cobalt(III) fluoride, are used for the vapor-phase fluorination of organic compounds, for example,

$$(CH_3)_3N \xrightarrow{\text{CoF}_3} (CF_3)_3N + (CF_3)_2NF + CF_3NF_2 + NF_3 \qquad (20\text{-}7.9)$$

4. Other methods for fluorination. A useful and selective fluorinating agent for oxygen compounds is SF_4 (Section 19-4); for example, ketones RR′CO may be converted to RR′CF$_2$, and carboxylate groups CO$_2$H to CF$_3$.

Cesium fluoride acts as a catalyst in various fluorination reactions, for example,

$$R_FCN + F_2 \xrightarrow{\text{CsF, }-78\,°C} R_FCF_2NF_2 \qquad (R_F = \text{perfluoralkyl})\ (20\text{-}7.10)$$

The F^- ion is very nucleophilic toward unsaturated fluorocarbons and adds to the positive center of a polarized multiple bond. The carbanion so produced may then undergo double-bond migration or may act as a nucleophile leading to the elimination of F^- or another ion by an S_N2 mechanism. Fluoride-initiated reactions of these types have wide scope. The reactions can be carried out in DMF or diglyme by using either the sparingly soluble CsF or the more soluble $(C_2H_5)_4NF$.

An example is:

$$CF_2{=}CFCF_3 \xrightarrow{\text{F}^-} (CF_3)_2CF^- \xrightarrow{\text{I}_2} (CF_3)_2CFI + I^- \qquad (20\text{-}7.11)$$

Properties of Organofluorine Compounds

The C—F bond energy is very high (486 kJ mol^{-1}; cf. C—H 415, and C—Cl 332 kJ mol^{-1}), but organic fluorides are not necessarily particularly stable thermodynamically. The low reactivities of fluorine derivatives can be attributed to the impossibility of expansion of the octet of fluorine and the inability of, say, water to coordinate to fluorine or carbon as the first step in hydrolysis. With chlorine this may be possible using outer *d* orbitals. Because of the small size of the F atom, H can be replaced by F with the least amount of strain or distortion, as compared with replacement by other halogen atoms. The F atoms also effectively shield the

C atoms from attack. Finally, since C bonded to F can be considered to be effectively oxidized (whereas in C—H it is reduced), there is no tendency for oxidation by oxygen. Fluorocarbons are attacked only by hot metals, for example molten Na. When pyrolyzed, they split at C—C rather than C—F bonds.

The replacement of H by F leads to increased density, but less than by other halogens. Completely fluorinated (called perfluoro) derivatives, C_nF_{2n+2}, have very low boiling points for their molecular weights and low intermolecular forces; the weakness of these forces is also shown by the very low coefficient of friction for polytetrafluoroethylene, $(CF_2-CF_2)_n$.

Chlorofluorocarbons are used as nontoxic, inert refrigerants, aerosol bomb propellants, and heat-transfer agents. Fluoroolefins are used as monomers for free radical initiated polymerizations to give oils, greases, and the like, and also as chemical intermediates. The compound $CF_3CHBrCl$ is a safe anaesthetic and the compound $CHClF_2$ is used for making tetrafluoroethylene.

$$2\ CHClF_2 \xrightarrow{\ 500-1000\ °C\ } CF_2{=}CF_2 + 2\ HCl \qquad (20\text{-}7.12)$$

Tetrafluoroethylene (bp −76.6 °C) can be polymerized thermally or in aqueous emulsion; the polymer is used for coating frying pans, resistant gaskets, and the like. Chlorofluorocarbons are now being phased out of use because they are photochemically decomposed in the upper atmosphere to give chlorine atoms, which catalyze ozone decomposition. Since destruction of any further significant percentage of this atmospheric ozone could have adverse effects, the problem of "ozone depletion" has been given serious study in recent years. It is not yet known to what extent permanent damage has already been done, nor is it clear what other events (namely, the increasing CO_2 and SO_2 concentrations in the atmosphere) will contribute to the complicated pattern of O_3 concentrations in the upper atmosphere.

Fluorinated carboxylic acids are strong acids. For example, CF_3CO_2H has $K_a = 5.9 \times 10^{-1}$, while for the parent acetic acid, CH_3CO_2H, $K_a = 1.8 \times 10^{-5}$. Many reactions of fluorocarboxylic acids leave the fluoroalkyl group intact. Consider, for example, the sequence of esterification by Reaction 20-7.13:

$$C_3F_7CO_2H \xrightarrow[C_2H_5OH]{H_2SO_4} C_3F_7CO_2C_2H_5 \qquad (20\text{-}7.13)$$

ammonolysis according to Reaction 20-7.14:

$$C_3F_7CO_2C_2H_5 \xrightarrow{\ NH_3\ } C_3F_7CONH_2 \qquad (20\text{-}7.14)$$

followed either by dehydration:

$$C_3F_7CONH_2 \xrightarrow{\ P_2O_5\ } C_3F_7CN \qquad (20\text{-}7.15)$$

or by reduction:

$$C_3F_7CONH_2 \xrightarrow{\ LiAlH_4\ } C_3F_7CH_2NH_2 \qquad (20\text{-}7.16)$$

all of which leaves the fluoroalkyl group untouched.

Perfluoroalkyl halides are made by Reaction 20-7.17.

$$R_FCO_2Ag + I_2 \xrightarrow{\ heat\ } R_FI + CO_2 + AgI \qquad (20\text{-}7.17)$$

Perfluoroalkyl halides are relatively reactive, undergoing free radical reactions when heated or irradiated. Because of the very strong electron-withdrawing nature of perfluoroalkyl groups, they do not undergo most of the nucleophilic reactions typical of the alkyl halides. Trifluoromethyl iodide is readily cleaved homolytically according to Reaction 20-7.18:

$$CF_3I = CF_3{}^{\cdot} + I^{\cdot} \qquad \Delta H = 115 \text{ kJ mol}^{-1} \qquad (20\text{-}7.18)$$

and radical reactions of CF_3I give CF_3 derivatives, an example being Reaction 20-7.19.

$$CF_3I + P \xrightarrow{\text{heat}} (CF_3)_n PI_{3-n} \qquad (20\text{-}7.19)$$

STUDY GUIDE

Study Questions

A. Review

1. Where, and in what chemical form, are the halogens found in nature?
2. How are the free halogens prepared from their halide salts?
3. List the main methods for the preparations of various anhydrous compounds of chlorine.
4. Give balanced equations for preparations of the following:
   (a) $CrCl_3$ from $[Cr(H_2O)_6]Cl_3$
   (b) $FeCl_3$ from Fe
   (c) PBr_3 from red P
   (d) CuI from aqueous $CuSO_4$
   (e) $FeCl_2$ from Fe
   (f) $GdCl_3$ from Gd_2O_3
5. Why is it impossible to make iodides of elements in high oxidation states, whereas corresponding bromides and chlorides are known?
6. Which elements give chlorides that are essentially insoluble in water or dilute HNO_3?
7. How may halides act as bridging ligands?
8. Give balanced equations for the preparations of the following oxo halogen compounds:
   (a) ClO_2 (b) I_2O_5
   (c) NaOCl(aq)   (d) $NaClO_2$
   (e) $NaClO_3$ (f) $NaClO_4$
9. What are the general formulas and names of the four types of oxo acids of the halogens and their anions? In the case of iodine, there is one of unique stoichiometry. What is its formula?
10. Name one cationic, one neutral, and one anionic interhalogen compound. In those consisting of three or more atoms, state the rule that predicts which atom will be the central atom.
11. Iodine has a very low solubility in water, but dissolves readily in KI(aq). Why?
12. Describe two methods for making fluoroorganic compounds.

B. Additional Exercises

1. The compound F_2O_2 has a very short O—O bond (1.217 Å) compared with those in H_2O_2 (1.48 Å) and O_2^{2-} (1.49 Å). It also has relatively long O—F bonds (1.575 Å) compared with those in OF_2. Why?

2. ClO_2 is a free radical with one unpaired electron, and it has less tendency to dimerize than does NO_2. Why?

3. Suggest a geometry for SbF_3 and $SbCl_5$. Classify each Sb atom according to the AB_xE_y scheme of Chapter 3.

4. Draw the shapes of the following molecules and ions, giving the AB_xE_y classification and the hybridization for each central atom. (a) ClF   (b) BrF_3 (c) IF_5 (d) IF_7 (e) ClF_4^- (f) I_3^- (g) BrF_4^+ (h) ICl_2^+

5. What is the order of acid strength for the following: HClO, $HClO_2$, $HClO_3$, and $HClO_4$? Why?

6. Why can F_2 not be obtained by electrolysis of aqueous solutions of NaF?

7. Predict the details of the structures of (a) O_2F_2 (b) ClO_2 (c) BrO_3^- (d) $H_4IO_6^-$ (e) BrO_4^-

8. Write balanced equations for each of the following:

(a) The oxidation of aqueous HCl by MnO_2.

(b) The oxidation of aqueous HI by MnO_4^-.

(c) Hydrolysis of SeF_6.

(d) Reduction of $KClO_3$ by oxalic acid.

(e) Reaction of aqueous barium chlorite with sulfuric acid.

(f) Oxidation of Mn^{2+} to MnO_4^- by periodic acid.

9. How might you obtain CF_3NO from CF_3I?

10. An unknown metal carbonyl (1.86 g) was heated with excess iodine dissolved in pyridine, liberating CO. The gas was passed over I_2O_5, and the resulting I_2 was extracted with CCl_4. The amount of I_2 in the CCl_4 solution was determined by reaction with sodium thiosulfate, 20.0 mL of a 1.00 M solution being required. Write balanced equations for each step in the analysis, and calculate the formula of the unknown metal carbonyl.

11. Describe the bonding in I_3^- and I_3^+.

12. Describe the three-center, four-electron (3c–4e) bond system of the molecular halide M_2Cl_{10}, Structure 20-II.

C. Problems from the Literature of Inorganic Chemistry

1. Consider the properties of the perbromate ion as reported by E. H. Appleman, *Inorg. Chem.*, **1969**, *8*, 223–227.

(a) Perbromic acid in aqueous solution and alkali perbromates were shown to contain the same tetrahedral BrO_4^- ion. On what basis was this conclusion reached?

(b) Periodates are rapidly hydrated to $H_4IO_6^-$. How was it demonstrated that this does not happen for perbromate?

2. The IF_4^- ion is featured in the work by K. O. Christe and D. Nauman, *Inorg, Chem.*, **1973**, *12*, 59–62.

(a) How was the square planar geometry of IF_4^- established?

(b) Discuss the geometries of IF_4^- and XeF_4 in terms of VSEPR theory.

3. The BrF_4^+ cation was studied by M. D. Lind and K. O. Christe, *Inorg. Chem.*, **1972**, *11*, 608–612.

(a) Discuss the structure in the solid state of $[BrF_4^+][Sb_2F_{11}^-]$ by taking the view that it is constructed through Lewis acid-base interactions between $[BrF_4^+]$, $[SbF_6^-]$, and SbF_5. Identify all donor–acceptor interactions in Fig. 1 of this paper.

(b) What would be the geometries of $[BrF_4^+]$, $[SbF_6^-]$, and SbF_5 in the absence of these solid state interactions?

4. The compound ClF_3O was described in a series of papers by K. O. Christe et al., *Inorg. Chem.*, **1972**, *11*, 2189, 2192, 2196, 2201, 2205, 2209, 2212.

(a) Write equations representing the synthesis of ClF_3O (i) from Cl_2O—note the precautions! (ii) from $NaClO_2$, and (iii) from $ClONO_2$.

(b) Write equations for the thermal decomposition of ClF_3, ClF_5, IOF_5, $FClO_2$, and ClF_3O.

(c) What reactions may be used in photochemical syntheses of ClF_3O?

(d) What is the structure of ClF_3O? Classify it according to the AB_xE_y system.

(e) List two reactions in which ClF_3O serves as a Lewis acid.

(f) List two reactions in which ClF_3O serves as a Lewis base.

(g) What are the structures of the ions ClF_4O^- and ClF_2O^+?

5. Make a list of all of the reasons given recently by Gillespie and Robinson (R. J. Gillespie and E. A. Robinson, *Inorg. Chem.*, **1992**, *31*, 1960–1963) for proposing a new value (0.54 Å) for the covalent radius of fluorine.

SUPPLEMENTARY READING

Banks, R. E., Ed. *Preparation, Properties, and Industrial Applications of Organofluorine Compounds*, Wiley-Horwood, Chichester, 1982.

Banks, R. E., Sharp, D. W. A., and Tatlow, J. C., *Fluorine—The First Hundred Years (1886–1986)*, Elsevier-Sequoia, Lausanne, 1986.

Christe, K. O. and Schack, C. J., "Chlorine Oxyfluorides," *Adv. Inorg. Chem. Radiochem.*, **1976**, *18*, 319.

Downs, A. J. and Adams, C. J., *The Chemistry of Chlorine, Bromine, Iodine, and Astatine*, Pergamon Press, New York, 1975.

Emeleus, H. J., *The Chemistry of Fluorine and its Compounds*, Academic, New York, 1969.

Foster, R., *Organic Charge Transfer Complexes*, Academic, New York, 1969.

German, L. and Zemskoreds, S., *New Fluorinating Agents in Organic Synthesis*, Springer-Verlag, Berlin, 1989.

Gillespie, R. J. and Morton, M. J., "Halogen and Interhalogen Cations," *Q. Rev.*, **1971**, *25*, 553.

Gutmann, V., *Halogen Chemistry*, Academic, New York, 1967.

Hagenmuller, P., Ed., *Inorganic Solid Fluorides: Physics and Chemistry*, Academic, New York, 1985.

Holloway, J. H. and Laycock, D., "Preparations and Reactions of Inorganic Main-Group Oxide Fluorides," *Adv. Inorg. Chem. Radiochem.*, **1983**, *27*, 157.

Naumann, D., *Fluorine and Fluorine Compounds; Special Inorganic Chemistry*, Vol. 2, Steinkopff, Darmstadt, 1980.

O'Donnell, T. A., *The Chemistry of Fluorine*, Pergamon Press, Oxford, 1975.

Schafer, H., "Gaseous Chloride Complexes Containing Halogen Bridges," *Adv. Inorg. Chem. Radiochem.*, **1983**, *26*, 201.

Sheppard, W. A. and Sharts, C. M., *Organo Fluorine Chemistry*, Benjamin, New York, 1970.

Solymosi, F., *Structure and Stability of Salts of Halogen Oxoacids in the Solid Phase*, Wiley, New York, 1977.

Tatlow, J. C. et al., Eds., *Advances in Fluorine Chemistry*, Vols. 1–7, Butterworths, London, 1966–1973.

Chapter 21

THE NOBLE GASES

21-1 Occurrence, Isolation, and Applications

The noble gases (Table 21-1) are minor constituents of the atmosphere, from which Sir William Ramsay was first able to isolate the elements Ne, Ar, Kr, and Xe. William F. Hillebrand had isolated helium gas from uranium minerals, and Ramsay was able to demonstrate that the gas has the same spectrum as the element identified spectroscopically in the sun by Sir J. Norman Lockyer and Sir E. Frankland in 1868.

Helium occurs in radioactive minerals and, notably, in some natural gases in the United States. Its origin is entirely from the decay of uranium or thorium isotopes that emit α-particles. These α-particles are helium nuclei that acquire electrons from surrounding elements, and if the rock is sufficiently impermeable, the helium remains trapped. The gas radon, all of whose isotopes are radioactive with short half-lives, was characterized in the decay series from uranium and thorium.

The elements Ne, Ar, Kr, and Xe are obtained from fractionation of liquid air. The gases were originally termed inert, and thought to have no chemical reactivity at all. They provided the key to the problem of valency, the interpretation of the periodic table, and the concept of the closed-electron shell configuration. Although we now know that some of the noble gases can form compounds, they still provide a point of reference in these respects.

A main use of helium is as the liquid in cryoscopy. Argon may be used to provide an inert atmosphere in laboratory apparatus, in welding, and in gas-filled electric light bulbs. Neon is used for discharge lighting tubes, giving the familiar red glow of "neon" signs.

Radon, formed from other elements by radioactive decay sequences, is a health hazard in houses in certain granite areas. It is taken into the lungs, where by-products from its decay sequences cause cancer. Thorough ventilation of the houses is important in such areas.

21-2 The Chemistry of Xenon

During studies with the very reactive gas PtF_6, Bartlett found that a crystalline solid, $[O_2^+][PtF_6^-]$, was formed with oxygen. He noted that since the ionization enthalpy of Xe is almost identical with that of O_2, an analogous reaction might be expected and, indeed, in 1962 he reported the first compound containing a noble gas, a red crystalline solid first believed to be $[Xe^+][PtF_6^-]$, but now known to be more complex.

Table 21-1 Some Properties of the Noble Gases

Element	Outer Configuration	First Ionization Enthalpy ($kJ\ mol^{-1}$)	Normal bp(K)	Vol. % in atmosphere ($\times 10^4$)
He	$1s^2$	2369	4.2	5.2
Ne	$2s^2 2p^6$	2078	27.1	18.2
Ar	$3s^2 3p^6$	1519	87.3	9340.0
Kr	$4s^2 4p^6$	1349	120.3	11.4
Xe	$5s^2 5p^6$	1169	166.1	0.08
Rn	$6s^2 6p^6$	1036	208.2	

There is now an extensive chemistry of xenon with bonds to F and O; one compound with a Xe—N bond is known, but compounds with bonds to other elements are highly unstable. A few krypton compounds exist, but while there should be an extensive chemistry of Rn, the short lifetimes of the isotopes make study impossible. Xenon only reacts directly with fluorine, but oxygen compounds can be obtained from the fluorides. Certain compounds are very stable and can be made in large quantities. Table 21-2 lists some of the more important compounds and their properties.

Table 21-2 Some Xenon Compounds

Oxidation State	Compound	Form	mp(°C)	Structure	Remarks
II	XeF_2	Colorless crystals	129	Linear	Hydrolyzed to $Xe + O_2$; very soluble in HF(ℓ)
IV	XeF_4	Colorless crystals	117	Square	Stable
VI	XeF_6	Colorless crystals	49.6	Complex, see text	Stable
	Cs_2XeF_8	Yellow solid		Archimedean antiprism	Stable to 400 °C
	$XeOF_4$	Colorless liquid	−46	Square pyramid	Stable
	XeO_2F_2	Colorless crystals	31	Seesaw F-axial	Stable
	XeO_3	Colorless crystals		Pyramidal	Explosive, hygroscopic; stable in solution
VIII	XeO_4	Colorless gas	−35.9	Tetrahedral	Explosive
	XeO_6^{4-}	Colorless salts		Octahedral	Anions $HXeO_6^{3-}$, $H_2XeO_6^{2-}$, $H_3XeO_6^{-}$ also exist

Fluorides

Thermodynamic studies of Reactions 21-2.1 to 21-2.3

$$Xe + F_2 = XeF_2 \qquad (21\text{-}2.1)$$

$$XeF_2 + F_2 = XeF_4 \qquad (21\text{-}2.2)$$

$$XeF_4 + F_2 = XeF_6 \qquad (21\text{-}2.3)$$

show that only these three fluorides exist. The three equilibria are established rapidly only above 250 °C, and the synthesis of one fluoride either from the others or instead of the others must be performed above this temperature. The three fluorides are volatile substances, subliming readily at 25 °C. They can be stored in nickel vessels, but XeF_4 and XeF_6 are exceptionally readily hydrolyzed, and even traces of water must be excluded.

Xenon difluoride (XeF_2) is best made by interaction of Xe with a deficiency of F_2 at high pressures. The deficiency of F_2 insures exclusive formation of the difluoride. It dissolves in water to give solutions with a pungent odor of XeF_2. Hydrolysis is slow in acid solution, but rapid in the presence of bases, due to Reaction 21-2.4.

$$XeF_2 + 2\ OH^- = Xe + \tfrac{1}{2} O_2 + 2\ F^- + H_2O \qquad (21\text{-}2.4)$$

Such aqueous solutions are strong oxidizers, converting HCl to Cl_2 and Ce^{3+} to Ce^{4+}. Xenon difluoride is also a mild fluorinating agent for organic compounds; for example, benzene forms C_6H_5F.

Xenon tetrafluoride (XeF_4) is the easiest of the three fluorides to prepare. On heating a 1:5 mixture of Xe and F_2 at 400 °C and about 6-atm pressure for a few hours, XeF_4 is formed quantitatively. It resembles XeF_2 except for its behavior on hydrolysis, as discussed later. Xenon tetrafluoride will fluorinate aromatic rings in compounds such as toluene.

Xenon hexafluoride (XeF_6) is obtained by the interaction of XeF_4 and F_2 under pressure or directly from Xe and F_2 at temperatures above 250 °C and pressures greater than 50 atm. Xenon hexafluoride is extremely reactive, attacking even quartz as in Reaction 21-2.5.

$$SiO_2 + 2\ XeF_6 \longrightarrow 2\ XeOF_4 + SiF_4 \qquad (21\text{-}2.5)$$

Xenon hexafluoride is a strong acid according to the Lux–Flood definition that was discussed in Chapter 7. It accepts oxide ion from other compounds and inserts fluoride ion in its place. The order of decreasing Lux–Flood acidity is

$$XeF_6 > XeO_2F_4 > XeO_4 > XeOF_4 > XeF_4 > XeO_2F_2 > XeO_3 > XeF_2$$

Any acid reacts by accepting oxide from any base beneath it in this series, and replacing it with fluoride. This can be useful in synthesis, and Reaction 21-2.6 is an example.

$$XeOF_4 + XeO_3 \longrightarrow 2\ XeO_2F_2 \qquad (21\text{-}2.6)$$

The colorless crystals of XeF_6 contain both tetramers and hexamers, each made up of XeF_5^+ units linked by unsymmetrical and bent F^- bridges, as shown in Fig. 21-1. Monomeric XeF_6 in the liquid or the vapor has a distorted octahedral structure because of a lone pair of electrons at Xe.

Xenon Fluoride Complexes

The xenon fluorides will react with strong Lewis acids such as SbF_5 or IrF_5 to give adducts. The three types of adducts formed by XeF_2 are $XeF_2 \cdot MF_5$, $2XeF_2 \cdot MF_5$, and $XeF_2 \cdot 2MF_5$, where M = Ru, Ir, Pt, and so on. Although $XeF_2 \cdot IF_5$ has a molecular rather than ionic structure, in most cases adduct formation involves fluoride ion transfer to give structures that contain ions, such as XeF^+ (formed by loss of F^- from XeF_2), $Xe_2F_3^+$ (which has a planar Structure 21-I),

$$Xe \overset{\overset{\displaystyle F}{\diagup \diagdown} 2.14\,Å}{\underset{151°}{}} Xe \diagdown 1.90\,Å$$

$$F \diagdown   F$$

21-I

and XeF_5^+ (formed by transfer of F^- from XeF_6). Examples include $[XeF_5^+][PtF_6^-]$ and Reaction 21-2.7.

$$2\ XeF_2 + AsF_5 \longrightarrow [Xe_2F_3^+]\ [AsF_6^-] \tag{21-2.7}$$

Xenon hexafluoride can act as a Lewis acid toward F^- and can be converted to heptafluoro or octafluoro xenates as in Reactions 21-2.8 and 21-2.9.

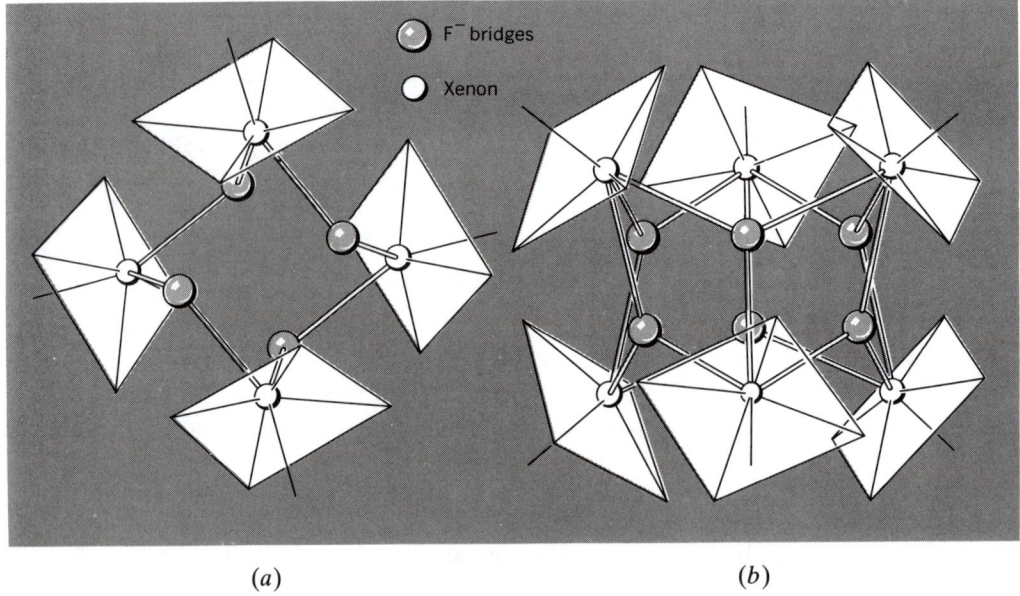

  ◉ F^- bridges

  ○ Xenon

(a)   (b)

Figure 21-1   The tetrameric (a) and the hexameric (b) units that make up the crystal structure of XeF_6. Each is built up of XeF_5^+ units bridged by F^- ions. There are at least four crystalline forms of the substance, three of which are built up of tetramers and one which includes both tetramers and hexamers.

$$XeF_6 + RbF \longrightarrow RbXeF_7 \qquad (21\text{-}2.8)$$

$$2\ RbXeF_7 \longrightarrow XeF_6 + Rb_2XeF_8 \qquad (21\text{-}2.9)$$

These rubidium octafluoroxenates are among the most stable xenon compounds known and decompose only above 400 °C.

Xenon–Oxygen Compounds

Xenon trioxide is formed in the hydrolysis of XeF_4 and XeF_6 according to Reactions 21-2.10 and 21-2.11.

$$3\ XeF_4 + 6\ H_2O \longrightarrow XeO_3 + 2\ Xe + \tfrac{3}{2}\ O_2 + 12\ HF \qquad (21\text{-}2.10)$$

$$XeF_6 + 3\ H_2O \longrightarrow XeO_3 + 6\ HF \qquad (21\text{-}2.11)$$

The colorless, odorless, and stable aqueous solutions of XeO_3 appear to contain XeO_3 molecules. On evaporation of water, XeO_3 is obtained as a white deliquescent solid that is dangerously explosive. In basic solution, a xenate(VI) ion $(HXeO_4^-)$ is formed, as in Reaction 21-2.12.

$$XeO_3 + OH^- \longrightarrow HXeO_4^- \qquad (21\text{-}2.12)$$

The ion $HXeO_4^-$ slowly disproportionates to give a xenate(VIII) (or perxenate ion, XeO_6^{4-}), as in Reaction 21-2.13.

$$2\ HXeO_4^- + 2\ OH^- \longrightarrow XeO_6^{4-} + Xe + O_2 + 2\ H_2O \qquad (21\text{-}2.13)$$

Perxenates are also formed by oxidation of $HXeO_4^-$ with ozone. The perxenate ions are yellow and are both powerful and rapid oxidizing agents. Salts such as $Na_4XeO_6 \cdot 8H_2O$ are stable and sparingly soluble in water.

In alkaline solution, the main form is the ion $HXeO_6^{3-}$, and perxenates are only slowly reduced by water. However, in acid solution, reduction by water according to Reaction 21-2.14 is almost instantaneous, and the hydroxyl radical is involved as an intermediate.

$$H_2XeO_6^{2-} + H^+ \longrightarrow HXeO_4^- + H_2O + \tfrac{1}{2}\ O_2 \qquad (21\text{-}2.14)$$

When barium perxenate is heated with concentrated sulfuric acid, xenon tetroxide (XeO_4) is formed as an explosive and unstable gas.

The aqueous chemistry of xenon is summarized by the potentials:

$$\text{Acid solution} \qquad H_4XeO_6 \xrightarrow{\ 2.36\ V\ } XeO_3 \xrightarrow{\ 2.12\ V\ } Xe$$

$$XeF_2 \xrightarrow{\ 2.64\ V\ } Xe$$

$$\text{Alkaline solution} \qquad HXeO_6^{3-} \xrightarrow{\ 0.94\ V\ } HXeO_4^- \xrightarrow{\ 1.26\ V\ } Xe$$

21-3   Other Noble Gas Chemistry

Radon might be expected to display even more chemistry than xenon, but because of the radioactivity of all radon isotopes, rather little has been learned about it. Apparently, at least one radon fluoride of uncertain composition does exist.

The other noble gas atoms have higher ionization energies than the xenon atom, and they therefore are much less reactive. In a consistent manner those compounds formed by the lighter noble gases are less stable than those of xenon.

Krypton difluoride (KrF_2) is obtained when an electric discharge is passed through a mixture of Kr and F_2 at -180 °C. It resembles XeF_2 being a volatile white solid constructed of linear FKrF molecules, but differs in that it is thermodynamically unstable, as indicated by Reactions 21-3.1 and 21-3.2.

$$KrF_2(g) = Kr(g) + F_2(g) \qquad \Delta H° = -63 \text{ kJ mol}^{-1} \qquad (21\text{-}3.1)$$

$$XeF_2(g) = Xe(g) + F_2(g) \qquad \Delta H° = 105 \text{ kJ mol}^{-1} \qquad (21\text{-}3.2)$$

Some compounds with Xe—C bonds are known. An example is $C_6F_5Xe^+$, which is made by Reaction 21-3.3.

$$XeF_2 + B(C_6F_5)_3 \longrightarrow [C_6F_5Xe^+][F_3BC_6F_5^-] \qquad (21\text{-}3.3)$$

STUDY GUIDE

Study Questions

A. Review

1. What is the origin of terrestrial helium?
2. Why do the boiling points of the noble gases vary systematically with atomic number? What interatomic forces account for this variation?
3. How are XeF_2, XeF_4, and XeF_6 prepared?
4. Write balanced equations for the hydrolyses of XeF_2, XeF_4, and XeF_6.
5. How are xenates and perxenates made?

B. Additional Exercises

1. Write balanced equations for
   (a) The oxidation of $HXeO_4^-$ by ozone.
   (b) The reduction of XeO_3 by I^- in acid solution to give Xe.
   (c) Oxidation of HCl by XeF_2.
   (d) Oxidation of Ce^{3+} by XeF_2.
   (e) Synthesis of $[XeF^+][SbF_6^-]$.
2. Show the electron-pair geometry around each atom in $Xe_2F_3^+$ and classify each atom using the AB_xE_y system.
3. Draw the Lewis diagrams and show the electron-pair geometries around each atom in

(a) XeF_4 (b) XeO_3

(c) XeO_6^{4-} (d) $XeOF_4$

(e) XeF_5^+ (f) $HXeO_4^-$

(g) XeF_8^{2-} (h) XeF_7^-

4. Prepare a MO description of the bonding in XeF_2 using only a colinear set of p orbitals.

5. Discuss the following reactions in terms of the Lux–Flood definition of acids and bases:

(a) $XeF_6 + XeO_2F_2 \longrightarrow 2\ XeOF_4$

(b) $XeO_3F_2 + XeO_2F_2 \longrightarrow XeOF_4 + XeO_4$

(c) $XeF_2 + Na_4XeO_6 \longrightarrow$ no reaction

C. Questions from the Literature of Inorganic Chemistry

1. Consider the work by J. L. Huston, *Inorg. Chem.*, **1982**, *21*, 685–688.

(a) Write a plausible sequence of reactions for the hydrolysis of XeF_4 in excess water.

(b) Explain Reactions 1, 2, 4, 11, 12, 16, and 17 in terms of the Lux–Flood definition of acids and bases. Identify the acid and base in each reaction.

(c) Prepare an order of base strength for each of the bases featured in the reactions of (b).

2. Consider the series of papers by N. Bartlett et al., *Inorg. Chem.*, **1973**, *12*, 1713, 1717, 1722.

(a) Explain how each of the following adducts may be considered to arise from fluoride ion transfer to give ionic compounds with weak F^- bridges in the solid state: (i) $XeF_4 \cdot 2SbF_5$; (ii) $XeF_2 \cdot RuF_5$ and $XeF_4 \cdot RuF_5$; (iii) $XeF_4 \cdot SbF_5$ and $XeF_4 \cdot 2SbF_5$; and (iv) $XeOF_4 \cdot SbF_5$ and $XeOF_4 \cdot 2SbF_5$

(b) Describe the geometries (ignoring the weak F^- bridges) of the cations in the compounds of (a). Use the AB_xE_y classification, and pay close attention to the positions of the lone electron pairs.

(c) Do the oxygen atoms in $XeOF_3^+$ and XeO_2F_2 prefer equatorial or axial positions?

SUPPLEMENTARY READING

Bartlett, N., *The Chemistry of the Noble Gases,* Elsevier, Amsterdam, 1971.

Halloway, J. H., *Noble Gas Chemistry,* Methuen, New York, 1968.

Hawkins, D. T., Falconer, W. E., and Bartlett, N., *Noble Gas Compounds, A Bibliography 1962–1976,* Plenum, New York, 1978.

Hopke, P. K., Ed., *Radon and its Decay Products,* ACS Symposium Series, American Chemical Society, Washington, DC, 1987.

Lazlo, P. and Schrobilgen, G. J., "One or Several Pioneers? The Discovery of Noble-Gas Compounds," *Angew. Chem. Int. Ed. Eng.,* **1988**, *27*, 479–489.

Moody, G. J., "A Decade of Xenon Chemistry," *J. Chem. Educ.,* **1974**, *51*, 628.

Nazeroff, W. K., Ed., *Radon and its Decay Products in Indoor Areas,* Wiley, New York, 1988.

Ozima, M. and Podosek, F. A., *Noble Gas Geochemistry,* Cambridge University Press, Cambridge, UK, 1983.

Selig, H. and Halloway, J. H. *Topics in Current Chemistry No. 124,* F. L. Bosche, Ed., Springer-Verlag, Berlin, 1984.

Seppelt, K. and Lentz, D. *Progress in Inorganic Chemistry,* Vol. 29, Wiley-Interscience, New York, 1982.

Chapter 22

ZINC, CADMIUM, AND MERCURY

22-1 Introduction

The position of Zn, Cd, and Hg in the periodic table is discussed in Section 2-5, and some properties are given in Table 8-2. Mercury shows such unique behavior that it cannot be considered as homologous to Zn and Cd.

Although these elements characteristically form +2 cations, they do not have much in common with the Be, Mg, and Ca to Ra group except for some resemblances between Zn, Be, and Mg. Thus BeO, $Be(OH)_2$, and BeS have the same structures as ZnO, $Zn(OH)_2$, and ZnS, and there is some similarity in the solution and complex chemistry of Zn^{2+} and Mg^{2+}. The main cause of the differences between the Group IIA(2) and the Group IIB(12) ions arises because of the high polarizability of the filled d shell of the Group IIB(12) ions compared with the nonpolarizable, noble gas-like electron configurations of the Group IIA(2) ions.

22-2 Occurrence, Isolation, and Properties of the Elements

The elements have relatively low abundance in nature (of the order 10^{-6} of the earth's crust for Zn and Cd), but have long been known because they are easily obtained from their ores.

Zinc occurs widely, but the main source is *sphalerite,* (ZnFe)S, which commonly occurs with galena (PbS); cadmium minerals are scarce but, as a result of its similarity to Zn, Cd occurs by isomorphous replacement in almost all Zn ores. Methods of isolation involve flotation and roasting, which yield the oxides; the ZnO and PbO are then reduced with carbon. Cadmium is invariably a by/product and is usually separated from Zn by distillation or by precipitation from sulfate solutions by Zn dust.

$$Zn + Cd^{2+} = Zn^{2+} + Cd \quad E° = +0.36 \text{ V} \qquad (22\text{-}2.1)$$

The only important ore of mercury is *cinnabar* (HgS); this ore is roasted to give the oxide which, in turn, decomposes at about 500 °C, the mercury vaporizing.

Zinc and cadmium are white, lustrous, but tarnishable metals. Their structures deviate only slightly from perfect hcp. Mercury is a shiny liquid at ordinary temperatures. All of these elements are remarkably volatile for heavy metals, mercury uniquely so. Mercury gives a monatomic vapor and has an appreciable vapor pressure $(1.3 \times 10^{-3}$ mm) at 20 °C. It is also surprisingly soluble in both polar and nonpolar liquids; a saturated solution in water at 25 °C has 6×10^{-8} g of Hg per gram of H_2O. Because of its high volatility and moderate toxicity, mercury should always be kept in stoppered containers and handled in well-ventilated areas. It becomes extremely hazardous in the biosphere because there are bacteria that convert it to the exceedingly toxic CH_3Hg^+ ion. Mercury is readily lost from aqueous solutions of mercuric salts owing to reduction by traces of reducing materials and by disproportionation of Hg_2^{2+}.

Both Zn and Cd are very electropositive and react readily with nonoxidizing acids, releasing H_2 and giving the divalent ions; Hg is inert to nonoxidizing acids. Zinc also dissolves in strong bases because of its ability to form zincate ions (see Reaction 22-2.2), commonly written ZnO_2^{2-}.

$$Zn + 2\ OH^- \longrightarrow ZnO_2^{2-} + H_2 \qquad (22\text{-}2.2)$$

Cadmium does not dissolve in bases.

Both Zn and Cd react readily when heated in O_2, to give the oxides. Although Hg and O_2 are unstable with respect to HgO at 25 °C, their rate of combination is exceedingly slow; the reaction proceeds at a useful rate at 300–350 °C, but above about 400 °C the ΔG becomes positive and HgO decomposes rapidly into the elements.

$$HgO(s) = Hg(\ell) + \tfrac{1}{2} O_2 \quad \Delta H_{diss} = 90.4 \text{ kJ mol}^{-1} \qquad (22\text{-}2.3)$$

This ability of Hg to absorb O_2 from the air and regenerate it as O_2 was of considerable importance in the earliest studies of the element oxygen by A. L. Lavoisier and J. Priestley.

All three elements react with halogens and with nonmetals such as S, Se, and P.

The elements Zn and Cd form many alloys. Some, such as brass, which is a copper–zinc alloy, are of technical importance. Mercury combines with many other metals, sometimes with difficulty but sometimes, as with Na or K, very vigorously, giving *amalgams*. Many amalgams are of continuously variable compositions, while others are compounds, such as Hg_2Na. Some of the transition metals do not form amalgams, and iron is commonly used for containers of Hg. Sodium amalgams and amalgamated Zn are frequently used as reducing agents for aqueous solutions.

22-3   The Univalent State

The elements Zn, Cd, and Hg form the ions M_2^{2+}. The Zn_2^{2+} and Cd_2^{2+} ions are unstable, especially Zn_2^{2+}, and are known only in melts or solids. Thus addition of Zn to fused $ZnCl_2$ gives a yellow solution and, on cooling, a yellow glass that contains Zn_2^{2+}.

The ions have a metal–metal bond $(^+M—M^+)$; Raman spectra allow the estimation of force constants, and they show that the order of bond strength is $Zn_2^{2+} < Cd_2^{2+} < Hg_2^{2+}$.

The mercury(I) ion (Hg_2^{2+}) is formed on reduction of mercury(II) salts in aqueous solution. X-ray diffraction studies on many compounds such as Hg_2Cl_2, Hg_2SO_4, and $Hg_2(NO_3)_2 \cdot 2H_2O$ show that the Hg—Hg distances range from 2.50 to 2.70 Å, depending on the associated anions. The shortest distances are found with the least covalently bound anions (e.g., NO_3^-).

Hg^I–Hg^{II} Equilibria

An understanding of the thermodynamics of these equilibria is essential to an understanding of the chemistry of the mercury(I) state. The important values are the potentials.

$$Hg_2^{2+} + 2\,e^- = 2\,Hg \qquad E° = 0.789\ \text{V} \qquad\qquad (22\text{-}3.1)$$

$$2\,Hg^{2+} + 2\,e^- = Hg_2^{2+} \qquad E° = 0.920\ \text{V} \qquad\qquad (22\text{-}3.2)$$

$$Hg^{2+} + 2\,e^- = Hg \qquad E° = 0.854\ \text{V} \qquad\qquad (22\text{-}3.3)$$

For the disproportionation equilibrium

$$Hg_2^{2+} = Hg + Hg^{2+} \qquad E° = -0.131\ \text{V} \qquad\qquad (22\text{-}3.4)$$

from which it follows that

$$K = \frac{[Hg^{2+}]}{[Hg_2^{2+}]} = 6.0\times10^{-3} \qquad\qquad (22\text{-}3.5)$$

The implication of the standard potentials is clearly that only oxidizing agents with potentials in the range -0.79 to -0.85 V can oxidize mercury to Hg^I, but not to Hg^{II}. Since no common oxidizing agent meets this requirement, it is found that when mercury is treated with an excess of oxidizing agent it is entirely converted into Hg^{II}. However, when mercury is in at least 50% excess, only Hg^I is obtained since, according to Reaction 22-3.4, Hg readily reduces Hg^{2+} to Hg_2^{2+}.

The equilibrium constant for Reaction 22-3.4 shows that Hg_2^{2+} is stable with respect to disproportionation, but by only a small margin. Thus any reagents that reduce the activity (by precipitation or complexation) of Hg^{2+}, to a significantly greater extent than they lower the activity of Hg_2^{2+}, will cause *disproportionation* of Hg_2^{2+}. There are many such reagents, so that the number of stable Hg^I compounds is quite restricted.

Thus, when OH^- is added to a solution of Hg_2^{2+}, a dark precipitate consisting of Hg and HgO is formed; evidently mercury(I) hydroxide, if it could be isolated, would be a stronger base than HgO. Similarly, addition of sulfide ions to a solution of Hg_2^{2+} gives a mixture of Hg and the extremely insoluble HgS. Mercury(I) cyanide does not exist because $Hg(CN)_2$, although soluble, is so slightly dissociated. The reactions in these cited cases are

$$Hg_2^{2+} + 2\,OH^- \longrightarrow Hg + HgO(s) + H_2O \qquad\qquad (22\text{-}3.6)$$

$$Hg_2^{2+} + S^{2-} \longrightarrow Hg + HgS(s) \qquad\qquad (22\text{-}3.7)$$

$$Hg_2^{2+} + 2CN^- \longrightarrow Hg + Hg(CN)_2(aq) \qquad\qquad (22\text{-}3.8)$$

Dimercury(I) Compounds

As we indicated previously, no hydroxide, oxide, or sulfide can be obtained by addition of the appropriate anion to aqueous Hg_2^{2+}, nor have these compounds been otherwise made.

Among the best known dimercury(I) compounds are the *halides*. The fluoride is unstable toward water, being hydrolyzed to hydrofluoric acid and hydroxide (which immediately disproportionates as shown previously). The other halides are insoluble, which thus precludes the possibilities of hydrolysis or disproportionation to give Hg^{II} halide complexes. Mercury(I) nitrate and perchlorate are soluble in water, but Hg_2SO_4 is sparingly soluble.

22-4   Divalent Zinc and Cadmium Compounds

Binary Compounds

Oxides

The *oxides* (ZnO and CdO) are formed on burning the metals in air or by pyrolysis of the carbonates or nitrates; oxide smokes can be obtained by combustion of the alkyls. The cadmium oxide smokes are exceedingly toxic. Zinc oxide is normally white but turns yellow on heating. Cadmium oxide varies in color from greenish yellow through brown to nearly black, depending on its thermal history. These colors are the result of various kinds of lattice defects. Both oxides sublime at very high temperatures.

The hydroxides are precipitated from solutions of salts by addition of bases. The compound $Zn(OH)_2$ readily dissolves in an excess of alkali bases to give "zincate" ions and solid zincates such as $NaZn(OH)_3$ and $Na_2[Zn(OH)_4]$ can be crystallized from concentrated solutions. Cadmium hydroxide, $Cd(OH)_2$, is insoluble in bases. Both Zn and Cd hydroxide readily dissolve in an excess of strong ammonia to form the ammine complexes, for example, $[Zn(NH_3)_4]^{2+}$. The complete set of formation constants for the cadmium system was presented in Section 6-4.

Sulfides

The *sulfides* are obtained by direct interaction or by precipitation by H_2S from aqueous solutions, acidic for CdS and neutral or basic for ZnS. The sulfides, as well as the selenides and tellurides, all have the wurtzite or zinc blende structures shown in Chapter 4.

Halides

The fluorides are essentially ionic, high melting solids, whereas the other *halides* are more covalent in nature. The fluorides are sparingly soluble in water, a reflection of the high lattice energies of the ZnF_2 (rutile) and CdF_2 (fluorite) structures. The other halides are much more soluble, not only in water but in alcohols, ketones, and similar donor solvents. Aqueous solutions of cadmium halides contain all the species Cd^{2+}, CdX^+, CdX_2, and CdX_3^- in equilibrium.

Oxo Salts and Aqua Ions

Salts of oxo acids such as the nitrate, sulfate, sulfite, perchlorate, and acetate are soluble in water. The Zn^{2+} and Cd^{2+} ions are rather similar to Mg^{2+}, and many of their salts are isomorphous with magnesium salts, for example, $Zn(Mg)SO_4 \cdot 7H_2O$. The aqua ions are acidic, and aqueous solutions of salts are hydrolyzed. In perchlorate solution the only species for Zn, Cd, and Hg below 0.1 M are the MOH^+ ions, for example,

$$Zn^{2+}(aq) + H_2O \rightleftharpoons ZnOH^+(aq) + H^+ \qquad (22\text{-}4.1)$$

For more concentrated cadmium solutions, the principal species is Cd_2OH^{3+}.

$$2\,Cd^{2+}(aq) + H_2O \rightleftharpoons Cd_2OH^{3+}(aq) + H^+ \qquad (22\text{-}4.2)$$

In the presence of complexing anions (e.g., halide), species such as $Cd(OH)Cl$ or $CdNO_3^+$ may be obtained.

Complexes

All of the halide ions except F^- form complex halogeno anions when present in excess, but for Zn^{2+} and Cd^{2+} the formation constants are many orders of magnitude smaller than those for Hg^{2+}. The same applies to complex cations with NH_3 and amines, many of which can be isolated as crystalline salts.

Zinc dithiocarbamates (Section 14-6) are industrially important as accelerators in the vulcanization of rubber by sulfur. Zinc complexes are also of great importance biologically (Section 31-9). Zinc compounds, especially $ZnCO_3$ and ZnO, are used in ointments, since zinc apparently promotes healing processes.

By contrast, cadmium compounds are extremely poisonous, possibly because of the substitution of Cd for Zn in an enzyme system, and consequently they constitute a serious environmental hazard (e.g., in the neighborhood of Zn smelters).

22-5 Divalent Mercury Compounds

Binary Compounds

Red HgO is formed on gentle pyrolysis of mercury(I) or mercury(II) nitrate, by direct interaction at 300–350 °C, or as red crystals by heating of an alkaline solution of K_2HgI_4. Addition of OH^- to aqueous Hg^{2+} gives a yellow precipitate of HgO; the yellow form differs from the red only in particle size.

No hydroxide has been obtained, but the oxide is soluble in water (10^{-3}–10^{-4} mol L^{-1}), the exact solubility depending on particle size, to give a solution of what is commonly assumed to be the hydroxide, although there is no proof for such a species. This "hydroxide" is an extremely weak base:

$$K = \frac{[Hg^{2+}][OH^-]^2}{[Hg(OH)_2]} = 1.8 \times 10^{-22} \qquad (22\text{-}5.1)$$

and is somewhat amphoteric, though more basic than acidic.

Mercury(II) sulfide (HgS) is precipitated from aqueous solutions as a black, highly insoluble compound. The solubility product is 10^{-54}, but the sulfide is somewhat more soluble than this figure would imply because of hydrolysis of Hg^{2+} and S^{2-} ions. The black sulfide is unstable with respect to a red form identical with the mineral cinnabar and changes into it when heated or digested with alkali polysulfides or mercury(I) chloride.

Mercury(II) fluoride is essentially ionic and crystallizes in the fluorite structure; it is almost completely decomposed even by cold water, as would be expected for an ionic compound that is the salt of a weak acid and an extremely weak base.

In sharp contrast to the fluoride, the other halides show marked covalent character. *Mercury(II) chloride* crystallizes in an essentially molecular lattice. Relative to ionic HgF_2, the other halides have very low melting and boiling points, for example, $HgCl_2$, mp 280 °C. They also show marked solubility in many organic solvents. In aqueous solution they exist almost exclusively (~99%) as HgX_2 molecules, but some hydrolysis occurs, the principal equilibrium being, for example,

$$HgCl_2 + H_2O \rightleftharpoons Hg(OH)Cl + H^+ + Cl^- \qquad (22\text{-}5.2)$$

Mercury(II) Oxo Salts

Among the mercury(II) salts that are essentially ionic and, hence, highly dissociated in aqueous solution are the nitrate, sulfate, and perchlorate. Because of the great weakness of mercury(II) hydroxide, aqueous solutions of these salts tend to hydrolyze extensively and must be acidified to be stable.

In aqueous solutions of $Hg(NO_3)_2$ the main species are $Hg(NO_3)_2$, $HgNO_3^+$, and Hg^{2+}, but at high concentrations of NO_3^- the complex anions $[Hg(NO_3)_{3,4}]^{-,2-}$ are formed.

Mercury(II) carboxylates, especially the acetate and the trifluoroacetate, are of considerable importance because of their utility in attacking unsaturated hydrocarbons (Section 29-6). These compounds are made by dissolving HgO in the hot acid and crystallizing. The trifluoroacetate is also soluble in benzene, acetone, and THF, which increases its utility, while the acetate is soluble in water and alcohols.

Mercury(II) ions catalyze a number of reactions of complex compounds such as the aquation of $[Cr(NH_3)_5X]^{2+}$. Bridged transition states, for example,

$$[(H_2O)_5CrCl]^{2+} + Hg^{2+} = [(H_2O)_5Cr\text{—}Cl\text{—}Hg]^{4+} \qquad (22\text{-}5.3)$$

are believed to be involved.

The silver ion is similarly able to accelerate the rates of substitution of halo and other ligands. In both cases, the catalysis arises because Hg^{2+} or Ag^+ is able to enhance the breaking of the bond to the leaving group in what would otherwise be a slow, dissociative (D or I_d in the terminology of Section 6-5) mechanism.

Mercury(II) Complexes

The Hg^{2+} ion forms many strong complexes. The characteristic coordination numbers and stereochemical arrangements are two-coordinate (*linear*) and four-coordinate (*tetrahedral*). Octahedral coordination is less common; a few three-and five-coordinate complexes are also known. There appears to be considerable covalent character in the mercury–ligand bonds, especially in the two-coordinate compounds.

In addition to halide or pseudohalide complex ions, such as $[HgCl_4]^{2-}$ or $[Hg(CN)_4]^{2-}$, there are cationic species, such as $[Hg(NH_3)_4]^{2+}$ and $[Hg(en)_3]^{2+}$.

There are also a number of novel compounds in which —Hg— or —HgX is bound to a transition metal. Some of these compounds may be obtained by reaction of $HgCl_2$ with carbonylate anions (Section 28-9), for example,

$$2\ Na^+Co(CO)_4^- + HgCl_2 = 2\ NaCl + (CO)_4Co{-}Hg{-}Co(CO)_4 \quad (22\text{-}5.4)$$

Mercury(II) also forms many compounds with PR_3 ligands. The compounds $HgX_2(PR_3)$ and $HgX_2(PR_3)_2$ are examples, and are either dimeric or polymeric with halide bridges.

With thiols, one obtains the well-known neutral thiolates, $Hg(SR)_2$, and thiolate anions such as $[Hg(SC_6H_5)_3]^-$ and $[Hg_2(SCH_3)_6]^{2-}$. In fact, the name "mercaptans" for thiols (RSH) originated from the high affinity of mercury for sulfur. The neutral thiolates are most commonly linear [e.g., $Hg(SCH_3)_2$ and $Hg(SC_2H_5)_2$], although there are often secondary bonding interactions in the solid state between sulfur and mercury. That is, a third or fourth thiolate ligand is found near enough to a given mercury atom to be considered to be within the van der Waals distance but not so close as to be regarded as covalently bonded. The 1:1 mercury(II) thiolates such as $[Hg(S\text{-}i\text{-}Pr)Cl]_n$ are chain polymers with thiolate bridges, as is the neutral 2:1 compound $[Hg(S\text{-}t\text{-}Bu)_2]_n$.

Anionic trithiolates, $[Hg(SR)_3]^-$, and tetrathiolates, $[Hg(SR)_4]^{2-}$, are also known. Some examples are $[Hg(SC_6H_5)_3]^-$, Structure 22-I, and $[Hg_2(SCH_3)_6]^{2-}$, Structure 22-II. Of the four-coordinate thiolates, only three have mononuclear, distorted tetrahedral geometries. One example is $[Hg(p\text{-}SC_6H_4Cl)_4]^{2-}$. The rest are at least dimeric, and sometimes polymeric, with sulfur atom bridges.

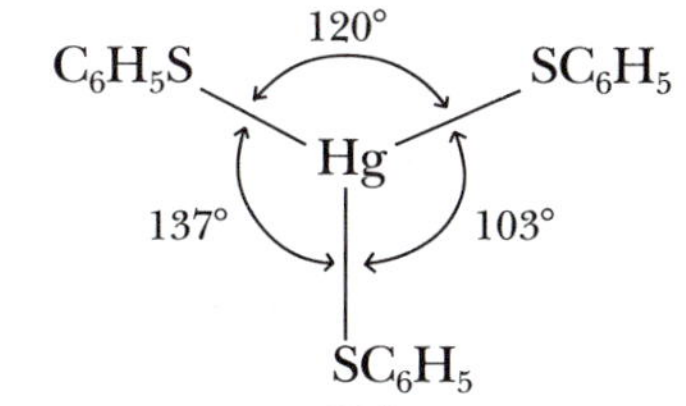

22-I

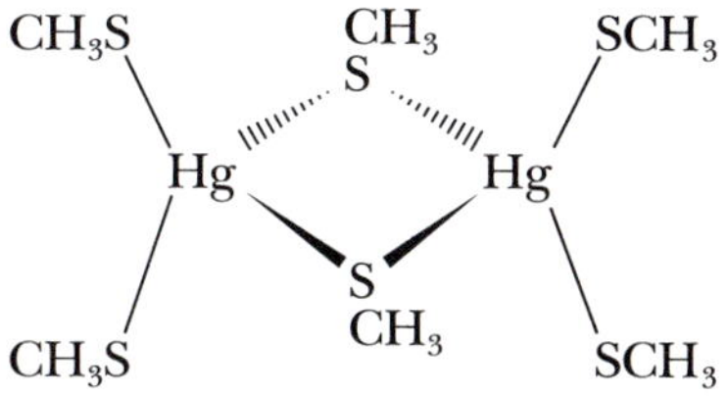

22-II

STUDY GUIDE

Study Questions

A. Review

1. Give the electronic structures of Zn, Cd, and Hg, and explain their position in the periodic table.
2. What are the electron configurations of the 2+ cations of Zn, Cd, and Hg?
3. Write balanced equations for the action on Zn of (a) 3 M HCl and (b) 3 M KOH.
4. Describe the interaction of Hg and O_2 and the properties of HgO.
5. What are the electron configurations of the 1+ cations?
6. What factors alter the ease of disproportionation of Hg^I?
7. Why do the hydroxide, oxide, or sulfide of Hg^I not exist?
8. Draw the structures of rutile, fluorite, and zinc blende.
9. What is the nature of $HgCl_2$ in solution and in the solid state?

B. Additional Exercises

1. Suggest the reason, in thermodynamic terms, why the sign of ΔG for the following reaction changes (from − to +) at about 400 °C

$$Hg(\ell) + \tfrac{1}{2} O_2(g) \longrightarrow HgO(s)$$

2. Why is it that when Hg is oxidized with an excess of oxidant only Hg^{II} is formed, yet when an excess of Hg is present during the oxidation, only Hg^I is formed?
3. By what methods can it be proved that the mercurous (mercury(I)) ion is the dimer Hg_2^{2+} in solution?
4. The zinc and cadmium dithiocarbamates are dimeric $[M(S_2CNR_2)_2]_2$. Draw a plausible structure.
5. Write balanced equations for
   (a) The disproportionation of mercury(I) hydroxide.
   (b) The hydrolysis of Hg^{II} fluoride.
   (c) The hydrolysis of zinc nitrate.
   (d) Thermal decomposition of mercury(II) oxide.
6. Calculate $\Delta G°$ for Reaction 22-3.4.

C. Questions from the Literature of Inorganic Chemistry

1. Based on the information concerning mercury(II) thiolate complexes presented by T. V. O'Halloran et al. (J. G. Wright, M. J. Natan, F. M. MacDonnell, D. M. Ralston, and T. V. O'Halloran, "Mercury(II)-Thiolate Chemistry and the Mechanism of Heavy Metal Biosensor MerR," in *Progress in Inorganic Chemistry*, Vol. 38, Wiley-Interscience, New York, 1990, 323, summarize the structural facts for the various Hg^{II}-thiolate complexes.
2. Two types of chloromercurate anions were found in the compounds reported by T. J. Kistenmacher, M. Rossi, C. C. Chiang, R. P. Van Duyne, and A. R. Siedle, "Crystal and Molecular Structure of an Unusual Salt formed from the Radical Cation of Tetrathiofulvalene (TTF) and the Trichloromercurate Anion ($HgCl_3^-$), (TTF)($HgCl_3$)," *Inorg. Chem.*, **1980**, *19*, 3604–3608.
   (a) What are the coordination geometries at Hg^{II} in the two types of halo-bridged ions reported in this article?

(b) What two types of chlorine atoms are there in these two ions, based on the Hg—Cl bond lengths?

(c) What are the relative numbers of terminal and bridging chlorine atoms in each ion reported here?

(d) If $Hg_2Cl_6^{2-}$ is described as an edge-sharing bitetrahedron, how might the sharing between adjacent units in polymeric $(HgCl_3^-)_n$ be described?

SUPPLEMENTARY READING

Dean, P. A. W., "The Coordination Chemistry of the Mercuric Halides," in *Progress in Inorganic Chemistry*, Vol. 24, Wiley-Interscience, New York, 1978, p. 109.

Larock, R. C., *Organomercury Compounds in Organic Synthesis*, Springer-Verlag, Berlin, 1985.

McAuliffe, C. A., Ed., *The Chemistry of Mercury*, Macmillan, New York, 1977.

Miller, M. W. and Clarkson, T. W., Eds., *Mercury, Mercurials, and Mercaptans*, Thomas, Springfield, IL, 1973.

Nriagu, J. O., *Zinc in the Environment*, Wiley, New York, 1980.

Roberts, H. L., "The Chemistry of Mercury," *Adv. Inorg. Chem. Radiochem.*, **1968,** *11,* 309.

Part 3

TRANSITION ELEMENTS

Chapter 23

INTRODUCTION TO THE TRANSITION ELEMENTS: LIGAND FIELD THEORY

23-1 Introduction

As we noted in Section 8-12, the transition elements are often defined as those which, *as elements,* have partly filled *d* or *f* shells. For practical purposes, however, we shall consider as transition elements all those elements that have partly filled *d* or *f* shells in any of their important compounds as well. Thus we include the coinage metals, Cu, Ag, and Au.

The transition elements are all metals, mostly hard strong ones that conduct heat and electricity well. They form many colored and paramagnetic compounds because of their partially filled shells.

In this part of the book we treat them in detail, beginning here with an account of their electronic structures, spectra, magnetic properties, and some other related matters. We then deal with the *d*-block elements, that is, those in which the partially filled shells are the 3*d*, 4*d*, or 5*d* shells. We shall then turn to the *lanthanides*, in which the 4*f* shell is partially filled, and, finally, the *actinides*, in which the 5*f* shell is partially filled.

23-2 Ligand Field Theory

The term *ligand field theory* refers to an entire body of theoretical apparatus used to understand the bonding and associated electronic (magnetic, spectroscopic, etc.) properties of complexes and other compounds formed by the transition elements.

There is nothing fundamentally different about the bonding in transition metal compounds as compared with that in compounds of the main group elements. All the usual forms of valence theory that are applied to the main group elements can be successfully applied to the transition elements. In general, the MO method applied to the transition metal compounds gives valid and useful results, the more so as the level of approximation is raised, just as in all other cases.

There are, however, two things that set the study of the electronic structures of transition metal compounds apart from the remaining body of valence theory. One is the presence of partly filled *d* and *f* shells. This leads to experimental observations not possible in most other cases: paramagnetism, visible absorption spectra, and apparently irregular variations in thermodynamic and structural properties. The second is that there is a crude but effective approximation, called *crystal field theory,* that provides a powerful yet simple method of understanding and correlating all of those properties that arise primarily from the presence of the partly filled shells.

The crystal field theory provides a way of determining, by simple electrostatic considerations, how the energies of the metal ion orbitals will be affected by the set of surrounding atoms or ligands. It works best when the symmetry is high but, with additional effort, can be applied more generally. Crystal field theory is a *model* and not a realistic description of the forces actually at work. However, its simplicity and convenience have earned it a place in the coordination chemist's "toolbox."

In the immediately following sections the crystal field theory is described and illustrated. Then the more complete MO method is outlined. After that, the electronic properties of transition metal complexes are discussed in terms of the "orbital splittings," which the crystal field theory enables us to work out relatively easily.

Our attention will be confined entirely to the *d*-block elements, and will be focused primarily on those of the 3*d* series. This is where the crystal field theory works best. The splittings of *f* orbitals are generally so small that they are not chemically important.

23-3   The Crystal Field Approach

Let us consider a metal ion (M^{m+}) lying at the center of an octahedral set of point charges, as is shown in Fig. 23-1. Let us suppose that this metal ion has a single *d* electron outside of closed shells; such an ion might be Ti^{3+} or V^{4+}. In the free ion, this *d* electron would have had equal probability of being in any one of the five *d* orbitals, since all are equivalent. Now, however, the *d* orbitals are not all equivalent. Some are concentrated in regions of space closer to the negative ions than are others, and the electron will obviously prefer to occupy the orbital(s) in which it can get as far as possible from the negative charges. Recalling the shapes

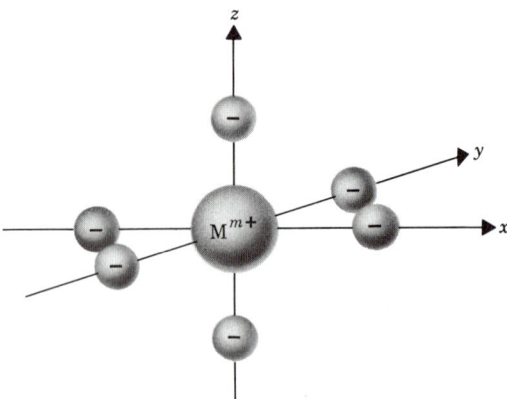

Figure 23-1   A sketch showing six negative charges arranged octahedrally around a central M^{m+} ion, with a set of Cartesian axes for reference.

of the d orbitals (Fig. 2-6) and comparing them with Fig. 23-1, we see that both the d_{z^2} and $d_{x^2-y^2}$ orbitals have lobes that are heavily concentrated in the vicinity of the charges, whereas the d_{xy}, d_{yz}, and d_{zx} orbitals have lobes that project between the charges. This is illustrated in Fig. 23-2. It can also be seen that each of the three orbitals in the latter group, namely, d_{xy}, d_{yz}, d_{zx}, is equally favorable for the electron; these three orbitals have entirely equivalent environments in the octahedral complex. The two relatively unfavorable orbitals, d_{z^2} and $d_{x^2-y^2}$, are also equivalent; this is not obvious from inspection of Fig. 23-2, but Fig. 23-3 shows why it is so. As indicated, the d_{z^2} orbital can be resolved into a linear combination of two orbitals, $d_{z^2-x^2}$ and $d_{z^2-y^2}$, each of which is obviously equivalent to the $d_{x^2-y^2}$ orbital. It is to be stressed, however, that these two orbitals do not have separate existences, and the resolution of the d_{z^2} orbital in this way is only a device to persuade the reader *pictorially* that d_{z^2} is equivalent to $d_{x^2-y^2}$ in relation to the octahedral distribution of charges.

Thus, in the octahedral environment of six negative charges, the metal ion now has two kinds of d orbitals: Three of one kind, equivalent to one another and labeled t_{2g}, and two of another kind, equivalent to each other, labeled e_g; furthermore, the e_g orbitals are of higher energy than the t_{2g} orbitals. These results may be expressed in an energy-level diagram as shown in Fig. 23-4(a).

In Fig. 23-4(a) it will be seen that we have designated the energy difference between the e_g and the t_{2g} orbitals as Δ_o, where the subscript o stands for octahedral. The additional feature of Fig. 23-4(a)—the indication that the e_g levels lie $\frac{3}{5}\Delta_o$ above and the t_{2g} levels lie $\frac{2}{5}\Delta_o$ below the energy of the unsplit d orbitals—will now be explained. Let us suppose that a cation containing ten d electrons, two in each of the d orbitals, is first placed at the center of a hollow sphere whose radius is equal to the M to X internuclear distance and that charge of total

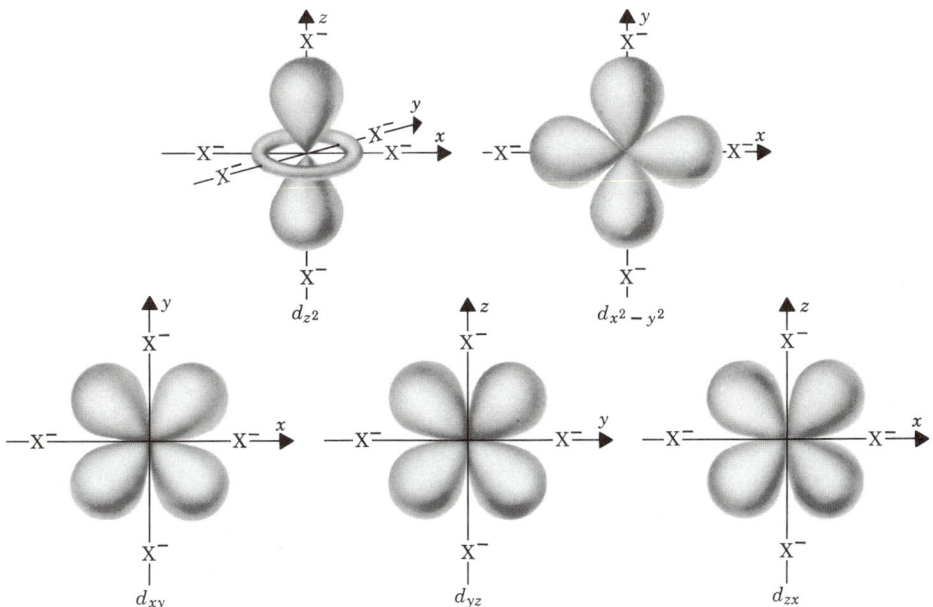

Figure 23-2   Sketches showing the distribution of electron density in the five d orbitals, and their orientation with respect to the set of six octahedrally arranged negative charges of Fig. 23-1.

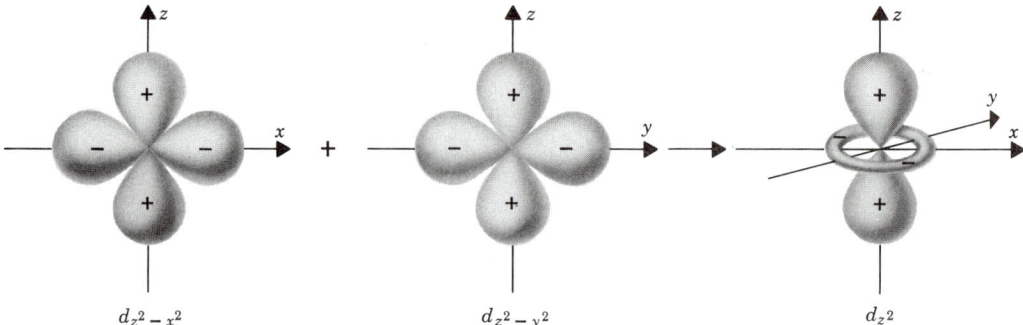

Figure 23-3   Sketches of the $d_{z^2-x^2}$ and the $d_{z^2-y^2}$ orbitals that are usually combined to make the d_{z^2} orbital.

quantity $6e$ is spread uniformly over the sphere. In this spherically symmetric environment the d orbitals are still fivefold degenerate. The energy of all orbitals is, of course, greatly raised when the charged sphere encloses the ion. The entire energy of the system, that is, the metal ion and the charged sphere, has a definite value. Now suppose the total charge on the sphere is caused to collect into six discrete point charges, each of magnitude e, and each lying at a vertex of an octahedron but still on the surface of the sphere. Merely redistributing the negative charge over the surface of the sphere in this manner cannot alter the total energy of the system when the metal ion consists entirely of spherically symmetrical electron shells, and yet we have already seen that, as a result of this redistribution, electrons in e_g orbitals now have higher energies than those in t_{2g} orbitals. It must therefore be that the total increase in energy of the four e_g electrons equals the total decrease in energy of the six t_{2g} electrons. This then implies that the rise in the energy of the e_g orbitals is $\frac{6}{4}$ times the drop in energy of the t_{2g} orbitals, which is equivalent to the $\frac{3}{5}:\frac{2}{5}$ ratio shown.

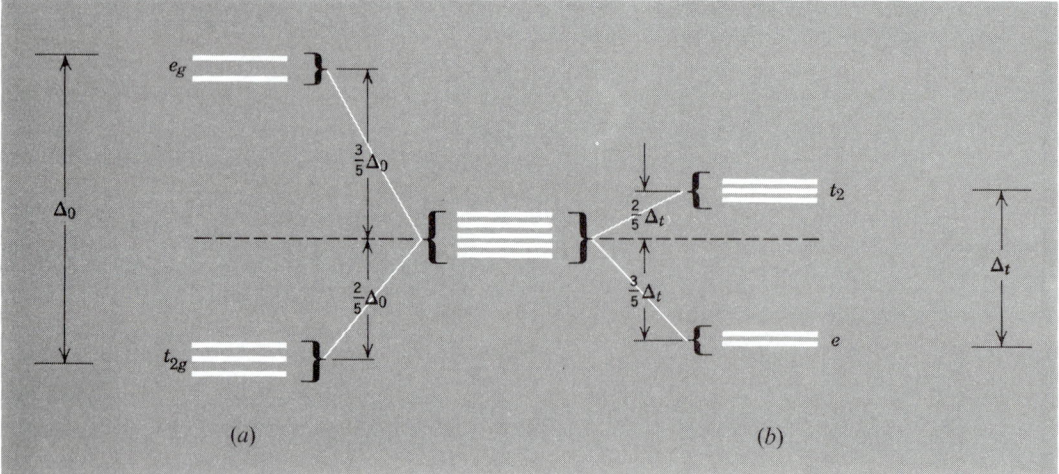

Figure 23-4   Energy-level diagrams showing the splitting of a set of d orbitals (degenerate in the uncoordinated ion at the center of the diagram) by octahedral and tetrahedral crystal fields. (*a*) The splitting caused by octahedral coordination of six ligands. (*b*) The splitting caused by tetrahedral coordination of four ligands. Brackets, } or {, designate orbitals that are degenerate.

   This pattern of splitting, in which the algebraic sum of all energy shifts of all orbitals is zero, is said to "preserve the center of gravity" of the set of levels. This center of gravity rule is quite general for any splitting pattern when the forces are purely electrostatic and where the set of levels being split is well removed in energy from all other sets with which they might be able to interact.

   By an analogous line of reasoning it can be shown that the electrostatic field of four charges surrounding an ion at the vertices of a tetrahedron causes the d shell to split up, as shown in Fig. 23-4(b). In this case the d_{xy}, d_{yz}, and d_{zx} orbitals are less stable than the d_{z^2} and $d_{x^2-y^2}$ orbitals. This may be appreciated qualitatively if the spatial properties of the d orbitals are considered with regard to the tetrahedral array of four negative charges, as depicted in Fig. 23-5. If the cation, the anions, and the cation–anion distance are the same in both the octahedral and tetrahedral cases, it can be shown that

$$\Delta_t = \tfrac{4}{9} \Delta_o$$

Since the pure electrostatic crystal field model is not quantitatively precise, the factor four ninths need not be taken literally. Rather, the practical interpretation of this result is that, other things being about equal, the crystal field splitting in a tetrahedral complex will be about one-half the magnitude of that in an octahedral complex.

   These results have been derived on the assumption that ionic ligands, such as F^-, Cl^-, or CN^-, may be represented by point negative charges. Ligands that are neutral, however, are dipolar (e.g., Structures 23-I and 23-II), and they approach

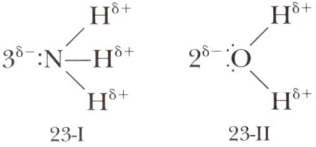

23-I   23-II

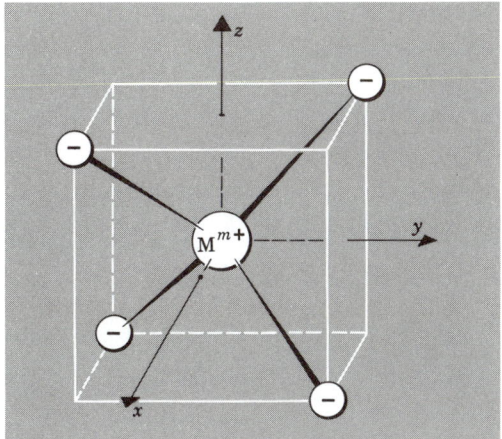

Figure 23-5   A sketch showing the tetrahedral arrangement of four negative charges around a cation, M^{m+}, with the Cartesian coordinate system oriented to identify the positions of the d orbitals.

the metal ion with their negative poles. Actually, in the field of the positive metal ion, such ligands are further polarized. Thus, in a complex such as a hexaammine, the metal ion is surrounded by six dipoles with their negative ends closest; this array has the same general effects on the d orbitals as an array of six anions, so that all of the above results are valid for complexes containing neutral, dipolar ligands.

We next consider the pattern of splitting of the d orbitals in tetragonally distorted octahedral complexes and in planar complexes. We begin with an octahedral complex (MX_6) from which we slowly withdraw two trans ligands. Let these be the two on the z axis. As soon as the distance from M^{m+} to these two ligands becomes greater than the distance to the other four, new energy differences among the d orbitals arise. First, the degeneracy of the e_g orbitals is lifted, the z^2 orbital becoming more stable than the $(x^2 - y^2)$ orbital. This occurs because the ligands on the z axis exert a much more direct repulsive effect on a d_{z^2} electron than upon a $d_{x^2-y^2}$ electron. At the same time the threefold degeneracy of the t_{2g} orbitals is lifted. As the ligands on the z axis move away, the yz and zx orbitals remain equivalent to one another, but they become more stable than the xy orbital because their spatial distribution makes them more sensitive to the charges along the z axis than is the xy orbital. Thus for a small tetragonal distortion of the type considered, we may draw the energy-level diagram shown in Fig. 23-6. It should be obvious that for the opposite type of tetragonal distortion, that is, one in which two trans ligands lie closer to the metal ion than the other four, the relative energies of the split components will be inverted.

As Fig. 23-6 shows, it is in general *possible* for the tetragonal distortion to become so large that the z^2 orbital eventually drops below the xy orbital. Whether this will actually happen for any particular case, even when the two trans ligands are completely removed so that we have the limiting case of a square, four-coordinated complex, depends on quantitative properties of the metal ion and the

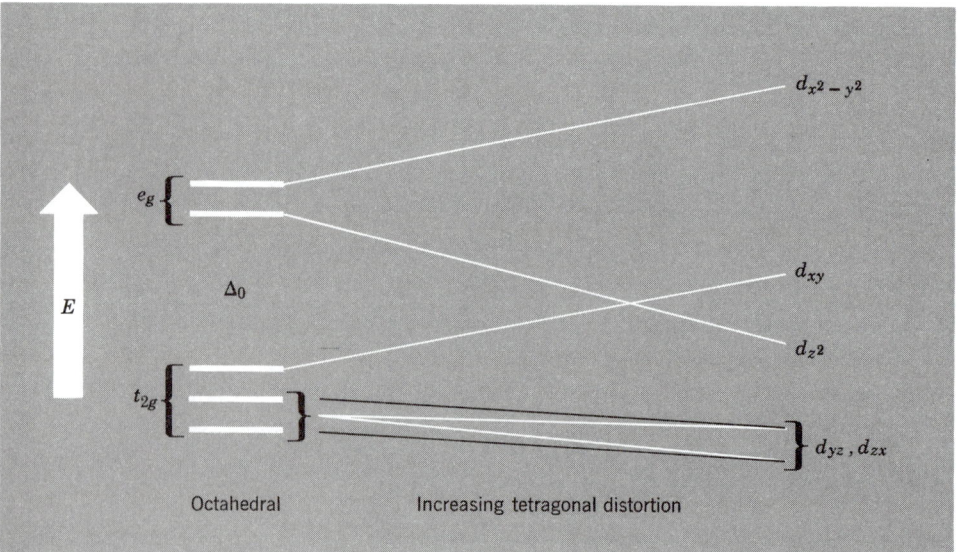

Figure 23-6 An energy-level diagram showing the further splitting of the d orbitals as an octahedral array of ligands becomes distorted by progressive withdrawal of two trans ligands along the z axis. Brackets, } or {, designate orbitals that are degenerate.

ligands concerned. Semiquantitative calculations with parameters appropriate for square complexes of Co^{II}, Ni^{II}, and Cu^{II} lead to the energy-level diagram shown in Fig. 23-7, in which the z^2 orbital has dropped so far below the xy orbital that it is nearly as stable as the (yz, zx) pair. As Fig. 23-6 indicates, the d_{z^2} level might even drop below the (d_{xz}, d_{yz}) levels and, in fact, experimental results suggest that in some cases (e.g., $PtCl_4^{2-}$) it does.

23-4   The Molecular Orbital Approach

The electrostatic crystal field theory is the simplest model that can account for the fact that the d orbitals split up into subsets in ligand environments. It is, of course, a physically unrealistic model in certain ways, and it is also incomplete as a treatment of metal–ligand bonding, since it deals only with the d orbitals. It is possible to treat the electronic structures of complexes from a MO point of view. This is more general, more complete, and potentially more accurate. It includes the crystal field model as a special case.

First, let us consider an octahedral complex MX_6, in which each ligand X has only a sigma orbital, directed toward the metal atom, and no π orbitals. The six σ orbitals of the ligands are designated σ_x and σ_{-x} (from those ligands along the x axis), σ_y and σ_{-y} (from those ligands along the y axis), and σ_z and σ_{-z} (from the ligands along the z axis). These six orbitals can be combined to make six distinct linear combinations, or ligand *group orbitals*, as shown in Fig. 23-8. Each ligand group orbital (Σ in Fig. 23-8) has a symmetry that is proper for overlap with only one of the metal s, p, or d orbitals. Each such overlap between one of the six ligand group orbitals and a metal valence orbital results in the formation of a bonding MO and an antibonding MO, according to the general principles of MO theory, as described in Chapter 3. Figure 23-9 gives an energy-level diagram that shows the formation of these bonding and anti-bonding MO's. Three of the d orbitals of the metal (d_{xy}, d_{yz}, and d_{zx}) are nonbonding, having zero overlap

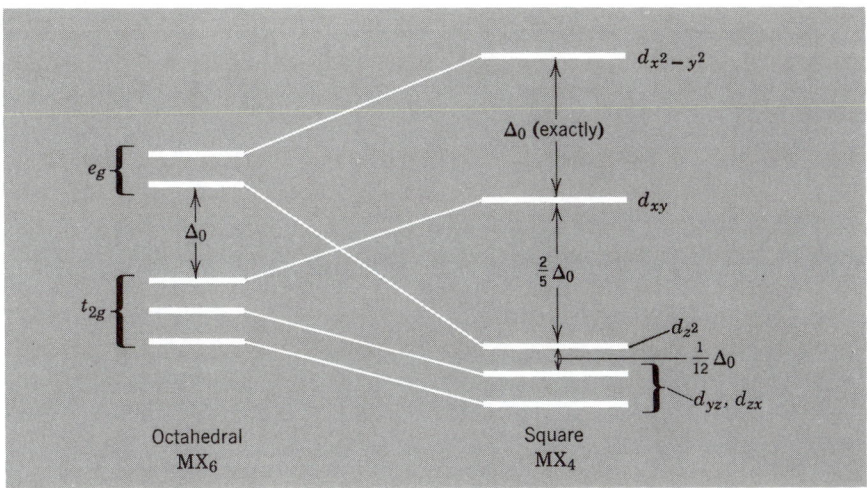

M = Co^{II}, Ni^{II}, Cu^{II}

Figure 23-7   The correspondence between the energy-level diagrams of octahedral ML_6 and square planar ML_4 complexes of some metal ions in the first transition series.

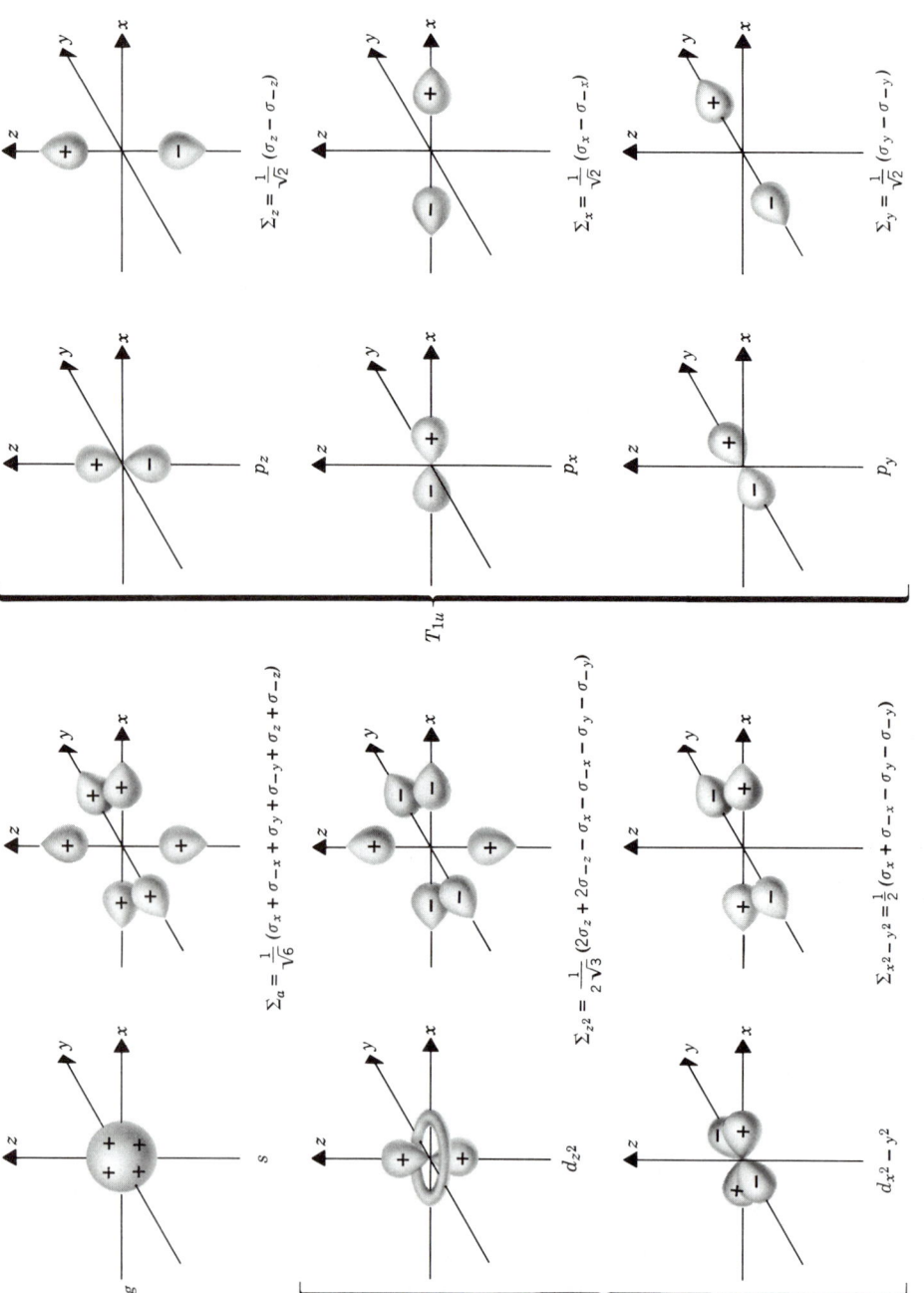

Figure 23-8 The six metal orbitals with σ symmetry and the ligand group orbitals (Σ) that overlap properly with these metal orbitals to form σ-type MO's in an ML_6 complex. Each ligand group orbital (Σ, with subscripts to designate the matching d orbital) is shown adjacent to the metal orbital with which it overlaps. The resulting MO's are listed in Fig. 23-9. The E_g set is doubly degenerate and the T_{1u} set is triply degenerate.

with each of the ligand group orbitals. These triply degenerate, nonbonding d orbitals are designated the t_{2g} set in Fig. 23-9.

The three MO's (bonding or antibonding) derived from the p orbitals have the same energy (they are degenerate), and are denoted t_{1u} (or t_{1u}^*). Similarly, the two MO's derived from the d_{z^2} and $d_{x^2-y^2}$ orbitals are degenerate and are denoted e_g (or e_g^*). The s orbital forms MO's denoted a_{1g} or a_{1g}^*. If each of the ligand σ orbitals originally contained an electron pair (which is the only situation of practical interest), these six electron pairs will then be found in the six (three t_{1u}, two e_g, a_{1g}) σ-bonding orbitals of the complex, as is also shown in Fig. 23-9.

It is evident that the MO discussion has lead to a result qualitatively the same as that from the crystal field theory with regard to the metal d orbitals: They are split into a set of two e_g^*, and a set of three t_{2g} orbitals, with the former having a higher energy than the latter. The MO picture also shows explicitly how the main binding energy of the complex arises, namely, by the formation of six two-electron bonds. The main difference between the MO and the crystal field results is that the e_g^* orbitals, as they are obtained in the MO treatment, *are not pure metal d orbitals.*

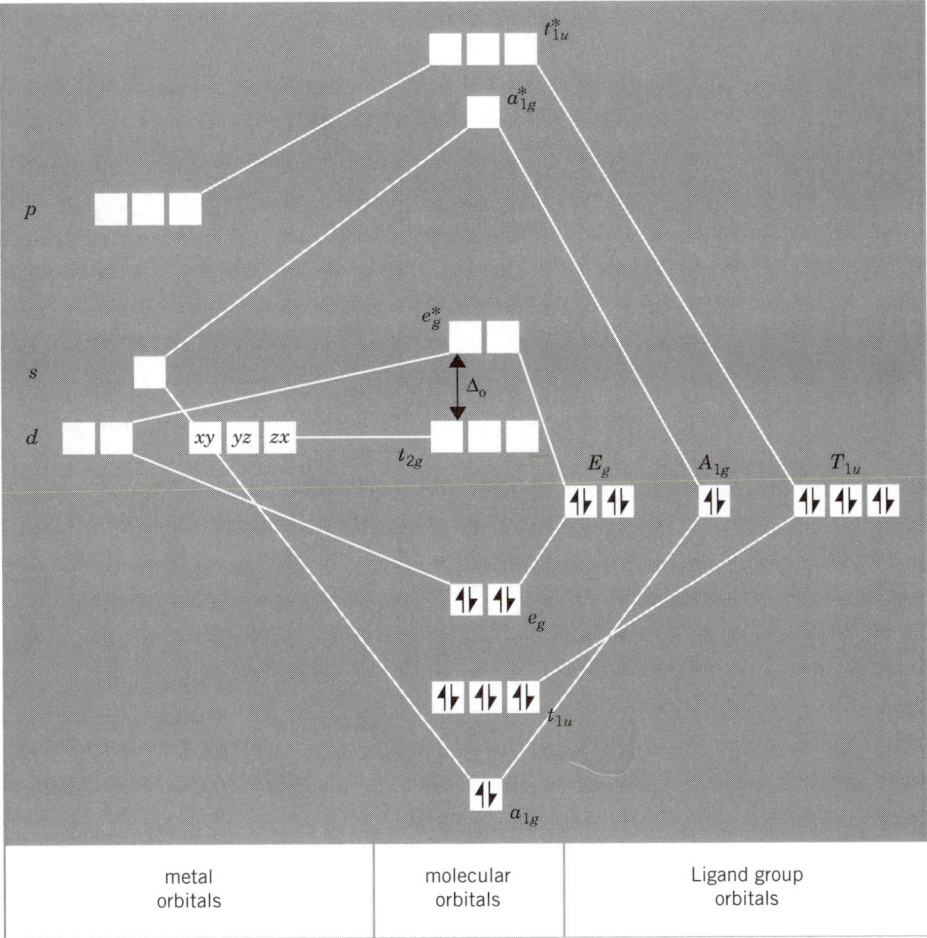

Figure 23-9   The MO energy-level diagram that arises from the σ-type MO's of Fig. 23-8. This diagram generally applies to an ML_6 complex with no π bonding.

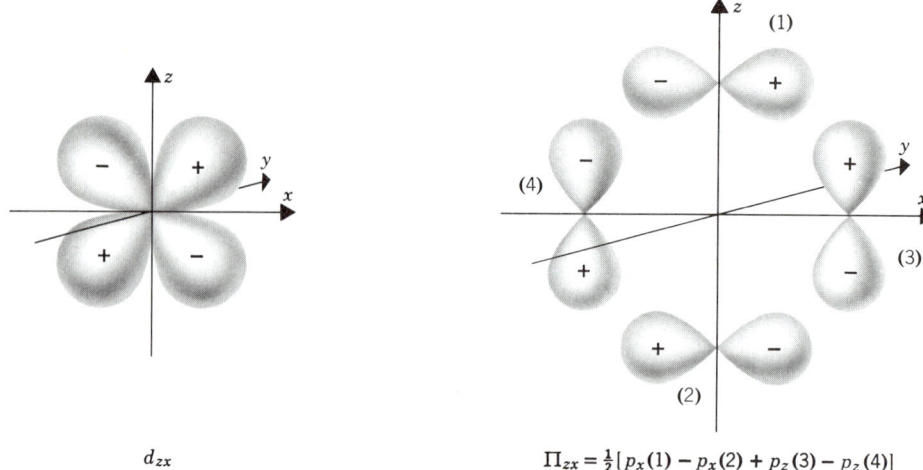

d_{zx} $\Pi_{zx} = \frac{1}{2}[p_x(1) - p_x(2) + p_z(3) - p_z(4)]$

Figure 23-10   At the right is a ligand group orbital Π_{zx} constructed by linear combination of ligand p orbitals and oriented to have optimum overlap with the metal d_{zx} orbital shown at the left. Analogous ligand group orbitals (Π_{xy} and Π_{yz}) overlap with the metal d_{xy} and d_{yz} orbitals, respectively.

We can generalize the MO treatment by supposing that the ligand atoms also possess π orbitals. Such π orbitals can overlap with the d_{xy}, d_{yz}, and d_{zx} orbitals, as is illustrated for the d_{zx} orbital in Fig. 23-10. Thus, instead of only one set of t_{2g} molecular orbitals, which are pure d orbitals, there will now be two sets. The positions of these sets of t_{2g} and t_{2g}^* orbitals in the MO energy-level diagram is quite variable depending on the nature of the ligand π orbitals. One case of rather general importance arises when the ligand π orbitals are empty and of higher energy than the metal d orbitals. Ligands that provide this situation include (1) phosphines, where the empty π orbitals are phosphorus $3d$ orbitals, and (2) CN^- and CO, where the empty π orbitals are antibonding $p\pi^*$ orbitals.

The interaction of the high-energy ligand π orbitals with the metal t_{2g} orbitals results in depressing the latter and thus increasing the separation between the t_{2g} and e_g^* orbitals, as shown in Fig. 23-11.

From the MO point of view, we see that a number of factors influence the ligand field splitting of the metal "d orbitals" and, further, that the "d orbitals" of crystal field theory are actually not pure d orbitals. It is remarkable, however, that the simple crystal field model is nevertheless a useful, qualitative working tool. In practice we do not try to use it to make quantitative predictions; that is, we do not try to calculate Δ_o (or Δ_t or any other d orbital splitting) from theory. Instead we derive these splittings from electronic spectra and use only the qualitative features of the d-orbital splitting patterns as given by crystal field theory.

23-5   Magnetic Properties of Transition Metal Compounds

One of the most useful applications of ligand field theory—whether in the simple electrostatic (crystal field) form or in a more sophisticated form—is to understand and correlate the magnetic properties of transition metal complexes.

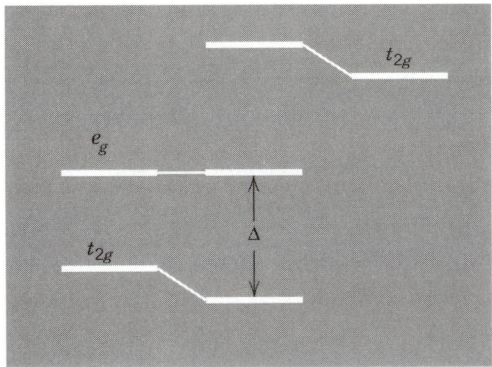

Figure 23-11 An energy-level diagram showing how π bonding such as that shown in Fig. 23-10 increases the value of Δ.

The magnetic properties of these compounds are important because, when properly interpreted, they are very useful in identifying and characterizing them.

The most basic question to ask concerning any paramagnetic ion is: How many unpaired electrons are present? We now see how this question may be handled in terms of the orbital splittings described in the preceding sections. We have already noted (Section 2-6) that according to Hund's first rule, if a group of n or fewer electrons (say n') occupy a set of n degenerate orbitals, they will spread themselves out among the orbitals and give n' unpaired spins. This is true because pairing of electrons is an unfavorable process; energy must be expended to make it occur. If two electrons are not only to have their spins paired but also to be placed in the same orbital, there is a further unfavorable energy contribution because of the increased electrostatic repulsion between electrons that are compelled to occupy the same regions of space. Let us suppose now that in some hypothetical molecule we have two orbitals separated by an energy ΔE and that two electrons are to occupy these orbitals. By referring to Fig. 23-12, we see that when we place one electron in each orbital, their spins will remain uncoupled and their combined energy will be $(2E_0 + \Delta E)$. If we place both of them in the lower orbital, their spins will have to be coupled to satisfy the exclusion principle, and the total energy will be $(2E_0 + P)$, where P stands for the energy required to cause pairing of two electrons in the same orbital. Thus, whether this

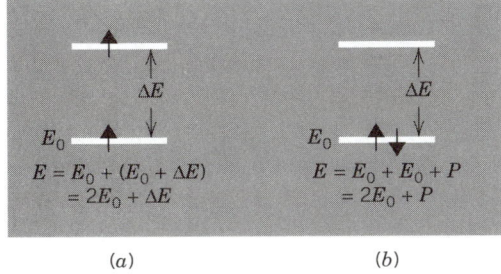

Figure 23-12 A hypothetical two-orbital system, showing the two possible distributions of two electrons. The resulting total energies are as shown.

system will have distribution (*a*) or (*b*) for its ground state depends on whether ΔE is greater or less than P.

Octahedral Complexes

An argument of this type can be applied to octahedral complexes, using the *d*-orbital splitting diagram previously deduced. As is indicated in Fig. 23-13, we may place one, two, or three electrons in the *d* orbitals without any possible uncertainty about how they will occupy the orbitals. They will follow Hund's first rule and enter the more stable t_{2g} orbitals with their spins all parallel, regardless of the strength of the crystal field (as measured by the magnitude of Δ_o). Furthermore, for ions with eight, nine, and ten *d* electrons, there is only one possible way in which the orbitals may be occupied to give the lowest energy, as shown in Fig. 23-13. For each of the remaining configurations, d^4, d^5, d^6, and d^7, two possibilities exist, and the question of which one represents the ground state can be answered only by comparing the values of Δ_o and P, an average pairing energy. The two configurations for each case, together with simple expressions for their energies, are set out in Fig. 23-14. The configurations with the maximum possible number of unpaired electrons are called the *high-spin* configura-

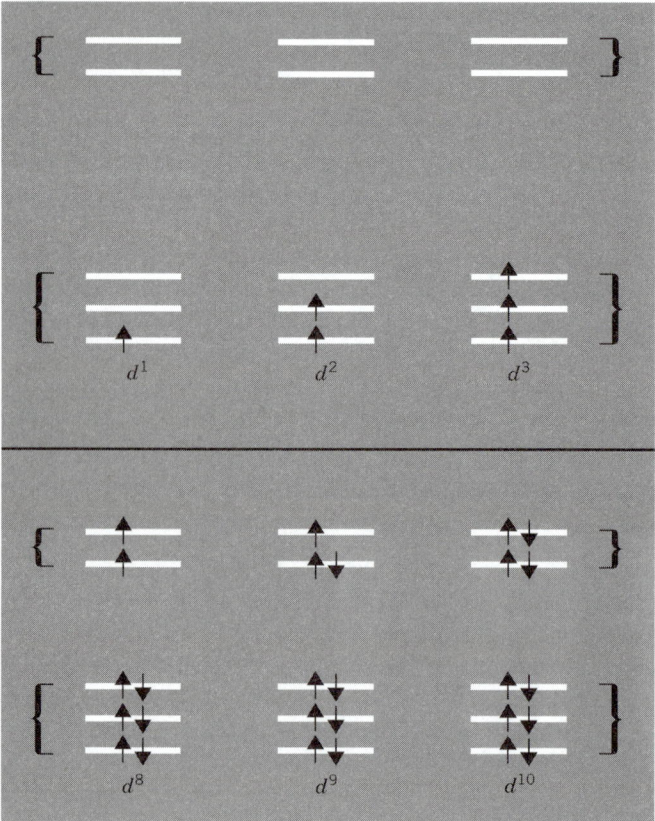

Figure 23-13 Sketches showing the unique ground-state electron configurations for *d* orbitals in octahedral fields with the *d* configurations d^1, d^2, d^3, d^8, d^9, and d^{10}.

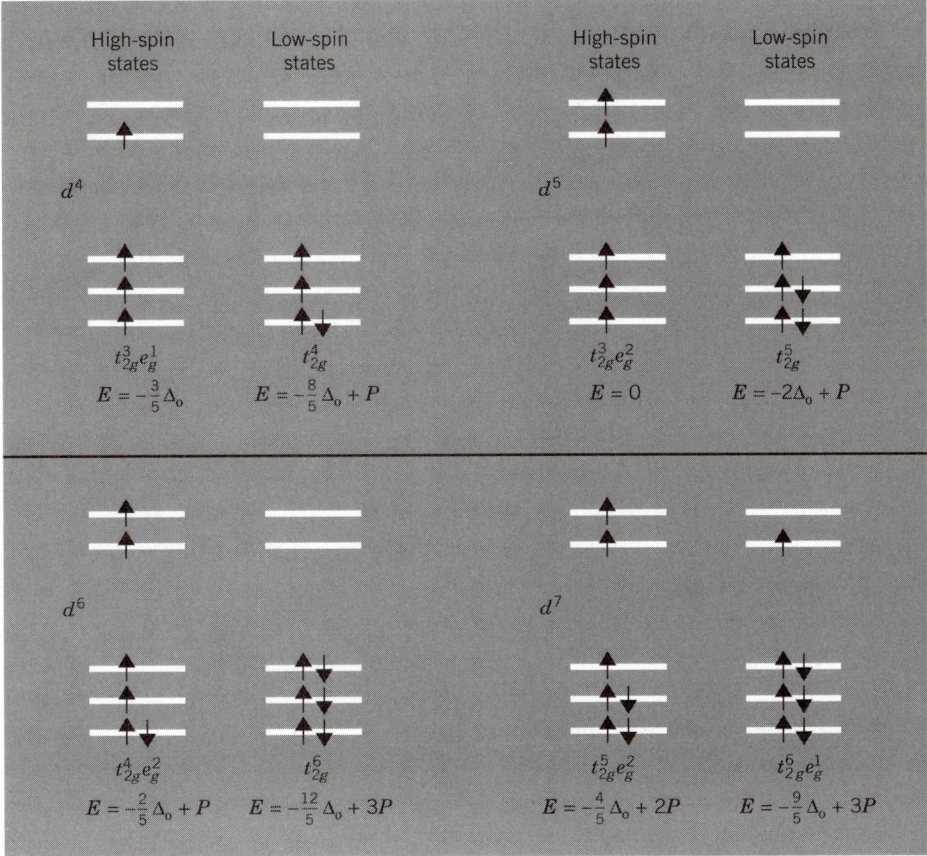

Figure 23-14 Diagrams showing the two possibilities (high spin and low spin) for the ground-state electron configurations of d^4, d^5, d^6, and d^7 ions in octahedral fields. Also shown is the notation for writing out the configurations and expressions for their energies, derived as explained in the text.

tions, and those with the minimum number of unpaired spins are called the *low-spin* or *spin-paired* configurations. These configurations can be written out in a notation similar to that used for electron configurations of free atoms, whereby we list each occupied orbital or set of orbitals, using a right superscript to show the number of electrons present. For example, the ground state for a d^3 ion in an octahedral field is t_{2g}^3; the two possible states for a d^5 ion in an octahedral field are t_{2g}^5 and $t_{2g}^3 e_g^2$. This notation is further illustrated in Fig. 23-14. The energies are referred to the energy of the unsplit configuration (the energy of the ion in a spherical shell of the same total charge) and are simply the sums of $-\frac{2}{5}\Delta_o$ for each t_{2g} electron, $+\frac{3}{5}\Delta_o$ for each e_g electron, and P for every pair of electrons occupying the same orbital.

For each of the four cases where high- and low-spin states are possible, we may obtain from the equations for the energies, which are given in Fig. 23-13, the following expression for the relation between Δ_o and P at which the high- and low-spin states have equal energies.

$$\Delta_o = P$$

The relationship is the same in all cases, and means that the spin state of any ion in an octahedral electrostatic field depends simply on whether the magnitude of the field, as measured by the splitting energy Δ_o, is greater or less than the mean pairing energy P for the particular ion. For a particular ion of the d^4, d^5, d^6, or d^7 type, the stronger the crystal field, the more likely it is that the electrons will crowd as much as possible into the more stable t_{2g} orbitals, whereas in the weaker crystal fields, where $P > \Delta_o$, the electrons will remain spread out over the entire set of d orbitals as they do in the free ion. For ions of the other types, d^1, d^2, d^3, d^8, d^9, and d^{10}, the number of unpaired electrons is fixed at the same number as in the free ion irrespective of how strong the crystal field may become.

Approximate theoretical estimates of the mean pairing energies for the relevant ions of the first transition series have been made from spectroscopic data. In Table 23-1 these energies, along with Δ_o values for some complexes (derived by methods to be described in Section 23-6), are listed. It is seen that this theory affords correct predictions in all cases. We note further that the mean pairing energies vary irregularly from one metal ion to another, as do the values of Δ_o, for a given set of ligands. Thus, as Table 23-1 shows, the d^5 systems should be exceptionally stable in their high-spin states, whereas the d^6 systems should be exceptionally stable in their low-spin states. These expectations are in excellent agreement with the experimental facts.

Tetrahedral Complexes

Metal ions in tetrahedral electrostatic fields may be treated by the same procedure outlined previously for the octahedral cases. For tetrahedral fields it is found that for the d^1, d^2, d^7, d^8, and d^9 cases only high-spin states are possible, whereas for d^3, d^4, d^5, and d^6 configurations both high-spin and low-spin states are in principle possible. Once again the existence of low-spin states requires that $\Delta_t > P$. Since Δ_t values are only about one half as great as Δ_o values, it is to be expected that low-spin tetrahedral complexes of first transition series ions with d^3, d^4, d^5, and d^6 configurations will be extremely rare, and that is the case.

Table 23-1 Crystal Field Splittings, Δ_o, and Mean Electron-Pairing Energies, P, for Several Transition Metal Ions (Energies in cm^{-1})

					Spin State	
Configuration	Ion	P	Ligands	Δ_o	Predicted	Observed
d^4	Cr^{2+}	23,500	6 H_2O	13,900	High	High
	Mn^{3+}	28,000	6 H_2O	21,000	High	High
d^5	Mn^{2+}	25,500	6 H_2O	7,800	High	High
	Fe^{3+}	30,000	6 H_2O	13,700	High	High
d^6	Fe^{2+}	17,600	6 H_2O	10,400	High	High
			6 CN^-	33,000	Low	Low
	Co^{3+}	21,000	6 F^-	13,000	High	High
			6 NH_3	23,000	Low	Low
d^7	Co^{2+}	22,500	6 H_2O	9,300	High	High

Square and Tetragonally Distorted Octahedral Complexes

The square and tetragonally distorted octahedral complexes must be considered together because, as we noted previously, they merge into one another.

Even when the strictly octahedral environment does not permit the existence of a low-spin state, as in the d^8 case, distortions of the octahedron will cause further splitting of degenerate orbitals that may become great enough to overcome pairing energies and cause electron pairing. Let us consider as an example the d^8 system in an octahedral environment that is then subjected to a tetragonal distortion. We have already seen (Fig. 23-6) how a decrease in the electrostatic field along the z axis may arise, by either moving the two z-axis ligands out to a greater distance than their otherwise identical neighbors in the xy plane, or by having two different ligands on the z axis that make an intrinsically smaller contribution to the electrostatic potential than the four in the xy plane. Irrespective of its origin, the result of a tetragonal distortion of an initially octahedral field is to split apart the (x^2-y^2) and z^2 orbitals. We have also seen that if the tetragonal distortion, that is, the disparity between the contributions to the electrostatic potential of the two z axis ligands and the other four, becomes sufficiently great, the z^2 orbital may fall below the xy orbital. In either case, the two least stable d orbitals are now no longer degenerate but are separated by some energy Q. Now the question of whether the tetragonally distorted d^8 complex will have high- or low-spin depends on whether the pairing energy P is greater or less than the energy Q. Figure 23-15(a) shows the situation for the case of a "weak" tetragonal distortion, that is, for one in which the second highest d orbital is still d_{z^2}.

Figure 23-15(b) shows a possible arrangement of levels for a strongly tetragonally distorted octahedron, or for the extreme case of a square, four-coordinate

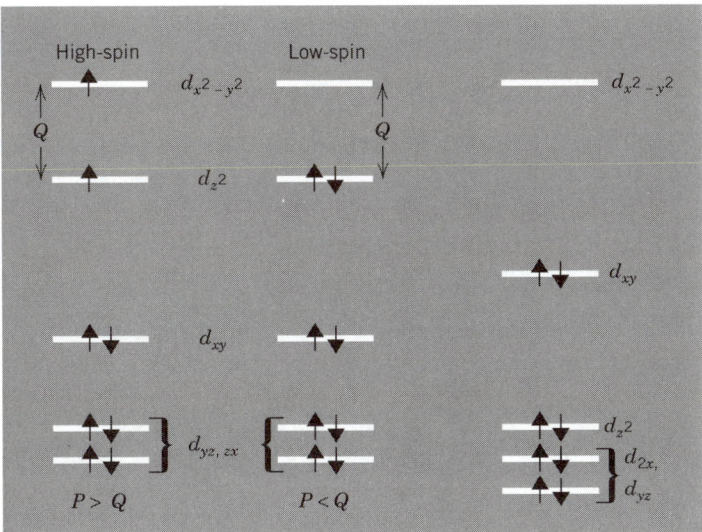

Figure 23-15   Energy-level diagrams showing the possible high-spin and low-spin ground states for a d^8 system (e.g., Ni^{2+}) in a tetragonally distorted octahedral field. (a) High-spin and low-spin possibilities for a weakly distorted system. (b) The low-spin result for a strongly distorted, or square complex.

complex (compared with Fig. 23-7), and the low-spin form of occupancy of these levels for a d^8 ion. In this case, due to the large separation between the highest and second highest orbitals, the high-spin configuration is impossible to attain considering the pairing energies of the real d^8 ions, e.g., Rh^I, Ir^I, Ni^{II}, Pd^{II}, Pt^{II}, and Au^{III}, which normally occur. All square complexes of these species are diamagnetic. Similarly, for a d^7 ion in a square complex, as exemplified by certain Co^{II} complexes, only the low-spin state with one unpaired electron should occur, and this is in accord with observation.

Other Forms of Magnetic Behavior

We have just indicated how the number of unpaired electrons on a transition metal ion in a complex, or other compound, can be understood in terms of the *d*-orbital splitting. The experimental method for determining the number of unpaired electrons has been discussed in Section 2-8; it is based on measuring the magnetic susceptibility of the substance. Here we must point out that certain additional factors must be considered in attempting to relate the magnetic moments of individual ions with the measured susceptibilities of bulk compounds.

Diamagnetism (which was briefly mentioned in Section 2-8) is a property of all forms of matter. All substances contain at least some if not all electrons in closed shells. In closed shells there is no net angular momentum, since the spin momenta cancel each other and so do the orbital momenta, and no net magnetic moment can result. However, when a substance is placed in a magnetic field, the closed shells are affected in such a way that the orbitals are all tipped and a small, net magnetic moment is set up in opposition to the applied field. This is called diamagnetism, and because the small induced moment is opposed to the applied field, the substance is repelled. In a substance that has no unpaired electrons, this will be the only response to the field. The substance will tend to move away from the strongest part of the field, and it is said to be diamagnetic. The susceptibility of a diamagnetic substance is negative and is independent of field strength and of temperature.

It is important to realize that even a substance that does have unpaired electrons also has diamagnetism because of whatever closed shells of electrons are also present. Thus the positive susceptibility measured is less than that expected for the unpaired electrons alone, because the diamagnetism partially cancels the paramagnetism. This is a small effect, typically amounting to less than 10% of the true paramagnetism, but in accurate work a correction for it must be applied.

Paramagnetism has already been discussed in Section 2-8. Simple paramagnetism occurs when the individual ions having the unpaired electrons are far enough apart to behave independently of one another. Curie's law (Eq. 2-8.1) is thus followed. The magnetic moment obtained can be directly, with allowance for small contributions (positive or negative) from orbital motion, interpreted in terms of the number of unpaired electrons.

Ferromagnetism and *antiferromagnetism* occur in substances where the individual paramagnetic atoms or ions are close together and each one is strongly influenced by the orientation of the magnetic moments of its neighbors. In ferromagnetism (so-called because it is very conspicuous in metallic iron) the interaction is such as to cause all moments to tend to point in the same direction. This enormously enhances the magnitude of the susceptibility of the substance as compared with what it would be if all the individual moments behaved

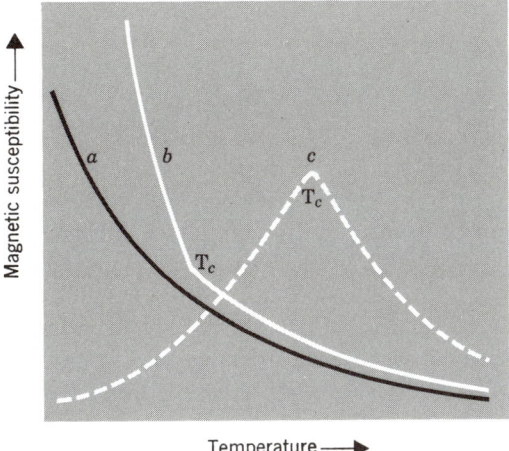

Figure 23-16   Magnetic susceptibility versus temperature plots for (*a*) a simple paramagnetic (Curie law) substance, (*b*) a ferromagnetic substance, and (*c*) an antiferromagnetic substance. The Curie temperatures are denoted by T_c in (*b*) and (*c*). For antiferromagnetism, the Curie point is also often called the Néel temperature.

independently. Ferromagnetism is generally found in the transition metals, and also in some of their compounds.

Antiferromagnetism occurs when the nature of the interaction between neighboring paramagnetic ions is such as to favor opposite orientations of their magnetic moments, thus causing partial cancellation. Antiferromagnetic substances thus have magnetic susceptibilities less than those expected for an array of independent magnetic ions. It occurs quite often among simple salts of ions, such as Mn^{2+}, Fe^{3+}, and Gd^{3+}, which have large intrinsic magnetic moments. The antiferromagnetic coupling involves interaction through the anions lying between the metal atoms in the crystal, and disappears in dilute solutions.

Ferro- or antiferromagnetic behavior causes deviations from the Curie law, as shown in Fig. 23-16. In each case there is a temperature at which the temperature dependence of the susceptibility changes abruptly. This is the Curie temperature (T_c), which is a characteristic property of the substance. Above T_c, the behavior is similar to that of the Curie law. Below T_c, the susceptibility either rises (ferromagnetism) or falls (antiferromagnetism) in a manner quite different from that implied by the Curie law. At the Curie temperature the effect of thermal energy in tending to randomize the individual spin orientations begins to get the upper hand over the ferro- or antiferromagnetic coupling interactions.

23-6   Electronic Absorption Spectroscopy

Ions with a Single *d* Electron

The simplest possible case of an ion with a single *d* electron is an ion with a d^1 configuration, lying at the center of an octahedral field, for example, the Ti^{III}

ion in $[Ti(H_2O)_6]^{3+}$. The d electron will occupy a t_{2g} orbital. On irradiation with light of frequency v, which is equal to Δ_o/h, where h is Planck's constant and Δ_o is the energy difference between the t_{2g} and e_g orbitals, it should be possible for such an ion to capture a quantum of radiation and convert that energy into energy of excitation of the electron from the t_{2g} to the e_g orbitals. The absorption band that results from this process is found in the visible spectrum of the hexa-aquotitanium(III) ion (shown in Fig. 23-17) and is responsible for its violet color. Two features of this absorption band are of importance here: its position and its intensity.

In discussing the positions of absorption bands in relation to the splittings of the d orbitals, it is convenient and common practice to use the same unit, the reciprocal centimeter or wave number, abbreviated cm^{-1}, for both the unit of frequency in the spectra and the unit of energy for the orbitals. With this convention, we see that the spectrum of Fig. 23-17 tells us that Δ_o in $[Ti(H_2O)_6]^{3+}$ is 20,000 cm^{-1}.

We note in Fig. 23-17 that the absorption band is very weak. Its molar absorbance at the maximum is five, whereas one-electron transitions that are theoretically "allowed" usually have absorbances of 10^4–10^5. This suggests that the transition in question is not "allowed" but is instead "forbidden" according to the quantum theory. That is indeed the case, for the following reason. It is a general rule of quantum mechanics that for any electronic transition to be "allowed" in a system that has a center of symmetry, it is a necessary (though not sufficient) condition that the electron move, as a result of the transition, from an orbital that is even with respect to inversion through the center of symmetry, to an orbital that is uneven with respect to inversion (or vice versa). Since all d orbitals are even with respect to inversion, this selection rule is not satisfied for electronic transitions that move an electron from one d orbital to another. Hence, d–d transitions of transition metal compounds are generally of low intensity (weakly absorbing).

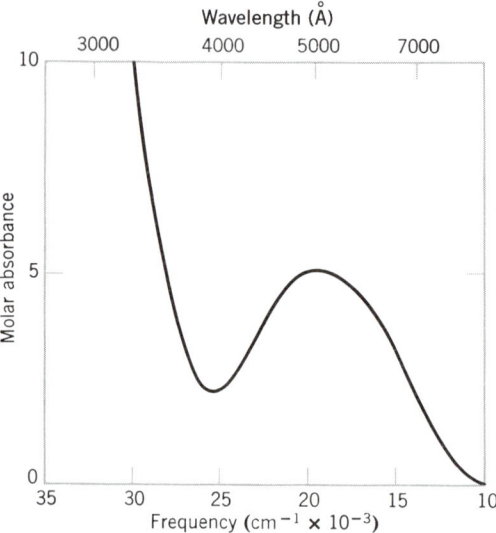

Figure 23-17   The electronic absorption spectrum of $[Ti(H_2O)_6]^{3+}$.

These d–d absorptions are measurable, though, and this selection rule evidently is relaxed in some way. For all transition metal compounds, there are vibrations of the ligands that slightly spoil the symmetry of the coordination sphere so as to remove the center of symmetry. This relaxes the rigorous requirement mentioned previously. In the case of tetrahedral compounds, the structures lack a center of symmetry. This makes the selection rule inapplicable, and the d–d transitions become not-forbidden. Consequently, it is a general observation that for tetrahedral complexes, the d–d absorption bands are considerably more intense than in octahedral complexes, often by a factor of 10 or more. This explains why, for example, the **pale** red color of the octahedral $[Co(H_2O)_6]^{2+}$ ion is changed by addition of chloride to the **intense** blue color of the tetrahedral $[CoCl_4]^{2-}$ ion.

In this discussion we have addressed the differences in absorption intensity that arise because one electronic transition may be more allowed than another. We must also address the question of absorption energy, as indicated by the wave number (cm^{-1}) of the electromagnetic radiation that is absorbed as a result of a particular electronic transition.

Ions with More than One *d* Electron

The majority of transition metal ions of practical interest have more than one d electron. An explanation of their electronic structures and electronic absorption spectra in terms of ligand field theory is considerably more complex, because there are now two forces to be considered; in addition to the repulsive forces exerted by the ligands on the electrons of the metal (the ligand field splitting), there are the forces between the electrons themselves. It is one of the great triumphs of modern physics that the methods for handling such complex problems in an accurate and useful way have been developed. Although it would be beyond the scope of this book to develop this methodology from first principles, it is important to provide a working sketch of how the electronic absorption spectra of coordination compounds may be interpreted. To do this, we shall first examine the case of a d^2 ion, where we shall be able to display all of the factors that are important for d^n ions in general. Once this basis is set down, it should be straightforward to apply the results to the remaining cases: d^3 to d^8.

Notations for the Electronic States of *d^n* Atoms and Ions

Certain symbols and terminologies are employed in examining the case of a d^n ion in an octahedral ligand field. The pertinent definitions are given below.

First, we must define two terms that will be used throughout the following discussion: electron configuration and electronic state. The term **electron configuration** refers to the way electrons occupy orbitals. We have frequently made use of this term, and have employed a shorthand notation for specifying electron configurations throughout the text. In Chapter 2, we wrote, for example, $1s^2 2s^2 2p^5$, as the electron configuration of the fluorine atom. For the vanadium ion (V^{3+}) we can say simply that the ground electron configuration is d^2, with the understanding that we are talking about the $3d$ orbitals and that all lower energy orbitals are fully occupied. For an ion in a ligand field (i.e., a coordination compound), where the d orbitals are split into subsets, a very similar notation for electron configuration is used. For example, in an octahedral field, the ground electron configuration of the V^{3+} ion would be written t_{2g}^2, and the two possible excited configurations would be written $t_{2g}^1 e_g^1$ and e_g^2. For a high-spin Mn^{2+} ion the electron configuration is $t_{2g}^3 e_g^2$, and for Ni^{2+} it is $t_{2g}^6 e_g^2$.

Having briefly mentioned electron configurations, it is now necessary to define electronic states. By the term **electronic state** we shall mean an energy level that is available to an ensemble of electrons. (It is, as such, something quite distinct from an energy for, say, a one-electron orbital, because the energy of an electronic state will be governed by the interactions of the electrons, as well as by the energies of the orbitals that house the electrons.) In general, more than one electronic state can arise from a given electron configuration. The only exceptions to this statement are a closed-shell configuration and, to a good approximation, the d^1 and d^9 configurations.

All other d^n electron configurations individually give rise to more than one electronic state because there are always several different ways that the electrons of a given electron configuration can interact with one another. Each different way of interacting among the electrons in any one electron configuration gives a different net energy for the ensemble of electrons. It is this energy that characterizes the resulting electronic state of the atom, molecule, or ion.

For example, if we have a p^2 electron configuration, the electron spins can be parallel (to give an electronic state with total spin equal to $2 \times \frac{1}{2} = 1$) or the two electrons in this electron configuration may be opposed (to give an electronic state with total spin equal to 0). There is a third electronic state that arises from this p^2 electron configuration, but we do not develop the details until later. For now it is sufficient to have demonstrated that a given electron configuration may give rise to a number of different electronic states. To illustrate the importance of this, consider carbon, which has the p^2 configuration. The states just mentioned for carbon differ in energy by about 125 kJ mol^{-1}.

Before proceeding, we need to designate the symbols to be used in specifying the different electronic states that we shall encounter. Just as lower case letters are used for orbitals of various degeneracies, so are capital letters used, as follows:

1. Singly degenerate orbitals: a   or   b
   Singly degenerate states: A   or   B
2. Doubly degenerate orbitals: e
   Doubly degenerate states: E
3. Triply degenerate orbitals: t
   Triply degenerate states: T

Subscripts 1 or 2 are used to distinguish among states of like degeneracy. In addition, subscripts g or u are employed for molecules that have a center of symmetry; subscript g designates states whose wave functions are even (from the German *gerade*), and subscript u designates states whose wave functions are uneven (from the German *ungerade*) with respect to inversion through the center of symmetry. For example, in an octahedral ligand environment, the set of p orbitals is triply degenerate and each p orbital is uneven with respect to inversion (i.e., the wave function changes sign upon inversion). The p orbitals in this environment are designated the t_{1u} set, and the electronic state arising from the electron configuration p^1 is designated T_{1u}. In an octahedral environment of ligands, three of the d orbitals are degenerate as well as even with respect to inversion. These three orbitals constitute the t_{2g} set in the octahedral ligand field, and the electronic state that arises from the t_{2g}^1 electron configuration is the T_{2g} state.

The symbols are further modified to designate the spin multiplicity of the electronic state, by adding a left superscript number. The left superscript, or spin multiplicity, of the electronic state is the value $(2S + 1)$, where S is the absolute value of the algebraic sum of the spins of the individual electrons. For example, for a d^1 electron configuration, $S = \frac{1}{2}$ and the spin multiplicity is two. For a d^2 electron configuration in which the electron spins are parallel, $S = \frac{1}{2} + \frac{1}{2} = 1$, and the spin multiplicity $(2S + 1)$ is three. In a d^5 electron configuration with four "up" spins and one "down" spin, $S = 4(\frac{1}{2}) + (-\frac{1}{2}) = \frac{3}{2}$, and the spin multiplicity $(2S + 1)$ is four.

The spin multiplicity is related to the total spin quantum number S by the relationship given previously, namely, spin multiplicity equals $2S + 1$ for the following reason. When an ion with a total spin quantum number S is placed in a magnetic field, the rules of quantization allow the total electron spin vector (whose length is $[S(S+1)]^{1/2}$) to take only those orientations relative to the magnetic field direction **H** that give projections in the field direction equal to $+S$, $S - 1$, $S - 2$, . . . , $-S$, as shown, for example, in Fig. 23-18. In general, the number of values from $+S$ to $-S$, in integral steps, is $(2S + 1)$, and that is where the relationship between S and spin multiplicity comes from. In verbal use, the spin multiplicities are pronounced as follows: 1 (singlet), 2 (doublet), 3 (triplet), 4 (quartet), 5 (quintet), 6 (sextet).

For a d^1 ion in an octahedral field of ligands, then, we have the two possible electron configurations e_g^1 and t_{2g}^1. The two states that arise are thus ^2E_g and $^2T_{2g}$, pronounced "doublet ee gee" and "doublet tee two gee," respectively. One electronic state that arises from the t_{2g}^2 electron configuration is the $^3T_{1g}$ state, pronounced "triplet tee one gee."

It remains only to point out that the **total degeneracy of a state** is the product of its spin and orbital degeneracies. The spin degeneracy is given by the spin multiplicity. The orbital degeneracy is designated by the letter scheme listed pre-

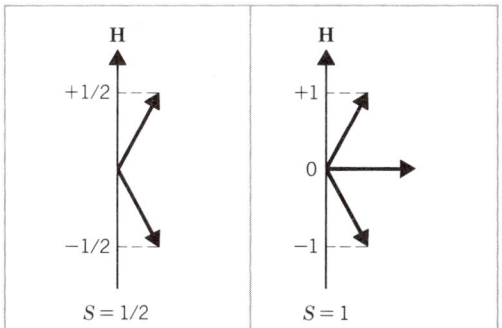

Figure 23-18   The allowed orientations of the electron spin vector in a magnetic field. Two cases are shown. The rules of quantization allow only those orientations of the vector that give projections equal to $+\frac{1}{2}$ or $-\frac{1}{2}$ (for the case where $S = \frac{1}{2}$) or which give projections of $+1$, 0, or -1 (for the case where $S = 1$).

viously. Thus, for the three states mentioned in the preceding paragraph, the total degeneracies are those given in Table 23-2.

The Electronic States Arising from a d^2 System

The procedure that we shall employ to work out the electronic states of a d^2 ion in an octahedral ligand field involves our considering two limiting cases, and then correlating the two. In one limit we have a ligand field so strong that the interactions between the two electrons are negligible in comparison with the energy differences between the various electronic states that arise from the electron configuration. This is called the **strong field** case. In the other extreme, we consider a ligand field that is so weak that the interaction among the electrons overshadows the ligand field, so that the various electronic states arising from the electron configuration have energies that are determined almost solely by interaction of the two electrons in the configuration. This is termed the **weak field** case.

The Strong Field Case. If we assume that the ligand field splitting is very large, then every electronic state arising from the electron configuration t_{2g}^2 will be of lower energy than every electronic state arising from the electron configuration $t_{2g}^1 e_g^1$, and similarly, every electronic state arising from the electron configuration e_g^2 will be higher in energy than those of the first two. This means that we can deal separately with each of the three possible electron configurations (t_{2g}^2, $t_{2g}^1 e_g^1$, and e_g^2), and we do so in the following way, beginning with the electron configuration e_g^2. The student may wish to preview the right side of Fig. 23-20, as this is the result towards which we are now working.

We have available a pair of e_g orbitals, and for each electron, two possible spin quantum numbers, $+\frac{1}{2}$ or $-\frac{1}{2}$. Thus we may assign the first electron of the set to the e_g orbitals in four different ways. We represent this by drawing a set of four boxes, each of which represents a distinct combination of orbital and spin quantum numbers for the single electron. We next ask how the second electron of the e_g^2 electron configuration may be assigned to the set of four boxes. Because of the Pauli exclusion principle, each box may hold only one electron. Then, there are six distinct and nonrepetitious ways of assigning two electrons to the four boxes, as illustrated in Fig. 23-19(a). This number six could have been anticipated by noting that we are simply asking for the number of ways to choose pairs from among four equivalent objects, and this is given by the product of the number of ways to choose the first one (4) times the remaining number of ways to choose the second (3), divided by two, since the order of choice for identical objects is immaterial. Thus we have for the e_g^2 electron configuration: $(4 \times 3)/2 = 6$. Each of these six distinct arrangements of electrons in the e_g^2 electron configuration is called a **microstate**.

Table 23-2 The Degeneracies of Three States

State	Spin Degeneracy	Orbital Degeneracy	Total Degeneracy
^2E_g	2	2	4
$^2T_{2g}$	2	3	6
$^3T_{1g}$	3	3	9

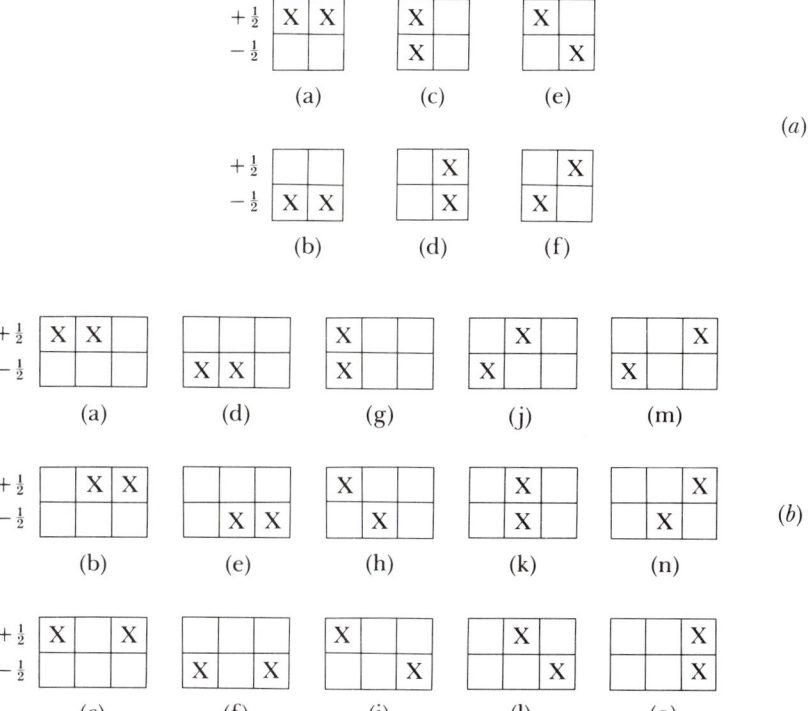

Figure 23-19 (a) The 6 ways (a–f) of arranging two electrons (with spin either $+\frac{1}{2}$ or $-\frac{1}{2}$) into two orbitals that are doubly degenerate. (b) The 15 ways (a–o) of arranging two electrons (with spin either $+\frac{1}{2}$ or $-\frac{1}{2}$) into three orbitals that are triply degenerate.

The arrangements of electrons in the boxes, as done in Fig. 23-19(a), represent the six microstates that are included in the electron configuration e_g^2. We shall now show, but not derive rigorously, the fact that these six microstates comprise three electronic states arising from the e_g^2 electron configuration. Two of the six microstates of Fig. 23-19(a) are spin triplets, because the electrons in each of the two microstates are parallel; these are microstates labeled (a) and (b) in Fig. 23-19(a). Together with one of the others, these constitute a spin-triplet state that is, orbitally, singly degenerate: $^3A_{2g}$. The remaining microstates of the e_g^2 electron configuration are the components of two singlet electronic states, $^1A_{1g}$ and ^1E_g. Note that when the orbital $(A + A + E)$ and spin (triplet + singlet + singlet) multiplicities of these three states are summed up, they correspond to a total of six: $(3 \times 1) + (1 \times 1) + (1 \times 2) = 6$, the total degeneracy of the e_g^2 electron configuration.

We now turn our attention to Fig. 23-19(b), where we display the 15 distinct and nonrepetitive microstates that are available to the electron configuration t_{2g}^2. Of these, three [(a), (b), and (c)] have $S = +1$ and three [(d), (e), and (f)] have $S = -1$. The other nine microstates all have $S = 0$. The three microstates with $S = +1$, the three microstates with $S = -1$, and three of the remaining nine microstates together constitute a $^3T_{1g}$ electronic state of the t_{2g}^2 electron configuration. From the remaining six microstates with $S = 0$, we can derive (without showing the details) the spin-singlet electronic states: $^1A_{1g}$, ^1E_g, and $^1T_{2g}$. Thus for the electron configuration t_{2g}^2 we find 4 electronic states with a total degeneracy of 15:

$$^3T_{1g} \qquad ^1T_{2g} \qquad ^1E_g \qquad ^1A_{1g}$$
$$(3 \times 3) + (1 \times 3) + (1 \times 2) + (1 \times 1) = 15$$

We may say this same thing in another way; out of the electron configuration t_{2g}^2 there arise 15 microstates that are grouped into 4 different electronic states, having 4 different energies.

Last we consider the third and only remaining possibility for a d^2 ion in an octahedral ligand field: The electron configuration $t_{2g}^1 e_g^1$. One electron is free to occupy one of four boxes (the electron in the e_g orbitals), whereas the other electron is free to occupy one of six boxes (the t_{2g} electron), leading to a total of 24 different microstates. Since each electron is in a different type of orbital, the Pauli exclusion principle is never a problem, no matter what spin assignments are made. Also, we need not divide by two even though the two electrons are indistinguishable. The result is that, among the 24 distinct and nonrepetitious microstates of the $t_{2g}^1 e_g^1$ electron configuration, there arise the electronic states $^1T_{1g}$, $^1T_{2g}$, $^3T_{1g}$, and $^3T_{2g}$, whose total degeneracy is, as required, 24.

The result of our analysis of the d^2 ion in the strong field limit is shown on the right of Fig. 23-20. We imply here that the energy difference among the electronic states is small (in the limit, zero) compared to the energy differences (Δ_o) between the three electron configurations.

The Weak Field Case. In the weak field limit (shown on the left in Fig. 23-20), we are dealing with a set of electronic states for which the energies are determined only by the interactions of the d electrons with one another. This is a problem that was solved by atomic spectroscopists quite independently of any work on either metal complexes or ligand field theory, because, in the weak field limit, there is necessarily no ligand set. For the metal ions of the first transition series (where the ligand field analysis of electronic absorption spectroscopy is most useful), there is, fortunately, a relatively convenient and more or less quantitative scheme for describing the electronic states that arise from a given $3d^n$ electron configuration. This same scheme is also reasonably satisfactory for $4d^n$ systems, but it has some inadequacies (from a quantitative point of view) for the $5d^n$ ions. We shall now describe this approach, which is called the Russell–Saunders or LS coupling scheme.

In this scheme, we use for the electronic states a set of quantum numbers and state symbols that closely parallel those used for a single electron. Just as an electron in a particular orbital has a certain orbital angular momentum quantum number ℓ {and an orbital angular momentum given by the quantity $[\ell(\ell + 1)]^{1/2}$}, so each electronic state arising from a given electron configuration d^n, has a total angular momentum quantum number L, and an orbital angular momentum given by the quantity $[L(L + 1)]^{1/2}$. Thus, just as we have $s, p, d, f, \ldots$ orbitals, we have $S, P, D, F, \ldots$ electronic states for a d^n ion. These letter designations for the electronic states correspond to total orbital angular momentum quantum numbers, $L = 0, 1, 2, \ldots$, respectively.

Each electronic state is also characterized by its spin quantum number S. Thus, even for a vanishing (or in the weak field limit, vanished) ligand field, ions with one, two, three, four, $\ldots$, unpaired electrons have spin quantum numbers,

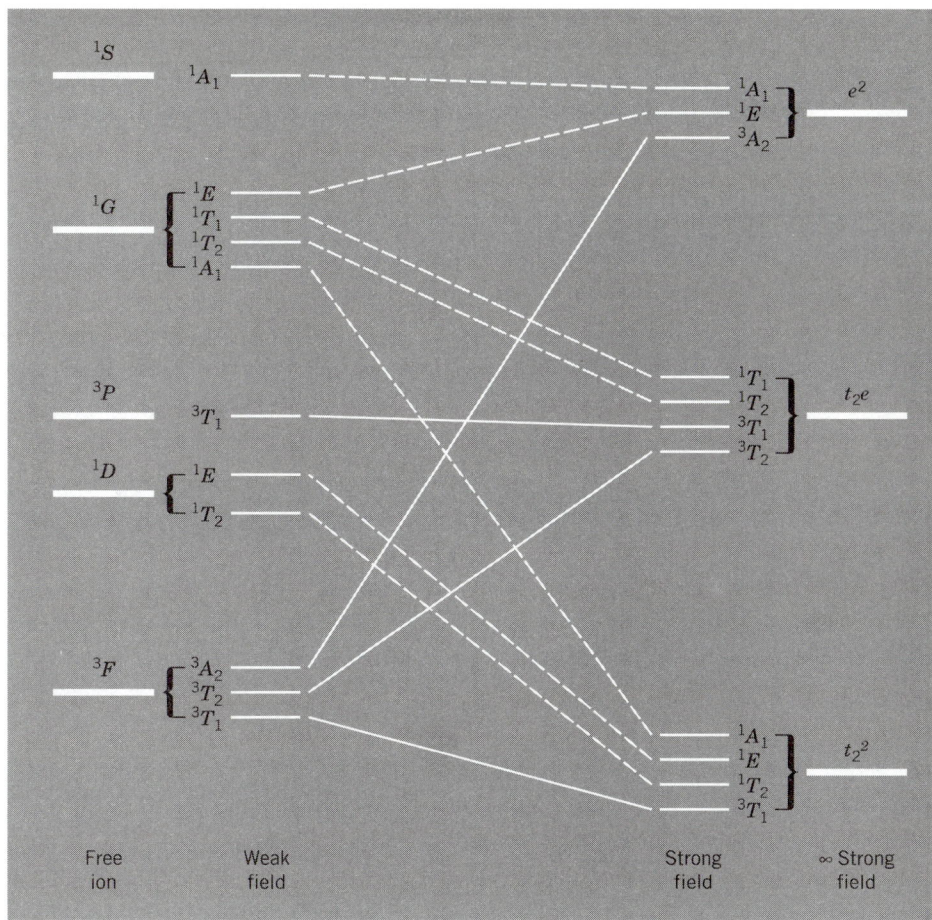

Figure 23-20 A correlation diagram for a d^2 ion in an octahedral environment. All states and orbitals are of the gerade type, and the subscript g has therefore been omitted.

$S = \frac{1}{2}, 1, \frac{3}{2}, 2, \ldots$, and, then, spin multiplicities (given by $2S + 1$) of 2 (doublet), 3 (triplet), 4 (quartet), 5 (quintet), $\ldots$, respectively.

We illustrate briefly with an ion having $L = 2$ and $S = \frac{3}{2}$. The electronic state would be designated ^4D. A few other examples are

$$L = 3, \, S = 1: \quad ^3F$$

$$L = 4, \, S = \tfrac{1}{2}: \quad ^2G$$

$$L = 0, \, S = 2: \quad ^5S$$

We note that for all d^1 ions, the electronic state must be solely ^2D, since $L = \ell = 2$.

For each electronic state that we identify using the Russell–Saunders coupling scheme, there is an orbital degeneracy as well as a spin degeneracy. The orbital degeneracies of the various electronic states are S (0), P (3), D (5), F (7), $\ldots$, and so on. These degeneracies correspond, of course, to those of the comparable orbitals having the same ℓ values in lowercase.

In summary, according to the Russell–Saunders coupling scheme, there are two quantum numbers L and S that characterize a given electronic state, just as we have ℓ and s quantum numbers that characterize a single electron. These two quantum numbers can be used to define and designate the various electronic states of a given d^n electron configuration. The Russell–Saunders scheme allows us to list all of the electronic states that arise from a d^n configuration and to estimate their energies. The scheme applies well to the $3d^n$ (and perhaps also to $4d^n$) cases involving weak field ligands. The procedures for applying the scheme demand a greater knowledge of quantum mechanics than is appropriate at this level, so we shall simply list the results of the analysis, along with some observations that will prove useful when applying the results to the interpretation of electronic absorption spectroscopy for the various coordination compounds. It is this latter purpose that remains the goal of Section 23-6.

For the d^2 case, the combinations of orbital and spin angular momenta (i.e., the couplings of quantum numbers L and S) that are consistent with the Pauli exclusion principle are represented by the following electronic state symbols:

$$^3F, \, ^1D, \, ^3P, \, ^1G, \, ^1S$$

It should be noted that a set of five d orbitals coupled with two choices for spin is equivalent to a set of 10 boxes. Thus a total of $(10 \times 9)/2 = 45$ microstates are possible in the d^2 electron configuration. As expected, the sum of the degeneracies of the states just listed will be found to be 45.

This list is given in order of increasing energy, as predicted by theory and established by experiment. Spectroscopic data are available for all of the d^2 ions that are of common occurrence, and the data provide not only the correct order of the electronic states, but also the exact value of the energy differences between the states.

The fact that the electronic state of lowest energy is the ^3F state is to be expected on the basis of Hund's rules. Hund's first rule, which was discussed in Chapter 2, says that for any partially filled shell, the most stable arrangement will be the one with the highest spin multiplicity: the arrangement with the maximum number of parallel electron spins. The second rule says that, among states of highest spin multiplicity (in this case, spin triplets), the state with highest orbital angular momentum is preferred. In this list for the states of a d^2 ion, and among those that are triplets, the one of lower energy is, then, the F state, for which $L = 3$.

In Fig. 23-20 we have placed these five free ion electronic states of the d^2 electron configuration in the correct order, on the vertical energy scale. We must next ask what happens to each of these states as the octahedral ligand field increases in strength from zero (for the free ion) to some small value. After doing this we shall need to connect the various states on the left and right sides of the diagram in order to trace the energies of the various states as the ligand field strength increases to the strong field case.

On the left portion of the diagram of Fig. 23-20 we have shown that a state with an orbital angular momentum quantum number L will split in the same way as a set or orbitals characterized by the corresponding value of the quantum number ℓ. Thus a D electronic state ($L = 2$) splits into E and T_2 states in an octahedral ligand field, just as a set of five d orbitals ($\ell = 2$) splits into the sets e_g and t_{2g} in an octahedral field. If, as here, the D electronic state is one derived from a d^n electron configuration, it and the states into which it splits will be of *gerade* character, just as the d orbitals are *gerade*. Also, weak ligand fields cannot alter the spin multiplicity of an electronic state, so that all states derived from a given free ion electronic state will retain the same spin multiplicity.

Table 23-3 lists the states into which the various free ion electronic states are split by the influence of an octahedral ligand field. Note that an S state, being nondegenerate, does not split, and a P state in an octahedral ligand field survives with its threefold orbital degeneracy. All the other free ion states are split to one extent or another into states by an octahedral ligand field. We say that the degeneracies of the various states are lifted, to one extent or another, by the ligand field. By employing the information of Table 23-3, we can complete the left side of Fig. 23-20 as shown for the weak field case.

We now have the two edges of a complete energy-level diagram for a d^2 ion in an octahedral ligand field. At the extreme left we see what states exist when the ligand field strength is zero and how these states are affected as a very weak field is applied. At the extreme right, we have information on what electronic states exist in the presence of a ligand field so strong that it completely overwhelms the electron–electron interactions. Our task now is to connect the two sides of this incomplete diagram so as to obtain a picture of how the electronic states of a d^2 configuration behave under realistic conditions, namely, at intermediate values of the ligand field strength.

We note first that the inventories of states on each side of Fig. 23-20 are the same; they must be if we are to be able to connect the two sides of the diagram completely. To carry this out, there are two rules that must be followed: First, only two states that are exactly alike may be connected. Second, connecting lines for states of the same type may never cross. With these rules, the connecting lines (or the state to state correlations) may be drawn in Fig. 23-20 in an absolutely unambiguous way. The correlations between triplet states have been drawn using solid lines, and those between singlet states have been drawn using broken lines, so that they are easy to distinguish.

Now that we have this diagram, what does it tell us? The following are the most important things as far as spectroscopy is concerned.

1. For all ligand field strengths, the ground electronic state of a d^2 system is a spin triplet, that is, the $^3T_{1g}$ state. Any d^2 ion in an octahedral field (however strong) of ligands will have two unpaired electrons in the ground state.

2. We have already pointed out that a general selection rule of quantum mechanics is that transitions between states of like parity (i.e., *gerade–gerade* or, sim-

Table 23-3   The Splitting of Free Ion Electronic States in an Octahedral Ligand Field

Electronic States of the Free Ion	Electronic States of the Ion in an Octahedral Field
S	A_1
P	T_1
D	$E + T_2$
F	$A_2 + T_1 + T_2$
G	$A_1 + E + T_1 + T_2$
H	$E + T_1 + T_1 + T_2$

ilarly, *ungerade–ungerade*) are discouraged. Recall, then, that any absorption bands assigned to transitions from the ground state of Fig. 23-20 to other electronic states of Fig. 23-20 are greatly discouraged by this selection rule.

3. It is another general selection rule of quantum mechanics that electronic transitions between states of different spin multiplicities are forbidden. As in most cases, the rigorous selection rule is relaxed, and we conclude more realistically that such transitions, though not completely unobserved, are at least greatly discouraged. Consequently, they are weak in absorption intensity when measured spectroscopically. Application of this selection rule to the d^2 case brings us to the conclusion that the only spin-allowed transitions are those from the triplet ground state to the triplet excited states. Thus three absorption bands for a d^2 system are predicted: those from the $^3T_{1g}$ ground state to the excited states $^3T_{2g}$, $^3A_{2g}$, and $^3T_{1g}$. The last excited electronic state originates from the ^3P free ion state.

4. Another general selection rule for electronic transitions is that two-electron transitions are much less probable than one-electron transitions. Thus the $^3A_{2g}$ state, which in the strong field limit correlates with the e_g^2 electron configuration, is not readily reached from the $^3T_{1g}$ ground state, which is derived from the t_{2g}^2 electron configuration. This transition should then be considerably weaker in absorption intensity than the other two spin-allowed transitions.

In summary, for a d^2 system, we expect to observe three electronic absorption bands, all of which are weak in absorption intensity. Each of the absorptions is spin allowed, but one is expected to be weaker than the other two because it is a transition involving two electrons. We shall make a direct comparison of these predictions with experiment as soon as we have introduced a more quantitative form of the energy-level diagram given in Fig. 23-20.

Quantitative Interpretation of Electronic Absorption Spectra for Various d^n Systems

The energy-level diagram (actually a simple correlation diagram) that is presented in Fig. 23-20 for the d^2 octahedral case is entirely correct as far as it goes, but it lacks some features necessary for practical, quantitative use. In fact, we have available to us calculations of electronic state energies as a function of ligand field strength, and we can use these results for a quantitative fitting of observed spectra. This has been done by Tanabe and Sugano. Tanabe–Sugano diagrams, which are presented in Fig. 23-21, are widely used to correlate and interpret spectra for ions of all types, from d^2 to d^8. To provide an understanding of these diagrams, we shall first take the d^2 case and compare it in detail with Fig. 23-20.

The energies of the various electronic states are given in the Tanabe–Sugano diagram on the vertical axis, and the ligand field strength increases from left to right on the horizontal axis. The symbols in the diagrams of Fig. 23-21 omit the subscript g, with the understanding that all states are *gerade* states. Also, in Tanabe–Sugano diagrams, the zero of energy for any particular d^n ion is taken to be the energy of the ground state. Regardless of the ligand field strength, then, the horizontal axis represents the energy of the ground state, and the energies of the excited electronic states are plotted against the energy of the ground state.

The unit of energy in a Tanabe–Sugano diagram is the parameter B, called

Racah's parameter. For different isoelectronic ions (i.e., the d^2 ions Ti^{2+} and V^{3+}, or the d^4 ions Cr^{2+}, Mn^{3+}, and Fe^{4+}), the values of B are different, as shown at the top of each diagram in Fig. 23-21. By plotting energies using the parameter B, one Tanabe–Sugano diagram may be used for all members of an isoelectronic group.

An Example of a d^2 Ion. Let us now take the V^{3+}(aq) ion as an example of an octahedral d^2 system. The well-known green color of this ion is caused by an electronic absorption spectrum that displays two weak absorption bands, one at

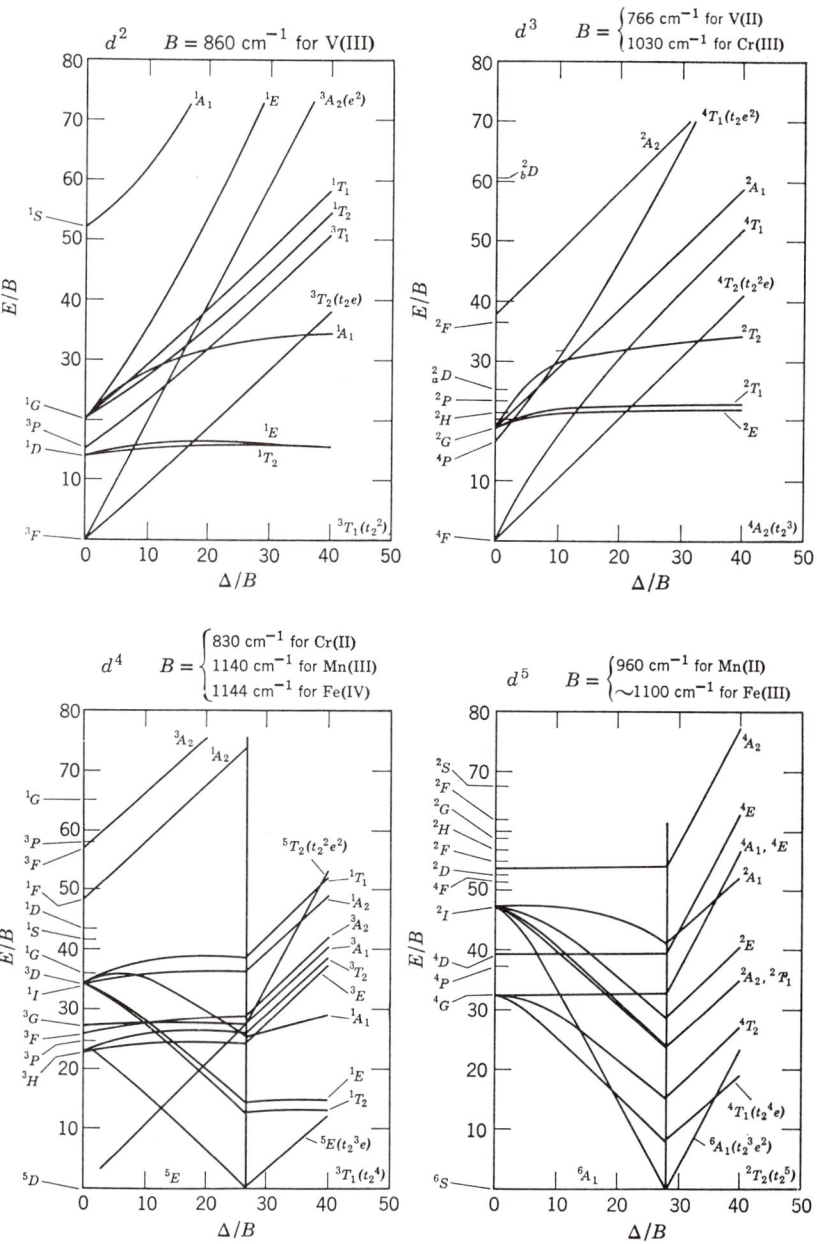

Figure 23-21 Energy-level diagrams [after Tanabe and Sugano, *J. Phys. Soc. Jpn.*, **1954**, *9*, 753.] for the d^2–d^8 configurations, in octahedral symmetry.

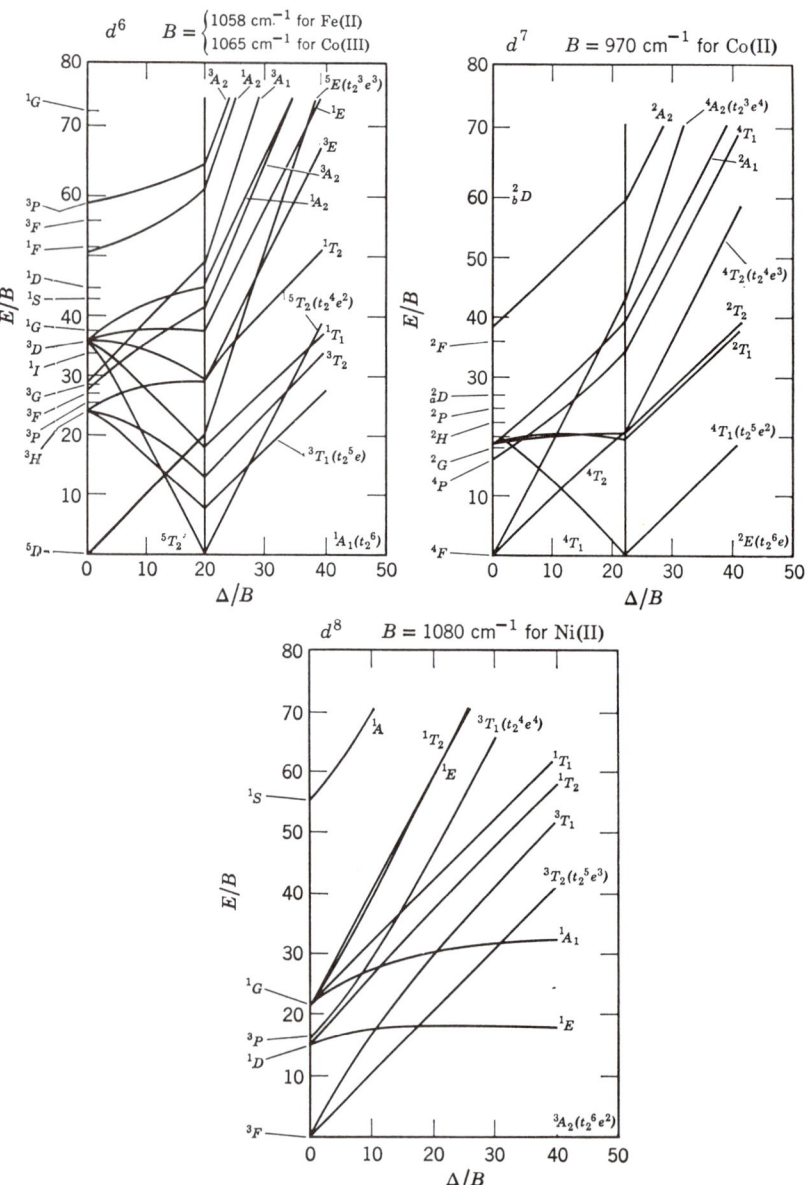

Figure 23-21 *(continued)*

17,800 and one at 25,700 cm^{-1}. The absorptions are weak in intensity, and have molar absorptivities (L mol^{-1} cm^{-1}) of 3.5 and 6.6, respectively. Such low values for molar absorptivity suggest that the electronic transitions that are responsible for the absorption of electromagnetic radiation are transitions that are forbidden, as we anticipated. Using the diagram for a d^2 ion from Fig. 23-21, we can discover only three spin allowed transitions, as we pointed out before using Fig. 23-20. Let us assign the first absorption band at 17,800 cm^{-1} to the transition $^3T_{1g}$ $\rightarrow$ $^3T_{2g}$ and the second band at 25,700 cm^{-1} to the transition $^3T_{1g} \rightarrow$ $^3T_{2g}(\text{P})$. We shall not assign one of the two observed absorptions to the transition to the $^3A_{2g}$ state, as that represents a two electron jump, as discussed previously. We can now proceed to use the Tanabe–Sugano diagram to try to get a "fit" of the energies for these two absorptions.

We now look at the d^2 Tanabe–Sugano diagram for the value on the horizontal axis of Δ/B that gives the best fit to these experimental absorption band energies. We find that the best value of Δ/B is about 29, as this is the point where the energies of the two states are in a ratio of 1.43 (as judged by the values of both states at this value of Δ/B), agreeing with the experimental ratio of 1.44. At Δ/B of 29.0, we find transition energies (differences in energy between the ground state and each of the two excited states) of 28.5 B and 40.5 B. Using the observed energies in reciprocal centimeters, we can then calculate that the value of B for the $[V(H_2O)_6]^{3+}$ ion must be 630 cm^{-1}. This is only 73% of the value of B for the free ion that is listed in the diagram for uncomplexed V^{3+}.

An Example of a d^8 Ion. The Ni^{2+} ion, like the vanadium ion discussed previously, exists in aqueous solutions of its salts as the hexaaqua nickel ion, $[Ni(H_2O)_6]^{2+}$. Aqueous solutions of salts such as $NiSO_4$, and $NiCl_2$ have a pale green color. Upon addition of aqueous NH_3 or the bidentate ethylenediamine, en ($H_2NCH_2CH_2NH_2$), the color of these aqueous solutions becomes deep blue or purple, respectively. We show in Fig. 23-22 the electronic absorption spectra of the ions $[Ni(H_2O)_6]^{2+}$ and $[Ni(en)_3]^{2+}$. The wavelengths involved cover the near-UV, through the visible, to the near-IR portions of the spectrum. The spectrum for each ion has three main absorption bands. For the hexaaqua ion, the lowest one is at about 8800 cm^{-1} and the uppermost one is at 24,000 cm^{-1}.

Let us turn now to the Tanabe–Sugano diagram for d^8 ions. At a value of Δ/B of about 11, we get good agreement between experiment and theory for the ratio of the energies of the two bands at 8800 and 24,000 cm^{-1} of the hexaaqua Ni^{2+} ion. Here we have assumed that the value of B in the complex is about 80% of that for the free ion. We are then able to predict that the middle absorption band in the spectrum, assigned to the transition $^3A_{2g} \rightarrow {}^3T_{1g}$, should be at $E/B =$

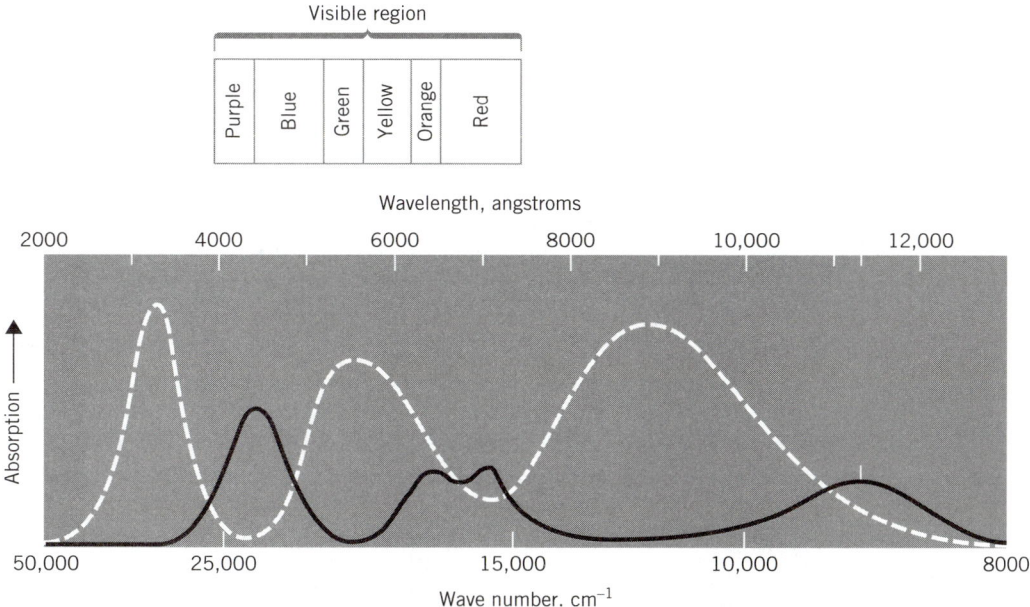

Figure 23-22   The electronic spectra of $[Ni(H_2O)_6]^{2+}$ (—) and $[Ni(en)_3]^{2+}$ (- - - -). Also shown is the correspondence between wavelength and the colors of the visible portion of the spectrum.

18, which becomes 16,000 cm^{-1}, if we let $B = 865$ (80% of 1080). This is in good agreement with the observed position of the middle "bands" in the spectrum of the hexaaqua ion. Similarly, for the Δ/B value of 13, and $B = 865$, we can fit all three bands in the spectrum of [Ni(en)$_3$]$^{2+}$.

We are now able to demonstrate that the ligand field strength of the three en ligands is greater than that of the six H$_2$O ligands at Ni^{2+}. In fact, it is greater by the amount $\frac{13}{11} = 1.18$, as based on the values of Δ/B deduced previously. This is what causes the spectrum to shift to higher energies (and the colors of the complexes to change) when the en ligand set replaces the water ligands.

Three other observations may be made about the nickel(II) ion spectra shown in Fig. 23-22. First we note that the middle absorption band in the spectrum of the ion [Ni(H$_2$O)$_6$]$^{2+}$ is really two close bands. The reason is that, at a ligand field strength of 11 B, there is a ^1E_g state that is practically degenerate with the $^3T_{1g}$ state. This can be seen in the d^8 Tanabe–Sugano diagram of Fig. 23-21. Because of an effect that has not been considered in our treatment, namely, "spin-orbit coupling," when two states of different spin multiplicity become nearly equal in energy, they get "mixed together," and there is then enough triplet character in both states to make transitions from the triplet ground state to both excited states become spin allowed. In other words, the transition $^3A_{2g} \rightarrow {}^1E_g$ becomes allowed because the spin selection rule is relaxed by spin-orbit coupling between the two excited states ^1E_g and $^3T_{1g}$.

Second, it can be seen that the spectrum of the [Ni(en)$_3$]$^{2+}$ ion is considerably more intense than that of the hexaaqua ion. This occurs because the [Ni(en)$_3$]$^{2+}$ ion lacks a center of symmetry, and the usual selection rule forbidding d–d transitions is relaxed.

Finally, on comparing the absorption spectra with the "color map" of the visible region of the spectrum, as is done in Fig. 23-22, it can be seen that the colors of the two ions are consistent with their spectra. The green [Ni(H$_2$O)$_6$]$^{2+}$ ion absorbs both the red and the blue ends of the spectrum, but transmits the central green region. The purple [Ni(en)$_3$]$^{2+}$ ion, however, absorbs the middle region of the visible spectrum (and some of the red), but transmits the purple and blue regions.

Examples of d^5 Ions. We now examine the spectra of the very pale pink, high-spin d^5 ions [Mn(H$_2$O)$_6$]$^{2+}$ and [Fe(H$_2$O)$_6$]$^{3+}$. It should be noted that the latter is obtained only in very strong noncomplexing acids, because the hexaaqua ion readily dissociates a proton to give [Fe(H$_2$O)$_5$OH]$^{2+}$, which is yellow brown. The reason both of these ions have extremely pale colors is that their d–d absorption spectra are even weaker (by a factor $\sim$100) than those we have looked at so far. The spectrum of the [Mn(H$_2$O)$_6$]$^{2+}$ ion is shown in Fig. 23-23, where it can be seen that the molar absorptivities (L mol^{-1} cm^{-1}) are extremely low. Why should these absorption bands be so extraordinarily weak? The answer must be given in terms of one or more selection rules that discourage the electronic transitions.

We can foresee the answer even before we look at the Tanabe–Sugano diagram. For a high-spin d^5 configuration, each d orbital is singly occupied, and all spins are parallel. This requires the ground electronic state to be a spin sextet ($^6A_{1g}$). There is, furthermore, no way to promote an electron from a t_{2g} orbital to an e_g orbital without reversing its spin. Thus, for the high-spin d^5 case, all possible d–d electronic transitions are spin forbidden. It is this extra degree of forbiddenness that decreases the absorption intensities by a factor of about 100

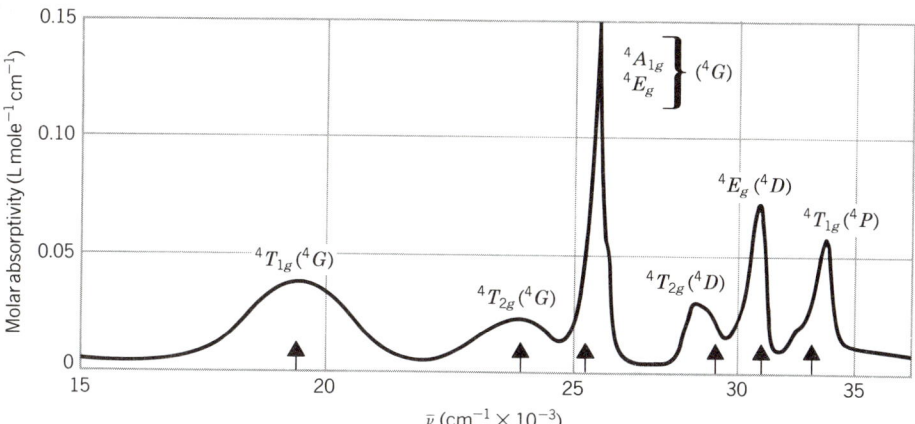

Figure 23-23   The electronic absorption spectrum of $[Mn(H_2O)_6]^{2+}$. Arrows indicate predicted band positions.

compared to the usual $d–d$ absorption band. Apart from this intensity question, the $[Mn(H_2O)_6]^{2+}$ absorption spectrum can be explained in the usual way, and the band assignments are shown in Fig. 23-23.

Charge-Transfer Spectra

While the $d–d$ absorption bands of many transition metal complexes are usually their most important electronic spectroscopic feature, there is another class of electronic transitions that can always occur and sometimes play a prominent role in the spectra of coordination compounds. A $d–d$ transition involves redistribution of electrons among orbitals that are mainly (in the crystal field model, entirely) localized on the metal atom. There are also electronic transitions in which an electron moves from an essentially ligand-based orbital to an essentially metal-based orbital, or vice versa. When this happens, charge is transferred from one part of the coordination sphere to another. The resulting spectroscopic features and the electronic transitions from which they arise are called, respectively, charge-transfer (CT) bands and charge-transfer (CT) transitions.

There are two broad classes of CT transitions. When an electron passes from a ligand-based orbital to a metal-based one, we have a ligand-to-metal charge-transfer (LMCT) absorption band, or transition. When the electron moves from an orbital that is largely metal based to one that is ligand based, we have a metal-to-ligand charge-transfer (MLCT) absorption, or transition. A few illustrations can now be given.

The most familiar CT transition may be the one that is responsible for the intense red color that identifies the Fe^{3+} ion upon addition of thiocyanate ion (SCN^-) to aqueous solutions of Fe^{3+}. The reason this color is so intense is because the LMCT transition of the complex is an allowed electronic transition in every aspect. No selection rule is violated by the transition. There is no change in spin multiplicity associated with the transition and the electron moves from a ligand orbital that is *ungerade* to a metal orbital that is *gerade*. Such a CT transition is allowed by these two important selection rules, and the absorption intensity (as measured by molar absorptivity) is about 1000 times greater than that of a typical $d–d$ transition. These are characteristic features of many CT transitions.

It may also be noted that the yellow-brown color of the $[Fe(H_2O)_5OH]^{2+}$ ion is due to an OH to metal LMCT band in the near UV, which is so intense that its tail, or low-energy edge, absorbs significantly in the blue end of the visible spectrum.

In general, LMCT bands are found in the visible or very near-UV region, for complexes having lone-pair electrons on anionic ligands and metal atoms in high oxidation states. These are just the sorts of complexes that should favor movement of electron density from the electron-rich ligands to metals with high positive charges. Furthermore, the high-spin Fe^{3+} ion has vacancies in both its t_{2g} and its e_g orbital sets, and both types of LMCT transition, that is, to either the metal e_g or t_{2g} orbitals, occur.

The other type of CT transition, namely, metal–ligand or MLCT transitions, are less common. They are likely to occur in the visible region of the spectrum. The MLCT absorptions are expected only in systems containing metals in low oxidation states and ligands with empty π^* orbitals. Organometallic compounds that we shall discuss in later chapters fall into this category. Thus, the Group VIA(6) metal hexacarbonyl molecules, $M(CO)_6$ (M = Cr, Mo, or W), all have a MLCT band around 35,000 cm^{-1}, which involves the transfer of a metal t_{2g} electron to the π^* orbitals of the CO ligands. These absorptions have molar absorptivities of about 15,000 L mol^{-1} cm^{-1}, as compared to values on the order of 1–100 for typical d–d absorption bands.

23-7   Some Generalizations Concerning Ligand Field Splittings and Spectra

Certain generalizations may be made about the dependence of the magnitudes of Δ values on the valence and atomic number of the metal ion, the symmetry of the coordination shell, and the nature of the ligands. For octahedral complexes containing high-spin metal ions, it may be inferred from the accumulated data for a large number of systems that

1. The Δ_o values for complexes of the first transition series are 7500–12,500 cm^{-1} for divalent ions and 14,000–25,000 cm^{-1} for trivalent ions.

2. The Δ_o values for corresponding complexes of metal ions in the same group and with the same valence increase by 30–50% on going from the first transition series to the second and by about this amount again from the second to the third. This is well illustrated by the Δ_o values for the complexes $[Co(NH_3)_6]^{3+}$, $[Rh(NH_3)_6]^{3+}$, and $[Ir(NH_3)_6]^{3+}$, which are, respectively, 23,000, 34,000, and 41,000 cm^{-1}.

3. The Δ_t values are about 40–50% of Δ_o values for complexes differing as little as possible except in the geometry of the coordination shell, in agreement with theoretical expectation.

4. The dependence of Δ values on the identity of the ligands follows a regular order known as the spectrochemical series, which will now be explained.

The Spectrochemical Series

Experimental study of the spectra of a large number of complexes containing various metal ions and various ligands led to the arrangement of ligands in a series according to their capacity to cause d-orbital splittings. This series, for the more common ligands, is $I^- < Br^- < Cl^- < F^- < OH^- < C_2O_4^{2-} < H_2O < —NCS^-$ $< py < NH_3 < en < bpy < o\text{-phen} < NO_2^- < CN^-$. The idea of this series is that the d-orbital splittings and, hence, the relative frequencies of visible absorption bands for two complexes containing the same metal ion but different ligands can be predicted from this series, whatever the particular metal may be. We have already seen a typical illustration of the relative positions of en and H_2O in the spectrochemical series when we examined the d–d spectra of the $[Ni(H_2O)_6]^{2+}$ and $[Ni(en)_3]^{2+}$ ions. Naturally, one cannot expect such a simple rule to be universally applicable. The following qualifications must be remembered in applying it.

1. The series is based on data for metal ions in common oxidation states. Because the nature of the metal–ligand interaction in an unusually high or unusually low oxidation state of the metal may be in certain respects qualitatively different from that for the metal in a normal oxidation state, striking violations of the order shown may occur for complexes in unusual oxidation states.

2. Inversions of the order of adjacent or nearly adjacent members of the series are sometimes found even for metal ions in their normal oxidation states.

23-8 Structural and Thermodynamic Effects of d-Orbital Splittings

Regardless of what type or level of theory is used to account for the existence of the d-orbital splittings, the fact that they do exist is of major importance. Their existence affects both structural and thermodynamic properties of the ions and their complexes.

Ionic Radii

Figure 23-24 shows a plot of the octahedral radii of the divalent ions of the first transition series. The points for Cr^{2+} and Cu^{2+} are indicated with open circles because the Jahn–Teller effect, to be discussed later, makes it difficult to obtain these ions in truly octahedral environments, thus rendering the assessment of their "octahedral" radii somewhat uncertain. A smooth curve has also been drawn through the points for Ca^{2+}, Mn^{2+}, and Zn^{2+} ions, which have the electron configurations $t_{2g}^0 e_g^0$, $t_{2g}^3 e_g^2$, and $t_{2g}^6 e_g^4$, respectively. In these three cases the distribution of d-electron density around the metal ion is spherical because all d orbitals are either unoccupied or equally occupied. Because the shielding of one d electron by another from the nuclear charge is imperfect, there is a steady contraction in these three ionic radii. It is seen that the radii of the other ions are all below the values expected from the curve passing through Ca^{2+}, Mn^{2+}, and Zn^{2+}. This occurs because the d electrons in these ions are not distributed uniformly (i.e., spherically) about the nuclei as we shall now explain.

The Ti^{2+} ion has the configuration t_{2g}^2. This means that the negative charge of two d electrons is concentrated in those regions of space away from the metal–ligand bond axes. Thus, compared with the effect that they would have if distributed spherically around the metal nucleus, these two electrons provide ab-

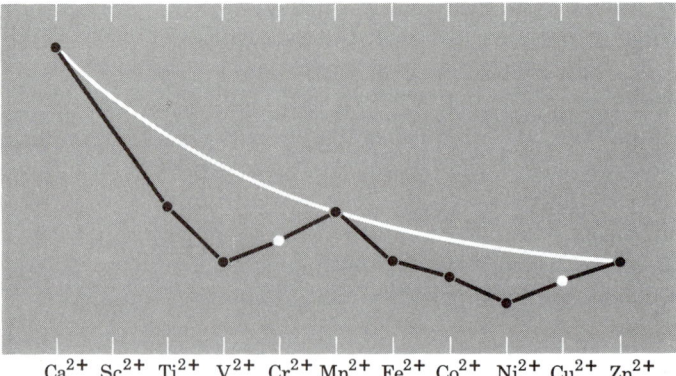

Ca^{2+} Sc^{2+} Ti^{2+} V^{2+} Cr^{2+} Mn^{2+} Fe^{2+} Co^{2+} Ni^{2+} Cu^{2+} Zn^{2+}

Figure 23-24   The relative ionic radii of divalent ions of the first transition series. The white line is a theoretical curve as explained in the text.

normally little shielding between the positive metal ion and the negative ligands; therefore, the ligand atoms are drawn in closer than they would be if the d electrons were spherically distributed. Thus, in effect, the radius of the metal ion is smaller than that for the hypothetical, isoelectronic spherical ion. In V^{2+} this same effect is found in even greater degree because there are now three t_{2g} electrons providing much less shielding between the metal ion and ligands than would three spherically distributed d electrons. For Cr^{2+} and Mn^{2+}, however, we have the configurations $t_{2g}^3 e_g^1$ and $t_{2g}^3 e_g^2$, in which the electrons added to the t_{2g}^3 configuration of V^{2+} go into orbitals that concentrate them mainly between the metal ion and the ligands. These e_g electrons thus provide a great deal more screening than would be provided by spherically distributed electrons, and indeed the effect is so great that the radii actually increase. The same sequence of events is repeated in the second half of the series. The first three electrons added to the spherical $t_{2g}^3 e_g^2$ configuration of Mn^{2+} go into the t_{2g} orbitals where the screening power is abnormally low, and the radii therefore decrease abnormally rapidly. On going from Ni^{2+}, with the configuration $t_{2g}^6 e_g^2$, to Cu^{2+} and Zn^{2+}, electrons are added to the e_g orbitals where their screening power is abnormally high, and the radii again cease to decrease and actually show small increases. Similar effects are found with trivalent ions, with ions of other transition series, and in tetrahedral complexes.

The Jahn–Teller Effect

In 1937 Jahn and Teller showed that in general no nonlinear molecule can be stable in a degenerate electronic state. The molecule must become distorted in such a way as to break the degeneracy. It develops that one of the most important areas of application of this Jahn–Teller theorem is the stereochemistry of the complexes of certain transition metal ions.

   To illustrate, we consider an octahedrally coordinated Cu^{2+} ion. There is one vacancy in the e_g orbitals, in either the $d_{x^2-y^2}$ or the d_{z^2} orbital. If the coordination is strictly octahedral, the two configurations $d_{x^2-y^2}^2 d_{z^2}^1$ and $d_{x^2-y^2}^1 d_{z^2}^2$, are of equal energy. This is the sense in which the electronic state of the Cu^{2+} ion is doubly degenerate. However, this is a state which, according to the Jahn–Teller theorem, cannot be stable, and the octahedron must distort so that the two configurations just mentioned are no longer of equal energy.

Actually, it is easy to see why this will happen. Suppose the actual configuration in the e_g orbitals is $d_{x^2-y^2}^1 d_{z^2}^2$. The ligands along the z axis are much more screened from the charge of the Cu^{2+} ion than are the four ligands along the x and y axes. The z-axis ligands will therefore tend to move further away. As they do so, however, the d_{z^2} orbital will become more stable than the $d_{x^2-y^2}$ orbital, thus removing the degeneracy, as is shown in Fig. 23-6. Of course, if we begin with a $d_{x^2-y^2}^2 d_{z^2}^1$ configuration, a distortion of the opposite kind would be expected. The question of which situation will actually occur is very difficult to predict, and there are, in fact, still other possibilities. However, it is the former type of distortion, the elongation on one axis, that is actually observed in a large number of Cu^{2+} complexes.

This is well illustrated by the copper(II) halides. In each case the Cu^{2+} ion has a coordination number of six, with four near neighbors in a plane and two more remote ones. The actual distances are shown in Fig. 23-25.

It is not difficult to see that the reasoning involved in the Cu^{2+} case will apply in all cases where an odd number (1 or 3) of electrons would occupy the e_g orbitals in an octahedral complex. In the case of a single electron, either the $d_{x^2-y^2}$ or the d_{z^2} orbital could be occupied, and the occupied orbital should "push away" the ligands toward which it is directed. The important cases in which this may be expected are

$$t_{2g}^3 e_g^1 \quad \text{high-spin } Cr^{2+} \text{ and } Mn^{3+}$$

$$t_{2g}^6 e_g^1 \quad \text{low-spin } Co^{2+} \text{ and } Ni^{3+}$$

Distortions similar to those for Cu^{2+} are, indeed, found for the "octahedral" complexes of these ions.

Ligand Field Stabilization Energies

We learned in Section 23-2 that the d orbitals of an ion in an octahedral field are split so that three of them become more stable (by $2\Delta_o/5$) and two of them less stable (by $3\Delta_o/5$) than they would be in the absence of the splitting. Thus, for example, a d^2 ion will have each of its two d electrons stabilized by $2\Delta_o/5$, giving a total stabilization of $4\Delta_o/5$. Recalling from Section 23-7 that Δ_o values run

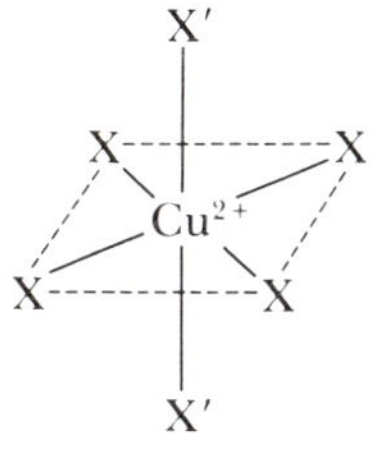

	Cu—X	Cu—X'
X = Cl	2.30	2.95
X = Br	2.40	3.18
X = F	1.93	2.27

Figure 23-25   The distorted six coordination found in the Cu^{II} halides, distances in angstroms. This elongation of the axial Cu—X' bonds constitutes an example of the Jahn–Teller effect.

about 10,000 and 20,000 cm^{-1} for di- and trivalent ions of the first transition series, we can see that these "extra" stabilization energies—extra in the sense that they would not exist if the d shells of the metal ions were symmetrical as are the other electron shells of the ions—will amount to about 100 and about 200 kJ mol^{-1}, respectively, for di- and trivalent d^2 ions. These *ligand field stabilization energies*, LFSEs, are of course of the same order of magnitude as the energies of most chemical changes, and will therefore play an important role in the thermodynamic properties of transition metal compounds.

Let us first consider high-spin octahedral complexes. Every t_{2g} electron represents a stability increase (i.e., energy lowering) of $2\Delta_o/5$, whereas every e_g electron represents a stability decrease of $3\Delta_o/5$. Thus, for any configuration $t_{2g}^p e_g^q$, the net stabilization will be given by $(2p/5 - 3q/5)\Delta_o$.

The results obtained for all of the ions, that is, d^0 to d^{10}, using this formula are collected in Table 23-4. Since the magnitude of Δ_o for any particular complex can be obtained from the spectrum, it is possible to determine the magnitudes of these crystal field stabilization energies independently of thermodynamic measurements and, thus, to determine what part they play in the thermodynamics of the transition metal compounds.

The enthalpies of hydration of the divalent ions of the first transition series are the energies of the processes:

$$M^{2+} (g) + \infty \, H_2O = [M(H_2O)_6]^{2+} (aq) \qquad (23\text{-}8.1)$$

They can be estimated by using thermodynamic cycles. The energies calculated are shown by the filled circles in Fig. 23-26. It will be seen that a smooth curve, which is nearly a straight line, passes through the points for the three ions, $Ca^{2+}(d^0)$, $Mn^{2+}(d^5)$, and $Zn^{2+}(d^{10})$, which have no LFSE, while the points for all other ions lie above this line. If we subtract the LFSE from each of the actual hydration energies, the values shown by open circles are obtained, and these fall on the smooth curve. It may be noted that, alternatively, LFSEs could have been estimated from Fig. 23-20 and used to calculate Δ_o values. Either way, the agreement between the spectroscopically and thermodynamically assessed Δ_o values provides evidence for the fundamental correctness of the idea of d-orbital splitting.

Table 23-4 Ligand Field Stabilization Energies, LFSEs, for Octahedrally and Tetrahedrally Coordinated High-Spin Ions

Number of d Electrons	Stabilization Energies		Difference, Octahedral-Tetrahedral[a]
	Octahedral	Tetrahedral	
1,6	$2\Delta_o/5$	$3\Delta_t/5$	$\Delta_o/10$
2,7[b]	$4\Delta_o/5$	$6\Delta_t/5$	$2\Delta_o/10$
3,8	$6\Delta_o/5$	$4\Delta_t/5$	$8\Delta_o/10$
4,9	$3\Delta_o/5$	$2\Delta_t/5$	$4\Delta_o/10$
0, 5, 10	0	0	0

[a]Assuming $\Delta_o = 2\Delta_t$.

[b]For the d^2 and d^7 ions, the figure obtained in this way and given above is not exactly correct because of the effect of configuration interaction.

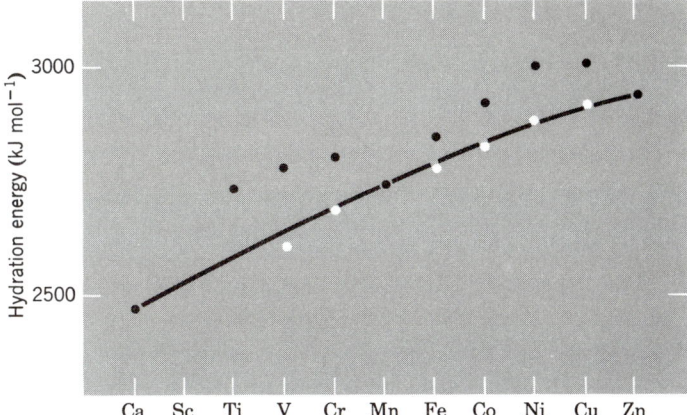

Figure 23-26 Hydration energies of some divalent ions of the first transition series. Solid circles are the experimentally obtained hydration energies. Open circles are energies corrected for LFSE.

Another important example of the thermodynamic consequences of ligand field splittings is shown in Fig. 23-27, where the lattice energies of the dichlorides of the metals from calcium to zinc are plotted versus atomic number. Once again they define a curve with two maxima, and a minimum at Mn^{2+}. As in previous cases, for all the ions having LFSEs, the energies lie above the curve passing through the energies of the three ions that do not have ligand field stabilization energy. Similar plots are obtained for the lattice energies of other halides and of the chalconides of di- and trivalent metals.

It is important to note that the LFSEs, critical as they may be in explaining the *difference* in energies between various ions in the series, make up only a small fraction, 5–10%, of the *total* energies of combination of the metal ions with the ligands. In other words, the LFSEs though crucially important in many ways, are not by any means major sources of the binding energies in complexes.

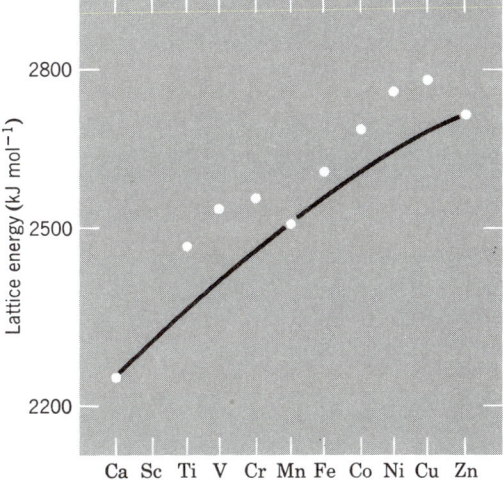

Figure 23-27 The lattice energies of the dichlorides of the elements from Ca to Zn.

Stability of Coordination Compounds

It is a fairly general observation that the equilibrium constants for the formation of analogous complexes of the divalent metal ions of Mn to Zn with ligands that contain nitrogen as the donor atom fall in the following order of the metal ions $Mn^{2+} < Fe^{2+} < Co^{2+} < Ni^{2+} < Cu^{2+} > Zn^{2+}$. The LFSEs are responsible for this general trend. If it is assumed that $\Delta S°$ values in the formation of a particular complex by the different metal ions will be essentially constant, then this order of formation constants is also the order of $-\Delta H°$ values for complex formation. Figure 23-26 shows that this order is the same as the order of hydration energies of gaseous ions. When an aqueous aqua ion, $[M(H_2O)_6]^{2+}$, reacts with a set of ligands to form a complex, LFSE in the complex is usually greater than that in the aqua ion. In each case, it will be greater by about the same fraction, say 20%. Thus, in each case, the replacement of water molecules by the new ligands will have a $-\Delta H°$ value that is proportional to the LFSE and the magnitudes of these $-\Delta H°$ values are in the same order as the LFSEs themselves.

STUDY GUIDE

Scope and Purpose

We have continued the discussion of coordination compounds that was begun in Chapter 6. The various theoretical treatments that have been presented are the electrostatic or crystal field model and the delocalized MO model. Each has its advantages, and the student should become comfortable with the language and the approach of both theories. The section on electronic absorption spectroscopy is certainly optional, but it does provide a concise sketch of a very important area of research into coordination compounds. The approaches to bonding theory that are presented here will be of great importance to the discussions in subsequent chapters.

Study Questions

A. Review

1. What is a practical definition of a transition element? What fraction of the approximately 109 known elements are of this type?

2. List some of the important characteristics of the transition elements.

3. Make drawings of the d orbitals, and state which fall into the e_g and which fall into the t_{2g} set in an octahedral ligand field.

4. What is the "center of gravity" rule and how does it apply to the splitting of the d orbitals in octahedral and in tetrahedral ligand fields?

5. Prepare a diagram that traces how the d-orbital splitting pattern changes as an octahedral complex is altered via a tetragonal distortion that is first weak and then reaches the extreme case where a square, four-coordinate complex is obtained.

6. According to the crystal field theory, the e_g and t_{2g} orbitals are purely metal d orbitals. How is this different from the approach of MO theory?

7. By using orbital splitting diagrams, show which d^n electron configurations are capa-

ble of giving both low-spin and high-spin configurations in an octahedral ligand field.

8. Calculate the spin-only magnetic moments that are expected in each case from Problem 7.

9. Why are $d-d$ electronic transitions weakly absorbing? Why are the absorptions observable at all, if they are forbidden?

10. Use the Tanabe–Sugano diagrams to show which d^n configurations that are high spin in the presence of a weak ligand field can become low spin at high values of the ligand field strength.

11. What is the spectrochemical series, and what limitations must be remembered in using it?

12. Which electronic transitions for a d^5 ion are spin allowed?

13. How does Δ_o change on going from one octahedral complex to another with the same ligand set, but (a) M^{3+} in place of M^{2+} or (b) a second series transition element in place of a first transition series element?

14. Explain the correlation between ion size and the number of d electrons.

15. Use Cr^{2+} to illustrate the influence of the Jahn–Teller effect on the ground-state structures of certain transition metal complexes.

16. Calculate, in units of Δ_o, the LFSEs of the following high-spin ions in their octahedral complexes Fe^{2+}, Mn^{2+}, Mn^{3+}, Co^{2+}.

B. Additional Exercises

1. How should the d-orbital splitting pattern for a tetrahedral complex be modified if the tetrahedron is flattened? Elongated?

2. What d-orbital splitting pattern would you expect for (a) a linear L—M—L complex. (b) a planar and triangular ML_3 complex, (c) a pyramidal ML_3 complex, (d) a trigonal bipyramidal ML_5 complex, (e) a square pyramidal ML_5 complex?

3. What d-orbital splitting pattern would you expect for an ML_8 complex with the eight ligands situated at the corners of a cube?

4. The complex $[NiCl_4]^{2-}$ is paramagnetic with two unpaired electrons, while $[Ni(CN)_4]^{2-}$ is diamagnetic. Deduce the structures of these two complexes and explain the observations in terms of ligand field theory.

5. Predict the relative positions of the absorption maximum in the spectra of $[Ti(CN)_6]^{3-}$, $[TiCl_6]^{3-}$, and $[Ti(H_2O)_6]^{3+}$.

6. What geometry do you expect for four-coordinate complexes of Zn^{2+}? Explain in terms of LFSEs.

7. Predict the magnetic properties and the LFSE for each of the following:
   (a) $[Fe(CN)_6]^{3-}$ (b) $[Ru(NH_3)_6]^{2+}$
   (c) $[Co(NH_3)_6]^{3+}$ (d) $[CoCl_4]^{2-}$
   (e) $[Fe(H_2O)_6]^{2+}$ (f) $[Mn(H_2O)_6]^{2+}$
   (g) $[CoF_6]^{3-}$ (h) $[Cr(H_2O)_6]^{2+}$

8. Why are tetrahedral complexes usually not low spin?

9. Prepare a drawing that shows the π-bond system that is responsible for the high position of CO in the spectrochemical series. Clearly show the donor orbital of the metal and the acceptor orbital of the CO ligand.

10. Consider the data of Fig. 23-24. For which ion is the effective nuclear charge highest? Why?

11. Use the appropriate Tanabe–Sugano diagram to estimate the positions of the three absorption bands for $[Ni(en)_3]^{2+}$.

12. Give the electronic state symbol for the ground electronic state arising from each of the octahedral d^n electron configurations, including high-spin and low-spin possibilities where appropriate.

SUPPLEMENTARY READING

Ballhausen, C. J. and Gray, H. B., "Electronic Structures of Metal Complexes," in *Coordination Chemistry*, A. E. Martell, Ed., Van Nostrand-Reinhold, New York, 1971.

Cotton, F. A., *Chemical Applications of Group Theory*, 3rd Edition, Wiley, New York, 1990.

Fackler, J. P., *Symmetry in Coordination Chemistry*, Academic, New York, 1971.

Figgis, B. N., *Introduction to Ligand Fields*, Wiley, New York, 1966.

Jørgensen, C. K., *Modern Aspects of Ligand Field Theory*, North-Holland, Amsterdam, 1971.

König, E. and Kremer, S., *Ligand Field Energy Level Diagrams*, Plenum, New York, 1977.

Lever, A. B. P., *Inorganic Electronic Spectroscopy*, 2nd ed., Elsevier, Amsterdam, 1984.

Mabbs, F. E. and Machin, D. J., *Magnetism and Transition Metal Complexes*, Wiley-Halsted, New York, 1973.

McClure, D. S. and Stephens, P. J., "Electronic Spectra of Coordination Compounds," in *Coordination Chemistry*, A. E. Martell, Ed., Van Nostrand-Reinhold, New York, 1971.

Schläfer, H. L. and Glieman, G., *Basic Principles of Ligand Field Theory*, Wiley, New York, 1969.

Chapter 24

THE ELEMENTS OF THE FIRST TRANSITION SERIES

As we have seen from their position in the periodic table (Section 2-5), the metals of the first transition series show variable valency. In this chapter we first discuss some of their common features and then consider the chemistry of individual elements.

24-1 The Metals

The metals are hard, refractory, electropositive, and good conductors of heat and electricity. The exception is copper, a soft and ductile metal, relatively noble, but second only to Ag as a conductor of heat and electricity. Some properties are given in Table 24-1. Manganese and iron are attacked fairly readily but the others are generally unreactive at room temperature. All react on heating with halogens, sulfur, and other nonmetals. The carbides, nitrides, and borides are commonly nonstoichiometric, interstitial, hard, and refractory.

24-2 The Lower Oxidation States

The oxidation states are given in Table 24-2, the most common and important (especially in aqueous chemistry) in bold type. Table 24-2 also gives the d electron configurations. Their chemistry can be classified on this basis; for example, the d^6 series is V^{-I}, Cr^0, Mn^I, Fe^{II}, Co^{III}, and Ni^{IV}. Comparisons of this kind can occasionally emphasize similarities in spectra and magnetic properties. However, the differences in properties of the d^n species due to differences in the nature of the metal, its energy levels, and especially the charge on the ion, often exceed the similarities.

    1. The oxidation states less than II. With the exception of copper, where copper(I) binary compounds and complexes and the Cu^+ ion are known, the chemistry of the I, 0, –I, and –II formal oxidation states is entirely concerned with:

    (a) π-Acid ligands such as CO, NO, PR_3, CN^-, and bpy.

    (b) Organometallic chemistry in which alkenes, acetylenes, or aromatic systems, such as benzene, are bound to the metal.

Table 24-1 Some Properties of the First Transition Series Metals

	Ti	V	Cr	Mn	Fe	Co	Ni	Cu
mp(°C)	1668	1890	1875	1244	1537	1493	1453	1083
Properties	Hard, corrosion resistant	Hard, corrosion resistant	Brittle, corrosion resistant	White, brittle, reactive	Lustrous, reactive	Hard, bluish color	Quite corrosion resistant	Soft and ductile, reddish color
Density (g cm^{-3})	4.51	6.11	7.19	7.18	7.87	8.90	8.91	8.94
E^0 (V^a)	—^{b}	-1.19	-0.91	-1.18	-0.44	-0.28	-0.24	+0.34
Solubility in acids	Hot HCl, HF	HNO$_3$, HF, concentrated H$_2$SO$_4$	dilute HCl, H$_2$SO$_4$	dilute HCl, H$_2$SO$_4$, and so on	dilute HCl, H$_2$SO$_4$, and so on	Slowly in dilute HCl	dilute HCl, H$_2$SO$_4$	HNO$_3$, hot concentrated H$_2$SO$_4$

^{a}For $M_{aq}^{2+} + 2 e^- = M(s)$.
^{b}No +2 ion in aqueous solution.

Table 24-2 Oxidation States of First Series Transition Elements[a]

Ti	V	Cr	Mn	Fe	Co	Ni	Cu
	0 d^5	0 d^6	0 d^7	0 d^8	0 d^9	**0** d^{10}	
	1 d^4	1 d^5	1 d^6		1 d^8	1 d^9	1 d^{10}
2 d^2	2 d^3	**2** d^4	2 d^5	2 d^6	2 d^7	**2** d^8	2 d^9
3 d^1	3 d^2	**3** d^3	3 d^4	3 d^5	3 d^6	3 d^7	3 d^8
4 d^0	4 d^1	4 d^2	4 d^3	4 d^4	4 d^5	4 d^6	
	5 d^0	5 d^1	5 d^2		5 d^4		
		6 d^0	6 d^1	6 d^2			
			7 d^0				

[a]Formal negative oxidation states are known in compounds of π-acid ligands, for example, Fe^{-II} in $[Fe(CO)_4]^{2-}$, Mn^{-I} in $[Mn(CO)_5]^-$, and so on.

There is an extensive chemistry of mixed compounds such as $(\eta^6\text{-}C_6H_6)Cr(CO)_3$ or $(\eta^4\text{-}C_4H_6)Fe(CO)_3$. These topics are described in Chapters 28 and 29. Some organometallic compounds in higher oxidation states are known, however, mainly for the cyclopentadienyl ligand as in $(\eta^5\text{-}C_5H_5)_2Ti^{IV}Cl_2$, $(\eta^5\text{-}C_5H_5)_2Fe^{II}$, and $[(\eta^5\text{-}C_5H_5)_2Co^{III}]^+$. With π-acid or organic ligands, transition metals also form many compounds with bonds to hydrogen, for example, $H_2Fe(PF_3)_4$. Compounds with M—H bonds are very important in certain catalytic reactions (Chapter 30).

2. The II oxidation state. The binary compounds in this state are usually ionic. The metal oxides are basic; they have the NaCl structure but are often nonstoichiometric, particularly for Ti, V, and Fe. The *aqua ions*, $[M(H_2O)_6]^{2+}$, except for the unknown Ti^{2+} ion, are well characterized in solution and in crystalline solids. The potentials and colors are given in Table 24-3. Note that the V^{2+}, Cr^{2+}, and Fe^{2+} ions are oxidized by air in acidic solution.

The aqua ions may be obtained by dissolution of the metals, oxides, carbonates, and so on, in acids and by electrolytic reduction of M^{3+} salts. Hydrated salts with noncomplexing anions usually contain $[M(H_2O)_6]^{2+}$; typical ones are

$$Cr(ClO_4)_2 \cdot 6H_2O \qquad Mn(ClO_4)_2 \cdot 6H_2O \qquad FeF_2 \cdot 8H_2O \qquad FeSO_4 \cdot 7H_2O$$

However, certain *halide hydrates* do *not* contain the aqua ion. Thus $VCl_2 \cdot 4H_2O$ is *trans*-$VCl_2(H_2O)_4$, and $MnCl_2 \cdot 4H_2O$ is a polymer with *cis*-$MnCl_2(H_2O)_4$ units; the diaqua species of Mn, Fe, Co, Ni, and Cu have a linear polymeric edge-shared chain structure with *trans*-$[MCl_4(H_2O)_2]$ octahedra. The $FeCl_2 \cdot 6H_2O$ compound contains *trans*-$FeCl_2(H_2O)_4$ units.

The water molecules of $[M(H_2O)_6]^{2+}$ can be displaced by ligands such as NH_3, en, $EDTA^{4-}$, CN^-, and acac. The resulting complexes may be cationic, neutral, or anionic depending on the charge of the ligands. For Mn^{2+} complexes, the formation constants in aqueous solution are low compared with those of the other ions, because of the absence of ligand field stabilization energy in the d^5 ion (Section 23-8). In complexes the ions are normally *octahedral*, but for the Cu^{2+} and Cr^{2+} ions two H_2O molecules in trans positions are much further from the metal than the other four equatorial ones, because of the Jahn–Teller effect (Section 23-8). For Mn, the complex $[Mn(edta)H_2O]$ is seven coordinate. With halide ions, SCN^-, and some other ligands, *tetrahedral* species MX_4^{2-} and MX_2L_2 may be formed, the tendency being greatest for Co, Ni, and Cu.

Addition of OH^- to the M^{2+} solutions gives *hydroxides,* some of which can be obtained as crystals. The compounds $Fe(OH)_2$ and $Ni(OH)_2$ have the brucite, $Mg(OH)_2$, structure. On addition of HCO_3^- the carbonates of Mn, Fe, Co, Ni, and Cu are precipitated.

24-3 The III Oxidation State

All of the elements form at least some compounds in this state but for Cu only a few complexes, not stable toward water, are known.

The fluorides (MF_3) and oxides (M_2O_3) are generally ionic but the chlorides, bromides, and iodides (where known), as well as sulfides and similar compounds, may have considerable covalent character.

The elements Ti to Co form octahedral ions, $[M(H_2O)_6]^{3+}$. The Co^{3+} and Mn^{3+} ions are very readily reduced by water (Table 24-3). The Ti^{3+} and V^{3+} ions are oxidized by air. In aqueous solution high acidities are required to prevent hydrolysis, for example,

$$[Ti(H_2O)_6]^{3+} = [Ti(H_2O)_5OH]^{2+} + H^+ \quad K = 1.3 \times 10^{-4} \quad (24\text{-}3.1)$$

Addition of OH^- to the solutions gives *hydrous oxides* rather than true hydroxides. In fairly concentrated halide solutions, complexes of the type $[MCl(H_2O)_5]^{2+}$, $[MCl_2(H_2O)_4]^+$, and so on, are commonly formed, and crystalline chlorides of V, Fe, and Cr are of the type *trans*-$[VCl_2(H_2O)_4]^+Cl^- \cdot 2H_2O$. The alums, such as $CsTi(SO_4)_2 \cdot 12H_2O$, or $KV(SO_4)_2 \cdot 12H_2O$ contain the hexaaqua ion as do certain hydrates like $Fe(ClO_4)_3 \cdot 10H_2O$.

There are many anionic, cationic, or neutral M^{III} complexes, which are mostly *octahedral.* For Cr^{III} and, especially for Co^{III}, hundreds of octahedral com-

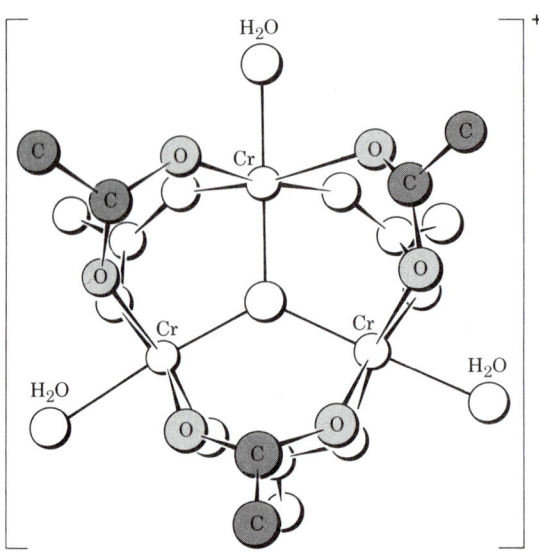

24-I

Table 24-3 Standard Potentials[a] (V, Acid Solution) and Colors for $[M(H_2O)_6]^{2+}$ and $[M(H_2O)_6]^{3+}$

	Ti	V	Cr	Mn	Fe	Co	Ni	Cu[b]
$M^{2+} + 2e^- = M$	—	-1.19	-0.91	-1.18	-0.44	-0.28	-0.24	+0.34
$M^{3+} + e^- = M^{2+}$	-0.37	-0.25	-0.41	+1.59	+0.77	+1.84	—	—
Color M^{2+}(aq)	—	Violet	Sky blue	Pale pink	Pale green	Pink	Green	Blue green
Color M^{3+}(aq)	Violet	Blue	Violet	Brown	v. Pale purple	Blue	—	—

[a]Some potentials depend on acidity and complexing anions, for example, for Fe^{3+}—Fe^{2+} in 1 M acids: HCl, +0.70; HClO$_4$, +0.75; H$_3$PO$_4$, +0.44; 0.5 M H$_2$SO$_4$, +0.68 V.

[b]$Cu^{2+} + e^- = Cu^+$; $E_0 = +0.15$ V; $Cu^+ + e^- = Cu$; $E_0 = +0.52$V.

plexes that are substitutionally inert are known. Representative octahedral complexes are $[TiF_6]^{3-}$, $[V(CN)_6]^{3-}$, $Cr(acac)_3$, and $[Co(NH_3)_6]^{3+}$.

The halides (MX_3) act as Lewis acids and form adducts, such as $VX_3(NMe_3)_2$, and $CrCl_3(thf)_3$, as well as the ionic species $[VCl_4]^-$, $[CrCl_4]^-$, and so on.

A special feature of the M^{3+} ions is the formation of basic carboxylates in which an O atom is in the center of a triangle of metal atoms (Structure 24-I). The latter are linked by carboxylate bridge groups, and the sixth coordination position is occupied by a water molecule or other ligand. This oxo-centered unit has been proved for carboxylates of V, Cr, Mn, Fe, Co, Ru, Rh, and Ir.

24-4 The IV and Higher Oxidation States

The IV state is the most important state for Ti, where the main chemistry is that of TiO_2 and $TiCl_4$ and derivatives. Although there are compounds like VCl_4, the main V^{IV} chemistry is that of the oxovanadium(IV) or vanadyl ion VO^{2+}. This ion can behave like an M^{2+} ion, and it forms many complexes that may be cationic, neutral, or anionic, depending on the ligand.

For the remaining elements, the IV oxidation state is not very common or well established except in fluorides, fluorocomplex ions, oxo anions, and a few complexes. Some tetrahedral compounds with —OR, —NR$_2$, or —CR$_3$ groups are known for a few elements, notably Cr; examples are $Cr(OCMe_3)_4$ and $Cr(1\text{-norbornyl})_4$.

The oxidation states V and above are known for V, Cr, Mn, and Fe in fluorides, fluorocomplexes or oxo anions (e.g., CrF_5, $KMnO_4$, and K_2FeO_4). All are powerful oxidizing agents.

TITANIUM

24-5 General Remarks: The Element

Titanium has the electronic structure $3d^24s^2$. The energy of removal of four electrons is so high that the Ti^{4+} ion may not exist and titanium(IV) compounds are covalent. There are some resemblances between Ti^{IV} and Sn^{IV} and their radii are similar. Thus TiO_2 (rutile) is isomorphous with SnO_2 (cassiterite) and is similarly yellow when hot. Titanium tetrachloride, like $SnCl_4$, is a distillable liquid readily hydrolyzed by water, behaving as a Lewis acid, and giving adducts with donor molecules. The bromide and iodide, which form crystalline molecular lattices, are also isomorphous with the corresponding Group IVB(14) halides.

Titanium is relatively abundant in the earth's crust (0.6%). The main ores are *ilmenite* $(FeTiO_3)$ and *rutile,* one of the several crystalline varieties of TiO_2. The metal cannot be made by reduction of TiO_2 with C because a very stable carbide is produced. The rather expensive Kroll process is used. Ilmenite or rutile is treated at red heat with C and Cl_2 to give $TiCl_4$, which is fractionated to free it from impurities, such as $FeCl_3$. The $TiCl_4$ is then reduced with molten Mg at about 800 °C in an atmosphere of argon. This gives Ti as a spongy mass from which the excess of Mg and $MgCl_2$ is removed by volatilization at 1000 °C. The sponge may then be fused in an electric arc and cast into ingots; an atmosphere of Ar or He must be used since Ti readily reacts with N_2 and O_2 when hot.

Titanium is lighter than other metals of similar mechanical and thermal properties and is unusually resistant to corrosion. It is used in turbine engines and industrial chemical, aircraft, and marine equipment. It is unattacked by dilute acids and bases. It dissolves in hot HCl giving Ti^{III} chloro complexes and in HF or $HNO_3 + HF$ to give fluoro complexes. Hot HNO_3 gives a hydrous oxide.

TITANIUM COMPOUNDS

The most important stereochemistries in titanium compounds are the following:

Ti^{II}	Octahedral	
Ti^{III}	Octahedral	in most compounds and in solution
Ti^{IV}	Tetrahedral	in $TiCl_4$, $Ti(CH_2C_6H_5)_4$, and so on
	Octahedral	in TiO_2 and Ti^{IV} complexes

24-6 Binary Compounds of Titanium

Titanium tetrachloride, a colorless liquid (bp 136 °C), has a pungent odor, fumes strongly in moist air, and is vigorously, though not violently, hydrolyzed by water.

$$TiCl_4 + 2\ H_2O = TiO_2 + 4\ HCl \qquad (24\text{-}6.1)$$

Partially hydrolyzed species are formed with a deficit of water or on addition of $TiCl_4$ to aqueous HCl.

Titanium oxide has three crystal forms—rutile (see Fig. 4-1), anatase, and brookite—all of which occur in nature. The dioxide that is used in large quantities as a white pigment in paints is made by vapor phase oxidation of $TiCl_4$ with oxygen. The precipitates obtained by addition of OH^- to Ti^{IV} solutions are best regarded as hydrous TiO_2, not a true hydroxide. This material is amphoteric and dissolves in concentrated NaOH.

Materials called "titanates" are of technical importance, for example, as ferroelectrics. Nearly all of them have one of the three major mixed metal oxide structures (Section 4-8). Indeed, the names of two of the structures are those of the titanium compounds that were the first found to possess them: $FeTiO_3$ (*ilmenite*) and $CaTiO_3$ (*perovskite*).

24-7 Titanium(IV) Complexes

Aqueous Chemistry: Oxo Salts

There is no firm evidence for the Ti^{4+} aqua ion. In aqueous solutions of Ti^{IV} there are only oxo species; basic oxo salts or hydrated oxides may be precipitated. These oxo salts have formulas such as $TiOSO_4\cdot H_2O$ and $(NH_4)_2TiO(C_2O_4)_2\cdot H_2O$, and have chains or rings, $(Ti-O-Ti-O-)_x$. There is spectroscopic evidence for TiO^{2+} only in 2 *M* $HClO_4$ solution, although some compounds with a $Ti=O$ group have been characterized.

Anionic Complexes

The solutions obtained by dissolving the metal or hydrous oxide in aqueous HF contain fluoro complex ions, mainly $[TiF_6]^{2-}$, which can be isolated as crystalline salts. In aqueous HCl, $TiCl_4$ gives yellow oxo complex anions but from solutions saturated with gaseous HCl, salts of the $[TiCl_6]^{2-}$ ion may be obtained.

Adducts of TiX_4

The halides form adducts, TiX_4L or TiX_4L_2, which are crystalline solids that are often soluble in organic solvents. These adducts are invariably *octahedral*. Thus $[TiCl_4(OPCl_3)]_2$ and $[TiCl_4(CH_3COOC_2H_5)]_2$ are dimeric, with two chlorine bridges, while $TiCl_4(OPCl_3)_2$ has octahedral coordination with *cis*-$OPCl_3$ groups.

Peroxo Complexes

One of the most characteristic reactions of aqueous Ti solutions is the development of an intense orange color on addition of H_2O_2. This reaction can be used for the colorimetric determination of either Ti or of H_2O_2. Below pH 1, the main species is $[Ti(O_2)(OH)]^+(aq)$.

Solvolytic Reactions of $TiCl_4$: Alkoxides and Related Compounds

Titanium tetrachloride reacts with compounds containing active hydrogen atoms with loss of HCl. The replacement of chloride is usually incomplete in the absence of an HCl acceptor such as an amine or alkoxide ion. The *alkoxides* are typical of other transition metal alkoxides, which we shall not discuss. They can be obtained by reactions such as

$$TiCl_4 + 4\ ROH + 4\ R'NH_2 \longrightarrow Ti(OR)_4 + 4\ R'NH_3Cl \qquad (24\text{-}7.1)$$

The alkoxides are liquids or solids that can be distilled or sublimed. They are soluble in organic solvents such as benzene, but are exceedingly readily hydrolyzed by even traces of water, to give polymeric species with —OH— or —O— bridges. The initial hydrolytic step probably involves coordination of water to the metal; a proton on H_2O could then interact with the oxygen of an OR group through hydrogen bonding, leading to hydrolysis:

$$\begin{array}{c}\text{H} \\ \diagdown \\ \text{O}-\text{M(OR)}_x \\ \diagup \\ \text{H}\end{array} \longrightarrow \begin{array}{c}\text{H} \\ \diagdown \\ \text{O}-\text{M(OR)}_{x-1} \\ \\ \text{H}\text{-----}:\text{O} \\ \diagdown \\ \text{R}\end{array} \qquad (24\text{-}7.2)$$

$$\text{M(OH)(OR)}_{x-1} + \text{ROH}$$

Although monomeric species can exist, for example, when made from secondary and tertiary alcohols, and in dilute solution, alkoxides are usually polymers. Solid $Ti(OC_2H_5)_4$ is a tetramer, with the structure shown in Fig. 24-1. The alkoxides are often referred to as "alkyltitanates" and under this name they are used in heat-resisting paints, where eventual hydrolysis to TiO_2 occurs.

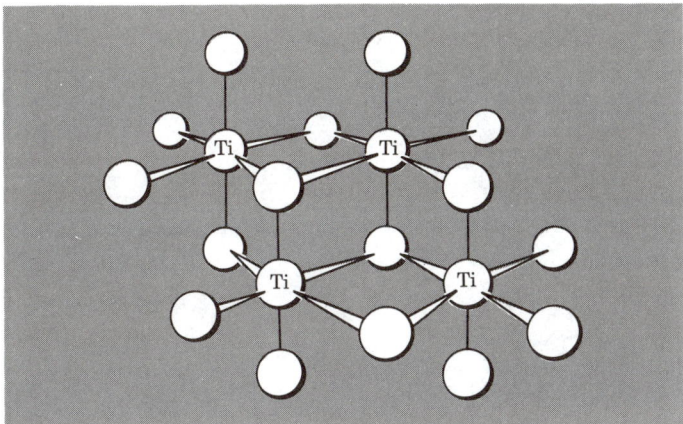

Figure 24-1   The tetrameric structure of $Ti(OC_2H_5)_4$. Only Ti and O atoms are shown. Each Ti is octahedrally coordinated.

Another class of titanium compounds, the *dialkylamides,* are also representative of similar compounds of other transition metals. These are liquids or volatile solids readily hydrolyzed by water. Unlike the alkoxides they are not polymeric. They are made by reaction of the metal halide with lithium dialkylamides:

$$TiCl_4 + 4\ LiNEt_2 = Ti(NEt_2)_4 + 4\ LiCl \qquad (24\text{-}7.3)$$

Such amides can undergo a wide range of "insertion" reactions (Section 30-3); thus with CS_2, the dithiocarbamates are obtained.

$$Ti(NEt_2)_4 + 4\ CS_2 = Ti(S_2CNEt_2)_4 \qquad (24\text{-}7.4)$$

24-8   The Chemistry of Titanium(III), d^1, and Titanium(II), d^2

Binary Compounds

Titanium trichloride ($TiCl_3$) has several crystalline forms. The violet α form is made by H_2 reduction of $TiCl_4$ vapor at 500–1200 °C. The reduction of $TiCl_4$ by aluminum alkyls (Section 30-10) in inert solvents gives a brown β form that is converted into the α form at 250–300 °C. The α form has a layer lattice containing $TiCl_6$ groups. The β-$TiCl_3$ is fibrous with single chains of $TiCl_6$ octahedra sharing edges. This structure is of particular importance for the stereospecific polymerization of propene using $TiCl_3$ as catalyst (Ziegler–Natta process) (Section 30-9).

The *dichloride* is obtained by high temperature syntheses:

$$TiCl_4 + Ti = 2\ TiCl_2 \qquad (24\text{-}8.1)$$

or

$$2\ TiCl_3 = TiCl_2 + TiCl_4 \qquad (24\text{-}8.2)$$

Aqueous Chemistry and Complexes

Aqueous solutions of the $[Ti(H_2O)_6]^{3+}$ ion are obtained by reducing aqueous Ti^{IV} either electrolytically or with zinc. The violet solutions reduce O_2 and, hence, must be handled in a N_2 or H_2 atmosphere.

$$\text{``TiO}^{2+}(\text{aq})\text{''} + 2\,H^+ + e^- = Ti^{3+} + H_2O \qquad E° = \text{about } 0.1\ V \quad (24\text{-}8.3)$$

The Ti^{3+} solutions are used as fairly rapid, mild reducing agents in volumetric analysis. In HCl solutions the main species is $[TiCl(H_2O)_5]^{2+}$.

There is no aqueous chemistry of Ti^{II} because of its ready oxidation, but a few Ti^{II} complexes, such as $[TiCl_4]^{2-}$, can be made in nonaqueous media.

VANADIUM

24-9 The Element

Vanadium is widely distributed but there are few concentrated deposits. It occurs in petroleum from Venezuela, and is recovered as V_2O_5 from flue dusts after combustion.

Very pure vanadium is rare because, like titanium, it is quite reactive toward O_2, N_2, and C at the elevated temperatures used in metallurgical processes. Since its chief commercial use is in alloy steels and cast iron, to which it lends ductility and shock resistance, commercial production is mainly as an iron alloy, *ferrovanadium*.

Vanadium metal is not attacked by air, alkalis, or nonoxidizing acids other than HF at room temperature. It dissolves in HNO_3, concentrated H_2SO_4, and aqua regia.

VANADIUM COMPOUNDS

The stereochemistries for the most important classes of vanadium compounds are the following:

$\left.\begin{array}{l} V^{II} \\ V^{III} \end{array}\right\}$ Octahedral as in $[V(H_2O)_6]^{2+}$
$VF_3(s)$ or $[V(ox)_3]^{3-}$

Tetrahedral as in VCl_4 or $V(CH_2SiMe_3)_4$

V^{IV} $\left\{\begin{array}{l} \text{Square pyramidal in } O{=}V(acac)_2 \\ \text{Octahedral in } VO_2, K_2VCl_6, O{=}V(acac)_2py, \text{ and so on} \end{array}\right.$

V^{V} $\left.\phantom{\begin{array}{l}a\\b\end{array}}\right\{$ Octahedral as in $[VO_2(ox)_2]^{3-}$, $VF_5(s)$

24-10 Binary Compounds

Halides

In the highest oxidation state only VF_5 is known. The colorless liquid (bp 48 °C) has a high viscosity (cf. SbF_5, Section 17-4) and has chains of VF_6 octahedra linked by *cis*-V—F—V bridges; it is monomeric in the vapor.

The *tetrachloride* is obtained from $V + Cl_2$ or from CCl_4 on red-hot V_2O_5. It is a dark red oil (bp 154 °C), which is violently hydrolyzed by water to give solutions of oxovanadium(IV) chloride. It has a high dissociation pressure and loses chlorine slowly when kept, but rapidly on boiling, leaving violet VCl_3. The latter may be decomposed to pale green VCl_2, which is then stable.

$$2 \, VCl_3(s) \longrightarrow VCl_2(s) + VCl_4(g) \qquad (24\text{-}10.1)$$

$$VCl_3(s) \longrightarrow VCl_2(s) + \tfrac{1}{2} Cl_2(g) \qquad (24\text{-}10.2)$$

Vanadium(V) Oxide

Addition of dilute H_2SO_4 to solutions of ammonium vanadate gives a brick-red precipitate of V_2O_5. This oxide is acidic and dissolves in NaOH to give colorless solutions containing the *vanadate* ion, $[VO_4]^{3-}$. On acidification, a complicated series of reactions occurs involving the formation of hydroxo anions and polyanions (cf. Section 5-4). In very strong acid solutions, the *dioxovanadium(V) ion* (VO_2^+) is formed.

24-11 Oxovanadium Ions and Complexes

The two oxo cations VO_2^+ and VO^{2+} have an extensive chemistry and form numerous complex compounds. All of the compounds show IR and Raman bands that are characteristic for M=O groups. The VO_2^+ group is angular. Examples of complexes are *cis*-$[VO_2Cl_4]^{3-}$, *cis*-$[VO_2edta]^{3-}$, and *cis*-$[VO_2(ox)_2]^{3-}$. The cis arrangement for dioxo compounds of metals with no *d* electrons is preferred over the trans arrangement that is found in some other metal dioxo systems (e.g., $[RuO_2]^{2+}$) because the strongly π-donating O ligands then have exclusive use of one $d\pi$ orbital each (d_{xz}, d_{yz}) and share a third one (d_{xy}), whereas in the trans configuration they would have to share two $d\pi$ orbitals and leave one unused.

The oxovanadium(IV) or vanadyl compounds are among the most stable and important of vanadium species, and the VO unit persists through a variety of chemical reactions. Solutions of V^{3+} are oxidized in air, while V^V is readily reduced by mild reducing agents to form the blue oxovanadium(IV) ion, $[VO(H_2O)_5]^{2+}$:

$$VO^{2+} + 2 \, H^+ + e^- = V^{3+} + H_2O \qquad E^\circ = 0.34 \, V \qquad (24\text{-}11.1)$$

$$VO_2^+ + 2 \, H^+ + e^- = VO^{2+} + H_2O \qquad E^\circ = 1.0 \, V \qquad (24\text{-}11.2)$$

Addition of base to $[VO(H_2O)_5]^{2+}$ gives the yellow hydrous oxide $VO(OH)_2$, which redissolves in acids giving the cation.

Oxovanadium(IV) compounds are usually blue green. They may be either five-coordinate square pyramidal (Structure 24-II) or six-coordinate with a distorted octahedron. Examples are $[VO(bpy)_2Cl]^+$, $VO(acac)_2$, and $[VO(NCS)_4]^{2-}$. The VO bonds are short (1.56–1.59 Å), and can properly be regarded as multiple ones, the π component arising from electron flow $O(p\pi) \rightarrow V(d\pi)$. Even in VO_2, which has a distorted rutile structure, one bond (1.76 Å) is conspicuously shorter than the others in the VO_6 unit (note that in TiO_2 all Ti—O bonds are substantially equal).

All of the five-coordinate complexes, such as Structure 24-II, take up a sixth ligand quite readily, becoming six coordinate.

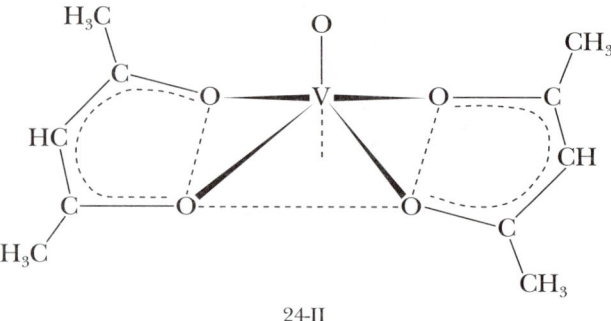

24-II

24-12 The Vanadium(III) Aqua Ion and Complexes

The electrolytic or chemical reduction of acid solutions of vanadates or V^{IV} solutions gives solutions of V^{III} that are quite readily reoxidized to VO^{2+}. Crystalline salts can be obtained. Addition of OH^- precipitates the hydrous oxide V_2O_3.

24-13 Vanadium(II)

When V^{III} solutions are reduced by Zn in acid, violet air-sensitive solutions of $[V(H_2O)_6]^{2+}$ are obtained. These are oxidized by water with evolution of hydrogen despite the fact that the V^{3+}/V^{2+} potential (Table 24-3) suggests otherwise. Vanadium(II) solutions are often used to remove traces of O_2 from inert gases.

The salt $[V(H_2O)_6]SO_4$ is obtained as violet crystals on addition of ethanol to reduced sulfate solutions. Because of its d^3 configuration the $[V(H_2O)_6]^{2+}$ ion like $[Cr(H_2O)_6]^{3+}$ is kinetically inert, and its substitution reactions are relatively slow.

CHROMIUM

24-14 The Element Chromium

Apart from stoichiometric similarities, chromium resembles the Group VIB(16) elements of the sulfur group only in the acidity of CrO_3 and the covalent nature and ready hydrolysis of CrO_2Cl_2 (cf. SO_3, SO_2Cl_2).

The chief ore is *chromite*, $FeCr_2O_4$, which is a spinel with Cr^{III} on octahedral sites and Fe^{II} on the tetrahedral ones. It is reduced by C to the carbon-containing alloy ferrochromium.

$$FeCr_2O_4 + 4\,C \xrightarrow{\text{heat}} Fe \cdot 2Cr + 4\,CO \qquad (24\text{-}14.1)$$

When pure Cr is required, the chromite is first treated with molten NaOH and O_2 to convert the Cr^{III} to CrO_4^{2-}. The melt is dissolved in water and sodium dichromate is precipitated. This precipitate is then reduced.

$$Na_2Cr_2O_7 + 2\,C \xrightarrow{\text{heat}} Cr_2O_3 + Na_2CO_3 + CO \qquad (24\text{-}14.2)$$

Next, the oxide is reduced.

$$Cr_2O_3 + 2\,Al \xrightarrow{\text{heat}} Al_2O_3 + 2\,Cr \qquad (24\text{-}14.3)$$

Chromium is resistant to corrosion, hence its use as an electroplated protective coating. It dissolves fairly readily in HCl, H_2SO_4, and $HClO_4$, but it is passivated by HNO_3.

CHROMIUM COMPOUNDS

The most common stereochemistries for chromium compounds are the following:

$$\left.\begin{array}{l} Cr^{II} \\ Cr^{III} \end{array}\right\} \quad \text{Octahedral as in } [Cr(H_2O)_6]^{2+} \text{ (distorted) or } [Cr(NH_3)_6]^{3+}$$

$$Cr^{IV} \quad \text{Tetrahedral as in } Cr(O\text{-}t\text{-}Bu)_4$$

$$\left.\begin{array}{l} Cr^{V} \\ Cr^{VI} \end{array}\right\} \quad \text{Tetrahedral as in } [CrO_4]^{3-}, [CrO_4]^{2-}, CrO_3$$

24-15 Binary Compounds

Halides

The anhydrous Cr^{II} halides are obtained by action of HCl, HBr, or I_2 on the metal at 600–700 °C or by reduction of the trihalides with H_2 at 500–600 °C. $CrCl_2$ dissolves in water to give a blue solution of the Cr^{2+} ion.

The red-violet trichloride, $CrCl_3$, is made by the action of $SOCl_2$ on the hydrated chloride. The flaky form of $CrCl_3$ is due to its layer structure.

Chromium(III) chloride forms adducts with donor ligands. The violet tetrahydrofuranate, fac-$CrCl_3(thf)_3$, which crystallizes from solutions formed by the action of a little zinc on $CrCl_3$ in thf, is a particularly useful material for the preparation of other chromium compounds, such as carbonyls or organometallic compounds.

Oxides

The green α-Cr_2O_3 (corundum structure) is formed on burning Cr in O_2, on thermal decomposition of CrO_3, or on roasting the hydrous oxide ($Cr_2O_3 \cdot n\,H_2O$). The latter, commonly called "chromic hydroxide," although its water content is variable, is precipitated on addition of OH^- to solutions of Cr^{III} salts. The hydrous oxide is amphoteric, dissolving readily in acid to give $[Cr(H_2O)_6]^{3+}$, and in concentrated alkali to form "chromites."

Chromium oxide and chromium supported on other oxides, such as Al_2O_3, are important catalysts for a wide variety of reactions.

Chromium(VI)oxide, CrO_3, is obtained as an orange-red precipitate on adding sulfuric acid to solutions of $Na_2Cr_2O_7$. It is thermally unstable above its melting point (197 °C), losing O_2 to give Cr_2O_3. The structure consists of infinite chains of CrO_4 tetrahedra sharing corners. This oxide is soluble in water and is highly poisonous.

Interaction of CrO_3 and organic substances is vigorous and may be explosive, but CrO_3 is used in organic chemistry as an oxidant, usually in acetic acid as solvent.

24-16 The Chemistry of Chromium(II), d^4

Aqueous solutions of the blue *chromium(II) ion* are best prepared by dissolving electrolytic Cr metal in dilute mineral acids. The solutions must be protected from air (Table 24-3)—even then, they decompose at rates varying with the acidity and the anions present, by reducing water with liberation of H_2.

The mechanisms of reductions of other ions by Cr^{2+} have been extensively studied, since the resulting Cr^{3+} complex ions are substitution inert. Much information regarding ligand-bridged transition states (Section 6-5) has been obtained in this way.

Chromium(II) *acetate*, $Cr_2(O_2CCH_3)_4(H_2O)_2$, is precipitated as a red solid when a Cr^{2+} solution is added to a solution of sodium acetate. Its structure is typical of carboxylate-bridged complexes with water end groups (Structure 24-X). The short Cr—Cr bond (2.36 Å) and diamagnetism are accounted for by the existence of a quadruple Cr—Cr bond, consisting of a σ, two π, and a δ component. This was the first compound containing a quadruple bond to be discovered (1844).

24-17 The Chemistry of Chromium(III), d^3

Chromium(III) Complexes

There are thousands of chromium(III) complexes which, with a few exceptions, are all six coordinate. The principal characteristic is their relative kinetic inertness in aqueous solutions. It is because of this that so many complex species can be isolated, and why much of the classical complex chemistry studied by early workers, notably S. M. Jørgensen and A. Werner, involved chromium. These complexes persist in solution, even where they are thermodynamically unstable.

The hexaaqua ion, $[Cr(H_2O)_6]^{3+}$, occurs in numerous salts, such as the violet hydrate, $[Cr(H_2O)_6]Cl_3$, and alums, $M^ICr(SO_4)_2 \cdot 12H_2O$. The chloride has three isomers, the others being the dark green *trans*-$[CrCl_2(H_2O)_4]Cl \cdot 2H_2O$, which is the usual form, and pale green $[CrCl(H_2O)_5]Cl_2 \cdot H_2O$. The ion is acidic and the hydroxo ion condenses to give a dimeric hydroxo bridged species.

$$[Cr(H_2O)_6]^{3+} \underset{H^+}{\overset{-H^+}{\rightleftharpoons}} [Cr(H_2O)_5OH]^{2+} \rightleftharpoons \left[(H_2O)_4Cr \underset{\underset{H}{O}}{\overset{\overset{H}{O}}{\diagup \diagdown}} Cr(H_2O)_4 \right]^{4+} \quad (24\text{-}17.1)$$

On addition of further base, soluble polymeric species of high-molecular weight and eventually dark green gels of the hydrous oxide are formed.

The most numerous complexes are those of amine ligands. These ligands provide examples of virtually all the kinds of isomerism possible in octahedral complexes. In addition to the mononuclear species, for example, $[Cr(NH_3)_5Cl]^{2+}$, there are many polynuclear complexes in which two or sometimes more metal atoms are bridged by hydroxo groups or, less commonly, oxygen in a linear Cr—O—Cr group. A representative example is $[(NH_3)_5Cr(OH)Cr(NH_3)_5]^{5+}$.

24-18 The Chemistry of Chromium(IV), d^2, and Chromium(V), d^1

The most readily accessible of these rare oxidation states are those with bonds to C, N, and O. A representative synthesis is

$$CrCl_3(thf)_3 + 4\ LiCH_2SiMe_3 \xrightarrow{\text{ether}} Li[Cr^{III}(CH_2SiMe_3)_4] + 3\ LiC \quad (24\text{-}18.1)$$

$$[Cr(CH_2SiMe_3)_4]^- \rightleftharpoons Cr(CH_2SiMe_3)_4 + e^- \quad (24\text{-}18.2)$$

The oxidation of the green Cr^{III} anion to the purple, petroleum-soluble Cr^{IV} compound can be made by air. The alkoxides and dialkylamides are similarly made from fac-$CrCl_3(thf)_3$; one example is the dark blue $Cr(OCMe_3)_4$.

For Cr^V some chromites containing CrO_4^{3-} are known. Reduction of CrO_3 with concentrated HCl in the presence of alkali ions at 0 °C gives salts $M_2[Cr^VOCl_5]$.

24-19 The Chemistry of Chromium(VI), d^0

Chromate and Dichromate Ions

In basic solutions above pH 6, CrO_3 forms the tetrahedral yellow *chromate* ion, CrO_4^{2-}. Between pH 2 and 6, $HCrO_4^-$ and the orange-red *dichromate* ion, $Cr_2O_7^{2-}$, are in equilibrium. At pH values below 1 the main species is H_2CrO_4. The equilibria are

$$HCrO_4^- \rightleftharpoons CrO_4^{2-} + H^+ \qquad K = 10^{-5.9} \quad (24\text{-}19.1)$$

$$H_2CrO_4 \rightleftharpoons HCrO_4^- + H^+ \qquad K = 4.1 \quad (24\text{-}19.2)$$

$$Cr_2O_7^{2-} + H_2O \rightleftharpoons 2\ HCrO_4^- \qquad K = 10^{-2.2} \quad (24\text{-}19.3)$$

In addition, there are the base-hydrolysis equilibria

$$Cr_2O_7^{2-} + OH^- \rightleftharpoons HCrO_4^- + CrO_4^{2-} \quad (24\text{-}19.4)$$

$$HCrO_4^- + OH^- \rightleftharpoons CrO_4^{2-} + H_2O \quad (24\text{-}19.5)$$

The CrO_4^{2-} ion is tetrahedral; $Cr_2O_7^{2-}$ has the structure shown in Fig. 24-2.

The pH-dependent equilibria are quite labile and on addition of cations that form insoluble chromates (e.g., Ba^{2+}, Pb^{2+}, and Ag^+), the chromates and not the dichromates are precipitated. Only for HNO_3 and $HClO_4$ are the equilibria as given previously. When HCl is used, there is essentially quantitative conversion into the *chlorochromate* ion, while with sulfuric acid a sulfato complex results.

$$CrO_3(OH)^- + H^+ + Cl^- \longrightarrow CrO_3Cl^- + H_2O \quad (24\text{-}19.6)$$

$$CrO_3(OH)^- + HSO_4^- \longrightarrow CrO_3(OSO_3)^{2-} + H_2O \quad (24\text{-}19.7)$$

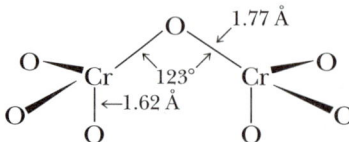

Figure 24-2   The structure of the dichromate ion as found in $Na_2Cr_2O_7$.

Acid solutions of dichromate are strong oxidants:

$$Cr_2O_7^{2-} + 14\ H^+ + 6\ e^- = 2\ Cr^{3+} + 7\ H_2O \qquad E° = 1.33\ V \quad (24\text{-}19.8)$$

In alkaline solution, the chromate ion is much less oxidizing:

$$CrO_4^{2-} + 4\ H_2O + 3\ e^- = Cr(OH)_3(s) + 5\ OH^- \qquad E° = 0.13\ V \quad (24\text{-}19.9)$$

Chromium(VI) does not give rise to the extensive and complex series of poly-acids and polyanions characteristic of the somewhat less acidic oxides of V^V, Mo^{VI}, and W^{VI}. The reason for this is perhaps the greater extent of multiple bonding (Cr=O) for the smaller chromium ion.

Chromyl chloride (CrO_2Cl_2), a deep-red liquid, is formed by the action of HCl on chromium(VI) oxide:

$$CrO_3 + 2\ HCl \longrightarrow CrO_2Cl_2 + H_2O \qquad\qquad (24\text{-}19.10)$$

or on warming dichromate with an alkali metal chloride in concentrated sulfuric acid:

$$K_2Cr_2O_7 + 4\ KCl + 3\ H_2SO_4 \longrightarrow 2\ CrO_2Cl_2 + 3\ K_2SO_4 + 3\ H_2O \quad (24\text{-}19.11)$$

It is photosensitive but otherwise rather stable, vigorously oxidizes organic matter, and is hydrolyzed by water to CrO_4^{2-} and HCl.

24-20 Peroxo Complexes of Chromium(IV), (V), and (VI)

Like other transition metals, notably Ti, V, Nb, Ta, Mo, and W, chromium forms peroxo compounds in the higher oxidation states. These complexes are all more or less unstable both in and out of solution, decomposing slowly with the evolution of O_2. Some are explosive or flammable in air.

When acid dichromate solutions are treated with H_2O_2, a deep blue color rapidly appears.

$$HCrO_4^- + 2\ H_2O_2 + H^+ \longrightarrow CrO(O_2)_2 + 3\ H_2O \qquad (24\text{-}20.1)$$

The blue species decomposes fairly readily, giving Cr^{3+}, but it may be extracted into ether where it is more stable. On addition of pyridine to the ether solution, the compound $(py)CrO_5$ is obtained. The pyridine complex has the bisperoxo structure shown in Fig. 24-3(a).

Treatment of alkaline chromate solutions with 30% H_2O_2 gives the red-brown peroxochromates, $M_3^I CrO_8$ [Fig. 24-3(b)], which are paramagnetic with one unpaired electron.

MANGANESE

24-21 The Element

The highest oxidation state of manganese corresponds to the total number of $3d$ and $4s$ electrons. It occurs in the oxo compounds MnO_4^-, Mn_2O_7, and MnO_3F and in amido complexes(Section 24-25). These compounds show some similarity to corresponding compounds of the halogens.

Manganese is relatively abundant, and occurs in substantial deposits, mainly oxides, hydrous oxides, or the carbonate. From them, or the Mn_3O_4 obtained by roasting them, the metal can be obtained by reduction with Al.

Manganese is quite electropositive, and readily dissolves in dilute, nonoxidizing acids.

MANGANESE COMPOUNDS

The most common stereochemistries of manganese compounds are the following:

$$\left.\begin{array}{c} Mn^{II} \\ Mn^{III} \\ Mn^{IV} \end{array}\right\}\ \text{Octahedral as in } [Mn(H_2O)_6]^{2+},\ [Mn(ox)_3]^{3-} \text{ and } [MnCl_6]^{2-}$$

$$\left.\begin{array}{c} Mn^{VI} \\ Mn^{VII} \end{array}\right\}\ \text{Tetrahedral as in } [MnO_4]^{2-} \text{and } MnO_4^-$$

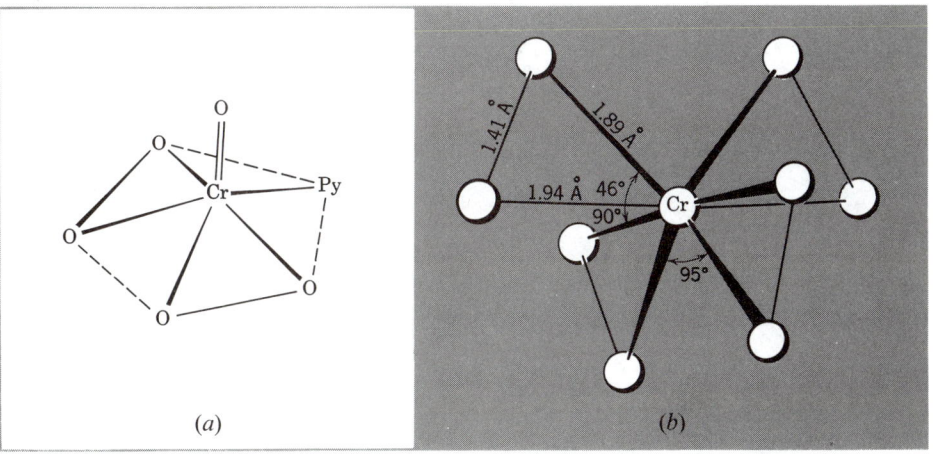

Figure 24-3   (a) The structure of $CrO(O_2)_2py$. The coordination polyhedron is approximately a pentagonal pyramid with the oxide ligand at the apex. (b) The dodecahedral structure of the CrO_8^{3-} ion, a tetraperoxide complex.

24-22 The Chemistry of Manganese(II), d^5

Manganese(II) salts are mostly water soluble. Addition of OH^- to Mn^{2+} solutions produces the gelatinous white *hydroxide*. This compound rapidly darkens in air due to oxidation, as shown by the base potentials:

$$MnO_2 \xrightarrow{-0.1\ V} Mn_2O_3 \xrightarrow{-0.2\ V} Mn(OH)_2 \qquad (24\text{-}22.1)$$

Addition of SH^- gives hydrous MnS, which also oxidizes becoming brown in air; on boiling in the absence of air the salmon pink material changes into green crystalline MnS. The sulfate, $MnSO_4$, is very stable and may be used for Mn analysis as it can be obtained on fuming down sulfuric acid solutions to dryness. The phosphate and carbonate are sparingly soluble. The equilibrium constants for the formation of manganese(II) complexes are relatively low as the Mn^{2+} ion has no ligand field stabilization energy (Section 23-8). However, chelating ligands, such as en, ox, or $EDTA^{4-}$, form complexes isolable from aqueous solution.

In aqueous solution the formation constants for halogeno complexes are very low, for example,

$$[Mn_{(aq)}]^{2+} + Cl^- \rightleftharpoons MnCl^+_{(aq)} \qquad K \approx 4 \qquad (24\text{-}22.2)$$

but in ethanol or acetic acid, salts of complex anions of varying types may be isolated, such as

MnX_3^-	Octahedral with perovskite structure
$[MnX_4]^{2-}$	Tetrahedral (green yellow) or polymeric octahedral with halide bridges (pink)
$[MnCl_6]^{4-}$	Only Na and K salts known; octahedral

The Mn^{2+} ions may occupy tetrahedral holes in certain glasses and substitute for Zn^{II} in ZnO. Tetrahedral Mn^{II} has a green-yellow color, far more intense than the pink of the octahedrally coordinated ion, and it very often exhibits intense yellow-green fluorescence. Most commercial phosphors are manganese-activated zinc compounds, wherein Mn^{II} ions are substituted for some of the Zn^{II} ions in tetrahedral surroundings as, for example, in Zn_2SiO_4.

Only the very strong ligand fields give rise to spin pairing as in the ions $[Mn(CN)_6]^{4-}$ and $[Mn(CNR)_6]^{2+}$, which have only one unpaired electron.

24-23 The Chemistry of Manganese(III), d^4

Oxides

When any manganese oxide or hydroxide is heated at 1000 °C, black crystals of Mn_3O_4 (*haussmannite*) are formed. This compound is a spinel, $Mn^{II}Mn_2^{III}O_4$. When $Mn(OH)_2$ is allowed to oxidize in air, a hydrous oxide is formed that gives $MnO(OH)$ on drying.

Manganese(III) Aqua Ion

The manganese(III) ion can be obtained by electrolytic or peroxodisulfate oxidation of Mn^{2+} solutions or by reduction of MnO_4^-. It cannot be obtained in high concentrations as it is reduced by water (Table 24-3). It also has a strong tendency to hydrolyze and to disproportionate in weakly acid solution.

$$2\,Mn^{3+} + 2\,H_2O = Mn^{2+} + MnO_2(s) + 4\,H^+ \qquad K \approx 10^9 \quad (24\text{-}23.1)$$

The dark brown crystalline *acetylacetonate* $Mn(acac)_3$ is readily obtained by oxidation of basic solutions of Mn^{2+} by O_2 or Cl_2 in the presence of acetylacetone.

The basic-oxo centered acetate (Structure 24-I), which is obtained by the action of $KMnO_4$ on Mn^{II} acetate in acetic acid, will oxidize alkenes to lactones. It is used industrially for oxidation of toluene to phenol.

Manganese(III) and (IV) complexes are probably important in photosynthesis, where oxygen evolution depends on manganese.

24-24 The Chemistry of Manganese(IV), d^3, and Manganese(V), d^2

The most common compound of Mn^{IV} is *manganese dioxide,* a gray to black solid found in nature as *pyrolusite.* When made by the action of oxygen on manganese at a high temperature it has the rutile structure found for many other dioxides, for example those of Ru, Mo, W, Re, Os, Ir, and Rh. However, as normally made, for example, by heating $Mn(NO_3)_2 \cdot 6H_2O$ in air, it is nonstoichiometric. A hydrated form is obtained by reduction of aqueous $KMnO_4$ in basic solution.

Manganese dioxide is inert to most acids except when heated, but it does not dissolve to give Mn^{IV} in solution; instead, it functions as an oxidizing agent, the exact manner of this depending on the acid. With HCl, chlorine is evolved.

$$MnO_2(s) + 4\,H^+ + 4\,Cl^- = Mn^{2+} + 2\,Cl^- + Cl_2 + 2\,H_2O \qquad (24\text{-}24.1)$$

With sulfuric acid at 110 °C, oxygen is evolved and an Mn^{III} acid sulfate is formed. Hydrated manganese dioxide is used in organic chemistry for the oxidation of alcohols and other compounds.

Manganese(IV) is known in MnF_6^{2-} and, like Mn^{VI}, in compounds with $Mn{=}O$ groups. Manganese(V) is little known except in bright blue "hypomanganates" that are formed by the reduction of permanganate with an excess of sulfite.

24-25 The Chemistry of Manganese(VI), d^1, and Manganese(VII), d^0

Manganese(VI) is found in the deep green *manganate* ion, MnO_4^{2-}. This ion is formed on oxidizing MnO_2 in fused KOH with KNO_3 or air.

The manganate ion is stable only in very basic solutions. In acid, neutral, or slightly basic solutions it readily disproportionates according to the equation

$$3\,MnO_4^{2-} + 4\,H^+ = 2\,MnO_4^- + MnO_2(s) + 2\,H_2O \qquad K \approx 10^{58} \quad (24\text{-}25.1)$$

Manganese(VII) is best known in the form of salts of the *permanganate ion.* The compound $KMnO_4$ is manufactured by electrolytic oxidation of a basic solution of K_2MnO_4. Aqueous solutions of MnO_4^- may be prepared by oxidation of solutions of the Mn^{II} ion with very powerful oxidizing agents such as PbO_2 or

$NaBiO_3$. The ion has an intense purple color, and crystalline salts appear almost black.

Solutions of permanganate are intrinsically unstable, decomposing slowly but observably in acid solution.

$$4\ MnO_4^- + 4\ H^+ \longrightarrow 3\ O_2(g) + 2\ H_2O + 4\ MnO_2(s) \qquad (24\text{-}25.2)$$

In neutral or slightly alkaline solutions in the dark, decomposition is immeasurably slow. It is catalyzed by light so that standard permanganate solutions should be stored in dark bottles.

In *basic* solution, permanganate is a powerful oxidant.

$$MnO_4^- + 2\ H_2O + 3\ e^- = MnO_2(s) + 4\ OH^- \qquad E° = +1.23\ V \quad (24\text{-}25.3)$$

In very strong base and with an excess of MnO_4^-, however, manganate ion is produced.

$$MnO_4^- + e^- = MnO_4^{2-} \qquad E° = +0.56\ V \qquad\qquad (24\text{-}25.4)$$

In *acid* solution permanganate is reduced to Mn^{2+} by an excess of reducing agent:

$$MnO_4^- + 8\ H^+ + 5\ e^- = Mn^{2+} + 4\ H_2O \qquad E° = +1.51\ V \quad (24\text{-}25.5)$$

but because MnO_4^- oxidizes Mn^{2+},

$$2\ MnO_4^- + 3\ Mn^{2+} + 2\ H_2O = 5\ MnO_2(s) + 4\ H^+ \qquad E° = +0.46\ V \quad (24\text{-}25.6)$$

the product in the presence of an excess of permanganate is MnO_2. The addition of $KMnO_4$ to concentrated H_2SO_4 gives stoichiometrically:

$$2\ KMnO_4 + 3\ H_2SO_4 \rightleftharpoons 2\ K^+ + Mn_2O_7 + H_3O^+ + 3\ HSO_4^- \quad (24\text{-}25.7)$$

the dangerous explosive oil, Mn_2O_7. This can be extracted into CCl_4 or chlorofluorocarbons in which it is reasonably stable and safe.

Until recently, only oxo compounds of Mn^{VI} and Mn^{VII} were known, but now the compound $Mn^{VII}(\text{N-}t\text{-Bu})_3Cl$ has been shown to be stable (mp 94–95 °C), whereas its oxo analog $MnClO_3$ detonates above 0 °C. A great many derivatives of $Mn(\text{N-}t\text{-Bu})_3Cl$ have also been prepared, examples being $Mn(\text{N-}t\text{-Bu})_3Br$ (mp 105–107 °C), $Mn(\text{N-}t\text{-Bu})_3(O_2CCH_3)$ (mp 49–59 °C), $Mn(\text{N-}t\text{-Bu})_3(OC_6F_5)$ (mp 95–97 °C), and $Mn(\text{N-}t\text{-Bu})_3(\text{NH-}t\text{-Bu})$, which is an unstable oil. Reduction of the chloride with sodium amalgam in THF gives the Mn^{VI} dimer $[(\text{N-}t\text{-Bu})_2Mn(\mu\text{-N-}t\text{-Bu})]_2$.

IRON

24-26 The Element Iron

Beginning with this element, there is no oxidation state equal to the total number of valence-shell electrons, which in this case is eight. The highest oxidation state is VI, and it is rare. Even the trivalent state, which rose to a peak of importance at chromium, now loses ground to the divalent state.

Iron is the second most abundant metal, after Al, and the fourth most abundant element in the earth's crust. The core of the earth is believed to consist mainly of Fe and Ni. The major ores are Fe_2O_3 (*hematite*), Fe_3O_4 (*magnetite*), FeO(OH) (*limonite*), and $FeCO_3$ (*siderite*).

Pure iron is quite reactive. In moist air it is rather rapidly oxidized to give a hydrous iron(III) oxide (rust) that affords no protection, since it flakes off, exposing fresh metal surfaces. Finely divided iron is pyrophoric.

The metal dissolves readily in dilute mineral acids. With nonoxidizing acids and in the absence of air, Fe^{II} is obtained. With air present or when warm dilute HNO_3 is used, some of the iron goes to Fe^{III}. Very strongly oxidizing media, such as concentrated HNO_3 or acids containing dichromate, passivate iron. Air-free water and dilute air-free solutions of OH^- have little effect, but iron is attacked by hot concentrated NaOH (see the following section).

IRON COMPOUNDS

The main stereochemistries of iron compounds are as follows:

Fe^{II} Octahedral as in $Fe(OH)_2$, $[Fe(H_2O)_6]^{2+}$, and $[Fe(CN)_6]^{4-}$

Fe^{III} Octahedral as in $[Fe(H_2O)_6]^{3+}$ and $Fe(acac)_3$

24-27 Binary Compounds

Oxides and Hydroxides

The addition of OH^- to Fe^{2+} solutions gives the pale green *hydroxide*, which is very readily oxidized by air to give red-brown hydrous iron(III) oxide.

The compound $Fe(OH)_2$, a true hydroxide with the $Mg(OH)_2$ structure, is somewhat amphoteric. Like Fe, it dissolves in hot concentrated NaOH, from which solutions, blue crystals of $Na_4[Fe^{II}(OH)_6]$ can be obtained.

The *oxide*, FeO, may be obtained as a black pyrophoric powder by ignition of Fe^{II} oxalate: It is usually nonstoichiometric, $Fe_{0.95}O$, which means that some Fe^{III} is present. The addition of OH^- to iron(III) solutions gives a red-brown gelatinous mass commonly called iron(III) hydroxide, but it is best described as a *hydrous oxide* ($Fe_2O_3 \cdot nH_2O$). This has several forms; one, FeO(OH), occurs in the mineral *lepidocrocite* and can be made by high-temperature hydrolysis of iron(III) chloride. On heating at 200 °C the hydrous oxides form red-brown $\alpha\text{-}Fe_2O_3$, which occurs as the mineral *hematite*. This has the corundum structure with an hcp array of O, with Fe^{3+} in the octahedral interstices.

The black crystalline oxide Fe_3O_4, a mixed Fe^{II}–Fe^{III} oxide, occurs in nature as *magnetite*. It can be made by ignition of Fe_2O_3 above 1400 °C. It has the inverse spinel structure (Section 4-8).

Chlorides are used as source material for the synthesis of other iron compounds. Anhydrous *iron(II) chloride* can be made by passing HCl gas over heated iron powder, by reducing $FeCl_3$ with Fe(s) in THF or by refluxing $FeCl_3$ in chlorobenzene. It is a very pale green, almost white, solid.

Iron(III) chloride is obtained by the action of chlorine on heated iron as almost black, red-brown crystals. Although in the gas phase there are dimers, Fe_2Cl_6, in the crystal the structure is nonmolecular and there are Fe^{III} ions occupying two thirds of the octahedral holes in alternate layers of Cl^- ions.

Iron(III) chloride quite readily hydrolyzes in moist air. It is soluble in ethers and other polar solvents.

Ferrites such as $M^{II}Fe_2O_4$ are important mixed oxide materials used in magnetic tapes for recording purposes.

24-28 Chemistry of Iron(II), d^6

The iron(II) ion, $[Fe(H_2O)_6]^{2+}$, gives many crystalline salts. Mohr's salt, $(NH_4)_2[Fe(H_2O)_6](SO_4)_2$, is reasonably stable toward air and loss of water, and is commonly used to prepare standard solutions of Fe^{2+} for volumetric analysis and as a calibration substance in magnetic measurements. By contrast, $FeSO_4 \cdot 7H_2O$ slowly effloresces and turns brown-yellow when kept in air.

Addition of HCO_3^- or SH^- to aqueous solutions of Fe^{2+} precipitates $FeCO_3$ and FeS, respectively. The Fe^{2+} ion is oxidized in acid solution by air to Fe^{3+}. With ligands other than water present, substantial changes in the potentials may occur, and the Fe^{II}–Fe^{III} system provides an excellent example of the effect of ligands on the relative stabilities of oxidation states.

$$[Fe(CN)_6]^{3-} + e^- = [Fe(CN)_6]^{4-} \qquad E° = 0.36 \text{ V} \qquad (24\text{-}28.1)$$

$$[Fe(H_2O)_6]^{3+} + e^- = [Fe(H_2O)_6]^{2+} \qquad E° = 0.77 \text{ V} \qquad (24\text{-}28.2)$$

$$[Fe(phen)_3]^{3+} + e^- = [Fe(phen)_3]^{2+} \qquad E° = 1.12 \text{ V} \qquad (24\text{-}28.3)$$

Complexes

Octahedral complexes are generally paramagnetic, and quite strong ligand fields are required to cause spin pairing. Diamagnetic complex ions are $[Fe(CN)_6]^{4-}$ and $[Fe(dipy)_3]^{2+}$. Formation of the red 2,2′-bipyridine and 1,10-phenanthroline complexes is used as a test for Fe^{2+}.

Some tetrahedral complexes like $[FeCl_4]^{2-}$ are known. Among the most important complexes are those involved in biological systems (Chapter 31) or models for them. An important iron(II) compound is ferrocene (Section 29-14).

24-29 The Chemistry of Iron(III), d^5

Iron(III) occurs in crystalline salts with most anions other than those, such as iodide, that are incompatible because of their reducing properties:

$$Fe^{3+} + I^- = Fe^{2+} + \tfrac{1}{2} I_2 \qquad (24\text{-}29.1)$$

Salts containing the ion $[Fe(H_2O)_6]^{3+}$, such as $Fe(ClO_4)_3 \cdot 10H_2O$, are pale pink to nearly white and the aquo ion is pale purple. Unless Fe^{3+} solutions are quite strongly acid, hydrolysis occurs and the solutions are commonly yellow because of the formation of hydroxo species that have charge-transfer bands in the UV region tailing into the visible region.

The initial hydrolysis equilibria are

$$[Fe(H_2O)_6]^{3+} = [Fe(H_2O)_5(OH)]^{2+} + H^+ \qquad K = 10^{-3.05} \qquad (24\text{-}29.2)$$

$$[Fe(H_2O)_5(OH)]^{2+} = [Fe(H_2O)_4(OH)_2]^+ + H^+ \qquad K = 10^{-3.26} \qquad (24\text{-}29.3)$$

$$2[Fe(H_2O)_6]^{3+} = [Fe(H_2O)_4(OH)_2Fe(H_2O)_4]^{4+} + 2\,H^+ \qquad (24\text{-}29.4)$$
$$K = 10^{-2.91}$$

The binuclear species in Reaction 24-29.4 probably has the Structure 24-III.

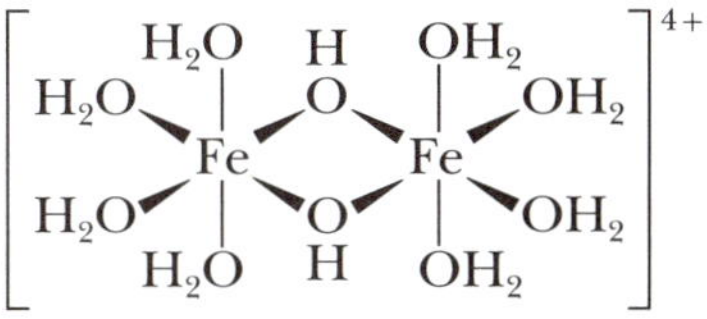

24-III

From the constants it is clear that, even at pH values of 2–3, hydrolysis is extensive. In order to have solutions containing say, about 99% $[Fe(H_2O)_6]^{3+}$ the pH must be around zero. As the pH is raised to about 2–3, more highly condensed species than the dinuclear one are formed, attainment of equilibrium becomes sluggish, and soon colloidal gels are formed. Ultimately, hydrous Fe_2O_3 is precipitated.

Iron(III) ion has a strong affinity for F^-

$$Fe^{3+} + F^- = FeF^{2+} \qquad K_1 \approx 10^5 \qquad (24\text{-}29.5)$$

$$FeF^{2+} + F^- = FeF_2^+ \qquad K_2 \approx 10^5 \qquad (24\text{-}29.6)$$

$$FeF_2^+ + F^- = FeF_3 \qquad K_3 \approx 10^3 \qquad (24\text{-}29.7)$$

The corresponding constants for chloro complexes are only about 10, 3, and 0.1, respectively. In very concentrated HCl the tetrahedral $FeCl_4^-$ ion is formed, and its salts with large cations may be isolated. Complexes with SCN^- are an intense red, and this serves as a sensitive qualitative and quantitative test for ferric ion; $Fe(SCN)_3$ and/or $Fe(SCN)_4^-$ may be extracted into ether. Fluoride ion, however, will discharge this color. In the solid state, $[FeF_6]^{3-}$ ions are known but in solutions only species with fewer F atoms occur.

The hexacyanoferrate ion, $[Fe(CN)_6]^{3-}$, in contrast to $[Fe(CN)_6]^{4-}$, is quite poisonous because the CN^- ions rapidly dissociate, whereas $[Fe(CN)_6]^{4-}$ is not labile.

The affinity of Fe^{III} for NH_3 and amines is low except for chelates, such as $EDTA^{4-}$; 2,2′-bipyridine and 1,10-phenanthroline, which produce ligand fields strong enough to cause spin pairing (cf. Fe^{II}) and form quite stable ions that can be isolated with large anions.

A number of hydroxo- and oxygen-bridged species, one of which has been mentioned previously, are of interest because they may show unusual magnetic properties due to coupling between the iron atoms via the bridges. One example is the complex of salen [bis(salicylaldehyde)ethylenediiminato],

[Fe(salen)]$_2$O, which has a nonlinear Fe—O—Fe group, while Fe(salen)Cl can form both mononuclear and binuclear complexes (Structure 24-IV). The latter has marked antiferromagnetic coupling between the Fe atoms.

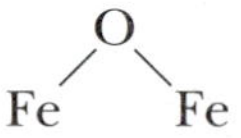

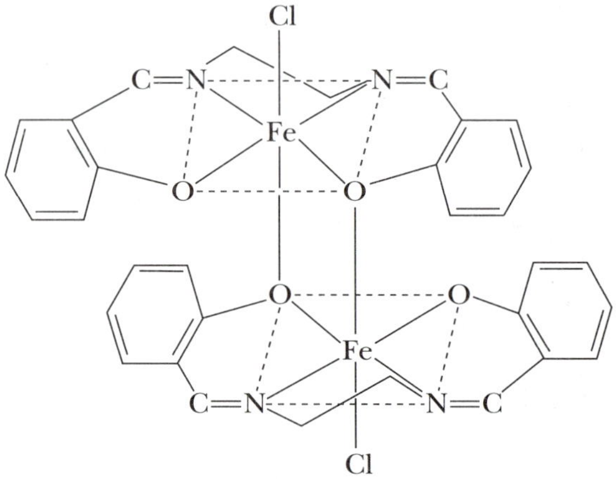

24-IV

24-30 The Chemistry of Iron(IV) and (VI)

There are only a few complexes, such as [Fe(S$_2$CNR$_2$)$_3$]$^+$ and [Fe(diars)$_2$Cl$_2$]$^{2+}$, for Fe^{IV}, and the unusual hydrocarbon soluble alkyl, Fe(1-norbornyl)$_4$ (Section 29-11). No stable Fe^V compounds are known.

The best known compound of iron(VI) is the oxo anion, [FeO$_4$]$^{2-}$, which is obtained by chlorine oxidation of suspensions of Fe$_2$O$_3$·nH$_2$O in concentrated NaOH or by fusing Fe powder with KNO$_3$. The red-purple ion is paramagnetic with two unpaired electrons. The Na and K salts are quite soluble but the Ba salt can be precipitated. The ion is relatively stable in basic solution but decomposes in neutral or acid solution according to the equation

$$2[\text{FeO}_4]^{2-} + 10\,\text{H}^+ = 2\,\text{Fe}^{3+} + \tfrac{3}{2}\,\text{O}_2 + 5\,\text{H}_2\text{O} \qquad (24\text{-}30.1)$$

It is an even stronger oxidizing agent than MnO$_4^-$ and can oxidize NH$_3$ to N$_2$, Cr^{II} to CrO$_4^{2-}$, and also primary amines and alcohols to aldehydes.

COBALT

24-31 The Element Cobalt

The trends toward decreased stability of the very high oxidation states and the increased stability of the II state relative to the III state, which occur through the series Ti, V, Cr, Mn, and Fe, persist with Co. The highest oxidation state is now IV, and only a few such compounds are known. Cobalt(III) is relatively unstable in simple compounds, but the low-spin complexes are exceedingly numerous and stable, especially where the donor atoms (usually N) make strong contributions to the ligand field. There are also numerous complexes of Co^I. This oxidation state is better known for cobalt than for any other element of the first transition series except copper. All Co^I complexes have π-acid ligands (Chapter 28).

Cobalt always occurs in association with Ni and will usually occur also with As. The chief sources of Co are "speisses," which are residues in the smelting of arsenical ores of Ni, Cu, and Pb.

Cobalt is relatively unreactive, although it dissolves slowly in dilute mineral acids.

COBALT COMPOUNDS

The main stereochemistries found in cobalt compounds are the following:

Co^{II}	Tetrahedral as in $[CoCl_4]^{2-}$ and $CoCl_2(PEt_3)_2$
	Octahedral as in $[Co(H_2O)_6]^{2+}$
Co^{III}	Octahedral as in $[Co(NH_3)_6]^{3+}$

24-32 Chemistry of Cobalt(II), d^7

The dissolution of Co, or the hydroxide or carbonate, in dilute acids gives the pink aqua ion, $[Co(H_2O)_6]^{2+}$, which forms many hydrated salts.

Addition of OH^- to Co^{2+} gives the *hydroxide*, which may be blue or pink depending on the conditions; it is weakly amphoteric dissolving in very concentrated OH^- to give a blue solution containing the $[Co(OH)_4]^{2-}$ ion.

Complexes

The most common Co^{II} complexes may be either octahedral or tetrahedral. There is only a small difference in stability and both types, with the same ligand, may be in equilibrium. Thus for water there is a very small but finite concentration of the tetrahedral ion:

$$[Co(H_2O)_6]^{2+} \rightleftharpoons [Co(H_2O)_4]^{2+} + 2\,H_2O \qquad (24\text{-}32.1)$$

Addition of excess Cl^- to pink solutions of the aqua ion readily gives the blue tetrahedral species:

$$[Co(H_2O)_6]^{2+} + 4\,Cl^- = [CoCl_4]^{2-} + 6\,H_2O \qquad (24\text{-}32.2)$$

Tetrahedral complexes, $[CoX_4]^{2-}$, are formed by halide, pseudohalide, and OH^- ions. Cobalt(II) forms tetrahedral complexes more readily than any other transition metal ion. The Co^{2+} ion is the only d^7 ion of common occurrence. For a d^7 ion, ligand field stabilization energies disfavor the tetrahedral configuration relative to the octahedral one to a smaller extent (see Table 23-4) than for any other $d^n (1 \le n \le 9)$ configuration. This argument is valid only in comparing the behavior of one metal ion with another and not for assessing the absolute stabilities of the configurations for any particular ion.

24-33 The Chemistry of Cobalt(III), d^6

In the absence of complexing agents, the oxidation of $[Co(H_2O)_6]^{2+}$ is very unfavorable (Table 24-3) and Co^{3+} is reduced by water. However, electrolytic or O_3 oxidation of cold acidic solutions of $Co(ClO_4)_2$ gives the aqua ion, $[Co(H_2O)_6]^{3+}$, in equilibrium with $[Co(OH)(H_2O)_5]^{2+}$. At 0 °C, the half-life of these diamagnetic ions is about a month. In the presence of complexing agents, such as NH_3, the stability of Co^{III} is greatly improved.

$$[Co(NH_3)_6]^{3+} + e^- = [Co(NH_3)_6]^{2+} \qquad E° = 0.1 \text{ V} \qquad (24\text{-}33.1)$$

In the presence of OH^- ion, cobalt(II) hydroxide is readily oxidized by air to a black hydrous oxide.

$$CoO(OH)(s) + H_2O + e^- = Co(OH)_2(s) + OH^- \qquad E° = 0.17 \text{ V} \qquad (24\text{-}33.2)$$

The Co^{3+} ion shows a particular affinity for N donors, such as NH_3, en, edta, and NCS^-, and complexes are exceedingly numerous. They generally undergo ligand-exchange reactions relatively slowly, like Cr^{3+} and Rh^{3+}. Hence, they have been extensively studied since the days of Werner and Jørgensen. A large part of our knowledge of the isomerism, modes of reaction, and general properties of octahedral complexes as a class is based on studies of Co^{III} complexes. Almost all Co^{III} complexes are octahedral.

Cobalt(III) complexes are synthesized by oxidation of Co^{2+} in solution in the presence of the ligands. Oxygen or hydrogen peroxide and a catalyst, such as activated charcoal, are used. For example,

$$4 Co^{2+} + 4 NH_4^+ + 20 NH_3 + O_2 \longrightarrow 4[Co(NH_3)_6]^{3+} + 2 H_2O \qquad (24\text{-}33.3)$$

$$4 Co^{2+} + 8en + 4enH^+ + O_2 = 4[Co(en)_3]^{3+} + 2 H_2O \qquad (24\text{-}33.4)$$

The green salt, *trans*-$[Co(en)_2Cl_2][H_5O_2]Cl_2$, is obtained from a reaction similar to Reaction 24-33.4 in the presence of HCl. This salt may be isomerized to the purple racemic cis isomer on evaporation of a neutral aqueous solution at 90–100 °C. Both the cis and the trans isomer are aquated when heated in water:

$$[Co(en)_2Cl_2]^+ + H_2O \longrightarrow [Co(en)_2Cl(H_2O)]^{2+} + Cl^- \qquad (24\text{-}33.5)$$

$$[Co(en)_2Cl(H_2O)]^{2+} + H_2O \longrightarrow [Co(en)_2(H_2O)_2]^{3+} + Cl^- \qquad (24\text{-}33.6)$$

and on treatment with solutions of other anions are converted into other $[Co(en)_2X_2]^+$ species, for example,

$$[Co(en)_2Cl_2]^- + 2 NCS^- \longrightarrow [Co(en)_2(NCS)_2]^+ + 2 Cl^- \qquad (24\text{-}33.7)$$

The initial reaction of Co^{II} with oxygen may involve oxidative–addition (Section 30-2) of O_2 to Co^{II} to give a transient Co^{IV} species that then reacts with another Co^{II} to produce a binuclear *peroxo-bridged* species:

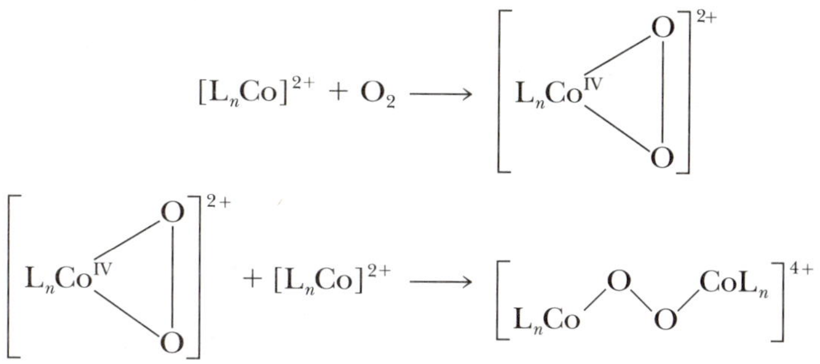

$$[L_nCo]^{2+} + O_2 \longrightarrow \left[L_nCo^{IV} \overset{O}{\underset{O}{\diagup\!\!\diagdown}} \right]^{2+}$$

$$\left[L_nCo^{IV} \overset{O}{\underset{O}{\diagup\!\!\diagdown}} \right]^{2+} + [L_nCo]^{2+} \longrightarrow \left[L_nCo \overset{O}{\diagdown}\underset{O}{\diagup} CoL_n \right]^{4+}$$

Complexes such as $[(NH_3)_5CoOOCo(NH_3)_5]^{4+}$ or $[(NC)_5CoOOCo(CN)_5]^{6-}$ have been isolated, although these ions decompose fairly readily in water or acids. The open-chain species $[(NH_3)_5CoOOCo(NH_3)_5]^{4+}$ can be cyclized in the presence of base to

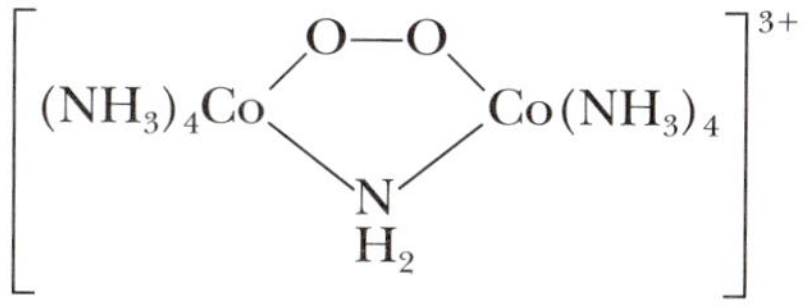

$$\left[(NH_3)_4Co \overset{O-O}{\underset{\underset{H_2}{N}}{\diagup\!\!\!\diagdown}} Co(NH_3)_4 \right]^{3+}$$

Such peroxo species, open chain or cyclic, contain low-spin Co^{III} and bridging peroxide (O_2^{2-}) ions.

The O_2-bridged binuclear complexes can often be oxidized in a one-electron step. The resulting ions were first prepared by Werner, who formulated them as peroxo-bridged complexes of Co^{III} and Co^{IV}. However, ESR data have shown that the single unpaired electron is distributed equally over *both* cobalt ions, and is best regarded as belonging formally to a superoxide (O_2^-) ion but delocalized over the planar Co^{III}—O—O—Co^{III} group. The structures (Fig. 24-4) show that the O—O distance is close to that for the O_2^- ion (1.28 Å) and much shorter than the distance (1.47 Å) in the peroxo complexes.

Although no cobalt-containing complex is known to be involved in oxygen metabolism, there are several that provide models for metal-to-oxygen binding in biological systems. Of greatest interest are those that undergo *reversible* oxygenation and deoxygenation in solution. The Schiff base complexes such as Co(acacen) (Structure 24-V) in DMF or py take up O_2 reversibly below 0 °C, for example,

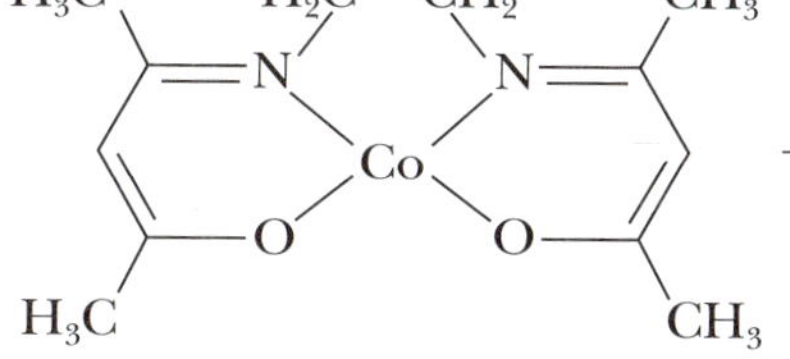

$$+ \, DMF + O_2 \longrightarrow Co(acacen)(dmf)O_2$$

$$K = 2.1 \text{ at } -10 \,°C$$

24-V

$$(24\text{-}33.10)$$

The initial complex has one unpaired electron, and so also do the oxygen adducts, but ESR data indicate that in the latter the electron is heavily localized on the oxygen atoms. There is also an IR absorption band due to an O—O stretching vibration. The adducts can be formulated as octahedral, low-spin Co^{III}

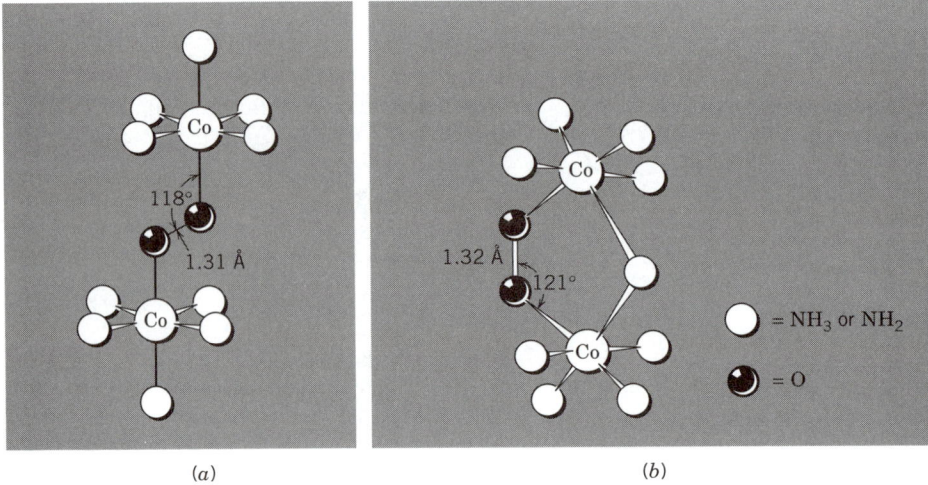

Figure 24-4   The structures of (a) $[(NH_3)_5CoO_2Co(NH_3)_5]^{5+}$, and

$$(b) \left[(NH_3)_4Co \overset{O_2}{\underset{\underset{H_2}{N}}{\diagup\diagdown}} Co(NH_3)_4 \right]^{4+}$$

There is octahedral coordination about each cobalt ion and the angles and distances shown are consistent with the assumption that there are bridging superoxo groups. The five-membered ring in (b) is essentially planar.

complexes containing a coordinated superoxide (O_2^-) ion. The Co—O—O chain is bent. A second type of complex involves the reversible formation of oxygen bridges (Co—O—O—Co), which are similar to those discussed previously.

Finally, we note in connection with oxidation that in acid solutions, cobalt(III) carboxylates catalyze the oxidation not only of alkyl side chains in aromatic hydrocarbons, but even of alkanes themselves. A cobalt catalyzed process is used commercially for oxidation of toluene to phenol. The actual nature of "cobaltic acetate," a green material made by ozone oxidation of Co^{2+} acetate in acetic acid is uncertain; it can, however, be converted by pyridine to an oxo centered species similar to those known for other M^{III} carboxylates (Structure 24-I).

24-34 Complexes of Cobalt(I), d^8

With the exception of reduced vitamin B_{12} and models for this system (Section 31-8), which appear to be Co^I species, all Co^I compounds involve ligands of the π-acid type (Chapter 28). The coordination is trigonal bipyramidal or tetrahedral. The compounds are usually made by reducing $CoCl_2$ in the presence of the ligand by agents such as N_2H_4, Zn, $S_2O_4^{2-}$, or Al alkyls.

Representative examples are $CoH(N_2)(PPh_3)_3$, $[Co(CNR)_5]^+$, $CoCl(PR_3)_3$.

NICKEL

24-35 The Element

The trend toward decreased stability of higher oxidation states continues, so that only Ni^{II} normally occurs with a few compounds *formally* containing Ni^{III} and Ni^{IV}. The relative simplicity of nickel chemistry in terms of oxidation number is balanced by considerable complexity in coordination numbers and geometries.

Nickel occurs in combination with arsenic, antimony, and sulfur as in *millerite* (NiS) and in *garnierite,* a magnesium-nickel silicate of variable composition. Nickel is also found alloyed with iron in meteors; the interior of the earth is believed to contain considerable quantities. In general, the ore is roasted in air to give NiO, which is reduced to Ni with C. Nickel is usually purified by electro-deposition but some high purity nickel is still made by the carbonyl process. Carbon monoxide reacts with impure nickel at 50 °C and ordinary pressure or with nickel–copper matte under more strenuous conditions, giving volatile $Ni(CO)_4$, from which metal of 99.90 to 99.99% purity is obtained on thermal decomposition at 200 °C.

Nickel is quite resistant to attack by air or water at ordinary temperatures when compact and is, therefore, often electroplated as a protective coating. It dissolves readily in dilute mineral acids. The metal or high Ni alloys are used to handle F_2 and other corrosive fluorides. The finely divided metal is reactive to air and may be pyrophoric. Nickel absorbs considerable amounts of hydrogen when finely divided and special forms of Ni (e.g., Raney nickel) are used for catalytic reductions.

NICKEL COMPOUNDS

24-36 The Chemistry of Nickel(II), d^8

The binary compounds, such as NiO and $NiCl_2$, need no special comment.

Nickel(II) forms a large number of *complexes* with coordination numbers six, five, and four having all the main structural types: octahedral, trigonal bipyramidal, square pyramidal, tetrahedral, and square. It is characteristic that complicated equilibria, which are generally temperature dependent and sometimes concentration dependent, often exist between these structural types.

Six-Coordinate Complexes

The commonest six-coordinate complex is the green aqua ion, $[Ni(H_2O)_6]^{2+}$, that is formed on dissolution of Ni, $NiCO_3$, and so on, in acids and gives salts like $NiSO_4 \cdot 7H_2O$.

The water molecules in the aqua ion can be readily displaced especially by amines to give complexes, such as *trans*-$[Ni(H_2O)_2(NH_3)_4]^{2+}$, $[Ni(NH_3)_6]^{2+}$, or $[Ni(en)_3]^{2+}$. These amine complexes are usually blue or purple because of shifts in absorption bands when H_2O is replaced by a stronger field ligand (Section 23-6).

Four-Coordinate Complexes

Most of the four-coordinate complexes are square. This is a consequence of the d^8 configuration, since the planar ligand set causes one of the d orbitals ($d_{x^2-y^2}$)

to be uniquely high in energy, and the eight electrons can occupy the other four *d* orbitals but leave this strongly antibonding one vacant. In tetrahedral coordination, on the other hand, occupation of antibonding orbitals is unavoidable. With the congeneric d^8 systems Pd^{II} and Pt^{II} this factor becomes so important that no tetrahedral complex is formed.

Planar complexes of Ni^{II} are thus invariably diamagnetic. They are frequently red, yellow, or brown owing to the presence of an absorption band of medium intensity ($\epsilon \approx 60$) in the range 450–600 nm.

Probably the best known example is the red *bis(dimethylglyoximato)nickel(II)*, $Ni(dmgH)_2$, which is used for the gravimetric determination of nickel; it is precipitated on addition of ethanolic $dmgH_2$ to ammoniacal nickel(II) solutions. It has Structure 24-VI, where the hydrogen bond is symmetrical, but these units are *stacked* one on top of the other in the crystal. Here, and in similar square compounds of Pd^{II} and Pt^{II} (Section 25-28), there is evidence of metal-to-metal interaction, even though the distance in the stack is too long for true bonding.

24-VI   24-VII

Similar square complexes are given by certain β-ketoenolates (e.g., Structure 24-VII), as well as by unidentate π-acid ligands [e.g., $NiBr_2(PEt_3)_2$], and by CN^- and SCN^-. The cyano complex, $[Ni(CN)_4]^{2-}$, is readily formed on addition of CN^- to Ni^{2+} (aq). The green $Ni(CN)_2$ which is first precipitated redissolves to give the yellow ion, which can be isolated as, for example, $Na_2[Ni(CN)_4] \cdot 3H_2O$. On addition of an excess of CN^- the red ion, $[Ni(CN)_5]^{3-}$, is formed, which can be precipitated only by use of large cations.

Tetrahedral Complexes

Tetrahedral complexes are less common than planar complexes, and are all paramagnetic. These complexes are of the types $[NiX_4]^{2-}$, NiX_3L^-, NiL_2X_2, and $Ni(L—L)_2$ where X is halogen, L is a neutral ligand (e.g., R_3P or R_3PO), and L—L is a bidentate uninegative ligand, $[NiL_4]^{2+}$, is known, where L = hexamethylphosphoramide.

Five-Coordinate Complexes

Five-coordinate complexes usually have trigonal-bipyramidal geometry but some are square pyramidal. Many contain the tetradentate "tripod" ligands, such as $N[CH_2CH_2N(CH_3)_2]_3$ (see Structures 6-XVa and 6-XVb).

24-37 Conformational Properties of Nickel(II) Complexes

The main structural and conformational changes that nickel(II) complexes undergo are the following:

1. *Formation of five- and six-coordinate complexes results from the addition of ligands to square complexes.* For *any* square complex NiL_4 the following equilibria with additional ligands (L') must in principle exist:

$$ML_4 + L' = ML_4L' \tag{24-37.1}$$

$$ML_4 + 2\,L' = ML_4L_2' \tag{24-37.2}$$

Where $L = L' = CN^-$, only the five-coordinate species is formed, but in most systems in which L' is a good donor such as py, H_2O and C_2H_5OH, the equilibria lie far in favor of the six-coordinate species. These have a trans structure and a high-spin electron configuration; many may be isolated as pure compounds. Thus, the β-diketone complex (Structure 24-VII) is normally prepared in the presence of water and/or alcohol and is first isolated as the green, paramagnetic dihydrate or dialcoholate, from which the red, square complex is then obtained by heating to drive off the solvent.

Another type of square complex that picks up H_2O, anions, or solvent is shown in Structure 24-VIII.

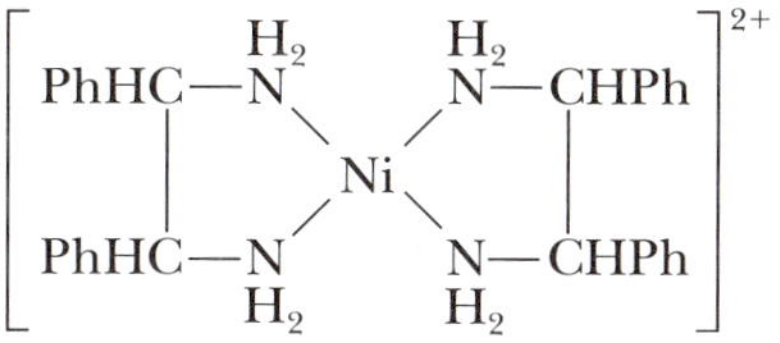

$$
\left[
\begin{array}{c}
\text{PhHC---N} \quad \overset{H_2}{\diagdown} \qquad \overset{H_2}{\diagup} \text{N---CHPh} \\
\mid \qquad\qquad \text{Ni} \qquad\qquad \mid \\
\text{PhHC---N} \quad \diagup \qquad \diagdown \text{N---CHPh} \\
\underset{H_2}{} \qquad\qquad \underset{H_2}{}
\end{array}
\right]^{2+}
$$

24-VIII

2. Monomer-polymer equilibria can occur. Four-coordinate complexes may associate or polymerize, to give five- or six-coordinate species. In some cases, the association is very strong and the four-coordinate monomers are observed only at high temperatures. In others the position of the equilibrium is such that both red, diamagnetic monomers, and green or blue, paramagnetic polymers, are present in a temperature- and concentration-dependent equilibrium around room temperature. A clear example of this situation is provided by the *acetylacetonate* (Fig. 24-5). As a result of the sharing of some oxygen atoms, each nickel atom achieves octahedral coordination. This trimer is very stable, and detectable quantities of monomer appear only at temperatures around 200 °C in a noncoordinating solvent. It is, however, readily cleaved by donors, such as H_2O or py, to give six-coordinate monomers. When the methyl groups of the acetylacetonate ligand are replaced by the very bulky $C(CH_3)_3$ groups, trimerization is completely prevented and the planar monomer (Structure 24-VII) results. When groups sterically intermediate between CH_3 and $C(CH_3)_3$ are used, temperature- and concentration-dependent monomer–trimer equilibria are observed in noncoordinating solvents.

3. Square-tetrahedral equilibria and isomerism can occur. Complexes, such as NiL_2X_2, where L represents a mixed alkylarylphosphine exist in solution in an

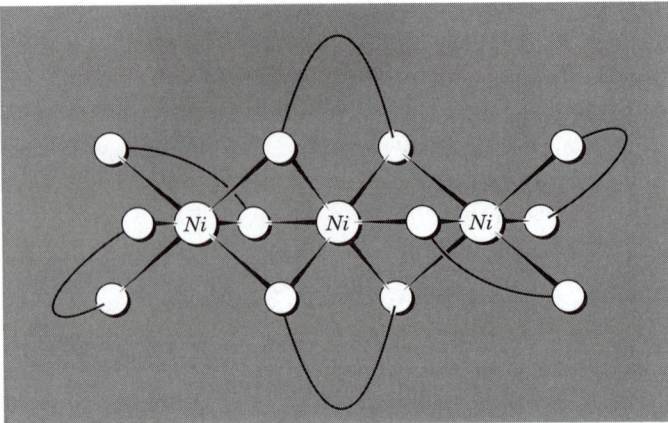

Figure 24-5 Sketch indicating the trimeric structure of nickel acetylacetonate. The unlabeled circles represent oxygen atoms, and the curved lines connecting them in pairs represent the remaining portions of the acetylacetonate rings. (Reproduced by permission from J. C. Bullen, R. Mason, and P. Pauling, *Inorg. Chem.*, **1965**, *4*, 456 © American Chemical Society.)

equilibrium distribution between the tetrahedral and square forms. In some cases it is possible to isolate two crystalline forms of the compound, one yellow-red and diamagnetic, the other green or blue with two unpaired electrons. There is even a case, $Ni[C_6H_5CH_2)(C_6H_5)_2P]_2Br_2$, in which *both* tetrahedral and square complexes are found together in the same crystalline substance.

24-38 Higher Oxidation States of Nickel

Oxides and Hydroxides

The action of Br_2 on alkaline solutions of Ni^{2+} gives a black hydrous oxide, $NiO(OH)$. Other black substances can be obtained by electrolytic oxidation; some of them contain alkali metal ions.

The Edison or nickel-iron battery, which uses KOH as the electrolyte, is based on the reaction

$$Fe + 2\,NiO(OH) + 2\,H_2O \underset{charge}{\overset{discharge}{\rightleftharpoons}}$$

$$Fe(OH)_2 + 2\,Ni(OH)_2 \qquad (\sim 1.3\ V)\quad (24\text{-}38.1)$$

but the mechanism and the true nature of the oxidized nickel species are not fully understood.

Complexes

There are several authentic complexes of *nickel(III)*. Oxidation of $NiX_2(PR_3)_2$ with the appropriate halogen gives $NiX_3(PR_3)_2$.

Nickel(IV) complexes are even rarer, and the dithiolene complexes (Section 28-18), which could formally be regarded as containing Ni^{4+} and $S_2CR_2^{2-}$ ligands, are best regarded as Ni^{II} complexes.

COPPER

24-39 The Element

Copper has a single *s* electron outside the filled $3d$ shell. It has little in common with the alkalis except formal stoichiometries in the +1 oxidation state. The filled *d* shell is much less effective than is a noble gas shell in shielding the *s* electron from the nuclear charge, so that the first ionization potential of Cu is higher than those of the alkalis. Since the electrons of the *d* shell are also involved in metallic bonding, the heat of sublimation and the melting point of copper are also much higher than those of the alkalis. These factors are responsible for the more noble character of copper. The effect is to make compounds more covalent and to give them higher lattice energies, which are not offset by the somewhat smaller radius of Cu^+ (0.93 Å) compared with Na^+ (0.95 Å) and K^+ (1.33 Å).

The second and third ionization potentials of Cu are very much lower than those of the alkalis and account in part for the transition metal character.

Copper is not abundant (55 ppm) but is widely distributed as a metal, in sulfides, arsenides, chlorides, and carbonates. The commonest mineral is chalcopyrite, $CuFeS_2$. Copper is extracted by oxidative roasting and smelting, or by microbial-assisted leaching, followed by electrodeposition from sulfate solutions.

Copper is used in alloys such as brass and is completely miscible with gold. It is very slowly superficially oxidized in moist air, sometimes giving a green coating of hydroxo carbonate and hydroxo sulfate (from SO_2 in the atmosphere).

Copper readily dissolves in nitric acid and in sulfuric acid in the presence of oxygen. It is also soluble in KCN or ammonia solutions in the presence of oxygen, as indicated by the potentials:

$$Cu(s) + 2\,NH_3 \xrightarrow{-0.12\ V} [Cu(NH_3)_2]^+ \xrightarrow{-0.01\ V} [Cu(NH_3)_4]^{2+} \quad (24\text{-}39.1)$$

COPPER COMPOUNDS

The stereochemistry of the more important copper compounds is as follows:

Cu^I Tetrahedral as in CuI(s) or $[Cu(CN)_4]^{3-}$

             Square as in CuO(s), $[Cu(py)_4]^{2+}$, or $[CuCl_4]^{2-}$

Cu^{II} Distorted octahedral with two longer trans bonds, for example, $[Cu(H_2O)_6]^{2+}$, $CuCl_2(s)$

24-40 The Chemistry of Copper(I), d^{10}

Copper(I) compounds are diamagnetic and, except where color results from the anion or charge-transfer bands, are colorless.

The relative stabilities of the Cu^I and Cu^{II} states are indicated by the potentials:

$$Cu^+ + e^- = Cu(s) \qquad E° = 0.52 \text{ V} \qquad\qquad (24\text{-}40.1)$$

$$Cu^{2+} + e^- = Cu^+ \qquad E° = 0.153 \text{ V} \qquad\qquad (24\text{-}40.2)$$

From these we have

$$Cu(s) + Cu^{2+} = 2\ Cu^+ \qquad E° = -0.37 \text{ V} \qquad\qquad (24\text{-}40.3)$$

$$\frac{[Cu^{2+}]}{[Cu^+]^2} = \sim\!10^6 \qquad\qquad (24\text{-}40.4)$$

The relative stabilities depend very strongly on the nature of anions or other ligands present, and vary considerably with solvent or the nature of neighboring atoms in a crystal.

In *aqueous* solution only low equilibrium concentrations of Cu^+ ($<10^{-2}$ M) can exist (see the following section). The only copper(I) compounds that are stable to water are the highly insoluble ones, such as $CuCl$ or $CuCN$. This instability toward water is due partly to the greater lattice and solvation energies and higher formation constants for complexes of the Cu^{2+} ion so that ionic Cu^I derivatives are unstable.

The equilibrium $2\ Cu^I \rightleftharpoons Cu + Cu^{II}$ can readily be displaced in either direction. Thus, with CN^-, I^-, and $(CH_3)_2S$, Cu^{II} reacts to give the Cu^I compound. The Cu^{II} state is favored by anions that cannot give covalent bonds or bridging groups (e.g., ClO_4^- and SO_4^{2-}) or by complexing agents that have their greater affinity for Cu^{II}. Thus ethylenediamine reacts with copper(I) chloride in aqueous potassium chloride solution.

$$2\ CuCl + 2en = [Cu(en)_2]^{2+} + 2\ Cl^- + Cu^0 \qquad\qquad (24\text{-}40.5)$$

The latter reaction also depends on the chelate nature of the ligand. Thus for ethylenediamine, K is about 10^7; for pentamethylenediamine (which does not chelate) K is 3×10^{-2}; and for ammonia K is 2×10^{-2}. Hence, in the last case the reaction is

$$[Cu(NH_3)_4]^{2+} + Cu^0 = 2[Cu(NH_3)_2]^+ \qquad\qquad (24\text{-}40.6)$$

The lifetime of the Cu^+ ion in water is usually very short (<1 s), but dilute solutions from reduction of Cu^{2+} with V^{2+} or Cr^{2+} may last for several hours in the absence of air.

An excellent illustration of how the stability of the Cu^I ion relative to that of the Cu^{II} ion may be affected by the solvent is the case of acetonitrile. The Cu^I ion is very effectively solvated by CH_3CN, and the halides have relatively high solubilities (e.g., CuI, 35 g/kg CH_3CN) versus negligible solubilities in H_2O. The Cu^I ion is more stable than Cu^{II} in CH_3CN, and Cu^{II} acts as a comparatively powerful oxidizing agent.

Copper(I) Binary Compounds

The *oxide* and *sulfide* are more stable than the corresponding Cu^{II} compounds at high temperatures. Cu_2O is made as a yellow powder by controlled reduction of an alkaline solution of a copper(II) salt with hydrazine or, as red crystals, by thermal decomposition of CuO. Copper(I) sulfide (Cu_2S) is a black crystalline solid prepared by heating copper and sulfur in the absence of air; it is, however, markedly nonstoichiometric.

Copper(I) chloride and *bromide* are made by boiling an acidic solution of the copper(II) salt with an excess of copper; on dilution, white CuCl or pale yellow CuBr is precipitated. Addition of I^- to a solution of Cu^{2+} forms a precipitate that rapidly and quantitatively decomposes to CuI and iodine. CuF is unknown. The halides have the zinc blende structure (tetrahedrally coordinated Cu^+). They are insoluble in water, but the solubility is enhanced by an excess of halide ions owing to formation of, for example, $[CuCl_2]^-$, $[CuCl_3]^{2-}$, and $[CuCl_4]^{3-}$.

Copper(I) Complexes

The most common types of Cu^I complexes are those of simple halide or amine ligands and are usually *tetrahedral*. Even those with stoichiometries such as K_2CuCl_3 still have tetrahedral coordination as there are chains sharing halide ions.

Copper(I) also forms several kinds of polynuclear complexes in which four Cu atoms lie at the vertices of a tetrahedron. In $Cu_4I_4L_4$ (L = R_3P or R_3As) species, there is a triply bridging I atom on each face of the Cu_4 tetrahedron and one ligand (L) is coordinated to a Cu atom at each vertex (Structure 24-IX).

24-41 The Chemistry of Copper(II), d^9

Most Cu^I compounds are fairly readily oxidized to Cu^{II}, but further oxidation to Cu^{III} is difficult. There is a well-defined aqueous chemistry of Cu^{2+}, and a large

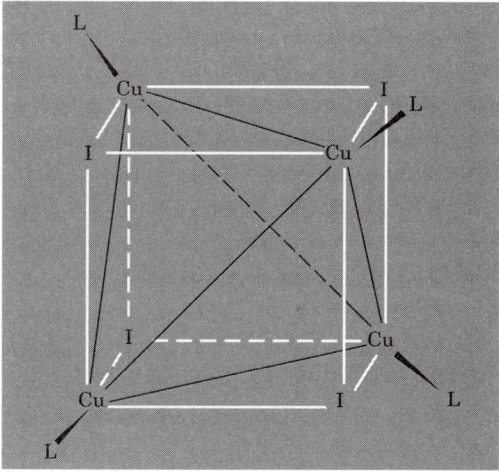

24-IX

number of salts of various anions, many of which are water soluble, that exist in addition to a wealth of complexes.

Before we discuss copper(II) chemistry, it is pertinent to note the stereo-chemical consequences of the d^9 configuration of Cu^{II}. This makes Cu^{II} subject to distortions (Section 23-8) if placed in an environment of cubic (i.e., regular octahedral or tetrahedral) symmetry. The result is that Cu^{II} is nearly always found in environments appreciably distorted from these regular symmetries. The characteristic distortion of the octahedron is such that there are *four short Cu—L bonds in the plane and two trans long ones.* In the limit, this elongation leads to a situation indistinguishable from square coordination, as found in CuO and many discrete complexes of Cu^{II}. Thus the cases of tetragonally distorted "octahedral" coordination and square coordination cannot be sharply differentiated.

Some distorted tetrahedral complexes, such as $M_2^I CuX_4$, are also known provided M is large like cesium. The compound $(NH_4)_2 CuCl_4$ has a planar anion.

Binary Compounds

Black crystalline CuO is obtained by pyrolysis of the nitrate or other oxo salts; above 800 °C it decomposes to Cu_2O. The *hydroxide* is obtained as a blue bulky precipitate on addition of $NaOH$ to Cu^{2+} solutions; warming an aqueous slurry dehydrates this to the oxide. The hydroxide is readily soluble in strong acids and also in concentrated $NaOH$, to give deep blue anions, probably of the type $[Cu_n(OH)_{2n-2}]^{2+}$. In ammoniacal solutions the deep blue tetraammine complex, $[Cu(NH_3)_4]^{2+}$, is formed.

The common *halides* are the yellow chloride and the almost black bromide, having structures with infinite parallel bands of square CuX_4 units sharing edges. The bands are arranged so that a tetragonally elongated octahedron is completed about each copper atom by halogen atoms of neighboring chains. Both $CuCl_2$ and $CuBr_2$ are readily soluble in water, from which hydrates may be crystallized, as well as in donor solvents, such as acetone, alcohol, and pyridine.

The Aqua Ion and Aqueous Chemistry

Dissolution of copper, the hydroxide, carbonate, and so on, in acids gives the blue-green aqua ion that may be written $[Cu(H_2O)_6]^{2+}$. Two of the H_2O molecules are further from the metal than the other four. Of the numerous crystalline hydrates the blue sulfate ($CuSO_4 \cdot 5H_2O$) is best known. It may be dehydrated to the virtually white anhydrous substance. Addition of ligands to aqueous solutions leads to the formation of complexes by successive displacement of water molecules. With NH_3, for example, the species $[Cu(NH_3)(H_2O)_5]^{2+} \cdots$ $[Cu(NH_3)_4(H_2O)_2]^{2+}$ are formed in the normal way, but addition of the fifth and sixth molecules of NH_3 is difficult. The sixth can be added only in liquid ammonia. The reason for this unusual behavior is connected with the Jahn-Teller effect. Because of it, the Cu^{II} ion does not bind the fifth and sixth ligands strongly (even the H_2O). When this intrinsic weak binding of the fifth and sixth ligands is added to the normally expected decrease in the stepwise formation constants (Section 6-4), the formation constants (K_5 and K_6) are very small indeed. Similarly, it is found with ethylenediamine that $[Cu(en)(H_2O)_4]^{2+}$ and

$[Cu(en)_2(H_2O)_2]^{2+}$ form readily, but $[Cu(en)_3]^{2+}$ is formed only at extremely high concentrations of en.

Multidentate ligands that coordinate through O or N, such as amino acids, form copper(II) complexes of considerable stability. The blue solutions formed by addition of tartrate to Cu^{2+} solutions (known as *Fehling's solution* when basic and when *meso*-tartrate is used) may contain monomeric, dimeric, or polymeric species at different pH values. The dimer, $Na_2[Cu\{(\pm)C_4O_6H_2\}]\cdot 5H_2O$, has square Cu^{II} coordination, two tartrate bridges, and a Cu—Cu distance of 2.99 Å.

Polynuclear Compounds with Magnetic Anomalies

Copper forms many compounds in which the Cu—Cu distances are short enough to indicate significant M-M interaction, but in no case is there an actual bond. Particular examples are the bridged *carboxylates* and the related 1,3-triazenido complexes (Structures 24-X and 24-XI). Although in other cases of carboxylates with the same structure (Cr_2^{II}, Mo_2^{II}, Rh_2^{II}, or $Ru_2^{II,III}$) there is a definite M—M bond, this is *not* so for Cu. However, there is weak coupling of the unpaired electrons, one on each Cu^{II} ion, giving rise to a singlet ground state with a triplet state lying only a few kilojoules per mole above it; the latter state is thus appreciably populated at normal temperatures and the compounds are paramagnetic. At 25 °C, μ_{eff} is typically about 1.4 BM per Cu atom and the temperature dependence is very pronounced. The interaction involves either the $d_{x^2-y^2}$ orbitals of the two metal atoms directly or transmission through the π orbitals of the bridge group, or both.

24-X   24-XI

Catalytic Properties of Copper Compounds

Copper compounds catalyze an exceedingly varied array of reactions, heterogeneously, homogeneously, in the vapor phase, in organic solvents, and in aqueous solutions. Many of these reactions, particularly if in aqueous solutions, involve oxidation–reduction systems and a Cu^I-Cu^{II} redox cycle. Molecular oxygen can often be utilized as an oxidant, for example in copper-catalyzed oxidations of ascorbic acid and in the Wacker process (Section 30-11).

The oxidation probably involves an initial oxidative addition reaction (Section 30-2):

$$Cu^+ + O_2 = CuO_2^+ \qquad (24\text{-}41.1)$$

$$CuO_2^+ + H^+ = Cu^{2+} + HO_2 \qquad (24\text{-}41.2)$$

$$Cu^+ + HO_2 = Cu^{2+} + HO_2^- \qquad (24\text{-}41.3)$$

$$H^+ + HO_2^- = H_2O_2 \qquad (24\text{-}41.4)$$

Copper compounds have many uses in organic chemistry for oxidations, for example of phenols by Cu^{2+}-amine complexes, halogenations, coupling reactions, and the like. Copper(II) has considerable biochemical importance (see Chapter 31).

STUDY GUIDE

Scope and Purpose

We have presented a rather large amount of information in a somewhat traditional and descriptive fashion, namely, a steady "march" through the metals of the first transition series and their compounds. For each element we have presented the important or interesting properties of the element and its inorganic compounds. The student should find it satisfying that the descriptions of the compounds and their reactivities are readily set down in the same "language" and using the same theories as those developed earlier in the text.
*SG For each metal we have organized the presentation in terms of important oxidation states, coordination numbers, geometries, number of
d electrons, and types of compounds. We also mention, where appropriate, the various thermodynamic stabilities of the derivatives of a particular metal ion, as well as the kinetic and mechanistic aspects of the reactions. The principal inorganic binary compounds are given first for most elements, followed by the more complex derivatives of an essentially inorganic nature, organized by the important oxidation states of the element. We anticipate covering the metalloorganic compounds in later chapters.

Study Questions

A. Review

1. Write down the ground-state electron configurations for the ions and atoms Ti^{4+}, V^{2+}, Cr^{5+}, Mn^{6+}, Fe^0, Co^+, Ni^{2+}, Cu^{3+}, and Ti^{3+}.

2. Which of the ions in Question 1 typically form octahedral complexes, tetrahedral complexes, or five-coordinate complexes?

3. Which of the complexes of Question 2 would you expect to show Jahn-Teller distortions?

4. What is the chief structural difference between $TiOSO_4 \cdot H_2O$ and $VOSO_4 \cdot 5\,H_2O$?

5. List two examples each where the transition metal compounds MCl_4 and MCl_3 behave as Lewis acids.

6. Give two examples of disproportionation reactions that were presented in this chapter.

7. Explain why the V—O stretching frequency changes when bis(acetylacetonato)oxo-vanadium(IV) is dissolved in pyridine.

8. What happens when a solution of $K_2Cr_2O_7$ is added to solutions of (a) F^- (b) Cl^- (c) Br^- (d) I^- (e) OH^- (f) NO_2^- (g) SO_4^{2-} (h) H_2O_2.

9. Give two examples each of the spinels, perovskites, and alums.

10. Explain why the trivalent ions give acid solutions in water. Write balanced equations to illustrate your answer.

11. Draw the structures of $Cr_2(CO_2CH_3)_4(H_2O)_2$ and $Cr_3O(CO_2CH_3)_6(H_2O)_3Cl$. Classify each atom in these structures according to the AB_xE_y scheme of Chapter 3, and choose a hybridization for each nonmetal, nonterminal atom.

12. What are the structures of $[Ni(acac)_2]_3$, $CrCl_3(thf)_3$, and CrO_5py?

13. Why is it that the freshly prepared hydroxide (a) of Mn^{2+} is white, but turns dark brown in air, (b) of Co^{2+} is blue, but turns pink on warming, and (c) of Cu^{2+} is blue, but turns black on warming?

14. What is the number of unpaired electrons in complexes of (a) spin-paired Mn^{2+}, (b) tetrahedral Cr^{4+}, (c) tetrahedral Co^{2+}, (d) octahedral V^{3+}, (e) octahedral Co^{3+}, (f) low-spin Fe^{2+}, and (g) high-spin Mn^{2+}?

15. Give an example of a complex representing each case in Problem 14.

16. How is oxygen bound in the complexes (a) $Cs_3[TiO_2F_5]$, (b) $K_3[CrO_8]$, and (c) $[Co_2O_2(NH_3)_{10}](SO_4)_2$?

17. Enumerate the possible isomers of $[Co(en)_2(SCN)_2]^+$, and name each one according to proper nomenclature.

B. Additional Exercises

1. Draw the structures of each reactant and product found in Reactions 24-7.4, 24-14.2, 24-18.1, 24-25.7, and 24-33.7.

2. Most M—O—M bonds are angular but some are linear, namely, that in $[(NH_3)_5Cr—O—Cr(NH_3)_5]^{4+}$. Why?

3. The densities of the metals Ca, Sc, and Zn are, respectively, 1.54, 3.00, and 7.13 g cm^{-3}. Make a plot of these data along with those given in Table 24-1 for the first transition series, and explain the various features and trends that arise.

4. Dimethyl sulfoxide (DMSO) reacts with $Co(ClO_4)_2$ in absolute ethanol to form a pink product that is a 1:2 electrolyte, and that has a magnetic moment of 4.9 BM. The compound $CoCl_2$, however, reacts with DMSO to form a dark blue product with a magnetic moment (per Co) of 4.6 BM. The latter is a 1:1 electrolyte that has an empirical formula of $Co(dmso)_3Cl_2$. Suggest a formula and a structure for each compound.

5. $Mn(acac)_3$ has axial Mn—O bond lengths (~ 1.94 Å) that are shorter than the *equatorial* ones (~ 2.00 Å). Explain.

6. Write balanced chemical equations for
   (a) Reaction of the aqua ion of Co^{2+} with the disodium salt of EDTA.
   (b) Addition of sodium bicarbonate to aqueous Fe^{2+}.
   (c) Reduction of Mn^{3+} by water.
   (d) Air oxidation of $Fe^{2+}(aq)$.
   (e) Hydrolysis of $TiCl_4$.
   (f) Oxidation of Ti^{3+} by H_2O_2.
   (g) Dissolution of the acidic V_2O_5 into aqueous NaOH.
   (h) Dissolution of the hydrous oxide $VO(OH)_2$ in aqueous HNO_3.
   (i) Burning of Cr in air.
   (j) A preparation of CrO_3.

(k) Hydrolysis of CrO_2Cl_2.

(l) Oxidation of aqueous Mn^{2+} with PbO_2.

(m) A preparation of $[Co(NH_3)_6]^{3+}$.

(n) Addition of I^- to aqueous Cu^{2+}.

(o) Reaction of aqueous Cu^{2+} with cyanide.

7. How is the preferred tetrahedral coordination obtained in complexes with an apparent nontetrahedral stoichiometry, such as K_2CuCl_3?

8. Predict the number of unpaired electrons in $[Fe(H_2O)_6]^{2+}$ and $[Fe(CN)_6]^{4-}$. Explain your reasoning.

9. Draw the structures of cis-$[VO_2Cl_4]^{3-}$ and cis-$[VO_2(ox)_2]^{3-}$.

10. Draw the structures of $[TiCl_4(OPCl_3)]_2$, $[TiCl_4(CH_3CO_2C_2H_5)]_2$, and $TiCl_4(OPCl_3)_2$. Classify each atom in the structures as AB_xE_y, and choose a hybridization for each nonterminal, nonmetal atom.

C. Problems from the Literature of Inorganic Chemistry

1. Consider the five-coordinate nickel(II) complex studied by K. N. Raymond, P. W. R. Corfield, and J. A. Ibers, *Inorg. Chem.*, **1968**, *7*, 1362–1372.

   (a) What geometries are reported for the $[Ni(CN)_5]^{3-}$ ion?

   (b) What hybridization should be chosen for each Ni^{II} ion? What crystal field diagrams should be drawn for each?

   (c) How big an energy difference is there, apparently, between these two coordination geometries?

   (d) What minimum sequence of atomic motions would be required to convert one geometry into the other?

   (e) Write balanced equations for the synthesis of this compound from cobalt(II) chloride, ethylenediamine, $[Ni(CN)_4]^{2-}$(aq), and KCN.

2. Titanium(IV) compounds are discussed in the article by T. J. Kistenmacher and G. D. Stucky, *Inorg. Chem.*, **1971**, *10*, 122–132.

   (a) Write balanced equations for the syntheses of $[PCl_4]_2[Ti_2Cl_{10}]$ and $[PCl_4][Ti_2Cl_9]$, as performed in this work.

   (b) Discuss the tendency for Ti^{IV} to be octahedrally coordinated, as illustrated by these two compounds.

   (c) Explain the two reactions from the viewpoint of chloride ion transfer.

   (d) How and why does the solvent (here either $SOCl_2$ or $POCl_3$) influence the formation of $[Ti_2Cl_{10}]^{2-}$ instead of $[Ti_2Cl_9]^-$?

3. Pentacoordinated copper(II) ions were reported by K. N. Raymond, D. W. Meek, and J. A. Ibers, *Inorg. Chem.*, **1968**, *7*, 1111–1117.

   (a) What geometry is reported for $[CuCl_5]^{3-}$ in this compound?

   (b) Compare the geometries and the crystal field diagrams of $[Ni(CN)_5]^{3-}$, $[CuCl_5]^{3-}$, and $[MnCl_5]^{2-}$.

   (c) Why are the axial Cu—Cl bond lengths in $[CuCl_5]^{3-}$ shorter than the equatorial ones?

4. Consider isomerism among nickel complexes as reported by R. G. Hayter and F. S. Humiec, *Inorg. Chem.*, **1965**, *4*, 1701–1706.

   (a) For the complexes $NiX_2(PRPh_2)_2$, state the trend that is observed for isomer preference when $X = Cl^-$, Br^-, or I^-. When does the system prefer square planar or tetrahedral geometry?

   (b) Which geometry should lead to paramagnetism and which should lead to diamagnetism? Explain with crystal field diagrams.

(c) How is isomer preference related to ligand field strength in the series of complexes with ligands $P(C_2H_5)_3$, $P(C_2H_5)_2C_6H_5$, $PC_2H_5(C_6H_5)_2$, and $P(C_6H_5)_3$ and for the series of complexes with ligands SCN^-, Cl^-, Br^-, and I^-?

5. Consider the adduct of oxovanadium(IV) dichloride as reported by J. E. Drake, J. Vekris, and J. S. Wood, *J. Chem. Soc. (A)*, **1968**, 1000–1005.

(a) Write an equation for the synthesis in ammonia of the title compound.

(b) What is the significance of the magnetic susceptibility ($\mu_{eff} = 1.74$ BM) found for this compound?

(c) Describe the V—O multiple bond by showing orbital-overlap diagrams.

(d) How strong is the V—O π bond as judged by the V—O distance?

(e) Why, according to the authors, is the coordination geometry around this oxovanadium(IV) compound a trigonal bipyramid and not the usual square pyramid?

6. Consider the redox reactions reported by A. J. Miralles, R. E. Armstrong, and A. Haim, *J. Am. Chem. Soc.*, **1977**, *99*, 1416–1420.

(a) Prepare crystal field diagrams (with electrons properly configured) for $[Ru(NH_3)_5py]^{3+}$, $[Co(NH_3)_5py]^{3+}$, $[Fe(CN)_6]^{4-}$, $[Fe(CN)_6]^{3-}$, and $[Ru(NH_3)_6]^{2+}$.

(b) How were these reactions shown to proceed via outer-sphere electron transfer mechanisms?

7. Consider the anation reactions studied by W. R. Muir and C. H. Langford, *Inorg. Chem.*, **1968**, *7*, 1032–1043.

(a) Why should exchange of dmso–ligand with DMSO-solvent be more rapid than anation? Explain by drawing the solvated activated complex along an I_d reaction pathway, and consider the probability of solvent versus anion entry into the first-coordination sphere.

(b) What evidence favoring an I_d mechanism for anation do the authors report?

(c) What evidence is cited in opposition to an associative mechanism?

SUPPLEMENTARY READING

Basolo, F., Bunnett, J. F., and Halpern, J. Eds., *Collected Accounts of Transition Metal Chemistry*, American Chemical Society, Washington, DC, 1973.

Clark, R. J. H., *The Chemistry of Titanium and Vanadium*, Elsevier, Amsterdam, 1968.

Colton, R. and Canterford, J. H., *Halides of the First Row Transition Metals*, Wiley, New York, 1969.

Cotton, F. A. and Wilkinson, G., *Advanced Inorganic Chemitsry*, 5th ed., Wiley-Interscience, New York, 1988.

Cotton, S. A., "Some Aspects of the Coordination Chemistry of Iron(III)," *Coord. Chem. Rev.*, **1972**, *8*, 185.

Hatfield, W. E. and Whyman, R., "Coordination Chemistry of Copper," *Transition Metal Chem.*, **1969**, *5*, 47.

Hathaway, B. J. and Billing, D. E., "The Electronic Properties and Stereochemistry of Mono-nuclear Complexes of the Copper(II) Ion,"

Coord. Chem. Rev., **1970**, *5*, 143.

Jardine, F. H., "Copper(I) Complexes," *Adv. Inorg. Chem. Radiochem.*, **1975**, *17*, 115.

Kepert, D. L., *The Early Transition Metals*, Academic, New York, 1972.

Levason, W. and McAuliffe, C. A., "Higher Oxidation State Chemistry of

Manganese," *Coord. Chem. Rev.,* **1972,** *7,* 353.

Levason, W. and McAuliffe, C. A., "Higher Oxidation State Chemistry of Iron, Cobalt, and Nickel," *Coord. Chem. Rev.,* **1974,** *12,* 151.

Parish, R. V., *The Metallic Elements,* Longman, London, 1977.

Smith, D. W., "Chlorocuprates(II)," *Coord. Chem. Rev.,*
1976, *21,* 93.

Toth, L. E., *Transition Metal Carbides and Nitrides,* Academic, New York, 1971.

Wells, A. F., *Structural Inorganic Chemistry,* 5th ed. Clarendon, Oxford, 1984.

Zordan, T. A. and Hepler, L. G., "Thermochemistry and Oxidation Potentials of Manganese and its Compounds," *Chem. Rev.,* **1968,** *68,* 737.

Chapter 25

THE ELEMENTS OF THE SECOND AND THIRD TRANSITION SERIES

25-1 General Remarks

Some important features of these elements compared with those of the first series are as follows:

Radii. The *radii* of the heavier metals and ions are larger than those of the first series. Because of the lanthanide contraction (Section 8-12 and Table 26-1) the radii of the third series show little difference from those of the second series, despite the increased atomic number and total number of electrons.

Oxidation States.

1. The higher oxidation states are significantly more stable than those of the first series. The oxo anions $[MO_4]^{n-}$ of Mo, W, Tc, Re, Ru, and Os are less readily reduced than those of Cr, Mn, and Fe. Some compounds, such as WCl_6, ReF_7, RuO_4, and PtF_6, have no analogs in the first series. The elements in Groups IVA(4) to VIA(6) prefer their highest oxidation state.
2. The II oxidation state is of relatively little importance except for Ru, Pd, and Pt. For Mo it is important, but is quite different (Mo_2^{4+}) from chromium (Cr^{2+}). Similarly the III oxidation state is relatively unimportant except for Rh, Ir, Ru, and Re.

Metal–Metal Bonding. For the first-row elements, M—M bonding occurs only in metal carbonyls and related compounds (Chapter 28) and in binuclear complexes of chromium(II), for example, $Cr_2(O_2CMe)_4(H_2O)_2$. The heavier elements are more prone to metal-to-metal bonding.

1. There are binuclear species of the type $M_2(O_2CCH_3)_4$, where M = Mo, Rh, and Ru, as well as binuclear halides of Mo, Tc, Re, and Os (e.g., $[Re_2Cl_8]^{2-}$) that have strong multiple M—M bonds.
2. There are halides of Nb, Ta, Mo, W, and Re that are cluster compounds (e.g., $[Ta_6Cl_{12}]^{2+}$ and $[Re_3Cl_{12}]^{3-}$). Some $[Au_{11}]^{3+}$ clusters are also known.

Magnetic Properties. The heavier elements tend to give *low-spin* compounds. Ions with an even number of electrons are often diamagnetic. Even where there is an odd number of d electrons, there is frequently only one unpaired electron. The simple interpretation of magnetic moments that is usually possible for first-row paramagnetic species can seldom be made because of complications due to spin-orbit coupling. The spin-pairing can be attributed to the greater spatial extension of the $4d$ and $5d$ orbitals. Double occupancy of an orbital produces less interelectronic repulsion than in the smaller $3d$ orbitals. The electronic absorption spectra are also more difficult to interpret in general. A given set of ligands produces splittings in the order $5d > 4d > 3d$.

Stereochemistries. For the early members of the second and third rows especially, higher coordination numbers of seven and eight are more common than in the first-row elements. However, for the platinum group metals, the maximum coordination number, with few exceptions, is six.

ZIRCONIUM AND HAFNIUM

25-2 General Remarks: The Elements

The atomic and ionic radii of Zr and Hf are virtually identical. Hence, their chemistry is remarkably similar. There are usually only small differences, for example, in solubilities or volatilities of compounds.

The significant differences from Ti are:

1. There are few compounds in oxidation states below IV.

2. The +4 ions have a high charge, there is no partly filled d shell that might give stereochemical preferences, and they are relatively large (0.74 and 0.75 Å). Thus they usually have coordination numbers in compounds and complexes of seven and eight rather than six, and various coordination polyhedra are formed, especially with O and F ligands. For example, we have

ZrF_6^{2-}	Octahedron in Li_2ZrF_6 and $CuZrF_6 \cdot 4H_2O$
ZrF_7^{3-}	Pentagonal bipyramid, in Na_3ZrF_7
ZrF_7^{3-}	Capped trigonal prism in $(NH_4)_3ZrF_7$
$Zr_2F_{12}^{4-}$	Pentagonal bipyramids sharing an edge in $K_2CuZr_2F_{12} \cdot 6H_2O$
ZrF_8^{4-}	Square antiprism, in $Cu_2ZrF_8 \cdot 12 H_2O$
$Zr_2F_{14}^{6-}$	Square antiprisms sharing an edge in $Cu_3Zr_2F_{14} \cdot 16H_2O$

3. Though there are a few compounds of Zr^{III}, particularly the halides, this is not nearly so important an oxidation state as for titanium.

Zirconium occurs as *baddeleyite* (a form of ZrO_2) and *zircon* ($ZrSiO_4$). Hafnium always accompanies Zr to the extent of fractions of the percentage of Zr. Separations are difficult, but solvent extraction or ion-exchange procedures are effective. The metals are made by the Kroll process (Section 24-5). They are similar to Ti both physically (being hard, resistant, and stainless steel-like in appearance) and chemically (being readily attacked only by HF, to give fluoro complexes).

COMPOUNDS OF ZIRCONIUM(IV), d^0

Since the chemistries of Zr and Hf are so similar, we shall refer only to Zr. All the compounds are white unless the anion is colored.

25-3 Binary Compounds

The *oxide* ZrO_2 is made by heating the *hydrous oxide,* which is precipitated on addition of OH^- to Zr^{IV} solutions. Zirconium dioxide is very refractory (mp 2700 °C) and exceptionally resistant to attack. It is used for crucibles, furnace linings, and so on.

The *tetrachloride* is made by direct interaction or by action of Cl_2 on a mixture of ZrO_2 and C. It is tetrahedral in the vapor phase but in crystalline form there are chains of $ZrCl_6$ octahedra.

Like $TiCl_4$, $ZrCl_4$ is a Lewis acid and forms adducts with donors such as Cl^-, $POCl_3$, and ethers. It is only partially hydrolyzed by water at room temperature to give the stable oxide chloride, $ZrOCl_2$.

25-4 Aqueous Chemistry and Complexes

Zirconium dioxide is more basic than titanium dioxide and is virtually insoluble in an excess of base. There is a more extensive aqueous chemistry of zirconium because of a lower tendency toward complete hydrolysis. Nevertheless, it is doubtful whether Zr^{4+} aqua ions exist even in strongly acid solutions. The hydrolyzed ion is often referred to as the "zirconyl" ion (ZrO^{2+}), but $Zr=O$ bonds do *not* exist. The complex $ZrOCl_2 \cdot 8H_2O$, which crystallizes from dilute HCl solutions, contains the ion $[Zr_4(OH)_8(H_2O)_{16}]^{8+}$. Here the Zr atoms lie in a distorted square, linked by pairs of hydroxo bridges, and also bound to four H_2O molecules so that each Zr atom is coordinated by eight oxygen atoms in a distorted dodecahedron.

Like Ce^{4+} and other +4 ions, Zr has an *iodate* insoluble in 6 M HNO_3. In fairly concentrated HF solutions, *only* $[ZrF_6]^{2-}$ is present in solution. The salts that crystallize *from* these solutions may contain $[ZrF_7]^{3-}$, $[ZrF_8]^{4-}$, and binuclear anions as listed in Section 25-2.

Other eight-coordinate complexes are the carboxylate, $Zr(O_2CR)_4$, the acetylacetonate, $Zr(acac)_4$, the oxalate, $Na_4[Zr(ox)_4]$, and the nitrate, $Zr(NO_3)_4$.

NIOBIUM AND TANTALUM
25-5 The Elements

Niobium and tantalum, though metallic, have chemistries in the V oxidation state that are similar to those of typical nonmetals. They have virtually no cationic chemistry but form numerous anionic complexes, most of which have coordination numbers of seven and eight. In their lower oxidation states they form many metal–atom cluster compounds. Only niobium forms lower states in aqueous solution.

Niobium is 10–12 times more abundant than Ta. The *columbite–tantalite* series of minerals have the general composition $(Fe/Mn)(Nb/Ta)_2O_6$, with variable ratios of (Fe/Mn) and (Nb/Ta). Niobium is also obtained from *pyrochlore,* a mixed calcium–sodium niobate. Separation and production of the metals is complex. Both metals are bright, high melting, and resistant to acids. They dissolve rapidly in an HNO_3–HF mixture, and very slowly in fused NaOH.

NIOBIUM AND TANTALUM COMPOUNDS

25-6   The Chemistry of Niobium and Tantalum (V), d^0

Binary Compounds

The *oxides* (M_2O_5) are dense white powders, commonly made by ignition of other Nb or Ta compounds in air. Addition of OH^- to halide solutions gives the gelatinous hydrous oxides. The oxides are scarcely attacked by acids other than HF but are dissolved by fused $NaHSO_4$ or NaOH. Alkali fusion gives oxo anions that are stable in aqueous solution only at high pH.

Halides and Their Complexes

The pentafluorides are volatile white solids obtained by direct interaction. These complexes have a tetrameric structure [Fig. 25-1(*a*)] that is characteristic for other pentafluorides. The *pentachlorides* are yellow solids obtained by direct interaction. They are hydrolyzed to the hydrous oxide. In the crystalline form and in solvents like CCl_4 they are dimeric [Fig. 25-1(*b*)]. Both halides abstract oxygen from donors like $(CH_3)_2SO$ or ether on heating to give oxochlorides $(MOCl_3)$. Niobium pentachloride, Nb_2Cl_{10}, is reduced by amines to give Nb^{IV} complexes e.g., $NbCl_4(py)_2$.

The halides are Lewis acids and give adducts with neutral donors and complex anions with halide ions.

The *fluoride solutions* contain $[NbOF_5]^{2-}$, $[NbF_6]^-$, and $[TaF_6]^-$ plus $[TaF_7]^{2-}$. However, from these solutions salts of different stoichiometry can be obtained: $[NbOF_6]^{3-}$, $[NbF_7]^{2-}$, and $[TaF_8]^{3-}$.

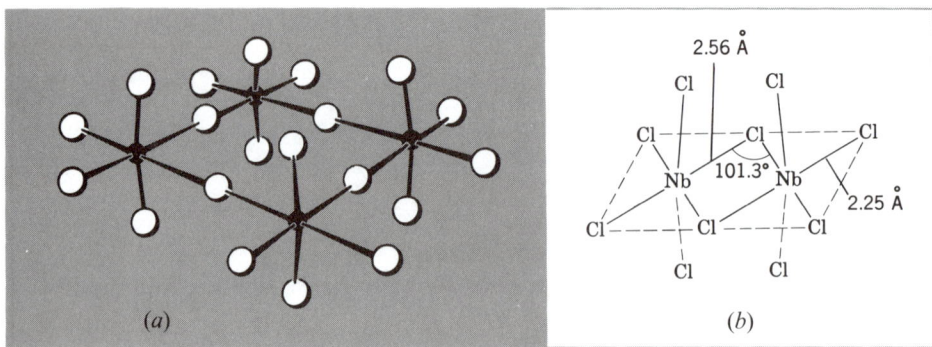

Figure 25-1   (*a*) The tetrameric structures of NbF_5 and TaF_5 (also MoF_5 and, with slight distortion, RuF_5 and OsF_5). The Nb—F bond lengths: 2.06 Å (bridging), 1.77 Å (nonbridging). [Adapted by permission from A. J. Edwards, *J. Chem. Soc.,* **1964**, 3714.] (*b*) The dinuclear structure of crystalline Nb_2Cl_{10}. The octahedra are distorted.

25-7 Lower Oxidation States of Niobium and Tantalum

The tetrahalides of niobium and tantalum are well known and readily form adducts, often with high coordination numbers. For example, with phosphines the following types of complexes are known:

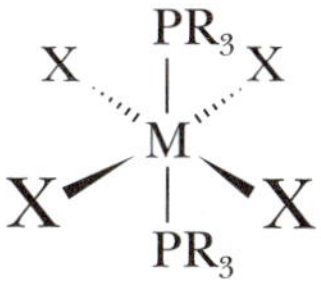

25-I

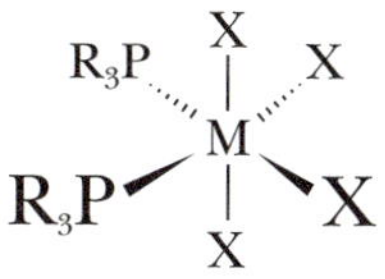

25-II

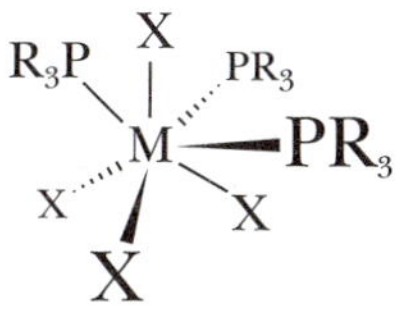

25-III

Coordination numbers of six (Structures 25-I and 25-II), seven (Structure 25-III), and eight (Structure 25-IV) occur frequently in complexes of Nb^{IV} and Ta^{IV}.

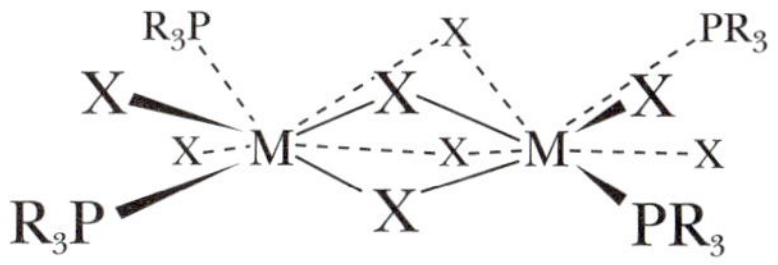

25-IV

The trivalent elements form some mononuclear octahedral complexes, but also form very stable dinuclear complexes containing metal–metal double bonds. Two of the important and typical ones, which involve edge-sharing and face-sharing octahedra are shown in Structures 25-V and 25-VI, respectively.

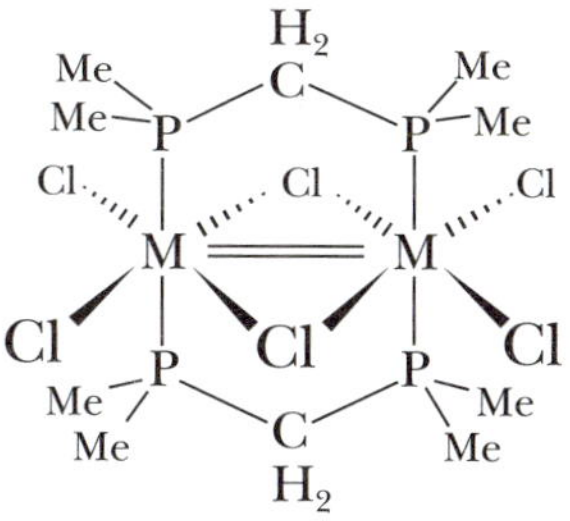

25-V

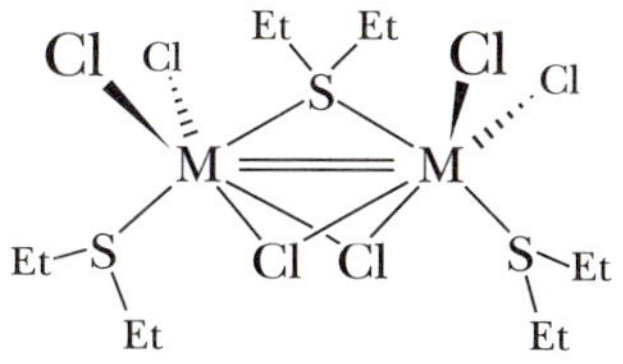

25-VI

There are also cluster halides that have stoichiometries $MX_{2.33}$ or $MX_{2.5}$. They contain the $[M_6X_{12}]^{n+}$ unit shown in Fig. 25-2. This has an octahedron of metal atoms with halogen bridges. In aqueous solution, redox reactions occur.

$$\underset{\text{Diamagnetic}}{[M_6Cl_{12}]^{2+}} \underset{}{\overset{-e^-}{\rightleftharpoons}} \underset{\substack{\text{1 Unpaired}\\ \text{electron}}}{[M_6Cl_{12}]^{3+}} \overset{-e^-}{\rightleftharpoons} \underset{\text{Diamagnetic}}{[M_6Cl_{12}]^{4+}} \tag{25-7.1}$$

Salts of these cations can be isolated.

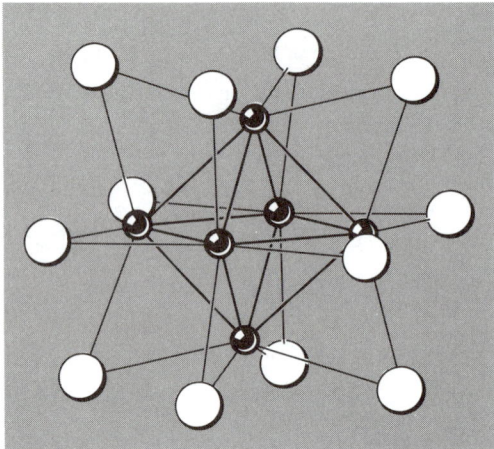

Figure 25-2   The structure of the $[M_6X_{12}]^{n+}$
units found in many halogen compounds of
lower valent niobium and tantalum.
[Reproduced by permission from L. Pauling,
The Nature of the Chemical Bond, 3rd ed.,
Cornell University Press, Ithaca, NY, 1960.
Used by permission of the publisher.]

MOLYBDENUM AND TUNGSTEN

25-8   The Elements

The elements Mo and W do not resemble Cr except in compounds with π-acid
ligands. Thus the +2 state is not well known except in compounds with quadru-
ply bonded Mo_2^{4+} units. The high stability of Cr^{III} in complexes has no counterpart in Mo or W chemistry. For Mo and W, the higher oxidation states are more
common and more stable against reduction.

Both Mo and W have a great range of stereochemistries in addition to the
variety of oxidation states, and their chemistry is among the most complex of the
transition elements.

Molybdenum occurs chiefly as molybdenite (MoS_2). Tungsten is found almost exclusively in tungstates such as $CaWO_4$ (*scheelite*) or $Fe(Mn)WO_4$ (*wol-
framite*).

Molybdenum is roasted to the oxide MoO_3. Tungsten is recovered after alkaline fusion and dissolution in water by precipitation of WO_3 with acids. The oxides are reduced with H_2 to give the metals as gray powders. These are readily attacked only by HF–HNO_3 mixtures or by oxidizing alkaline fusions with Na_2O_2
or KNO_3–NaOH.

The chief use of both metals is in alloy steels; even small amounts cause
tremendous increases in hardness and strength. "High-speed" steels, which are
used to make cutting tools that remain hard even at red heat, contain W and Cr.
Tungsten is also used for lamp filaments. The elements give hard, refractory, and
chemically inert interstitial compounds with B, C, N, or Si on direct reaction at
high temperatures. Tungsten carbide is used for tipping cutting tools, and the
like.

Molybdenum is used in oxide and other systems as a catalyst for a variety of reactions, one example being the "ammonoxidation" synthesis of acrylonitrile:

$$CH_2{=}CHCH_3 + NH_3 + \tfrac{3}{2} O_2 \longrightarrow CH_2{=}CHCN + 3\ H_2O \qquad (25\text{-}8.1)$$

Molybdenum is present in some enzymes, notably those that reduce N_2.

MOLYBDENUM AND TUNGSTEN COMPOUNDS

25-9 Oxides and Oxoanions

The *trioxides* are obtained on heating the metal or other compounds in air. Molybdenum trioxide (MoO_3) is white and tungsten trioxide (WO_3) is yellow. They are not attacked by acids other than HF but dissolve in bases to form molybdates or tungstates. Alkali metal or NH_4^+ salts that are water soluble contain the tetrahedral ions MoO_4^{2-} and WO_4^{2-}. Most other cations give insoluble salts; $PbMoO_4$ can be used for the gravimetric determination of Mo.

When solutions of molybdates or tungstates are made weakly acid, condensation occurs giving complicated polyanions. In more strongly acid solutions the *hydrated oxides*, $MoO_3{\cdot}2H_2O$ (yellow) and $WO_3{\cdot}2H_2O$ (white), are formed. These contain MoO_6 octahedra sharing corners.

Unlike chromates that are powerful oxidants (Section 24-19), the Mo and W anions are weak oxidants.

25-10 Halides

Interaction of Mo or W with F_2 gives the colorless *hexafluorides* MoF_6 (bp 35 °C) and WF_6 (bp 17 °C). Both are readily hydrolyzed.

Chlorination of hot Mo gives only the dimeric pentachloride (Mo_2Cl_{10}), which is a dark red solid with a structure in the crystal very similar to that of Nb_2Cl_{10} [Fig. 25-1(*b*)].

The compound Mo_2Cl_{10} is soluble in benzene and in polar organic solvents. It is monomeric in solution and is presumably solvated. It readily abstracts oxygen from oxygenated solvents to give the oxo species and is rapidly hydrolyzed by water. The preparation of other Mo chlorides is shown in Fig. 25-3.

Chlorination of hot W gives the dark blue-black monomeric *hexachloride* WCl_6. It is soluble in CS_2, CCl_4, alcohol, and ether. It reacts slowly with cold and rapidly with hot water to give tungstic acid. Both Mo_2Cl_{10} and WCl_6 are the usual starting materials for synthesis of a variety of compounds such as dialkylamides, alkoxides, organometallics, and carbonyls.

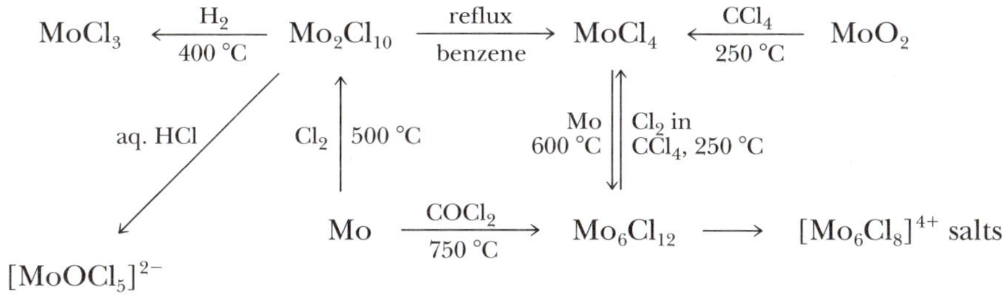

Figure 25-3 The various preparations of molybdenum chlorides.

The so-called "dihalides" (M_6Cl_{12}) contain $[M_6Cl_8]^{4+}$ clusters (Fig. 25-4) similar to those of Nb and Ta, but with 8 face-bridging rather than 12 edge-bridging chlorine atoms. The Mo clusters differ in that they do not undergo reversible oxidation, but W_6Cl_{12} can be oxidized by Cl_2 at high temperatures. The $[M_6X_8]^{4+}$ units can coordinate six electron-pair donors, one to each metal atom along a fourfold axis of the octahedron. Thus, in molybdenum dichloride, the $(Mo_6Cl_8)^{4+}$ units are connected by bridging Cl atoms (four per unit) and there are nonbridging Cl atoms in the remaining two coordination positions.

The bridging groups in the $[M_6X_8]^{4+}$ units can undergo replacement reaction only slowly, whereas the six outer ligands are labile. Thus mixed halides such as $Mo_6Cl_8Br_4$ and complexes such as $[Mo_6Cl_8(Me_2SO)_6]^{4+}$ and $Mo_6Cl_8Cl_4(PPh_3)_2$ can be made.

In aqueous solution $[M_6X_8]^{4+}$ units are unstable to strongly nucleophilic groups such as OH^-, CN^-, or SH^-.

25-11 Complexes of Molybdenum and Tungsten

There are very many complexes of all types in oxidation states from II to VI.

Mo^{II} Species with Mo—Mo Quadruple Bonds

Interaction of $Mo(CO)_6$ (Section 28-8) with carboxylic acids gives dimers, $Mo_2(CO_2R)_4$, that have the same tetrabridged structure as $Cr_2(O_2CR)_4$ (Section 24-16). Although $Cr_2(O_2CCH_3)_4$ with HCl gives only Cr^{2+}, the Mo—Mo bond is *much* stronger and persists giving chloro complexes with quadruple Mo—Mo bonds

$$Mo_2^{II}(O_2CCH_3)_4 \xrightarrow{\text{HCl(aq)}} [Mo_2^{II}Cl_8]^{4-} \xrightarrow{\text{4 L}} Mo_2Cl_4L_4 \quad (25\text{-}11.1)$$

where L is almost any neutral ligand.

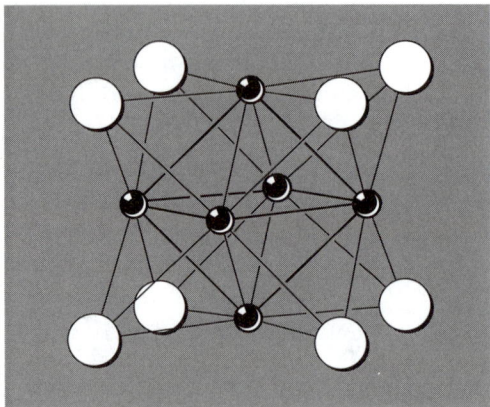

Figure 25-4   The structural unit $M_6X_8^{4+}$, as found frequently in halide cluster compounds.

Oxo Complexes

The most accessible complex used for the synthesis of other complexes is the emerald green pentachlorooxomolybdate(V) ion, $[MoOCl_5]^{2-}$. This is obtained by reduction of $[MoO_4]^{2-}$ in HCl solutions or by dissolving Mo_2Cl_{10} in concentrated aqueous HCl. Paramagnetic salts such as $K_2[MoOCl_5]$ can be isolated. On addition of NaOH to acid solutions, equilibria involving dimeric species occur.

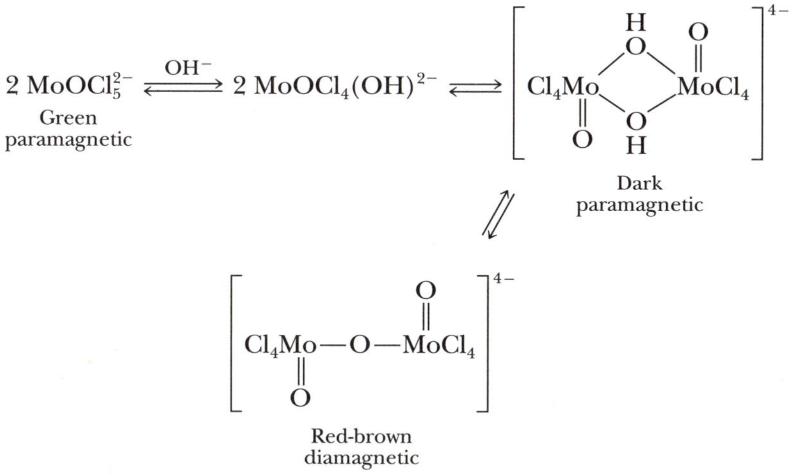

$$2\ MoOCl_5^{2-} \xrightleftharpoons{OH^-} 2\ MoOCl_4(OH)^{2-} \rightleftharpoons \left[\begin{array}{c} Cl_4Mo \underset{O}{\overset{H\,O\quad O}{\diamond}} MoCl_4 \\ \end{array} \right]^{4-}$$

Green
paramagnetic

Dark
paramagnetic

$$\left[\begin{array}{c} O \\ \| \\ Cl_4Mo{-}O{-}MoCl_4 \\ \| \\ O \end{array} \right]^{4-}$$

(25-11.2)

Red-brown
diamagnetic

The red-brown species represents a common type of Mo^V oxo species. These have an Mo_2O_3 unit with either a linear or a bent Mo—O—Mo bridge. Other types have dioxo or disulfur bridges.

In view of the interest in models for enzyme systems such as xanthine oxidase and sulfide oxidase, complexes with amino acids and organosulfur ligands have been studied. Two examples are the xanthate $Mo_2^V O_3(S_2COC_2H_5)_4$ (Structure 25-VII) and the oxalate complex, $[Mo_2^V O_4(C_2O_4)_2(H_2O)_2]^{2-}$ (Structure 25-VIII).

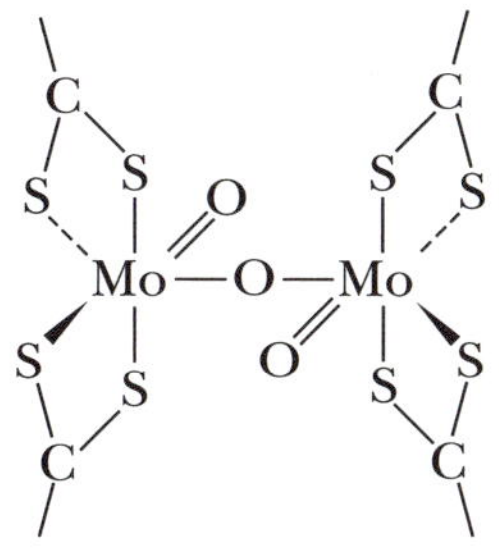

25-VII

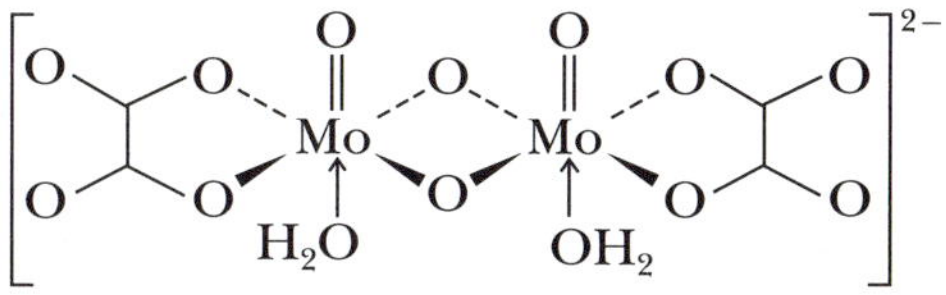

25-VIII

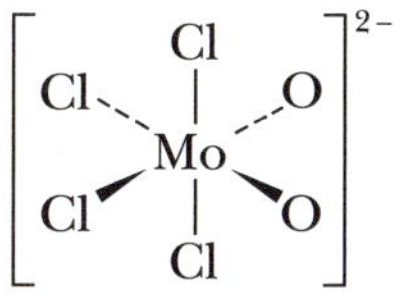

25-IX

Molybdenum(VI) commonly forms *dioxo* species in which the two Mo=O bonds are cis. Thus MoO_3 in 12 M HCl gives $[MoO_2Cl_4]^{2-}$ (Structure 25-IX).

Tungsten does not form a comparable variety of oxo complexes, although a few are known.

TECHNETIUM AND RHENIUM

25-12 The Elements

Technetium and rhenium differ considerably from Mn, the first-row element.

1. There is no analog of $Mn^{2+}(aq)$ and only a very few complexes are known in the II oxidation state.
2. There is little cationic chemistry in any oxidation state even in complexes.
3. Both elements have an extensive chemistry in the IV and V oxidation states, and especially as oxo compounds in the V oxidation state.
4. The oxo anions MO_4^- are much weaker oxidants than permanganate.
5. The formation of clusters and metal-to-metal bonds is a feature of the chemistry in the II to IV oxidation states.

Rhenium is recovered from flue dusts in the roasting of MoS_2 ores and from residues in the smelting of some Cu ores. It is usually left in solutions as the perrhenate ion, ReO_4^-. After concentration, the addition of KCl precipitates the sparingly soluble salt, $KReO_4$.

All isotopes of technetium are radioactive. ^{99}Tc (2.12×10^5 year) is recovered in kilogram quantities from fission product wastes. There may be more ^{99}Tc in existence on the earth than Re.

The metals are obtained by H_2 reduction of the oxides or the $(NH_4)MO_4$ compounds. They are very high melting and unreactive at room temperature, but they burn in O_2 at 400 °C to give the volatile oxides M_2O_7. They dissolve to give the oxo acid in warm aqueous Br_2 or hot HNO_3. Rhenium dissolves in 30% H_2O_2.

Rhenium is used mainly in a Pt—Re alloy supported on alumina for catalytic re-forming of petroleums. Technetium, because of its radioactivity, is used for radiographic scanning of the liver, the heart, and other organs.

RHENIUM COMPOUNDS

For present purposes, technetium compounds can be assumed to be similar to those of Re.

25-13 Binary Compounds

The yellow volatile oxide (Re_2O_7) is very hygroscopic and dissolves in water, from which the oxide hydrate, $O_3ReOReO_3(H_2O)_2$, can be obtained by evaporation. In NaOH, the perrhenate ion (ReO_4^-) is formed.

Saturation of HCl or H_2SO_4 solutions of ReO_4^- with H_2S gives the black *sulfide* Re_2S_7. This procedure is used for recovery of Re from residues.

The only *halides* in the VI- and VII-oxidation states are the volatile ReF_6 and ReF_7. Chlorination of Re at about 550 °C gives dark red-brown Re_2Cl_{10}. It is a dimer like Mo_2Cl_{10} or Ta_2Cl_{10}. On heating, the liquid decomposes to give the *trichloride*. This is a cluster compound whose structure is shown in Fig. 25-5. The Re_3Cl_9 units are linked into a polymer by sharing of Cl atoms. This unit is extremely stable, persists in the vapor at 600 °C, and forms the structural basis for much of Re^{III} chemistry.

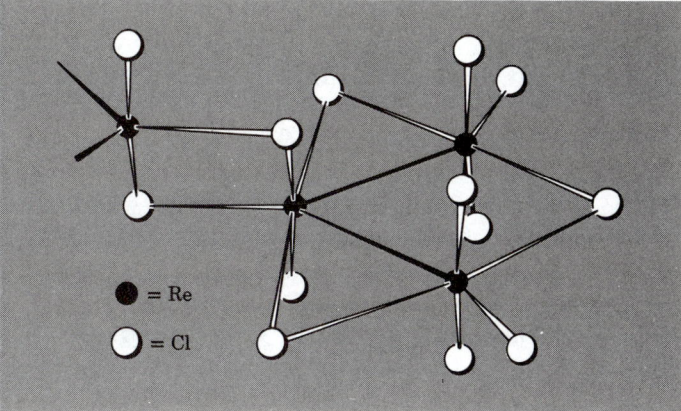

Figure 25-5   The cluster structure of Re_3Cl_9.

25-14 Oxo Compounds and Complexes

As with Mo and W, oxo compounds are important especially in the V and VII oxidation states.

The salts of the *perrhenate* ion (ReO_4^-) have solubilities similar to perchlorates, but salts of TcO_4^- are more soluble than either. An insoluble perrhenate suitable for gravimetric analysis, $(Ph_4As)(ReO_4)$, is given by the tetraphenylarsonium ion.

The ions are stable in water and are weak oxidants. In HCl solution ReO_4^- is reduced by hypophosphite, partially to the chloro complex ion, $[Re^{IV}Cl_6]^{2-}$, which forms stable salts such as K_2ReCl_6, and partially to the $[Re_2Cl_8]^{2-}$ ion, which is isoelectronic with the $[Mo_2Cl_8]^{4-}$ ion and contains a quadruple bond between the metal atoms.

Oxo Complexes

The oxo complexes are numerous, as with Mo. Re_2Cl_{10} dissolves in aqueous strong HCl to give $[ReOCl_5]^{2-}$. Oxo species may have the groups Re=O, Re—O—Re and *trans* O=Re=O (Mo^{VI} has cis dioxo groups) as well as linear O=Re—O—Re=O.

There is an extensive chemistry of oxorhenium(V) compounds containing phosphine ligands. The complexes $ReOCl_3(PR_3)_2$ are obtained by interaction of ReO_4^- with PR_3 in ethanol containing HCl. The halide ion (or other ligand) opposite to the Re=O bond is labile; in ethanol, for example, it is rapidly replaced, giving $ReOCl_2(OC_2H_5)(PR_3)_2$.

THE PLATINUM METALS

25-15 General Remarks

Ruthenium, osmium, rhodium, iridium, palladium, and platinum are the six heaviest members of those in Groups VIIIA(8), VIIIA(9), and VIIIA(10), commonly known as the platinum metals. They are rare elements, with platinum being the most common having an abundance of about $10^{-6}\%$, whereas the oth ers have abundances on the order of $10^{-7}\%$. These elements occur as metals, often as alloys such as osmiridium, and in arsenide, sulfide, and other ores. The elements are usually associated not only with one another but also with nickel, copper, silver, and gold.

The compositions of the ores and the extraction methods vary considerably.An important source is South African Ni–Cu sulfide. The ore is concentrated by gravitation and flotation, after which it is smelted with lime, coke, and sand and is bessemerized in a convertor. The resulting Ni–Cu sulfide "matte" is cast into anodes. On electrolysis in sulfuric acid solution, Cu is deposited at the cathode, and Ni remains in solution, from which it is subsequently recovered by electrodeposition, while the platinum metals, silver, and gold collect in the anode slimes. The subsequent procedures for separation of the elements are very complicated. Although most of the separations used to involve classical precipitations or crystallizations, ion-exchange and solvent-extraction procedures are now used.

The metals are grayish white and are obtained initially as powders by ignition of salts, such as $(NH_4)_2PtCl_6$. Almost all compounds of these elements give the metal when heated. However, Os is readily oxidized by air to the very volatile oxide OsO_4, while Ru gives RuO_2, so that reduction by hydrogen is necessary to recover the metals.

The metals can also be thrown out from acid solutions by the action of Zn, a common recovery procedure known as "footing."

The metals are chemically inert especially when massive. The elements Ru and Os are best attacked by an alkaline oxidizing fusion, Rh and Ir by HCl + $NaClO_3$ at 125 to 150 °C, and Pd and Pt by concentrated HCl + Cl_2 or aqua regia.

The metals, as gauze or foil and especially on supports such as alumina or charcoal, onto which the metal salts are absorbed and reduced *in situ,* are extensively used as catalysts in industry. One of the biggest uses of Pt is as Pt–Re or Pt–Ge on alumina catalysts in the re-forming or "platforming" of crude petroleum. The Pd and Rh compounds are used in homogeneous catalytic syntheses (Chapter 30). The catalytic "after burners" in use on automobile exhausts use a platinum metal catalyst.

Platinum or its alloys are used in electrical contacts.

Both Pd and Pt are capable of absorbing large volumes of molecular hydrogen, and Pd is used for the purification of H_2 by diffusion, since Pd metal is uniquely permeable to hydrogen.

25-16 General Remarks on the Chemistry of the Platinum Metals

The chemistries of these elements have some common features, but there are wide variations depending on differing stabilities of oxidation states, stereochemistry, and the like. There is little similarity to Fe, Co, and Ni except in some compounds of π-acid ligands such as CO and in stoichiometries of compounds. The important oxidation states are listed in Table 25-1. Some general points are as follows.

Binary Compounds. The halides, oxides, sulfides, and phosphides are not of great importance.

Aqueous Chemistry. This chemistry is almost exclusively that of complex compounds. Aqua ions of Ru^{II}, Ru^{III}, Rh^{III}, Ir^{III}, Pd^{II} and Pt^{II} exist in solutions of non-complexing anions, namely, ClO_4^-, BF_4^-, $CF_3SO_3^-$ or *p*-toluenesulfonate, but are not ordinarily of importance.

A vast array of complex ions, predominantly with halide or nitrogen donor ligands, are water soluble. Exchange and kinetic studies have been made with many of these because of interest in (a) trans effects, especially with square Pt^{II}, (b) differences in substitution mechanisms between the ions of the three transition metal series, and (c) the unusually rapid electron-transfer processes with heavy metal complex ions.

Compounds with π-Acid Ligands.

1. Binary carbonyls are formed by all but Pd and Pt, and the majority of them are polynuclear. Substituted polynuclear carbonyls are known for Pd and Pt, and all six elements give carbonyl halides and a variety of carbonyl complexes containing other ligands.

2. For Ru, nitrosyl (NO) complexes are a common feature of the chemistry, especially in solutions containing nitric acid.

3. There is an extensive chemistry of complexes with tertiary phosphines and phosphites, and to a lesser extent with R_3As and R_2S. Some of these are useful homogeneous catalysts (Chapter 30).

    Mixed complexes of PR_3 with CO, alkenes, halides, and hydride ligands in at least one oxidation state are common for all of the elements.

4. All the elements have a strong tendency to form bonds to carbon, especially with alkenes and alkynes; Pt^{II}, Pt^{IV}, and to a lesser extent Pd^{II} have a strong tendency to form σ bonds, while Pd^{II} very readily forms π-allyl species (Section 29-16).

5. A characteristic feature is the formation of complexes with M—H bonds when the metal halides in higher oxidation states are reduced, especially in the presence of tertiary phosphines or other ligands. Hydrogen abstraction from reaction media such as alcohols or DMF is common.

Oxidation States. The main *oxidation states* are given in Table 25-1.

Table 25-1   Oxidation States of Platinum Metals (Bold Type Shows Main States)

	Ru	Os	Rh	Ir	Pd	Pt
	0	**0**	0	0	**0**	**0**
	1			1	1	
	2	2	2	2	**2**	**2**
	3	3	3	3		3^c
	4	**4**	**4**	**4**	4	**4**
	$5^{a,b}$	$5^{a,b}$	5^a	5^a	5^a	5^a
	$6^{a,b}$	$6^{a,b}$	6^a	6^a	6^a	6^a
	$7^{a,b}$	$7^{a,b}$				
	$8^{a,b}$	$8^{a,b}$				

[a] In fluorides or fluoro complexes.

[b] In oxides or oxo anions.

[c] Usually in binuclear compounds with M—M bonds.

Stereochemistry. The coordination number exceeds six in only a few compounds, for example, $OsH_4(PR_3)_3$ and $IrH_5(Pr_3)_2$. Most complexes in the +3 and +4 oxidation states are octahedral. The d^8 species Rh^I, Ir^I, Pd^{II}, and Pt^{II} normally are square or five coordinate complexes.

The +2 oxidation states for Ru and Os form five or six coordinate complexes.

RUTHENIUM AND OSMIUM

25-17 Oxo Compounds of Ruthenium and Osmium

One of the most characteristic features of the chemistry of Ru and Os is the oxidation by aqueous oxidizing agents to give the volatile *tetraoxides.*

Orange-yellow RuO_4 (mp 25 °C) is formed when acid solutions containing Ru are oxidized by MnO_4^-, Cl_2, or hot $HClO_4$. It can be distilled from the solutions or swept out by a gas stream.

Colorless OsO_4 (mp 40 °C) is more easily obtained and HNO_3 is a sufficiently powerful oxidant. The distillation first of OsO_4 and then of RuO_4 is used in their separation from other platinum metals. The RuO_4 is collected in strong HCl solutions where it is reduced to a mixture of Ru^{III} and Ru^{IV} chloro complexes. The evaporated product is sold as $RuCl_3 \cdot 3H_2O$, the commonest starting material for syntheses of Ru compounds.

The tetraoxides consist of tetrahedral molecules. These compounds are extracted from aqueous solutions by CCl_4. Both are powerful oxidants. OsO_4 is used in organic chemistry since it oxidizes alkenes to cis diols. It is also used for biological staining as organic matter reduces it. Osmium tetraoxide presents an especial hazard to the eyes and must be handled carefully. Ruthenium tetraoxide is much more reactive and can react vigorously with organic matter; it is very toxic.

Dissolution of OsO_4 in base gives a colorless *oxo anion*

$$OsO_4 + 2\ OH^- = [OsO_4(OH)_2]^{2-} \qquad (25\text{-}17.1)$$

which can be reduced to $[OsO_2(OH)_4]^{2-}$.

For Ru, the most important oxo anion is orange $[RuO_4]^{2-}$, obtained by fusing Ru compounds with Na_2O_2 and dissolving the melt in water. The difference in stoichiometry may be due to the greater ability of the $5d$ anion to increase its coordination shell.

Reduction of RuO_4 by HCl in the presence of KCl gives $K_4[Ru_2OCl_{10}]$ as red crystals. This oxo species of $Ru^{IV}(d^4)$ is diamagnetic because the electrons become paired in a MO extending over the *linear* Ru—O—Ru bridge.

25-18 Ruthenium Chloro Complexes and Aqua Ions

The commercial product $RuCl_3 \cdot 3H_2O$, on evaporation with HCl, is reduced to ruthenium(III) chloro complexes. The ion $[RuCl_6]^{3-}$ may be obtained with high concentrations of Cl^-. The rate of replacement of Cl^- by H_2O decreases as the number of Cl^- ions decreases so that whereas the aquation of $[RuCl_6]^{3-}$ to $[RuCl_5(H_2O)]^{2-}$ occurs within seconds in water, the half-reaction time for conversion of $[RuCl(H_2O)_5]^{2-}$ to $[Ru(H_2O)_6]^{3+}$ is about 1 year. Intermediate species such as *trans*-$[RuCl_2(H_2O)_4]^+$ can be isolated by ion exchange procedures.

The Cl⁻ can be removed by $AgBF_4$ and the +3 ion electrolytically reduced to the easily oxidized +2 aqua ion.

$$[Ru(H_2O)_6]^{3+} + e^- = [Ru(H_2O)_6]^{2+} \qquad E° = 0.23 \text{ V} \qquad (25\text{-}18.1)$$

25-19 Ruthenium Amine Complexes

There is an extensive chemistry of Ru with nitrogen ligands. Some of the chemistry is summarized in Fig. 25-6. The $[Ru(NH_3)_5]^{2+}$ group has remarkable π-bonding properties. It forms complexes with CO, RNC, N_2O, and SO_2, and its dinitrogen complex was the first N_2 complex to be made.

25-20 Nitric Oxide Complexes

Both ruthenium and osmium form octahedral complexes (ML_5NO) that have an M—NO group. Depending on the nature of L, they may be cationic, anionic, or neutral. The MNO group can survive many chemical transformations of such complexes.

Ruthenium solutions that have at any time been treated with HNO_3 can, and usually do, contain nitrosyl species that are then difficult to remove. They are readily detected by their IR absorption in the region 1930–1845 cm⁻¹ (Section 28-14).

25-21 Tertiary Phosphine Complexes

Both ruthenium and osmium have an extensive chemistry with these π-acid ligands. Some representative reactions are shown for Ru in Fig. 25-7. The $RuHCl(PPh_3)_3$ and $RuH_2(PPh_3)_3$ complexes are of interest in that they are highly active catalysts for the selective homogeneous hydrogenation of alk-1-enes (Section 30-7).

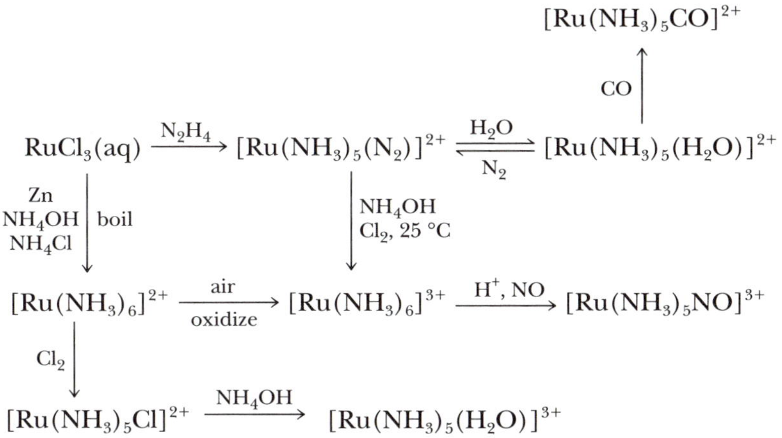

Figure 25-6 Some reactions of ruthenium ammines.

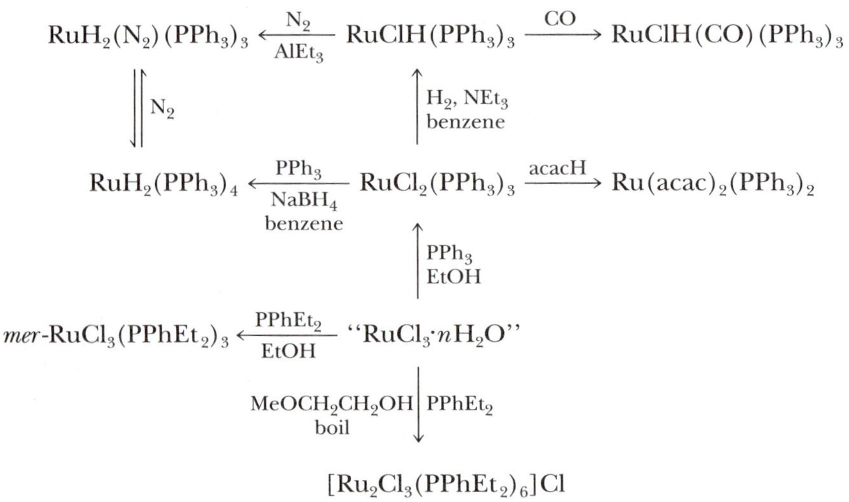

$$\text{RuH}_2(\text{N}_2)(\text{PPh}_3)_3 \xleftarrow[\text{AlEt}_3]{\text{N}_2} \text{RuClH}(\text{PPh}_3)_3 \xrightarrow{\text{CO}} \text{RuClH}(\text{CO})(\text{PPh}_3)_3$$

$$\Big\Updownarrow \text{N}_2 \qquad\qquad \Big\uparrow \begin{array}{l}\text{H}_2,\ \text{NEt}_3 \\ \text{benzene}\end{array}$$

$$\text{RuH}_2(\text{PPh}_3)_4 \xleftarrow[\substack{\text{NaBH}_4 \\ \text{benzene}}]{\text{PPh}_3} \text{RuCl}_2(\text{PPh}_3)_3 \xrightarrow{\text{acacH}} \text{Ru}(\text{acac})_2(\text{PPh}_3)_2$$

$$\Big\uparrow \begin{array}{l}\text{PPh}_3 \\ \text{EtOH}\end{array}$$

$$\textit{mer-}\text{RuCl}_3(\text{PPhEt}_2)_3 \xleftarrow[\text{EtOH}]{\text{PPhEt}_2} \text{``RuCl}_3{\cdot}n\text{H}_2\text{O''}$$

$$\Big\downarrow \begin{array}{l}\text{MeOCH}_2\text{CH}_2\text{OH} \mid \text{PPhEt}_2 \\ \text{boil}\end{array}$$

$$[\text{Ru}_2\text{Cl}_3(\text{PPhEt}_2)_6]\text{Cl}$$

Figure 25-7   Some reactions of tertiary phosphine complexes of ruthenium. Note that the use of different phosphines may lead to different products.

RHODIUM AND IRIDIUM

25-22 Complexes of Rhodium(III) and Iridium(III), d^6

There are many diamagnetic, kinetically inert octahedral complexes similar to those of Co^{III}. They differ from those of Co^{III}, first, in that octahedral halogeno complexes are readily formed, for example, $[\text{RhCl}_5(\text{H}_2\text{O})]^{2-}$ and $[\text{IrCl}_6]^{3-}$. Second, on reduction of the trivalent complexes the divalent complex is not obtained, except under special circumstances for Rh. When the ligands are halogens, amines, or water, reduction gives the metal, or under controlled conditions, a *hydride* complex like $[\text{Rh}(\text{NH}_3)_5\text{H}]\text{SO}_4$; when π-acid ligands are present, reduction to Rh^{I} or Ir^{I}, or to iridium(III) hydrido complexes occurs.

Chlororhodates: The Rh^{III} Aqua Ion

Fusion of Rh with NaCl in Cl_2 followed by dissolution in water and crystallization gives $\text{Na}_3[\text{RhCl}_6]$. Addition of OH^- to this pink ion gives the *hydrous oxide* Rh_2O_3. Dissolution of this in dilute HClO_4 gives $[\text{Rh}(\text{H}_2\text{O})_6]^{3+}$, yellow salts of which can be crystallized.

When Rh_2O_3 is dissolved in HCl and the solutions are evaporated, a dark red deliquescent material ($\text{RhCl}_3{\cdot}n\text{H}_2\text{O}$) is obtained. This is the usual starting material for synthesis of Rh compounds. It is soluble in alcohols as well as water. Fresh solutions do not give AgCl with Ag^+ ion, but on boiling, the red-brown solutions do turn to the yellow of $[\text{Rh}(\text{H}_2\text{O})_6]^{3+}$. Some of its reactions are shown in Fig. 25-8.

25-23 Complexes of Rhodium(IV) and Iridium(IV), d^5

It is very difficult to oxidize Rh^{III} and only a few unstable compounds of Rh^{IV} are known. Octahedral complexes of Ir^{IV} are stable; they have one unpaired electron (t_{2g}^5).

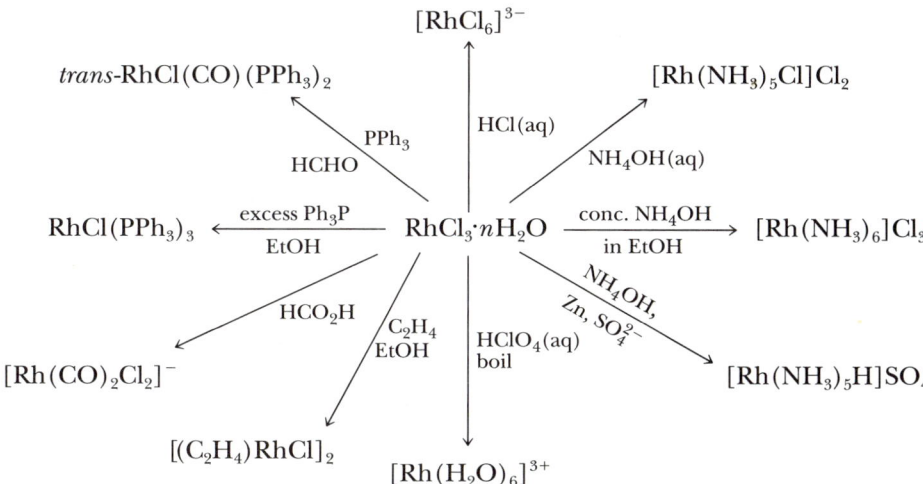

Figure 25-8   Some reactions of rhodium trichloride.

Hexachloroiridates

Hexachloroiridates are made by heating Ir + NaCl in Cl_2. The black salt Na_2IrCl_6 is very soluble in water; the so-called "chloroiridic acid" (cf. chloroplatinic acid Section 25-29) is an oxonium salt $(H_3O)_2IrCl_6 \cdot 4H_2O$. These materials are used to prepare other Ir complexes.

The dark red-brown $[Ir^{IV}Cl_6]^{2-}$ ion is rapidly and quantitatively reduced in strong OH^- solution to give yellow-green $[Ir^{III}Cl_6]^{3-}$:

$$2[IrCl_6]^{2-} + 2\ OH^- \rightleftharpoons 2[IrCl_6]^{3-} + \tfrac{1}{2}O_2 + H_2O \qquad (25\text{-}23.1)$$

The $[IrCl_6]^{2-}$ ion will oxidize many organic compounds, and it is also quantitatively reduced by KI and $[C_2O_4]^{2-}$.

In *acid* solution we have

$$2[IrCl_6]^{2-} + H_2O = 2[IrCl_6]^{3-} + \tfrac{1}{2}O_2 + 2\ H^+$$
$$K = 7 \times 10^{-8}\ atm^{1/2}\ mol^3\ L^{-2}\ (25\ °C) \qquad (25\text{-}23.2)$$

so that in 12 *M* HCl, oxidation of $[Ir^{III}Cl_6]^{3-}$ occurs partially at 25 °C and completely on boiling.

25-24 Complexes of Rhodium(II), d^7

Only a few of these complexes are known, the major ones being the diamagnetic binuclear *carboxylates* that have the common tetrabridged structure. The end positions may be occupied by solvent molecules; with oxygen donors the complexes are green or blue, but with π acids such as $P(C_6H_5)_3$, they are orange red. The carboxylates are made by boiling $RhCl_3(aq)$ with NaO_2CR in methanol. Action of very strong noncomplexing acids gives the Rh_2^{4+} aqua ion that also has a Rh—Rh bond.

25-25 Complexes of Rhodium(I) and Iridium(I), d^8

These square or five-coordinate diamagnetic complexes all have π-acid ligands. They are formed by reduction of Rh^{III} or Ir^{III} in the presence of the ligand. There have been many studies on these complexes because they provide the best systems for the study of the oxidative–addition reaction (Section 30-2) that is a characteristic feature of square d^8 complexes. For $trans$-$IrX(CO)(PR_3)_2$ the equilibria, for example,

$$trans\text{-}Ir^I Cl(CO)(PPh_3)_2 + HCl \rightleftharpoons Ir^{III}Cl_2H(CO)(PPh_3)_2 \qquad (25\text{-}25.1)$$

lie well to the Ir^{III} side and the Ir^{III} complexes can be readily characterized. For Rh, the Rh^{III} complexes are much less stable.

The two yellow compounds, $trans$-*chlorocarbonylbis(triphenylphosphine)rhodium* or -*iridium, i.e. trans*-$MCl(CO)(PPh_3)_2$, are obtained by reducing the halides in alcohols containing PPh_3 by HCHO, which acts as a reductant and source of CO.

Bis[(dicarbonyl)chlororhodium], $[Rh(CO)_2Cl]_2$, is obtained by passing CO saturated with ethanol over $RhCl_3 \cdot 3H_2O$ at about 100 °C, when it sublimes as red needles. It has the structure shown in Fig. 25-9, where the coordination around each Rh atom is planar, and there are bridging chlorides with a marked dihedral angle, along the Cl—Cl line. There appears to be some direct interaction between electrons in rhodium orbitals perpendicular to the planes of coordination.

This carbonyl chloride is a useful source of other rhodium(I) species, and it is cleaved by donor ligands to give cis-dicarbonyl complexes, for example.

$$[Rh(CO)_2Cl]_2 + 2\,L \longrightarrow 2\,RhCl(CO)_2L \qquad (25\text{-}25.2)$$

$$[Rh(CO)_2Cl]_2 + 2\,Cl^- \longrightarrow 2\,[Rh(CO)_2Cl_2]^- \qquad (25\text{-}25.3)$$

$$[Rh(CO)_2Cl]_2 + 2(acac)^- \longrightarrow 2\,Rh(CO)_2(acac) + 2\,Cl^- \quad (25\text{-}25.4)$$

Hydridocarbonyltris(triphenylphosphine)rhodium is a yellow crystalline solid with a trigonal bipyramidal structure with equatorial phosphine groups. It is prepared by the reaction

$$trans - RhCl(CO)(PPh_3)_2 + PPh_3 \xrightarrow[\text{EtOH}]{\text{NaBH}_4} RhH(CO)(PPh_3)_3 \qquad (25\text{-}25.5)$$

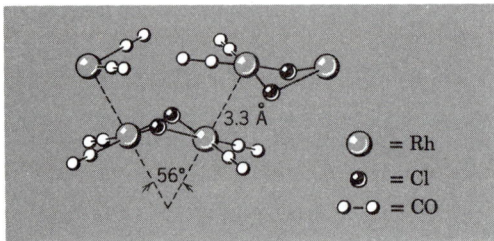

Figure 25-9   The structure of crystalline $[Rh(CO)_2Cl]_2$. The chloride bridges are readily cleaved by nucleophiles.

but it is also formed by the action of $CO + H_2$ under pressure with virtually any rhodium compound in the presence of an excess of PPh_3. Its main importance is as a hydroformylation catalyst for alkenes (Section 30-9).

Chlorotris(triphenylphosphine)rhodium, $RhCl(PPh_3)_3$, is a red-violet crystalline solid formed by reduction of ethanolic solutions of $RhCl_3 \cdot 3H_2O$ with an excess of $P(C_6H_5)_3$. It is a catalyst for hydrogenation of alkenes and other unsaturated substances (Section 30-7). It undergoes many oxidative addition reactions (Section 30-2), and it abstracts CO readily from metal carbonyl complexes and from organic compounds such as acyl chlorides and aldehydes, often at room temperature, to give $RhCl(CO)(PPh_3)_2$.

PALLADIUM AND PLATINUM

25-26 Chlorides

Palladous chloride, $PdCl_2$, is obtained by chlorination of Pd. Above 550 °C an unstable α form is produced, while below 550 °C it is in a β form. There are also α and β forms of $PtCl_2$. The β forms have a molecular structure with M_6Cl_{12} units (Structure 25-X); the stabilization is due to halogen bridges rather than metal–metal bonds. Although the structure of α-$PtCl_2$ is not certain, it differs from that of α-$PdCl_2$, which has a flat chain (Structure 25-XI). In both structures, the metal has the square coordination characteristic of Pd^{II} and Pt^{II}.

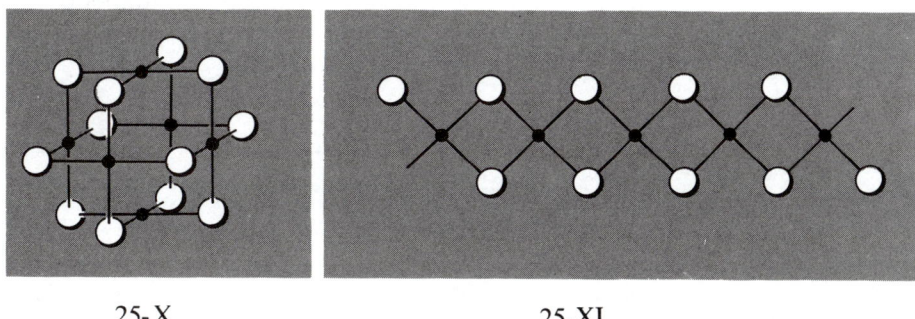

25-X   25-XI

Platinum(IV) chloride, $PtCl_4$, is obtained as red-brown crystals by heating chloroplatinic acid, $(H_3O)_2PtCl_6$, in chlorine. It is soluble in water and in HCl. The analogous chloride of Pd^{IV} does not exist.

25-27 Complexes of Palladium(II) and Platinum(II), d^8

The palladium(II) ion, Pd^{2+}, occurs in PdF_2 and is paramagnetic. However, the *aqua ion*, $[Pd(H_2O)_4]^{2+}$, is spin paired and all Pd and Pt complexes are diamagnetic. Brown deliquescent salts like $[Pd(H_2O)_4](ClO_4)_2$ can be obtained when Pd is dissolved in HNO_3 or PdO in $HClO_4$.

Palladium(II) acetate is obtained as brown crystals when Pd sponge is dissolved in acetic acid containing HNO_3. It is a trimer, $[Pd(CO_2CH_3)_2]_3$. The metal atoms form a triangle with bridging acetate groups. The acetate acts like Pb^{IV} and Hg^{II} acetates (Section 15-6) in attacking aromatic hydrocarbons; such "palladation" reactions are involved in many catalytic processes (cf. Chapter 30).

Palladium(II) and platinum(II) complexes are square or five-coordinate complexes with the formulas ML_4^{2+}, ML_5^{2+}, ML_3X^+, *cis*- and *trans*-ML_2X_2, MX_4^-, and ML_3X_2, where L is a neutral ligand and X a uninegative ion. The palladium complexes are thermodynamically and kinetically less stable than their Pt^{II} analogs. Otherwise the two series of complexes are similar. The kinetic inertness of the Pt^{II} (and also Pt^{IV}) complexes has allowed them to play a very important role in the development of coordination chemistry. Many studies of geometrical isomerism and reaction mechanisms using platinum complexes have had a profound influence on our understanding of complex chemistry (cf. also Cr^{III}, Co^{III}, and Rh^{III}).

There is a preference for amine ligands, halogens, CN^-, tertiary phosphines, and sulfides (R_2S), but little affinity for oxygen ligands and F^-. The concepts of hard and soft acids and bases, or class A and B metals, are clearly shown here (Section 7-9). The strong binding of heavy donor atoms, such as P, is due in part to π bonding.

Many complexes have halide or other bridges, for example,

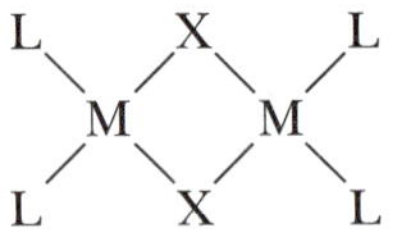

25-XII 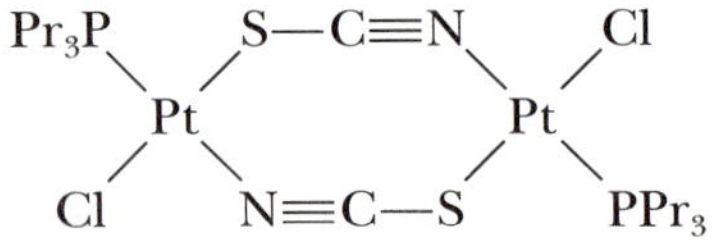   25-XIII 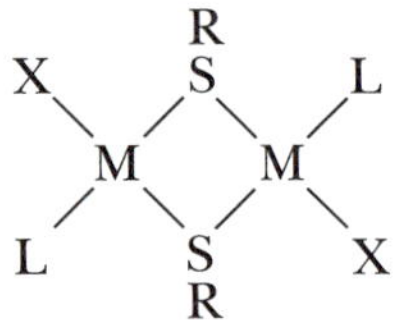   25-XIV

Bridged complexes can be cleaved by donors to give mononuclear species, for example,

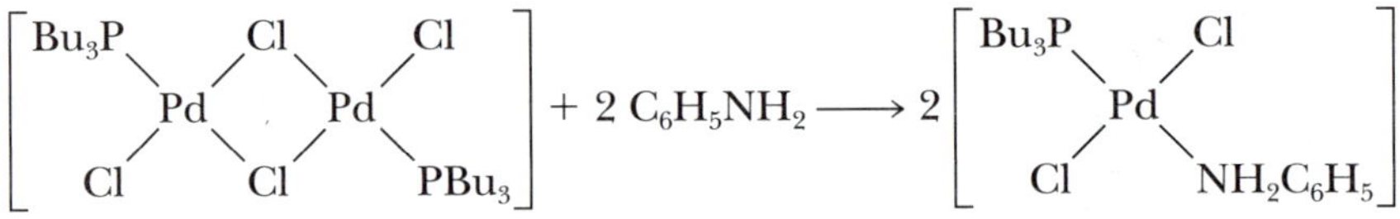

$$(25\text{-}27.1)$$

Salts of the halogeno anions, $[MCl_4]^{2-}$, are common source materials. The yellowish $[PdCl_4]^{2-}$ ion is obtained when $PdCl_2$ is dissolved in HCl. The red $[PtCl_4]^{2-}$ ion is made by reduction of $[PtCl_6]^{2-}$ with oxalic acid or N_2H_5Cl.

25-28 Metal–Metal Interactions in Square Complexes

In crystals the square complexes are often stacked one above the other. Even though the metal-to-metal distances may be too long for true bonding, weak interactions can occur between d orbitals on adjacent metal atoms. An example is $Pt(en)Cl_2$, which is shown in Fig. 25-10(a); others are Ni and Pd dimethylglyoximates.

Salts such as $[Pt(NH_3)_4][PtCl_4]$, $[Pd(NH_3)_4][Pd(SCN)_4]$, or $[Cu(NH_3)_4][PtCl_4]$ also have stacked cations and anions so that there are chains of metal atoms. When *both* metal atoms are Pt^{II}, the crystal is green, although the constituent cations are colorless or pale yellow and the anions are red. There is also (a) marked dichroism with high absorption of light polarized in the direction of the metal chains and (b) increased electrical conductivity along the chain. If steric

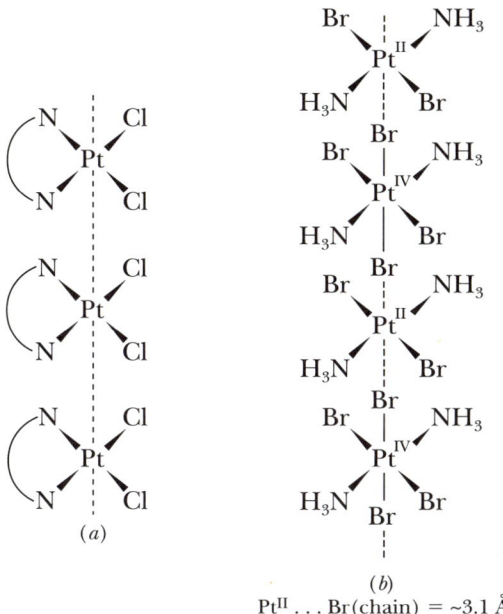

Figure 25-10 (*a*) Linear stacks of planar Pt(en)Cl$_2$ molecules. (*b*) Chains of alternating Pt^{II} and Pt^{IV} atoms, with bridging bromide ions, in Pt(NH$_3$)$_2$Br$_3$.

hindrance is too large, as in [Pt(EtNH$_2$)$_4$][PtCl$_4$], the structure is different and the crystal has a pink color, the sum of the colors of the constituents.

A related class of compounds with chainlike structures differ in that the metals are linked by halide bridges [Fig. 25-10(*b*)]. Again, there is high electrical conductivity along the —X—M^{II}—X—M^{IV}—X— chain.

Five-coordinate complexes are important in substitution and isomerization of square Pd^{II} and Pt^{II} complexes, which proceed by an associative pathway. Some stable complexes have multifunctional ligands such as tris[*o*-(diphenylarsino)phenylarsine] (QAS), which gives salts, for example, [Pd(QAS)I]$^+$. Platinum gives the salts (R$_4$N)$_3$[Pt(SnCl$_3$)$_5$].

25-29 Complexes of Platinum(IV), d^6

There are few complexes of Pd^{IV}. When Pd is dissolved in concentrated HNO$_3$, a nitrato complex is formed. In contrast, platinum(IV) forms many thermally stable and kinetically inert octahedral complexes, ranging from cationic such as [Pt(NH$_3$)$_6$]Cl$_4$ to anionic like K$_2$[PtCl$_6$].

The most important are the sodium or potassium *hexachloroplatinates,* which are starting materials for synthesis of other compounds. The *acid* called "chloroplatinic acid" is an oxonium salt, (H$_3$O)$_2$PtCl$_6$. It is formed as orange crystals when the solution of Pt in aqua regia or in HCl saturated with chlorine is evaporated.

25-30 Complexes of Palladium(0) and Platinum(0), d^{10}

All of these involve π-acid ligands, mainly tertiary phosphines. The complex $M(PPh_3)_4$ is obtained when K_2PdCl_4 or K_2PtCl_4 is reduced by N_2H_4 in ethanol containing PPh_3. These complexes readily undergo oxidative addition reactions (Section 30-2) in which two PPh_3 molecules are lost, for example,

$$Pt(PPh_3)_4 + CH_3I = PtI(CH_3)(PPh_3)_2 + 2\ PPh_3 \qquad (25\text{-}30.1)$$

These elements also give complexes with O_2, alkenes, and alkynes (Chapter 29).

SILVER AND GOLD

25-31 General Remarks

In spite of the similarity in electronic structures, with an s electron outside a completed d shell and high ionization potentials, there are only limited resemblances between Ag, Au, and Cu. These are as follows:

1. The metals crystallize with the same face-centered cubic (ccp) lattice.
2. Both Cu_2O and Ag_2O have the same body-centered cubic (bcc) structure, where the metal atom has two close O neighbors and every O is tetrahedrally surrounded by four metal atoms.
3. Although the stability constant sequence for halogeno complexes of many metals is F > Cl > Br > I, Cu^I and Ag^I belong to the group of ions of the more noble metals for which it is the reverse.
4. Both Cu^I and Ag^I (and to a lesser extent Au^I) form similar types of complexes, such as $[MCl_2]^-$, $[Et_3AsMI]_4$, and K_2MCl_3.
5. Certain complexes of Cu^{II} and Ag^{II} are isomorphous, and Ag^{III}, Au^{III}, and Cu^{III} also give similar complexes.

The only stable cation, apart from complex ions, is Ag^+. The Au^+ ion is exceedingly unstable with respect to the disproportionation:

$$3\ Au^+(aq) = Au^{3+}(aq) + 2\ Au(s) \qquad K \approx 10^{10} \qquad (25\text{-}31.1)$$

Gold(III) is *invariably* complexed in all solutions, usually as anionic species such as $[AuCl_3OH]^-$. The other oxidation states, Ag^{II}, Ag^{III}, and Au^I, are either unstable to water or exist only in insoluble compounds or complex species. Intercomparisons of the standard potentials are of limited utility, particularly since these strongly depend on the nature of the anion; some useful ones are:

$$Ag^{2+} \xrightarrow{\ 2.0\ V\ } Ag^+ \xrightarrow{\ 0.799\ V\ } Ag \qquad (25\text{-}31.2)$$

$$Ag(CN)_2^- \xrightarrow{\ -0.31\ V\ } Ag + 2\ CN^- \qquad (25\text{-}31.3)$$

$$AuCl_4^- \xrightarrow{\ 1.0\ V\ } Au + 4\ Cl^- \qquad (25\text{-}31.4)$$

$$Au(CN)_2^- \xrightarrow{\ -0.6\ V\ } Au + 2\ CN^- \qquad (25\text{-}31.5)$$

25-32 The Elements

The elements are widely distributed as metals, in sulfides and arsenides, and as AgCl. Silver is usually recovered from the work-up of other ores, for example, of lead, the platinum metals, and particularly copper. The elements are extracted by treatment with cyanide solutions in the presence of air, whereby the cyano complexes, $[M(CN)_2]^-$, are formed, and are recovered from them by addition of zinc. They are purified by electrodeposition.

Silver is white, lustrous, soft, and malleable (mp 961 °C) with the highest known electrical and thermal conductivities. It is less reactive than copper, except toward sulfur and hydrogen sulfide, which rapidly blacken silver surfaces. Silver dissolves in oxidizing acids and in cyanide solutions in the presence of oxygen or peroxide.

Gold is soft and yellow (mp 1063 °C) with the highest ductility and malleability of any element. It is unreactive and is not attacked by oxygen or sulfur but reacts readily with halogens or with solutions containing or generating chlorine, such as aqua regia. It dissolves in cyanide solutions in the presence of air or H_2O_2 to form $[Au(CN)_2]^-$.

Both silver and gold form many useful alloys.

SILVER AND GOLD COMPOUNDS

25-33 Silver(I), d^{10}, Compounds

The silver(I) ion (Ag^+) is evidently solvated in aqueous solution but an aqua ion does *not* occur in salts, practically all of which are anhydrous. The compounds $AgNO_3$, $AgClO_3$, and $AgClO_4$ are water soluble but Ag_2SO_4 and AgO_2CCH_3 are sparingly so. The salts of oxo anions are ionic. Although the water-insoluble halides AgCl and AgBr have the NaCl structure, there appears to be appreciable covalent character in the $Ag \cdots X$ interactions. The addition of NaOH to Ag^+ solutions produces a dark brown *oxide* that is difficult to free from alkali ions. It is basic, and its aqueous suspensions are alkaline:

$$\tfrac{1}{2} Ag_2O(s) + \tfrac{1}{2} H_2O = Ag^+ + OH^- \qquad \log K = -7.42 \qquad (25\text{-}33.1)$$

$$\tfrac{1}{2} Ag_2O(s) + \tfrac{1}{2} H_2O = AgOH \qquad \log K = -5.75 \qquad (25\text{-}33.2)$$

They absorb CO_2 from the air to give Ag_2CO_3. The oxide decomposes above about 160 °C and is reduced to the metal by hydrogen. The treatment of water-soluble halides with a suspension of silver oxide is a useful way of preparing hydroxides, since the silver halides are insoluble.

The action of H_2S on Ag^+ solutions gives the black sulfide Ag_2S. The coating often found on silver articles is Ag_2S; this can be readily reduced by contact with aluminum in dilute Na_2CO_3 solution.

Silver fluoride is unique in forming hydrates, such as $AgF \cdot 4H_2O$. The other halides are precipitated by the addition of X^- to Ag^+ solutions; the color and insolubility in water increase Cl < Br < I. *Silver chloride* can be obtained as rather tough sheets that are transparent over much of the IR region and have been used for cell materials. Silver chloride and bromide are light sensitive and have been intensively studied because of their importance in photography. For monodentate ligands, the *complex ions* AgL^+, AgL_2^+, AgL_3^+, and AgL_4^+ exist. The constants K_1 and K_2 are usually high, whereas K_3 and K_4 are relatively small. The main species are, hence, AgL_2^+, which are linear. Because of this, chelating ligands cannot form simple ions, and they give polynuclear complexes instead. The commonest complexes are those such as $[Ag(NH_3)_2]^+$, which are formed by dissolving silver chloride in NH_3, $[Ag(CN)_2]^-$, and $[Ag(S_2O_3)_2]^{3-}$. Silver halides also dissolve in solutions with excess halide ion and excess Ag^+, for example,

$$AgI + n\,I^- \rightleftharpoons [AgI_{n+1}]^{n-} \qquad (25\text{-}33.3)$$

$$AgI + n\,Ag^+ \rightleftharpoons Ag_{n+1}I^{n+} \qquad (25\text{-}33.4)$$

25-34 Silver(II), d^9, and Silver(III), d^8, Compounds

Silver(II) fluoride is a brown solid formed on heating Ag in F_2; it is a useful fluorinating agent. A black oxide ($Ag^IAg^{III}O_2$) is obtained by oxidation of Ag_2O in alkaline solution.

Both Ag^{II} and Ag^{III} occur in complexes with appropriate ligands; the usual procedure is to oxidize Ag^+ in the presence of the ligand. Thus oxidation by $S_2O_8^{2-}$ in the presence of pyridine gives the red ion $[Ag(py)_4]^{2+}$ while in alkaline periodate solution the ion $[Ag(IO_6)_2]^{7-}$ is obtained.

25-35 Gold Compounds

The *oxide* Au_2O_3 decomposes to Au and O_2 at about 150 °C. Chlorination of gold at 200 °C gives *gold(III) chloride*, Au_2Cl_6, as red crystals; on heating at 160 °C this in turn gives *gold(I) chloride*, AuCl.

Complexes

The *dicyanoaurate ion*, $[Au(CN)_2]^-$, is readily formed by dissolving gold in cyanide solutions in the presence of air or H_2O_2.

The interaction of Au_2Cl_6 in ether with tertiary phosphines gives gold(I) complexes (R_3PAuCl); Cl^- can be replaced by I^- or SCN^-. On reduction with $NaBH_4$, these complexes give *gold cluster compounds* with a stoichiometry $Au_{11}X_3(PR_3)_7$. The cluster is an incomplete icosahedron with a central Au atom.

Gold alkylsulfides, $[Au(SR)]_n$, and similar compounds made from sulfurized terpenes are very soluble in organic solvents and are also probably cluster compounds. They are used as "liquid gold" for decorating ceramic and glass articles, which are then fired leaving a gold film.

Gold(III) d^8 is isoelectronic with Pt^{II}, and thus its compounds are *square*. Dissolution of Au in aqua regia or of Au_2Cl_6 in HCl gives a solution that on evaporation deposits yellow crystals of $[H_3O][AuCl_4] \cdot 3H_2O$. The tetrachloroaurate(III) ion quite readily hydrolyzes to $[AuCl_3OH]^-$.

From dilute HCl solutions Au^{III} can be extracted with a very high partition coefficient into ethyl acetate or diethyl ether. The yellow species in the organic layer is probably $[H_3O][AuCl_3OH]$.

STUDY GUIDE

Scope and Purpose

The scope and purpose in this chapter are the same as those for Chapter 24. The student should note the differences that arise between transition elements of the first and subsequent transition series.

Study Questions

A. Review

1. State the chief differences between the second- and third-row transition elements on the one hand and those of the first series on the other with respect to (a) atomic and ionic radii, (b) oxidation states, (c) formation of metal to metal bonds, (d) stereo-chemistry, and (e) magnetic properties.

2. Why are the chemical and physical properties of hafnium and zirconium compounds so similar?

3. What elements characteristically form cluster compounds in their lower oxidation states? Give examples of the three major types, two of which have six metal atoms, and the other three.

4. Draw the structures of the following: $Mo_2(O_2CCF_3)_4$, $[Re_2Cl_8]^{2-}$, $TaCl_5$, NbF_5, $Mo_2O_3(S_2COC_2H_5)_4$, and $Rh_2Cl_2(CO)_4$.

5. Describe the chemical and physical properties of RuO_4 and OsO_4, including preparations and toxicology.

6. List all the elements in the group called the "platinum metals" and show how and where they are arranged in the periodic table. Indicate the relative importance of oxidation states I–VI for each.

7. What is the true nature of the so-called "dihalides" of molybdenum and tungsten?

8. Discuss the terrestrial abundance and commercial availability of technetium.

9. What evidence is there for metal-to-metal interactions in compounds containing square complexes of Ni^{II}, Pd^{II}, and Pt^{II} stacked so the metal atoms form chains perpendicular to the parallel planes of the complexes?

10. Show with sketches the structures of the α and β forms of $PdCl_2$. What role is direct metal–metal bonding thought to play in each?

11. What is the structure of Pd^{II} acetate?

12. How is $Pt(Ph_3P)_4$ prepared? What product is formed when it reacts with methyl iodide?

13. Contrast the chemistry of Cu with that of Ag and Au. First mention the important similarities, and then several important differences.

14. Compare the chemistries of Ag^I and Au^I.

15. Write balanced equations for the following processes: (a) leaching of metallic gold by CN^- in the presence of oxygen. (b) The reaction of AgI with a solution of thiosulfate (photographer's "hypo"). (c) The reaction of aqueous $AgNO_3$ with $S_2O_8^{2-}$ in the presence of excess pyridine.

16. Name the most important silver salts that are (a) soluble in water, and (b) insoluble in water.

17. Starting with a Ni–Cu sulfide ore containing significant amounts of the platinum metals, what are the main steps by which the latter, as a group, are isolated?

B. Additional Exercises

1. What is the lanthanide contraction and what effect does it have on the chemistry of the heavier elements?

2. How would you most easily (a) dissolve tantalum metal, (b) precipitate zirconium from aqueous solution in the presence of aluminum, (c) prepare molybdenum(V) chloride from MoO_3, (d) prepare rhenium(III) chloride, (e) dissolve WO_3, (f) prepare $Rh(CO)H(PPh_3)_3$, (g) make $K_2[MoOCl_5]$ from MoO_3?

3. A number of different ions can be precipitated from solutions of $[ZrF_6]^{2-}$. These include $[ZrF_6]^{2-}$, $[ZrF_7]^{3-}$, $[Zr_2F_{12}]^{4-}$, $[ZrF_8]^{4-}$, and $[Zr_2F_{14}]^{6-}$, as discussed in Sections 25-2 and 25-4. Make careful drawings of each of these ions (from the information provided in Section 25-2).

4. How would you dissolve an alloy of Au and Ag and obtain the metals separately?

5. Give two examples of "bridge cleaving reactions" of either Rh or Pt complexes.

6. How is commercial $RuCl_3 \cdot 3H_2O$ made? What is its actual composition and structure? Suggest products when it is (a) dissolved in conc. HCl and evaporated carefully to dryness, (b) heated with aqueous hydrazine, (c) boiled in aqueous NH_4Cl/NH_4OH with zinc powder, and (d) heated with triphenylphosphine in ethanol.

7. How is commercial $RhCl_3 \cdot 3H_2O$ prepared and what is its composition and structure? What happens when it is (a) boiled with aqueous HCl, (b) warmed with excess triphenylphosphine in ethanol, (c) heated with ammonia in ethanol, or (d) boiled with sodium acetate in ethanol?

8. Suggest explanations for the following:

(a) The aqua nickel(II) ion is paramagnetic, but the aqua palladium(II) ion is diamagnetic.

(b) The contrast noted in (a) is not observed when NiF_2 and PdF_2 are compared; the latter are isostructural and both are paramagnetic.

(c) There is important metal–metal bonding in the $[M_6Cl_{12}]^{n+}$ systems when M = Nb or Ta, but not when M = Pd or Pt.

9. Write balanced chemical equations representing

(a) Preparation of ZrO_2 from aqueous Zr^{IV} solutions using hydroxide.

(b) Hydrolysis of zirconium tetrachloride.

(c) Reaction of molybdenum(II) chloride with chlorine.

(d) Oxidation of Mo with Cl_2.

(e) Three different preparations of $MoCl_4$.

(f) Reaction of $Cr_2(O_2CCH_3)_4$ with HCl.

(g) Reaction of $Mo_2(O_2CCH_3)_4$ with HCl.

(h) A preparation of $[MoOCl_5]^{2-}$.

(i) Treatment of MoO_3 with 12 M HCl.

(j) Recovery of Re as the sulfide.

(k) Dissolution of Re_2Cl_{10} in aqueous HCl.

(l) A preparation of $ReOCl_2(OEt)(PPh_3)_2$.

(m) Reduction of RuO_4 in aqueous HCl.

(n) Reduction of $[IrCl_6]^{2-}$ by KI.

(o) Preparation of the hexaaquarhodium(III) ion beginning with Rh metal.

(p) Cleavage of the $[Rh(CO)_2Cl]_2$ dimer by pyridine.

(q) Dissolution of $PdCl_2$ in HCl.

10. One would expect an octahedral complex of Ru^{IV} to be paramagnetic. Explain this with a crystal field diagram. *Octahedral* Ru^{IV} in the linear, oxo-bridged dimer $[Ru_2OCl_{10}]^{4-}$ is, however, diamagnetic. Show the orbital overlap that takes place in this system to allow spin pairing, as discussed in Section 25-17.

11. Draw the structure of $Mo_2(O_2CH_3)_4$.

12. Draw the structure of $Zr(acac)_4$.

13. Draw the structure of Mo_2Cl_{10}.

14. Draw the structure of *trans*-$RhCl(CO)(PPh_3)_2$.

15. Determine the oxidation state of the metal in each compound of Problems 11–14, above.

16. Draw the structure of each reactant and product in Eq. 25-30.1. Explain why the reaction is called an "oxidative addition." What gets oxidized? What gets reduced?

17. Describe the π-bond system in (a) $[Ru(NH_3)_5(N_2)]^{2+}$, (b) $[Ru_2OCl_{10}]^{4-}$, and (c) $RuCl(H)(CO)(PPh_3)_3$.

C. Questions from the Literature of Inorganic Chemistry

1. Consider $NbCl_4$ as reported by D. R. Taylor, J. C. Calabrese, and E. M. Larsen, *Inorg. Chem.*, **1977**, *16*, 721–722.

(a) Write a balanced chemical equation for the synthesis as reported here of $NbCl_4$.

(b) What structural features suggest metal–metal bonding?

(c) What magnetic feature indicates a metal–metal interaction in this (formally) d^1 system.

(d) There is no metal–metal bonding in the dimeric $[NbCl_5]_2$. Account for this difference.

(e) Draw the structure of $[NbCl_5]_2$.

2. Rh complexes are reported in the article by M. J. Bennett and P. B. Donaldson, *Inorg. Chem.*, **1977**, *16*, 655–660.

(a) What is the geometry about Rh in the complexes $Rh(PPh_3)_3Cl$ as reported here? What is the oxidation state of Rh?

(b) Show orbital overlap diagrams of the metal to ligand π back-bond that is involved in the "π acidity" of the unique phosphine ligand.

(c) Why is metal to ligand π back-bonding strongest in the bond to the $P(C_6H_5)_3$ ligand that is uniquely trans to Cl^-?

3. Consider the paper by A. J. Edwards, *J. Chem. Soc. Dalton Trans.*, **1972**, 582–584.

(a) $WOCl_4$ has an oxygen-bridged infinite chain structure, with a melting point of 209 °C. What does the melting point of $ReOCl_4$ (30 °C) suggest about its structure?

(b) Draw the structure of the $ReOCl_4$ dimers reported here. How strong is the bridging Re···Cl interaction? How do you know?

(c) What does the author suggest is the geometry in that vapor state of (i) $MoCl_5$, (ii) $WOCl_4$, (iii) $WSCl_4$, and (iv) $ReOCl_4$? Explain these geometries.

4. Consider the work by K. G. Caulton and F. A. Cotton, *J. Am. Chem. Soc.*, **1969**, *91*, 6517–6518.

(a) What evidence do the authors present for the presence of a Rh—Rh single bond?

(b) Show the orbital overlap that is responsible for the formation of this Rh—Rh bond.

(c) Should the molecule be paramagnetic or diamagnetic? Explain your reasoning.

SUPPLEMENTARY READING

Bottomley, F., "Nitrosyl Complexes of Ruthenium," *Coord. Chem. Rev.*, **1978**, *26*, 7.

Burgmayer, S. J. N. and Stiefel, E. I., "Molybdenum Enzymes, Cofactors, and Model Systems," *J. Chem. Educ.*, **1985**, *62*, 943.

Canterford, J. H. and Colton, R., *Halides of the Second and Third Transition Series*, Wiley-Interscience, New York, 1968.

Colton, R., *The Chemistry of Rhenium and Technetium*, Wiley, New York, 1965.

Cotton, F. A., "Compounds with Multiple Metal to Metal Bonds," *Chem. Soc. Rev.*, **1975**, *4*, 27.

Dellien, I., Hall, F. M., and Hepler, L. G., "Chromium, Molybdenum, and Tungsten: Thermodynamic Properties, Chemical Equilibria, and Standard Potentials," *Chem. Rev.*, **1976**, *76*, 283.

Fairbrother, F., *The Chemistry of Niobium and Tantalum*, Elsevier, Amsterdam, 1967.

Griffith, W. P., *The Chemistry of the Rarer Platinum Metals*, Wiley-Interscience, New York, 1967.

Hartley, F. R., *The Chemistry of Palladium and Platinum*, Wiley, New York, 1973.

Hartley, F. R., Ed., *Chemistry of the Platinum Group Metals: Recent Developments*, Elsevier, New York, 1991.

Hill, J. O., Worsley, I. G., and Hepler, L. G., "Thermochemistry and Oxidation Potentials of Vanadium, Niobium, and Tantalum," *Chem. Rev.*, **1971**, *71*, 127.

Larsen, E. M., "Zirconium and Hafnium Chemistry," *Adv. Inorg. Chem. Radiochem.*, **1970**, *13*, 1.

MacDermott, T. E., "The Structural Chemistry of Zirconium Compounds," *Coord. Chem. Rev.*, **1973**, *11*, 1.

Miller, D. A. and Bereman, R. D., "The Chemistry of the d^1 Complexes of Niobium, Tantalum, Zirconium, and Hafnium," *Coord. Chem. Rev.*, **1973**, *9*, 107.

Mitchell, P. C. H., Ed., "The Chemistry and Uses of Molybdenum," *J. Less-Common Met.*, **1974**, *36*.

Puddephat, R. J., *The Chemistry of Gold*, Elsevier, Amsterdam, 1978.

Rard, J. A. "Inorganic Aspects of Ruthenium Chemistry," *Chem. Rev.*, **1985**, 81, 1.

Rouschias, G., "Recent Advances in the Chemistry of Rhenium," *Chem. Rev.*, **1974**, *74*, 531.

Seddon, E. A. and Seddon, K. R. *The Chemistry of Ruthenium*, Elsevier, Amsterdam, 1984.

Stiefel, E. I., "The Coordination and Bioinorganic Chemistry of Molybdenum," in *Progress in Inorganic Chemistry*, Vol. 22, Wiley-Interscience, New York, 1977.

Walton, R. A., "Halides and Oxyhalides of the Early Transition Series and Their Stability and Reactivity in Nonaqueous Media," in *Progress in Inorganic Chemistry*, Vol. 16, Wiley-Interscience, New York, 1972.

Chapter 26

SCANDIUM, YTTRIUM, LANTHANUM, AND THE LANTHANIDES

26-1 General Features

The position of these elements in the periodic table is discussed in Section 2-5. Note that actinium, although it is the first member of the actinide elements (Chapter 27), is a true member of the Group IIIA(3) series, Sc, Y, La, and Ac. Except for some similarities in the chemistry of Sc and Al, little resemblance exists between these elements and the Group IIIB(13) elements (Al to Tl).

The elements and some of their properties are given in Table 26-1. Strictly speaking, the lanthanide elements are the 14 that follow the element La, and in which the $4f$ electrons are successively added to the La configuration. However, the term lanthanide is usually taken to include lanthanum itself, since this element is the prototype for the succeeding 14 elements. The progressive decrease in the radii of the atoms and ions of these elements, which when summed is called the *lanthanide contraction,* has been discussed (Section 8-12).

The elements are all highly electropositive with the M^{3+}/M potential varying from -2.25 V (Lu) to -2.52 V (La). The chemistry is predominantly ionic and of the M^{3+} ions.

Yttrium, which lies above La in Group IIIA(3) has a similar $+3$ ion with a noble gas core; because of the effect of the lanthanide contraction, the Y^{3+} radius is close to the values for Tb^{3+} and Dy^{3+}. Consequently, Y occurs in lanthanide minerals. The lighter element in Group IIIA(3), *scandium,* is also considered here, although it has a smaller ionic radius and shows chemical behavior intermediate between that of Al and that of Y and the lanthanides.

Variable Valency

Certain lanthanides (Table 26-1) form $+2$ or $+4$ ions. The $+2$ ions are readily oxidized and the $+4$ ions are readily reduced to the $+3$ ion. A simplified explanation for the occurrence of these valences is that empty, half-filled or filled f shells are especially stable. A similar phenomenon has been noted concerning the ionization enthalpies of the elements of the first transition series (Section 2-7), and half-filling of the $3d$ shell accounts for the stability of manganese(II). For the lanthanides, the oxidation state IV for cerium gives Ce^{4+} with the empty f shell configuration of La^{3+}. Similarly, the formation of Yb^{2+} gives this ion an f^{14} configuration. The half-filled f^7 configuration of Gd^{3+} is formed by reduction to give

Table 26-1   Some Properties of Scandium, Yttrium, and the Lanthanides

Z	Name	Symbol	Electron Configuration	Valences	M^{3+} Radius (Å)	M^{3+} Color
21	Scandium	Sc	$[Ar]3d^14s^2$	3	0.68	Colorless
39	Yttrium	Y	$[Kr]4d^15s^2$	3	0.88	Colorless
57	Lanthanum	La	$[Xe]5d^16s^2$	3	1.06	Colorless
58	Cerium	Ce	$[Xe]4f^15d^16s^2$	3, 4	1.03	Colorless
59	Praseodymium	Pr	$[Xe]4f^36s^2$	3, 4	1.01	Green
60	Neodymium	Nd	$[Xe]4f^46s^2$	3	0.99	Lilac
61	Promethium	Pm	$[Xe]4f^56s^2$	3	0.98	Pink
62	Samarium	Sm	$[Xe]4f^66s^2$	2, 3	0.96	Yellow
63	Europium	Eu	$[Xe]4f^76s^2$	2, 3	0.95	Pale pink
64	Gadolinium	Gd	$[Xe]4f^75d6s^2$	3	0.94	Colorless
65	Terbium	Tb	$[Xe]4f^96s^2$	3, 4	0.92	Pale pink
66	Dysprosium	Dy	$[Xe]4f^{10}6s^2$	3	0.91	Yellow
67	Holmium	Ho	$[Xe]4f^{11}6s^2$	3	0.89	Yellow
68	Erbium	Er	$[Xe]4f^{12}6s^2$	3	0.88	Lilac
69	Thulium	Tm	$[Xe]4f^{13}6s^2$	3	0.87	Green
70	Ytterbium	Yb	$[Xe]4f^{14}6s^2$	2, 3	0.86	Colorless
71	Lutetium	Lu	$[Xe]4f^{14}5d6s^2$	3	0.85	Colorless

Eu^{2+} or by oxidation to give Tb^{4+}. That other factors are involved, however, is shown by the existence of many +2 ions stabilized in CaF_2 lattices and of Pr^{4+} and Nd^{4+} fluoride complexes.

Magnetic and Spectral Properties

The lanthanide ions that have unpaired electrons are colored and are paramagnetic. There is a fundamental difference from the d-block elements in that the $4f$ electrons are inner electrons and are very effectively shielded from the influence of external forces by the overlying $5s^2$ and $5p^6$ shells. Hence, there are essentially only very weak effects of ligand fields. As a result, electronic transitions between f orbitals give rise to extremely narrow absorption bands, quite unlike the broad bands resulting from d–d transitions, and the magnetic properties of the ions are little affected by their chemical surroundings.

Coordination Numbers and Stereochemistry

It is characteristic of the M^{3+} ions that *coordination numbers exceeding six are common*. Very few six-coordinate species are known but coordination numbers of seven, eight, and nine are important. In the ion $[Ce(NO_3)_6]^{2-}$, the Ce is surrounded by 12 oxygen atoms of chelate NO_3 groups.

The decrease in radii from La to Lu also has the effect that different crystal structures and coordination numbers may occur for different parts of the lanthanide group. For example, the metal atoms in the trichlorides La to Gd are nine-coordinate, whereas the chlorides of Tb to Lu have an $AlCl_3$ type structure with the metal being octahedrally coordinated. Similar differences in coordination numbers occur for ions in solution.

26-2   Occurrence and Isolation

Scandium is quite a common element being as abundant as As and twice as abundant as B. However, it is not readily available, partly owing to a lack of rich ores, and partly due to the difficulty of separation. It may be separated from Y and the lanthanides, which may be associated with Sc minerals, by cation exchange procedures using oxalic acid as elutant.

The lanthanide elements, including La and Y, were originally known as the rare earths—from their occurrence in oxide (or in old usage, earth) mixtures. They are *not* rare elements and their absolute abundances are relatively high. Thus even the scarcest, Tm, is as common as Bi and more common than As, Cd, Hg, or Se. The major source is *monazite,* a heavy dark sand of variable composition. Monazite is essentially a lanthanide orthophosphate, but may contain up to 30% thorium. The elements La, Ce, Pr, and Nd usually account for about 90% of the lanthanide content of minerals, with Y and the heavier elements accounting for the rest. Minerals carrying lanthanides in the +3 oxidation state are usually poor in Eu which, because of its tendency to give the +2 state, is often concentrated in minerals of the Ca group.

Promethium occurs only in traces in U ores as a spontaneous fission fragment of ^{238}U. Milligram quantities of pink ^{147}Pm^{3+} salts can be isolated by ion exchange methods from fission products in spent fuel of nuclear reactors where ^{147}Pm(β^-, 2.64 year) is formed.

The lanthanides are separated from most other elements by precipitation of oxalates or fluorides from HNO_3 solution, and from each other by ion exchange on resins. Cerium and europium are normally first removed. Cerium is oxidized to Ce^{IV} and is then precipitated from 6 M HNO$_3$ as Ce^{IV} iodate or separated by solvent extraction. Europium is reduced to Eu^{2+} and is removed by precipitation as $EuSO_4$.

The ion-exchange behavior depends primarily on the hydrated ionic radius. As with the alkalis, the smallest ion crystallographically (Lu) has the largest hydrated radius, while La has the smallest hydrated radius. Hence, La is the most tightly bound and Lu is the least, and the elution order is Lu $\rightarrow$ La (Fig. 27-3). This trend is accentuated by use of complexing agents at an appropriate pH; the ion of smallest radius also forms the strongest complexes and, hence, the preference for the aqueous phase is enhanced. Typical complexing ligands are α-hydroxyisobutyric acid, $(CH_3)_2CH(OH)CO_2H$, EDTAH$_4$, and other hydroxo or amino carboxylic acids. From the eluates the M^{3+} ions are recovered by acidification with dilute HNO_3 and addition of oxalate ion, which precipitates the oxalates essentially quantitatively. These are then ignited to the oxides.

Cerium(IV) (also Zr^{IV}, Th^{IV}, and Pu^{IV}) is readily extracted from HNO$_3$ solutions by tributyl phosphate dissolved in kerosene or other inert solvents and can be separated from the +3 lanthanide ions. The +3 lanthanide nitrates can also be extracted under suitable conditions with various phosphate esters or acids. Extractability under given conditions increases with increasing atomic number; it is higher in strong acid or high NO_3^- concentrations.

The Metals

The lighter metals (La to Gd) are obtained by reduction of the trichlorides with Ca at 1000 °C or more. For Tb, Dy, Ho, Er, Tm, and also Y the trifluorides are used because the chlorides are too volatile. Promethium is made by reduction of

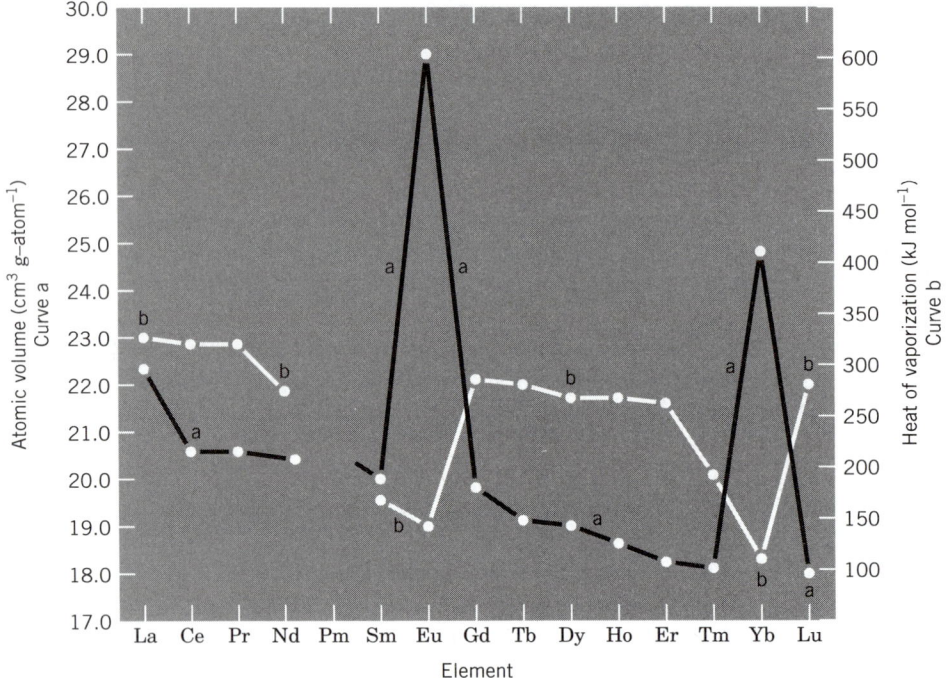

Figure 26-1 The atomic volumes (curve a) and heats of vaporization (curve b) of the lanthanide metals.

PmF_3 with lithium. The Eu, Sm, and Yb trichlorides are reduced only to the dihalides by Ca. Reduction of the +3 oxides with La at high temperatures gives the metals.

The metals are silvery white and highly electropositive. They react with water, slowly in the cold, rapidly on heating, to liberate hydrogen. These metals tarnish in air and burn easily to give the oxides M_2O_3; cerium is the exception giving CeO_2. Lighter "flints" are mixed metals containing mostly cerium. Yttrium is resistant to air even up to 1000 °C owing to formation of a protective oxide coating. The metals react with H_2, C, N_2, Si, P, S, halogens, and other nonmetals at elevated temperatures.

Many physical properties of the metals change smoothly along the series, except for Eu and Yb, and occasionally Sm and Tm (cf. Fig. 26-1). The deviations occur with those lanthanides that have the greatest tendency to exist in the +2 state; presumably these elements tend to donate only two electrons to the conduction bands of the metal, thus leaving larger cores and affording lower binding forces. Note, too, that Eu and Yb dissolve in ammonia (Section 10-3).

LANTHANIDE COMPOUNDS

26-3 The Trivalent State

Oxides and Hydroxides

The oxide Sc_2O_3 is less basic than the other oxides and closely resembles Al_2O_3; it is similarly amphoteric, dissolving in NaOH to give a "scandate" ion, $[Sc(OH)_6]^{3-}$.

The oxides of the remaining elements resemble CaO and absorb CO_2 and H_2O from the air to form carbonates and hydroxides, respectively. The *hydroxides*, $M(OH)_3$, are true compounds whose basicities decrease with increasing Z, as would be expected from the decrease in ionic radius. They are precipitated from aqueous solutions by bases as gelatinous masses. These oxides are not amphoteric.

Halides

Scandium is again exceptional. Its fluoride resembles AlF_3, being soluble in excess HF to give the $[ScF_6]^{3-}$ ion; Na_3ScF_6 is like cryolite (Fig. 13-2). However, $ScCl_3$ is not a Friedel–Crafts catalyst like $AlCl_3$ and does not behave as a Lewis acid; its structure is like that of $FeCl_3$ (Section 24-27).

Lanthanide fluorides are of importance because of their insolubility. Addition of HF or F^- precipitates MF_3 from solutions even 3 *M* in HNO_3 and is a characteristic test for lanthanide ions. The fluorides of the heavier lanthanides are slightly soluble in an excess of HF owing to complex formation. Fluorides may be redissolved in 3 *M* HNO_3 saturated with H_3BO_3, which removes F^- as BF_4^-.

The *chlorides* are soluble in water, from which they crystallize as hydrates. The anhydrous chlorides are best made by the reaction:

$$M_2O_3 + 6\,NH_4Cl \xrightarrow{\sim 300\,°C} 2\,MCl_3 + 3\,H_2O + 6\,NH_3 \qquad (26\text{-}3.1)$$

Aqua Ions, Oxo Salts, and Complexes

Scandium forms a hexaaqua ion $[Sc(H_2O)_6]^{3+}$ that is readily hydrolyzed. Scandium β-diketonates are also octahedral like those of Al and unlike those of the lanthanides.

For the lanthanides and yttrium, the aqua ions $[M(H_2O)_n]^{3+}$ have coordination numbers exceeding six, as in $[Nd(H_2O)_9]^{3+}$. They are hydrolyzed in water.

$$[M(H_2O)_n]^{3+} + H_2O \rightleftharpoons [M(OH)(H_2O)_{n-1}]^{2+} + H_3O^+ \qquad (26\text{-}3.2)$$

The tendency to hydrolyze increases from La to Lu, which is consistent with the decrease in the ionic radii. Yttrium also gives predominantly $Y(OH)^{2+}$. For Ce^{3+}, however, only about 1% of the metal ion is hydrolyzed without forming a precipitate, and the main equilibrium appears to be

$$3\,Ce^{3+} + 5\,H_2O \rightleftharpoons [Ce_3(OH)_5]^{4+} + 5\,H^+ \qquad (26\text{-}3.3)$$

In aqueous solutions, rather weak *fluoride* complexes, MF^{2+}, are formed. Complex anions are *not* formed, a feature that distinguishes the +3 lanthanides as a group from the +3 actinide elements that *do* form anionic complexes in strong HCl solutions.

The most stable and common complexes are those with *chelating oxygen ligands*. The formation of water-soluble complexes by citric and other hydroxo acids is utilized in ion-exchange separations, as we noted previously. The complexes usually have coordination numbers greater than 6.

Beta-diketone (β-dik) ligands, such as acetylacetone, are especially important, since some of the fluorinated β-diketones give complexes that are volatile and suitable for gas chromatographic separation. The preparation of β-diketonates by conventional methods invariably gives hydrated or solvated species, such as $M(acac)_3 \cdot C_2H_5OH \cdot 3H_2O$, that have coordination numbers greater than 6. Prolonged drying over $MgClO_4$ gives the very hygroscopic $M(\beta\text{-dik})_3$.

An interesting use of Eu and Pr β-diketonate complexes, which are soluble in organic solvents, such as those derived from 1,1,1,2,2,3,3-heptafluoro-7,7-dimethyl-4,6-octanedione, is as shift reagents in NMR spectrometry. The paramagnetic complex deshields the protons of complicated molecules, and vastly improved separation of the resonance lines may be obtained.

Other uses for lanthanide compounds depend on their spectroscopic properties. The elements Y and Eu in oxide or silicate lattices have fluorescent or luminescent behavior and the phosphors are used in color television tubes. In CaF_2 lattices the +2 ions show laser activity as do salts of $[Eu(\beta\text{-dik})_4]^-$.

26-4   The Tetravalent State

Cerium(IV)

This is the only +4 lanthanide that exists in aqueous solution as well as in solids. The dioxide (CeO_2) is obtained by heating $Ce(OH)_3$ or oxo salts in air. It is unreactive and is dissolved by acids only in the presence of reducing agents $(H_2O_2,$ Sn^{II}, etc.) to give Ce^{3+} solutions. Hydrous cerium(IV) oxide, $CeO_2 \cdot nH_2O$, is a yellow, gelatinous precipitate obtained on treating Ce^{IV} solutions with OH^-; it redissolves in acids.

The *cerium(IV) ion* in solution is obtained by oxidation of Ce^{3+} in HNO_3 or H_2SO_4 with $S_2O_8^{2-}$ or bismuthate. Its chemistry is similar to that of Zr^{4+} and +4 actinides. Thus Ce^{4+} gives phosphates insoluble in 4 M HNO_3 and iodates insoluble in 6 M HNO_3, as well as an insoluble oxalate. The phosphate and iodate precipitations can be used to separate Ce^{4+} from the trivalent lanthanides.

The yellow-orange hydrated ion $[Ce(H_2O)_n]^{4+}$ is a fairly strong acid, hydrolyzes readily, and probably exists only in strong $HClO_4$ solution. In other acids complex formation accounts for the acid dependence of the potential.

$$Ce^{IV} + e^- = Ce^{III} \qquad E° = + 1.28 \text{ V } (2 \text{ } M \text{ HCl})$$
$$+ 1.44 \text{ V } (1 \text{ } M \text{ H}_2\text{SO}_4)$$
$$+ 1.61 \text{ V } (1 \text{ } M \text{ HNO}_3) \qquad (26\text{-}4.1)$$
$$+ 1.70 \text{ V } (1 \text{ } M \text{ HClO}_4)$$

Comparison of the potential in H_2SO_4, where at high SO_4^{2-} concentrations the major species is $[Ce(SO_4)_3]^{2-}$, with that for the oxidation of water

$$O_2 + 4 \text{ H}^+ + 4 \text{ e}^- = 2 \text{ H}_2\text{O} \qquad E° = +1.229 \text{ V} \qquad (26\text{-}4.2)$$

shows that the acid Ce^{IV} solutions commonly used in analysis are metastable.

Cerium(IV) is used as an oxidant in analysis and in organic chemistry, where it is commonly used in acetic acid. The solutions oxidize aldehydes and ketones at the α-carbon atom. Benzaldehyde yields benzoin.

Table 26-2   Properties of the Lanthanide +2 Ions

Ion	Color	$E°$ (V)^a	Crystal Radius, (Å)^b
Sm^{2+}	Blood red	−1.55	1.11
Eu^{2+}	Colorless	−0.43	1.10
Yb^{2+}	Yellow	−1.15	0.93

^{a}For $M^{3+} + e^- = M^{2+}$.
^{b}Pauling radii: Ca^{2+}, 0.99; Sr^{2+}, 1.13; Ba^{2+}, 1.35.

Complex anions are formed quite readily. The analytical standard "ceric ammonium nitrate," which can be crystallized from HNO_3, contains the hexanitratocerate anion, $[Ce(NO_3)_6]^{2-}$.

Praseodymium(IV) and Terbium(IV)

These exist only in oxides and fluorides. The oxide systems are very complex and nonstoichiometric. The potential Pr^{IV}/Pr^{III} is estimated to be +2.9 V so that it is not surprising that Pr^{IV} does not exist in aqueous solution.

26-5   The Divalent State

The +2 state is known in both solutions and solid compounds of Sm, Eu, and Yb (Table 26-2). Less well established are the ions Tm^{2+} and Nd^{2+}, but the +2 ions of all the lanthanides can be prepared and stabilized in CaF_2 or BaF_2 lattices by reduction of, for example, MF_3 in CaF_2 with Ca.

The *europium(II)* ion can be made by reducing aqueous Eu^{3+} solutions with Zn or Mg. The other ions require the use of Na amalgam. All three can be prepared by electrolytic reduction in aqueous solution or in halide melts.

The ions Sm^{2+} and Yb^{2+} are quite rapidly oxidized by water. The Eu^{2+} ion is oxidized by air.

The Eu^{2+} ion resembles Ba^{2+}. Thus the sulfate and carbonate are insoluble, whereas the hydroxide is soluble. The stability of the Eu^{2+} complex with $EDTA^{4-}$ is intermediate between those of Ca^{2+} and Sr^{2+}.

Crystalline compounds of Sm, Eu, and Yb are usually isostructural with the Sr^{2+} or Ba^{2+} analogs.

STUDY GUIDE

Study Questions

A. Review

1. Name the lanthanide elements and give their electron configurations.
2. Explain the position of the lanthanides in the periodic table and their relation to the Al, Ga, In, and Tl group.
3. What is the "lanthanide contraction"? What effect does it have on the chemistry of later elements?

4. Compare the main features of the chemistry of ions of highly electropositive elements with charges +1, +2, and +3.

5. Why are scandium and yttrium usually considered along with the lanthanide elements?

6. Which lanthanide elements show departure from the usual +3 oxidation state? Give the electron configurations of these ions.

7. What is characteristic about the coordination numbers of lanthanide ions? Give examples.

8. How are the lanthanide ions separated from each other?

9. What are the characteristic precipitation reactions of lanthanide +2, +3, and +4 ions?

10. How are anhydrous lanthanide chlorides made?

11. What are the interesting features of lanthanide β-diketonates?

B. Additional Exercises

1. Work out the number of unpaired electrons in the ions (a) Pr^{3+}, (b) Pm^{3+}, (c) Sm^{2+}, (d) Gd^{3+}, (e) Tb^{4+}, (f) Tm^{3+}, and (g) Lu^{2+}.

2. Why do the electronic absorption spectra of lanthanide ions have sharp bands unlike the broad bands in the spectra of the $3d$ elements?

3. Write balanced chemical equations representing

   (a) Preparation of anhydrous $PrCl_3$.

   (b) Reduction of CeO_2 in aqueous HCl solution by Sn^{2+}.

   (c) Dissolution of $CeO_2 \cdot nH_2O$ in aqueous HCl.

   (d) Oxidation of Ce^{3+} in aqueous HNO_3 by $S_2O_8^{2-}$.

   (e) Reduction of aqueous Eu^{3+} with Zn.

4. Discuss the pH and anion dependence of the Ce^{III}–Ce^{IV} couple.

5. Why is Pr^{IV} not stable in aqueous solution? Write a balanced equation for its reaction with water. What is $E°$ for this reaction?

6. Explain the increase in hydrolysis that takes place from La to Lu, as the size of the ions M^{3+} decreases through the lanthanide contraction.

C. Questions from the Literature of Inorganic Chemistry

1. Consider the Nd complex reported by R. A. Anderson, D. H. Templeton, and A. Zalkin, *Inorg. Chem.*, **1978**, *17*, 1962–1965.

   (a) Write balanced chemical equations for the synthesis (two steps) of the title compound.

   (b) Calculate the percentage yield.

   (c) Prepare diagrams of each class of oxygen as found in this structure: (i) terminal, (ii) edge bridging, and (iii) trigonal face bridging.

   (d) What geometry is defined by the Nd_6 group?

   (e) Based on the reported magnetic susceptibility data, what is the number of unpaired electrons per Nd atom? What is the formal charge on each Nd? What is the ground-state electron configuration for each Nd?

2. Two types of lanthanide compounds are described by D. C. Bradley, J. S. Ghotra, F. A. Hart, M. B. Hursthouse, and P. R. Raithby, *J. Chem. Soc. Dalton Trans.*, **1977**, 1166–1172.

   (a) Write balanced chemical equations for the preparations, as reported here, of the adducts $[M\{N(SiMe_3)_2\}_3(PPh_3O)]$, where M = La, Eu, or Lu. What is the oxidation state of the metal in these complexes?

(b) These complexes contain the monoanionic bis(trimethylsilyl)amido ligands, $[N(SiMe_3)_2]^-$, and the neutral triphenylphosphine oxide ligand (Ph_3PO). Draw a Lewis diagram for each of these ligands, and classify each nonhydrogen atom in these ligands according to the AB_xE_y system. What hybridization is appropriate for each nonterminal atom in these ligands?

(c) What is the coordination geometry about La in the complex from (a)?

(d) Write balanced chemical equations for the syntheses, as reported here, of the μ-peroxo dimers $[M_2\{N(SiMe_3)_2\}_4(O_2)(Ph_3PO)_2]$, where M = La, Pr, Sm, or Eu.

(e) What is unusual about the peroxo bridge that is reported here? (Compare the O—O distance with those found in Table 18-2.) Should this be considered to be an O_2^{2-} ligand?

SUPPLEMENTARY READING

Asprey, L. B. and Cunningham, B. B., "Unusual Oxidation States of Some Actinide and Lanthanide Elements," *Progress in Inorganic Chemistry*, Vol. 2, Wiley-Interscience, New York, 1960.

Bagnall, K. W., Ed., *Lanthanides and Actinides*, Butterworths, London, 1972.

Brown, D., *Halides of the Lanthanides and Actinides*, Wiley-Interscience, New York, 1968.

Bunzli, J. G. and Wessner, D., "Rare Earth Complexes with Neutral Macrocyclic Ligands," *Coord. Chem. Rev.,* **1984,** *60,* 191.

Callow, R. J., *The Industrial Chemistry of Lanthanons, Yttrium, Thorium, and Uranium*, Pergamon Press, New York, 1967.

Cotton, S. A. and Hart, F. A., *The Heavy Transition Elements*, Macmillan, New York, 1975.

Horowitz, C. T., *Scandium*, Academic, New York, 1975.

Koppikar, D. K., Sivapullaiah, P. V., Ramakrishnan, L., and Soundararajan, S., "Complexes of the Lanthanides with Neutral Oxygen Donor Ligands," *Struct. Bonding,* **1978,** *34* 135.

Melson, G. A. and Stotz, R. W., "The Coordination Chemistry of Scandium," *Coord. Chem. Rev.,* **1971,** *7,* 133.

Morss, L. R., "Thermochemical Properties of Yttrium, Lanthanum, and the Lanthanide Elements and Ions," *Chem. Rev.,* **1974,** *74,* 827.

Sinha, S. P., *Europium*, Springer-Verlag, Berlin, 1968.

Topp, N. E., *The Chemistry of the Rare Earth Elements*, Elsevier, Amsterdam, 1965.

Chapter 27

THE ACTINIDE ELEMENTS

27-1 General Features

The actinide elements and the electronic structures of the atoms are given in Table 27-1. Their position in the periodic table and their relation to the lanthanide elements are discussed in Chapter 8. It will be evident in the following pages that the term *actinides* is not as apt for these elements as is the term lanthanides for elements 59–72. The elements immediately following Ac, which is similar to La and has only the +3 state, do not resemble it very closely at all. Thorium, protactinium and, to a lesser extent, uranium are homologous with their vertical groups in the periodic table, that is, Hf, Ta, and W. However, beginning with Am, there is pronounced lanthanide-like behavior. This, coupled with the existence of the +3 state for all the elements, justifies the term actinide.

The atomic spectra of these heavy elements are very complex, and it is difficult to identify levels in terms of quantum numbers and configurations. The energies of the $5f$, $6d$, $7s$, and $7p$ levels are comparable, and the energies involved in an electron moving from one level to another may lie within the range of chemical binding energies. Thus the electronic structure of an ion in a given oxidation state may be different in different compounds, and in solution it may be dependent on the nature of the ligands. It is thus often impossible to say which orbitals are being used in bonding or to decide whether the bonding is covalent or ionic.

One difference from the $4f$ group is that the $5f$ orbitals have a greater spatial extension relative to the $7s$ and $7p$ orbitals than the $4f$ orbitals have relative to the $6s$ and $6p$. Thus $5f$ orbitals can, and do, participate in bonding to a far greater extent than the $4f$ orbitals. A reflection of this potential for covalent bonding is shown by the formation of organometallic compounds similar to those formed by the d-block elements. Examples are di-η^8-cyclooctatetraenyl uranium, $(\eta^8\text{-}C_8H_5)_2U$, and tri-$\eta^5$-cyclopentadienyl uranium benzyl, $(\eta^5\text{-}C_5H_5)_3UCH_2C_6H_5$.

Ionic Radii

The ionic radii of actinide and lanthanide ions are compared in Fig. 27-1. Notice that there is an "actinide contraction" similar to the lanthanide contraction.

Magnetic and Spectroscopic Properties

The magnetic properties of the actinide ions are complicated and difficult to interpret. The electronic absorption spectra that result from f–f transitions consist, like those of the lanthanides, of quite narrow bands.

Table 27-1   The Actinide Elements and Some of Their Properties

| | | | Electronic Structure[a] | Radii (Å) | |
| | | | | M^{3+} | M^{4+} |
Z	Name	Symbol	of Atom		
89	Actinium	Ac	$6d7s^2$	1.11	
90	Thorium	Th	$6d^27s^2$		0.90
91	Protactinium	Pa	$5f^26d7s^2$ or $5f^16d^27s^2$		0.96
92	Uranium	U	$5f^36d7s^2$	1.03	0.93
93	Neptunium	Np	$5f^57s^2$	1.01	0.92
94	Plutonium	Pu	$5f^67s^2$	1.00	0.90
95	Americium	Am	$5f^77s^2$	0.99	0.89
96	Curium	Cm	$5f^76d7s^2$	0.985	0.88
97	Berkelium	Bk	$5f^86d7s^2$ or $5f^97s^2$	0.98	
98	Californium	Cf	$5f^{10}7s^2$	0.977	
99	Einsteinium	Es	$5f^{11}7s^2$		
100	Fermium	Fm	$5f^{12}7s^2$		
101	Mendelevium	Md	$5f^{13}7s^2$		
102	Nobelium	No	$5f^{14}7s^2$		
103	Lawrencium	Lr	$5f^{14}6d7s^2$		
104	Rutherfordium	Rf			

[a]Outside Rn structure.

Oxidation States

There is a far greater range of oxidation states compared with the lanthanides, which is in part attributable to the fact that the $5f$, $6d$, and $7s$ levels are of comparable energies. The known states are given in Table 27-2.

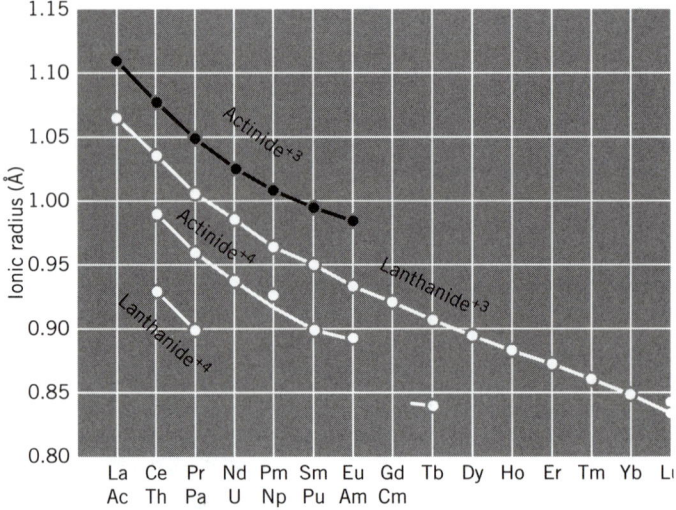

Figure 27-1   Radii of actinide and lanthanide ions. [Reproduced by permission from D. Brown, *Halides of Lanthanides and Actinides,* Wiley-Interscience Publishers, New York, 1968]

Table 27-2   Oxidation States of Actinides with Corresponding Members of Lanthanides

Ac	Th	Pa	U	Np	Pu	Am	Cm	Bk	Cf	Es	Fm	Md	No	Lr
						2			2	2	2	2	**2**	
3	3	3	3	3	3	**3**	3	3	**3**	**3**	**3**	**3**	3	3
	4	4	4	**4**	4	4	4	4						
		5	5	5	5	5			5?					
			6	6	6	6								
				7	7									

La	Ce	Pr	Nd	Pm	Sm	Eu	Gd	Tb	Dy	Ho	Er	Tm	Yb	Lu
f^0							f^7							f^{14}

27-2   Occurrence and Properties of the Elements

All of the elements are radioactive. The terrestrial occurrence of U and Th is due to the half-lives of ^{235}U, ^{238}U, and ^{232}Th, which are sufficiently long to have enabled them to persist since genesis. These isotopes are the ones formed in the radioactive series and found in U and Th minerals. The half-lives of even the most stable of the elements following U are so short that any amounts formed at genesis would have disappeared quite quickly.

The first new elements, neptunium and plutonium, named like uranium after the planets, were made in 1940 by E. M. McMillan and P. Abelson and by G. T. Seaborg, E. M. McMillan, J. W. Kennedy, and A. Wahl, respectively, by bombardments of uranium using particles from the cyclotron in Berkeley. Both are now obtained from spent uranium fuel elements of nuclear reactors where they are formed by capture of neutrons produced in the fission of ^{235}U fuel:

$$^{238}U \xrightarrow{n\gamma} {}^{239}U \xrightarrow[23.5\ m]{\beta^-} {}^{239}Np \xrightarrow[2.35\ d]{\beta^-} {}^{239}Pu\,(24,360\ year) \quad (27\text{-}2.1)$$

$$^{235}U \xrightarrow{2\ n\gamma} \;\;\; {}^{237}U \xrightarrow[6.75\ d]{\beta^-} {}^{237}Np\,(2.2 \times 10^6\ year) \quad (27\text{-}2.2)$$
$$^{238}U \xrightarrow{n,\,2n} $$

Only Pu is normally recovered since ^{239}Pu has fission properties similar to ^{235}U and can be used as a fuel or in nuclear weapons. Some ^{237}Np is used to prepare ^{238}Pu (86.4 year), which is used as a power source for satellites.

Isotopes of elements following Pu are made by successive neutron capture in ^{239}Pu in nuclear reactors. Examples are

$$^{239}Pu \xrightarrow{n\gamma} {}^{240}Pu \xrightarrow{n\gamma} {}^{241}Pu \xrightarrow[13.2\ year]{\beta^-} {}^{241}Am\,(433\ year) \quad (27\text{-}2.3)$$

$$^{239}Pu \xrightarrow{4\ n\gamma} {}^{243}Pu \xrightarrow[5\ h]{\beta^-} {}^{243}Am \xrightarrow{n\gamma} {}^{244}Am$$
$$\qquad\qquad\qquad\qquad\qquad\qquad 26\ min \downarrow \beta^-$$
$$\qquad\qquad\qquad\qquad\qquad {}^{244}Cm\,(7.6\ year) \quad (27\text{-}2.4)$$

The elements 100–104 are made by bombardment of Pu, Am, or Cm with accelerated ions of B, C, or N.

The isotopes ^{237}Np and ^{239}Pu can be obtained in multikilograms; Am and Cm in greater than 100-g amounts; Bk, Cf, and Es in milligrams; and Fm in 10^{-6} g quantities. The isotopes of elements above Fm are short-lived and only tracer quantities are yet accessible. The metals are all chemically very reactive. The intense radiation from the elements with short half-lives can cause rapid decomposition of compounds. Both Ac and Cm glow in the dark.

27-3   General Chemistry of the Actinides

The chemistry of the actinides is very complicated, especially in solutions. It has been studied in great detail because of its relevance to nuclear energy, and the chemistry of plutonium is better known than that of many natural elements.

The principal features of the actinides, all of which are electropositive metals, are the following:

1. *Actinium* has only the +3 state and is entirely lanthanide-like.

2. *Thorium* and *protactinium* show limited resemblance to the other elements. They can perhaps best be regarded as the heaviest members of the Ti, Zr, Hf and V, Nb, and Ta groups, respectively.

3. *Uranium, neptunium, plutonium,* and *americium* are all quite similar, differing mainly in the relative stabilities of their oxidation states, which range from +3 to +6.

4. *Curium* is lanthanide-like and corresponds to gadolinium in that at Cm the 5f shelf is half full. It differs from Gd in having +4 compounds. By comparison with the lanthanides the previous element *americium* should show the +2 state, like Eu, and the succeeding element, berkelium the +4 state, like Tb. This is the case.

5. The elements Cm and Lr are lanthanide-like. *Lawrencium,* like Lu, has a filled f shell so that element 104 should and, as far as is yet known, does have hafnium-like behavior. The elements from 104 onward should be analogs of Hf, Ta, W, and so on. For example, element 112, for which an unsubstantiated claim was made, should resemble Hg. It is uncertain how many more elements can be synthesized. The observation of element 109 was recently claimed (cf. Chapter 8) but only those up to 106 have been confirmed.

6. A characteristic feature of the compounds and complexes of actinides, like the lanthanides, is the occurrence of *high coordination numbers* up to 12 as in $[Th(NO_3)_6]^{2-}$. Coordination geometries in solids are especially complicated.

7. The various cations of U, Np, Pu, and Am have a very complex solution chemistry. The free energies of various oxidation states differ little, and for Pu the +3, +4, +5, and +6 states can actually coexist. The chemistry is complicated by hydrolysis, polymerization, complexing, and disproportionation reactions. Also, for the most radioactive species, chemical reactions are induced by the intense radiation.

The Metals

The metals are prepared by the reduction of anhydrous fluorides, chlorides, or oxides by Li, Mg, or Ca at 1100–1400 °C. They are silvery white and reactive, tarnishing in air, and pyrophoric when finely divided. They are soluble in common acids; HNO_3 or HCl are the best solvents.

Uranium normally has a black oxidized film. When enriched in ^{235}U, the metal can initiate a nuclear explosion above a certain critical mass, and this is true also for plutonium. The metals U, Np, and Pu are similar and are the densest of metals.

Americium and *Cm* are much lighter metals with higher melting points than U, Np, and Pu and resemble the lanthanides. The metallic radius of *californium* indicates that it is divalent like Eu and Yb.

Oxidation States

The oxidation states have been summarized in Table 27-2.

The +3 state is the one common to all actinides except for Th and Pa. It is the preferred state for Ac, Am, and all the elements following Am. The most readily oxidized +3 ion is U^{3+}, which is oxidized by air or more slowly by water.

The chemistry is similar to that of the lanthanides. For example, the fluorides are precipitated from dilute HNO_3 solutions. Since the ionic sizes of both series are comparable, there is considerable similarity in the formation of complex ions, such as citrates, and in the magnitude of the formation constants. The separation of +3 lanthanides and actinides into groups and from each other requires ion-exchange methods (Section 26-2).

The +4 state is the principal state for thorium. For Pa, U, Np, Pu, and Bk, +4 cations are known in solution, but for Am and Cm in solution there are only complex fluoroanions. All form solid +4 compounds. Element 104 has been found only in the +4 state.

The +4 cations in acid solution can be precipitated by iodate, oxalate, phosphate, and fluoride. The *dioxides* (MO_2) from Th to Bk have the fluorite structure. The tetrafluorides (MF_4) for both actinides and lanthanides are isostructural.

The +5 state is the preferred state for Pa, in which it resembles Ta. For U to Am only a few solid compounds are known. For these elements the *dioxo ions*, $[MO_2]^+(aq)$, are of importance (discussed shortly).

In the +6 state the only simple compounds are the hexafluorides (MF_6) of U, Np, and Pu. The principal chemistry is that of the *dioxo ions*, $[MO_2]^{2+}$, of U, Np, Pu, and Am (discussed shortly).

The +2 and +7 states are quite rare. The +2 state is confined to Am (the 5f analog of Eu), where the +2 ion can occur in CaF_2 lattices, and to Cf, Es, Fm, Md, and No, which have +2 ions in solution. These are chemically similar to Ba^{2+}. The Md^{2+} ion is less readily oxidized than Eu^{2+} ($E° = -0.15$ vs -0.43 V).

The +7 state is known only in oxoanions of Np and Pu when alkaline solutions are oxidized by O_3 or PuO_2 and Li_2O are heated in oxygen. Representative oxo anions are $[NpO_4(OH)_2]^{3-}$ and $[PuO_6]^{5-}$.

The dioxo ions $[MO_2]^+$ and $[MO_2]^{2+}$ are both formed. The stabilities of the $[MO_2]^+$ ions are determined by the ease of disproportionation, for example,

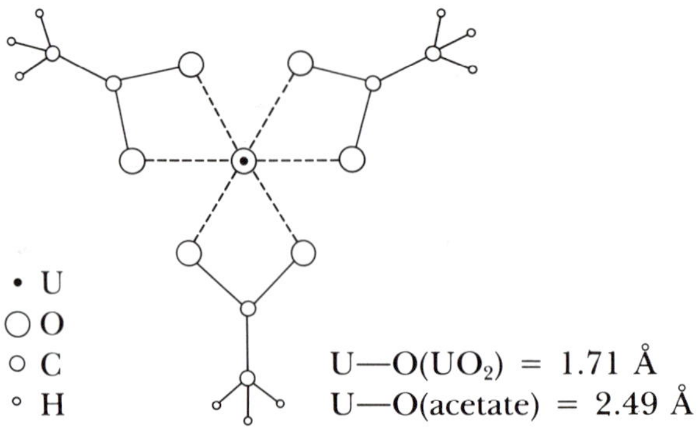

• U
○ O
○ C
° H

$U—O(UO_2) = 1.71$ Å
$U—O(acetate) = 2.49$ Å

Figure 27-2   The structure of the anion in $Na[UO_2(O_2CCH_3)_3]$ viewed along the linear UO_2 group. The carboxylate groups are bidentate and equivalent. The U—O distance in UO_2 is much shorter than the U—O distances in the equatorial plane.

$$2\,UO_2^+ + 4\,H^+ = U^{4+} + UO_2^{2+} + 2\,H_2O \qquad (27\text{-}3.1)$$

The stability order is Np > Am > Pu > U but, of course, there is dependence on the acid concentration. The UO_2^+ ion has only a transient existence in solution but is most stable in the pH range 2–4.

The $[MO_2]^{2+}$ ions are quite stable; $[AmO_2]^{2+}$ is most easily reduced, the stability order being U > Pu > Np > Am.

The $[AmO_2]^+$ and $[AmO_2]^{2+}$ ions undergo reduction at a few percent per hour by the products of their own α-radiation.

The linear dioxo ions can persist through a variety of chemical changes. They also appear as structural units in crystalline higher oxides. The ions are normally coordinated by solvent molecules or anions with four, or most often, five or six ligand atoms in or near the equatorial plane of the linear O—M—O group. These equatorial ligands are often not exactly coplanar. An example is the anion in sodium uranyl acetate shown in Fig. 27-2. Similar structures occur in $UO_2(NO_3)_2(H_2O)_2$, $Rb[UO_2(NO_3)_3]$, and so on.

27-4   Actinium

Actinium occurs in traces in U minerals, but can be made on a milligram scale by the neutron reaction

$$^{226}Ra\,(n\gamma)\ ^{227}Ra \xrightarrow{\ \beta^-\ } \ ^{227}Ac\,(\alpha,\ 21.7\ \text{year}) \qquad (27\text{-}4.1)$$

It is lanthanum-like in its chemistry, which is difficult to study because of the intense radiation of the decay products.

27-5   Thorium

Thorium is widely distributed, but the chief mineral is *monazite* sand, a complex phosphate that also contains lanthanides. The sand is digested with sodium hydroxide and the insoluble hydroxides are dissolved in hydrochloric acid. When the pH of the solution is adjusted to 5.8, the thorium, uranium, and about 3% of the lanthanides are precipitated as hydroxides. The thorium is recovered by extraction from greater than 6 M hydrochloric acid solution by tributyl phosphate in kerosene.

The commonest thorium compound is the *nitrate,* $Th(NO_3)_4 \cdot 5H_2O$. This is soluble in water and alcohols, ketones, and esters. In aqueous solution the Th^{4+} ion is hydrolyzed at a pH higher than about 3. It forms complex salts such as $K_4[Th(ox)_4] \cdot 4H_2O$ and $M^{II}[Th(NO_3)_6]$. On heating, the nitrate gives the white refractory dioxide ThO_2. Action of CCl_4 on this at 600 °C gives the white crystalline $ThCl_4$, which acts as a Lewis acid.

27-6 Protactinium

Protactinium can be isolated from residues after the extraction of uranium from pitchblende. It is exceedingly difficult to handle, except in fluoride solutions where it forms complexes (cf. Ta). In most other acid solutions it hydrolyzes to give polymeric species and colloids that are adsorbed on vessels and precipitates. Only a few compounds, some of Pa^{IV} but mostly Pa^V, are known; they generally resemble those of Ta. For example, the chloride is Pa_2Cl_{10}, the oxide is Pa_2O_5, and the fluoroanions $[PaF_6]^-$, $[PaF_7]^{2-}$, and $[PaF_8]^{3-}$ are formed.

27-7 Uranium

Until the discovery of nuclear fission by Lise Meitner, Otto Hahn, and Fritz Strassman in 1939, uranium was used only for coloring glass and ceramics, and the main reason for working its ores was to recover radium for use in cancer therapy. The isotope ^{235}U (0.72% abundance) is the prime nuclear fuel; although natural uranium can be used in nuclear reactors moderated by D_2O, most reactors and nuclear weapons use enriched uranium. Large-scale separation of ^{235}U employs gaseous diffusion of UF_6, but a gas centrifuge method now appears more economical.

Uranium is widely distributed and is more abundant than Ag, Hg, Cd, or Bi. It has few economic ores, the main one being *uraninite* (one form is *pitchblende*) an oxide of approximate composition UO_2. Uranium is recovered from nitric acid solutions by

1. Extraction of uranyl nitrate into diethyl ether or isobutylmethylketone; a salt such as NH_4^+, Ca^{2+}, or Al^{3+} nitrate is added as a "salting-out" agent to increase the extraction ratio to technically usable values. If tributyl phosphate in kerosene is used, no salting-out agent is necessary.
2. Removal from the organic solvent by washing with dilute HNO_3.
3. Recovery as U_3O_8 or UO_3 (see next subsection) by precipitation with ammonia.

Oxides

The U–O system is extremely complex. The main oxides are orange-yellow UO_3, black U_3O_8, and brown UO_2. Uranium trioxide (UO_3) is made by heating the hydrous oxide, mainly $UO_2(OH)_2 \cdot H_2O$, which is obtained by adding NH_4OH to $[UO_2]^{2+}$ solutions. The other oxides are obtained by the reactions

$$3\,UO_3 \xrightarrow{\;700\;°C\;} U_3O_8 + \tfrac{1}{2}O_2 \tag{27-7.1}$$

$$UO_3 + CO \xrightarrow{\;350\;°C\;} UO_2 + CO_2 \tag{27-7.2}$$

All oxides dissolve in HNO_3 to give uranyl nitrate, $UO_2(NO_3)_2 \cdot nH_2O$.

Halides

The *hexafluoride* UF_6 is obtained as colorless volatile crystals (mp 64 °C) by fluorination at 400 °C of UF_3 or UF_4. It is a very powerful oxidizing and fluorinating agent and is vigorously hydrolyzed by water.

The green *tetrachloride* is obtained on refluxing UO_3 with hexachloropropene. It is soluble in polar organic solvents and in water. The action of Cl_2 on UCl_4 gives U_2Cl_{10} and, under controlled conditions, the rather unstable UCl_6.

Hydride

Uranium reacts with dihydrogen even at 25 °C to give a pyrophoric black powder.

$$U + \tfrac{3}{2}H_2 \underset{\text{heat}}{\overset{250\ °C}{\rightleftharpoons}} UH_3 \tag{27-7.3}$$

This hydride is often more suitable for the preparation of uranium compounds than is the massive metal. Some typical reactions are

$$UH_3 + \begin{Bmatrix} H_2O & 350\ °C \\ Cl_2 & 200\ °C \\ H_2S & 450\ °C \\ HF & 400\ °C \\ HCl & 250\ °C \end{Bmatrix} = \begin{Bmatrix} UO_2 \\ UCl_4 \\ US_2 \\ UF_4 \\ UCl_3 \end{Bmatrix} \begin{matrix} (27\text{-}7.4) \\ (27\text{-}7.5) \\ (27\text{-}7.6) \\ (27\text{-}7.7) \\ (27\text{-}7.8) \end{matrix}$$

Dioxouranium(VI) or Uranyl Salts

The most common uranium salt is the yellow uranyl nitrate, which may have two, three, or six molecules of water depending on whether it is crystallized from fuming, concentrated, or dilute nitric acid. When extracted from aqueous solution into organic solvents uranyl nitrate is accompanied by four H_2O molecules, and the NO_3^- ions and water are coordinated in the equatorial plane.

On addition of an excess of sodium acetate to UO_2^{2+} solutions in dilute acetic acid, the insoluble salt $Na[UO_2(O_2CCH_3)_3]$ is precipitated. The uranyl ion is reduced to red-brown U^{3+} by Na/Hg or zinc, and U^{3+} is oxidizable by air to green U^{4+}. The potentials (1 M $HClO_4$) are

$$\mathrm{UO_2^{2+}} \xrightarrow{0.06\,\mathrm{V}} \mathrm{UO_2^+} \xrightarrow{0.58\,\mathrm{V}} \mathrm{U^{4+}} \xrightarrow{-0.63\,\mathrm{V}} \mathrm{U^{3+}} \xrightarrow{-1.8\,\mathrm{V}} \mathrm{U}$$

$$\underset{0.32\,\mathrm{V}}{\underbrace{}}$$

27-8   Neptunium, Plutonium, and Americium

The extraction of plutonium from uranium fuel elements involves (a) removal of the highly radioactive fission products that are produced simultaneously in comparable amounts, (b) recovery of the uranium for reprocessing, (c) remote control of all the chemical operations because of the radiation hazard. An additional hazard is the extreme toxicity of Pu, 10^{-6} g of which is potentially lethal; a particle of $^{239}\mathrm{PuO_2}$ only 1-μm in diameter can give a very high dose of radiation, enough to be strongly carcinogenic.

The separation methods of Np, Pu, and Am from U are based on the following chemistry.

1. Stabilities of oxidation states.   The stabilities of the major ions involved are $\mathrm{UO_2^{2+}} > \mathrm{NpO_2^{2+}} > \mathrm{PuO_2^{2+}} > \mathrm{AmO_2^{2+}}$; $\mathrm{Am^{3+}} > \mathrm{Pu^{3+}} \gg \mathrm{Np^{3+}}$, $\mathrm{U^{4+}}$. It is thus possible by choice of suitable oxidizing or reducing agents to obtain a solution containing the elements in different oxidation states; they can then be separated by precipitation or solvent extraction. For example, Pu can be oxidized to $\mathrm{PuO_2^{2+}}$ while Am remains as $\mathrm{Am^{3+}}$. The former can then be removed by solvent extraction or the latter by precipitation of $\mathrm{AmF_3}$.

2. Extractability into organic solvents.   The $\mathrm{MO_2^{2+}}$ ions are extracted from nitrate solutions into ethers. The $\mathrm{M^{4+}}$ ions are extracted into tributyl phosphate in kerosene from 6 M nitric acid solutions; the $\mathrm{M^{3+}}$ ions are similarly extracted from 10 to 16 M nitric acid, and neighboring actinides can be separated by a choice of conditions.

3. Precipitation reactions.   Only $\mathrm{M^{3+}}$ and $\mathrm{M^{4+}}$ give insoluble fluorides or phosphates from acid solutions. The higher oxidation states give either no precipitate or can be prevented from precipitation by complex formation with sulfate or other ions.

4. Ion exchange methods.   These are used mainly for small amounts of material as in the separation of Am and the following elements, as discussed later.

The following are examples of the separation of Pu from a nitric acid solution of the uranium fuel (plus its aluminum or other protective jacket).

The combination of oxidation–reduction cycles coupled with solvent extraction and/or precipitation methods removes the bulk of fission products (FP's). Certain elements—notably Ru, which forms cationic, neutral, and anionic nitrosyl complexes—may require special elimination steps. The initial uranyl nitrate solution contains $\mathrm{Pu^{4+}}$, since nitric acid cannot oxidize this to $\mathrm{Pu^V}$ or $\mathrm{Pu^{VI}}$.

Lanthanum Fluoride Cycle

This classical procedure was first developed by McMillan and Abelson for the isolation of neptunium, and is still of great utility. For the U to Pu separation, the cycle in Scheme 27-1 is repeated, with progressively smaller amounts of

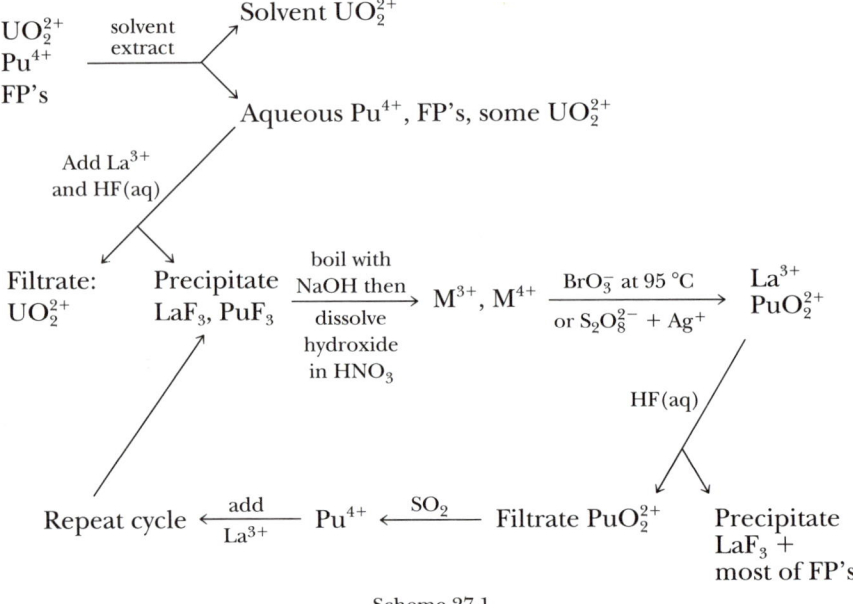

Scheme 27-1

lanthanum carrier and smaller volumes of solution, until plutonium becomes the bulk phase.

Tributyl Phosphate Solvent Extraction Cycle

The extraction coefficients from 6 N nitric acid solutions into 30% tributyl phosphate (TBP) in kerosene are $Pu^{4+} > PuO_2^{2+}$; $Np^{4+} \sim NpO_2^{2+} \gg Pu^{3+}$; $UO_2^{2+} > NpO_2^+ > PuO_2^{2+}$. The M^{3+} ions have very low extraction coefficients in 6 M acid, but from 12 M hydrochloric acid or 16 M nitric acid the extraction increases and the order is Np < Pu < Am < Cm < Bk.

Thus in the U to Pu separation, after addition of NO_2^- to adjust all of the plutonium to Pu^{4+}, we have Scheme 27-2.

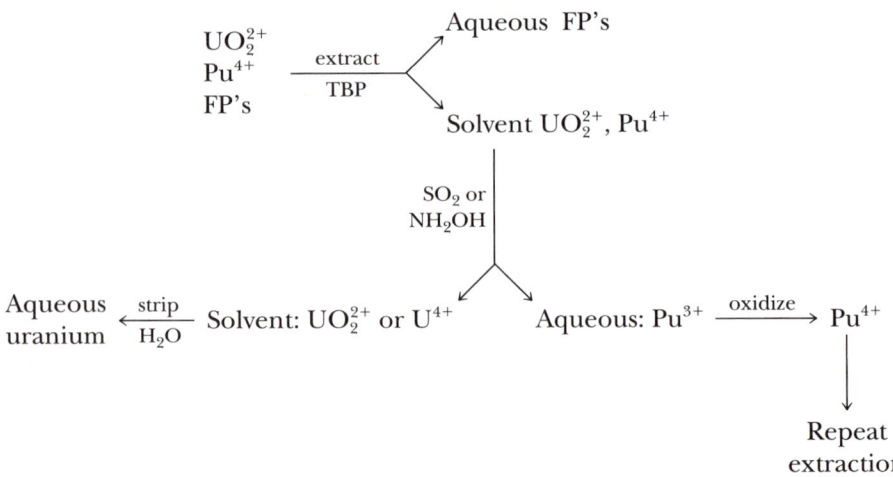

Scheme 27-2

The extraction of ^{237}Np involves similar principles of adjustment of oxidation state and solvent extraction; Pu is reduced by iron(II) sulfamate plus hydrazine to unextractable Pu^{III}, while Np^{IV} remains in the solvent from which it is differentially stripped by water to separate it from U.

The chemistries of U, Np, Pu, and Am are quite similar and solid compounds are usually isomorphous. The main differences are in stabilities of oxidation states in solution.

For Np, the oxidation states are well separated, but by contrast to UO$_2^+$, NpO$_2^+$ is reasonably stable. Plutonium chemistry is complicated because the potentials are not well separated and, indeed, in 1 M HClO$_4$ all four oxidation states can coexist.

For Am, the normal state is Am^{3+} and powerful oxidants are required to reach the higher states.

The cations all tend to hydrolyze in water, the ease of hydrolysis being Am > Pu > Np > U and M^{4+} > MO$_2^{2+}$ > M^{3+} > MO$_2^+$. The tendency to complexing also decreases Am > Pu > Np > U.

27-9 The Elements Following Americium

The isotope ^{242}Cm was first isolated among the products of α bombardment of ^{239}Pu, and its discovery actually preceded that of americium. Isotopes of the other elements were first identified in products from the first hydrogen bomb explosion (1952) or in cyclotron bombardments.

Ion-exchange methods have been indispensable in the separation of the elements following americium (often called the trans-americium elements) and also for tracer quantities of Np, Pu, and Am. By comparison with the elution of lanthanide ions, where La is eluted first and Lu last (Section 26-2), and by extrapolating data for Np^{3+} and Pu^{3+} the order of elution of the ions can be predicted accurately. Even a few *atoms* of an element can be identified because of the characteristic nuclear radiation.

The actinides as a group may be separated from lanthanides (always present as fission products from irradiations that produce the actinides) by use of concentrated HCl or 10 M LiCl, because the actinide ions more readily form chloro complex anions than lanthanides. Hence, actinides can be removed from cation exchange resins, or conversely, absorbed on anion exchange resins. There is also, in addition to the group separation, some separation of Pu, Am, Cm, Bk, and Cf to Es.

The actinide ions are usually separated from each other by elution with citrate or a similar elutant; some typical elution curves in which the relative positions of the corresponding lanthanides are given are shown in Fig. 27-3. Observe that a striking similarity occurs in the spacings of corresponding elements in the two series. There is a distinct break between Gd and Tb and between Cm and Bk, which can be attributed to the small change in ionic radius occasioned by the half-filling of the 4f and 5f shells, respectively. The elution order is not always as regular as that in Fig. 27-3.

After separation by ion exchange, macro amounts of the actinides can be precipitated by F$^-$ or oxalate; tracer quantities can be collected by using a La^{3+} carrier.

Solid compounds of Cm, Bk, Cf, and Es, mainly oxides and halides, have been characterized.

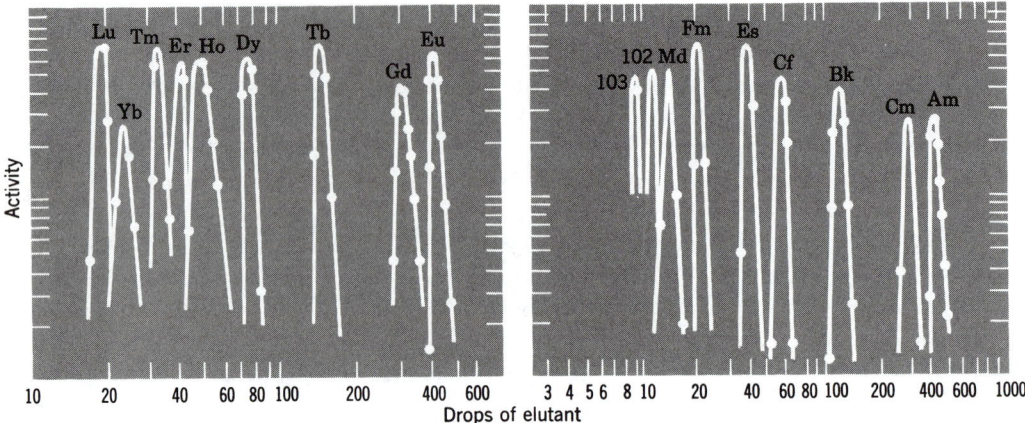

Figure 27-3   Elution of lanthanide +3 ions (left) and actinide +3 ions (right) from Dowex 50 cation-exchange resin. Buffered ammonium 2-hydroxybutyrate was the elutant. The predicted positions of No (102) and Lr (103) (unobserved here) are also shown. [Reproduced by permission from J. J. Katz and G. T. Seaborg, *The Chemistry of the Actinide Elements*, Methuen, London, 1957.]

STUDY GUIDE

Study Questions

A. Review

1. Name the actinide elements and list their electron configurations.
2. List the oxidation states for actinide elements.
3. Which actinide isotopes can be obtained in macroscopic amounts?
4. What are characteristic reactions of actinide +3 and +4 ions?
5. Which +3 ion has its $5f$ shell half full? What oxidation states do the preceding and succeeding elements show?
6. Which actinide element corresponds to Lu?
7. How are actinide metals made? What are their main features?
8. What is the structure of the dioxo ions MO_2^{2+} in, for example, uranyl nitrate hydrate?
9. How is actinium isolated? Which element does it most resemble?
10. What are the main sources of (a) thorium, and (b) uranium?
11. Uranium is usually recovered as uranyl nitrate. How is this converted to the metal?
12. What are the properties and main use of UF_6?
13. How is uranium hydride obtained? What are its uses?
14. What elements would the elements 105, 107, 112, and 118 be expected to resemble?

B. Additional Exercises

1. What are the main principles upon which the separations of Np, Pu, and Am from U are made?
2. Describe the lanthanum fluoride cycle for separation of Np or Pu from U.
3. Describe the tributyl phosphate extraction separation of Np and Pu from U.

4. How are the elements Am to Lr usually separated? Why is it first necessary to separate lanthanides as a group from the actinides as a group and how is this done?

5. Compare and contrast the chemistry of the dioxo ions of U, Np, Pu, and Am.

C. Questions from the Literature of Inorganic Chemistry

1. An actinide metallocarborane is described by F. R. Fronczek, G. W. Halstead, and K. N. Raymond, *J. Am. Chem. Soc.,* **1977,** *99,* 1769–1775.

    (a) What reasons do the authors offer (in the introductory paragraphs of this paper) for anticipating that the complex reported here should be stable?

    (b) What is the coordination geometry of the Li^+ counterion for this complex? What are the ligands?

    (c) What is the oxidation state of the uranium in this dicarbollide complex?

    (d) Why can actinide ions (namely, the uranium ion in this complex) accommodate more ligands than typical *d*-block elements?

    (e) What is the coordination number of U in this dicarbollide complex? Is this to be considered a coordinatively saturated system?

2. A protactinium–oxygen bond was examined by D. Brown, C. T. Reynolds, and P. T. Moseley, *J. Chem. Soc. Dalton Trans.,* **1972,** 857–859.

    (a) What evidence(s) have the authors given to suggest the presence of a Pa=O double bond?

    (b) With what other mono–oxo or dioxo–halogeno complexes does this Pa—O distance compare?

3. Consider the work by R. T. Paine, R. R. Ryan, and L. B. Asprey, *Inorg. Chem.,* **1975,** *14,* 1113–1117.

    (a) Write balanced chemical equations for the hydrolytic preparations reported here of UOF_4 and UO_2F_2.

    (b) How is water slowly obtained in the stoichiometric amounts necessary for limited hydrolysis of UF_6?

    (c) How is the controlled hydrolysis reported here different from total hydrolysis in the presence of excess water?

    (d) What evidence do the authors cite for an oxygen in a "terminal, axial position"? For what other MOF_4 systems is terminal–axial placement of oxygen found?

    (e) How many different coordination environments are realized by F^- groups in this structure?

SUPPLEMENTARY READING

Bagnall, K. W., Ed., *Lanthanides and Actinides,* Butterworths, London, 1972.

Bagnall, K. W., *The Actinide Elements,* Elsevier, Amsterdam, 1972.

Brown, D., "Some Recent Preparative Chemistry of Protactinium," *Adv. Inorg. Chem. Radiochem.,* **1969,** *12,* 1.

Casellato, U., Vigato, P. A., and Vidali, M., "Actinide Complexes with Carboxylic Acids," *Coord. Chem. Rev.,* **1978,** *26,* 85.

Cleveland, J. M., *The Chemistry of Plutonium,* Gordon and Breach, New York, 1970.

Cordfunke, E. H. P., *The Chemistry of Uranium,* Elsevier, Amsterdam, 1969.

Edelstein, M. M., Ed., *Actinides in Perspective,* Pergamon Press, New York, 1982.

Fields, P. R. and Moeller, T., "Lanthanide/Actinide Chemistry," in *Advances in*

Chemistry Series, ACS Monograph No. 71, American Chemical Society, Washington, DC, 1971.

Katz, J. J., Seaborg, G. T., and Morss, L. R., *The Chemistry of the Actininde Elements,* 2nd ed., Chapman & Hall, New York, 1986.

Lodhi, M. A. K., *Superheavy Elements,* Pergamon Press, New York, 1978.

Meyer, G. and Morss, L. R., *Synthesis of Lanthanide and Actinide Compounds,* Kluwer, Dordrecht, 1991.

Morss, L. R. and Fuger, J., *Transuranium Elements—A Half Century,* American Chemical Society, Washington DC, 1992.

Seaborg, G. T., *Man-Made Transuranium Elements,* Prentice-Hall, New York, 1963.

Seaborg, G. T., *Transuranium Elements, Products of Modern Alchemy,* Academic, New York, 1978.

Taube, M., *Plutonium: A General Survey,* Verlag Chemie, Weinheim, 1974.

Taylor, J. C., "Systematic Features in the Structural Chemistry of the Uranium Halides, Oxyhalides, and Related Transition Metal and Lanthanide Halides," *Coord. Chem. Rev.,* **1976,** *20,* 197.

Part 4

SOME SPECIAL TOPICS

Chapter 28

METAL CARBONYLS AND OTHER TRANSITION METAL COMPLEXES WITH π-ACCEPTOR (π-ACID) LIGANDS

28-1 Introduction

A characteristic feature of the d-block transition metal atoms is their ability to form complexes with a variety of neutral molecules (e.g., carbon monoxide, isocyanides, substituted phosphines, arsines and stibines, and nitric oxide) and various molecules with delocalized π orbitals such as pyridine (py), 2,2′-bipyridine (bpy) and 1,10-phenanthroline (phen). Very diverse types of complexes exist, ranging from binary molecular compounds such as $Cr(CO)_6$ or $Ni(PF_3)_4$, to complex ions such as $[Fe(CN)_5CO]^{3-}$, $[Mo(CO)_5I]^-$, $[Mn(CNR)_6]^+$, and $[V(phen)_3]^+$.

In many of these complexes, the metal atoms are in low-positive, zero, or even negative *formal* oxidation states. It is a characteristic of the ligands now under discussion that they can stabilize low oxidation states. This property is associated with the fact that these ligands have vacant π orbitals in addition to lone pairs. These vacant orbitals accept electron density from filled metal orbitals to form a type of π bonding that supplements the σ bonding arising from lone-pair donation. High electron density on the metal atom—of necessity in low oxidation states—can thus be *delocalized onto the ligands*. The ability of ligands to accept electron density into low-lying empty π orbitals is called π acidity. The word acidity is used in the Lewis sense.

The stoichiometries of most complexes of π-acid ligands can be predicted by use of the noble gas formalism. This formalism requires that the number of valence electrons possessed by the metal atom plus the number of pairs of σ electrons contributed by the ligands be equal to the number of electrons in the succeeding noble gas atom. The basis for this rule is the tendency of the metal atom to use its valence orbitals, nd, $(n + 1)s$, and $(n + 1)p$, as fully as possible, in forming bonds to ligands. Although it is of considerable utility in the design of new compounds, particularly of metal carbonyls, nitrosyls and isocyanides, and their substitution products, it is by no means infallible. It fails altogether for the bipyridine and dithiolene type of ligand, and there are significant exceptions even among carbonyls such as $V(CO)_6$ and $[Mo(CO)_2(diphos)_2]^+$, where diphos = 1,2-bis(diphenylphosphino)ethane.

641

CARBON MONOXIDE COMPLEXES

The most important π-acceptor ligand is carbon monoxide. Many carbonyl complexes are of considerable structural interest as well as being important industrially and in catalytic and other reactions. Carbonyl derivatives of at least one type are known for all of the transition metals. The first metal carbonyls, $Ni(CO)_4$ and $Fe(CO)_5$, were discovered by A. Mond in 1890 and 1891; he developed an industrial process for the isolation of pure nickel based on the formation and subsequent thermal decomposition of the volatile $Ni(CO)_4$.

28-2   Mononuclear Metal Carbonyls

The simplest carbonyls are of the type $M(CO)_x$ (Table 28-1A). The compounds are all hydrophobic, volatile, and soluble to varying degrees in nonpolar solvents. Of the d-block metals, the ones that form stable mononuclear carbonyls are principally those that require an integral number of carbonyl ligands to attain the number of valence electrons in the succeeding noble gas atom. The only important exception is vanadium, which forms the $V(CO)_6$ molecule. Since the number of valence electrons for the noble gases is 18, the noble gas formalism may be simplified to the *18-electron rule*—stable metal complexes will be those which, in acquiring electrons from ligands, attain a total of 18 electrons (metal valence electrons + donated ligand electrons) in their valence shell. It obviously becomes necessary to know how to count ligand electrons properly in applying this formalism.

Although there are exceptions, for the majority of simpler transition metal organometallics, and especially for the mononuclear and binuclear metal carbonyls and their derivatives, the 18-electron formalism is useful. We start with the mononuclear binary carbonyls of Table 28-1.

The 18-Electron Rule as Applied to Mononuclear Metal Carbonyls

Group VIA(6) Metals

The stable binary carbonyls are the hexacarbonyls, $M(CO)_6$, because the valence electrons of the metal (6 valence electrons for Cr, Mo, or W) plus 12 electrons from the ligands (each of the six CO ligands is considered to be a 2-electron donor) brings the total to 18. Stable derivatives of the mononuclear carbonyls include those where one or more CO groups have been replaced by an equal number of 2-electron donors, so that the total number of electrons provided by ligands remains 12. Two examples are shown in Reactions 28-2.1 and 28-2.2:

$$W(CO)_6 + Cl^- \longrightarrow W(CO)_5Cl^- + CO \qquad (28\text{-}2.1)$$

$$Cr(CO)_6 + R_2S \longrightarrow Cr(CO)_5SR_2 + CO \qquad (28\text{-}2.2)$$

where the chloride anion or the thioether are considered to be 2-electron σ donors. Substitution reactions of the Group VIA(6) hexacarbonyls proceed by dissociative mechanisms because loss of a carbonyl ligand to give a 16-electron intermediate is more favorable than the gain of an extra ligand (associative mechanism).

Table 28-1   Some Representative Metal Carbonyls and Carbonyl Hydrides

Compound	Color and Form	Structure	Comments
A. Mononuclear Carbonyls			
$V(CO)_6$	Black crystals; decomposes 70 °C; sublimes in vacuum	Octahedral	Yellow orange in solution; paramagnetic (1 e^-)
$Cr(CO)_6$ $Mo(CO)_6$ $W(CO)_6$	Colorless crystals; all sublime in vacuum	Octahedral	Stable to air; decompose 180–200 °C
$Fe(CO)_5$	Yellow liquid; mp −20 °C bp 103 °C	tbp	Action of UV gives $Fe_2(CO)_9$
$Ru(CO)_5$	Colorless liquid; mp −22 °C	tbp (by IR)	Very volatile and difficult to prepare
$Ni(CO)_4$	Colorless liquid; mp −25 °C bp 43 °C	Tetrahedral	Very toxic; musty smell; flammable; decomposes readily to metal
B. Polynuclear Carbonyls			
$Mn_2(CO)_{10}$[a]	Yellow solid mp 151 °C sublimes 50 °C (10^{-2} mm)	See Fig. 28-2	The Mn—Mn bond is long (2.93 Å) and $Mn_2(CO)_{10}$ is reactive
$Fe_2(CO)_9$	Gold solid mp 100 °C decomposes	See Fig. 28-2	Very insoluble and nonvolatile
$Fe_3(CO)_{12}$	Green-black solid mp 140–150 °C decomposes	See Fig. 28-2	Moderately soluble
$Rh_4(CO)_{12}$	Brick red solid mp 150 °C decomposes sublimes 65 °C (10^{-2} mm)	See Fig. 28-2	Useful reagent for many carbonyl rhodium compounds
C. Carbonyl Hydrides[b]			
$HMn(CO)_5$[a]	Colorless liquid mp −25 °C	Octahedral	Stable at 25 °C, weak acid $\delta = -7.5$ ppm[b]
$H_2Fe(CO)_4$	Yellow liquid, colorless gas mp −70 °C	v. distorted octahedron	Decomposes −10 °C Weak acid $\delta = -10.1$ ppm[b]
$H_2Fe_3(CO)_{11}$	Dark red liquid	Uncertain	
$HCo(CO)_4$	Yellow liquid, colorless gas, mp −20 °C	Distorted tbp	Decomposes above mp, strong acid $\delta = -10$ ppm[b]

[a]Very similar Tc and Re analogs are known.

[b]δ value is the position of the high-resolution proton magnetic resonance line in parts per million referred to tetramethylsilane reference as 0.0 ppm. Negative values to high field.

Fe(CO)₅

Here it is the pentacarbonyl that is favored. Eight valence electrons from the metal plus 10 from the five CO groups give the stable 18-electron configuration. This occurs similarly for the other members of the group, Ru and Os, although the monomers are unstable towards formation of the polynuclear systems to be discussed shortly. Replacement of a CO ligand by another 2-electron donor is a common reaction of iron pentacarbonyl.

$$Fe(CO)_5 + py \longrightarrow Fe(CO)_4py + CO \qquad (28\text{-}2.3)$$

Again, reactions such as 28-2.3 proceed by dissociative mechanisms because of the greater likelihood of dissociation of a CO group to give a 16-electron intermediate (or transition state) than of ligand gain (associative mechanism), which would exceed the 18-electron configuration. The 18-electron rule requires that 2-electron reduction of $Fe(CO)_5$ as in Reaction 28-2.4:

$$Fe(CO)_5 + 2 Na \longrightarrow Na_2[Fe(CO)_4] + CO \qquad (28\text{-}2.4)$$

be accompanied by loss of one CO ligand. Accordingly, the product of Reaction 28-2.4 is the tetracarbonyl dianion. Here we consider that 8 electrons from the four CO ligands, 8 from the Fe atom, and 2 electrons that are added to provide the 2− charge give the stable 18-electron total.

Ni(CO)₄

This nickel compound achieves the 18-electron total by coordination of four CO ligands to the 10-electron nickel center.

28-3 Polynuclear Metal Carbonyls

In each of the mononuclear metal carbonyls metioned in Section 28-2, an even number of metal valence electrons allowed the 18-electron formalism to be satisfied by coordination from an integral number of 2-electron donor ligands. Where the metal brings an odd number of valence electrons to the structure (Mn, Tc, Re; Co, Rh, or Ir) or where condensation to polynuclear metal carbonyls is thermodynamically favorable (Fe, Ru, or Os), an understanding of how the 18-electron configuration is achieved requires consideration of metal-to-metal bonds.

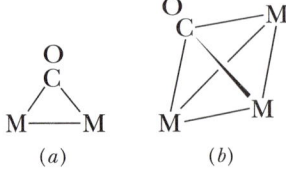

Figure 28-1 The two main types of bridging CO groups: (*a*) doubly bridging and (*b*) triply bridging.

There are numerous polynuclear carbonyls that may be homonuclear, for example $Fe_3(CO)_{12}$ or heteronuclear $MnRe(CO)_{10}$. In these compounds there are not only linear M—C—O groups but also either M—M bonds alone or M—M bonds plus *bridging carbonyl* groups. The two principal types of bridging group are depicted in Fig. 28-1. The doubly bridging type occurs fairly frequently and practically always in conjunction with an M—M bond.

Some important polynuclear carbonyls are listed in Table 28-1B and their structures and those of others are shown in Fig. 28-2.

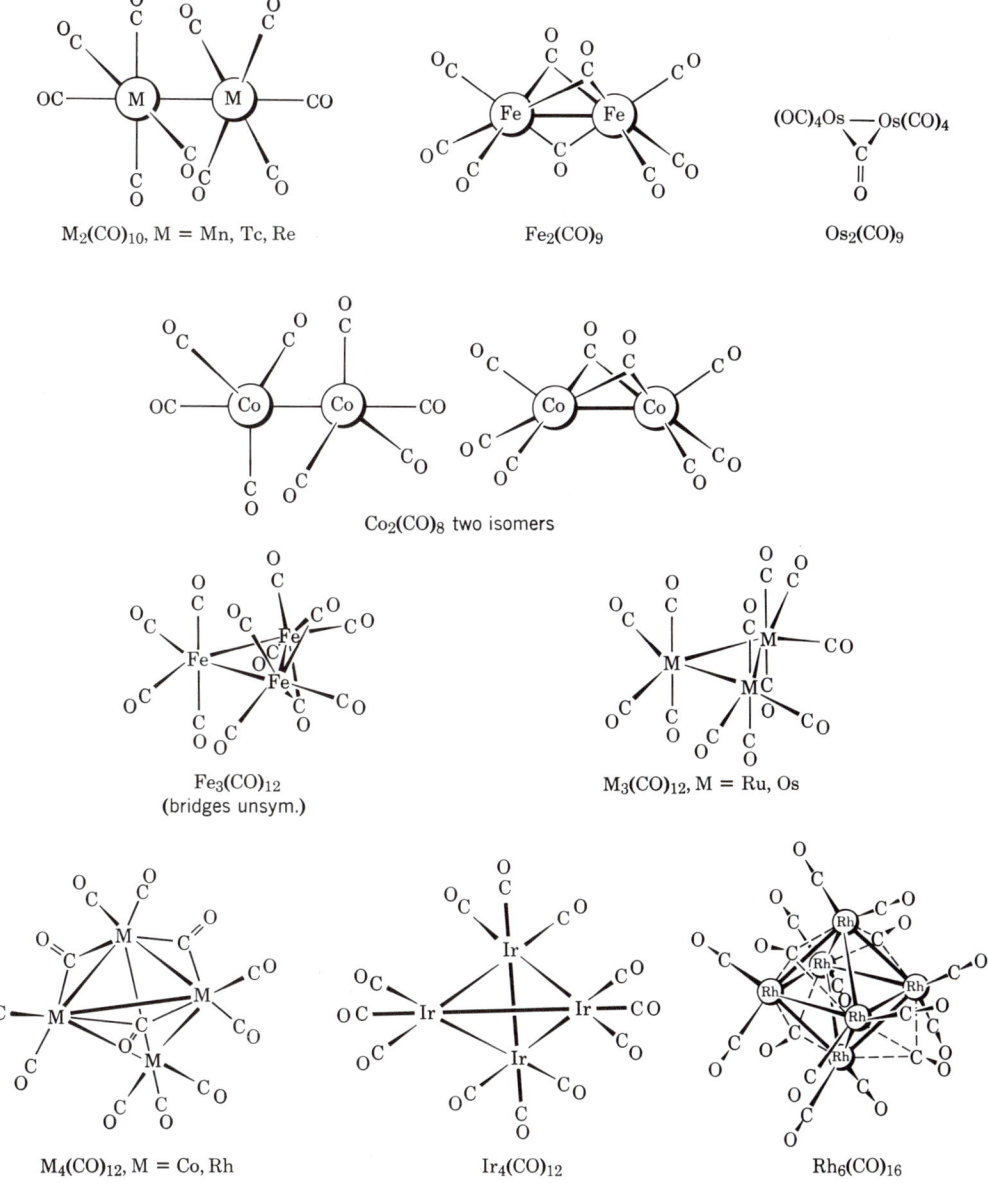

Figure 28-2 The structures of some polynuclear metal carbonyls.

Bridging CO groups very often occur in pairs, as in Structure 28-Ia. Any pair of bridging CO groups can only be regarded as an alternative to a nonbridged arrangement with two terminal groups, as in Structure 28-Ib.

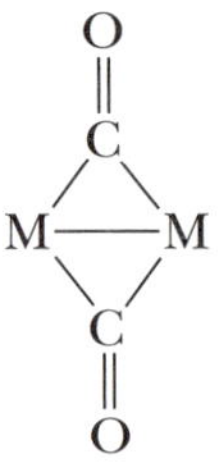

28-Ia   28-Ib

The relative stabilities of the alternatives appear to depend markedly on the size of the metal atoms. The larger the metal atoms the greater is the preference for a nonbridged structure. Thus, in any group the relative stability of nonbridged structures increases as the group is descended. For example, $Fe_3(CO)_{12}$ has two bridging CO groups while $Ru_3(CO)_{12}$ and $Os_3(CO)_{12}$ have none. The generalization concerning metal atom size also covers the trend horizontally in the periodic table. Thus, the large Mn atoms form only the nonbridged $(OC)_5Mn—Mn(CO)_5$ molecule, whereas the dinuclear cobalt carbonyl, $Co_2(CO)_8$, exists as an equilibrium mixture of the bridged and nonbridged structures.

Carbonyl groups less commonly bridge triangular arrays of three metal atoms [Fig. 28-1(*b*)] as in $Rh_6(CO)_{16}$ (Fig. 28-2).

The presence of bridging CO groups can often be recognized from the IR spectra of the compounds (see Section 28-7).

The 18-Electron Rule as Applied to Binuclear Metal Carbonyls

The counting of electrons in binuclear metal carbonyls should obey the following conventions:

1. Electrons in metal–metal bonds should be assigned homolytically (divided evenly) among the two metals.
2. Terminal CO groups are considered to be 2-electron donors, as usual; doubly bridging CO groups contribute 1 electron to each metal.
3. Where two isomers arise because of terminal-bridging tautomerism of CO groups, the total number of valence electrons, in either case, should be found to be 18, because the number of valence electrons is unaffected by tautomerism. (Compare Structures 28-Ia and 28-Ib, where in each one, each metal atom receives 2 electrons from CO ligands.)

The counting of electrons at each metal develops as follows:

$Mn_2(CO)_{10}$:

Mn	7 valence electrons
Terminal CO groups	$2 \times 5 = 10$ electrons
Mn—Mn bond	1 electron
Total	18 electrons

$Fe_2(CO)_9$:

Fe	8 valence electrons
Terminal CO groups	$2 \times 3 = 6$ electrons
Bridging CO groups	$1 \times 3 = 3$ electrons
Fe—Fe bond	1 electron
Total	18 electrons

$Os_2(CO)_9$:

Os	8 valence electrons
	(exclusive of f electrons)
Terminal CO groups	$2 \times 4 = 8$ electrons
Bridging CO group	$1 \times 1 = 1$ electron
Os—Os bond	1 electron
Total	18 electrons

$Co_2(CO)_8$, the nonbridged isomer:

Co	9 valence electrons
Terminal CO groups	$2 \times 4 = 8$ electrons
Co—Co bond	1 electron
Total	18 electrons

$Co_2(CO)_8$, the bridged isomer:

Co	9 valence electrons
Terminal CO groups	$2 \times 3 = 6$ electrons
Bridging CO groups	$1 \times 2 = 2$ electrons
Co—Co bond	1 electron
Total	18 electrons

The counting of electrons in clusters containing three or more metals is not always such a straightforward affair. Many clusters are found to be "formally" unsaturated, and we shall not pursue the topic here in detail. We may, however, mention two cases where the procedure for a given metal atom is quite easy:

$Ru_3(CO)_{12}$:

Ru	8 valence electrons
Terminal CO groups	$2 \times 4 = 8$ electrons
Two Ru—Ru bonds	$1 \times 2 = 2$ electrons
Total	18 electrons

$Ir_4(CO)_{12}$:

Ir	9 valence electrons
Terminal CO groups	$2 \times 3 = 6$ electrons
Three Ir—Ir bonds	$1 \times 3 = 3$ electrons
Total	18 electrons

28-4 Stereochemical Nonrigidity in Carbonyls

It is very common for bi- and polynuclear metal carbonyls to undergo rapid intramolecular rearrangements in which CO ligands are scrambled over the two or more metal atoms. These scrambling processes are observed and studied by NMR spectroscopy.

In many binuclear compounds the mechanism of scrambling has as its key steps the opening and closing of pairs of bridges, as is illustrated in the following two cases, where Cp represents the C_5H_5 group, which we discuss in detail in Chapter 29. For now, it is sufficient to note that CO groups that are labeled with *, a, or b are scrambled by the processes shown here.

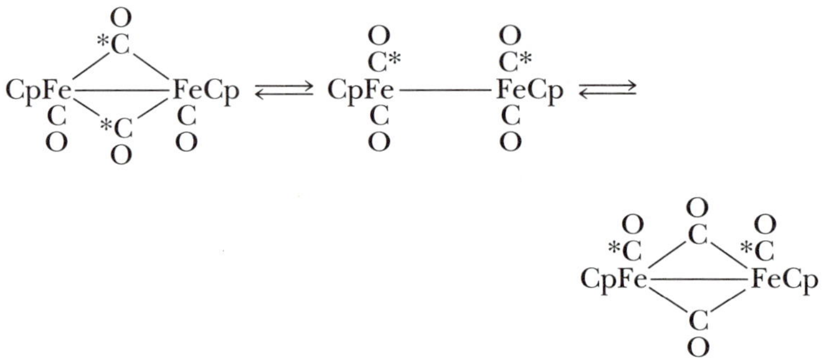

(28-4.1)

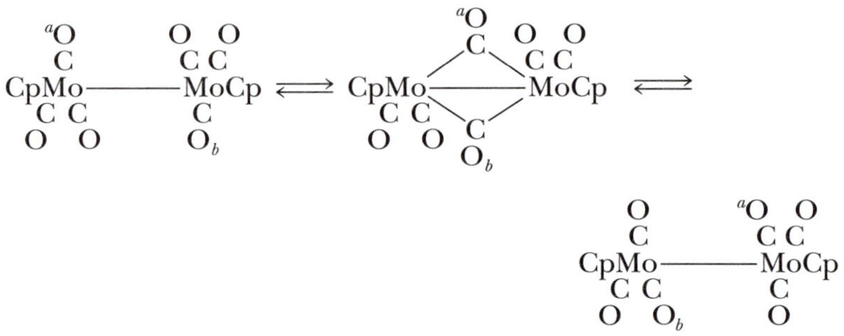

(28-4.2)

A more elaborate example is presented by $Rh_4(CO)_{12}$ in which the 12 CO ligands move rapidly over the entire tetrahedral skeleton in a series of steps, each involving the concerted opening or closing of a set of three bridges, as shown here

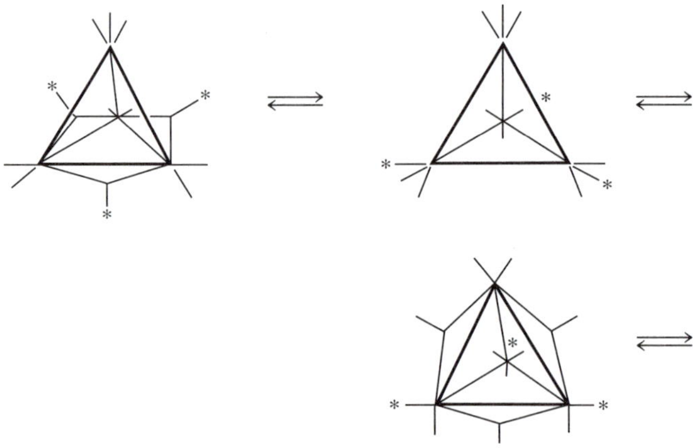

⇌ and so on   (28-4.3)

The ease with which these processes proceed in nearly all cases is attributable to the fact that in most polynuclear carbonyls the bridged and non-bridged structures differ very little in energy and, thus, whichever one is the ground state [cf. the $Cp_2Fe_2(CO)_4$ and $Cp_2Mo_2(CO)_6$ cases in reactions 28-4.1 and 28-4.2] the other provides an energetically accessible intermediate for the scrambling. In the examples cited, the rates at which the individual steps occur at room temperature are in the range of $10–10^3$ times per second. Thus, in the course of any ordinary chemical reaction, complete scrambling will occur—many times over.

28-5   Preparation of Metal Carbonyls

Although many metals, when prepared in a highly dispersed form, will react with CO, only $Ni(CO)_4$ and $Fe(CO)_5$ are normally made this way. Finely divided nickel will react at room temperature; an appreciable rate of reaction with iron requires elevated temperatures and pressures.

In general, carbonyls are formed when metal compounds are reduced in the presence of CO. Usually high pressures (200–300 atm) of CO are required. In some cases, CO itself serves as the only necessary reducing agent, for example,

$$Re_2O_7 + 17\,CO \longrightarrow Re_2(CO)_{10} + 7\,CO_2 \qquad (28\text{-}5.1)$$

but usually an additional reducing agent is needed. Typical reducing agents are H_2, metals (e.g., Na, Al, Mg, or Cu), or compounds such as trialkylaluminum or $(C_6H_5)_2CO^-Na^+$:

$$2\,CoCO_3 + 2\,H_2 + 8\,CO \xrightarrow{\text{250–300 atm}} Co_2(CO)_8 + 2\,CO_2 + 2\,H_2O \qquad (28\text{-}5.2)$$

$$2\,Mn(acac)_3 + 10\,CO \xrightarrow{(C_2H_5)_3Al} Mn_2(CO)_{10} \qquad (28\text{-}5.3)$$

$$CrCl_3 + 6\,CO \xrightarrow{C_6H_5MgBr} Cr(CO)_6 \qquad (28\text{-}5.4)$$

The reaction mechanisms are obscure but when Na, Mg, or Al are used, reduction to metal probably occurs. When organometallic reducing agents are employed, unstable organo derivatives of the transition metal may be formed as intermediates.

28-6   Bonding in Linear M—C—O Groups

The fact that refractory metals, with high heats of atomization (~ 400 kJ mol^{-1}), and an inert molecule like CO are capable of uniting to form stable, molecular compounds is quite surprising, especially when the CO molecules retain their individuality. Moreover, the Lewis basicity of CO is *negligible*. However, the explanation lies in the multiple nature of the M—CO bond, for which there is much evidence, some of it semiquantitative.

Although we can formulate the bonding in terms of a resonance hybrid of Structures 28-IIa and 28-IIb, a MO formulation is more detailed and accurate.

$$\overset{-}{M}-\overset{+}{C}\equiv O: \longleftrightarrow M=C=\overset{..}{\underset{..}{O}}:$$

28-IIa   28-IIb

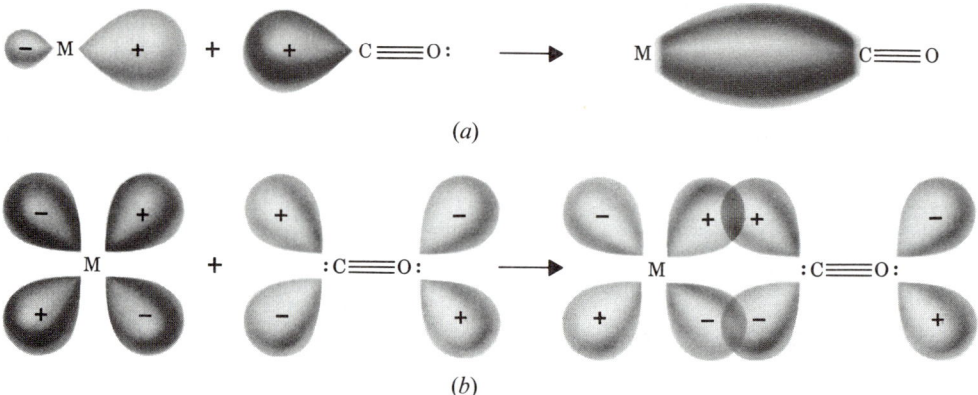

Figure 28-3   (*a*) The formation of the metal←CO σ bond using a lone electron pair (most likely in an *sp* hybrid) on carbon. (*b*) The formation of the metal→CO π back-bond. Other orbitals on the CO ligand are omitted for clarity.

First, there is a dative overlap of the filled carbon σ orbital [Fig. 28-3(*a*)] and, second, a dative overlap of a filled *d*π or hybrid *dp*π metal orbital with an empty antibonding *p*π orbital of the CO [Fig. 28-3(*b*)]. This bonding mechanism is synergic, since the drift of metal electrons into CO orbitals will tend to make the CO as a whole negative and, hence, will increase its basicity via the σ orbital of carbon; also the drift of electrons to the metal in the σ bond tends to make the CO positive, thus enhancing the acceptor strength of the π* orbitals. Thus, the effects of σ-bond formation strengthen the π bonding and vice versa.

The main lines of physical evidence showing the multiple nature of the M—CO bonds are bond lengths and vibrational spectra. According to the preceding description of the bonding, as the extent of back-donation from M to CO increases, the M—C bond becomes stronger and the C≡O bond becomes weaker. Thus the multiple bonding should be evidenced by shorter M—C and longer C—O bonds as compared with M—C single bonds and C≡O triple bonds, respectively. Although C—O bond lengths are rather insensitive to bond order, for M—C bonds in selected compounds there *is* appreciable shortening consistent with the π-bonding concept.

28-7   Vibrational Spectra of Metal Carbonyls

Infrared spectra have been widely used in the study of metal carbonyls since the C—O stretching frequencies give very strong sharp bands that are well separated from other vibrational modes of any other ligands also present.

The CO molecule has a stretching frequency of 2143 cm^{-1}. Terminal CO groups in neutral metal carbonyl molecules are found in the range 2125–1850 cm^{-1}, showing the reduction in CO bond orders. Moreover, when changes are made that should increase the extent of M—C back-bonding, the CO frequencies are shifted to even lower values. Thus, if some CO groups are replaced by ligands with low or negligible back-accepting ability, those CO groups that remain must accept more *d*π electrons from the metal to prevent the accumulation of negative charge on the metal atom. Hence, the frequency for $Cr(CO)_6$ is about 2000 cm^{-1} (exact values vary with phase and solvent) whereas, when three CO groups are replaced by amine groups that have essentially no ability to back-accept, as in $Cr(CO)_3(dien)$, where dien = $NH(CH_2CH_2NH_2)_2$, there are two CO stretching modes with frequencies of about 1900 and 1760 cm^{-1}. Similarly, in $V(CO)_6^-$, where more negative charge must be taken from the metal atom, a band is found at about 1860 cm^{-1} corresponding to the one found at about 2000 cm^{-1} in $Cr(CO)_6$. Conversely, a change that would tend to inhibit the shift of electrons from metal to CO π orbitals, such as placing a positive charge on the metal, should cause the CO frequencies to rise, for example,

$$Mn(CO)_6^+ \sim 2090 \qquad Mn(dien)(CO)_3^+ \sim 2020, \sim 1900$$

$$Cr(CO)_6 \sim 2000 \qquad Cr(dien)(CO)_3 \sim 1900, \sim 1760$$

$$V(CO)_6^- \sim 1860$$

The most important use of IR spectra of CO compounds is in *structural* diagnosis, whereby bridging and terminal CO groups can be recognized.

For terminal M—CO the frequencies of C—O stretches range from 1850 to 2125 cm^{-1}, but for bridging CO groups the range is from 1750 to 1850 cm^{-1}. Figure 28-4 shows how these facts may be used to infer structures. Observe that $Fe_2(CO)_9$ has strong bands in both the terminal and the bridging regions. From this alone it could be inferred that the structure must contain both types of CO groups; X-ray study shows that this is true. For $Os_3(CO)_{12}$ several structures consistent with the general rules of valence can be envisioned; some of these would have bridging CO groups, while the actual one (Fig. 28-2) does not. The IR spectrum alone [Fig. 28-4(b)] shows that no structure with bridging CO groups is acceptable, since there is no absorption band below 2000 cm^{-1}.

In using the *positions* of CO stretching bands to infer the presence of bridging CO groups, certain conditions must be remembered. The frequencies of terminal CO stretches can be quite low if (a) there are ligands present that are good

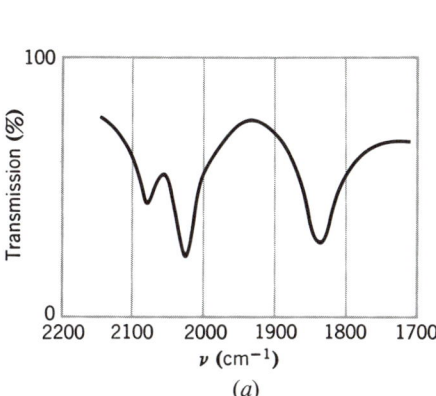

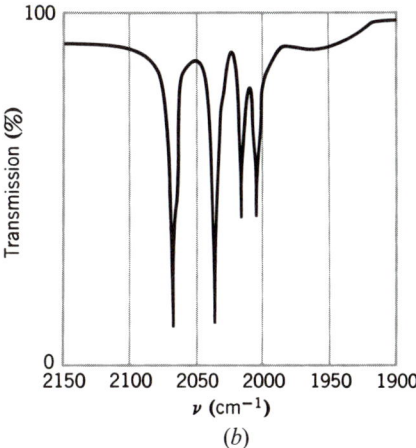

Figure 28-4   The IR spectra in the CO stretching region of (a) solid $Fe_2(CO)_9$, and (b) $Os_3(CO)_{12}$ in solution. Notice the greater sharpness of the solution spectrum. The most desirable spectra are those obtained in nonpolar solvents or in the gas phase.

donors but poor π acceptors, or (b) there is a net negative charge on the species. In either case, back-donation to the CO groups becomes very extensive, thus increasing the M—C bond orders, decreasing the C—O bond orders, and driving the CO stretching frequencies down.

28-8   Reactions of Metal Carbonyls

The variety of reactions of the various carbonyls is so large that only a few types can be mentioned. For $Mo(CO)_6$ and $Fe(CO)_5$, Fig. 28-5 gives an indication of the extensive chemistry that is typical for any individual carbonyl.

The most important general reactions of carbonyls are those in which CO groups are displaced by ligands such as PX_3, PR_3, $P(OR)_3$, SR_2, NR_3, OR_2, and RNC, or unsaturated organic molecules such as C_6H_6 or cycloheptatriene. Derivatives of organic molecules are discussed in Chapter 29.

Another important general reaction is that with bases (OH^-, H^-, NH_2^-), leading to carbonylate anions (discussed in Section 28-9).

Substitution reactions may proceed by either thermal or photochemical activation. In some instances, only the photochemical reaction is practical. Generally, the photochemical process first involves expulsion of a CO group after absorption of a photon, followed by entry of the substituent into the coordination sphere. For example,

$$Cr(CO)_6 \xrightarrow[-CO]{h\nu} Cr(CO)_5 \xrightarrow{+L} Cr(CO)_5 L \qquad (28\text{-}8.1)$$

The advantage offered by the photochemical route of Reaction 28-8.1 is that di- and trisubstituted products can be avoided.

If we further consider the reactions of $Fe(CO)_5$ as shown in Fig. 28-5, we find that four of these involve simple substitution

$$Fe(CO)_5 + C_7H_8 \longrightarrow C_7H_8Fe(CO)_3 + 2\ CO \qquad (28\text{-}8.2)$$

$$Fe(CO)_5 + C_8H_8 \longrightarrow C_8H_8Fe(CO)_3 + 2\ CO \qquad (28\text{-}8.3)$$

$$Fe(CO)_5 + RNC \longrightarrow RNCFe(CO)_4 + CO \qquad (28\text{-}8.4)$$

$$Fe(CO)_5 + n\ PPh_3 \longrightarrow (PPh_3)_nFe(CO)_{5-n} + n\ CO \qquad (28\text{-}8.5)$$

Other reactions of $Fe(CO)_5$, as shown in Fig. 28-5, include reduction to carbonylate anions or carbonyl hydrides, as discussed in Section 28-9.

28-9   Carbonylate Anions and Carbonyl Hydrides

Carbonylate anions and carbonyl hydrides are formed in a number of ways. The anionic hydride $[HFe(CO)_4]^-$ is obtained when $Fe(CO)_5$ is treated with aqueous hydroxide, as in Reaction 28-9.1

$$Fe(CO)_5 + 3\ NaOH(aq) \longrightarrow$$
$$Na[HFe(CO)_4](aq) + Na_2CO_3(aq) + H_2O \qquad (28\text{-}9.1)$$

or when the dianion $[Fe(CO)_4]^{2-}$ is protonated:

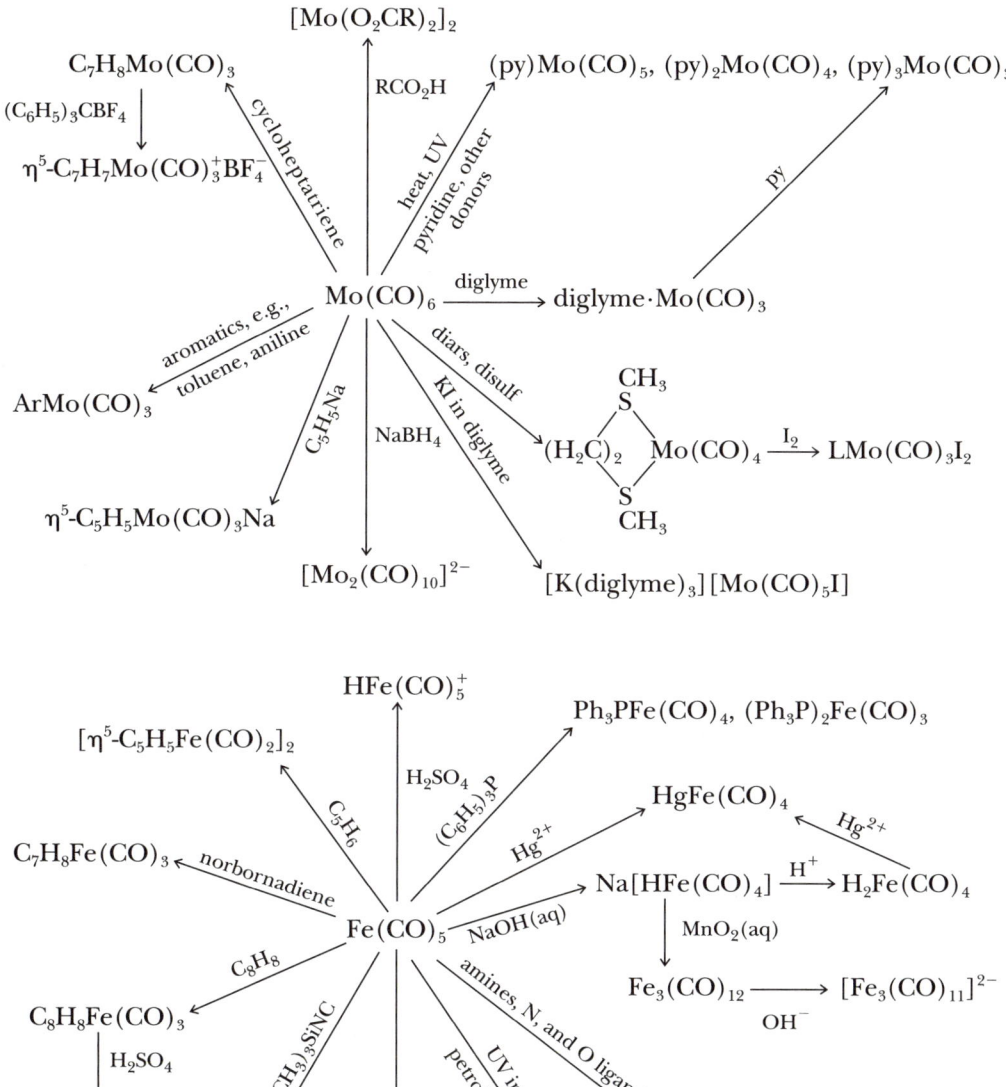

Figure 28-5   Some reactions of the carbonyls $Mo(CO)_6$ and $Fe(CO)_5$. Further discussion is given in the text.

$$[Fe(CO)_4]^{2-} + H^+ \longrightarrow [HFe(CO)_4]^- \qquad (28\text{-}9.2)$$

Carbonylate anions may be prepared by reduction with sodium, as in Reactions 28-9.3 and 28-9.4.

$$Co_2(CO)_8 + 2 \; Na \; Hg \xrightarrow{\text{THF}} 2 \, Na[Co(CO)_4] \qquad (28\text{-}9.3)$$

$$Cr(CO)_6 + 2 \, Na \longrightarrow Na_2[Cr(CO)_5] + CO \qquad (28\text{-}9.4)$$

Reaction 28-9.3 involves cleavage of a metal-to-metal bond by Na. Similarly, Li cleaves a metal-to-metal bond in Reaction 28-9.5.

$$Mn_2(CO)_{10} + 2 \, Li \xrightarrow{\text{THF}} 2 \, Li[Mn(CO)_5] \qquad (28\text{-}9.5)$$

The cobalt tetracarbonyl anion may also be prepared by Reaction 28-9.6.

$$2 \, Co^{2+}(aq) + 11 \, CO + 12 \, OH^- \xrightarrow{\text{KCN(aq)}} 2 \, [Co(CO)_4]^- + 3 \, CO_3^{2-} + 6 \, H_2O \qquad (28\text{-}9.6)$$

The stoichiometries of the simpler carbonylate anions obey the 18-electron rule (noble gas formalism). Most of them are readily oxidized by air. The alkali metal salts are soluble in water, from which they can be precipitated by large cations such as $[(C_6H_5)_4As]^+$. In the presence of water and other weak acids, though, many of the carbonylate anions can be protonated to give hydrides.

The general reaction of carbonylate anions with halogen compounds is important. Thus with alkyl halides and with acyl halides we have Reactions 28-9.7 and 28-9.8:

$$[Fe(CO)_4]^{2-} + RX \longrightarrow [RFe(CO)_4]^- + X^- \qquad (28\text{-}9.7)$$

$$[Fe(CO)_4]^{2-} + RC(O)Cl \longrightarrow [RC(O)Fe(CO)_4]^- + X^- \qquad (28\text{-}9.8)$$

which proceed by classic S_N2 mechanisms to give metal alkyls and metal acyls, respectively. As another example, consider the formation of a metal-to-carbon bond as in Reaction 28-9.9.

$$Mn(CO)_5^- + ClCH_2CH{=}CH_2 = (CO)_5MnCH_2CH{=}CH_2 + Cl^- \qquad (28\text{-}9.9)$$

In addition, metal-to-metal bonds may be formed, as in Reactions 28-9.10 and 28-9.11.

$$[Fe(CO)_4]^{2-} + 2 \, Ph_3PAuCl \longrightarrow (Ph_3PAu)_2Fe(CO)_4 + 2 \, Cl^- \qquad (28\text{-}9.10)$$

$$Co(CO)_4^- + Mn(CO)_5Br \longrightarrow (OC)_4CoMn(CO)_5 + Br^- \qquad (28\text{-}9.11)$$

As has already been mentioned, hydrides corresponding to carbonylate anions can be isolated. A few of the neutral ones are listed in Table 28-1C, along with their properties. These neutral carbonyl hydrides, which are usually rather unstable, can be obtained by acidification of the appropriate alkali carbonylates, as in Reactions 28-9.12 or 28-9.13.

$$Na[Co(CO)_4] + H^+(aq) \longrightarrow HCo(CO)_4 + Na^+(aq) \qquad (28\text{-}9.12)$$

$$Na_2[Fe(CO)_4] + 2\,H^+ \longrightarrow H_2Fe(CO)_4 + 2\,Na^+ \qquad (28\text{-}9.13)$$

They may also be obtained by reduction of metal carbonyl halides (Section 28-10) as in Reaction 28-9.14:

$$Fe(CO)_4I_2 \xrightarrow{\text{NaBH}_4 \text{ in THF}} H_2Fe(CO)_4 \qquad (28\text{-}9.14)$$

or by cleavage of metal-to-metal bonds by H_2:

$$Mn_2(CO)_{10} + H_2 \xrightarrow[200\,°C]{200\,\text{atm}} 2\,HMn(CO)_5 \qquad (28\text{-}9.15)$$

$$Co_2(CO)_8 + H_2 \longrightarrow 2\,HCo(CO)_4 \qquad (28\text{-}9.16)$$

Reaction 28-9.17 is another route to the neutral carbonyl hydride of cobalt.

$$Co + 4\,CO + \tfrac{1}{2}H_2 \xrightarrow[150\,°C]{50\,\text{atm}} HCo(CO)_4 \qquad (28\text{-}9.17)$$

The neutral carbonyl hydrides are slightly soluble in water where they behave as acids, ionizing to give carbonylate anions, as in Reactions 28-9.18 to 28-9.20:

$$HMn(CO)_5 = H^+ + [Mn(CO)_5]^- \qquad pK \sim 7 \qquad (28\text{-}9.18)$$

$$H_2Fe(CO)_4 = H^+ + [HFe(CO)_4]^- \qquad pK_1 \sim 4 \qquad (28\text{-}9.19)$$

$$HCo(CO)_4 = H^+ + [Co(CO)_4]^- \qquad \text{Strong acid} \qquad (28\text{-}9.20)$$

The neutral carbonyl hydrides have structures in which the hydrogen atom occupies a regular place in the coordination polyhedron, and the M—H distances are approximately equal to the values expected from the sum of the single-bond covalent radii. A good example is afforded by the structure of $HMn(CO)_5$, shown in Fig. 28-6. For purposes of electron counting, the hydrogen atom can be considered to add one electron to the $M(CO)_n$ entity to which it is attached.

In contrast to the neutral carbonyl hydrides, in the anionic hydrido carbonyls such as $[HM(CO)_5]^-$ (where M = Cr or W) and the previously mentioned $[HFe(CO)_4]^-$, the hydrogen atoms have much more hydridic character. Consequently, the anionic hydrides are not proton donors, but can be hydride (H^-) donors. In this way they find application as reducing agents for alkyl halides or acid chlorides, as in Reactions 28-9.21 and 28-9.22.

$$RX + [HFe(CO)_4]^- \longrightarrow RH + [XFe(CO)_4]^- \qquad (28\text{-}9.21)$$

$$RC(O)Cl + [HCr(CO)_5]^- \longrightarrow RC(O)H + [ClCr(CO)_5]^- \qquad (28\text{-}9.22)$$

In nonpolar solvents, and in the presence of acids (e.g., acetic acid), the anionic hydrides are also useful as reducing agents for aldehydes and ketones, giving alcohols, as in Reactions 28-9.23 and 28-9.24:

$$RC(O)H + [HCr(CO)_5]^- + HOAc \longrightarrow$$
$$RCH_2OH + [(OAc)Cr(CO)_5]^- \qquad (28\text{-}9.23)$$

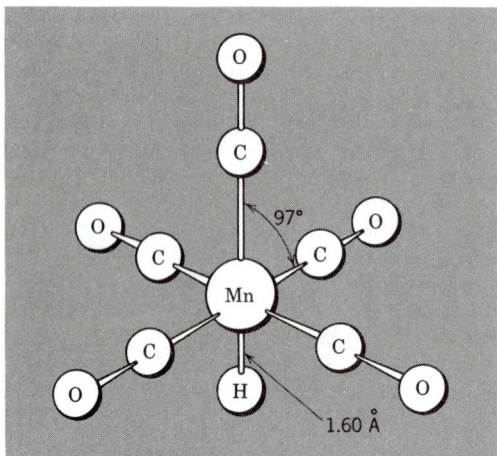

Figure 28-6   The structure of the neutral hydride $HMn(CO)_5$, showing the stereochemical activity of the hydrogen atom and the metal–hydrogen distance (which approximates the sum of the normal covalent radii).

$$RC(O)R' + [HCr(CO)_5]^- + HOAc \longrightarrow$$
$$RCH(OH)R' + [(OAc)Cr(CO)_5]^- \quad (28\text{-}9.24)$$

In the presence of strong acids, or with time in the presence of weak acids, most anionic hydrido carbonyls decompose via loss of H_2.

28-10 Carbonyl Halides and Related Compounds

Carbonyl halides, $M_x(CO)_yX_z$, are known for most of the elements forming binary carbonyls and also for Pd, Pt, Au, Cu^I, and Ag^I. They are obtained either by the direct interaction of metal halides and carbon monoxide, usually at high pressure, or by the cleavage of polynuclear carbonyls by halogens:

$$Mn_2(CO)_{10} + Br_2 \xrightarrow{\,40\,°C\,} 2\,Mn(CO)_5Br \underset{\substack{CO \\ 150\ atm}}{\overset{\substack{in\ petrol \\ at\ 120\ °C}}{\rightleftharpoons}}$$

$$[Mn(CO)_4Br]_2 + 2\,CO \quad (28\text{-}10.1)$$

$$RuI_3 + 2\,CO \xrightarrow{\,220\,°C\,} [Ru(CO)_2I_2]_n + \tfrac{1}{2}I_2 \quad (28\text{-}10.2)$$

$$2\,PtCl_2 + 2\,CO \longrightarrow [Pt(CO)Cl_2]_2 \quad (28\text{-}10.3)$$

Examples of the halides and some of their properties are listed in Table 28-2. Carbonyl halide anions are also known; they are often derived by reaction of ionic halides with metal carbonyls or substituted carbonyls.

$$M(CO)_6 + R_4N^+X^- \xrightarrow{\text{diglyme}} R_4N^+[M(CO)_5X]^- + CO$$

$$M = \text{Cr, Mo, or W} \quad (28\text{-}10.4)$$

$$Mn_2(CO)_{10} + 2\,R_4N^+X^- \longrightarrow (R_4N^+)_2[Mn_2(CO)_8X_2]^{2-} + 2\,CO \quad (28\text{-}10.5)$$

Dimeric or polymeric carbonyl halides are invariably bridged through the halogen atoms and not by carbonyl bridges, for example, in Structures 28-III and 28-IV.

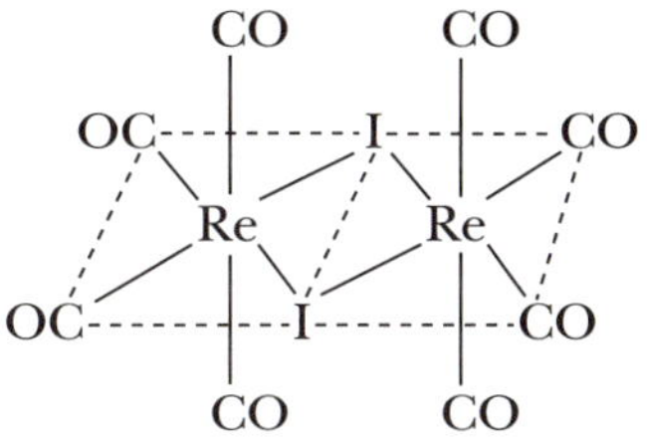

28-III 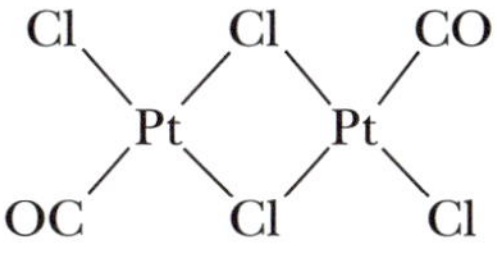   28-IV

The halogen bridges can be broken by numerous donor ligands such as pyridine, substituted phosphines, and isocyanides, as in the following reaction:

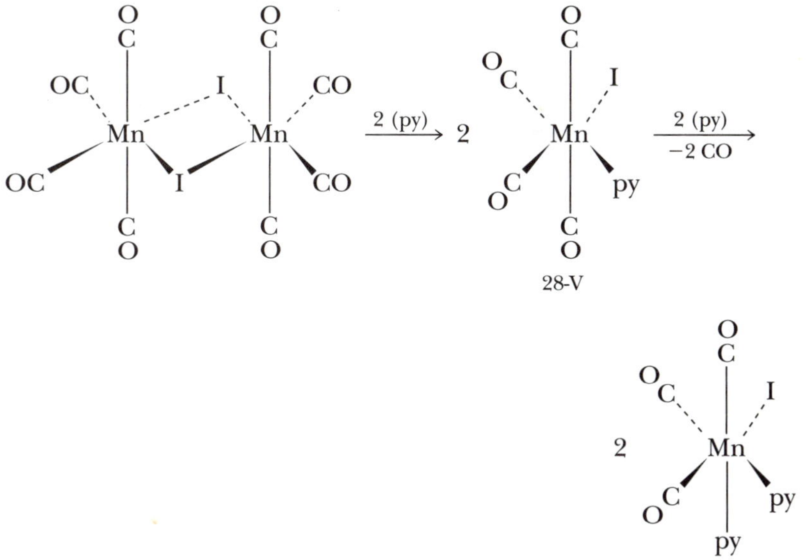

28-V

$$(28\text{-}10.6)$$

28-VI

CARBON MONOXIDE ANALOGS

28-11 Isocyanide Complexes

An isocyanide ($R-N\equiv C:$) is very similar electronically to $:O\equiv C:$, and there are many isocyanide complexes stoichiometrically analogous to metal carbonyls. Isocyanides can occupy bridging as well as terminal positions. Examples are such crystalline air-stable compounds as red $Cr(CNC_6H_5)_6$, white $[Mn(CNCH_3)_6]I$, and orange $Co(CO)(NO)(CNC_7H_7)_2$, all of which are soluble in benzene.

Isocyanides generally appear to be stronger σ donors than CO, and various complexes, such as $[Ag(CNR)_4]^+$, $[Fe(CNR)_6]^{2+}$, and $[Mn(CNR)_6]^{2+}$, are known

Table 28-2   Some Examples of Carbonyl Halide Complexes

Compound	Form	mp °C	Comment
$Mn(CO)_5Cl$	Pale yellow crystals	Sublimes	Loses CO at 120 °C in organic solvents; can be substituted by pyridine, and so on
$[Re(CO)_4Cl]_2$	White crystals	Decomp. >250	Halogen bridges cleavable by donor ligands or by CO (pressure)
$[Ru(CO)_2I_2]_n$	Orange powder	Stable >200	Halide bridges cleavable by ligands
$[Pt(CO)Cl_2]_2$	Yellow crystals	195; sublimes	Hydrolyzed by H_2O; PCl_3 replaces CO

where π bonding is of relatively little importance; derivatives of this type are not known for CO. However, the isocyanides are capable of extensive back-acceptance of π electrons from metal atoms in low oxidation states. This is indicated qualitatively by their ability to form compounds such as $Cr(CNR)_6$ and $Ni(CNR)_4$, analogous to the carbonyls and more quantitatively by C≡N stretching frequencies which, like CO stretching frequencies, are markedly lowered when the ligand acts as a π acid.

28-12 Dinitrogen (N_2) Complexes

The fact that CO and N_2 are isoelectronic had for years led to speculation as to the possible existence of M—NN bonds analogous to M—CO bonds, but it was only in 1965 that the first example, $[Ru(NH_3)_5N_2]Cl_2$, was reported. Subsequent work has shown that the $[Ru(NH_3)_5N_2]^{2+}$ cation can be obtained in a number of ways, for example,

by reaction of N_2H_4 with aqueous $RuCl_3$
by reaction of NaN_3 with $[Ru(NH_3)_5(H_2O)]^{3+}$
by reaction of N_2 with $[Ru(NH_3)_5H_2O]^{2+}$
by reaction of $RuCl_3(aq)$ with Zn in $NH_3(aq)$

Of these the direct reaction with N_2 to displace H_2O is perhaps most notable. Despite much study, no effective way of reducing coordinated N_2 to NH_3 has yet been found. However, there are several systems in which reduction of N_2 to NH_3 and/or N_2H_4 is catalyzed by low-valent metal compounds, presumably via transient M—N_2 complexes.

A *bridging* N_2 ligand, of the M—N—N—M type, is formed in the reaction

$$[Ru(NH_3)_5Cl]^{2+} \xrightarrow[N_2]{Zn/Hg} \{[Ru(NH_3)_5]_2N_2\}^{4+} \qquad (28\text{-}12.1)$$

The terminal-type N_2 ligands have strong IR bands in the range 1930–2230 cm^{-1} (100–400 cm^{-1} below that of free N_2, 2331 cm^{-1}) that may be used diagnostically.

The formation of N_2 complexes by direct uptake of N_2 gas at 1 atm has been observed, especially with tertiary phosphine ligands in reactions such as:

$$Co(acac)_3 + 3\,Ph_3P + N_2 \xrightarrow{\ Al(i\text{–}C_4H_9)_3\ } Co(H)(N_2)(Ph_3P)_3 \quad (28\text{-}12.2)$$

$$FeCl_2 + 3\,PEtPh_2 + N_2 \xrightarrow{\ NaBH_4,\ EtOH\ } FeH_2(N_2)(PEtPh_2)_3 \quad (28\text{-}12.3)$$

$$MoCl_4(PPhMe_2)_2 + N_2 + 2\,PPhMe_2 \xrightarrow{\ Na\ Hg\ in\ THF\ }$$
$$cis\text{-}Mo(N_2)_2(PPhMe_2)_4 \quad (28\text{-}12.4)$$

Several typical compounds containing M—NN groups have been structurally characterized. The three atom chains are essentially linear, the N—N distances are slightly longer than that in the N_2 molecule, and the M—N distances are short enough to indicate some multiple bond character.

The bonding in M—N_2 groups is similar to that in terminal M—CO groups. The same two basic components, M←N_2 σ donation and M→N_2 π acceptance, are involved. The major quantitative differences, which account for the lower stability of N_2 complexes, arise from small differences in the energies of the MO's of CO and N_2. It appears that N_2 is weaker than CO in both its σ-donor and π-acceptor functions, which accounts for the poor stability of N_2 complexes in general.

28-13 Thiocarbonyl Complexes

The CS molecule, unlike CO, does not exist under ordinary conditions, although it can be made in dilute gas streams by photolysis of CS_2. Nevertheless, CS can be stabilized by complexing and a few compounds are known. Thus $RhCl(PPh_3)_3$ reacts with CS_2 to give $RhCl(\eta^1\text{-}CS_2)(\eta^2\text{-}CS_2)(PPh_3)_2$, which in methanol gives $trans\text{-}RhCl(CS)(PPh_3)_2$.

Thiocarbonyl complexes have CS stretches in the region 1270-1360 cm^{-1}, depending on the oxidation state of the metal, charge on the complex, and the like, whereas the stretch for CS trapped in a matrix at -190 °C is at 1274 cm^{-1}. The $d\pi$–$p\pi$ bonding is similar to that for the carbonyls.

28-14 Nitrogen Monoxide Complexes

The NO molecule is similar to CO except that it contains one more electron, which occupies a π^* orbital (cf. Section 3-6). Consistent with this similarity, CO and NO form many comparable complexes although, as a result of the presence of the additional electron, NO also forms a class (bent MNO) with no carbonyl analogs.

Linear, Terminal MNO Groups

We have seen that the CO group reacts with a metal atom that presents an empty σ orbital and a pair of filled $d\pi$ orbitals, as illustrated in Fig. 28-3, to give a linear MCO grouping with a C → M σ bond and a significant degree of M → C π bonding. The NO group engages in an entirely analogous interaction with a metal atom that may be considered, at least formally, to present an empty σ orbital and a pair of $d\pi$ orbitals containing only three electrons. The full set of four electrons for the $Md\pi \rightarrow \pi^*(NO)$ interactions is thus made up of three electrons from M and one from NO. In effect, NO contributes three electrons to the total bonding configuration under circumstances where CO contributes only two. Thus, for purposes of formal electron "bookkeeping," the ligand NO can be regarded as a three-electron donor in the same sense as the ligand CO is considered a two-electron donor. This leads to the following very useful general rules concerning stoichiometry, which may be applied without specifically allocating the difference in the number of electrons to any particular (i.e., σ or π) orbitals:

1. Compounds isoelectronic with one containing an $M(CO)_n$ grouping are those containing $M'(CO)_{n-1}(NO)$, $M''(CO)_{n-2}(NO)_2$, and so on, where M', M'', and so on, have atomic numbers that are 1, 2, . . . , and so on, less than M. Some examples are

(a) $(\eta^5\text{-}C_5H_5)Cu(CO)$ and $(\eta^5\text{-}C_5H_5)Ni(NO)$
(b) $Fe(CO)_5$ and $Mn(CO)_4(NO)$
(c) $Ni(CO)_4$, $Co(CO)_3(NO)$, $Fe(CO)_2(NO)_2$, $Mn(CO)(NO)_3$, and $Cr(NO)_4$.

The isoelectronic and isostructural series in (c) is the longest one known. All are, like $Ni(CO)_4$, tetrahedral molecules.

2. Three CO groups can be replaced by two NO groups. Examples of pairs of compounds so related are

$$Fe(CO)_5 \qquad\qquad Fe(CO)_2(NO)_2$$
$$Mn(CO)_4NO \qquad\qquad Mn(CO)(NO)_3$$

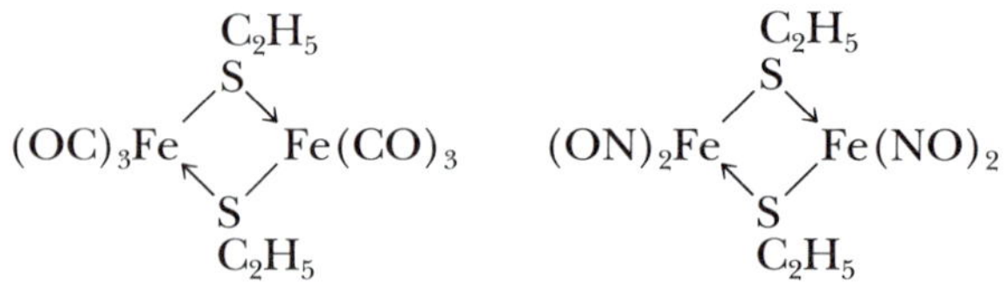

Structural data suggest that under comparable circumstances M—CO and M—NO bonds are about equally strong, but in a chemical sense the M—N bonds appear to be stronger, since substitution reactions on mixed carbonyl–nitrosyl compounds typically result in displacement of CO in preference to NO. For example, $Co(CO)_3NO$ reacts with R_3P, X_3P, amine, and RNC ligands, invariably to yield the $Co(CO)_2(NO)L$ product.

The NO vibration frequencies for linear MNO groups substantiate the idea of extensive M to N π bonding, leading to appreciable population of NO π^* orbitals. Nitrogen monoxide has its unpaired electron in a π^* orbital; the N—O stretching frequency is 1860 cm^{-1}. For typical linear MNO groups in molecules with small or zero charge, the observed frequencies are in the range 1800–1900 cm^{-1}. This indicates the presence of approximately one electron pair shared between metal $d\pi$ and NO π^* orbitals.

Bent, Terminal MNO Groups

It has long been known that NO can form single bonds to univalent groups such as halogens and alkyl radicals, affording the bent species

$$\underset{X}{\diagup}\ddot{N}{=}\ddot{O}\colon \quad \text{and} \quad \underset{R}{\diagup}\ddot{N}{=}\ddot{O}\colon$$

Metal atoms with suitable electron configurations and partial coordination shells may bind NO in a similar way. This type of NO complex is formed when the incompletely coordinated metal ion (L_nM) would have a $t_{2g}^6e_g^1$ configuration, thus being prepared to form one more single σ bond. The M—N—O angles are in, or near, the range 120–140°. Typical compounds are $[Co(NH_3)_5NO]Br_2$ and $IrCl_2(PPh_3)_2NO$.

Bridging NO Groups

These are less common than bridging CO groups, but well-established cases of both double and triple bridges are known. As in carbonyls, the bridging NO frequencies are at lower frequencies than terminal ones.

Bridging NO groups are also to be regarded as three-electron donors. The doubly bridging ones may be represented as

$$\ddot{N}\colon\colon\ddot{O}$$

where the additional electron required to form two metal-to-nitrogen single bonds is supplied by one of the metal atoms.

28-15 Complexes of Group VB(15) and Group VIB(16) Ligands

Trivalent phosphorus, arsenic, antimony, and bismuth compounds, as well as divalent sulfur and selenium compounds, can give complexes with transition metals. These donors are, of course, quite strong Lewis bases and give complexes with Lewis acids such as BR_3 compounds, where d orbitals are not involved. However, the donor atoms do also have empty $d\pi$ orbitals and back-acceptance into these orbitals is possible, as shown in Fig. 28-7.

Based on IR data an extensive series of ligands involving Groups VB(15) and VIB(16) donor atoms can be arranged in the following order of decreasing π acidity:

$$CO \sim PF_3 > PCl_3 \sim AsCl_3 \sim SbCl_3 > PCl_2(OR) > PCl_2R > PCl(OR)_2$$
$$> PClR_2 \sim P(OR)_3 > PR_3 \sim AsR_3 \sim SbR_3 \sim SR_2$$

It is noteworthy that IR spectral evidence, as well as photoelectron spectroscopy, shows that PF_3 is as good or better than CO as a π acid. Thus, it is not surprising that PF_3 forms an extensive group of $M_x(PF_3)_y$ compounds, many of which are analogs of corresponding $M_x(CO)_y$ compounds and some of which, for example, $Pd(PF_3)_4$ and $Pt(PF_3)_4$, are more stable than their carbonyl analogs, which can be observed only at very low temperatures.

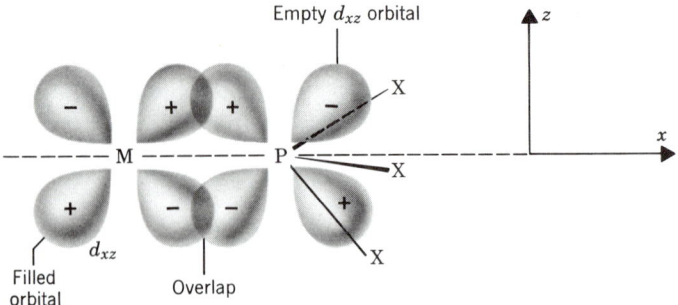

Figure 28-7   Diagram showing the π back-bonding from a filled metal d orbital to an empty phosphorus $3d$ orbital of the PX_3 ligand, taking the z axis as the M—P bond axis. A similar overlap occurs in the yz plane using the d_{yz} orbital.

The other Group VB(15) and VIB(16) ligands are all capable of replacing some CO groups, to form compounds such as $(R_3P)_3Mo(CO)_3$ or even $(R_3P)_4Mo(CO)_2$, but rarely can they replace all CO groups starting from a carbonyl. However, by special methods (such as cocondensation of metal atoms with $(CH_3)_3P$, or by treating metal compounds with diphosphines under strongly reducing conditions), products such as $Mo(PMe_3)_6$ and $Mo(Me_2PCH_2CH_2PMe_2)_3$ can be obtained.

28-16 Cyanide Complexes

The formation of cyanide complexes is restricted almost entirely to the transition metals of the d block and their near neighbors Zn, Cd, and Hg. This suggests that metal—CN π bonding is of importance in the stability of cyanide complexes, and there is evidence of various types to support this. However, the π-accepting tendency of CN^- is much lower than for CO, NO, or RNC. This is, of course, reasonable in view of its negative charge. Cyanide ion is a strong σ donor so that back-bonding does not have to be invoked to explain the stability of its complexes with metals in normal (i.e., II or III) oxidation states. Nonetheless, because of the formal similarity of CN^- to CO, NO, and RNC, it is convenient to discuss its complexes in this chapter.

The majority of cyano complexes have the general formula $[M^{n+}(CN)_x]^{(x-n)-}$ and are anionic, such as $[Fe(CN)_6]^{4-}$, $[Ni(CN)_4]^{2-}$, and $[Mo(CN)_8]^{3-}$. Mixed complexes, particularly of the type $[M(CN)_5X]^{n-}$, where X may be H_2O, NH_3, CO, NO, H, or a halogen, are also well known.

Although bridging cyanide groups might be expected in analogy with those formed by CO, none has been definitely proved. However, linear bridges (M—CN—M) are well known and play an important part in the structures of many crystalline cyanides and cyano complexes. Thus AuCN, $Zn(CN)_2$, and $Cd(CN)_2$ are all polymeric with infinite chains.

The free anhydrous acids corresponding to many cyano anions can be isolated; $H_3[Rh(CN)_6]$ and $H_4[Fe(CN)_6]$ are examples. These acids are different from those corresponding to many other complex ions, such as $[PtCl_6]^{2-}$ or

$[BF_4]^-$, which cannot be isolated except as hydroxonium (H_3O^+) salts. These compounds are also different from metal carbonyl hydrides in that they do not contain metal-to-hydrogen bonds. Instead, the hydrogen atoms are situated in hydrogen bonds between anions, that is, MCN- - -H- - -NCM.

STUDY GUIDE

Study Questions

A. Review

1. For each of the π-acceptor ligands mentioned in this chapter, state the nature of the acceptor orbital(s).

2. Write the formulas for the mononuclear metal carbonyl molecules formed by V, Cr, Fe, and Ni. Which ones satisfy the noble gas formalism?

3. Why are the simplest carbonyls of the metals Mn, Tc, Re and Co, Rh, Ir groups polynuclear?

4. Explain, with necessary orbital diagrams, how CO, which has negligible donor properties toward simple acceptors such as BF_3, can form strong bonds to transition metal atoms.

5. In what ways can CO be bound to a metal atom?

6. Discuss and explain the trend in CO stretching frequencies in the series $V(CO)_6^-$, $Cr(CO)_6$, $Mn(CO)_6^+$.

7. Draw the structures of $Fe_2(CO)_9$, $Ru_3(CO)_{12}$, and $Rh_4(CO)_{12}$.

8. Which are the only two metals to react directly with CO under conditions suitable for practical syntheses?

9. What is the general type of reaction used to prepare metal carbonyls? State the main ingredients, the function of each, and some examples.

10. How are the following compounds made? What are their principal physical characteristics?
    (a) $Fe(CO)_5$ from iron powder
    (b) $Co_2(CO)_8$ from hydrated cobalt(II) sulfate
    (c) $Cr(CO)_6$ from hydrated chromium(III) chloride
    (d) $Mn_2(CO)_{10}$ from hydrated manganese(II) chloride
    (e) $Fe_3(CO)_{12}$ from $Fe(CO)_5$

11. Explain why $Mo(py)_2(CO)_4$ has two forms, one having a single CO stretching band in the IR spectrum, the other four.

12. Give the formulas of some simple carbonylate ions and carbonyl hydrides. Do they follow the noble gas rule?

13. In a carbonyl complex with a linear OC—M—CO group, how will the CO stretching frequency change when (a) one CO is replaced by triethylamine, (b) a positive charge is put on the complex, and (c) a negative charge is put on the complex?

14. How is N_2 related to CO? Are N_2 complexes more or less stable than CO complexes? What was the first N_2 complex discovered and when?

15. Describe the bonding of NO to a metal in the case where the M—N—O chain is essentially linear; specifically contrast it with the analogous M—C—O bonding in terms of how many electrons are involved.

16. Besides linear M—N—O bonding, what other kinds are there?

17. Explain why nitric oxide can be regarded as a three-electron donor ligand. Would you expect the stoichiometries of compounds with the following chelate ligands to be similar to those formed by NO?

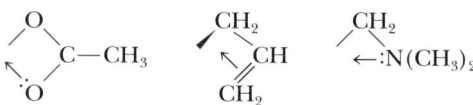

Would there be any difference in the formal oxidation state of the metal, for example, in $Mn(CO)_4NO$ and $Mn(CO)_4[CH_2N(CH_3)_2]$ (shown on the right)?

18. Explain how trialkyl or aryl phosphines can bind to a metal.

19. Which PX_3 ligand is most similar in its bonding ability to CO?

20. Discuss the similarities and differences between CN^- and CO as ligands.

B. Additional Exercises

1. In order to have a vanadium carbonyl that satisfies the noble gas formalism, what would be the simplest formula? Why do you think this fails to occur?

2. It is known that in $Mn_2(CO)_{10}$ the carbonyl groups move rapidly from one manganese atom to the other. On the basis of what you find in Sections 28-3 and 28-4 suggest a plausible intermediate for this process.

3. Do you think that carbonyls of the lanthanides are likely to be stable? Whatever your answer, give reasons.

4. Write both bridged and nonbridged structures for $Mn_2(CO)_{10}$ and $Co_2(CO)_8$. The former has CO stretching bands only in the range 2044–1980 cm^{-1}, while the latter has bands in the range 2071–2022 cm^{-1}, as well as two at 1860 and 1858 cm^{-1}. Which structure is indicated to be correct in each case?

5. What are the formulas of the metal carbonyls that are isoelectronic with $Cr(NO)_4$, $Mn(CO)(NO)_3$, $Mn(CO)_4NO$, $Fe(CO)_2(NO)_2$?

6. Write balanced equations for the following reactions
   (a) $Mn_2(CO)_{10}$ is heated with I_2
   (b) $Mo(CO)_6$ is refluxed with KI in THF
   (c) $Fe(CO)_5$ is shaken with aqueous KOH
   (d) $Ni(CO)_4$ is treated with PCl_3
   (e) $Co_2(CO)_8$ is treated with NO in hexane

7. What is the difference between a π-acid ligand like RNC and a ligand like C_2H_4 that forms π complexes?

8. In a linear group $R_3P—M—CO$, how would the CO frequency change when

$$R = F, CH_3, C_6H_5, \text{—⬡—} CH_3, \text{ or } \text{—⬡—} F$$

9. Why is pK_2 for $H_2Fe(CO)_4$ smaller than pK_1 by nine units? What does this tell us?

10. Put the following ligands in decreasing order of π acidity
    CH_3CN $(C_2H_5)_2O$ PCl_3 $As(C_6H_5)_3$ CH_3NC $(C_2H_5)_3N$

11. Determine whether or not the following structures obey the 18-electron rule:
    (a) $H_2Fe(CO)_4$ (b) $V(CO)_6$ (c) $[V(CO)_6]^-$
    (d) $W(CO)_5P(C_6H_5)_3$ (e) $Mn(CO)_4NO$ (f) $[Cr(CO)_5]^-$
    (g) $Co(H)(N_2)(PPh_3)_3$ (h) $Mn_2(I)_2(CO)_8$ (i) $[W(CO)_5Cl]^-$
    (j) $HMn(CO)_4$ (k) $[V(CO)_5H]^{2-}$

12. Explain the relative position of the IR stretching absorptions in $[V(CO)_6]^-$ versus $Cr(CO)_6$.

13. Explain why $V(CO)_6$ is readily reduced to the monoanion.

14. Describe the bonding of a doubly bridging CO group between two metals. Use the three-center, two-electron MO approach.

15. Use the MO approach to describe the bonding of a triply bridging CO group to three metal centers.

16. Suggest preparations of the following products:
    (a) $W(CO)_5(py)$
    (b) $Na[HCr(CO)_5]$
    (c) $MeNCFe(CO)_4$
    (d) $K[Mo(CO)_5I]$
    (e) $Mo(CO)_3(py)_3$
    (f) $HCo(CO)_4$
    (g) $K[Co(CO)_4]$
    (h) $Mn(CO)_4(I)(py)$
    (i) $Mn(CO)_3(I)(py)_2$
    (j) $Re(CO)_4Cl(py)$
    (k) $trans$-$RhCl(CS)(PPh_3)_2$

17. Write balanced equations for each of the preparations of $[Ru(NH_3)_5N_2]^{2+}$.

18. Write balanced equations for three separate routes to the anion $[Co(CO)_4]^-$. What do you expect its geometry to be?

C. Questions from the Literature of Inorganic Chemistry

1. Consider the paper by F. A. Cotton, D. J. Darensbourg, and B. W. S. Kolthammer, *Inorg. Chem.*, **1981**, *20*, 4440–4442.
    (a) Write a balanced equation for the synthesis of the title compound. [*Note:* The PPN$^+$ counterion is a large organic cation, bis(triphenylphosphine)iminium, which may be considered to be noninvolved in the essential chemistry of this system.] What is the role of the methanol reagent in this preparation?
    (b) In which compounds mentioned in this paper do steric effects preclude a short M—P bond?
    (c) What significant inter- or intramolecular contacts (or lack of contacts) lead the authors to propose that the M—P bond reported here is not influenced by steric effects?
    (d) What mechanism for substitution has been assumed in predicting that the $P(CH_3)_3$ ligand in $W(CO)_5P(CH_3)_3$ should be less labile than the $P(t\text{-Bu})_3$ ligand in $W(CO)_5P(t\text{-Bu})_3$? Explain.

2. Consider the work by R. J. Dennenberg and D. J. Darensbourg, *Inorg. Chem.*, **1972**, *11*, 72–77.
    (a) Summarize the evidence that is presented in favor of a dissociative mechanism for the substitution (decomposition) reactions reported here.
    (b) Why is cleavage of a M—N bond easier than cleavage of a M—CO bond?
    (c) Show with orbital overlap diagrams the π bond between M and py that is responsible for the slower substitution of this unsaturated amine than of the saturated amines.

3. Consider triphenyltris(thf)chromium(III) as reported by S. I. Khan and R. Bau, *Organometallics*, **1983**, *2*, 1896–1897.
    (a) Assign oxidation states to the ligands and to the metal. How many electrons should each ligand be considered to donate to chromium?
    (b) What evidence is there for the $d\pi$–$p\pi$ bonding to the phenyl ligands? Show the orbitals that would be involved in such a π-bond system.
    (c) The dianion $[Cr(C_6H_5)_5]^{2-}$ is mentioned. Does it satisfy the 18-electron rule?

4. Consider the hydrido pentacarbonyl of chromium as reported by M. Y. Darensbourg and J. C. Deaton, *Inorg. Chem.*, **1981**, *20*, 1644–1646.

   (a) Write equations for each of the routes (outlined in the introduction), which involve the monoanion $[HCr(CO)_5]^-$. Why do these routes not serve as useful methods for isolation of $[HCr(CO)_5]^-$?

   (b) Should the dimer $[(CO)_5Cr—H—Cr(CO)_5]^-$ be considered to be saturated from the standpoint of the 18-electron rule? Answer also for the monomeric $[HCr(CO)_5]^-$.

   (c) Write an equation for each step in the successful synthesis, as reported here, of the tetraethylammonium salt of $[HCr(CO)_5]^-$.

5. Consider the anionic hydrido carbonyl dimer $[HFe_2(CO)_8]^-$ as reported by H. B. Chin and R. Bau, *Inorg. Chem.*, **1978**, *17*, 2314–2317.

   (a) How is the structure of this anion related to that of $Fe_2(CO)_9$?

   (b) What bonds in the anion should be described by the three-center, two-electron formalism?

   (c) What arguments do the authors present that M—Cl—M systems should not be formulated as electron deficient (three-center, two-electron) bonds?

   (d) What IR evidence suggests the presence of bridging CO ligands in this compound?

SUPPLEMENTARY READING

Abel, E. W. and Stone, F. G. A., "The Chemistry of Transition-Metal Carbonyls: Structural Considerations," *Q. Rev.*, **1969**, *23*, 325.

Abel, E. W. and Stone, F. G. A., "The Chemistry of Transition-Metal Carbonyls: Synthesis and Reactivity," *Q. Rev.*, **1970**, *24*, 498.

Allen, A. D., "Complexes of Dinitrogen," *Chem. Rev.*, **1973**, *73*, 11.

Braterman, P. S., *Metal Carbonyl Spectra*, Academic, New York, 1975.

Coates, G. E., Green, M. L. H., Powell, P., and Wade, K., *Principles of Organometallic Chemistry*, Chapman & Hall, London, 1977.

Darensbourg, M. Y., "Ion Pairing Effects on Transition Metal Carbonyl Anions," in *Progress in Inorgance Chemistry*, Vol. 33, Wiley-Interscience, New York, 1985.

Enemark, J. H. and Feltham, R. D., "Nitric Oxide Complexes," *Coord. Chem. Rev.*, **1974**, *13*, 339.

Hoffmann, R., "Theoretical Organometallic Chemistry," *Science*, **1981**, *211*, 995.

Malatesta, L. and Cenini, S., *Zerovalent Complexes of Metals*, Academic, New York, 1974.

McAuliffe, C. A., Ed., *Transition Metal Complexes of Phosphorus, Arsenic, and Antimony Ligands*, Macmillan, New York, 1973.

Pearson, A. J., *Metallo-Organic Chemistry*, Wiley-Interscience, New York, 1985.

Pearson, R. G., "The Transition-Metal-Hydrogen Bond," *Chem. Rev.*, **1985**, *85*, 41.

Singleton, E. and Oosthuizen, H. E., "Metal Isocyanide Complexes," *Advances in Organometallic Chemistry*, Vol. 22, F. G. A. Stone and R. West, Eds., Academic, New York, 1983.

Wender, I. and Pino, P., Eds., "Organic Syntheses via Metal Carbonyls," Wiley, New York, 1968.

Wilkinson, G., Stone, F. G. A., and Abel, E. W., Eds., *Comprehensive Organometallic Chemistry*, Pergamon Press, New York, 1982.

Chapter 29

ORGANOMETALLIC COMPOUNDS

29-1 General Survey of Types

Organometallic compounds are those in which the *carbon* atoms of organic groups are bound to metal atoms. For example, an alkoxide such as $(C_3H_7O)_4Ti$ is not considered to be an organometallic compound because the organic group is bound to Ti by oxygen, whereas $C_6H_5Ti(OC_3H_7)_3$ is, because a metal-to-carbon bond is present. The term organometallic is usually rather loosely defined and compounds of elements such as boron, phosphorus, and silicon, which are at best scarcely metallic, are included in the category. A few general comments on the various types of compounds can be made first.

Ionic Compounds of Electropositive Metals. The organometallic compounds of highly electropositive metals are usually ionic, insoluble in hydrocarbon solvents, and are very reactive toward air, water, and the like. The stability and reactivity of ionic compounds are determined in part by the stability of the carbanion. Compounds containing unstable anions (e.g., $C_nH_{2n+1}^-$) are generally highly reactive and often unstable and difficult to isolate. Metal salts of carbanions whose stability is enhanced by delocalization of electron density are more stable, although still quite reactive; examples are $(C_6H_5)_3C^-Na^+$ and $(C_5H_5^-)_2Ca^{2+}$.

σ-Bonded Compounds. Organometallic compounds in which the organic residue is bound to a metal atom by a normal two-electron covalent bond (albeit in some cases with appreciable ionic character) are formed by most metals of lower electropositivity and, of course, by nonmetallic elements. The normal valence rules apply in these cases, and partial substitution of halides, hydroxides, and so on, by organic groups occurs as in $(CH_3)_3SnCl$, $(CH_3)SnCl_3$, and so on. In most of these compounds, the bonding is predominantly covalent and the chemistry is organic-like, although there are many differences from carbon chemistry due to the following factors:

1. The possibility of using higher d orbitals in SiR_4, for example, which is not feasible in CR_4.
2. Donor ability of alkyls or aryls with lone pairs, as in $P(C_2H_5)_3$, $S(CH_3)_2$, and so on.
3. Lewis acidity due to incomplete valence shells as in BR_3 or coordinative unsaturation as in ZnR_2.
4. Effects of electronegativity differences between M—C and C—C bonds.

Transition metals may form simple alkyls or aryls but these are normally less stable than those of main group elements for reasons that we discuss later (Section 29-11). There are numerous compounds in which additional ligands such as CO or PR_3 are present.

Nonclassically Bonded Compounds. In many organometallic compounds there is a type of metal-to-carbon bonding that cannot be explained in terms of ionic or electron-pair σ bonds. One class comprises the alkyls of Li, Be, and Al that have *bridging* alkyl groups. Here, there is electron deficiency as in boron hydrides, and the bonding is of a similar multicenter type. A second, much larger class comprises compounds of transition metals with alkenes, alkynes, benzene, and other ring systems such as $C_5H_5^-$.

First, we consider the organometallic compounds of the main group elements, including the nonclassically bonded ones, and then turn to the transition metal compounds.

29-2   Synthetic Methods

There are many ways of generating metal-to-carbon bonds that are useful for both main group and transition metals. Some of the more important ways are as follows:

1. Direct reactions of metals.   The earliest synthesis, by the English chemist Sir Edward Frankland in 1845, was the interaction of Zn and an alkyl halide. Frankland was, in fact, attempting to synthesize alkyl radicals; his discovery played a decisive part in the development of modern ideas of chemical bonds. Much more useful, however, was the discovery by the French chemist, V. Grignard, of what are now called Grignard reagents by interaction of Mg with alkyl or aryl halides in ether.

$$Mg + CH_3I \xrightarrow{\text{ether}} CH_3MgI \qquad (29\text{-}2.1)$$

Direct reactions of alkyl or aryl halides occur also with Li, Na, K, Mg, Ca, Zn, and Cd.

2. Use of alkylating agents.   The previously mentioned compounds can be utilized to make other organometallic compounds. The most important and widely used are Grignard and lithium reagents. Aluminum and mercury alkyls and certain sodium derivatives, especially $Na^+C_5H_5^-$, are also useful alkylating agents.

Most nonmetal and metal halides or halide derivatives can be alkylated in ethers, or hydrocarbon solvents, for example,

$$PCl_3 + 3\ C_6H_5MgCl = P(C_6H_5)_3 + 3\ MgCl_2 \qquad (29\text{-}2.2)$$

$$VOCl_3 + 3\ (Me)_3SiCH_2MgCl = VO(CH_2SiMe_3)_3 + 3\ MgCl_2 \quad (29\text{-}2.3)$$

$$PtCl_2(PEt_3)_2 + MeMgCl = PtClMe(PEt_2)_2 + MgCl_2 \qquad (29\text{-}2.4)$$

3. Interaction of metal or nonmetal hydrides with alkenes or alkynes.
One of the best examples for nonmetals, and one that finds wide use in synthe-

sis, is the hydroboration shown in Reaction 29-2.5.

$$\tfrac{1}{2} \, B_2H_6 + 3 \; \diagdown\!C\!\!=\!\!C\diagup \; \xrightarrow{\text{ether}} \; B\!\!\left(\!\!-\overset{|}{\underset{|}{C}}-\overset{|}{\underset{|}{C}}-\!\!\right)_{\!3} \qquad (29\text{-}2.5)$$

Reaction 29-2.5 may be regarded as an addition across the double bond of the alkene. The intermediate trialkyl borane may be protonated as in Reaction 29-2.6:

$$\left(\!\!R\!-\!\!\overset{H}{\underset{H}{\overset{|}{C}}}\!-\!\!\overset{H}{\underset{H}{\overset{|}{C}}}\!-\!\!\right)_{\!3}\!\!\!B \; \xrightarrow{H_3O^+} \; 3R\!-\!\!\overset{H}{\underset{H}{\overset{|}{C}}}\!-\!\!\overset{H}{\underset{H}{\overset{|}{C}}}\!-\!H + B(OH)_3 \qquad (29\text{-}2.6)$$

and the net result is hydrogenation of the original alkene. Oxidation of the trialkyl borane molecule as in Reaction 29-2.7

$$\left(\!\!R\!-\!\!\overset{H}{\underset{H}{\overset{|}{C}}}\!-\!\!\overset{H}{\underset{H}{\overset{|}{C}}}\!-\!\!\right)_{\!3}\!\!\!B \; \xrightarrow{H_2O_2} \; 3R\!-\!\!\overset{H}{\underset{H}{\overset{|}{C}}}\!-\!\!\overset{H}{\underset{H}{\overset{|}{C}}}\!-\!OH + B(OH)_3 \qquad (29\text{-}2.7)$$

gives the alcohol.

For transition metals and hydride complexes, such reactions are of prime importance in that many catalytic syntheses involving transition metals (Chapter 30) have, as an early step, the reaction:

$$L_nMH + \; \diagdown\!C\!\!=\!\!C\diagup \; = L_nM\!-\!\overset{H}{\underset{|}{\overset{|}{C}}}\!-\!\overset{|}{\underset{|}{C}}\!- \qquad (29\text{-}2.8)$$

4. Oxidative addition reactions.   The so-called oxad reactions (Section 30-2), where alkyl or aryl halides are added to coordinatively unsaturated transition metal compounds, generate metal–carbon bonds; for example,

$$RhCl(PPh_3)_3 + MeI = RhClMe(PPh_3)_2 + PPh_3 \qquad (29\text{-}2.9)$$

5. Insertion reactions. Certain "insertion" reactions (Section 30-3) may also allow the generation of bonds to carbon, for example,

$$[(CN)_5Co\!-\!Co(CN)_5]^{4-} + HC\!\!\equiv\!\!CH = \left[(CN)_5Co\!-\!\overset{H}{\underset{H}{\overset{|}{C}}}\!\!=\!\!C\!-\!Co(CN)_5\right]^{4-}$$
$$(29\text{-}2.10)$$

$$SbCl_5 + 2\,HC\!\!\equiv\!\!CH = Cl_3Sb(CH\!\!=\!\!CHCl)_2 \qquad (29\text{-}2.11)$$

The reactions in number 3 above can also be regarded as "insertions" into the M—H bond.

PART A

MAIN GROUP ELEMENTS

29-3   Lithium Alkyls and Aryls

Organometallic compounds of lithium have been discussed in a preliminary fashion in Section 10-8. One of the major uses of metallic lithium, industrially and in the laboratory, is for the preparation of organolithium compounds which, in their reactions, generally resemble Grignard reagents, although the lithium reagents are usually more reactive. Organolithium compounds are prepared by interaction of the metal with an organic halide, usually the chloride, in benzene or alkanes. Ethers can be used as solvents, but they are attacked slowly by the lithium compounds. Examples of typical preparations include reaction with ethyl chloride as in Reaction 29-3.1:

$$C_2H_5Cl + 2 \text{ Li} \longrightarrow C_2H_5Li + LiCl \qquad (29\text{-}3.1)$$

metal–hydrogen exchange as in Reaction 29-3.2:

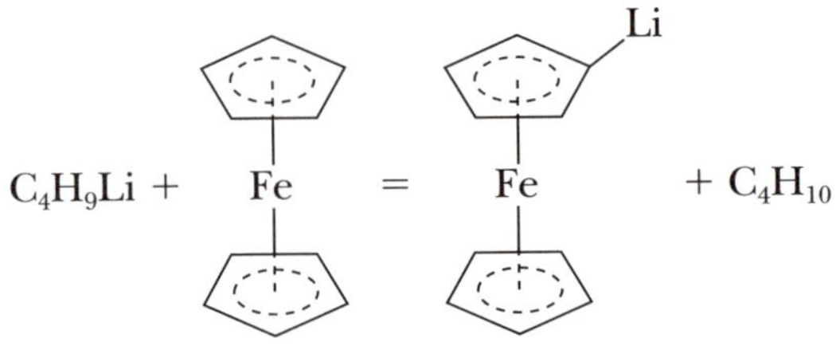

$$C_4H_9Li + \quad = \quad + C_4H_{10} \qquad (29\text{-}3.2)$$

metal–halogen exchange as in Reaction 29-3.3:

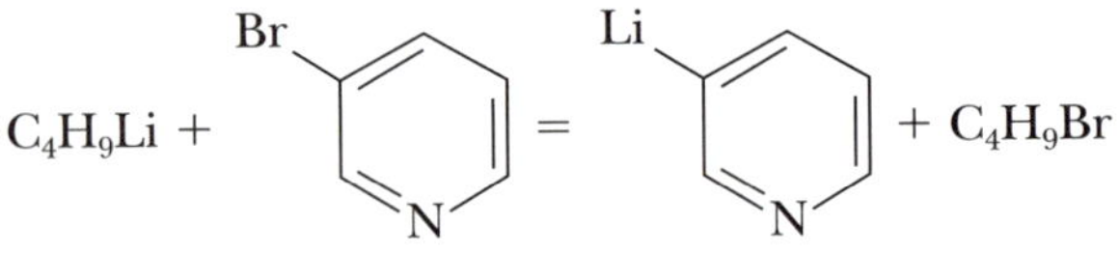

$$C_4H_9Li + \quad = \quad + C_4H_9Br \qquad (29\text{-}3.3)$$

and metal–metal exchange as in Reaction 29-3.4:

$$2 \text{ Li} + R_2Hg \longrightarrow 2 \text{ RLi} + Hg \qquad (29\text{-}3.4)$$

Butyllithium in hexane, benzene, or ethers is commonly used for such reactions. Methyllithium is also prepared by exchange through the interaction of C_4H_9Li and CH_3I in hexane at low temperatures, at which point it precipitates as insoluble white crystals.

Organolithium compounds all react rapidly with oxygen, being usually spontaneously flammable in air, and also with liquid water and with water vapor. However, lithium bromide and iodide form solid complexes of stoichiometry $RLi(LiX)_{1-6}$ with the alkyls, and these solids are stable in air.

Organolithium compounds are among the very few alkali metal compounds that have properties—solubility in hydrocarbons or other nonpolar liquids and high volatility—typical of covalent substances. They are generally liquids or low-melting solids. Molecular association is an important feature of the alkyls in both

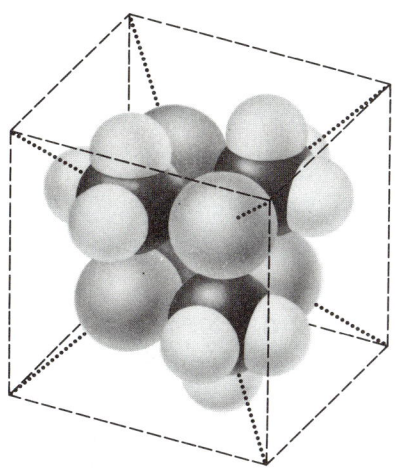

Figure 29-1   The structure of
$(CH_3Li)_4$ showing the tetrahedral
Li_4 unit with the methyl groups located symmetrically above each
face of the tetrahedron. [Adapted
from E. Weiss and E. A. C. Lucken,
J. Organomet. Chem., **1964**, *2*, 197.]
See also Fig. 10-4.

crystals and solutions. Thus in methyllithium (Fig. 29-1) the Li atoms are at the corners of a tetrahedron with the alkyl groups centered over the facial planes. Each CH_3 group is thus symmetrically bound to three Li atoms, and this alkyl bridge bonding is of the electron-deficient multicenter type (Section 3-7). Aggregate formation is due principally to the Li—C—Li rather than to Li—Li bonding interactions.

In solutions the nature of the polymerized species depends on the solvent, the steric nature of the organic radical, and temperature. In hydrocarbons CH_3Li, C_2H_5Li, C_3H_7Li, and some others are hexamers, but *tert*-butyllithium, which presumably is too bulky, is only tetrameric. In ethers or amines solvated tetramers are formed. There are no aggregates smaller than tetramers. However, when chelating ditertiary amines, notably tetramethylethylenediamine (TMED), $(CH_3)_2NCH_2CH_2N(CH_3)_2$, are used, comparatively stable monomeric alkyl-lithium complexes are obtained.

The alkyls and aryls also form complexes with other metal alkyls such as those of Mg, Cd, and Zn. For example,

$$2\,LiC_6H_5 + Mg(C_6H_5)_2 = Li_2[Mg(C_6H_5)_4] \qquad (29\text{-}3.5)$$

It is not surprising that there are wide variations in the comparative reactivities of Li alkyls depending on the differences in aggregation and ion-pair inter-actions. An example is benzyllithium, which is monomeric in THF and reacts with a given substrate more than 10^4 times as fast as the tetrameric methyl-lithium. The monomeric TMED complexes mentioned previously are also very much more reactive than the corresponding aggregated alkyls. Alkyllithiums can polylithiate acetylenes, acetonitrile, and other compounds; thus, $CH_3C{\equiv}CH$ gives Li_4C_3, which can be regarded as a derivative of C_3^{4-}.

Reactions of lithium alkyls are generally considered to be carbanionic in nature. Lithium alkyls are widely employed as stereospecific catalysts for the polymerization of alkenes, notably isoprene, which gives up to 90% of 1,4-*cis*-polyisoprene; numerous other reactions with alkenes have been studied. The TMED complexes again are especially active. Not only will they polymerize ethylene, but they will even metallate benzene and aromatic compounds, as well as reacting with hydrogen at 1 atm to give LiH and alkane.

29-4 Organosodium and Organopotassium Compounds

These compounds have been discussed in Section 10-8, are all essentially ionic, and are not soluble to any appreciable extent in hydrocarbons. They are exceedingly reactive, sensitive to air, and are hydrolyzed vigorously by water.

Most important are the sodium compounds from acidic hydrocarbons such as cyclopentadiene, indene, and acetylenes. These compounds are obtained by reaction of the hydrocarbon with metallic sodium or sodium dispersed in THF or DMF.

29-5 Magnesium

The organic compounds of Ca, Sr, and Ba are highly ionic and reactive and are not useful, but the magnesium compounds are probably the most widely used of all organometallic compounds, finding application extensively in organic chemistry, as well as in the synthesis of alkyl and aryl compounds of other elements. Magnesium compounds are of the types RMgX (the Grignard reagents) and MgR_2. The former are made by direct interaction of the metal with an organic halide (RX) in a suitable solvent, usually an ether such as diethyl ether or THF. The reaction is normally most rapid with iodides (RI) and iodine may be used as an initiator. For most purposes, RMgX reagents are used *in situ*. The species MgR_2 are best made by the dry reaction:

$$HgR_2 + Mg(\text{excess}) \longrightarrow Hg + MgR_2 \qquad (29\text{-}5.1)$$

The dialkyl or diaryl is then extracted with an organic solvent. Both RMgX, as solvates, and R_2Mg are reactive, being sensitive to oxidation by air and to hydrolysis by water.

The nature of Grignard reagents *in solution* is complex and depends on the nature of the alkyl and halide groups, and on the solvent, concentration, and temperature. Generally, the equilibria involved are of the type:

$$
RMg\overset{\displaystyle X}{\underset{\displaystyle X}{\diagup\diagdown}}MgR \rightleftharpoons 2\,RMgX \rightleftharpoons R_2Mg + MgX_2
$$

$$
\begin{array}{c}
R\diagdown\;\;\diagup X \\
Mg\qquad Mg \\
R\diagup\;\;\diagdown X
\end{array}
\qquad (29\text{-}5.2)
$$

Solvation (not shown) occurs and association is predominantly by halide rather than by carbon bridges, except for methyl compounds, where bridging by CH_3 groups may occur.

In dilute solutions and in more strongly donating solvents the monomeric species normally predominate, but in diethyl ether at concentrations greater than 0.1 M association gives linear or cyclic polymers. For crystalline Grignard reagents both of the structures $RMgX \cdot nS$, where the number (n) of solvent (S) molecules depends on the nature of R, and $R(S)Mg(\mu\text{-}X)_2Mg(S)R$ have been found. The Mg atom is usually tetrahedrally coordinated.

Zinc and *cadmium* compounds are similar to those of magnesium but differ in their reactivities. The lower alkyls of zinc are liquids spontaneously flammable in air. These compounds react vigorously with water.

29-6 Mercury

A vast number of organomercury compounds are known, some of which have useful physiological properties. They are of the types RHgX and R_2Hg. These compounds are commonly made by the interaction of $HgCl_2$ and RMgX, but Hg—C bonds can also be made in other ways which are discussed below.

The *RHgX compounds* are crystalline solids. When X can form covalent bonds to mercury, for example, Cl, Br, I, CN, SCN, or OH, the compound is a covalent nonpolar substance more soluble in organic liquids than in water. When X is SO_4^{2-} or NO_3^-, the substance is saltlike and presumably quite ionic, for instance, $[RHg]^+[NO_3]^-$.

The *dialkyls* and *diaryls* are nonpolar, volatile, toxic, colorless liquids, or low-melting solids. They are unaffected by air or water, presumably because of the low polarity of the Hg—C bond and the low affinity of mercury for oxygen. However, they are photochemically and thermally unstable, as would be expected from the low bond strengths (50–200 kJ mol^{-1}). In the dark, mercury compounds can be kept for months. The decomposition generally proceeds by homolysis of the Hg—C bond and ensuing free radical reactions.

All RHgX and R_2Hg molecules are linear. The principal utility of dialkyl- and diarylmercury compounds, and a very valuable one, is in the preparation of other organometallic compounds by *interchange* reactions. For example,

$$\tfrac{n}{2} R_2Hg + M \longrightarrow R_nM + \tfrac{n}{2} Hg \tag{29-6.1}$$

This reaction proceeds essentially to completion with the Li and Ca groups, and with Zn, Al, Ga, Sn, Pb, Sb, Bi, Se, and Te; but with In, Tl, and Cd, reversible equilibria are established. Partial alkylation of reactive halides can be achieved, for example,

$$AsCl_3 + (C_2H_5)_2Hg \longrightarrow C_2H_5HgCl + C_2H_5AsCl_2 \tag{29-6.2}$$

Mercury released to the environment (e.g., as metal) by losses from electrolytic cells used for NaOH and Cl_2 production, or as compounds such as alkylmercury seed dressings or fungicides, constitutes a serious hazard. This is a result of biological methylation to give highly toxic $(CH_3)_2Hg$ or CH_3Hg^+. Models for vitamin B_{12}, such as methylcobaloximes (Section 31-8), which have

Co—CH_3 bonds, will transfer CH_3 to Hg^{2+}. There are a number of microorganisms that can perform the same function, possibly by similar routes.

Mercuration and Oxomercuration

An important reaction for the formation of Hg—C bonds, and one that can be adapted to the synthesis of a wide variety of organic compounds, is the addition of mercuric salts, notably the acetate, trifluoroacetate, or nitrate to unsaturated compounds.

Mercuration of aromatic compounds occurs as follows:

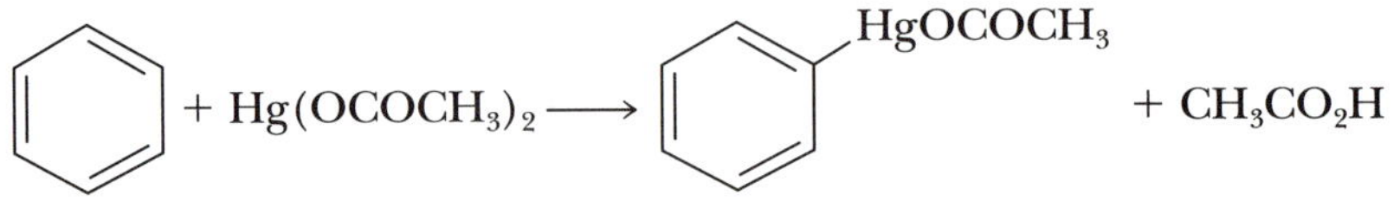

$$\text{C}_6\text{H}_6 + \text{Hg(OCOCH}_3)_2 \longrightarrow \text{C}_6\text{H}_5\text{HgOCOCH}_3 + CH_3CO_2H \quad (29\text{-}6.3)$$

Mercuric salts also react with alkenes in a reversible reaction:

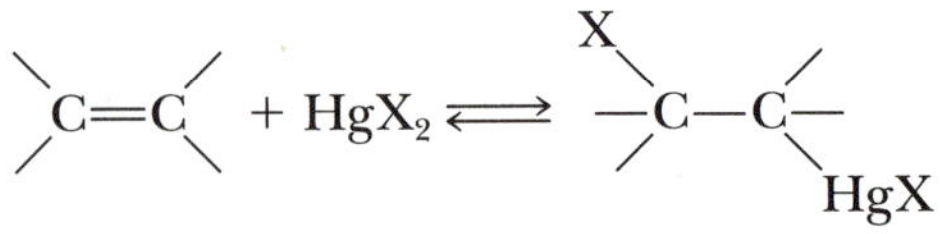

$$\text{C}{=}\text{C} + \text{HgX}_2 \rightleftharpoons \overset{X}{-}C{-}C\underset{HgX}{-} \quad (29\text{-}6.4)$$

The reversibility is readily shown by using $Hg(OCOCF_3)_2$, since the latter is soluble in nonpolar solvents; the equilibrium constants for Reaction 29-6.4 can be measured. In most instances, the reactions must be carried out in an alcohol or other protic medium, where further reaction with the solvent occurs. The reaction is then called *oxomercuration*. For example,

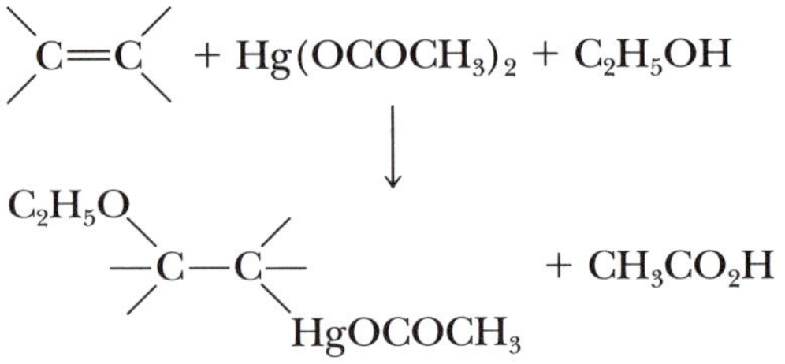

$$\text{C}{=}\text{C} + \text{Hg(OCOCH}_3)_2 + \text{C}_2\text{H}_5\text{OH}$$

$$\downarrow$$

$$\overset{C_2H_5O}{-}C{-}C\underset{HgOCOCH_3}{-} + CH_3CO_2H \quad (29\text{-}6.5)$$

The evidence that HgX_2 adds across the double bond is usually indirect, e.g., by observing the products on hydrolysis.

$$CH_2{=}CH_2 + Hg(NO_3)_2 \xrightarrow{\ OH^-\ } HOCH_2CH_2Hg^+ + 2\,NO_3^- \quad (29\text{-}6.6)$$

In these reactions, *mercurinium ions* of the type found in Structures 29-I and 29-II are believed to be intermediates. In FSO_3H—SbF_5—SO_2 at -70 °C long-

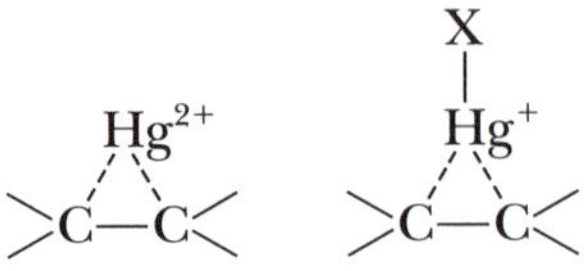

$$\underset{\text{29-I}}{\overset{Hg^{2+}}{>\!C\!-\!C\!<}} \qquad \underset{\text{29-II}}{\overset{\overset{X}{|}\ \ }{\overset{Hg^+}{>\!C\!-\!C\!<}}}$$

lived mercurinium ions have been obtained by reactions such as:

$$CH_3OCH_2CH_2HgCl \xrightarrow{H^+} CH_3OH_2^+ + CH_2CH_2Hg^{2+} + HCl \qquad (29\text{-}6.7)$$

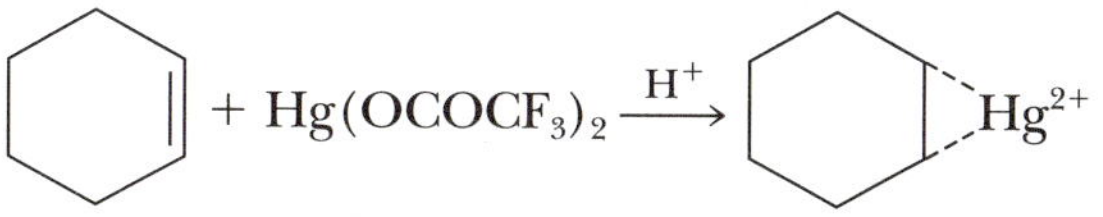

$$\text{(cyclohexene)} + Hg(OCOCF_3)_2 \xrightarrow{H^+} \text{(ring)} Hg^{2+} \qquad (29\text{-}6.8)$$

This type of addition has been used for the synthesis of alcohols, ethers, and amines from alkenes and other unsaturated substances. The additions of HgX_2 are carried out in water, alcohols, or acetonitrile, respectively. The mercury is removed from the intermediate by reduction with sodium borohydride. An example is:

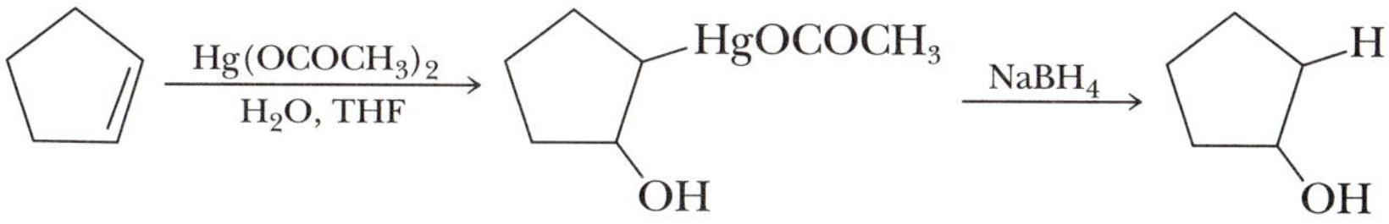

$$\text{(cyclopentene)} \xrightarrow[H_2O, THF]{Hg(OCOCH_3)_2} \text{(ring with HgOCOCH}_3\text{, OH)} \xrightarrow{NaBH_4} \text{(ring with H, OH)} \qquad (29\text{-}6.9)$$

29-7 Boron

There is a very extensive chemistry of organoboron compounds.

The *trialkyl-* and *triarylboranes* are made from the halides by lithium or Grignard reagents, and by hydroboration. The lower alkyls inflame in air, but the aryls are stable. Like other BX_3 compounds alkylboranes behave as Lewis acids giving adducts (e.g., $R_3B:NR_3'$). Furthermore, when boron halides are treated with four equivalents of an alkylating agent, the trialkyl or triaryl gives an anion BR_4^-. The most important compound is *sodium tetraphenylborate*, $Na[B(C_6H_5)_4]$. This is soluble in water and is stable in weakly acid solution; it gives insoluble precipitates with larger cations (such as K^+, Rb^+, or $(CH_3)_4N^+$) that are suitable for gravimetric analysis. There are also di- and monoalkyl compounds such as R_2BX or RBX_2, where X may be a halogen, OH, H, and so on.

29-8 Aluminum

The alkyls of Al are important because of their industrial use as catalysts for the polymerization of ethylene and propylene (Section 30-10). They are also widely used as reducing and alkylating agents for transition metal complexes.

The alkyls may be prepared by the reactions

$$2\,Al + 3\,R_2Hg \longrightarrow 2\,R_3Al \text{ or } [R_3Al]_2 + 3\,Hg \qquad (29\text{-}8.1)$$

$$RMgCl + AlCl_3 \longrightarrow RAlCl_2, R_2AlCl, R_3Al \qquad (29\text{-}8.2)$$

More direct methods suitable for large-scale use are

$$AlH_3 + 3\,C_nH_{2n} \longrightarrow Al(C_nH_{2n+1})_3 \qquad (29\text{-}8.3)$$

$$LiAlH_4 + 4\,C_nH_{2n} \longrightarrow Li[Al(C_nH_{2n+1})_4] \qquad (29\text{-}8.4)$$

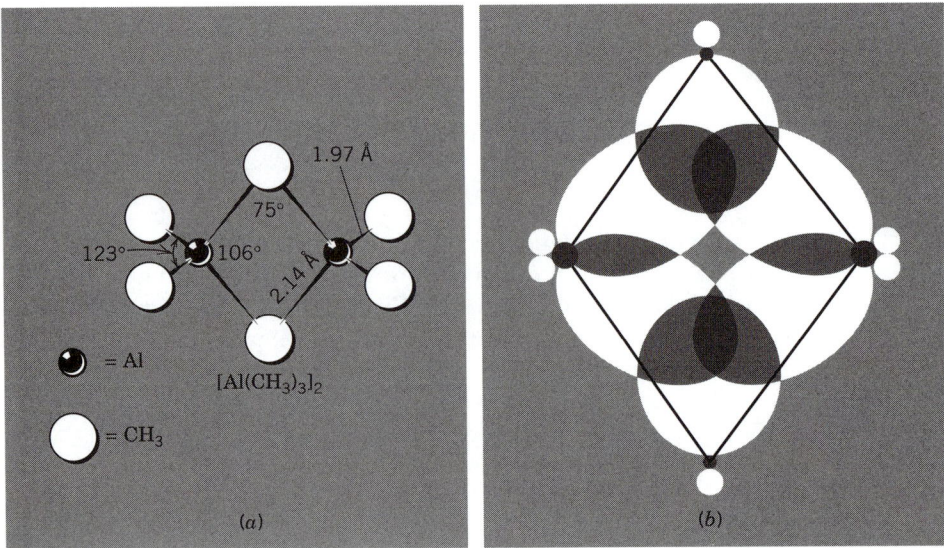

Figure 29-2   (a) The structure of $Al_2(CH_3)_6$. (b) The orbital overlap in the plane of the Al—C—Al bridge bonds.

Although $(AlH_3)_n$ cannot be made by direct interaction of Al and H_2, nevertheless, in the presence of aluminum alkyl, the following reaction can occur to give the dialkyl hydride:

$$Al + \tfrac{3}{2} H_2 + 2\, AlR_3 \longrightarrow 3\, AlR_2H \qquad (29\text{-}8.5)$$

This hydride will then react with alkenes.

$$AlR_2H + C_nH_{2n} \longrightarrow AlR_2(C_nH_{2n+1}) \qquad (29\text{-}8.6)$$

Thus the direct interaction of Al, H_2, and alkene can be used to give either the dialkyl hydrides or the trialkyls.

Other technically important compounds are the "sesquichlorides," such as $(CH_3)_3Al_2Cl_3$ or $(C_2H_5)_3Al_2Cl_3$. These compounds can be made by direct interaction of Al or Mg–Al alloy with the alkyl chloride.

The aluminum lower alkyls are reactive liquids, inflaming in air and exploding with water. All other derivatives are similarly sensitive to air and moisture though not all are spontaneously flammable. Certain aluminum alkyls form reasonably stable dimers. The structure of trimethylaluminum is shown in Fig. 29-2(a). The alkyl bridge is formed by multicenter bonding, that is, Al—C—Al, (3c–2e) bonds (Section 3-7). Each Al atom supplies an sp^3 hybrid orbital and so does the C atom. The bonding situation is shown in Fig. 29-2(b). A similar description holds true for bridging in $[Be(CH_3)_2]_n$, which is a linear polymer, each Be atom being tetrahedral.

There is no simple explanation why boron trialkyls do not dimerize in a similar way, especially since hydrogen bridges are very important in the boranes (Sections 3–7 and 12-5). The coordinative unsaturation of Al alkyls also means that they behave as Lewis acids, giving adducts such as R_3AlNR_3 or anionic species like $Li[Al(C_2H_5)_4]$. In this respect all of the coordinatively unsaturated alkyls of Groups IIIB(13) and IIB(12) elements are similar.

29-9   Silicon, Ge, Sn, and Pb

There is an extensive chemistry of the Group IVB(14) elements bound to carbon. Some of the compounds, notably silicon—oxygen polymers and alkyltin and alkyllead compounds, are of commercial importance.

Essentially all the compounds are of the tetravalent elements. In the divalent state the trimethylsilylmethyl derivatives, $M[CH(SiMe_3)_2]_2$, are the only well-established compounds with σ bonds. Other tin compounds that might appear to contain Sn^{II} are linear or cyclic polymers of Sn^{IV}.

For all four elements the compounds can generally be designated $R_{4-n}MX_n$, where R is an alkyl or aryl and X can be H, Cl, O, COR′, OR′, NR_2', SR′, $Mn(CO)_5$, $W(CO)_3(\eta^5\text{-}C_5H_5)$, and so on. The elements can also be incorporated into heterocyclic rings of various types.

For a given class of compound, those with C—Si and C—Ge bonds have higher thermal stability and lower reactivity than those with bonds to Sn and Pb. In catenated compounds, similarly, Si—Si and Ge—Ge bonds are more stable and less reactive than Sn—Sn and Pb—Pb bonds. For example, $Si_2(CH_3)_6$ is very stable, but $Pb_2(CH_3)_6$ blackens in air and decomposes rapidly in CCl_4 although it is farily stable in benzene.

The bonds to carbon are usually made via interaction of Li, Hg, or Al alkyls or RMgX and the Group IVB(14) halide, but there are some special synthetic methods noted in the following sections.

Silicon

The organometallic compounds of Si and Ge are very similar in their properties.

Silicon–carbon bond energies are less than those of C—C bonds but are still quite high, in the region 250–335 kJ mol^{-1}. Hence, the tetraalkyls and tetraaryls are thermally quite stable; $Si(C_6H_5)_4$, for example, boils unchanged at 530 °C.

The chemical reactivity of Si—C bonds is generally greater than that of C—C bonds because (a) the greater polarity of the bond ($Si^{\delta+}$—$C^{\delta-}$) allows easier nucleophilic attack on Si and electrophilic attack on C than for C—C compounds, and (b) displacement reactions at silicon are facilitated by its ability to form five-coordinate transition states by utilization of d orbitals.

Alkyl- and Arylsilicon Halides

These halides are of special importance because of their hydrolytic reactions. They may be obtained by normal Grignard procedures from $SiCl_4$ or, in the case of the methyl derivatives, by the *Rochow process* in which methyl chloride is passed over heated, copper-activated silicon.

$$CH_3Cl + Si(Cu) \longrightarrow (CH_3)_nSiCl_{4-n} \qquad (29\text{-}9.1)$$

The halides are liquids that are readily hydrolyzed by water, usually in an inert solvent. The silanol intermediates R_3SiOH, $R_2Si(OH)_2$, and $RSi(OH)_3$ can sometimes be isolated, but the diols and triols usually condense under the hydrolysis conditions to *siloxanes* that have Si—O—Si bonds. The exact nature of the products depends on the hydrolysis conditions, and linear, cyclic, and complex cross-linked polymers of varying molecular weights can be obtained. These compounds are often referred to as *silicones;* the commercial polymers usually have R = CH_3, but other groups may be incorporated for special purposes.

Controlled hydrolysis of the alkyl halides in suitable ratios can give products of particular physical characteristics. The polymers may be liquids, rubbers, or solids, which have in general high thermal stability, high dielectric strength, and resistance to oxidation and chemical attack.

Examples of simple siloxanes are $(C_6H_5)_3SiOSi(C_6H_5)_3$ and the cyclic trimer or tetramer $(Et_2SiO)_{3(or\ 4)}$; linear polymers contain —SiR_2—O—SiR_2—O— chains, whereas the cross-linked sheets have the basic unit

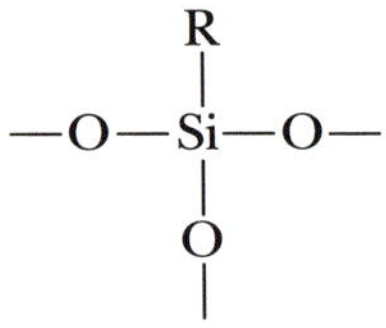

$$-O-\underset{\underset{O}{|}}{\overset{\overset{R}{|}}{Si}}-O-$$
|

Tin

Where the compounds of tin differ from those of Si and Ge, they do so mainly because of a greater tendency of Sn^{IV} to show coordination numbers higher than four and because of ionization to give cationic species.

Trialkyltin compounds (R_3SnX) are always associated in the solid by anion bridging (Structures 29-III and 29-IV). The coordination of the tin atom is close to tbp with planar $Sn(CH_3)_3$ groups. In water the perchlorate and some other compounds ionize to give cationic species, for example, $[(CH_3)_3Sn(H_2O)_2]^+$.

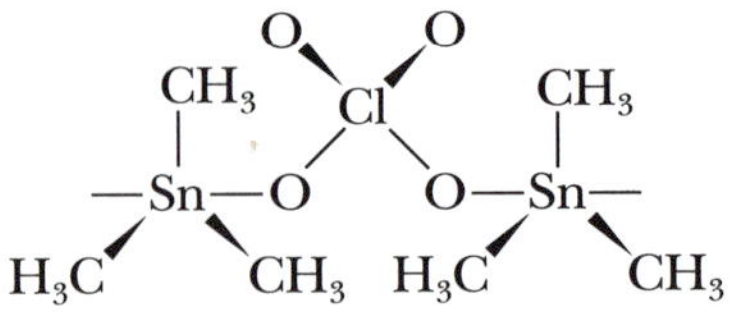

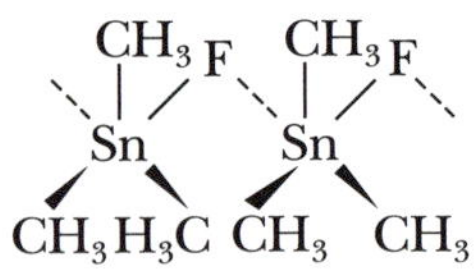

Dialkyltin compounds (R_2SnX_2) have a behavior similar to that of the trialkyl compounds. Thus the fluoride $(CH_3)_2SnF_2$ is again polymeric, with bridging F atoms, but Sn is octahedral and the group CH_3—Sn—CH_3 is linear. However, the chloride and bromide have low melting points (90 and 74 °C) and are essentially molecular compounds. The halides also give conducting solutions in water, and the aqua ion has the linear C—Sn—C group characteristic of the dialkyl species [cf. the linear species $(CH_3)_2Hg$, $(CH_3)_2Tl^+$, $(CH_3)_2Cd$, or $(CH_3)_2Pb^{2+}$], probably with four water molecules completing octahedral coordination. The linearity in these species appears to result from maximizing of *s* character in the bonding orbitals of the metal atoms. Organotin *hydrides* are useful reducing agents in organic chemistry and can add to alkenes by free radical reactions to generate other organotin compounds.

Organotin compounds have a number of uses in marine antifouling paints, fungicides, wood preserving, and as catalysts for curing silicone and epoxy resins.

Lead

The most important compounds of lead are $(CH_3)_4Pb$ and $(C_2H_5)_4Pb$, which were made in huge quantities for use as antiknock agents in gasoline. The environmental increase in lead is largely due to the burning of leaded gasolines, and their use has been largely phased out in the USA and elsewhere.

The commercial synthesis is the interaction of a sodium–lead alloy with CH_3Cl or C_2H_5Cl in an autoclave at 87–100 °C, without solvent for C_2H_5Cl but in toluene at a higher temperature for CH_3Cl. The reaction is complicated and not fully understood, and only a quarter of the lead appears in the desired product.

$$4\ NaPb + 4\ RCl \longrightarrow R_4Pb + 3\ Pb + 4\ NaCl \qquad (29\text{-}9.2)$$

The required recycling of the lead is disadvantageous and electrolytic procedures have been developed.

The lower alkyls are nonpolar highly toxic liquids. Tetramethyllead decomposes around 200 °C and tetraethyllead around 110 °C by free radical processes.

29-10 Phosphorus, As, Sb, and Bi

The chemistry of organometallic compounds is extensive, especially that of phosphorus and arsenic. This chemistry was developed largely because of the physiological properties of these compounds. Thus, one of the first chemotherapeutic agents, salvarsan, which was discovered by P. Ehrlich, led to a wide study of arylarsenic compounds.

The so-called "organophosphorus" compounds that have anticholinesterase activity and are widely used as insecticides do *not* contain P—C bonds, but are P^V derivatives such as phosphates or thionates. For example, parathion is $(C_2H_5O)_2P(S)(OC_6H_4NO_2)$.

Most of the genuine organometallic derivatives are compounds with only three or four bonds to the central atom, although a few R_5M compounds are known. The simplest synthesis is the reaction

$$(O)MX_3 + 3\ RMgX \longrightarrow (O)MR_3 + 3\ MgX_2 \qquad (29\text{-}10.1)$$

Trimethylphosphine is spontaneously flammable in air, but the higher trialkyls are oxidized more slowly. The R_3MO compounds, which may be obtained from the oxo halides as shown in Reaction 29-10.1 or by oxidation of the corresponding R_3M compounds, are all very stable. The trialkyl- or alkylarylphosphines are usually liquids with an unpleasant odor. The triarylphosphines are white crystalline solids reasonably stable in air. Tertiary phosphines, arsines, and stibines are all good π-acid ligands for *d*-group transition metals (Section 28-15). The oxides (R_3MO) also form many complexes, but they function as simple donors. Trialkyl- and triarylphosphines, -arsines, and -stibines generally react with alkyl and aryl halides to form quaternary salts.

$$R_3M + R'X \longrightarrow [R_3R'M]^+X^- \qquad (29\text{-}10.2)$$

The tetraphenylphosphonium and tetraphenylarsonium ions are useful for precipitating large anions such as ReO_4^-, ClO_4^- and complex anions of metals.

An important phosphonium compound is obtained by the reaction

$$PH_3 + 4\ HCHO + HCl(aq) = [P(CH_2OH)_4]^+Cl^- \qquad (29\text{-}10.3)$$

It is a white crystalline solid, soluble in water, and is used in the flameproofing of fabrics.

Triphenylphosphine is an important ligand, and is also utilized in the Wittig reaction for alkene synthesis. This reaction involves the formation of alkylidenetriphenylphosphoranes from the action of butyllithium or other base on the quarternary halide, for example,

$$[(C_6H_5)_3PCH_3]^+Br^- \xrightarrow{\ n\text{-}butyllithium\ } (C_6H_5)_3P{=}CH_2 \qquad (29\text{-}10.4)$$

This intermediate reacts very rapidly with aldehydes and ketones to give zwitterionic compounds (Structure 29-V), which eliminate triphenylphosphine oxides under mild conditions to give alkenes (Structure 29-VI).

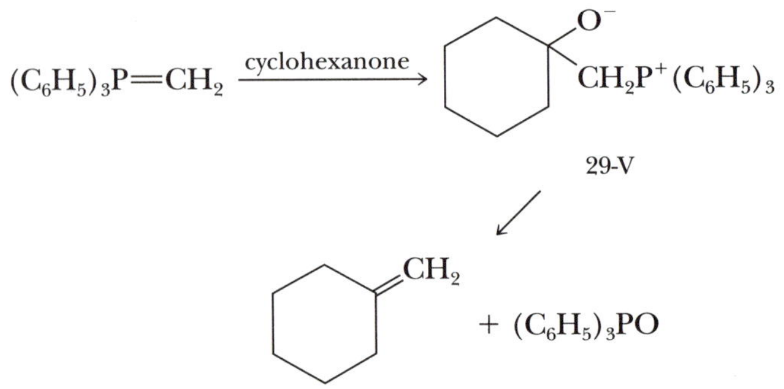

$$(C_6H_5)_3P{=}CH_2 \xrightarrow{\ cyclohexanone\ }$$

29-V

$$+\ (C_6H_5)_3PO \qquad (26\text{-}10.5)$$

29-VI

Alkylidenephosphoranes such as $(CH_3)_3P{=}CH_2$, $(C_2H_5)_3P{=}CH_2$, $(CH_3)_2C_2H_5P{=}CH_2$, and $(C_2H_5)_3P{=}CHCH_3$ are all colorless liquids, stable for long periods in an inert atmosphere.

PART B

TRANSITION METALS

Sigma-bonded alkyls or aryls of the transition metals are stable only under special circumstances. Unstable or labile species with σ bonds to carbon are of great significance, particularly in catalytic reactions of alkenes and alkanes induced by transition metals or metal complexes. Transition metal to carbon σ bonds also exist in nature in Vitamin B_{12} derivatives (Section 31-8).

The unique characteristics of d orbitals allow the binding to metal atoms of unsaturated hydrocarbons and other molecules. The bonding is nonclassical and the metal complexes of alkenes, alkynes, arenes, and the like, have no counterparts elsewhere in chemistry.

29-11 Transition Metal to Carbon σ Bonds

Although the compound $[(CH_3)_3PtI]_4$, which has a structure based on a cube with Pt and I atoms at alternative corners and each Pt bound to three CH_3 groups, was made in 1909 by Pope and Peachy, attempts to prepare compounds such as $(C_2H_5)_3Fe$ by reactions of Grignard reagents with metal halides failed. Although evidence indicated the alkyls were present in solution at low temperatures, complicated decomposition and coupling reactions occurred at ambient temperatures.

It was found over 30 years ago that, provided other ligands such as the η^5-cyclopentadienyl group described later in this chapter, or those of the π-acid type (Chapter 28) were present, alkyl compounds could be isolated; one example is $CH_3Mn(CO)_5$. It now appears that the principal reason for the stability of these compounds is that the coordination sites required for decomposition reactions to proceed are blocked. The main reason for the instability of most binary alkyls or aryls is that they are coordinatively unsaturated, and there are easy pathways for thermodynamically possible decomposition reactions to occur. Possible decomposition pathways include homolysis of the M—C bond, which generates free radicals, as well as the transfer of a hydrogen atom from carbon to the metal. A particularly common reaction is the transfer from the β carbon of the alkyl chain (Reaction 29-11.1)

$$M-CH_2-CH_2-R \rightleftharpoons \left[\begin{array}{c} H \\ | \\ M \longleftarrow \begin{array}{c} CHR \\ \| \\ CH_2 \end{array} \end{array} \right] \longrightarrow$$

$$MH + CHR = CH_2 \qquad (29\text{-}11.1)$$

resulting in the elimination of alkene and formation of an M—H bond. The reverse of this reaction, that is, the formation of alkyls by addition of alkenes to M—H bonds (cf. Section 30-6) is of very great importance in catalytic reactions discussed in Chapter 30. Once the hydrogen atom has been transferred to the metal, further reaction can occur to give the metal and hydrogen, or the hydrogen can be transferred to the alkene to form alkane. Thus it has been shown that the copper alkyl $(Bu_3P)CuCH_2C(Me)_2Ph$ decomposes largely by a free radical pathway but that the similar alkyl, $(Bu_3P)CuCH_2CH_2CH_2CH_3$, decomposes by a nonradical pathway involving Cu—H bond formation. The difference is that the latter, but not the former, has a hydrogen atom on the β-carbon atom.

There are a number of reasonably thermally stable alkyls that cannot undergo the β-hydride-transfer, alkene-elimination sequence. These have groups such as $-CH_2C_6H_5$, $-CH_2Si(CH_3)_3$, $-CH_2C(CH_3)_3$ $-CH_2P^+(CH_3)_3$, and 1-norbornyl (Structure 29-VII).

M

29-VII

Although hydrogen transfer from an α-carbon atom to produce a hydrido carbene intermediate as the first step in decomposition:

Table 29-1   Some Binary Transition Metal Alkyls

Compound	Properties
$Ti(CH_2Ph)_4$	Red crystals, mp 70 °C; tetrahedral
$VO(CH_2SiMe_3)_3$	Yellow needles, mp 75 °C; has V=O bond
$Cr(1\text{-norbornyl})_4$	Red-brown crystals; tetrahedral; d^2 paramagnetic
$Mo_2(CH_2SiMe_3)_6$	Yellow plates, decomp. 135 °C; has Mo≡Mo bond
$Re(CH_3)_6$	Green crystals; octahedral; d^1 paramagnetic

$$M\text{—CHRR'} \longrightarrow M=CRR' \quad \overset{H}{|} \qquad (29\text{-}11.2)$$

is possible, this is evidently less favorable than the β transfer and is rarely observed. Methyl metal compounds such as $(CH_3)_4Ti$ or the $[(CH_3)_3PtI]_4$ already mentioned are accordingly much more stable than the homologous ethyl metal compounds. However, even $Ti(CH_3)_4$ decomposes at about −80 °C, but on addition of ligands such as bipyridine which leads to coordinative saturation as in $Ti(bpy)(CH_3)_4$, a substantial increase in thermal stability results. This again shows the necessity of having coordination sites on the metal available in order to allow decomposition reactions to proceed. Another striking example of this principle is that substitution-inert complexes (Section 6-5) of Cr^{III}, Co^{III}, and Rh^{III} may have M—C bonds even when H_2O or NH_3 are ligands; one example is $[Rh(NH_3)_5C_2H_5]^{2+}$. Particularly important are the cobalt complexes of the vitamin B_{12} type and their synthetic analogs, which are discussed in Chapter 31. One example is the dimethylglyoxime complex shown in Structure 29-VIII.

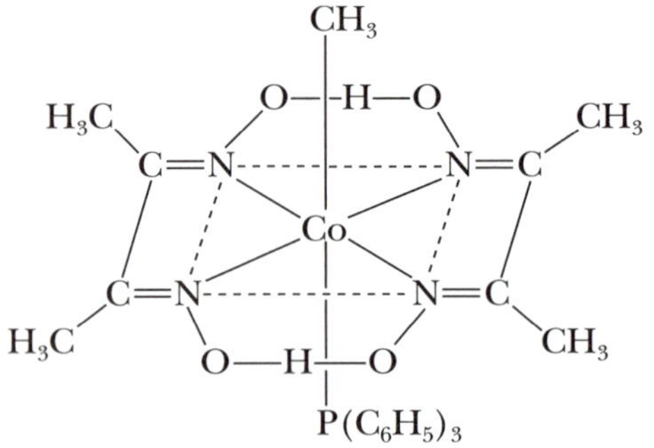

29-VIII

Some representative examples of alkyls are given in Table 29-1.

29-12 Alkene Complexes

About 1830, W. C. Zeise, a Danish pharmacist, characterized compounds that had stoichiometries $PtCl_2C_2H_4$ and $K[PtCl_3C_2H_4]$. Although these were the first organometallic derivatives of transition metals to be prepared, their true nature was fully established only around 1953.

Ethylene and most other alkenes can be bound to transition metals in a wide variety of complexes. The structures of two such complexes are shown in Fig. 29-3. The fact that the plane of the alkene and the C=C axis are perpendicular to one of the expected bond directions from the central metal atom is of key significance. In addition, the expected line of a bond orbital from the metal atom strikes the C=C bond at its midpoint.

The most useful description of the bonding in alkene complexes is illustrated in Fig. 29-4. The bonding consists of two interdependent components: (a) overlap of the π-electron density of the alkene with a σ-type acceptor orbital on the metal atom; and (b) a "back-bond" resulting from flow of electron density from filled metal d_{xz} or other $d\pi-p\pi$ hybrid orbitals into the π*-*antibonding* orbital on the carbon atoms. It is thus similar to that discussed for the bonding of CO and other π-acid ligands (Chapter 28) and implies the retention of appreciable "double-bond" character in the alkene. Of course, the donation of π-bonding electrons to the metal σ orbital and the introduction of electrons into the π*-antibonding orbital both weaken the π bonding in the alkene, and in every case except the anion of Zeise's salt there is significant lengthening of the alkene C—C bond. There appears to be some correlation between lengthening of the bond and the electron-withdrawing power of the substituents of the alkene. This is exemplified by the structures shown in Fig. 29-3 where the $C_2(CN)_4$ complex has a C—C bond about as long as a normal single bond.

In the extreme of a very long C—C distance the bonding could be formulated as a kind of metallocyclopropane ring, involving two (2c–2e) M—C bonds and a C—C single bond. In Fig. 29-3(*b*) the bond angles at the two alkene carbon atoms are consistent with this view. This representation of the bonding and the MO description are complementary, and there is a smooth gradation of one description into the other.

Alkenes with unconjugated double bonds can form independent linkages to a metal atom. Two representative complexes, of 1,5-cyclooctadiene (cod) and norbornadiene, are Structures 29-IX and 29-X, respectively. Three unconjugated

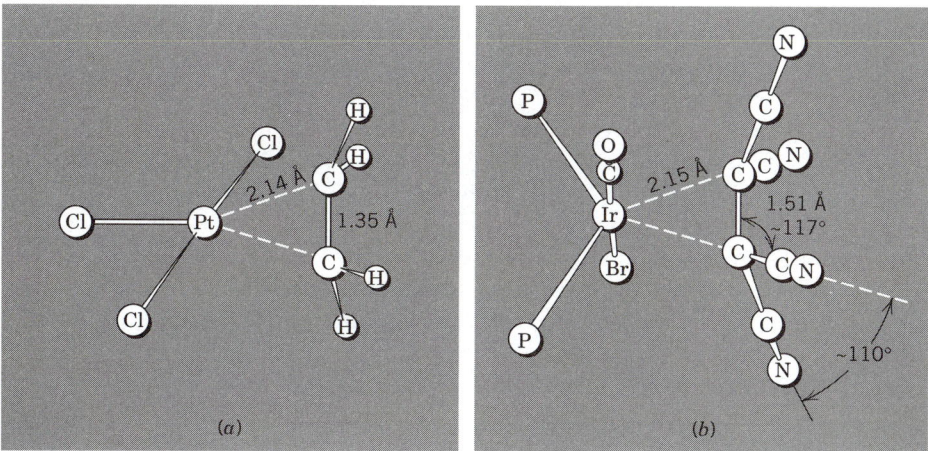

Figure 29-3   The structures of two monoalkene complexes. (*a*) The $[PtCl_3C_2H_4]^-$ anion of Zeise's salt. (*b*) The molecule $(Ph_3P)_2(CO)(Br)Ir[C_2(CN)_4]$, with phenyl groups of the triphenylphosphine ligands omitted for clarity. The monoalkene $C_2(CN)_4$ is tetracyanoethylene.

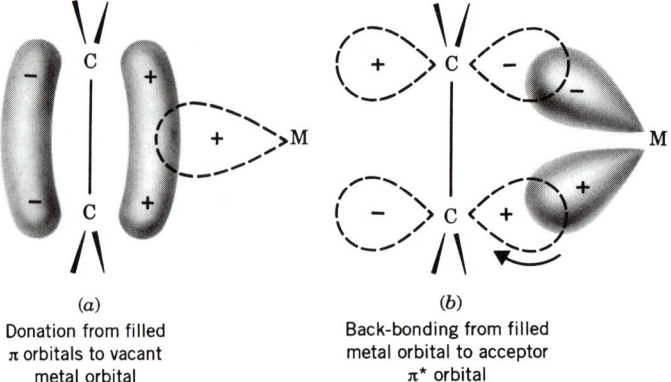

(a)   (b)
Donation from filled   Back-bonding from filled
π orbitals to vacant   metal orbital to acceptor
metal orbital   π* orbital

Figure 29-4   Diagrams showing the MO view of alkene–metal bonding. (*a*) The donation of π-electron density from the alkene to the metal. (*b*) The donation of *d*π-electron density from the metal to the alkene.

double bonds may be coordinated to one metal atom as in the *trans, trans, trans*-cyclododecatriene complex (Structure 29-XI).

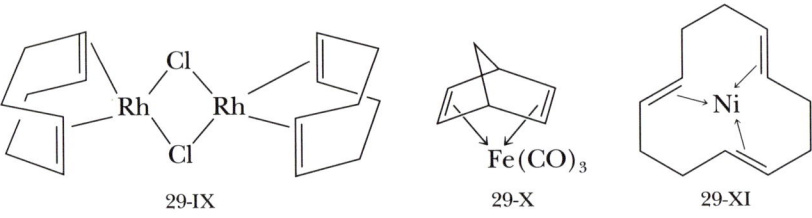

29-IX   29-X   29-XI

When two or more *conjugated* double bonds are engaged in bonding to a metal atom the interactions become more complex, though qualitatively the two types of basic, synergic components are involved. Buta-1,3-diene is an important case and shows why it is an oversimplification to treat the bonding as simply collections of separate monoalkene-metal interactions.

Two extreme formal representations of the bonding of a buta-1,3-diene group to a metal atom are shown in Fig. 29-5. The degree to which individual structures approach either of these extremes can be judged by the lengths of the C—C bonds. A short–long–short pattern is indicative of Fig. 29-5(*a*) while a long–short–long pattern is indicative of Fig. 29-5(*b*). In no case has a pronounced short–long–short pattern been established and the actual variation seems to lie between approximate equality of all three bond lengths and the long–short–long pattern.

From a purely formal point of view each double bond in any alkene can be considered as a two-electron donor. If we have a polyalkene involved, the metal atom usually reacts so as to complete its normal coordination. For example, $Mo(CO)_6$ and $Fe(CO)_5$ react with cyclohepta-1,3,5-triene to give Structures 29-XII and 29-XIII, respectively. In Structure 29-XIII there is one uncoordinated double bond.

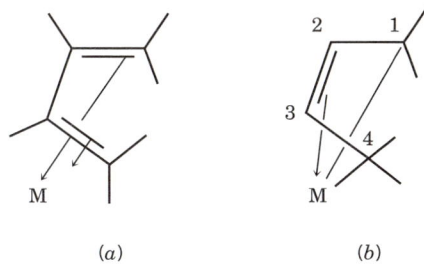

Figure 29-5 Two extreme formal representations of the bonding of a *cis*-buta-1,3-diene
group to a metal atom. Part (*a*) implies that
there are two substantially independent mon-
alkene-metal interactions, while (*b*) depicts σ
bonds to carbon atoms 1 and 4 coupled with
an alkene–metal interaction at carbon atoms 2
and 3.

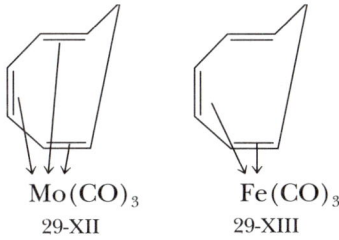

$$Mo(CO)_3$$
29-XII

$$Fe(CO)_3$$
29-XIII

Cyclooctatetraene with four essentially unconjugated double bonds can bind
in several ways depending on the metal system. With $PtCl_2$, it uses its 1- and
5-olefinic linkages, as in Structure 29-XIV, and with $Fe(CO)_3$, which has a
predilection for binding 1,3-dialkenes, it is bound as in Structure 29-XV.

29-XIV

29-XV

Synthesis

Alkene complexes are usually synthesized by the interaction of metal carbonyls,
halides, or occasionally other complexes with the alkene. Some representative
examples are:

$$Mo(CO)_6 + C_7H_8 \xrightarrow{\text{reflux}} Mo(CO)_3C_7H_8 + 3\,CO \qquad (29\text{-}12.1)$$

$$RhCl_3(aq) + C_2H_4 \xrightarrow{25\,°C \text{ in ethanol}} [(C_2H_4)_2RhCl]_2 \qquad (29\text{-}12.2)$$

Some of the earliest studies were made with the Ag^+ ion and in solutions we have equilibria of the type:

$$Ag^+(aq) + alkene = [Ag\ alkene]^+ \qquad (29\text{-}12.3)$$

The interaction of hydrocarbons with Ag^+ ions sometimes gives crystalline precipitates that are useful for purification of the alkene. Thus cyclooctatetraene or bicyclo-2,5-heptadiene, when shaken with aqueous silver perchlorate (or nitrate), give white crystals of stoichiometry alkene·$AgClO_4$ or 2 alkene·$AgClO_4$, depending on the conditions. Benzene also gives crystalline complexes with $AgNO_3$, $AgClO_4$, or $AgBF_4$. In $[C_6H_6 \cdot Ag]^+ClO_4^-$ the metal ion is asymmetrically located with respect to the ring.

29-13 Notation and Electron Counting in Alkene and Related Complexes

In addition to alkene complexes, there are more complicated systems (such as the allyls to be discussed in Section 29-16) in which delocalized π electrons are bound to metals. Some systematic notation is required to designate the number of carbon atoms that are bound to the metal. This is done by use of the term *hapto-* (from the Greek, to fasten). Prefixes designate the number of carbon atoms that are fastened to the metal: *trihapto-, tetrahapto-, pentahapto-,* and so on. An equivalent designation uses the Greek η, with superscripts: η^3, η^4, η^5, and so on. Structures 29-XVI, 29-XVII, and 29-XVIII in Fig. 29-6 should help to make clear the use of the notation.

Electron Counting Rules as Applied to Alkene Complexes

We take the convention as stated previously that each double bond in a neutral alkene is a two-electron donor. The following cases are representative.

$[PtCl_3C_2H_4]^-$, Fig. 29-3(a):

Pt(II)		8 electrons
Cl^-	$2 \times 3 =$	6 electrons
C_2H_4	$2 \times 1 =$	2 electrons
	Total =	16 electrons

$Fe(CO)_3(C_7H_8)$, Structure 29-X:

Fe		8 electrons
Norbornadiene	$2 \times 2 =$	4 electrons
CO	$2 \times 3 =$	6 electrons
	Total =	18 electrons

Ni(cyclododecatriene), Structure 29-XI:

Ni	10 electrons
Cyclododecatriene	$2 \times 3 =$ 6 electrons

Total = 16 electrons

Mo(CO)$_3$(η^6-cyclohepta-1,3,5-triene),
 Structure 29-XII:

Mo		6 electrons
η^6-C$_7$H$_8$	$2 \times 3 =$	6 electrons
CO	$2 \times 3 =$	6 electrons

Total = 18 electrons

Fe(CO)$_3$(η^4-cyclohepta-1,3,5-triene),
 Structure 29-XIII:

Fe		8 electrons
η^4-C$_7$H$_8$	$2 \times 2 =$	4 electrons
CO	$2 \times 3 =$	6 electrons

Total = 18 electrons

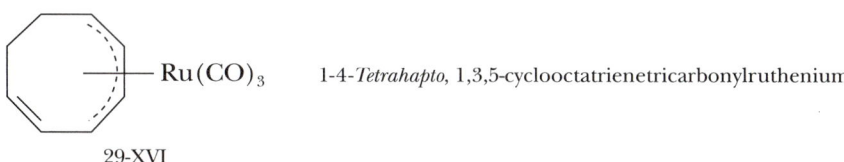

Ru(CO)$_3$ 1-4-*Tetrahapto*, 1,3,5-cyclooctatrienetricarbonylruthenium

29-XVI

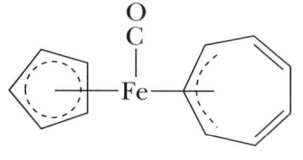

(*Pentahapto*cyclopentadienyl)(1-3-*trihapto*cycloheptatrienyl)carbonyliron

29-XVII

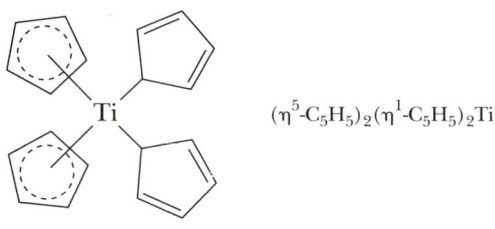

(η^5-C$_5$H$_5$)$_2$(η^1-C$_5$H$_5$)$_2$Ti

29-XVIII

Figure 29-6 Illustrations of the *hapto* notation. Structure 29-XVI contains an η^4 ligand. Structure 29-XVII contains both an η^3 and an η^5 ligand. Structure 29-XVIII contains two η^5 and two η^1 cyclopentadienyl ligands.

29-14 Other π-Donor Ligands; Delocalized Carbocyclic Groups

In 1951 a compound with the formula $(C_5H_5)_2Fe$ was reported and was subsequently shown to have a unique "sandwich" Structure (29-XIX) in which the metal lies between two planar cyclopentadienyl rings. Many $\eta^5\text{-}C_5H_5$ compounds are now known. Some have only one $\eta^5\text{-}C_5H_5$ ring, as in Structure 29-XX, while others have two rings but with these at an angle as in Structure 29-XXI.

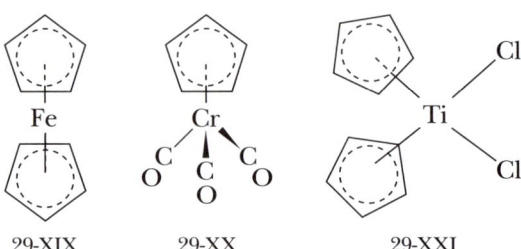

29-XIX   29-XX   29-XXI

Other symmetric ring systems now known to form complexes are $C_3(C_6H_5)_3$, C_4H_4, C_6H_6, C_7H_7, and C_8H_8. There is a formalism of describing these ring systems as if they assume the charge required to achieve an aromatic electron configuration. The "magic numbers" for aromaticity are 2, 6, and 10 so that these carbocycles can be written as

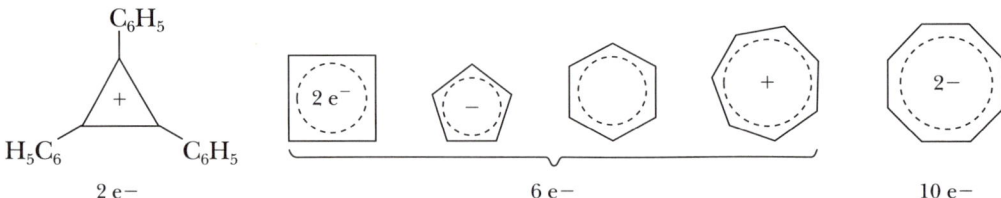

The charges may be used in assigning formal oxidation numbers to the metal atoms in the complexes. Thus $(\eta^5\text{-}C_5H_5)_2Fe$ can be regarded as formed from two cyclopentadienide ions, $C_5H_5^-$, and Fe^{2+}, so that the compound contains Fe^{II}. In the benzene compound $C_6H_6Cr(CO)_3$ chromium has the formal oxidation state 0 as in $Cr(CO)_6$.

Examples of carbocyclic complexes are Structures 29-XXII to 29-XXV.

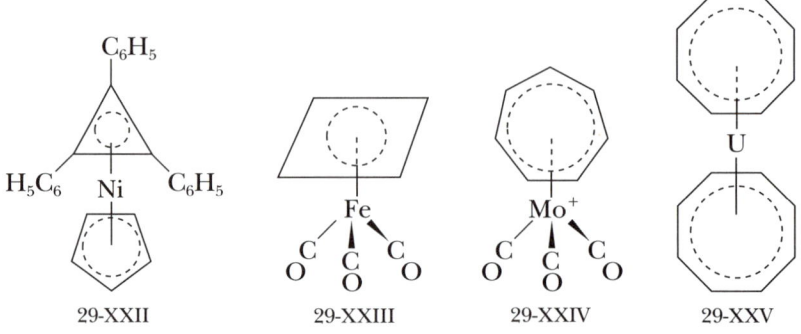

29-XXII   29-XXIII   29-XXIV   29-XXV

Cyclopentadienyls

Cyclopentadiene is a weak acid ($pK_a \sim 20$) and with strong bases forms the cyclopentadienide ion $C_5H_5^-$. The general method for synthesizing metal complexes is the reaction of this ion with a metal halide or other complex, for example,

$$C_5H_6 + Na \xrightarrow{\text{THF}} C_5H_5^- + Na^+ + \tfrac{1}{2}H_2 \qquad \text{(main reaction)} \qquad (29\text{-}14.1)$$

$$2\,C_5H_5^- + NiCl_2 \xrightarrow{\text{THF}} (\eta^5\text{-}C_5H_5)_2Ni + 2\,Cl^- \qquad\qquad (29\text{-}14.2)$$

Two other methods are (a) the use of C_5H_5Tl, which is insoluble in water, stable, and easily stored:

$$C_5H_6 + TlOH \xrightarrow{\text{H}_2\text{O}} C_5H_5Tl(s) + H_2O \qquad\qquad (29\text{-}14.3)$$

$$FeCl_2 + 2\,TlC_5H_5 \xrightarrow{\text{THF}} 2\,TlCl + (\eta^5\text{-}C_5H_5)_2Fe \qquad\qquad (29\text{-}14.4)$$

and (b), the use of a strong organic base as a proton acceptor,

$$2\,C_5H_6 + CoCl_2 + 2(C_2H_5)_2NH \xrightarrow[\text{amine}]{\text{in excess}}$$

$$(\eta^5\text{-}C_5H_5)_2Co + 2(C_2H_5)_2NH_2Cl \quad (29\text{-}14.5)$$

Since the C_5H_5 anion acts as a uninegative ligand, the dicyclopentadienyl compounds are of the type $(\eta^5\text{-}C_5H_5)_2MX_{n-2}$, where the oxidation state of the metal (M) is n and X is a uninegative ion. When $n = 2$ we obtain neutral molecules like $(\eta^5\text{-}C_5H_5)_2Fe^{II}$. When $n = 3$, we may obtain a cation like $[(\eta^5\text{-}C_5H_5)_2Co^{III}]^+$ or, when $n = 4$, a halide like $(\eta^5\text{-}C_5H_5)_2Ti^{IV}Cl_2$. Some typical η^5-cyclopentadienyl compounds are given in Table 29-2. The C—C distance and bond order in η^5-C_5H_5 rings are similar to those in benzene. Aromatic-like reactions can be carried out for two compounds, $(\eta^5\text{-}C_5H_5)_2Fe$, which has been given the trivial name *ferrocene,* and $(\eta^5\text{-}C_5H_5)Mn(CO)_3$ or *cymantrene.* These compounds will survive the reaction conditions, but other η^5-C_5H_5 compounds are decomposed. Typical reactions are Friedel-Crafts acylation, metalation by butyllithium, sulfonation, and so on. Indeed, there is an extensive "organic chemistry" of these molecules.

The bonding in η^5-C_5H_5 complexes is well described by an MO approach involving overlap of metal d orbitals (principally the d_{z^2}, d_{xz} and d_{yz} orbitals) with various π molecular orbitals of the cyclopentadienyl rings. For instance, if the z axis passes through the ring centers and the metal atom, then a bonding π molecular orbital of each cyclopentadienyl ring overlaps well with the corresponding lobe of the d_{z^2} orbital. In this fashion, bonding between the central metal atom and the "π-cloud" of each cyclopentadienyl ring is established.

In the compounds with only one η^5-C_5H_5 ring, the lobes of the d orbitals not involved in bonding to the ring can overlap with suitable orbitals in other ligands such as CO, NO, and R_3P. Observe that only in the neutral compounds and $(\eta^5\text{-}C_5H_5)_2M^+$ are the rings parallel; in other compounds, such as Structure 29-XXI, the rings are at an angle.

Table 29-2   Some Di-η^5-cyclopentadienyl Metal Compounds[a]

Compound	Appearance; mp (°C)	Unpaired Electrons	Other Properties
$(\eta^5\text{-}C_5H_5)_2Fe$	Orange crystals; 174	0	Oxidized by Ag^+(aq) or dilute HNO_3 to blue cation $\eta^5\text{-}Cp_2Fe^+$. Stable thermally to >500 °C
$(\eta^5\text{-}C_5H_5)_2Cr$	Scarlet crystals; 173	2	Very air sensitive
$(\eta^5\text{-}C_5H_5)_2Co^+$	Yellow ion and salts	0	Forms numerous salts and a stable strong base (absorbs CO_2 from air); thermally stable to ~400 °C
$(\eta^5\text{-}C_5H_5)_2TiCl_2$	Bright red crystals; 230	0	C_6H_5Li gives $\eta^5\text{-}Cp_2Ti(C_6H_5)_2$; reducible to $\eta^5\text{-}Cp_2TiCl$
$(\eta^5\text{-}C_5H_5)_2WH_2$	Yellow crystals; 163	0	Moderately stable in air, soluble in benzene, and so on; soluble in acids giving $\eta^5\text{-}Cp_2WH_3^+$ ion

[a]$Cp = C_5H_5$. Note that many substituted derivatives are known and that pentamethylcyclopentadienyls (*Cp) have been especially well studied. Their properties sometimes differ from those of Cp analogs. For example, Cp_2Mn is high spin when dilute, whereas *Cp_2Mn is low spin.

Benzenoid–Metal Complexes

Of other carbocycles, those containing benzene and substituted benzenes are the most important. Curiously, the first $(\eta^6\text{-}C_6H_6)M$ compounds were prepared as long ago as 1919, but their true identities were recognized only in 1954. A series of chromium compounds was obtained by F. Hein in 1919 from the reaction of $CrCl_3$ with C_6H_5MgBr; these compounds were formulated as "polyphenylchromium" compounds, namely, $(C_6H_5)_nCr^{0,1+}$, where $n = 2$, 3, or 4. They actually contain "sandwich"-bonded C_6H_6 and C_6H_5—C_6H_5 groups as, for example, in Structure 29-XXVI.

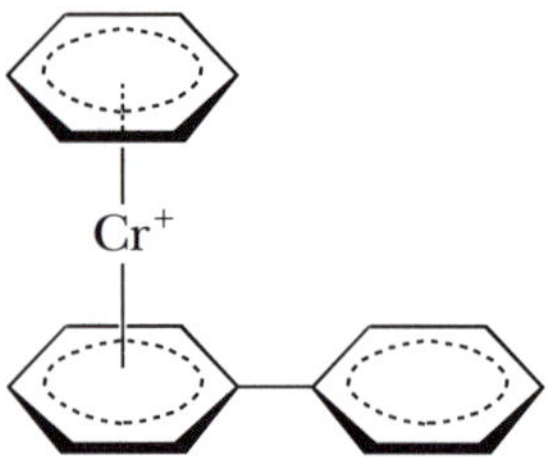

29-XXVI 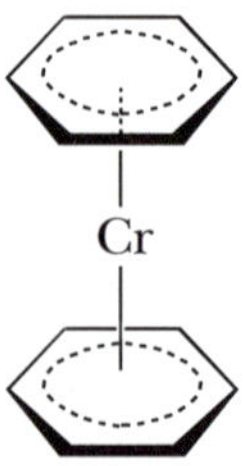   29-XXVII

The prototype neutral compound, *dibenzenechromium,* $(C_6H_6)_2Cr$ (Structure

29-XXVII), has also been obtained from the Grignard reaction of $CrCl_3$. A more effective method, discovered by E. O. Fischer, is the direct interaction of an aromatic hydrocarbon and a metal halide in the presence of Al powder as a reducing agent and halogen acceptor, plus $AlCl_3$ as a Friedel–Crafts-type activator. Although the neutral species are formed directly in the case of chromium, the usual procedure is to hydrolyze the reaction mixture with dilute acid to give the cations $(C_6H_6)_2Cr^+$, (mesitylene)$_2Ru^{2+}$, and so on. These cations may then be reduced to the neutral molecules.

Dibenzenechromium, which forms dark brown crystals, is much more sensitive to air than is ferrocene, with which it is isoelectronic; it does not survive the reaction conditions of aromatic substitution. As with the η^5-C_5H_5 compounds, complexes with only one arene ring can be prepared:

$$C_6H_5CH_3 + Mo(CO)_6 \xrightarrow{\text{reflux}} \eta^6\text{-}C_6H_5CH_3Mo(CO)_3 + 3\,CO \quad (29\text{-}14.6)$$

$$C_6H_6 + Mn(CO)_5Cl + AlCl_3 \longrightarrow [\eta^6\text{-}C_6H_6Mn(CO)_3]^+[AlCl_4]^- \quad (29\text{-}14.7)$$

The cyclooctatetraenyl ion ($C_8H_8^{2-}$) forms similar sandwich compounds with actinides, for example, $(\eta^8\text{-}C_8H_8)_2U^{IV}$ (Structure 29-XXV). It appears that f orbitals are involved in the bonding here.

29-15 Alkyne Complexes

In alkynes there are *two* π bonds at 90° to each other and both can be bound to a metal as in Structure 29-XXVIII. The Co atoms and the alkyne carbon atoms form a distorted tetrahedron, and the C_6H_5 (or other groups) on the alkyne are bent away as shown.

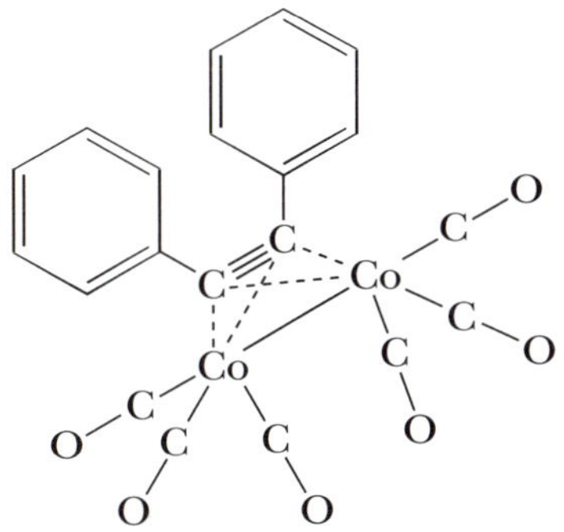

29-XXVIII

There are also complexes where the alkyne is coordinated to only one metal atom and serves simply as the equivalent of an alkene or carbon monoxide ligand. Thus we have the reactions:

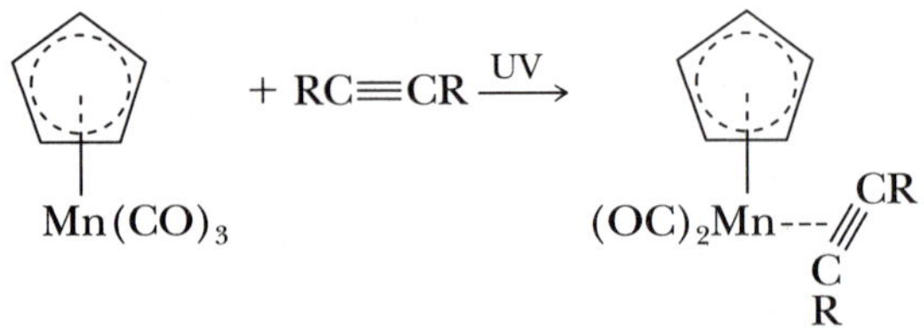

$$\text{(29-15.1)}$$

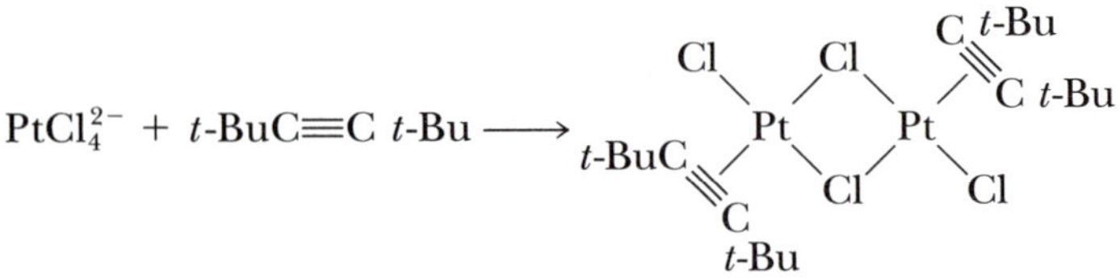

$$\text{(29-15.2)}$$

A third way of bonding, notably in Pt, Pd, and Ir complexes, is that shown in Fig. 29-7. In these the C—C stretching frequency is lowered considerably, to the range 1750–1770 cm^{-1}, indicative of a C—C double bond. The C—C bond length of 1.32 Å is consistent with this view, as is the large distortion from linearity.

Finally, many important reactions of alkynes, especially with metal carbonyls, involve incorporation of the alkynes into rings, thus producing species with new organic ligands bound to the metals. Some examples are the following:

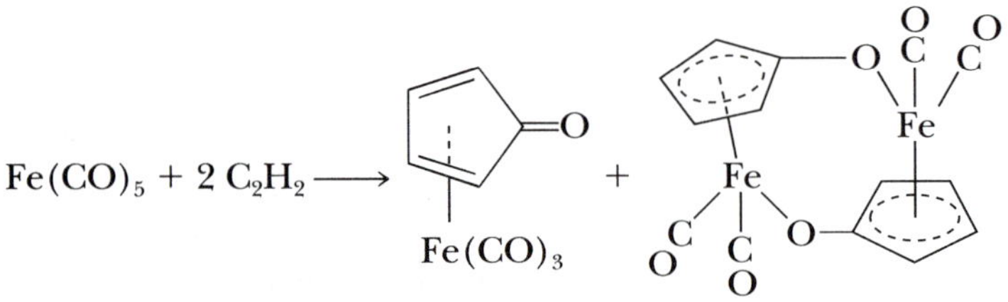

$$\text{(29-15.3)}$$

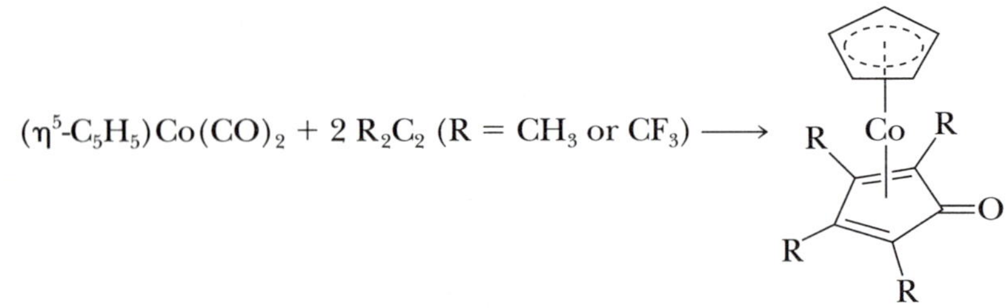

$$\text{(29-15.4)}$$

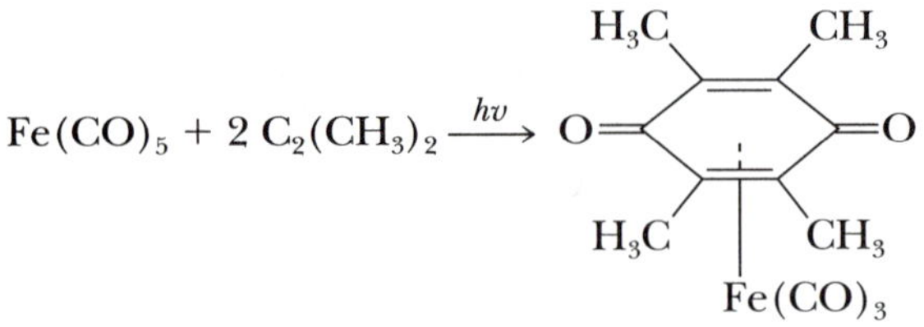

$$\text{(29-15.5)}$$

29-16 Allyl Complexes

The unsaturated, bent, three-carbon allylic group can be bound to a metal atom in either of two ways, as shown in Reaction 29-16.1. The first is as a trihapto (or

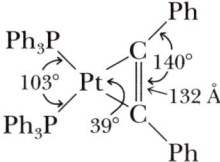

Figure 29-7 The structure of $(Ph_3P)_2Pt(PhC_2Ph)$, in which diphenylacetylene is most simply formulated as a divalent, bidentate ligand.

π-allyl) radical, which serves as a three-electron π donor to the metal. A representative example is shown in Fig. 29-8, where it can be seen that the plane of the allyl carbon atoms neither coincides with, nor is perpendicular to, the central $Pd(\mu$-Cl$)_2$Pd plane. The second type of metal allyl is the *monohapto*, or σ-allyl, which, as shown in Reaction 29-16.1, is often in equilibrium with the π-allyl. Monohapto allyls are best considered to be a special type of alkyl ligand.

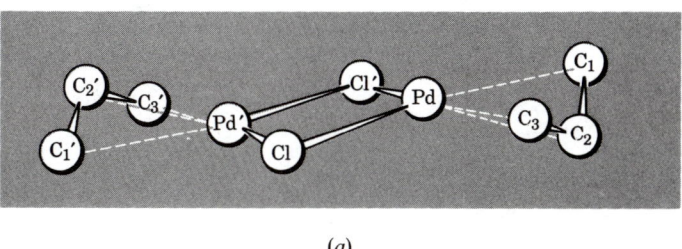

(a)

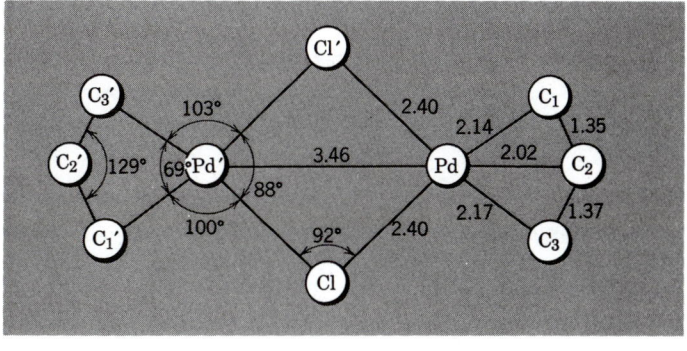

(b)

Figure 29-8 Structure of the allylpalladium(II) chloride dimer: (*a*) side view and (*b*) top view. [Reprinted by permission from W. E. Oberhansli and L. F. Dahl, *J. Organomet. Chem.*, **1965**, *3*, 43.]

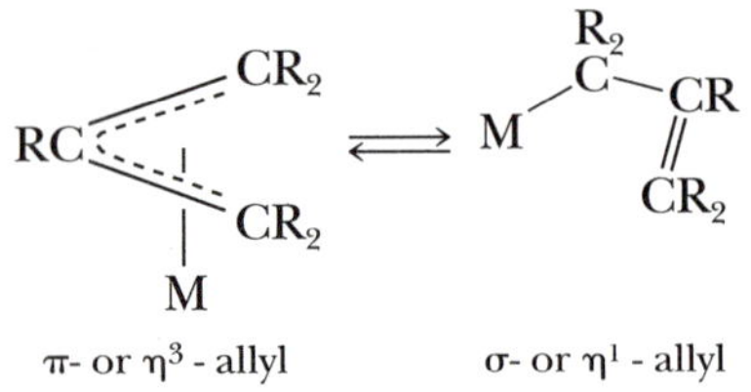

$$RC \underset{\underset{M}{\overset{|}{\underset{}{}}}}{\overset{CR_2}{\underset{CR_2}{\rlap{}}}} \rightleftharpoons M \overset{R_2}{\overset{C}{\underset{CR_2}{\overset{CR}{\parallel}}}} \tag{29-16.1}$$

π- or η³ - allyl   σ- or η¹ - allyl

Allyl complexes may be obtained from allyl Grignard reagents or from allyl chloride as in the reactions:

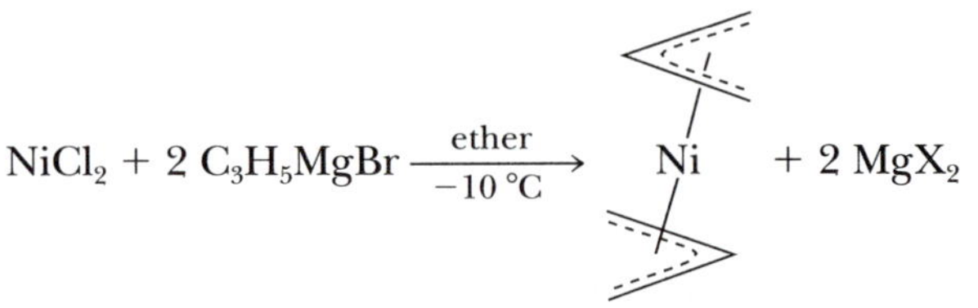

$$NiCl_2 + 2\ C_3H_5MgBr \xrightarrow[-10\,°C]{ether} Ni + 2\ MgX_2 \tag{29-16.2}$$

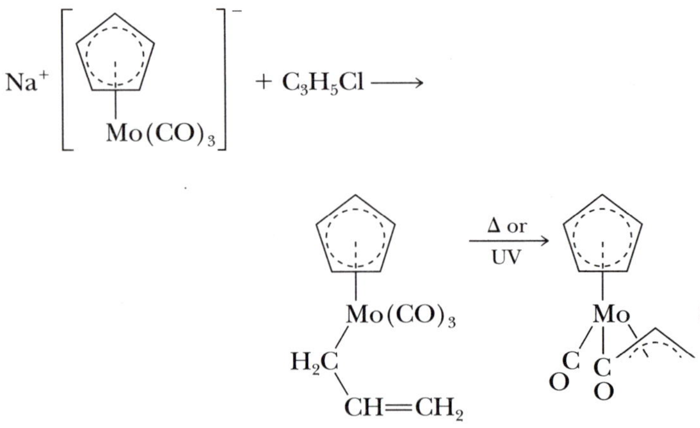

$$Na^+ \left[\underset{Mo(CO)_3}{\boxed{}} \right]^- + C_3H_5Cl \longrightarrow \tag{29-16.3}$$

They can also be obtained by protonation of butadiene complexes, for example,

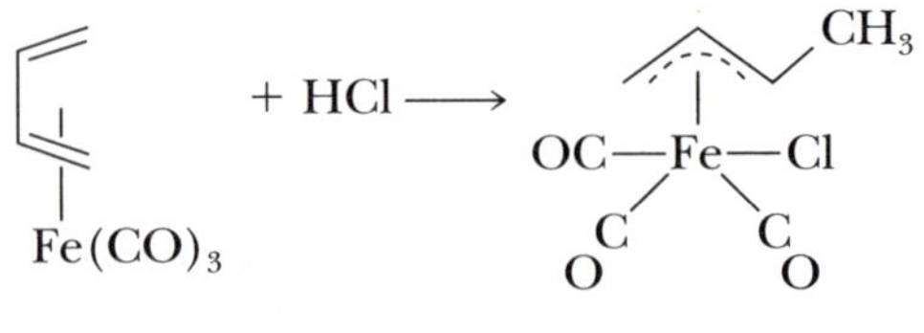

$$+ HCl \longrightarrow \tag{29-16.4}$$

Allyl complexes play an important role in many catalytic reactions, particularly those involving conjugated alkenes.

29-17 Carbene and Alkylidene Complexes

Although carbenes ($:CR_2$) are short lived in the free state, many organic reactions proceed by way of carbene intermediates. An increasing number of compounds are known in which carbenes are "stabilized" by binding to a transition metal. A carbene could be regarded as a two-electron donor comparable to CO, since there is a lone pair of electrons present on what is formally a divalent carbon atom. This view is depicted in Structure 29-XXIX. An alternative view is that there is a metal-to-carbon double bond (similar to a C=C double bond), as in Structure 29-XXX.

$$M \longleftarrow :CR_2 \qquad M{=}CR_2$$

$$\text{29-XXIX} \qquad\qquad \text{29-XXX}$$

Indeed there is evidence to support both formal types of bonding, since there exist two main classes of carbene complexes: one where the metal is in a formally low oxidation state (Structure 29-XXIX), the other where the metal is in a high oxidation state (Structure 29-XXX). The compounds with metals in low oxidation states are properly called **carbene complexes**, and they are sometimes referred to as Fischer-type complexes after their discoverer, E. O. Fischer. General methods of synthesis involve attack of nucleophilic reagents on coordinated CO or RNC, followed by an electrophilic attack, as shown in the two steps of Reaction 29-17.1.

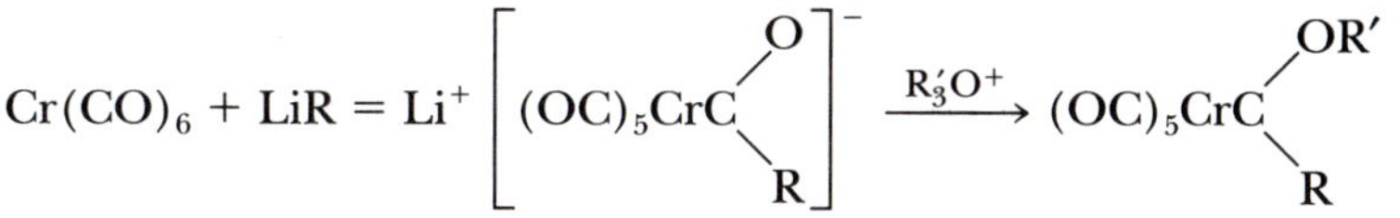

$$Cr(CO)_6 + LiR = Li^+ \left[(OC)_5CrC\begin{matrix} O \\ \diagup \\ \diagdown \\ R \end{matrix} \right]^- \xrightarrow{R_3'O^+} (OC)_5CrC\begin{matrix} OR' \\ \diagup \\ \diagdown \\ R \end{matrix} \qquad (29\text{-}17.1)$$

Carbene complexes can also be formed by cleavage of a C=C double bond in certain electron-rich alkenes, as in Reaction 29-17.2.

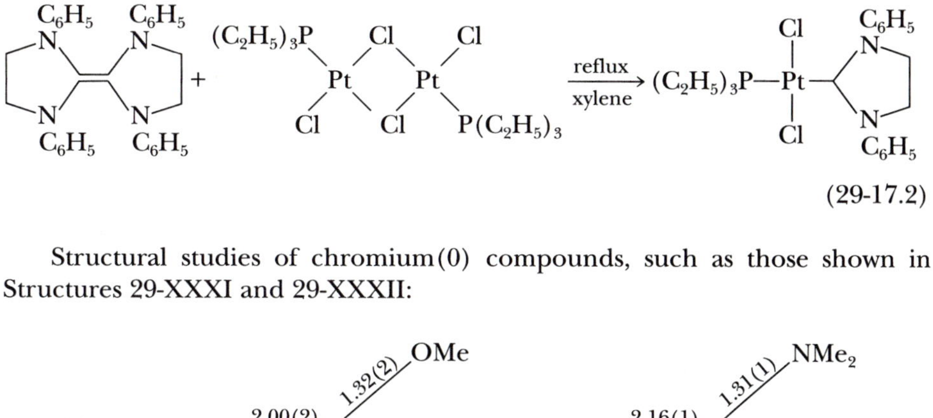

$$(29\text{-}17.2)$$

Structural studies of chromium(0) compounds, such as those shown in Structures 29-XXXI and 29-XXXII:

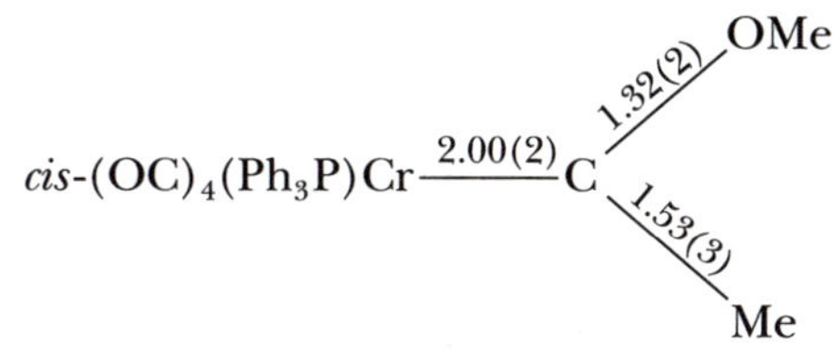

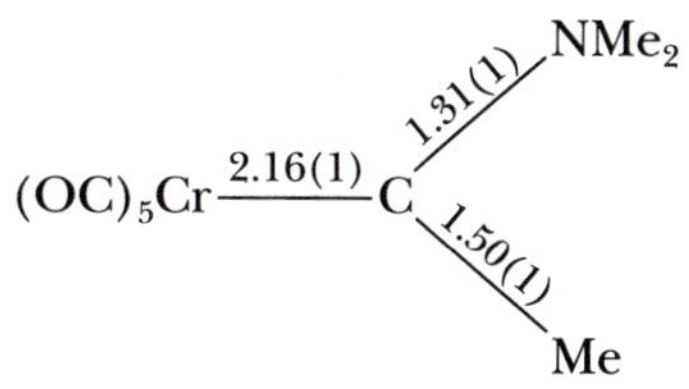

show that the M—CXY skeleton is always planar, while the M—C distances indicate some $d\pi$–$p\pi$ bonding to the carbene carbon, as in the bonding of π-acid ligands such as CO.

Complexes of the second type, where the metal is in a high oxidation state (i.e., Structure 29-XXX), are properly termed **alkylidene complexes.** They are sometimes also called Schrock-type complexes, after their discoverer, R. R. Schrock. These are commonly obtained by deprotonation reactions of alkyl groups, as in Reaction 29-17.3.

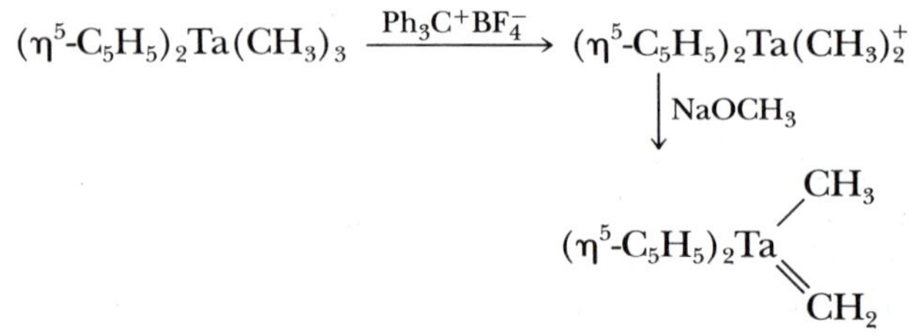

$$(\eta^5\text{-}C_5H_5)_2Ta(CH_3)_3 \xrightarrow{Ph_3C^+BF_4^-} (\eta^5\text{-}C_5H_5)_2Ta(CH_3)_2^+$$

$$\downarrow NaOCH_3$$

$$(\eta^5\text{-}C_5H_5)_2Ta\diagdown\genfrac{}{}{0pt}{}{CH_3}{CH_2} \qquad (29\text{-}17.3)$$

The tantalum(V) product of Reaction 29-17.3 provides a unique comparison of a metal-to-carbon single bond (2.246 Å) and a metal-to-carbon double bond (2.026 Å).

Alkylidene complexes are known to be intermediates in a number of organic reactions that are catalyzed by transition metal complexes, especially the alkene metathesis reaction. This involves exchange of groups on an alkene, as shown generally in Reaction 29-17.4.

$$\begin{array}{c} RCH{=}CH_2 \\ + \\ RCH{=}CH_2 \end{array} \longrightarrow \begin{array}{c} RCH \\ \| \\ RCH \end{array} + \begin{array}{c} CH_2 \\ \| \\ CH_2 \end{array} \qquad (29\text{-}17.4)$$

Alkene metathesis is catalyzed by Mo, W, and Re oxohalides in the presence of an alkylating agent, and is considered to proceed by way of the sequence shown in Reaction 29-17.5,

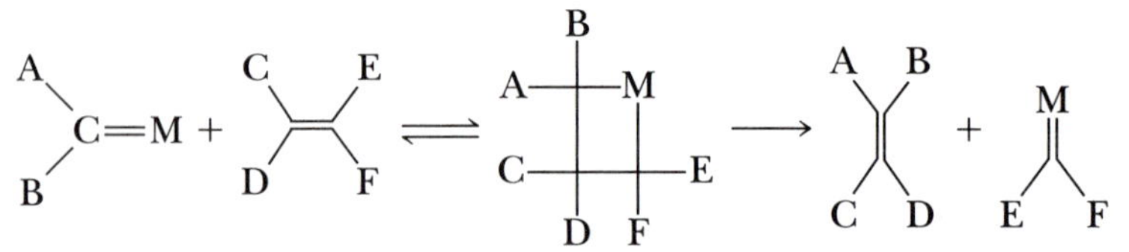

$$\genfrac{}{}{0pt}{}{A}{B}{C}{=}M + \genfrac{}{}{0pt}{}{C}{D}{=}\genfrac{}{}{0pt}{}{E}{F} \rightleftharpoons \begin{array}{cc} B \\ A{-}{\vert}{-}M \\ C{-}{\vert}{-}E \\ D\quad F \end{array} \longrightarrow \genfrac{}{}{0pt}{}{A}{C}{\;}\genfrac{}{}{0pt}{}{B}{D} + \genfrac{}{}{0pt}{}{M}{\;}\genfrac{}{}{0pt}{}{E}{F} \qquad (29\text{-}17.5)$$

where there is a metallacyclic intermediate.

The decomposition of certain metal alkyl compounds can also give alkylidene complexes by α-hydrogen transfer, as in Reaction 29-17.6.

$$M{-}CH_3 \longrightarrow \overset{\overset{\displaystyle H}{\displaystyle |}}{M}{=}CH_2 \qquad (29\text{-}17.6)$$

The chemical reactivities of carbene and alkylidene complexes differ considerably. In carbene complexes the carbon bound to the metal is electrophilic, and it is readily attacked by nucleophiles as in Reaction 29-17.7.

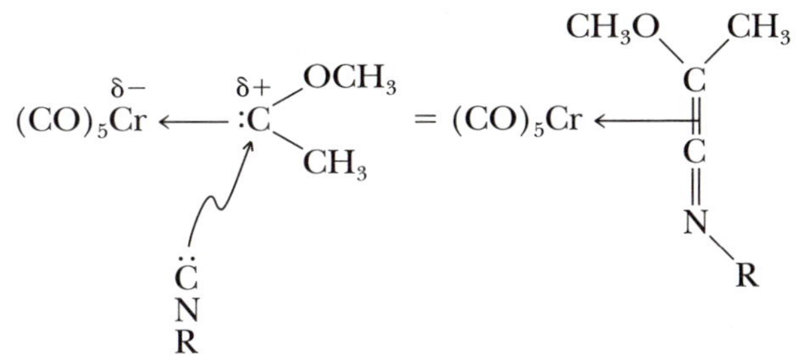

$$(CO)_5 \overset{\delta-}{Cr} \longleftarrow \overset{\delta+}{:C} \overset{OCH_3}{\underset{CH_3}{\Big\langle}} = (CO)_5 Cr \longleftarrow \overset{\overset{CH_3O \quad CH_3}{\underset{}{C}}}{\underset{\underset{R}{\overset{\parallel}{N}}}{\overset{\parallel}{C}}} \qquad (29\text{-}17.7)$$

By contrast, the carbon atom bound to the metal in alkylidenes is nucleophilic.

29-18 Carbyne and Alkylidyne Complexes

These have M—CR groupings, and, in the same way as carbenes, the metals fall into two classes: those in either high or low oxidation states. **Carbyne compounds** can be regarded as derived from the four-electron donor CR. The metal —C—R groups are usually linear or nearly so. Carbyne compounds can be obtained from carbene compounds, for example by Reaction 29-18.1.

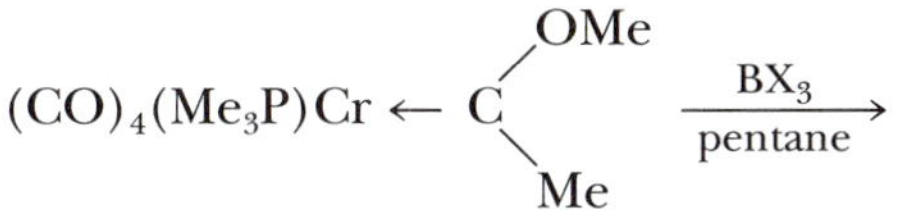

$$(CO)_4(Me_3P)Cr \leftarrow C \overset{OMe}{\underset{Me}{\Big\langle}} \xrightarrow[\text{pentane}]{BX_3}$$

$$(CO)_3(Me_3P)XCr\text{—}CMe + CO + BX_2OMe \qquad (29\text{-}18.1)$$

Alkylidyne compounds of metals in high oxidation states can be formulated with M≡C triple bonds. They can be obtained by deprotonation of alkylidenes as in Reaction 29-18.2.

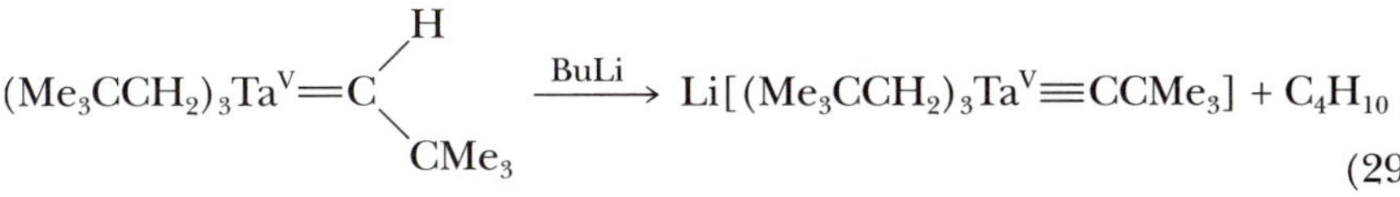

$$(Me_3CCH_2)_3Ta^V{=}C\overset{H}{\underset{CMe_3}{\Big\langle}} \xrightarrow{BuLi} Li[(Me_3CCH_2)_3Ta^V{\equiv}CCMe_3] + C_4H_{10}$$
$$(29\text{-}18.2)$$

Some remarkable compounds have been made that have single, double, and triple MC bonds, for example, the tungsten(VI) complex shown in Structure 29-XXXIII.

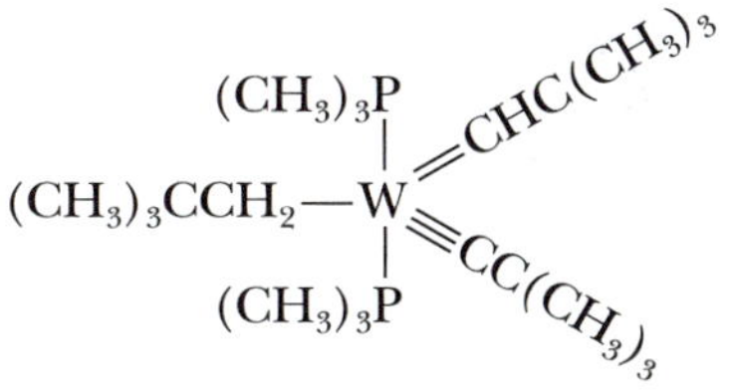

$$(CH_3)_3CCH_2\text{—}\underset{\underset{(CH_3)_3P}{\overset{|}{\underset{|}{\underset{}{}}}}}{\overset{(CH_3)_3P}{\overset{|}{\underset{}{}}}}W\overset{\overset{=CHC(CH_3)_3}{}}{\underset{\equiv CC(CH_3)_3}{}}$$

29-XXXIII

The analogy between alkynes ($RC{\equiv}CR$) and alkylidynes ($LM{\equiv}CR$) suggested to F. G. A. Stone that similar behavior towards other transition metal complexes could be expected. This concept has led to a wide variety of metal cluster compounds of the type shown in Structure 29-XXXIV.

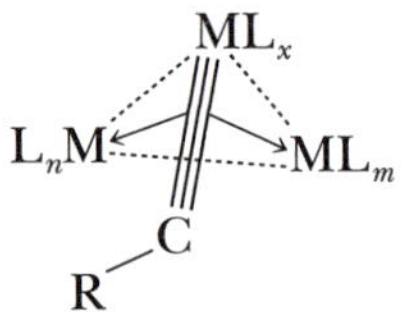

$$L_nM \cdots \overset{\displaystyle ML_x}{\underset{\underset{R}{\overset{|}{C}}}{\big\|}} \cdots ML_m$$

29-XXXIV

The latter can be compared with the alkyne complex shown in Structure 29-XXVIII, which has a Co_2C_2 core. The alkylidyne-derived complexes (Structure 29-XXXIV) have M_3C cores.

STUDY GUIDE

Scope and Purpose

This chapter is an indispensable prerequisite to the study of Chapter 30. It is also a natural continuation of topics covered in earlier chapters.

Although organometallic chemistry is a subarea of both inorganic and organic chemistry, it is such a large field that it could be considered a full-fledged branch of chemistry in itself. It draws on both inorganic and organic chemistry; yet, the whole is greater than the sum of the parts.

We have covered in this one chapter only some of the salient points. Familiar as well as new and novel types of structures are encountered, and some new concepts of bonding, structure, and the like are presented. Each type of structure and each class of substance deserves careful study.

We have somewhat relaxed the normal expectation concerning the need for full, completely balanced chemical equations. This is done in some cases for clarity, and in other cases because the nature of all the products is not clearly established.

STUDY QUESTIONS

A. Review

1. Give a definition of an organometallic compound.
2. What are the three broad classes of organometallic compounds? Cite an example of each.
3. Describe at least three important general methods for preparing organometallic compounds.
4. What is the most characteristic structural feature of lithium alkyls?
5. Why are the tetramethylethylenediamine complexes of the lithium alkyls more reactive than the alkyls themselves?

6. What is a formula for a Grignard reagent and how are Grignard reagents prepared? What is the other general type of organomagnesium compound?

7. What sorts of species are believed actually to exist, in equilibrium, in a Grignard reagent in diethyl ether solution?

8. Give an example of a metal interchange reaction involving an organomercury compound.

9. Indicate with sketches the structures of the following:
   $LiCH_3$ $MgCH_3Br[O(C_2H_5)_2]_2$ $Hg(CH_3)_2$ $Al(CH_3)_3$ $(CH_3)_3SnF$

10. Write an equation illustrating each of the terms: hydroboration reaction, mercuration, and oxymercuration.

11. How would you prepare each of the following: $NaB(C_6H_5)_4$ (from BCl_3); cyclopropyl mercury bromide (from mercury); diethylzinc (from zinc); trimethylaluminum (from aluminum)?

12. What are siloxanes? Silicones? How are they made?

13. What is an alkylidenetriphenylphosphorane (Wittig reagent)? How are they made and what are they used for?

14. Why is $Ti(bpy)(CH_3)_4$ much more thermally stable than $Ti(CH_3)_4$?

15. Explain mechanistically why transition metal alkyls that have a β-hydrogen atom are usually unstable, whereas analogous compounds in which the alkyls do not have β-hydrogen atoms generally are stable.

16. Besides β-hydrogen transfer, what is another important mode of decomposition of some metal alkyls?

17. Describe the structure of the anion $[PtCl_3C_2H_4]^-$, in Zeise's salt, emphasizing the significant features on which an understanding of the metal–alkene bonding must be based.

18. Show with drawings the two important types of orbital overlap that explain the metal–alkene bonding in $[PtCl_3C_2H_4]^-$.

19. Show with drawings the expected structures of the following cyclooctatetraene (cot) complexes: $(cot)Cr(CO)_3$, $(cot)Fe(CO)_3$, $(cot)PtCl_2$.

20. Give the formal names for the following:

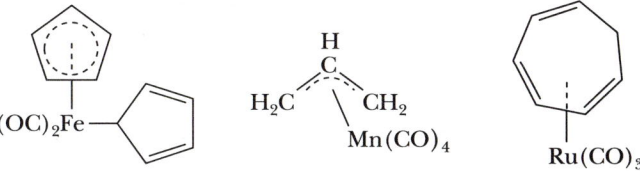

21. Write equations for a two-step preparation of $(\eta^5\text{-}C_5H_5)_2Ni$ from C_5H_6, Na, and $NiCl_2$.

22. List the five symmetric ring systems that are known to form carbocyclic complexes of the type $[(RC)_n]ML_x$.

23. Enumerate the differences between (a) carbenes and alkylidenes and (b) carbynes and alkylidynes.

24. State two ways of obtaining η^3-allyl complexes.

25. Show how carbene complexes can be obtained from metal carbonyls.

B. Additional Exercises

1. What are mercurinium ions and what part do they play in oxomercuration reactions?

2. Describe the bonding in (a) dimethylberyllium, and (b) trimethylaluminum in terms

of multicenter bonds.

3. Discuss the mechanism of the synthesis of alkenes from aldehydes or ketones by use of the Wittig reaction.

4. Explain the following observations

    (a) Although the M—C force constants and presumably bond strengths are comparable, $PbMe_4$ begins to decompose by radical formation at about 200 °C, while $TiMe_4$ is unstable above −80 °C.

    (b) Alkyl halides, R′X, react with phosphate esters, $P(OR)_3$, to give dialkylphosphonates, $O{=}P(OR)_2R′$ and RX.

    (c) The compound $(CH_3)_2PBH_2$ is trimeric and is extraordinarily stable and inert.

    (d) At −75 °C the proton resonance spectrum of trimethylaluminum shows two resonances in the ratio 2:1, but at 25 °C only one peak at an average position is found.

5. Consider the interaction of a hydrido species L_nMH with hex-1-ene in benzene solution at 25 °C.

    (a) Why and how are *cis*- and *trans*-2-enes formed?

    (b) If L_nMD was used, where would the deuterium finish up?

6. Assuming H transfer from CH_3 to metal is a plausible first step in decomposition of a methyl compound, write a mechanism for decomposition of $Ti(CH_3)_4$ in alkane solution.

7. The interaction of $Na_2Fe(CO)_4$ with $(CH_3)_2NCH_2I$ gives the carbene complex $(CO)_4FeCHN(CH_3)_2$. Write a plausible reaction sequence.

8. The interaction of $(\eta^5\text{-}C_5H_5)_2Co^+$ salts with sodium borohydride as H^- source gives a red diamagnetic hydrocarbon-soluble product, $C_{10}H_{11}Co$, whose NMR spectrum is quite complex. The interaction of $(\eta^5\text{-}C_5H_5)_2Co$ with CH_3I gives $(\eta^5\text{-}C_5H_5)_2CoI$ and $C_{11}H_{13}Co$. Explain these reactions.

9. Compare the bonding of C_2H_4 and O_2 in the compounds $(Ph_3P)_2Pt(O_2)$ and $(Ph_3P)_2Pt(C_2H_4)$.

10. Compare the bond distances in Structures 29-XXXI and 29-XXXII with the sum of the appropriate single-bond covalent radii to determine the extent of π bonding to each of the three substituents of these carbene ligands. Draw pictures representing the π-bond network about a planar carbene carbon.

11. Apply the 18-electron formalism to the reactant and product of Reaction 29-16.4, and give the *hapto* notation that is appropriate for each organic ligand.

12. Apply the 18-electron formalism to each of the five compounds found in Table 29-2.

13. Draw the structure of each of the substances found in Reactions 29-12.1 and 29-12.2.

14. Explain why the cyclohepta-1,3,5-triene ligand is *hexahapto* towards the $Cr(CO)_3$ fragment but only *tetrahapto* towards the $Fe(CO)_3$ fragment.

15. Would you expect Zeise's salt to add other ligands readily? Explain.

16. Why are there two different Al—C bond lengths in $Al_2(CH_3)_6$?

17. Suggest a synthesis for:

    (a) $Na[B(C_2H_5)_4]$ (b) $Li[Al(C_2H_5)_4]$ (c) C_2H_5MgBr (d) $P(C_2H_5)_3$

C. Questions from the Literature of Inorganic Chemistry

1. Consider the two alkene complexes of iron(0) as reported by M. V. R. Strainer and J. Takats, *Inorg. Chem.*, **1982**, *21*, 4044–4049.

    (a) Of the two ligands def and dem, which is the stronger π acceptor? Support your conclusion with data from the article.

(b) Apply the 18-electron formalism to the title compounds.

(c) Why do alkenes prefer the equatorial position in five-coordinate complexes such as these?

(d) Arrange the ligands CO, dem, def, and PPh_3 in order of π acidity as found in these five-coordinate complexes. Justify your order based on data found in this article.

2. Consider the alkyne complexes of tantalum reported by G. Smith, R. R. Schrock, M. R. Churchill, and W. J. Youngs, *Inorg. Chem.,* **1981**, *20,* 387–393.

(a) Write balanced chemical equations for the preparations of the four alkyne complexes reported here.

(b) Apply the 18-electron formalism to the two complexes $Ta(\eta^5\text{-}C_5Me_5)$(cyclooctene)$(Cl)_2$ and $Ta(\eta^5\text{-}C_5Me_5)(PhC\equiv CPh)(Cl)_2$.

(c) Why do you suppose the two-electron donor alkenes are readily replaced by the four-electron donor alkynes in these Ta compounds?

3. Consider the η^2-alkyne ligands found in the report by F. W. B. Einstein, K. G. Tyers, and D. Sutton, *Organometallics,* **1985**, *4,* 489–493.

(a) Apply the 18-electron formalism to Compounds 1*a* and 1*b* of this paper.

(b) Enumerate the differences between this diphenylacetylene ligand and the one featured in the article of Question 2 above. What accounts for the differences between a four-electron donor alkyne and a two-electron donor alkyne ligand?

4. Consider the alkyl–alkenyl–alkynylboranes of H. C. Brown, D. Basavaiah, and N. G. Bhat, *Organometallics,* **1983**, *2,* 1468–1470.

(a) Write a balanced chemical equation for each step in the synthesis of Compound *4a,* starting with Compound *3a,* ethylbromoborane: $S(CH_3)_2$.

(b) Describe each step in (a) as either hydroboration, methanolysis, nucleophilic displacement, or adduct formation.

SUPPLEMENTARY READING

Alper, H., *Transition Metal Organometallics in Organic Synthesis,* Vol. 2, Academic, New York, 1978.

Atwood, J. D., *Inorganic and Organometallic Reaction Mechanisms,* Brooks/Cole, Monterey, CA, 1985.

Becker, E. I. and Tsutsui, M., *Organometallic Reactions,* Vols. 1–4, Wiley-Interscience, New York, 1970–1972.

Coates, G. E., Green, M. L. H., Powell, P., and Wade, K., *Principles of Organometallic Chemistry,* Chapman & Hall, London, 1977.

Collman, J. P., Hegedus, L. S., Norton, J. R., and Finke, R. G., *Principles and Applications of Organometallic Chemistry,* 2nd ed., University Science Books, Mill Valley, CA, 1987.

Davies, S. G., *Organotransition Metal Chemistry, Applications to Organic Synthesis,* Pergamon Press, New York, 1982.

Dotz, K. H. et al., *Transition Metal Carbene Complexes,* Verlag Chemie, Weinheim, 1983.

Elsenbroich, C. and Salzer, A., *Organometallics,* 2nd ed., VCH, New York 1990.

Haiduc, I., and Zuckermann, J. J., *Basic Organometallic Chemistry,* Walter de Gruyter, New York, 1985.

Hartley, F. R., and Patai, S., Eds., *The Chemistry of the Metal–Carbon Bond,* Vol. 3, Wiley,

New York, 1985.

Herberhold, M., *Metal Pi Complexes,* Elsevier, New York, 1972.

Kochi, J. K., *Organometallic Mechanisms and Catalysis,* Academic, New York, 1978.

Korchmiedes, S. U. and Wilkinson, G., "Homoleptic and Related Aryls of Transition Metals," *Polyhedron,* **1991,** *10,* 135.

Kosolapoff, G. M. and Maier, L., *Organophosphorus Compounds,* Wiley, New York, 1972.

Lukehart, C. M., *Fundamental Transition Metal Organometallic Chemistry,* Brooks/Cole, Monterey, CA, 1985.

Matteson, D. S., *Organometallic Reaction Mechanisms of the Nontransition Elements,* Academic, New York, 1974.

Mole, T. and Jeffery, E. A., *Organoaluminum Compounds,* Elsevier, New York, 1972.

Pruchnick, F. P., *Organometallic Chemistry of the Transition Elements,* Plenum Press, New York, 1990.

Sawyer, A. K., *Organotin Compounds,* Vols. 1–3, Dekker, New York, 1974.

Wakefield, B. J., *Organolithium Compounds,* Pergamon Press, New York, 1974.

Wilkinson, G., Stone, F. G. A., and Abel, E. W., Eds., *Comprehensive Organometallic Chemistry,* Pergamon Press, New York, 1982.

Yamamoto, A., *Organotransition Metal Chemistry,* Wiley-Interscience, New York, 1986.

Chapter 30

STOICHIOMETRIC AND CATALYTIC REACTIONS OF ORGANOMETALLIC COMPOUNDS

The production of most bulk organic chemicals, including gasoline and fuel oils, ultimately rests on the abundance of oil, natural gas, and coal. Most of the processes used to convert these raw materials into useful products depend on catalytic reactions involving transition metals. Heterogeneous processes where one or more metals (with other additives) are supported on zeolites, alumina, silica, or graphite are most commonly used. This is especially true where vast quantities of materials are involved, as in the processing and reforming of petroleums. Heterogeneous catalysis has the advantage that the catalyst can often operate at high temperatures and that the catalyst is readily separated from feed stock and products. There are also some disadvantages, such as mass transfer problems in solids, limited variability, lack of high selectivity, and the fact that only a small part of the metal content may be in "active sites." Homogeneous catalysis in solution has advantages in that high activities are possible and alteration of electronic and steric factors through ligand substitution may allow for design of high selectivity. The disadvantages of homogeneous catalysis are that such systems are commonly decomposed at high temperatures, and most important, that there can be serious problems in separating feedstocks, products, and catalysts. Nevertheless, a number of important commercial homogeneous catalytic processes are in use. Among the most important are the hydroformylation of alkenes, the Monsanto acetic acid process, and the Wacker process for making acetaldehyde from ethylene. Homogeneous reactions can also provide a greater insight into the mechanisms of catalyzed reactions since they are more amenable to study, especially by spectroscopic methods.

Before we discuss specific catalytic reactions, we must consider a number of stoichiometric ones that are important for themselves, as well as for their relevance to catalysis. Although the principles discussed in this chapter have some applicability to heterogeneous catalysis, we shall not discuss these processes.

STOICHIOMETRIC REACTIONS

30-1 Coordinative Unsaturation

If two substances A and B are to react at the central metal of a complex in solution, then there must be vacant sites for their coordination to the metal. In heterogeneous reactions, the surface atoms of metals and metal oxides, halides, and so on, are necessarily coordinatively unsaturated. In solution, even intrinsically coordinatively unsaturated complexes, such as square planar d^8 systems, are solvated, and coordinated solvent molecules will have to be replaced by reacting molecules, as in Reaction 30-1.1.

$$\begin{array}{c} \text{Solvent} \\ L_1 \underset{\diagdown}{\overset{|}{\underset{M}{\diagup}}} L_3 \\ L_2 \overset{|}{\underset{\text{Solvent}}{\diagup}} L_4 \end{array} + A + B \longrightarrow \begin{array}{c} L_1 \underset{\diagdown}{\overset{A}{\underset{M}{\diagup}}} L_3 \\ L_2 \overset{|}{\underset{B}{\diagup}} L_4 \end{array} + 2\,\text{Solvent} \qquad (30\text{-}1.1)$$

In five- or six-coordinated metal complexes, coordination sites must be made available, either by thermal or photochemical dissociation of one or more ligands. For example, up to two phosphine ligands can be dissociated in the Rh complex of Reaction 30-1.2, allowing the Rh complex to be useful as a catalyst at 25 °C. The iridium analog, namely, $IrH(CO)(PPh_3)_3$, is not useful as a catalyst unless the dissociation of li-gands (a slow process at 25 °C) is induced by heat or UV light.

$$RhH(CO)(PPh_3)_3 \underset{+PPh_3}{\overset{-PPh_3}{\rightleftharpoons}} RhH(CO)(PPh_3)_2 \underset{+PPh_3}{\overset{-PPh_3}{\rightleftharpoons}} RhH(CO)(PPh_3) \quad (30\text{-}1.2)$$

The ready substitution of ligands that is necessary for catalytic activity may also be prompted by a change of oxidation state at the metal, as in the oxidative addition reactions discussed in Section 30-2.

30-2 Metal Atoms as Centers of Acid–Base Behavior in Complexes

Protonation and Lewis Base Behavior of Metals

In electron-rich complexes, the metal may have substantial electron density located on it and, consequently, may be attacked by protons or other electrophilic reagents. An example is the bis(cyclopentadienyl)rhenium hydride, which may be protonated as in Reaction 30-2.1,

$$(\eta^5\text{-}C_5H_5)_2HRe + H^+ \rightleftharpoons (\eta^5\text{-}C_5H_5)_2H_2Re^+ \qquad (30\text{-}2.1)$$

having a pK_b comparable to that of ammonia. Many carbonyl, phosphine, or phosphite complexes may be similarly protonated. Some examples are Reactions 30-2.2 to 30-2.5.

$$Fe(CO)_5 + H^+ \rightleftharpoons FeH(CO)_5^+ \qquad (30\text{-}2.2)$$

$$Ni[P(OEt)_3]_4 + H^+ \rightleftharpoons NiH[P(OEt)_3]_4^+ \qquad (30\text{-}2.3)$$

$$Ru(CO)_3(PPh_3)_2 + H^+ \rightleftharpoons [RuH(CO)_3(PPh_3)_2]^+ \qquad (30\text{-}2.4)$$

$$Os_3(CO)_{12} + H^+ \rightleftharpoons [HOs_3(CO)_{12}]^+ \qquad (30\text{-}2.5)$$

The protonation of carbonylate anions (Section 28-9) may similarly be regarded as addition of H^+ to an electron-rich metal center, as in Reaction 30-2.6.

$$Mn(CO)_5^- + H^+ \rightleftharpoons HMn(CO)_5 \tag{30-2.6}$$

The Oxidative Addition Reaction

Coordinatively unsaturated compounds, whether transition metal or not, can generally add neutral or anionic nucleophiles. Such reactions are simple additions. Two examples are given in Reactions 30-2.7 and 30-2.8.

$$PF_5 + F^- \rightleftharpoons PF_6^- \tag{30-2.7}$$

$$TiCl_4 + 2\ OPCl_3 \rightleftharpoons TiCl_4(OPCl_3)_2 \tag{30-2.8}$$

Coordinatively unsaturated complexes may still readily add nucleophiles even when the metal is formally electron-rich. Consider Reactions 30-2.9 and 30-2.10.

$$trans\text{-}IrCl(CO)(PPh_3)_2 + CO \rightleftharpoons IrCl(CO)_2(PPh_3)_2 \tag{30-2.9}$$

$$[PdCl_4]^{2-} + Cl^- \rightleftharpoons [PdCl_5]^{3-} \tag{30-2.10}$$

A special circumstance arises when addition of a ligand is accompanied by oxidation of the metal. The oxidative addition reaction may be written generally as in Eq. 30-2.11.

$$L_nM + XY \longrightarrow L_n(X)(Y)M \tag{30-2.11}$$

It is a reaction in which the formal oxidation state of the metal increases by two. Also, the formal oxidation state of the group XY is reduced upon addition, often with cleavage of an intraligand bond.

For an oxidative addition reaction to proceed, we must have (a) nonbonding electron density on the metal, (b) two vacant coordination sites on the reacting complex L_nM (in order to allow formation of two new bonds to X and Y), and (c) a metal with stable oxidation states separated by two units.

Many reactions of even nonmetal compounds may be regarded as oxidative additions. Consider, for example, oxidative addition reactions of the halogens represented in Reactions 30-2.12 to 30-2.14.

$$(CH_3)_2S + I_2 \rightleftharpoons (CH_3)_2SI_2 \tag{30-2.12}$$

$$PF_3 + F_2 \rightleftharpoons PF_5 \tag{30-2.13}$$

$$SnCl_2 + Cl_2 \rightleftharpoons SnCl_4 \tag{30-2.14}$$

Formally, at least, we consider that the zero-valent halogens have oxidized the central atoms in these reactions, and that they are coordinated in the products as halide (X^-) groups. For transition metals, the most common oxidative addition reactions involve complexes of the metals with d^8 and d^{10} electron configurations, notably Fe^0, Ru^0, Os^0; Rh^I, Ir^I; Ni^0, Pd^0, Pt^0 and Pd^{II} and Pt^{II}. An especially well-studied complex is the square-planar $trans\text{-}IrCl(CO)(PPh_3)_2$, which undergoes reactions such as Reaction 30-2.15.

$$trans\text{-}Ir^{I}Cl(CO)(PPh_3)_2 + HCl \rightleftharpoons Ir^{III}HCl_2(CO)(PPh_3)_2 \quad (30\text{-}2.15)$$

It should be noted that in oxidative additions of molecules such as H_2, HCl, or Cl_2, two new bonds to the metal are formed, and the X—Y bond is broken. However, molecules that contain multiple bonds may be added oxidatively to a metal complex without cleavage. In such cases, three-membered rings are formed with the metal. Reactions 30-2.16 and 30-2.17 are two examples.

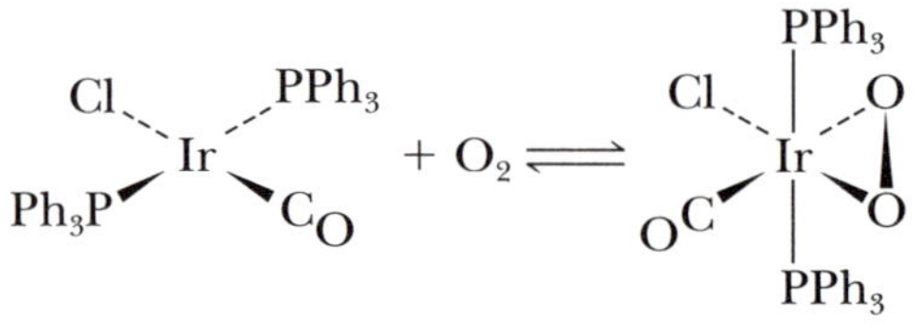

$$(30\text{-}2.16)$$

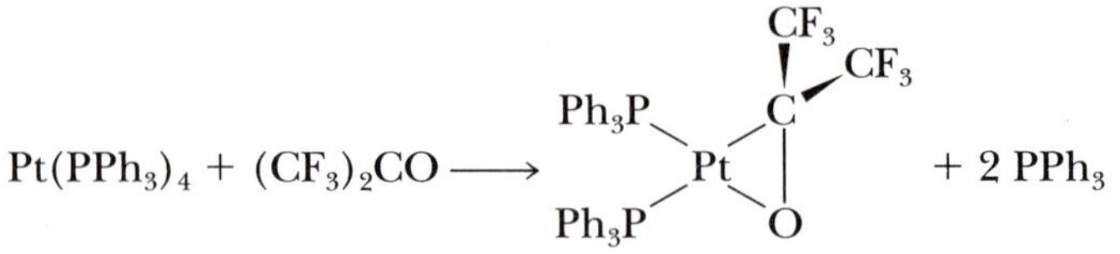

$$(30\text{-}2.17)$$

The latter reaction also provides an example of the situation where the most stable coordination number in the oxidized state would be exceeded, so that expulsion of ligands (in this case two PPh_3) attends the process.

In Table 30-1 we list the types of molecules that are known to add oxidatively to a complex in at least one instance.

Oxidative addition reactions can be regarded as equilibria of the type shown in Reaction 30-2.18.

$$L_mM^n + XY \rightleftharpoons L_mM^{n+2}XY \quad (30\text{-}2.18)$$

Whether the equilibrium lies on the reduced-metal or the oxidized-metal side depends very critically on (a) the nature of the metal and its ligands, (b) the nature of the addend XY and of the M—X and M—Y bonds so formed, and (c) the medium in which the reaction is conducted. When the molecule XY adds without severance of X from Y, the two new bonds to the metal are necessarily in cis positions. When X and Y are separated by addition to the metal, the product may be one of several isomers with cis or trans MX and MY groups, as shown in Reaction 30-2.19.

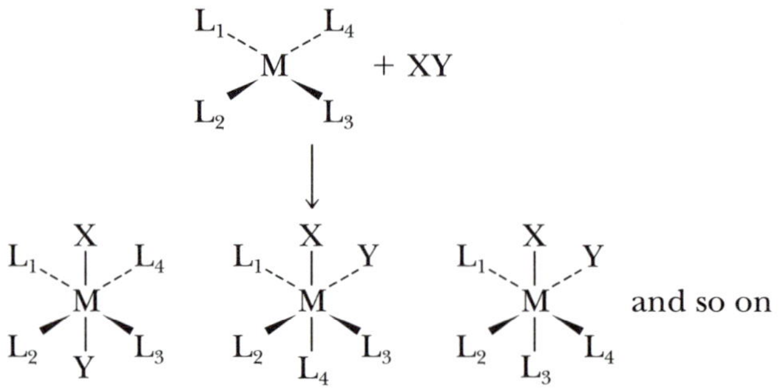

$$(30\text{-}2.19)$$

The final product will be the isomer or mixture of isomers that is the most stable

Table 30-1 Some Substances that Can Be Added
to Complexes in Oxidative–Addition Reactions

Atoms Become Separated		Atoms Remain Connected
H_2		O_2
HX (X = Cl, Br, I, CN, RCO_2, ClO_4)		SO_2
H_2S, C_6H_5SH		$CF_2{=}CF_2$, $(CN)_2C{=}C(CN)_2$
RX		$RC{\equiv}CR'$
RCOX	R = CH_3, C_6H_5, CF_3, and so on	RNCS
RSO_2X		RNCO
R_3SnX	X = Cl, Br, I	$RN{=}C{=}NR'$
R_3SiX		$RCON_3$
Cl_3SiH		$R_2C{=}C{=}O$
$(C_6H_5)_3PAuCl$		CS_2
HgX_2, CH_3HgX (X = Cl, Br, I)		$(CF_3)_2CO$, $(CF_3)_2CS$, CF_3CN
C_6H_6		

thermodynamically under the reaction conditions. The ligands, solvent, temperature, pressure, and the like, will have a decisive influence on this. Consequently, the nature of the final product does not necessarily give a guide to the initial product of the addition, since isomerization of the initial product may occur under the conditions of the addition.

Mechanisms of Oxidative Addition Reactions

Studies indicate that the following types of pathways are possible:

1. A purely *ionic mechanism* is favored by a polar medium. In polar media, HCl or HBr would be dissociated, and protonation of a square complex would first produce a five-coordinate intermediate.

$$MXL_3 + H^+(\text{solvated}) \rightarrow MHXL_3^+ \qquad (30\text{-}2.20)$$

Intramolecular isomerization followed by coordination of X^- would then give the final product:

$$MHXL_3^+ + Cl^-(\text{solvated}) \rightarrow MHClXL_3 \qquad (30\text{-}2.21)$$

2. In an S_N2 attack of the type common in organic chemistry, the transition metal complex attacks an alkyl halide, as in Reactions 30-2.22 and 30-2.23.

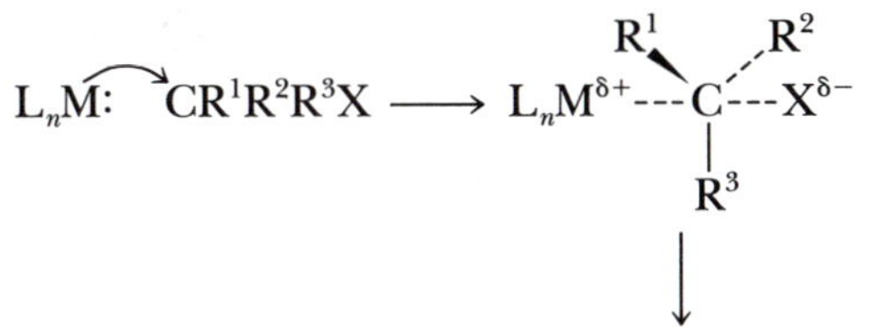

$$L_n\overset{\frown}{M:} \ \ CR^1R^2R^3X \longrightarrow L_nM^{\delta+}\text{---}\overset{\overset{\displaystyle R^1 \diagdown \ \diagup R^2}{\displaystyle C}}{\underset{\displaystyle R^3}{|}}\text{---}X^{\delta-} \qquad (30\text{-}2.22)$$

$$L_nMX(CR^1R^2R^3) \longleftarrow [L_nM\text{---}CR^1R^2R^3]^+ + X^- \qquad (30\text{-}2.23)$$

3. Several oxidative addition reactions are *free radical* in nature and can be initiated by free radical sources such as peroxides.

4. Under nonpolar conditions, particularly with molecules that have little or no polarity (such as hydrogen or oxygen) *one step, concerted processes* give products with the new bonds formed cis to one another as in Reaction 30-2.24.

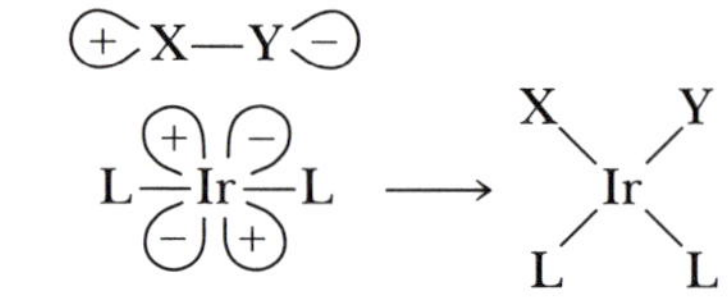

$$(+) X—Y (-)$$

$$L—Ir—L \longrightarrow \begin{matrix} X & Y \\ & Ir & \\ L & & L \end{matrix}$$ (30-2.24)

Notice that the filled nonbonding *d* orbitals of the metal in Reaction 30-2.24 have the proper symmetry to populate (i.e., reduce) the antibonding orbital of X—Y.

Consider the following observations:

1. When solid *trans*-IrCl(CO)(PPh$_3$)$_2$ reacts with HCl(g), the product has H and Cl in cis positions.

2. The addition of HCl or HBr to *trans*-IrCl(CO)(PPh$_3$)$_2$ in nonpolar solvents (such as benzene) also gives cis-addition products. If wet or polar solvents (such as DMF) are used, the addition products are cis and trans mixtures.

The concerted mechanism seems appropriate for the reactions in nonpolar media, and the ionic mechanism in polar media.

The fact that many of the d^8 complexes react with molecular hydrogen might seem surprising in light of the high energy (~ 450 kJ mol^{-1}) of the H—H bond. The attack on the hydrogen molecule probably places electron density from the metal into the σ-antibonding MO of dihydrogen, helping to weaken the H—H bond. Homolysis of the H—H bond results. This mechanism implies initial coordination of molecular hydrogen to the metal and indeed some complexes of this type are now characterized, for example, *trans*-(R$_3$P)$_2$(CO)$_3$W(H$_2$). It has even been possible to detect the following series of steps in certain cases with the use of NMR:

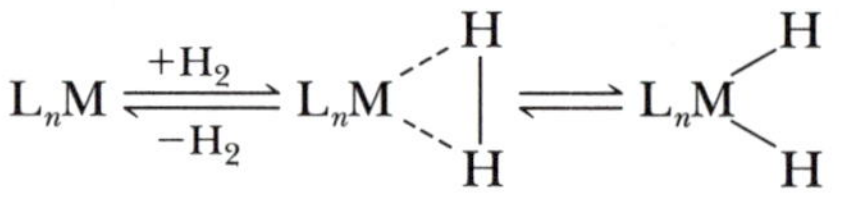

$$L_nM \underset{-H_2}{\overset{+H_2}{\rightleftharpoons}} L_nM \overset{H}{\underset{H}{\big|}} \rightleftharpoons L_nM \overset{H}{\underset{H}{\diagup}}$$ (30-2.25)

An alternate mechanism, which operates in other cases, is heterolytic cleavage of the H—H bond by base-promoted removal of H$^+$, as in Reaction 30-2.26.

$$RuCl_2(PPh_3)_3 + H_2 + Et_3N \rightarrow RuHCl(PPh_3)_3 + Et_3NH^+Cl^-$$ (30-2.26)

30-3 Migration of Atoms or Groups from Metal to Ligand: The Insertion Reaction

"Insertion" is a broadly applicable description of a reaction in which any atom or group is inserted between two other atoms initially bound together. In the general case, Reaction 30-3.1, the group YZ is inserted into the M—X bond. Some representative examples are given in Reactions 30-3.2 to 30-3.6.

$$L_nM—X + YZ \longrightarrow L_nM—(YZ)—X \tag{30-3.1}$$

$$R_3SnNR_2 + CO_2 \longrightarrow R_3SnOC(O)NR_2 \tag{30-3.2}$$

$$Ti(NR_2)_4 + 4 CS_2 \longrightarrow Ti(S_2CNR_2)_4 \tag{30-3.3}$$

$$R_3PbR' + SO_2 \longrightarrow R_3PbOS(O)R' \tag{30-3.4}$$

$$[(NH_3)_5RhH]^{2+} + O_2 \longrightarrow [(NH_3)_5RhO_2H]^{2+} \tag{30-3.5}$$

$$(CO)_5MnCH_3 + CO \longrightarrow (CO)_5MnCOCH_3 \tag{30-3.6}$$

For transition metals, detailed studies have been made on the insertion of CO and SO_2 into metal to carbon bonds. Insertions of CO_2 into M—H and M—O bonds are also known. We shall consider only insertions of CO into M—C bonds.

Mechanistic studies using ^{14}CO-labeled $CH_3Mn(CO)_5$ have shown that (a) the CO molecule that becomes the acyl ligand is *not* derived from the external CO, but from one that was already coordinated to the metal atom; that (b) the incoming CO is added cis to the acyl group, as in Reaction 30-3.7,

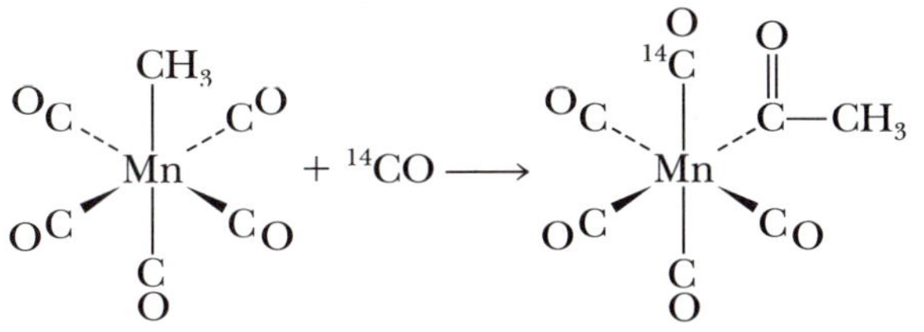

$$\tag{30-3.7}$$

and that (c) the conversion of the alkyl ligand into the acyl ligand can be promoted by the addition of ligands other than CO, for instance excess $P(C_6H_5)_3$ as in Reaction 30-3.8.

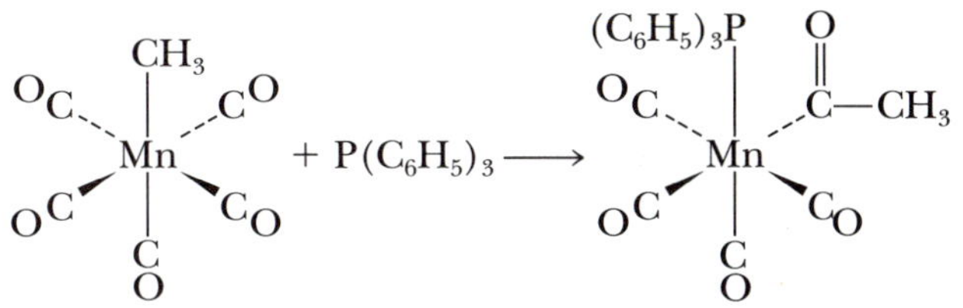

$$\tag{30-3.8}$$

Kinetic studies of such insertion reactions show that the first step involves an equilibrium between the octahedral alkyl and a five-coordinate acyl intermediate.

$$CH_3Mn(CO)_5 \rightleftharpoons CH_3C(O)Mn(CO)_4 \tag{30-3.9}$$

The in-coming ligand (whether CO, $P(C_6H_5)_3$ or the like) then adds to the five-coordinate intermediate, as in Reaction 30-3.10.

$$CH_3C(O)Mn(CO)_4 + L \longrightarrow CH_3C(O)Mn(CO)_4L \tag{30-3.10}$$

The insertion of CO is thus best considered to be an alkyl migration to a coordinated carbon monoxide. The migration is to a cis CO ligand and probably proceeds through a three-centered transition state, as in Reaction 30-3.11.

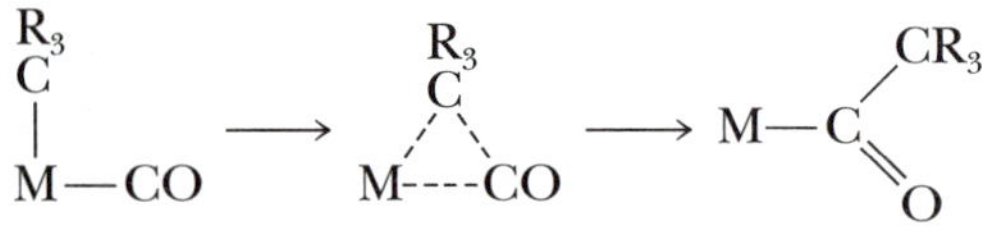

$$(30\text{-}3.11)$$

Since five-coordinate species can undergo intramolecular rearrangements, more than one isomer of the final product *may* be formed.

Multiple insertions may occur under certain circumstances. Consider Reaction 30-3.12,

$$W(CH_3)_6 + 9\,CO \longrightarrow W(CO)_6 + 3(CH_3)_2CO \qquad (30\text{-}3.12)$$

which involves initial transfer to give an acetyl ligand, followed by methyl migration to give acetone.

One further important example that can be regarded as an insertion is Reaction 30-3.13,

$$M\text{—}H + CH_2\text{==}CH_2 \rightleftharpoons M\text{—}CH_2CH_3 \qquad (30\text{-}3.13)$$

which will be discussed in Section 30-6.

30-4 Reactions of Coordinated Ligands

Nucleophilic Attack at Coordinated Ligands

This is a broadly general type of reaction of which there are numerous examples involving nucleophiles such as OH^-, RO^-, RCO_2^-, N_3^-, NR_3, and so on. Ligands that are generally susceptible to attack by nucleophiles include CO, NO, RCN, RNC, alkenes, and so on. It is not always apparent that attack on the ligand is direct, since prior coordination of the nucleophile to the metal center may occur. Then the reactions should be regarded as intramolecular transfers.

Examples long known to involve reaction directly at the ligand are the attack of OH^- at coordinated NO:

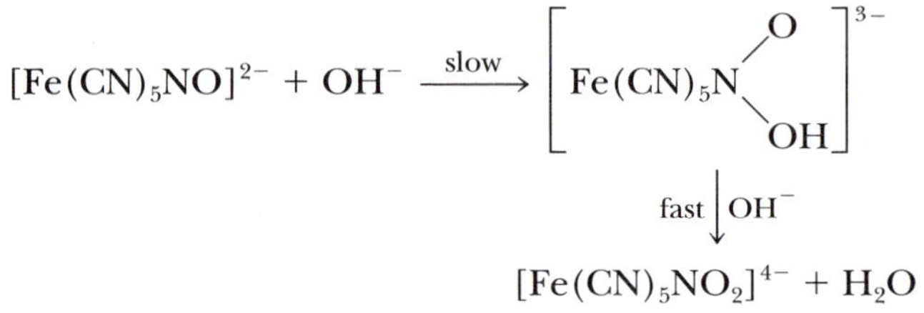

$$(30\text{-}4.1)$$

and attack of OH^- at coordinated CO:

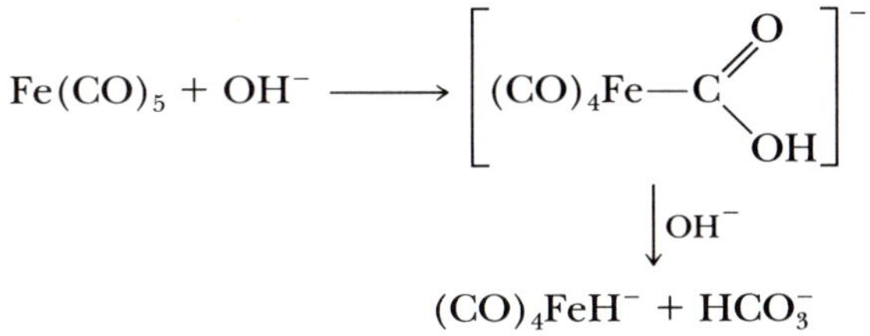

$$Fe(CO)_5 + OH^- \longrightarrow \left[(CO)_4Fe-C \begin{matrix} O \\ \diagup\diagdown \\ OH \end{matrix} \right]^-$$

$$(30\text{-}4.2)$$

$$\Big\downarrow OH^-$$

$$(CO)_4FeH^- + HCO_3^-$$

The attack of alkoxide ions RO^- on CO gives the M—C(O)OR group as in Reaction 30-4.3.

$$[Ir(CO)_3(PPh_3)_2]^+ \underset{H^+}{\overset{MeO^-}{\rightleftharpoons}} Ir(CO)_2(\overset{\overset{O}{\|}}{C}OMe)(PPh_3)_2 \qquad (30\text{-}4.3)$$

Such reactions are important in the catalyzed syntheses of carboxylic acids and esters from alkenes, carbon monoxide, and water or alcohols.

Coordinated CO can also be attacked by lithium alkyls, as in Reaction 30-4.4.

$$LiCH_3 + W(CO)_6 \longrightarrow Li^+[(CO)_5W-C(O)CH_3]^- \qquad (30\text{-}4.4)$$

Alkene and dienyl complexes are also attacked by nucleophiles, as in Reaction 30-4.5.

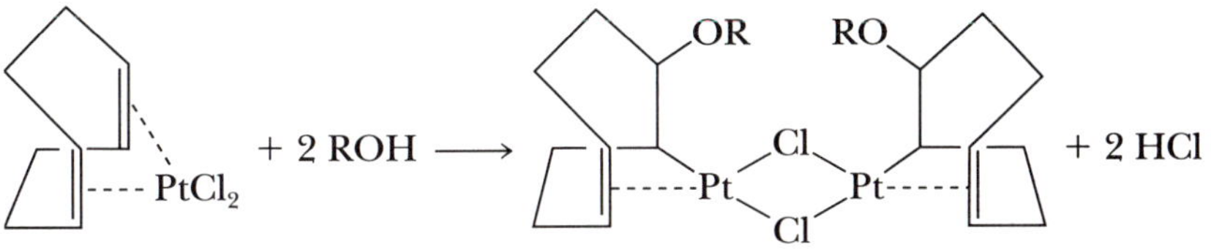

$$(30\text{-}4.5)$$

Isocyanide complexes, on the other hand, are attacked by alcohols to form "carbene" complexes, as in Reaction 30-4.6.

$$(Et_3P)Cl_2PtCNPh + EtOH \longrightarrow (Et_3P)Cl_2Pt\!=\!\!C\begin{matrix} \diagup OEt \\ \diagdown NHPh \end{matrix} \qquad (30\text{-}4.6)$$

Attack Involving Hydride Ion

Reduction by H^- of certain η^5-C_5H_5 rings produces η^4-cyclopentadiene ligands, as in Reaction 30-4.7, while hydride reduction of η^6-C_6H_6 (arene) ligands gives

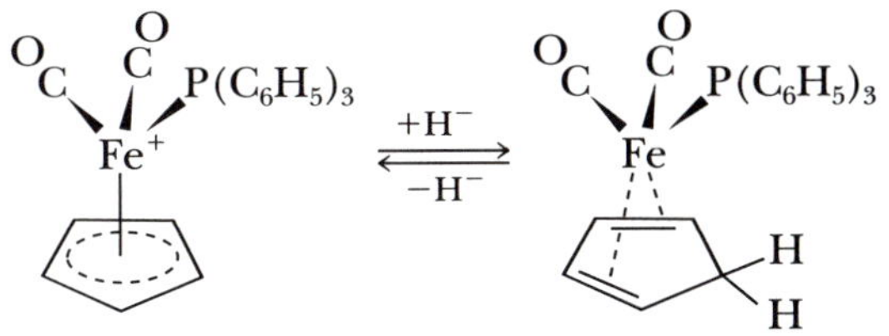

$$(30\text{-}4.7)$$

η^5-cyclohexadienyls:

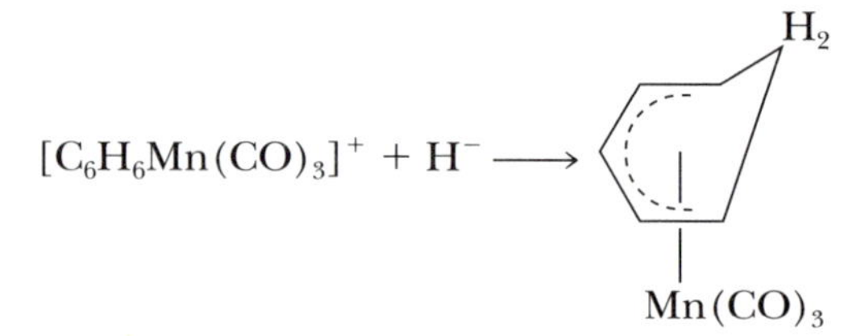

$[C_6H_6Mn(CO)_3]^+ + H^- \longrightarrow$
H_2
$Mn(CO)_3$

(30-4.8)

Hydride transfer also occurs with certain complex alkyls, where conversion into an alkene complex, as in Reaction 30-4.9, can be achieved by abstraction of H^- from the alkyl ligand, using triphenylmethyl tetrafluoroborate.

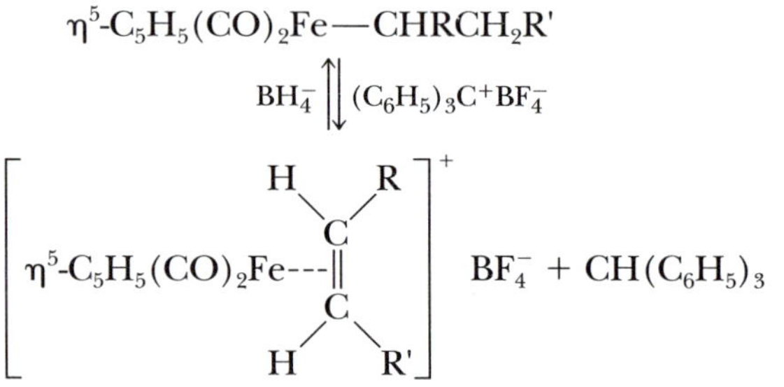

η^5-$C_5H_5(CO)_2Fe$—$CHRCH_2R'$
BH_4^-
$(C_6H_5)_3C^+BF_4^-$
H
R
C
η^5-$C_5H_5(CO)_2Fe$
C
H
R'
+
$BF_4^- + CH(C_6H_5)_3$

(30-4.9)

Intramolecular Hydrogen Transfer

A special case of transfer reactions is one in which a hydrogen atom is initially transferred to the metal giving an intermediate hydride ligand. Such reactions are especially important for triarylphosphines and triarylphosphites. An example is Reaction 30-4.10.

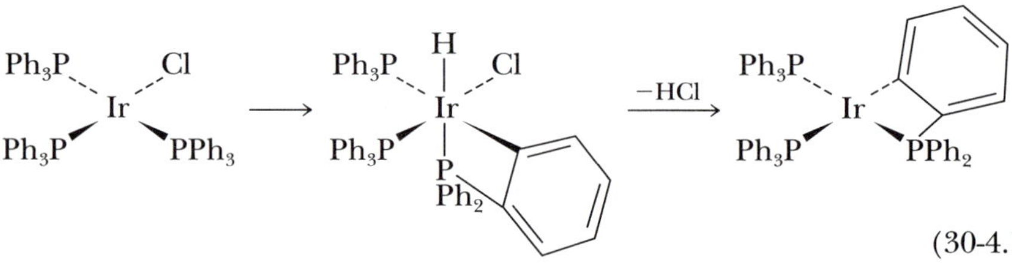

Ph_3P
Cl
Ir
Ph_3P
PPh_3
H
Ph_3P
Cl
Ir
Ph_3P
P
Ph_2
−HCl
Ph_3P
Ir
Ph_3P
PPh_2

(30-4.10)

Such reactions are termed ortho- or cyclometalations.

30-5 Reactions of Coordinated Molecular Oxygen

It has been observed that molecular oxygen can add to certain metal complexes without full cleavage of the O—O bond, and that there is some correlation between O—O distances in the O_2 ligands and the reversibility with which an O_2 complex is formed. In the extreme case, full oxidative addition of O_2 gives the O_2^{2-} ligand with long O—O distances in the ligand. Less severe oxidation upon addition leads to reversible adduct formation, as in hemoglobin.

Coordinated oxygen may be more reactive than free oxygen because of the O—O bond weakening that accompanies even weak addition to metals. Thus coordinated O_2 may be attacked more successfully. The mechanisms of attack on coordinated oxygen are poorly understood, but in many cases free radicals are involved. However, for some phosphine complexes, the reaction proceeds through peroxo intermediates that may be isolated. An example is the platinum peroxocarbonate of Reaction 30-5.1.

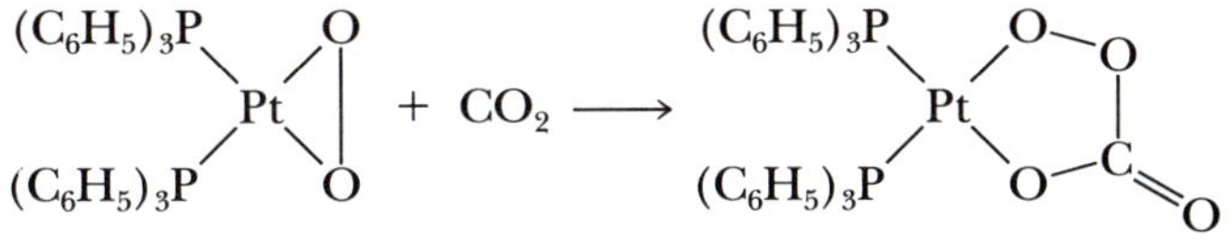

$$(C_6H_5)_3P \diagdown_{Pt}\diagup^O + CO_2 \longrightarrow (C_6H_5)_3P\diagdown_{Pt}\diagup^{O-O} \tag{30-5.1}$$

The mechanism of oxidation of SO_2 by the oxygen complex $IrCl(CO)(O_2)$-$(PPh_3)_2$ to give a sulfato complex,

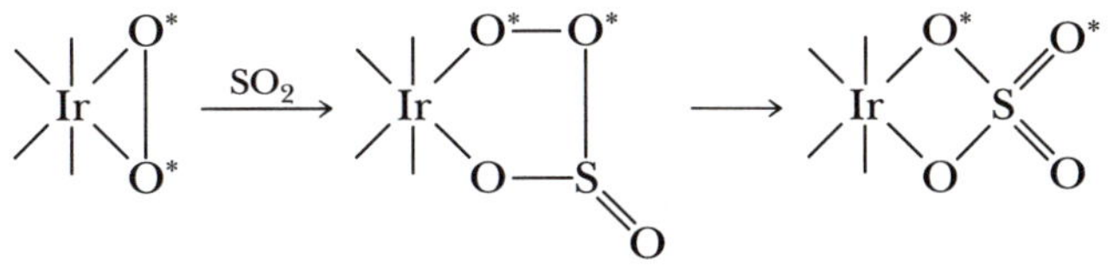

$$\ce{Ir} \overset{SO_2}{\longrightarrow} \ce{Ir} \longrightarrow \ce{Ir} \tag{30-5.2}$$

has been studied with an ^{18}O tracer, designated O^*.

CATALYTIC REACTIONS

The term catalyst requires careful use. The term—meaning a substance added to accelerate a reaction—may have some application in heterogeneous systems. In such systems, where, for example, a gas mixture is passed over a solid catalyst, the solid catalyst may be recovered unchanged. Homogeneous catalytic systems, however, proceed in solution by way of a series of linked chemical reactions involving different metal complexes at each point in the process. What one adds initially to the reaction mixture is quickly engaged in a number of reactions and equilibria. The concept of one particular species being "the catalyst" then has no validity. Instead, one speaks in terms of intermediates involved in the various steps of a catalytic cycle.

The catalytic cycles to be described all involve the same sorts of changes in oxidation state and coordination number that have been delineated for stoichiometric reactions in the preceding sections. In the catalytic systems, however, the metal complex is returned to its original state at the end of one cycle.

30-6 Isomerization

Many transition metal complexes, especially those of the metals from Groups VIIIA(8)–VIIIA(10), promote double-bond migration (isomerization) in alkenes. The products are the thermodynamically most stable isomeric mixtures. Thus 1-alkenes give mixtures of cis and trans 2-alkenes. This reaction is characteristic for many transition metal hydrido complexes; the isomerization involves transfer of H from the transition metal hydride to a coordinated alkene, giving a metal–alkyl complex. In addition, many complexes that do not have M—H bonds, for example, $(Et_3P)_2NiCl_2$, will catalyze the isomerization of alkenes provided a source of hydride (such as molecular hydrogen) is present.

The first step in the catalytic cycle must be the coordination of the alkene.

$$L_nMH + RCH{=}CH_2 \rightleftharpoons L_nMH(RCH{=}CH_2) \qquad (30\text{-}6.1)$$

Reaction 30-6.1 is followed by hydride transfer to form an alkyl ligand.

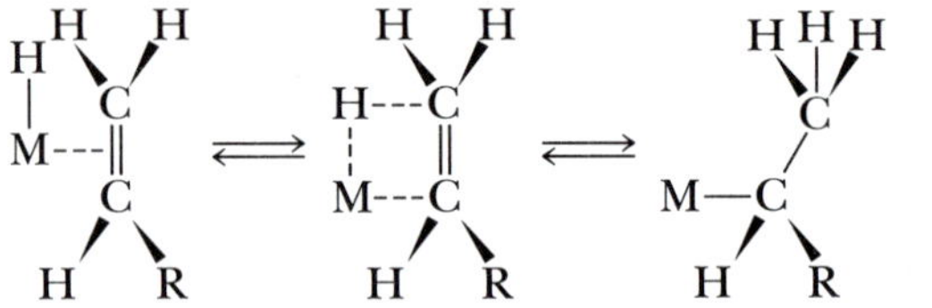

$$(30\text{-}6.2)$$

Reaction 30-6.2, in the reverse direction, has already been identified as the primary route for the decomposition of many alkyls. (The reverse of Reaction 30-6.2 is termed β elimination; the alkene is eliminated from the metal alkyl by transfer to the metal of a hydrogen atom from the β carbon of the alkyl ligand.) Thus we expect that Reaction 30-6.2 is readily reversible, and that there should be a scrambling of all hydrogen atoms during the catalytic process. When fluoroalkenes are used, as in Reaction 30-6.3,

$$RhH(CO)(PPh_3)_3 + C_2F_4 \longrightarrow Rh(CF_2CF_2H)(CO)(PPh_3)_2 + PPh_3 \quad (30\text{-}6.3)$$

stable metal alkyls are obtained, and it can be demonstrated that the hydride of the original metal complex is attached to the β carbon of the alkyl ligand. This is required if the four-centered transition state of Reaction 30-6.2 is involved.

With alkenes other than ethylene, there is the possibility of addition of M—H across the double bond in either Markovnikov or anti-Markovnikov fashion. Thus we may have Reactions 30-6.4 or 30-6.5.

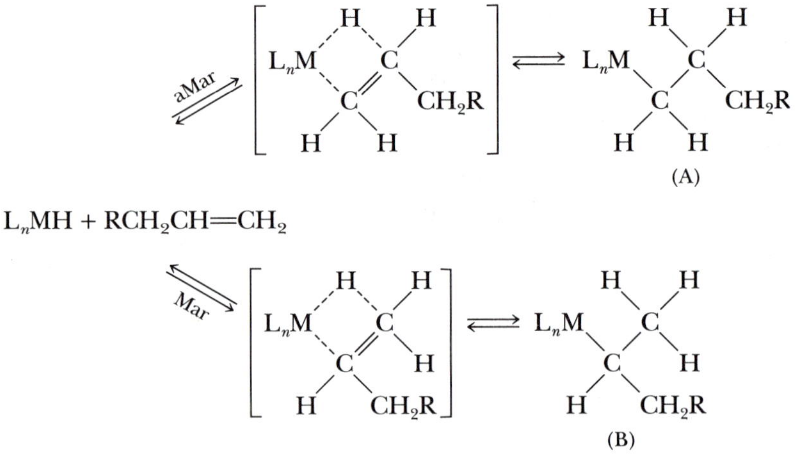

$$(30\text{-}6.4)$$

$$L_nMH + RCH_2CH{=}CH_2$$

$$(30\text{-}6.5)$$

aMar = anti-Markovnikov
Mar = Markovnikov

In the anti-Markovnikov addition (Reaction 30-6.4), the hydrogen atom is transferred from the metal to the β carbon of the chain, giving the primary alkyl derivative (A). The reverse of Reaction 30-6.4 requires β elimination from compound A. Since only one carbon is positioned β to the metal, the original alkene must re-form upon reversal of Reaction 30-6.4. Observe, however, that because of free rotation about C—C bonds, the same hydrogen need not be eliminated to reform the alkene as was given to the alkene in making the alkyl ligand of compound A. Thus in anti-Markovnikov addition there will be scrambling of hydrogen atoms at the β position, but not isomerization of the alkene. On the other hand, there are two possibilities for β elimination from the secondary alkyl derivative B that is formed by Markovnikov addition, as in Reaction 30-6.5. If the H atom is β eliminated from the CH_3 group of Compound B, the original alkene is reformed. If the H atom is β eliminated from the methylene carbon of the CH_2R group in Compound B, then a 2-alkene is reformed. Thus isomerization may occur through initial Markovnikov addition of the metal hydride across the alkene. Note that either cis or trans 2-alkenes, or both, may be formed.

30-7 Hydrogenation

The fact that molecular hydrogen reacts with many substances at room temperature has allowed the design of useful catalysts for the reduction of unsaturated compounds, such as alkenes or alkynes, by H_2. The most successful catalytic cycles use complexes such as $RhCl(PPh_3)_3$, in benzene or ethanol–benzene solution. The rates of hydrogenation depend on the nature of the groups at the site of reduction, and selectivity is, therefore, possible. In Reaction 30-7.1,

for example, only one of two C=C double bonds is actually reduced. Furthermore, in contrast to heterogeneous catalysis, where scattering of hydrogen (traced by use of D_2) throughout the molecule usually results, homogeneous catalysis leads to selective addition of H_2 (D_2) to one C=C site.

The mechanism of hydrogenations using $RhCl(PPh_3)_3$ appears to involve the cycle shown in Fig. 30-1, where P = PPh_3.

There are a number of similar catalytic systems that can hydrogenate not only C=C and C≡C groups but also $\diagdown$C=O, —N=N—, and —CH=N—.

Another catalytic system for hydrogenation employs $RhCl_3(py)_3$ in DMF plus $NaBH_4$.

One of the most important developments has been the use of optically active phosphine ligands to achieve highly selective hydrogenation of prochiral unsaturated compounds. Rhodium cationic complexes of the type [RhLL(solvent)]$^+$ (where LL is a chelating phosphine ligand) are the most commonly used. A wide variety of ligands, LL, with chirality at carbon or phosphorus have been investigated. An important example is (+ or −)-2,3-*O*-isopropylidene-2,3-di-

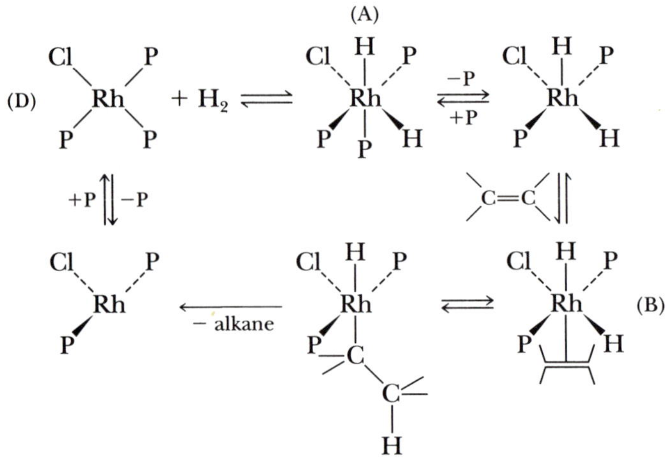

Figure 30-1 A mechanism for the catalytic hydrogenation of an alkene. Clockwise from the top left: oxidative addition of H_2 to form the dihydride, A; dissociation of a phosphine ligand to give a free coordination site for alkene addition; formation of the alkene complex, B; insertion of the alkene ligand into the M—H bond to give an alkyl ligand, C; reductive elimination of the alkane and regeneration of the catalyst precursor, D. (P = PPh_3)

hydroxy-1,4-bis(diphenylphosphino)butane, usually called (+ or −)diop, as shown in Structure 30-I.

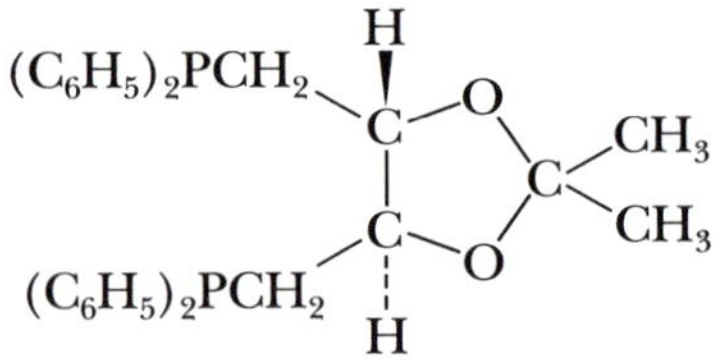

30-I

An important application has been the synthesis of the chiral drug L-DOPA (dihydroxyphenylalanine, used in the treatment of Parkinson's disease) by the Monsanto Company. Prochiral compounds of the type $R'CH{=}C(NHR_2)CO_2H$ can be reduced to chiral amino acids, and the optical purity may be greater than 95%.

It is important to note that, unlike the catalytic cycle shown in Fig. 30-1, which uses monodentate phosphine ligands, with catalytic systems involving chelating phosphine ligands, a complex with substrate forms first. It is then that oxidative addition of H_2 takes place. The enantiomeric selectivity arises from preferential complexation of the prochiral substrate at the metal containing the chiral LL. In some cases intermediate complexes can be characterized by ^{31}P NMR spectra; one example is shown in Structure 30-II,

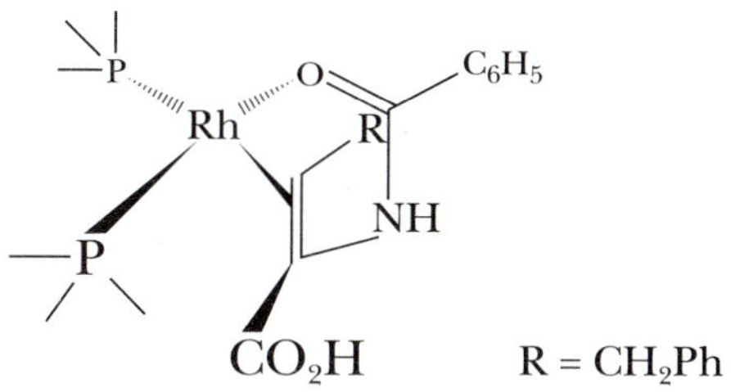

30-II   R = CH$_2$Ph

which is the rhodium complex of α-benzamidocinnamic acid. Two stereoisomers, differing in the configuration at the bound alkene, exist in a 10:1 ratio at room temperature.

30-8 Other Catalyzed Additions to Alkenes

There are two addition reactions of alkenes that are commercially important. These are the hydrosilylation and the hydrocyanation reactions.

Hydrosilylation of Alkenes

The hydrosilylation of alkenes is similar to hydrogenation except that H and SiR$_3$ from a silane (R$_3$SiH) are added across the double bond, as in Reaction 30-8.1.

$$RCH{=}CH_2 + HSiR_3 \longrightarrow RCH_2CH_2SiR_3 \qquad (30\text{-}8.1)$$

The commercial process uses hexachloroplatinic acid as the catalyst, but phosphine complexes of cobalt, rhodium, palladium, or nickel may also be used. The ready addition of silanes to *trans*-IrCl(CO)(PPh$_3$)$_2$, as in Reaction 30-8.2,

$$IrCl(CO)(PPh_3)_2 + R_3SiH \longrightarrow IrHCl(CO)(PPh_3)_2SiR_3 \qquad (30\text{-}8.2)$$

suggests that the first step in hydrosilylations is oxidative addition of the Si—H group to the metal center. In the case of Reaction 30-8.2 the product is coordinatively saturated; there is no further available site at the metal center, and the process ends here. In the actual catalytic systems, there must be a further coordination site available, because the next step is addition of the M—H group across the double bond of the alkene to form an alkyl group. Reductive elimination of the newly formed alkyl with the SiR$_3$ group yields the product and regenerates the catalyst.

Hydrocyanation of Alkenes

The DuPont Company has patented a process using nickel phosphite complexes for the addition of HCN to alkenes. The process also employs Lewis acid co-catalyst and yields high percentages of adiponitrile, an important nylon precursor. The process works because HCN, although only a weak acid, adds oxidatively to nickel phosphite compounds (NiL$_4$), as in Reaction 30-8.3.

$$NiL_4 + HCN = NiH(CN)L_2 + 2\,L \qquad (30\text{-}8.3)$$

Likely further steps in the catalytic cycle include addition of the alkene as in Reaction 30-8.4:

$$NiH(CN)L_2 + RCH{=\!\!=}CH_2 \longrightarrow (RCH{=\!\!=}CH_2)NiH(CN)L_2 \quad (30\text{-}8.4)$$

insertion of the olefinic ligand into the Ni—H bond to form an alkyl ligand:

$$(RCH{=\!\!=}CH_2)NiH(CN)L_2 \longrightarrow RCH_2CH_2{-}Ni(CN)L_2 \quad (30\text{-}8.5)$$

reductive elimination of the nitrile, as in Reaction 30-8.6:

$$RCH_2CH_2{-}Ni(CN)L_2 \longrightarrow RCH_2CH_2CN + NiL_2 \quad (30\text{-}8.6)$$

and, finally, regeneration of the active catalytic species by oxidative addition of a second equivalent of HCN, as in Reaction 30-8.7.

$$NiL_2 + HCN \longrightarrow NiH(CN)L_2 \quad (30\text{-}8.7)$$

This proposal provides a good example of the general requirement that such a catalyst readily undergo additions, oxidative additions, and reductive eliminations—the same sequence that is apparently involved in the hydrosilylations.

30-9   Hydroformylation

The hydroformylation reaction is the addition of H_2 and CO (or formally of H and the formyl group, HCO) to an alkene, usually a terminal, or 1-alkene, as in Reaction 30-9.1:

$$RCH{=\!\!=}CH_2 + H_2 + CO \longrightarrow RCH_2CH_2CHO \quad (30\text{-}9.1)$$

The aldehyde product may be further reduced under the reaction conditions to an alcohol, as in Reaction 30-9.2:

$$RCH_2CH_2CHO + H_2 \longrightarrow RCH_2CH_2CH_2OH \quad (30\text{-}9.2)$$

Originally, cobalt compounds were used as catalysts at temperatures of about 150 °C and greater than 200-atm pressure, and some 3 million tons/year of alcohols, usually C_7–C_9, have been produced in this way. The original processes gave mixtures of straight- and branched-chain products in the ratio of about 3:1, but considerable effort has been made to improve the yield of the straight-chain products. Cobalt catalysts also gave, undesirably, reduction of the feedstock alkenes to alkanes. They have been superseded by rhodium catalysts.

Extensive information is available on the catalytic cycle employing $RhH(CO)(PPh_3)_3$, which is active even at 25 °C and 1-atm pressure. In addition, the rhodium catalyst produces only an aldehyde, making it ideally suited to study. On use of high concentrations of PPh_3, high yields of linear aldehydes can be obtained, and little of the alkene reactant is lost as the simple alkane reduction products. The reaction cycle is shown in Fig. 30-2. The initial step is addition of the alkene to $RhH(CO)_2(PPh_3)_2$ (compound A in Fig. 30-2), followed by

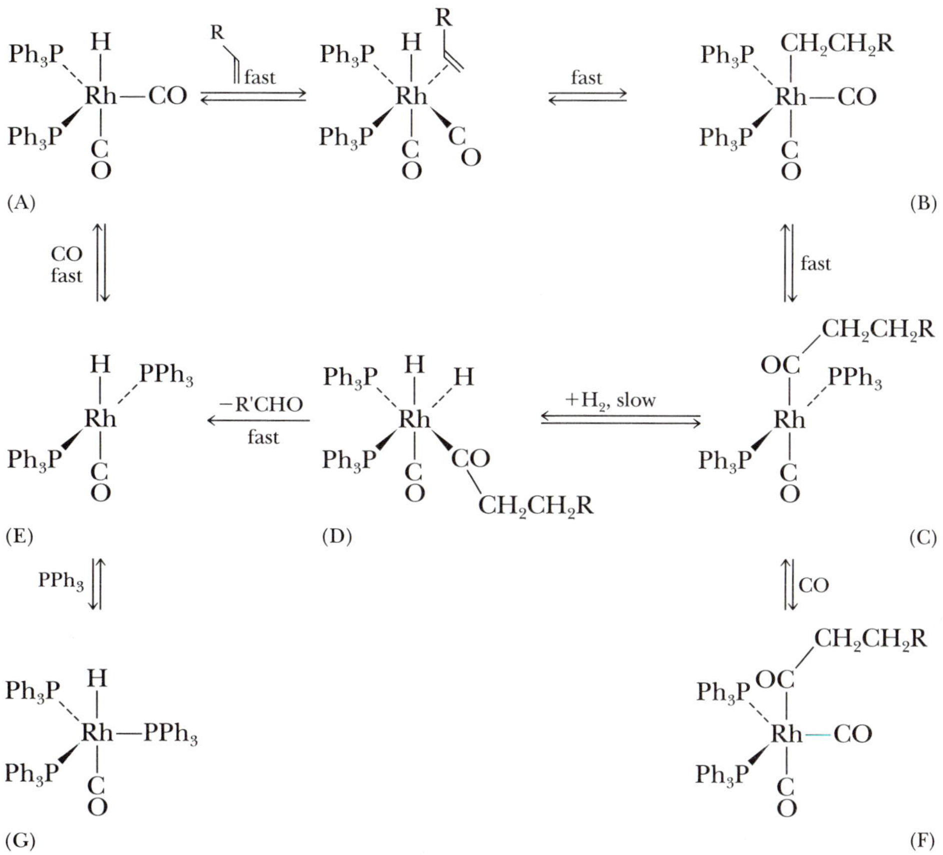

(A)

(B)

(E)

(D)

(C)

(G)

(F)

−R'CHO
fast

+H₂, slow

CO
fast

fast

fast

CO

PPh₃

Figure 30-2   A catalytic cycle for the hydroformylation of alkenes using triphenylphosphine complexes. The configurations of the complexes are not known with certainty. The equilibria involving F and G are "nonproductive."

insertion of the olefinic ligand into the Rh—H bond to give the alkyl complex, B. The latter then undergoes migratory insertion of CO into the Rh—C bond to give the acyl derivative, C. Oxidative addition of H_2 then gives the dihydrido acyl, D. It is this last step, the only one involving a change in oxidation state for the Rh, that is most likely the rate-determining step in the cycle. The final steps are reductive elimination of the aldehyde to give E and reformation of A by addition of CO.

The high PPh_3 concentrations that are essential in providing high yields (> 95%) of linear aldehydes are probably required to suppress dissociative formation of monophosphine species, and thus to force the attack of alkene on bis(phosphine) species such as Compound A. The bis(phosphine) complexes favor anti-Markovnikov addition to the alkene, and thereby lead to linear aldehydes.

30-10 Ziegler–Natta Polymerization

Hydrocarbon solutions of $TiCl_4$ in the presence of triethylaluminum polymerize ethylene at 1-atm pressure. An extension of this Ziegler–Natta polymerization of

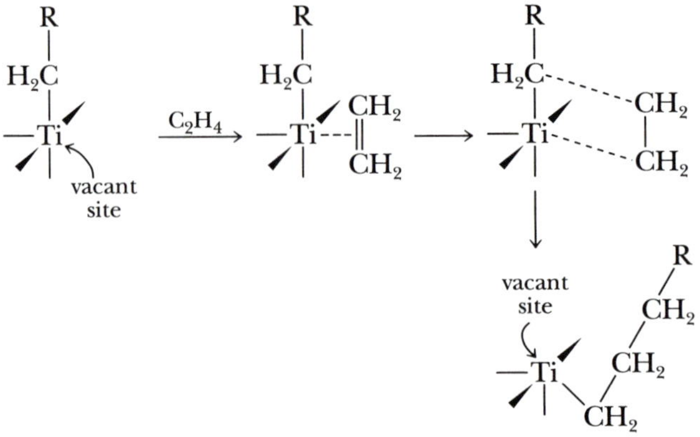

Figure 30-3   Steps in the Ziegler–Natta, $TiCl_3$-catalyzed polymerization of ethylene.

ethylene is the copolymerization of styrene, butadiene, and a third component (usually dicyclopentadiene or 1,4-hexadiene) to give synthetic rubbers. Vanadyl halides instead of titanium halides are then the preferred catalysts.

The Ziegler–Natta system is heterogeneous, and the active species is a fibrous form of $TiCl_3$ that is formed *in situ* from $TiCl_4$ and $Al(C_2H_5)_3$. Preformed $TiCl_3$ may also be used. During the course of the polymerization, many different alkyl groups become available, and it appears that a second role of the aluminum species, in addition to that of forming $TiCl_3$, is replacement of a chloride at the $TiCl_3$ surface by an alkyl group. Thus the catalytic process as in Fig. 30-3 begins with addition of ethylene to the vacant coordination site of a surface Ti atom. The alkyl group is transferred to the coordinated ethylene, another ethylene is bound to the newly created vacant coordination site, and the process of polymerization continues.

30-11 Palladium-Catalyzed Oxidations

It was long known that ethylene compounds of palladium, $[(C_2H_4)PdCl_2]_2$ for example, are rapidly decomposed in aqueous solution to form acetaldehyde (an oxidation product of ethylene) and Pd metal, according to the stoichiometry of Reaction 30-11.1.

$$C_2H_4 + PdCl_2 + H_2O \longrightarrow CH_3CHO + Pd + 2\,HCl \qquad (30\text{-}11.1)$$

The conversion of this stoichiometric reaction into a catalytic one (the Wacker process) required the linking together of Reaction 30-11.1 with the following known reactions:

$$Pd + 2\,CuCl_2 \longrightarrow PdCl_2 + 2\,CuCl \qquad (30\text{-}11.2)$$

$$2\,CuCl + 2\,HCl + \tfrac{1}{2}\,O_2 \longrightarrow 2\,CuCl_2 + H_2O \qquad (30\text{-}11.3)$$

The sum of Reactions 30-11.1, 30-11.2, and 30-11.3 is the desired oxidation of ethylene (Reaction 30-11.4).

$$C_2H_4 + \tfrac{1}{2}O_2 \longrightarrow CH_3CHO \tag{30-11.4}$$

The catalytic oxidation of ethylene by Pd^{II}–Cu^{II} chloride solutions is essentially quantitative, and only low Pd concentrations are required.

Since the reaction proceeds in Pd^{II} solutions with a chloride concentration greater than 0.2 M, the metal is most likely present as $[PdCl_4]^{2-}$. Reactions 30-11.5 and 30-11.6 then occur.

$$[PdCl_4]^{2-} + C_2H_4 \rightleftharpoons [PdCl_3(C_2H_4)]^- + Cl^- \qquad \text{(fast)} \tag{30-11.5}$$

$$[PdCl_3(C_2H_4)]^- + H_2O \rightleftharpoons [PdCl_2(H_2O)(C_2H_4)] + Cl^- \tag{30-11.6}$$

The neutral product of Reaction 30-11.6 is attacked nucleophilically by water giving the hydroxy–alkyl ligand shown in Structure 30-III

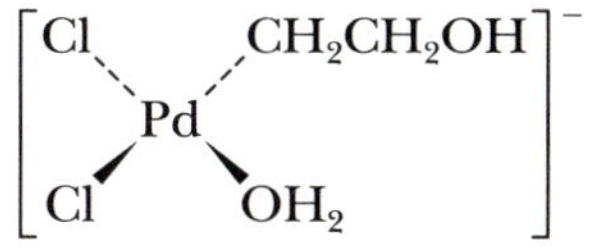

30-III

which eventually leads to products by the sequence shown in Reactions 30-11.7 and 30-11.8.

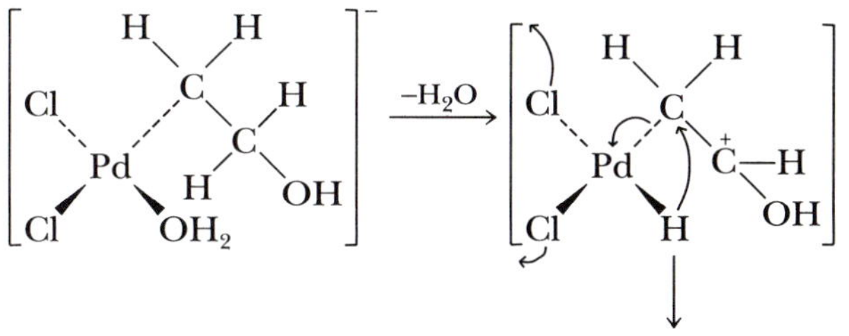

$$\tag{30-11.7}$$

$$CH_3CHO + H^+ \longleftarrow CH_3CHOH^+ + Pd^0 + 2\,Cl^- \tag{30-11.8}$$

The mechanism for the required reoxidation of Pd metal by Cu(II) chloro complexes (Reaction 30-11.2) is not well understood, but inner-sphere electron transfer by chloro bridges is probably involved.

The reactivity of palladium complexes in other systems has been extensively studied, and there now are many catalytic processes involving alkenes, arenes, carbon monoxide, alkynes, and the like. Extensions of the Wacker process using media other than water are known; thus in acetic acid, vinyl acetate is obtained, while in alcohols, vinyl ethers are obtained. Also, with alkenes other than ethylene, ketones may be obtained. Propene, for example, gives acetone.

30-12 Catalytic Reactions of Carbon Monoxide

We have already considered the hydroformylation reaction: an addition of H_2 and CO to alkenes. There are other important reactions involving CO, two of which we consider here.

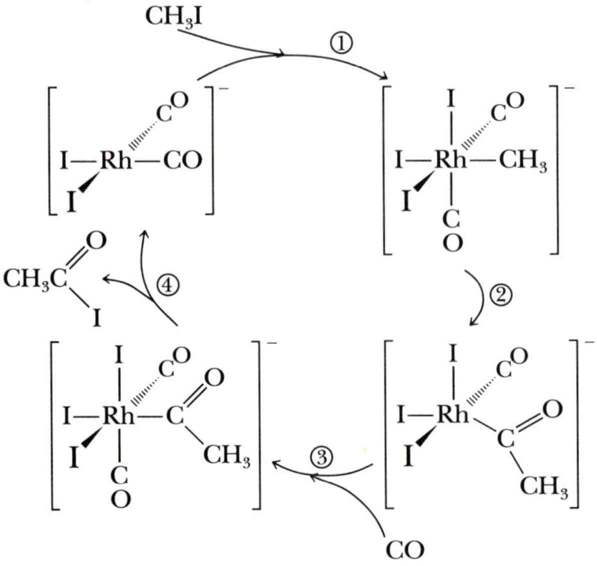

Figure 30-4   Catalytic cycle for the synthesis of acetic acid from methanol. The steps are (1) oxidative addition of CH_3I, (2) migratory insertion of CO, (3) addition of CO, and (4) reductive

elimination of $H_3C-\overset{\overset{O}{\|}}{C}-I$. Subsequent hydrolysis

of $H_3C-\overset{\overset{O}{\|}}{C}-I$ gives CH_3CO_2H.

Acetic Acid Synthesis

Acetic acid can be made by carbonylation of methanol. Originally, a high-temperature and high-pressure reaction using cobalt iodide was used. In the 1960s, the Monsanto Chemical Company introduced a process using rhodium that operates under milder conditions: about 180 °C and 40 atm. The key to this reaction (and to other carbonylations, e.g., of methyl acetate, to give acetic anhydride) involves the use of methyl iodide, which can oxidatively add to Rh^I as in Fig. 30-4. Carbon monoxide insertion gives an acyl intermediate that undergoes

reductive elimination of acetyl iodide, $H_3C-\overset{\overset{O}{\|}}{C}-I$. Hydrolysis of the latter by water in the aqueous–methanol feed then gives acetic acid and HI, as in Reaction 30-12.1.

$$H_3C-\overset{\overset{O}{\|}}{C}-I + H_2O \longrightarrow CH_3CO_2H + HI \qquad (30\text{-}12.1)$$

HI then reacts with methanol to regenerate methyl iodide, as in Reaction 30-12.2.

$$HI + CH_3OH \longrightarrow CH_3I + H_2O \tag{30-12.2}$$

In the absence of water and in the presence of lithium acetate, carbonylations of methanol or methyl acetate give acetic anhydride via the reaction:

$$CH_3C(O)I + CO_2CH_3^- \longrightarrow CH_3C(O)O_2CCH_3 \tag{30-12.3}$$

The cycle is similar to that in Fig. 30-4.

Fischer–Tropsch Reactions

These reactions were discovered by F. Fischer and H. Tropsch in Germany, in the late 1920s. The reactions make use of iron or other oxide catalysts to reduce CO by hydrogen, giving hydrocarbons, the simplest representative example being Reaction 30-12.4.

$$CO + 3\, H_2 \longrightarrow CH_4 + H_2O \tag{30-12.4}$$

Under selected conditions, petroleum or fuel oils can be made. The process is not very economical, even with very cheap coal as a source of feedstock synthesis gas, but is operated commercially in South Africa. A much more important reaction is methanol synthesis (Reaction 30-12.5),

$$CO + 2\, H_2 \longrightarrow CH_3OH \qquad \Delta H° = -92 \text{ kJ mol}^{-1} \tag{30-12.5}$$

which may be accomplished by using heterogeneous, copper-promoted, zinc oxide catalysts at 250 °C and 50 atm. This process provides a means of using methane waste gases from oil wells by oxidation to synthesis gas, as in Reaction 30-12.6, followed by conversion to methanol by Reaction 30-12.5. The methanol can then be converted to acetic acid and acetic anhydride as discussed previously, bringing the formerly wasted methane usefully into the petrochemicals market.

$$CH_4 + H_2O \longrightarrow CO + 3\, H_2 \tag{30-12.6}$$

It is possible to make other alcohols directly from CO and H_2, notably ethylene glycol by use of homogeneous rhodium catalysts. Such processes are, however, not yet economical.

Finally, it should be noted that the water–gas shift reaction (Reaction 30-12.7)

$$CO + H_2O = CO_2 + H_2 \tag{30-12.7}$$

can be catalyzed both heterogeneously and homogeneously. The reaction is used to remove CO from synthesis gas, thereby increasing the amount of H_2 that is available for ammonia synthesis, as mentioned in Section 9-1.

STUDY GUIDE

Study Questions

A. Review

1. What is meant by a coordinatively unsaturated species? Give two examples, and explain how these species may arise in solutions beginning with coordinatively saturated ones.

2. Define the term oxidative addition (oxad) reaction. What conditions must be met for such a reaction to occur? What is the reverse of such a reaction called?

3. Draw plausible structures for the reaction products of $IrCl(CO)(PR_3)_2$ with H_2, CH_3I, C_6H_5NCS, CF_3CN, $(CF_3)_2CO$, O_2.

4. How can one account for the low activation energy for oxidative addition of H_2 with its very strong H—H bond?

5. What is an insertion reaction? Give two real examples.

6. Describe the actual pathway for the reaction of $P(C_2H_5)_3$ with $CH_3Mn(CO)_5$ to give $CH_3COMn(CO)_4P(C_2H_5)_3$.

7. Complete the following equations and show with diagrams the structures of the principal products:

   (a) $Ru(CO)_3(PPh_3)_2 + HBF_4 \longrightarrow$

   (b) $Ir(CO)_3(PPh_3)_2^+ + CH_3O^- \longrightarrow$

   (c) $W(CO)_6 \xrightarrow{\text{LiCH}_3} [A] \xrightarrow{(CH_3)_3O^+} [B]$

   (d) $[Fe(CN)_5NO]^{2-} + 2\ OH^- \longrightarrow$

8. Show the steps by which a hydrido complex can cause isomerization of 1-alkenes to 2-alkenes? Is this generally stereospecific?

9. Write a balanced equation showing the overall (net) reaction in each of the following processes: hydroformylation; hydrosilylation; the Ziegler–Natta process; the Wacker process for synthesis of acetaldehyde.

10. Outline the main steps by which Ziegler–Natta polymerization proceeds.

11. Outline the mechanism of the Wacker process.

B. Additional Exercises

1. Write a plausible mechanism for the reaction of $Ti(NEt_2)_4$ with CS_2 to give $Ti(S_2CNEt_2)_4$.

2. Give a plausible catalytic cycle to account for the conversion of ethylene to propionaldehyde employing $RhH(CO)(PPh_3)_3$ as the catalyst.

3. Suggest a catalytic cycle to account for the action of

$$[Rh(PEtPh_2)_2(CH_3OH)_2]^+PF_6^-$$

   in methanol as a catalyst for hydrogenation of but-1-ene by H_2 at 25 °C and 1-atm pressure.

4. The complex $Ni[P(OEt)_3]_4$ in acidic solution is used in the synthesis of hexa-1,4-diene from ethylene and butadiene. Suggest a plausible catalytic cycle.

5. $Ni[P(OEt)_3]_4$ is also used to catalyze the process

$$CH_2{=}CH{-}CH{=}CH_2 + HCN \longrightarrow NC(CH_2)_4CN$$

   Again, suggest a sensible sequence of steps.

6. Suggest a mechanism for the following so-called 1,3-insertion reaction.

$$1 \quad 2 \quad 3\,4$$
$$(\eta^5\text{-}C_5H_5)(CO)_2Fe\text{—}CH_2C\equiv CCH_3 + SO_2$$

$$\begin{array}{c} 4\\ CH_3\\ | \\ 3\\ C \end{array}$$

$$(\eta^5\text{-}C_5H_5)(CO)_2Fe\text{—}\overset{2}{C}\underset{H_2C\overset{1}{\text{—}}O}{\Big\diagup}\overset{}{\diagdown}S{=}O$$

7. It has been proved that the alkyl group retains its configuration when CO insertion to produce the acyl occurs for $(\eta^5\text{-}C_5H_5)(CO)_2Fe\text{—}CHD\text{—}CHD\text{—}C(CH_3)_3$. Propose a mechanism that accounts for this.

8. Write a mechanism for the conversion of butadiene to *trans-trans-trans*-cyclododecatriene, using a Ni^0 species.

9. $RhH(CO)(PR_3)_3$ in benzene under pressure of ethylene reacts with benzoyl chloride to give propiophenone, $C_6H_5C(O)C_2H_5$. Suggest a mechanism.

10. A catalytic process for making propionic acid from acetic acid has been developed. It uses HI, H_2, CO, and H_2O as stoichiometric reagents and a patented $Ru(CO)_xI_y$ compound as catalyst. HI and H_2O are regenerated by the full catalytic process.

Plausible steps in the catalytic cycle include (a) oxidative addition of $H_3C\overset{\overset{\displaystyle O}{\|}}{—C}—I$; (b) addition of H_2 to Ru, and migration to give an α-hydroxyethyl ligand; (c) hydrogenation of the latter, with elimination of water to give an ethyl ligand; (d) insertion of CO to give an acyl ligand; and (e) hydrolysis to propionic acid. Write out the catalytic cycle in the style of Fig. 30-4.

11. The complex $[\eta^5\text{-}C_5H_5Re(CO)_2NO]^+$ can be reduced using $NaBH_4$ in THF–water mixtures giving first a formyl complex, second a hydroxymethyl complex, and third a methyl complex. Draw structures of the four compounds, apply the 18-electron formalism to each, and discuss the relevance of these reactions to our understanding of Fischer–Tropsch chemistry.

C. Problems from the Literature of Inorganic Chemistry

1. Consider the report (of a Rh catalyst for olefin hydrogenation) by C. O'Connor and G. Wilkinson, *J. Chem. Soc. (A)*, **1968**, 2665–2671.

   (a) Prepare sketches of each rhodium compound in the hydrogenation cycle proposed here.

   (b) Describe (addition, oxidative addition, reductive elimination, etc.) each step in the process.

   (c) Enumerate the data or reasoning favoring each structure in (a) and each step in (b).

   (d) Why, according to the authors, is hydrogenation with this catalyst only possible, apparently, for terminal alkenes of the formula $RHC{=}CH_2$?

2. Read the paper on reductions using $CO + H_2O$ in place of H_2 by H. Kang, C. H. Mauldin, T. Cole, W. Slegeir, K. Cann, and R. Pettit, *J. Am. Chem. Soc.*, **1977**, *99*, 8323–8325.

   (a) Propose reactions and mechanisms for the formation from CO, H_2O and $Fe(CO)_5$ of $HFe(CO)_4^-$, $H_2Fe(CO)_4$, and H_2.

(b) Propose a mechanism for the pH-dependent reduction (actually hydroformylation) of ethylene to propanol. How do the authors account for the pH dependence?

(c) Explain how $Fe(CO)_5$ serves as a catalyst for the water–gas shift reaction.

3. Consider the study reported by E. L. Muetterties and P. L. Watson, *J. Am. Chem. Soc.*, **1976**, *98*, 4665–4667.

(a) Write balanced equations for two separate preparations of the hydrogen adduct, or dihydride, $H_2Co[P(OCH_3)_3]_4^+$.

(b) Draw the structure of the dihydride in (a). What data indicate a predominantly cis arrangement of the two (formally) H^- ligands? What is the difference between a "dihydrogen adduct" and a "dihydride"?

(c) What experiments suggest that reductive elimination of H_2 from the dihydride in (a) is a unimolecular process?

(d) How has reductive elimination of CH_4 been studied here? How does CH_4 elimination compare with H_2 elimination?

4. Consider the paper by E. M. Hyde and B. L. Shaw, *J. Chem. Soc. Dalton Trans.*, **1975**, 765–767.

(a) Enumerate the differences between oxidative addition of H_2 and CH_3I to *trans*-$[IrX(CO)L_2]$ complexes in general.

(b) How were rate constants obtained in this study for the addition of H_2 to $IrCl(CO)(PR_3)_2$ complexes?

(c) How does H_2 addition to $[IrCl(CO)(PMe_2Ph)_2]$ compare with H_2 addition to $[IrCl(CO)\{PMe_2(C_6H_4OMe)_2\}_2]$?

(d) How does CH_3I oxidative addition compare with H_2 oxidative addition towards both of the complexes in (c)?

(e) What mechanistic interpretation do the authors give for the differences noted in (c) and (d)?

5. Consider the paper by K. L. Brown, G. R. Clark, C. E. L. Headford, K. Mardson, and W. R. Roper, *J. Am. Chem. Soc.*, **1979**, *101*, 503–505.

(a) Apply the 18-electron formalism to explain the various preparations and reactions of this η^2-formyl complex of Os^0.

(b) The "hydrido-formyl" $Os(CHO)H(CO)_2(PPh_3)_2$ eliminates H_2, while the "chloroformyl" $Os(CHO)Cl(CO)_2(PPh_3)_2$ eliminates CO, not HCl. Why? Explain.

6. Read the article by C. P. Casey, M. W. Meszaros, S. M. Neuman, I. G. Cesa, and K. J. Haller, *Organometallics*, **1985**, *4*, 143–149.

(a) Compare the syntheses and structures of the analogous acetyl and formyl complexes as reported here.

(b) Apply the 18-electron formalism to each reactant and product in your answer to (a).

(c) What does a comparison of these two structures seem to indicate about the reason why CO insertion into a M—C bond occurs more readily than CO insertion into a M—H bond?

SUPPLEMENTARY READING

Alper, H., Ed., *Transition Metal Organometallics in Organic Synthesis*, Vol. 1, 1976, Vol. 2, 1978, Academic, New York.

Collman, J. P. and Hegedus, L. S., *Principles and Applications of Organotransition Metal Chemistry*, University Science Books, Mill Valley, CA, 1987.

Gates, B. C., Katzer, J. R., and Schuit, G. C. A., *Chemistry of Catalytic Processes*, McGraw-Hill, New York, 1979.

James, B. R., *Homogeneous Hydrogenation*, Wiley-Interscience, New York, 1973.

Kochi, J. K., *Organometallic Mechanisms and Catalysis*, Academic, New York, 1978.

Masters, C., *Homogeneous Transition Metal Catalysis, A Gentle Art*, Chapman & Hall, London, 1981.

Parshall, G. W., *Homogeneous Catalysis*, Second Edition Wiley-Interscience, New York, 1992.

Tolman, C. A., "The 16- and 18-Electron Rule in Organometallic Chemistry and Homogeneous Catalysis," *Chem. Soc. Rev.*, **1972,** *11,* 337.

Wilkinson, G., Abel, E. W., and Stone, F. G. A., Eds., *Comprehensive Organometallic Chemistry*, Pergamon Press, New York, 1982.

Yamamoto, A., *Organotransition Metal Chemistry*, Wiley-Interscience, New York, 1986.

Chapter 31

BIOINORGANIC CHEMISTRY

31-1 Overview

Biochemistry is not merely an elaboration of organic chemistry. The chemistry of life involves, in essential and indispensible ways, at least 25 elements. In addition to the "organic" elements C, H, N, and O, there are 9 other elements that are required in relatively large quantities, and called, therefore, **macronutrients.** These elements are Na, K, Mg, Ca, S, P, Cl, Si, and Fe. There are also many other elements, **micronutrients,** that are required in small amounts by at least some forms of life: V, Cr, Mn, Co, Ni, Cu, Zn, Mo, W, Se, F, and I. As research activity intensifies, and as instrumental methods of analysis and detection become more sophisticated and sensitive, it is likely that other elements will be added to the list of micronutrients. The elements Cr, Ni, W, and Se have been added only within the last few years.

The metallic elements play a variety of roles in biochemistry. Several of the most important roles are the following:

1. *Regulatory action* is exercised by Na^+, K^+, Mg^{2+}, and Ca^{2+}. The flux of these ions through cell membranes and other boundary layers sends signals that turn metabolic reactions on and off.

2. *The structural role* of calcium in bones and teeth is well known, but many proteins owe their structural integrity to the presence of metal ions that tie together and make rigid certain portions of these large molecules, portions that would otherwise be only loosely linked. Metal ions particularly known to do this are Ca^{2+} and Zn^{2+}.

3. An enormous amount of *electron-transfer* chemistry goes on in biological systems, and nearly all of it critically depends on metal-containing electron-transfer agents. These include cytochromes (Fe), ferredoxins (Fe), and a number of copper-containing "blue proteins," such as azurin, plastocyanin, and stellacyanin.

4. *Metalloenzymes or metallocoenzymes* are involved in a great deal of enzymatic activity, which depends on the presence of metal ions at the active site of the enzyme or in a key coenzyme. Of the latter, the best known is vitamin B_{12}, which contains Co. Important metalloenzymes include carboxypeptidase (Zn), alcohol dehydrogenase (Zn), superoxide dismutase (Cu, Zn), urease (Ni), and cytochrome P-450 (Fe).

5. All aerobic forms of life depend on *oxygen carriers,* molecules that carry oxygen from the point of intake (such as the lungs) to tissues where O_2 is

used in oxidative processes that generate energy. There are three major types of oxygen carriers, and all of them contain metal ions that provide the actual binding sites for the O_2 molecules. These types are

Hemoglobins (Fe), found in all mammals.

Hemerythrins (Fe), found in various marine invertebrates.

Hemocyanins (Cu), found in arthropods and molluscs.

Each of these roles will be discussed in this chapter.

31-2 The Role of Model Systems

Because of the size and complexity of most biochemical molecules and processes, it is often advantageous to find smaller and simpler models upon which controlled experiments can be more easily performed, and with which hypotheses can be tested. Bioinorganic chemistry has been an especially fruitful area for the use of model systems, particularly where transition metals are involved. Of course it is not always possible to find or develop suitable models, and it can be dangerously misleading should overly simplistic models be used naively. Even in the best of circumstances, a model can give only a partial view of how the real system works. If these limitations are recognized, the model system approach can provide valuable guidance to eventual study of the real systems.

The broad and detailed knowledge that we have of coordination chemistry sets the stage for an understanding of the role of metal ions in biological systems. Fundamental principles and generalizations about the behavior of metal complexes are valid whether the metal is coordinated by some relatively simple set of man-made ligands or by a gigantic protein molecule, where the coordinating groups are often carboxyl oxygen atoms, thiol sulfur atoms, or amine nitrogen atoms. Moreover, the optical spectra, magnetic moments, and EPR spectra of transition metal ions afford the same powerful methods of study as when applied to the simpler complexes. Thus we have methods for checking the models against the real systems.

Throughout this chapter we shall frequently refer to model systems that have played a role in understanding real bioinorganic systems. Among these are iron–porphyrin compounds relevant to the understanding of hemoglobin, myoglobin, cytochromes, and enzyme P-450; models for hemerythrin; the cobaloxime model for vitamin B_{12}; iron–sulfur cluster compounds as models for ferredoxins; and a number of copper complexes that serve as models for a variety of copper-containing enzymes.

31-3 The Alkali and Alkaline Earth Metals

The elements Na, K, Mg, and Ca are ubiquitous in living systems and play an assortment of vital roles. Inorganic chemists who were interested in coordination chemistry used to have a tendency to regard these elements as relatively uninteresting. Nothing could be further from the truth if one is seeking an understanding of life processes.

Sodium and Potassium

Sodium and potassium ion concentrations and the balance (or ratio) of their concentrations in various parts of an organism are controlled by a number of special complexing agents. These generally are **cavitands,** that is, macrocyclic molecules with polar interior groups for binding the ions and nonpolar (hydrophobic) exterior groups that enable the cavitands to carry the Na^+ or K^+ ions across cell boundaries. An example is the cyclic dodecapeptide **valinomycin,** shown as Structure 31-I, and in Chapter 10 as Structure 10-III.

31-I

The Na^+ concentration within animal cells has to be kept about 10 times lower than that in the extracellular fluids, whereas an opposite **gradient** (by a factor of 30) must be maintained for the K^+ ion. The maintenance of these balances requires energy, and when such balances are abruptly changed, electrical potentials responsible for the transmission of nerve impulses are created.

Calcium

Calcium serves in a staggering variety of roles, the most obvious being in structural materials such as teeth, bones, shells, and a number of other less well-known calcium-rich deposits. It is important to note that none of these calciferous biological materials is an inert "mineral." Bone, for example, though consisting largely of calcium carbonate and phosphate, is continually being deposited and reabsorbed, and it acts as a buffer for body Ca^{2+} and phosphate ions, as controlled by hormonal action. The form of calcium phosphate that occurs in bone and teeth has the same composition as the mineral apatite, $Ca_{10}(PO_4)_6X_2$, where X represents F, Cl, or OH, or a mixture of these.

Calcium is essential to the action of extracellular enzymes, and it participates in many regulatory processes. It is generally complexed by the side-chain carboxyl groups of proteins, with additional bonding sometimes to peptide carbonyl groups and hydroxyl groups.

Magnesium

Magnesium, because of its high charge/radius ratio and consequent strong hydration {as $[Mg(H_2O)_6]^{2+}$}, plays biological roles that are very different from those of calcium. One of its major roles is as a counterion to the negatively charged $ROPO_3H^-$ groups in nucleotides and polynucleotides. Sometimes it approaches the phosphate anions as $[Mg(H_2O)_6]^{2+}$, but it is also found as $[Mg(H_2O)_5]^{2+}$ or $[Mg(H_2O)_4]^{2+}$ with one or two phosphate oxygen atoms, respectively, completing its first coordination sphere. The magnesium ion helps to stabilize the three-dimensional structure of ribonucleic acid (RNA) and deoxyribonucleic acid (DNA) and is thus crucial to the proper functioning of the genetic machinery of the cell. Also, adenosine diphosphate (ADP) and adenosine triphosphate (ATP, shown in Structure 31-II) exist mainly as 1:1 complexes with the magnesium ion. Magnesium also has a unique role in the plant kingdom as the central atom of chlorophyll, which will be discussed in Section 31-4.

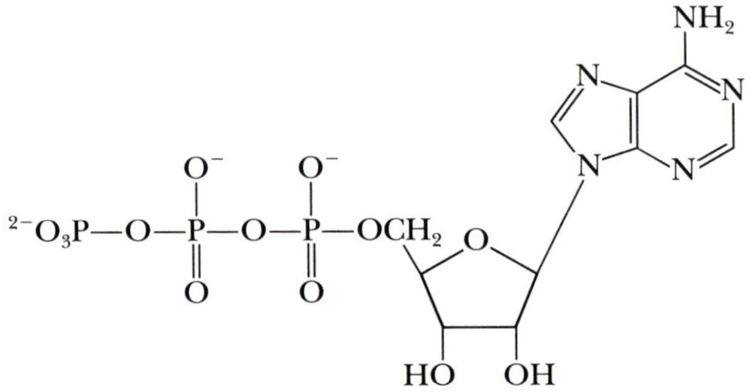

31-II

31-4 Metalloporphyrins

One of the most important ways in which metal ions are involved in biochemistry is in complexes with a type of macrocyclic ligand called a *porphyrin*. Porphyrins are derivatives of *porphine*. They differ in the arrangement of substituents around the periphery. The porphine molecule is shown in Fig. 31-1(a), and the two most important metal complexes of porphyrins, *chlorophyll* and *heme*, are shown in Fig. 31-1(b) and (c). In these complexes the inner hydrogen atoms have been displaced by the metal ions.

Chlorophyll

There are several very similar but not identical chlorophyll molecules. Green plants contain two and various algae contain others. Notice that in Fig. 31-1(b) the basic porphine system has been modified in two ways. In pyrrole ring IV, one of the double bonds has been trans-hydrogenated, and a cyclopentanone ring has been fused to the side of pyrrole ring III. Nevertheless, the fundamental properties of the porphine system are retained.

Photosynthesis is a complex sequence of processes in which solar energy is first absorbed and ultimately—in a series of redox reactions, some of which proceed in the dark—used to drive the overall endothermic process of combining water and carbon dioxide to give glucose; molecular oxygen is released simultaneously:

$$6\ CO_2 + 6\ H_2O = C_6H_{12}O_6 + 6\ O_2 \qquad (31\text{-}4.1)$$

The function of the chlorophyll molecules in the chloroplast is to absorb photons in the red part of the visible spectrum (near 700 nm) and pass this energy of excitation on to other species in the reaction chain. The ability to absorb the light is due basically to the conjugated polyene structure of the porphine ring system. The role of the magnesium ion is, at least, twofold. (1) It helps to

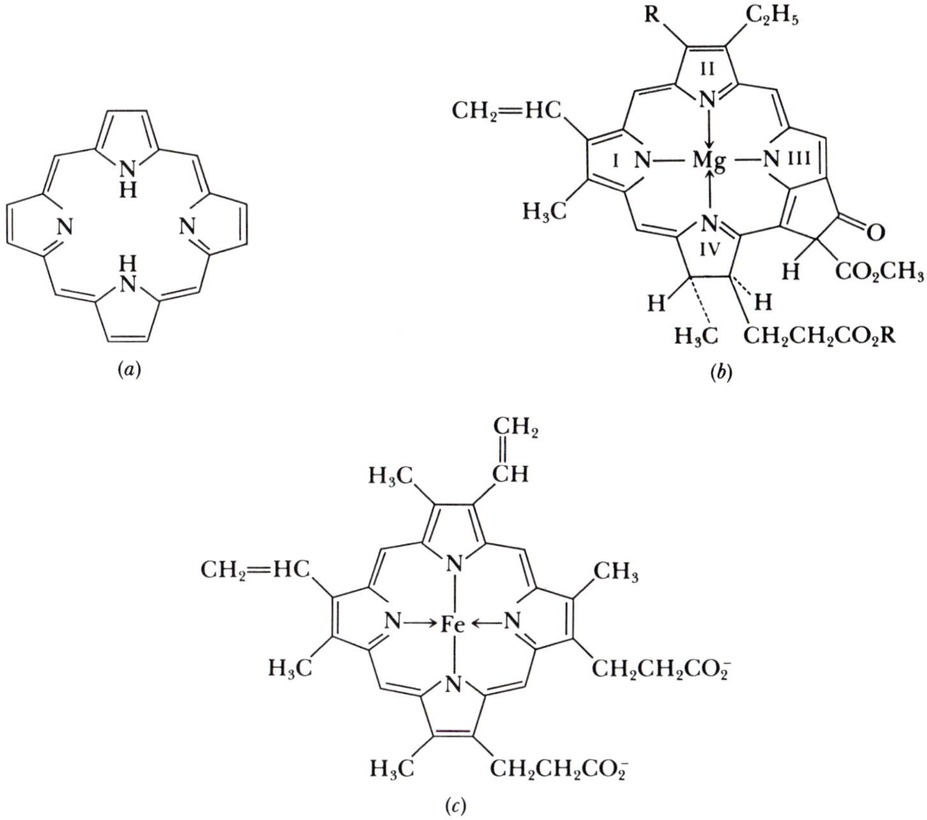

Figure 31-1 (*a*) The prototype porphine molecule. (*b*) One of the chlorophyll molecules. (*c*) The heme group.

make the entire molecule rigid so that energy is not too easily lost thermally, that is, degraded to molecular vibrations. (2) It enhances the rate at which the short-lived singlet excited state initially formed by photon absorption is transformed into the corresponding triplet state, which has a longer lifetime and thus can transfer its excitation energy into the redox chain.

At an early stage of the electron-transfer sequence that leads ultimately to the release of molecular oxygen, a polynuclear manganese complex, of unknown composition, undergoes reversible redox reactions. At still other stages, iron-containing substances, called cytochromes and ferredoxins, and a copper-containing substance, called plastocyanin, also participate. Thus, photosynthesis requires the participation of complexes of no less than four metallic elements.

Heme Proteins

Iron is certainly the most widespread of the transition metals in living systems. Its compounds participate in a variety of activities. The two main functions of iron-containing materials are (1) transport of oxygen, and (2) mediation in electron-transfer chains. So much iron is required for these purposes that there is also a chemical system to store and transport iron. We turn first to compounds in which the iron is present as heme, the porphyrin complex depicted in Fig. 31-1(c). The heme group functions in all cases in intimate association with a protein molecule. The chief heme proteins are

1. Hemoglobins
2. Myoglobins
3. Cytochromes, including a special type, P-450
4. Enzymes such as catalase and peroxidase

Hemoglobin and Myoglobin

These are closely related. Hemoglobin has a molecular weight of 64,500 and consists of four subunits, each containing one heme group. Myoglobin is very similar to one of the subunits of hemoglobin, one of which is shown in Fig. 31-2. Hemoglobin has two functions. (1) It binds oxygen molecules to its iron atoms and transports them from the lungs to muscles where they are delivered to myoglobin molecules. These store the oxygen until it is required for metabolic ac-

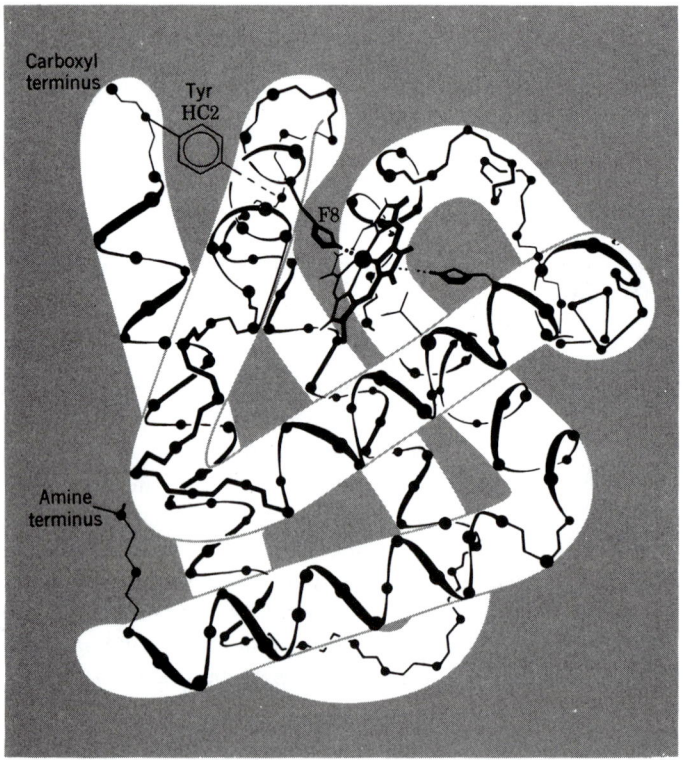

Figure 31-2   A representation of one of the four subunits of hemoglobin. The continuous black band represents the peptide chain and the various sections of the helix. Dots on the helical chain represent α-carbon atoms. The heme group is near the top of the diagram (just to the right of center), with the iron atom represented as a large dot. The coordinated histidine is labeled F8, meaning the 8th residue of the F helix. [This diagram was adapted from one kindly provided by M. Perutz.]

tion. (2) The hemoglobin then uses certain amino groups to bind carbon dioxide and carry it back to the lungs.

The heme group is attached to the protein in both hemoglobin and myoglobin through a coordinated histidine–nitrogen atom (F8), shown in Fig. 31-2. The position trans to the histidine–nitrogen atom is occupied by a water molecule in the deoxy species or O_2 in the oxygenated species. The structure of the $Fe—O_2$ grouping is still uncertain, but changes in the oxidation state of iron and the introduction of O_2 (and other ligands) cause important changes in the structure of heme, as we describe here.

Hemoglobin is not simply a passive container for oxygen but an intricate molecular machine. This may be appreciated by comparing its affinity for O_2 to that of myoglobin. For myoglobin (Mb) we have the following simple equilibrium:

$$Mb + O_2 = MbO_2 \qquad K = \frac{[MbO_2]}{[Mb][O_2]} \qquad (31\text{-}4.2)$$

If f represents the fraction of myoglobin molecules bearing oxygen and P represents the equilibrium partial pressure of oxygen, then

$$K = \frac{f}{(1-f)P} \qquad \text{and} \qquad f = \frac{KP}{1+KP} \qquad (31\text{-}4.3)$$

This is the equation for the hyperbolic curve labeled Mb in Fig. 31-3. Hemoglobin with its four subunits has more complex behavior; it approximately follows the equation

$$f = \frac{KP^n}{1+KP^n} \qquad n \approx 2.8 \qquad (31\text{-}4.4)$$

where the exact value of n depends on pH. Thus, for hemoglobin (Hb) the oxygen-binding curves are sigmoidal, as is shown in Fig. 31-3. The fact that n exceeds unity can be ascribed physically to the fact that attachment of O_2 to one heme group increases the binding constant for the next O_2, which in turn increases the constant for the next one, and so on.

Although Hb is about as good an O_2 binder as Mb at high O_2 pressure, it is much poorer at the lower pressures prevailing in muscle and, hence, passes on its oxygen to the Mb as required. Moreover, the need for O_2 will be greatest in tissues that have already consumed oxygen and simultaneously have produced CO_2. The CO_2 lowers the pH, thus causing the Hb to release even more oxygen to the Mb. The pH-sensitivity (called the Bohr effect), as well as the progressive increase of the O_2 binding constants in Hb, is due to interactions between the subunits; Mb behaves more simply because it consists of only one unit. It is clear that each of the two is essential in the complete oxygen-transport process. Carbon monoxide, PF_3, and a few other substances are toxic because they become bound to the iron atoms of Hb more strongly than O_2; their effect is one of competitive inhibition.

The way in which interactions between the four subunits in Hb give rise to both the cooperativity in oxygen binding and to the Bohr effect (pH dependence), both of which are so essential to the role played by Hb, is now partly un-

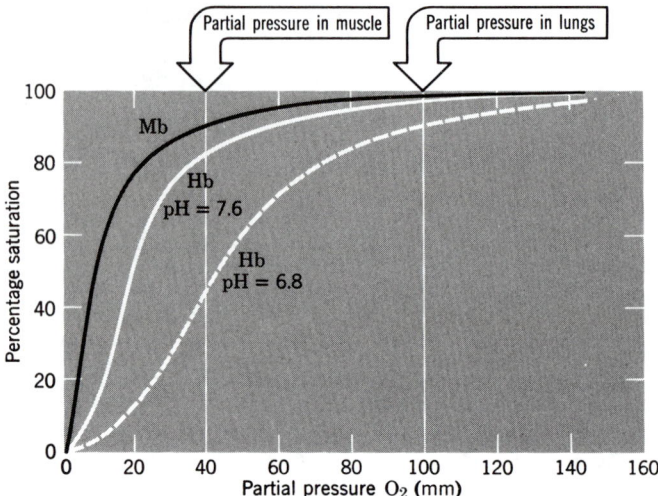

Figure 31-3   The oxygen-binding curves for myoglobin (Mb) and hemoglobin (Hb), showing also the pH dependence for the latter.

derstood. The mechanism is very intricate, but one essential feature depends directly on the coordination chemistry involved. Deoxyhemoglobin has a high-spin distribution of electrons, with one electron occupying the $d_{x^2-y^2}$ orbital that points directly toward the four porphyrin nitrogen atoms. The presence of this electron in effect increases the radius of the iron atom in these directions by repelling the lone-pair electrons of the nitrogen atoms. The result is that the iron atom actually lies about 0.7–0.8 Å out of the plane of these nitrogen atoms, in order that it not be in too close contact with them. The iron atom is also coordinated by a nitrogen atom on the imidazole ring of the amino acid histidine, labeled F8 in Fig. 31-2. Thus the iron atom in deoxyhemoglobin has square pyramidal coordination, as is shown in Fig. 31-4(a).

When an oxygen molecule becomes bound to the iron atom, it occupies a position opposite to the imidazole–nitrogen atom. The presence of this sixth ligand alters the strength of the ligand field, and the iron atom goes into a low-spin state, in which the six d electrons occupy the d_{xy}, d_{yz}, and d_{zx} orbitals. The $d_{x^2-y^2}$ orbital is then empty and the previous effect of an electron occupying this orbital in repelling the porphyrin nitrogen atoms vanishes. The iron atom is thus able to slip into the center of an approximately planar porphyrin ring and an essentially octahedral complex is formed, as shown in Fig. 31-4(b).

As the iron atom moves, it pulls the imidazole side chain of histidine F8 with it, thus moving that ring about 0.75 Å. This shift is then transmitted to other parts of the protein chain to which F8 belongs and, in particular, a large movement of the phenolic side chain of tyrosine HC2 is produced. From here various shifts of atoms in the neighboring subunit are caused, and these shifts influence the oxygen-binding capability of the heme group in that subunit. Thus the movement of the iron atom of the heme group in one subunit of hemoglobin acts as a kind of "trigger," which sets into motion extensive structural changes in other subunits.

One of the interesting problems about oxygen binding by hemoglobin concerns the structure of the Fe–O₂ grouping. Three possibilities are shown in Fig.

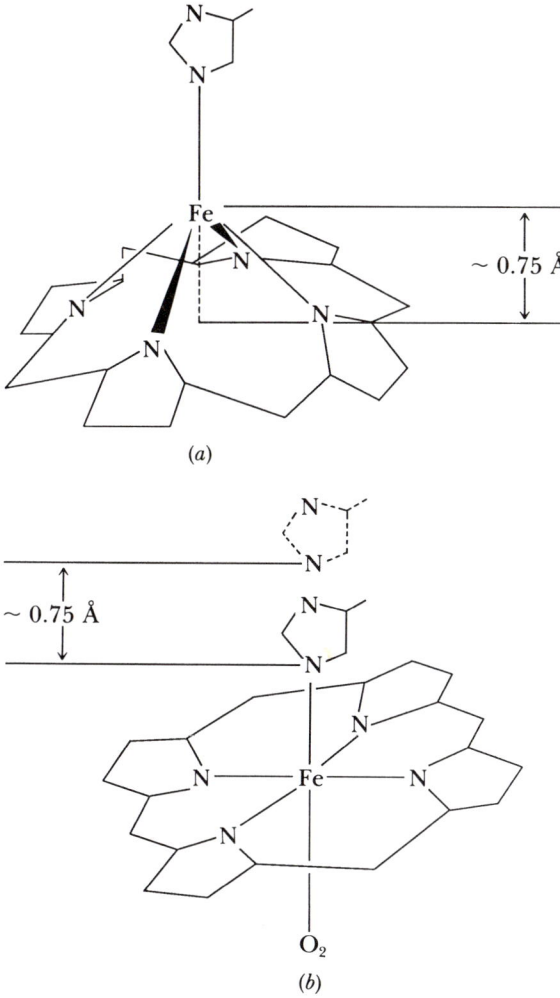

(a)

(b)

Figure 31-4 (*a*) The five-coordinate, high-spin Fe^{II} in deoxyhemoglobin. (*b*) The six-coordinate, low-spin iron in oxyhemoglobin, showing the distance that the side-chain histidine residue F8 has moved upon oxygenation.

31-5. The linear geometry has no precedent and is least probable. The side-on arrangement is found in some simple O_2 complexes involving other metals, such as $(PPh_3)_2ClIrO_2$, but is very unlikely for hemoglobin. The bent chain appears most probable, since O_2 is isoelectronic with NO^+, and since the latter forms complexes with bent Co^{III}—N—O chains. Also, there is one fairly good model compound, an iron(II)porphyrin complex of O_2, in which the bent arrangement has been found.

Recent X-ray studies on both $Mb(O_2)$ and $Hb(O_2)_4$ indicated that O_2 binds in the bent end-on fashion, with an Fe—O—O angle of approximately 130°.

Hemoglobin Modeling

The ability of the heme in hemoglobin or myoglobin to bind an O_2 molecule and later release it without the iron atom becoming permanently oxidized

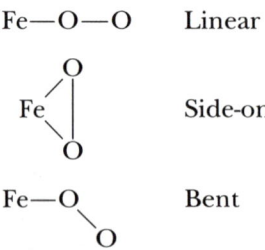

Fe—O—O   Linear

Fe⟨O... Side-on

Fe—O⟨O Bent

Figure 31-5 Three conceivable O_2–iron bonding geometries for hemoglobin or myoglobin.

to the iron(III) state is obviously essential to the functioning of these oxygen carriers. This remarkable ability has been taken for granted in the preceding discussion, but it merits further discussion. It is the reversibility of the hemoglobin and myoglobin reactions with O_2 that must be matched by any useful model.

Early attempts to employ simple Fe^{II}–porphyrin complexes, or even free heme itself, plus an aromatic amine molecule (to take the place of the histidine F8) were not successful. On exposing such a "model" to O_2, oxidation (rather than oxygenation) occurred promptly and irreversibly. Oxygen was absorbed, but not released. The reason for this is now understood: Dioxygen reacts to produce an O-bridged dinuclear complex of iron(III), as in Reaction 31-4.5.

$$2\,(\text{Amine})\text{Fe} + \tfrac{1}{2}\,O_2 \longrightarrow (\text{Amine})\text{Fe}—O—\text{Fe}(\text{Amine}) \qquad (31\text{-}4.5)$$

In hemoglobin and myoglobin, the bulk of the protein surrounding the heme unit assures that each heme unit remains isolated. To have an effective model, something must be added to the simple iron–porphyrin to accomplish this same degree of bulk. The two ways in which this has been done are represented schematically in Fig. 31-6 and then more realistically in Fig. 31-7. Model compounds such as those shown in Fig. 31-7 do engage in reversible oxygen binding, quite similar to the behavior of myoglobin. In the two examples shown in Fig. 31-7, a suitable amine (such as pyridine) is bound on the unprotected side, and the O_2 molecule then enters either between the "pickets" or under the "cap," where it is bound end-on to the iron atom.

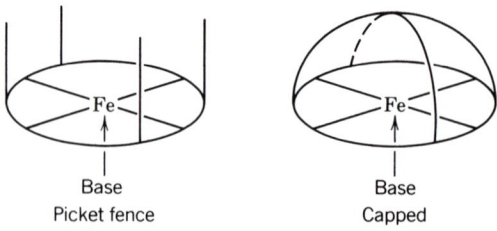

Base
Picket fence

Base
Capped

Figure 31-6 Schematic representations of two ways in which hemelike models may be modified to preclude dimerization via μ–O bridging.

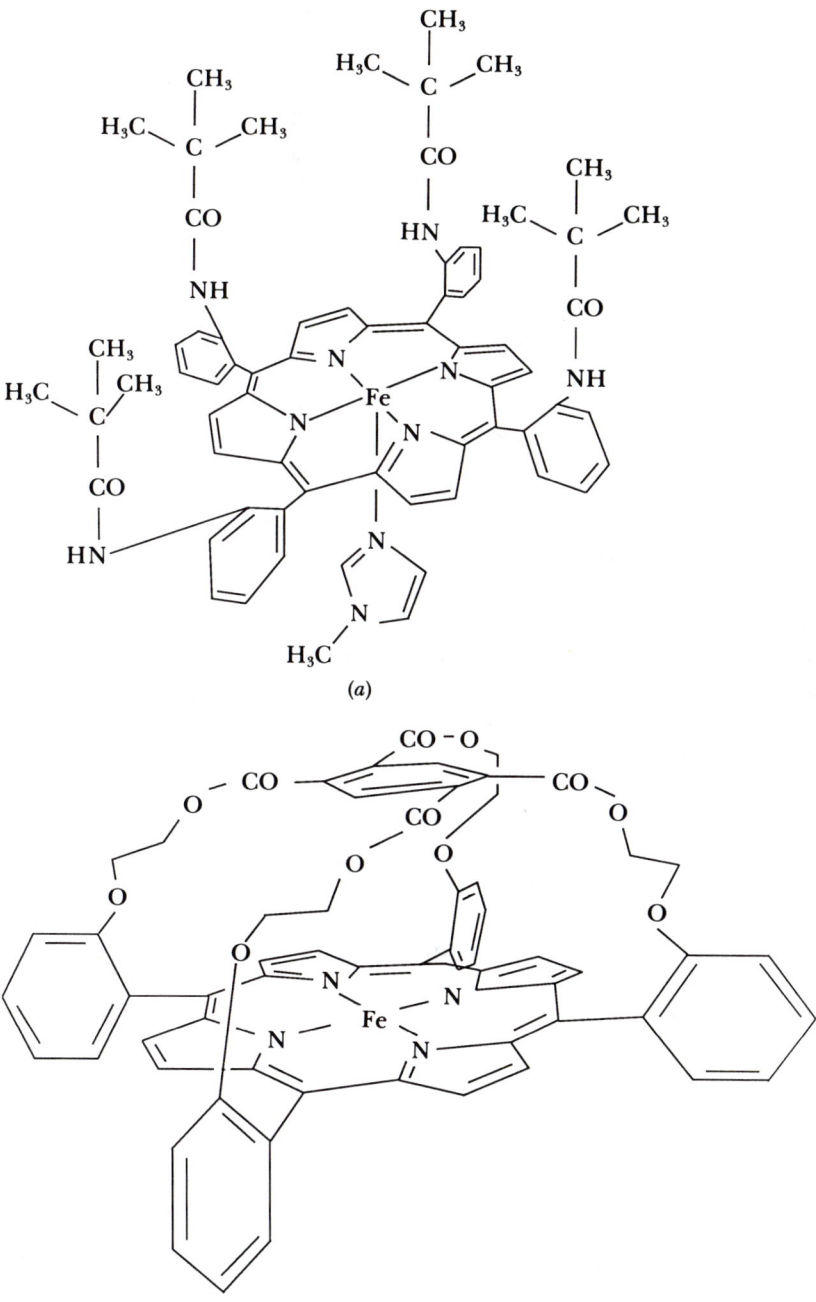

(a)

(b)

Figure 31-7   Actual examples of (*a*) the "picket fence" and (*b*) the "capped" types of heme models.

Other Heme Proteins

It is a fascinating fact that heme, the iron–porphyrin complex shown in Fig. 31-1(*c*), functions in Nature for a number of other tasks in addition to carrying oxygen. We shall not go into any of these in detail, but they should at least be mentioned since inorganic chemists have contributed to our understanding of all of them, both through research on the natural materials themselves, and through fabrication and study of model systems.

Cytochromes. Cytochromes are found in both plants and animals and serve as electron carriers. They accept an electron from a slightly better reducing agent and pass it on to a slightly better oxidizing agent. In the cytochromes, the heme iron is coordinated by a nitrogen atom of an imidazole ring on one side of the porphyrin plane, and it is coordinated on the other side of the porphyrin plane by the sulfur atom of a methionine residue from a different part of the protein backbone. Thus the potential oxygen-carrying capacity of the heme in cytochromes is blocked.

Cytochrome P–450 Enzymes. These enzymes are heme-containing oxygenases that catalyze the introduction of oxygen atoms into substrates. Of the many possible substrates, the most important are molecules in which C—H groups are converted into C—OH groups. The catalytic cycle entails a substance in which the iron atom attains a high (IV or V) oxidation state. The coordination sphere of the iron atom includes, in addition to the porphyrin ring, one sulfur atom, but whether the sixth coordination position is occupied by a water ligand or is vacant in the resting enzyme is uncertain.

Peroxidases and Catalases. Peroxidases catalyze the oxidation of a variety of substances by peroxides, mainly H_2O_2. Catalases catalyze decomposition of H_2O_2 (and some other peroxides) to H_2O and O_2. They have many similarities both in structure and in aspects of their mechanisms. They both have high-spin ferric heme groups lodged deeply in large protein molecules, with a histidine nitrogen atom occupying the fifth coordination site. The sixth coordination position may be occupied by a water ligand in the resting enzyme. There is growing evidence that a porphyrin—Fe^{IV}=O substance is the key intermediate in the function of peroxidases and catalases.

31-5   Iron–Sulfur Proteins

Iron–sulfur proteins contain strongly bound, functional iron atoms, but not porphyrins. The iron atoms are bound by sulfur atoms. These proteins all participate in electron-transfer sequences.

Rubredoxins

These are found in anaerobic bacteria where they are believed to participate in biological redox reactions. They are relatively low-molecular weight proteins (~ 6000), and usually contain only one iron atom. In the best characterized rubredoxin, from the bacterium *Clostridium pasteurianum,* the iron atom, which is normally in the III oxidation state, is surrounded by a distorted tetrahedron of cysteinyl sulfur atoms. The Fe—S distances range from 2.24 to 2.33 Å, and the S—Fe—S angles from 104 to 114°. A schematic representation of this is given in Fig. 31-8. When the Fe^{III} is reduced to Fe^{II}, there is a slight (0.05 Å) increase in the Fe—S distances, but the essentially tetrahedral coordination is maintained. Mössbauer spectroscopy has shown that the iron is in the high-spin condition in both oxidation states. Inorganic chemists have prepared and studied $[Fe(SR)_4]^{2-}$ and $[Fe(SR)_4]^-$ complexes as models to help understand the properties of the rubredoxins.

Ferredoxins

Ferredoxins are also relatively small proteins (6000–12,000) that contain iron–sulfur redox centers that are held in place by bonds from cysteine sulfur atoms to iron. The difference from rubredoxin is that here the redox centers are clusters of two, three, or four iron atoms, together with several sulfur atoms (so-called inorganic sulfur). In each case, an approximate tetrahedron of sulfur atoms is completed about each iron atom by the sulfur atoms of cysteine residues of the peptide. These systems are generally called ferredoxins and are often abbreviated Fd.

The two-iron Fd's, complete with their attached cysteine sulfur atoms, can be described as two tetrahedral FeS_4 units sharing an edge. In a convenient notation, the two-iron clusters can be represented as $[2Fe–2S]^{n+}$. They have relatively simple behavior. Their normal state is $[2Fe–2S]^{2+}$ [meaning that both iron atoms are iron(III)], but they can be reduced at potentials similar to that of the standard H^+/H_2 electrode (i.e., –0.4 V on the hydrogen scale) to $[2Fe–2S]^+$.

Several kinds of spectroscopic evidence indicate that in the reduced $[2Fe–2S]^+$ cluster, the added electron is localized on one iron atom, so that one Fe^{II} and one Fe^{III} atom are present. In the $[2Fe–2S]^{2+}$ cluster, the Fe–Fe distance is only 2.72 Å, and the two formally high-spin (d^5) iron atoms have their magnetic moments so strongly coupled antiferromagnetically that the cluster is diamagnetic. Upon reduction to give $[2Fe–2S]^+$, this coupling persists, and the $[2Fe–2S]^+$ cluster has only one unpaired electron. This has been very helpful, since it means that ESR detection of the reduced cluster is quite easy.

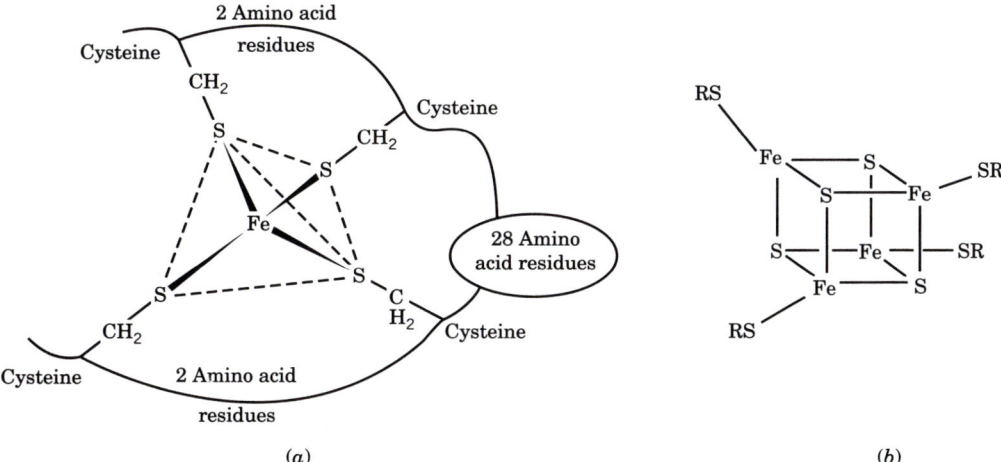

(a)   (b)

Figure 31-8   (a) The environment of the iron atom in the rubredoxin molecule. (b) The Fe_4S_4 cluster found in the four-iron ferredoxins and HiPiPs. The thiolate side-chains of cysteines are represented by RS.

In recent years it has become known that there are important Fd's that contain three-iron clusters, which have the general Structure 31-III. The structure is a fragment of the four-iron cluster structure (31-IV) and the $[3Fe-4S]^{n+}$ unit can have oxidation states corresponding to $n = +1, 0,$ and -1.

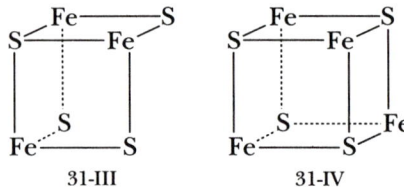

The four-iron Fd's, which contain $[4Fe-4S]^{n+}$ clusters appear to be more common than the two-iron or three-iron ones, and they have quite complex behavior. In biological systems they have three oxidation levels, giving the charges of +3, +2, or +1. In any given system, though, only one pair of these charge types is employed. For many of these the normally isolated substance contains a diamagnetic $[4Fe-4S]^{2+}$ cluster; this can be reversibly reduced at about -0.4 V (vs. the hydrogen electrode) to give $[4Fe-4S]^+$, which has one unpaired electron.

One particularly important class of four-iron Fd's are sometimes called *high-potential iron–sulfur proteins* (abbreviated HiPIP). Here the operative redox couple, at about $+0.75$ V is between the clusters $[4Fe-4S]^{3+}$ and $[4Fe-4S]^{2+}$. The redox behavior of both HiPIPs and other Fd's is summarized in Reaction 31-5.1,

$$[4Fe-4S]^{3+} \underset{-e^-}{\overset{+e^-}{\rightleftharpoons}} [4Fe-4S]^{2+} \underset{-e^-}{\overset{+e^-}{\rightleftharpoons}} [4Fe-4S]^{+} \qquad (31.5.1)$$

$$\begin{array}{ccc} S = \tfrac{1}{2} & S = 0 & S = \tfrac{1}{2} \text{ or } \tfrac{3}{2} \\ \sim +0.35 \text{ V} & \sim +0.40 \text{ V} & \\ \text{HiPIP couples} & \text{other Fd couples} & \end{array}$$

where the redox potentials are given in volts (V) against the standard hydrogen electrode.

Let us now emphasize a very important point, for which there is not yet a generally accepted explanation. Both HiPIP and the other Fd's are normally isolated with the $[4Fe-4S]^{2+}$ cluster. For the latter, a reversible, one electron reduction occurs at about -0.40 V, but reversible oxidation to the 3+ level has never been accomplished. Conversely, reversible one-electron oxidation readily occurs for HiPIPs, but reduction can be achieved only under forcing conditions having no relevance to the biological situation. Unquestionably, however, the $[4Fe-4S]^{2+}$ clusters in the two types of compounds are the same. What, then, causes the marked difference in their redox behavior?

Two hypotheses are under consideration. One focuses on the number of hydrogen bonds from surrounding protein NH groups to cysteinyl sulfur atoms. There appear to be about twice as many of these for the usual Fd than for a HiPIP; thus reduction (the introduction of negative charge) would be preferred for a usual Fd. A second hypothesis is that oxidation and reduction of the $[4Fe-4S]^{2+}$ cluster lead to different sorts of structural deformations, and that the protein conformations about the cluster in usual Fd's and HiPIPs differ so as to favor the reductively induced changes in the Fd case and the oxidatively induced ones in the HiPIP case. This is a fascinating question which will, no doubt, be resolved as better structural data are obtained and, perhaps, as model systems become better characterized.

The study of ferredoxin biochemistry provides a classic example of how inorganic chemists can use model systems to investigate complex biological processes. It has been possible to synthesize compounds containing $[Fe_4S_4(SR)_4]^{x-}$ anions that are very similar in many aspects (especially structural ones) to the $[4Fe-4S]^{n+}$ clusters that are bonded to the four cysteinyl sulfur atoms of the peptide chain. By treating ferredoxins with solutions of mercaptides, RS^-, it is even possible to extract the $[4Fe-4S]^{2+}$ clusters from the protein and capture them as $[Fe_4S_4(SR)_4]^{2-}$ anions.

31-6 Hemerythrin

In a number of marine worms, there exists a different solution to the oxygen carrying problem. Again, the active metal is iron, but the rest of the picture is quite different: no porphyrin ligand is involved, and two iron atoms are required to bind one molecule of O_2. The full details of how the active site of a hemerythrin actually works are still incomplete, but there is good evidence (not conclusive, however) that the process goes according to the scheme shown in Fig. 31-9.

The two-iron active site has the iron atoms connected by three bridging groups, two of which are carboxyl anions from the side chains of glutamic and aspartic acid residues. The other bridging ligand is either O^{2-} or OH^-, but probably OH^-. All of the remaining ligands (which complete an octahedron about one iron atom and a type of five-coordination about the other iron atom) are imidazole nitrogen atoms from histidine residues. The possibility that a sixth ligand (very weakly held) may be present at the second iron atom cannot be entirely ruled out.

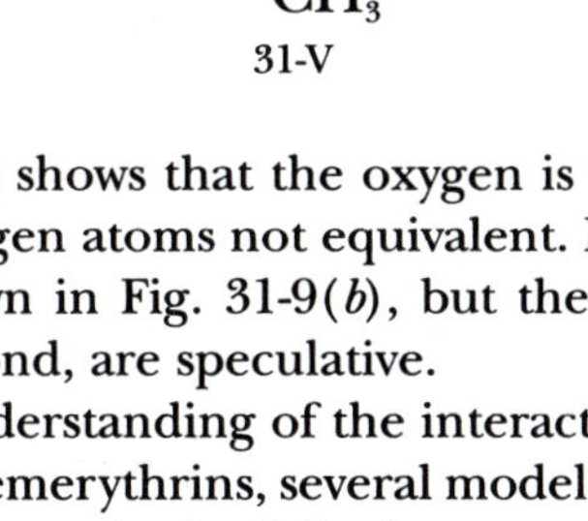

31-V

Spectroscopic evidence shows that the oxygen is definitely bound in a peroxo form, with the two oxygen atoms not equivalent. It is virtually certain that it occupies the position shown in Fig. 31-9(*b*), but the finer details, such as the OO—H · · · O hydrogen bond, are speculative.

To develop a better understanding of the interactions between the two iron atoms in the active site of hemerythrins, several model systems, of the type shown in structure 31-V, have been synthesized. In these models, the bridging carboxyl groups are derived from acetic acid, and the nitrogen atoms are supplied by tridentate triamines whose conformations cause them naturally to occupy three mutually *cis* positions.

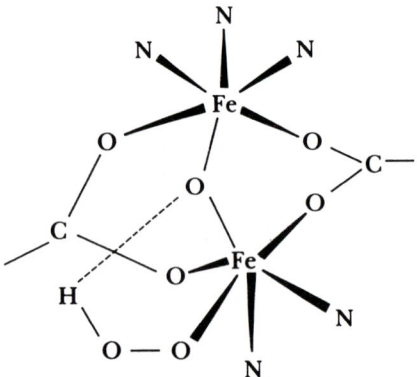

Figure 31-9 A possible description of the mode of oxygen binding by hemerythrin.

31-7 Iron Supply and Transport

Iron metabolism requires provision for storing and transporting iron. In humans and in many other higher animals the storage materials are *ferritin* and *hemosiderin*. These are present in liver, spleen, and bone marrow. Ferritin is a water soluble, crystalline substance consisting of a shell, or sheath, of protein surrounding a spherical core that contains the iron. The diameter of the core varies from 40 to 88 Å, and may contain up to 4500 iron atoms, having a composition closely approximating $(FeOOH)_8 \cdot FeO \cdot H_2PO_4$. The diffraction pattern of this core is similar to that of the substance ferrihydrite, $5Fe_2O_3 \cdot 9H_2O$, which is formed when NH_4OH is added slowly to a solution of ferric nitrate at 80–90 °C. The phosphate is not a part of the bulk structure of the core, but appears to play some role in covering the iron particles of the core and perhaps attaching them to each other and to the protein sheath. Up to 23% of the dry weight may be iron. The protein portion alone, called *apoferritin*, is stable, forms crystals suitable for X-ray diffraction, and has a molecular weight of about 45,000. *Hemosiderin* contains larger proportions of hydrous metal oxide, but is rather variable in composition and properties. It is poorly understood compared to ferritin.

The manner in which the iron enters and leaves ferritin is not well understood. The core can be formed only from aqueous iron(II), so that oxidation to give the correct proportion of iron(III) must accompany, or follow, incorporation in the core. Iron release is controlled by the protein sheath and can occur very rapidly when necessary.

Transferrin is a protein that binds iron(III) very strongly, and transports it from the stomach to the iron metabolic processes of the body. As iron passes from the stomach (which is acidic) into the blood (pH $\cong$ 7.4), it is oxidized to Fe^{III} in a process catalyzed by the copper metalloenzyme ceruloplasmin, after which it is picked up by transferrin molecules. These are proteins with a molecular weight of about 80,000, and they contain two similar but not identical sites that bind iron tightly but reversibly in the presence of certain anions such as CO_3^{2-} and HCO_3^-. The binding constant is approximately 10^{26}, making transferrin an extremely efficient scavenger of iron. Eventually transferrin becomes bound to the cell wall of an immature red cell, which utilizes the iron. Transferrin also carries iron to ferritin, the process of iron(II) transfer being a complex one requiring ATP and ascorbic acid.

In microorganisms, iron is transported by substances called *ferrichromes* and *ferrioxamines*. The former are trihydroxamic acids in which the three hydroxamate groups are on three side chains of a cyclic hexapeptide. The latter have the three hydroxamate groups as part of the peptide chain, which may be cyclic or acyclic. Typical structures are shown in Fig. 31-10.

The importance of these compounds derives from their exceptional ability to chelate iron(III) and then pass through cell membranes, thus carrying iron from inorganic sources, such as $Fe_2O_3 \cdot x\, H_2O$, to points of need in the cells.

31-8 The Bioinorganic Chemistry of Cobalt: Vitamin B₁₂

The best-known biological function of cobalt is its intimate involvement in the coenzymes related to vitamin B₁₂, the structure of which is shown in Fig. 31-11. This structure is not as overwhelming as it might seem at first glance. It consists of four principal components:

1. A cobalt atom.

2. A macrocyclic ligand called the *corrin* ring, which bears various substituents. The essential corrin ring system is shown in bold lines. It resembles the porphine ring, but differs in various ways, notably in the absence of one methine ($=$CH—) bridge between a pair of pyrrole rings.

3. A complex organic portion consisting of a phosphate group, a sugar, and an organic base, the latter being coordinated to the cobalt atom.

4. A sixth ligand may be coordinated to the cobalt atom. This ligand can be varied, and when the cobalt atom is reduced to the oxidation state +1, it is evidently absent.

The entire entity shown in Fig. 31-11, but neglecting the ligand X, is called cobalamin.

The term vitamin B₁₂ refers to cyanocobalamin, which has cobalt in the +3 oxidation state and CN^- as the ligand X. The cyanide ligand is introduced dur-

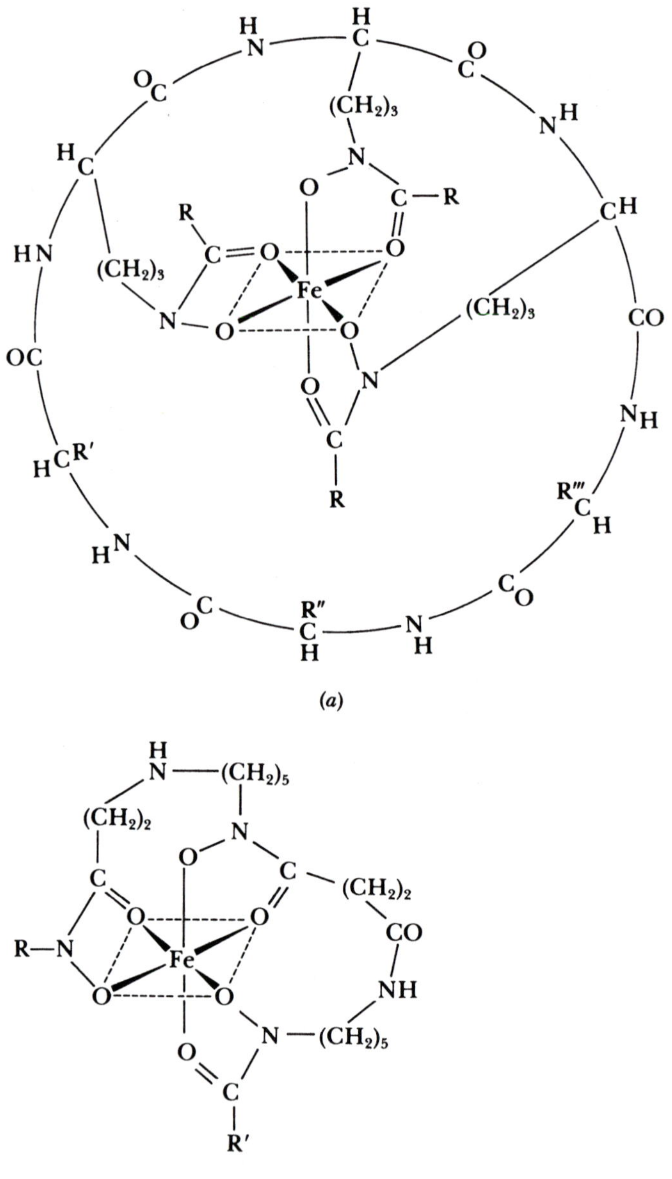

Figure 31-10 (*a*) A typical ferrichrome. (*b*) Typical structure of an acyclic ferrioxamine.

ing the isolation procedure and is not present in any active form of the vitamin. In the biological system, the ligand X is likely to be H_2O much of the time, but another possibility, which has been identified by actual isolation of the complex, is the 5′-deoxyadenosyl radical, as shown in Fig. 31-12. The particular coenzyme in which this is found was the first organometallic compound to be observed in a living system.

The B_{12} coenzymes act in concert with a number of enzymes, but the best studied systems involve the dioldehydrases, where reactions such as 31-8.1 are catalyzed.

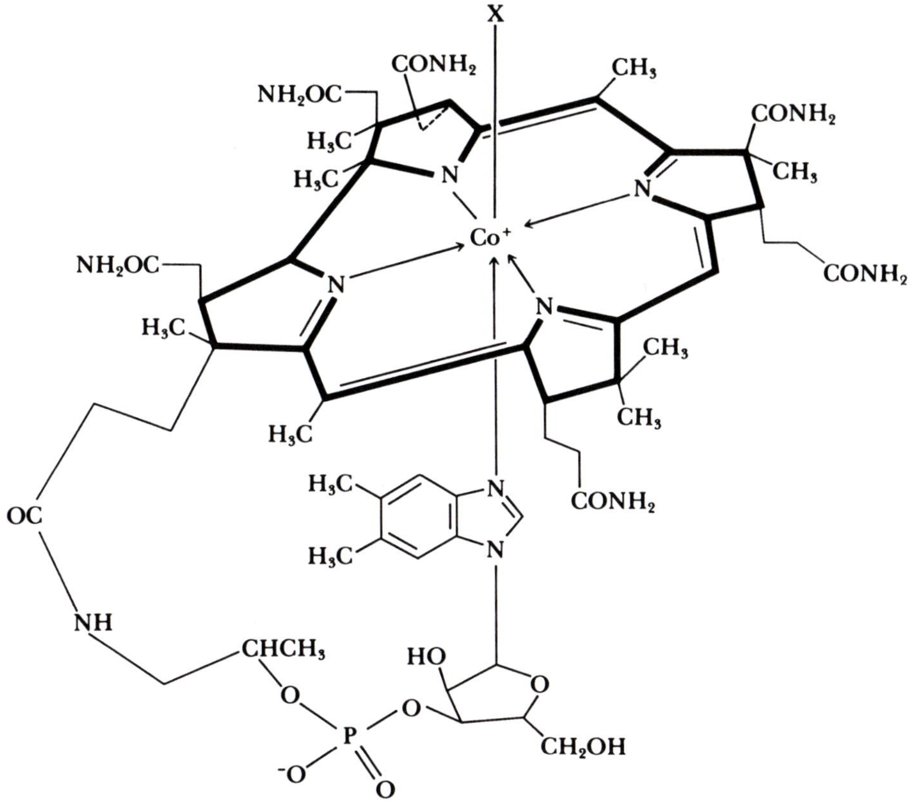

Figure 31-11 The structure of cobalamin. The corrin ring is shown in heavy lines.

$$RCHOHCH_2OH \longrightarrow RCH_2CHO + H_2O \qquad (R = CH_3 \text{ or } H) \quad (31\text{-}8.1)$$

From studies of the nonenzymic chemistry of B$_{12}$ coenzymes and of model systems noted below, a body of knowledge about fundamental B$_{12}$ chemistry has been built up. Some of this chemistry undoubtedly plays a role in its activities as a coenzyme. The cobalamins can be reduced in neutral or alkaline solution to give cobalt(II) and cobalt(I) species, often called B$_{12r}$ and B$_{12s}$, respectively. The latter is a powerful reducing agent, decomposing water to give hydrogen and B$_{12r}$. These reductions can apparently be carried out *in vivo* by reduced ferredoxin. When cyano- or hydroxocobalamin is reduced, the ligand (CN$^-$ or OH$^-$) is lost, and the resulting five-coordinate cobalt(I) species reacts with ATP in the presence of a suitable enzyme to generate the B$_{12}$ coenzyme.

In nonenzymic systems, rapid reaction of B$_{12s}$ occurs with alkyl halides, alkynes, and the like, as shown in Reactions 31-8.2 to 31-8.4, where [Cb] represents the cobalamin group. Methylcobalamin has an extensive chemistry, some of which is involved in the metabolism of methane-producing bacteria. It transfers CH$_3$ groups to Hg^{II}, Tl^{III}, Pt^{II}, and Au^{I}. It is, evidently, in this way that certain bacteria accomplish their unfortunate feat of converting relatively harmless elemental mercury, which collects in sea or lake bottoms, into the exceeding toxic methylmercury ion CH$_3$Hg$^+$.

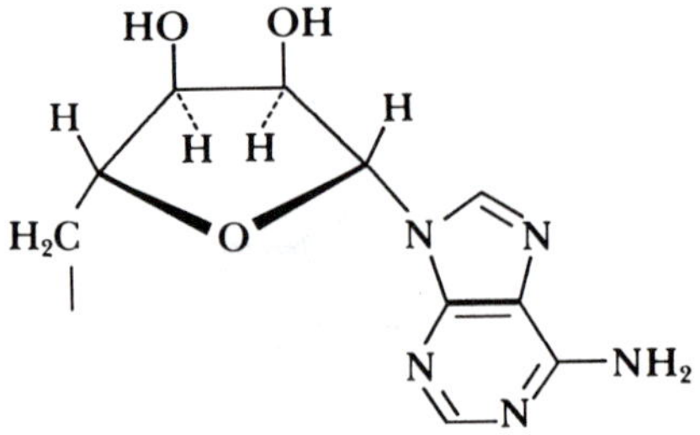

Figure 31-12   The 5'-deoxyadenosyl group that may constitute the ligand X in Fig. 31-11.

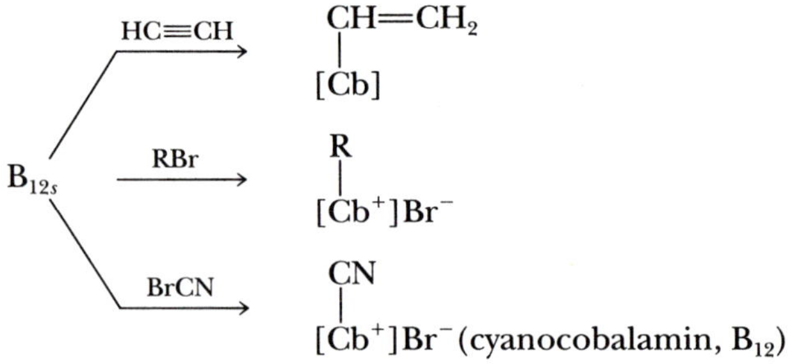

$$
B_{12s}
\begin{cases}
\xrightarrow{HC\equiv CH} &
\begin{array}{c} CH=CH_2 \\ | \\ [Cb] \end{array} & (31\text{-}8.2) \\[2ex]
\xrightarrow{RBr} &
\begin{array}{c} R \\ | \\ [Cb^+]Br^- \end{array} & (31\text{-}8.3) \\[2ex]
\xrightarrow{BrCN} &
\begin{array}{c} CN \\ | \\ [Cb^+]Br^- \,(\text{cyanocobalamin, } B_{12}) \end{array} & (31\text{-}8.4)
\end{cases}
$$

A number of models for vitamin B_{12} have been synthesized and studied. The best known are the bis(dimethylglyoximato) complexes, an example of which is shown in Fig. 31-13. This and other models have as their essential feature a planar tetradentate ligand with amido-type nitrogen atoms. Many of these quite successfully model the reducibility to the cobalt(I) state, as well as the formation and reactions of the key cobalt–carbon bonds.

It is interesting that cobalt porphyrins are not very good models for B_{12} since they cannot be reduced to the cobalt(I) state under conditions where vitamin B_{12s} is obtained. This inability of the porphyrin ligand to stabilize the cobalt(I) species may be a reason why the corrin ring system was evolved.

31-9   Metalloenzymes

Enzymes are large protein molecules so built that they can bind at least one reactant (called the substrate) and catalyze an important biochemical reaction. These compounds are extremely efficient as catalysts, typically causing rates to increase 10^6 times or more compared to the uncatalyzed rate. They are also usually highly specific, catalyzing only one, or a few reactions, rather than all those of a given class.

Some enzymes incorporate one or more metal atoms in their normal structure. The metal ion does not merely participate during the time that the enzyme–substrate complex exists, but is a permanent part of the enzyme. The metal atom, or at least one of the metal atoms when two or more are present, occurs at or very near to the active site (the locus of the bound, reacting substrate) and plays a role in the activity of the enzyme. Such enzymes are called *metalloenzymes,* and at least 100 have been identified.

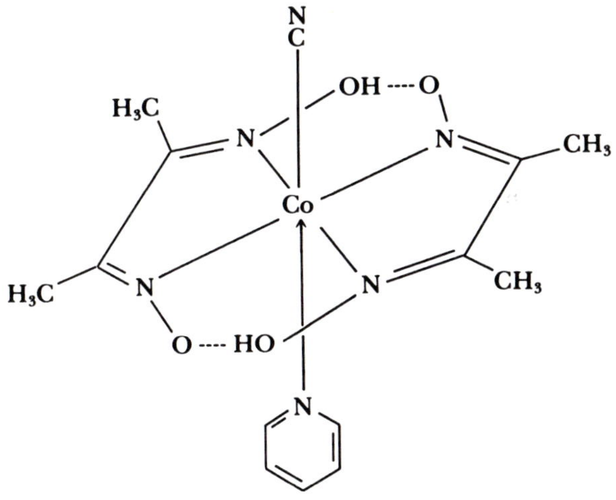

Figure 31-13 A cobaloxime, or bis(dimethylglyoximato)cobalt complex, which is a model for
cyanocobalamin, vitamin B_{12}.

The following metals are most often found in metalloenzymes, especially the
last three: Mo, Ca, Mn, Fe, Cu, and Zn. Although Co^{2+} can often be made to replace Zn^{2+} in zinc metalloenzymes, with retention or even enhancement of activity, the actual presence of Co^{2+} in the native enzymes is rare.

Zinc Metalloenzymes

No less than 30 zinc metalloenzymes are known. Two of the most important, or
at least best studied, are the following:

Carbonic anhydrase (MW = 30,000; 1 Zn):

This enzyme occurs in red blood cells and catalyzes the dehydration of the
bicarbonate ion and the hydration of CO_2 according to Reaction 31-9.1.

$$OH^- + CO_2 = HCO_3^- \tag{31-9.1}$$

These reactions would otherwise proceed too slowly to be compatible with physiological requirements.

Carboxypeptidase (MW = 34,300; 1 Zn):

This enzyme in the pancreas of mammals catalyzes the hydrolysis of the peptide bound at the carboxyl end of a peptide chain, as in Reaction 31-9.2.

$$-R''CH-C(O)NH-CHR'-C(O)NH-CHRCO_2^- + H_2O \longrightarrow$$
$$-R''CH-C(O)NH-CHR'-CO_2^- + H_3N^+CHRCO_2^- \tag{31-9.2}$$

The enzyme has a particular preference for substrates in which the side chain R
is aromatic, that is, $-CH_2C_6H_5$ or $-CH_2C_6H_4OH$.

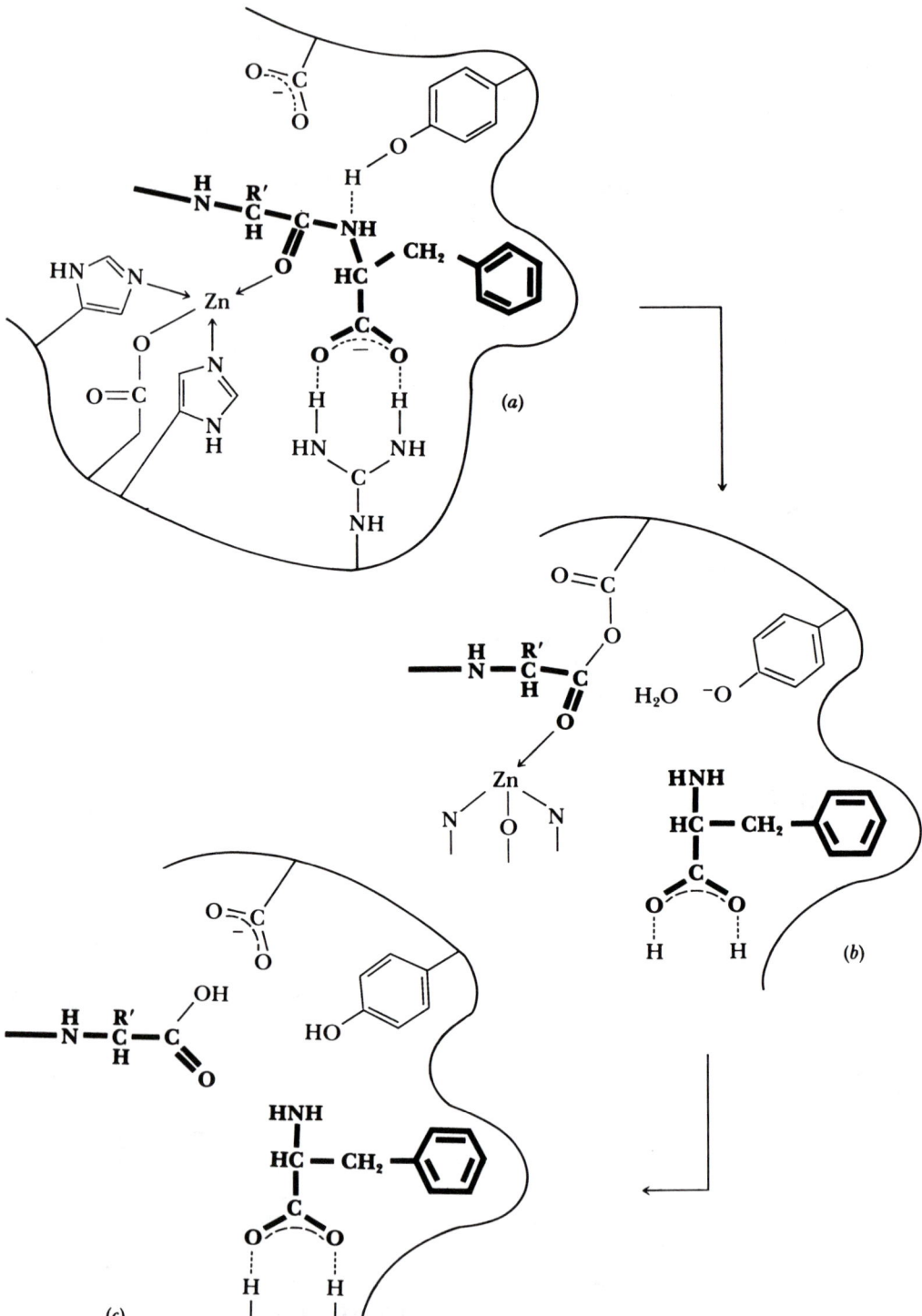

Figure 31-14 (*a*) A proposed mode of binding of the substrate in carboxypeptidase. The substrate is shown in heavy type and lines. The curved line schematically defines the "surface" of the enzyme molecule. (*b*) A possible first step in the mechanism, wherein a carboxyl side chain attacks the carbonyl carbon atom, forming an anhydride. (*c*) Subsequent steps in the proposed mechanism, including hydrolysis of the intermediate anhydride and dissociation of the products from the active site.

The structure and main mechanistic features of carboxypeptidase have been elucidated. The zinc ion is bound in a distorted tetrahedral environment, with two histidine nitrogen atoms, one glutamate carboxyl oxygen atom and a water molecule as ligands. The binding of the substrate probably occurs as shown in Fig. 31-14(a). Notice that the carbonyl oxygen atom of the peptide linkage that is to be broken has replaced the water molecule in the coordination sphere of the zinc ion.

The key step in a possible, but speculative, mechanism is shown in Fig. 31-14(b). Once the peptide bond has been broken with formation of the acid anhydride, rapid hydrolysis of the anhydride would occur, as in Fig. 31-14(c). The products would then vacate the active site, leaving it ready to bind another molecule of substrate and repeat the cycle.

Copper Metalloenzymes

More than 20 of these have been isolated, but in no case is structure or function well understood. The copper enzymes are mostly oxidases, that is, enzymes that catalyze oxidations. Examples are (1) *Ascorbic acid oxidase* (MW = 140,000; 8 Cu), which is widely distributed in plants and microorganisms. It catalyzes oxidation of ascorbic acid (vitamin C) to dehydroascorbic acid. (2) *Cytochrome oxidase,* the terminal electron acceptor in the oxidative pathway of cell mitochondria. This enzyme also contains heme. (3) Various *tyrosinases,* which catalyze the formation of pigments (melanins) in a host of plants and animals.

In many lower animals, such as crabs and snails, the oxygen-carrying molecule is a copper-containing protein *hemocyanin,* which despite the name, contains no heme group. The hemocyanins represent the third system in Nature (besides hemoglobins and hemerythrins) for oxygen carrying from the point of intake to those tissues where O_2 is required. Like hemoglobin, hemocyanins have many subunits in the complete molecule and, therefore, exhibit cooperativity in O_2 binding. The active sites consist of two copper atoms (~ 3.8 Å apart) that jointly bind one O_2 molecule. The way they do this apparently involves the conversion of the colorless Cu^I . . . Cu^I deoxy center to a peroxide-bridged Cu^{II}—O—O—Cu^{II}, which is bright blue.

31-10 Nitrogen Fixation

Elemental nitrogen (N_2) is relatively unreactive. In order to "fix" nitrogen, that is, make nitrogen react with other substances to produce nitrogen compounds, it is generally necessary to use energy-rich conditions. High temperatures or electrical discharges can supply the necessary activation energy. However, primitive bacteria and some blue-green algae can fix nitrogen under mild conditions, that is, ambient temperature and pressure. Metalloenzymes play a key role in this process.

Bacterial Nitrogenase Systems

Our more detailed information about nitrogen fixation comes mainly from studies of free-living soil bacteria. These can be cultured in the laboratory and essential components can be isolated and purified. Biological nitrogen fixation is a reductive process. An important fact, which was established by using $^{15}N_2$, is that the first recognizable product is always NH_3. Apparently, all intermediates remain bound to the enzyme system.

It has been known since 1930 that molybdenum is essential for bacterial nitrogen fixation, since this function can be turned off and on by removing and then restoring molybdenum to the environment. Magnesium and iron are also essential components.

In 1960, the first active cell-free extracts were prepared, and since then, *nitrogenases,* as the enzymes are called, have been obtained in fairly pure condition from several bacteria. In each case the nitrogenase can be separated into two proteins, one with molecular weight of about 260,000 (the Fe–protein) and the other around 240,000 (the MoFe protein). Neither of these proteins is separately active, but on mixing them activity is obtained immediately. The Fe–protein consists of two identical subunits that clasp a ferredoxin unit (Fe_4S_4) between them by forming Fe—S bonds to two cysteine residues in each subunit. It is believed that the Fe–protein plays its role by coupling electron transfer and hydrolysis of ATP, but that the actual conversion of N_2 to NH_3 is carried out at the active site of the larger protein, the MoFe–protein, so-called because it contains both molybdenum and iron.

Until very recently, there has been no direct indication of how the iron and molybdenum atoms are arranged in the MoFe protein, nor did we have any completely reliable knowledge of exactly how many of each type of metal atom is present. However, in late 1992 an X-ray crystallographic study revealed a metal cluster arrangement, as shown in Fig. 31-15.

This structure is still somewhat inaccurate and one of the bridging groups (Y) has not yet been conclusively identified. Overall, the structure has had an enormous impact. Previously, it had been correctly assumed that some sort of mixed iron–molybdenum–sulfur species was present, but it was also assumed that the molybdenum atom was the seat of reactivity, that is, the atom to which N_2 would first become attached and then reduced. In view of the apparent coordinative saturation of the Mo atom and the possibility that the middle part of the cluster, where the two halves are joined by the μ-S, μ-S, and μ-Y linkages, might be capable of accepting the N_2 molecule and retaining the various intermediates, the mechanism of action might be quite different from what was previously imagined.

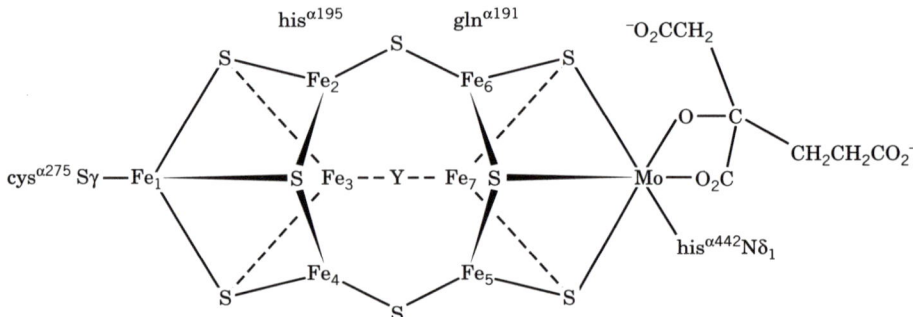

Fig. 31-15   The Fe_7Mo–sulfur cluster system, and its immediate surroundings, found in the MoFe–protein of nitrogenase.

STUDY GUIDE

Scope and Purpose

We have sketched some important inorganic aspects of the chemistry of life. This has been an area of great recent interest among researchers, and new understandings develop so frequently, that the reader should expect to consult recent journal articles for more up-to-date information. Continued study in the references provided under "Supplementary Reading" is highly encouraged.

This chapter's major message is that the chemistry of life involves more than 20 elements besides those traditionally treated in organic chemistry. Though these other elements tend to have limited roles, life processes require them just as surely as they require proteins, carbohydrates, and lipids.

Study Questions

A. Review

1. Name four transition metals and two nontransition metals that play important roles in biological processes.

2. Draw the structure of porphine and explain how the structures of heme and chlorophyll are related to it.

3. What role does the magnesium ion play in the functioning of chlorophyll?

4. What constitutes a heme protein? Name three of them.

5. What are the functions of hemoglobin and myoglobin? What are the principal similarities in their structures?

6. What changes occur in the heme groups of hemoglobin on going from deoxy- to oxy-hemoglobin?

7. What is the structure of the redox center of HiPIP and of the 4-Fe and 8-Fe ferredoxins?

8. What functions do ferrichromes and ferrioxamines have? What are their chief chemical features?

9. State the main components of cobalamin. How do B_{12}, B_{12r}, and B_{12s} differ?

10. What role does the zinc ion play in the action of carboxypeptidase?

11. What is the principal function of nitrogenase?

12. List the ways in which the cobaloximes resemble cobalamin.

C. Questions from the Literature of Inorganic Chemistry

1. Consider the paper by J. Halpern, "Mechanisms of Coenzyme B_{12}-Dependent Rearrangements," *Science*, **1985**, *227*, 869.

    (a) What is the significance of the observation that reactions involving the coenzyme B_{12} give scrambling of the methylene hydrogens from the 5'-deoxyadenosine of the coenzyme with the hydrogen atom involved in the migration [e.g., Eq. (1)] at the substrate?

    (b) Through what various spin states does the cobalt atom of the coenzyme B_{12} progress during the operation of the mechanism shown in Fig. 2 of this article? What is the difference in the number of *d* electrons on B_{12} and B_{12r}?

    (c) What factors are said to influence the critical cobalt–carbon bond dissociation energies?

(d) What features do the "DH" and the "saloph" cobalt complex model systems have in common with the coenzyme B_{12}?

(e) What analogy does the author draw between the reversible cobalt–carbon bond dissociation of coenzyme B_{12} and the reversible binding of dioxygen as in Eqs. 23 and 24?

2. Consider the extensive work by J. P. Collman and students, represented by the following paper, and the references therein: J. P. Collman, J. I. Brauman, B. I. Iverson, J. L. Sessler, R. M. Morris, and Q. H. Gibson, *J. Am. Chem. Soc.*, **1983,** *105,* 3052.

(a) What are the main similarities and differences, structurally, between the "picket fence" and "pocket" porphyrins that are described in this article?

(b) How is solvation thought to reduce affinities for O_2 of the unprotected iron(II) porphyrins?

(c) What advantages in O_2 binding do the "picket fence" and "pocket" porphyrins have over those iron(II) porphyrins that are "unprotected" from solvation effects?

(d) How do the O_2 and CO affinities of the "picket fence" porphyrins compare with those of the "pocket" porphyrins?

(e) What geometries for M—O_2 and M—CO groups seem to make sense in explaining the observations in (d)?

3. Consider the work by J. Chatt on nitrogen fixation analogs: J. Chatt, A. J. Pearman, and R. L. Richards, *J. Chem. Soc. Dalton Trans.*, **1977,** 1852.

(a) The N_2 complexes reported here are protonated to give ammonia. How is this reaction of interest to the molybdenum nitrogenase systems?

(b) In other studies mentioned in the introduction to this paper, other complexes were protonated to give not ammonia, but intermediate reduction products. Enumerate the findings concerning the formation of diazenido, diazine, and hydrazido ligands.

(c) What is the difference between protonation of the N_2 ligand in complexes containing two bidentate dppe ligands and protonation of N_2 ligand in complexes containing four monodentate $P(CH_3)_2C_6H_5$ ligands? What bonding arguments do the authors present to account for these differences?

(d) At what stage do the authors propose a splitting of the N—N bond? When is this likely to occur in the overall stepwise process that is proposed?

(e) How is the oxidation state of the metal at the end of reaction sequence (5) different from the oxidation state that is likely in the enzymic system? How do the authors propose that the enzyme avoids this high an oxidation state?

SUPPLEMENTARY READING

Bertini, I., Gray, H. B., Lippard, S. J., and Valentine, J. S., Eds., *Bioinorganic Chemistry,* University Science Books, Mill Valley, CA, 1994.

Brill, A. S., *Transition Metals in Biochemistry,* Springer-Verlag, Berlin, 1977.

Chatt, J., Dilworth, J. R., and Richards, R. L., "Recent Advances in the Chemistry of Nitrogen Fixation," *Chem. Rev.,* **1978,** *78,* 589.

da Silva, J. R. F. and Williams, R. J. P, *The Biological Chemistry of the Elements—The Inorganic Chemistry of Life,* Clarendon Press, Oxford, 1991.

Dickerson, R. E. and Geis, I., *The Structure and Action of Proteins,* Harper & Row, New York, 1969.

Dickerson, R. E. and Geis, I., *Hemoglobin: Structure, Function, Evolution, and Pathology,* Benjamin-Cummings, Menlo Park, CA, 1983.

Eichhorn, G. L. and Marzilli, L. G., *Advances in Inorganic Biochemistry,* Vols. 1–6, Elsevier, New York.

Harrison, P. M., Ed., *Metalloproteins,* Parts 1 and 2, Macmillan, New York, 1985.

Henderson, R. A., Leigh, G. J., and Pickett, C. J., "The Chemistry of Nitrogen Fixation and Models for the Reactions of Nitrogenase," *Adv. Inorg. Chem. Radiochem.,* **1983,** *27,* 197.

Hughes, M. N., *The Inorganic Chemistry of Biological Processes,* Wiley-Interscience, New York, 1981.

Lippard, S. J., Ed., "Bioinorganic Chemistry," a special issue of *Progress in Inorganic Chemistry,* Vol. 38, Wiley-Interscience, New York, 1990.

Lippard, S. J. and Berg, J. M., "Principles of Bioinorganic Chemistry," University Science Books, Mill Valley, CA, 1994.

McMillin, D. R., Ed., "Bioinorganic Chemistry—The State of the Art," *J. Chem. Educ.,* **1985,** *62,* 916–1011. An excellent series of articles.

Niederhoffer, E. C., Timmons, J. H., and Martell, A. E., "Thermodynamics of Oxygen Binding in Natural and Synthetic Dioxygen Complexes," *Chem. Rev.,* **1984,** *84,* 137–203.

Ochiai, E. I., *Bioinorganic Chemistry,* Allyn and Bacon, Boston, 1977.

Peisach, J., Aisen, P., and Blumberg, W. E., Eds., *The Biochemistry of Copper,* Academic, New York, 1966.

Postgate, B., Ed., *The Chemistry and Biochemistry of Nitrogen Fixation,* Plenum, New York, 1971.

Pratt, J. M., "The B_{12}-Dependent Isomerase Enzymes; How the Protein Controls the Active Site," *Chem. Soc. Rev.,* **1985,** *14,* 161.

Siegel, H. and Sigel, A., Eds., *Metal Ions in Biological Systems,* Vols. 1–27, Marcel-Dekker, New York.

Stiefel, E. I. and Cramer, S. P., "Chemistry and Biology of the Iron–Molybdenum Cofactor of Nitrogenase," in *Molybdenum Enzymes,* T. G. Spiro, Ed., Wiley-Interscience, New York, 1985.

Stiefel, E. I., Coucouvanis, D., and Newton, W. E., Eds., *Molybdenum Enzymes, Cofactors, and Model Systems,* ACS Symposium Series, American Chemical Society, Washington, DC, 1994.

Chapter 32

THE INORGANIC
SOLID STATE

32-1 Introduction

Why do we need a special chapter on the subject of solid substances? Because many solid substances differ *fundamentally* from gases and most liquids. Gases and most liquids consist of molecules (or atoms in the case of the noble gases). In gases the molecules are practically independent and the properties of a gas (except at extremely high pressures) are predictable from the properties of its constituent molecules. For many liquids this is also at least approximately true, although the closer contacts between molecules in a liquid do introduce additional factors. There are, of course, some liquids, for example molten salts or strongly hydrogen-bonded liquids such as water, that cannot be treated simply as collections of loosely interacting molecules. Finally, it is quite true that many solids consist exclusively of ordered (crystalline) arrays of molecules that make only van der Waals contacts with each other. These *molecular solids* pose no special problems. Indeed, they are in some ways easier to understand than molecular liquids because of the long-range order displayed.

The special properties we need to deal with here are those found in solids that do *not* consist simply of ordered arrays of small molecules, loosely touching each other. Instead, when the constituents of the solid are atoms or ions that make very strong contacts, either ionic or covalent, and have extended interactions, the properties can only be understood as properties of a large array as a whole. We have already discussed in Chapter 4 certain aspects of one such class, the *ionic solids*. There is, however, more to say about them and there are other major types of nonmolecular solids with properties of both theoretical and practical importance. These other materials include polymers, metals, alloys, and infinite covalent materials such as silicon, graphite, or ceramics.

While it is true that inorganic solids have many useful properties of a mechanical nature (think, for example, of portland cement or tungsten carbide), the more interesting important properties of solids are electrical, magnetic, and optical. The purpose of this chapter is to survey this vast area from the point of view of the chemist, and particularly the inorganic chemist. Indeed, most solid state chemistry is inorganic chemistry because organic solids are nearly all molecular. Organic polymers, of course, are an exception to this generalization.

32-2 Preparation of Inorganic Nonmolecular Solids

Molecular compounds are mostly prepared by a reaction in solution, with the solid product precipitating, either immediately, after cooling, or by evaporation of the solvent. Such preparations are thus carried out, generally, under mild thermal conditions, limited by the boiling point of the solvent.

Of course, a number of nonmolecular solids, for example, oxides and sulfides of the transition metals, can also be prepared by precipitating them from a solvent—usually water. However, a large number of the most interesting inorganic compounds in the solid state, and especially in crystalline form, are prepared by high-temperature (>500 °C) reactions.

One of the simplest, but most widely used, procedures (often dismissively called "shake and bake") involves intimately mixing two or more finely powdered starting materials, placing the mixture in a sealed inert container, and heating the entire container in an oven. The shortcomings of this approach are that it is often difficult to predict the stoichiometry of the product(s) and homogeneity is often difficult to achieve. Of course, once an interesting product has been identified, it is usually possible to prepare it efficiently by mixing components in the exact corresponding proportions. Homogeneity can be increased by grinding the product of one reaction step and repeating the heating process.

Many important and useful solids are made by the "shake and bake" technique. The recently discovered *high-temperature superconductors* provide excellent examples. A typical material of this class is made by the following reaction, in which it is also necessary to control the partial pressure of oxygen.

$$Y_2(CO_3)_3 + 4\,BaCO_3 + 6\,CuCO_3 + \tfrac{1}{2}\,O_2 \longrightarrow 2\,YBa_2Cu_3O_7 + 13\,CO_2 \qquad (32\text{-}2.1)$$

An intimate mixture of carbonates is made by coprecipitating them from an aqueous solution of the three cations, Y^{3+}, Ba^{2+}, and Cu^{2+}.

In many cases the formation of a product can be expedited by using a *flux*. A flux is a substance that does not participate in the net reaction, but "lubricates" the process by increasing mobilities of the reactants. Shorter times, lower temperatures, and greater homogeneity can thus result. Sometimes traces of water have this effect, while in other cases substantial amounts of the flux (which is then in a sense a solvent) are used. For example, a "shake and bake" preparation of $LiFe_5O_8$ by Reaction 32-2.2 requires much regrinding and refiring, but the addition of a eutectic mixture of Li_2SO_4/Na_2SO_4 leads to a smooth, one-step reaction at about 800 °C.

$$Li_2CO_3(s) + 5\,Fe_2O_3(s) \longrightarrow 2\,LiFe_5O_8(s) + CO_2 \qquad (32\text{-}2.2)$$

Other techniques that are important in synthesizing solid, nonmolecular inorganic compounds include *hydrothermal synthesis* and *vapor-phase transport*. The former often employs supercritical water, that is, water contained in a closed, high-pressure reactor and heated above its triple point temperature (>373 °C). One of the most important applications of hydrothermal synthesis is the manufacture of zeolitic aluminosilicates, although in many of these processes the temperatures are not supercritical. For example, the important zeolite mordenite is made by first preparing a precipitated gel formed from sodium aluminate, sodium carbonate, and silicic acid, with the desired ratio of Al to Si (about 1:5). This gel is then heated with water in a closed autoclave at a temperature of about 300 °C. On cooling, crystalline, hydrated mordenite, $Na_2O{\cdot}Al_2O_3{\cdot}10SiO_2{\cdot}6H_2O$, is obtained. The water can be driven off to give the anhydrous material that is employed industrially.

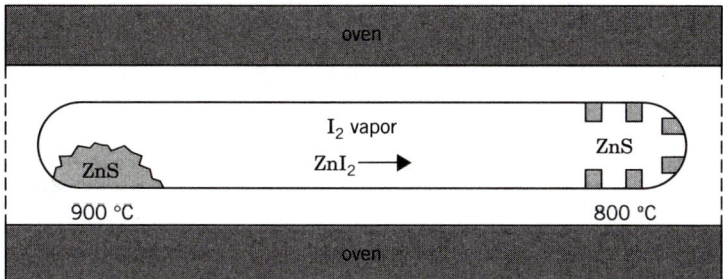

Figure 32-1 An example of vapor phase transport to produce a crystalline solid. Amorphous ZnS is transported as $ZnI_2 + S$ to redeposit ZnS in crystalline form.

Vapor-phase transport is a very useful technique for converting an amorphous solid to a crystalline one. A classic example is provided by Reaction (32-2.3)

$$ZnS(s) + I_2 = ZnI_2 + \tfrac{1}{2}\, S_2 \qquad\qquad (32\text{-}2.3)$$

Amorphous ZnS and I_2 are placed in a sealed tube and the end of the tube containing the ZnS is placed in a zone of the furnace where the temperature is 900 °C, while the other end of the tube is in a zone where the temperature, ramped down along the length of the tube, is 800 °C (Fig. 32-1). Because of the above reaction the original ZnS is transported as $ZnI_2 + \tfrac{1}{2}\, S_2$ in the vapor phase to the cooler end of the tube where crystalline ZnS is deposited.

32-3 Bonding in Infinite Arrays

In Chapter 3 we have seen how the increasing overlap of two orbitals, one on each of two atoms, as these atoms approach each other closely, gives rise to a bonding molecular orbital (MO) and an antibonding MO. We have also seen that if we take three atoms we obtain three MO's: one bonding, one antibonding, and a third one, lying between these in energy, that is approximately nonbonding. These two cases are simply the beginning of a potentially infinite series in which the linear chain of atoms becomes ever longer. Let us suppose in these first two cases all atoms are H atoms, each of which has only a $1s$ orbital to use for bonding, and that the chain of three, as well as all longer chains, are linear and have uniform spacing. When there are four atoms in the chain we will obtain four MO's; the results for H_2, H_3, and H_4 are shown in Fig. 32-2. *In general,* a chain of n atoms will give n MO's. The most stable one will always be the one in which all $1s$ orbitals have the same sign, since this gives the greatest total positive overlap. The least stable MO will always be the one in which the signs alternate, since this gives the greatest total negative overlap. As the chain lengthens, these greatest + and − values increase slightly because greater numbers of second, third, and so on, nearest neighbor contributions must be counted. However, since these longer range interactions are small and decay very rapidly with distances, the bottom and top energies approach asymptotic limits. It should be obvious that the distribution of energy levels as n increases will be as shown in Fig. 32-3. When n becomes very large, the large number of orbitals

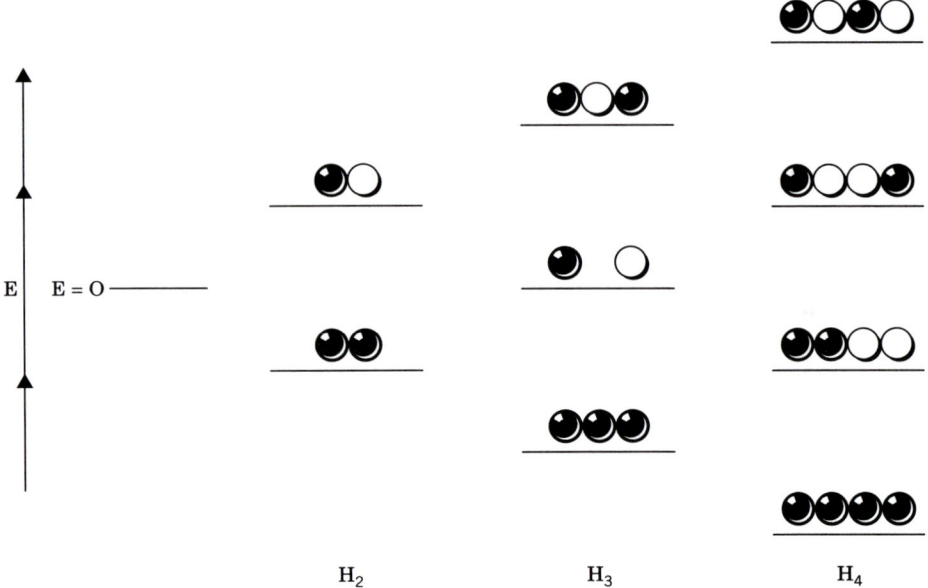

Figure 32-2   The molecular orbitals formed by linear chains of H atoms (H_2, H_3, and H_4).

packed between the upper and lower limits get so close together that they merge in the limit of $n \to \infty$. We then refer to them collectively as a *band*. This is illustrated in Fig. 32-3.

The case of a very long H_n chain is the simplest possible example of energy band formation: there is only one dimension and only one type of band (based on one type of atomic orbital). To deal with real solids, this simple model must be generalized in two ways: (1) Because, generally, atoms have more than one kind of valence shell orbital, there will be more than one sort of band, each with its own width and energy. (2) The one-dimensional picture must be developed into a three-dimensional picture.

With regard to the first point, the alkali metals illustrate the first step that may be taken to include more than one type of band, because here the valence shells include both ns and np orbitals, which have different energies. If the s–p energy gap is sufficiently large, the result will be a separation between the s and the p band, but if not, the two bands may overlap. The two possibilities are shown schematically in Fig. 32-4. In fact, the bands do overlap in metallic sodium.

With regard to developing a band picture in three dimensions, this would require a mathematical derivation beyond the scope of this book. For our purposes, an explicit development is not really necessary. The general idea that what we have just examined in a 1D structure will also happen in a 3D structure is sufficient.

Incompletely Filled Bands: The Fermi Level

Just as in the filling of discrete orbitals in individual atoms, electrons occupy bands from the bottom up. If the solid is at the absolute zero, there will be a sharp cutoff when all the available electrons have been added. The highest occupied level is called the *Fermi level*. This is illustrated in Fig. 32-5(*a*). What is

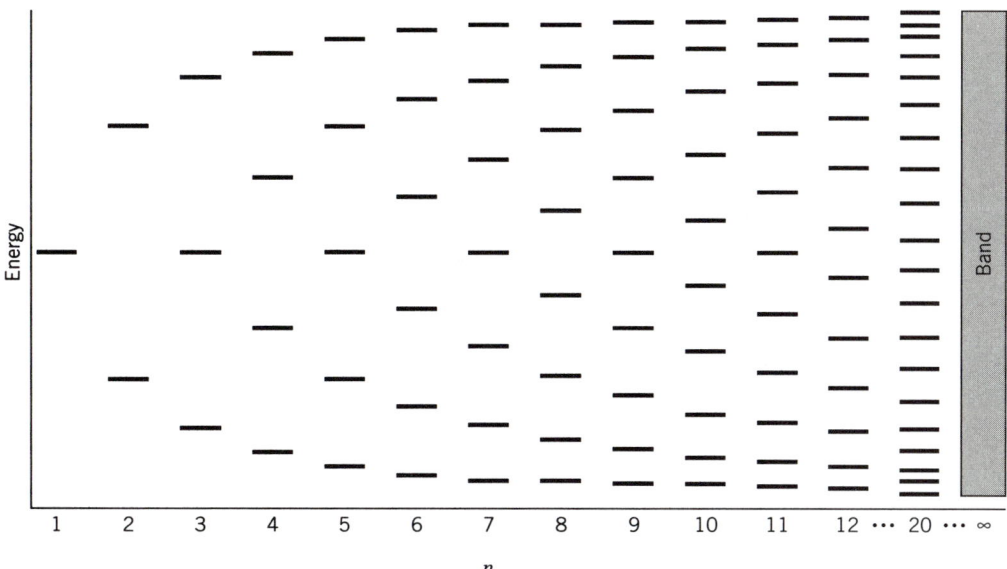

Figure 32-3   A diagram showing how the distribution of energy levels in linear chains (H_n) develops as $n \to \infty$.

shown in this figure corresponds to the situation we would expect for the *s* band in hypothetical metallic hydrogen. Since there is only one electron per atom, only one-half of the band can be filled. At temperatures above 0 K there will be a blurring of the electron energy distribution about the Fermi level, as shown in Fig. 32-5(b).

More Realistic Bands: Density of States; Band Gaps

So far, in our illustrations bands have been depicted in an extremely simple way. For real energy bands the capacity to hold electrons is not uniform from top to bottom. This is indicated in realistic diagrams by employing the horizontal axis as a measure of the *density of states* (DOS). By this we mean the number of energy levels per unit of energy. Typically, for a 3D band formed from only one type of atomic orbital on each metal atom, the density of states is greatest at the center. Thus, a more realistic diagram for hypothetical metallic hydrogen might be that shown in Fig. 32-6. When bands are completely nonoverlapping, as shown, there is an energy range in which the density of states is zero. This energy between the highest energy of one band and the lowest energy of the next is called a *band gap*.

In most real substances, there are so many different kinds of orbitals that can overlap to form bands, that the energy band diagram becomes very elaborate, with many peaks and valleys in the DOS profile. A representative example is shown in Fig. 32-7.

Metallic Conduction

The situation we have just seen for hypothetical metallic hydrogen is characteristic of that for metals in general. The highest occupied energy band is not fully occupied and, hence, electrons are free to flow when a potential difference is applied.

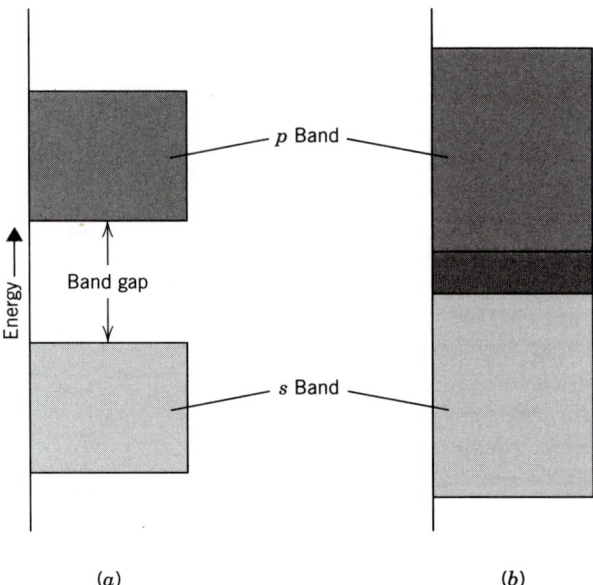

Figure 32-4   Two possibilities for the *s* and *p* bands of an alkali metal: (*a*) narrow separated bands (with a band gap) and (*b*) wide overlapping bands.

It is a well-known characteristic of metals, however, that their electrical conductance decreases as the temperature increases. Indeed, this type of temperature dependence of the electrical conductance is often taken as the major defining experimental criterion of a metal. Why should metals behave in this way? Naively, one might have guessed that since the electrons would become more mobile when thermal excitation is increased, increasing the temperature would cause greater conductance. There is, however, a much more important countervailing effect. The ability of an electron to move through a solid in a partially filled band depends on the uniformity of the structure from which the band arises. In a perfectly ordered structure in which the atoms did not vibrate about

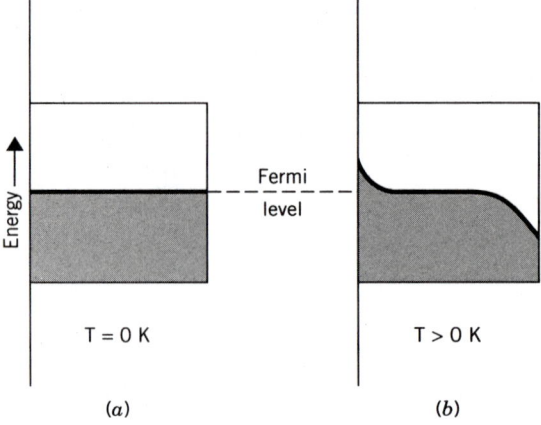

Figure 32-5   The Fermi level in a half-filled band at (*a*) the absolute zero and (*b*) at a higher temperature.

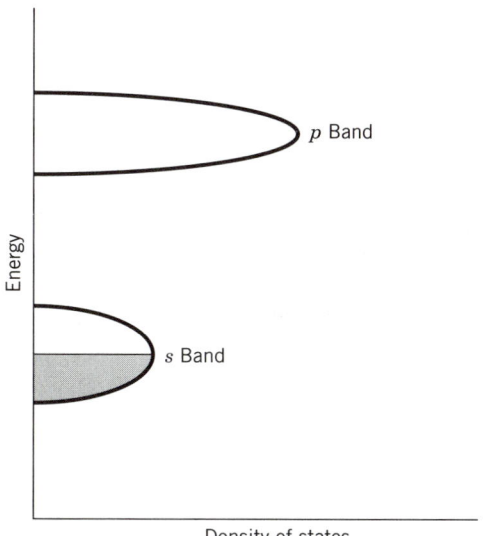

Figure 32-6   A Diagram of the *s* and *p* bands of hypothetical metallic hydrogen showing a variation in the density of states over the band energies.

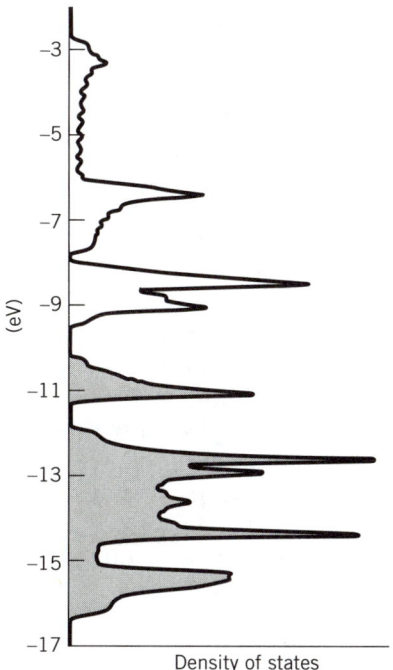

Figure 32-7   An example of the calculated band structure for a real substance (MoS_2). Filled bands are shaded. Note that there is only a small band gap, making MoS_2 an intrinsic semiconductor.

their mean positions, the maximum mobility would occur. Actually, the atoms vibrate about their mean positions even at 0 K, but as the temperature rises these vibrations become more and more violent and this disrupts the band structure. The net result is that it becomes harder for electrons to move as the temperature rises.

Nonmetals: Insulators and Semiconductors

Any substance in which there are only filled and empty bands, with a large energy gap between the highest filled and the lowest unfilled band will be a nonconductor of electricity, that is, an *insulator*. Application of a potential difference to a filled band does not cause net electron flow and with a large band gap no significant number of electrons can be thermally excited at ordinary temperatures.

Insulators are not usually discussed in terms of band theory because the presence of a neatly filled band and a large band gap is equivalent to a more familiar picture of the electronic structure. The more familiar picture is either that pertaining to an ionic solid, such as NaCl, or a localized covalent bond description as applied to substances such as diamond, silicon, or B_2O_3. In NaCl, for example, the components each have closed shells and their valence shell orbitals are very different in energy. Because like ions are well separated from each other there is little tendency to form bands in the first place, and, to the extent that they do form, we expect that there will be a lower chlorine band that is completely filled and that the lowest empty sodium band will lie above it by many electron volts.

In a covalent extended solid such as silicon (with the diamond structure), the complete set of doubly occupied bonding orbitals corresponds to a filled band and the lowest empty band, corresponding to the set of Si—Si antibonding orbitals, is again so far above it as to be thermally inaccessible.

A *semiconductor* is a substance that has an electrical conductance that is small compared to those of metals, but which increases with increasing temperature; that is, it has the opposite temperature dependence to that of metals. There are two types of semiconductors: *intrinsic* and *extrinsic*.

An intrinsic semiconductor is a pure material that resembles an insulator except that the band gap is sufficiently small that at normal temperatures a significant number of electrons are thermally excited from the filled to the empty band, as shown in Fig. 32-8(a). Because the magnitude of the conductance depends on how many electrons have enough thermal energy to cross the band gap, the temperature dependence of the conductance follows an exponential law, in the same way as the rate of a chemical reaction, where thermal energy is required to get some molecules over an energy barrier (there called an activation energy). Thus, the conductance σ of a semiconductor obeys an equation of the form:

$$\sigma = Ce^{-E/kt} \tag{32-3.1}$$

Here C is a constant characteristic of the material and the energy E can be shown to be approximately equal to one-half of the band gap.

Extrinsic semiconductors are actually far more important than intrinsic ones. Very few pure substances have suitable band gaps, but, by suitably doping a pure substance, that is, introducing very low levels of a suitable impurity, semicon-

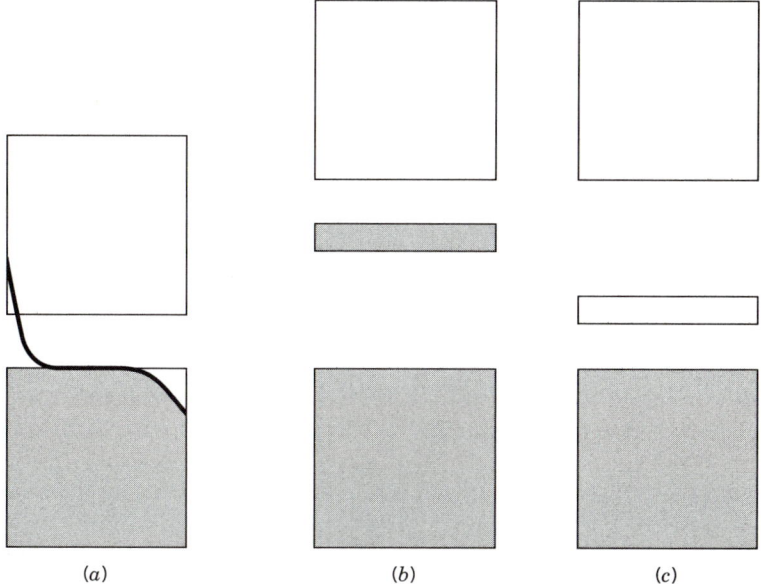

Figure 32-8   Energy bands for (*a*) an intrinsic semiconductor, (*b*) an
n-type, doped, extrinsic semiconductor, and (*c*) a **p**-type, doped, extrinsic semiconductor.

ductance can be engendered in an otherwise poorly semiconducting or insulating material. These doped materials are called *extrinsic* semiconductors. Classic examples of extrinsic semiconductors are silicon or germanium doped with gallium or arsenic.

Let us consider what happens when a few As atoms (as little as one As per 10^8 Si) are doped into pure silicon. Each As atom replaces a Si atom in the silicon structure, but after forming four bonds to its four Si neighbors, the As atom still has one electron, which is forced to occupy an orbital higher in energy than those used in the As—Si bonds. The effect that this has on the band structure is shown in Fig. 32-8(*b*). A new, narrow filled band is introduced, much closer in energy to the upper unfilled band. From here, it is much easier to have electrons thermally excited into this empty band, where they will be able to migrate through the crystal under the influence of a potential difference. Thus, the essentially insulating silicon becomes a semiconductor in which the conductance is attributable to the movement of the excess negative charges. It is thus called an *n-type* semiconductor.

If, on the other hand, pure silicon is doped with some gallium atoms, which have only three valence electrons, and also have their valence shell orbitals at slightly higher energy than those of the silicon atoms, the net result is as shown in Fig. 32-8(*c*). A narrow empty band not far above the filled band is introduced, and electrons from the filled band can be thermally excited to this new, narrow band, leaving positive holes in the filled band. Since the conductance in this case can be regarded as due to the migration of positive holes in the now incomplete lower band, this type of semiconductivity is called *p-type* semiconductivity.

32-4   Defects in Solids

All solid substances, even when very pure, have defects in their structures, that is, faults, absences, excesses, or misalignments relative to the idealized crystal structure. These defects may have an important influence on the properties of the substance. The presence of some defects is a consequence of thermodynamics. The presence of defects causes disorder in the substance and this means that the entropy S is increased. The most stable state of any system occurs when its free energy G is minimal, and it is well known that G is related to enthalpy H and S by the equation:

$$G = H - TS \qquad (32\text{-}4.1)$$

where T is the absolute temperature. The introduction of defects into an initially perfect crystal costs energy, meaning that H increases. However, the term TS will also increase with the number of defects, provided the crystal is not at the absolute zero ($T = 0$). Hence, at a given temperature, the thermodynamic picture is as shown in Fig. 32-9. Clearly, the existence of a certain number of defects is thermodynamically required, except at $T = 0$; the higher the temperature the more defects there will be.

We now ask in more detail what we mean by defects. The majority fall into one of the following two classes:

1. Point defects.
2. Extended defects.

Point defects are of two main types: *vacancies* and *interstitials.* *Vacancies* (sometimes called Shottky defects) at some lattice sites are very common. Generally speaking, if we are dealing with an ionic substance (e.g., a halide), there will be equal numbers of cation and anion vacancies, so as to preserve electroneutrality. While the number of vacancies is usually so small that they are not easily de-

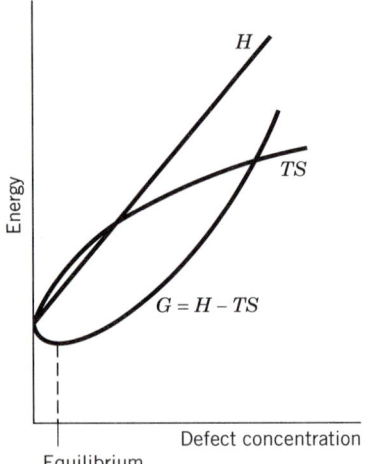

Figure 32-9   The dependence of H, TS, and $G = H - TS$ on temperature as the number of defects in a solid increases, at a given temperature.

tected, some substances, of which TiO is a well-known example, have a sizeable number at room temperature. Thus it is not uncommon for TiO to have a measured density 10%, or more, below that calculated for the perfect rock salt structure, thus showing that the corresponding percentage of vacancies is present.

Interstitials (sometimes called Frenkel defects) are atoms or ions that are displaced from their normal site in a crystal to a position in between the normal sites. This situation is most likely to arise in substances with relatively open structures and with metal ions that do not have any marked preference for octahedral versus tetrahedral coordination. Thus, there may be vacant octahedral sites but metal ions in nearly tetrahedral ones. This, of course, causes no change in composition and has no first-order effect on the density, but it can be detected in other ways, namely, by spectroscopic and electrical properties or by very sensitive X-ray diffraction measurements. The wurtzite form of ZnS (see Fig. 4-1) is prone to have Frenkel defects.

Still another type of point defect worth mentioning, not because it is very common but because it is easily noticed when it occurs, is the *color center* defect. It can occur, for example, in an alkali halide crystal that has been heated in the vapor of the alkali metal. This introduces additional metal ions, but no additional halide ions. Electroneutrality is maintained by having some anion sites occupied by an electron, which is trapped at the site. Within its "box" the electron has access to a ladder of quantum states and by absorption of visible light it can be promoted from the lowest quantum state to one or more higher ones, thus giving rise to the color. For example, strongly heated cadmium sulfide readily forms color center defects by loss of sulfur.

Extended defects often occur when a relatively large number of potential point defects become associated or clustered so as to allow a shift of one entire portion of the structure relative to an adjacent portion. This introduces *shear planes* that can actually be directly seen in high-resolution electron micrographs of crystals. Shear plane defects are very common in the higher oxides of titanium, vanadium, molybdenum, and tungsten. For example, WO_3 can actually exist in a range of compositions from the ideal WO_3 down to about $WO_{2.9}$. Electroneutrality is conserved because a certain number of W atoms are W^V instead of W^{VI}.

Instead of the oxygen vacancies being randomly distributed, in an otherwise unchanged structure, these vacancies are actually eliminated by a closing up of the structure, as shown in Fig. 32-10. In the perfect WO_3 structure we have a perfect checkerboard of WO_6 octahedra in which only corners are shared; after the shear plane has formed there are a number of edges shared between oxygen octahedra. The movement of the lower (light) section of the structure relative to the upper (dark) section [Fig. 32-10(a)] leads to the arrangement shown in Fig. 32-10(b), and entails the formation of a series of units of the type shown in Fig. 32-10(c) running in a diagonal direction. For each such unit formed, one less oxygen atom is required. Thus, if a small number of these shear planes are formed randomly, the overall structure may undergo little change but the composition will be WO_n, where $n < 3$. As already noted, shear planes can often be directly observed by electron microscopy, and an example is shown in Fig. 32-11.

It turns out that if very many of these shear planes form they begin to influence each other and become organized in a periodic manner. A detailed analysis shows that these periodic arrangements should correspond to compositions that can be expressed as W_nO_{3n-2}, with $n = 20, 24, 25,$ and 40 (i.e., $W_{20}O_{58}$, etc.).

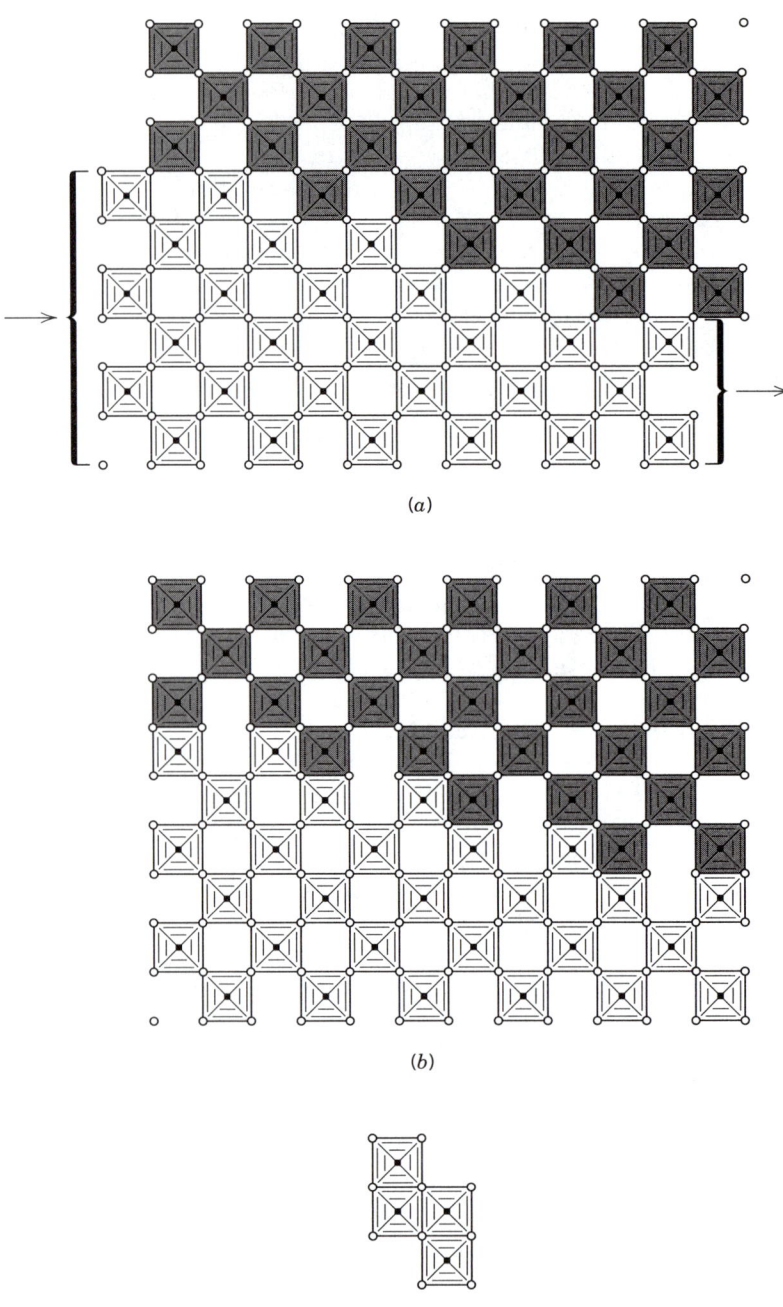

(a)

(b)

(c)

Figure 32-10   Diagrams showing how a perfect WO_3 structure (*a*) can respond to a deficiency of oxygen atoms by developing a shear plane (*b*).

Diffusion in Solids. A discussion of defects in solids leads naturally into the topic of diffusion. In most cases, although we shall later consider some important exceptions, diffusion takes place, albeit slowly, because of the presence of defects. If holes are present, adjacent atoms or ions may slide over into them,

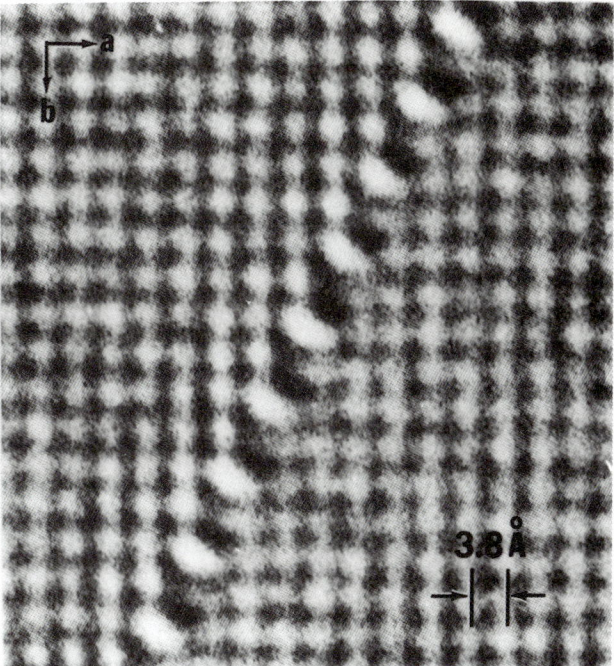

Figure 32-11   A high-resolution electron microscope image of a shear plane in a WO_{3-x} crystal. This can be compared with the diagram in Fig. 32-10. [Reproduced by permission from S. Iijima, *J. Solid State Chem.*, **1975**, *14*, 52.]

making a new hole. Interstitial atoms can migrate from one site to an adjacent one. The rate of defect-controlled diffusion is highly temperature dependent for two reasons: (1) The higher the temperature the more defects there will be, and (2) there is an energy barrier to be surmounted for each step and thus an Arrhenius-like dependence of the rate per defect is to be expected. We can write Equation (32-4.2)

$$D = D_0 e^{-E/RT} \tag{32-4.2}$$

to express the diffusion rate as a function of temperature. The exponential factor arises from the mean barrier (per mole) for the hopping process, while D_0 is proportional to the number of defects (of the type that facilitate diffusion). Log D is about linearly proportional to $1/T$.

Solid Electrolytes. While we normally think of ionic conduction as a process that occurs in solutions or molten salts, some solids are so constituted that they permit the diffusion of ions (generally cations) without the need of defects. The general requirements for a good solid electrolyte are

1. The presence of a large number of mobile ions.
2. The presence of many empty sites that the mobile ions can jump into.
3. The empty sites should be of similar or the same energy as the filled ones, with only a small energy barrier between them.

4. There should be an anionic framework within which there are either open channels, or else the framework should be soft, polarizable, and deformable.

Let us consider two examples of materials that have "hard" frameworks but permit ionic conduction because of the presence of channels through which small cations present can move relatively easily. The best known example of this is provided by β-alumina. The name β-alumina is a misnomer, assigned many years ago when the substance was believed to be a polymorph of pure Al_2O_3 (of which there are two genuine examples, α-Al_2O_3 and γ-Al_2O_3). In reality, β-alumina contains a nonstoichiometric amount of sodium and can be formulated as $Na_{1+x}Al_{11}O_{17+x/2}$.

The reason β-alumina has mobile Na^+ ions, and hence conducts electricity, is because of its structure, shown schematically in Fig. 32-12. It consists of sheets of hard, rigid γ-Al_2O_3 held together by thin layers of Na^+ and O^{2-} ions. In these thin layers the Na^{2+} ions are fairly mobile. Similar materials with K^+ or other mobile cations are also known. The possible use of β-alumina as a solid electrolyte in Na/S batteries has made it the object of very detailed study.

Another well-established solid ionic conductor is a material called NASICON (an acronym for sodium superionic conductor). The formula is $Na_3Zr_2PSi_2O_{12}$, and it consists of a framework built of corner-sharing ZrO_6 octahedra and PO_4/SiO_4 tetrahedra. In this framework there is a network of tunnels in which the Na^+ ions reside, but they occupy only a fraction of all the available sites. Thus, in an electric field, the Na^+ ions can hop from one site to the next, much as they do in the β-alumina structure.

A prominent class of "soft" solid electrolytes is provided by silver iodide (AgI) itself and, better yet, ternary silver iodides such as $HgAg_2I_4$ and $RbAg_4I_5$. In all of these, the Ag^+ ions can move fairly easily even though there are no large channels, because of the low lattice energies of these substances.

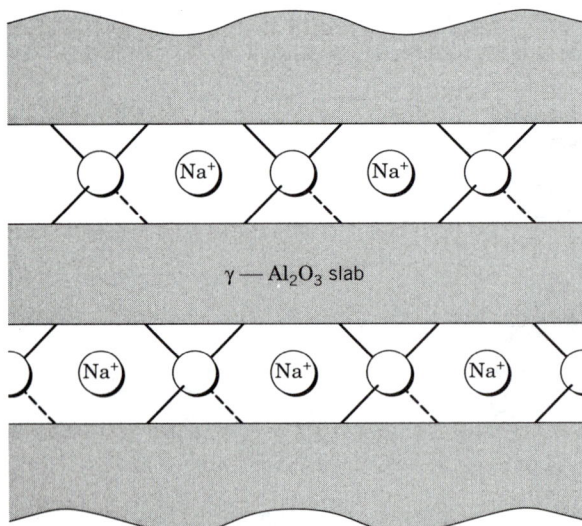

Figure 32-12   A schematic representation of the β-alumina structure. Slabs of γ-Al_2O_3 are held together by tetrahedral oxygen atoms with mobile Na^+ ions between them.

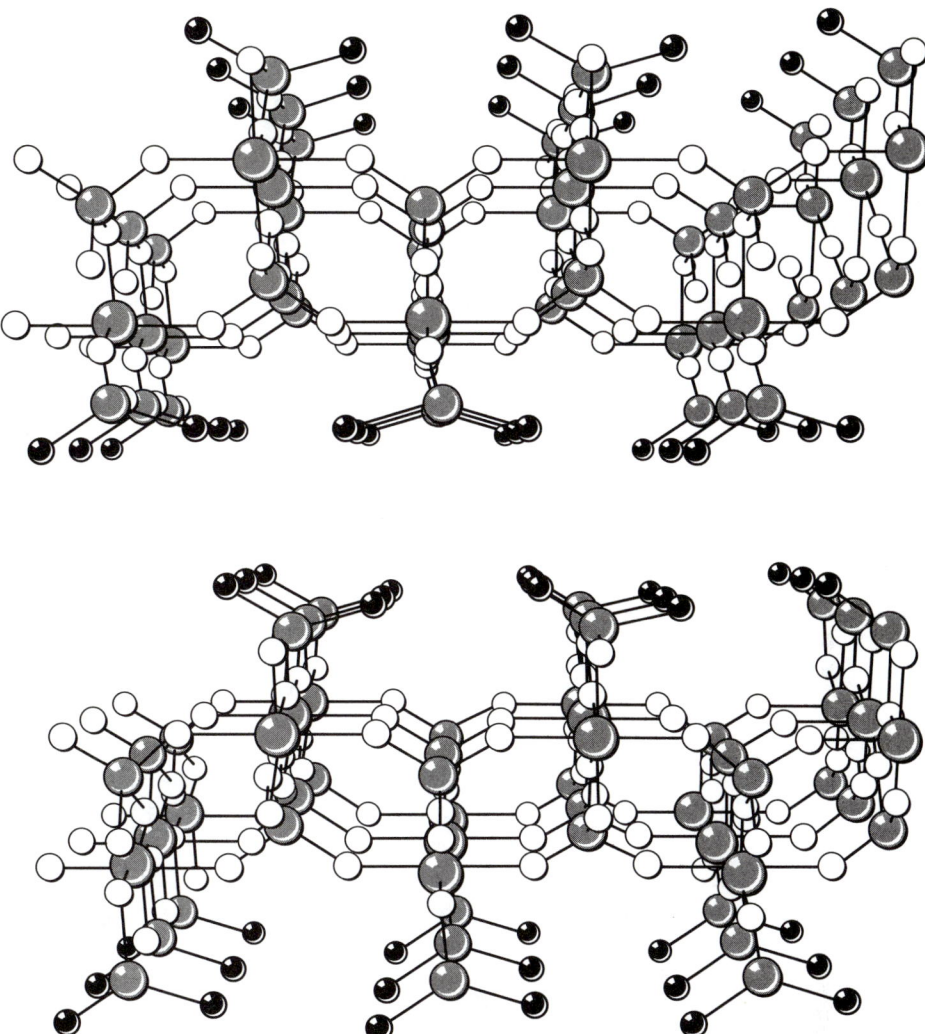

Figure 32-13 The layer structure of a γ-zirconium phosphate, $Zr(PO_4)(H_2PO_4)\cdot 2H_2O$. The two OH groups are represented by black circles and are located between the layers. Within each layer the Zr atoms are octahedrally coordinated by phosphate oxygen atoms, giving infinite chains of octahedra sharing opposite edges.

32-5 Nonstoichiometry

In molecular chemistry we expect, and generally find, that substances have precise and unvarying compositions. Thus, whether liquid or solid, water always has the composition H_2O within the accuracy of the most sensitive analytical tools known. This is to be expected since, within every molecule, the valences of each atom must be exactly satisfied by the other atoms. One cannot just "leave out one atom" from time to time. In a nonmolecular solid, on the other hand, it is possible to do exactly that, and in many cases such inexact or *nonstoichiometric* substances are known. It is characteristic of most nonstoichiometric substances that the same basic structure is retained while the composition varies.

A very good and well-known example is provided by "FeO." Rarely does one find a specimen that has exactly that composition. Usually, there are fewer iron atoms than oxygen atoms and typical compositions range from $Fe_{0.9}O$ to $Fe_{0.96}O$. However, all of them display an X-ray diffraction pattern corresponding to the rock salt structure, albeit with some slight changes. This variability can be attributed to the tendency of a few iron atoms to be Fe^{III} so that the total number of iron atoms required to counterbalance the charges of n O^{2-} ions is less than n. There are then some vacant cation sites in the structure.

We have already discussed a nonstoichiometric compound, WO_n, $2.9 < n < 3$, where the structure does not remain *entirely* unaltered but accommodates to a deficiency of one component by forming shear planes or some other type of stacking fault.

A class of compounds that are very prone to being nonstoichiometric are the transition metal hydrides. These are substances in which the metal atoms retain the same spatial arrangement as they have in the metal itself while interstices become occupied by hydrogen atoms. The limiting compositions MH, MH_2, or MH_3 are never (or almost never) attained. Instead, the substances have compositions such as $NbH_{0.7}$, $ZrH_{1.6}$, and $LuH_{2.2}$.

32-6 Some Important Structures

Many of the important types of structures for nonmolecular inorganic solids have been mentioned earlier in this book. It will be useful, however, to give an overview here.

Ionic 3D Structures. In Chapter 4 were depicted six of the most important structure types for ionic (or partially ionic) substances of a binary nature, that is, substances containing only one type of cation and one type of anion. These should be reexamined at this point. It will be noted that all of these six structures are three dimensional, in the sense that no subdivision of the arrangement into layers or chains is possible.

We have also mentioned three of the most important structure types for mixed, usually ternary, oxides (see pages 141–142). These were the spinel ($MgAl_2O_4$), ilmenite ($FeTiO_3$), and perovskite ($CaTiO_3$) structures.

Nonionic 3D Structures. Unquestionably, the diamond structure is the most important. It is shown on page 245. It will be evident to the observant reader that the diamond structure is simply the zinc blende (or sphalerite) structure with all the atoms identical instead of being of two types. In both cases, the key feature is that each atom is tetrahedrally surrounded by four nearest neighbors, and the tetrahedra are linked so as to give the overall network cubic symmetry.

The diamond structure is also adopted by silicon and germanium and the sphalerite structure is adopted by numerous nonionic binary compounds (e.g., GaAs and CdS), which are very important in solid state electronics.

Layer Structures. A layer structure is one in which a network of covalent bonds extends throughout the structure in only two directions. In the third dimension (i.e., perpendicular to the layers) there are only van der Waals forces. The most venerable example of a layer structure is provided by graphite (Chapter 8). However, there are many chemical compounds that have layer structures. All of those silicate minerals called micas have layer structures, as is evident in their macroscopic appearance. There are somewhat similar zirconium phosphate compounds, an example of which is shown in Fig. 32-13.

The oxohalides, MOCl, MOBr, and MOI, of the elements Al, Ga, La, Ti, V, and Fe, as well as the remaining lanthanides, form layer structures of a type shown schematically below. The central layer consists of oxygen atoms. Above and below this are layers of metal atoms, and finally on the top and on the bottom are layers of halogen atoms. These five-ply sheets are then stacked, with only van der Waals contacts (Cl $\cdots$ Cl).

$$
\begin{array}{ccccc}
\text{X} & \text{X} & \text{X} & \text{X} & \text{X} \\
& \text{M} & \text{M} & \text{M} & \text{M} \\
\text{O} & \text{O} & \text{O} & \text{O} & \text{O} \\
& \text{M} & \text{M} & \text{M} & \text{M} \\
\text{X} & \text{X} & \text{X} & \text{X} & \text{X}
\end{array}
$$

One other notable class of layer compounds is typified by molybdenite (MoS_2). This is the most common compound of molybdenum in Nature. It consists of infinite sandwiches with hexagonal close-packed layers of sulfur atoms on the outside and Mo atoms in trigonal (not octahedral) interstices between them. A portion of the structure is shown in Fig. 32-14. The compounds WS_2, $MoSe_2$, and several other disulfides also have this structure. Closely related is the layer structure displayed by TiS_2, ZrS_2, and several other compounds in which the two outer sulfur layers are related in such a way as to create octahedral holes for the metal atoms.

Intercalation. One of the most important properties of layer structures is their ability to allow other species, atoms, ions, or molecules, to penetrate be-

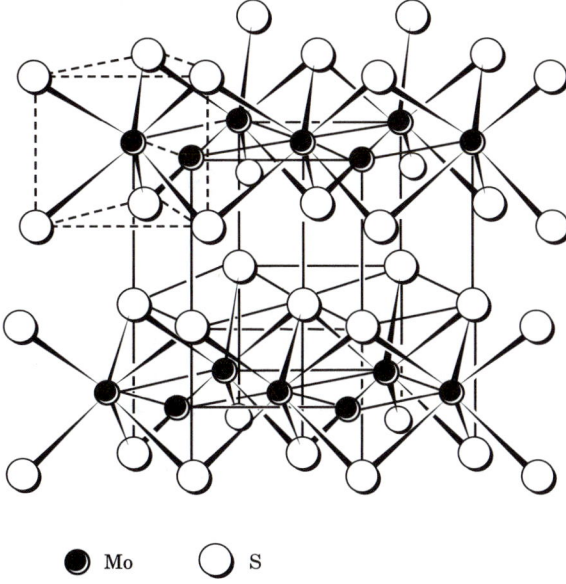

● Mo   ◯ S

Figure 32-14   A portion of the MoS_2 structure showing how trigonal prismatic sites for the Mo atoms are formed.

tween the layers, thus forming intercalation compounds. The process of doing this is called an insertion reaction, or an intercalation reaction, and the resulting "compounds" are not usually stoichiometric.

The earliest observations of intercalation were made with graphite using alkali metals. When intercalation occurs, the alkali metal atoms transfer their valence electrons to the graphite sheets and become cations. Indeed, this electron transfer is necessary in order for the intercalation to occur. More recently, there have been extensive studies of this sort of intercalation with the layered MX_2 compounds.

In some cases the alkali metal may be introduced, and also removed, by thermochemical processes, as in the following equations:

$$MS_2 + x\,Na \xrightarrow{\sim 800\,°C} Na_x MS_2 \qquad x = 0.4 - 0.7 \qquad (32\text{-}6.1)$$

$$LiVS_2 + \tfrac{1}{2} I_2 \xrightarrow{\hspace{2cm}} VS_2 + LiI \qquad\qquad (32\text{-}6.2)$$

In other cases insertion and deinsertion are driven electrochemically, a process that is often preferred because the extent of intercalation can be controlled by measuring the number of coulombs of electricity used. Certain metallocenes, $(C_5H_5)_2M$ (where M = Co or Cr), can also intercalate and they, too, transfer an electron. This is consistent with the fact that $(C_5H_5)_2Fe$, which has a higher oxidation potential, does not intercalate.

There has been much interest in such intercalation compounds because they are potentially useful in batteries. In fact, there are practical batteries now made of which the elements are a lithium anode, an MS_2 cathode, and a polar but nonprotonic solvent; they can give voltages exceeding 2 V.

Amorphous Solids: Ceramics and Glasses.   The term ceramic refers to any nonmetallic inorganic material that is nonmolecular and usually obtained in an amorphous (i.e., noncrystalline) condition. It is characteristic of ceramics that they are made into their final useful form by strong heating (firing), generally in the range of 1000–2000 °C. The vast majority of ceramics are made of silicate clays (kaolins) to which other substances, such as feldspars, are added. However, there are many specialized ceramics, such as ferrites, which contain iron oxide (Fe_3O_4) together with other metal oxides (MgO or ZnO), and are used in a variety of applications where ferromagnetism is required. Another specialized application of ceramics is as abrasives, and here alumina, SiC, and boron carbide are among the most common.

Porcelain, which is the most useful ceramic for pottery, dishes, and coatings on iron vessels, is made from kaolin mixed with a finely powdered feldspar. On heating to about 1450 °C the feldspar becomes vitreous (see below) and binds the mixture together.

Ceramics are valued because of their resistance to heat and abrasion, but their chief drawback, brittleness, has limited their range of application.

A *glass* is a special type of ceramic that can be regarded as a rigid liquid. A glassy substance is also said to be in a *vitreous state*. In general, a glass is obtained when a liquid is cooled so quickly that crystallization does not have time to occur; the disordered or tangled molecules in the liquid are trapped in that condition. Therefore, in a glass there is no long-range order.

The substances that most commonly and easily form glasses are those consisting of oligomers or polymers (in one or two dimensions) that become randomly arranged in the liquid state and require very long periods of very slow cooling to arrange themselves in a pattern with long-range order. Silica and silicates are perhaps the commonest and most important examples.

Returning in more detail to the question of how a glass is formed, let us look at Fig. 32-15. First, this figure shows that as the liquid substance is cooled, its volume decreases until, at the temperature of freezing, T_f, (or, alternatively, if we imagine starting with the crystalline solid, the melting point), there is a discontinuous change in volume as the crystalline solid forms. After this, further cooling causes only a very gradual contraction in volume. However, if the liquid is cooled rapidly enough, it will continue to contract but no crystals will form. It is now a *supercooled liquid*. Finally, however, a temperature will be reached when it will become rigid. This temperature (T_g) is called the *glass temperature,* and it usually corresponds to only a slight kink in the cooling curve.

Fused quartz (SiO_2) has a very high T_g, and thus the temperature required to work it are inconvenient. Ordinary glass, often called soft glass, contains Na_2O and CaO, and has a much lower T_g, thus making it easier to fabricate into sheets, bottles, and so on. When B_2O_3 is added to quartz, a glass commonly called *pyrex* is obtained. This has a higher T_g than soft glass, but lower than quartz. In addition, it is less sensitive to thermal shock and has a lower coefficient of thermal expansion. It is therefore a preferred glass for cookware and for laboratory apparatus.

32-7 Electrical and Magnetic Properties

The electrical and magnetic properties of inorganic materials have long been employed in technological applications, and even today, the use of metals as

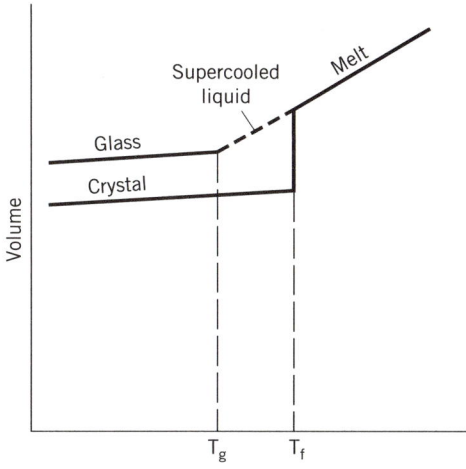

Figure 32-15 Phase diagram showing the relation between liquid, solid, and glassy forms. *Temperatures T_g and T_f are the* glass transition temperature and the melting temperature of the crystalline material, respectively.

electrical conductors and the use of iron, Fe_3O_4 and various metallic alloys as magnets continues. In recent years, however, a host of new types of substances with remarkable and highly useful electrical and magnetic properties have opened a completely new era. In this section we shall draw attention to three of these phenomena and to the types of materials that are used in technological applications.

Solid State Electronics. Modern electronic technology is totally dependent on the electrical properties of semiconductors. All solid state devices, such as transistors, silicon chips, photocells, and others, employ semiconductors. As an illustrative example we shall discuss the simplest example of a transistor, namely, the type with one **pn** junction, which functions as the equivalent of a vacuum tube diode.

Recall from Section 32-3 that there are two ways to dope silicon so as to make it an extrinsic semiconductor. In one case we have an **n** type and in the other case a **p** type. Suppose, however, that we take a small single crystal of very pure silicon and dope it so that one half is **n** type and the other half is **p** type. If we now consider the situation at the junction of these two halves, we will find something like what is shown in Fig. 32-16. There will be a spontaneous blurring of the boundary since some electrons will drift across it, from the higher energy **n**-type region into the holes in the **p**-type region. However, the extent of this drift is limited by the fact that a back voltage is built up.

Let us now apply an external voltage, even a low one, to the silicon crystal, in a direction perpendicular to the junction, so that there is a source of electrons (the negative terminal) in contact with the **n**-type portion and a sink for electrons (the positive terminal) in contact with the **p**-type portion. Current will now flow continuously, entering the **n**-type portion, crossing the interface to the **p**-type portion, and exiting at the positive terminal of the voltage source. However, if a low voltage were applied the other way around, no current would flow because the electrons cannot be driven up from the **p**-type into the **n**-type portion. A **pn** junction thus serves as the equivalent of a vacuum tube diode. It is a rectifier and can be used to convert ac current into dc current.

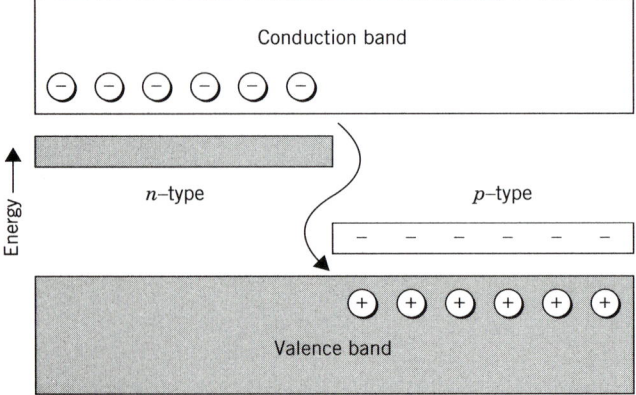

Figure 32-16   A schematic representation of the junction between the **n**-type and **p**-type halves of a simple **pn** transistor, which functions as a rectifier.

We shall not pursue the point in detail but by making **pnp** or **npn** junctions one can create the equivalent of vacuum tube triodes, which serve as voltage or current amplifiers.

Superconductivity. Superconductors are characterized by two extraordinary properties. Below some critical temperature T_c, their electrical resistance goes to zero. Above T_c, most superconductors display ordinary metallic conductance. Second, when in the superconducting state, a superconductor excludes the lines of force of a magnetic field and is thus repelled by such a field. This behavior is called the Meissner effect.

Superconductivity was first discovered in 1911 for the element mercury, which has a T_c of only 4.2 K. For many years the majority of known superconducting substances were metals or alloys and the T_c values were very low. Until 1986 the highest known T_c was 23 K, which was found in a metallic alloy Nb_3Ge.

In general, superconductivity is suppressed by the application of a magnetic field. In some materials this occurs abruptly at some critical field H_c, which is characteristic of the material. In others, conductivity declines gradually above H_c. Since one of the most important applications of superconducting materials is to carry the electric current required to generate high magnetic fields in electromagnets, this antagonism between the superconductivity and the magnetic field poses a practical problem. Thus, research aimed at inventing better superconductors has two goals: to raise both T_c and H_c. While there has recently been spectacular progress in raising T_c (to be discussed shortly), it is interesting to note that one of the first nonmetallic type superconductors to be discovered, the Chevrel phases, were particularly interesting because of their high H_c values, even though their T_c values ($\sim$ 15 K) are not exceptional.

The Chevrel phases, discovered by R. Chevrel, a French inorganic chemist, in 1971, have Mo_6S_8 units packed with other metal atoms in the manner shown in Fig. 32-17 for the particular compound $PbMo_6S_8$. Other Chevrel phases may contain Li, Mn, or Cd in place of Pb, and Se or Te in place of S. Unfortunately, the Chevrel phases have one practical problem, as compared to metallic superconductors, namely, that they are brittle and thus not easy to fabricate into useful shapes, such as wire. However, progress is being made in solving this problem.

The search for superconductors with higher T_c values is driven by the fact that it is very difficult, and expensive, to maintain materials in their superconducting state when they must be kept in containers surrounded by liquid helium or liquid hydrogen. For many years, the dream of finding superconducting materials with T_c values above the boiling point of liquid nitrogen (77 K) was as attractive as it was elusive, but at last, in 1987, it was achieved. Let us briefly recount those exciting developments.

In January of 1986 two scientists working at the IBM European Research Laboratory, Bednorz and Müller (Nobel prize in physics, 1987), found that materials entirely different from the best previously used metals and alloys, namely, mixed-metal oxides of the type $BaLa_xCu_yO_z$, such that the average oxidation state of the copper is between +2 and +3, could have T_c as high as 35 K, which is a considerable jump. This work immediately drew attention to this class of compounds in general and an enormous research effort suddenly arose all over the world.

The story now switches to the laboratory of Paul Chu, a physicist at the University of Houston, who found that by applying pressure to a Bednorz and Müller type compound, T_c could be raised to 57 K. This observation led Chu, in

"PbMo$_6$S$_8$"

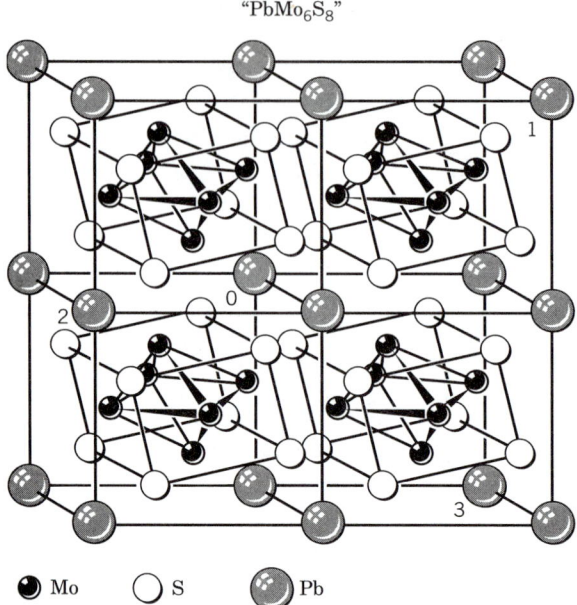

⬤ Mo   ◯ S   ◉ Pb

Figure 32-17   The structure of Chevrel phases. The packing of Mo$_6$S$_8$ units and Pb atoms in PbMo$_6$S$_8$ is shown.

cooperation with a group led by M. K. Wu at the University of Alabama at Huntsville, to consider the possibility of simulating the effect of external pressure by replacing the lanthanum atoms with smaller trivalent ions, and only a little over a year after Bednorz and Müller's initial discovery, Chu and Wu reported a material (Ba$_{0.8}$Y$_{1.2}$CuO$_x$), with $T_c \geq 90$ K. Other laboratories almost immediately reported similar results. The highest confirmed T_c is 122 K, which has been obtained for a material of composition Tl$_2$Ba$_2$Ca$_2$Cu$_3$O$_{10}$.

Since these exciting developments, there have been no further increases in T_c; scattered reports claiming higher values have occasionally appeared, but were not reproducible. Research in the field of high-temperature superconductivity has subsequently been heavily concentrated on understanding the properties of the Ba—Y—Cu—O systems. A particularly well investigated substance is the so-called 1-2-3 material YBa$_2$Cu$_3$O$_7$, which has $T_c = 95$ K.

The 1-2-3 compound is typical of all those in this general class of high T_c materials in having a layer structure, a portion of which is shown in Fig. 32-18. It can be seen that there are two environments for the copper atoms. Some have five oxygen atom neighbors and these CuO$_5$ units are linked by shared basal corners into infinite sheets. The other copper atoms are only four coordinate, and are linked into infinite chains by the sharing of some oxygen atoms. From the formula YBa$_2$Cu$_3$O$_7$, with the assumption that Y and Ba have their normal valences (+3 and +2, respectively), the average valence of copper is +2⅓. This means that both Cu^{2+} and Cu^{3+} ions are present, but discrete sites for them cannot be identified. There is general agreement that it is the copper ions and the way in which their oxidation states vary that is the key to the superconducting behavior, but a precise explanation has yet to be found.

It is necessary to recognize that these 1-2-3 materials are easily obtained in

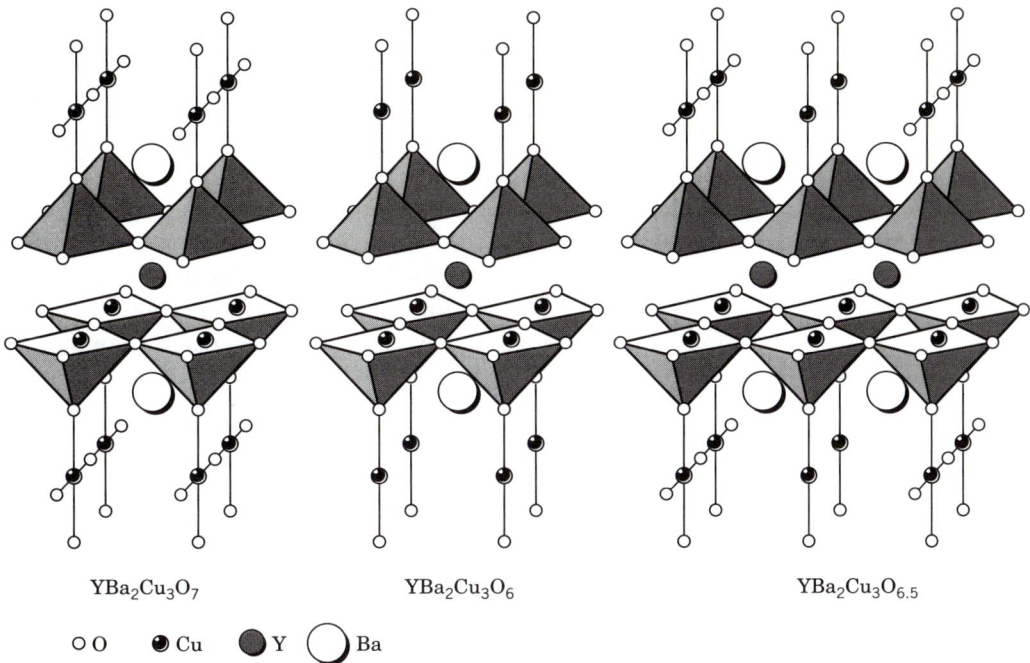

YBa$_2$Cu$_3$O$_7$ YBa$_2$Cu$_3$O$_6$ YBa$_2$Cu$_3$O$_{6.5}$

○ O   ◑ Cu   ● Y   ◯ Ba

Figure 32-18 The layer structure of compounds of the type YBa$_2$Cu$_3$O$_x$ (x = 7, 6, and 6.5). [Figure courtesy of Professor T. R. Hughbanks.]

nonstoichiometric condition in which there is a deficit of oxygen. One way in which this can occur is by loss of some oxygen atoms from the chains, whereby some of the copper atoms become two coordinate and are presumed to become Cu$^+$ ions. If oxygen loss occurs from one half the chains, in an ordered way, the composition becomes YBaCu$_3$O$_{6.5}$ (as shown in Fig. 32-18) and if all such oxygen atoms are lost, the composition becomes YBa$_2$Cu$_3$O$_6$. These latter two composi- tions are not high-temperature superconductors. Because of this tendency to nonstoichiometry and consequent loss of superconductivity, the thermal history of a 1-2-3 compound is of crucial importance.

Needless to say, right from the beginning a theoretical understanding of su- perconductivity posed a great problem. However, Bardeen, Cooper, and Schrieffer (Nobel prize in physics, 1972) developed a theory (the BCS theory) that is considered satisfactory for all the low-temperature superconductors. In this theory, the central concept is the existence of pairs of electrons called *Cooper pairs*. The electrons in a Cooper pair interact with each other indirectly by way of displacements of the atoms from their mean lattice positions. The presence of one electron at a given position in the crystal will distort its immediate environ- ment so as to make it attractive for a second electron to come into the same re- gion, creating the Cooper pair. However, the persistence of the distortion, which is responsible for the "virtual attraction" between the two electrons, is opposed by the thermal vibrations of the atoms, and thus the Cooper pairs can persist only at low temperature.

A Cooper pair can travel through a solid much more easily than a single elec- tron because it is not scattered by collisions with the atoms as much as is a single electron. This means that when large numbers of Cooper pairs exist, conduc- tance becomes so great that the material becomes a superconductor.

It is not clear whether the basic concepts of the BCS theory can be adapted to explain the high-temperature superconductors, or whether new ideas will be required. In view of the theoretical uncertainties, there is no reason to believe that superconductors with even higher values of T_c may not exist. The ultimate dream, of course, would be a substance that is superconducting at ice temperature, or even room temperature.

Cooperative Magnetic Properties. We have earlier (Chapter 23) discussed the magnetic properties of individual metal ions, and briefly mentioned the occurrence of more complicated magnetic behavior such as ferromagnetism and antiferromagnetism. These more complicated phenomena occur only in the solid state where interactions between the magnetic moments of adjacent metal ions in the infinite nonmolecular structure can occur. In this section more will be said about these cooperative magnetic interactions, especially ferromagnetism.

The major types of cooperative magnetic interactions in the solid state are those leading to ferromagnetism, antiferromagnetism, and ferrimagnetism.

Ferromagnetism results when the spins on the metal centers communicate with one another—usually by an electronic coupling through the anions that lie between them—so as to favor the alignment of all of them in the same direction. This sort of cooperation can be sustained over very large *domains,* consisting of thousands of neighboring metal ions, and in this way the degree of magnetization of the material will be enormously greater than for a normal paramagnetic substance, where the magnetic moments of the individual metal ions each interact with the magnetic field independently of the others. However, there is some temperature above which the tendency to thermal randomization becomes so great that the collective interaction of the spins on different metal ions is overcome. Above this critical temperature, called the Curie temperature (T_C) the substance behaves like an ordinary paramagnetic material.

Below the Curie temperature, however, ferromagnetic substances are not only able to develop very large magnetizations in a magnetic field, but they *remain permanently magnetized* when removed from the field. The extent to which a ferromagnetic material retains its magnetization determines the use to which it can be put. All ferromagnetic materials respond nonlinearly to the applied magnetic field, giving a hysteresis loop, as shown in Fig. 32-19. If the loop is broad [as shown in Fig. 32-19(a)] the substance retains a very high degree of magnetization when the field is reduced to zero and can only be demagnetized by applying a high field in the opposite direction. Such substances, called hard ferromagnets, are suitable for making permanent magnets. On the other hand, soft ferromagnets, which have narrow hysteresis loops like that in Fig. 32-19(b), can follow a rapidly oscillating magnetic field and are used in transformer cores.

Antiferromagnetic substances behave in a manner opposite to that of ferromagnetics. Below a critical temperature, here called the Néel temperature (T_N) they have a cooperative interaction between the spins on adjacent metal ions that leads to an alternating pattern of alignment. In this way, the magnetization is very low below T_N and tends to zero as the temperature approaches 0 K.

Ferrimagnetism is displayed when there are two types of magnetic ions present. Those of each type tend to align with the magnetic field but in opposite directions. However, because the two types have different inherent moments, one

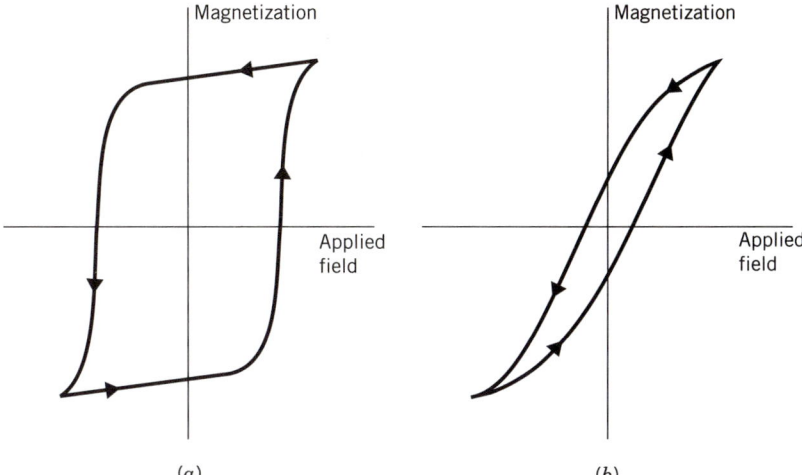

Figure 32-19   Hysteresis loops for the magnetization of ferromagnetic materials, showing (*a*) large hysteresis and (*b*) small hysteresis.

set does not completely cancel the other set, and there is a net magnetization, even at 0 K.

A particularly important class of substances from the point of view of cooperative magnetic properties are the *ferrites*. These are substances of the spinel type (see chapter 4) that contain iron (i.e., MFe_2O_4). In fact, they are so-called inverse spinels, in which half the Fe^{3+} ions are in tetrahedral sites. Depending on the composition, ferrites may be either ferromagnetic or antiferromagnetic.

32-8   Concluding Remarks

The field of solid state inorganic chemistry and physics is very large and of enormous practical as well as scientific importance. In this brief chapter we have been able to do little more than introduce some of the key phenomena and concepts. Along with polymers, pharmaceuticals, and catalysts, the many inorganic substances with exceptional physical, electrical and magnetic properties play one of the most important roles in our contemporary high-tech world.

There are exciting new areas for research and development. Some that would follow directly from the topics covered here are (1) still better and more useful high-temperature superconductors, (2) ferrimagnetic materials with properties more precisely suited to specific applications, and (3) ceramics with better mechanical properties.

A very important new frontier is that of composite materials. These are formed when different substances are interleaved or interwoven at a microscopic level, thus giving rise to macroscopic properties that combine the virtues of all of the components.

There is, without question, a wide range of important and interesting challenges awaiting inorganic chemists in the field of solid state science.

Study Questions

A. Review

1. List the important ways in which solids differ from liquids and gases.
2. State the difference between molecular solids and nonmolecular solids.
3. Describe and contrast the methods of preparing solid compounds. Consider specifically:
   (a) the so-called "shake and bake" method.
   (b) the vapor-phase transport process.
   (c) the use of a flux.
   (d) hydrothermal synthesis.
4. What is a flux? Give an example and state how it is used.
5. Prepare a drawing that depicts vapor-phase transport as used in the synthesis of crystalline ZnS from amorphous ZnS.
6. Explain the formation of an energy band in a solid.
7. Describe the energy bands in an alkali metal.
8. What is a fermi level?
9. What are band gaps in solids?
10. What are the characteristic temperature dependences of electrical conductance in metals and in semiconductors?
11. How does an insulator differ from a semiconductor?
12. What is the difference between an intrinsic and an extrinsic semiconductor? Give examples of each.
13. What is an **n**-type semiconductor? Prepare a diagram of its band structure as a part of your answer.
14. Answer question 13 for a **p**-type semiconductor.
15. Why do all solids have some defects at temperatures above 0 K?
16. What are point defects in solids?
17. What are vacancies in solids?
18. What are interstitial defects in solids?
19. Name a substance in which one commonly finds:
    (a) vacancies
    (b) Frenkel defects
    (c) color center defects
    (d) shear plane defects
20. Why are samples of TiO commonly less dense than ideal?
21. Explain in general terms how defects in solids give rise to diffusion in solids.
22. What is a solid electrolyte? Give examples.
23. What is NASICON? How does it function as a solid electrolyte?
24. What are nonstoichiometric solids? Give examples.
25. Give the compostion of some nonstoichiometric hydrides.
26. What is intercalation? Does it give rise to stoichiometric or nonstoichiometric compounds?
27. Describe the intercalation of alkali metals into graphite.
28. What are amorphous solids?
29. What is kaolin? How is it used in the production of porcelain?
30. What constitutes a ceramic material?

31. What is a glass? A glass temperature?

32. What are the characteristic properties of superconductors?

33. Make a list of modern superconducting materials, beginning with the substance in which superconductivity was first discovered.

34. How is β-alumina employed in making a new type of battery?

35. Explain why "ferrous oxide" is nearly always nonstoichiometric.

36. Describe the structure of FeOCl.

37. Explain how a **pn** junction functions as a solid state diode.

38. When were the first high-temperature superconductors discovered? What is the highest T_c that has been reproducibly observed in such a material?

39. What effect does a strong magnetic field have on a superconductor?

B. Additional Exercises

1. What are the practical advantages of extrinsic (i.e., doped) semiconductors over intrinsic ones? Give examples involving the doping of pure silicon.

2. The formula for β-alumina was stated to be $Na_{1+x}Al_{11}O_{17+x/2}$. Do you think this is distinguishable analytically from $Na_{1+x}Al_9O_{14+x/2}$?

3. When strongly heated, cadmium sulfide becomes bright orange. Explain.

4. Compare the structure of diamond with those of silicon and wurtzite.

5. Even though WO_n with n < 3 is oxygen-deficient compared to the ideal formula (i.e., WO_3), there are no vacant oxygen atom sites. Explain.

6. What is a hysteresis loop as this term is applied to the behavior of a ferromagnetic solid?

7. True of false: The BCS theory of superconductivity straightforwardly explains the behavior of $YBa_2Cu_3O_7$.

8. What are the major structural changes that occur to the host substance when an intercalation compound is formed?

9. Do you think the level of defects in an ionic solid will depend on the lattice energy? If so, how?

10. How are the Arrhenius equation and the temperature dependence of a semiconductor related?

SUPPLEMENTARY READING

Adams, D. M., *Inorganic Solids*, Wiley, New York, 1984.

Borg, R. J., *The Physical Chemistry of Solids*, Academic Press, Boston, 1992.

Cheetham, A. K. and Day, P., Eds., *Solid State Chemistry*, Clarendon Press, New York, 1992.

Cox, P. A., *The Electronic Structure and Chemistry of Solids*, Oxford University Press, Oxford, 1987.

Duffy, J. A., *Bonding, Energy Levels, and Bonds in Inorganic Solids*, Longman, Essex, 1990.

Hagenmuller, N. B. and Van Gool, W., *Solid Electrolytes, General Principles, Characterization, Materials, and Applications*, Academic Press, New York, 1978.

Hannay, N. B., *Solid State Chemistry*, Prentice-Hall, 1967.

Hoffmann, R., *Solids and Surfaces, A Chemist's View of Bonding in Extended Structures*, VCH Publishers, New York, 1988.

Ladd, M. C. F., *Structure and Bonding in Solid State Chemistry*, Wiley, New York, 1979.

Rao, C. N. R. and Gopalakrishnan, J., *New Directions in Solid State Chemistry*, Cambridge University Press, 1986.

Smart, L. and Moore, E., *Solid State Chemistry*, Chapman and Hall, London, 1992.

Sutton, A. P., *Electronic Structure of Materials*, Oxford University Press, Oxford, England, 1993.

West, A., *Solid State Chemistry and Its Applications*, Wiley, New York, 1984.

Appendix I

ASPECTS OF SYMMETRY AND POINT GROUPS

The material for Appendix I was taken from F. A. Cotton and G. Wilkinson, *Advanced Inorganic Chemistry*, 5th ed., Wiley-Interscience, New York, 1988, pp. 1389–1409 (used with permission) and F. A. Cotton, *Chemical Applications of Group Theory*, 2nd ed., Wiley-Interscience, New York, 1971, pp. 45–52 (used with permission).

AI-1 Symmetry Operations and Elements

When we say that a molecule has *symmetry,* we mean that *certain parts of it can be interchanged with others without altering either the identity or the orientation of the molecule.* The interchangeable parts are said to be equivalent to one another by symmetry. Consider, for example, a trigonal bipyramidal molecule such as PF_5 (AI-I). The three equatorial P—F bonds to F_1, F_2, and F_3, are equivalent. They

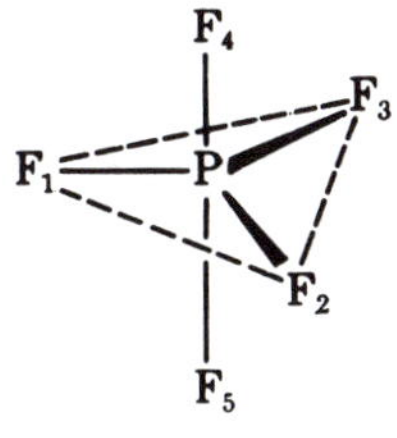

(A-I)

have the same length, the same strength, and the same type of spatial relation to the remainder of the molecule. Any permutation of these three bonds among themselves leads to a molecule indistinguishable from the original. Similarly, the axial P—F bonds to F_4 and F_5 are equivalent. *But,* axial and equatorial bonds are different types (e.g., they have different lengths), and if one of each were to be interchanged, the molecule would be noticeably perturbed. These statements are probably self-evident, or at least readily acceptable, on an intuitive basis; but for systematic and detailed consideration of symmetry, certain formal tools are needed. The first set of tools is a set of *symmetry operations.*

Symmetry operations are geometrically defined ways of exchanging equivalent parts of a molecule. There are four kinds that are used conventionally and these are sufficient for all our purposes.

1. Simple rotation about an axis passing through the molecule by an angle $2\pi/n$. This operation is called a *proper rotation* and is symbolized C_n. If it

is repeated n times, of course, the molecule comes all the way back to the original orientation.

2. Reflection of all atoms through a plane that passes through the molecule. This operation is called *reflection* and is symbolized σ.

3. Reflection of all atoms through a point in the molecule. This operation is called *inversion* and is symbolized **i**.

4. The combination, in either order, of rotating the molecule about an axis passing through it by $2\pi/n$ and reflecting all atoms through a plane that is perpendicular to the axis of rotation is called *improper rotation*. The symbol for improper rotation is S_n.

These operations are *symmetry operations if, and only if,* the appearance of the molecule is *exactly* the same after one of them is carried out as it was before. For instance, consider rotation of the molecule H_2S by $2\pi/2$ about an axis passing through S and bisecting the line between the H atoms. As shown in Fig. AI-1, this operation interchanges the H atoms and interchanges the S—H bonds. Since these atoms and bonds are equivalent, there is no physical (i.e., physically meaningful or detectable) difference after the operation. For HSD, however, the corresponding operation replaces the S—H bond by the S—D bond, and vice versa, and one can see that a change has occurred. Therefore, for H_2S, the operation C_2 is a symmetry operation; for HSD it is not.

These types of symmetry operation are graphically explained by the diagrams in Fig. AI-2, where it is shown how an arbitrary point (0) in space is affected in each case. Filled dots represent points above the xy plane and open dots represent points below it. Let us examine first the action of proper rotations, illustrated here by the C_4 rotations, that is, rotations by $2\pi/4 = 90°$. The operation C_4 is seen to take the point 0 to the point 1. The application of C_4 twice, designated C_4^2, generates point 2. Operation C_4^3 gives point 3 and, of course, C_4^4, which is a rotation by $4 \times 2\pi/4 = 2\pi$, regenerates the original point. The set of four points, 0, 1, 2, 3 are permutable, cyclically, by repeated C_4 proper rotations and are equivalent points. It will be obvious that in general repetition of a C_n operation will generate a set of n equivalent points from an arbitrary initial point, provided that point lies off the axis of rotation.

The effect of reflection through symmetry planes perpendicular to the xy plane, specifically, σ_{xz} and σ_{yz}, is also illustrated in Fig. AI-2. The point 0 is re-

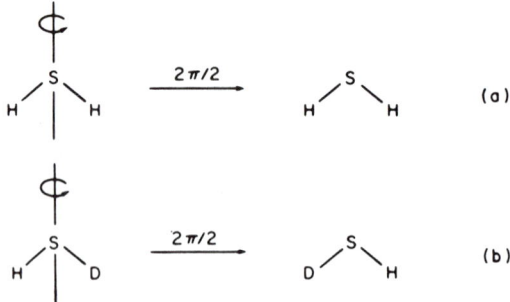

Figure AI-1   The operation C_2 carries H_2S into an orientation indistinguishable from the original, but HSD goes into an observably different orientation.

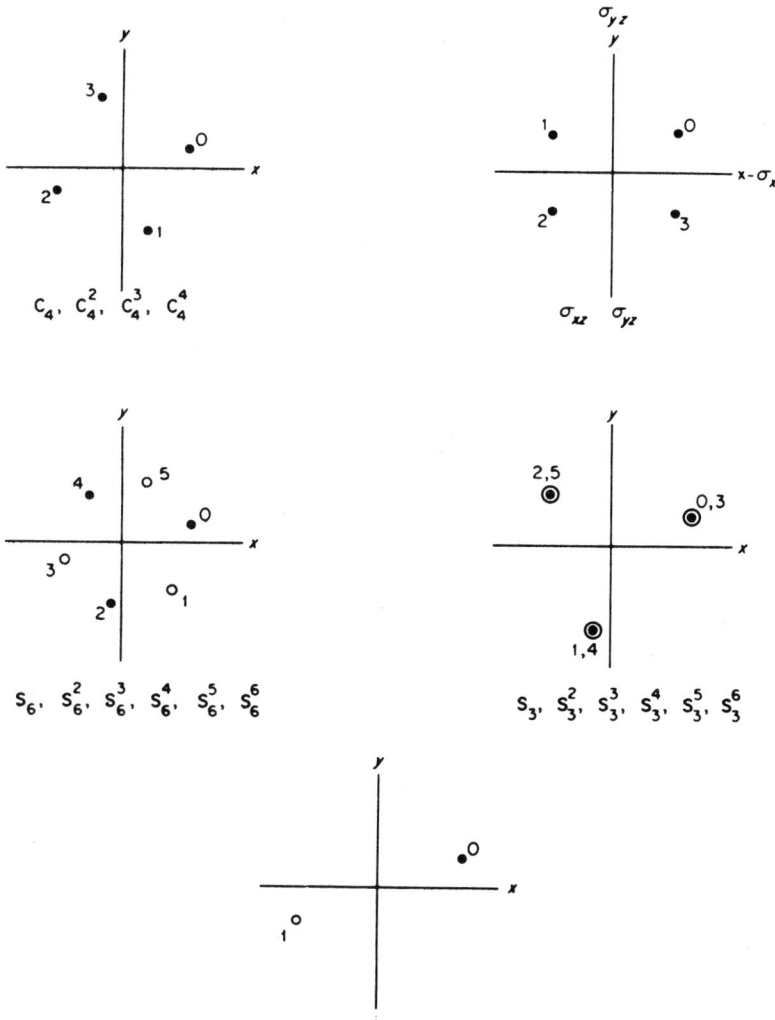

Figure AI-2 The effects of symmetry operations on an arbitrary point, designated 0, thus generating sets of points.

lated to point 1 by the $\boldsymbol{\sigma}_{yz}$ operation and to the point 3 by the $\boldsymbol{\sigma}_{xz}$ operation. By reflecting either point 1 or point 3 through the second plane, point 2 is obtained.

The set of points generated by the repeated application of an improper rotation will vary in appearance depending on whether the order of the operation $\mathbf{S}_n$ is even or odd, order being the number n. A crown of n points, alternately up and down, is produced for n even, as illustrated for $\mathbf{S}_6$. For n odd there is generated a set of $2n$ points, which form a right n-sided prism, as shown for $\mathbf{S}_3$.

Finally, the operation $\mathbf{i}$ is seen to generate from point 0 a second point, 1, lying on the opposite side of the origin.

Let us now illustrate the symmetry operations for various familiar molecules as examples. As this is done it will also be convenient to employ the concept of *symmetry elements*. A symmetry element is an *axis* (line), *plane*, or *point* about which symmetry operations are performed. The existence of a certain symmetry operation implies the existence of a corresponding symmetry element, and conversely, the presence of a symmetry element means that a certain symmetry operation or set of operations is possible.

Consider the ammonia molecule (Fig. AI-3). The three equivalent hydrogen atoms may be exchanged among themselves in two ways: by proper rotations and by reflections. The molecule has an axis of threefold proper rotation; this is called a C_3 axis. It passes through the N atom and through the center of the equilateral triangle defined by the H atoms. When the molecule is rotated by $2\pi/3$ in a clockwise direction H_1 replaces H_2, H_2 replaces H_3, and H_3 replaces H_1. Since the three H atoms are physically indistinguishable, the numbering having no physical reality, the molecule after rotation is indistinguishable from the molecule before rotation. This rotation, called a C_3 or threefold proper rotation, is a symmetry operation. Rotation by $2 \times 2\pi/3$ also produces a configuration different, but physically indistinguishable, from the original and is likewise a symmetry operation; it is designated $\mathbf{C}_3^2$. Finally, rotation by $3 \times 2\pi/3$ carries each atom all the way around and returns it to its initial position. This operation ($\mathbf{C}_3^3$) has the same net effect as performing no operation at all, but for mathematical reasons it must be considered as an operation generated by the C_3 axis. This, and other operations, which have no net effect, are called *identity* operations and are symbolized by $\mathbf{E}$. Thus, we may write $\mathbf{C}_3^3 = \mathbf{E}$.

The interchange of hydrogen atoms in NH_3 by reflections may be carried out in three ways; that is, there are three planes of symmetry. Each plane passes through the N atom and one of the H atoms, and bisects the line connecting the other two H atoms. Reflection through the symmetry plane containing N and H_1 interchanges H_2 and H_3; the other two reflections interchange H_1 with H_3, and H_1 with H_2.

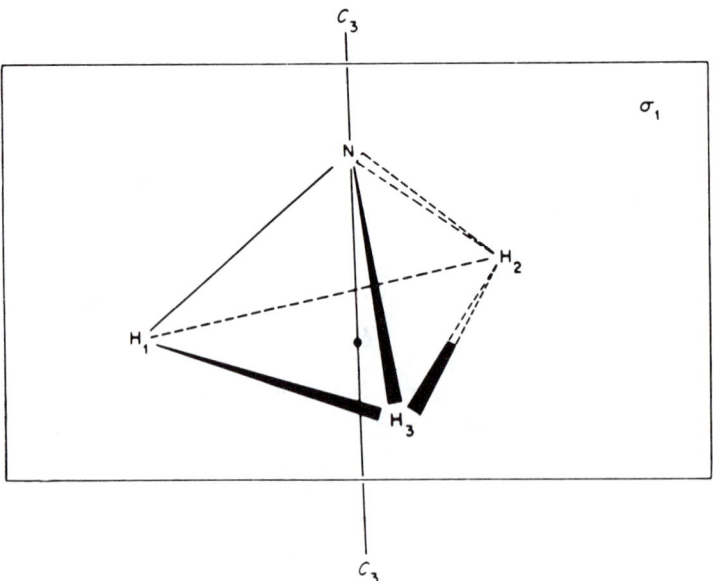

Figure AI-3   The ammonia molecule, showing its threefold symmetry axis C_3, and one of its three planes of symmetry σ_1, which passes through H_1 and N and bisects the H_2—H_3 line.

Inspection of the NH_3 molecule shows that no other symmetry operations besides these six (three rotations, $\mathbf{C}_3$, $\mathbf{C}_3^2$, $\mathbf{C}_3^3 = \mathbf{E}$, and three reflections, $\boldsymbol{\sigma}_1$, $\boldsymbol{\sigma}_2$, $\boldsymbol{\sigma}_3$) are possible. Put another way, the only symmetry elements the molecule possesses are C_3 and the three planes that we may designate σ_1, σ_2, and σ_3. Specifically, it will be obvious that no sort of improper rotation is possible, nor is there a center of symmetry.

As a more complex example, in which all four types of symmetry operation and element are represented, let us take the $Re_2Cl_8^{2-}$ ion, which has the shape of a square parallepiped or right square prism (Fig. AI-4). Altogether this ion has six axes of proper rotation, of four different kinds. First, the Re_1—Re_2 line is an axis of fourfold proper rotation C_4, and four operations ($\mathbf{C}_4$, $\mathbf{C}_4^2$, $\mathbf{C}_4^3$, $\mathbf{C}_4^4 \equiv \mathbf{E}$) may be carried out. This same line is also a C_2 axis, generating the operation $\mathbf{C}_2$. It will be noted that the $\mathbf{C}_4^2$ operation means rotation by $2 \times 2\pi/4$, which is equivalent to rotation by $2\pi/2$, that is, to the $\mathbf{C}_2$ operation. Thus the C_2 axis and the $\mathbf{C}_2$ operation are implied by, not independent of, the C_4 axis. There are, however, two other types of C_2 axis that exist independently. There are two of the type that passes through the centers of opposite vertical edges of the prism (C_2' axes) and two more that pass through the centers of opposite vertical faces of the prism (C_2'' axes).

The $Re_2Cl_8^{2-}$ ion has three different kinds of symmetry plane [see Fig. AI-4(b)]. There is a unique plane that bisects the Re—Re bond and all the vertical edges of the prism. Since it is customary to define the direction of the highest proper axis of symmetry, C_4 in this case, as the vertical direction, this symmetry plane is horizontal and the subscript h is used to identify it, σ_h. There are then two types of vertical symmetry plane, namely, the two that contain opposite vertical edges, and two others that cut the centers of the opposite vertical faces. One of these two sets may be designated $\sigma_v^{(1)}$ and $\sigma_v^{(2)}$, the v implying that they are vertical. Since those of the second vertical set bisect the dihedral angles between those of the first set, they are then designated $\sigma_d^{(1)}$ and $\sigma_d^{(2)}$, the d standing for dihedral. Both pairs of planes are vertical and it is actually arbitrary which are labeled σ_v and which σ_d.

Continuing with $Re_2Cl_8^{2-}$, we see that an axis of improper rotation is present. This is coincident with the C_4 axis and is an S_4 axis. The $\mathbf{S}_4$ operation about this axis proceeds as follows. The rotational part, through an angle of $2\pi/4$, in the clockwise direction has the same effect as the $\mathbf{C}_4$ operation. When this is coupled with a reflection in the horizontal plane ($\boldsymbol{\sigma}_h$) the following shifts of atoms occur:

$$Re_1 \longrightarrow Re_2 \qquad Cl_1 \longrightarrow Cl_6 \qquad Cl_5 \longrightarrow Cl_2$$
$$Re_2 \longrightarrow Re_1 \qquad Cl_2 \longrightarrow Cl_7 \qquad Cl_6 \longrightarrow Cl_3$$
$$Cl_3 \longrightarrow Cl_8 \qquad Cl_7 \longrightarrow Cl_4$$
$$Cl_4 \longrightarrow Cl_5 \qquad Cl_8 \longrightarrow Cl_1$$

Finally, the $Re_2Cl_8^{2-}$ ion has a center of symmetry i and the inversion operation $\mathbf{i}$ can be performed.

In the case of $Re_2Cl_8^{2-}$ the improper axis S_4 might be considered as merely the inevitable consequence of the existence of the C_4 axis and the σ_h, and, indeed, this is a perfectly correct way to look at it. However, it is important to emphasize that there are cases in which an improper axis S_n exists without inde-

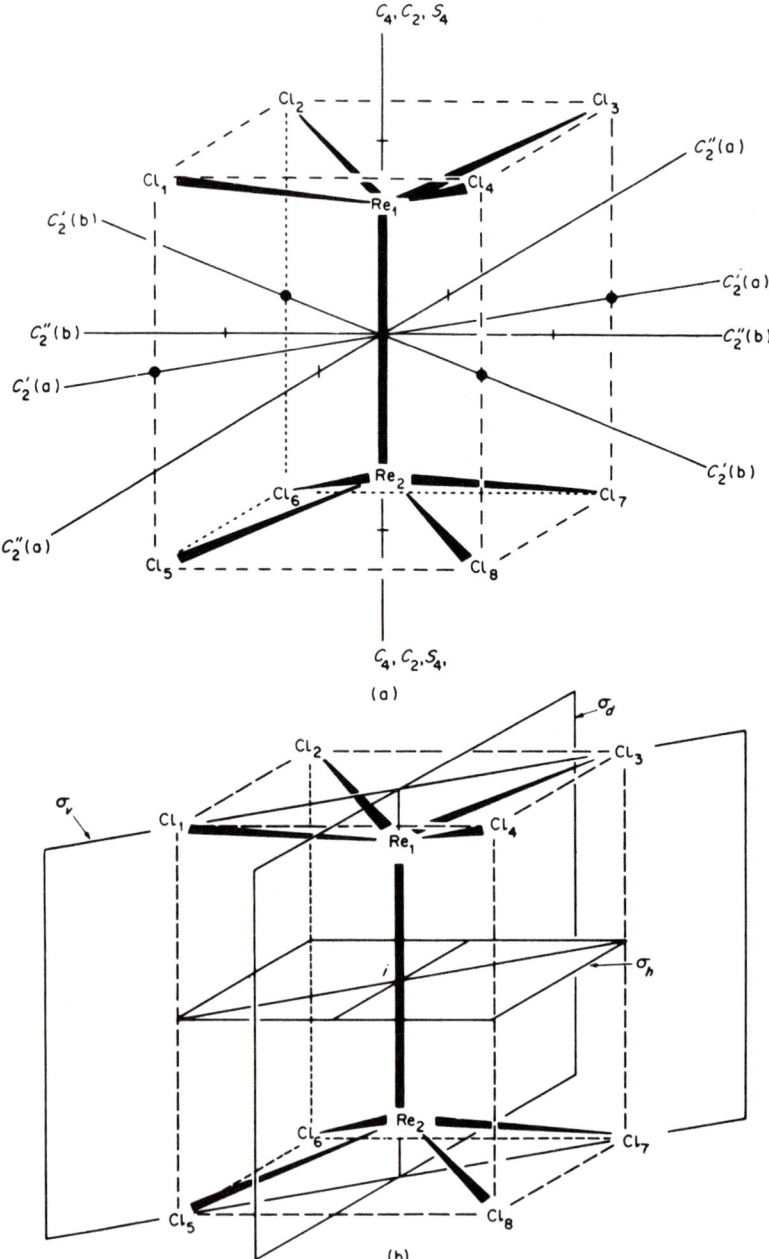

Figure AI-4   The symmetry elements of the $Re_2Cl_8^{2-}$ ion. (*a*) The axes of symmetry. (*b*) One of each type of plane and the center of symmetry.

pendent existence of either C_n or σ_h. Consider, for example, a tetrahedral molecule as depicted in Structure AI-II, where the $TiCl_4$ molecule is shown inscribed

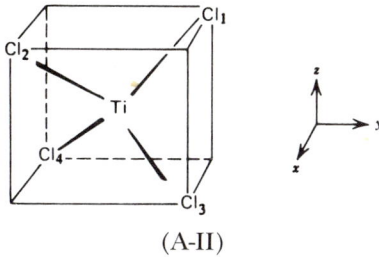

(A-II)

in a cube and Cartesian axes, x, y, and z are indicated. Each of these axes is an S_4 axis. For example, rotation by $2\pi/4$ about z followed by reflection in the xy plane shifts the Cl atoms as follows:

$$Cl_1 \longrightarrow Cl_3 \qquad Cl_3 \longrightarrow Cl_2$$
$$Cl_2 \longrightarrow Cl_4 \qquad Cl_4 \longrightarrow Cl_1$$

Note, however, that the Cartesian axes are not C_4 axes (though they are C_2 axes) and the principal planes (namely, xy, xz, and yz) are not symmetry planes. Thus we have here an example of the existence of the S_n axis without C_n or σ_h having any independent existence. The ethane molecule in its staggered configuration has an S_6 axis and provides another example.

AI-2   Symmetry Groups

The complete set of symmetry operations that can be performed on a molecule is called the *symmetry group* for that molecule. The word "group" is used here not as a mere synonym for "set" or "collection," but in a technical, mathematical sense, and this meaning must first be explained.

Introduction to Multiplying Symmetry Operations

We have already seen in passing that if a proper rotation $\mathbf{C}_n$ and a horizontal reflection $\boldsymbol{\sigma}_h$ can be performed, there is also an operation that results from the combination of the two which we call the improper rotation $\mathbf{S}_n$. We may say that $\mathbf{S}_n$ is the product of $\mathbf{C}_n$ and $\boldsymbol{\sigma}_h$. Noting also that the order in which we perform $\boldsymbol{\sigma}_h$ and $\mathbf{C}_n$ is immaterial,* we can write

$$\mathbf{C}_n \times \boldsymbol{\sigma}_h = \boldsymbol{\sigma}_h \times \mathbf{C}_n = \mathbf{S}_n$$

This is an algebraic way of expressing the fact that successive application of the two operations shown has the same effect as applying the third one. For obvious reasons, it is convenient to speak of the third operation as being the product obtained by multiplication of the other two.

This example is not unusual. Quite generally, any two symmetry operations can be multiplied to give a third. For example, in Fig. AI-2 the effects of reflec-

*This is, however, a special case; in general, order of multiplication matters as noted later.

tions in two mutually perpendicular symmetry planes are illustrated. It can be seen that one of the reflections carries point 0 to point 1. The other reflection carries point 1 to point 2. Point 0 can also be taken to point 2 by way of point 3 if the two reflection operations are performed in the opposite order. But a moment's thought will show that a direct transfer of point 0 to point 2 can be achieved by a C_2 operation about the axis defined by the line of intersection of the two planes. If we call the two reflections $\sigma(xz)$ and $\sigma(yz)$ and the rotation $C_2(z)$, we can write

$$\sigma(xz) \times \sigma(yz) = \sigma(yz) \times \sigma(xz) = C_2(z)$$

It is also evident that

$$\sigma(yz) \times C_2(z) = C_2(z) \times \sigma(yz) = \sigma(xz)$$

and

$$\sigma(xz) \times C_2(z) = C_2(z) \times \sigma(xz) = \sigma(yz)$$

Also note that if any one of these three operations is applied twice in succession, we get no net result or, in other words, an identity operation, namely,

$$\sigma(xz) \times \sigma(xz) = \mathbf{E}$$

$$\sigma(yz) \times \sigma(yz) = \mathbf{E}$$

$$C_2(z) \times C_2(z) = \mathbf{E}$$

Introduction to a Group

If we pause here and review what has just been done with the three operations $\sigma(xz)$, $\sigma(yz)$, and $C_2(z)$, we see that we have formed all the nine possible products. To summarize the results systematically, we can arrange them in the annexed tabular form. Note that we have added seven more multiplications, namely, all those in which the identity operation $\mathbf{E}$ is a factor. The results of these are trivial, since the product of any other, nontrivial operation with $\mathbf{E}$ must be just the nontrivial operation itself, as indicated.

	$\mathbf{E}$	$C_2(z)$	$\sigma(xz)$	$\sigma(yz)$
$\mathbf{E}$	$\mathbf{E}$	$C_2(z)$	$\sigma(xz)$	$\sigma(yz)$
$C_2(z)$	$C_2(z)$	$\mathbf{E}$	$\sigma(yz)$	$\sigma(xz)$
$\sigma(xz)$	$\sigma(xz)$	$\sigma(yz)$	$\mathbf{E}$	$C_2(z)$
$\sigma(yz)$	$\sigma(yz)$	$\sigma(xz)$	$C_2(z)$	$\mathbf{E}$

The set of operations $\mathbf{E}$, $C_2(z)$, $\sigma(xz)$, and $\sigma(yz)$ evidently has the following four interesting properties:

1. There is one operation $\mathbf{E}$, the identity, that is the trivial one of making no change. Its product with any other operation is simply the other operation.

2. There is a definition of how to multiply operations: we apply them successively. The product of any two is one of the remaining ones. In other words, this collection of operations is self-sufficient, all its possible products being already within itself. This is sometimes called the property of *closure*.

3. Each of the operations has an *inverse*, that is, an operation by which it may be multiplied to give **E** as the product. In this case, each operation is its own inverse, as shown by the occurrence of **E** in all diagonal positions of the table.

4. It can also be shown that if we form a triple product, this may be subdivided in any way we like without changing the result, thus

$$\boldsymbol{\sigma}(xz) \times \boldsymbol{\sigma}(yz) \times \mathbf{C}_2(z)$$

$$= [\boldsymbol{\sigma}(xz) \times \boldsymbol{\sigma}(yz)] \times \mathbf{C}_2(z) = \mathbf{C}_2(z) \times \mathbf{C}_2(z)$$

$$= \boldsymbol{\sigma}(xz) \times [\boldsymbol{\sigma}(yz) \times \mathbf{C}_2(z)] = \boldsymbol{\sigma}(xz) \times \boldsymbol{\sigma}(xz)$$

$$= \mathbf{E}$$

Products that have this property are said to obey the *associative law* of multiplication.

The four properties just enumerated are of fundamental importance. They are the properties—and the *only* properties—that any collection of symmetry operations must have to constitute a *mathematical group*. Groups consisting of symmetry operations are called *symmetry groups* or sometimes *point groups*. The latter term arises because all the operations leave the molecule fixed at a certain point in space. This is in contrast to other groups of symmetry operations, such as those that may be applied to crystal structures in which individual molecules move from one location to another.

The symmetry group we have just been examining is one of the simpler groups, but nonetheless, an important one. It is represented by the symbol C_{2v}; the origin of this and other symbols is discussed later. It is not an entirely representative group in that it has some properties that are *not* necessarily found in other groups. We have already called attention to one, namely, that each operation in this group is its own inverse; this is actually true of only three kinds of operation: reflections, twofold proper rotations, and inversion **i**. Another special property of the group C_{2v} is that all multiplications in it are *commutative;* that is, every multiplication is equal to the multiplication of the same two operations in the opposite order. It can be seen that the group multiplication table is symmetrical about its main diagonal, which is another way of saying that all possible multiplications commute. In general, multiplication of symmetry operations is *not* commutative, as subsequent discussion will illustrate.

For another simple, but more general, example of a symmetry group, let us recall our earlier examination of the ammonia molecule. We were able to discover six and only six symmetry operations that could be performed on this molecule. If this is indeed a complete list, they should constitute a group. The easiest way to see if they do is to attempt to write a multiplication table. This will contain 36 products, some of which we already know how to write. Thus we know the result of all multiplications involving **E**, and we know that

$$\mathbf{C}_3 \times \mathbf{C}_3 = \mathbf{C}_3^2$$
$$\mathbf{C}_3 \times \mathbf{C}_3^2 = \mathbf{C}_3^2 \times \mathbf{C}_3 = \mathbf{E}$$

It will be noted that the second of these statements means that $\mathbf{C}_3$ is the inverse of $\mathbf{C}_3^2$ and vice versa. We also know that $\mathbf{E}$ and each of the $\boldsymbol{\sigma}$'s is its own inverse. So all operations have inverses, thus satisfying requirement 3.

To continue, we may next consider the products when one $\boldsymbol{\sigma}_v$ is multiplied by another. A typical example is shown in Fig. AI-5(a). When point 0 is reflected first through $\boldsymbol{\sigma}^{(1)}$ and then through $\boldsymbol{\sigma}^{(2)}$, it becomes point 2. But point 2 can obviously also be reached by a clockwise rotation through $2\pi/3$, that is, by the operation $\mathbf{C}_3$. Thus we can write

$$\boldsymbol{\sigma}^{(1)} \times \boldsymbol{\sigma}^{(2)} = \mathbf{C}_3$$

If, however, we reflect first through $\boldsymbol{\sigma}^{(2)}$ and then through $\boldsymbol{\sigma}^{(1)}$, point 0 becomes point 4, which can also be reached by $\mathbf{C}_3 \times \mathbf{C}_3 = \mathbf{C}_3^2$. Thus we write

$$\boldsymbol{\sigma}^{(2)} \times \boldsymbol{\sigma}^{(1)} = \mathbf{C}_3^2$$

Clearly, the reflections $\boldsymbol{\sigma}^{(1)}$ and $\boldsymbol{\sigma}^{(2)}$ do not commute. The reader should be able to make the obvious extension of the geometrical arguments just used to obtain the following additional products:

$$\boldsymbol{\sigma}^{(1)} \times \boldsymbol{\sigma}^{(3)} = \mathbf{C}_3^2$$
$$\boldsymbol{\sigma}^{(3)} \times \boldsymbol{\sigma}^{(1)} = \mathbf{C}_3$$
$$\boldsymbol{\sigma}^{(2)} \times \boldsymbol{\sigma}^{(3)} = \mathbf{C}_3$$
$$\boldsymbol{\sigma}^{(3)} \times \boldsymbol{\sigma}^{(2)} = \mathbf{C}_3^2$$

There remain, now, the products of $\mathbf{C}_3$ and $\mathbf{C}_3^2$ with $\boldsymbol{\sigma}^{(1)}$, $\boldsymbol{\sigma}^{(2)}$ and $\boldsymbol{\sigma}^{(3)}$. Figure AI-5(b) shows a type of geometric construction that yields these products. For example, we can see that the reflection $\boldsymbol{\sigma}^{(1)}$ followed by the rotation $\mathbf{C}_3$ carries

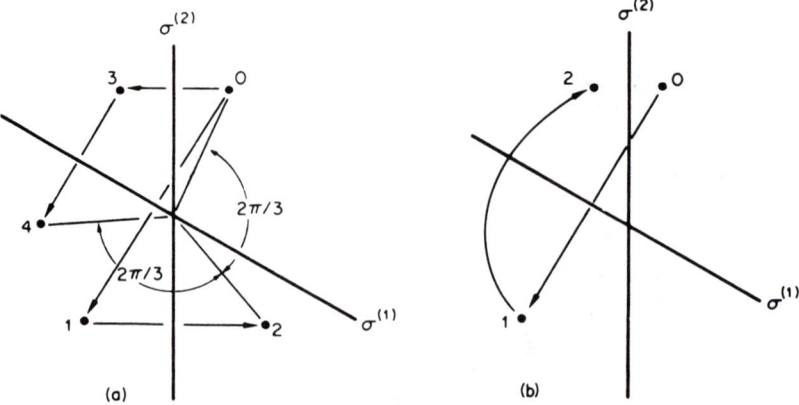

Figure AI-5   The multiplication of symmetry operations: (*a*) reflection times reflection and (*b*) reflection followed by $\mathbf{C}_3$.

point 0 to point 2, which could have been reached directly by the operation $\sigma^{(2)}$. By similar procedures all the remaining products can be easily determined. The complete multiplication table for this set of operations is given here.

	E	C_3	C_3^2	$\sigma^{(1)}$	$\sigma^{(2)}$	$\sigma^{(3)}$
E	E	C_3	C_3^2	$\sigma^{(1)}$	$\sigma^{(2)}$	$\sigma^{(3)}$
C_3	C_3	C_3^2	E	$\sigma^{(3)}$	$\sigma^{(1)}$	$\sigma^{(2)}$
C_3^2	C_3^2	E	C_3	$\sigma^{(2)}$	$\sigma^{(3)}$	$\sigma^{(1)}$
$\sigma^{(1)}$	$\sigma^{(1)}$	$\sigma^{(2)}$	$\sigma^{(3)}$	E	C_3	C_3^2
$\sigma^{(2)}$	$\sigma^{(2)}$	$\sigma^{(3)}$	$\sigma^{(1)}$	C_3^2	E	C_3
$\sigma^{(3)}$	$\sigma^{(3)}$	$\sigma^{(1)}$	$\sigma^{(2)}$	C_3	C_3^2	E

The successful construction of this table demonstrates that the set of six operations does indeed form a group. This group is represented by the symbol C_{3v}. The table shows that its characteristics are more general than those of the group C_{2v}. Thus it contains some operations that are not, as well as some which are, their own inverse. It also involves a number of multiplications that are not commutative.

AI-3   Some General Rules for Multiplication of Symmetry Operations

In the preceding section several specific examples of multiplication of symmetry operations have been worked out. On the basis of this experience, the following general rules should not be difficult to accept:

1. The product of two proper rotations must be another proper rotation. Thus, although rotations can be created by combining reflections [recall: $\sigma(xz) \times \sigma(yz) = C_2(z)$], the reverse is not possible.

2. The product of two reflections in planes meeting at an angle θ is a rotation by 2θ about the axis formed by the line of intersection of the planes (recall: $\sigma^{(1)} \times \sigma^{(2)} = C_3$ for the ammonia molecule).

3. When there is a rotation operation C_n and a reflection in a plane containing the axis, there must be altogether n such reflections in a set of n planes separated by angles of $2\pi/2n$, intersecting along the C_n axis [recall: $\sigma^{(1)} \times C_3 = \sigma^{(2)}$ for the ammonia molecule].

4. The product of two C_2 operations about axes that intersect at an angle θ is a rotation by 2θ about an axis perpendicular to the plane containing the two C_2 axes.

5. The following pairs of operations always commute:
   (a) Two rotations about the same axis.
   (b) Reflections through planes perpendicular to each other.
   (c) The inversion and any other operation.
   (d) Two C_2 operations about perpendicular axes.
   (e) C_n and σ_h, where the C_n axis is vertical.

AI-4 A Systematic Listing of Symmetry Groups, with Examples

The symmetry groups to which real molecules may belong are very numerous. However, they may be systematically classified by considering how to build them up using increasingly more elaborate combinations of symmetry operations. The outline that follows, though neither unique in its approach nor rigorous in its procedure, affords a practical scheme for use by most chemists.

The simplest nontrivial groups are those of order 2, that is, those containing but one operation in addition to $\mathbf{E}$. The additional operation must be one that is its own inverse; thus the only groups of order 2 are

$$C_s: \mathbf{E}, \boldsymbol{\sigma}$$

$$C_i: \mathbf{E}, \mathbf{i}$$

$$C_2: \mathbf{E}, \mathbf{C}_2$$

The symbols for these groups are rather arbitrary, except for C_2 which, we shall soon see, forms part of a pattern.

Molecules with C_s symmetry are fairly numerous. Examples are the thionyl halides and sulfoxides (AI-III), and secondary amines (AI-IV). Molecules having a center of symmetry as their *only* symmetry element are quite rare; two types are shown as Structures AI-V and AI-VI. The reader should find it very challenging, though not impossible, to think of others. Molecules of C_2 symmetry are fairly common, two examples being Structures AI-VII and AI-VIII.

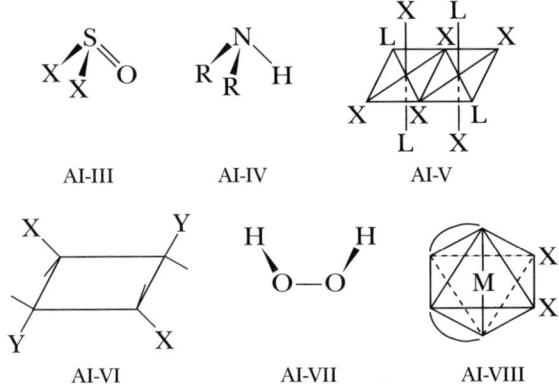

AI-III   AI-IV   AI-V

AI-VI   AI-VII   AI-VIII

The Uniaxial or C_n Groups

These are the groups in which all operations are due to the presence of a proper axis as the sole symmetry element. The general symbol for such a group, and the operations in it, are

$$C_n: \mathbf{C}_n, \mathbf{C}_n^2, \mathbf{C}_n^3 \cdots \mathbf{C}_n^{n-1}, \mathbf{C}_n^n \equiv \mathbf{E}$$

A C_n group is thus of order n. We have already mentioned the group C_2. Molecules with pure axial symmetry other than C_2 are rare. Two examples of the group C_3 are shown in Structures AI-IX and AI-X.

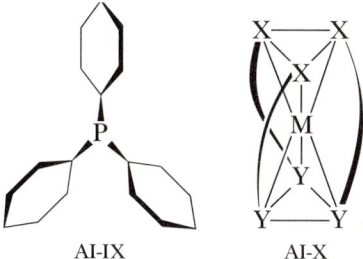

AI-IX   AI-X

The C_{nv} Groups

If, in addition to a proper axis of order n, there is also a set of n vertical planes, we have a group of order $2n$, designated C_{nv}. This type of symmetry is found quite frequently and is illustrated in Structures AI-XI to AI-XV, where the values of n are 2–6.

AI-XI   AI-XII   AI-XIII

AI-XIV   AI-XV

The C_{nh} Groups

If in addition to a proper axis of order n there is also a horizontal plane of symmetry, we have a group of order $2n$, designated C_{nh}. The $2n$ operations include S_n^m operations that are products of C_n^m and σ_h for n odd, to make the total of $2n$. Thus for C_{3h} the operations are

$$C_3, C_3^2, C_3^3 \equiv E$$

$$\sigma_h$$

$$\sigma_h \times C_3 = C_3 \times \sigma_h = S_3$$

$$\sigma_h \times C_3^2 = C_3^2 \times \sigma_h = S_3^5$$

Molecules of C_{nh} symmetry with $n > 2$ are relatively rare; examples with $n = 2, 3$, and 4 are shown in Structures AI-XVI to AI-XVIII.

AI-XVI   AI-XVII   AI-XVIII

The D_n Groups

When a vertical C_n axis is accompanied by a set of n C_2 axes perpendicular to it, the group is D_n. Molecules of D_n symmetry are, in general, rare, but there is one very important type, namely, the trischelates (AI-XIX) of D_3 symmetry.

AI-XIX

The D_{nh} Groups

If we add reflection in a horizontal plane of symmetry to the operations making up a D_n group, the group D_{nh} is obtained. It should be noted that products of the type $\mathbf{C}_2 \times \boldsymbol{\sigma}_h$ will give rise to a set of reflections in vertical planes. These planes *contain* the C_2 axes; this point is important in regard to the distinction between D_{nh} and D_{nd}, mentioned next. The D_{nh} symmetry is found in a number of important molecules, a few of which are benzene (D_{6h}), ferrocene in an eclipsed configuration (D_{5h}), $Re_2Cl_8^{2-}$, which we examined previously, (D_{4h}), $PtCl_4^{2-}$ (D_{4h}), and the boron halides (D_{3h}) and $PF_5(D_{3h})$. All right prisms with regular polygons for bases as illustrated in Structures AI-XX and AI-XXI, and all bipyramids, as illustrated in Structures AI-XXII and AI-XXIII, have D_{nh}-type symmetry.

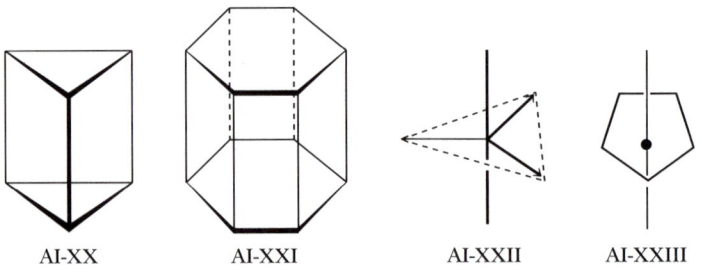

AI-XX   AI-XXI   AI-XXII   AI-XXIII

The D_{nd} Groups

If to the operations making up a D_n group we add a set of vertical planes that bisect the angles between pairs of C_2 axes (note the distinction from the vertical planes in D_{nh}), we have a group called D_{nd}. The D_{nd} groups have no σ_h. Perhaps the most celebrated examples of D_{nd} symmetry are the D_{3d} and D_{5d} symmetries of $R_3W \equiv WR_3$ and ferrocene in their staggered configurations, Structures AI-XXIV and AI-XXV.

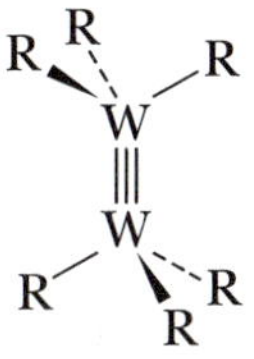

AI-XXIV 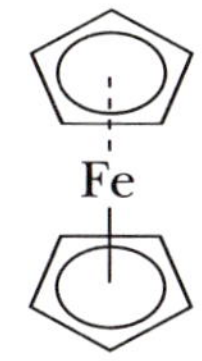   AI-XXV

Two comments about the scheme so far outlined may be helpful. First, the reader may have wondered why we did not consider the result of adding to the operations of C_n *both* a set of n σ_v's and a σ_h. The answer is that this is simply another way of getting to D_{nh}, since a set of C_2 axes is formed along the lines of intersection of the σ_h with each of the σ_v's. By convention, and in accord with the symbols used to designate the groups, it is preferable to proceed as we did. Second, in dealing with the D_{nh}-type groups, if a horizontal plane is found, there must be only the n vertical planes *containing* the C_2 axis. If dihedral planes were also present, there would be, in all $2n$ planes and hence, as shown previously, a principal axis of order $2n$, thus vitiating the assumption of a D_n type of group.

The S_n Groups

So far, our scheme has overlooked one possibility, namely, that a molecule might contain an S_n axis as its only symmetry element (except for others that are directly subservient to it). It can be shown that for n odd, the groups of operations arising would actually be those forming the group C_{nh}. For example, take the operations generated by an S_3 axis:

$$S_3$$
$$S_3^2 \equiv C_3^2$$
$$S_3^3 \equiv \sigma_h$$
$$S_3^4 \equiv C_3$$
$$S_3^5$$
$$S_3^6 \equiv E$$

Comparison with the list of operations in the group C_{3h} shows that the two lists are identical.

It is only when n is an even number that new groups can arise that are not already in the scheme. For instance, consider the set of operations generated by an S_4 axis:

$$\mathbf{S}_4$$

$$\mathbf{S}_4^2 \equiv \mathbf{C}_2$$

$$\mathbf{S}_4^3$$

$$\mathbf{S}_4^4 \equiv \mathbf{E}$$

This set of operations satisfies the four requirements for a group and is not a set that can be obtained by any procedure previously described. Thus S_4, S_6, and so on, are new groups. These groups are distinguished by the fact that they contain no operation that is not an $\mathbf{S}_n^m$ operation, even though it may be written in another way, as with $\mathbf{S}_4^2 \equiv \mathbf{C}_2$ above.

Note that the group S_2 is not new. A little thought will show that the operation $\mathbf{S}_2$ is identical with the operation $\mathbf{i}$. Hence, the group that could be called S_2 is the one we have already called C_i.

An example of a molecule with S_4 symmetry is shown in Structure AI-XXVI. Molecules with S_n symmetries are not very common.

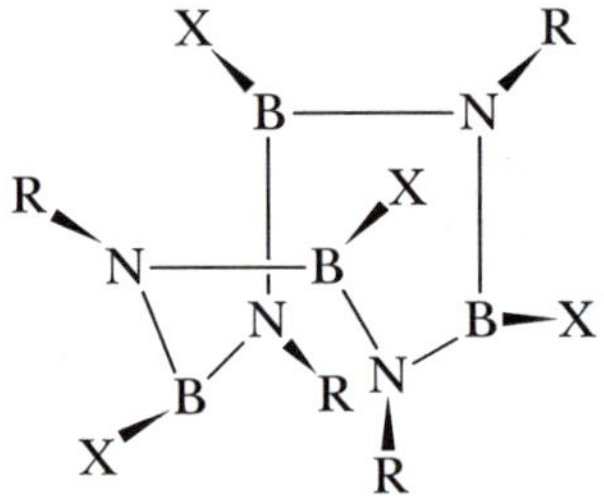

AI-XXVI

Linear Molecules

There are only two kinds of symmetry for linear molecules. There are those represented by structure AI-XXVII, which have identical ends. Thus, in addition to an infinitefold rotation axis C_∞, coinciding with the molecular axis, and an infinite number of vertical symmetry planes, they have a horizontal plane of symmetry and an infinite number of C_2 axes perpendicular to C_∞. The group of these operations is $D_{\infty h}$. A linear molecule with different ends (Structure AI-XXVIII) has only C_∞ and the σ_v's as symmetry elements. The group of operations generated by these is called $C_{\infty v}$.

$$\text{A—B—C—B—A} \qquad \text{A—B—C—D}$$
$$\text{(AI-XXVII)} \qquad\qquad \text{(AI-XXVIII)}$$

AI-5 The Groups of Very High Symmetry

The scheme followed in the preceding section has considered only cases in which there is a single axis of order equal to or greater than 3. It is possible to have symmetry groups in which there are several such axes. There are, in fact, seven such groups, and several of them are of paramount importance.

The Tetrahedron

We consider first a regular tetrahedron. Figure AI-6 shows some of the symmetry elements of the tetrahedron, including at least one of each kind. From this it can be seen that the tetrahedron has altogether 24 symmetry operations, which are as follows:

There are three S_4 axes, each of which gives rise to the operations $\mathbf{S}_4$, $\mathbf{S}_4^2 \equiv \mathbf{C}_2$, $\mathbf{S}_4^3$ and $\mathbf{S}_4^4 \equiv \mathbf{E}$. Neglecting the $\mathbf{S}_4^4$'s, this makes $3 \times 3 = 9$.

There are four C_3 axes, each giving rise to $\mathbf{C}_3$, $\mathbf{C}_3^2$, and $\mathbf{C}_3^3 \equiv \mathbf{E}$. Again omitting the identity operations, this makes $4 \times 2 = 8$.

There are six reflection planes, only one of which is shown in Fig. AI-6, giving rise to six $\boldsymbol{\sigma}_d$ operations.

Thus there are $9 + 8 + 6 +$ one identity operation $= 24$ operations. This group is called T_d. It is worth emphasizing that despite the considerable amount of symmetry, there is no inversion center in T_d symmetry. There are, of course, numerous molecules having full T_d symmetry, such as CH_4, SiF_4, ClO_4^-, $Ni(CO)_4$, and $Ir_4(CO)_{12}$, and many others where the symmetry is less but approximates to it.

If we remove the reflections from the T_d group, it turns out that $\mathbf{S}_4$ and $\mathbf{S}_4^3$ operations are also lost. The remaining 12 operations ($\mathbf{E}$, four $\mathbf{C}_3$ operations, four $\mathbf{C}_3^2$ operations and three $\mathbf{C}_2$ operations) form a group, designated T. This group in itself has little importance, since it is very rarely, if ever, encountered in real molecules. However, if we then add to the operations in the group T a different set of reflections in the three planes defined so that each one contains two of the C_2 axes, and work out all products of operations, we get a new group of 24 operations ($\mathbf{E}$, four $\mathbf{C}_3$, four $\mathbf{C}_3^2$, three $\mathbf{C}_2$, three $\boldsymbol{\sigma}_h$, $\mathbf{i}$, four $\mathbf{S}_6$, four $\mathbf{S}_6^5$) denoted T_h. This, too, is rare, but it occurs in some "octahedral" complexes in which the ligands are planar and arranged as in Structure AI-XXIX. The important feature

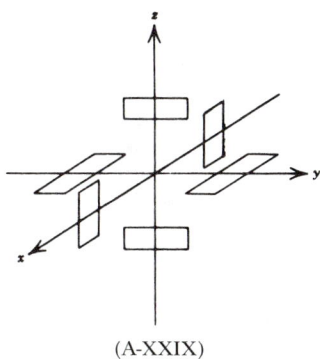

(A-XXIX)

here is that each pair of ligands on each of the Cartesian axes is in a different one of the three mutually perpendicular planes, xy, xz, and yz. Real cases are provided by $W(NMe_2)_6$ and several $M(NO_3)_6^{n-}$ ions in which the NO_3^- ions are bidentate.

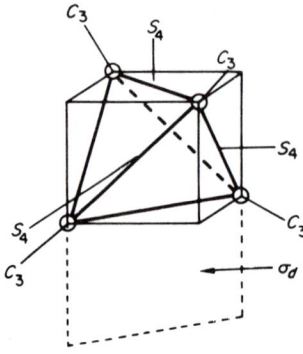

Figure AI-6 The tetrahedron, showing some of its essential symmetry elements. All S_4 and C_3 axes are shown, but only one of the six dihedral planes σ_d.

The Octahedron and the Cube

These two bodies have the same elements, as shown in Fig. AI-7, where the octahedron is inscribed in a cube, and the centers of the six cube faces form the vertices of the octahedron. Conversely, the centers of the eight faces of the octahedron form the vertices of a cube. Figure AI-7 shows one of each of the types of symmetry element that these two polyhedra possess. The list of symmetry operations is as follows:

There are three C_4 axes, each generating $\mathbf{C_4}$, $\mathbf{C_4^2} \equiv \mathbf{C_2}$, $\mathbf{C_4^3}$, $\mathbf{C_4^4} \equiv \mathbf{E}$. Thus there are $3 \times 3 = 9$ rotations, excluding $\mathbf{C_4^4}$'s.

There are four C_3 axes giving four $\mathbf{C_3}$'s and four $\mathbf{C_3^2}$'s.

There are six C_2' axes bisecting opposite edges, giving six $\mathbf{C_2''}$'s.

There are three planes of the type σ_h and six of the type σ_d, giving rise to nine reflection's.

The C_4 axes are also S_4 axes and each of these generates the operations $\mathbf{S_4}$, $\mathbf{S_4^2} \equiv \mathbf{C_2}$ and $\mathbf{S_4^3}$, the first and last of which are not yet listed, thus adding $3 \times 2 = 6$ more to the list.

The C_3 axes are also S_6 axes and each of these generates the new operations $\mathbf{S_6}$, $\mathbf{S_6^3} \equiv \mathbf{i}$, and $\mathbf{S_6^5}$. The $\mathbf{i}$ counts only once, so there are then $(4 \times 2) + 1 = 9$ more new operations.

The entire group thus consists of the identity $+ 9 + 8 + 6 + 9 + 6 + 9 = 48$ operations. This group is denoted O_h. It is, of course, a very important type of symmetry since octahedral molecules (e.g., SF_6), octahedral complexes [$Co(NH_3)_6^{3+}$ and $IrCl_6^{3-}$], and octahedral interstices in solid arrays are very common. There is a group O, which consists of only the 24 proper rotations from O_h, but this, like T, is rarely if ever encountered in Nature.

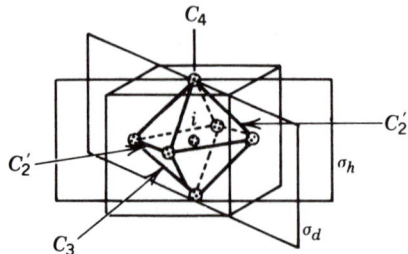

Figure AI-7 The octahedron and the cube, showing one of each of their essential types of symmetry element.

The Pentagonal Dodecahedron and the Icoshedron

These bodies (Fig. AI-8) are related to each other in the same way as are the octahedron and the cube, the vertices of one defining the face centers of the other, and vice versa. Both have the same symmetry operations, a total of 120! We shall not list them in detail but merely mention the basic symmetry elements: six C_5 axes, ten C_3 axes, fifteen C_2 axes, and fifteen planes of symmetry. The group of 120 operations is designated I_h and is often called the icosahedral group.

There is one known example of a molecule that is a pentagonal dodecahedron, namely, dodecahedrane ($C_{12}H_{12}$). The icosahedron is a key structural unit in boron chemistry, occurring in all forms of elemental boron as well as in the $B_{12}H_{12}^{2-}$ ion.

If the symmetry planes are omitted, a group called I consisting of only proper rotations remains. This is mentioned purely for the sake of completeness, since no example of its occurrence in Nature is known.

AI-6   Molecular Dissymmetry and Optical Activity

Optical activity, that is, rotation of the plane of polarized light coupled with unequal absorption of the right- and left-circularly polarized components, is a property of a molecule (or an entire 3D array of atoms or molecules) that is not superposable on its mirror image. When the number of molecules of one type exceeds the number of those that are their nonsuperposable mirror images, a net optical activity results. To predict when optical activity will be possible, it is necessary to have a criterion to determine when a molecule and its mirror image will not be identical, that is, superposable.

Molecules that are not superposable on their mirror images are called *dissymmetric*. This term is preferable to "asymmetric," which means "without symmetry," whereas dissymmetric molecules can and often do possess some symmetry, as will be seen.*

A compact statement of the relation between molecular symmetry properties and dissymmetric character is: *A molecule that has no axis of improper rotation is dissymmetric.*

This statement includes and extends the usual one to the effect that optical isomerism exists when a molecule has neither a plane nor a center of symmetry. It has already been noted that the inversion operation **i** is equivalent to the improper rotation S_2. Similarly, S_1 is a correct although unused way of representing

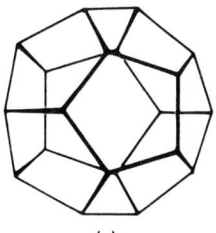

(a)   (b)

Figure AI-8   The two regular polyhedra having I_h symmetry. (a) The pentagonal dodecahedron and (b) the icosahedron.

*Synonyms for dissymmetric and dissymmetry are chiral and chirality. Chiral comes from the Greek word for hand and thus refers to the nonsuperimposable mirror relationship of the right and left hands.

σ, since it implies rotation by $2\pi/1$, equivalent to no net rotation, in conjunction with the reflection. Thus σ and i are simply special cases of improper rotations, namely S_1 and S_2.

However, even when σ and i are absent, a molecule may still be identical with its mirror image if it possesses an S_n axis of some higher order. A good example of this is provided by the $(-RNBX-)_4$ molecule shown in Structure AI-XXVI. This molecule has neither a plane nor a center of symmetry, but inspection shows that it can be superposed on its mirror image. As we have noted, it belongs to the symmetry group S_4.

Dissymmetric molecules either have no symmetry at all, or they belong to one of the groups consisting only of proper rotation operations, that is, the C_n or D_n groups. (Groups T, O, and I are, in practice, not encountered, though molecules in these groups must also be dissymmetric.) Important examples are the bischelate and trischelate octahedral complexes (Structures AI-VIII, AI-X, and AI-XIX).

AI-7   A Systematic Procedure for Symmetry Classification of Molecules

In this section we shall describe a systematic procedure for deciding to what point group any molecule belongs. This will be done in a practical, "how-to-do-it" manner, but the close relationship of this procedure to the arguments used in deriving the various groups should be evident. The following sequence of steps will lead systematically to a correct classification.

1. We determine whether the molecule belongs to one of the "special" groups, that is $C_{\infty v}$, $D_{\infty h}$, or one of those with multiple high-order axes. Only linear molecules can belong to $C_{\infty v}$ or $D_{\infty h}$, so these cannot possibly involve any uncertainty. The specially high symmetry of the others is usually obvious. All of the cubic groups, T, T_h, T_d, O, and O_h, require four C_3 axes, while I and I_h require ten C_3 and six C_5 axes. These multiple C_3 and C_5 axes are the key things to look for. In practice only molecules built on a central tetrahedron, octahedron, cuboctahedron, cube, or icosahedron will qualify, and these figures are usually very conspicuous.

2. If the molecule belongs to none of the special groups, we search for proper or improper axes of rotation. If no axes of either type can be found, we look for a plane or center of symmetry. If a plane only is found, the group is C_s. If a center only is found (this is *very* rare), the group is C_i. If no symmetry element at all is present, the group is the trivial one containing only the identity operation and designated C_1.

3. If an *even*-order improper axis (in practice only S_4, S_6, and S_8 are common) is found but no planes of symmetry or any proper axis except a colinear one (or more), whose presence is automatically required by the improper axis, the group is S_4, S_6, S_8, ... An S_4 axis requires a C_2 axis; an S_6 axis requires a C_3 axis; an S_8 axis requires C_4 and C_2 axes. The important point here is that the S_n (n even) groups consist exclusively of the operations generated by the S_n axis. If any additional operation is possible, we are dealing with a D_n, D_{nd}, or D_{nh} type of group. Molecules belonging to these S_n groups are relatively rare, and the conclusion that a molecule belongs to one of them should be checked thoroughly before it is accepted.

4. Once it is certain that the molecule belongs to none of the groups so far considered, we look for the highest order proper axis. It is possible that there will be no one axis of uniquely high order but instead three C_2 axes. In such a case, we look to see whether one of them is geometrically unique in some sense, for example, in being colinear with a unique molecular axis. This occurs with the molecule allene, which is one of the examples to be worked through later. If all of the axes appear quite similar to one another, then any one may be selected at random as the axis to which the vertical or horizontal character of planes will be referred. Suppose that C_n is our reference or principal axis. The crucial question now is whether there exists a set of n C_2 axes perpendicular to the C_n axis. If so, we proceed to Step 5. If not, the molecule belongs to one of the groups C_n, C_{nv}, and C_{nh}. If there are no symmetry elements except the C_n axis, the group is C_n. If there are n vertical planes, the group is C_{nv}. If there is a horizontal plane, the group is C_{nh}.

5. If in addition to the principal C_n axis there are n C_2 axes lying in a plane perpendicular to the C_n axis, the molecule belongs to one of the groups D_n, D_{nh}, and D_{nd}. If there are no symmetry elements besides C_n and the n C_2 axes, the group is D_n. If there is also a horizontal plane of symmetry, the group is D_{nh}. A D_{nh} group will also, necessarily, contain n vertical planes; these planes

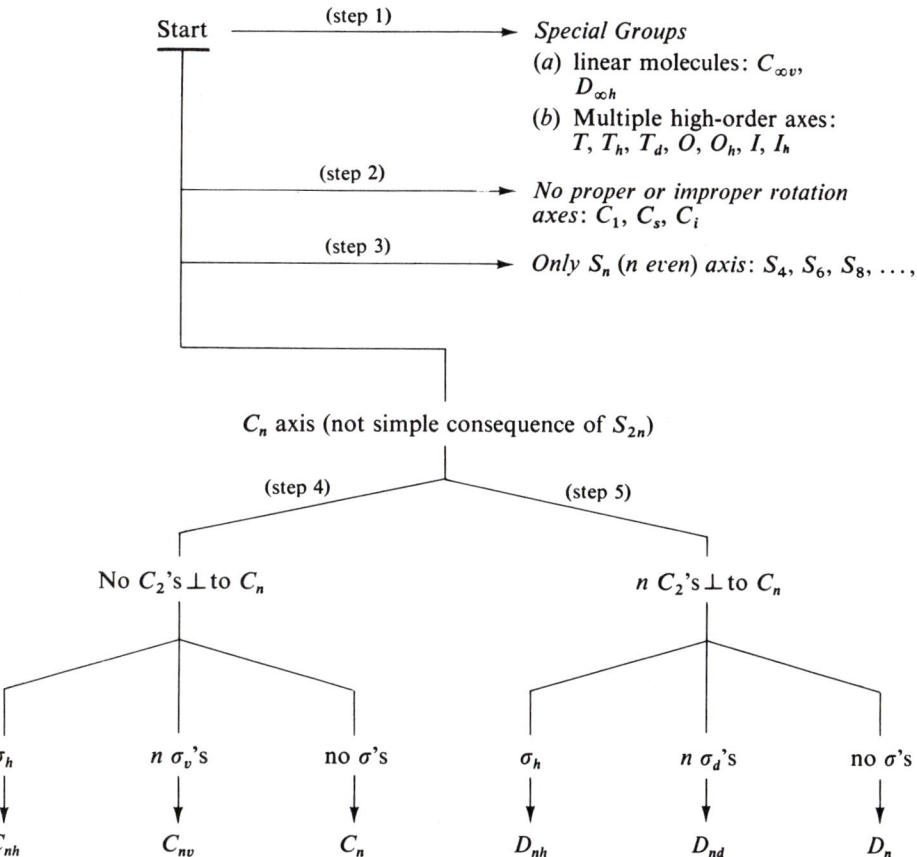

Figure Al-9   A five-stage procedure for the symmetry classification of molecules.

contain the C_2 axes. If there is no σ_h but there is a set of n vertical planes that *pass between* the C_2 axes, the group is D_{nd}.

The five-step procedure just explained is summarized in the flow sheet of Fig. AI-9.

AI-8   Illustrative Examples

The scheme just outlined for allocating molecules to their point groups will now be illustrated. We shall deal throughout with molecules that do not belong to any of the special groups, and we shall also omit molecules belonging to C_1, C_s, and C_i. Thus, each illustration will begin at Step 3, the search for an even-order S_n axis.

Example 1: H$_2$O

3. Water possesses no improper axis.
4. The highest order proper axis is a C_2 axis passing through the oxygen atom and bisecting a line between the hydrogen atoms. There are no other C_2 axes. Therefore H$_2$O must belong to C_2, C_{2v}, or C_{2h}. Since it has two vertical planes, one of which is the molecular plane, it belongs to the group C_{2v}.

Example 2: NH$_3$

3. There is no improper axis.
4. The only proper axis is a C_3 axis; there are no C_2 axes at all. Hence, the point group must be C_3, C_{3v}, or C_{3h}. There are three vertical planes, one passing through each hydrogen atom. The group is thus C_{3v}.

Example 3: Allene

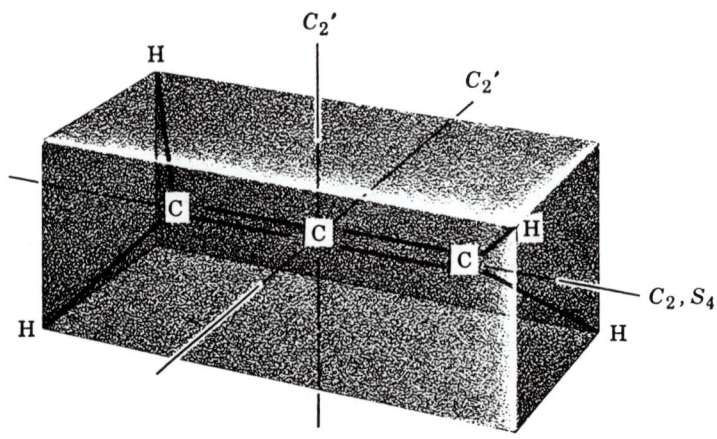

3. There is an S_4 axis coinciding with the main, molecular (C=C=C) axis. However, there are also other symmetry elements besides the C_2 axis which is a necessary consequence of the S_4. Most obvious, perhaps, are the planes of symmetry passing through the H_2C=C=C and C=C=CH$_2$ sets of atoms. Thus, although an S_4 axis is present, the additional symmetry rules out the point group S_4.

4. As noted, there is a C_2 axis lying along the C=C=C axis. There is no higher order proper axis. There are two more C_2 axes perpendicular to this one, as shown in the sketch. Thus, the group must be a D type, and we proceed to Step 5.

5. Taking the C_2 axis lying along the C=C=C axis of the molecule as the reference axis, we look for a σ_h. There is none, so the group D_{2h} is eliminated. There are, however, two vertical planes (which lie between C_2' axes), so the group is D_{2d}.

Example 4: H₂O₂

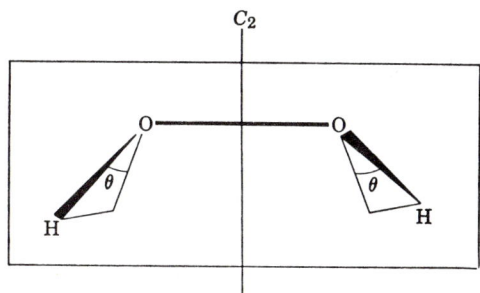

A. The nonplanar equilibrium configuration

3. There is no improper axis.

4. As indicated in the sketch, there is a C_2 axis and no other proper axis. There are no planes of symmetry. The group is therefore C_2. Note that the C_2 symmetry is in no way related to the value of the angle θ except when θ equals 0° or 90°, in which case the symmetry is higher. We shall next examine these two nonequilibrium configurations of the molecule.

B. The cis-planar configuration (θ = 0°)

3. Again there is no even-order S_n axis.

4. The C_2 axis, of course, remains. There are still no other proper axes. The molecule now lies in a plane, which is a plane of symmetry, and there is another plane of symmetry intersecting the molecular plane along the C_2 axis. The group is C_{2v}.

C. The trans-planar configuration (θ = 90°)

3. Again, there is no even-order S_n axis. (except $S_2 \equiv i$).

4. The C_2 axis is still present, and there are no other proper axes. There is now a σ_h, which is the molecular plane. The group is C_{2h}.

Example 5: 1,3,5,7 - Tetramethylcyclooctatetraene

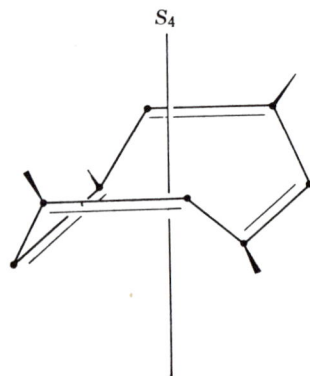

3. There is an S_4 axis. There are no additional independent symmetry elements; the set of methyl groups destroys all the vertical planes and horizontal C_2 axes that exist in C_8H_8 itself. The group is therefore S_4.

   It may be noted that this molecule contains no center of symmetry or any plane of symmetry and yet it is *not* dissymmetric. It thus provides an excellent illustration of the rule developed in Section AI-6.

Example 6: Cyclooctatetraene

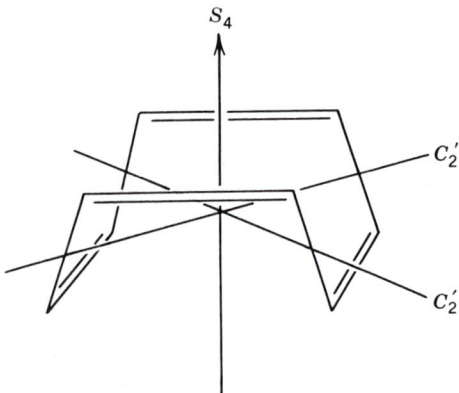

3. There is an S_4 axis. However, there are also numerous other symmetry elements that are independent of the S_4 axis. We thus proceed to Step 4.

4. Coincident with the S_4 axis there is (by necessity) a C_2 axis. No proper axis of higher order can be found, but there are two more, equivalent C_2' axes in a plane perpendicular to the S_4–C_2 axis. Thus we are dealing with a D_2 type of group.

5. There is no σ_h, thus ruling out D_{2h}. There are however, vertical planes of symmetry bisecting opposite double bonds. These pass between the C_2' axes, and the point group is D_{2d}.

Example 7: Benzene

3. There is an S_6 axis, perpendicular to the ring plane, but there are also other symmetry elements independent of the S_6 axis.
4. There is a C_6 axis perpendicular to the ring plane and six C_2 axes lying in the ring plane. Hence, the group is a D_6 type.
5. Since there is a σ_h, the group is D_{6h}. Note that there are vertical planes of symmetry, but they contain the C_2 axes.

Example 8: PF$_5$ (Trigonal Bipyramidal)

3. There is no even-order S_n axis.
4. There is a unique C_3 axis, and there are three C_2 axes perpendicular to it.
5. There is a σ_h; the group is D_{3h}.

Example 9: Ferrocene

A. The staggered configuration

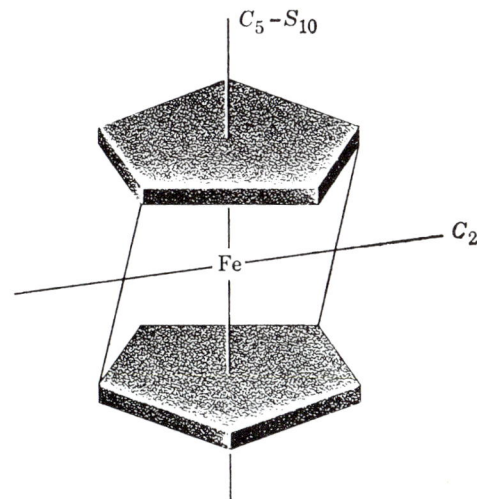

3. There is an even-order improper axis, S_{10}, as indicated in the sketch, but there are also other unrelated symmetry elements, so the group is not S_{10}.
4. The unique, high-order, proper axis is a C_5 axis, as shown. Perpendicular to this there are five C_2 axes.
5. Because of the staggered relationship of the rings there is no σ_h. There are, however, five vertical planes of symmetry that pass between the C_2 axes. The group is thus D_{5d}.

B. The eclipsed configuration

3. There is no even-order S_n axis.

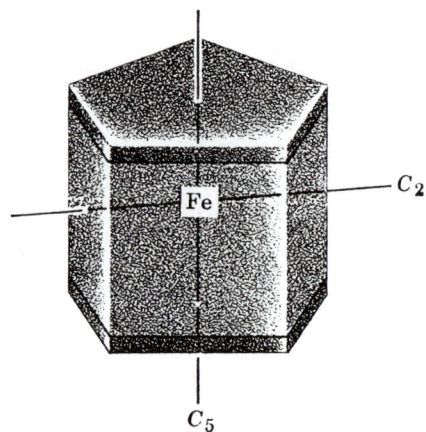

4. There is a C_5 axis as shown. There are five C_2 axes perpendicular to the C_5 axis.
5. There is a σ_h, so the group is D_{5h}.

Appendix IIA

Table AII-A The Hydrogen-Like Atomic Orbital Wave Functions, Factored Into the Radial $[R(r)]$ and Angular $[\Theta(\theta)\Phi(\phi)]$ Components.[a]

Orbital Designation	$\Theta(\theta)\Phi(\phi)$	$R(r)$
$1s$	$\Psi(s) = \left(\dfrac{1}{4\pi}\right)^{1/2}$	$\Psi(1s) = 2\left(\dfrac{Z}{a_0}\right)^{3/2} e^{-\sigma/2}$
$2s$	$\Psi(s) = \left(\dfrac{1}{4\pi}\right)^{1/2}$	$\Psi(2s) = \dfrac{1}{2\sqrt{2}}\left(\dfrac{Z}{2a_0}\right)^{3/2}(2-\sigma)e^{-\sigma/2}$
$2p_z$	$\Psi(p_z) = \left(\dfrac{3}{4\pi}\right)^{1/2}\cos\theta$	
$2p_x$	$\Psi(p_x) = \left(\dfrac{3}{4\pi}\right)^{1/2}\sin\theta\cos\phi$	$\Psi(2p) = \dfrac{1}{2\sqrt{6}}\left(\dfrac{Z}{a_0}\right)^{3/2}\sigma e^{-\sigma/2}$
$2p_y$	$\Psi(p_y) = \left(\dfrac{3}{4\pi}\right)^{1/2}\sin\theta\sin\phi$	
$3s$	$\Psi(s) = \left(\dfrac{1}{4\pi}\right)^{1/2}$	$\Psi(3s) = \dfrac{1}{9\sqrt{3}}\left(\dfrac{Z}{a_0}\right)^{3/2}(6-6\sigma+\sigma^2)e^{-\sigma/2}$
$3p_z$	$\Psi(p_z) = \left(\dfrac{3}{4\pi}\right)^{1/2}\cos\theta$	
$3p_x$	$\Psi(p_x) = \left(\dfrac{3}{4\pi}\right)^{1/2}\sin\theta\cos\phi$	$\Psi(3p) = \dfrac{1}{9\sqrt{6}}\left(\dfrac{Z}{a_0}\right)^{3/2}(4-\sigma)\sigma e^{-\sigma/2}$
$3p_y$	$\Psi(p_y) = \left(\dfrac{3}{4\pi}\right)^{1/2}\sin\theta\sin\phi$	
$3d_{z^2}$	$\Psi(d_{z^2}) = \left(\dfrac{5}{16\pi}\right)^{1/2}(3\cos^2\theta-1)$	
$3d_{xz}$	$\Psi(d_{xz}) = \left(\dfrac{15}{4\pi}\right)^{1/2}\sin\theta\cos\theta\cos\phi$	
$3d_{yz}$	$\Psi(d_{yz}) = \left(\dfrac{15}{4\pi}\right)^{1/2}\sin\theta\cos\theta\sin\phi$	$\Psi(3d) = \dfrac{1}{9\sqrt{30}}\left(\dfrac{Z}{a_0}\right)^{3/2}\sigma^2 e^{-\sigma/2}$
$3d_{x^2-y^2}$	$\Psi(d_{x^2-y^2}) = \left(\dfrac{15}{4\pi}\right)^{1/2}\sin\theta\cos 2\phi$	
$3d_{xy}$	$\Psi(d_{xy}) = \left(\dfrac{15}{4\pi}\right)^{1/2}\sin^2\theta\sin 2\phi$	

[a] The quantity $a_0 = 0.529$ Å is the first Bohr radius and the quantity σ is defined below

$$\sigma = \frac{2\,Zr}{na_0}$$

Appendix IIB

IONIZATION ENTHALPIES OF THE ELEMENTS

The values given below are the standard enthalpies (in kJ mol^{-1}) of stepwise electron removal (ionization), as in the following equations:

First Ionization

$$X(g) = X^+(g) + e^-(g) \qquad \Delta H°_{ion}(1)$$

Second Ionization

$$X^+(g) = X^{2+}(g) + e^-(g) \qquad \Delta H°_{ion}(2)$$

Third Ionization

$$X^{2+}(g) = X^{3+}(g) + e^-(g) \qquad \Delta H°_{ion}(3)$$

Fourth Ionization

$$X^{3+}(g) = X^{4+}(g) + e^-(g) \qquad \Delta H°_{ion}(4)$$

Values of $\Delta H°_{ion}(n)$ (kJ mol^{-1})

Z	First	Second	Third	Fourth
1 H	1311			
2 He	2372	5249		
3 Li	520.0	7297	11,810	
4 Be	899.1	1758	14,850	21,000
5 B	800.5	2428	2,394	25,020
6 C	1086	2353	4,618	6,512
7 N	1403	2855	4,577	7,473
8 O	1410	3388	5,297	7,450
9 F	1681	3375	6,045	8,409
10 Ne	2080	3963	6,130	9,363
11 Na	495.8	4561	6,913	9,543
12 Mg	737.5	1450	7,731	10,540
13 Al	577.5	1817	2,745	11,580
14 Si	786.3	1577	3,228	4,355
15 P	1012	1903	2,910	4,955
16 S	999.3	2260	3,380	4,562
17 Cl	1255	2297	3,850	5,160
18 Ar	1520	2665	3,950	5,771

19 K	418.7	3069	4,400	5,876
20 Ca	589.6	1146	4,942	6,500
21 Sc	631	1235	2,389	7,130
22 Ti	656	1309	2,650	4,173
23 V	650	1414	2,828	4,600
24 Cr	652.5	1592	3,056	4,900
25 Mn	717.1	1509	3,251	
26 Fe	762	1561	2,956	
27 Co	758	1644	3,231	
28 Ni	736.5	1752	3,489	
29 Cu	745.2	1958	3,545	
30 Zn	906.1	1734	3,831	
31 Ga	579	1979	2962	6190
32 Ge	760	1537	3301	4410
33 As	947	1798	2735	4830
34 Se	941	2070	3090	4140
35 Br	1142	2080	3460	4560
36 Kr	1351	2370	3560	
37 Rb	402.9	2650	3900	
38 Sr	549.3	1064		5500
39 Y	616	1180	1979	
40 Zr	674.1	1268	2217	3313
41 Nb	664	1381	2416	3700
42 Mo	685	1558	2618	4480
43 Tc	703	1472	2850	
44 Ru	710.6	1617	2746	
45 Rh	720	1744	2996	
46 Pd	804	1874	3177	
47 Ag	730.8	2072	3360	
48 Cd	876.4	1630	3615	
49 In	558.1	1820	2705	5250
50 Sn	708.2	1411	2942	3928
51 Sb	833.5	1590	2440	4250
52 Te	869	1800	3000	3600
53 I	1191	1842		
54 Xe	1169	2050	3100	
55 Cs	375.5	2420		
56 Ba	502.5	964		
57 La	541	1103	1849	
72 Hf	760	1440	2250	3210
73 Ta	760	1560		
74 W	770	1710		
75 Re	759	1600		
76 Os	840	1640		
77 Ir	900			
78 Pt	870	1791		
79 Au	889	1980		
80 Hg	1007	1809	3300	
81 Tl	588.9	1970	2880	4890
82 Pb	715.3	1450	3080	4082
83 Bi	702.9	1609	2465	4370
84 Po	813			
86 Rn	1037			
88 Ra	509.1	978.6		
89 Ac	670	1170		

Appendix IIC

IONIC RADII

Selected ionic radii (in Å) determined by the method of Shannon and Prewitt (*Acta Crystallogr.* **1976,** *A32,* 751)[a]

Ion	C.N.	Radius	Ion	C.N.	Radius
A. Alkali and Alkaline Earth Cations					
Li^+	4	0.73	Ba^{2+}	6	1.49
	6	0.90		8	1.56
	8	1.06		10	1.66
Na^+	4	1.13		12	1.75
	6	1.16	Ra^{2+}	8	1.62
	8	1.32		12	1.84
	12	1.53	*B. Group IB (11)*		
K^+	4	1.51	Cu^+	2	0.60
	6	1.52		4	0.74
	8	1.65		6	0.91
	10	1.73	Ag^+	2	0.81
	12	1.78		4	1.14
Rb^+	6	1.66		4 (sq)	1.16
	8	1.75		6	1.29
	10	1.80		8	1.42
	12	1.86	Au^+	6	1.51
	14	1.97	Cu^{2+}	4	0.71
Cs^+	6	1.81		4 (sq)	0.71
	8	1.88		6	0.87
	10	1.95	Au^{3+}	4 (sq)	0.82
	12	2.02		6	0.99
Fr^+	6	1.94	*C. Group IIB (12)*		
Be^{2+}	4	0.41	Zn^{2+}	4	0.74
	6	0.59		6	0.88
Mg^{2+}	4	0.71		8	1.04
	6	0.86	Cd^{2+}	4	0.92
	8	1.03		6	1.09
Ca^{2+}	6	1.14		8	1.24
	8	1.26		12	1.45
	10	1.37	Hg^{2+}	2	0.83
	12	1.48		4	1.10
Sr^{2+}	6	1.32		6	1.16
	8	1.40		8	1.28
	10	1.50			
	12	1.58			

Ion	C.N.	Radius	Ion	C.N.	Radius
D. Other Non-Transition Metal Ions			Ni^{2+}	4	0.69
NH_4^+	6	1.61		4 (sq)	0.63
Tl^+	6	1.64		6	0.83
	8	1.73	Ti^{3+}	6	0.81
	12	1.84	V^{3+}	6	0.78
Pb^{2+}	6	1.33	Cr^{3+}	6	0.76
	8	1.43	Mn^{3+}	6 (LS)	0.72
	10	1.54		6 (HS)	0.79
	12	1.63	Fe^{3+}	4 (HS)	0.63
B^{3+}	4	0.25		6 (LS)	0.69
Al^{3+}	4	0.53		6 (HS)	0.79
	6	0.68	Co^{3+}	6 (LS)	0.69
Ga^{3+}	4	0.61		6 (HS)	0.75
	6	0.76	Ni^{3+}	6 (LS)	0.70
In^{3+}	4	0.76		6 (HS)	0.74
	6	0.94	Ti^{4+}	6	0.75
	8	1.06	**F. Second Transition Series Elements**		
Tl^{3+}	4	0.89	Pd^{2+}	4 (sq)	0.78
	6	1.03		6	1.00
	8	1.12	Nb^{3+}	6	0.86
Sb^{3+}	6	0.90	Mo^{3+}	6	0.83
Bi^{3+}	6	1.17	Ru^{3+}	6	0.82
Sc^{3+}	6	0.89	Rh^{3+}	6	0.81
	8	1.01	Nb^{4+}	6	0.82
Y^{3+}	6	1.04	Mo^{4+}	6	0.79
C^{4+}	4	0.29	Ru^{4+}	6	0.76
Si^{4+}	4	0.40	Rh^{4+}	6	0.74
	6	0.54	**G. Third Transition Series Elements**		
Ge^{4+}	4	0.53			
	6	0.67	Pt^{3+}	4 (sq)	0.74
Sn^{4+}	4	0.69		6	0.94
	6	0.83	Ta^{3+}	6	0.86
	8	0.95	Ir^{3+}	6	0.82
Pb^{4+}	4	0.79	Hf^{4+}	6	0.85
	6	0.92	Ta^{4+}	6	0.82
	8	1.08	W^{4+}	6	0.80
			Re^{4+}	6	0.77
E. First Transition Series Metals			Os^{4+}	6	0.78
Ti^{2+}	6	1.00	Ir^{4+}	6	0.77
V^{2+}	6	0.93	Pt^{4+}	6	0.77
Cr^{2+}	6 (LS)	0.87	Th^{4+}	6	1.08
	6 (HS)	0.94			
Mn^{2+}	4 (HS)	0.80	**H. Anions**		
	6 (LS)	0.81	F^-	2	1.15
	6 (HS)	0.97		4	1.17
Fe^{2+}	4 (HS)	0.77		6	1.19
	6 (LS)	0.75	Cl^-	6	1.67
	6 (HS)	0.92	Br^-	6	1.82
Co^{2+}	4 (HS)	0.72	I^-	6	2.06
	6 (LS)	0.79	O^{2-}	2	1.21
	6 (HS)	0.89		3	1.22

Ion	C.N.	Radius	Ion	C.N.	Radius
	4	1.24	OH^-	2	1.18
	6	1.26		3	1.20
O^{2-}	8	1.28		4	1.21
S^{2-}	6	1.70		6	1.23
Se^{2-}	6	1.84	N^{3-}	4	1.32
Te^{2-}	6	2.07			

[a]Values are given for each of an ion's important coordination number (C.N.) as well as for both high spin (HS) and low spin (LS) cases for the appropriate transition metal ions. Six-coordinate ions are octahedral, and "sq" designates a square coordination geometry. Otherwise no particular coordination geometry is implied.

ELECTRON ATTACHMENT ENTHALPIES OF SELECTED ELEMENTS

The values given below are the standard enthalpies of electron attachment, ΔH_{EA} (in kJ mol^{-1}), as defined by the following process:

$$X(g) + e^- = X^-(g) \qquad \Delta H^{\circ}{}_{EA}$$

Note: All alkaline earth and noble gas elements have $\Delta H^{\circ}_{EA} > 0$.

Z	Element	ΔH°_{EA} (kJ mol^{-1})
1	H	−72.77
3	Li	−59.8
5	B	−27
6	C	−122.3
7	N	~0
8	O	−141
9	F	−328
11	Na	−53
13	Al	−45
14	Si	−132.2
15	P	−71
16	S	−200
17	Cl	−349
19	K	−48.4
33	As	−77
34	Se	−194
35	Br	−324
37	Rb	−46.9
51	Sb	−101
52	Te	−190.1
53	I	−295.3
55	Cs	−45.5
79	Au	−222.8
83	Bi	−97

Appendix IIE

A COMPARISON OF ELECTRONEGATIVITY VALUES (PAULING UNITS) FROM FOUR SOURCES

Atom	χ_{spec}[a]	χ_P[b]	$\chi_{A\&R}$[c]	χ_M[d]
H	2.300	2.20	2.20	3.059
Li	0.912	0.98	0.97	1.282
Be	1.576	1.57	1.47	1.987
B	2.051	2.04	2.01	1.828
C	2.544	2.55	2.50	2.671
N	3.066	3.04	3.07	3.083
O	3.610	3.44	3.50	3.215
F	4.193	3.98	4.10	4.438
Ne	4.787			4.597
Na	0.869	0.93	1.01	1.212
Mg	1.293	1.31	1.23	1.630
Al	1.613	1.61	1.47	1.373
Si	1.916	1.90	1.74	2.033
P	2.253	2.19	2.06	2.394
S	2.589	2.58	2.44	2.651
Cl	2.869	3.16	2.83	3.535
Ar	3.242			3.359
K	0.734	0.82	0.91	1.032
Ca	1.034	1.00	1.04	1.303
Ga	1.756	1.81	1.82	1.343
Ge	1.944	2.01	2.02	1.949
As	2.211	2.18	2.20	2.256
Se	2.424	2.55	2.48	2.509
Br	2.685	2.96	2.74	3.236
Kr	2.966			2.984

Rb	0.706	0.82	0.89	0.994
Sr	0.963	0.95	0.99	1.214
In	1.656	1.78	1.49	1.298
Sn	1.824	1.96	1.72	1.833
Sb	1.984	2.05	1.82	2.061
Te	2.158	2.10	2.01	2.341
I	2.359	2.66	2.21	2.880
Xe	2.582			2.586

[a] L.C. Allen, *J. Am. Chem. Soc.* **1989,** *111*, 9003.

[b] Pauling's values, taken from A. L. Allred, *J. Inorg. Nucl. Chem.,* **1961,** *17,* 215.

[c] A.L. Allred and E. G. Rochow, *J. Inorg. Nucl. Chem.,* **1958,** *5,* 264.

[d] Mulliken's values, taken from H. Hotop and W. C. Lineberger, *J. Phys. Chem. Ref. Data,* **1985,** *14,* 731.

Glossary[*]

Absorbance (n). The $\log_{10}$ of the ratio I_0/I, where I_0 is the intensity of incident light and I is the intensity of the transmitted light. It is usually denoted A and is equal to the unitless quantity in Beer's law: $A = \epsilon \, cd$, where ϵ is the molar absorptivity, c is the concentration, and d is the path length.

Absorption (n). The process by which the intensity of radiation is reduced as it passes through a material.

Absorptivity, molar (n). The constant ϵ (in $L \, mol^{-1} \, cm^{-1}$) in Beer's law: $A = \epsilon cd$. It is also called the extinction coefficient.

Acid–base reaction (n). (a) According to the definition of Brønsted-Lowry, the neutralization of a proton donor by a proton acceptor; (b) according to the definition of Lewis, the formation of an adduct between an electron-pair donor and an electron-pair acceptor; (c) according to the Lux–Flood definition, the reaction of an oxide ion acceptor with an oxide ion donor.

Actinide elements (n). The elements ^{90}Th through ^{103}Lr, which follow actinium, ^{89}Ac.

Activated complex (n). That arrangement of atoms, groups, molecules, or ions that has the highest free energy along the reaction coordinate (free energy profile) for a reaction. This is also known as the transition state and corresponds to the minimum free energy that must be possessed by the reactive ensemble in order to consummate the reaction.

Addition reaction (n). A reaction in which a group, molecule, or ion combines with another. Common examples are additions across a multiple bond and addition to an atom that is able to undergo coordination sphere expansion (an increase in occupancy). This reaction is the converse of elimination.

Adduct (n). The product of the addition of a Lewis acid to a Lewis base.

Adiabatic (adj). Without heat transfer.

Adsorption (n). The adhering or retention of a substance (usually a gas, liquid, or a mixture of these) on the surface of a material.

Alkene (n). A substance containing a $C{=}C$ double bond.

Allotrope (n). One of the two or more distinct forms or structures adopted by an element, for example, O_2 and O_3.

Alloy (n). A solid solution of two or more metals.

Alum (n). An ionic sulfate containing a trivalent cation (nominally Al^{3+}) and any of a number of monovalent cations such as K^+. This class of substances is named after the parent potassium alum, $KAl(SO_4)_2(H_2O)_{12}$, and may contain practically any combination of monovalent and trivalent cations.

Amalgam (n). An alloy of a metal with mercury.

Amorphous (adj). Having a random or disordered arrangement in the solid state, that is, the antithesis of crystalline.

Amphoteric (adj). Capable of reacting either as an acid or as a base.

Angular momentum (n). A property associated with angular motion, equal to the product of angular velocity (ω) and the moment of inertia (mr).

Angular wave function (n). That portion of the total wave function $\psi(r, \theta, \phi) = R(r)\Theta(\theta)\Phi(\phi)$ that is factorable from $R(r)$, namely, $\Theta(\theta)\Phi(\phi)$.

Anhydride, acidic (n). An oxide that reacts with water to form an acid. Acidic anhydrides are usually nonmetal oxides such as SO_3 and P_2O_5. Thus acidic anhydrides are acids with water removed.

---

[*]n = noun, adj = adjective, v = verb.

Anhydride, basic (n). An oxide of a metal that reacts with water to give an aqueous hydroxide, or that reacts with protic acids to give aqueous metal salts. These can sometimes be formed by thorough dehydration of hydrous oxides or hydroxides.

Anhydrous (adj). Lacking water.

Antibonding orbital (n). A molecular orbital at higher energy than the orbitals from which it was formed, resulting from negative overlap of atomic orbitals, and having less electron density (or electron probability) between the nuclei than would be true of the simple sum of the electron densities from the combining orbitals of the separate atoms.

Aqua ion (n). A metal ion that is exclusively coordinated by a given number of water molecules. (This is sometimes spelled aquo.)

Autooxidation (n). The apparently spontaneous oxidation of a substance that is exposed to the atmosphere, hence, oxidation by gaseous dioxygen.

Azimuthal quantum number (n). The orbital angular momentum quantum number $l = 0, 1, 2, \ldots, (n-1)$.

Band gap (n). In a solid, any energy level with a density of states equal to zero.

Bidentate (adj). Twice attached; used to describe a ligand.

Bond, covalent (n). The strong attractive force that holds together atoms within a molecule or complex ion, and that arises between a pair of bonded atoms through the sharing of a pair of electrons, one electron of the pair being contributed by each atom.

Bond, coordinate covalent (also called a dative bond) (n). A covalent bond in which both electrons originate from the same atom. This type of bond arises from the addition of a Lewis base to a Lewis acid. It is the electron-pair donor–acceptor bond of an adduct.

Bond, ionic (n). A bond consisting of the electrostatic attraction between a cation and an anion.

Bond, polar covalent (n). An electron-pair or covalent bond in which the electron density is not distributed equally or shared evenly between the two atoms because of a difference in electronegativity. The electron density in the bond is shifted (polarized) towards the more electronegative atom.

Bonding orbital (n). A molecular orbital at lower energy than the orbitals from which it was formed, resulting from positive overlap of orbitals from separate atoms, and having more electron density (or electron probability) between the nuclei than would be true of the simple sum of the electron densities from the combining orbitals of the separate nuclei.

Borate (n). A compound containing polynuclear oxo anions of boron, which are ring or chain polymeric anions containing planar BO_3 or tetrahedral BO_4 units. Also, neutral borate esters that may be considered to be derived from boric acid.

Borax (n). The sodium salt of the ring anion "$[B_4O_7]$," of composition $Na_2B_4O_7 \cdot 10H_2O$, more properly written $Na_2[B_4O_5(OH)_4] \cdot 8 H_2O$ since the tetraborate dianion is dihydrated and contains four B—OH groups.

Brass (n). Any of the alloys variously composed of copper and zinc, sometimes also containing small amounts of other constituents, often tin or lead.

Bronze (n). Any of the alloys composed of at least 88% copper, tin (8–10%), and sometimes zinc.

Buckminsterfullerene (n). The C_{60} molecule. See also fullerene.

Calcination (n). Strong heating of a material (usually an ore) in a furnace to achieve some desired decomposition or change, e.g. of calcium carbonate to give calcium oxide and carbon dioxide.

Canonical form (n). One of the contributing resonance structures of a substance.

Carbonyl (n). A compound containing a CO group.

Catalyst (n). A substance that increases the rate (lowers the activation barrier) for attainment of equilibrium, and that, ideally, can be recovered at the end of the reaction.

Catenation (n). The self-linking of an element in its compounds; the forming of chains.

Center of inversion (n). A symmetry element (point) in a structure through which inversion [changing every location $(+x, +y, +z)$ to $(-x, -y, -z)$] leaves the structure indistinguishable from the original.

Center of symmetry (n). Another term for center of inversion.

Chalcogen (n). An element of Group VIB(16), i.e. O, S, Se, Te, or Po.

Chelate (n). A ligand that is able to bond to a central metal atom simultaneously through more than one donor atom.

Chiral (a). Not superimposable on its mirror image.

Clathrate (n). A solid in which are trapped one or more substances (usually gases or volatile liquids) within various interstices of the structure. These are not compounds with definite and fixed compositions, although there are limiting compositions in which all the interstices appropriate for holding the trapped substance are occupied.

Coke (n). A porous residue of carbon and mineral ash that is obtained from furnaces where coal is heated in a deficiency or absence of oxygen to drive off volatile materials.

Colloidal system (n). A dispersion of one substance within another such that one (the colloid) is uniformly distributed throughout the other (the dispersing medium) in a manner that is intermediate between a true solution and a suspension. The colloidal particle sizes may vary from 10 to 10,000 Å, and although they are too small to be seen with the unaided eye, they can be illuminated with a beam of light.

Combustion (n). Any reaction that is sufficiently exothermic to be self-sustaining; usually used for reactions with O_2.

Conformation (n). One of the various arrangements of atoms in a molecule or complex ion, based on differences in angle(s) of rotation about single bonds.

Conformer (n). One of a number of conformational isomers or rotamers.

Conjugate acid–base pair (n). Substances related to one another by proton transfer.

Corundum (n). The hard, α-form of Al_2O_3, containing aluminum ions in two-thirds of the octahedral holes in a hexagonally close-packed array of oxide ions.

Coulomb (n). The SI unit of charge, that of an electron being $1.6021892 \times 10^{-19}$ coulombs.

Cryolite (n). The mineral Na_3AlF_6 which, as the melt, finds application as a medium for the electrolytic production of aluminum from bauxite.

Crystalline (adj). Having a regular and continuous three-dimensional arrangement of atoms in the solid state.

Cyclization reaction (n). A reaction that leads to the formation of rings.

Dative bond (n). See Bond, coordinate covalent.

Degenerate orbitals (n). Orbitals having the same energy.

Delta bond (n). A bond formed from the face-to-face overlap of d orbitals from separate atoms, such that the internuclear axis coincides with the intersection of the two nodal planes that divide the electron density of the bond.

Diastereomers (n). Isomers that are individually chiral, but that are not mirror images of one another.

Diastereotopic (n). Leading to diastereomers.

Dielectric constant (n). The constant k in the equation for the force f between two unit charges q separated by a distance r: $f = q^2/kr^2$.

Diffusion (n). A movement of molecules (or particles) throughout a solvent (or within a mixture) so as to make the system uniform or homogeneous.

Disproportionation reaction (n). Self-reaction of one substance to give simultaneously two or more dissimilar substances.

Ductile (adj). Able to be drawn to a longer length without breaking.

Effective nuclear charge (n). That portion of the total nuclear charge that is experienced by a given electron. It is equal to the total, or formal nuclear charge, less the amount by which other electrons shield the given electron from the nucleus.

Effusion (n). The escape of a gas through a small hole in its container.

Elastomer (n). A macromolecular substance that can be stretched to at least twice its unstressed length and will return on release to nearly its original length. The elastic character of rubber is improved through vulcanization.

Electronegativity (n). The ability of an atom in a molecule to attract electrons to itself.

Electron affinity (n). The energy that is released when an electron is added to the valence shell of an atom. It is the negative of the electron attachment enthalpy (ΔH_{EA}).

Electron configuration (n). A listing of the electrons of an atom or ion according to their distribution within the various available orbitals.

Electron spin (n). The characteristic angular momentum associated with an electron, and that is independent of orbital angular momentum or motion.

Electron transfer reaction (n). An oxidation-reduction reaction in which electrons pass from one reactant to another.

Elimination reaction (n). A reaction in which a group, molecule, or ion is separated from another. Examples are dehydrohalogenations, and eliminations from metal centers that are able to undergo coordination number reduction (a decrease in occupancy). This is the converse of an addition reaction.

Emulsion (n). A type of colloidal mixture or system.

Enantiomer (n). Enantiomorph.

Enantiomorph (n). One of a pair of optical isomers, that is, one of a pair of chiral isomers, each of which is the nonsuperimposable mirror image of the other.

Enantiotopic (adj). Leading to enantiomorphs.

Enclosure system (n). A boundary representation of an orbital such that some arbitrary fraction (usually large) of total electron density is distributed between the surface and the origin.

Eutectic mixture (n). The lowest melting composition obtainable from a given set of components that form a solid solution.

Exchange reaction (n). A reaction in which two atoms, ions, or groups switch places either between two different molecules or ions (intermolecular exchange) or within the same molecule or ion (intramolecular exchange).

Fermi level (n). The highest occupied energy level in the electronic structure of a solid.

Fixation (n). Any process by which otherwise inert dinitrogen is combined with other elements, most notably of dinitrogen with H_2 to give ammonia.

Fluorescence (n). The emission, immediately following excitation, of electromagnetic radiation (at longer wavelength than that necessary to accomplish the excitation) from a substance in an excited electronic state. The electronic process that accounts for the release of the electromagnetic radiation is characteristically a spin-allowed process. This and phosphorescence together constitute the general behavior known as luminescence.

Flux (n). An additive that aids in the fusion of a material, and that often imparts to the melt a resistance to oxide formation.

Frenkel defect (n). A point defect in a solid, arising when atoms or ions are displaced from their normal sites in a crystal to a position between their normal sites.

Frequency (n). In any periodic motion, the number of cycles completed in a unit of time, that is, cycles per second, or s^{-1}. In electromagnetic radiation, frequency v (in s^{-1}) equals the speed c (cm s^{-1}) divided by the wavelength λ (cm), or $v = c/\lambda$.

Friedel–Crafts reaction (n). A reaction catalyzed by Al_2Cl_6 and resulting in the condensation of alkyl or aryl halides with benzene or its derivatives, and in which an alkyl (R) or acyl

(RCO) group is substituted for a hydrogen atom of the aromatic ring to give, respectively, an alkyl or a ketone.

Fullerene (n). An allotrope of carbon composed of large spheroidal C_n molecules, where n is equal to 60 or more. The molecule C_{60} has been termed buckminsterfullerene. Less formally, these allotropes are also known as buckyballs, after their obvious structural similarity to a soccer ball.

Fusion (n). Melting.

Galvanized (adj). Coated with zinc, either by dipping in molten zinc (hot dipping), sherardizing (rolling in powdered zinc at ~ 300 °C), or electrodeposition.

Gas hydrate (n). A clathrate compound in which water is the host (a pentagonal dodecahedral arrangement of water molecules being common) and gases are trapped.

Geometrical isomers (n). Molecules or complex ions having the same empirical formulae, and the same atomic linkages, but differing in the spatial orientation of like groups.

Glass (n). An amorphous solid (formed from a supercooled liquid) in which there is the same type of arrangement as in the liquid but without appreciable translational energy. A glass should be regarded as a metastable material because the corresponding crystalline material would have a lower free energy, but the glassy structure cannot rearrange to the preferred crystalline lattice. One normally thinks of the silica glasses but, technically, even metals may form glasses.

Grignard reagent (n). An organomagnesium halide, of formal composition RMgX, usually prepared in anhydrous ether solution and used to transfer R groups.

Group orbitals (n). Linear combinations of orbitals from separate atoms and conforming to molecular geometry so that, as a group, they can overlap with orbitals of other groups or atoms, leading to bonding.

Halogen (n). An element of Group VIIB(17), that is, F, Cl, Br, I, and At.

Hapto (adj). A prefix used (in conjunction with a designation mono-, di-, tri-, tetra-, penta-, hexa-, etc.) to specify the number of atoms within a ligand that are attached to a metal atom.

Heterogeneous (adj). Consisting of dissimilar components (antonym: homogeneous).

Heterolysis (n). Cleavage of an electron-pair bond in an unsymmetrical fashion so that one atom of the pair retains both electrons of the bond.

Heterolytic (adj). Leading to or pertaining to heterolysis.

Homogeneous (adj). Having uniform composition, structure, and properties throughout (antonym: heterogeneous).

Homologous (adj). Consisting of the same general class, but differing by the addition of various numbers of some common unit, i.e. the homologous series of alkanes built up of CH_2 units.

Homolysis (n). Cleavage of an electron-pair bond symmetrically so as to allow each atom of the pair to retain one electron.

Homolytic (adj). Leading to or pertaining to homolysis.

Hybrid orbital (n). A combination of two or more atomic orbitals of like energy on the same atom.

Hydrated (adj). Containing in the crystalline form a distinct, fixed, and reproducible number of water molecules incorporated into the crystal structure, for example, $CuSO_4 \cdot 5H_2O$ or $CoCl_2 \cdot 6 H_2O$.

Hydrolysis (n). A decomposition with water, the hydrogen and hydroxyl of which are found in separate products of the reaction. A more general definition—any reaction with water—is sometimes also used.

Hydrolytic (adj). Involving hydrolysis.

Hydrous (adj). Generally containing water, but not in the distinct, fixed, and reproducible proportions typical of a hydrated substance. Examples: A hydrous metal oxide, $MO \cdot n H_2O$ as opposed to a distinct hydroxide, $M(OH)_2$.

I_a. The associative interchange mechanism of ligand substitution.

I_d. The dissociative interchange mechanism of ligand substitution.

Inert (adj). Slow to react, and, in particular, having a half-life for reaction of a minute or longer.

Insertion reaction (n). The interposition of a new molecule, group, or ion between atoms in a structure such that the added molecule, group, or ion separates the two parts of the structure that were formerly bonded together.

Insulator (n). A substance having a low electrical conductivity, due to a large band gap between the highest filled band and the lowest unfilled band.

Intercalation (v). Penetration of other atoms, ions, or molecules into the regions between layers in solids having a layer structure.

Intermediate (n). A structure that occurs along the reaction pathway and that has a lower free energy than the two transition states that bracket it. Also, a precursor to some desired product. These are generally postulated rather than actually isolable.

Inversion center. See center of inversion.

Inversion reaction (n). A change of the chirality at a single center of chirality.

Ion pair (n). A pair of oppositely charged ions that are associated by electrostatic forces.

Isoelectronic (adj). Having the same electron configuration.

Isomer (n). One of two or more substances that are structurally and physically different but which have the same elemental composition.

Isomerization reaction (n). A conversion of one isomer into another.

Isomorphous (adj). Having the same shape, form, or structure; generally said of crystals.

Isostructural (adj). Having the same structure.

Isotope (n). One of two or more forms of an atom, all of which have the same atomic number (i.e., the same number of protons), but which differ in the number of neutrons in the nucleus of the atom. The atomic mass and the mass number of the various isotopes of an element are different.

Jahn–Teller Effect (n). The requirement that a nonlinear molecule with degenerate orbitals must undergo a structural change in order to lift the degeneracy; in other words, the requirement that molecules adopt geometries that do not lead to a degeneracy in valence-level, or d electron, configurations.

Kieselguhr (n). Infusorial or diatomaceous earth, which is a fine powder used as an absorbent or clarifying agent. In its most useful form it is a soft, white, porous powder, mostly of hydrated silica.

Labile (adj). Quick to react, in particular having a half-life for reaction of less than a minute.

Lanthanides (n). The elements ^{58}Ce to ^{71}Lu, which follow lanthanum, ^{57}La.

Ligand (n). A molecule, group, or anion that bonds to a metal atom or cation.

Ligand field (n). The electrostatic field created by a set of ligands arranged in some particular geometry about a metal.

Lime (n). Calcium oxide.

Limestone (n). Naturally occurring calcium carbonate.

Luminescence (n). Either fluorescence or phosphorescence.

Malleable (adj). Extensible or deformable in all dimensions, without loss in character, by hammering or rolling.

Magnesite (n). The mineral $MgCO_3$.

Magnetite (n). The mineral Fe_3O_4 or lodestone.

Metal (n). A solid whose electrical conductivity decreases with increasing temperature. See also metallic.

Metallic (adj). Having the properties of a metal: luster (surface sheen), high thermal conductivity, and high electrical conductivity. Malleability and ductility also characterize many metals.

Metalloid (n). An element that exhibits some metallic characteristics together with some non-metallic ones; examples are Ge and Te.

Metathesis (n). An exchange of comparable groups, such that two compounds form two new ones: $AX + BY \longrightarrow AY + BX$.

Mineral (n). A naturally occurring inorganic substance, which has a characteristic elemental composition and structure, and which is found in pure crystalline form, or in a composite rock in the earth's crust.

Moderator (n). A substance that reduces the speed of neutrons created in nuclear fusion reactors.

Molar absorptivity. See absorptivity, molar.

Molecular sieves (n). Sodium or calcium aluminosilicates with porous cavities generally 3–4 Å in size, which are able to discriminate among substances based on molecular sizes.

Momentum (n). Classically, the product of mass and velocity.

Multiplicity, spin (n). For an ensemble of electrons, the spin multiplicity equals the quantity $(2S + 1)$, where S is the spin quantum number.

Node (n). A point, line, or surface where electron density is zero, caused by a change in the sign of the wave function. Also, in stationary waves, a point, line, or surface at which there is no displacement.

Nonbonding (adj). Neither bonding nor antibonding.

Nuclear reaction (n). A reaction that changes the atomic number or mass number of an atom.

Occupancy (n). A spatial designation for an atom in a molecule or complex ion. When an atom A resides in a molecule AB_xE_y, with x being the number of other atoms B bound to A, and y being the number of lone pairs of electrons (E) at atom A, the occupancy at A is $x + y$.

Octet (n). A set of eight.

Octet rule (n). A rule whose application is limited to elements of the first short period, stating that a set of eight electrons (as various combinations of shared and lone electrons) at an atom in its compounds is most stable.

Olefin (n). See alkene.

Oleum (n). Fuming sulfuric acid; 100% H_2SO_4 containing some dissolved SO_3.

Oligomer (n). A polymer made up of only a few (usually less than four) monomers.

Optical activity (n). The ability of a substance to rotate the plane of plane-polarized light, and a characteristic property of individual enantiomorphs or chiral substances.

Optical isomerism (n). The isomerism associated with chirality.

Orbit (n). The path followed by the electron in Bohr's theory of the hydrogen atom, characterized by the quantity r, the distance from the nucleus of the electron.

Orbital (n). The space-dependent (time-independent) portion of the wave function for an electron in an atom, molecule, or ion.

Ore (n). A natural inorganic material from which important metals or nonmetals may be extracted.

Oxidation (n). A loss of electrons, corresponding to an increase in oxidation state.

Oxidizing agent (n). A substance that causes oxidation, and which is, as a consequence, reduced.

Oxo process (n). The general catalytic reaction of alkenes with carbon monoxide and molecular hydrogen to give, variously, alcohols, aldehydes, and so on.

Oxo acid (n). A protic acid containing an element combined with oxygen; the ionizable hydrogen atoms are attached to O atoms.

Passivation (of metals) (n). A treatment that renders the metal surface inert, and which usually results from the action of strong oxidizing agents.

Pauli exclusion principle (n). A postulate that no single atom may possess two electrons that have the same four quantum numbers.

Perovskite (n). The mineral $CaTiO_3$, a mixed-metal oxide.

Phosphorescence (n). The delayed emission of electromagnetic radiation from a substance in an excited electronic state. As in fluorescence, the emitted wavelength is longer than that necessary to accomplish the excitation. Phosphorescence is different in that the emission is delayed because the electronic process that accounts for the emission is a spin-forbidden process by the rules of quantum mechanics. Phosphorescence and fluorescence together constitute the general behavior known as luminescence.

Pi bond or π bond (n). A bond formed from side-to-side overlap of p or d orbitals from separate atoms, such that the internuclear axis lies in a single nodal plane that divides the electron density of the bond.

Polar (adj). Having a full or partial separation of opposite electrical charges by some distance.

Polar coordinates (n). A coordinate system employing a length r and two angles θ and ϕ for representing the location of a point in space. The location of a point in space by the Cartesian coordinate system (x, y, z) is related to its location by the polar coordinate system as follows: $x = r \sin \theta \cos \phi$, $y = r \sin \theta \sin \phi$, $z = r \cos \theta$.

Polarizable (adj). Able to be induced to have polar character.

Polarizability (n). The ease with which polar character can be induced. Technically this is the size of the electric dipole moment that is induced by a given electric field.

Polymorphic (adj). Having multiple shapes, forms, or structures (said normally of crystalline solids).

Polyphosphate (n). A condensed phosphate containing P—O—P linkages, an example being diphosphate, $[O_3P\text{—}O\text{—}PO_3]^{4-}$. The general formula is $[P_nO_{3n+1}]^{(n+2)-}$.

Precursor (n). An intermediate preceding the desired product.

Promoted catalyst (n). A catalyst that has been altered so as to increase its activity.

Pyrolysis (n). Breaking down by heat.

Pyrophoric (n). Spontaneously flammable in air.

Quantum (n). The indivisible or most elementary amount of electromagnetic radiation. One quantum is involved per electron in electronic transitions in atoms and the like. Amounts (intensities) of electromagnetic radiation are restricted to integral numbers of the quantum. The energy of a particular quantum is given by Planck's equation $E = hv$, where v is frequency (in cycles s^{-1}) and h is Planck's constant (in erg s).

Quartz (n). The most stable and most dense of the normal crystalline forms of SiO_2.

Quicklime (n). Calcium oxide.

Quicksilver (n). Mercury.

Racemic (adj). Containing equal amounts of each of a pair of enantiomorphs such that the mixture is not optically active, although both enantiomorphs are individually chiral.

Racemization (n). Production of a mixture that is racemic, through an interconversion among enantiomorphs.

Radial wave function (n). That portion of the total wave function $\Psi(r, \theta, \phi)$ that is dependent only on r, the radial distance from the nucleus.

Rare earth element. See lanthanides.

Rearrangement (n). A change in geometry.

Reducing agent (n). A substance that causes reduction, and which is, as a consequence, oxidized.

Reduction (n). A gain of electrons, corresponding to a decrease in oxidation state.

Rock (n). A hard and compacted aggregate of various minerals, sometimes having a uniform and characteristic composition (e.g., marble), but often being visibly heterogeneous (e.g., granite).

Rutile (n). One of the mineral forms of titanium dioxide.

S_N1 (adj). Nucleophilic substitution that has a unimolecular rate-determining step.

S_N2 (adj). Nucleophilic substitution that has a bimolecular rate-determining step.

Saturated (adj). Having sufficient electrons to allow two-center, two-electron single bonds throughout.

Selection rule (n). A statement about the quantum mechanical allowedness of a process.

Semiconductor (n). A solid whose electrical conductivity increases with increasing temperature.

Sequester (v). To draw aside, remove, or bind up a substance so as to influence its freedom of movement or independent action; the process by which a polydentate ligand can surround a metal ion and render it unreactive.

Sigma bond (n). A bond formed by the end-to-end overlap of orbitals from separate atoms, such that there is no nodal plane that includes the internuclear axis.

Silane (n). A hydride of silicon, e.g., SiH_4; In general, Si_nH_{2n+2} or cyclic Si_nH_{2n}.

Silica (n). Silicon dioxide (SiO_2), which, in its most stable crystalline form is quartz, but which exists in numerous other variations including, but not limited to, (a) other crystalline materials formed from quartz by the aid of fluxes (e.g., tridymite, cristobalite, or keatite), (b) amorphous solids, (c) hydrated silicas including aqueous colloidal systems (sols and gels), (d) silica glass, and (e) biogenic silicas such as diatomaceous earth.

Silica gel (n). A solid network of spherical colloidal silica particles.

Silica sol (n). A colloidal suspension of amorphous silica in water.

Silicates (n). Compounds containing polynuclear oxoanions of silicon that contain SiO_4 tetrahedra, which are variously linked to give chain, ring, sheet, cage, or framework structures.

Silicone (n). A linear or chain polymer containing —R_2Si—O—SiR_2—O— repeating units, sometimes cross-linked by $RSiO_3$ units.

Solvation (n). An association with solvent.

Solvolysis (n). A reaction with solvent; when the solvent is water, it is called hydrolysis.

Sphalerite (n). Zinc blende, the denser or β-form of ZnS, the α-form being wurtzite.

Spin angular momentum. That angular momentum due to the vector sum of all electron spins in an atom or molecule.

Steam reforming (n). A thermal and catalyzed degradation in the presence of steam.

Superconductor (n). A substance that has no electrical resistance below a certain characteristic temperature T_c, called the critical temperature.

Symmetry (n). The property of having two or more identical parts that are related to each other by rotation, reflection, or inversion.

Symmetry, center of. See center of inversion.

Tautomer (n). One of a set of isomers that are readily interconverted by rearrangements of atoms.

Tautomerism (n). The occurrence of tautomers.

Tautomerization (n). The (usually reversible) interconversion of tautomers.

Transition state (n). The point (on the reaction profile for a chemical reaction) at which the activated complex has been reached. See activated complex.

Unsaturated (adj). (a) Having one or more multiple bonds formed by electrons that might also be used to bond to additional atoms, to achieve saturation. (b) Any compound containing an atom that may yet add more groups, that is, may undergo an increase in coordination number or occupancy is said to be coordinatively unsaturated.

Valence (n). The capacity of an atom to form bonds to other atoms.

Viscosity (n). Resistance to flow.

Volatile (adj). Easily vaporized.

Vulcanization (n). Irreversible treatment of a rubber compound so that the substance is made less plastic and more elastic. The process is accomplished by certain chemical changes, such as an increase in the extent of cross linking, often by sulfur atoms, in rubber compounds.

Wave number (n). The reciprocal of wavelength, generally stated in units of reciprocal centimeters (cm^{-1}). It can be used as a unit of energy because, according to Planck, $E = h\nu = hc/\lambda$.

Wurtzite (n). The less dense α-form of ZnS, the other form being sphalerite.

Zeolites (n). Framework aluminosilicates that contain cavities, into which ions and molecules of various sizes are more or less free to move, and be retained.

Zinc blende. See sphalerite.

Index